43 uic 110
lbf 36303

Ausgeschieden im Jahr 2025

Fast and Efficient Algorithms in Computational Electromagnetics

For a listing of recent titles in the *Artech House Antennas and Propagation Library*, turn to the back of this book.

Fast and Efficient Algorithms in Computational Electromagnetics

Weng Cho Chew
Jian-Ming Jin
Eric Michielssen
Jiming Song

Editors

Artech House
Boston • London
www.artechhouse.com

Library of Congress Cataloging-in-Publication Data
Fast and efficient algorithms in computational electromagnetics / Weng Cho Chew ... [et al.], editors.
        p. cm. — (Artech House antennas and propagation library)
    Includes bibliographical references and index.
    ISBN 1-58053-152-0 (alk. paper)
      1. Electromagnetism—Data processing. 2. Computer algorithms. I. Chew, Weng Cho.
    II. Series.

QC760.54 .F38 2001
537'.0285—dc21   2001022860

British Library Cataloguing in Publication Data
Fast and efficient algorithms in computational
    electromagnetics. — (Artech House antennas and propagation
    library)
    1. Electromagnetism—Data processing 2. Electromagnetism—
    Mathematics 3. Algorithms
    I. Chew, Weng Cho
    537

ISBN 1-58053-152-0

Cover design by Gary Ragaglia

© 2001 ARTECH HOUSE, INC.
685 Canton Street
Norwood, MA 02062

All rights reserved. Printed and bound in the United States of America. No part of this book may be reproduced or utilized in any form or by any means, electronic or mechanical, including photocopying, recording, or by any information storage and retrieval system, without permission in writing from the publisher.
    All terms mentioned in this book that are known to be trademarks or service marks have been appropriately capitalized. Artech House cannot attest to the accuracy of this information. Use of a term in this book should not be regarded as affecting the validity of any trademark or service mark.

International Standard Book Number: 1-58053-152-0
Library of Congress Catalog Card Number: 2001022860

10 9 8 7 6 5 4 3 2 1

To our wives and our parents

Contents

Preface   xix

Acknowledgments   xxv

1 Introduction to Electromagnetic Analysis and Computational Electromagnetics   1
  1.1 Introduction   1
  1.2 A Bit of History   4
  1.3 More on Differential Equation Solvers   9
      1.3.1 Convergence Rate of Iterative Differential Equation Solvers   10
  1.4 Integral Equation Solvers   13
      1.4.1 Surface Integral Equations   13
      1.4.2 The Internal Resonance Problem   15
      1.4.3 Volume Integral Equation   19
      1.4.4 Green's Function   19
      1.4.5 Method of Moments   20
      1.4.6 Fast Integral Equation Solvers   21
  1.5 A Simplified View of the Multilevel Fast Multipole Algorithm   22
  1.6 Conclusion   26
      References   26

2 Fast Multipole Method and Multilevel Fast Multipole Algorithm in 2D   39
  2.1 Introduction   39
  2.2 Introduction to Fast Multipole in 2D   39
      2.2.1 A 2D MOM Problem   40
      2.2.2 Addition Theorem for Bessel Functions   42
      2.2.3 An Inefficient Factorization of the Green's Function   44
      2.2.4 Diagonalization of the Translation Operator   46
      2.2.5 Summary and Hindsight   49
      2.2.6 An Alternative Derivation of the Diagonalized Translator   49
      2.2.7 Physical Interpretation of Aggregation, Translation, and Disaggregation   50
      2.2.8 Bandwidth of the Radiation Pattern   51
      2.2.9 Error Control   52

2.3	Motivation for Multilevel Method		53
	2.3.1	Factorization of the Green's Function	54
2.4	The Multilevel Fast Multipole Algorithm		55
	2.4.1	The Aggregation Process	56
	2.4.2	$\overline{\alpha}$ Translation and Disaggregation	58
	2.4.3	More on Interpolation and Anterpolation	59
	2.4.4	Computational Complexity of MLFMA	61
2.5	Interpolation Error		63
	2.5.1	Global Interpolation (Exact)	63
	2.5.2	Local Interpolation (Exponentially Accurate)	65
2.6	FMM and Group Theory		67
	2.6.1	Groups	68
	2.6.2	Example of a Group	68
	2.6.3	Representation of a Group	68
	2.6.4	Green's Function	71
	2.6.5	Plane-Wave Representation of the Green's Function	73
2.7	Conclusion		74
	References		74

3 FMM and MLFMA in 3D and Fast Illinois Solver Code — 77

3.1	Introduction		77
3.2	Three-Dimensional FMM and MLFMA		78
	3.2.1	Integral Equations and the Method of Moments	78
	3.2.2	Three-Dimensional FMM	80
3.3	Multilevel Fast Multipole Algorithm		83
3.4	Error Analysis in FMM and MLFMA		85
	3.4.1	Truncation Error in Scalar Green's Function	86
	3.4.2	Truncation Error in the Vector Green's Function	88
	3.4.3	Error in Numerical Integral	92
	3.4.4	Error in Local Interpolation	92
3.5	Large-Scale Computing		92
	3.5.1	Block Diagonal Preconditioner	92
	3.5.2	Initial Guess	93
	3.5.3	Approximation of Bistatic RCS to Monostatic RCS	94
	3.5.4	Interpolation of Translation Matrix in MLFMA	96
	3.5.5	MLFMA for Calculating Radiation Fields	102

3.6	Fast Illinois Solver Code (FISC)		103
	3.6.1	Capabilities	104
	3.6.2	Complexity and Accuracy	106
	3.6.3	Scaling of Memory Requirements	109
	3.6.4	CPU Time Scaling	110
	3.6.5	More Results and Summary	112
3.7	Conclusions		114
	References		114

4 Parallelization of Multilevel Fast Multipole Algorithm on Distributed Memory Computers — 119

4.1	Introduction		119
4.2	The MPI Programming Model		120
4.3	Mathematical Preliminaries		122
	4.3.1	Notations	122
	4.3.2	Essentials of Diagonal Forms of Radiation Fields	122
	4.3.3	FMM Representation of Matrix Elements in Method of Moments	124
4.4	The Parallel MLFMA		125
	4.4.1	The Algorithm	127
4.5	Implementation Issues		132
	4.5.1	Storage of the Tree	132
	4.5.2	Construction of the Tree and Domain Decomposition	133
	4.5.3	Scaling of Translation Matrices	134
	4.5.4	A Costzone Scheme for Load Balancing	135
4.6	ScaleME : A Brief Description		136
4.7	Numerical Experiments		138
	4.7.1	The Integral Equation Formulation	138
	4.7.2	The TRIMOM+ScaleME Code	140
	4.7.3	Single Processor Performance	141
	4.7.4	Processor Scaling	142
	4.7.5	Large-Scale Problems	142
4.8	ScaleME-2: An Improved Parallel MLFMA		144
4.9	Conclusions		147
	References		148

5 Multilevel Fast Multipole Algorithm at Very Low Frequencies — 151
- 5.1 Introduction — 151
- 5.2 Two-Dimensional Multilevel Fast Multipole Algorithm at Very Low Frequencies — 153
  - 5.2.1 Core Equation of the 2D Undiagonalized Dynamic MLFMA — 153
  - 5.2.2 Core Equation of the 2D Diagonalized Dynamic MLFMA — 155
  - 5.2.3 2D Uniformly Normalized LF-MLFMA — 156
  - 5.2.4 Nonuniformly Normalized Form of 2D LF-MLFMA — 158
  - 5.2.5 Computational Complexity of 2D LF-MLFMA — 158
  - 5.2.6 Applying 2D Dynamic MLFMA and LF-MLFMA to CFIE for PEC Structures — 160
- 5.3 3D Multilevel Fast Multipole Algorithm at Very Low Frequencies — 174
  - 5.3.1 General Formulations for the 3D Dynamic MLFMA — 174
  - 5.3.2 Core Equation for 3D Diagonalized Dynamic MLFMA — 176
  - 5.3.3 Core Equation for 3D LF-MLFMA — 177
  - 5.3.4 Computational Complexity of 3D LF-MLFMA — 180
  - 5.3.5 Core Equation for 3D Static MLFMA — 181
  - 5.3.6 Rotation of the Translation Matrices for 3D LF-MLFMA and 3D Static MLFMA — 184
  - 5.3.7 3D LF-MLFMA Based on RWG Basis — 189
- 5.4 Conclusions — 197
- References — 199

6 Error Analysis of Surface Integral Equation Methods — 203
- 6.1 Introduction — 203
  - 6.1.1 Surface Integral Equations and the Method of Moments — 204
  - 6.1.2 Error Measures — 206
  - 6.1.3 Approaches to Error Analysis — 208
  - 6.1.4 Spectral Convergence Theory — 211
- 6.2 Spectral Convergence Theory—2D — 212
  - 6.2.1 Circular Cylinder—TM — 213
  - 6.2.2 Circular Cylinder—TE — 222
  - 6.2.3 Flat Strip—TM — 225
  - 6.2.4 Flat Strip—TE — 234
  - 6.2.5 Flat Strip—Edge Error — 239
  - 6.2.6 Rectangular Cavity — 242
  - 6.2.7 Higher-Order Basis Functions — 247

	6.2.8	Summary	255
6.3	Spectral Convergence Theory—3D		256
	6.3.1	Flat Plate	256
	6.3.2	Rooftop Basis Functions	258
6.4	Iterative Solution Methods		261
	6.4.1	Iteration Count Estimates	262
	6.4.2	Condition Number Estimates	263
6.5	Conclusion		277
	References		278

7 Advances in the Theory of Perfectly Matched Layers — 283

7.1	Introduction		283
7.2	PML via Complex Space Coordinates		284
	7.2.1	Frequency Domain Analysis	284
	7.2.2	Time Domain Analysis	287
7.3	PML-FDTD for Dispersive Media with Conductive Loss		288
	7.3.1	Time Domain Analysis	288
	7.3.2	Dispersive Medium Models	288
	7.3.3	Incorporation into FDTD Update	291
7.4	Maxwellian PML		293
7.5	Extension to (Bi)Anisotropic Media		294
	7.5.1	Non-Maxwellian Formulation	295
	7.5.2	Maxwellian Formulation	295
7.6	PML for Inhomogeneous Media		298
7.7	Curvilinear PML		299
	7.7.1	Cylindrical PML-FDTD	299
	7.7.2	Spherical PML-FDTD	305
	7.7.3	Maxwellian PML in Cylindrical and Spherical Coordinates	307
	7.7.4	Conformal (Doubly Curved) PML	310
7.8	Stability Issues		317
	7.8.1	Cartesian PML Analysis	320
	7.8.2	Cylindrical PML Analysis	323
	7.8.3	Spherical PML Analysis	327
	7.8.4	Imposing Stability a Posteriori: The Quasi-PML	330
7.9	Generalized PML-FDTD Schemes		331
	7.9.1	Cylindrical PML-PLRC-FDTD: Split-Field Formulation	332

		7.9.2 Cylindrical PML-PLRC-FDTD: Maxwellian Formulation	334

    7.10 Unified Theory: Brief Discussion   336
        7.10.1 PML as a Change on the Metric of Space   336
        7.10.2 Metric and Topological Structure of Maxwell's Equations   339
        7.10.3 Hybrid PMLs   341
    References   342

8 Fast Forward and Inverse Methods for Buried Objects   347

    8.1 Introduction   347
    8.2 Green's Functions   349
        8.2.1 Green's Function in Integral Equation   350
        8.2.2 Green's Function for Incident Field   351
        8.2.3 Green's Function for Scattered Field   352
        8.2.4 Reduction to 2D Case   353
    8.3 Fast Forward Scattering Methods   354
        8.3.1 2D Buried Dielectric Cylinders   354
        8.3.2 3D Buried Conducting Plates   360
        8.3.3 3D Buried Dielectric Objects   364
    8.4 Detection of Buried Objects Using Forward Method   370
        8.4.1 VETEM System   370
        8.4.2 Numerical Modeling: Loop-Antenna Model   372
        8.4.3 Numerical Modeling: Magnetic-Dipole Model   374
        8.4.4 Simulation Results   375
    8.5 Fast Inverse Scattering Methods   380
        8.5.1 Single-Frequency DBIM Algorithm for 2D Objects   382
        8.5.2 Multifrequency DBIM Algorithm for 2D Objects   392
        8.5.3 DT Algorithm for 2D Objects   396
        8.5.4 DT Algorithm for 3D Objects   407
    References   418

9 Low-Frequency Scattering from Penetrable Bodies   425

    9.1 Introduction   425
    9.2 Low-Frequency Scattering from a Single Penetrable Body   428
        9.2.1 A Brief Review of Basis Functions   428
        9.2.2 General Equations and Frequency Normalization   432
        9.2.3 Interpretations   436

9.3	Scattering from a Multibody		440
	9.3.1	PMCHWT Formulation for Multibody Problem	440
	9.3.2	Number-of-Unknowns Reduction Scheme for RWG Basis	443
	9.3.3	Number-of-Unknowns Reduction Scheme for Loop-Tree Basis	449
	References		458

10 Efficient Analysis of Waveguiding Structures — 461

10.1	Introduction		461
10.2	Finite Difference Formulation		462
	10.2.1	Boundary Conditions	465
10.3	Solution to the Sparse Matrix Equation		467
	10.3.1	Complexity and Storage Issues	468
10.4	Waveguide Discontinuities		469
	10.4.1	The Single Junction Problem	471
	10.4.2	The n-Junction Problem	474
10.5	Numerical Examples		477
10.6	Conclusions		480
	References		484

11 Volume-Surface Integral Equation — 487

11.1	Introduction		487
11.2	The Formulation of the Integral Equations		488
	11.2.1	Volume Integral Equation (VIE)	489
	11.2.2	Hybrid Volume-Surface Integral Equation (VSIE)	491
11.3	Numerical Solution of the Hybrid VSIE		492
	11.3.1	Mesh Generating	492
	11.3.2	Discretization of the Hybrid Integral Equation	495
	11.3.3	Mesh Termination	499
	11.3.4	Enforcing the Continuity Condition	501
	11.3.5	Other Cell Shapes	502
11.4	Combined Field Integral Equation		506
11.5	Singular Integral Treatments		508
11.6	Solution of VSIE by Fast Multipole Method		513
11.7	Numerical Examples		514
	11.7.1	Results by Volume Integral Equation	515

	11.7.2 Results by Hybrid Volume-Surface Integral Equation	517
11.8	Other Applications	518
	11.8.1 Indoor Radio Wave Propagation Simulation	518
	11.8.2 Microwave Thermal Effect Simulation	519
	11.8.3 Antenna Radome Modeling	519
	References	537

12 Finite Element Analysis of Complex Axisymmetric Problems — 541

12.1	Introduction	541
12.2	Formulation	542
	12.2.1 Problem Definition	543
	12.2.2 Variational Formulation	543
	12.2.3 Solution of the Equations	549
	12.2.4 Far-Field Calculations	551
12.3	Cylindrical PML	552
	12.3.1 Parameter Definitions	553
	12.3.2 Systematic Error Reduction	554
12.4	Numerical Results	554
	12.4.1 Scattering	554
	12.4.2 Radiation	560
12.5	BOR with Appendages	566
12.6	Conclusion	569
	References	571

13 Hybridization in Computational Electromagnetics — 575

13.1	Introduction	575
13.2	Hybrid FEM/ABC Technique	577
	13.2.1 Problem Statement	578
	13.2.2 Finite Element Analysis	578
	13.2.3 Numerical Results	582
13.3	Hybrid FEM/BIE Technique	582
	13.3.1 Formulation	584
	13.3.2 Application of MLFMA	589
	13.3.3 Numerical Results	590
13.4	Hybrid FEM/AABC Technique	590
	13.4.1 Formulation	593

		13.4.2 Numerical Results	598
	13.5	Hybrid FEM/SBR Technique	602
		13.5.1 Formulation	604
		13.5.2 Scattered Field Calculation	608
		13.5.3 Analysis of the Hybrid Technique	609
		13.5.4 Numerical Results	610
	13.6	Hybrid MOM/SBR Technique	614
		13.6.1 Formulation	615
		13.6.2 Scattered Field Calculation	618
		13.6.3 Iterative Improvement	619
		13.6.4 Numerical Results	620
	13.7	Summary	621
		References	628
14	**High-Order Methods in Computational Electromagnetics**		**637**
	14.1	Introduction	637
	14.2	Higher-Order MOM and MLFMA	638
		14.2.1 Formulation	639
		14.2.2 Numerical Examples	645
	14.3	Point-Based Implementation of Higher-Order MLFMA	652
		14.3.1 Formulation	654
		14.3.2 Complexity Analysis	657
		14.3.3 Numerical Results	658
	14.4	Higher-Order FEM	658
		14.4.1 Higher-Order Tetrahedral Elements	661
		14.4.2 Application to Cavity Scattering	664
	14.5	Mixed-Order Prism Elements	672
	14.6	Point-Based Grid-Robust Higher-Order Bases	680
		14.6.1 Vector Basis Functions	682
		14.6.2 MOM Formulation	686
	14.7	Numerical Results	688
	14.8	Summary	693
		References	694
15	**Asymptotic Waveform Evaluation for Broadband Calculations**		**699**
	15.1	Introduction	699

15.2	The AWE Method	700
15.3	Analysis of Metallic Antennas	702
	15.3.1 Formulation	703
	15.3.2 Numerical Examples	706
15.4	Analysis of Metallic Scatterers	708
	15.4.1 Formulation	708
	15.4.2 Numerical Examples	713
15.5	Analysis of Dielectric Scatterers	713
	15.5.1 Formulation	714
	15.5.2 Numerical Examples	717
15.6	Analysis of Microstrip Antennas	718
	15.6.1 Formulation	718
	15.6.2 Numerical Examples	720
15.7	Summary	725
	References	725

16 Full-Wave Analysis of Multilayer Microstrip Problems — 729

16.1	Introduction	729
16.2	Green's Functions for Multilayer Media	730
16.3	The Method-of-Moments Solution	737
16.4	Fast Frequency-Sweep Calculation	746
16.5	The Conjugate Gradient–FFT Method	752
16.6	The Adaptive Integral Method	759
16.7	The Multilevel Fast Multipole Algorithm	764
16.8	Summary	770
	References	772

17 The Steepest-Descent Fast Multipole Method — 781

17.1	Introduction	781
17.2	Field Evaluation on Quasi-Planar Surfaces	782
	17.2.1 The Scalar Case	782
	17.2.2 The Vector Case	787
17.3	Computational Complexity Estimates	788
17.4	Scattering from Random Rough Surfaces	791
	17.4.1 Model Development	791
	17.4.2 Integral Equation Formulations	792

	17.4.3 SDFMM-Based Solutions	793
	17.4.4 Simulation Results	795
17.5	Quantum Well Grating Analysis	798
	17.5.1 Introduction and Formulation	798
	17.5.2 Periodic and Quasi-Random Grating Analysis	802
	17.5.3 Random Rough Surface Couplers	804
17.6	Analysis of Microstrip Antenna Arrays on Finite Substrates	804
	17.6.1 Introduction	804
	17.6.2 Integral Equation Formulation and SDFMM Solution	805
	17.6.3 MOM Formulation	807
	17.6.4 Simulation Results	808
17.7	Conclusion	810
	References	810

18 Plane-Wave Time-Domain Algorithms — 815

18.1	Introduction	815
18.2	The Marching-on-in-Time Method	817
18.3	The Plane-Wave Time-Domain Algorithm	819
	18.3.1 Plane Wave Decomposition	820
	18.3.2 Implementation Issues	826
18.4	Implementation of the PWTD-Enhanced MOT Schemes	830
	18.4.1 A Two-Level PWTD-Enhanced MOT Algorithm	831
	18.4.2 A Multilevel PWTD-Enhanced MOT Algorithm	836
18.5	The Windowed Plane-Wave Time-Domain Algorithm	842
	18.5.1 Windowed Plane-Wave Decomposition	843
	18.5.2 Implementation Using Sampled Field Representations	844
18.6	Implementation of the Windowed PWTD-Enhanced MOT Schemes	849
	18.6.1 Sphere-to-Sphere Translation	850
	18.6.2 A Two-Level Windowed PWTD-Enhanced MOT Algorithm	853
	18.6.3 A Multilevel Windowed PWTD-Enhanced MOT Algorithm	854
18.7	Summary	855
	References	856

19 Plane-Wave Time-Domain Algorithm Enhanced Time-Domain Integral Equation Solvers — 859

19.1	Introduction	859

19.2	Formulation	861
	19.2.1 Integral Equations	862
	19.2.2 Marching-on-in-Time Formulation	865
19.3	Plane-Wave Time-Domain Algorithm	867
	19.3.1 Plane Wave Representations	868
	19.3.2 Implementation of Two-Level PWTD Enhanced MOT Solvers	873
	19.3.3 Complexity Analysis	875
19.4	Numerical Results	875
	19.4.1 Efficacy of the CFIE	876
	19.4.2 Validating the PWTD-Augmented MOT Solver	881
	19.4.3 Efficacy of the PWTD-Augmented MOT Scheme for Large-Scale Analysis	883
19.5	Summary	888
	References	889
About the Authors		**893**
Index		**899**

Preface

This book documents recent advances in computational electromagnetics performed under the auspices of the Center for Computational Electromagnetics at the University of Illinois, funded mainly by the Multidisciplinary University Research Initiative (MURI), a program administered by the Air Force Office of Scientific Research. Other funding agencies also contributed to the success of the Center, such as the National Science Foundation, Office of Naval Research, Army Research Office, and Department of Energy.

There is a tremendous need to bring the science of electromagnetic simulation, also known as computational electromagnetics, to the same confidence level as that achieved by circuit simulation. However, computational electromagnetics involves solving Maxwell's equations, which are more complex than circuit equations. It is hoped that one day electromagnetic simulation will master this complexity and enjoy the same pervasiveness in engineering design as does circuit simulation. We are grateful for the foresight of these funding agencies who share our passion for developing this technology.

This book does not pretend to be complete, as it reflects our viewpoint of computational electromagnetics. However, we believe that the knowledge required to support electromagnetic simulation in a sophisticated manner has to come from physicists, engineers, mathematicians, and computer scientists. Since electrical engineering is an offshoot of applied physics, we play the role of applied physicists in the development of this technology: we develop this technology based on our physical insight into the problems, while drawing on knowledge from mathematicians and computer scientists. The presentation style of most of the chapters of this book is in the manner of applied physicists or of traditional electromagneticists—hopefully, we sacrifice mathematical rigor for physical clarity.

This book is not an introduction to computational electromagnetics. It documents recent advances in computational electromagnetics in the manner of a monograph. A seasoned researcher in the area of computational electromagnetics should have little difficulty reading the material. It is also hoped that a graduate student or a professional with some preliminary background in computational electromagnetics or a classicist in electromagnetics who has done some rapid background reading, can easily digest the work reported in this book. For one who intends to perform research in this area, this book will be an excellent starting point. The variety of topics covered is sufficient to nourish many different research directions in this very interesting field.

Even though this book deals only with linear problems associated with Maxwell's equations, it can be gleaned from a cursory reading that such problems are rich; they are amenable to different mathematical analyses, and allow for different and interesting algorithm designs. Because of the linearity of the problems, both differential equation and integral equation solvers can be developed. Moreover, the problems can be solved in the frequency domain as well as the time domain, enhancing the efficiency and enriching the variety of these methods.

Solutions to Maxwell's equations have been sought since the very early days of the equations' discovery. Electromagnetic analysis has always played an important role in understanding many scientific and engineering problems.

Chapter 1 gives an introduction to electromagnetic analysis and explains how the field has evolved into computational electromagnetics in the last few decades. It also introduces, in a very simplified manner, the recent fast algorithms developed to solve Maxwell's equations. The chapter also attempts to give a historical perspective on electromagnetic analysis and to describe how far we have come since the advent of Maxwell's equations.

Chapter 2 presents an introduction to the fast multipole method (FMM) and the multilevel fast multipole algorithm (MLFMA) in two dimensions. Interpolation, truncation, and integration errors are discussed. An attempt is also made to relate FMM to group theory, and to the inherent symmetry of space.

Chapter 3 describes the three-dimensional version of FMM and MLFMA and demonstrates the application of the fast algorithm to real-world problems. The algorithm has also been parallelized on a shared-memory machine, and tour-de-force computation involving close to 10 million unknowns is the most important achievement of this work.

Chapter 4 outlines the distributed-memory parallelization of MLFMA, encapsulated in a code called ScaleME (Scaleable Multipole Engine). The parallelization of MLFMA on a distributed memory machine is not an easy task, because different parts of the computation may reside on different processors. The increased communication cost with more processors can be an issue here. A 10-million-unknown problem has also been solved with ScaleME.

Chapter 5 reports on the low-frequency solution of Maxwell's equations using fast algorithms. This chapter describes the treatment needed for FMM and MLFMA to prevent their catastrophic breakdown at low frequencies. It also describes a method to apply the LF-MLFMA based on Rao-Willon-Glisson (RWG), wire, and wire-surface bases while the intrinsic expansion bases are still the loop-tree-star bases. These bases are designed for low-frequency problems to make the LF-MLFMA efficient for problems with global loops.

Chapter 6 delves into different error issues involved when solving surface integral equations related to Maxwell's theory. Discretization error due to the use of basis functions, and integration error by replacing integrals with summation are discussed. Errors result from solving the matrix equation, and deconditioning of the matrix equation by MOM and its impact on errors are studied. This chapter also discusses deconditioning due to the near-resonance problem and the low-frequency breakdown problem.

Chapter 7 deals with a recent topic of intense interest in differential equation solvers—the theory of perfectly matched layers (PML). The concept of complex coordinate stretching is discussed. PML is generalized to curvilinear coordinates as well as to complex media. In this chapter, stability issues related to PML are studied, and a unified analysis of various PML formulations using differential forms is included.

Chapter 8 addresses the issue of efficiently solving the forward and inverse problems for buried objects using FFT-based methods. The detection of buried objects usually involves loop antennas, and the forward problem involving the solution of loop antennas over a buried object is discussed in great detail. Moreover, recent advances in different inversion algorithms are also described.

Chapter 9 touches upon solving the penetrable problem at very low frequencies. The low-frequency problem encountered in Chapter 5 for metallic objects also occurs for dielectric and lossy material objects. This chapter describes a way to solve this problem so that the solution of integral equations remains stable all the way from zero frequency to microwave frequencies.

Chapter 10 describes an algorithm to solve three-dimensional waveguide structures using numerical mode matching, but using the finite difference method. The spectral Lanczos decomposition method is used to find the modes. An algorithm with $O(N)$ memory complexity and $O(N^{1.5})$ computational complexity is achieved.

Chapter 11 addresses the problem of solving the volume integral equation concurrently with the surface integral equation. This is particularly important when dealing with structures having metals as well as dielectric materials. The solutions are also accelerated with MLFMA as demonstrated in the chapter. Many practical illustrations of the use of this solution technique are given in this chapter.

Chapter 12 deals with solving axially symmetric, body-of-revolution (BOR) geometry using the finite element method (FEM). This reduces a three-dimensional problem to two dimensions, greatly enhancing the efficiency of the solution. Both material-coated and metallic objects are considered. The chapter also shows the practical use of cylindrical PML for truncating the FEM mesh. Treatment of BOR geometry with appendages is also considered.

Chapter 13 reports on the hybridization in computational electromagnetics. Hybridization between FEM and the absorbing boundary condition (ABC) is discussed alongside the boundary integral equation (BIE), MLFMA, adaptive absorbing boundary condition (AABC), and shooting and bouncing ray (SBR). Hybridization between MOM and SBR is also considered. AABC is a promising method of hybridizing FEM with fast solvers in the future.

Chapter 14 presents different higher-order methods in computational electromagnetics. Higher-order methods for the surface integral equation as well as for FEM are considered. Also, the efficient coupling of higher-order methods to fast solvers such as MLFMA is discussed. In particular, the use of point-based MLFMA is illustrated. Moreover, a higher-order grid-robust method is also studied in this chapter.

Chapter 15 touches on the topic of asymptotic waveform evaluation (AWE) for broadband calculation in electromagnetics. Illustrations of this acceleration technique for broadband calculation are given for metallic antennas, wire antennas, dielectric scatterers, and microstrip antennas.

Chapter 16 details the analysis of microstrip structure on top of a layered medium. The derivation of the layered medium Green's function together with its numerical approximation by the complex images is discussed. The use of the fast frequency sweep method, adaptive integral method, and MLFMA to accelerate solution speed is studied. A higher-order method to improve solution accuracy is also demonstrated.

Chapter 17 reviews the steepest-descent FMM (SDFMM) to accelerate the solution speed of quasi-planar structures. For this class of structures, this method reduces both the computational and memory complexity of MLFMA from $O(N \log N)$ to $O(N)$. Applications to scattering from random rough surfaces, quantum-well gratings, and microstrip antennas are demonstrated with this analysis method.

Chapter 18 elaborates on the plane-wave time-domain (PWTD) algorithm, which is an ingenious way of arriving at the time-domain equivalent of FMM and MLFMA. The integral equation is solved using the marching-on-in-time (MOT) method. Stability and accuracy issues are carefully analyzed in this chapter. Both the two-level and multilevel algorithms are presented and demonstrated with examples.

Chapter 19 further develops PWTD for large-scale and real-world applications. The use of PWTD with the magnetic field integral equation (MFIE), electric field integral equation (EFIE), and combined field integral equation (CFIE) is illustrated. Furthermore, scattering and error analysis from complex targets such as aircraft, almond shapes, and cone-spheres are considered.

Even though a large variety of topics is covered here, we do feel that there is still a myriad of problems in computational electromagnetics begging to be solved. Due to the complex nature of computational electromagnetics compared to circuit simulation, the robustness and stability of these algorithms are still issues to be addressed.

Another issue is the computational labor associated with these algorithms—more research needs to be done to enhance their speed. We hope, however, that the work at our Center marks a new beginning in the era of fast algorithms in computational electromagnetics.

During the MURI support, we have demonstrated our ability to solve problems involving 10 million unknowns using the supercomputing facilities of the University of Illinois. With continued support in this field, together with improvements in computer technology, we predict that a decade from now, solving a problem of this size will be routine for many applications.

If only electromagnetic fields can talk, they will speak volumes!

WENG CHO CHEW

Urbana-Champaign, Illinois, June 2001

Acknowledgments

First, we are indebted to many colleagues who have contributed to this field. We are also indebted to many from whom we have learned this material, making this work possible. We also owe much to many graduate students, postdoctoral associates, and research scientists who have worked tirelessly to make this work possible. Many of them are coauthors of chapters in this book.

Financial support from the following organizations in the course of our research is gratefully acknowledged:

- Air Force Office of Scientific Research under MURI grant F49620-96-1-0025;

- National Science Foundation;

- Office of Naval Research;

- Department of Energy.

In addition, a number of industrial organizations have contributed to our research—notably, HRL, Intel, Lockheed-Martin, Mobil-Exxon, MRC, Northrop-Grumman, Raytheon, and Schlumberger. At the University of Illinois, we thank NCSA for the use of the supercomputing facilities and support from the CSE Program. One of the authors of Chapter 19, AAE, would like to thank Gebze Institute of Technology for their support.

We thank Jamie Hutchinson of our Publications Office, who painstakingly read all chapters and checked for editorial corrections. For their willingness to serve and provide feedback during the course of this program, many thanks also go to the advisory board members of our Center: Tom Blalock, Bill Hall, Kristopher Kim, Charlie Liang, Bob Mailloux, Louis Medgyesi-Mitschang, Don Pflug, Maurice

Sancer, Joe Shang, Anthony Terzuoli, Steve Wandzura, Arthur Yaghjian, and Long Yu. We owe Dennis Andersh and S. W. Lee of SAIC (formerly of UIUC) a word of gratitude for their moral support and interest during this program. Finally, we thank Arje Nachman of AFOSR, whose no-nonsense approach to monitoring our program provides constant impetus.

Weng Cho Chew thanks several students—Alaeddin Aydiner, Yunhui Chu, Eric Forgy, Larkin Hastriter, and Lijun Jiang—for proofreading and providing useful feedback on some of the chapters. Thanks is also due to Sanjay Velamparambil, who spent hours perfecting the template to meet the requirements of the publisher. Last but not least, we are grateful to our wives for their support despite the trying times they went through during the course of this research.

Weng Cho Chew, Jian-Ming Jin, Eric Michielssen, and Jiming Song

1

Introduction to Electromagnetic Analysis and Computational Electromagnetics

Weng Cho Chew

1.1 INTRODUCTION

Electromagnetic analysis, a discipline whereby one solves Maxwell's equations to obtain a better understanding of a complex system, is becoming increasingly important in electrical engineering. One reason is that Maxwell's theory is essential for the manipulation of electricity and hence is indispensable. Another reason is that Maxwell's theory has proven to have strong predictive power. This strong predictive power, together with the advent of computer technology, has changed the practice of electrical engineering in recent years. A complete solution to Maxwell's equations can expedite many electrical engineering design processes.

A notable example is the integrated circuits industry where computer-aided design predictive tools, using circuit theory that is partly based on low-frequency Maxwell's theory, have completely changed how electrical engineering design processes are performed.

In the beginning, electromagnetic analyses were performed with pencil and paper, solving for closed form and approximate solutions. However, this has changed with the advent of numerical methods and computers, which have spurred the field of "computational electromagnetics."

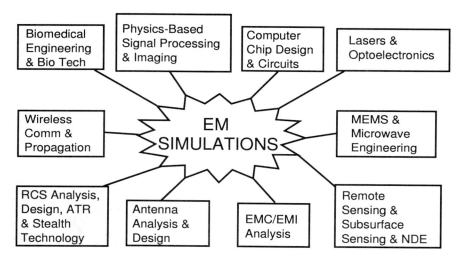

Figure 1.1 Because solving Maxwell's equations is fundamental for the manipulation of electricity, the ability to solve Maxwell's equations to perform electromagnetic simulations impacts many different areas.

Traditional numerical methods are inefficient, but fast algorithms in computational electromagnetics will alter the use of many of these electromagnetic analysis methods in the future. For a problem with N degrees of freedom, or N unknowns, many of these fast algorithms use $O(N \log N)$ memory, and close to $O(N \log N)$ time, as opposed to the traditional methods requiring $O(N^2)$ memory and $O(N^2)$ time. As N becomes very large, there will be a tremendous discrepancy in memory and time usage between the fast solvers and traditional solvers.

The ability to perform computer simulations with Maxwell's equations portends great changes in many areas in electrical engineering and related fields such as computer chip design and circuit analysis, lasers and optoelectronics, microelectro-mechanical sensors (MEMS) and microwave engineering, remote sensing and subsurface sensing, electromagnetic compatibility and electromagnetic interference (EMC/EMI), antenna analysis and design, radar cross-section (RCS) analysis and design, stealth technology, automatic target recognition (ATR), wireless communication and propagation, bioengineering and biotechnology, and physics-based signal processing and imaging (see Figure 1.1).

Despite its arcane sounding name, "computational electromagnetics" is actually highly interdisciplinary, drawing knowledge from physics, mathematics, and computer science (Figure 1.2). An electromagnetic problem that begs to be solved nowadays is orders of magnitude more complex than those at the end of the 19th century. As such, a good understanding of the physics of the problem is essential to

Introduction to Electromagnetic Analysis and CEM 3

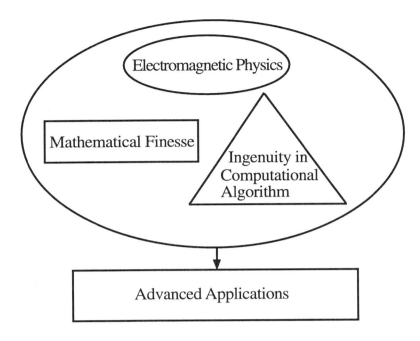

Figure 1.2 Computational electromagnetics is a highly interdiscplinary field drawing expertise from mathematics, physics, and computer science.

avoid developing methods that will lead us on a primrose path. Good physical insight is also important in understanding the physical mechanism within a calculation, and arriving at an improved algorithm.

However, the use of numerical methods to simulate physical phenomena cannot be done without finesse in the underlying mathematical analysis. Mathematical analysis is needed for understanding error analysis and error control within an algorithm. It is necessary for the understanding of the convergence and conditioning of iterative solvers when applied to a system of linear algebraic equations. More recently, mathematical analysis is necessary for the factorization of operators into factors, which can be computed very efficiently.

The final test of an algorithm is its ability to harness the power of a parallel computer to its fullest. Issues such as memory latency, cache hit and miss, interprocessor communication, and shared memory versus local memory architecture have to be considered. Software engineering issues such as encapsulation, reusability, and ease of software development and maintenance have to be addressed. The synergism between different disciplines is described in Figure 1.2.

1.2 A BIT OF HISTORY[1]

Maxwell's theory was completed over a century ago in 1864 [1]. At the time of its completion, it was regarded as a triumph by mathematicians and physicists alike. Maxwell's theory unifies both the theory of light and the theory of electromagnetism, which were thought to be two different theories. The motivation to add displacement current to Ampere's Law to complete Maxwell's equations was both a mathematical and a physical one. It was noted that—assuming the existence of the displacement current, which, unlike conduction current, could flow through vacuum—wave phenomena could be mathematically predicted by Maxwell's equations. The existence of electromagnetic fields as waves was finally confirmed by Heinrich Hertz in 1888, and later the propagation of radio wave across the Atlantic Ocean was demonstrated by Guglielmo Marconi in 1901. We can view those two historical experiments as very early uses of Maxwell's theory to predict the outcome of experiments.

Maxwell's theory has been regarded by most physicists to be one of the most elegant theories in physics. Maxwell's theory also suggests the theory of special relativity and Lorentz transformation. Most physicists believe that if you lock a graduate student in a room and have him perform an electromagnetic calculation correctly, and if you perform an experiment that does not agree with the graduate student's calculations, then you better check your experiment (C. N. Yang, private communication).

Maxwell's equations are given by

$$\nabla \times \mathbf{E} = -\frac{\partial \mathbf{B}}{\partial t} \tag{1.1}$$

$$\nabla \times \mathbf{H} = \frac{\partial \mathbf{D}}{\partial t} + \mathbf{J} \tag{1.2}$$

$$\nabla \cdot \mathbf{D} = \rho \tag{1.3}$$

$$\nabla \cdot \mathbf{B} = 0 \tag{1.4}$$

They have been used to predict electromagnetic phenomena from the subatomic lengthscale to the intergalactic lengthscale. At the subatomic level, the quantized version of electromagnetic theory—namely, quantum electrodynamics—is often used. In this case, electromagnetic fields are treated as quantum mechanical operators. On the galactic scale, a supernova explosion that happened 168,000 light years away was observed in 1987 [2]. At the time of this supernova explosion (SN1978A, Large Magellanic Cloud), there was no civilization on Earth, only primitive men, such as Cro-Magnon men and Neanderthal men who roamed the surface of the Earth. Despite that, we have been able to observe such electromagnetic radiation that has been traversing the galaxy for 168,000 years through the immensity of the universe. More

---

[1] Due to the long history of electromagnetics, the history here is necessarily incomplete.

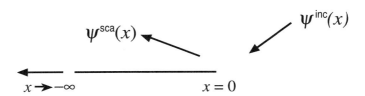

Figure 1.3 The Sommerfeld half-plane problem of a diffraction by a half screen was solved by Sommerfeld in 1896.

mind boggling still is the remnant of the electromagnetic radiation from the big bang that happened 10 billion years ago. This electromagnetic radiation has been traveling at the speed of light for over 10 billion years, and its remnants were detected by two Bell Lab electrical engineers, Arno Penzias and Robert Wilson, who won the Nobel prize in physics in 1978. The cosmic background microwave radiation they detected corresponds to the three-degree black body radiation from the universe.

As a consequence of their predictive power and their validity over a vast range of lengthscales, solutions for Maxwell's equations have been used for analysis and prediction in engineering design. Shortly after the advent of Maxwell's theory, closed form solutions to Maxwell's equations were sought by the separation of variables. For instance, the Mie scattering solution for a sphere using the method of the separation of variables was found around the end of the 19th century [3, 4]. The guided wave solution in a hollow waveguide was given by Rayleigh in 1897 [5]. The famous Sommerfeld half-plane solution for diffraction of waves by a semi-infinite half plane was presented by Sommerfeld in 1896 [6] (see Figure 1.3). Moreover, the Sommerfeld problem of a dipole source over a half-space Earth was obtained in 1909 [7] (see Figure 1.4). In addition to the above, many solutions were sought for scatterer shapes that conform to a separable coordinate systems. There are reasons for the rapid growth of solutions to Maxwell's equations during the end of the 19th century. The precursors to electromagnetic theory were the theory of fluid, commonly expressed in the Navier-Stokes equation and the theory of sound, which was expressed in terms of the scalar wave equation. Many mathematical techniques were developed to solve fluid and acoustic problems related to simple shapes. Such problems were solvable with harmonic functions and harmonic analysis, as in the case of low-Reynold's-number flow and inviscid flow [8]. Hence, we often hear names like Lagrange, Stokes, Gauss, and Euler in electromagnetics literature. For instance, Lord Rayleigh was well familiar with theory of fluid and sound, and he wrote two volumes on the theory of sound during his voyage on the Nile river recuperating from a pneumatic fever [9]. Many of these closed form solutions were generalized to more complex geometries during the first half of the 20th century [10–12]. Many of these solutions to simple shapes are documented in [13].

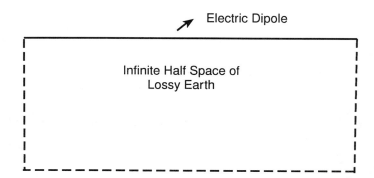

Figure 1.4 The Sommerfeld half-space solution of an electric dipole radiating on top of the lossy Earth was solved by Sommerfeld in 1909.

It was said that Sommerfeld found the solution to the half-plane problem by inspection. The Sommerfeld half-plane solution is usually taught to students nowadays using the Wiener-Hopf technique [14]. The Wiener-Hopf technique is a "solution in search of a problem," a term used by some scientists to describe the narrow application of some analysis technique (L. Felsen, private communication). The Wiener-Hopf technique is a technique to deconvolve a convolutional integral of the form

$$\phi_{sca}(x) = -\phi_{inc}(x) = \int_{-\infty}^{0} g(x - x')j(x')dx', \qquad -\infty < x < 0 \qquad (1.5)$$

by Wiener-Hopf factorization. In the above $g(x - x')$ is a Green's function, while $j(x)$ describes the current on the half-plane.

Sommerfeld's solution to the other classical problem for the radiation of an electric dipole source on top of the half-space Earth involves an integral and often expressed as

$$\phi(\mathbf{r}) = \int_{0}^{\infty} dk_\rho J_n(k_\rho \rho) \left[e^{ik_z|z-z'|} + R(k_\rho)e^{ik_z z} \right] \qquad (1.6)$$

where $k_z = \sqrt{k^2 - k_\rho^2}$, $J_n(x)$ is a Bessel function of nth order, and $R(k_\rho)$ is a reflection coefficient. Such integrals are known as Sommerfeld integrals for an obvious reason, and they were hard to evaluate at the time of Sommerfeld. The integrals involve branch cuts and poles that can easily confound the uninitiated. But evaluating such integrals is a routine task these days with the use of the digital computer. To date, the class of closed form integral solutions has expanded to more complex geometry. For instance, the solution of a dipole source in layered media has been generalized [12, 15–22]. These Sommerfeld solutions have also been generalized to the time domain [23–25].

During Sommerfeld's time, asymptotic methods such as the method of stationary phase and the method of steepest descent were developed to evaluate such integrals approximately [15–18, 20, 21, 26–29]. Also, the method of Watson transformation can be applied to the Mie series solution to obtain approximate solutions when the frequency is high [28]. For instance, the physics of a creeping surface wave can be elucidated as such [17, 29]. In these asymptotic methods, a large parameter is often assumed, where the approximation becomes increasingly good when the parameter is large. For electromagnetics, the large parameter is usually the wavenumber that becomes large as the wavelength becomes small or when the frequency is high.

These approximate solutions, derived with mathematical elegance, often entail nice physical meanings, lending further insight into these problems that the closed form solutions, such as the Sommerfeld half-plane and the Sommerfeld half-space solution, do not offer. Moreover, they illustrate that many of the physical mechanisms of scattering and interaction with complex geometries are local in nature, such as edge diffraction and specular reflection.

However, the availability of closed form solutions for simple geometries in separable coordinate systems did not meet the need of engineers and scientists alike. There are only a finite number of separable coordinate systems [30], and the solutions are often very complex. Very soon, innovative methods were sought for solutions of more complex geometries using perturbation and approximate methods. Such methods parallel the perturbation methods that appeared in the fluid community [8]. Like the asymptotic evaluation of integrals, many of these perturbation methods are asymptotic methods where a large or a small parameter has to be assumed.

Mathematicians and engineers capitalized on the physics that can be elucidated from the asymptotic solutions and proposed an ansatz-based solution technique to finding the solution from complex geometries as the frequency becomes very high. Approximate solutions are proposed as ansatz with an asymptotic series that resembles [31]

$$\phi(\mathbf{r}) \sim \frac{e^{ikr}}{(kr)^\alpha} \left[a_0 + a_1 \frac{1}{kr} + a_2 \frac{1}{(kr)^2} + \cdots \right], \quad kr \to \infty \quad (1.7)$$

The amplitude a_0 and the exponent α are found by solving canonical problems with a closed form solution, whose asymptotic approximations are obtained with the aforementioned methods. These theories are variously known as the geometrical optics theory, physical optics theory, geometrical theory of diffraction, physical theory of diffraction, and so on. However, these high frequency theories usually become invalid at ray caustics (caustics are defined as points in space where rays meet or bunch together), and shadow boundaries (the boundary between the illuminated region and the shadow region of a wave). In such cases, the uniform asymptotic theory or uniform theory of diffraction has to be used. Many of these works are reported in [16, 32–38]. Recently, these methods have been numerically implemented as a shooting and bouncing ray [39] method, available in a code called XPATCH.

Another approximation method is the use of matched asymptotic expansions, which are used in fluid theory, but rarely in electromagnetics [40]. The solution of scattering by small particles by Rayleigh [41] is an excellent example of a solution by matched asymptotic expansion—the far field is expressed in terms of a dipole scattering field, while the near field is expressed in terms of the Laplacian field. The boundary condition on the particle surface is matched using the Laplacian field, but the far field radiation condition is met with the radiation field. In this manner, Rayleigh was able to predict that the scattering cross-section of small particles grows as the fourth power of frequency—the reason why the sky is blue, as high frequency lights are scattered more intensely by air molecules. However, it does not explain why the sky is not purple! To understand that, one will have to study Mie scattering [3] or Thompson scattering [42], both of which indicate that the scattering cross-section of a particle does not grow without bound with increasing frequency.

Though greatly enlarging the class of solvable problems, it was clear that even approximate methods prove insufficient in providing solutions for many engineering and scientific analyses. Soon after the advent of computers in the 1940s [43], scientists and engineers delved into numerical methods for solving fluid equations. The intractable problems found in fluid theory presage the need for electronic computers. Before the advent of electronic computers, human "computers" were used to design aerodynamic vehicles and calculate bomb trajectories. During the Second World War, "computers" meant a bevy of intelligent women with slide rules at hand, doing countless calculations that would advance the frontier in the science of developing the best aerodynamic vehicle. Due to the inherent nonlinearity in the equation of fluid, it was clear that the fluid theorists could not advance their solution technique without the help of the computers, while electromagnetic theorists, largely due to the linearity of the problem, still could sustain themselves with analytic tools.

The 1960s saw the development of many numerical methods to solve Maxwell's equations. The earliest of these were the finite-difference time-domain (FDTD) method developed by Yee [44] and the method of moments (MOM) developed by Harrington [45]. FDTD proposes to solve the partial differential equation (PDE) associated with Maxwell's equations directly, whereas MOM converts Maxwell's equations into integral equations using the principle of linear superposition and the Green's function. The ability to convert Maxwell's equations to integral equations assumes the linearity of Maxwell's equations, which may not be the case in some media. MOM evolves from Galerkin's method, and it is also similar to the boundary element method (BEM) developed in the structure community [46]. Many works have been devoted to extending MOM for solving 3D electromagnetic problems [47–51].

The finite element method, which was popular in the structure community, was soon ported over to the electromagnetics community for solving problems [52–54]. Many methods popular in the fluid community for solving PDEs are also being brought into the electromagnetics community for solving Maxwell's equations [55, 56]. In the past few decades, differential equation-based methods have become immensely popular because of their ease of implementation and their efficient use of memory. All numerical solutions to Maxwell's equations can be cast into solving an equivalent matrix equation. For the differential equation-based methods, these equivalent matrices are sparse, and for the case of FDTD, the method is matrix-free and mesh-free, as the equivalent matrix-vector product can be generated with some very simple computer operations. The mesh, being rectilinear, need not be stored. Moreover, it is an optimal algorithm in the sense that it generated $O(N^\alpha)$ numbers with $O(N^\alpha)$ operations. As a consequence, FDTD has become immensely popular, primarily by the effort of Taflove and many others [57–59]. The growth in the literature of the FDTD method has been tremendous and is too voluminous for us to report here. A Web site has been maintained to track the work in this area [60].

In contrast, the integral equation-based method gives rise to a smaller number of unknowns and is very versatile for geometry handling. The integral equation method gives rise to dense matrices that are expensive to store and to solve; however, the recent advent in fast solvers for integral equations has opened a new horizon for integral equation methods that did not previously exist.

1.3 MORE ON DIFFERENTIAL EQUATION SOLVERS

There is a distinct difference between differential equation solvers and integral equation solvers. In differential equation solvers, one solves for the field that permeates all of space. In contrast, integral equation solvers solve for the sources. For surface scattering phenomena, the equivalent currents reside on the surface or at the interface between regions, reducing the dimensionality of the problem. Hence, the unknown count in integral equation solvers can be much smaller than differential equation solvers for a problem of equivalent size in wavelengths.

Since the computer has only a finite amount of memory, to mimic an infinite space, differential equation solvers require the use of the absorbing boundary condition to emulate the radiation condition at infinity [18, 57]. The recent advent of a perfectly matched layer by Berenger [61] has spurred much activity in this area [62, 63], much of which is reported in [64–69, 71, 90].

In contrast, an integral equation can be derived with Green's functions satisfying the radiation condition, and hence providing solutions that satisfy the radiation condition. An alternative way to enforce the radiation condition in differential equation solvers is to truncate the solution region with boundary integrals, which in turn are solved with integral equation techniques [72–75].

In solving differential equations, one constructs a numerical grid on which a field is propagated—hence, a field propagates from point A to point B via a numerical grid. A small amount of error is often committed in this mode of field propagation, giving rise to a phase error in the field. This phase error is cumulative and becomes larger with larger simulation size. Consequently, one has to increase the grid density to mitigate this error. Therefore, the larger the problem size, the higher the grid density required in differential equation solvers [76–78]. In contrast, an integral equation solver propagates a field from point A to point B using an exact, closed-form Green's function. Hence, the problem of grid dispersion error is often less severe in integral equation solvers.

For a second-order accurate scheme such as the Yee scheme, the grid density in one dimension (number of grid points per wavelength) has to grow as $(kd)^{0.5}$ where d is the "diameter" of the simulation region, and k is the wavenumber [77]. Therefore, the total number of unknowns scales as $(kd)^{1.5}$ in one dimension. In two dimensions, it scales as $(kd)^3$, and in three dimensions, it scales as $(kd)^{4.5}$. A way to mitigate this growth in the unknown count is to use higher-order methods, or to hybridize a differential equation solver with an integral equation solver when a large region of homogeneity exists [79–81].

Fast direct solvers also exist for solving the matrix system resulting from differential equations. The pertinent sparse matrices can be solved in $O(N^2)$ operations in 2D, and $O(N^{1.67})$ operations in 3D. The nested dissection ordering method [82] permits the direct solution of the matrix system in $O(N^{1.5})$ operations in 2D and $O(N^{1.33})$ operations in 3D [83]. The inverse of a sparse matrix is usually a dense matrix. But the frontal algorithm can be used to greatly reduce the memory storage and computer time in some applications.

1.3.1 Convergence Rate of Iterative Differential Equation Solvers

Since differential equation solvers are associated with sparse matrices with $O(N)$ elements, a matrix-vector product is achieved in $O(N)$ operations. When the matrix system is solved iteratively, the only remaining avenue for accelerating differential equation solvers is to reduce the number of iterations needed or the number of time steps needed in time-stepping methods. The convergence rate in a matrix equation is determined by the condition number of the matrix, which is the ratio of its largest eigenvalue to its smallest eigenvalue. For the Laplace equation, the equivalent matrix can be always positive definite because the Laplacian operator is a negative definite operator with either the homogeneous Dirichlet boundary condition ($\phi(\mathbf{r}) = 0$) or the homogeneous Neumann boundary condition ($\hat{n} \cdot \nabla \phi(\mathbf{r}) = 0$). This can easily be shown using integration by parts:

$$\int_V d\mathbf{r} \phi^*(\mathbf{r}) \nabla^2 \phi(\mathbf{r}) = -\int_V d\mathbf{r} \nabla \phi^*(\mathbf{r}) \cdot \nabla \phi(\mathbf{r}) < 0 \tag{1.8}$$

Finding the eigenvalues of a Laplacian operator, which is equivalent to solving the equation:

$$-\nabla^2 \phi(\mathbf{r}) = \lambda \phi(\mathbf{r}) \tag{1.9}$$

is the same as solving a resonant problem involving the Helmholtz equation. For instance, in the 2D case, the Helmholtz equation is

$$\left[\nabla^2 + k^2\right] \phi(\mathbf{r}) = 0 \tag{1.10}$$

Hence, the largest eigenvalue corresponds to the highest spatial frequency supported by the mesh, which is proportional to $(1/\Delta)^2$ where Δ is the discretization size of the mesh. The lowest eigenvalue is the lowest spatial frequency supported by the region, which is proportional to $(1/L)^2$ where L is the length of the simulation region. Therefore, the condition number of the matrix system is proportional to $(L/\Delta)^2$. Since N is proportional to $(L/\Delta)^2$, the condition number is proportional to N.

In a symmetric, positive definite matrix system, the number of iterations is proportional to the square-root of the condition number. Therefore, the number of iterations is proportional to $O(N^{0.5})$. The computational complexity of solving a simple Laplace equation problem is then of $O(N^{1.5})$ [83].

The above simple result has a very nice physical interpretation for the fact that the number of iterations needed to solve a Laplace problem is roughly proportional to $O(N^{0.5})$. This follows from the fact that the differential equation operator gives rise to a sparse matrix system, which accounts for only near-neighbor interactions. Therefore, each matrix-vector product will send the information $O(1)$ grid points away. Therefore, it takes $O(N^{0.5})$ steps to send the information completely through the simulation region. In 3D, this translates to $O(N^{0.33})$ steps, yielding a computational complexity of $O(N^{1.33})$.

The above matrix system for 2D can be preconditioned to give a condition number proportional to $O(N^{0.5})$ yielding a final computational complexity of $O(N^{1.25})$. In 3D, with preconditioning, this reduces to $O(N^{1.17})$ [83].

The convergence of solving Helmholtz equation is harder to ascertain because we are solving a system in which

$$\left[\nabla^2 + k^2\right] \phi(\mathbf{r}) = s(\mathbf{r}) \tag{1.11}$$

In the above, since k^2 is positive definite and the Laplacian operator is negative definite, the resultant operator is indefinite, if the above is solved in a closed region with Neumann or Dirichlet boundary conditions. The indefiniteness comes about because of the possible resonances in the closed surface at which the eigenvalue of the above system is zero. As such, solving the Helmholtz equation iteratively for a closed cavity may not converge at all at the resonant frequencies. When translated to the FDTD method, this nonconvergence implies that it takes infinitely many time steps to solve the wave equation in a closed cavity.

However, this problem of resonance can be mitigated by letting the energy radiate to infinity which, on a computer, is mimicked by putting absorbing boundary conditions to absorb the energy at the edge of the simulation region. In this manner, the system will not have a zero eigenvalue, mitigating the ill convergence problem.

It is clear from the above that it is not possible to estimate the asymptotic computational complexity for solving a Helmholtz problem due to the possibility of encountering a resonance structure. However, we can regard the asymptotic complexity derived for the Laplace problem to be the lower bound for the Helmholtz problem.

The above physical picture of the convergence rate lends insight into designing methods to accelerate the convergence rate for the Laplace equation. Also, a notable feature of the Laplace equation is its scale-invariant nature. Therefore, one can design a coarser grid to propagate the information rapidly across the simulation region, and interpolate from a coarse grid to a fine grid, and vice versa, to capture high spatial frequency information for accuracy. A hierarchy of meshes of different coarseness can be built for this purpose. This is known as a multigrid method, and it can be shown that the number of iterations can be made independent of the size of the problem. In this manner, the Laplace problem can then be solved in $O(N)$ operations [83].

Unfortunately, there is a limitation on accelerating the solution time of the Helmholtz problem using the multigrid method. The solution to the Helmholtz equation is oscillatory, and the mesh cannot be made arbitrarily coarse due to the Nyquist sampling theorem, which says that at least two points per wavelength is needed to fully capture the information in an oscillatory function. We shall call this barrier to coarsification the "Nyquist barrier."

An iterative solver is also considered a Krylov subspace method. A series of matrix-vector products after M times, starting with a seed vector $\mathbf{x}_0$, generates a subspace spanned by these vectors:

$$\mathcal{K} = \text{span}\left[\mathbf{x}_0, \overline{\mathbf{A}} \cdot \mathbf{x}_0, (\overline{\mathbf{A}})^2 \cdot \mathbf{x}_0, \cdots, (\overline{\mathbf{A}})^M \cdot \mathbf{x}_0\right] \quad (1.12)$$

The numerical solution to the equation $\overline{\mathbf{A}} \cdot \mathbf{x} = \mathbf{b}$ is then sought approximately in this subspace [84, 85].

A notable Krylov subspace method is the spectral Lanczos decomposition method (SLDM) [86]. While it is of the same complexity as the conjugate gradient method, it can be used to generate solutions for all frequencies without additional work. This method is also particularly suitable for solving differential equations where sections of uniformity exist, and the method of lines [87] or the numerical mode matching method can be used [88, 89].

Yet another differential equation solver is the spectral method, in which the solution is expanded in terms of Fourier bases, or orthogonal bases [90, 91]. Due to the natural smoothness of the solution to the Helmholtz equation, this method attains higher-order convergence rapidly, and has been shown to reduce grid-dispersion error.

The above discussions were for Laplace and Helmholtz equations, but we can regard the property of static electromagnetic fields to be Laplacian in nature, and that of dynamic electromagnetic fields to be Helmholtz in nature. The Laplace equation is termed elliptic by mathematicians [83], while the Helmholtz equation is related to the wave equation which is hyperbolic. A diffusion equation, which is midway between elliptic and hyperbolic, is termed parabolic.

Introduction to Electromagnetic Analysis and CEM   13

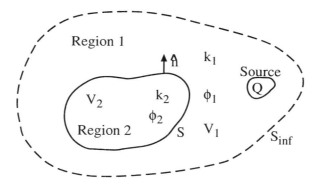

Figure 1.5 The geometry for deriving the integral equation of scattering.

1.4 INTEGRAL EQUATION SOLVERS

The integral equation of scattering can be derived using the Green's function. For electromagnetics, the integral equation is called the Stratton-Chu integral equation [92]. For the sake of clarity, we will present the derivation of surface integral equations for scalar waves. Figure 1.5 is an illustration of the problem.

1.4.1 Surface Integral Equations

Given two equations

$$[\nabla^2 + k_1^2]\phi_1(\mathbf{r}) = Q(\mathbf{r}) \tag{1.13}$$

$$[\nabla^2 + k_1^2]g_1(\mathbf{r}, \mathbf{r}') = -\delta(\mathbf{r} - \mathbf{r}') \tag{1.14}$$

one can multiply the first by $g_1(\mathbf{r}, \mathbf{r}')$ and the second by $\phi_1(\mathbf{r})$ and subtract the two equations. Upon doing so, and integrating the result over V_1, we obtain

$$\int_{V_1} dV \left[g_1(\mathbf{r}, \mathbf{r}')\nabla^2 \phi_1(\mathbf{r}) - \phi_1(\mathbf{r})\nabla^2 g_1(\mathbf{r}, \mathbf{r}')\right]$$

$$= \int_{V_1} dV\, g_1(\mathbf{r}, \mathbf{r}')Q(\mathbf{r}) + \phi_1(\mathbf{r}'), \quad \mathbf{r}' \in V_1 \tag{1.15}$$

We notice that $\nabla \cdot (g\nabla\phi - \phi\nabla g) = g\nabla^2\phi - \phi\nabla^2 g$. Hence, Gauss' divergence theorem can be used to convert the volume integral to a surface integral yielding

$$-\int_{S+S_{inf}} dS\, \hat{n} \cdot [g_1(\mathbf{r}, \mathbf{r}')\nabla\phi_1(\mathbf{r}) - \phi_1(\mathbf{r})\nabla g_1(\mathbf{r}, \mathbf{r}')]$$

$$= -\phi_{inc}(\mathbf{r}') + \phi_1(\mathbf{r}'), \quad \mathbf{r}' \in V_1 \tag{1.16}$$

where
$$\phi_{inc}(\mathbf{r}') = -\int_{V_1} dV\, g_1(\mathbf{r},\mathbf{r}')Q(\mathbf{r}) \tag{1.17}$$

The surface integral at infinity can be shown to vanish, finally yielding (after swapping $\mathbf{r}$ and $\mathbf{r}'$)

$$\phi_1(\mathbf{r}) = \phi_{inc}(\mathbf{r}) - \int_S dS'\, [g_1(\mathbf{r},\mathbf{r}')\partial_{n'}\phi_1(\mathbf{r}') - \phi_1(\mathbf{r}')\partial_{n'} g_1(\mathbf{r},\mathbf{r}')], \quad \mathbf{r} \in V_1 \tag{1.18}$$

where $\partial_{n'} = \hat{n}' \cdot \nabla'$. If the observation point is inside V_2, the left-hand side would have been zero, and we obtain

$$\left. \begin{array}{c} \mathbf{r} \in V_1, \quad \phi_1(\mathbf{r}) \\ \mathbf{r} \in V_2, \quad 0 \end{array} \right\} = \phi_{inc}(\mathbf{r}) - \int_S dS'\, [g_1(\mathbf{r},\mathbf{r}')\partial_{n'}\phi_1(\mathbf{r}') - \phi_1(\mathbf{r}')\partial_{n'} g_1(\mathbf{r},\mathbf{r}')] \tag{1.19}$$

The above implies that the field outside the scatterer consists of the incident field $\phi_{inc}(\mathbf{r})$ and a scattered field generated by equivalent sources $\partial_{n'}\phi_1(\mathbf{r}')$ and $\phi_1(\mathbf{r}')$, where $\phi_1(\mathbf{r})$ satisfies the Helmholtz wave equation. The former is a charge source while the latter corresponds to a double layer (dipole layer) source. However, if the observation point is placed in V_2, then the field is identically zero—this is known as the *extinction theorem* [93].

A similar equation can be derived in Region 2 to obtain

$$\left. \begin{array}{c} \mathbf{r} \in V_2, \quad \phi_2(\mathbf{r}) \\ \mathbf{r} \in V_1, \quad 0 \end{array} \right\} = \int_S dS'\, [g_2(\mathbf{r},\mathbf{r}')\partial_{n'}\phi_2(\mathbf{r}') - \phi_2(\mathbf{r}')\partial_{n'} g_2(\mathbf{r},\mathbf{r}')] \tag{1.20}$$

By imposing the above equations in S^+ and S^-, where the former is just outside S and the latter is just inside S, we obtain the integral equation of scattering for a penetrable body:

$$\mathbf{r} \in S^-, \quad \phi_{inc}(\mathbf{r}) = \int_S dS'\, [g_1(\mathbf{r},\mathbf{r}')\partial_{n'}\phi_1(\mathbf{r}') - \phi_1(\mathbf{r}')\partial_{n'} g_1(\mathbf{r},\mathbf{r}')] \tag{1.21}$$

$$\mathbf{r} \in S^+, \quad 0 = \int_S dS'\, [g_2(\mathbf{r},\mathbf{r}')\partial_{n'}\phi_2(\mathbf{r}') - \phi_2(\mathbf{r}')\partial_{n'} g_2(\mathbf{r},\mathbf{r}')] \tag{1.22}$$

When a homogeneous Dirichlet boundary condition or homogeneous Neumann boundary condition is imposed on the surface S, the integral equation of scattering reduces to

$$\phi_{inc}(\mathbf{r}) = \int_S dS'\, g_1(\mathbf{r},\mathbf{r}')\partial_{n'}\phi_1(\mathbf{r}'), \quad \mathbf{r} \in S \tag{1.23}$$

for the homogeneous Dirichlet case, and

$$\phi_{inc}(\mathbf{r}) = -\int_S dS' \, \phi_1(\mathbf{r}') \partial_{n'} g_1(\mathbf{r}, \mathbf{r}'), \quad \mathbf{r} \in S^- \quad (1.24)$$

for the homogeneous Neumann case.

1.4.2 The Internal Resonance Problem[2]

The integral equations (1.23) and (1.24) suffer from the nonuniqueness problem at the internal resonant frequencies of a cavity formed by the surface S. This fact can be proven by deriving the integral equation for the internal resonance of the cavity formed by S.

Let us consider the field inside the surface S due to sources on S. Assume that the volume inside S is now filled with a medium with wavenumber k_1. Then by the same reasoning as with (1.20), we have

$$\phi_1(\mathbf{r}) = \int_S dS' \, [g_1(\mathbf{r}, \mathbf{r}') \partial_{n'} \phi(\mathbf{r}') - \phi_1(\mathbf{r}') \partial_{n'} g_1(\mathbf{r}, \mathbf{r}')], \quad \mathbf{r} \in V_2 \quad (1.25)$$

The field inside V_2 is now termed $\phi_1(\mathbf{r})$ since the wavenumber is k_1 and the corresponding Green's function is $g_1(\mathbf{r}, \mathbf{r}')$. If the cavity has a boundary condition such that $\phi_1(\mathbf{r}) = 0, \mathbf{r} \in S$, then the above becomes

$$0 = \int_S dS' g_1(\mathbf{r}, \mathbf{r}') \partial_{n'} \phi_1(\mathbf{r}'), \quad \mathbf{r} \in S \quad (1.26)$$

At the internal resonant frequencies of the cavity, the above integral equation has a nontrivial solution. The integral operator in (1.26) is precisely that in (1.23). Hence, the integral operator in (1.23) has a null space at the internal resonant frequencies of the cavity. Therefore, the null-space solutions of the exterior Dirichlet problem are due to the internal resonance of the interior Dirichlet problem.

Strangely enough, the null-space solution of the exterior Neumann problem (1.24) is still due to the null-space solution of the interior Dirichlet problem. By the extinction theorem, both (1.25) and (1.26) evaluate to zero for $\mathbf{r} \in V_1$. Keeping $\mathbf{r} \in V_1$, we can take the normal derivative of (1.26) with respect to $\mathbf{r}$ to obtain

$$0 = \int_S dS' \partial_n g_1(\mathbf{r}, \mathbf{r}') \partial_{n'} \phi_1(\mathbf{r}'), \quad \mathbf{r} \in S^+ \quad (1.27)$$

where $\mathbf{r} \in S^+$ so that $\mathbf{r} \in V_1$. The above has a nontrivial solution at the same resonant frequencies as in (1.26). Moreover, it can be shown that the integral operator in (1.27)

---

[2]This section can be skipped on first reading.

is the transpose of the integral operator in (1.24). Therefore, when (1.27) is discretized into a matrix operator, for example by Galerkin's method, its null space is the left null space of the corresponding matrix equation derived from (1.24). A matrix operator with a left null space is ill-conditioned, and hence (1.24) becomes ill-conditioned at the same internal resonant frequencies as (1.23).

The remedy to the nonuniqueness due to the internal resonances is to use the combined field formulation known in the electromagnetics literature as the combined field integral equation (CFIE) [94]. A list of papers on the topic can be found in [18, 95]. In this case, one imposes (1.23) for $\mathbf{r} \in S^-$ and takes its normal derivative to obtain a new equation:

$$\partial_n \phi_{inc}(\mathbf{r}) = \int_S dS' \partial_n g_1(\mathbf{r}, \mathbf{r}') \partial_{n'} \phi_1(\mathbf{r}'), \qquad \mathbf{r} \in S^- \qquad (1.28)$$

Then (1.23) is combined with (1.28) to obtain

$$\mathcal{P} \phi_{inc}(\mathbf{r}) = \int_S dS' \mathcal{P} g_1(\mathbf{r}, \mathbf{r}') \partial_{n'} \phi_1(\mathbf{r}'), \qquad \mathbf{r} \in S^- \qquad (1.29)$$

where $\mathcal{P} = 1 + \lambda \partial_n$, and λ is an appropriately chosen combination parameter to eliminate the null-space solution of (1.29).

It can be proven that with an appropriate choice of λ, (1.29) does not have a null-space solution at the internal resonances of the cavity. To show this, we return to the interior cavity problem (1.25) and impose an impedance boundary condition

$$\phi_1(\mathbf{r}') = -\lambda \partial_{n'} \phi_1(\mathbf{r}'), \quad \mathbf{r}' \in S \qquad (1.30)$$

By imposing the above boundary condition for $\mathbf{r} \in S^+$, we have

$$0 = \int_S dS' \left[g_1(\mathbf{r}, \mathbf{r}') + \lambda \partial_{n'} g_1(\mathbf{r}, \mathbf{r}') \right] \partial_{n'} \phi_1(\mathbf{r}'), \quad \mathbf{r} \in S^+ \qquad (1.31)$$

The above does not have a nontrivial solution when λ is chosen to be a complex or imaginary number. This can be seen by integrating the following equation:

$$\phi_1^* \left(\nabla^2 + k_1^2 \right) \phi_1 = 0 \qquad (1.32)$$

over the volume V_2 with the boundary condition (1.30) to obtain

$$-\lambda^{-1} \int_S dS |\phi_1|^2 - \int_{V_2} d\mathbf{r} |\nabla \phi_1|^2 + k_1^2 \int_{V_2} d\mathbf{r} |\phi_1|^2 = 0 \qquad (1.33)$$

When λ is a complex number, the only way the above can be satisfied is for k_1^2 to be complex. Hence, the equation $\left(\nabla^2 + k_1^2 \right) \phi_1(\mathbf{r}) = 0$, $\mathbf{r} \in V_2$, with the boundary condition (1.30) can only admit solutions with complex resonant frequencies. Equation (1.30) with a complex λ implies either a dissipative or active impedance boundary condition, giving rise to complex resonances.

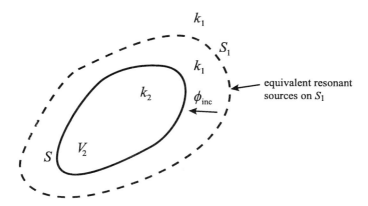

Figure 1.6 In the Gedanken experiment to prove the existence of a null-space solution for penetrable scatterer case, we assume that the scatterer is impinged upon by an incident wave generated by a resonant source impressed on S_1.

1.4.2.1 The Penetrable Scatterer Case

Surprisingly enough, even the integral equation (1.21) and (1.22) for a penetrable scatterer is not immune to the internal resonance problem. This can be seen by the following *Gedanken* experiment (thought experiment), which indicates that (1.21) and (1.22) has a nontrivial solution even when $\phi_{inc}(\mathbf{r}) = 0$ (Figure 1.6).

Let us first enclose S within a larger surface S_1. The surface S_1 has an internal resonance for the interior Dirichlet problem where $\phi(\mathbf{r}) = 0$, $\mathbf{r} \in S_1$. We first assume that the scatterer V_2 is absent. By the equivalence principle, we can put equivalent resonant sources on S_1 at its resonant frequency that generates a field within S_1 and a zero field everywhere outside V_2. Next, we insert scatterer V_2 into S_1, and the scatterer will scatter the quiescent field inside S_1, generating a scattered field everywhere outside V_2. We then shrink S_1 gradually so that it becomes S, also adjusting the frequency and the equivalent sources so that the sources are always equivalent to the internal resonance problem of S_1. During this shrinkage, the field thus generated will always have $\phi_{inc}(\mathbf{r}) = 0$, $\mathbf{r} \in S_1$. There will always be a scattered field outside the scatterer V_2.

When S_1 collapses exactly onto S, $\phi_{inc}(\mathbf{r}) = 0$, $\mathbf{r} \in S$, and $\mathbf{r} \in S_1$ simultaneously, but there will still be a scattered field since $\phi_{inc}(\mathbf{r}) \neq 0$, $\mathbf{r} \in V_2$. Therefore, we have a case of having a scattered field solution when $\phi_{inc}(\mathbf{r}) = 0$, $\mathbf{r} \in S$, in (1.21) and (1.22). This proves that (1.21) and (1.22) have a nontrivial solution even when $\phi_{inc}(\mathbf{r}) = 0$, $\mathbf{r} \in S$. The remedy for this problem is again to use a combined field integral equation by taking the normal derivative of (1.21) and (1.22) and combining the result with itself.

1.4.2.2 PMCHWT Formulation

Another way to avoid the internal resonance problem is to use the PMCHWT formulation [96–98]. Commonly named the PMCHW formulation to credit the authors Poggio, Miller, Chang, Harrington, Wu, and Tsai, it should rightfully be called PMCHWT formulation. It was originally developed for electromagnetic scattering problems, but we will present it using the scalar wave theory to be consistent with the other sections.

In this method, (1.19) is first evaluated for $\mathbf{r} \in S$ by taking its limit from $\mathbf{r} \in V_1$ to obtain

$$\frac{1}{2}\phi_1(\mathbf{r}) = \phi_{inc}(\mathbf{r}) - \fint_S dS' \left[g_1(\mathbf{r},\mathbf{r}')\partial_{n'}\phi_1(\mathbf{r}') - \phi_1(\mathbf{r}')\partial_{n'}g_1(\mathbf{r},\mathbf{r}')\right] \quad (1.34)$$

and similarly, from (1.20) we have

$$\frac{1}{2}\phi_2(\mathbf{r}) = \fint_S dS' \left[g_2(\mathbf{r},\mathbf{r}')\partial_{n'}\phi_2(\mathbf{r}') - \phi_2(\mathbf{r}')\partial_{n'}g_2(\mathbf{r},\mathbf{r}')\right] \quad (1.35)$$

where $\fint$ implies a principal value integral.

By assuming that $\phi_1(\mathbf{r}) = \phi_2(\mathbf{r}), p_1\partial_n\phi_1(\mathbf{r}) = p_2\partial_n\phi_2(\mathbf{r}), \mathbf{r} \in S$, is the interface boundary condition, we have

$$\phi_{inc}(\mathbf{r}) = \fint_S dS' \left\{ \left[g_1(\mathbf{r},\mathbf{r}') + \frac{p_2}{p_1}g_2(\mathbf{r},\mathbf{r}')\right] \partial_{n'}\phi_1(\mathbf{r}') \right. \\ \left. - \left[\partial_{n'}g_1(\mathbf{r},\mathbf{r}') + \partial_{n'}g_2(\mathbf{r},\mathbf{r}')\right] \phi_1(\mathbf{r}') \right\} \quad (1.36)$$

By taking the normal derivative of (1.19) and (1.20) and letting $\mathbf{r} \to S$, we obtain a second equation:

$$p_1\partial_n\phi_{inc}(\mathbf{r}) = \fint_S dS' \left\{ \left[\partial_n g_1(\mathbf{r},\mathbf{r}') + \partial_n g_2(\mathbf{r},\mathbf{r}')\right] p_1\partial_{n'}\phi_1(\mathbf{r}') \right. \\ \left. - \left[p_1\partial_n\partial_{n'}g_1(\mathbf{r},\mathbf{r}') + p_2\partial_n\partial_{n'}g_2(\mathbf{r},\mathbf{r}')\right] \phi_1(\mathbf{r}') \right\} \quad (1.37)$$

The above are two equations for the two surface unknowns $\phi_1(\mathbf{r})$ and $\partial_n\phi_1(\mathbf{r})$. If we go through the same Gedanken experiment as we did in in the previous subsection, the left-hand side does not go to zero anymore, and hence, the proof that a null-space solution exists does not hold here. Experience has shown that the above equations are well conditioned even at the internal resonance frequencies of a cavity formed by S.

1.4.3 Volume Integral Equation

The volume integral equation is for the inhomogeneous medium problem. Its derivation is best illustrated with the scalar wave equation. Given a scalar wave equation

$$\left[\nabla^2 + k^2(\mathbf{r})\right]\phi(\mathbf{r}) = s(\mathbf{r}) \tag{1.38}$$

where $k^2(\mathbf{r})$ is a function of position, we can rewrite the above as

$$\left[\nabla^2 + k_0^2\right]\phi(\mathbf{r}) = \left[k_0^2 - k^2(\mathbf{r})\right]\phi(\mathbf{r}) + s(\mathbf{r}) \tag{1.39}$$

Using the Green's function, which is a solution to

$$\left[\nabla^2 + k_0^2\right]g(\mathbf{r}, \mathbf{r}') = -\delta(\mathbf{r} - \mathbf{r}') \tag{1.40}$$

we can write, by the principle of linear superposition, the solution to (1.39), namely,

$$\phi(\mathbf{r}) = \phi_{inc}(\mathbf{r}) + \int_V d\mathbf{r}' g(\mathbf{r}, \mathbf{r}')\left[k^2(\mathbf{r}) - k_0^2\right]\phi(\mathbf{r}) \tag{1.41}$$

where $\phi_{inc}(\mathbf{r}) = -\int_V d\mathbf{r}' g(\mathbf{r}, \mathbf{r}')s(\mathbf{r}')$. The above is known as the volume integral equation, or the volume current integral equation. A similar electromagnetic volume integral equation can be derived.

1.4.4 Green's Function

The Green's function is essential for the derivation of the integral equation, warranting its discussion here. The Green's function for the scalar Helmholtz equation satisfies

$$[\nabla^2 + k^2]g(\mathbf{r}, \mathbf{r}') = -\delta(\mathbf{r} - \mathbf{r}') \tag{1.42}$$

which is equivalent to the response to a point source at $\mathbf{r}'$. The Green's function in 3D can be derived to be

$$g(\mathbf{r}, \mathbf{r}') = \frac{e^{ik|\mathbf{r}-\mathbf{r}'|}}{4\pi|\mathbf{r} - \mathbf{r}'|} \tag{1.43}$$

In general, for a homogeneous medium Green's function, $g(\mathbf{r}, \mathbf{r}') = g(\mathbf{r} - \mathbf{r}')$, and $g(\mathbf{r}, \mathbf{r}') = g(\mathbf{r}', \mathbf{r})$.

The Green's function in 2D is

$$g(\boldsymbol{\rho}, \boldsymbol{\rho}') = \frac{i}{4}H_0^{(1)}(k|\boldsymbol{\rho} - \boldsymbol{\rho}'|) \tag{1.44}$$

where $\boldsymbol{\rho} = \hat{x}x + \hat{y}y$ and $\boldsymbol{\rho}' = \hat{x}x' + \hat{y}y'$.

For the Laplace equation, the Green's function in 3D is

$$g(\mathbf{r}, \mathbf{r}') = \frac{1}{4\pi|\mathbf{r} - \mathbf{r}'|} \tag{1.45}$$

and in 2D, it is
$$g(\rho, \rho') = \frac{1}{2\pi} \ln(|\rho - \rho'|) \tag{1.46}$$

The above Green's function appears as the kernel in the integral equation of scattering. Notice that the Green's function is oscillatory for the Helmholtz equation, while it is nonoscillatory for the Laplace equation. This has both a mathematical and a physical significance. The oscillatory nature of the Green's function implies that the higher derivatives of the Green's function become larger, implying the nonsmoothness of the kernel even away from the source point. In contrast, the higher derivatives of the Laplace Green's function becomes smaller as one recedes from the source point. Consequently, the Laplacian kernel loses information content away from the source point, but this is not the case for the Helmholtz kernel, which carries information to infinity.

Physically, the above implies that the Laplace kernel gives rise to short range interaction whereas the Helmholtz kernel yields long range interaction. Consequently, a vast amount of information can be carried by electromagnetic waves. For instance, we can see the craters on the surface of the moon using optical signals, but we cannot delineate details of the moon if we have only the knowledge of the gravitational field of the moon—which is a static field.

1.4.5 Method of Moments

The method of moments (MOM) was developed to discretize the integral equation, or in general to convert a linear operator equation to a matrix equation. This method is also known as the Petrov-Galerkin method [99]. Given a linear operator equation

$$\mathcal{L}f = g \tag{1.47}$$

where $\mathcal{L}$ is a linear operator, mapping functions in the domain space D to the range space R. We can first approximate f with a set of basis functions that can approximate the function in the domain D, namely,

$$f = \sum_{n=1}^{N} a_n f_n \tag{1.48}$$

Substituting the above into (1.47), we have

$$\sum_{n=1}^{N} a_n \mathcal{L} f_n = g \tag{1.49}$$

Multiplying the above by t_m that approximately spans the range R and integrating over the the range space, we have

$$\sum_{n=1}^{N} a_n \langle t_m, \mathcal{L} f_n \rangle = \langle t_m, g \rangle \quad m = 1, \cdots, N \tag{1.50}$$

where t_m is known as the testing function. In the above, we define the inner product as

$$\langle t, f \rangle = \int_S dS\, t(\mathbf{r}) f(\mathbf{r}) \tag{1.51}$$

1.4.6 Fast Integral Equation Solvers

There are essentially two ways to solve the ensuing matrix equation (1.50) that result from the integral equation by the method of moments: (1) a direct solver that seeks the inversion of the matrix equation, and (2) an iterative solver for the solution of the matrix equation. A direct solver of the matrix equation can be Gaussian elimination, or lower-upper-triangular decomposition (LUD). The number of operations is proportional to $O(N^3)$, while the matrix storage requirement is of $O(N^2)$. An iterative method to solve a matrix equation can either be Gauss-Seidel, Jacobi relaxation, conjugate gradient, or the derivatives of the conjugate gradient algorithm [83]. All iterative methods require the performance of a matrix-vector product, which usually is the bottleneck of the computation. For dense matrices, such a matrix-vector product requires operations of $O(N^2)$. If the matrix equation is solved in N_{iter} iterations, the computational cost is proportional to $N_{iter}N^2$. Moreover, traditional iterative solvers require the generation and storage of the matrix itself.

For large scale computing, it is imperative that the memory requirements have a low complexity. Therefore, the $O(N)$ or $O(N \log N)$ memory requirement of iterative fast solvers is the best path to large scale computing at this point. Historically, fast solvers for static problems have been proposed for simulating interstellar interaction in astronomy, since stars interact via gravitational fields, which are Laplacian [100, 101]. Rokhlin and Greengard [102] have proposed a fast multipole method in 1987 which has caught on in popularity in a number of communities. The method is matrix-free, allowing a matrix-vector product to be effected in $O(N)$ operations with $O(N)$ memory requirement.

For dynamic problems, one needs to solve either the Helmholtz or the Maxwell problem. Due to the oscillatory nature of the solution because it describes a wave phenomenon, fast methods to solve dynamic problems did not surface until later. During 1989–1993, some fast direct solvers such as RATMA and NEPAL [103, 104, 106] were proposed by our research group, but they were abandoned in favor of iterative solvers that consume less memory. In 1990, Rokhlin proposed a fast algorithm for the Helmholtz equation [107], which was followed by other works [108–110] and later our group plus others also developed a multilevel version of it during 1994–1995, now commonly known as the multilevel fast multipole algorithm (MLFMA) [111–113]. MLFMA allows a matrix-vector product to be effected in $O(N \log N)$ operations for matrix equations resulting from the method of moments. Moreover, it is almost matrix-free, requiring a memory of $O(N \log N)$. Recently, MLFMA has been demonstrated to solve an equivalent dense matrix system resulting from MOM with close to 10 million unknowns, and a full-size aircraft at 8 GHz with a length of 400 wavelengths. A scattering solution using MOM with 9.6 million unknowns has been compared to a Mie series solution for a sphere of 120 wavelengths in diameter [114]. It seems to have been the only path for large-scale computing for the Helmholtz and the Maxwell problems. Moreover, MLFMA has recently been extended by Michielssen, et al. to the time domain called plane-wave time-domain (PWTD) method [115]. A variant of it has also been developed by Brennan and Cullen [116, 117].

Other fast solvers for the Helmholtz equation are the impedance matrix localization (IML) method proposed by Canning [118] and the matrix decomposition method proposed by Michielssen and Boag [119]. The wavelet decomposition method has been proposed to compress or sparsify matrices resulting from solving the Laplace, Helmholtz, or Maxwell problems [120–129]. For Laplace problems, due to the infinitely smooth nature of the static Green's function which forms the kernel of the integral equation, the resultant matrix can be compressed or sparsified to having $O(N \log N)$ elements. As a consequence, a matrix-vector product of the compressed matrix can be effected in $O(N \log N)$ operations. However, the wavelet compression technique applied to integral equations for the Helmholtz problem faces the same problem as the multigrid method applied to the Helmholtz equation for the differential equation case. The Nyquist barrier or the oscillatory nature of the Green's function, which forms the kernel for the integral equation, precludes the compression of the matrix equation beyond a certain limit. Hence, a matrix sparsity is still of $O(N^2)$, failing to reduce the computational complexity of a matrix-vector multiply. Recently, the adaptive wavelet packets have been specially designed to further reduce the complexity of a matrix-vector product [128, 129].

Other fast solvers are FFT-based where FFT is used to perform a matrix-vector product in $O(N \log N)$ operations [130–144]. This method exploits the convolutional nature of the integral equation operator, implying that a fast convolution can be performed in the Fourier space by a volumetric Fourier transform. This method is particularly suited for the volume integral equation as the unknowns are tightly packed in space. For surface scatterers, the unknown currents are sparsely distributed through space, making the FFT-based method inefficient.

1.5 A SIMPLIFIED VIEW OF THE MULTILEVEL FAST MULTIPOLE ALGORITHM

In this section, we will give a very simplistic and pedestrian view of the multilevel fast multipole algorithm. However, many details are sacrificed in order to elucidate the salient features of the algorithm—the accurate description of the algorithm is compromised in order to reduce the description of the algorithm to its simplest.

Introduction to Electromagnetic Analysis and CEM

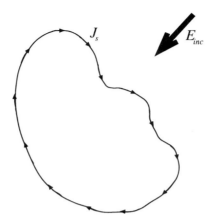

Figure 1.7 When an incident field impinges on a scatterer, currents are induced on the scatterer. The induced currents cooperate with each other to generate a secondary field that cancels the incident field on the surface of the scatterer to satisfy the boundary condition.

When solving an integral equation of scattering, the pertinent equation is

$$\phi_{inc}(\mathbf{r}) = \int_S dS'\, g(\mathbf{r},\mathbf{r}')j(\mathbf{r}'), \quad \mathbf{r} \in S \tag{1.52}$$

where we have replaced the $\partial_n \phi(\mathbf{r})$ in 1.24 with $j(\mathbf{r})$. The physical essence of the above equation says that the current $j(\mathbf{r})$ produces a field via the Green's function $g(\mathbf{r},\mathbf{r}')$ that cancels the incident field on the surface and in the scatterer (Figure 1.7). The Green's function is not a local function, but a function that connects two points irrespective of their separation. Therefore, all the current sources on the scatterer cooperate with each other to produce a field that cancels the incident field inside the scatterer. In a numerical discretization of this problem, the current is described by discrete elements on the surface of the scatterer. The cooperative behavior of the current elements can be likened to a telephone network consisting of N telephones where each telephone is connected to every other telephone by direct wire connections. As illustrated in Figure 1.8, the number of wires needed to set up such a telephone network requires $O(N^2)$ telephone lines.

However, the telephone company knows better. The number of telephone lines needed to connect N telephones can be greatly reduced using the hub system. As illustrated in Figure 1.9, telephones are first divided into groups according to their proximity to each other, and those within the same group are connected to a single hub. Then, wires are used to connect the hubs together. In this manner, the number of telephone lines needed to connect all telephones together is greatly reduced.

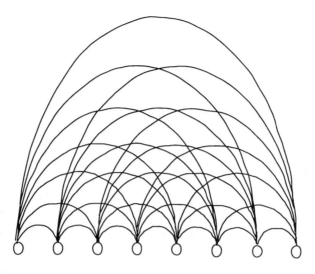

Figure 1.8 A one-level matrix-vector multiply where all current elements talk directly to each other. The number of "links" is proportional to N^2 where N is the number of current elements.

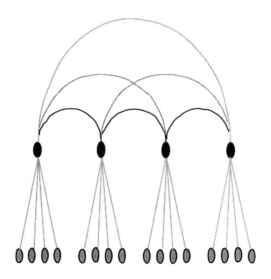

Figure 1.9 A two-level matrix-vector multiply where "hubs" are established to reduce the number of direct "links" between the current elements. This could potentially reduce the complexity of a matrix-vector multiply.

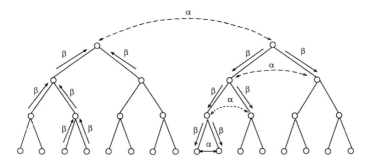

Figure 1.10 A tree structure showing the aggregation and the disaggregation procedure to form a multilevel algorithm.

A matrix element A_{ij} is a conduit to transmit field information from current element j to current element i in a single step. In order to facilitate a two-step field information transmission procedure, a matrix element A_{ij} has to be factorized according to

$$A_{ij} = \mathbf{V}_{il}^t \cdot \overline{\mathbf{T}}_{ll'} \cdot \mathbf{V}_{l'j} \tag{1.53}$$

This factorization can be achieved using the translational addition theorem to factorize the Green's function. The first factor $\mathbf{V}_{l'j}$ is responsible for information transmission between point j and hub l'. The factor $\overline{\mathbf{T}}_{ll'}$ transmits information between hubs l' and l. Finally, the factor $\mathbf{V}_{il}^t$ disseminates the information from hub l to point j. Consequently, a matrix-vector product is replaced by

$$\sum_{j=1}^N A_{ij} x_j = \mathbf{V}_{il}^t \cdot \left(\sum_{l'=1}^{N_g} \overline{\mathbf{T}}_{ll'} \cdot \left(\sum_{j \in G_{l'}} \mathbf{V}_{l'j} x_j \right) \right), \ i \in G_{l'}, \ l = 1, \cdots, N_g \tag{1.54}$$

Notice that in the above, a scalar number A_{ij} has been exploded to a vector times a matrix followed by a multiplication of another vector. It seems that a simple information transmission method has been replaced by a more cumbersome one. However, if the matrix $\overline{\mathbf{T}}_{ll'}$ is made diagonal, making the information transmission between two groups a lot more efficient, a reduction in computational labor is possible. This is because the matrix $\overline{\mathbf{T}}_{ll'}$ is reused for information transmission between all elements residing in two groups. The diagonal factorization of the Green's function for the Helmholtz problem was first achieved by Rokhlin in 1990 [107].

The above idea can be logically extended to multilevel just as a telephone network:

$$A_{ij} = \mathbf{V}_{il_1}^t \cdot \overline{\beta}_{l_1 l_2} \cdots \overline{\beta}_{l_{L-1} l_L} \cdot \overline{\mathbf{T}}_{l_L, l'_L} \cdot \overline{\beta}_{l'_L l'_{L-1}} \cdots \overline{\beta}_{l'_2 l'_1} \cdot \mathbf{V}_{l'_1 j} \tag{1.55}$$

In the above, the $\overline{\beta}$ matrices are also diagonal matrices. As such, the above matrices are all the same size. However, a straightforward application of the above factorization does not yield an algorithm of $O(N \log N)$ complexity. Interpolation and anterpolation operators [145] have to be added between levels in order to arrive at an $O(N \log N)$ for fast matrix-vector product [111, 112].

1.6 CONCLUSION

This book documents recent advances in fast and efficient methods in computational electromagnetics. There are methods for both differential equation solvers as well as integral equation solvers. Methods for both frequency and time domain will be presented. Hybridization methods will be discussed, as well as methods to solve Maxwell's equations stably from static to microwave frequencies.

Despite the voluminous amount of work presented here, we believe that much work still needs to be done in computational electromagnetics so that some day, it will provide the same confidence level in electromagnetic simulation as circuit theory has provided in the microchip industry. Hopefully, this book will serve as an inspiration for workers in the field to advance the frontier of knowledge in this area.

Other literature on the subject matter can be found in [146–151].

REFERENCES

1. P. M. Harman (ed.), *The Scientific Letters and Papers of James Clerk Maxwell, Vol. II, 1862-1873*, Cambridge, U.K.: Cambridge University Press, 1995.

2. I. Shelton, O. Duhalde, and A. Jones, *Int. Astro. Union Circ.*, p. 4316, 1987.

3. G. Mie, "Beiträge zur optik trüber medien speziel kolloidaler metallösungen," *Ann. Phys. (Leipzig)*, vol. 25, p. 377, 1908.

4. P. Debye, *Ann. d. Physik*, vol. 4, no. 30, p. 57, 1909.

5. J. W. Strutt Rayleigh (Lord Rayleigh), "On the passage of electric waves through tubes, or the vibra cylinder," *Phi. Mag.*, vol. 43, pp. 125–132, 1897.

6. A. Sommerfeld, "Mathematische Theorie der Diffraction," *Mathematische Annalen*, vol. 47, no. s319, pp. 317–374, 1896.

7. A. Sommerfeld, "Uber die Ausbreitung der Wellen in der drahtlosen Telegraphie," *Ann. Physik*, vol. 28, pp. 665–737, 1909.

8. R. S. Brodkey, *The Phenomena of Fluid Motions*, New York: Dover Publ., 1995.

9. J. W. Strutt Rayleigh (Lord Rayleigh), *Theory of Sound*, New York: Dover Publ., 1976. (Originally published 1877.)

10. L. J. Chu and J. A. Stratton, "Forced oscillations of a prolate spheroid," *J. Appl. Phys.*, vol. 12, pp. 241–248, 1941.

11. N. Marcuvitz, "Field representations in spherically stratified regions," *Comm. Pure and Appl. Math.*, vol. 4, pp. 263–315, 1951.

12. J. R. Wait, "Radiation from a vertical antenna over a curved stratified ground," *J. Res. NBS*, vol. 56, p. 237, 1956.

13. J. J. Bowman, T. B. A. Senior, and P. L. E. Uslenghi, *Electromagnetic and Acoustic Scattering by Simple Shapes*, Amsterdam: North-Holland, 1969.

14. B. Noble, *Methods Based on the Wiener-Hopf Technique for the Solution of Partial Differential Equations*, New York: AMS Chelsea Publ., 1988.

15. L. M. Brekhovskikh, *Waves in Layered Media*, New York: Academic Press, 1960.

16. L. B. Felsen and N. Marcuvitz, *Radiation and Scattering of Electromagnetic Waves*, Englewood Cliffs, New Jersey: Prentice-Hall, 1973.

17. J. A. Kong, *Electromagnetic Wave Theory*, New York: John Wiley & Sons, 1986.

18. W. C. Chew, *Waves and Fields in Inhomogeneous Media*, New York: Van Nostrand Reinhold, 1990, reprinted by Piscataway, New Jersey: IEEE Press, 1995.

19. J. A. Kong, "Electromagnetic field due to dipole antennas over stratified anisotropic media," *Geophysics*, vol. 38, pp. 985–96, 1972.

20. L. Tsang and J. A. Kong, "Interference patterns of a horizontal electric dipole over layered dielectric media," *J. Geophys. Res.*, vol. 78, pp. 3287–3300, 1973.

21. W. C. Chew and J. A. Kong, "Asymptotic approximation of waves due to a dipole on a two-layer medium," *Radio Science*, vol. 17, pp. 509–513, 1982.

22. K. A. Michalski and D. Zheng, "Electromagnetic scattering and radiation by surfaces of arbitrary shape in layered media, Part I: Theory," *IEEE Trans. Antennas Propagat.*, vol. AP-38, pp. 335–344, March 1990.

23. L. Cagniard, *Réflection et Réfraction des Ondes Séi smiques Progressives*, Gauthier-Villars, 1939. Translated and revised by E. A. Flinn and C. H. Dix, *Reflection and Refraction of Progressive Seismic Waves*, New York: McGraw-Hill, 1962.

24. A. T. de Hoop, "A modification of Cagniard's method for solving seismic pulse problems," *Appl. Sci. Res.*, vol. B8, pp. 349–356, 1960.

25. M. Tygel and P. Hubral, *Transient Waves in Layered Media,* New York: Elsevier, 1987.

26. A. Baños, Jr., *Dipole Radiation in the Presence of a Conducting Half-Space,* New York: Pergamon Press, 1966.

27. N. Bleistein, "Uniform asymptotic expansions of integrals with many nearby stationary points and algebraic singularities," *J. Math. Mech.*, vol. 17, pp. 533–59, 1967.

28. G. N. Watson, "The diffraction of electric waves by the earth," *Proc. Roy. Soc.* (London), vol. A95, pp. 83–99, 1918.

29. D. S. Jones, *The Theory of Electromagnetism*, New York: Macmillan, 1964.

30. P. H. Moon and D. E. Spencer, *Field Theory Handbook: Including Coordinate Systems, Differential Equations and Their Solutions*, Berlin: Springer-Verlag, 1961.

31. B. D. Seckler and J. B. Keller, "Geometrical theory of diffraction in inhomogeneous media," *J. Acoust. Soc. Am.*, vol. 31, pp. 192–205, 1959.

32. J. Boersma, "Ray-optical analysis of reflection in an open-ended parallel-plane waveguide: II-TE case," *Proc. IEEE,* vol. 62, pp. 1475–1481, 1974.

33. H. Bremmer, "The WKB approximation as the first term of a geometric-optical series," *Comm. Pure and Appl. Math.*, vol. 4, p. 105, 1951.

34. D. S. Jones and M. Kline, "Asymptotic expansion of multiple integrals and the method of stationary phase," *J. Math. Phys.*, vol. 37, pp. 1–28, 1958.

35. V. A. Fock, *Electromagnetic Diffraction and Propagation Problems*, New York: Pergamon Press, 1965.

36. R. C. Hansen, (ed.), *Geometric Theory of Diffraction*, Piscataway, New Jersey: IEEE Press, 1981.

37. R. G. Kouyoumjian and P. H. Pathak, "A uniform geometrical theory of diffraction for an edge in a perfectly conducting surface," *Proc. IEEE,* vol. 62, pp. 1448–1461, 1974.

38. S. W. Lee and G. A. Deschamps, "A uniform asymptotic theory of electromagnetic diffraction by a curved wedge," *IEEE Trans. Ant. Progat.*, vol. 24, pp. 25–35, 1976.

39. S. W. Lee, H. Ling, and R. C. Chou, "Ray tube integration in shooting and bouncing ray method," *Micro. Opt. Tech. Lett.*, vol. 1, pp. 285–289, Oct. 1988.

40. M. Van Dyke, *Perturbation Methods in Fluid Mechanics,* Stanford, CA: Parabolic Press, Inc., 1975.

41. R. B. Lindsay, *Lord Rayleigh, the Man and His Works,* London: Oxford, 1970.

42. J. J. Sakurai, *Advanced Quantum Mechanics,* Reading, MA: Addison-Wesley Pub., 1967.

43. P. E. Ceruzzi, *A History of Modern Computing,* Cambridge, MA: MIT Press, 1998.

44. K. S. Yee, "Numerical solution of initial boundary value problems involving Maxwell's equations in isotropic media," *IEEE Trans. Ant. Propag.,* vol. 14, pp. 302–307, 1966.

45. R. F. Harrington, *Field Computation by Moment Method,* Malabar, FL: Krieger Publ., 1982.

46. C. A. Brebbia and J. Dominguez, *Boundary Elements: An Introductory Course,* Boston: Computational Mechanics Publ., 1992.

47. S. M. Rao, D. R. Wilton, and A. W. Glisson, "Electromagnetic scattering by surfaces of arbitrary shape," *IEEE Trans. Antennas Propagat.,* vol. 30, no. 3, pp. 409–418, 1982.

48. M. I. Sancer, R. L. McClary, and K. J. Glover, "Electromagnetic computation using parametric geometry," *Electromagnetics,* vol. 10, pp. 85–103, 1990.

49. E. H. Newman, "Polygonal plate modeling," *Electromagnetics,* vol. 10, pp. 65–83, 1990.

50. S. Wandzura, "Electric current basis functions for curved surfaces," *Electromagnetics,* vol. 12, no. 1, pp. 77–91, 1992.

51. R. D. Graglia, D. R. Wilton, and A. F. Peterson, "Higher order interpolatory vector bases on prism elements," *IEEE Trans. Ant. Propag.,* vol. 46, pp. 442–450, 1998.

52. P. P. Silvester and R. L. Ferrari, *Finite Elements for Electrical Engineers,* Second Edition, Cambridge, U.K.: Cambridge University Press, 1990.

53. J. M. Jin, *The Finite Element Method in Electromagnetics,* New York: John Wiley & Sons, 1993.

54. J. L. Volakis, A. Chatterjee, and L. C. Kempel, *Finite Element Method for Electromagnetics: Antennas, Microwave Circuits, and Scattering Applications,* Piscataway, New Jersey: IEEE Press, 1998.

55. V. Shankar, W. F. Hall, A. H. Mohammadian, "A time-domain differential solver for electromagnetic scattering problems," *Proc. IEEE.*, vol. 77, pp. 709–721, May 1989.

56. J. S. Shang, "Characteristic-based algorithms for solving the Maxwell equations in the time domain," *IEEE Ant. Propag. Magazine,* vol. 37, pp. 15–25, June 1995.

57. A. Taflove, *Computational Electrodynamics: The Finite-Difference Time-Domain Method,* Norwood, MA: Artech House, 1995.

58. K. S. Kunz and R. J. Luebbers, *The Finite Difference Time Domain Method for Electromagnetics,* Boca Raton: CRC Press, 1993.

59. J. A. Svigelj and R. Mittra, "The dispersive boundary condition applied to nonuniform orthogonal meshes," *IEEE Trans. Micro. Theory Tech.*, vol. 47, no. 3, pp. 257–264, 1999.

60. J. B. Schneider, http://www.fdtd.org/.

61. J.-P. Berenger, "A perfectly matched layer for the absorption of electromagnetic waves," *J. Comp. Phys.*, vol. 114, pp. 185–200, 1994.

62. W. C. Chew and W. H. Weedon, "A 3-D perfectly matched medium from modified Maxwell's equations with stretched coordinates," *Micro. Opt. Tech. Lett.*, vol. 7, no. 13, pp. 599–604, 1994.

63. D. S. Katz, E. T. Thiele, and A. Taflove, "Validation and extension to three dimensions of the Berenger PML absorbing boundary condition for FD-TD meshes," *IEEE Micro. Guided Wave Lett.*, vol. 4, no. 8, pp. 268–270, 1994.

64. R. Mittra and Ü. Pekel, "A new look at the perfectly matched layer (PML) concept for the reflectionless absorption of electromagnetic waves," *IEEE Micro. Guided Wave Lett.*, vol. 5, no. 3, pp. 84–86, 1995.

65. E. A. Navarro, C. Wu, P. Y. Chung, and J. Litva, "Application of PML superabsorbing boundary condition to non-orthogonal FDTD method," *Electronics Lett.*, vol. 30, no. 20, pp. 1654–1656, 1994.

66. Z. S. Sacks, D. M. Kingsland, R. Lee, and J.-F. Lee, "A perfectly matched anisotropic absorber for use as an absorbing boundary condition," *IEEE Trans. Ant. Prop,* vol. 43, no. 12, pp. 1460–1463, 1995.

67. M. Gribbons, S. K. Lee, and A. C. Cangellaris, "Modification of Berenger's perfectly matched layer for the absorption of electromagnetic waves in layered media," *11th Ann. Rev. of Progress in ACES,* pp. 498–503, 1995.

68. W. V. Andrew, C. A. Balanis, and P. A. Tirkas, "A comparison of the Berenger perfectly matched layer and the Lindman higher-order ABC's for the FDTD method," *IEEE Microwave Guided Wave Lett.*, vol. 5, no. 6, pp. 192–194, 1995.

69. J. Fang and Z. Wu, "Generalized perfectly matched layer—an extension of Berenger's perfectly matched layer boundary condition," *IEEE Microwave Guided Wave Lett.*, vol. 5, no. 12, pp. 451–453, 1995.

70. S. Abarbanel and D. Gottlieb, "On the construction and analysis of absorbing layers in CEM," *App. Num. Math.*, vol. 27, pp. 331–340, 1998.

71. F. L. Teixeira and W. C. Chew, "PML-FDTD in cylindrical and spherical grids," *Micro. Guided Wave Lett.*, vol. 7, no. 9, pp. 285–287, 1997.

72. B. McDonald and A. Wexler, "Finite element solution of unbounded field problems," *IEEE Trans. Microwave Theory Tech.*, vol. 20, pp. 841–847, 1972.

73. K. K. Mei, "Unimoment method of solving antenna and scattering problems," *IEEE Trans. Antennas Propagat.*, vol. 22, pp. 760–766, Nov. 1974.

74. J. M. Jin and V. V. Liepa, "A note on hybrid finite element method for solving scattering problems," *IEEE Trans. Antennas Propagat.*, vol. 36, no. 10, pp. 1486–1489, 1988.

75. J. M. Jin, J. L. Volakis, and J. D. Collins, "A finite element-boundary integral method for scattering and radiation by two- and three-dimensional structures," *IEEE Antennas Propagat. Mag.*, vol. 33, no. 3, pp. 22–32, June 1991.

76. A. Bayliss, C. I. Goldstein, and E. Turkel, "On accuracy conditions for the numerical computation of waves," *J. Computational Phys.*, vol. 59, pp. 396–404, 1985.

77. R. Lee and A. C. Cangellaris, "A study of discretization error in the finite element approximation of wave solution," *IEEE Trans. Ant. Propag.*, vol. 40, no. 5, pp. 542–549, 1992.

78. W. R. Scott, Jr., "Errors due to spatial discretization and numerical precision in the finite-element method," *IEEE Trans. Ant. Propag.*, vol. 42, no. 11, pp. 1565–1569, 1994.

79. T. Deveze, L. Beaulieu, and W. Tabbara, "A fourth-order scheme for the FDTD algorithm applied to Maxwell's equations," *IEEE APS Int. Symp. Dig.*, pp. 346–349, 1992.

80. C. W. Manry, S. L. Broschat, and J. B. Schneider, "Higher-order FDTD methods for large problems," *J. Appl. Comput. Electromag. Soc.*, vol. 10, no. 2, pp. 17–29, 1995.

81. D. J. Riley, C. D. Turner, and J. D. Kotulski, "Transient application of sub-cell wire and slot models to unstructured, tetrahedral-element meshes," *IEEE Antennas and Propagat. Soc. Int. Symposium*, Atlanta, GA, vol. 1, pp. 592–595, June 1998.

82. A. George, "Nested dissection of a regular finite element mesh," *SIAM J. Numer. Anal.*, vol. 10, pp. 345–363, 1973.

83. O. Axelsson and V. A. Barker, *Finite Element Solution of Boundary Value Problems: Theory and Computation*, New York: Academic Press, 1984.

84. M. R. Hestenes and E. Stiefel, "Methods of conjugate gradients for solving linear systems," *J. Res. Nat. Bur. Stand.*, Sect. B, vol. 49, pp. 409–436, 1952.

85. T. K. Sarkar, "On the application of the generalized biconjugate gradient method," *J. Electromag. Waves Appl.*, vol. 1, no. 3, pp. 223–242, 1987.

86. V. Druskin and L. Knizhnerman, "A spectral semidiscrete method for the numerical solution of 3-D non-stationary problems in electrical prospecting," *Izv. Acad. Sci. US Phys. Solid Earth*, no. 8, pp. 63–74, 1988.

87. U. Schulz and R. Pregla, "A new technique for the analysis of the dispersion characteristics of planar waveguides," *Arch. Elek. Übertragung. (AEÜ)*, vol. 34, pp. 169–173, 1980.

88. W. C. Chew, S. Barone, C. Hennessy, and B. Anderson, "Diffraction of axisymmetric waves in a borehole by bed boundary discontinuities," *Geophysics*, vol. 49, no. 10, pp. 1586–1595, 1984.

89. Q. H. Liu and W. C. Chew, "Analysis of discontinuities in planar dielectric waveguides: results and applications," *IEEE Trans. Microwave Theory Tech.*, vol. 39, no. 3, pp. 422–430, 1991.

90. B. Yang, D. Gottlieb and J. S. Hesthaven, "Spectral Simulation of Electromagnetic Wave Scattering," *J. Comput. Phys.*, vol. 134, no. 2, pp. 216–230, 1997.

91. Q. H. Liu, "Large-scale simulations of electromagnetic and acoustic measurements using the pseudospectral time-domain (PSTD) algorithm," *IEEE Trans. Geosci. Remote Sensing*, vol. 37, no. 2, pp. 917–926, 1999.

92. J. A. Stratton, *Electromagnetic Theory*, New York: McGraw-Hill, 1941.

93. M. Born and E. Wolf, *Principles of Optics*, New York: Pergammon Press, 1980.

94. J. R. Mautz and R. F. Harrington, "H-field, E-field, and combined field solution for conducting bodies of revolution," *Arch. Elektron. Übertragungstech (AEÜ)*, vol. 32, no. 4, pp. 157–164, 1978.

95. A. F. Peterson, "The interior resonance problem associated with surface integral equations of electromagnetics: Numerical consequences and a survey of remedies," *Electromagnetics,* vol. 10, pp. 293–312, 1990.

96. A. J. Poggio and E. K. Miller, "Integral equation solution of three-dimensional scattering problems," in *Computer Techniques for Electromagnetics,* R. Mittra (ed.), New York: Pergamon Press, 1973; reprint ed., New York: Hemisphere Publishing, 1987.

97. Y. Chang and R. F. Harrington, "A surface formulation for characteristic modes of material bodies," *IEEE Trans. Antennas Propag.,* vol. #AP25, pp. 789–795, 1977.

98. T. K. Wu and L. L. Tsai, "Scattering from arbitrarily-shaped lossy dielectric bodies of revolution," *Radio Science,* vol. 12, pp. 709–718, 1977.

99. G. Petrov, "Application of Galerkin's method to a problem of the stability of the flow of a viscous liquid," (Russian), *Prikad. Matem. i Mekh.,* vol. 4, no. 3, pp. 3–12, 1940.

100. A. W. Appel, "An efficient program for many-body simulation," *SIAM J. Sci. Stat. Comput.,* vol. 6, no. 1, pp. 85–103, 1985.

101. J. Barnes and P. Hut, "A hierarchical $O(N \log N)$ force calculation algorithm," *Nature,* vol. 324, pp. 446–449, 1986.

102. L. Greengard and V. Rokhlin, "A Fast Algorithm for Particle Simulations," *Journal of Computational Physics,* vol. 73, pp. 325–348, 1987.

103. W. C. Chew and Y. M. Wang, "A fast algorithm for solution of a scattering problem using a recursive aggregate tau matrix method," *Micro. Opt. Tech. Lett.,* vol. 3, no. 5, pp. 164–169, 1990.

104. W. C. Chew and C. C. Lu, "NEPAL–An algorithm for solving the volume integral equation," *Micro. Opt. Tech. Lett.,* vol. 6, no. 3, pp. 185–188, 1993.

105. W. C. Chew and C. C. Lu, "A fast recursive algorithm to compute the wave scattering solution of a large strip," *J. Comput. Phys.,* vol. 107, pp. 378–387, 1993.

106. C. C. Lu and W. C. Chew, "The use of Huygens' equivalence principle for solving 3D volume integral equation of scattering," *IEEE Trans. Ant. Propag.,* vol. 43, no. 5, pp. 500–507, 1995.

107. V. Rokhlin, "Rapid solution of integral equations of scattering theory in two dimensions," *J. Comput. Phys.,* vol. 36, no. 2, pp. 414–439, 1990.

108. R. Coifman, V. Rokhlin, and S. Wandzura, "The fast multipole method for the wave equation: A pedestrian prescription," *IEEE Ant. Propag. Mag.*, vol. 35, no. 3, pp. 7–12, June 1993.

109. C. C. Lu and W. C. Chew, "A fast algorithm for solving hybrid integral equation," *IEE Proceedings-H*, vol. 140, no. 6, pp. 455–460, 1993.

110. L. R. Hamilton, P. A. Macdonald, M. A. Stalzer, R. S. Turley, J. L. Visher, and S. M. Wandzura, "3D method of moments scattering computations using the fast multipole method," *IEEE Antennas Propagat. Int. Symp. Dig.*, vol. 1, pp. 435–438, July 1994.

111. C. C. Lu and W. C. Chew, "A multilevel algorithm for solving boundary-value scattering," *Micro. Opt. Tech. Lett.*, vol. 7, no. 10, pp. 466–470, July 1994.

112. J. M. Song and W. C. Chew, "Multilevel fast-multipole algorithm for solving combined field integral equations of electromagnetic scattering," *Micro. Opt. Tech. Lett.* vol. 10, no. 1, pp. 14–19, Sept. 1995.

113. B. Dembart and E. Yip, "A 3D fast multipole method for electromagnetics with multiple levels," *Ann. Rev. Prog. Appl. Computat. Electromag.*, vol. 1, pp. 621–628, 1995.

114. J. M. Song and W. C. Chew, "Large scale computations using FISC," *IEEE Antennas Propag. Soc. Int. Symp.*, Salt Lake City, Utah, vol. 4, pp. 1856–1859, July 16-21, 2000.

115. A. A. Ergin, B. Shanker, and E. Michielssen, "The plane-wave time-domain algorithm for the fast analysis of transient wave phenomena," *IEEE Ant. Propag. Magazine*, vol. 41, no. 4, pp. 39–52, Aug. 1999.

116. C. Brennan and P. J. Cullen, "Tabulated interaction method for UHF terrain propagation problems," *IEEE Trans. Ant. Propag.*, vol. 46, no. 5, pp. 738–739, May 1998.

117. C. Brennan and P. J. Cullen, "Application of the fast far-field approximation to the computation of UHF path-loss over irregular terrain," *IEEE Trans. Ant. Propag.*, vol. 46, no. 6, pp. 881–890, June 1998.

118. F. X. Canning, "Transformations that produce a sparse moment matrix," *J. Electromag. Waves Appl.*, vol. 4, pp. 983–993, 1990.

119. E. Michielssen and A. Boag, "Multilevel evaluation of electromagnetic fields for the rapid solution of scattering problems," *Micro. Opt. Tech. Lett.*, vol. 7, no. 17, pp. 790–795, Dec. 1994.

120. S. Jaffard, "Wavelet methods for fast resolution of elliptic problems," *SIAM J. Numer. Anal.*, vol. 29, no. 4, pp. 965–986, 1992.

121. B. Alpert, G. Beylkin, R. R. Coifman, and V. Rokhlin, "Wavelet-like bases for the fast solution of second-kind integral equations," *SIAM J. Sci. Comput.*, vol. 14, no. 1, pp. 159–184, 1993.

122. H. Kim and H. Ling, "On the application of fast wavelet transform to the integral-equation solution of electromagnetic scattering problems," *Micro. Opt. Tech. Lett.*, vol. 6, no. 3, pp. 168–173, 1993.

123. B. Z. Steinberg and Y. Leviatan, "On the use of wavelet expansions in the method of moments," *IEEE Trans. Antennas Propagat.*, vol. 41, no. 5, pp. 610–619, 1993.

124. R. L. Wagner, G. P. Otto, and W. C. Chew, "Fast waveguide mode computation using wavelet-like basis functions," *IEEE Micro. Guid. Wave Lett.*, vol. 3, no. 7, pp. 208–210, 1993.

125. K. Sabetfakhri and L. P. B. Katehi, "Analysis of integrated millimeter-wave and submillimeter-wave waveguides using orthonormal wavelet expansion," *IEEE Trans. Microwave Theory Tech.*, vol. 42, pp. 2412–2422, 1994.

126. G. Wang, G. Pan, and B. K. Gilbert, "A hybrid wavelet expansion and boundary element analysis for multiconductor transmission lines in multilayered dielectric media," *IEEE Trans. Microwave Theory Tech.*, vol. 43, pp. 664–674, 1995.

127. R. L. Wagner and W. C. Chew, "A study of wavelets for the solution of electromagnetic integral equations," *IEEE Trans. Ant. Propag.*, vol. 43, no. 8, pp. 802–810, 1995.

128. W. L. Golik, "Sparsity and conditioning of impedance matrices obtained with semi-orthogonal and bi-orthogonal wavelet bases," *IEEE Trans. Ant. Propag.*, vol. 48, no. 4, pp. 473–481, 2000.

129. H. Deng and H. Ling, "On a class of predefined wavelet packet bases for efficient representation of electromagnetic integral equations," *IEEE Trans. Ant. Propag.*, vol. 47, no. 12, pp. 1772–1779, 1999.

130. D. T. Borup and O. P. Gandhi, "Fast-Fourier transform method for calculation of SAR distributions in finely discretized inhomogeneous models of biological bodies," *IEEE Trans. Microwave Theory Tech.*, vol. 32, no. 4, pp. 355–360, 1984.

131. C. Y. Shen, K. J. Glover, M. I. Sancer, and A. D. Varvatsis, "The discrete Fourier transform method of solving differential-integral equations in scattering theory," *IEEE Trans. Ant. Propag.*, vol. 37, no. 8, pp. 1032–1041, 1989.

132. P. Zwamborn and P. M. van den Berg, "The three-dimensional weak form of the conjugate gradient FFT method for solving scattering problems," *IEEE Trans. Microwave Theory Tech.*, vol. 40, no. 9, pp. 1757–1766, 1992.

133. M. F. Catedra, E. Gago, and L. Nuno, "A numerical scheme to obtain the RCS of three-dimensional bodies of size using the conjugate gradient method and the fast Fourier transform," *IEEE Trans. Ant. Propag.*, vol. 37, no. 5, pp. 528–537, 1989.

134. J. R. Phillips and J. K. White, "Efficient capacitance computation of 3D structures using generalized pre-corrected FFT methods," *Proceedings of the 3rd Topical Meeting on Electric Performance of Electronic Packaging*, Nov. 2-4, Monterey, CA, 1994.

135. K. Lumme and J. Rahola, "Light scattering by porous dust particles in the discrete-dipole approximation," *Astrophys. J.*, vol. 425, pp. 653–667, 1994.

136. E. Bleszynski, M. Bleszynski, and T. Jaroszewicz, "A fast integral-equation solver for electromagnetic scattering problems," *IEEE APS Int. Symp. Dig.*, pp. 416–419, 1994.

137. C. H. Chan and L. Tsang, "A sparse-matrix canonical-grid method for scattering by many scatterers," *Micro. Opt. Tech. Lett.*, vol. 8, no. 2, pp. 114–118, Feb. 1995.

138. H. Gan and W. C. Chew, "A discrete BCG-FFT algorithm for solving 3D inhomogeneous scatterer problems," *J. Electromag. Waves Appl.*, vol. 9, no. 10, pp. 1339–1357, 1995.

139. W. C. Chew, J. H. Lin, and X. G. Yang, "An FFT T-matrix method for 3D microwave scattering solution from random discrete scatterers," *Micro. Opt. Tech. Lett.*, vol. 9, no. 4, pp. 194–196, July 1995.

140. J. H. Lin and W. C. Chew, "BiCG-FFT T-matrix method for solving for the scattering solution from inhomogeneous bodies," *IEEE Trans. Microwave Theory Tech.*, vol. 44, no. 7, pp. 1150–1155, July 1996.

141. J. J. Mallorqui and M. Rodrigeuz, "GRATMA method for biomedical applications: comparison with the CGM-FFT," *IEEE APS Int. Symp. Dig.*, pp. 1581–1583, 1995.

142. C. F. Wang and J. M. Jin, "Simple and efficient computation of electromagnetic fields in arbitrarily shaped inhomogeneous dielectric bodies using transpose-free QMR-and FFT," *IEEE Trans. Micro. Theory Tech.*, vol. 46, no. 5, pt. 1, pp. 553–558, 1998.

143. F. Ling, D. Jiao, and J. M. Jin, "Efficient electromagnetic modeling of microstrip structures in multilayer media," *IEEE Trans. Micro. Theory Tech.*, vol. 47, no. 9, pt. 2, pp. 1810–1818, 1999.

144. S. S. Bindiganavale, J. L. Volakis, and H. Anastassiu, "Scattering from planar structures containing small features using the adaptive integral method (AIM)," *IEEE Trans. Ant. Propag.*, vol. 46, no. 12, pp. 1867–1878, 1998.

145. A. Brandt, "Multilevel computations of integral transforms and particle interactions with oscillatory kernels," *Comp. Phys. Comm.*, vol. 65, pp. 24-38, 1991.

146. S. S. Bindiganavale and J. L. Volakis, "Guidelines for using the fast multipole method to calculate the RCS of large objects," *Micro. Opt. Tech. Lett.*, vol. 11, no. 4, pp. 190–194, 1996.

147. L. S. Canino, J. J. Ottusch, M. A. Stalzer, J. L. Visher, and S. Wandzura, "Numerical solution of the Helmholtz equation in 2D and 3D using a higher-order Nyström discretization," *J. Comput. Phys.*, vol. 146, pp. 627–663, 1998.

148. N. Geng, A. Sullivan, and L. Carin, "Multilevel fast-multipole algorithm for scattering from conducting targets above or embedded in a lossy half space," *IEEE Trans. Geosci. Remote Sensing,* vol. 38, pp. 1561–1573, 2000.

149. W. C. Chew, "Fast algorithms for wave scattering developed at the Electromagnetics Laboratory, University of Illinois," *IEEE Ant. Propag. Mag.*, vol. 35, no. 4, pp. 22–32, 1994.

150. W. C. Chew, C. C. Lu, and Y. M. Wang, "Review of efficient computation of three-dimensional scattering of vector electromagnetic waves," *J. Opt. Soc. Am. A*, vol. 11, pp. 1528–1537, 1994.

151. W. C. Chew, J. M. Jin, C. C. Lu, E. Michielssen, and J. M. Song, "Fast solution methods in electromagnetics," *IEEE Trans. Ant. Propag.*, vol. 45, pp. 533–543, 1997.

2

Fast Multipole Method and Multilevel Fast Multipole Algorithm in 2D

Weng Cho Chew and Jiming Song

2.1 INTRODUCTION

The fast multipole method (FMM) was suggested by Rokhlin to seek rapid solution to integral equation of scattering [1] for Helmholtz problems. Earlier, Rokhlin and Greengard had proposed a fast multipole algorithm for Laplace problems [2] suitable for solving static problems. However, the static algorithm needs to be modified for Helmholtz and electromagnetic problems. These dynamic algorithms have been recently generalized to electromagnetics and to multilevel algorithms [3–7]. FMM and its multilevel generalization, the multilevel fast multipole algorithm (MLFMA), are promising in terms of providing a path to large-scale computing in electromagnetics.

2.2 INTRODUCTION TO FAST MULTIPOLE IN 2D

To obviate many of the derivations without the complexity of 3D expansion functions, it is best to illustrate first the derivation of the FMM in 2D. In this manner, we can see the simplicity of the mathematics and gain insight into why FMM works. As

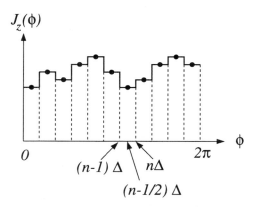

Figure 2.1 Approximation of the current on the surface of a scatterer using the pulse basis functions.

we said before, the key feature of the FMM is to diagonalize the translation matrix. This diagonalization was first given by Rokhlin [1]. The diagonalized form of the translation matrix can be derived in many ways. We will illustrate one of the derivations here.

2.2.1 A 2D MOM Problem

Before we do that, we demonstrate a very simple 2D method of moments (MOM) to discretize the 2D integral equation of scattering [8]. In electromagnetics, the 2D scattering problem can be cast entirely into a scalar problem; therefore, the mathematics of 2D electromagnetic scattering is the same as that of scalar acoustic wave scattering. Consider the following scattering problem of a metallic cylinder by an E_z polarized electromagnetic wave:

$$i\omega\mu_0 \int_S dS' g_0(\boldsymbol{\rho} - \boldsymbol{\rho}') J_z(\boldsymbol{\rho}') = -E_z^{inc}(\boldsymbol{\rho}), \quad \boldsymbol{\rho} \in S \tag{2.1}$$

In the above,

$$g_0(\boldsymbol{\rho} - \boldsymbol{\rho}') = \frac{i}{4} H_0^{(1)}(k|\boldsymbol{\rho} - \boldsymbol{\rho}'|) \tag{2.2}$$

is the 2D Green's function. $J_z(\boldsymbol{\rho})$ is the induced current on the surface of the metallic scatterer. The above equation says that the induced current J_z generates an electromagnetic field E_z polarized in the z direction that cancels the incident electric field on the surface of the scatterer.

We illustrate a very simple discretization of the above integral equation by using the pulse basis function by the method of moments (see Figure 2.1). In other words, letting $J_z(\rho) = \sum_{i=1}^{N} a_i j_i(\rho)$, where $j_i(\rho)$ is a basis function, which is the pulse basis here, and substituting into the above integral equation, we have, after testing by a Dirac delta function located at the center of the pulse basis,

$$\sum_{i=1}^{N} A_{ji} a_i = b_j, \qquad j = 1, \ldots, N \tag{2.3}$$

where

$$\begin{aligned} A_{ji} &= \langle \delta(\rho - \rho_j), i\omega\mu_0 g_0(\rho - \rho'), j_i(\rho) \rangle \\ &= i\omega\mu_0 \langle g_0(\rho_j - \rho'), j_i(\rho) \rangle \end{aligned} \tag{2.4}$$

In the above,

$$\begin{aligned} \langle f(\rho), g_0(\rho, \rho'), h(\rho) \rangle \\ = \int d\rho f(\rho) \int d\rho' g_0(\rho - \rho') h(\rho') \end{aligned} \tag{2.5}$$

The commas indicate an integration over the surface of the scatterer, and ρ_j is located at the midpoint of the jth pulse basis. After performing the integration for the self term ($i = j$) by approximating the Hankel function with its small argument approximation [9], and approximating the nonself term by a one-point integration rule for simplicity, we have

$$A_{ji} = \begin{cases} \frac{\omega\mu_0}{4} \Delta_i \left[1 + \frac{2i}{\pi} \ln\left(\frac{\gamma k \Delta_i}{4e}\right) \right], & i = j \\ \frac{\omega\mu_0}{4} \Delta_i H_0^{(1)}(k\rho_{ji}), & i \neq j \end{cases} \tag{2.6}$$

where Δ_i is the discretization size that may not be uniform. The above shows that the 2D Green's function (i.e., the Hankel function) still plays an important role in the matrix equation. Furthermore,

$$b_j = -E_z^{inc}(\rho_j) \tag{2.7}$$

which is obtained by testing the incident field at the location ρ_j. In addition, we can think of

$$a_i = J_z(\rho_i) \tag{2.8}$$

which is an approximation of the current at the location ρ_i. In the above, $\rho_{ji} = |\rho_j - \rho_i|$ and $\frac{\gamma}{4e} = 0.163805$.

The matrix-vector multiplication $\sum_{i=1}^{N} A_{ji} a_i$ is the bottleneck in the speed of an iterative solver such as the conjugate gradient algorithm [10, 11]. The fast multipole method expedites this matrix-vector product by a divide-and-conquer scheme followed by a three-step transmission procedure, rather than a one-step procedure (see Figure 2.2). This essentially emulates the telephone network connection we discussed in Chapter 1.

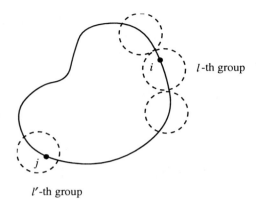

Figure 2.2 In FMM the adjacent current elements on the surface of the scatterer are grouped together in groups of size M.

2.2.2 Addition Theorem for Bessel Functions

Understanding the translational addition theorem is imperative for understanding the fast multipole algorithm. We will review the 2D translational addition theorem here. The theorem arises out of addition of angular momentum in quantum mechanics, from whence its name is derived [12]. In the mathematics literature, it is known as the Graf's formula [9].

The addition theorem for Hankel functions is [13]

$$H_m^{(1)}(k|\boldsymbol{\rho} - \boldsymbol{\rho}'|)e^{im\phi''} = \begin{cases} \sum_{n=-\infty}^{\infty} J_{n-m}(k\rho')e^{-i(n-m)\phi'} H_n^{(1)}(k\rho)e^{in\phi}, \\ \qquad\qquad\qquad\qquad\qquad\qquad \rho > \rho' \\ \sum_{n=-\infty}^{\infty} H_{n-m}^{(1)}(k\rho')e^{-i(n-m)\phi'} J_n(k\rho)e^{in\phi}, \\ \qquad\qquad\qquad\qquad\qquad\qquad \rho < \rho' \end{cases} \qquad (2.9)$$

Here, ϕ, ϕ', and ϕ'' are the angles that $\boldsymbol{\rho}$, $\boldsymbol{\rho}'$, and $\boldsymbol{\rho} - \boldsymbol{\rho}'$ make with the x-axis, respectively (Figure 2.3). Physically, the above says that for a cylindrical wave emanating from the origin O'', it can be expressed in terms of outgoing Hankel cylindrical wave harmonics emanating from the origin O when $\rho > \rho'$. This is necessary to satisfy the radiation condition at infinity. When $\rho < \rho'$, the wave can be expressed in terms of an "incoming" Bessel cylindrical wave harmonics about the origin O. The Bessel wave is actually a standing wave consisting of an incoming Hankel wave plus an outgoing Hankel wave as obviated by the identity $J_n(x) = 1/2(H_n^{(1)}(x) + H_n^{(2)}(x))$, where $H_n^{(1)}(x)$ represents an outgoing wave and

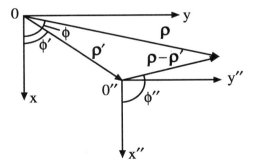

Figure 2.3 Translation in the cylindrical coordinate system.

$H_n^{(2)}(x)$ represents an incoming wave. We term the Bessel wave as incoming wave because an incoming Hankel wave will naturally reflect to an outgoing Hankel wave at the origin, yielding a standing Bessel wave.

For the Bessel function of the first kind, we have

$$J_m(k|\boldsymbol{\rho} - \boldsymbol{\rho}'|)e^{im\phi''} = \sum_{n=-\infty}^{\infty} J_{n-m}(k\rho')J_n(k\rho)e^{in\phi - i(n-m)\phi'} \quad (2.10)$$

Alternatively, in vector notation [13], the above can be written more concisely as

$$\psi^t(\boldsymbol{\rho} - \boldsymbol{\rho}') = \begin{cases} \psi^t(\boldsymbol{\rho}) \cdot \overline{\boldsymbol{\beta}}(\boldsymbol{\rho}'), & \rho > \rho' \\ \Re g\psi^t(\boldsymbol{\rho}) \cdot \overline{\boldsymbol{\alpha}}(\boldsymbol{\rho}'), & \rho < \rho' \end{cases} \quad (2.11)$$

$$\Re g\psi^t(\boldsymbol{\rho} - \boldsymbol{\rho}') = \Re g\psi^t(\boldsymbol{\rho}) \cdot \overline{\boldsymbol{\beta}}(\boldsymbol{\rho}'), \quad \forall \rho, \rho' \quad (2.12)$$

In the above, $\psi^t(\boldsymbol{\rho})$ is an infinitely long row vector containing $H_n^{(1)}(k\rho)$ and $\Re g$ implies the "regular part" or the part of Hankel function that contains the Bessel function. Moreover, $\overline{\alpha}$ and $\overline{\beta}$ are infinite dimensional matrices. Consequently,

$$\psi^t(\boldsymbol{\rho}_j) = \begin{cases} \psi^t(\boldsymbol{\rho}_i) \cdot \overline{\boldsymbol{\beta}}_{ij}, & \rho_i > |\boldsymbol{\rho}_i - \boldsymbol{\rho}_j| \\ \Re g\psi^t(\boldsymbol{\rho}_i) \cdot \overline{\boldsymbol{\alpha}}_{ij}, & \rho_i < |\boldsymbol{\rho}_i - \boldsymbol{\rho}_j| \end{cases} \quad (2.13)$$

$$\Re g\psi^t(\boldsymbol{\rho}_j) = \Re g\psi^t(\boldsymbol{\rho}_i) \cdot \overline{\boldsymbol{\beta}}_{ij}, \quad \forall \rho_i, \rho_j \quad (2.14)$$

The $\overline{\alpha}$ and $\overline{\beta}$ matrices are known as translation matrices or operators. They are used to translate the coordinate system of a wave function to another coordinate system. The $\overline{\alpha}$ matrix translates an outgoing wave from one coordinate system to an incoming wave at another coordinate system. The $\overline{\beta}$ matrix translates an outgoing wave from one coordinate system to an outgoing wave at another coordinate system. It also translates an incoming wave from one coordinate system to an incoming wave at another coordinate system.

The use of the addition theorem allows us to express the field in one coordinate system in terms of the field in another coordinate system. Notice the dichotomous behavior of the expansion, namely, that one expansion is valid outside a circle, while the other expansion is only valid inside the same circle. We term this the violation of the addition theorem. Also,

$$\begin{aligned}\psi^t(\rho_{ji}) &= \psi^t(\rho_{jl}) \cdot \overline{\beta}_{li} \\ &= \Re g \psi^t(\rho_{jl'}) \cdot \overline{\alpha}_{l'l} \cdot \overline{\beta}_{li} \\ &= \Re g \psi^t(0) \cdot \overline{\beta}_{jl'} \cdot \overline{\alpha}_{l'l} \cdot \overline{\beta}_{li} \\ &= \beta_{jl'}^t \cdot \overline{\alpha}_{l'l} \cdot \overline{\beta}_{li}\end{aligned} \quad (2.15)$$

where ρ_{ji} points from point i to point j. For a monopole source, since only one column of $\overline{\beta}_{li}$ is needed, this becomes

$$H_0^{(1)}(k\rho_{ji}) = \beta_{jl'}^t \cdot \overline{\alpha}_{l'l} \cdot \beta_{li} \quad (2.16)$$

2.2.3 An Inefficient Factorization of the Green's Function

We illustrate first an alternative way to compute interactions between all the current sources by using the above factorization of the Green's function, but it does not give rise to a more efficient algorithm for Helmholtz problems. However, it does yield a more efficient algorithm for Laplace problems. Nevertheless, this factorization illustrates the role the multipoles play in the algorithm. It also illustrates why there is a need to diagonalize the translation operators for Helmholtz and electromagnetic problems.

To this end, we divide the N subscatterers into groups, each of which contains M subscatterers (current elements). Hence, there are N/M groups altogether. Furthermore, because A_{ji} contains the Hankel function when $i \ne j$, the translational addition theorem for Hankel function can be used to rewrite A_{ji}. Hence, it follows from (2.16) that

$$H_0^{(1)}(k\rho_{ji}) = \overbrace{\beta_{jl'}^t}^{1 \times P} \cdot \overbrace{\overline{\alpha}_{l'l}}^{P \times P} \cdot \overbrace{\beta_{li}}^{P \times 1} \quad (2.17)$$

where l' and l are the centers of the l'th and lth groups respectively (see Figure 2.2), and matrix $\overline{\alpha}$ and vector β are defined as

$$[\overline{\alpha}_{l'l}]_{nm} = H_{n-m}^{(1)}(k_0 \rho_{l'l}) e^{-i(n-m)\phi_{l'l}} \quad (2.18)$$

$$[\beta_{jl'}]_n = J_n(k_0 \rho_{jl'}) e^{in(\phi_{jl'} - \pi)} \qquad [\beta_{li}]_n = J_n(k_0 \rho_{li}) e^{-in\phi_{li}} \quad (2.19)$$

where $\phi_{\alpha\beta}$ is the angle a vector (that points from point α to point β) makes with the x axis. The above allows us to factorize the Green's function from the Helmholtz equation into a vector times a matrix times another vector.

The addition theorem allows one to expand the field due to a source at point i about a new coordinate system with the center at l by using multipole expansions. In (2.17), P indicates the number of multipoles needed to make the expansion accurate. Consequently, the vector β_{li} converts a monopole to a P term multipole. Moreover, the translation matrix $\overline{\alpha}_{l'l}$ converts the outgoing wave multipoles with coordinate center at l to incoming wave multipoles with coordinate center at l'. Finally, the vector β_{jl}^t represents the evaluation of these incoming multipoles at the field point j.

Using (2.17), we can rewrite the far interaction (2.3) as

$$b'_j = \frac{\omega\mu_0}{4} \beta_{jl'}^t \cdot \sum_{\substack{l \neq l'+NN \\ l=1}}^{N/M} \overline{\alpha}_{l'l} \cdot \sum_{i \in \mathcal{G}_l} \beta_{li}\Delta_i a_i, \quad j \in \mathcal{G}_{l'}, \quad l' = 1,\ldots,N/M \quad (2.20)$$

The above illustrates a three-stage transmission of field information from a source point i to a field point. The first stage is the aggregation stage, the second stage is the translation stage, and the third stage is the disaggregation stage. Due to the violation of the addition theorem, the above mode of calculation is only good when the source point and the field point belong to different groups that are far apart. Hence $l \neq l' + NN$ implies when l is not the self group l' or the near neighbor groups.

In order to maintain the accuracy of (2.17), P is proportional to M. Here, P indicates the number of multipoles needed to make the factorization accurate. If the scatterer is discretized such that the discretization size Δ is about 0.1λ, then the size of the group in terms of wavelength is linearly proportional to M. However, the larger the group size in terms of wavelength, the more the number of multipoles needed to approximate the outgoing waves from the group center l. It can be seen easily that P is also proportional to the electrical size (size in terms of wavelength) of the group, and hence is also proportional to M. This fact can be proven also by looking at the relevant mathematical equation for the convergence of the addition theorem, and we will revisit this point again when we study the error analysis and control of this algorithm.

Hence, the cost of first stage, the aggregation stage, which requires computing $\mathbf{c}_l = \sum_{i \in \mathcal{G}_l} \beta_{li}\Delta_i a_i, l = 1,\ldots,N/M$, is

$$T_1 = \frac{N}{M}M^2 = NM \quad (2.21)$$

The cost of the second stage, the translation stage, in computing $\mathbf{d}'_l = \sum_{l=1}^{N/M} \overline{\alpha}_{l'l} \cdot \mathbf{c}_l$, $l = 1, \ldots, N/M$ is

$$T_2 = \left(\frac{N}{M}\right)^2 M^2 = N^2 \tag{2.22}$$

The cost of the last stage, the disaggregation stage, in computing $\beta^t_{jl'} \cdot \mathbf{d}_{l'}$, $j \in \mathcal{G}_{l'}, l', \ldots, N/M$ is

$$T_3 = \left(\frac{N}{M}\right) M^2 = NM \tag{2.23}$$

But alas, the cost of the translation stage using the $\overline{\alpha}_{l'l}$ is still $O(N^2)$. Therefore, there is little advantage at this point in rewriting (2.3) as (2.20) as now the cost of computing via (2.20) is still of the same computational complexity as before.

2.2.4 Diagonalization of the Translation Operator

It is clear why the above algorithm is inefficient—we have exploded a scalar matrix element A_{ij} into a vector times a matrix times another vector. The $\overline{\alpha}_{l'l}$ matrix, which translates outgoing multipoles to incoming multipoles, is a dense matrix making the aforementioned algorithm inefficient.

Therefore, another factorization where $\overline{\alpha}_{l'l}$ is a diagonal matrix is needed. The diagonal factorization was first presented by Rokhlin [1, 14] and it has since been studied by other workers [15–18]. We will first derive a diagonalized factorization from the above; and later we present another, simpler way of deriving the diagonalized factorization.

To this end, we substitute in the definition of $\overline{\alpha}$ and β in (2.17) so that it can be written as

$$H_0^{(1)}(k\rho_{ji}) = \sum_{m=-\infty}^{\infty} J_m(k\rho_{jl'}) e^{im(\phi_{jl'} - \pi)}$$
$$\cdot \sum_{n=-\infty}^{\infty} H_{m-n}^{(1)}(k\rho_{l'l}) e^{-i(m-n)\phi_{l'l}} J_n(k\rho_{li}) e^{-in\phi_{li}} \tag{2.24}$$

where $\phi_{\alpha\beta}$ is the angle a vector that points from point α to point β makes with the x axis.

Even though $H_{m-n}^{(1)}(x) \to \infty$ when $|m - n| \to \infty$, the inner summation above converges because $J_n(x) \to 0$ more rapidly when $|n| \to \infty$. Notice that the inner summation in the above is the convolution of two discrete Fourier series. However, we can truncate the inner summation since it converges and express (2.24) as [1]

$$H_0^{(1)}(k\rho_{ji}) = \sum_{m=-\infty}^{\infty} J_m(k\rho_{jl'})e^{im(\phi_{jl'}-\pi)} \\ \cdot \sum_{n=m-P}^{m+P} H_{m-n}^{(1)}(k\rho_{l'l})e^{-i(m-n)\phi_{l'l}} J_n(k\rho_{li})e^{-in\phi_{li}} \quad (2.25)$$

Equation (2.25) can be expressed in the Fourier space by finding the Fourier transforms of the Fourier coefficients. In the Fourier space, the above discrete convolution becomes a multiplication. The last summation in (2.25) can be expressed in terms of Fourier transforms using Parseval's theorem. The Fourier transforms of the Fourier coefficients involving Bessel functions can be found by using the integral representation of the Bessel function:

$$J_m(k\rho_{jl'})e^{im\phi_{jl'}} = \frac{1}{2\pi} \int_0^{2\pi} d\alpha \, e^{ik\rho_{jl'}\cos(\alpha-\phi_{jl'})+im(\alpha-\frac{\pi}{2})} \quad (2.26)$$

$$J_n(k\rho_{li})e^{-in\phi_{li}} = \frac{1}{2\pi} \int_0^{2\pi} d\alpha' \, e^{ik\rho_{li}\cos(\alpha'+\phi_{li})+in(\alpha'-\frac{\pi}{2})} \quad (2.27)$$

Using (2.26) and (2.27) in (2.25), invoking the convolutional property of Fourier series and using Parseval's theorem for discrete Fourier series, we have

$$H_0^{(1)}(k\rho_{ji}) = \frac{1}{2\pi} \int_0^{2\pi} d\alpha \, \tilde{\beta}_{jl'}(\alpha) \tilde{\alpha}_{l'l}(\alpha) \tilde{\beta}_{li}(\alpha) \quad (2.28)$$

where

$$\tilde{\alpha}_{l'l}(\alpha) = \sum_{p=-P}^{P} H_p^{(1)}(k\rho_{l'l})e^{-ip(\phi_{l'l}-\alpha+\frac{\pi}{2})} \quad (2.29)$$

and

$$\tilde{\beta}_{jl'}(\alpha) = e^{-ik\rho_{jl'}\cos(\alpha-\phi_{jl'})} \qquad \tilde{\beta}_{li}(\alpha) = e^{-ik\rho_{li}\cos(\alpha-\phi_{li})} \quad (2.30)$$

---

[1] We will revisit the issue of error introduced by this truncation later.

Using (2.28) in replacement of (2.17) in (2.20), we have a new formula that allows us to transmit information by a three-stage computation, namely,

$$b'_j = \frac{\omega\mu_0}{8\pi} \int_0^{2\pi} d\alpha \tilde{\beta}_{jl'}(\alpha) \sum_{\substack{l \neq l'+NN \\ l=1}}^{N/M} \tilde{\alpha}_{l'l}(\alpha) \sum_{i \in \mathcal{G}_l} \tilde{\beta}_{li}(\alpha) \Delta_i a_i,$$

$$j \in \mathcal{G}_{l'}, \quad l' = 1, \ldots, N/M \quad (2.31)$$

for intergroup interactions. The integral in (2.31) can be replaced by a Q-point summation yielding

$$\frac{\omega\mu_0}{4Q} \sum_{q=1}^{Q} \tilde{\beta}_{jl'}(\alpha_q) \sum_{\substack{l \neq l'+NN \\ l=1}}^{N/M} \tilde{\alpha}_{l'l}(\alpha_q) \sum_{i \in \mathcal{G}_l} \tilde{\beta}_{li}(\alpha_q) \Delta_i a_i = b'_j, \quad j \in \mathcal{G}_{l'} \quad (2.32)$$

As shall be shown later, Q is linearly proportional to M. We can rewrite the summation over q above in terms of a matrix notation as

$$\frac{\omega\mu_0}{4Q} \tilde{\beta}_{jl'} \cdot \sum_{\substack{l \neq l'+NN \\ l=1}}^{N/M} \tilde{\overline{\alpha}}_{l'l} \cdot \sum_{i \in \mathcal{G}_l} \tilde{\beta}_{li} \Delta_i a_i = b'_j, \quad j \in \mathcal{G}_{l'} \quad (2.33)$$

Clearly, $\tilde{\overline{\alpha}}_{l'l}$ is a diagonal matrix in the above compared to (2.20).

The replacement of $\overline{\alpha}_{l'l}$ with a diagonal operator $\tilde{\overline{\alpha}}_{l'l}$ reduces the cost in the second stage to

$$T_2 = \frac{N^2}{M} \quad (2.34)$$

The cost in the first and third stages is still the same. Therefore, the total cost is

$$T = C_1 \frac{N^2}{M} + C_2 NM \quad (2.35)$$

Optimizing (2.35) with respect to M yields $M = \sqrt{\frac{C_1}{C_2}} N$. Therefore,

$$T = 2\sqrt{C_1 C_2} N^{1.5} \quad (2.36)$$

The above exercise illustrates the importance of diagonalizing the translation operator to arrive at a two-level algorithm with reduced computational complexity.

2.2.5 Summary and Hindsight

The diagonalization of the translation operator can be regarded as a change of basis or a similarity transform of the translation matrix.

The β's were originally expressed in terms of multipoles, and the wave used to express the outgoing waves were multipoles. The $\widetilde{\beta}$'s are actually plane waves. So the relationship between the new $\widetilde{\beta}$'s and the old β's, as given by (2.26) and (2.27), can be regarded as a change of basis which is expressible as a similarity transform:

$$\beta = \overline{\mathbf{S}}^t \cdot \widetilde{\beta} \tag{2.37}$$

The above is just the inverse of (2.26) and (2.27). Then,

$$H_0^{(1)}(k\rho_{ji}) = \widetilde{\beta}_{jl'}^t \cdot \overline{\mathbf{S}} \cdot \overline{\alpha}_{l'l} \cdot \overline{\mathbf{S}}^t \cdot \widetilde{\beta}_{li} \tag{2.38}$$

$$H_0^{(1)}(k\rho_{ji}) = \widetilde{\beta}_{jl'}^t \cdot \widetilde{\overline{\alpha}}_{l'l} \cdot \widetilde{\beta}_{li} \tag{2.39}$$

and

$$H_0^{(1)}(k\rho_{ji}) = \frac{1}{2\pi} \int_0^{2\pi} d\alpha \, \widetilde{\beta}_{jl'}(\alpha) \widetilde{\alpha}_{l'l}(\alpha) \widetilde{\beta}_{li}(\alpha) \tag{2.40}$$

where

$$\widetilde{\alpha}_{l'l}(\alpha) = \sum_{p=-P}^{P} H_p^{(1)}(k\rho_{l'l}) e^{-ip(\phi_{l'l} - \alpha + \frac{\pi}{2})} \tag{2.41}$$

and

$$\begin{aligned}\widetilde{\beta}_{jl'}(\alpha) &= e^{-ik\rho_{jl'}\cos(\alpha - \phi_{jl'})} = e^{-i\mathbf{k}\cdot\boldsymbol{\rho}_{jl'}} \\ \widetilde{\beta}_{li}(\alpha) &= e^{-ik\rho_{li}\cos(\alpha - \phi_{li})} = e^{-i\mathbf{k}\cdot\boldsymbol{\rho}_{li}}\end{aligned} \tag{2.42}$$

In general, for a MOM matrix element, when i and j refer to field and source points not within the same group plus the near neighbor groups, it can be factorized as

$$A_{ij} = \mathbf{V}_{il}^t \cdot \widetilde{\overline{\alpha}}_{ll'} \cdot \mathbf{V}_{l'j} \tag{2.43}$$

2.2.6 An Alternative Derivation of the Diagonalized Translator

The derivation of (2.25) can follow an alternative path. We can first write

$$\rho_{jl} = |\boldsymbol{\rho}_j - \boldsymbol{\rho}_l| = |\boldsymbol{\rho}_{jl'} + \boldsymbol{\rho}_{l'l}| \tag{2.44}$$

(see Figure 2.4). Using this in the translational addition theorem [13], we have

$$H_0^{(1)}(k\rho_{jl}) = \sum_{p=-P}^{P} H_p^{(1)}(k\rho_{l'l}) e^{-ip\phi_{l'l}} J_p(k\rho_{jl'}) e^{ip(\phi_{jl'} - \pi)} \tag{2.45}$$

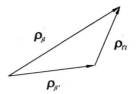

Figure 2.4 The addition of vectors for forming ρ_{jl}.

We have truncated the series *a priorily* since the series is exponentially convergent when $\rho_{jl'} < \rho_{l'l}$. Using the integral representation of the Bessel function as in (2.26) and (2.27), we arrive at

$$H_0^{(1)}(k\rho_{jl}) = \frac{1}{2\pi} \int_0^{2\pi} d\alpha\, e^{-i\mathbf{k}\cdot\boldsymbol{\rho}_{jl'}} \tilde{\alpha}_{l'l}(\alpha) \qquad (2.46)$$

The condition for the validity of the above is that $\rho_{l'l} > \rho_{jl'}$. This constraint comes from the use of the addition theorem in the derivation of the above—it prevents the violation of the addition theorem. When the above formula is applied to the case

$$\rho_{ji} = |\boldsymbol{\rho}_j - \boldsymbol{\rho}_i| = |\boldsymbol{\rho}_{jl'} + \boldsymbol{\rho}_{l'l} + \boldsymbol{\rho}_{li}| \qquad (2.47)$$

where $\rho_{l'l} > |\boldsymbol{\rho}_{jl'} + \boldsymbol{\rho}_{li}|$, we have

$$H_0^{(1)}(k\rho_{ji}) = \frac{1}{2\pi} \int_0^{2\pi} d\alpha\, e^{-i\mathbf{k}\cdot\boldsymbol{\rho}_{jl'}} \cdot \tilde{\alpha}_{l'l}(\alpha) e^{-i\mathbf{k}\cdot\boldsymbol{\rho}_{li}} \qquad (2.48)$$

This yields a diagonalization that is the same as the previous case in (2.28).

2.2.7 Physical Interpretation of Aggregation, Translation, and Disaggregation

Notice that the factor $e^{-i\mathbf{k}\cdot\boldsymbol{\rho}_{jl'}}$ occurs when we calculate the far field pattern of a radiation source as in

$$\phi(\boldsymbol{\rho}) = i \int_V d\boldsymbol{\rho}' \frac{i}{4} H_0^{(1)}(k_\rho|\boldsymbol{\rho} - \boldsymbol{\rho}'|) j(\boldsymbol{\rho}')$$
$$\sim \frac{1}{\sqrt{2\pi\rho}} \int_V d\boldsymbol{\rho}'\, e^{-i\mathbf{k}\cdot\boldsymbol{\rho}'} j(\boldsymbol{\rho}'), \quad \rho \to \infty \qquad (2.49)$$

Therefore, the aggregation stage in (2.32) is equivalent to calculating the far field radiation pattern function of a group. The function $e^{-i\mathbf{k}\cdot\boldsymbol{\rho}_{li}}$ is the contribution to this radiation function centered at ρ_l due to a point source at ρ_i. The translation stage takes this far field radiation pattern function of each group and converts it into an incoming wave pattern. Therefore, it is a far-field-to-near-field transform. The disaggregation stage takes the incoming field pattern and multiplies it by the receiving pattern function of an observation point. The function $e^{-i\mathbf{k}\cdot\boldsymbol{\rho}_{jl'}}$ can be thought of as the receiving pattern function of the observation point at ρ_j with respect to the center $\rho_{l'}$. Notice that the radition pattern function of a point source is the same as the receiving pattern function of an observation point due to reciprocity.

The richness of the radiation pattern is dependent on the size of the group with respect to the wavelength. Therefore, the larger the group size, the richer the radiation pattern, and the more the multipole terms needed to capture this far field pattern.

2.2.8 Bandwidth of the Radiation Pattern

The radiation pattern generated by a source has a finite bandwidth because the far field does not have a singularity and is smooth. The far field pattern about the group center due to a group of sources is obtained by the following summation, which gives the far field radiation function:

$$F(\alpha) = \sum_{i=1}^{M} a_i e^{-i\mathbf{k}\cdot\boldsymbol{\rho}_{li}} \qquad (2.50)$$

where ρ_{il} is the distance of the source point from the group center. The expression $e^{-i\mathbf{k}\cdot\boldsymbol{\rho}_{il}} = e^{-ik\rho_{il}\cos(\alpha-\phi_{il})}$ is a periodic function. Notice that it oscillates more rapidly as a function of α when ρ_{il} becomes larger. Therefore, the largest frequency component of the far field function is determined by $R = \max[\rho_{il}]$, which we shall call the radius of the group. To establish a tighter relationship between the bandwidth of the radiation function and R, we expand:

$$e^{-ik\rho_{il}\cos(\alpha-\phi_{il})} = \sum_{n=-\infty}^{\infty} J_n(k\rho_{il})e^{in(\alpha-\phi_{il})}i^{-n} \qquad (2.51)$$

The above identity is just the Fourier series expansion of the function $e^{-ik\rho_{il}\cos(\alpha-\phi_{il})}$. Because $J_n(x)$ tends to zero exponentially fast when $n \to \infty$, the above series has exponential convergence.

We can perform a simple analysis to estimate how many terms in the series are needed before the error in the summation is exponentially small. The number of terms also indicates the bandwidth of the function in (2.51). It can be shown that when $n \to \infty$ and $x \sim O(n)$, then [9]

$$J_n(x) \sim \frac{e^{\sqrt{n^2-x^2}-n\cosh^{-1}(n/x)}}{\sqrt{2\pi(n^2-x^2)}} \qquad (2.52)$$

The above decays exponentially fast when $n > x$. Hence, we first let $n/x \approx 1 + \delta$ where $\delta \ll 1$, and it is a measure of how much n should be in excess of x for the above to be exponentially small. In this case, $\cosh^{-1}(n/x) \sim \sqrt{2\delta}$, $\sqrt{n^2 - x^2} \sim x\sqrt{2\delta}$. Therefore, the above becomes

$$J_n(x) \sim \frac{e^{x(1-n/x)\sqrt{2\delta}}}{2x\sqrt{\pi\delta}} \approx \frac{e^{-x\sqrt{2}\delta^{3/2}}}{2x\sqrt{\pi\delta}} \tag{2.53}$$

Hence, if $x\delta^{3/2} \gg 1$, or $\delta \gg x^{-2/3}$, or if $\delta \approx Cx^{-2/3}$, where $C \gg 1$, then $J_n(x)$ is exponentially small. This happens when

$$\frac{n}{x} \approx 1 + Cx^{-2/3} \tag{2.54}$$

The above analysis is self-consistent since when x is large, δ can still be small even when the above inequalities are satisfied.

The above shows that for the Fourier series in (2.51) to converge, the number of terms n needed is given by

$$n \approx kR + C(kR)^{1/3} \tag{2.55}$$

where R is the radius of the group size. The above formula has also been derived in [19–21].

2.2.9 Error Control

There are two sources of errors in FMM:

- Truncation of the addition theorem;
- Integration error at the final stage.

All these errors can be controlled to be exponentially small. Hence, in principle, FMM has exponential convergence. However, there is often a need to trade off between speed and accuracy.

The truncation of the summation in (2.45) gives rise to a truncation error. The summation in (2.29) is a divergent summation when $P \to \infty$ because the Hankel function grows larger when P is larger than its argument [9]. This unstable behavior comes about because we have exchanged the order of integration and summation when the integral representation was used in (2.45). In principle, we should let P be as large as possible in (2.45) to reduce the truncation error, but this is not possible because of the instability in the resultant formula (2.46) and the associated formula (2.29) due to the finite machine precision. Therefore, if we can truncate the summation before this divergent behavior sets in, we will have a more stable formula. To achieve this stability, we make the argument of the Hankel function $k\rho_{ll'}$ large so that P can be sufficiently large before it is larger than the argument of the Hankel function precluding the divergent behavior of the series. This is achieved by using the factorized form for computation only for groups that are buffered by at least one group size between them making $k\rho_{ll'}$ large. Higher stability and accuracy can be achieved by using more buffer boxes.

Referring to (2.45), to obtain an estimate for the truncation error, we can replace the Hankel function $H_p^{(1)}(k\rho_{l'l})$ by its large argument approximation, namely, [22]

$$H_p^{(1)}(x) \sim \sqrt{\frac{2}{\pi x}} e^{i(x - \frac{p\pi}{2} - \frac{\pi}{4})} \qquad (2.56)$$

when $p < x$. Therefore, the truncation error in (2.45) can be estimated to be

$$\text{error}_{\text{trun}} = \sqrt{\frac{2}{\pi k \rho_{l'l}}} \left| \sum_{|p|>P} e^{-ip(\frac{\pi}{2} + \phi_{l'l} + \phi_{jl'})} J_p(k\rho_{jl'}) \right| \qquad (2.57)$$

Using the large-order approximation for the Bessel function in the above, we can see that the error is exponentially small since a Bessel function becomes exponentially small when its order is larger than its argument [9].

The above argument strikes a delicate balance because we have required that p be small to arrive at (2.56), and in (2.57), the summation is over $|p| > P$. However, since the series in (2.57) converges exponentially fast, only a few terms are needed to estimate the truncation error.

Going from (2.45) to (2.46) requires the use of Parseval's theorem. Notice that $\tilde{\alpha}_{l'l}(\alpha)$ is a function with a finite bandwidth of $2P$. If we choose this bandwidth to be larger than the bandwidth of $e^{-i\mathbf{k}\cdot\boldsymbol{\rho}_{jl'}}$ (see Section 2.2.8), then the error in (2.46) will be exponentially small. In other words, the bandwidth of the integrand in (2.46) and (2.48) is $4P$. A band-limited function can be integrated to exponential precision with a finite number of integration points Q [22, 23] where

$$Q \sim O(P) \qquad (2.58)$$

Consequently, both the truncation error and integration error can be made exponentially small, making these errors controllable.

2.3 MOTIVATION FOR MULTILEVEL METHOD

A natural extension of the two-level FMM is to a multilevel one. As we have seen in Chapter 1, a two-level algorithm reduces the computational complexity by reducing the number of "links" between the current elements. Naturally, a multilevel algorithm will further reduce the number of "links." We call this the multilevel fast multipole algorithm (MLFMA) [4–6]. A simple nesting of the two-level algorithm to the multilevel one will not yield an $O(N \log N)$ algorithm. Interpolators and anterpolators have to be inserted between levels to arrive at an $N \log N$ algorithm [24].

2.3.1 Factorization of the Green's Function

A simple way to have a multistage multilevel implementation of the fast multipole algorithm is to factorize the Green's function into multifactors. Using the addition theorem repeatedly, the Hankel function can be written as the following matrix products:

$$H_0^{(1)}(kr_{ji}) = \beta_{jJ_1} \cdot \overline{\beta}_{J_1 J_2} \cdot \overline{\alpha}_{J_2 I_2} \cdot \overline{\beta}_{I_2 I_1} \cdot \beta_{I_1 i} \qquad (2.59)$$

where

$$[\beta_{jJ_1}]_{1,l} = J_l(k\rho_{jJ_1})e^{il(\phi_{jJ_1} - \pi)} \qquad (2.60)$$

$$[\beta_{J_1 J_2}]_{l,m} = J_{l-m}(k\rho_{J_1 J_2})e^{-i(l-m)\phi_{J_1 J_2}} \qquad (2.61)$$

$$[\alpha_{J_2 I_2}]_{m,n} = H_{m-n}^{(1)}(k\rho_{J_2 I_2})e^{-i(m-n)\phi_{J_2 I_2}} \qquad (2.62)$$

$$[\beta_{I_2 I_1}]_{n,k'} = J_{n-k'}(k\rho_{I_2 I_1})e^{-i(n-k')\phi_{I_2 I_1}} \qquad (2.63)$$

$$[\beta_{I_1 i}]_{k',1} = J_{k'}(k\rho_{I_1 i})e^{-ik'\phi_{I_1 i}} \qquad (2.64)$$

The above equation can be diagonalized just as in the two-level case:

$$H_0^{(1)}(kr_{ji}) = \int_0^{2\pi} d\alpha' \, \tilde{\beta}_{jJ_1}(\alpha')\tilde{\beta}_{J_1 J_2}(\alpha')\tilde{\alpha}_{J_2 I_2}(\alpha')\tilde{\beta}_{I_2 I_1}(\alpha')\tilde{\beta}_{I_1 i}(\alpha) \qquad (2.65)$$

where

$$\tilde{\beta}_{\alpha\beta}(\alpha) = e^{-i\mathbf{k}\cdot\boldsymbol{\rho}_{\alpha\beta}} \qquad (2.66)$$

and $\boldsymbol{\rho}_{\alpha\beta}$ is a vector that points from $\boldsymbol{\rho}_\alpha$ to $\boldsymbol{\rho}_\beta$. Equation (2.65) can also be easily derived from (2.46).

The integral above can be discretized and rewritten as products of matrices and vector:

$$H_0^{(1)}(kr_{ji}) = \tilde{\beta}_{jJ_1}^t \cdot \tilde{\overline{\beta}}_{J_1 J_2} \cdot \tilde{\overline{\alpha}}_{J_2 I_2} \cdot \tilde{\overline{\beta}}_{I_2 I_1} \cdot \tilde{\beta}_{I_1 i} \qquad (2.67)$$

In the above, the $\tilde{\overline{\beta}}_{J_1 J_2}$, $\tilde{\overline{\alpha}}_{J_2 I_2}$, and $\tilde{\overline{\beta}}_{I_2 I_1}$ are diagonal matrices. They are all translation matrices represented by a new basis—the $\overline{\beta}$ matrices translate an outgoing wave from one coordinate center to another outgoing wave from another coordinate center. They also translate an incoming wave at one coordinate center to another incoming wave at another coordinate center. The $\overline{\alpha}$ matrix is as before, a matrix that translates an outgoing wave to incoming waves from one coordinate to another.

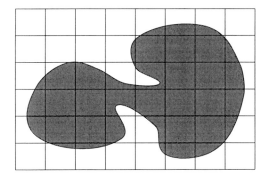

Figure 2.5 A scatterer is immersed in a gridded space.

2.4 THE MULTILEVEL FAST MULTIPOLE ALGORITHM

The result of a matrix vector product is to compute the field due to the ith current element at the jth current element for $i = 1, \ldots, N$ and $j = 1, \ldots, N$. It involves N^2 operations for MOM matrices because they are dense. MLFMA facilitates this product in $O(N \log N)$ operations for sparse scatterers, and in $O(N)$ operations for densely packed scatterers. It is an algorithm for Helmholtz and electromagnetic problems where the matrix-vector product associated with a matrix from a surface integral equation can be effected in $O(N \log N)$ operations. We summarize the important points of MLFMA below:

- First, a scatterer is discretized into finite elements by using the method of moments. Each element is about 0.1λ to 0.2λ in size.

- Then the space that contains the scatterer is gridded into little boxes so that the smallest boxes, at the lowest level, contain at most a few current elements (Figure 2.5).

- In 2D, the boxes are organized into a quad-tree structure with the smallest boxes at the bottom of the inverted tree, and the largest box at the root of the inverted tree. The smallest boxes are also known as the leafy branches of the tree. Not all leafy branches are filled, and the empty boxes with no current element in them are pruned.

- When the element i is far from element j, the field due to current element i at the current element j is computed indirectly using the factorized Green's function in a multilevel multistage fashion. Multistage here implies the aggregation stages followed by translation, and then the disaggregation stages.

- The field due to the near neighbor current elements are computed by the traditional method using the unfactorized MOM matrix element A_{ji}.

- The far element calculation is divided into (1) the aggregation process, and (2) the $\overline{\alpha}$ translation-disaggregation process. The aggregation process is to compute the radiation patterns (outgoing fields) of the sources at different levels starting from the lowest level in the inverted tree using the $\overline{\beta}$ translations. The $\overline{\alpha}$ translation converts outgoing waves into incoming waves, and next, the disaggregation process converts these incoming waves from a higher level into incoming waves at a lower level, and finally, into received fields at the desired field points.

2.4.1 The Aggregation Process

In the aggregation process, we compute the radiation patterns of the current sources. The radiation patterns due to the sources contained in each and every box are computed including the smallest box to the largest box. Instead of computing the radiation patterns of the larger boxes directly, their radiation patterns are computed using the radiation patterns of the smaller boxes contained in them using $\overline{\beta}$ translation. In this manner, we reuse computations and save computer time. To save more computer time, we introduce interpolation operators between levels in the radiation pattern computation. The important points of the aggregation process are summarized as follows (Figure 2.6):

- The radiation functions due to the sources contained in the boxes in the lowest level are first computed using the formula

$$\mathbf{b}_{I_1} = \sum_i \tilde{\beta}_{I_1 i} J_i \qquad (2.68)$$

and the radiation functions of boxes at the next level up are computed using

$$\mathbf{b}_{I_2} = \sum_{I_1} \tilde{\overline{\beta}}_{I_2 I_1} \cdot \mathbf{b}_{I_1} \qquad (2.69)$$

and so on.

- This is done until the radiation function of the boxes at the third highest level is obtained.

- The above is the same as achieving the calculation

$$\mathbf{b}_{I_L} = \sum_{I_{L-1}} \tilde{\overline{\beta}}_{I_L I_{L-1}} \cdots \sum_{I_1} \tilde{\overline{\beta}}_{I_2 I_1} \cdot \sum_i \tilde{\beta}_{I_1 i} J_i \qquad (2.70)$$

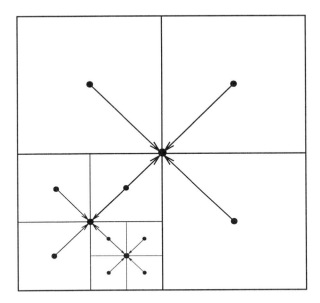

Figure 2.6 The aggregation procedure.

- The radiation patterns of the boxes at the lowest level have narrower spectral content (smaller bandwidth) than the radiation patterns of the boxes at the higher level. Hence, it is not necessary to store all the radiation patterns at different levels with the same sampling rate.

- The radiation patterns at the lower levels are stored with a coarser sampling rate than those at the higher levels. This is imperative to reduce the memory and the CPU time requirements and eventually the memory complexity and computational complexity of the algorithm.

- Hence, it is necessary to interpolate the radiation pattern of the boxes at a given level to a higher sampling rate before it is used to aggregate to build the radiation pattern of the boxes at the next level up.

- Mathematically, this is equivalent to adding interpolation matrices at the dot product in the above formula:

$$\mathbf{b}_{I_L} = \sum_{I_{L-1}} \hat{\bar{\beta}}_{I_L I_{L-1}} \cdot \bar{\mathbf{I}}_{L-1} \cdots \bar{\mathbf{I}}_2 \cdot \sum_{I_1} \hat{\bar{\beta}}_{I_2 I_1} \cdot \bar{\mathbf{I}}_1 \cdot \sum_i \hat{\beta}_{I_1 i} J_i \quad (2.71)$$

- The interpolation matrices are nonsquare and can be made sparse. They will reduce the cost of the aggregation process significantly.

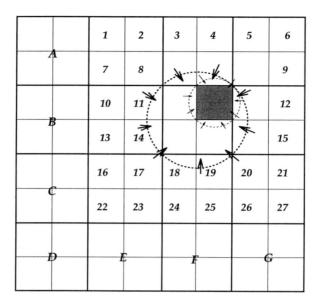

Figure 2.7 The translation-disaggregation procedure.

2.4.2 $\overline{\alpha}$ Translation and Disaggregation

In the aggregation process, the radiation patterns of all the boxes are computed and stored. This sets the stage for the $\overline{\alpha}$ translation and disaggregation process. In MLFMA, the $\overline{\alpha}$ translation and disaggregation procedures are performed simultaneously. This $\overline{\alpha}$ translation process converts the radiation patterns (outgoing fields) into incoming fields. The disaggregation process converts the incoming wave into the received field using the receiving pattern corresponding to the jth field point. We summarize the important points of the translation-disaggregation process below:

- The $\overline{\alpha}$ translation is complicated by the violation of the addition theorem. Therefore, the incoming waves from the outgoing waves cannot be calculated for near neighbor boxes. To improve the numerical accuracy of the calculation, only outgoing waves from boxes separated by at least one box size are converted into incoming waves using the $\overline{\alpha}$ translation.

- Hence, by an inductive procedure, we first assume that the incoming field at a given box is computed and is only due to sources in boxes one box removed from the given box (see Figure 2.7).

- When we want to calculate the incoming field for one of the children of this given box, we will first shift the precomputed incoming field of this box to the center of the child box (by using diagonalized $\overline{\beta}$ translators).

- But this resultant incoming field at the child box then comes from the field generated by sources outside the parent box excluding the neighbors of the parent boxes (boxes A to G at the parent's level).

- Hence, we have to include the field generated by the 27 boxes as shown, which are at the same level as the child box. These 27 boxes are known as the interaction list boxes.

- This is achieved by multiplying the vector that contains the radiation pattern of every one of the interaction list boxes by the diagonalized $\overline{\alpha}$ matrix for the corresponding translation.

- In this manner, the incoming wave at the child box (shaded box in Figure 2.7) will be receiving the incoming field from all the boxes one box removed from itself. We can inductively continue this process, and at the lowest level of the inverted tree, these smallest boxes will be receiving fields from all the other boxes, except for sources in its near neighbor boxes and in the same box.

- When shifting the incoming field from the parent box by the $\overline{\beta}$ translation in the disaggregation process, one needs to down-interpolate, because the spectral content of the receiving pattern of field points inside the parent box is higher than the spectral content of the receiving pattern of field points inside the child box. This is just the converse of the aggregation process, where one needs to up-interpolate.

- This down-interpolation is a smoothing process, or a low-pass filtering process. It is also known as the adjoint interpolation, transpose interpolation, or anterpolation [24]. It is facilitated by the transpose of the interpolation matrices described above.

2.4.3 More on Interpolation and Anterpolation

Interpolation and anterpolation are intimately related to each other. Consider that we are integrating the product of two band-limited functions:

$$I = \int_0^{2\pi} d\alpha\, f_2(\alpha) f_1(\alpha) \tag{2.72}$$

where $f_2(\alpha)$ has twice the bandwidth of $f_1(\alpha)$ (i.e., $f_1(\alpha)$ has a bandwidth of W and $f_2(\alpha)$ has a bandwidth of $2W$). The whole integrand has a bandwidth of $3W$.

Since the integrand is band-limited, we can replace the above with a quadrature rule committing only exponentially small error [22, 23]:

$$I \doteq \sum_{i=1}^{Q} w_i f_2(\alpha_i) f_1(\alpha_i) \qquad (2.73)$$

Moreover, if the functions are exactly band-limited, the use of the Nyquist theorem and evenly spaced sampling allows the above integral to be evaluated exactly.

By Nyquist theorem, we need to store $f_1(\alpha)$ at half the sampling rate required to store $f_2(\alpha)$. Therefore, to reduce storage, we can design an interpolation matrix $\overline{\mathbf{I}}^{\text{int}}$ such that

$$f_1(\alpha_i) = \sum_{j=1}^{M} I_{ji}^{\text{int}} \hat{f}_1(\alpha_j), \quad i = 1, \cdots Q \qquad (2.74)$$

In the above, $M < Q$, and hence, $\overline{\mathbf{I}}^{\text{int}}$ is nonsquare. Consequently, the sum in (2.73) can be written as

$$I = \mathbf{f}_2^t \cdot \overline{\mathbf{W}} \cdot \overline{\mathbf{I}}^{\text{int}} \cdot \hat{\mathbf{f}}_1 \qquad (2.75)$$

where $\overline{\mathbf{W}}$ is a diagonal matrix that contains the weights from a quadrature rule. (For an evenly spaced sample, $\overline{\mathbf{W}}$ is just a constant times an identity matrix, and it can be ignored in the following discussion.) When we read (2.75) from right to left, it says that the interpolator first interpolates a function stored in $\hat{\mathbf{f}}_1$ at a lower sampling rate to a vector with a higher sampling rate. Then the inner product of the resultant vector with the $\mathbf{f}_2$ at a higher sampling rate is formed (ignoring $\overline{\mathbf{W}}$).

If we take the transpose of the above equation, it takes on a new meaning without changing its value:

$$I = \hat{\mathbf{f}}_1^t \cdot \left(\overline{\mathbf{I}}^{\text{int}} \right)^t \cdot \overline{\mathbf{W}} \cdot \mathbf{f}_2 \qquad (2.76)$$

The above says that we first take a function stored in a vector $\mathbf{f}_2$ at a higher sampling rate, and anterpolate (also known as transpose interpolate) the vector, which is the same as multiplying the vector by the transpose of the interpolation operator. The resultant vector then forms an inner product with a function stored in the vector at a lower sampling rate.

Anterpolation here can be interpreted as a filtering, smoothing, or down-sampling process. If the interpolation matrix is a sparse matrix with $O(Q)$ elements, it is achieved in $O(Q)$ operations without the effort of an FFT. From the above, it is seen that the interpolation and anterpolation errors are the same. For band-limited functions, interpolation and anterpolation can be performed with exponential accuracy as we shall see.

For MLFMA, the use of interpolation and anterpolation between levels is essential in arriving at an $O(N \log N)$ algorithm for surface scatterers solved with surface integral equations. We will emphasize the following points about interpolation and anterpolation:

FMM and MLFMA in 2D

- Interpolation and anterpolation are crucial because they allow us to work with a lower sampling rate near the leafy branches of the tree (lower levels in an inverted tree) and thereby to reduce the workload. Since the number of leafy branches is a lot higher, cutting down this workload is important.
- This is essential for achieving $O(N \log N)$ or $O(N)$ computational and memory complexity. Less memory is needed because radiation patterns for smaller box sizes can be stored in shorter vectors. Computing with smaller vectors also reduces the computer time.
- Mathematically, this is equivalent to a new factorization of the Green's operator:

$$H_0^{(1)}(kr_{ji}) = \tilde{\beta}_{jJ_1}^t \cdot \bar{\mathbf{I}}_1^t \cdot \tilde{\bar{\beta}}_{J_1 J_2} \cdot \bar{\mathbf{I}}_2^t \cdots \tilde{\bar{\beta}}_{J_{M-1} J_M} \cdot \tilde{\bar{\alpha}}_{J_M I_M} \cdot \tilde{\bar{\beta}}_{I_M I_{M-1}} \\ \cdots \bar{\mathbf{I}}_2 \cdots \tilde{\bar{\beta}}_{I_2 I_1} \cdot \bar{\mathbf{I}}_1 \cdot \tilde{\beta}_{I_1 i} \tag{2.77}$$

- Since the box size doubles as we ascend the inverted tree, the bandwidth of the radiation functions doubles.
- The interpolation matrices are not square. Asymptotically, they interpolate functions from a given sampling rate to twice its sampling rate.
- The use of interpolation and anterpolation matrices allows us to work with smaller matrices at the lower levels, thereby greatly reducing the computational cost and memory requirements.
- It can be proven that for band-limited signals, interpolation can be achieved with exponential convergence even with a sparse matrix. Alternatively, interpolation and anterpolation can be achieved with FFTs, albeit with added cost.
- The spectral bandwidth or content of the radiation function is proportional to the size of the group, or in this case, the size of the pertinent box. If the box has a diameter D, the spectral content is proportional to kD.

2.4.4 Computational Complexity of MLFMA

In this section, we will analyze the computational and memory complexity of MLFMA when applied to surface and volume scatterers. The MLFMA offers an $O(N \log N)$ computational and memory complexity for performing a matrix-vector product for surface scatterers, and $O(N)$ computational and memory complexity for volume scatterers.

- Let us assume the scatter to be a surface scatterer first, so that the current elements sparsely populate the gridded boxes. Only nonempty boxes are involved in the computation.

- If the nonempty boxes at the lowest level have on the average M current elements, there are about N/M nonempty boxes. If the diameter of the box at the lowest level is d_0, then the spectral bandwidth of the radiation pattern is proportional to kd_0. Moreover, if Q_0 discrete samples are used to capture their radiation pattern, then $Q_0 \sim kd_0$.

- The aggregation process at the lowest level involves matrix-vector products of a $Q_0 \times M$ matrix with a length M vector, repeated N/M times. Therefore, the workload is proportional to $Q_0 N$.

- In MOM, the smallest element size is assumed fixed when N increases. Therefore, the smallest box size and hence Q_0, are fixed when N increases. As a result, the workload at the lowest level is $O(N)$ when $N \to \infty$.

- The boxes at the next higher level are two times bigger than the boxes at the lowest level. Hence, the bandwidth of their radiation pattern is two times wider. Therefore, the sampling rate at the next level up, Q_1, is about two times Q_0.

- However, the number of boxes reduces by a factor of two or more when one ascends the inverted tree. Therefore, the memory requirement for storing the radiation pattern function does not increase compared to the previous level.

- Before shifting the radiation pattern function to the next level, we need to interpolate the sampling rate to twice its rate. This again is an $O(N)$ operation.

- Next, we shift these up-sampled radiation patterns to the center of the parent boxes. Even though we have to work with a radiation pattern with twice the sampling rate, the number of boxes decreases by a factor of two in the worst case as we move up to the next level. Hence, the workload of aggregation remains $O(N)$ at the next level.

- To understand the worst-case scenario, we consider scattering by a horizontal strip. This problem can be solved using MLFMA with a binary tree instead of a quad tree. Then all the leafy branches of the binary tree are filled. From this, we see that the number of boxes decreases by a factor of two as we ascend the inverted tree. For a sinuous scatterer winding through a 2D space, it is better than the worst case, and the number of boxes decreases by more than a factor of two.

- Hence, by induction, the workload at every level of the aggregation process is $O(N)$, and the memory requirement at every level is also of $O(N)$.

- Since there are $\log N$ levels, the whole aggregation process takes $O(N \log N)$ operations with $O(N \log N)$ memory requirement.

- If the scatterers are densely packed as in volume scattering, then the number of boxes will decrease by a factor of four, while the sampling rate will increase by a factor of two when one ascends the inverted tree.

- Hence, the workload is halved every time one ascends the inverted tree. Consequently, the total workload is $N + \frac{N}{2} + \frac{N}{4} + \frac{N}{8} + \cdots \approx 2N$. Therefore, for densely packed scatterers, the whole aggregation process can be completed in $O(N)$ operations.

- The aggregation process is the same as finding the Fourier transform of a function defined at a set of nonuniformly spaced points. It is also a nonuniformly spaced FFT algorithm.

- The disaggregation process is the transpose of the aggregation process. The translation process requires an extra step to address the violation of the addition theorem by introducing a buffer zone. The contribution from the sources in the buffer zone, which are in boxes called the interaction list, introduces an addition $O(N)$ workload at each level to the translation-disaggregation process because the number of boxes in the interaction list is a constant independent of N. Hence, it is also $O(N \log N)$ for surface scatterers and $O(N)$ for dense scatterers.

- MLFMA is used for far-neighbor calculations, and traditional MOM is used for near-neighbor calculations and calculation within the same box at the leafy branches or boxes at the lowest level. Since the size of the smallest boxes at the leafy branches are independent of N, the near-neighbor calculation and the self-box calculation are $O(N)$ also.

- Therefore, the whole matrix-vector multiply can be effected in $O(N \log N)$ operations for surface scatterers, and $O(N)$ for densely packed volume scatterers!

2.5 INTERPOLATION ERROR

The interpolation error in MLFMA can be controlled to be exponentially small by using local interpolation. We briefly describe the theory of local interpolation so that this fact can be understood. Before doing so, we review the theory of global interpolation.

2.5.1 Global Interpolation (Exact)

Assume that $f(t)$ is a band-limited function as shown in Figure 2.8. Nyquist theorem says that $f(t)$ can be recovered from $f(n\Delta t)$, $-\infty < n < \infty$, where $\Delta t \leq \pi/\omega_{max}$,

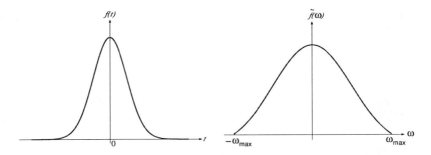

Figure 2.8 A function $f(t)$ and its Fourier transform $\tilde{f}(\omega)$, which is band-limited.

and ω_{max} is the maximum frequency content of $f(t)$. That is, if $\tilde{f}(\omega)$ is the Fourier transform of $f(t)$, then $\tilde{f}(\omega) = 0$ for $|\omega| > \omega_{max}$. In fact, we can write exactly that

$$f(t) = \left(\sum_{n=-\infty}^{\infty} f(n\Delta t)\delta(t - n\Delta t) \right) \otimes \text{sinc}\left(\frac{\pi t}{\Delta t}\right) \quad (2.78)$$

where $\otimes$ implies a convolution. The above assertion can be proven by taking the Fourier transform of (2.78) so that

$$\tilde{f}(\omega) = \tilde{f}_\Delta(\omega)\tilde{B}(\omega) \quad (2.79)$$

where

$$\tilde{f}_\Delta(\omega) = FT\left[\sum_n f(n\Delta t)\delta(t - n\Delta t) \right] \quad (2.80)$$

$$\tilde{B}(\omega) = FT\left[\text{sinc}\left(\frac{\pi t}{\Delta t}\right) \right] \quad (2.81)$$

From (2.80), we see that

$$\tilde{f}_\Delta(\omega) = FT\left[\sum_n f(t)\delta(t - n\Delta t) \right]$$
$$= FT\left[f(t) \sum_n \delta(t - n\Delta t) \right] \quad (2.82)$$
$$= \tilde{f}(\omega) \otimes FT\left[\sum_n \delta(t - n\Delta t) \right] \frac{1}{2\pi}$$

The series $\sum_{n=-\infty}^{\infty} \delta(t - n\Delta t)$ is known as an impulse train. Its Fourier transform is also an impulse train, which can be easily proven. Hence the Fourier transform of the impulse train is:

$$\tilde{T}(\omega) = \sum_{p=-\infty}^{\infty} \frac{2\pi}{\Delta t} \delta\left(\omega - \frac{2\pi pt}{\Delta t}\right) \quad (2.83)$$

Consequently, (2.82) becomes, after using (2.83),

$$\tilde{f}_\Delta(\omega) = \tilde{f}(\omega) \otimes \sum_{p=-\infty}^{\infty} \frac{1}{\Delta t} \delta\left(\omega - \frac{2\pi pt}{\Delta t}\right) = \sum_{p=-\infty}^{\infty} \tilde{f}\left(\omega - \frac{2\pi pt}{\Delta t}\right) \frac{1}{\Delta t} \quad (2.84)$$

The Fourier transform of a sinc function is $\tilde{B}(\omega)$ which is given by

$$\tilde{B}(\omega) = \begin{cases} \Delta t, & |\omega| \le \frac{\pi}{\Delta t} \\ 0, & |\omega| \ge \frac{\pi}{\Delta t} \end{cases} \quad (2.85)$$

The right-hand side in (2.79) is a product of (2.84) and (2.85). Since $\Delta t \le \frac{\pi}{\omega_{max}}$, $\tilde{f}(\omega)$ is exactly reproduced on the right-hand side, and hence, is equal to the left-hand side.

The above is an exact interpolation. If $f(t)$ is sampled at $f(n\Delta t)$, $f(t)$ can be exactly reproduced if it is band-limited and $\Delta t \le \frac{\pi}{\omega_{max}}$. However, evaluating (2.78) is expensive since it uses all $f(n\Delta t)$ to find $f(t)$. If $f(n\Delta t)$ is stored at N points, and N values of $f(t)$ is needed, it is an N^2 algorithm. The next question is whether it is possible to form a local interpolation using fewer points.

2.5.2 Local Interpolation (Exponentially Accurate)

We shall outline the theory for a simple local interpolator [25,26]. The generalization of this theory to other local interpolators should be straightforward. To this end, we replace (2.78) with

$$f(t) \doteq \left(\sum_{n=-\infty}^{\infty} f(n\Delta t)\delta(t - n\Delta t)\right) \otimes \Omega(t) \quad (2.86)$$

For the discussion here, we assume that $\Omega(t) = \text{sinc}(\frac{\pi t}{\Delta t})e^{-(t/W)^2}$ even though other functions are possible. In this case, we choose W to be one or two Δt, so that $\Omega(t)$ is negligibly small after one or two Δt, or $t > W$. In this case, when the convolution (2.86) is performed for a given t on the left-hand side, only a few values of $f(n\Delta t)$ in the neighborhood of t are needed, hence the term *local interpolation*.

To study the error in (2.86), we take the Fourier transform of (2.86) to obtain

$$\tilde{f}(\omega) \doteq \tilde{f}_\Delta(\omega)\tilde{B}_\Omega(\omega) \quad (2.87)$$

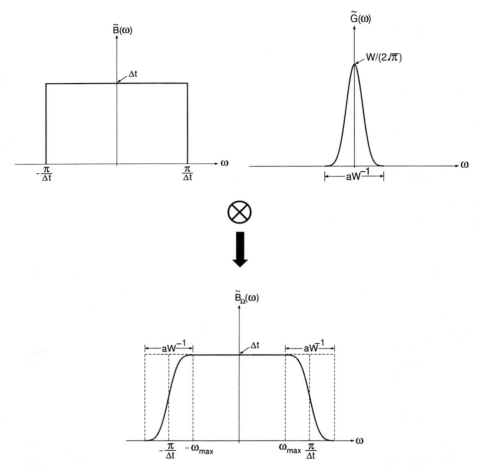

Figure 2.9 The local interpolator is a product of the sinc function with a window function whose Fourier transform is also band-limited. Hence, the Fourier transform of the local interpolator is a convolution of the box function with a band-limited function, producing a band-limited function as shown.

where $\tilde{f}_\Delta(\omega)$ is given by (2.84), and $\tilde{B}_\Omega(\omega)$ is the Fourier transform of $\Omega(t)$, and it is clear that

$$\tilde{B}_\Omega(\omega) = \tilde{B}(\omega) \otimes \tilde{G}(\omega) \tag{2.88}$$

where

$$\tilde{G}(\omega) = e^{-(\omega W)^2/4} \frac{W}{2\sqrt{\pi}} \tag{2.89}$$

and we have made use of the integral identity

$$\int_{-\infty}^{\infty} ds\, e^{-s^2} = \sqrt{\pi} \tag{2.90}$$

Moreover, using (2.90), it is easy to show that

$$\int_{-\infty}^{\infty} d\omega\, \tilde{G}(\omega) = 1 \tag{2.91}$$

Since $\tilde{B}_\Omega(\omega)$ is the convolution of $\tilde{B}(\omega)$ with $\tilde{G}(\omega)$ with unit area, the sketch of $\tilde{B}_\Omega(\omega)$ is as shown in Figure 2.9. We assume $\tilde{B}_\Omega(\omega)$ to be negligibly small when $\omega = 0.5aW^{-1}$, where $a > 4$.

Notice that $\tilde{B}_\Omega(\omega)$ is similar to $\tilde{B}(\omega)$ except in the vicinity of $\pm\frac{\pi}{\Delta t}$. Therefore, if $f(t)$ is band-limited, and if we pick Δt and W such that

$$\frac{\pi}{\Delta t} - 0.5aW^{-1} > \omega_{max} \tag{2.92}$$

then the error in (2.86) is exponentially small. Therefore, for local interpolation, the sampling rate Δt has to be higher than the Nyquist rate. Moreover, if (2.92) is satisfied, the difference between $\tilde{B}(\omega)$ and $\tilde{B}_\Omega(\omega)$ is exponentially small and hence, the exponential convergence of this local interpolator.

The local interpolator in (2.86) can be replaced with other local interpolators such as the approximate prolate spheroid, the Lagrange, or bicubic spline interpolators. In some of these interpolators, their Fourier transforms may not be a band-limited function with an exponentially small tail as the example shown here, but with an algebraically decaying tail. However, the amplitude of the algebraically decaying tail can be made exponentially smaller by increasing the order of the interpolator. In this case, exponential convergence is still achieved.

2.6 FMM AND GROUP THEORY

There is an intimate relationship between FMM and group theory. The success of FMM relies on the factorization of the Green's function. This is intimately related to the translational invariance of the physics of free-space—physics experiment performed in one part of space is the same as the same experiment performed in another part of space. The solutions of the Helmholtz equation or Maxwell's equations in free space are translationally invariant, exhibiting translational symmetry. Translational symmetry is often expressed by the translation group in group theory. To facilitate the discussion, we give a brief overview of groups.

2.6.1 Groups

A group is a set of elements with an operation defined such that the following properties hold:

- *Closure*: For all a and b in the group $\mathcal{G}$, ab is also in $\mathcal{G}$.
- *Associativity*: For all a, b, c in $\mathcal{G}$, $(ab)c = a(bc)$.
- *Unit Element*: For all a in $\mathcal{G}$, there exists an element e such that $ea = a$.
- *Inverse Element*: For all a in $\mathcal{G}$, there exists an element a^{-1} such that $a^{-1}a = e$.

If a group has the property that $ab = ba$ (i.e., the elements are commutative), the group is Abelian. There are many interesting properties associated with groups, but for the lack of space, we will confine ourselves to simple properties so as not to encumber the discussion here [27, 28].

2.6.2 Example of a Group

The translation operator $T_\mathbf{a}$ has the property of

$$T_\mathbf{a} f(\mathbf{r}) = f(\mathbf{r} - \mathbf{a}) \qquad (2.93)$$

All possible elements of the translation operator $T_\mathbf{a}$ form the translation group. If we call this group $\mathcal{T}$, it is easy to verify the following:

- If $T_\mathbf{a} \in \mathcal{T}$, $T_\mathbf{b} \in \mathcal{T}$, then $T_\mathbf{a} T_\mathbf{b} \in \mathcal{T}$.
- $(T_\mathbf{a} T_\mathbf{b}) T_\mathbf{c} = T_\mathbf{a}(T_\mathbf{b} T_\mathbf{c})$.
- $T_\mathbf{a} I = T_\mathbf{a}$, where I is the identity operator.
- $T_{-\mathbf{a}} T_\mathbf{a} = I$.

Furthermore, it can be shown that $T_\mathbf{a} T_\mathbf{b} = T_\mathbf{b} T_\mathbf{a}$. Hence, the translation group is Abelian. The translation group in 2D is usually called $\mathcal{T}_2$ while that in 3D is called $\mathcal{T}_3$.

2.6.3 Representation of a Group

The translation operator above can be regarded as an abstract notion. However, integral, differential, or matrix operators can be derived to explicitly represent it. Consider a function in 1D which is $f(x)$. Assuming that its derivative exists to infinite order, then its shift can be written by Taylor series expansion as:

$$T_a f(x) = f(x - a) = \sum_{n=0}^{\infty} \frac{(-a)^n}{n!} \left(\frac{d}{dx}\right)^n f(x) = \exp\left(-a \frac{d}{dx}\right) f(x) \qquad (2.94)$$

In two and three dimensions, this can be generalized to

$$T_{\mathbf{a}} f(\mathbf{r}) = f(\mathbf{r} - \mathbf{a}) = \exp(-\mathbf{a} \cdot \nabla) f(\mathbf{r}) \tag{2.95}$$

Hence, a representation of the translation operator in coordinate space is that

$$T_{\mathbf{a}} = \exp(-\mathbf{a} \cdot \nabla) \tag{2.96}$$

In (2.95), if we replace

$$f(\mathbf{r}) = \frac{1}{(2\pi)^3} \int_{-\infty}^{\infty} d\mathbf{k}\, \tilde{f}(\mathbf{k}) e^{i\mathbf{k} \cdot \mathbf{r}} \tag{2.97}$$

then it becomes

$$T_{\mathbf{a}} f(\mathbf{r}) = f(\mathbf{r} - \mathbf{a}) = \frac{1}{(2\pi)^3} \int_{-\infty}^{\infty} d\mathbf{k}\, \tilde{f}(\mathbf{k}) e^{i\mathbf{k} \cdot \mathbf{r}} e^{-i\mathbf{k} \cdot \mathbf{a}} \tag{2.98}$$

Hence, the representation of the translation operator becomes very simple in the Fourier space (momentum space). Its effect can be represented by multiplying the Fourier transform $\tilde{f}(\mathbf{k})$ by a scalar number $e^{-i\mathbf{k} \cdot \mathbf{a}}$.

2.6.3.1 Cylindrical Harmonics

We can also find the representation of the translation operator in subspaces. The subspace we will first consider will be the space of functions spanned by the solution of the source-free Helmholtz equation. In this space, the translation operators can be represented by a matrix operator.

For example, we know that

$$\phi_m(\boldsymbol{\rho}) = J_m(k\rho) e^{im\phi} \tag{2.99}$$

is a solution of the Helmholtz equation

$$(\nabla^2 + k^2) \phi_m(\boldsymbol{\rho}) = 0 \tag{2.100}$$

in 2D. Since the Helmholtz operator $\nabla^2 + k^2$ commutes with the translation operator, then

$$T_{\boldsymbol{\rho}'} \phi_m(\boldsymbol{\rho}) = \phi_m(\boldsymbol{\rho} - \boldsymbol{\rho}') = J_m(k|\boldsymbol{\rho} - \boldsymbol{\rho}'|) e^{im\phi''} \tag{2.101}$$

is also a solution to the Helmholtz equation (see Figure 2.3 for the definition of variables). We can use the addition theorem to expand the right-hand side of (2.101) to obtain

$$J_m(k|\boldsymbol{\rho} - \boldsymbol{\rho}'|) e^{im\phi''} = \sum_{n=-\infty}^{\infty} J_{n-m}(k\rho') e^{-i(n-m)\phi'} J_n(k\rho) e^{in\phi} \tag{2.102}$$

The above can be put in a vector notation:

$$\Re g \Psi^t(\rho - \rho') = \Re g \Psi^t(\rho) \cdot \overline{\beta}(\rho') \tag{2.103}$$

where

$$[\Psi(\rho)]_n = H_n^{(1)}(k\rho)e^{in\phi} \tag{2.104}$$

$$[\overline{\beta}(\rho')]_{nm} = J_{n-m}(k\rho')e^{-i(n-m)\phi'} \tag{2.105}$$

and $\Re g H_n^{(1)}(x) = J_n(x)$. The above addition theorem can also be derived by the application of the translation operator (2.96) to (2.99) [27].

Consequently, we can put (2.99) into a column vector and rewrite (2.101) as

$$T_{\rho'} \Re g \Psi^t(\rho) = \Re g \Psi^t(\rho - \rho') = \Re g \Psi^t(\rho) \cdot \overline{\beta}(\rho') \tag{2.106}$$

Therefore, the effect of $T_{\rho'}$ on $\Re g \Psi^t(\rho)$ is the same as multiplying or transforming $\Re g \Psi^t(\rho)$ with the infinite dimensional matrix $\overline{\beta}(\rho')$. We call $\overline{\beta}(\rho')$ the matrix representation of the translation operator $T_{\rho'}$.

It follows that

$$\begin{aligned} T_{\rho_1} T_{\rho_2} \Re g \Psi^t(\rho) &= T_{\rho_1} \Re g \Psi^t(\rho) \cdot \overline{\beta}(\rho_2) \\ &= \Re g \Psi^t(\rho) \cdot \overline{\beta}(\rho_1) \cdot \overline{\beta}(\rho_2) \end{aligned} \tag{2.107}$$

In this notation, the matrix representation of $T_{\rho_1} T_{\rho_2}$ is

$$\overline{\beta}(\rho_1) \cdot \overline{\beta}(\rho_2) \tag{2.108}$$

A general solution of the source-free Helmholtz equation can be written as

$$\phi(\rho) = \Re g \Psi^t(\rho) \cdot \mathbf{a} \tag{2.109}$$

By using the translation operator, we can easily factorize the solution; that is,

$$T_{\rho'} \phi(\rho) = \Re g \Psi^t(\rho) \cdot \overline{\beta}(\rho') \cdot \mathbf{a} \tag{2.110}$$

Repeated application of the translation operator allows us to factorize the solution of the Helmholtz equation. The $\overline{\beta}$ matrices also faithfully reproduce the properties of the translation group. They are also commutative just as the translation operators.

2.6.3.2 Plane Waves

Now we replace (2.99) by

$$\phi_m(\rho) = e^{i\mathbf{k}_m \cdot \rho} \tag{2.111}$$

Here, the number of plane waves can be chosen to be independent of each other, and they span a subspace of functions which can be expressed as a sum of these plane waves. Then,

$$T_{\rho'} \phi_m(\rho) = \phi_m(\rho - \rho') = e^{i\mathbf{k}_m \cdot \rho} e^{-i\mathbf{k}_m \cdot \rho'} = e^{i\mathbf{k}_m \cdot \rho} \tilde{\beta}_m(\rho') \tag{2.112}$$

where $\tilde{\beta}_m(\rho') = e^{-i\mathbf{k}_m \cdot \rho'}$ is a scalar number instead of a matrix in (2.103). If we stack (2.111) into a column vector, we can rewrite (2.112) as

$$T_{\rho'}\phi^t(\rho) = \phi^t(\rho - \rho') = \phi^t(\rho) \cdot \tilde{\overline{\beta}}(\rho') \qquad (2.113)$$

where $\tilde{\overline{\beta}}(\rho')$ is now a diagonal matrix whose element on the diagonal is given by the scalar number $\tilde{\beta}_m(\rho')$. It is still a matrix representation of the translation operator $T_{\rho'}$, but it is a diagonal representation. The difference between (2.99) and (2.111) is that (2.99) uses cylindrical harmonics as a basis, while (2.111) uses plane waves as a basis. Hence, the matrix representation of the translation operator under a plane wave basis is diagonal.

A general solution of the source-free Helmholtz equation can be expressed as

$$\phi(\rho) = \sum_m e^{i\mathbf{k}_m \cdot \rho} a_m = \phi^t(\rho) \cdot \mathbf{a} \qquad (2.114)$$

where $\mathbf{k}_m$ corresponds to plane waves propagating in different directions. Equation (2.114) can be made as accurately as one wishes by including more plane wave directions, including inhomogeneous plane waves. As in (2.110), we can easily factorize (2.114) as is done in (2.110) by

$$T_{\rho'}\phi(\rho) = \phi^t(\rho) \cdot \tilde{\overline{\beta}}(\rho') \cdot \mathbf{a} \qquad (2.115)$$

The deeper reason for the existence of a diagonal representation of the translation operator is that the translation group is Abelian. Hence, the translation operators commute, and subsequently, they can be represented by a scalar number as in (2.112), or a diagonal matrix as in (2.113).

2.6.4 Green's Function

The Green's function is a solution of the Helmholtz equation when the source is a point source in space. When the source is at the origin, the Green's function in 2D is a solution of

$$\left(\nabla^2 + k^2\right) g(\rho) = -\delta(\rho) \qquad (2.116)$$

where $g(\rho) = \frac{i}{4} H_0^{(1)}(k\rho)$. The Green's function satisfies the radiation condition at infinity. Since the Helmholtz operator commutes with the translation operator, when the point source is being translated to a location ρ', the Green's function is given by

$$g(\rho - \rho') = \frac{i}{4} H_0^{(1)}(k|\rho - \rho'|) \qquad (2.117)$$

It is the solution of

$$\left(\nabla^2 + k^2\right) g(\rho - \rho') = -\delta(\rho - \rho') \qquad (2.118)$$

Since the variables ρ and ρ' are interchangeable, the Green's function is also a solution of
$$\left(\nabla'^2 + k^2\right) g(\rho - \rho') = -\delta(\rho - \rho') \tag{2.119}$$
where ∇' is expressed in terms of x', y', z' variables. In other words, the Green's function is symmetric with respect to the ρ and ρ' variables. For the following discussion, we will ignore the factor $\frac{i}{4}$ in (2.117), and the Green's function can then be written as

$$g(\rho - \rho') = \begin{cases} \Re g \boldsymbol{\Psi}^t(\rho) \cdot \hat{\boldsymbol{\Psi}}(\rho'), & |\rho| < |\rho'| \\ \boldsymbol{\Psi}^t(\rho) \cdot \Re g \boldsymbol{\Psi}^*(\rho'), & |\rho'| < |\rho| \end{cases} \tag{2.120}$$

We assume k is real in the above, and that $\left[\hat{\boldsymbol{\Psi}}(\rho')\right]_n = H_n^{(1)}(k\rho')e^{-in\phi'}$.

The bottom part of the equation can be derived by applying the translation operator (2.96) to $g(\rho)$ with $\mathbf{a} = \rho'$. The rationale for this is that a translated solution of the Helmholtz equation satisfying the radiation condition still satisfies the Helmholtz equation with the radiation condition. The top part of the equation can be derived by using the symmetry property of the Green's function. The dichotomous behavior of the expansion is necessary for the convergence of the resultant expansion.

In the lower part of (2.120), $\boldsymbol{\Psi}^t(\rho)$ is singular when $|\rho| = 0$, but for the region $|\rho| > |\rho'|$, we can apply $T_\mathbf{a}$ to (2.120) to obtain

$$T_\mathbf{a} g(\rho - \rho') = g(\rho - \rho' - \mathbf{a}) = \boldsymbol{\Psi}^t(\rho - \mathbf{a}) \cdot \Re g \boldsymbol{\Psi}^*(\rho') \tag{2.121}$$

where we ensure that $|\rho - \mathbf{a}| > |\rho'|$ after translation for the convergence of the resultant expression. We can apply the addition theorem to express

$$\boldsymbol{\Psi}^t(\rho - \mathbf{a}) = \Re g \boldsymbol{\Psi}^t(\rho) \cdot \overline{\alpha}(\mathbf{a}), \qquad |\rho| < |\mathbf{a}| \tag{2.122}$$

where

$$[\overline{\alpha}(\mathbf{a})]_{nm} = H_{n-m}^{(1)}(ka) e^{-i(n-m)\phi_a} \tag{2.123}$$

The above addition theorem can also be derived by applying the raising operator to the prime coordinates in the upper part of (2.120) [13, 27].

Consequently, we have

$$g(\rho - \rho' - \mathbf{a}) = \Re g \boldsymbol{\Psi}^t(\rho) \cdot \overline{\alpha}(\mathbf{a}) \cdot \Re g \boldsymbol{\Psi}^*(\rho') \tag{2.124}$$

Next, we want to show that this $\overline{\alpha}$ matrix commutes with the $\overline{\beta}$ matrix. To this end, we apply $T_\mathbf{b}$ to the above to further obtain

$$g(\rho - \rho' - \mathbf{a} - \mathbf{b}) = \Re g \boldsymbol{\Psi}^t(\rho) \cdot \overline{\beta}(\mathbf{b}) \cdot \overline{\alpha}(\mathbf{a}) \cdot \Re g \boldsymbol{\Psi}^*(\rho') \tag{2.125}$$

Equation (2.124) is also a solution of the source-free Helmholtz equation in the ρ' variable. We can rewrite it as

$$g(\rho - \rho' - \mathbf{a}) = \Re g \boldsymbol{\Psi}^\dagger(\rho') \cdot \overline{\alpha}^t(\mathbf{a}) \cdot \Re g \boldsymbol{\Psi}(\rho) \tag{2.126}$$

Applying $T_{-\mathbf{b}}$ to the ρ' variable, we have

$$\begin{aligned}g(\rho - \rho' - \mathbf{a} - \mathbf{b}) &= \Re g \mathbf{\Psi}^\dagger(\rho') \cdot \overline{\boldsymbol{\beta}}^*(-\mathbf{b}) \cdot \overline{\boldsymbol{\alpha}}^t(\mathbf{a}) \cdot \Re g \mathbf{\Psi}(\rho) \\ &= \Re g \mathbf{\Psi}^t(\rho) \cdot \overline{\boldsymbol{\alpha}}(\mathbf{a}) \cdot \overline{\boldsymbol{\beta}}^\dagger(-\mathbf{b}) \cdot \Re g \mathbf{\Psi}^*(\rho')\end{aligned} \quad (2.127)$$

Comparing (2.125) and (2.127), we have

$$\overline{\boldsymbol{\beta}}(\mathbf{b}) \cdot \overline{\boldsymbol{\alpha}}(\mathbf{a}) = \overline{\boldsymbol{\alpha}}(\mathbf{a}) \cdot \overline{\boldsymbol{\beta}}^\dagger(-\mathbf{b}) \quad (2.128)$$

But from (2.105), it can be shown that $\overline{\boldsymbol{\beta}}(\mathbf{b}) = \overline{\boldsymbol{\beta}}^\dagger(-\mathbf{b})$; hence, the above implies that

$$\overline{\boldsymbol{\beta}}(\mathbf{b}) \cdot \overline{\boldsymbol{\alpha}}(\mathbf{a}) = \overline{\boldsymbol{\alpha}}(\mathbf{a}) \cdot \overline{\boldsymbol{\beta}}(\mathbf{b}) \quad (2.129)$$

The fact that $\overline{\boldsymbol{\beta}}$ and $\overline{\boldsymbol{\alpha}}$ commute implies that they can be diagonalized by the same basis. It has been shown that the plane-wave basis diagonalizes both $\overline{\boldsymbol{\alpha}}$ and $\overline{\boldsymbol{\beta}}$ matrices simultaneously.

2.6.5 Plane-Wave Representation of the Green's Function

The Green's function given by (2.117) can be expanded in terms of plane waves, namely,

$$g(\boldsymbol{\rho}_i - \boldsymbol{\rho}_j) = \frac{i}{8\pi} \int_{-\infty}^{\infty} dk_x \frac{e^{ik_x(x_i-x_j) \pm ik_y(y_i-y_j)}}{k_y}, \quad \begin{matrix} y_i > y_j \\ y_i < y_j \end{matrix} \quad (2.130)$$

When $y_i > y_j$ or $y_i < y_j$, the integrand represents plane waves including inhomogeneous waves. If we confine ourselves to the region where $y_i > y_j$ always, (2.130) can be written as

$$g(\boldsymbol{\rho}_i - \boldsymbol{\rho}_j) = \frac{i}{8\pi} \int_{-\infty}^{\infty} dk_x \frac{e^{i\mathbf{k}_\rho \cdot (\boldsymbol{\rho}_i - \boldsymbol{\rho}_j)}}{k_y} \quad (2.131)$$

As such, we can write

$$\begin{aligned}\boldsymbol{\rho}_i - \boldsymbol{\rho}_j = \boldsymbol{\rho}_{ij} &= (\boldsymbol{\rho}_i - \boldsymbol{\rho}_l) + (\boldsymbol{\rho}_l - \boldsymbol{\rho}_{l'}) + (\boldsymbol{\rho}_{l'} - \boldsymbol{\rho}_j) \\ &= \boldsymbol{\rho}_{il} + \boldsymbol{\rho}_{ll'} + \boldsymbol{\rho}_{l'j}\end{aligned} \quad (2.132)$$

where $\boldsymbol{\rho}_{ab} = \boldsymbol{\rho}_a - \boldsymbol{\rho}_b$ in the above. Using (2.132) in (2.131), we have

$$g(\boldsymbol{\rho}_{ij}) = \frac{i}{8\pi} \int_{-\infty}^{\infty} dk_x \frac{1}{k_y} e^{i\mathbf{k}_\rho \cdot \boldsymbol{\rho}_{il}} e^{i\mathbf{k}_\rho \cdot \boldsymbol{\rho}_{ll'}} e^{i\mathbf{k}_\rho \cdot \boldsymbol{\rho}_{l'j}} \quad (2.133)$$

The above can be approximated by a quadrature rule to give

$$g(\boldsymbol{\rho}_{ij}) = \sum_m w_m e^{i\mathbf{k}_\rho^m \cdot \boldsymbol{\rho}_{il}} e^{i\mathbf{k}_\rho^m \cdot \boldsymbol{\rho}_{ll'}} e^{i\mathbf{k}_\rho^m \cdot \boldsymbol{\rho}_{l'j}} = \tilde{\boldsymbol{\beta}}_{il}^t \cdot \tilde{\overline{\boldsymbol{\alpha}}}_{ll'} \cdot \tilde{\boldsymbol{\beta}}_{l'j} \quad (2.134)$$

where $\tilde{\bar{\alpha}}_{ll'}$ is a diagonal matrix whose elements are

$$\left[\tilde{\bar{\alpha}}_{ll'}\right]_{mm'} = \delta_{mm'} w_m e^{i\mathbf{k}_\rho^m \cdot \boldsymbol{\rho}_{ll'}} \qquad (2.135)$$

and

$$\left[\tilde{\beta}_{\gamma\beta}\right]_{mm'} = e^{i\mathbf{k}_\rho^m \cdot \boldsymbol{\rho}_{\gamma\beta}} \qquad (2.136)$$

Such plane-wave representation of the translation operator has been used to design fast algorithms in computational electromagnetics [17, 18, 29].

2.7 CONCLUSION

This chapter describes the details of the fast multipole method and the multilevel fast multipole algorithm in 2D. The important point is that MLFMA allows a matrix-vector product to be effected in $O(N \log N)$ operations, where the matrix is associated with an integral equation of the Helmholtz and electromagnetic problem. The key to obtaining an $O(N \log N)$ algorithm in both CPU time and memory requirement is the use of interpolation and anterpolation between levels. We further show that the error in the interpolation and anterpolation are controllable. We also discuss the relationship of FMM to group theory, and explain the deeper reason for the diagonalizability of the translation operators that exist in FMM. This method can be generalized to 3D and will be discussed in the next chapter. More references can be found on this topic in [30].

REFERENCES

1. V. Rokhlin, "Rapid solution of integral equations of scattering theory in two dimensions," *J. Comput. Phys.*, vol. 86, pp. 414–439, 1990.

2. L. Greengard and V. Rokhlin, "A fast algorithm for particle simulations," *Journal of Computational Physics,* vol. 73, pp. 325–348, 1987.

3. C. C. Lu and W. C. Chew, "A fast algorithm for solving hybrid integral equation," *IEE Proceedings-H*, vol. 140, no. 6, pp. 455–460, 1993.

4. C. C. Lu and W. C. Chew, "A multilevel algorithm for solving boundary-value scattering," *Micro. Opt. Tech. Lett.*, vol. 7, no. 10, pp. 466–470, July 1994.

5. J. M. Song and W. C. Chew, "Multilevel fast-multipole algorithm for solving combined field integral equations of electromagnetic scattering," *Micro. Opt. Tech. Lett.*, vol. 10, no. 1, pp. 14–19, Sept. 1995.

6. J. M. Song, C.-C. Lu, and W. C. Chew, "Multilevel fast multiple algorithm for electromagnetic scattering by large complex objects," *IEEE Trans. Ant. Propag.*, vol. 45, no. 10, pp. 1488–1493, Oct. 1997.

7. B. Dembart and E. Yip, "A 3D fast multipole method for electromagnetics with multiple levels," *Ann. Rev. Prog. Appl. Computat. Electromag.*, vol. 1, pp. 621–628, 1995.

8. R. F. Harrington, *Field Computation by Moment Method*, Malabar, FL: Krieger Publ., 1982.

9. M. Abramowitz and I. A. Stegun, *Handbook of Mathematical Functions*, New York: Dover Publ., 1965.

10. M. R. Hestenes and E. Stiefel, "Methods of conjugate gradients for solving linear systems," *J. Res. Nat. Bur. Stand.*, Sect. B, vol. 49, pp. 409–436, 1952.

11. T. K. Sarkar, "On the application of the generalized biconjugate gradient method," *J. Electromag. Waves Appl.*, vol. 1, no. 3, pp. 223–242, 1987.

12. A. Messiah, *Quantum Mechanics*, Amsterdam: North-Holland, Amsterdam, 1958.

13. W. C. Chew, *Waves and Fields in Inhomogeneous Media*, New York: Van Nostrand Reinhold, 1990; reprinted by IEEE Press, 1995.

14. V. Rokhlin, "Diagonal forms of translation operators for the Helmholtz equation in three dimensions," *Appl. Comput. Harmon. Anal.*, vol. 1, no. 1, pp. 82–93, 1993.

15. M. Epton, B. Dembart, "Multipole translation theory for the three-dimensional Laplace and Helmholtz equations," *SIAM J. Sci. Comput.*, vol. 16, no. 4, pp. 865-897, 1995.

16. W. C. Chew, S. Koc, J. M. Song, C. C. Lu, and E. Michielssen, "A succinct way to diagonalize the translation matrix in three dimensions," *Micro. Opt. Tech. Lett.*, vol. 15, no. 3, pp. 144–147, June 1997.

17. E. Michielssen and W. C. Chew, "Fast steepest descent path algorithm for analyzing scattering from two-dimensional objects," *Radio Science*, vol. 31, no. 5, pp. 1215–1224, Sept.-Oct. 1996.

18. B. Hu and W. C. Chew, "Fast inhomogeneous plane wave algorithm for electromagnetic solutions in layered medium structures - 2D case," *Radio Science*, vol. 35, no. 1, 2000.

19. W. J. Wiscombe, "Improved Mie scattering algorithms," *Applied Optics*, vol. 19, no. 9, May 1980.

20. O. M. Bucci and G. Franceschetti, "On the spatial bandwidth of scattered fields," *IEEE Trans. Antennas Propag.*, vol. 35, no. 12, pp. 1445–1455, Dec. 1987.

21. V. Rokhlin, "Sparse diagonal forms for translation operations for the Helmholtz equation in two dimensions," *Research Report YALEU/DCS/RR-1095,* Dept. Comp. Sci. Yale University, Dec. 1995.

22. S. Koc, J. Song, and W. C. Chew, "Error analysis for the numerical evaluation of the diagonal forms of the scalar spherical addition theorem," *SIAM J. Numerical Analysis,* vol. 36, no. 3, pp. 906-921, 1999.

23. F. Hildebrand, *Introduction to Numerical Analysis*, New York: McGraw-Hill, 1874.

24. A. Brandt, "Multilevel computations of integral transforms and particle interactions with oscillatory kernels," *Comp. Phys. Comm.*, vol. 65, pp. 24–38, 1991.

25. J. J. Knab, "Interpolation of band-limited functions using the approximate prolate series," *IEEE Trans. Inform. Theory*, vol. 25, no. 6, pp. 717–720, 1979.

26. O. M. Bucci, C. Gennareli, and C. Savarese, "Optimal interpolation of radiated fields over a sphere," *IEEE Trans. Antennas Propag.*, vol. 39, no. 11, pp. 1633–1643, 1991.

27. W. K. Tung, *Group Theory in Physics,* Philadelphia, PA: World Scientific Publ., 1985.

28. L. M. Falicov, *Group Theory and its Physical Applications,* Chicago: University of Chicago Press, 1966.

29. W. C. Chew and C. C. Lu, "A fast recursive algorithm to compute the wave scattering solution of a large strip," *J. Comput. Phys.*, vol. 107, pp. 378–387, 1993.

30. W. C. Chew, J. M. Jin, C. C. Lu, E. Michielssen, J. M. Song, "Fast solution methods in electromagnetics," *IEEE Trans. Ant. Propag.*, vol. 45, pp. 533–543, 1997.

3
FMM and MLFMA in 3D and Fast Illinois Solver Code

Jiming Song and Weng Cho Chew

3.1 INTRODUCTION

We have discussed the 2D fast multipole method (FMM) and the multilevel fast multipole algorithm (MLFMA) in the last chapter. FMM was extended to 3D by Coifman, Rokhlin, and Wandzura [1] and was implemented in solving the electric field integral equation (EFIE) [2, 3] and combined field integral equation (CFIE) [4] for scattering from conducting objects, reducing the complexity of a matrix-vector multiply and memory requirement from $O(N^2)$ to $O(N^{3/2})$. Later, Song, Lu, and Chew [5, 6] generalized to MLFMA with $O(N\log N)$ complexity and memory requirement using translation, interpolation, anterpolation (adjoint interpolation), and a grid-tree data structure. Dembart and Yip [7, 8] have implemented MLFMA using signature function, interpolation, and filtering, with a complexity of $O(N\log^2 N)$. Gyure and Stalzer [9] also implemented MLFMA using FFT and 1D FMM with a complexity of $O(N\log^2 N)$. FMM and MLFMA have been applied to problems in a number of areas in electromagnetics such as impedance boundary condition (IBC) [10], homogeneous media and inhomogeneous media using hybridization with the finite element method (FEM) [11, 12], impedance of surface-wire structures [13], electromagnetic compatibility analysis [14], and complex images in microstrip structures [15]. MLFMA has been combined with the higher-order Galerkin's method [16, 17] (see also Chapter 14) and Nystrom method [18].

Since the basic concepts in 3D FMM and MLFMA are similar to those discussed in the last chapter, we will give a brief description of FMM and MLFMA in 3D in the next section. The differences between the 2D and 3D FMM and MLFMA will be emphasized. Then we will give a detailed discussion of error analysis in the 3D FMM and MLFMA. Later, we will discuss some implementations for large-scale computing, such as the block diagonal preconditioner, good initial guess, bistatic to monostatic RCS approximation, interpolation of translation matrix in MLFMA, and calculating radiated fields using MLFMA. Finally, we will present the Fast Illinois Solver Code (FISC) and discuss its capabilities, complexity, accuracy, and scaling of memory requirements and CPU time. More results will be shown in the end of this chapter.

3.2 THREE-DIMENSIONAL FMM AND MLFMA

Before we derive the formulations of FMM and MLFMA in 3D, we will briefly introduce the integral equations and the method of moments (MOM). As presented in the last chapter, one way to derive FMM formulations is to use the diagonalized form of the translation matrix, which has been discussed by some researchers [19–21]. In this section, we will derive it in a simple way by using the addition theorem and an elementary identity.

3.2.1 Integral Equations and the Method of Moments

The electromagnetic field scattering by a 3D arbitrarily shaped conductor can be obtained by finding the solution of an integral equation where the unknown function is the induced current distribution. For conducting objects, the electric field integral equation (EFIE) is given by

$$\hat{t} \cdot \int_S \overline{\mathbf{G}}(\mathbf{r}, \mathbf{r}') \cdot \mathbf{J}(\mathbf{r}') dS' = \frac{4\pi i}{k\eta} \hat{t} \cdot \mathbf{E}^i(\mathbf{r}) \tag{3.1}$$

for $\mathbf{r}$ on surface S, where $\mathbf{J}(\mathbf{r})$ is the unknown current distribution, $\hat{t}$ is any unit tangent vector on S, and

$$\overline{\mathbf{G}}(\mathbf{r}, \mathbf{r}') = (\overline{\mathbf{I}} - \frac{1}{k^2} \nabla \nabla') g(\mathbf{r}, \mathbf{r}') \tag{3.2}$$

$$g(\mathbf{r}, \mathbf{r}') = \frac{e^{ikR}}{R}, \qquad R = |\mathbf{r} - \mathbf{r}'| \tag{3.3}$$

For closed conducting objects, the magnetic field integral equation (MFIE) is given by

$$2\pi \hat{t} \cdot \mathbf{J}(\mathbf{r}) - \hat{t} \cdot \hat{n} \times \nabla \times \int_S dS' g(\mathbf{r},\mathbf{r}') \mathbf{J}(\mathbf{r}') = 4\pi \hat{t} \cdot \hat{n} \times \mathbf{H}^i(\mathbf{r}) \quad (3.4)$$

for $\mathbf{r}$ approaching to S from the outside, where $\hat{n}$ is an outwardly directed normal.

At the resonant frequencies of the cavity formed by the closed objects, both EFIE and MFIE have null-space solutions (nonzero solutions without excitations). The currents from the null-space solutions of EFIE will not radiate, and null-space solutions of MFIE will radiate. Therefore, neither EFIE nor MFIE can give correct current solutions, while EFIE gives a correct RCS but MFIE does not. The combined field integral equation (CFIE) [22] is used to solve this problem, which is simply a linear combination of (3.1) and (3.4) and is of the form

$$\alpha \text{EFIE} + (1-\alpha)\frac{i}{k}\text{MFIE} \quad (3.5)$$

The combination parameter α ranges from 0 to 1 and can be chosen to be any value within this range. Some studies have been done on choosing an optimum α, and they have found $\alpha = 0.2$ to be an overall good choice [23].

Let us apply the method of moments to CFIE. The unknown current $\mathbf{J}(\mathbf{r})$ is first expanded in an appropriately chosen set of basis functions $\{\mathbf{j}_i\}$,

$$\mathbf{J}(\mathbf{r}) = \sum_{i=1}^N a_i \mathbf{j}_i(\mathbf{r}) \quad (3.6)$$

where a_i are the unknown expansion coefficients. The CFIE is discretized by substituting the above expansion in terms of unknowns a_i. Then, rather than forcing CFIE to be satisfied for $\mathbf{r}$ approaching S from the outside, it is multiplied by a set of N testing functions $\{\mathbf{t}_j\}$ and the inner products are taken to yield

$$\sum_{i=1}^N A_{ji} a_i = F_j, \qquad j = 1, 2, \cdots, N \quad (3.7)$$

where

$$F_j = \frac{4\pi i}{k} \left[\frac{\alpha}{\eta} \int_S dS \mathbf{t}_j(\mathbf{r}) \cdot \mathbf{E}^i(\mathbf{r}) \right.$$
$$\left. + (1-\alpha) \int_S \mathbf{t}_j(\mathbf{r}) \cdot \hat{n} \times \mathbf{H}^i(\mathbf{r}) dS \right] \quad (3.8)$$

$$A_{ji} = \alpha A_{ji}^E + (1-\alpha) A_{ji}^M \quad (3.9)$$

$$A_{ji}^E = \int_S dS \mathbf{t}_j(\mathbf{r}) \cdot \int_S dS' \overline{\mathbf{G}}(\mathbf{r},\mathbf{r}') \cdot \mathbf{j}_i(\mathbf{r}') \qquad (3.10)$$

$$A_{ji}^M = \frac{2\pi i}{k} \int_S dS \mathbf{t}_j(\mathbf{r}) \cdot \mathbf{j}_i(\mathbf{r})$$
$$- \frac{i}{k} \int_S dS \mathbf{t}_j(\mathbf{r}) \cdot \hat{n} \times \nabla \times \int_S dS' g(\mathbf{r},\mathbf{r}') \mathbf{j}_i(\mathbf{r}') \qquad (3.11)$$

Consequently, the integral equations are approximated by matrix equation (3.7) using the MOM. When we use iterative methods like the conjugate gradient (CG) to solve the matrix equation (3.7), a matrix-vector multiply must be computed in each iteration. In the next two sections, we will apply FMM and MLFMA to accelerate the calculations.

3.2.2 Three-Dimensional FMM

The addition theorem [1, 24] has the form

$$\frac{e^{ik|\mathbf{D}+\mathbf{d}|}}{|\mathbf{D}+\mathbf{d}|} = ik \sum_{l=0}^{\infty} (-1)^l (2l+1) j_l(kd) h_l^{(1)}(kD) P_l(\hat{d} \cdot \hat{D}) \qquad (3.12)$$

where $j_l(x)$ is a spherical Bessel function of the first kind, $h_l^{(1)}(x)$ is a spherical Hankel function of the first kind, $P_l(x)$ is a Legendre polynomial, and $d < D$.

Substituting the elementary identity [25] (p. 410)

$$4\pi i^l j_l(kd) P_l(\hat{d} \cdot \hat{D}) = \int d^2 \hat{k} e^{i\mathbf{k} \cdot \mathbf{d}} P_l(\hat{k} \cdot \hat{D}) \qquad (3.13)$$

into (3.12) yields, after truncating the summation series and exchanging the order of integration and summation,

$$\frac{e^{ik|\mathbf{D}+\mathbf{d}|}}{|\mathbf{D}+\mathbf{d}|} \approx \frac{ik}{4\pi} \int d^2 \hat{k} e^{i\mathbf{k}\cdot\mathbf{d}} \sum_{l=0}^{L} i^l (2l+1) h_l^{(1)}(kD) P_l(\hat{k} \cdot \hat{D})$$
$$= \frac{ik}{4\pi} \int d^2 \hat{k} e^{i\mathbf{k}\cdot\mathbf{d}} T_L(\hat{k} \cdot \hat{D}) \qquad (3.14)$$

We have truncated the summation of infinite series, where $\int d^2 \hat{k}$ represents the integrals over the unit sphere, and

$$T_L(\cos\theta) = \sum_{l=0}^{L} i^l (2l+1) h_l^{(1)}(kD) P_l(\cos\theta) \qquad (3.15)$$

To apply FMM in the matrix-vector multiply, we divide the N basis functions into G localized groups, labeled by an index m, each supporting about $M = N/G$ basis functions. For the nearby group pairs (m, m'), we calculate the matrix elements by direct numerical computation. Letting $\mathbf{r}_j$ and $\mathbf{r}_i$ be the field point and source point, respectively, we have

$$\begin{aligned} \mathbf{r}_{ji} = \mathbf{r}_j - \mathbf{r}_i &= \mathbf{r}_j - \mathbf{r}_m + \mathbf{r}_m - \mathbf{r}_{m'} + \mathbf{r}_{m'} - \mathbf{r}_i \\ &= \mathbf{r}_{jm} + \mathbf{r}_{mm'} - \mathbf{r}_{im'} \end{aligned} \quad (3.16)$$

where $\mathbf{r}_m$ and $\mathbf{r}_{m'}$ are the centers of the mth and m'th groups, to which $\mathbf{r}_j$ and $\mathbf{r}_i$ belong, respectively. For nonnearby group pairs (m, m'), using (3.14) with $\mathbf{D} = \mathbf{r}_{mm'}$ and $\mathbf{d} = \mathbf{r}_{jm} - \mathbf{r}_{im'}$, we approximate the scalar Green's function as

$$\frac{e^{ikr_{ji}}}{r_{ji}} = \int d^2 \hat{k} e^{i\mathbf{k}\cdot(\mathbf{r}_{jm}-\mathbf{r}_{im'})} \alpha_{mm'}(\mathbf{k}, \mathbf{r}_{mm'}) \quad (3.17)$$

where

$$\alpha_{mm'}(\mathbf{k}, \mathbf{r}_{mm'}) = \frac{ik}{4\pi} \sum_{l=0}^{L} i^l (2l+1) h_l^{(1)}(kr_{mm'}) P_l(\hat{r}_{mm'} \cdot \hat{k}) \quad (3.18)$$

The integration in (3.17) will be evaluated by Gaussian quadratures with $K = 2L^2$ points.

Substituting (3.17) into (3.2) yields

$$\overline{\mathbf{G}}(\mathbf{r}_j - \mathbf{r}_i) = \int d^2 \hat{k} (\overline{\mathbf{I}} - \hat{k}\hat{k}) e^{i\mathbf{k}\cdot(\mathbf{r}_{jm}-\mathbf{r}_{im'})} \alpha_{mm'}(\mathbf{k}, \mathbf{r}_{mm'}) \quad (3.19)$$

By substituting (3.19) into (3.10), we have the EFIE matrix elements in the form

$$A_{ji}^E = \int d^2 \hat{k} \mathbf{V}_{fmj}^E(\hat{k}) \cdot \alpha_{mm'}(\mathbf{k}, \mathbf{r}_{mm'}) \mathbf{V}_{sm'i}^E(\hat{k}) \quad (3.20)$$

where

$$\mathbf{V}_{fmj}^E(\hat{k}) = \int_S dS e^{i\mathbf{k}\cdot\mathbf{r}_{jm}} (\overline{\mathbf{I}} - \hat{k}\hat{k}) \cdot \mathbf{t}_j(\mathbf{r}_{jm}) \quad (3.21)$$

$$\mathbf{V}_{sm'i}^E(\hat{k}) = \int_S dS e^{-i\mathbf{k}\cdot\mathbf{r}_{im'}} (\overline{\mathbf{I}} - \hat{k}\hat{k}) \cdot \mathbf{j}_i(\mathbf{r}_{im'}) \quad (3.22)$$

Since $\overline{\mathbf{I}} - \hat{k}\hat{k} = \hat{\theta}\hat{\theta} + \hat{\phi}\hat{\phi}$, then $(\overline{\mathbf{I}} - \hat{k}\hat{k}) \cdot \mathbf{V} = V_\theta \hat{\theta} + V_\phi \hat{\phi}$, $\mathbf{V}_{fmj}^E$ and $\mathbf{V}_{sm'i}^E$ have only θ and ϕ components.

We have derived FMM formulation solution for EFIE. Let us derive the formula for MFIE. Substituting (3.17) into (3.11) and applying vector operations yield

$$A_{ji}^M = \int d^2 \hat{k} \mathbf{V}_{fmj}^M(\hat{k}) \cdot \alpha_{mm'}(\mathbf{k}, \mathbf{r}_{mm'}) \mathbf{V}_{sm'i}^M(\hat{k}) \quad (3.23)$$

where
$$\mathbf{V}^M_{fmj}(\hat{k}) = -\hat{k} \times \int_S dS e^{i\mathbf{k}\cdot\mathbf{r}_{jm}} \mathbf{t}_j(\mathbf{r}_{jm}) \times \hat{n} \tag{3.24}$$

$$\mathbf{V}^M_{sm'i}(\hat{k}) = \int_S dS e^{-i\mathbf{k}\cdot\mathbf{r}_{im'}} \mathbf{j}_i(\mathbf{r}_{im'}) \tag{3.25}$$

Since $\mathbf{V}^M_{fmj}(\hat{k})$ has $\hat{\theta}$ and $\hat{\phi}$ components only, then only the $\hat{\theta}$ and $\hat{\phi}$ components of $\mathbf{V}^M_{sm'i}(\hat{k})$ are needed. Therefore, $\mathbf{V}^M_{sm'i}(\hat{k})$ can be replaced by $\mathbf{V}^E_{sm'i}$, which will be called $\mathbf{V}_{sm'i}$. Consequently, the matrix elements in nonnearby group pairs for CFIE are reduced to

$$A_{ji} = \int d^2\hat{k} \mathbf{V}_{fmj}(\hat{k}) \cdot \alpha_{mm'}(\mathbf{k}, \mathbf{r}_{mm'}) \mathbf{V}_{sm'i}(\hat{k}) \tag{3.26}$$

$$\mathbf{V}_{fmj}(\hat{k}) = \alpha \mathbf{V}^E_{fmj} + (1-\alpha) \mathbf{V}^M_{fmj} \tag{3.27}$$

Finally, we can rewrite the matrix-vector multiply as

$$\sum_{i=1}^N A_{ji}a_i = \sum_{m'\in B_m} \sum_{i\in G_{m'}} A_{ji}a_i + \int d^2\hat{k} \mathbf{V}_{fmj}(\hat{k}) \\ \cdot \sum_{m'\notin B_m} \alpha_{mm'}(\mathbf{k}, \mathbf{r}_{mm'}) \sum_{i\in G_{m'}} \mathbf{V}_{sm'i}(\hat{k})a_i \tag{3.28}$$

for $j \in G_m$, where G_m denotes all elements in the mth group, and B_m denotes all nearby groups of the mth group (including itself). The first term is the contribution from nearby groups, and the second term is the far interaction calculated by FMM. The cost of computing the first term is

$$T_1 = c_1 BNM \tag{3.29}$$

where B is the average number of nearby groups, and c_1 is machine and implementation dependent. The cost of computing $\mathbf{S}_m(\hat{k}) = \sum_{i\in G_m} \mathbf{V}_{smi}(\hat{k})a_i$, $m = 1, 2, ..., G$ is

$$T_2 = c_2 KN \tag{3.30}$$

where K is defined just after (3.18). The cost of computing $\mathbf{g}_m(\hat{k}) = \sum_{m'} \alpha_{mm'} \mathbf{S}_{m'}(\hat{k})$ is

$$T_3 = c_3 KG(G - B) \tag{3.31}$$

And finally, the cost of computing $\int d^2\hat{k} \mathbf{V}_{fmj}(\hat{k}) \cdot \mathbf{g}_m(\hat{k})$ is

$$T_4 = c_4 KN \tag{3.32}$$

In Section 3.3, we will show that $L = kd + \beta(kd)^{1/3}$, where β is dependent on the accuracy and d is the maximum diameter of a group size ($d = \max(|\mathbf{r}_{jm} - \mathbf{r}_{im'}|)$). Thus L is proportional to the size d, and K is proportional to the surface area of the group. Since the number of unknowns in each group (M) is proportional to the surface area for a fixed discretization density, $K \sim M$. The computational cost of the matrix-vector multiply is $T = C_1 NG + C_2 N^2/G$. The total cost is minimized by choosing $G = \sqrt{C_2 N/C_1}$. Therefore, $T = 2\sqrt{C_1 C_2} N^{3/2}$, the same as for the 2D FMM. Our numerical simulations also show that the computational complexity is of order $O(N^{3/2})$ [3, 4].

3.3 MULTILEVEL FAST MULTIPOLE ALGORITHM

To further reduce the computational complexity of the matrix-vector multiply, the FMM should be extended to a multilevel algorithm, which yields the multilevel fast multipole algorithm, as discussed in the last chapter.

To implement the 3D MLFMA, the entire object is first enclosed in a large cube, which is partitioned into eight smaller cubes. Each subcube is then recursively subdivided into smaller cubes until the edge length of the finest cube is about a quarter wavelength or until each finest cube includes a few basis functions. Cubes at all levels are indexed. At the finest level, we find the cube in which each basis function resides by comparing the coordinates of the center of the basis function with the center of cube. We further find nonempty cubes by sorting. Only nonempty cubes are recorded using tree-structured data at all levels [26, 27]. Thus, the computational cost depends only on the nonempty cubes.

We use L to denote the number of levels in MLFMA. Level L refers to the finest level and level 0 is the coarsest level. At level 0, there is one and only one nonempty cube. At level 1, there are at most eight nonempty cubes and there are at most 64 at level 2. Although we describe the algorithm using cubic structures, this does not mean that the algorithm is efficient only for cubic objects like cubes and spheres. In fact, we use cubes at the finest level. Then the cubes will form arbitrary shapes at all other levels. Hence, this algorithm is very efficient in handling various types of objects, such as aircraft, long cylinders, 3D strips, and linear antennas.

The basic algorithm for matrix-vector multiply is broken down into two sweeps [27]: the first sweep consists of constructing outgoing wave expansions (represented by the radiation pattern at a number of directions) for each nonempty cube from level L to 2. At level L, the outgoing wave expansions for a cube are calculated by combining all sources in the cube as shown in (3.28). Then from level $L - 1$ to 2, the expansions for each cube are computed from the expansions of its children cubes. The second sweep consists of constructing local incoming wave expansions contributed from the well-separated cubes from level 2 to L. The well-separated cubes are defined as the cubes separated by at least one cube. At level 2, the local incoming wave expansions for a cube are constructed by translating the outgoing wave expansions from all the well-separated cubes at the same level. From levels 3 to L, the local expansions for each cube consist of two parts: incoming waves received by its parent cube from all well-separated cubes of the parent cube, and incoming waves from all its well-separated cubes whose parent cubes are not well-separated because the contribution is already included in the first part. When the cube becomes larger as one progresses from a finer level to a coarser level, the number of plane wave directions needed to represent the radiation pattern should increase. In the first sweep, the outgoing wave expansions are computed at the finest level, and then the expansions for larger cubes are obtained using shifting and interpolation. Let $\mathbf{r}_{m'_l}$ and $\mathbf{r}_{m'_{l-1}}$ be the cube centers at levels l and $l-1$, respectively; then the outgoing wave expansions for coarser level $l-1$ should be

$$\mathbf{V}_{sm'_{l-1}i}(\hat{k}) = e^{-i\mathbf{k}\cdot\mathbf{r}_{m'_l m'_{l-1}}} \mathbf{V}_{sm'_l i}(\hat{k}) \qquad (3.33)$$

But $\mathbf{V}_{sm'_l i}(\hat{k})$ has only K_l values, and we need K_{l-1} values of $\mathbf{V}_{sm'_{l-1}i}(\hat{k})$. Therefore, we will interpolate $\mathbf{V}_{sm'_l i}(\hat{k})$ to K_{l-1} values first. That is,

$$\mathbf{V}_{sm'_{l-1}i}(\hat{k}_{(l-1)n'}) = e^{-i\mathbf{k}_{(l-1)n'}\cdot\mathbf{r}_{m'_l m'_{l-1}}} \sum_{n=1}^{K_l} W_{n'n} \mathbf{V}_{sm'_l i}(\hat{k}_{ln}) \qquad (3.34)$$

At the second sweep, the local expansions for smaller cubes include the contributions from parent cubes using shifting and anterpolation as discussed in the last chapter, and from the cubes that are well-separated at this level but not at the parent level. If the local incoming wave expansions received by a cube center at level $l-1$ are $\mathbf{B}(\hat{k})$, then the contribution from all well-separated cubes can be written as

$$I = \int d^2\hat{k}\, \mathbf{V}_{fm_{l-1}j}(\hat{k}) \cdot \mathbf{B}(\hat{k}) = \sum_{n'=1}^{K_{l-1}} \mathbf{V}_{fm_{l-1}j}(\hat{k}_{(l-1)n'}) \cdot \mathbf{B}(\hat{k}_{(l-1)n'}) \qquad (3.35)$$

where weighting factors for the numerical integral are included in $\mathbf{B}(\hat{k}_{(l-1)n'})$. Substituting the interpolation expression for $\mathbf{V}_{fm_{l-1}j}(\hat{k}_{(l-1)n'})$ into (3.35) and changing the order of two summations lead to

$$I = \sum_{n=1}^{K_l} \mathbf{V}_{fm_l j}(\hat{k}_{ln}) \cdot \sum_{n'=1}^{K_{l-1}} W_{n'n} \mathbf{B}(\hat{k}_{(l-1)n'}) e^{i\mathbf{k}_{(l-1)n'}\cdot\mathbf{r}_{m'_l m'_{l-1}}} \qquad (3.36)$$

The above operation is called anterpolation [28] as discussed in Chapter 2. At the finest level, the contributions from non-well-separated cubes are calculated directly.

There are two approaches for the interpolation: global interpolation and local interpolation. In the global interpolation, the interpolation matrix is full. To reduce the complexity in memory and CPU time requirements, the matrix should never be formed explicitly and the interpolation should be performed by oversampling. Since 3D radiated waves are a periodic function of ϕ, the interpolation over ϕ is achieved with the effort of an FFT [29]. The interpolation over θ can be achieved using FFT by forming a periodic function of θ, or 1D FMM over θ, or other algorithms [7, 9]. The global anterpolation can be interpreted as a filtering, smoothing, or down-sampling. For N_s samples in the radiation pattern, the global interpolation/anterpolation needs $O(N_s \log N_s)$ operations.

In the local interpolation, the radiation pattern at a given point can be interpolated using values at the neighbor points. The interpolation matrix is sparse, and each row has a finite number of nonzero elements. It can be precalculated and used in both interpolation and anterpolation. Therefore, the local interpolation is easy to implement and has $O(N_s)$ operations. In the next section, we will show that for the band-limited radiation pattern, local interpolation/anterpolation can be performed with exponential accuracy.

Since only nonempty cubes are considered, the number of nonempty cubes is reduced by a factor of four from a finer level to a coarser level for a surface scatterer. But the number of samples for each cube increases by a factor of four, so the number of levels is proportional to $\log_4 N$ and each level needs $O(N)$ operations if the local interpolation/anterpolation is used. Therefore, the complexity for MLFMA is $O(N \log N)$, and the memory requirements for MLFMA are also of order $O(N \log N)$ as we discussed in the last section.

3.4 ERROR ANALYSIS IN FMM AND MLFMA

One of the most important mathematical formulas in FMM is the addition theorem. In the numerical implementation of the addition theorem, the infinite series should be truncated. The error analysis for the truncation error in the scalar Green's functions has been done by many researchers [1, 3, 6, 8, 9, 30]. In this section, the number of terms needed for the scalar Green's function is derived, and the error analysis for the truncation error in the multipole expansion of the vector Green's functions is given. We also discuss the error in the numerical integral and the local interpolation and anterpolation.

3.4.1 Truncation Error in Scalar Green's Function

We rewrite the addition theorem (3.12) as

$$g(R) = \frac{e^{ikR}}{kR} = \frac{e^{ik|\mathbf{D}+\mathbf{d}|}}{k|\mathbf{D}+\mathbf{d}|}$$

$$= i\sum_{l=0}^{\infty}(-1)^l(2l+1)j_l(kd)h_l^{(1)}(kD)P_l(\hat{d}\cdot\hat{D}) \qquad (3.37)$$

$$\approx i\sum_{l=0}^{L}(-1)^l(2l+1)j_l(kd)h_l^{(1)}(kD)P_l(\hat{d}\cdot\hat{D})$$

where $R = |\mathbf{r}-\mathbf{r}'| = |\mathbf{D}+\mathbf{d}|$, $\mathbf{r}$ and $\mathbf{r}'$ are for field and source points, $\mathbf{D}$ is a vector between two group centers, $\mathbf{d}$ is the summation of two local vectors, and $d < D$.

The infinite series of (3.37) is truncated at the Lth term. The leading error term is $(2L+3)j_{L+1}(kd)/(kD)$ (for $L > kd$). Some researchers [1, 3, 6, 8, 9, 30] have given the semi-empirical formula:

$$L \approx kd + \alpha \ln(\pi + kd) \qquad (3.38)$$

where α is dependent on the accuracy; for example, $\alpha = 1$ gives the accuracy of 0.1 and $\alpha = 5$ results in 10^{-6} accuracy. Recently, Rokhlin derived a new formula for 2D [31]:

$$L \approx kd + \beta(kd)^{1/3} \qquad (3.39)$$

where β is also dependent on the accuracy. The same formula was used in calculating the Mie series [32] and the optimal sampling of scattered fields [33]. For a given accuracy, we calculate the true L needed in (3.37), and then compare the true L with the values given by (3.38) and (3.39). In Figure 3.1, we plot the differences between the approximated L and true L for an accuracy 10^{-6}. It is found that $kd + 5\ln(\pi + kd)$ is a good approximation for kd up to 40. But $kd + 6(kd)^{1/3}$ is always a good approximation.

Let us derive β as a function of the accuracy requirement. The relative error in the truncated scalar Green's function of (3.37) can be written as

$$\epsilon = kR \left| \sum_{l=L+1}^{\infty}(-1)^l(2l+1)j_l(kd)h_l^{(1)}(kD)P_l(\hat{d}\cdot\hat{D}) \right| \qquad (3.40)$$

The error in each term of (3.40) is maximum when $\mathbf{D}$ and $\mathbf{d}$ are collinear [30]. Applying the large argument approximation of the spherical Hankel function, we have

$$\epsilon \approx \frac{R}{D}\left|\sum_{l=L+1}^{\infty}i^l(2l+1)j_l(kd)\right| \approx (2L+3)j_{L+1}(kd) \qquad (3.41)$$

FMM and MLFMA in 3D and Fast Illinois Solver Code

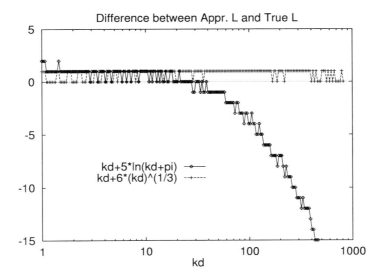

Figure 3.1 The differences between the L calculated using $kd + 5\ln(\pi + kd)$ or $kd + 6(kd)^{1/3}$ and true L for an accuracy 10^{-6}.

where only the leading error term is kept. With large order and argument, $j_{L+1}(kd)$ can be approximated as [24]

$$j_{L+1}(x) \approx \frac{1}{2\sqrt{f(L,x)x}} e^{f(L,x)-(L+3/2)\ln\{[L+3/2+f(L,x)]/x\}} \quad (3.42)$$

where $x = kd$, and $f(L,x) = [(L+3/2)^2 - x^2]^{1/2}$. Let us change the variable using

$$L + 3/2 = x(1+\delta) \quad (3.43)$$

Since the spherical Bessel function decreases very fast when the order is larger than the argument, δ is very small compared to x. So we have the approximations

$$f(L,x) \approx x\sqrt{2\delta}(1+\delta/4) \quad (3.44)$$

$$\epsilon \approx (2\delta)^{-1/4} e^{-x(2\delta)^{3/2}/3} \quad (3.45)$$

The second factor in (3.45) is much smaller than the first and hence dominates when the log of (3.45) is taken. Therefore, we have

$$\epsilon \approx e^{-x(2\delta)^{3/2}/3} \qquad \delta \approx \frac{1}{2}\left[\frac{3\ln(1/\epsilon)}{x}\right]^{2/3} = 1.8\left[\frac{\log(1/\epsilon)}{x}\right]^{2/3} \quad (3.46)$$

Finally, we have a more refined formula:

$$L \approx kd + 1.8 d_0^{2/3}(kd)^{1/3} \quad (3.47)$$

where $d_0 = \log(1/\epsilon)$, the number of digits of accuracy. Equation (3.47) is a very good approximation. For kd varying from 1 to 500, ϵ varies from 10^{-1} to 10^{-10}, the difference between the approximated L and true L is between -1 and 2.

3.4.2 Truncation Error in the Vector Green's Function

The dyadic Green's function is given by

$$\overline{\mathbf{G}}(\mathbf{r},\mathbf{r}') = \left[\overline{\mathbf{I}} - \frac{\nabla\nabla'}{k^2}\right] g(R) = \\ \left\{\left[-\frac{2i}{kR} + \frac{2}{(kR)^2}\right]\hat{R}\hat{R} + \left[1 + \frac{i}{kR} - \frac{1}{(kR)^2}\right](\hat{\Theta}\hat{\Theta} + \hat{\Phi}\hat{\Phi})\right\} g(R) \quad (3.48)$$

where $\hat{R}$ is the unit vector in the radial directions of $\mathbf{r} - \mathbf{r}'$, and $\hat{\Theta}$ and $\hat{\Phi}$ are two unit vectors in transverse directions. It is noted that the dyadic Green's function has three diagonal components only. The transverse components decrease as $1/R$, but the radial component decreases as $1/R^2$.

Applying the addition theorem (3.37) to (3.48) and expanding the spherical wave in plane waves yield [1, 3]

$$\overline{\mathbf{G}}(\mathbf{r},\mathbf{r}') \approx \int d^2\hat{k}(\hat{\mathbf{I}} - \hat{k}\hat{k})e^{i\mathbf{k}\cdot\mathbf{d}}\alpha_L(\hat{D}\cdot\hat{k}) \quad (3.49)$$

where $\alpha_L(\hat{D}\cdot\hat{k}) = \frac{i}{4\pi}\sum_{l=0}^{L} i^l(2l+1)h_l^{(1)}(kD)P_l(\hat{D}\cdot\hat{k})$. Expanding $\overline{\mathbf{I}} - \hat{k}\hat{k}$ in Cartesian coordinates, we have

$$\begin{aligned}\overline{\mathbf{I}} - \hat{k}\hat{k} =& (1 - \sin^2\theta\cos^2\phi)(\hat{x}\hat{x} + \hat{y}\hat{y}) + \sin^2\theta\hat{z}\hat{z} \\ & - \sin^2\theta\sin\phi\cos\phi(\hat{x}\hat{y} + \hat{y}\hat{x}) - \sin\theta\cos\theta\cos\phi(\hat{x}\hat{z} + \hat{z}\hat{x}) \\ & - \sin\theta\cos\theta\sin\phi(\hat{y}\hat{z} + \hat{z}\hat{y})\end{aligned} \quad (3.50)$$

The error in (3.37) and (3.49) is maximum when $\mathbf{D}$ and $\mathbf{d}$ are collinear [30]. To simplify our analysis, we assume that both $\mathbf{D}$ and $\mathbf{d}$ are along the z-axis and replace $\cos\theta$ with x. So the components of the dyadic Green's function in Cartesian coordinates are given by

$$\begin{aligned} G_{zz} &= \int d^2\hat{k}\sin^2\theta\alpha_L(\cos\theta)e^{ikd\cos\theta} = 2\pi\int_{-1}^{1}dx(1-x^2)\alpha_L(x)e^{ixkd} \\ G_{yy} &= G_{xx} = \int d^2\hat{k}(1-\sin^2\theta\cos^2\phi)\alpha_L(\cos\theta)e^{ikd\cos\theta} = g(R) - G_{zz}/2 \\ G_{xy} &= G_{yx} = G_{xz} = G_{zx} = G_{yz} = G_{zy} = 0\end{aligned}$$

$$(3.51)$$

There are three nonzero diagonal components. G_{zz} is the radial component and should not have the $1/R$ term. G_{xx} and G_{yy} are the transverse components. To check the error term in G_{zz} of (3.51), we use the expansion of a plane wave to spherical waves [$e^{ixkd} = \sum_{n=0}^{\infty} i^n(2n+1)j_n(kd)P_n(x)$], the translation $\alpha_L(x) = \frac{i}{4\pi}\sum_{l=0}^{L} i^l(2l+1)h_l^{(1)}(kD)P_l(x)$, the recurrence relations for Legendre polynomials [$xP_l(x)$ and $x^2P_l(x)$], and the asymptotic form of $h_l^{(1)}(kD)$ for large kD [24]. Then we have

$$G_{zz} \approx \frac{e^{ikD}}{2kD} \sum_{n=0}^{\infty} i^n(2n+1)j_n(kd) \sum_{l=0}^{L}$$
$$\int_{-1}^{1} dx P_n(x) \left\{ \frac{l(l-1)}{2l-1}[P_l(x) - P_{l-2}(x)] \right. \quad (3.52)$$
$$\left. + \frac{(l+1)(l+2)}{2l+3}[P_l(x) - P_{l+2}(x)] \right\}$$

Using the orthogonality of Legendre polynomials, we find that $G_{zz} = 0$ (no $1/D \approx 1/R$ term) when $L \to \infty$. But when FMM truncates the series at the Lth term, the leading order error term in G_{zz} is $\frac{L(L+1)}{2L+1}j_{L-1}(kd)/(kD)$. Similarly, the leading order error term in G_{xx} and G_{yy} is found to be

$$\left[(2L+3)j_{L+1}(kd) - \frac{L(L+1)}{2(2L+1)}j_{L-1}(kd)\right]/(kD) \quad (3.53)$$

For small kd, L is much larger than kd, and hence, and the second term is dominant. So, if the scalar Green's function is truncated at the Lth term and the relative error is ϵ [see (3.41)], then the relative error of the transverse components of the dyadic Green's function is approximately $\epsilon/8$ if it is truncated at the $(L+2)$th term. For the radial component, the relative error normalized by the transverse component is $\epsilon/4$ if it is truncated at the $(L+2)$th term. The relative error normalized by itself is $kR\epsilon/8$. Therefore, the relative error of the dyadic Green's function is $\epsilon/4$.

The vector Green's function for the MFIE is just the gradient of the scalar Green's function,

$$\nabla g(R) = k[i - 1/(kR)]g(R)\hat{R} \quad (3.54)$$

Its FMM factorization can be written as [4]

$$\nabla g(R) \approx \int d^2\hat{k}\mathbf{k}e^{i\mathbf{k}\cdot\mathbf{d}}\alpha_L(\hat{D}\cdot\hat{k}) \quad (3.55)$$

Using the above analysis, we find that the leading error term in $\nabla g(R)$ is $(L+1)j_L(kd)/(kD)$. So the relative accuracy is $\epsilon/2$ if it is truncated at the $(L+1)$th term.

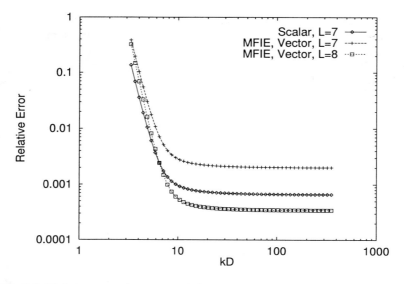

Figure 3.2 Relative truncation errors of the scalar Green's function and vector Green's function for MFIE [$\nabla g(R)$] for $\mathbf{D} = D\hat{z}$ and $\mathbf{d} = 0.4\lambda\hat{z}$.

In Figure 3.2, we plot the relative truncation errors of the scalar Green's function and the vector Green's function for MFIE [$\nabla g(R)$]. It is found that if the vector Green's function retains one more term than the scalar Green's function, the truncation error of the vector Green's function is one half the error of the scalar Green's function. Corresponding to the above analysis, the relative errors of the transverse and radial components of the dyadic Green's function are plotted in Figures 3.3 and 3.4. The error for the radial component is normalized by the transverse component. If the dyadic Green's function retains two more terms than the scalar Green's function, the radial component has $1/4$ of the error of the scalar Green's function and the transverse component has $1/8$ of the error, as indicated by the above analysis.

In summary, if the number of digits of accuracy in the truncated Green's function is d_0, the number of terms is given by $L = kd + 1.8d_0^{2/3}(kd)^{1/3}$. We have also analyzed the truncation error in FMM expansion of the vector Green's function. The error term in the vector Green's function is proportional to $1/R$. If the scalar Green's function is truncated at the Lth term and the relative error is ϵ, then the relative error in the dyadic Green's function is $\epsilon/4$ if it is truncated at the $(L+2)$th term. For the vector Green's function related to MFIE, the relative error is $\epsilon/2$ if it is truncated at the $(L+1)$th term.

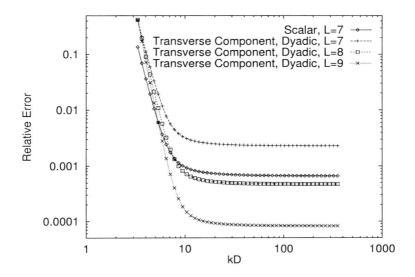

Figure 3.3 Relative truncation errors of the scalar Green's function and the transverse component of the dyadic Green's function for $\mathbf{D} = D\hat{z}$ and $\mathbf{d} = 0.4\lambda\hat{z}$.

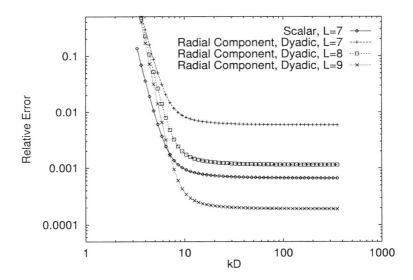

Figure 3.4 Relative truncation errors of the scalar Green's function and the radial component (normalized by the transverse component) of the dyadic Green's function for $\mathbf{D} = D\hat{z}$ and $\mathbf{d} = 0.4\lambda\hat{z}$.

3.4.3 Error in Numerical Integral

The integrand in (3.28) can be expanded as a summation of band-limited spherical functions up to order $2L$ with a leading error term proportional to $j_{L+1}(kd)$. Since it is a periodic function of ϕ with a period of 2π, the $2L+1$ point trapezoidal quadrature rule on the interval $[0, 2\pi]$ over ϕ and the $L+1$ point Gaussian-Legendre quadrature rule on the interval $[0, \pi]$ over θ give accurate results for the spherical functions up to order $2L$. Therefore, the error in the numerical integral decreases exponentially.

3.4.4 Error in Local Interpolation

Since the bandwidth of the radiation pattern is about $kr + \beta(kr)^{1/3}$, where r is the radius of the group, and we sample it at $kd + \beta(kd)^{1/3}$, where d is the diameter, we oversample it by a factor of two. In [30], we proved that a local interpolation (n points along ϕ and n points along θ, including 0 and π, totally n^2 points are used) has a relative error

$$\delta \approx \sqrt{\frac{2}{n\pi}} \left(\frac{\pi}{4}\right)^n \tag{3.56}$$

This result reveals that the interpolation error decreases exponentially as the number of interpolation points is increased.

In this section, we have shown that truncation in multipole expansions, numerical integral, and local interpolation and anterpolation are three major error sources in MLFMA and all of them are controllable to decrease exponentially.

3.5 LARGE-SCALE COMPUTING

In the last section, we presented the use of MLFMA to reduce the complexity of the matrix-vector multiply to $O(N\log N)$. But some computations in implementing MLFMA to the MOM code may have higher complexity than $O(N\log N)$. When the translation matrix in MLFMA is calculated directly, the complexity is $O(N^{3/2})$. For calculating the bistatic RCS or the radiation pattern on a one-plane cut, the number of observation angles M is proportional to the frequency for a surface scatterer. So the complexity of calculating the bistatic RCS or the radiation pattern is $O(NM)$ or $O(N^{3/2})$. In this section, we use interpolation to calculate the translation matrix, and the complexity is reduced to $O(N)$. MLFMA has been implemented to calculate the bistatic RCS or radiation pattern reducing its complexity to $O(M) + O(N\log N)$.

3.5.1 Block Diagonal Preconditioner

The CPU time for iterative methods is proportional to the number of iterations needed to get the desired accuracy. The convergence rate depends on the spectral properties of the MOM matrix. Hence, one may want to transform the original matrix equation $\overline{\mathbf{A}} \cdot \mathbf{x} = \mathbf{b}$ into $\overline{\mathbf{M}}^{-1} \cdot \overline{\mathbf{A}} \cdot \mathbf{x} = \overline{\mathbf{M}}^{-1} \cdot \mathbf{b}$, which has the same solution, but with a more favorable spectral property, where $\overline{\mathbf{M}}^{-1}$ is called a preconditioner.

If basis functions in one of the finest cubes are considered as one group, the matrix $\overline{\mathbf{A}}$ has block structure and can be further divided as

$$\overline{\mathbf{A}} \cdot \mathbf{x} = (\overline{\mathbf{A}}_0 + \overline{\mathbf{A}}_1) \cdot \mathbf{x} + \overline{\mathbf{A}}_2 \cdot \mathbf{x} \qquad (3.57)$$

where matrices $\overline{\mathbf{A}}_0$ and $\overline{\mathbf{A}}_1$ account for near interactions and $\overline{\mathbf{A}}_0$ is the block diagonal part. The matrix $\overline{\mathbf{A}}_2$ accounts for far interactions, and we perform $\overline{\mathbf{A}}_2 \cdot \mathbf{x}$ using MLFMA. Choosing $\overline{\mathbf{A}}_0^{-1}$ as a preconditioner, we have

$$\overline{\mathbf{A}}_0^{-1} \cdot \overline{\mathbf{A}} \cdot \mathbf{x} = \mathbf{x} + \overline{\mathbf{A}}_0^{-1} \cdot (\overline{\mathbf{A}}_1 \cdot \mathbf{x} + \overline{\mathbf{A}}_2 \cdot \mathbf{x}) \qquad (3.58)$$

Because we replace $\overline{\mathbf{A}}_0$ by its LUD form for $\overline{\mathbf{A}}_0^{-1}$, the block diagonal preconditioner needs no extra memory and no extra CPU time in each matrix-vector multiply. $\overline{\mathbf{A}}_0$ is a block diagonal matrix with a block size of M, which is the number of unknowns in one cube. When M is a constant, the LUD of $\overline{\mathbf{A}}_0$ takes $O(M^3 N/M) = O(N)$ operations. Our numerical results show that the block diagonal preconditioner reduces the number of iterations significantly [6], especially when we use CFIE for the closed target.

3.5.2 Initial Guess

For iterative solutions of monostatic RCS, different incident angles require different iterative solutions. Since a small change in the incident angle corresponds to a small change in the current, we can use the current solution from the previous angle as the initial guess for the next angle. Because the right-hand side of the MOM matrix equation has a term $e^{i\mathbf{k}_i \cdot \mathbf{r}}$, where $\mathbf{k}_i$ is the incident direction vector, we can shift the phase of the current solution as

$$\tilde{\mathbf{J}}(\mathbf{r}) = \mathbf{J}(\mathbf{r}) e^{-i\mathbf{k}_i \cdot \mathbf{r}} \qquad (3.59)$$

In general, $\tilde{\mathbf{J}}(\mathbf{r})$ changes more slowly than $\mathbf{J}(\mathbf{r})$ when $\mathbf{k}_i$ changes. So we can use the current solution from the previous angle with phase correction as the initial guess for the next angle. We assume that $\mathbf{J}_1(\mathbf{r})$ is a solution for an incident direction $\mathbf{k}_1$, then we use $\mathbf{J}_1(\mathbf{r}) e^{-i\mathbf{k}_1 \cdot \mathbf{r}} e^{i\mathbf{k}_2 \cdot \mathbf{r}}$ as the initial guess for the solution $\mathbf{J}_2$ at an incident direction $\mathbf{k}_2$. This technique significantly reduces the number of iterations. As an illustration, we calculate the monostatic RCS from the VFY218 at 100 MHz for VV polarization. Zero degree corresponds to the incidence angle on the nose. In Figure 3.5, we plot the number of iterations for different incident angles using three kinds of initial guesses. The first case, which uses zero as the initial guess for all angles, needs about 85 iterations on average for each angle. The second case, which uses the solution of the previous angle (2° step size) as the initial guess for the next

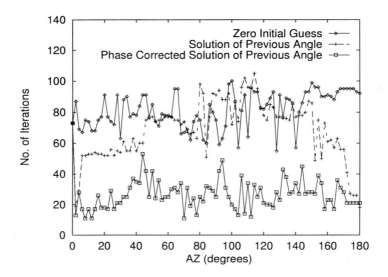

Figure 3.5 Number of iterations as functions of incident angles for different initial guesses using zero initial guess for all angles, and using the solution of the previous angle for the next angle with/without phase corrections.

angle, needs about 65 iterations per angle. The third case, which uses the phase-corrected solution of the previous angle as the initial guess for the next angle, needs only about 30 iterations per angle.

We also apply the linear extrapolation of the current solutions from the previous two angles with the phase correction as the initial guess for the third angle, and further reduce the number of iterations. The extrapolation from more than two previous angles does not show more improvements.

3.5.3 Approximation of Bistatic RCS to Monostatic RCS

For iterative solutions of monostatic RCS, different incident angles require different iterative solutions. Once the current distributions are found, calculating the RCS for one angle requires only $O(N)$ operations. Calculating the bistatic RCS is much less time-consuming than calculating the monostatic RCS. Thus, we calculate the monostatic RCS using the bistatic RCS [34, 35].

The bistatic RCS σ of a target illuminated by a plane wave with frequency f and electric field vector amplitude $\mathbf{E}_0$ can be expressed in terms of the scattered electric field $\mathbf{E}_s$ as

$$\sigma(f, \alpha_v, \alpha_i) = \lim_{r \to \infty} 4\pi r^2 \frac{|\mathbf{E}_s(f, \alpha_v)|^2}{|\mathbf{E}_i(f, \alpha_i)|^2} \tag{3.60}$$

where α_i is the incident angle and α_v is the view angle. Using the concept of reradiation lobe patterns of the individual scattering centers, Kell [34] showed that the bistatic RCS is very closely approximated by the monostatic RCS measured on the bisector of the bistatic angle and measured at a frequency lower than the true frequency by the factor $\cos\frac{\beta}{2}$

$$\sigma\left(f, \alpha - \frac{\beta}{2}, \alpha + \frac{\beta}{2}\right) \approx \sigma\left(f \cos\frac{\beta}{2}, \alpha, \alpha\right) \quad (3.61)$$

We need to find how small β must be to give a similar RCS at frequencies f and $f \cos\frac{\beta}{2}$. For a target with length L, the maximum phase delay P between the scattered and incident fields at frequency f is given by

$$P = \frac{4\pi L}{c} f \quad (3.62)$$

where c is the speed of light and a round trip is considered. If the maximum phase difference between frequencies f and $f \cos\frac{\beta}{2}$ is ΔP, then

$$\frac{4\pi f L}{c}\left(1 - \cos\frac{\beta}{2}\right) < \Delta P \quad (3.63)$$

Using the small argument approximation of $\cos\frac{\beta}{2}$, we have

$$\beta < \sqrt{\frac{2c\Delta P}{\pi L f}} \quad (3.64)$$

If $\Delta P = 11°$ is chosen, then $\beta \approx \frac{20}{\sqrt{L/\lambda}}$ (degrees). For example, β should be $10°$ and $5°$ for $L/\lambda = 4$ and 16, respectively. If (3.64) is satisfied, we can approximate the monostatic RCS using bistatic RCS:

$$\sigma(f, \alpha, \alpha) \approx \sigma\left(f, \alpha - \frac{\beta}{2}, \alpha + \frac{\beta}{2}\right) \quad (3.65)$$

In calculating the monostatic RCS, when we double the frequency, the number of incident/observation angles should be doubled too. But if we use the bistatic RCS to monostatic RCS approximation, the maximum bistatic (sweeping) angle is inversely proportional to the square root of the frequency. So the number of incident angles increases more slowly than the number of observation angles when the frequency increases.

We have calculated the monostatic RCS of the VFY218 as an example using the approximation. At 100 MHz, we need the monostatic RCS at 181 angles on the horizontal plane for the azimuthal (AZ) angle from $0°$ to $180°$ for the VFY218. We find the current distribution for 19 angles, and then use the approximation of bistatic

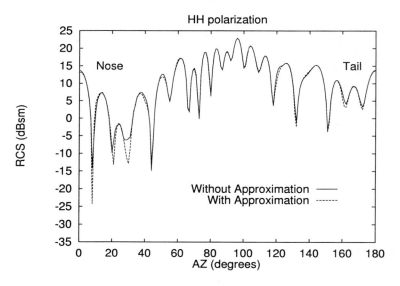

Figure 3.6 The monostatic RCS of the VFY218 at 100 MHz. The numerical results are calculated with/without the approximation of bistatic RCS to monostatic RCS (HH polarization).

RCS to monostatic RCS to calculate the monostatic RCS at all 181 points. In Figure 3.6, we plot the monostatic RCS of the VFY218 at 100 MHz for HH polarization. The solid line is the monostatic RCS calculated without the approximation of bistatic RCS to monostatic RCS, and the dashed line is the monostatic RCS calculated with the approximation, which needs only 1.5 hours on a DEC Alpha workstation.

3.5.4 Interpolation of Translation Matrix in MLFMA

In MLFMA, the translation matrix T is used for translating the outgoing wave (radiation pattern) from a group to an incoming wave at another group, the matrix is given by (3.15) and can be rewritten as

$$T(\hat{X} \cdot \hat{k}) = \sum_{l=0}^{L} i^l (2l+1) h_l^{(1)}(kX) P_l(\hat{X} \cdot \hat{k}) \qquad (3.66)$$

where $\mathbf{X} = X\hat{X}$ is the distance vector between two group centers. L is the number of multipole expansions needed in the addition theorem and is given by $L \approx kd + \beta(kd)^{1/3}$ from the last section, where $\beta = 1.8 d_0^{2/3}$ and d_0 is the number of digits of accuracy.

For a pair of intergroup interactions, the number of directions $\hat{k}$ in (3.66) needed is $K = 2L^2$. If we directly calculate $T(\hat{X} \cdot \hat{k})$ at K directions, this computation requires $O(KL \approx K^{3/2})$ operations. In MLFMA, the number of pairs of intergroup calculations at each level is a constant, and K increases by a factor of four from a finer level to a coarser level. At the coarsest level in MLFMA (level 2), K is proportional to N, the number of unknowns. Then, the complexity for filling all the translation matrices for all pairs of intergroup calculations and for all levels is $O(N^{3/2})$, being dominated by the coarsest level. For a problem with a moderate electrical size, the CPU time for calculating the translation matrices can be negligible because of the small constant factor in its complexity.

For example, in the 2,408,448-unknown problem [10], the CPU time spent on filling the translation matrices is only 5 minutes without parallelization. But for larger problems, this CPU time increases significantly. In Table 3.1, the computation time for filling the translation matrices is given for the VFY218 at different frequencies. It is found that when the frequency is doubled, the number of unknowns N increases by a factor of four, but the CPU time for filling the translation matrices increases by a factor of eight. The last two rows of Table 3.1 will be explained later.

Table 3.1 CPU Time for Filling Translation Matrices for the VFY218

Frequency (GHz)	0.25	0.5	1	2	4
Levels in MLFMA	5	6	7	8	9
Unknowns (N)	10,698	41,832	163,344	625,638	2,206,278
Direct Calculation (sec)	2.67	20.11	156.04	1,251.38	9,880.52
Interpolation (sec)	4.04	15.83	64.36	251.63	987.71
Inpl. with 4 nodes (sec)	1.17	4.31	16.29	65.06	257.73

The fast Legendre expansion [36] was proposed in [1] to reduce the complexity of calculating the translation matrix. But it has a large crossover point. In this section, we will use the interpolation technique. First the spectrum of the translation matrix is discussed. Then, two interpolation methods are introduced and the advantages and disadvantages of these two methods are discussed. Finally, the speed-up factors using the interpolation will be given.

3.5.4.1 Spectrum of the Translation Matrix

The translation matrix in (3.66) depends on $\hat{r} \cdot \hat{k}$ only. Letting $\cos\theta = \hat{r} \cdot \hat{k}$, we have

$$T(\theta) = \sum_{l=0}^{L} i^l (2l+1) h_l^{(1)}(kr) P_l(\cos\theta) \tag{3.67}$$

This is a band-limited function of θ. Using the large argument approximation for the spherical Hankel function of the first kind, we have

$$T(\theta) \approx \frac{e^{ikr}}{ikr} \sum_{l=0}^{L} (2l+1) P_l(\cos\theta) = \frac{e^{ikr}}{ikr} \sum_{j=0}^{L} b_j \cos(j\theta) \quad (3.68)$$

Using the formula given in [37], we can write the Legendre polynomial as

$$P_l(\cos\theta) = \frac{1}{2^l} \sum_{j=0}^{l} \frac{(2j-1)!!}{j!} \frac{(2l-2j-1)!!}{(l-j)!} \cos[(l-2j)\theta] = \sum_{j=0}^{l} a_{lj} \cos(j\theta) \quad (3.69)$$

where the double factorial notation is defined as $(2n-1)!! = 1 \cdot 3 \cdot 5 \cdots (2n-1)$. Using Stirling's formula [24], a_{lj} can be approximated as

$$a_{lj} \approx \frac{4\delta_j}{\pi\sqrt{(l+2/\pi)^2 - j^2}} \quad (3.70)$$

where

$$\delta_j = \begin{cases} 0.5, & j=0 \\ 1, & j>0 \end{cases} \quad (3.71)$$

We look at a_{lj} as a function of j when $l = 80$ [38]. It is found that (3.70) is a good approximation, where even for small l, the relative error is less than 4%.

Substituting (3.70) into (3.68) leads to

$$b_j = \sum_{l=j}^{L} (2l+1) a_{lj} \approx \frac{4\delta_j}{\pi} \sum_{l=j}^{L} \frac{2l+1}{\sqrt{(l+2/\pi)^2 - j^2}} \approx \frac{4\delta_j}{\pi} \sqrt{(L+1)^2 - j^2} \quad (3.72)$$

It is seen that $T(\theta)$ is a band-limited function of θ with order up to L. When L increases, the coefficient b_j increases almost linearly with respect to L. But we know that after the integration over K $\hat{k}$-directions, the calculated Green's function becomes more accurate and changes little when L is greater than the one given in (3.39). Therefore, the high frequency components in $T(\theta)$ should be evaluated very accurately, in order to evaluate the accurate (but small) contributions in the $\hat{k}$ integration. This is especially so when r in (3.67) is not too large. The calculation of the translation matrix is different from the evaluation of the outgoing plane wave coefficients for each basis and group, where the amplitudes of the high frequency components decrease faster than exponential decay.

The approximations used in this section are for understanding the spectrum of the translation matrix in order to find an accurate, efficient interpolation algorithm to reduce the computation time.

3.5.4.2 Interpolation of Translation Matrix

In this section, two interpolation methods are introduced first, then the advantages and disadvantages of these two methods are discussed. Finally, the speed-up factors using interpolations will be given.

Lagrange Polynomial Interpolation. Lagrange polynomial interpolation [39] is a simple and efficient interpolation method. If the number of sampling points for θ between 0 and π is $M = sL$ and $n = 2p$ interpolation points are used, the interpolation formula is given by

$$\tilde{T}(\theta) = \sum_{m=m_0-p+1}^{m_0+p} T(m\Delta\theta) \prod_{\substack{j=m_0-p+1 \\ j \neq m}}^{j=m_0+p} \frac{j\Delta\theta - \theta}{j\Delta\theta - m\Delta\theta}$$

$$= \sum_{m=m_0-p+1}^{m_0+p} T(m\Delta\theta) \prod_{\substack{j'=m_0-m-p+1 \\ j' \neq 0}}^{j'=m_0-m+p} \frac{j'\Delta\theta - (\theta - m\Delta\theta)}{j'\Delta\theta} \quad (3.73)$$

$$= \sum_{m=m_0-p+1}^{m_0+p} T(m\Delta\theta) Q_p(\theta - m\Delta\theta)$$

where $\Delta\theta = 2\pi/(2M+1)$, $m_0 = \text{Int}[\theta/\Delta\theta]$, with Int denoting the integer part; and since $m_0 - m = \text{Int}[\theta/\Delta\theta] - m = \text{Int}\lfloor(\theta - m\Delta\theta)/\Delta\theta\rfloor$, we have

$$Q_p(\theta) = \begin{cases} \prod_{\substack{j=j_0-p+1 \\ j \neq 0}}^{j=j_0+p} \frac{j\Delta\theta - \theta}{j\Delta\theta}, & |\theta| \leq p\Delta\theta \\ 0, & |\theta| > p\Delta\theta \end{cases} \quad (3.74)$$

where $j_0 = \text{Int}\lfloor\theta/\Delta\theta\rfloor$. The interpolation error is found to be [30]

$$|R_n| \leq T_0 \frac{(2p)!}{(p!)^2} \left(\frac{\pi}{4s}\right)^{2p} \quad (3.75)$$

where T_0 is the maximum of $|T(\theta)|$ and s is a measurement of the oversampling factor. Using Stirling's formula to approximate the factors leads to

$$|R_n| \leq \frac{T_0}{\sqrt{2n\pi}} \left(\frac{\pi}{2s}\right)^n \quad (3.76)$$

The more the sampling and interpolation points, the smaller the error.

Optimal Interpolation—Approximate Prolate Spheroidal (APS). The optimal interpolation [40] is very useful for the interpolation of the radiation pattern. It uses fewer sampling points and has lower upper-bounds on the error. One of the approximations to the optimal interpolation is the approximate prolate spheroidal (APS) function, whose interpolation formula is given by [40]

$$\tilde{T}(\theta) = \sum_{m=m_0-p+1}^{m_0+p} T(m\Delta\theta) S_N(\theta - m\Delta\theta, \theta_0) D_M(\theta - m\Delta\theta) \quad (3.77)$$

where

$$S_N(\theta, \theta_0) = \frac{R_N(\theta, \theta_0)}{R_N(0, \theta_0)} \quad (3.78)$$

$$R_N(\theta, \theta_0) = \frac{\sinh\left[(2N+1)\sinh^{-1}\sqrt{\sin^2(\theta_0/2) - \sin^2(\theta/2)}\right]}{\sqrt{\sin^2(\theta_0/2) - \sin^2(\theta/2)}} \quad (3.79)$$

$$D_M(\theta) = \frac{\sin[(2M+1)\theta/2]}{(2M+1)\sin(\theta/2)} \quad (3.80)$$

where $M = sL$ is the total number of sampling points, $\Delta\theta = 2\pi/(2M+1)$, $n = 2p$ is total number of interpolation points, $\theta_0 = p\Delta\theta$, $N = M - L = (s-1)L$ is the number of oversampling points, $m_0 = \text{Int}[\theta/\Delta\theta]$, and $S_N(\theta, \theta_0)$ is a windowing function, while $D_M(\theta)$ is the periodic sinc function. The error's upper bound can be found in [40].

Another optimal interpolation function is the Tschebysceff sampling (TS) series [40]. We found that its accuracy and CPU time requirements are similar to those of the APS function.

Now, we need to interpolate $T(\theta)$ from $M = sL$ sampling points $T(m\Delta\theta)$ to $K = 2L^2$ points. The number of operations for computing these sampling points can be written as

$$T_{\text{fill}} = c_0 s L^2 \quad (3.81)$$

where c_0 is the number of operations needed to evaluate one term in the summation of (3.67). The number of operations for the interpolation can be written as

$$T_{\text{inpl}} = 2c_1 L^2 \quad (3.82)$$

where c_1 is the number of operations for one interpolation. Since (3.78) and (3.80) contain some complicated functions, we find that the Lagrange polynomial interpolation needs less operation count than the optimal interpolation.

Since T_{inpl} is larger than T_{fill} for a moderately small n, we will use a large sampling rate and a small number of interpolation points to reduce the total CPU time for filling translation matrices. Next, we need to decide the parameters s and n by examining the spectrum of interpolation functions. In general, the interpolation formula can be written as

$$\tilde{T}(\theta) = \sum_{m=m_0-p+1}^{m_0+p} T(m\Delta\theta) P(\theta - m\Delta\theta) = T(\theta) \otimes P(\theta) \quad (3.83)$$

where $\otimes$ denotes the discrete convolution with sampling step $\Delta\theta$ and $P(\theta)$ is the interpolation function, which is $Q_n(\theta)$ for Lagrange polynomial interpolation and $S_N(\theta)D_M(\theta)$ for the optimal interpolation. In the spectral domain, the above relation can be written as

$$\mathcal{F}(\tilde{T}) = \mathcal{DFT}(T)\mathcal{F}(P) \quad (3.84)$$

where $\mathcal{F}$ denotes the Fourier transform and $\mathcal{DFT}$ denotes the discrete Fourier transform with sampling step $\Delta\theta$. Since $T(\theta)$ is a band-limited (up to L) function, the interpolated translation matrix $[\tilde{T}(\theta)]$ is the same as the original one $[T(\theta)]$ if $\mathcal{F}(P)$ is a rectangular window function and is wider than the bandwidth. We know that a sinc function gives exact results, but it needs all points ($n = 2M + 1$) to calculate any $\tilde{T}(\theta)$.

Our numerical results [38] showed that for a small oversampling rate and large number of interpolation points, APS gives better results than the Lagrange polynomial. Unfortunately, it requires longer CPU time. On the other hand, for a large oversampling rate and small number of interpolation points, the Lagrange polynomial gives better results than APS, but both need longer CPU time for T_{fill}. After testing a few more cases and considering the accuracy and CPU time requirements, we found that Lagrange polynomial interpolation with $s = 5$ and $p = 3$ is the best, and the relative error is less than 10^{-4}.

Using Lagrange polynomial interpolation with $s = 5$ and $p = 3$, we recalculate the translation matrices for the example shown in Table 3.1. The new CPU time is also shown in Table 3.1. We found that the CPU time for filling the translation matrices increases linearly as the number of unknowns. At 4 GHz, the CPU time is reduced from 3 hours to 17 minutes. For a smaller L, the CPU time for calculating the translation matrices using interpolation may be more than calculating them directly. It is found that $L = 50$ is the crossover point, we use interpolation when $L > 50$.

We parallelize the subroutine for calculating the translation matrices, the CPU time for four nodes is shown in the last row of Table 3.1. A speed-up factor of 3.8 is obtained.

We have also tested cubic spline interpolation [39] and found that it needs more memory, longer CPU time for T_{fill}, shorter CPU time for T_{inpl}, and that its accuracy is not as good as Lagrange polynomial interpolation with six points for five times the required sampling rates.

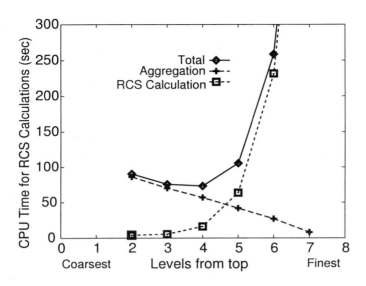

Figure 3.7 CPU time for 1,201 points of bistatic RCS as a function of the level exited for calculating RCS with interpolation.

3.5.5 MLFMA for Calculating Radiation Fields

Once the electric current distributions are solved, we can calculate the scattered or radiation patterns. For each observation point, the operation count needed to calculate the radiated field is proportional to N, which is proportional to the square of the frequency for surface scatterers and radiators. For calculating the bistatic RCS or radiation pattern on a one-plane cut, the number of observation angles M is proportional to the frequency. So the complexity of calculating the bistatic RCS or radiation pattern is $O(NM)$ or $O(N^{3/2})$. MLFMA has been implemented to calculate the bistatic RCS or radiation pattern to reduce the complexity to $O(M) + O(N\log N)$.

There are two steps in this approach: aggregating the radiation pattern for all nonempty boxes from the finest level to a level and calculating the bistatic RCS using interpolation from the radiation patterns at the level. The CPU time increases for the first step and decreases for the second step from the finest level to a coarser level. So there is an optimal level as shown in Figure 3.7, in which we plot the CPU time for calculating 1,201 points of bistatic RCS of a 24λ sphere on one CPU of an SGI R8000 (90 MHz) machine. In Figure 3.8, we plot the CPU time for the RCS calculations using the direct method or using MLFMA for a 2m sphere from 0.225 to 3.6 GHz. The number of points of bistatic RCS has increased from 181 to 1,201. At 3.6 GHz, the CPU time is reduced from 3 hours to 1.5 minutes.

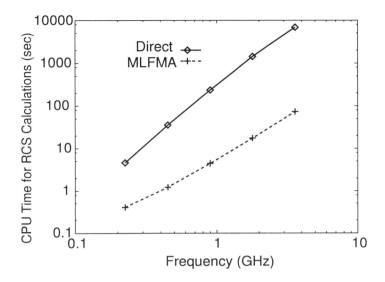

Figure 3.8 CPU time for RCS calculation using direct method or using MLFMA on 1 CPU of an SGI R8000 (90 MHz) machine.

3.6 FAST ILLINOIS SOLVER CODE (FISC)

Because MLFMA expedites matrix-vector multiplies, it can be used to speed up iterative solutions of scattering problems. We have combined MLFMA with the conjugate gradient (CG) and the biconjugate gradient (BCG) methods to arrive at efficient solvers for matrix equations arising from the integral equation of scattering. The computational and memory-requirement complexities of this algorithm are both $O(N \log N)$ for surface scatterers. The total CPU time of the matrix solver is hence proportional to $N_{\text{iter}} N \log N$. Compared to traditional matrix solvers requiring $O(N^2)$ memory, as well as $N_{\text{iter}} N^2$ CPU time for iterative solvers and $O(N^3)$ CPU time for LUD, this is a vast improvement, especially for large problems. Therefore, large problems that previously required the resources of a supercomputer to be solved, can now be solved on a workstation-size computer.

FISC (Fast Illinois Solver Code) [10,41,42] is designed to be an industrial-strength code using the most current technology from the method of moments (MOM) [6,43] and MLFMA. The method of moments is used carefully to develop a matrix equation from the integral equation. Both curvilinear quad-patch and tri-patch are used. Careful integrations are performed to account for singularities and near singularities in the evaluation for the matrix elements.

The code was first designed to model the scattering solution from a complex metallic target like an aircraft, so that its RCS can be ascertained. Impedance boundary conditions, thin dielectric sheets, and resistive boundary conditions can be modeled by the code. We here applied the hybridization of finite element and boundary-integral (FE-BI) methods to solve scattering from material-coated targets or inhomogeneous dielectric objects. This hybrid method divides the problem into interior and exterior regions, and employs the finite-element method (FEM) to deal with the interior problem. The exterior problem is formulated using the boundary-integral method which, when coupled with the interior fields, provides an efficient solution to the original problem. But the capability of the technique is limited mainly by the full matrix generated by the discretization of the boundary-integral using MOM. FMM and MLFMA are used to speed up the matrix-vector multiply related to boundary integrals at a reduced complexity. Please read Chapter 13 for more details about the hybridization.

3.6.1 Capabilities

In the released version of FISC, the triangular facet format is used to describe the geometry. The original geometry file should describe the target accurately. FISC will refine it for different frequencies when the maximum edge length (in wavelengths) is given. For a closed target (used for nonzero thickness objects), all the sequential right-hand normal (RHN) points "outward." FISC checks the normal directions and automatically rearranges facets with wrong normal directions.

3.6.1.1 Integral Equations

Three kinds of integral equations [4,22] are used in FISC: EFIE, MFIE, and CFIE. We implement Galerkin's method and the line matching method in FISC. In Galerkin's method, the testing functions are the same as the basis function, the Rao-Wilton-Glisson (RWG) basis [43]. In line matching, the testing functions are constant along the line joining the centers of two adjacent patches.

3.6.1.2 Boundary Conditions

In addition to perfect electrical conductor (PEC), FISC can deal with impedance boundary conditions (IBC), resistive sheets (R-card), and thin dielectric sheets (TDS). When the coating material has a large refractive index ($|\sqrt{\epsilon\mu/(\epsilon_0\mu_0)}|$, where ϵ and μ are the permittivity and permeability of the coating material), the wave inside the coating sheet propagates approximately in the direction normal to the sheet. Then, the incident angle dependence of the impedance can be neglected. If the material is lossy or the coating sheet is very thin, it can be modeled as IBC with

$$\eta_s = -i\sqrt{\frac{\mu}{\epsilon}} \tan\left(\sqrt{\frac{\epsilon\mu}{\epsilon_0\mu_0}} k_0 d\right) \qquad (3.85)$$

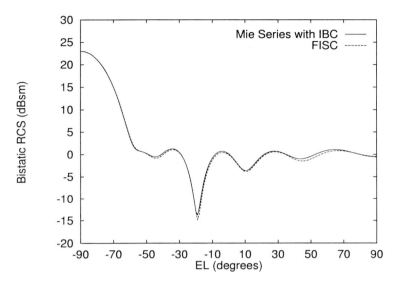

Figure 3.9 Bistatic RCS for a sphere with 1m radius at 300 MHz as a function of elevation angle $(90 - \theta)$. The impedance of the IBC is $100 - i37.7 \, \Omega$.

where k_0 is the wavenumber in the free space, and d is the thickness. The IBC also can be used for lossy, electrically large scatterers, such as the Earth and the ocean surface. The scatterer is modeled with IBC using

$$\eta_s = \sqrt{\frac{\mu}{\epsilon}} \qquad (3.86)$$

In Figure 3.9, we plot the bistatic RCS for a sphere with 1m radius at 300 MHz as a function of elevation angle $(90 - \theta)$. The impedance of the IBC is $100 - i37.7 \, \Omega$. The sphere is described by 1,568 flat triangular facets, and the three-level MLFMA is used. A good agreement between Mie series and numerical results by FISC is observed.

A thin dielectric sheet can be approximated by an impedance sheet

$$Z = \frac{i}{\omega d \Delta \epsilon} \qquad (3.87)$$

where d is its thickness, and $\Delta \epsilon = \epsilon - \epsilon_b$, where ϵ_b is the permittivity of the background.

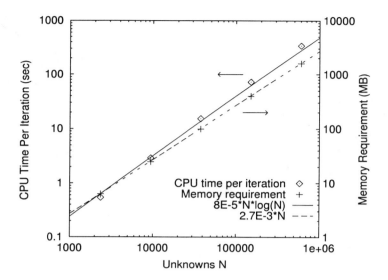

Figure 3.10 CPU time per iteration and memory requirements (points) as functions of number of unknowns in FISC. Two curves, $8 \times 10^{-5} N \log N$ and $2.7 \times 10^{-3} N$, are also plotted for comparison.

3.6.2 Complexity and Accuracy

To test the complexity, we calculate the electromagnetic scattering from a conducting sphere solving the CFIE. The machine used is an SGI Power Challenge with four processors (R8000 with 90 MHz) and 2 GB of memory. But only one processor is used for this simulation. The radius of the sphere is from 0.75 to 12λ, the number of unknowns N is from 2,352 to 602,112, and the number of levels in MLFMA is from 3 to 7. The CPU time per iteration and memory requirement are plotted in Figure 3.10 as functions of the number of unknowns. Two curves, $8 \times 10^{-5} N \log N$ and $2.7 \times 10^{-3} N$, are also plotted on the same figure for comparison.

To obtain an accurate solution from the method of moments, the matrix elements should be evaluated accurately. We apply different orders of Gaussian quadrature rules to the integral over triangles based on the distance between the field triangle and source triangle. For finite thickness objects, only the self terms have a singularity, and only self singularity extraction [44] is needed. For very thin objects, both self and near singularity extractions [45] are required to obtain correct matrix elements. One numerical example is shown in Figure 6 of [6].

In Figure 3.11, we plot the RMS error between Mie series and FISC results for a sphere from 0.75 to 100λ. The number of unknowns is 588–9,633,792, and the number of levels in MLFMA is from 2 to 9. It is observed that the error is almost a

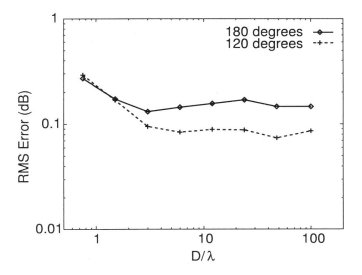

Figure 3.11 RMS error for a sphere from 0.75 to 100λ in diameter. Number of levels in MLFMA is from 2 to 9.

constant except for the first two points, where the geometry error is dominant. The 9,633,792 unknown model is also used to calculate the RCS for a 120λ sphere as shown in Figure 3.12. This problem uses 32 processors of an Origin2000, 26.7 GB of memory, 1.5 hours for filling the matrix, 13.0 hours for 43 iterations in GMRES (restart after 15 iterations) to reach 0.001 residual error, and 3 minutes for 1,800 points of RCS (MLFMA is used to calculate the bistatic RCS). The RMS error is 0.20 dB for all 1,801 points for elevation angle ranging from 90° to −90° with the incident angle at 90°, and 0.11 dB for elevation angle ranging from 90° to −30°. Because of the memory limit, the accuracy setting is lower than the run for 100λ.

Figure 3.13 shows the monostatic RCS for the aircraft (VFY218) at 100 MHz as a function of azimuth angle in the horizontal plane using the Rockwell flat triangular patch model for HH polarization. The wings of the VFY218 are on the x-y plane (horizontal plane). Zero degree corresponds to the incidence angle on the nose. A five-level MLFMA is used. The measurement data are from Wang, Sanders, and Woo at the Naval Air Warfare Center [46]. Good agreement between the numerical results and the measurements is observed. For this 9,747 unknown problem, FISC needs 102 MB of memory for this single-precision code and requires 3 hours of CPU time on a DEC Alpha workstation for 181 incident angles. In contrast, the LUD solution is estimated to need 800 MB of memory and 10 hours of CPU time for the LUD, and $O(N^2)$ calculations for each incident angle. We estimate that FISC would only need the same CPU time as the LUD solution for 1,000 incident angles. But FISC needs

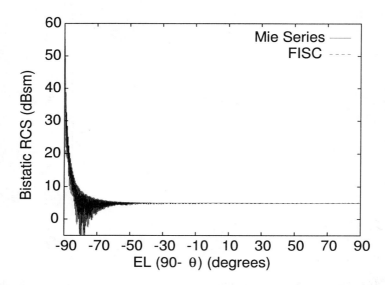

Figure 3.12 The Mie series and FISC result for a sphere with 120λ. A total of 9,633,792 unknowns and 9-level MLFMA are used.

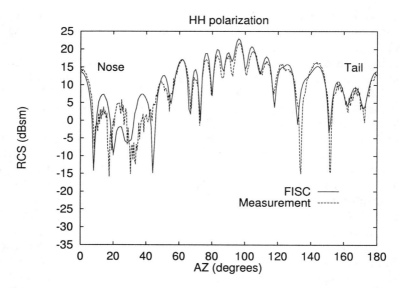

Figure 3.13 The monostatic RCS of the VFY218 at 100 MHz. The measurement data are from Naval Air Warfare Center [46] (HH polarization).

much less memory (102 MB) than the LUD solution (800 MB). The comparison is more in favor of FISC when N becomes larger.

3.6.3 Scaling of Memory Requirements

FISC solves the surface integral equation and the surface is described by a facet model. The RWG basis function [43] is assigned to each interior edge. The total number of unknowns N is the number of interior edges, which is about one-and-a-half (1.5) times the number of triangular facets. In most cases, average edge length a should be about 0.1λ. Therefore, we can express the number of unknowns N as

$$N \simeq 350 S \left(\frac{0.1\lambda}{a}\right)^2 \tag{3.88}$$

where S is the total surface area in the unit of square wavelength (λ^2). For a given target, N is proportional to the square of the frequency. If we refine the facet size by half, N increases by a factor of four.

Each single precision complex number needs 8 bytes. Hence, for a full matrix, the memory (in MB) is

$$MEM_{\text{full}} = 8 \left(\frac{N}{1000}\right)^2 \simeq 0.98 S^2 \left(\frac{0.1\lambda}{a}\right)^4 \tag{3.89}$$

The memory needed for a full matrix is proportional to the fourth power of the frequency. This increases drastically if we use a finer facet model (that is, smaller a).

We can empirically approximate FISC's memory requirement as

$$MEM_{\text{FISC}} \simeq 0.75 S \left[0.2 + 0.8 \left(\frac{0.1\lambda}{a}\right)^2\right] \tag{3.90}$$

The memory requirement increases linearly as the number of unknowns increases and is only proportional to the square of the frequency. If we refine the facet size by half, FISC only needs about three times as much memory—not 16 times as much as in the case of a full matrix. The complexity of MLFMA's memory requirement is $O(N \log N)$ or $O(S \log S)$. But for the problems we have tested (up to 10 million unknowns), the linear term in the memory requirement (the storage for near-interaction elements in the MOM matrix and each basis's radiation pattern) is dominant.

Figure 3.14 shows FISC's estimated memory requirement as a function of surface area with three different discretization densities (lines) and the memory used running FISC (points).

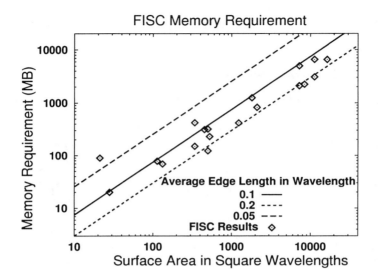

Figure 3.14 FISC's estimated memory requirement as a function of surface area (lines) and the memory used running FISC (points).

3.6.4 CPU Time Scaling

We divide the total CPU time into two parts: setup time (matrix filling) and solution time. The setup time is for one frequency point and is not a function of numbers of incident angles, observation angles, and polarization. If we use a full matrix, the setup time is proportional to N^2.

If we use LUD to solve the full matrix, it needs $O(N^3)$ operations to perform LUD, and then $O(N^2)$ for each incident angle. If we use an iterative solver with a full matrix, the CPU time for each incident angle is proportional to $N_{\text{iter}}N^2$, where N_{iter} is the number of iterations.

The FISC CPU time requirement is based on one CPU of an SGI CRAY Origin2000. Because the number of near-interaction elements in the MOM matrix is a constant for a basis in MLFMA, we can empirically approximate the setup time (filling the MOM matrix for near interactions) T_{fill} (sec) for one CPU of an SGI CRAY Origin2000 as

$$T_{\text{fill}} \simeq 5S \tag{3.91}$$

The setup time is linearly proportional to the problem size. However, the matrix-filling time for a full matrix is proportional to N^2.

We can approximate the solution time T_{solve} (sec) for each angle as

$$T_{\text{solve}} \simeq 0.05 S \log(S)(S^{1/4} + I_{\text{open}} S^{1/2}) \log\left(\frac{1}{\epsilon}\right) \frac{0.1\lambda}{a} \left(\frac{1}{r_{\min}}\right)^{1/2} \quad (3.92)$$

where I_{open} is 1 for open targets and 0 for closed targets, ϵ is the relative residual error for iterative solvers, and $r_{\min}$ is the minimum aspect ratio (the aspect ratio is always less than one).

In (3.92), we use the fact that the CPU time per matrix-vector multiply using MLFMA is $O(N \log N)$ or $O(S \log S)$. For closed targets, CFIE is used to reduce the number of iterations. The number of iterations needed for CFIE increases slowly when the problem size increases, and can be approximated by $O(N^{1/4})$ or $O(S^{1/4})$. For open targets, we can only use EFIE. It converges much more slowly than CFIE for the same problem size, where the required number of iterations is proportional to $N^{1/2}$ or $S^{1/2}$.

CFIE is mandatory for closed surfaces because it removes the nonuniqueness problem at the internal resonant frequencies of the closed structure. In addition, it provides a second-kind integral equation, which is well known to be better conditioned than EFIE's first-kind integral equation [47].

When $\epsilon = 0.01$, $a = 0.1\lambda$, and $r_{\min} \simeq 1$, the solution time T_{solve} can be simplified as

$$T_{\text{solve}} \simeq 0.1 S \log(S)(S^{1/4} + I_{\text{open}} S^{1/2}) \quad (3.93)$$

In Figure 3.15, we plot three curves as functions of S. The first curve (solid line) is for setup time calculated using (3.91), and the second and third curves (the dashed lines) are for the solution time calculated using (3.93) with $I_{\text{open}} = 1$ (using the EFIE for open target) and 0 (using CFIE for closed target), respectively.

Therefore, the total CPU time for FISC is

$$T_{\text{total}} = N_{\text{freq}}(T_{\text{fill}} + N_{\text{inc}} N_{\text{pol}} T_{\text{solve}}) \quad (3.94)$$

where N_{freq} is the number of frequency points, N_{inc} is the number of incident angles, and N_{pol} is the number of polarizations. Because we use the previous solution with phase correction as the initial guess for the next incident angle in the FISC, the CPU time in the second part of (3.94) increases more slowly as the number of incident angles increases. When we use the approximation of bistatic RCS to monostatic RCS available in FISC [41], we can further reduce the number of incident angles used for iterative solutions. For example, we used 30 incident angles to calculate the monostatic RCS of VFY218 at 1 GHz from 0° to 180° with 900 points. Similarly, we use the frequency interpolation to reduce the number of frequency points.

Figures 3.14 and 3.15 also plot FISC numerical results used to obtain the approximated FISC requirements and scaling. Our FISC results come from solving spheres with different electric sizes, VFY218 at different frequencies, bicones, squares, closed cylinders, trihedrons, and so on.

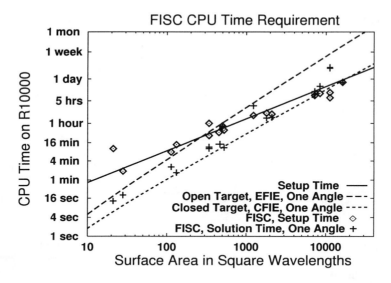

Figure 3.15 Estimated setup and solution time requirements as functions of surface area (lines) and the CPU time used running FISC (points).

3.6.5 More Results and Summary

With the features introduced in the last two sections, we have solved the problem of scattering from a 120λ sphere with 9,633,792 unknowns and VFY218 at 8 GHz with 9,990,918 unknowns. In Figure 3.16, we plot the bistatic RCS of the VFY218 at 8 GHz. The incident direction is from broadside with H-polarization (EL=0°, AZ=90°). This problem used 64 processors of the Origin2000, 46 GB of memory, and 19 hours. Numerical results calculated by Xpatch [48], a high-frequency approximation code, are also plotted for comparison.

MLFMA is also implemented for targets with a PEC ground plane. Our approach uses the mirror image of the radiation pattern of the targets in free space. The aggregation and disaggregation are unchanged: only the translation needs to include the mirror image term. It needs only 10% more memory and 35% more CPU time than MLFMA for the same targets in free space. In [49] we show the synthetic aperture radar (SAR) image of a 2m sphere located 2m above a PEC ground plane. The 64 frequencies are from 0.9 to 1.5 GHz and the 64 incident angles are around 30° from the normal direction. A total of 37,632 unknowns are used and the run takes only 1 minute per frequency and incident angle on a DEC personal workstation.

Since FISC v1.0 was released in March 1997, more than 400 copies of Xpatch/FISC have been distributed to governmental and industrial users. FISC is designed to compute the RCS of a target described by a triangular facet file. MLFMA has been

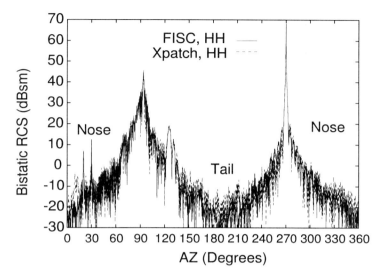

Figure 3.16 The bistatic RCS of the VFY218 at 8 GHz (H-pol.). A total of 9,990,918 unknowns and 10-level MLFMA are used.

implemented to speed up the matrix-vector multiplies. Both the memory requirements and the CPU time per iteration are of $O(N \log N)$. We have developed many features such as a block diagonal preconditioner, near singularity extraction, phase corrected previous solution for the initial guess, frequency interpolation/extrapolation of electric currents [50] or scattered fields, multiple right-hand-side (RHS) solvers, the approximation of bistatic RCS to monostatic RCS, calculating radiated fields using MLFMA, and PEC ground plane. Using upgraded FISC on a small computer, we can solve for the electromagnetic scattering by large complex 3D objects like aircraft (VFY218) and automobiles.

We also give the scaling properties of the CPU and memory requirements of FISC. The number of unknowns scales approximately as the surface area of the scatterer, or $N \simeq 350 S \left(\frac{0.1\lambda}{a}\right)^2$, where S is the target surface area in the unit of square wavelength, and a is the average edge length.

The memory requirement scales approximately as the number of unknowns, and hence, the surface area of the scatterer. The memory requirement is approximately $0.75 S \left[0.2 + 0.8 \left(\frac{0.1\lambda}{a}\right)^2\right]$ MB.

If we use one CPU of an SGI CRAY Origin2000, for each frequency, FISC needs the setup time $T_{\text{fill}} \simeq 5S$ (sec), and solution time $T_{\text{solve}} = 0.1 S \log(S)(S^{1/4} + I_{\text{open}} S^{1/2})$ (sec) for each angle. This formula is true for nonresonant structures.

CFIE is de rigueur for scatterers with closed surfaces. It yields a formulation with the second-kind integral equation with better condition numbers, and requiring fewer iterations. The number of iterations required, N_iter, scales as $N^{0.25}$ in this case.

For scatterers with open surfaces, our only choice is to use EFIE, which is a first-kind integral equation with a worse condition number than a second-kind integral equation. The number of iterations required, N_iter, scales as $N^{1/2}$ in this case.

The scaling properties of N_iter at this point are empirically derived via numerical experimentation of very large scale computing problems. A more theoretical analysis is necessary to arrive at these scaling properties.

3.7 CONCLUSIONS

In this chapter, we have described 3D FMM and MLFMA and their implementations. The interpolation and anterpolation are important in achieving $O(N\log N)$ complexity in MLFMA for a matrix-vector multiply. We have discussed the errors in FMM and MLFMA and show that all of them are controllable and decrease exponentially. Low complexity in a matrix-vector multiply is not good enough for efficient, large-scale computing with many iterations, multiple incident angles, and frequencies. We present a number of features in FISC for efficient calculating. Using FISC, we show that both memory and CPU time requirements are of $O(N\log N)$, that the accuracy for a problem ranging from 588 to 9,633,792 with 2 to 9 levels in MLFMA is almost a constant, and that scattering from a VFY218 at 8 GHz can be solved using the MOM with MLFMA. In the next few chapters, we will present more applications using FMM and MLFMA.

REFERENCES

1. R. Coifman, V. Rokhlin, and S. Wandzura, "The fast multipole method for the wave equation: A pedestrian prescription," *IEEE Ant. Propag. Mag.*, vol. 35, pp. 7–12, June 1993.

2. L. R. Hamilton, P. A. Macdonald, M. A. Stalzer, R. S. Turley, J. L. Visher, and S. M. Wandzura, "3D method of moments scattering computations using the fast multipole method," *IEEE Antennas Propagat. Int. Symp. Dig.*, vol. 1, pp. 435–438, July 1994.

3. J. M. Song and W. C. Chew, "Fast multipole method solution using parametric geometry," *Micro. Opt. Tech. Lett.*, vol. 7, pp. 760–765, Nov. 1994.

4. J. M. Song and W. C. Chew, "Fast multipole method solution of combined field integral equation," *11th Annual Review of Progress in Applied Computational Electromagnetics*, vol. 1, pp. 629–636, March 1995.

5. J. M. Song and W. C. Chew, "Multilevel fast-multipole algorithm for solving combined field integral equations of electromagnetic scattering," *Micro. Opt. Tech. Lett.*, vol. 10, pp. 14–19, Sept. 1995.

6. J. M. Song, C. C. Lu, and W. C. Chew, "MLFMA for electromagnetic scattering from large complex objects," *IEEE Trans. Ant. Propag.*, vol. 45, pp. 1488–1493, Oct. 1997.

7. B. Dembart and E. Yip, "A 3D fast multipole method for electromagnetics with multiple level," *11th Annual Review of Progress in Applied Computational Electromagnetics*, vol. 1, pp. 621–628, March 1995.

8. B. Dembart and E. Yip, "The accuracy of fast multipole methods for Maxwell's equations," *IEEE Computational Science and Engineering*, vol. 5, pp. 48–56, July-Sept. 1995.

9. M. F. Gyure and M. A. Stalzer, "A prescription for the multilevel Helmholtz FMM," *IEEE Computational Science and Engineering*, vol. 5, pp. 39–47, July-Sept. 1998.

10. J. M. Song, C. C. Lu, W. C. Chew, and S. W. Lee, "Fast Illinois Solver Code (FISC)," *IEEE Ant. Propag. Mag.*, vol. 48, pp. 27–34, June 1998.

11. X. Q. Sheng, J. M. Jin, J. M. Song, C. C. Lu, and W. C. Chew, "On the formulation of hybrid finite-element and boundary-integral methods for 3-D scattering," *IEEE Trans. Ant. Propag.*, vol. 46, pp. 303–311, March 1998.

12. X. Q. Sheng, J. M. Jin, J. M. Song, W. C. Chew, and C. C. Lu, "Solution of combined-field integral equation using multilevel fast multipole algorithm for scattering by homogeneous bodies," *IEEE Trans. Ant. Propag.*, vol. 46, pp. 1718–1726, Nov. 1998.

13. H. Y. Chao, W. C. Chew, J. M. Song, and E. Michielssen, "Impedance calculation of complex surfaces-wire structures with multilevel fast multipole algorithm and variational formulation," *IEEE AP-S Symposium*, vol. 3, pp. 1848–1851, July 1999.

14. H. Y. Chao, W. C. Chew, E. Michielssen, and J. M. Song, "The multilevel fast multipole algorithm for electromagnetic compatibility analysis," *IEEE-EMC Int. Symp.*, vol. 2, pp. 844–847, 1999.

15. F. Ling, J. M. Jin, and J. M. Song, "Multilevel fast multipole algorithm for analysis of large-scale microstrip structures," *Micro. Opt. Tech. Lett.*, vol. 9, pp. 508–510, Dec. 1999.

16. K. Donepudi, J. M. Song, J. M. Jin, G. Kang, and W. C. Chew, "Point-based MLFMA for Galerkin's method," *IEEE Trans. Ant. Propag.*, vol. 48, pp. 1192–1197, Aug. 2000.

17. K. Donepudi, J. M. Jin, S. Velamparambil, J. M. Song, and W. C. Chew, "A higher-order parallelized fast multipole algorithm for 3D scattering," *IEEE Trans. Ant. Propag.*, vol. 49, May 2001.

18. L. S. Canino, J. J. Ottusch, M. A. Stalzer, J. L. Visher, and S. Wandzura, "Numerical solution of the Helmholtz equation in 2D and 3D using a high-order Nystrom discretization," *Comput. Phys.*, vol. 146, pp. 627–663, 1998.

19. V. Rokhlin, "Diagonal forms of translation operators for the Helmholtz equation in three dimensions," *Applied and Comp. Harmonic Analysis*, vol. 1, pp. 82–93, Jan. 1993.

20. M. A. Epton and B. Dembart, "Multipole translation theory for the three dimensional Laplace and Helmholtz equations," *SIAM J. Scientific Computing*, vol. 16, pp. 865–897, July 1995.

21. W. C. Chew, S. Koc, J. M. Song, C. C. Lu, and E. Michielssen, "A succinct way to diagonalize the translation matrix in three dimensions," *Micro. Opt. Tech. Lett.*, vol. 15, pp. 144–147, June 1997.

22. J. R. Mautz and R. F. Harrington, "H-field, E-field, and combined field solutions for conducting bodies of revolution," *A.E.U.*, vol. 32, pp. 159–164, April 1978.

23. G. Antilla and N. G. Alexopoulos, "Scattering from complex three-dimensional geometries using a curvilinear hybrid finite-element-integral equation approach," *Optical Soc. America*, vol. 11, pp. 1445–1457, April 1994.

24. M. Abromowitz and I. A. Stegun, *Handbook of Mathematical Functions*, New York: Dover Publication, 1972.

25. J. Stratton, *Electromagnetic Theory*, New York: McGraw-Hill, 1941.

26. L. Hernquist, "Performance characteristics of tree codes," *Astrophysical Journal Supp.*, vol. 64, pp. 715–734, 1987.

27. C. R. Anderson, "An implementation of the fast multipole method without multipoles," *SIAM Journal of Scientific and Statistical Computing*, vol. 13, pp. 923–947, July 1992.

28. A. Brandt, "Multilevel computations of integral transforms and particle interactions with oscillatory kernels," *Proceedings IMACS 1st. Int. Conf. on Comp. Phys*, vol. 65, pp. 24–38, June 1990.

29. C. C. Lu and W. C. Chew, "A multilevel algorithm for solving boundary integral equations of wave scattering," *Micro. Opt. Tech. Lett.*, vol. 7, pp. 466–470, July 1994.

30. S. Koc, J. M. Song, and W. C. Chew, "Error analysis for the numerical evaluation of the diagonal forms of the scalar spherical addition theorem," *SIAM J. Numer. Anal.*, vol. 36, no. 3, pp. 906–921, 1999.

31. V. Rokhlin, *Sparse Diagonal Forms for Translation Operations for the Helmholtz Equation in Two Dimensions*, Research Report YALEU/DCS/RR-1095, Dec. 1995.

32. W. J. Wiscombe, "Improved Mie scattering algorithms," *Applied Optics*, vol. 19, May 1980.

33. O. M. Bucci and G. Franceschetti, "On the spatial bandwidth of scattered fields," *IEEE Trans. Ant. Propag*, vol. 35, pp. 1445–1455, Dec. 1987.

34. R. E. Kell, "On the derivation of bistatic RCS from monostatic measurements," *Proc. IEEE*, vol. 53, pp. 983–988, Aug. 1965.

35. M. J. Schuh, A. C. Woo, and M. P. Simon, "The monostatic/bistatic approximation," *IEEE Antennas Propagat. Mag.*, vol. 36, pp. 76–78, Aug. 1994.

36. B. K. Alpert and V. Rokhlin, "A fast algorithm for the evaluation of Legendre expansions," *SIAM J. Sci. Stat. Comput.*, vol. 12, pp. 158–179, Jan. 1991.

37. I. S. Gradshteyn and I. M. Ryzhik, *Table of Integrals, Series, and Products*, New York: Academic Press, 1965.

38. J. M. Song and W. C. Chew, "Interpolation of translation matrix in MLFMA," *Micro. Opt. Tech. Lett.*, July 20, 2001.

39. W. H. Press, S. A. Teukolsky, W. T. Vetterling, and B. P. Flannery, *Numerical Recipes in FORTRAN*, Second Edition, New York: Cambridge University Press, 1992.

40. O. M. Bucci, C. Gennareli, and C. Savarese, "Optimal interpolation of radiated fields over a sphere," *IEEE Trans. Ant. Propag*, vol. 39, pp. 1633–1643, Nov. 1991.

41. J. M. Song, C. C. Lu, W. C. Chew, and S. W. Lee, *User's Manual for FISC (Fast Illinois Solver Code)*, Champaign, IL: Center for Computational Electromagnetics, University of Illinois and DEMACO, Inc., Jan. 1997.

42. J. M. Song and W. C. Chew, "The Fast Illinois Solver Code: requirements and scaling properties," *IEEE Computational Science and Engineering*, vol. 5, pp. 19–23, July–Sept. 1998.

43. S. M. Rao, D. R. Wilton, and A. W. Glisson, "Electromagnetic scattering by surfaces of arbitrary shape," *IEEE Trans. Ant. Propag.*, vol. 30, pp. 409–418, May 1982.

44. J. M. Song and W. C. Chew, "Moment method solutions using parametric geometry," *J. Electromagn. Waves and Appl.*, vol. 9, pp. 71–83, Jan.–Feb. 1995.

45. G. Antilla, Y. C. Ma, P. V. Alstine, and M. I. Sancer, *Hybrid Finite Element-Method of Moments for Electromagnetic Prediction of Complex 3D Geometries, Course Note*, Newport Beach, CA: IEEE AP-S Int. Symp. and URSI Radio Science Meeting, June 1995.

46. H. T. G. Wang, M. L. Sanders, and A. C. Woo, "Radar cross section measurement data of the VFY 218 configuration," *NAWCWPNS TM 7621*, Jan. 1994.

47. G. C. Hsiao and R. E. Kleinman, "Error analysis in numerical solution of acoustic integral equations," *Int. J. Numer. Meth. Engg.*, vol. 37, pp. 2921–2933, 1994.

48. S. W. Lee, *User's Manual for Xpatch*, Champaign, IL: DEMACO, Inc., Jan. 1995.

49. J. M. Song and W. C. Chew, "Large scale computations using FISC," *IEEE Antennas Propagat. Int. Symp. Dig.*, vol. 4, pp. 1856–1859, July 2000.

50. Y. X. Wang, H. Ling, J. M. Song, and W. C. Chew, "A frequency extrapolation algorithm for FISC," *IEEE Trans. Antennas Propag.*, vol. 45, pp. 1891–1893, Dec. 1997.

4

Parallelization of Multilevel Fast Multipole Algorithm on Distributed Memory Computers

Sanjay Velamparambil, Jiming Song, and Weng Cho Chew

4.1 INTRODUCTION

Integral equation methods are widely used for the solution of electromagnetic scattering problems. In this approach, the problem is first formulated in terms of an appropriate integral equation and then reduced to a system of linear equations using the method of moments (MOM). The resulting linear system is then solved by either a direct method such as LU-decomposition or by an iterative method such as the conjugate gradient method (CGM). It is well known that the number of operations necessary for the direct method is $O(N^3)$, where N is the number of unknowns; and hence, the maximum size that can be solved is limited to a few thousand unknowns even with modern supercomputers. On the other hand, iterative methods require the application of the coefficient matrix to a sequence of vectors in the solution process. Each of these evaluations requires $O(N^2)$ operations, making it prohibitively expensive for large problems.

In recent years, a number of techniques have been proposed for the rapid application of the coefficient matrix to a given vector. Among these, the fast multipole method [1] and its multilevel or recursive variant, the multilevel fast multipole algorithm (MLFMA) [2], have established themselves as the most powerful. It has been shown that MLFMA reduces the computational complexity of a matrix-vector multiply to $O(N \log N)$. Using MLFMA-accelerated MOM codes, researchers have solved very large-scale problems [2, 3].

Advances in supercomputing and low-cost cluster computing have demonstrated that the future of high-performance computing lies in distributed memory parallel computing. This includes the possibility of heterogeneous computing over geographically separated supercomputers linked through fast networks. It is clear that in order to exploit this new technology for solving extremely large-scale problems in electromagnetics, distributed memory versions of MLFMA have to be developed.

Recently, in an attempt to meet this demand, we have developed a parallel version of dynamic MLFMA. This new portable, scalable, parallel implementation of the dynamic MLFMA is called the ScaleME (*Scale*able *M*ultipole *E*ngine). ScaleME is a portable implementation of the dynamic MLFMA using the message passing paradigm. It uses the Message Passing Interface (MPI) for communication. It is implemented as an *application-independent kernel*, which can be retrofitted into existing applications and is targeted at distributed memory machines and networked cluster of workstations. Using ScaleME we have been able to solve very large scale scattering problems, including that of a full scale aircraft at 8GHz involving more than 10 million unknowns.

In this chapter, we discuss the various aspects of parallelizing the dynamic MLFMA on a distributed memory platform. In particular, we shall discuss the characteristic features that are unique to the dynamic MLFMA, as opposed to the static case, and their effects on the parallel algorithm design. We shall also discuss some features of ScaleME , our implementation of MLFMA for distributed memory computers.

This chapter is organized as follows. In the next section, we briefly discuss the MPI. In Section 4.3 we summarize the relevant facts from the multipole translation theory and establish our notations. In Section 4.4 we describe the parallel MLFMA in detail. Section 4.5 describes certain optimizations we have developed for accelerating and reducing the memory requirements of the distributed memory implementation. In Section 4.6 we briefly discuss the features of ScaleME from a user's point of view. In Section 4.7 we report numerical results demonstrating the performance of ScaleME as applied to a specific problem. In Section 4.8 we report a preliminary result from ScaleME-2, an improved implementation of the parallel MLFMA, followed by conclusions in Section 4.9.

4.2 THE MPI PROGRAMMING MODEL

Message passing is a programming paradigm used widely on parallel computers. The message passing model can be explained with respect to Figure 4.1. In this model,

Parallelization of MLFMA on Distributed Memory Computers

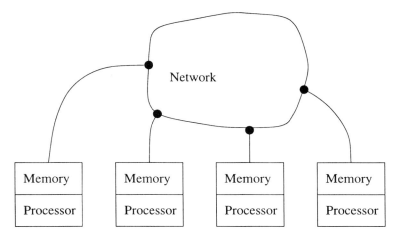

Figure 4.1 Schematic of the computing model used in MPI.

each computing node essentially consists of a processor and a certain amount of RAM. The processing elements (PEs) or *nodes* are connected through a network. Thus, a processor can only access the memory local to the node it is associated with, and each node is capable of executing programs independent of the other. Exchange of data between nodes is made through explicit messages. Such a computer can be a cluster of independent workstations linked through a TCP/IP network or a high-performance parallel computer with distributed memory.

MPI is a standardized and portable message passing system designed to function on a wide variety of parallel computers. The MPI standard defines the user interface, syntax, and semantics of a set of message passing functionalities. Examples include point-to-point communication operations such as send and recieve, and collective communications such as broadcast. It is important to note that MPI standard only defines an interface and not a "program" by itself. Several implementations are available, some of them free and some supplied by vendors. An excellent resource for MPI related information is the Web site [4].

We note that although MPI provides a standardized interface for communication operations, the operating environment varies from one implementation to another. These variations generally reflect the way processes are created, compilation commands, and interface to the standard input and output devices.

In developing the parallel MLFMA, we have assumed the single program multiple data (SPMD) paradigm of parallel computing. This means that the same executable runs on all the processors, but the execution graphs of the processors may vary.

4.3 MATHEMATICAL PRELIMINARIES

The basic principle behind FMM is to decompose the computation of matrix-vector products into two parts: one involving the interaction between nearby sources and the other involving those between well separated ones. The nearby interactions are computed using the usual technique adopted in the method of moments approach. The interaction between far away sources is evaluated using an asymptotic representation of the radiated fields. In this section, we summarize the essential results from the multipole translation theory. The discussion is very brief, by no means original; for detailed descriptions the reader is referred to [1, 5–7].

4.3.1 Notations

We shall use the following notations in the discussion below. Let $D = \{\mathbf{r} \in \mathbf{R}^d : |\mathbf{r} - \mathbf{c}| < R, \mathbf{c} \in \mathbf{R}^d, R > 0\}, d = 2, 3$. If $\mathbf{r} \in \mathbf{R}^d$, the length of the vector will be denoted by $|\mathbf{r}|$ and the polar/spherical coordinates by the ordered pair $(|\mathbf{r}|, \hat{r})$.

We shall consider three open disks D_1, D_2, D_3 in $\mathbf{R}^d$ such that $D_2 \subset D_1$ and $D_3 \cap D_1 = \emptyset$. The respective closures of these disks will be denoted by $\bar{D}_1, \bar{D}_2$, and $\bar{D}_3$. The centers and radii of these disks will be denoted by $\mathbf{c}_1, \mathbf{c}_2, \mathbf{c}_3$ and R_1, R_2, R_3, respectively.

We shall denote the unit sphere centered at the origin by the symbol $S_0 = \{\mathbf{r} \in \mathbf{R}^d : |\mathbf{r}| = 1\}$ and any element of S_0 by the symbol $\hat{s}$. Note that our definition of the sphere does not include its "interior."

4.3.2 Essentials of Diagonal Forms of Radiation Fields

If a function $\psi : \mathbf{R}^d \mapsto \mathbf{C}$ satisfies the Helmholtz equation

$$\nabla^2 \psi + k^2 \psi = 0 \tag{4.1}$$

outside D and the radiation condition

$$\lim_{r \to \infty} \psi(r\hat{r}) = C e^{ikr} r^{\frac{1-d}{2}} + O(r^{1-d}) \tag{4.2}$$

for some $C < \infty$, then we define an asymptotic representation to ψ, or the radiation pattern of ψ, as a function $F_{\psi, \mathbf{c}} : S_0 \mapsto \mathbf{C}$ for a given $\mathbf{c} \in \mathbf{R}^d$, defined by the formula

$$F_{\psi, \mathbf{c}}(\hat{s}) = \lim_{r \to \infty} (kr)^{\frac{d-1}{2}} \psi(\mathbf{c} + r\hat{s}) e^{-ikr} \tag{4.3}$$

We note that this definition is in tandem with that of [1] but not with that in [5]. We adopt this definition because it provides a uniform set of definitions and notations.

Example 1 Let $\psi(x) = H_0(k|\mathbf{r} - \mathbf{r}'|)$. Then $F_{\psi, \mathbf{c}}(\hat{s}) = -ie^{-ik(\mathbf{c} - \mathbf{r}') \cdot \hat{s}}$.

On the other hand, if $\phi\colon \mathbf{R}^d \mapsto \mathbf{C}$ is a function satisfying the Helmholtz equation inside D, then under certain conditions, there exist unique numbers $U(\hat{s}), V(\hat{s})$ such that
$$\lim_{kr\to\infty}\left\{(kr)^{\frac{d-1}{2}}\phi(\mathbf{c}+r\hat{s}) - a_1 U(\hat{s})e^{ikr} - a_2 V(\hat{s})e^{-ikr}\right\} = 0$$
where a_1, a_2 are constants that depend only on the dimension d. We define the asymptotic representation to the function ϕ as the function $L_{\phi,\mathbf{c}}\colon S_0 \mapsto \mathbf{C}$ by the formula:
$$L_{\phi,\mathbf{c}}(\hat{s}) = U(\hat{s}) \tag{4.4}$$

In the following, we shall refer to the asymptotic forms $F_{\psi,\mathbf{c}}(\hat{s})$ and $L_{\psi,\mathbf{c}}(\hat{s})$ as the radiation patterns of ψ centered at $\mathbf{c}$.

Lemma 1 If $\psi : \mathbf{R}^d \mapsto \mathbf{C}$ is a radiation field analytical outside $\bar{D}_2$ and satisfying the radiation condition given by (4.2) at ∞, then for any $\hat{s} \in S_0$
$$F_{\psi,\mathbf{c}_1}(\hat{s}) = \lambda_\infty(\hat{s}) F_{\psi,\mathbf{c}_2}(\hat{s}) \tag{4.5}$$
with $\lambda_\infty(\hat{s})$ defined as
$$\lambda_\infty(\hat{s}) = e^{-ik(\mathbf{c}_1-\mathbf{c}_2).\hat{s}} \tag{4.6}$$

Lemma 2 If $\phi : \mathbf{R}^d \mapsto \mathbf{C}$ is a radiation field analytical inside D_1, then for any $\hat{s} \in S_0$,
$$L_{\phi,\mathbf{c}_2}(\hat{s}) = \mu_\infty(\hat{s}) L_{\phi,\mathbf{c}_1}(\hat{s})$$
where
$$\mu_\infty(\hat{s}) = e^{-ik(\mathbf{c}_2-\mathbf{c}_1).\hat{s}} \tag{4.7}$$

Remark 1 If k is real, then $\mu_\infty(\hat{s}) = \lambda_\infty^*(\hat{s})$.

Lemma 3 If $\psi : \mathbf{R}^d \mapsto \mathbf{C}$ is a radiation field analytical outside $\bar{D}_1$ and satisfying the radiation condition given in (4.2) at ∞, then for any $\epsilon > 0$ there exists $N > 0$ such that for any $n \geq N$,
$$|\psi(\mathbf{r}) - \phi_n(\mathbf{r})| < \epsilon$$
for any $\mathbf{r} \in D_3$, where $\phi_n\colon \mathbf{R}^d \mapsto \mathbf{C}$ is a radiation field analytical inside D_3 with the asymptotic representation defined by the formula
$$L_{\phi_n,\mathbf{c}_3}(\hat{s}) = \nu_n(\hat{s}) F_{\psi,\mathbf{c}_1}(\hat{s})$$
and the function $\nu_n(\hat{s})$ defined as
$$\nu_n(\hat{s}) = \begin{cases} \sqrt{\frac{\pi}{2}} e^{\frac{i\pi}{4}} \sum_{j=-n}^{n} H_j(k|\mathbf{c}_3-\mathbf{c}_1|) e^{-ij(\theta_{13}-\pi-\theta)} & \text{if } d = 2 \\ \sum_{m=0}^{n} i^m (2m+1) P_m(\hat{s}.\hat{s}_{13}) h_m(k|\mathbf{c}_3-\mathbf{c}_1|) & \text{if } d = 3 \end{cases} \tag{4.8}$$

where θ_{13} is the polar angle in two dimensions between the vector $\mathbf{c}_3 - \mathbf{c}_1$ and the x axis. Also $\hat{s}_{13} = \frac{\mathbf{c}_3-\mathbf{c}_1}{|\mathbf{c}_3-\mathbf{c}_1|}$.

These results provide a tool for shifting the center of asymptotic expansions. The following lemma gives a tool for computing the actual field and its derivative at any point from its asymptotic representation [5].

Lemma 4 Suppose that under conditions of Lemma 2, the field $\phi(\mathbf{r}), \mathbf{r} \in D_1$ has the asymptotic representation $L_{\phi,\mathbf{c}_1} : S_0 \mapsto \mathbf{C}$. Then for any $x \in D_1$,

$$\phi(\mathbf{r}) = \frac{i^{d-2}}{|S_0|} \int_{S_0} L_{\phi,\mathbf{c}_1}(\hat{s}) \cdot e^{ik(\mathbf{r}-\mathbf{c}_1)\cdot\hat{s}} \, d\omega_s \quad (4.9)$$

where $|S_0|$ denotes the surface area of the d-dimensional unit sphere and $d\omega_s$ represents the elemental area over the unit sphere.

Proofs of these results or related ones can be found in [1, 5, 8].

Although not directly related to the theory of diagonal forms, we state a useful result involving the diagonal form of the gradient of a scalar field.

Lemma 5 Let $\psi : \mathbf{R}^d \setminus \bar{D} \mapsto \mathbf{C}$ be an analytic function satisfying the Helmholtz equation (4.1) everywhere outside $\bar{D}$. Let $F_{\psi,c}(\hat{s})$ denote its far field signature as defined in (4.3). Then,

$$\lim_{r \to \infty} (kr)^{\frac{d-1}{2}} e^{-ikr} \nabla \psi(\mathbf{c} + r\hat{s}) = ik F_{\psi,\mathbf{c}}(\hat{s})\hat{s}$$

Remark 2 (Numerical Approximation) For the purpose of implementation, we approximate these asymptotic forms with a set of n samples over the unit sphere S_0 for a suitably chosen positive integer $n \geq 2$. Let these sampling points be denoted by $\{s_1, \cdots, s_n\}$. Furthermore, these finite dimensional approximations will be denoted by $F_{\psi,\mathbf{c}_1}^n$ and $L_{\phi,\mathbf{c}_1}^n$, respectively.

4.3.3 FMM Representation of Matrix Elements in Method of Moments

When we use MOM for the solution of integral equations, each element in the resultant matrix takes the form $a_{ij} = \langle t_i, Lf_j \rangle$ with the usual notations. We shall further assume that the basis and testing functions used in MOM have compact support. In this section, we provide an explicit representation of these elements using the diagonal forms. Of course, we assume that the kernel of the operator L is some form of the free space Green's function to the Helmholtz equation (4.1). That is,

$$\psi_j(\mathbf{r}) = (Tf_j)(\mathbf{r}) = \int_G f_j K(\mathbf{r}, \mathbf{r}') \, d\mathbf{r}'$$

such that the function $\psi_j(\mathbf{r})$ satisfies the Helmholtz equation outside the suppf_j. In the case of functions satisfying the vector Helmholtz equation, we shall consider the rectangular components separately.

If $\text{supp} f_j \subset D_1$ from under the conditions of Lemma 3 and Lemma 4, we have the following representation for $\psi_j(\mathbf{r})$:

$$\psi_j(\mathbf{r}) \approx \frac{i^{d-2}}{|S_0|} \int_{S_0} L^m_{\psi,\mathbf{c}_3}(\hat{s}) \cdot e^{ik(\mathbf{r}-\mathbf{c}_3)\cdot\hat{s}} \, d\omega_s \quad (4.10)$$

where $L^n_{\psi,\mathbf{c}_3}(\hat{s}) = \nu_n(\hat{s}) F_{\psi,\mathbf{c}_1}(\hat{s})$. Substituting (4.10) into the expression for the matrix elements, we obtain the following result:

$$a_{ij} \approx \int_{\text{supp } t_i} d\mathbf{r} \, t_i(\mathbf{r}) \frac{i^{d-2}}{|S_0|} \int_{S_0} L^n_{\psi,\mathbf{c}_3}(\hat{s}) \cdot e^{ik(\mathbf{r}-\mathbf{c}_3)\cdot\hat{s}} \, d\omega_s$$

Definition 1 (Receiving Pattern) Under the conditions of Lemma 3 and Lemma 4 we define the receiving pattern of a testing function by the formula:

$$R_{t_i,\mathbf{c}_3}(\hat{s}) = \int_{\text{supp } t_i} t_i(\mathbf{r}) e^{ik(\mathbf{r}-\mathbf{c}_3)\cdot\hat{s}} \, d\mathbf{r}$$

Using the above definition, we can rewrite the representation for a_{ij} as

$$a_{ij} \approx \frac{i^{d-2}}{|S_0|} \int_{S_0} L^n_{\psi,\mathbf{c}_3}(\hat{s}) \cdot R_{t_i,\mathbf{c}_3}(\hat{s}) \, d\omega_s \quad (4.11)$$

which is of fundamental importance in the design of the ScaleME.

4.4 THE PARALLEL MLFMA

In this section, we shall describe the parallel multilevel fast multipole algorithm as implemented in ScaleME. The general strategy of an MLFMA is that of clustering basis functions or particles at various spatial lengths and computing the interactions with the testing functions from sufficiently distant clusters using multipole expansions. For nearby interactions, direct computation is used.

By B_0, we denote the smallest d-cell that encompasses the domain $G \subset \mathbf{R}^d$ containing the scatterer and refer to it as the *computational box*. S_l is the set of all boxes at level l, S_0 consisting only of B_0. S_{l+1} is obtained by subdividing each nonempty box in S_l into 2^d boxes. The boxes obtained by dividing a given box b are called the children of b, and b is called the parent. The parent of a box $b \in S$ will be denoted by parent(b) and the ith child of $b \in S$ will be denoted by child$_i(b)$. The level of refinement of any box is denoted by the integer $nlev$ such that $nlev = 0$ for the computational box B_0. The set of all boxes is denoted by S. The maximum number of levels is denoted by maxlev. Further, for any box $b \in S$, the center and length of the side are denoted by $\mathbf{c}_b$ and w_b, respectively. If $b \in S$, then the set of all basis functions in b will be denoted by Π_b and number of basis functions in box b

will be denoted by $N_\pi(b)$. If $i \in \Pi_b$ for any $b \in S_{\text{maxlev}}$, we shall denote the strength of the ith source by the notation q_i.

For describing parallel algorithms, we assume that each processor is uniquely identified by an integer in the range $0, 1, \ldots, P-1$, where P is the total number of processors. This identifier will be referred to as the *rank* of the processor. For convenience, if A is a set that is distributed across P processors, by $A(p) \subset A$ we denote the subset which belongs to the processor with rank p. For example, the set of all nonempty boxes at level l belonging to processor p will be denoted by $S_l(p)$.

If $A \subset G$, where G is the domain enclosing the particles, then the radiation field outside A due to all the sources in A is denoted by ψ_A. Further, the asymptotic form of ψ_A with respect to a center $\mathbf{c}_A \in \mathbf{R}^d$, defined by (4.3), and its finite-dimensional approximation will be denoted by $F_A(\hat{s})$ and F_A^n, respectively. Also, $F_i(\hat{s})$ will denote the radiation pattern of the ith particle. Similarly, if $B \subset G$ and if there exist two disks D_1, and, D_2 with centers $\mathbf{c}_1$ and $\mathbf{c}_2$, respectively, such that $A \subset D_1$, $B \subset D_2$ with $D_1 \cap D_2 = \emptyset$, then under conditions of Lemma 3 we set

$$L_A^r(\hat{s}) = \nu_r(\hat{s}) F_A(\hat{s}) \tag{4.12}$$

for any integer $r > 0$, where we have omitted the explicit dependence of ν_r on the centers of A and B for simplicity. In particular, for any box $b \in S_l$, by $L_b^r(\hat{s})$ we denote the finite-dimensional receiving pattern of b, due to all the sources sufficiently well separated from b.

Definition 2 (Distance Between Boxes) The distance between two boxes $b, b' \in S$ is denoted by $d_B(b, b')$ and is defined by

$$d_B(b, b') = \inf\{|\mathbf{x} - \mathbf{y}| \,|\, \mathbf{x} \in b, \mathbf{y} \in b'\}$$

For each nonempty box $b \in S$, and a given positive integer m, we shall define two lists of other boxes for b.

Definition 3 (Near Neighbor List) For the computational box B_0, we shall define the list of all near neighbors to contain the the box B_0 itself. Given integers $0 < l \leq$ maxlev and $m > 0$ and a box $b \in S_l$, we shall define the list of all near neighbors of b, denoted by near(b), as the set all boxes $b' \in S_l$ such that:

1. Parent of b' is a near neighbor to parent of b;

2. $d_B(b, b') < m w_b$.

Definition 4 (Interaction List) Given integers $0 \leq l \leq$ maxlev and $m > 0$ and a box $b \in S_l$, we shall define the interaction list of b, denoted by Interaction_List(b), as the set all boxes $b' \in S_l$ such that:

1. Parent of b' is a near neighbor to parent of b;

2. $d_B(b, b') \geq mw_b$.

In the case of dynamic fast multipole methods, it is well known that the number of samples required at each level is proportional to the size of the box at that level, in terms of the wavelength. For each level l we shall denote the number of samples used in the finite dimensional approximation to the asymptotic fields by $N_s(l)$. Further, because of this, during the aggregation phase before translating from a finer level to a coarser level, the field samples are to be *interpolated*. Similarly, during the disaggregation phase, before translating the receiving pattern of a parent to its children, it has to be filtered down to a smaller number of samples. We assume that there exist suitably defined *interpolation* and *filtering* operators for doing these operations. For more details on these aspects, please refer to [9, 10]. In the following we shall denote the interpolation operator from level $l+1$ to level l by I_{l+1}^l and the filter from level l to $l+1$ by M_l^{l+1}. In the current implementation, we use polynomial interpolation and anterpolation suggested in [9], and we note that these operators are sparse matrices and that $M_l^{l+1} = (I_{l+1}^l)^T$.

We now describe the parallel MLFMA in broad strokes. We shall later take up the issues associated in implementing it.

4.4.1 The Algorithm

We shall begin by assuming that the tree has been constructed and is distributed across the processors, and that *no* processor has the complete information about the tree. We shall also not assume any preferences to tree construction or its distribution. In other words, we shall assume that the nodes of the tree are assigned to processors arbitrarily.

The evaluation of the matrix-vector product can be divided into three phases. In the aggregation (F2F) phase the far fields of each box are computed by shifting and combining the far fields of its children. In the next, called the translation (F2L) phase, the far fields of every box in the interaction list of a given box are translated to the incoming field or the local asymptotic representations ($L_b(\hat{s})$). In the third phase, called the disaggregation (L2L) phase, the local fields $L_b(\hat{s})$ of every parent box are shifted down to the center of each of its children. There are also two minor phases associated with the F2F and L2L phases which do not differ from from the sequential algorithm.

In the following discussion, for simplicity, we shall use the notations for the continuous forms of radiation fields for denoting the finite-dimensional approximation. For example, we shall denote the radiation pattern of a box b with respect to a center $\mathbf{c}$ by the symbol $F_{b,\mathbf{c}}(\hat{s})$ instead of $F_{b,\mathbf{c}}^n$, where n is the number of samples at the level of refinement of b.

First, we shall describe the aggregation phase (F2F) of the parallel MLFMA in detail. We begin by observing that since the tree is distributed across several processors, some of the nodes may have either their parent or children (or both) in different processors. In such a situation, the processors need to communicate to complete the aggregation phase. In order to handle this situation, we split the aggregation phase into two sections. In the first section, each processor traverses the tree from the finest level and completes as much computation as possible with locally available data. After that, each processor communicates with others to complete the computation involving nonlocal data. This is summarized below:

4.4.1.1 Aggregation Phase for Processor p

[**Comment:** At the finest level, combine the radiation pattern of basis functions belonging to a box to obtain the radiation pattern of the box.]

Step 1: for $b \in S_{\text{maxlev}}(p)$ do
$$F_b(\hat{s}) = \sum_{i \in \Pi_b} q_i F_i(\hat{s})$$
end for

[**Comment:** Starting from the finest level, combine the radiation patterns of *locally available* children to get a *partial* radiation pattern of parent boxes. For a given box, if every child is locally available, set a status flag, b.status, to F2F_COMPLETE, indicating that the box is ready for translation phase and for any required communication.]

Step 2: for $l = \text{maxlev} - 1$ down to 2 do
    for $b \in S_l(p)$ do
        for each locally available child(b) of b do
$$F_b(\hat{s}) \leftarrow F_b(\hat{s}) + \lambda_\infty(\hat{s}) I_{l+1}^l (F_{\text{child}(b)}(\hat{s}))$$
        end for
        If every non-empty child of b has been accounted for
            set b.status = F2F_COMPLETE;
    end for
end for

[**Comment:** Communicate with other processors and complete the rest of the computations. We assume that we have constructed a list of boxes which are to be *sent* to other processors and let nsnd be the number of boxes to be sent. Further, we assume that we know the number of boxes to be received, nrcv.]

Step 3: while there are boxes to be communicated {
    Select the next box b to be communicated, if any,
    if b.status = F2F_COMPLETE

            Queue up the radiation pattern of b for sending.
            Mark the box as SENT.
        end if
        if there is any incoming message
            Receive the incoming box b;
            $F_{\text{parent}(b)}(\hat{s}) \leftarrow F_{\text{parent}(b)}(\hat{s}) + \lambda_\infty(\hat{s}) I^l_{l+1}(F_b(\hat{s}))$
            if the radiation pattern of every child of parent(b) been translated
                Set parent(b).status = F2F_COMPLETE;
                UpdateSubTree(parent(b));
            end if
        end if
} end while

Two things require explanation: the significance of the status flag and Update-SubTree routine. In order to do this, refer to Figure 4.2. In this figure, P_1, P_2, P_3 refer to the processors, and n_0 to n_5 refer to the nodes of the MLFMA tree. Let us examine what happens during the aggregation phase. Processor P_1 can complete all its computations in Step 1 itself. In Step 2, since the node n_4 does not have any children (or trivially has all the children locally available) P_1 sets the status flag to F2F_COMPLETE. It then gets into Step 3 and sends the data, $F_{n_4}(\hat{s})$, to processor P_2. In the mean time, processor P_2 computes the radiation pattern of the node n_5, shifts, and combines with that of node n_2. However, P_2 cannot proceed further until $F_{n_4}(\hat{s})$ from P_1 arrives. So it starts executing Step 3 and waits for the data.

As soon as the data arrives, it is combined with the radiation pattern of node n_2 and sets the status flag to F2F_COMPLETE. It then calls the routine UpdateSubTree which recursively traverses the tree upwards from the given box, updating the radiation pattern of each node it visits, by shifting the radiation pattern of the node it visited immediately before to the center of the current node and by adding the shifted pattern with that of the current node. However, it can proceed from a given node n_g to its parent if and only if the radiation pattern of n_g is *complete*. That is, only if the radiation pattern of the current node n_g contains information from every child. So after updating the current node, the routine UpdateSubTree checks if the radiation pattern of every child has been translated. If this condition is met, it sets the status flag of the current node to F2F_COMPLETE. Otherwise, it exits.

This can be best explained with respect to Figure 4.2. When the data of node n_4 arrives in P_2, it is shifted and combined with the radiation pattern of node n_2. Then a call is made to UpdateSubTree, which will recursively traverse to node n_1. However, at node n_1, it cannot proceed further, since the radiation pattern $F_{n_3}(\hat{s})$ of node n_3 is necessary for completing the computation for node n_1. So it exits. Thus, UpdateSubTree makes it possible to have the computation to proceed asynchronously.

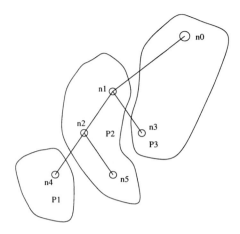

Figure 4.2 Illustration of the significance of UpdateSubTree.

4.4.1.2 The Translation (F2L) Phase

In this phase, the radiation pattern of each box in the interaction list of a given box is translated to an incoming pattern centered around the given box. As in the case of the aggregation phase, this phase first loops over every locally available box and completes as much computation as possible using locally available data. Once that is over, it communicates with other processors to gather data and updates the local patterns of each local box.

To improve the efficiency, we exploit the symmetry of interaction lists of boxes. It is easy to see that if $a \in$ Interaction_List(b) then $b \in$ Interaction_List(a). This property is exploited in the Step 2 below. We note that this approach is a special case of the *inverse interaction list* discussed in [11].

> [**Comment:** For each local box, translate the radiation pattern of every locally available box in the interaction list and combine them to get the partial local patterns.]

Step 1: for $l = 2$ to **maxlev** do
    for $b \in S_l(p)$ do
        for each locally available $b' \in$ Interaction_List(b) do
            $L_b(\hat{s}) \leftarrow L_b(\hat{s}) + \nu_r(\hat{s}) F_{b'}(\hat{s})$
        end for
    end for
end for

Parallelization of MLFMA on Distributed Memory Computers

[**Comment:** Communicate with other processors and gather data for the nonlocal boxes in the interaction lists. When a box arrives, compute its interaction list and use the symmetry. As in the case of aggregation phase, each processor keeps a list of boxes to be sent.]

Step 2: while there are boxes to be communicated {
    if there is any more boxes left to be sent,
        queue it up for sending.
    if there is any incoming message {
        Receive the incoming box b;
        Create the Interaction_List(b);
        for each locally available $b' \in$ Interaction_List(b) do
            $L_{b'}(\hat{s}) \leftarrow L_{b'}(\hat{s}) + \nu_r(\hat{s}) F_b(\hat{s})$
        end for
    end if
} end while

A word of explanation: Note the reversal of targets and sources from Step 1 to Step 2. In Step 1, the far field pattern of each box in the interaction list of the given box is converted to a local pattern centered at the given box. However, in Step 2, the far field pattern of the given incoming box is converted to a local expansion centered at each box in its interaction list. That is, the interaction list is now "inverted." See Figure 4.3 for a pictorial illustration.

The boxes marked S belong to the Interaction List of the box marked D.

The boxes marked D belong to the Inverse Interaction List of the box marked S.

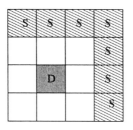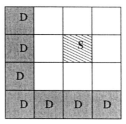

S : Source Box
D : Destination Box

The direction of translation is from a source box to a destination box.

Figure 4.3 Illustration of forward and inverse interaction lists.

In a later stage, we shall also discuss what are called *ghost boxes* used in conjunction with a certain compressed form of the translation operators ν_r.

4.4.1.3 Disaggregation (L2L) Phase

This phase is similar to the aggregation phase, except that it traverses the tree from top to bottom. Analogous to the UpdateSubTree, there is a corresponding update stage for disaggregation phase. We leave out the details.

The final evaluation phase contains two parts. First, the local pattern $L_b(\hat{s})$ of each box is evaluated at every observation point of the receiver using Lemma 4, Definition 1, and (4.11). In the second part, the interactions with particles in the nearby boxes are evaluated directly. Note that the symmetry of the near_list, like that of the interaction list, can be used to improve the efficiency.

4.5 IMPLEMENTATION ISSUES

Going from the abstract algorithm described in the previous section to the actual implementation is a long way as several issues emerge. We discuss some of the important ones in this section.

4.5.1 Storage of the Tree

In the sequential algorithm, a tree can be elegantly stored and accessed using pointers. This approach can also work with shared memory machines where a global memory is available. However, in the case of truly distributed memory machines, such a direct addressing scheme is not possible without considerable hardware support. Therefore, we need to develop a unique identifier for each box in the tree. Salmon and Warren [12] have developed an elegant scheme based on Morton ordering for this problem. They have used this scheme along with a hash table for storing the tree.

Following [12], in our implementation, we use the Morton ordering for providing a unique global id (uid) for each box. However, although the hash table provides $O(1)$ access on the average, a careful choice of the hashing function is necessary for good performance. Also, since the geometric data remain the same throughout the solution of the integral equation, dynamic regeneration of the tree is not necessary.

In our implementation, boxes are stored in a two-dimensional array with variable length rows. That is, the pointers to the boxes at a given level are stored in a distinct one dimensional array, sorted with respect to their Morton keys. If N_b is the number of nonempty boxes at the given level, then accessing any box b' at this level at random requires $O(\log N_b)$ operations using a binary search. Since the number of boxes at any given level l is bounded by $(2^d)^l$, where d is the number of dimensions, it follows that accessing any box at any given level l has a complexity $O(l)$. Furthermore, if we tacitly agree (see [13] for a discussion on the theoretical bounds) that the number of levels is bounded by $\log N$, where N is the number of particles, it follows that the random access requires $O(\log N)$ operations. In addition, when the boxes are accessed in sequence, such as during the translation phase, each box can be fetched in $O(1)$ operations.

4.5.2 Construction of the Tree and Domain Decomposition

Several approaches have been discussed in the literature for tree construction [12, 14] for particle simulation applications. However, in the case of integral equation solvers, there exist two important geometrical properties that influence the choice of the tree construction algorithm.

The first property of the geometric data arising from the discretization of integral equations is that they are stationary in time. That is, the geometry does not change with time, in contrast to the particle simulation applications. This property can be used to generate a good load balancing strategy during the initialization time itself.

However, the second property is not so helpful. It refers to the fact that most of the discretization procedures use basis functions with nontrivial support. That is, the sources are no longer located at a single point, but spread over a region. In such problems, the region could be modeled using polygons or polyhedra, specified in terms of vertices, edges, and facets; and the support of these basis functions could extend over multiple elements. An important example of such a discretization is Galerkin's method with the Rao-Wilton-Glisson (RWG) functions for solving electromagnetic scattering problems [15]. The RWG functions have support over two triangular facets.

Such discretizations make the so called "connection matrix" nondiagonal. The connection matrix relates the basis functions to the geometric description of the object. For instance, in the case of RWG basis functions, the connection matrix relates each basis function with the indices of the triangular facets on which it is defined. Therefore, any decomposition of basis functions across multiple processors will have to ensure *completeness* of local data. That is, one needs to make sure that if a particular basis function is assigned to a processor, all the geometric data associated with that basis function is also available at the same processor. To give an example, with RWG basis functions, if a particular basis function is assigned to a processor, we also have to keep the two facets associated with that basis function in the same processor. But this in turn implies that we also need to keep the vertices and the edges of the facets in the same processor. Such a scheme would ensure that the processors need not communicate with each other to get relevant geometric data. We call such an assignment a *locally complete decomposition*. Note that a locally complete decomposition may still have replication to some extent.

One way to handle this problem is to simply replicate the full geometric structures in every processor. This is the simplest approach and lets existing programs use

ScaleME without major modifications. However, memory for geometric data structures is $O(N)$, where N is the number of unknowns, and a full replication has the undesirable scaling of $O(Np)$ where p is the total number of processors. Nevertheless, for small- to medium-scale problems, ranging from a few thousand unknowns to a few hundreds of thousands, this approach is very practical.

Obviously, for very-large-scale problems such a simplistic approach is unacceptable and any serious implementation of MLFMA on distributed memory computers will have to support an appropriate domain decomposition.

Fortunately, we can construct a good decomposition of geometric data, which requires $O(N)$ memory independent of the number of processors, provided we *already have a distributed tree*. It can be shown that once we have the tree, all we need to store is the connectivity and geometric information of every basis function in the near lists of local boxes. Since the number of direct interactions is $O(N)$, it follows that the total storage requirement still remains $O(N)$. However, note that the actual storage will be more than that required for one copy of the geometric data structures because of a few replications. This can be accomplished through the use of a *skeleton* of the tree in the preprocessing stage [7].

4.5.3 Scaling of Translation Matrices

In the dynamic MLFMA, the speed-up in computing the matrix-vector products is achieved through the use of *diagonalized translation operators*. Usually, in any practical implementation of MLFMA, the diagonalized translation operators are precomputed at the setup stage and stored. In the straightforward implementation of MLFMA for distributed memory computers using the message passing paradigm, the translation operators are replicated in each processor in order to reduce communication costs and to accelerate the matrix-vector multiply. However, this requires an $O(N)$ storage in each processor. Thus the memory requirements for translation operators scale as $O(Np)$, where p is the number of processors. This is unacceptable for large-scale problems.

A naive approach to solving this problem is to compute the translation operators "on the fly." In other words, compute them as they are required. However, it can be shown that such an approach would increase the computational complexity of the algorithm to at least $O(N^{3/2})$ [7, 16].

In order to solve this problem, we have developed a new *compressed* representation for the translation matrices which can be evaluated rapidly when they are required [16]. To explain this, we recall that the expression for the diagonal translation operator is given by

$$\nu_L(z, \hat{s}_0, \hat{s}) = \sum_{m=0}^{L-1} i^m (2m+1) h_m^{(1)}(z) P_m(\hat{s} \cdot \hat{s}_0) \qquad (4.13)$$

where $\hat{s}$ and $\hat{s}_0$ are unit vectors, $z \in \mathbf{C}$, $P_m(x)$ are Legendre polynomials of order m, $h_m^{(1)}(z)$ are the spherical Hankel functions of order m, and L is a positive integer.

The new representation is based on the observation that for a given $\hat{s}_0$, the product $\hat{s} \cdot \hat{s}_0 \in [-1, 1]$ and that the function $\nu_L(\hat{s})$ is thus a polynomial of degree $L - 1$ in the interval $[-1, 1]$. Given this, a rather obvious modification of the one-dimensional fast multipole algorithm [17] can be used to construct a fast, polynomial representation which requires only $O(\sqrt{N})$ storage and $O(N)$ evaluation time, thereby bringing down the scaling of storage requirements of the translation operator to $O(p\sqrt{N})$. (See [16] for more details of this work.)

However, the introduction of the compressed translation operators increases the time for matrix-vector products significantly. This is because, although the asymptotic complexity of the evaluation of translation operator has been reduced, the constant associated with the overall MLFMA goes up. Fortunately, this can be avoided by rearranging the translation phase in such a way that each translation matrix is evaluated once and only once during each matrix-vector computation. For this, instead of associating an interaction list with every box, the interaction lists can be defined with respect to the translation operators. See [7] for a complete discussion of this issue.

On a single-processor implementation, this scheme ensures that each translation operator is accessed or computed only once. However, in a distributed memory implementation it may happen that some of the boxes in the interaction list of a translation operator may not be locally available. Hence, a direct implementation of the above scheme cannot be used. However, this problem can be solved by using the concept of *ghost boxes* [18].

4.5.4 A Costzone Scheme for Load Balancing

In this section we describe a static costzone scheme for balancing the CPU load across the processors. The approach we employ here is similar to the one discussed in [19]. We note that, in the case of acoustic and electromagnetic scattering problems, the positions of the particles remain the same throughout the simulation. As a consequence, the MLFMA tree used for computing matrix-vector products also remains the same. Thus it is feasible to look for a load balancing scheme which can be incorporated as an inexpensive preprocessing scheme for tree construction, since it is a one-time affair.

For simplicity, we ignore the communication overhead associated with the computations. We shall also assume that the costs of transmitting data across any two processors are the same. Although this assumption is not always true, it simplifies the situation considerably.

We begin by mapping each particle to a box at the finest level, thereby assigning a Morton key, not necessarily unique, to each of them. After sorting the particles according to their keys, the finest level of the MLFMA tree is obtained. We shall keep this sorted list of basis functions and their respective keys in each processor and use it throughout the computations.

Corresponding to each box, we shall define a data structure, `node_work`, which can be represented as a 4-tuple: (`key`, `nparticle`, `first_particle`, `work`), where `key` is the Morton key to the box, `nparticle` is the number of particles in the box, `first_particle` is a pointer to the first particle in the box in the sorted list, and `work` is the total work required for the box. The variable `work` comprises the work required for all the F2L, F2F, L2L translations and direct interactions for the given box.

An important fact to be noted is that computing the work required for each node is not the same as doing the actual computations. Therefore, under the assumption that the total number of nonempty boxes is $O(N)$, it can be shown that this process has a complexity of $O(N \log N)$, with a very small constant of proportionality.

Once dummy tree is constructed, we compute the total work W required for each matrix-vector product. The ith processor is then assigned all the nodes in the tree which contribute from $\frac{i-1}{p}W$ to $\frac{i}{p}W$ segment of the total work, where p is the number of processors. This can be readily done by traversing the tree in postorder and summing the contributions from each node as the traversal progresses.

4.6 ScaleME : A BRIEF DESCRIPTION

`ScaleME` is a portable implementation of the dynamic MLFMA for distributed memory computers, using the message passing paradigm. It uses the Message Passing Interface (MPI) for communication. As mentioned earlier, it is implemented as an *application-independent kernel*, which can be retrofitted into existing applications.

One of the motivations of the present work stems from the fact that there exists a large number of research and commercial electromagnetic/acoustic integral equation solvers which can benefit from a parallelized MLFMA and parallel processing power offered by an existing cluster of workstations.

However, implementing MLFMA on a distributed memory machine requires considerable expertise and time, which are often unavailable. Therefore, `ScaleME` aims at simplifying the effort required in exploiting the power of a parallel MLFMA by carefully hiding away the MLFMA-specific and the parallel-computer-specific technical details from the user. Such a design lets the user concentrate on the application specific details necessary to couple it with MLFMA through a fairly simple interface.

`ScaleME` uses the single program multiple data (SPMD) paradigm of parallel computing. That is, the same program runs on all the nodes. For a complete discussion of various parallel programming paradigms, we refer the reader to any standard textbook on parallel computing, for instance [21]. In the following, we shall refer to the program which is to be embedded with `ScaleME` as a *host*. There are two interfaces to ScaleME : (1) the sequential host interface, and (2) the parallel host interface.

The sequential host interface is aimed at already existing sequential programs that need to be quickly retrofitted with ScaleME. It is essentially a master-slave strategy. To explain this further, assume that we have an integral equation solver that uses some iterative technique, say the conjugate gradient method (CGM). We wish to replace the current routine for matrix-vector products with the parallel MLFMA-based routine. Suppose that we are interested in solving only those problems for which the computational cost is dominated by matrix-vector products and we can neglect the cost of the rest of the computations, such as dot products. In such a situation, we would like to have our CGM routine running on one processor alone and we would like to use the rest of the processors for computing the matrix-vector products. The sequential host interface of ScaleME is designed precisely for this situation, in which the iterative solver runs one processor, but *all* the processors are used for computing the matrix-vector products. Note that in this case, the original code requires very minimal modification and the coupling between the host program and ScaleME is rather loose.

In contrast, the parallel host interface is directed at problems where the iterative solver itself is *parallelized*, such as a parallel GMRES solver, and runs on a multitude of processors. In this case, we need to *embed* ScaleME into the iterative solver. This requires a tighter coupling between the host program and ScaleME and hence may require considerable tuning of the original host program. However, it may be noted that the parallel host interface is really aimed at MOM codes that are already parallelized.

The current stable version of ScaleME has about 40 user-callable functions. These routines can be broadly divided into four categories:

1. Configuration routines;

2. Initializing and finalizing routines;

3. Execution routines or routines for the evaluation of matrix-vector products;

4. Routines for assisting external fine tuning.

Configuration routines are usually for setting the parameters that are invariant for the class of problems the application sets about to solve. Initialization routines, as the name implies, initialize the data structures and parameters for the given instance of a problem. The third class consists of the routines for controlling the evaluation of matrix-vector products. The fourth class of routines provide some assistance in exploiting some of the internal structure of MLFMA and allow an experienced user to fine tune the performance of ScaleME. In addition to these, the user has to supply ScaleME with three subroutines which describe the physics of the problem. Although the presence of such a large number of routines may sound intimidating, it is not so.

This is because ScaleME has been designed so that while offering a simple, intuitive and yet flexible interface, it also tries to follow the philosophy that *what you don't know should not hurt you!* For a complete discussion of the features and interface to ScaleME we refer the reader to the user's guide [22].

4.7 NUMERICAL EXPERIMENTS

As we mentioned earlier, ScaleME is an application-independent kernel which can be retrofitted into independently developed integral equation solvers. We have interfaced it with several such codes developed at the Center for Computational Electromagnetics of the University of Illinois. In this section, we report some results from one such code for demonstration purposes. For more details on this specific application, see [3].

The application considered is that of electromagnetic scattering from perfectly conducting obstacles. This is a classical and very important problem having a wide variety of applications. We shall first formulate the problem in terms of appropriate integral equations and describe the code in detail. We shall then report various results from this code. In the following discussion we shall use the following notations:

- k: the free space wave number, m^{-1};
- η: impedance of free space, Ω;
- S: surface bounding the scatterer;
- $\hat{t}$, $\hat{n}$: unit tangent and normal vectors at any given point on S;
- $\mathbf{J}$: induced surface electric current;
- $\mathbf{H}^i$, $\mathbf{E}^i$: incident magnetic and electric field vectors;
- $\bar{\mathbf{I}}$: the unit dyad;
- α: combination coefficient of CFIE, $\alpha \in [0, 1]$;
- $\mathbf{r}$, $\mathbf{r}'$: position vectors in $\mathbf{R}^3$.

4.7.1 The Integral Equation Formulation

The scattering of electromagnetic waves from perfectly conducting objects can be formulated in terms of EFIE, given by

$$\frac{ik\eta}{4\pi}\hat{t} \cdot \int_S \bar{\mathbf{G}}(\mathbf{r}, \mathbf{r}') \cdot \mathbf{J}(\mathbf{r}') \, dS' = -\hat{t} \cdot \mathbf{E}^i(\mathbf{r}), \quad \mathbf{r} \in S \qquad (4.14)$$

with
$$\bar{G}(r,r') = (\bar{I} - \frac{1}{k^2}\nabla\nabla')g(r,r'), \text{ and } g(r,r') = \frac{e^{ik|r-r'|}}{|r-r'|}$$

If the surface bounding the object is closed, it can also be described using MFIE:

$$-\hat{t}\cdot\frac{J(r)}{2} + \frac{1}{4\pi}\hat{t}\cdot\hat{n}\times\nabla\times\int_S g(r,r')J(r')\,dS' = -\hat{t}\cdot\hat{n}\times H^i(r) \quad (4.15)$$

However, both EFIE and MFIE suffer from nonunique solutions at resonant frequencies. To alleviate this problem, electromagneticists often use CFIE, which is defined by the relation:

$$\alpha\,\text{EFIE} + \eta(1-\alpha)\,\text{MFIE}$$

In order to numerically solve these equations, we model the surface using flat triangular patches and then expand the current in terms of the standard RWG basis functions [15]. That is, if $f_m(r)$ denotes the mth basis function, we expand the current J as

$$J = \sum_{m=1}^{N} I_m f_m$$

Using Galerkin's method, the approximate integral equation is then reduced to a system of linear equations. The matrix element corresponding to the EFIE is given by

$$Z_{mn}^E = \frac{ik\eta}{4\pi}\int_S dS f_m(r) \cdot \int_S \bar{G}(r,r')\cdot f_n(r')\,dS' \quad (4.16)$$

and that corresponding to the MFIE is given by

$$Z_{mn}^M = -\int_S dS\,f_m(r)\cdot\frac{f_n(r)}{2}$$
$$+ \frac{1}{4\pi}\int_S dS\,f_m(r)\cdot\hat{n}\times\nabla\times\int_S g(r,r')f_n(r')\,dS' \quad (4.17)$$

The matrix element for the CFIE can then be derived as $Z_{mn} = \alpha Z_{mn}^E + \eta(1-\alpha)Z_{mn}^M$. Then, the integral equations reduce to the matrix equation:

$$\bar{Z}\cdot I = V$$

where

$$V_m = -\left(\alpha\int_S f_m(r)\cdot E^i(r)\,dS + \eta(1-\alpha)\int_S f_m(r)\cdot\hat{n}H^i(r)\,dS\right)$$

In order to use MLFMA for accelerating the matrix-vector computations, we need to define the radiation and receiving patterns associated with the basis and testing functions [6, 9]. Starting with (4.3), it can be shown that the *radiation pattern* for the CFIE corresponding to the nth basis function, with respect to a given center of expansion $\mathbf{c}$, is given by

$$\mathbf{F}_m(\hat{s}) = k(\bar{\mathbf{I}} - \hat{s}\hat{s}) \int_S \mathbf{f}_m(\mathbf{r}') e^{ik\hat{s}\cdot(\mathbf{c}-\mathbf{r}')} \quad (4.18)$$

Similarly, the *receiving pattern* corresponding to the mth testing function can be shown to be

$$\begin{aligned} \mathbf{R}_m(\hat{s}) = \frac{ik\eta}{4\pi} &\left(\alpha(\bar{\mathbf{I}} - \hat{s}\hat{s}) \cdot \int_S \mathbf{f}_m(\mathbf{r}) e^{ik\hat{s}\cdot(\mathbf{r}-\mathbf{c}')} \, dS \right. \\ &\left. - (1-\alpha)\hat{s} \times \int_S \mathbf{f}_m(\mathbf{r}) \times \hat{n} e^{ik\hat{s}\cdot(\mathbf{r}-\mathbf{c}')} \, dS \right) \end{aligned} \quad (4.19)$$

We note that the matrix elements corresponding to sufficiently separated basis functions have a representation of the form

$$Z_{mn} = \frac{i}{4\pi} \int_{S_0} \mu_{\mathbf{c},\mathbf{c}'}(\hat{s}) \mathbf{F}_m(\hat{s}) \cdot \mathbf{R}_m(\hat{s}) \, d\Omega$$

where $\mu_{\mathbf{c},\mathbf{c}'}(\hat{s})$ is a complex function with a branch cut along the negative real axis and $\hat{s} \in S_0$ with $S_0 = \{\mathbf{r} \in R^3 \mid |\mathbf{r}| = 1\}$.

4.7.2 The TRIMOM+ScaleME Code

This code was developed at the University of Illinois by combining an existing MOM code, called **TRIMOM** with ScaleME. **TRIMOM** was originaly developed in FORTRAN by Lu and Chew at the University of Illinois [23]. This code is also the backbone of the Fast Illinois Solver Code developed at the center [24]. In order to retrofit **TRIMOM** to ScaleME, three subroutines were added. The first function returns the center of a basis function given its index. The second implements the radiation pattern given by (4.18), taking three arguments, the center of expansion $\mathbf{c}$, the direction $\hat{s}$ in which the radiation pattern is desired, and the index of the basis function. The third function implements the receiving pattern given by (4.19), whose semantics are similar to those for the radiation pattern. We note that this is typical of the work required for using ScaleME in an existing code [25]. After having verified the accuracy of the code, the domain decomposition (Section 4.5.2) was introduced and the higher levels of the code were rewritten in C.

A few general remarks are in order. First, all the results reported in this paper were computed with a low accuracy setting for MLFMA. This means that the number of multipole terms at each level was computed using the formula $L = kD + 1.0 \log(kD + \pi)$, where kD is the diameter of the smallest sphere enclosing any box at a given level [26]. Furthermore, we have used a four-point Lagrange interpolation during the aggregation phase and a corresponding anterpolation scheme during the downward pass.

4.7.3 Single Processor Performance

The first set of results we report is that of comparison of single processor performance of the new code to the well-tested code called FISC [24]. In order to do this, we choose the classical example of scattering from perfectly conducting spheres of various sizes. This comparison was carried out on an SGI Power Challenge series multiprocessor with 2-GB RAM, running at 90 MHz. During this test, the parameters of ScaleME, such as number of terms in the multipole expansions and the order of the interpolation, are chosen to match exactly with that of FISC. The results are plotted in Figure 4.4. It is seen that the ScaleME is about 23% slower than that of FISC. This is a very good performance since the MLFMA as implemented in FISC is tightly coupled to the MOM code and hence uses several specialized optimizations for efficiency. On the other hand, we see that the memory requirements for ScaleME are almost the same as those for FISC. This test demonstrates that the single processor performance of ScaleME is comparable to that of the best known sequential code.

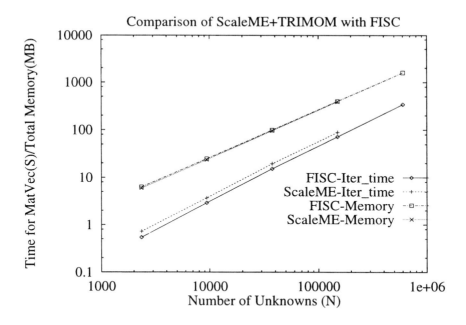

Figure 4.4 Comparison of speed and the memory requirements for TRIMOM+ScaleME with FISC on a single processor. In the figure, the time for each iteration, denoted by "iter," is in seconds and the memory, denoted by "Memory," is in megabytes.

4.7.4 Processor Scaling

We now demonstrate the processor scaling of the dynamic MLFMA as implemented in ScaleME. Before we present the results, two remarks are in order. Note that in the case of dynamic MLFMA the number of terms in the multipole expansion—or its equivalent, the number of samples over the unit sphere—does not remain constant for all levels. As mentioned earlier, the number of samples of the unit sphere is directly proportional to the surface area of the cube. Hence, the cost of communicating multipole expansions between two processors depends heavily on the level of the expansion. As a consequence of this, the cost of communication is much higher for the dynamic MLFMA.

Another factor which makes the performance of dynamic MLFMA different from that of static codes is the interpolation/filtering operations. These operations tend to make the upward and downward passes more expensive.

With these observations in mind, we now present the results of a scaling study. We consider the case of the scattering of electromagnetic waves from a perfectly conducting sphere of diameter 12λ where λ is the free-space wavelength. The sphere was modeled using 100,352 triangular facets. The resulting number of unknowns is 150,528. The code was run on an SGI Origin2000 at National Center for Supercomputing Applications (NCSA). The results are shown in Figure 4.5.

4.7.5 Large-Scale Problems

In order to demonstrate the ability to solve very-large-scale problems, we consider two cases. First, we solve the problem of scattering from a sphere of diameter 80λ. The sphere was modeled using 3,981,312 unknowns. This problem was solved on 16 nodes of an SGI Origin2000 at the NCSA using a nine-level MLFMA. The total memory used was 15.12 GB. The time for setting up the MLFMA kernel was about 30 minutes and the average wall clock time for each matrix-vector product was about 552 seconds. The total solution time was 26 hours and 34 minutes, using a conjugate gradient solver, which took 78 iterations to converge to a residual error of 1.0×10^{-3}. We plot the bistatic RCS of the sphere and compare it with the analytical solution in Figure 4.6. Also in Figure 4.7 we show the *average kernel time* spent in each processor for each matrix-vector product, as an indicator of the load balance obtained.

For this problem, we have used the compressed translation operators and ghost boxes (Section 4.5.3) for the levels 2–5. We have used a "one buffer box" criterion to determine two boxes are far apart, that is $m = 1$ in Definition 4. With this condition, there are 316 translation matrices at each level. Table 4.1 shows the breakdown of the memory required for a single copy of the translation operators for the top four levels and the savings obtained using compressed operators and ghost boxes. In the table, the third column gives the memory required for a single translation operator

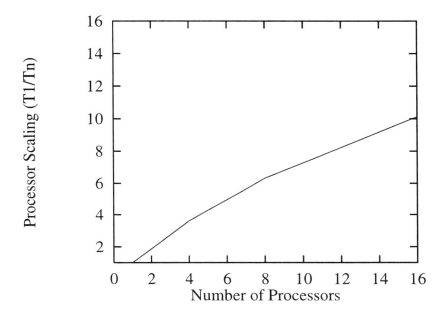

Figure 4.5 Time for matrix-vector product as a function of the number of processors. The number of unknowns was 150,528 and the test was conducted on an SGI Origin2000 at the NCSA.

and the fourth for all the translation matrices at the given level. The advantages of the technique are clearly seen from the table.

In the second example, we compute the RCS of a full-scale aircraft, namely a VFY218, at 500 MHz. The aircraft was modeled using 54,990 unknowns. This problem was solved on a Beowulf [27] class cluster, called *Orion*, of PCs running Linux. Each node of this cluster has a 350 MHz AMD K6-2 processor and 128 MB of RAM. The nodes are connected through a 100 Mbps ethernet and a 24 port, Xyplex Network's Megaswitch which gives a maximum throughput of 3.2 Gbps. More details on the cluster can be found in [28].

The total memory used by ScaleME was 696 MB. For this problem we have used a built-in block diagonal preconditioner along with a CGNR solver. The setup time was about 265 seconds and each matrix vector product took 34.31 seconds. The solution to the vertical polarization required 68 iterations and that for the horizontal was 69 iterations. The bistatic RCS is presented in Figure 4.8.

An important issue that we addressed earlier was the domain decomposition. In TRIMOM+ScaleME, we use the MLFMA induced domain decomposition. Since the MLFMA induced decomposition uses only the information from the centers of

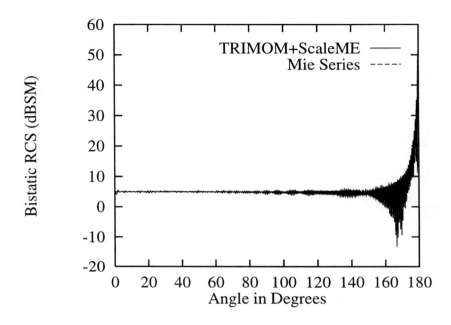

Figure 4.6 The Bistatic radar cross-section of the 80λ sphere for the vertical polarization. "Mie Series" refers to the analytical solution.

basis functions, this can be done in two steps in the case of the triangular patch modeling and RWG basis function-based discretization. First, the host program collects the list of all basis functions that are needed in each processor. Then, using the connection matrix relating the facets, edges, and nodes, the stored geometric data in each processor is trimmed to the bare minimum. We report the results of this in Table 4.2. The savings are obvious.

4.8 ScaleME-2: AN IMPROVED PARALLEL MLFMA

We have seen that as the problem size increases, the size of the radiation and receiving patterns also increases in proportion. In fact, we have noted earlier that the number of samples at the coarser levels of the tree is $O(N)$, where N is the number of unknowns. This causes a deterioration in performance, owing to excessively long messages, in addition to many other memory bottlenecks.

ScaleME-2 is a new scheme that exploits certain insights gained while solving large-scale problems on distributed memory environments. It can be shown that better communication performance can be achieved by distributing the radiation and

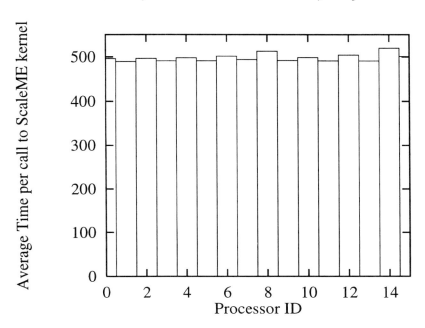

Figure 4.7 The average time spent in the kernel for each matrix-vector product plotted against the processor ID. We use this as a measure of the load balance achieved.

Table 4.1 Memory Required at Each Level for the Translation Operators for the 80λ Sphere Problem and the Savings Obtained Using Compressed Translation Operators and Ghost Boxes

Level	N_s	Memory (1) (MB)	Memory(Full) (MB)
2	170532	1.3	411.13
3	45004	0.343	108.50
4	12172	0.093	29.35
5	3532	27.6 (KB)	8.52
Memory required for one processor			557.5
Total memory for 16 processors (MB)			8,912
Memory for compressed operators (MB)			12.34
Total memory for ghost boxes (MB)			2,048
Savings in memory (MB)			6,674.6

Note: N_s is the number of samples over the unit sphere corresponding to a given level (see Remark 2).

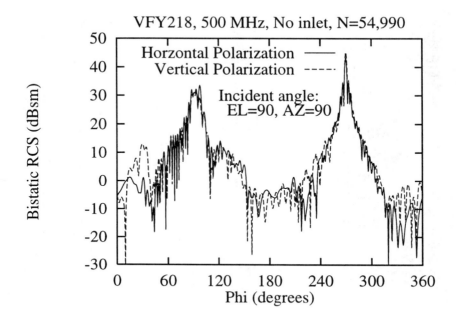

Figure 4.8 The Bistatic radar cross-section of the aircraft VFY218 at 500 MHz.

Table 4.2 Influence of Domain Decomposition on Memory Requirement for Geometric Data Structures

Problem	Memory (before)	Memory (after)
VFY-218	45.92 MB	11.29 MB
Sphere-80	3.24 GB	1.37 GB

receiving patterns of larger boxes across processors, instead of the boxes themselves. This is a consequence of the diagonalized translations and the excellent data locality of the interpolation and anterpolation schemes used in our MLFMA. A detailed discussion of this technique is beyond the scope of this chapter.

We have implemented this algorithm and using the new scheme we have been able to efficiently solve very large scale problems. In particular, we have solved the bistatic scattering from an aircraft (VFY218) at 8GHz. This simulation involved nearly **10.186 million** unknowns. The computations were done on a 128 processor SGI Origin 2000 supercomputer at the NCSA, University of Illinois. The total

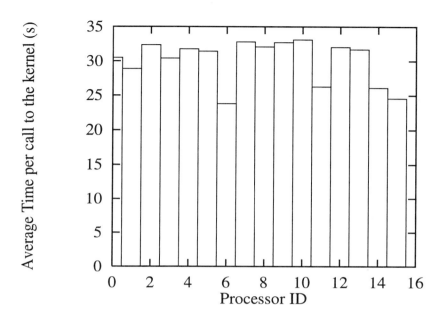

Figure 4.9 The average time spent in the kernel for each matrix-vector product for the VFY218 plotted against the processor ID for the aircraft.

memory used was about 69 GB. The initialization phase required approximately one hour and the evaluation of a matrix-vector product required about 106s of wall clock time, and approximately the same amount of CPU time. Using a GMRES iterative solver with restart after 20 iterations, convergence to 1.0×10^{-2} was achieved in 81 iterations for the V-polarization and in 77 iterations for the H-polarization. The total run time was about 7 hours and 15 minutes.

4.9 CONCLUSIONS

In this paper, we have presented the design and implementation details of a portable, distributed memory multilevel fast multipole kernel for electromagnetic and acoustic problems. The MLFMA was implemented as a library of user-callable functions. The Message Passing Interface was used for communications. Several efficient techniques have been developed for enhancing the parallel performance. Furthermore, the library can be retrofitted into existing electromagnetic codes without difficulty. We have incorporated this into an existing method of moments code for electromagnetic problems, and we have presented some representative results from this new code. Work is currently in progress to incorporate new techniques to improve the communication algorithms.

REFERENCES

1. V. Rokhlin, "Diagonal forms of translation operators for the Helmholtz equation in three dimensions," *Applied and Comp. Harmonic Analysis*, vol. 1, pp. 82–93, 1993.

2. J. Song, C. C. Lu, and W. C. Chew, "MLFMA for electromagnetic scattering from large complex objects," *IEEE Trans. Ant. Propag.*, vol. 45, pp. 1488–1493, Oct. 1997.

3. S. V. Velamparambil, J. M. Song, and W. C. Chew, "A portable parallel multilevel fast multipole solver for scattering from perfectly conducting bodies," *IEEE Antennas Propagat. Symp.*, vol. 1, pp. 648–651, July 1999.

4. http://www-unix.mcs.anl.gov/mpi/.

5. V. Rokhlin, "Rapid solution of integral equations of scattering theory in two dimensions," *Journal of Computational Physics*, vol. 86, pp. 414–439, 1990.

6. M. A. Epton and B. Dembart, "Multipole translation theory for the three dimensional Laplace and Helmholtz equations," *SIAM J. of Scientific Computing*, vol. 16, pp. 865–897, July 1995.

7. S. Velamparambil, J. Song, and W. C. Chew, "ScaleME: A portable, distributed memory multilevel fast mulipole kernel for electromagnetic and acoustic integral equation solvers," Tech. Rep. CCEM-23-99, Center for Computational Electromagnetics, Dept. of Electrical and Computer Engineering, University of Illinois at Urbana-Champaign, Urbana, IL, 1999.

8. S. V. Velamparambil, *A Fast Adaptive Multipole Algorithm for Scattering from Inhomogeneous Dielectric Cylinders of Arbitrary Cross-section*, Ph.D. dissertation, Indian Institute of Science, Bangalore, July 1994.

9. J. Song and W. C. Chew, "Multilevel fast multipole algorithm for solving combined field integral equations of electromagnetic scattering," *Micro. Opt. Tech. Lett.*, vol. 10, pp. 14–19, Sept. 1995.

10. B. Dembart and E. Yip, "A 3D fast multipole method for electromagnetics with multiple levels," *Applied Computational Electromagnetics Society*, pp. 621–628, 1995.

11. W. T. Rankine and J. A. Board, "A portable distributed implementation of the parallel fast multipole tree algorithm," Tech. Rep. 95-002, Dept. of Electrical Engineering, Duke University, Durham, NC, August 1994.

12. M. S. Warren and J. K. Salmon, "A parallel hashed oct-tree N-body algorithm," in *Supercomputing '93 Proceedings*, (Washington, D.C.), pp. 12–21, IEEE Comp. Soc. Press, 1993.

13. P. B. Callahan and S. R. Kosaraju, "A decomposition of multidimensional point sets with applications to k-nearest neighbors and n-body potential fields," *J. of Association of Computing Machinery*, vol. 42, pp. 67–90, Jan. 1995.

14. A. Y. Grama, V. Kumar, and A. Sameh, "Scalable parallel formulations of the barnes-hut method for n-body simulations," in *Supercomputing '94 Proceedings*, (Washington, D.C.), pp. 439–448, IEEE Comp. Soc. Press, 1994.

15. S. M. Rao, D. R. Wilton, and A. W. Glisson, "Electromagnetic scattering by surfaces of arbitrary shape," *IEEE Trans. Antennas Propagat.*, vol. 30, pp. 409–418, May 1982.

16. S. Velamparambil and W. C. Chew, "Fast polynomial representations for the diagonal translation operators of the three dimensional Helmholtz equation," Tech. Rep. CCEM-17-99, Center for Computational Electromagnetics, University of Illinois at Urbana-Champaign, Urbana, IL, 1999.

17. A. Dutt, M. Gu, and V. Rokhlin, "Fast algorithms for polynomial interpolation, integration, and differentiation," *SIAM J. Numerical Analysis*, vol. 33, pp. 1689–1711, Oct. 1996.

18. E. J. Lu and D. I. Okunbor, "A massively parallel fast multipole algorithm in three dimensions," in *The Proceedings of Fifth IEEE International Symposium on Parallel and Distributed Processing*, pp. 40–48, 1996.

19. A. Grama, V. Kumar, and A. Sameh, "Scalable parallel formulations of the Barnes-Hut method for n-body simulations," *Parallel Computing*, vol. 24, pp. 797–822, June 1998.

20. M. Snir, S. W. Otto, S. Huss-Lederman, D. W. Walker, and J. Dongarra, *MPI: The Complete Reference*, Scientific and Engineering Computation Series, Cambridge, MA: MIT Press, 1996.

21. V. Kumar, A. Grama, A. Gupta, and G. Karypis, *Introduction to Parallel Computing: Design and Analysis of Algorithms*, Redwood City, CA: The Benjamin/Cummings Publishing Company, Inc., 1994.

22. S. Velamparambil and W. C. Chew, "ScaleME: Application interface; a programmer's guide and reference," Tech. Rep. CCEM-27-99, Center for Computational Electromagnetics, University of Illinois, Urbana, 1999.

23. C.-C. Lu and W. C. Chew, "Calculation of EM scattering from conducting surfaces of arbitrary shape using the method of moments: Triangular patch formulation," Tech. Rep., Center for Computational Electromagnetics, University of Illinois, Urbana, 1996.

24. J. Song, C. Lu, W. Chew, and S. Lee, "Fast Illinois Solver Code," *IEEE Antennas and Propagation Magazine*, vol. 40, pp. 27–34, June 1998.

25. S. Velamparambil, J. Song, W. Chew, and K. Gallivan, "ScaleME: A portable, scalable multipole engine for electromagnetic and acoustic integral equation solvers," *IEEE Antennas Propagat. Symp.*, vol. 3, pp. 1774–1777, 1998.

26. J. Song and W. C. Chew, "Error analysis for the truncation of multipole expansion of vector Green's function," *IEEE Antennas and Propagation Symposium, Orlando, Florida*, vol. 1, pp. 628–631, 1999.

27. http://www.beowulf.org/.

28. S. V. Velamparambil, J. E. Schutt-Aine, J. G. Nickel, J. M. Song, and W. C. Chew, "Solving large scale electromagnetic problems using a Linux cluster and parallel MLFMA," *IEEE Antennas Propagat. Symp.*, vol. 1, pp. 636–639, July 1999.

5

Multilevel Fast Multipole Algorithm at Very Low Frequencies

Junsheng Zhao and Weng Cho Chew

5.1 INTRODUCTION

For solving general two-dimensional (2D) and three-dimensional (3D) electromagnetic problems, a straightforward and accurate approach is to use the method of moments (MOM) to solve the integral equation where the current density or the charge density distribution is the unknown [1]. MOM generates a linear equation corresponding to a full matrix. For a conventional matrix solution, the required computer storage is $O(N^2)$, where N is the number of unknowns, and the number of floating-point operations is $O(N^3)$ for direct inversion methods or $O(N^2)$ per iteration for plain iterative solvers. To expedite the matrix-vector multiplication, which is the most time-consuming part in plain iterative solvers, and to decrease the storage, a number of techniques have been proposed. There are the application of wavelet basis functions [2], the impedance matrix localization (IML) technique [3], the multilevel matrix decomposition algorithm (MLMDA) [4], the complex multipole beam approach (CMBA) [5], the fast inhomogeneous plane wave algorithm [6,7], and the multilevel fast multipole algorithm (MLFMA) [8–24]. MLFMA is efficient, error controllable, and flexible. It is developed by combining the fast multipole method (FMM) and multilevel method with interpolation and anterpolation techniques.

FMM was originally proposed to evaluate particle simulations and to solve static integral equations [8], and then extended to solve acoustic wave scattering problems [9] and electromagnetic scattering problems by many researchers in both 2D and 3D cases [10–18]. By combining FMM and the multilevel method and applying the interpolation and anterpolation techniques [19] for the dynamic problems, the MLFMA was developed with $O(N)$ computational complexity for static problems [16, 24] and with $O(N\log N)$ computational complexity for dynamic problems [20–23].

The development of MLFMA makes it possible to calculate very large linear systems corresponding to an arbitrary structure without compromising numerical precision. The diagonalized-form dynamic 2D MLFMA [20] and 3D MLFMA [21–23] are excellent for the dynamic case, but cannot be used at low frequencies. The instability of the divergent series in the diagonalized translation matrix incurs a large error at low frequencies. The originally undiagonalized-form 2D and 3D dynamic MLFMA, with no reduction in the computational complexity in the dynamic case, can reduce the computational complexity at low frequencies, but cannot be applied directly. The arguments of the Bessel-type functions, Hankel and Bessel functions for 2D MLFMA and spherical Hankel and spherical Bessel functions for 3D MLFMA, appearing in the translation matrices are very small. Due to the asymptotic behaviors of Bessel-type functions with small arguments, the multipole amplitudes of cylindrical waves for 2D cases and spherical waves for 3D cases are either very large or very small at any level. If the magnitudes of these numbers are not suppressed, this algorithm is useless, because the values of elements in translation matrices can easily yield floating-point overflows halting the computation.

In the simulation of electromagnetic phenomena in microwave integrated circuits where the simulated object can be a tiny fraction of the wavelength but the number of unknowns may be very large, neither the static MLFMA nor the dynamic MLFMA can be used. The original 3D undiagonalized dynamic FMM is applied to simulate the electromagnetic compatibility (EMC) problems [25]. Due to the intrinsic drawbacks of the 3D undiagonalized dynamic FMM discussed above, it cannot work at very low frequencies. The fine details cannot be simulated accurately. Therefore, fast 2D and 3D algorithms which can work at very low frequencies must be developed to bridge the gap from static to electrodynamic.

Based on the 2D and 3D originally undiagonalized dynamic MLFMA, 2D and 3D low-frequency multilevel fast multipole algorithms (LF-MLFMA) with a computational complexity of $O(N)$ are developed for low-frequency problems [26–28]. The memory requirements and the workload per iteration for the normalized LF-MLFMA all scale as $O(N)$. From the 3D LF-MLFMA, a more explicit and succinct representation of the static MLFMA is also derived [28]. The matrix rotation technique [29] is applied to the normalized 3D LF-MLFMA and the static MLFMA to save storage without increasing the order of the workload [30]. Some symmetries of the translation matrices along the z direction for the very low frequency case and the

MLFMA at Very Low Frequencies 153

static case are also derived to further reduce storage. In this chapter, we will describe the 2D and 3D LF-MLFMA and the matrix rotation techniques.

5.2 TWO-DIMENSIONAL MULTILEVEL FAST MULTIPOLE ALGORITHM AT VERY LOW FREQUENCIES

In this section, we will first repeat the review of the undiagonalized and the diagonalized dynamic MLFMA as was done in Chapter 2. Then based on the 2D undiagonalized dynamic MLFMA, by applying the normalization idea [31] for 3D problems at low frequencies, we derive the 2D LF-MLFMA. The efficiency test and some numerical results will also be given.

5.2.1 Core Equation of the 2D Undiagonalized Dynamic MLFMA

To implement the 2D MLFMA, the entire cross-section of the object is first enclosed in a large square, which is then partitioned into four smaller subsquares. Each subsquare is then recursively subdivided into smaller subsquares until the smallest squares are obtained. The dimensions of the smallest squares depend on the structure of the object, or more precisely, on the property of the current distribution. The center of each square is the new center for outgoing or incoming waves.

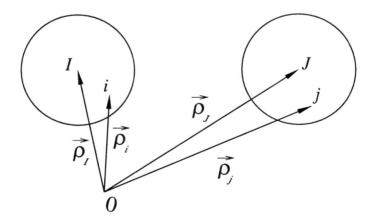

Figure 5.1 Source and field positions and their relative groups.

As shown in Figure 5.1, a source point ρ_j is located in group J whose center is at ρ_J, and a field point ρ_i is located in group I whose center is at ρ_I. There is no overlap between group I and group J. The vector from the source point ρ_j to the

field point ρ_i is
$$\boldsymbol{\rho}_{ij} = \boldsymbol{\rho}_i - \boldsymbol{\rho}_j \qquad (5.1)$$
The amplitude and the angle of vector $\boldsymbol{\rho}_{ij}$ are ρ_{ij} and ϕ_{ij}, respectively. Vector $\boldsymbol{\rho}_{ij}$ can be represented as
$$\boldsymbol{\rho}_{ij} = \boldsymbol{\rho}_{iI} - (\boldsymbol{\rho}_{jJ} - \boldsymbol{\rho}_{IJ}) \qquad (5.2)$$
By applying the addition theorem for Hankel's function of high order, we have the core equation of the 2D undiagonalized dynamic MLFMA:
$$\alpha_{nm}(\boldsymbol{\rho}_{ij}) = \sum_{N=-\infty}^{+\infty} \sum_{M=-\infty}^{+\infty} \beta_{nN}(\boldsymbol{\rho}_{iI}) \alpha_{NM}(\boldsymbol{\rho}_{IJ}) \beta_{Mm}(\boldsymbol{\rho}_{Jj}) \qquad (5.3)$$
where
$$\alpha_{nm}(\boldsymbol{\rho}_{ij}) = H_{n-m}^{(1)}(k_\rho \rho_{ij}) e^{-i(n-m)\phi_{ij}} \qquad (5.4)$$
$$\beta_{nm}(\boldsymbol{\rho}_{ij}) = J_{n-m}(k_\rho \rho_{ij}) e^{-i(n-m)\phi_{ij}} \qquad (5.5)$$
Equation (5.3) can also be written in matrix form as [27]
$$\overline{\alpha}_{ij} = \overline{\beta}_{iI} \cdot \overline{\alpha}_{IJ} \cdot \overline{\beta}_{Jj} \qquad (5.6)$$

As discussed in Chapter 2, MLFMA expedites the matrix-vector multiplication by dividing the calculation into two parts: the aggregation and the translation-disaggregation. The aggregation process can be described as
$$\mathbf{b}_{k'_{(l+1)}} = \sum_{k \in G_{k'_{(l+1)}}} \overline{\beta}_{k'k} \cdot \mathbf{b}_{k_{(l)}} \qquad (5.7)$$
where $\mathbf{b}_{k'_{(l+1)}}$ and $\mathbf{b}_{k_{(l)}}$, respectively, are amplitude vectors of the outgoing cylindrical harmonics for group k' at level $l+1$ and group k at level l, $G_{k'_{(l+1)}}$ represents all the child groups at level l of group k' at level $l+1$, and $\overline{\beta}_{k'k}$ is the aggregation translator from group k to group k'. The translation-disaggregation process can be described as
$$\mathbf{S}_{k_{(l)}} = \sum_{m \in g_{k_{(l)}}} \overline{\alpha}_{mk} \cdot \mathbf{b}_{m_{(l)}} + \overline{\beta}_{kk'} \cdot \mathbf{S}_{k'_{(l+1)}} \qquad (5.8)$$
where $\mathbf{S}_{k_{(l)}}$ and $\mathbf{S}_{k'_{(l+1)}}$, respectively, are amplitude vectors of the incoming cylindrical harmonics for group k at level l and group k' at level $l+1$, $g_{k_{(l)}}$ represents the groups which are separated from group k at level l but within the nearby groups of group k' at level $l+1$, $\overline{\alpha}_{mk}$ and $\overline{\beta}_{kk'}$ are relative translators.

Equations (5.3) and (5.6) cannot be used directly in either the dynamic case or the quasi-static case (very low frequency case). For the dynamic case, the number of multipole cylindrical harmonics of a group is proportional to the group size at any level. Hence, the computational complexity is still $O(N^2)$ [10–12, 20–22]. No reduction in the order of the computational complexity is achieved.

For very low frequency cases, the number of the multipole cylindrical harmonics can be chosen the same at different levels due to the scale invariance of the Laplace equation. Consequently, the computational complexity is reduced to $O(N)$. However, other problems appear.

When the frequency becomes very low compared to the dimensions, the arguments of the Bessel and Hankel functions in the translation matrices $\overline{\beta}$ and $\overline{\alpha}$ are very small. Therefore, the elements in $\overline{\beta}$ are very small, tending to zero, whereas the elements in $\overline{\alpha}$ are very large, tending to infinity. It can be seen from (5.3) that a moderate value is formed by a summation of a series where every term is formed by the product of a very large value and a very small value. If the magnitudes of these numbers are not suppressed, this formula is useless, because the values of elements in translation matrices easily exceed the floating number range resulting in floating-point underflows or overflows.

5.2.2 Core Equation of the 2D Diagonalized Dynamic MLFMA

The factorization in (5.3) can be used potentially to reduce computational labor in a matrix vector multiplication because it allows the grouping of source points and receiving points. Unfortunately, the $\overline{\alpha}$ matrices are dense matrices. Hence, there is no reduction in computational cost for dynamic cases. We need to diagonalize the $\overline{\alpha}$ matrix for dynamic cases. To achieve the diagonalization of the $\overline{\alpha}$ matrix, by representing the elements in $\overline{\beta}$ in plane waves by using its integral representation, we have the core equation of the 2D diagonalized dynamic MLFMA [20, 22], (see Chapter 2):

$$\alpha_{nm}\left(\boldsymbol{\rho}_{ij}\right) = \frac{1}{2\pi}\int_0^{2\pi} d\alpha\, e^{in\left(\alpha-\frac{\pi}{2}\right)} e^{ik_\rho \rho_{iI} \cos(\alpha+\phi_{iI})} \tilde{\alpha}_{IJ}(\alpha)$$
$$e^{ik_\rho \rho_{Jj} \cos(\alpha+\phi_{Jj})} e^{-im\left(\alpha-\frac{\pi}{2}\right)} \quad (5.9)$$

where

$$\tilde{\alpha}_{IJ}(\alpha) = \sum_p H_p^{(1)}\left(k_\rho \rho_{IJ}\right) e^{-ip\left(\phi_{IJ}+\alpha-\frac{\pi}{2}\right)} \quad (5.10)$$

Equation (5.9) above represents the diagonal form of (5.3), where the middle factor $\tilde{\alpha}_{IJ}(\alpha)$ is diagonal, so that instead of a double summation in (5.3), (5.9) has only one single integral. The calculation of $\alpha_{nm}\left(\boldsymbol{\rho}_{ij}\right)$ via (5.9) is much more efficient than via (5.3).

The summation of (5.10) is divergent if taken over an infinite range. The reason is that $H_p^{(1)}\left(k_\rho \rho_{IJ}\right)$ becomes infinitely large when $|p| \to \infty$. Therefore, the summation in (5.10) does not converge. To prevent this from happening, the summation in (5.10) should be truncated (see Chapter 2).

From (5.9) we can see the relations between strengths of cylindrical harmonics and plane waves are [20]

$$\tilde{b}_{k_{(l)}}(\alpha) = \sum_{n=-N_l}^{N_l} e^{-in(\alpha-\pi/2)} \left[\mathbf{b}_{k_{(l)}}\right]_n \qquad (5.11)$$

$$\left[\mathbf{b}_{k_{(l)}}\right]_n = \frac{1}{2\pi} \int_0^{2\pi} d\alpha \, e^{in(\alpha-\pi/2)} \tilde{b}_{k_{(l)}}(\alpha) \qquad (5.12)$$

where $\mathbf{b}_{k_{(l)}}$ and $\tilde{b}_{k_{(l)}}$ are, respectively, the amplitude vectors of the cylindrical harmonics and the plane waves of the $k_{(l)}$th group at level l. The interpolation and anterpolation techniques are used in aggregation and disaggregation separately to solve the problem of unequal sampling rates at different levels. With the application of these techniques, the computational complexity is reduced to $O(N\log N)$.

5.2.3 2D Uniformly Normalized LF-MLFMA

Although the derivation of the dynamic algorithm is completely valid, when the frequency becomes very low compared to the relevant dimensions, the divergent series in the diagonalized translation matrix $\overline{\alpha}$ in (5.10) is unstable. Therefore, the 2D diagonalized dynamic algorithm (5.9) cannot be used for the quasi-static case.

As mentioned above and will be proved later, the computational complexity is $O(N)$ without diagonalizing the translation matrices for the quasi-static case. The only problems are that the elements in $\overline{\beta}$ tend to zero and the elements in $\overline{\alpha}$ tend to infinity, making floating-point overflows appear during computing. To solve these problems, we apply the normalization idea for the 3D problem at low frequencies [31] to keep moderate all the values of the amplitudes of outgoing harmonics $\mathbf{b}$ and the amplitudes of the incoming harmonics $\mathbf{S}$ throughout the computation, and meanwhile $\overline{\beta}$ and $\overline{\alpha}$ are floating-point-overflow controllable during computing [26].

By considering the asymptotic behaviors of the Bessel-type functions when the argument tends to zero [32, 33], we can define normalized Bessel, Neumann, and Hankel functions with moderate values as follows [26]:

$$J_n^N(x) = \frac{1}{t^{|n|}} J_n(x) \qquad (5.13)$$

$$Y_n^N(x) = t^{|n|} Y_n(x) \qquad (5.14)$$

$$H_n^{(i)N}(x) = t^{|n|} H_n^{(i)}(x) \qquad (5.15)$$

where the superscript N denotes the normalized terms. This notation is used all through this chapter. The term t is a normalization factor that satisfies $\frac{x}{t} \sim O(1)$. All these normalized Bessel-type functions can be calculated by recurrence method.

By considering the asymptotic property of matrix β, we can see from (5.6) that the amplitudes of the multipole cylindrical harmonics of a group $k'_{(1)}$ at the finest level $\left[\mathbf{b}_{k'_{(1)}}\right]_m$ scales $O(t^{|m|})$ when $t \to 0$, where t is the normalization factor that satisfies $\frac{k_\rho \bar{\rho}_{k'_{(1)}}}{t} \sim O(1)$, and $\bar{\rho}_{k'_{(1)}}$ is the average distance from the subscatterers in group $k'_{(1)}$ to their group center. Therefore, $\frac{1}{|t|^m}\left[\mathbf{b}_{k'_{(1)}}\right]_m \sim O(1)$ when $t \to 0$.

Since the interaction should be of moderate value, the amplitude of the incoming multipole cylindrical harmonics $\left[\mathbf{S}_{k_{(1)}}\right]_m \sim O(t^{-|m|})$ when $t \to 0$. Therefore, $t^{|m|}\left[\mathbf{S}_{k_{(1)}}\right]_m \sim O(1)$, when $t \to 0$.

The multipole amplitude $\left[\mathbf{b}_{k_{(l)}}\right]_m$ at a coarser level l ($l > 1$) obtained by multilevel aggregation must be the same as those obtained by assuming this level to be the finest level. Applying the discussion above, we derive that $\left[\mathbf{b}_{k_{(l)}}\right]_m \sim O(t^{|m|})$ when $t \to 0$. Performing the same analysis on $\left[\mathbf{S}_{k_{(l)}}\right]_m$, we derive that $\left[\mathbf{S}_{k_{(l)}}\right]_m \sim O(t^{-|m|})$ when $t \to 0$. Therefore, we can normalize $\left[\mathbf{b}_{k_{(l)}}\right]_m$ and $\left[\mathbf{S}_{k_{(l)}}\right]_m$ as follows [26]:

$$\left[\mathbf{b}^N_{k_{(l)}}\right]_m = \frac{1}{t^{|m|}}\left[\mathbf{b}_{k_{(l)}}\right]_m \tag{5.16}$$

$$\left[\mathbf{S}^N_{k_{(l)}}\right]_m = t^{|m|}\left[\mathbf{S}_{k_{(l)}}\right]_m \tag{5.17}$$

All of the normalized amplitudes of the outgoing and incoming multipole cylindrical harmonics are of moderate value. Comparing (5.4) to (5.5) and applying (5.13) and (5.15), we derive the translation matrices for the normalized multipole amplitude in terms of the normalized Bessel and Hankel functions as follows [26]:

$$\begin{aligned}\left[\overline{\alpha}^N_{ij}(\boldsymbol{\rho}_{ij},t)\right]_{mn} &= t^{|m|+|n|-|m-n|}H^{(1)N}_{m-n}(k\rho_{ij})\,e^{-i(m-n)\phi_{ij}}\\ &= t^{|m|+|n|}\left[\overline{\alpha}_{ij}(\boldsymbol{\rho}_{ij})\right]_{mn}\end{aligned} \tag{5.18}$$

$$\begin{aligned}\left[\overline{\beta}^N_{ij}(\boldsymbol{\rho}_{ij},t)\right]_{mn} &= t^{|l^<|-|l^>|+|l^<-l^>|}J^N_{m-n}(k\rho_{ij})\,e^{-i(m-n)\phi_{ij}}\\ &= t^{|l^<|-|l^>|}\left[\overline{\beta}_{ij}(\boldsymbol{\rho}_{ij})\right]_{mn}\end{aligned} \tag{5.19}$$

where $l^<$ and $l^>$ are the index number of the cylindrical harmonics at the lower level and the higher level between m and n, separately. The term t is the normalization coefficient. With these normalizations, the core equation of the 2D LF-MLFMA for the quasi-static case is

$$\overline{\alpha}^N_{ij}(\boldsymbol{\rho}_{ij},t) = \overline{\beta}^N_{iI}(\boldsymbol{\rho}_{iI},t)\cdot\overline{\alpha}^N_{IJ}(\boldsymbol{\rho}_{IJ},t)\cdot\overline{\beta}^N_{Jj}(\boldsymbol{\rho}_{Jj},t) \tag{5.20}$$

In (5.18) and (5.19), because $t \leq 1$,

$$t^{|m|+|n|-|m-n|} \leq 1 \tag{5.21}$$

$$t^{||l^<|-|l^>|+|l^<-l^>||} \leq 1 \tag{5.22}$$

Only floating-point underflow may appear during computing, but it does not pose a problem as the possible floating-point underflow terms are exponential functions. Also because the multipole amplitudes at any level are of moderate value, $\left[\overline{\beta}^N\right]_{mn}$ or $\left[\overline{\alpha}^N\right]_{mn}$ with very small value plays a trivial role. Ignoring the elements that may cause a floating-point underflow will not decrease the precision of the final results. Therefore, the normalized algorithm is precise and the floating-point overflow during computing is avoided.

5.2.4 Nonuniformly Normalized Form of 2D LF-MLFMA

The uniformly normalized 2D LF-MLFMA (5.20) can be applied directly at very low frequencies or for very small structures compared to the wavelength without a significant variation in discretization size. For problems whose geometry property or material properties vary significantly in some subareas, we need to choose different normalization coefficients at different subareas because adaptive meshing techniques may be used to describe the geometry. In this case, we modify the uniformly normalized 2D LF-MLFMA to the nonuniformly normalized form as

$$\alpha_{nm}^N\left(\boldsymbol{\rho}_{ij},t_{ij}\right) = \sum_{N=-\infty}^{+\infty}\sum_{M=-\infty}^{+\infty}\left(\frac{t_{ij}}{t_{iI}}\right)^{|n|}\beta_{nN}^N\left(\boldsymbol{\rho}_{iI},t_{iI}\right)\left(\frac{t_{iI}}{t_{IJ}}\right)^{|N|}$$
$$\alpha_{NM}^N\left(\boldsymbol{\rho}_{IJ},t_{IJ}\right)\left(\frac{t_{Jj}}{t_{IJ}}\right)^{|M|}\beta_{Mm}^N\left(\boldsymbol{\rho}_{Jj},t_{Jj}\right)\left(\frac{t_{ij}}{t_{Jj}}\right)^{|m|} \tag{5.23}$$

where t_{ij}, t_{IJ}, t_{Jj}, and t_{iI} are, respectively, normalization coefficients for $\alpha_{nm}\left(\boldsymbol{\rho}_{ij},t_{ij}\right)$, $\alpha_{NM}\left(\boldsymbol{\rho}_{IJ},t_{IJ}\right)$, $\beta_{nN}\left(\boldsymbol{\rho}_{iI},t_{iI}\right)$, and $\beta_{Mm}\left(\boldsymbol{\rho}_{Jj},t_{Jj}\right)$ and are indicated in parentheses of these translators.

This normalized algorithm is easily merged with the dynamic algorithm to solve the large-scale problem with dense subgridded areas. Figure 5.2 shows the group structure for a conducting cylinder whose cross-section is an ogive. In the dense subgridded areas, more levels are needed to reach the small group size required in these areas. We can first use the normalized algorithm and then switch to the dynamic algorithm when the group size is large enough to meet the requirement of the dynamic algorithm. The switch between these two algorithms can be easily done with the help of (5.11) and (5.12) and the relative normalization coefficients.

5.2.5 Computational Complexity of 2D LF-MLFMA

In the implementation of the 2D LF-MLFMA, only nonempty groups are stored and calculated at any level. Assuming there are on the average M subscatterers

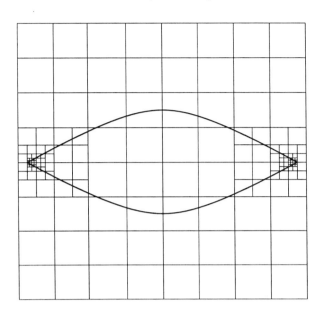

Figure 5.2 Group structure for a conducting cylinder whose cross-section is an ogive.

in each group at the finest level, then the floating-point operations that are relevant directly with subscatterers are the aggregation from subscatterers to the finest level, the disaggregation from the finest level to subscatterers, and the interactions of the neighbor terms. The workload is

$$T_0 = C'_0 MN + C''_0 KN = (C'_0 M + C''_0 K)N \tag{5.24}$$

where N is the number of unknowns, and K is the number of the multipole cylindrical harmonics at the finest level.

The number of the groups at the next coarser level is roughly $\frac{1}{I}$ of that at the current level. I is smaller than 4 for 2D cases. The number of levels needed in the 2D LF-MLFMA is roughly

$$L = \log_I \frac{N}{M} - 1 \tag{5.25}$$

For the low-frequency case, or when all the groups at different levels are much smaller than the wavelength, the number of the multipole cylindrical waves can be chosen the same at different levels due to the scale invariance of the low-frequency case. We index the levels from the finest level to the coarsest level successively with $l = 1, 2, \ldots, L$. The workload for the aggregation from level l to level $l + 1$ and the disaggregation from level $l+1$ to level l are

$$T'_l = \begin{cases} C' \left(\frac{1}{I}\right)^l \frac{N}{M} K^2, & l = 1, 2, \ldots, L-1 \\ 0, & l = L \end{cases} \quad (5.26)$$

The workload for the translations at level l is

$$T''_l = C'' \left(\frac{1}{I}\right)^l \frac{N}{M} K^2, \quad l = 1, 2, \ldots, L \quad (5.27)$$

The total workload per iteration needed in the normalized 2D MLFMA is

$$\begin{aligned} T &= T_0 + \sum_{l=1}^{L} (T'_l + T''_l) \\ &= \left\{ (C'_0 M + C''_0 K) + \left[C' \left(1 - \frac{1}{I^{L-1}}\right) + C'' \left(1 - \frac{1}{I^L}\right) \right] \frac{K^2}{M} \right\} N \quad (5.28) \\ &= \left\{ (C'_0 M + C''_0 K) + (C' + C'') \frac{K^2}{M} \right\} N - (C'I + C'')IK^2 \end{aligned}$$

Therefore, the workload per iteration is clearly $O(N)$ when $N \to \infty$ such that the scatterer size is still in the low-frequency regime. The memory requirement is also $O(N)$ because only nonempty groups are stored at any level.

Figures 5.3 and 5.4, respectively, show the comparison of the CPU time per iteration and the CPU time for setup between the 2D LF-MLFMA and the plain conjugate gradient (CG) method. Figure 5.5 shows the memory requirements for the 2D LF-MLFMA and the plain CG versus the number of unknowns. The 2D LF-MLFMA is more efficient even for a small number of unknowns. It is clear that the computational complexity of the 2D LF-MLFMA is $O(N)$, where N is the number of unknowns.

5.2.6 Applying 2D Dynamic MLFMA and LF-MLFMA to CFIE for PEC Structures

In this section, we will show how to apply 2D dynamic MLFMA and LF-MLFMA to solve the combined fields integral equation (CFIE) for TM polarization and TE polarization, respectively. For simplicity, we only describe the two-level case. The general multilevel cases are straightforward extensions from the two-level case.

5.2.6.1 CFIE for TM Polarization

For TM polarization, the incident electric fields are parallel to the axis of the scattering cylinder. Assuming the axis of the cylinder is z, the induced current is z-directed and

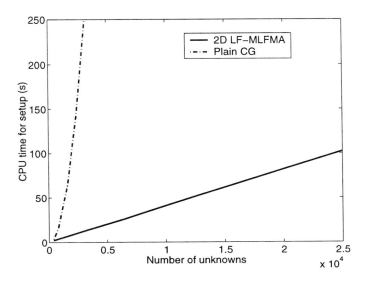

Figure 5.3 Comparison of the CPU time for setup versus the number of unknowns between normalized 2D MLFMA and plain CG method.

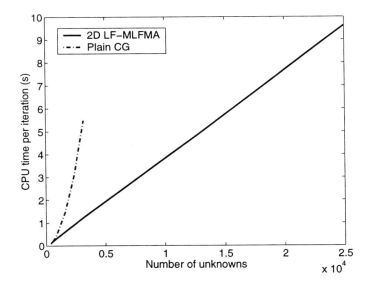

Figure 5.4 Comparison of the CPU time per iteration versus the number of unknowns between normalized 2D MLFMA and plain CG method.

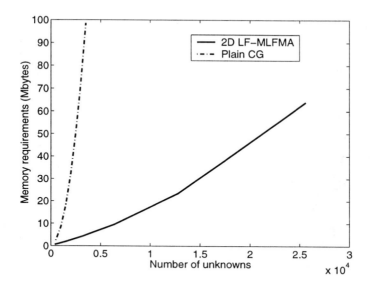

Figure 5.5 Comparison of the memory requirements versus the number of unknowns between normalized 2D MLFMA and plain CG method.

z-independent. The EFIE is

$$\frac{\omega\mu}{4}\int_C H_0^{(1)}(k\,|\,\rho-\rho'\,|)J_z(\rho')\,dl' = E_z^{inc}(\rho), \qquad \rho \text{ on } C \qquad (5.29)$$

where $E_z^{inc}(\rho)$ is the incident electric field, $J_z(\rho)$ is the induced current, and C is the boundary of the cylinder. The MFIE is

$$-\frac{i}{4}\int_C \hat{n}\cdot\nabla H_0^{(1)}(k\,|\,\rho-\rho'\,|)\,J_z(\rho')\,dl' = \frac{1}{\eta}(\hat{n}\cdot\hat{k})E_z^{inc}(\rho),$$
$$\rho \to C^- \qquad (5.30)$$

where $\hat{n}$ is the outward normal unit vector of the cylinder. The CFIE is the linear combination of the EFIE and MFIE. It can be constructed as

$$\text{CFIE} = \xi\text{EFIE} + (1-\xi)\eta\text{MFIE} \qquad (5.31)$$

By expanding $J_z(\rho)$ in terms of a set of local basis functions

$$J_z(\rho) = \sum_{j=1}^{L} I_j J_j(\rho)$$

and testing (5.31) by a set of local testing functions $J_{ti}(\boldsymbol{\rho})$, $i = 1, 2, \ldots, L$, we obtain the matrix equation of the CFIE:

$$\overline{\mathbf{Z}}^{CE} \cdot \mathbf{I} = \mathbf{V}^{CE} \tag{5.32}$$

where $\mathbf{I}$ represents the current, and

$$Z_{ij}^{CE} = \int_{C_i} dl_i \int_{C_j} dl_j \, J_{ti}(\boldsymbol{\rho}_i) g_{ij}^{CEK} J_j(\boldsymbol{\rho}_j) \tag{5.33}$$

$$g_{ij}^{CEK} = \frac{k\eta}{4} \left[\xi - \frac{i(1-\xi)}{k} \hat{n}_i \cdot \nabla_i \right] H_0^{(1)}(k|\boldsymbol{\rho}_i - \boldsymbol{\rho}_j|) \tag{5.34}$$

$$V_i^{CE} = \int_{C_i} J_{ti}(\boldsymbol{\rho}_i) E_i^{CEK} dl_i \tag{5.35}$$

$$E_i^{CEK} = \left[\xi + (1-\xi)(\hat{n}_i \cdot \hat{k}) \right] E_z^{inc}(\boldsymbol{\rho}_i) \tag{5.36}$$

First, we describe how to apply the 2D dynamic MLFMA to expedite the matrix-vector multiplication. From the core equation of the 2D diagonalized dynamic MLFMA (5.9), we have

$$H_0^{(1)}(k|\boldsymbol{\rho}_i - \boldsymbol{\rho}_j|) = \frac{1}{2\pi} \int_0^{2\pi} d\alpha \, e^{ik\rho_{iI} \cos(\alpha + \phi_{iI})} \tilde{\alpha}_{IJ}(\alpha) e^{ik\rho_{Jj} \cos(\alpha + \phi_{Jj})} \tag{5.37}$$

where I and J are the groups in which i and j reside, respectively. Substituting (5.37) in (5.34) leads to

$$g_{ij}^{CEK} = \frac{k\eta}{4} \left[\xi - \frac{i(1-\xi)}{k} \hat{n}_i \cdot \nabla_i \right]$$
$$\frac{1}{2\pi} \int_0^{2\pi} d\alpha \, e^{ik\rho_{iI} \cos(\alpha + \phi_{iI})} \tilde{\alpha}_{IJ}(\alpha) e^{ik\rho_{Jj} \cos(\alpha + \phi_{Jj})} \tag{5.38}$$

In (5.38),

$$\hat{n}_i \cdot \nabla_i e^{ik\rho_{iI} \cos(\alpha + \phi_{iI})}$$
$$= ike^{ik\rho_{iI} \cos(\alpha + \phi_{iI})} \hat{n}_i \cdot \nabla_i \left\{ \rho_{l'm} \cos(\alpha + \phi_{l'm}) \right\}$$
$$= ike^{ik\rho_{iI} \cos(\alpha + \phi_{iI})} \hat{n}_i \cdot \nabla_i \left\{ (x_i - x_I) \cos \alpha - (y_i - y_I) \sin \alpha \right\} \tag{5.39}$$
$$= ike^{ik\rho_{iI} \cos(\alpha + \phi_{iI})} (n_{ix} \cos \alpha - n_{iy} \sin \alpha)$$

By applying (5.38) and (5.39) in (5.33), we have

$$Z_{ij}^{CE} = \frac{k\eta}{4} \frac{1}{2\pi} \int_0^{2\pi} d\alpha$$
$$\int_{C_i} dl_i \, J_{ti}(\boldsymbol{\rho}_i) \left[\xi + (1-\xi)(n_{ix} \cos \alpha - n_{iy} \sin \alpha) \right] e^{ik\rho_{iI} \cos(\alpha + \phi_{iI})} \tag{5.40}$$
$$\tilde{\alpha}_{IJ}(\alpha) \int_{C_j} dl_j \, J_j(\boldsymbol{\rho}_j) e^{ik\rho_{Jj} \cos(\alpha + \phi_{Jj})}$$

If the CG method is used to solve the CFIE, the transpose of the impedance matrix is also needed. Similarly, we have

$$\begin{aligned}
Z_{ji}^{CE} &= \int_{C_j} dl_j \int_{C_i} dl_i \, J_{tj}(\boldsymbol{\rho}_i) g_{ji}^{CEK} J_i(\boldsymbol{\rho}_i) \\
&= \int_{C_j} dl_j \int_{C_i} dl_i \, J_{tj}(\boldsymbol{\rho}_i) J_i(\boldsymbol{\rho}_i) \\
&\quad \frac{k\eta}{4} \left[\xi - \frac{i(1-\xi)}{k} \hat{n}_j \cdot \nabla_j \right] H_0^{(1)}(k|\boldsymbol{\rho}_j - \boldsymbol{\rho}_i|) \\
&= \int_{C_j} dl_j \int_{C_i} dl_i \, J_{tj}(\boldsymbol{\rho}_i) J_i(\boldsymbol{\rho}_i) \\
&\quad \frac{k\eta}{4} \left[\xi - \frac{i(1-\xi)}{k} \hat{n}_j \cdot \nabla_j \right] H_0^{(1)}(k|\boldsymbol{\rho}_i - \boldsymbol{\rho}_j|) \\
&= \frac{k\eta}{4} \frac{1}{2\pi} \int_0^{2\pi} d\alpha \\
&\quad \int_{C_i} dl_i \, J_i(\boldsymbol{\rho}_i) \left[\xi + (1-\xi)(-n_{jx}\cos\alpha + n_{jy}\sin\alpha) \right] e^{ik\rho_{iI}\cos(\alpha + \phi_{iI})} \\
&\quad \tilde{\alpha}_{IJ}(\alpha) \int_{C_j} dl_j \, J_{tj}(\boldsymbol{\rho}_j) e^{ik\rho_{Jj}\cos(\alpha + \phi_{Jj})}
\end{aligned} \tag{5.41}$$

Second, we apply the 2D LF-MLFMA to expedite the matrix-vector multiplication. Equation (5.34) can be rewritten as

$$\begin{aligned}
g_{ij}^{CEK} &= \frac{k\eta}{4} \left[\xi - \frac{i(1-\xi)}{k} \hat{n}_i \cdot \nabla_i \right] H_0^{(1)}(k|\boldsymbol{\rho}_i - \boldsymbol{\rho}_j|) \\
&= \frac{k\eta}{4} \left\{ \xi H_0^{(1)}(k\rho_{ij}) + \frac{i(1-\xi)}{2} \right. \\
&\quad \left. \left[\tilde{\tilde{n}}_i H_1^{(1)}(k\rho_{ij}) e^{-i\phi_{ij}} - \tilde{\tilde{n}}_i^* H_{-1}^{(1)}1(k\rho_{ij}) e^{i\phi_{ij}} \right] \right\}
\end{aligned} \tag{5.42}$$

where
$$\tilde{\tilde{n}}_i = n_{ix} + i n_{iy}$$

By applying the core equation of the 2D LF-MLFMA (5.23), we have

$$\begin{aligned}
g_{ij}^{CEK} = \frac{k\eta}{4} \sum_{N=-\infty}^{+\infty} \sum_{M=-\infty}^{+\infty} \beta_{0N}^{ND}(\boldsymbol{\rho}_{iI}, t_{iI}) \left(\frac{t_{iI}}{t_{IJ}} \right)^{|N|} \\
\alpha_{NM}^{N}(\boldsymbol{\rho}_{IJ}, t_{IJ}) \left(\frac{t_{Jj}}{t_{IJ}} \right)^{|M|} \beta_{M0}^{N}(\boldsymbol{\rho}_{Jj}, t_{Jj})
\end{aligned} \tag{5.43}$$

where

$$\beta_{0N}^{ND}(\boldsymbol{\rho}_{iI}, t_{iI}) = \xi \beta_{0N}^{N}(\boldsymbol{\rho}_{iI}, t_{iI}) \\ + \frac{i(1-\xi)}{2t_{iI}} \left(\tilde{\hat{n}}_i \beta_{1N}^{N}(\boldsymbol{\rho}_{iI}, t_{iI}) - \tilde{\hat{n}}_i^* \beta_{(-1)N}^{N}(\boldsymbol{\rho}_{iI}, t_{iI}) \right) \quad (5.44)$$

By applying (5.43) in (5.33), we have

$$Z_{ij}^{CE} = \frac{k\eta}{4} \sum_{N=-\infty}^{+\infty} \sum_{M=-\infty}^{+\infty} \int_{C_i} dl_i \, J_{ti}(\boldsymbol{\rho}_i) \beta_{0N}^{ND}(\boldsymbol{\rho}_{iI}, t_{iI}) \left(\frac{t_{iI}}{t_{IJ}} \right)^{|N|} \\ \alpha_{NM}^{N}(\boldsymbol{\rho}_{IJ}, t_{IJ}) \left(\frac{t_{Jj}}{t_{IJ}} \right)^{|M|} \int_{C_j} dl_j \, J_j(\boldsymbol{\rho}_j) \beta_{M0}^{N}(\boldsymbol{\rho}_{Jj}, t_{Jj}) \quad (5.45)$$

Similarly, we have

$$g_{ji}^{CEK} = \frac{k\eta}{4} \left[\xi - \frac{i(1-\xi)}{k} \hat{n}_j \cdot \nabla_j \right] H_0^{(1)}(k|\boldsymbol{\rho}_j - \boldsymbol{\rho}_i|) \\ = \frac{k\eta}{4} \left[\xi - \frac{i(1-\xi)}{k} \hat{n}_j \cdot \nabla_j \right] H_0^{(1)}(k|\boldsymbol{\rho}_i - \boldsymbol{\rho}_j|) \\ = \frac{k\eta}{4} \left\{ \xi H_0^{(1)}(k\rho_{ij}) - \frac{i(1-\xi)}{2} \\ \left[\tilde{\hat{n}}_i H_1^{(1)}(k\rho_{ij})e^{-i\phi_{ij}} - \tilde{\hat{n}}_i^* H_{-1}^{(1)}1(k\rho_{ij})e^{i\phi_{ij}} \right] \right\} \\ = \frac{k\eta}{4} \sum_{N=-\infty}^{+\infty} \sum_{M=-\infty}^{+\infty} \beta_{0N}^{N}(\boldsymbol{\rho}_{iI}, t_{iI}) \left(\frac{t_{iI}}{t_{IJ}} \right)^{|N|} \\ \alpha_{NM}^{N}(\boldsymbol{\rho}_{IJ}, t_{IJ}) \left(\frac{t_{Jj}}{t_{IJ}} \right)^{|M|} \beta_{M0}^{ND}(\boldsymbol{\rho}_{Jj}, t_{Jj}) \quad (5.46)$$

where

$$\beta_{M0}^{ND}(\boldsymbol{\rho}_{Jj}, t_{Jj}) = \xi \beta_{M0}^{N}(\boldsymbol{\rho}_{Jj}, t_{Jj}) \\ - \frac{i(1-\xi)}{2t_{Jj}} \left(\tilde{\hat{n}}_i \beta_{M(-1)}^{N}(\boldsymbol{\rho}_{Jj}, t_{Jj}) - \tilde{\hat{n}}_i^* \beta_{M1}^{N}(\boldsymbol{\rho}_{Jj}, t_{Jj}) \right) \quad (5.47)$$

By applying (5.46) in (5.33), we have

$$Z_{ji}^{CE} = \frac{k\eta}{4} \sum_{N=-\infty}^{+\infty} \sum_{M=-\infty}^{+\infty} \int_{C_i} dl_i \, J_i(\boldsymbol{\rho}_i) \beta_{0N}^{N}(\boldsymbol{\rho}_{iI}, t_{iI}) \left(\frac{t_{iI}}{t_{IJ}} \right)^{|N|} \\ \alpha_{NM}^{N}(\boldsymbol{\rho}_{IJ}, t_{IJ}) \left(\frac{t_{Jj}}{t_{IJ}} \right)^{|M|} \int_{C_j} dl_j \, J_{tj}(\boldsymbol{\rho}_j) \beta_{M0}^{ND}(\boldsymbol{\rho}_{Jj}, t_{Jj}) \quad (5.48)$$

5.2.6.2 CFIE for TE Polarization

For TE polarization to z, the incident and scattered magnetic fields have only a z component:

$$\mathbf{H}^{inc} = H_z^{inc}\hat{z}$$

$$\mathbf{H}^s = H_z^s\hat{z}$$

The induced current has only a transverse component:

$$\mathbf{J}(\boldsymbol{\rho}) = J(\boldsymbol{\rho})\hat{n} \times \hat{z}$$

From the condition that the tangential component of the electric field vanishes on the surface C

$$-\hat{n} \times \mathbf{E}^s = \hat{n} \times \mathbf{E}^{inc} \tag{5.49}$$

we derive the EFIE for TE polarization:

$$-\frac{k}{4}(\hat{n}\cdot\hat{n}')\int_C H_0^{(1)}(k|\boldsymbol{\rho}-\boldsymbol{\rho}'|)J(\boldsymbol{\rho}')\,dl'$$
$$+\frac{1}{4}\int_C \frac{(\hat{n}\times\hat{z})\cdot(\boldsymbol{\rho}-\boldsymbol{\rho}')}{|\boldsymbol{\rho}-\boldsymbol{\rho}'|}H_1^{(1)}(k|\boldsymbol{\rho}-\boldsymbol{\rho}'|)\nabla'\cdot\mathbf{J}(\boldsymbol{\rho}')\,dl' = (\hat{k}\cdot\hat{n})H_z^{inc} \tag{5.50}$$

MLFMA only applies to separated terms which have no singularity. For the convenience of applying MLFMA, we use another condition:

$$-\hat{n} \times (\nabla \times \hat{z}H_z^s) = \hat{n} \times (\nabla \times \hat{z}H_z^{inc}) \tag{5.51}$$

By using the identity of

$$\hat{n} \times \nabla \times (f\hat{z}) = -\hat{z}\hat{n}\cdot\nabla f \tag{5.52}$$

we derive another expression of EFIE for TE polarization:

$$-\frac{1}{4k}\int_C (\hat{n}\cdot\nabla)(\hat{n}'\cdot\nabla')H_0^{(1)}(k|\boldsymbol{\rho}-\boldsymbol{\rho}'|)J(\boldsymbol{\rho}')\,dl' = (\hat{n}\cdot\hat{k})H_z^{inc}(\boldsymbol{\rho}) \tag{5.53}$$

Based on the fact that the magnetic fields just on the inside of the closed contour C are zero, we derive the MFIE for TE polarization:

$$-\frac{i}{4}\int_C \hat{n}'\cdot\nabla' H_0^{(1)}(k|\boldsymbol{\rho}-\boldsymbol{\rho}'|)J(\boldsymbol{\rho}')\,dl' = H_z^{inc}(\boldsymbol{\rho}), \qquad \boldsymbol{\rho} \to C^- \tag{5.54}$$

Using (5.31) and MOM leads to the matrix equation of the CFIE for TE polarization:

$$\overline{\mathbf{Z}}^{CH} \cdot \mathbf{I} = \mathbf{V}^{CH} \tag{5.55}$$

where

$$Z_{ij}^{CH} = \int_{C_i} dl_i \int_{C_j} dl_j\, J_{ti}(\boldsymbol{\rho}_i) g_{ij}^{CHK} J_j(\boldsymbol{\rho}_j) \tag{5.56}$$

MLFMA at Very Low Frequencies

$$g_{ij}^{CHK} = -\frac{1}{4}\left\{\xi\frac{1}{k}(\hat{n}_i \cdot \nabla_i) + i(1-\xi)\right\}(\hat{n}_j \cdot \nabla_j)H_0^{(1)}(k\rho_{ij}) \tag{5.57}$$

$$V_i^{CH} = \int_{C_n} J_{ti}(\boldsymbol{\rho}_i)H_i^{CHK}dl_i \tag{5.58}$$

$$H_i^{CHK} = \left[\xi(\hat{n}_i \cdot \hat{k}) + (1-\xi)\right]H_z^{inc}(\boldsymbol{\rho}_i) \tag{5.59}$$

Substituting (5.37) in (5.57) leads to

$$\begin{aligned}g_{ij}^{CHK} = &-\frac{1}{4}\left\{\xi\frac{1}{k}(\hat{n}_i \cdot \nabla_i) + i(1-\xi)\right\}(\hat{n}_j \cdot \nabla_j)\\ &\frac{1}{2\pi}\int_0^{2\pi} d\alpha\, e^{ik\rho_{iI}\cos(\alpha+\phi_{iI})}\tilde{\alpha}_{IJ}(\alpha)e^{ik\rho_{Jj}\cos(\alpha+\phi_{Jj})}\end{aligned} \tag{5.60}$$

As in (5.39), we have

$$\begin{aligned}\hat{n}_j \cdot \nabla_j &e^{ik\rho_{Jj}\cos(\alpha+\phi_{Jj})}\\ &= ike^{ik\rho_{Jj}\cos(\alpha+\phi_{Jj})}(-n_{jx}\cos\alpha + n_{jy}\sin\alpha)\end{aligned} \tag{5.61}$$

Substituting (5.39) and (5.61) in (5.60) leads to

$$g_{ij}^{CHK} =$$

$$\frac{1}{2\pi}\int_0^{2\pi} d\alpha\left[i\xi(n_{ix}\cos\alpha - n_{iy}\sin\alpha) + i(1-\xi)\right]e^{ik\rho_{iI}\cos(\alpha+\phi_{iI})} \tag{5.62}$$

$$\tilde{\alpha}_{IJ}(\alpha)e^{ik\rho_{Jj}\cos(\alpha+\phi_{Jj})}\frac{-ik}{4}(-n_{jx}\cos\alpha + n_{jy}\sin\alpha)$$

By substituting (5.62) in (5.56), we have

$$Z_{ij}^{CH} =$$

$$\frac{1}{2\pi}\int_0^{2\pi} d\alpha \int_{C_i} dl_i\, J_{ti}(\boldsymbol{\rho}_i)\left[i\xi(n_{ix}\cos\alpha - n_{iy}\sin\alpha) + i(1-\xi)\right]e^{ik\rho_{iI}\cos(\alpha+\phi_{iI})}$$

$$\tilde{\alpha}_{IJ}(\alpha)\int_{C_j} dl_j\, J_j(\boldsymbol{\rho}_j)e^{ik\rho_{Jj}\cos(\alpha+\phi_{Jj})}\frac{-ik}{4}(-n_{jx}\cos\alpha + n_{jy}\sin\alpha) \tag{5.63}$$

Similarly, we have

$$\begin{aligned}g_{ji}^{CHK} &= -\frac{1}{4}\left\{\xi\frac{1}{k}(\hat{n}_j \cdot \nabla_j) + i(1-\xi)\right\}(\hat{n}_i \cdot \nabla_i)H_0^{(1)}(k\rho_{ji})\\ &= -\frac{1}{4}\left\{\xi\frac{1}{k}(\hat{n}_j \cdot \nabla_j) + i(1-\xi)\right\}(\hat{n}_i \cdot \nabla_i)H_0^{(1)}(k\rho_{ij})\\ &= \frac{1}{2\pi}\int_0^{2\pi} d\alpha\,\frac{-ik}{4}(n_{ix}\cos\alpha - n_{iy}\sin\alpha)e^{ik\rho_{iI}\cos(\alpha+\phi_{iI})}\tilde{\alpha}_{IJ}(\alpha)\\ &\quad e^{ik\rho_{Jj}\cos(\alpha+\phi_{Jj})}\left[i\xi(-n_{jx}\cos\alpha + n_{jy}\sin\alpha) + i(1-\xi)\right]\end{aligned} \tag{5.64}$$

$$Z_{ij}^{CH} = \frac{1}{2\pi} \int_0^{2\pi} d\alpha$$
$$\int_{C_i} dl_i\, J_{ti}(\boldsymbol{\rho}_i) \frac{-ik}{4}(n_{ix}\cos\alpha - n_{iy}\sin\alpha)e^{ik\rho_{iI}\cos(\alpha+\phi_{iI})}\tilde{\alpha}_{IJ}(\alpha) \quad (5.65)$$
$$\int_{C_j} dl_j\, J_j(\boldsymbol{\rho}_j) e^{ik\rho_{Jj}\cos(\alpha+\phi_{Jj})}\left[i\xi(-n_{jx}\cos\alpha + n_{jy}\sin\alpha) + i(1-\xi)\right]$$

Now we describe how to apply the 2D LF-MLFMA to CFIE for TE polarization. From the core equation of the 2D LF-MLFMA (5.23), we have

$$H_0^{(1)}(k|\boldsymbol{\rho}_i - \boldsymbol{\rho}_j|) = \sum_{N=-\infty}^{+\infty} \sum_{M=-\infty}^{+\infty} \beta_{0N}^N(\boldsymbol{\rho}_{iI}, t_{iI}) \left(\frac{t_{iI}}{t_{IJ}}\right)^{|N|}$$
$$\alpha_{NM}^N(\boldsymbol{\rho}_{IJ}, t_{IJ}) \left(\frac{t_{Jj}}{t_{IJ}}\right)^{|M|} \beta_{M0}^N(\boldsymbol{\rho}_{Jj}, t_{Jj}) \quad (5.66)$$

By applying (5.66) in (5.57), we have

$$g_{ij}^{CHK} = -\frac{1}{4}\left\{\xi\frac{1}{k}(\hat{n}_i \cdot \nabla_i) + i(1-\xi)\right\}(\hat{n}_j \cdot \nabla_j)$$
$$\sum_{N=-\infty}^{+\infty} \sum_{M=-\infty}^{+\infty} \beta_{0N}^N(\boldsymbol{\rho}_{iI}, t_{iI}) \left(\frac{t_{iI}}{t_{IJ}}\right)^{|N|} \quad (5.67)$$
$$\alpha_{NM}^N(\boldsymbol{\rho}_{IJ}, t_{IJ}) \left(\frac{t_{Jj}}{t_{IJ}}\right)^{|M|} \beta_{M0}^N(\boldsymbol{\rho}_{Jj}, t_{Jj})$$

In (5.67),

$$\hat{n}_i \cdot \nabla_i \beta_{0N}^N(\boldsymbol{\rho}_{iI}, t_{iI})$$
$$= t_{iI}^{-|N|} \hat{n}_i \cdot \nabla_i \left\{J_{-N}(k\rho_{iI})e^{iN\phi_{iI}}\right\} \quad (5.68)$$
$$= (-1)^N t_{iI}^{-|N|} \hat{n}_i \cdot \nabla_i \left\{J_N(k\rho_{iI})e^{iN\phi_{iI}}\right\}$$

and

$$\hat{n}_i \cdot \nabla_i \left\{J_N(k\rho_{iI})e^{iN\phi_{iI}}\right\}$$
$$= e^{iN\phi_{iI}} \hat{n}_i \cdot \nabla_i J_N(k\rho_{iI}) + J_N(k\rho_{iI})\hat{n}_i \cdot \nabla_i e^{iN\phi_{iI}} \quad (5.69)$$
$$= kJ_N'(k\rho_{iI})e^{iN\phi_{iI}} \hat{n}_i \cdot \hat{\rho}_{iI} + iNJ_N(k\rho_{iI})e^{iN\phi_{iI}} \hat{n}_i \cdot \nabla_i \phi_{iI}$$

where

$$\nabla_i \phi_{iI} = \nabla_i \tan^{-1}\frac{y_i - y_I}{x_i - x_I}$$
$$= \frac{(x_i - x_I)^2}{\rho_{iI}^2}\left\{-\frac{y_i - y_I}{(x_i - x_I)^2}\hat{x} + \frac{1}{x_i - x_I}\hat{y}\right\} \quad (5.70)$$
$$= \frac{1}{\rho_{iI}}\left\{-\sin\phi_{iI}\hat{x} + \cos\phi_{iI}\hat{y}\right\}$$

MLFMA at Very Low Frequencies

$$J'_N(k\rho_{iI}) = J_{N-1}(k\rho_{iI}) - \frac{N}{k\rho_{iI}} J_N(k\rho_{iI}) \quad (5.71)$$

By applying (5.70) and (5.71) in (5.69), we have

$$\hat{n}_i \cdot \nabla_i \left\{ J_N(k\rho_{iI}) e^{iN\phi_{iI}} \right\}$$

$$= k J_{N-1}(k\rho_{iI}) e^{iN\phi_{iI}} \hat{n}_i \cdot \hat{\rho}_{iI} - \frac{N}{\rho_{iI}} J_N(k\rho_{iI}) e^{iN\phi_{iI}} \hat{n}_i \cdot \hat{\rho}_{iI}$$

$$+ \frac{iN}{\rho_{iI}} J_N(k\rho_{iI}) e^{iN\phi_{iI}} \hat{n}_i \cdot \{-\sin\phi_{iI}\hat{x} + \cos\phi_{iI}\hat{y}\}$$

$$= k J_{N-1}(k\rho_{iI}) e^{iN\phi_{iI}} \hat{n}_i \cdot \hat{\rho}_{iI} - \frac{N}{\rho_{iI}} J_N(k\rho_{iI}) e^{iN\phi_{iI}}$$

$$\{n_{ix}\cos\phi_{iI} + n_{iy}\sin\phi_{iI} + i n_{ix}\sin\phi_{iI} - i n_{iy}\cos\phi_{iI}\}$$

$$= k J_{N-1}(k\rho_{iI}) e^{iN\phi_{iI}} \hat{n}_i \cdot \hat{\rho}_{iI} - \frac{N}{\rho_{iI}} J_N(k\rho_{iI}) e^{iN\phi_{iI}}$$

$$\left\{ \tilde{\hat{n}}_i^* \cos\phi_{iI} + i\tilde{\hat{n}}_i^* \sin\phi_{iI} \right\}$$

$$= k J_{N-1}(k\rho_{iI}) e^{iN\phi_{iI}} \hat{n}_i \cdot \hat{\rho}_{iI} - \frac{N}{\rho_{iI}} J_N(k\rho_{iI}) \tilde{\hat{n}}_i^* e^{i(N+1)\phi_{iI}}$$

$$= k J_{N-1}(k\rho_{iI}) e^{iN\phi_{iI}} \left\{ n_{ix} \frac{e^{i\phi_{iI}} + e^{-i\phi_{iI}}}{2} + n_{iy} \frac{e^{i\phi_{iI}} - e^{-i\phi_{iI}}}{2i} \right\}$$

$$- \frac{N}{\rho_{iI}} J_N(k\rho_{iI}) \tilde{\hat{n}}_i^* e^{i(N+1)\phi_{iI}}$$

$$= \frac{1}{2} k J_{N-1}(k\rho_{iI}) e^{iN\phi_{iI}} \left\{ \tilde{\hat{n}}_i^* e^{i\phi_{iI}} + \tilde{\hat{n}}_i e^{-i\phi_{iI}} \right\}$$

$$- \frac{N}{\rho_{iI}} J_N(k\rho_{iI}) \tilde{\hat{n}}_i^* e^{i(N+1)\phi_{iI}}$$

$$= \frac{1}{2} k J_{N-1}(k\rho_{iI}) e^{i(N-1)\phi_{iI}} \tilde{\hat{n}}_i$$

$$- \frac{k}{2} \tilde{\hat{n}}_i^* e^{i(N+1)\phi_{iI}} \left\{ \frac{2N}{k\rho_{iI}} J_N(k\rho_{iI}) - J_{N-1}(k\rho_{iI}) \right\}$$

$$= \frac{k}{2} \tilde{\hat{n}}_i J_{N-1}(k\rho_{iI}) e^{i(N-1)\phi_{iI}} - \frac{k}{2} \tilde{\hat{n}}_i^* J_{N+1}(k\rho_{iI}) e^{i(N+1)\phi_{iI}} \quad (5.72)$$

By substituting (5.72) in (5.68), we have

$$\hat{n}_i \cdot \nabla_i \beta_{0N}^N (\boldsymbol{\rho}_{iI}, t_{iI})$$

$$= t_{iI}^{-|N|} \frac{k}{2} \left\{ \tilde{\hat{n}}_i J_{N-1}(k\rho_{iI}) e^{i(N-1)\phi_{iI}} - \tilde{\hat{n}}_i^* J_{N+1}(k\rho_{iI}) e^{i(N+1)\phi_{iI}} \right\}$$

$$= \frac{k}{2} t_{iI}^{-|N|} \left\{ \tilde{\hat{n}}_i^* J_{-1-N}(k\rho_{iI}) e^{i(N+1)\phi_{iI}} - \tilde{\hat{n}}_i J_{1-N}(k\rho_{iI}) e^{i(N-1)\phi_{iI}} \right\} \quad (5.73)$$

$$= \frac{k}{2} \left\{ \tilde{\hat{n}}_i^* t_{iI}^{|N+1|-|N|} \beta_{(-1)N}^N (\boldsymbol{\rho}_{iI}, t_{iI}) - \tilde{\hat{n}}_i t_{iI}^{|N-1|-|N|} \beta_{1N}^N (\boldsymbol{\rho}_{iI}, t_{iI}) \right\}$$

Similarly,

$$\begin{aligned}
&\hat{n}_j \cdot \nabla_j \beta_{M0}^N (\boldsymbol{\rho}_{Jj}, t_{Jj}) \\
&= t_{Jj}^{-|M|} \hat{n}_j \cdot \nabla_j \left\{ J_M (k\rho_{Jj}) e^{-iM\phi_{Jj}} \right\} \\
&= t_{Jj}^{-|M|} \hat{n}_j \cdot \nabla_j \left\{ J_{-M} (k\rho_{jJ}) e^{-iM\phi_{jJ}} \right\} \\
&= t_{Jj}^{-|M|} \frac{k}{2} \left\{ \tilde{\hat{n}}_j J_{-M-1}(k\rho_{jJ})e^{i(-M-1)\phi_{jJ}} - \tilde{\hat{n}}_j^* J_{-M+1}(k\rho_{jJ})e^{i(-M+1)\phi_{jJ}} \right\} \\
&= t_{Jj}^{-|M|} \frac{k}{2} \left\{ \tilde{\hat{n}}_j J_{M+1}(k\rho_{Jj})e^{-i(M+1)\phi_{Jj}} - \tilde{\hat{n}}_j^* J_{M-1}(k\rho_{Jj})e^{-i(M-1)\phi_{Jj}} \right\} \\
&= \frac{k}{2} \left\{ \tilde{\hat{n}}_j t_{Jj}^{|M+1|-|M|} \beta_{M(-1)}^N (\boldsymbol{\rho}_{Jj}, t_{Jj}) - \tilde{\hat{n}}_i^* t_{Jj}^{|M-1|-|M|} \beta_{M1}^N (\boldsymbol{\rho}_{Jj}, t_{Jj}) \right\}
\end{aligned}$$
(5.74)

By substituting (5.73) and (5.74) in (5.67), we have

$$\begin{aligned}
g_{ij}^{CHK} &= -\frac{1}{4} \left\{ \xi \frac{1}{k} (\hat{n}_i \cdot \nabla_i) + i(1-\xi) \right\} (\hat{n}_j \cdot \nabla_j) \\
&\sum_{N=-\infty}^{+\infty} \sum_{M=-\infty}^{+\infty} \beta_{0N}^N (\boldsymbol{\rho}_{iI}, t_{iI}) \left(\frac{t_{iI}}{t_{IJ}} \right)^{|N|} \\
&\alpha_{NM}^N (\boldsymbol{\rho}_{IJ}, t_{IJ}) \left(\frac{t_{Jj}}{t_{IJ}} \right)^{|M|} \beta_{M0}^N (\boldsymbol{\rho}_{Jj}, t_{Jj}) \\
&= \sum_{N=-\infty}^{+\infty} \sum_{M=-\infty}^{+\infty} \beta_{0N}^{NHD} (\boldsymbol{\rho}_{iI}, t_{iI}) \left(\frac{t_{iI}}{t_{IJ}} \right)^{|N|} \\
&\alpha_{NM}^N (\boldsymbol{\rho}_{IJ}, t_{IJ}) \left(\frac{t_{Jj}}{t_{IJ}} \right)^{|M|} \beta_{M0}^{NHA} (\boldsymbol{\rho}_{Jj}, t_{Jj})
\end{aligned}$$
(5.75)

where

$$\begin{aligned}
&\beta_{0N}^{NHD} (\boldsymbol{\rho}_{iI}, t_{iI}) \\
&= \frac{\xi}{2} \left\{ \tilde{\hat{n}}_i^* t_{iI}^{|N+1|-|N|} \beta_{(-1)N}^N (\boldsymbol{\rho}_{iI}, t_{iI}) - \tilde{\hat{n}}_i t_{iI}^{|N-1|-|N|} \beta_{1N}^N (\boldsymbol{\rho}_{iI}, t_{iI}) \right\} \\
&\quad + i(1-\xi)\beta_{0N}^N (\boldsymbol{\rho}_{iI}, t_{iI})
\end{aligned}$$
(5.76)

$$\begin{aligned}
&\beta_{M0}^{NHA} (\boldsymbol{\rho}_{Jj}, t_{Jj}) \\
&= -\frac{k}{8} \left\{ \tilde{\hat{n}}_j t_{Jj}^{|M+1|-|M|} \beta_{M(-1)}^N (\boldsymbol{\rho}_{Jj}, t_{Jj}) - \tilde{\hat{n}}_i^* t_{Jj}^{|M-1|-|M|} \beta_{M1}^N (\boldsymbol{\rho}_{Jj}, t_{Jj}) \right\}
\end{aligned}$$
(5.77)

By applying (5.75) in (5.33), we have

$$Z_{ij}^{CH} = \sum_{N=-\infty}^{+\infty} \sum_{M=-\infty}^{+\infty} \int_{C_i} dl_i \, J_{ti}(\boldsymbol{\rho}_i) \beta_{0N}^{NHD}(\boldsymbol{\rho}_{iI}, t_{iI}) \left(\frac{t_{iI}}{t_{IJ}}\right)^{|N|}$$
$$\alpha_{NM}^N (\boldsymbol{\rho}_{IJ}, t_{IJ}) \left(\frac{t_{Jj}}{t_{IJ}}\right)^{|M|} \int_{C_j} dl_j \, J_j(\boldsymbol{\rho}_j) \beta_{M0}^{NHA}(\boldsymbol{\rho}_{Jj}, t_{Jj})$$

(5.78)

As in (5.75), we have

$$\begin{aligned}
g_{ji}^{CHK} &= -\frac{1}{4} \left\{ \xi \frac{1}{k} (\hat{n}_j \cdot \nabla_j) + i(1-\xi) \right\} (\hat{n}_i \cdot \nabla_i) H_0^{(1)}(k\rho_{ji}) \\
&= -\frac{1}{4} \left\{ \xi \frac{1}{k} (\hat{n}_j \cdot \nabla_j) + i(1-\xi) \right\} (\hat{n}_i \cdot \nabla_i) H_0^{(1)}(k\rho_{ij}) \\
&= -\frac{1}{4} \left\{ \xi \frac{1}{k} (\hat{n}_j \cdot \nabla_j) + i(1-\xi) \right\} (\hat{n}_i \cdot \nabla_i) \\
&\quad \sum_{N=-\infty}^{+\infty} \sum_{M=-\infty}^{+\infty} \beta_{0N}^N (\boldsymbol{\rho}_{iI}, t_{iI}) \left(\frac{t_{iI}}{t_{IJ}}\right)^{|N|} \\
&\quad \alpha_{NM}^N (\boldsymbol{\rho}_{IJ}, t_{IJ}) \left(\frac{t_{Jj}}{t_{IJ}}\right)^{|M|} \beta_{M0}^N (\boldsymbol{\rho}_{Jj}, t_{Jj}) \\
&= \sum_{N=-\infty}^{+\infty} \sum_{M=-\infty}^{+\infty} \beta_{0N}^{NHDT} (\boldsymbol{\rho}_{iI}, t_{iI}) \left(\frac{t_{iI}}{t_{IJ}}\right)^{|N|} \\
&\quad \alpha_{NM}^N (\boldsymbol{\rho}_{IJ}, t_{IJ}) \left(\frac{t_{Jj}}{t_{IJ}}\right)^{|M|} \beta_{M0}^{NHAT} (\boldsymbol{\rho}_{Jj}, t_{Jj})
\end{aligned}$$

(5.79)

where

$$\beta_{0N}^{NHDT}(\boldsymbol{\rho}_{iI}, t_{iI}) = \\ -\frac{k}{8} \left\{ \tilde{n}_i^* t_{iI}^{|N+1|-|N|} \beta_{(-1)N}^N (\boldsymbol{\rho}_{iI}, t_{iI}) - \tilde{n}_i t_{iI}^{|N-1|-|N|} \beta_{1N}^N (\boldsymbol{\rho}_{iI}, t_{iI}) \right\}$$

(5.80)

$$\beta_{M0}^{NHAT}(\boldsymbol{\rho}_{Jj}, t_{Jj}) = \frac{\xi}{2} \left\{ \tilde{n}_j t_{Jj}^{|M+1|-|M|} \beta_{M(-1)}^N (\boldsymbol{\rho}_{Jj}, t_{Jj}) \right. \\ \left. - \tilde{n}_i^* t_{Jj}^{|M-1|-|M|} \beta_{M1}^N (\boldsymbol{\rho}_{Jj}, t_{Jj}) \right\} + i(1-\xi) \beta_{M0}^N (\boldsymbol{\rho}_{Jj}, t_{Jj})$$

(5.81)

By applying (5.79) in (5.33), we have

$$Z_{ji}^{CH} = \sum_{N=-\infty}^{+\infty} \sum_{M=-\infty}^{+\infty} \int_{C_i} dl_i\, J_i(\boldsymbol{\rho}_i) \beta_{0N}^{NHDT}(\boldsymbol{\rho}_{iI}, t_{iI}) \left(\frac{t_{iI}}{t_{IJ}}\right)^{|N|}$$
$$\alpha_{NM}^{N}(\boldsymbol{\rho}_{IJ}, t_{IJ}) \left(\frac{t_{Jj}}{t_{IJ}}\right)^{|M|} \int_{C_j} dl_j\, J_{tj}(\boldsymbol{\rho}_j) \beta_{M0}^{NHAT}(\boldsymbol{\rho}_{Jj}, t_{Jj})$$
(5.82)

Figure 5.6 shows the normalized induced current density on a square conducting cylinder at various frequencies by using the 2D LF-MLFMA and plain CG. The agreement between the 2D LF-MLFMA and the plain CG is excellent.

Figures 5.7 and 5.8 show the bistatic RCS of a conducting cylinder whose cross-section is an ogive for TE and TM polarization, respectively. Here, 50 segments per wavelength around the singular areas and 10 segments per wavelength in smooth areas are chosen to discretize the integral equation, and CFIE is used. The 2D LF-MLFMA is used first at lower levels in the singular areas and is then switched to the 2D diagonalized dynamic MLFMA for the dynamic case when the group sized is big enough to meet the requirement of the 2D diagonalized dynamic MLFMA. It is convenient to solve large-scale structures with nonuniform segments by combining the 2D LF-MLFMA and the 2D diagonalized dynamic MLFMA.

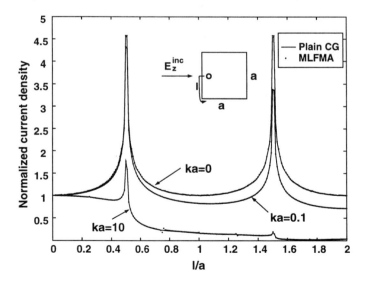

Figure 5.6 Normalized induced current on a square conducting cylinder immersed in a TM wave at different frequencies.

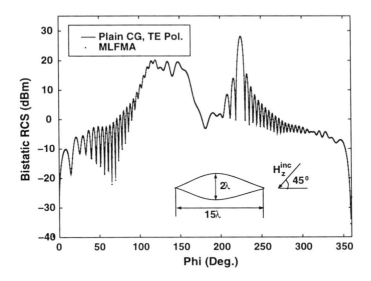

Figure 5.7 Comparison of the bistatic RCS between 2D MLFMA and plain CG method. The cross-section is an ogive, and CFIE is used with TE polarization.

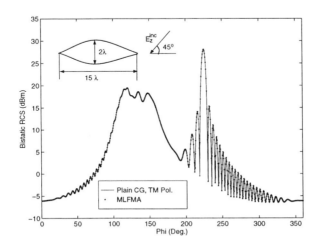

Figure 5.8 Comparison of the bistatic RCS between 2D MLFMA and plain CG method. The cross-section is an ogive, and CFIE is used with TM polarization.

5.3 3D MULTILEVEL FAST MULTIPOLE ALGORITHM AT VERY LOW FREQUENCIES

5.3.1 General Formulations for the 3D Dynamic MLFMA

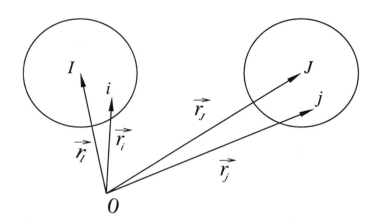

Figure 5.9 Source and field positions and their relative groups.

As shown in Figure 5.9, the core equation for the 3D undiagonalized dynamic MLFMA is [22, 27]

$$\alpha_{L,L'}(\mathbf{r}_{ji}) = \sum_{L_1} \sum_{L_2} \beta_{L,L_1}(\mathbf{r}_{jJ}) \alpha_{L_1,L_2}(\mathbf{r}_{JI}) \beta_{L_2,L'}(\mathbf{r}_{Ii}) \quad (5.83)$$

or in matrix form

$$\overline{\alpha}_{ji} = \overline{\beta}_{jJ} \cdot \overline{\alpha}_{JI} \cdot \overline{\beta}_{Ii} \quad (5.84)$$

where

$$\alpha_{L',L}(\mathbf{r}_{Ji}) = \sum_{L''} 4\pi i^{(l'+l''-l)} \Psi_{L''}(k, \mathbf{r}_{Ji}) A_{L,L',L''} \quad (5.85)$$

$$\beta_{L',L}(\mathbf{r}_{Ii}) = \sum_{L''} 4\pi i^{(l'+l''-l)} \Re g \Psi_{L''}(k, \mathbf{r}_{Ii}) A_{L,L',L''} \quad (5.86)$$

$$A_{L,L',L''} = \int d\Omega_k \, Y_L(\Omega_k) Y_{L'}^*(\Omega_k) Y_{L''}^*(\Omega_k) \quad (5.87)$$

$$L = (l, m), L' = (l', m'), L'' = (l'', m'') \quad (5.88)$$

$$\mathbf{r}_{ji} = \mathbf{r}_j - \mathbf{r}_i \quad (5.89)$$

$$\int d\Omega_k = \int_0^{2\pi} \int_0^{\pi} \sin\theta_k d\theta_k \, d\phi_k \tag{5.90}$$

and $\Psi_L(k, \mathbf{r})$ is a scalar wave function given by

$$\Psi_L(k, \mathbf{r}) = h_l^{(1)}(kr) Y_{lm}(\theta, \phi) \tag{5.91}$$

Here, k is the wavenumber in the Helmholtz equation, $h_l^{(1)}(x)$ is the spherical Hankel function of first kind, and $Y_{lm}(\theta, \phi)$ is the spherical harmonic function involving the Legendre function $P_l^m(\cos\theta)$ and $e^{im\phi}$ and can be expressed as [27]

$$Y_{lm}(\theta, \phi) = \sqrt{\frac{2l+1}{4\pi} \frac{(l-m)!}{(l+m)!}} P_l^m(\cos\theta) e^{im\phi}, \qquad m \geq 0 \tag{5.92}$$

The spherical harmonic function with minus order is defined as

$$Y_{l,(-m)}(\theta, \phi) = (-1)^m Y_{lm}(\theta, \phi) \tag{5.93}$$

The range of the summation on L' can be characterized as

$$l' > 0, \qquad -l' \leq m' \leq l' \tag{5.94}$$

The integral $A_{L,L',L''}$ appearing in (5.85) and (5.86) can be expressed in terms of Wigner 3-j symbols by applying the Gaunt coefficient [27]

$$\begin{aligned} A_{L,L',L''} &= \int d\Omega_k \, Y_L(\Omega_k) Y_{L'}^*(\Omega_k) Y_{L''}^*(\Omega_k) \\ &= (-1)^m \sqrt{\frac{1}{4\pi}(2l+1)(2l'+1)(2l''+1)} \\ &\quad \begin{pmatrix} l & l' & l'' \\ 0 & 0 & 0 \end{pmatrix} \begin{pmatrix} l & l' & l'' \\ -m & m' & m'' \end{pmatrix} \end{aligned} \tag{5.95}$$

where $L = (l, m)$, $L' = (l', m')$, $L'' = (l'', m'')$. Only considering the nonzero terms of the 3-j symbols, we can see that in (5.85) and (5.86),

$$|l - l'| \leq l'' \leq l + l' \tag{5.96}$$

where l'' increments by 2 and $m'' = m - m'$.

From similar discussions as the 2D case, we can see that the core equation (5.83) cannot be used directly in either the dynamic case or the very low frequency case. For the dynamic case, no reduction in the order of the computational complexity is achieved. For the very low frequency case, the computational complexity is reduced to $O(N)$. But the values of elements in translation matrices can easily exceed the floating point number range, which results in floating-point underflows or overflows.

5.3.2 Core Equation for 3D Diagonalized Dynamic MLFMA

By expressing spherical harmonics in terms of plane waves, the core equation of the 3D undiagnoalized dynamic MLFMA (5.83) can be diagonalized [22]:

$$\alpha_{L,L'}(\mathbf{r}_{ji}) = \int d\Omega_k \, i^l Y_L^*(\Omega_k) e^{i\mathbf{k}\cdot\mathbf{r}_{jJ}} \tilde{\alpha}_{JI}(\Omega_k) e^{i\mathbf{k}\cdot\mathbf{r}_{Ii}} Y_{L'}(\Omega_k) i^{-l'} \quad (5.97)$$

where

$$\tilde{\alpha}_{JI}(\Omega_k) = \sum_{L''} \Psi_{L''}(k, \mathbf{r}_{JI}) 4\pi i^{l''} Y_{L''}(\Omega_k) \quad (5.98)$$

After expressing the outgoing and incoming spherical waves in terms of plane waves, all translation matrices are diagonalized. The interpolation and anterpolation techniques are used in aggregation and disaggregation separately to solve the problem of unequal sampling rates at different levels. With the application of these techniques, the computational complexity is reduced to $O(N\log N)$ [21, 22] (see Chapter 3).

In (5.98), the double summation inherent in the index $L'' = (l'', m'')$ can be removed if one selects the z-axis to coincide with the direction of $\mathbf{r}_{JI}$ [10, 22]:

$$\tilde{\alpha}_{JI}(\Omega_k) = \sum_{l=0}^{L_{max}} i^l (2l+1) h_l^{(1)}(kr_{JI}) P_l(\hat{k} \cdot \hat{r}_{JI}) \quad (5.99)$$

The above form is expressed in a coordinate-independent form, and the argument of the Legendre polynomial depends on the angle between $\hat{k}$ and $\hat{r}_{JI}$.

From (5.97), we derive the transform pair between the amplitudes of the multipole plane waves and the amplitudes of the multipole spherical waves:

$$\begin{cases} \tilde{b}_i(\Omega_k) = \sum_{L_i} i^{-l_i} Y_{L_i}(\Omega_k) [b_i]_{L_i} \\ [b_i]_{L_i} = \int d\Omega_k \, i^{l_i} Y_{L_i}^*(\Omega_k) \tilde{b}_i(\Omega_k) \end{cases} \quad (5.100)$$

where $\tilde{b}_i(\Omega_k)$ are the amplitudes of the multipole plane waves, and $[b_i]_{L_i}$ are the amplitudes of the multipole spherical waves. Equation (5.100) can also be derived by the following procedures. The aggregation for spherical waves is

$$[b_I]_{L_I} = \sum_{L_i} \beta_{L_I, L_i}(\mathbf{r}_{Ii}) [b_i]_{L_i} \quad (5.101)$$

Substituting the plane wave expression of β_{L_I, L_i} into (5.101) leads to

$$\begin{aligned} [b_I]_{L_I} &= \sum_{L_i} i^{l_i - l_i} \int d\Omega_k \, Y_{L_I}^*(\Omega_k) Y_{L_i}(\Omega_k) e^{i\mathbf{k}\cdot\mathbf{r}_{Ii}} [b_i]_{L_i} \\ &= \int d\Omega_k \, i^{l_I} Y_{L_I}^*(\Omega_k) e^{i\mathbf{k}\cdot\mathbf{r}_{Ii}} \sum_{L_i} i^{-l_i} Y_{L_i}(\Omega_k) [b_i]_{L_i} \end{aligned} \quad (5.102)$$

If we define the transform pairs between the amplitudes of the plane waves and the spherical waves as in (5.100), the amplitudes of the plane waves have the following relationship:

$$\tilde{b}_I(\Omega_k) = e^{i\mathbf{k}\cdot\mathbf{r}_{Ii}}\tilde{b}_i(\Omega_k) \qquad (5.103)$$

This is just the diagonalized aggregation that we seek for the plane waves.

5.3.3 Core Equation for 3D LF-MLFMA

The summation in (5.98) is divergent if taken over an infinite range. The reason is that α_{L_1,L_2} becomes infinitely large when $|l_1 - l_2| \to \infty$. Therefore, the inner summation in (5.98) does not converge. To prevent this from happening, the summation in (5.98) should be truncated. But when the frequency is very low, this truncated summation is unstable. This makes the diagonalized form of FMM no longer useful at very low frequencies.

In the undiagonalized equation, (5.85) is used for α_{L_1,L_2}. In this case, $\alpha_{L_1,L_2} = 0$ when $|l_1 - l_2| > l''$. Therefore, the summations over L'' in (5.83) are automatically truncated.

The only problems are that the elements in $\overline{\beta}$ tend to zero and the elements in $\overline{\alpha}$ tend to infinity, making floating-point overflows appear during computing. To solve these problems, the normalization method for the 2D case can be applied directly here [26–28, 31].

For low frequencies ($k \to 0$) or small structures ($kr_{ji} \to 0$),

$$\alpha_{L,L'}(\mathbf{r}_{ji}) \approx O\left(\frac{1}{t^{l'+l+1}}\right) \qquad (5.104)$$

where t is a dimensionless parameter that satisfies $\frac{t}{kr_{ji}} \approx O(1)$. From (5.104), we can define the normalized $\alpha_{L,L'}(\mathbf{r}_{ji})$ as follows:

$$\alpha^N_{L,L'}(\mathbf{r}_{ji}) = t^{l'+l+1}\alpha_{L,L'}(\mathbf{r}_{ji}) \qquad (5.105)$$

which is of moderate value. Substituting (5.105) into (1), we have

$$\alpha^N_{L,L'}(\mathbf{r}_{ji}) = \sum_{L_1}\sum_{L_2} t^{l-l_1}\beta_{L,L_1}(\mathbf{r}_{jJ})\alpha^N_{L_1,L_2}(\mathbf{r}_{JI})t^{l'-l_2}\beta_{L_2,L'}(\mathbf{r}_{Ii}) \qquad (5.106)$$

According to (5.106), we define the normalized β as

$$\beta^N_{L^i,L^j}(\mathbf{r}) = t^{l^<-l^>}\beta_{L^i,L^j}(\mathbf{r}) \qquad (5.107)$$

where the superscript $<$ denotes the term at the lower level, and the superscript $>$ denotes the term at the higher level. With definitions in (5.105) and (5.107), we have

$$\alpha^N_{L,L'}(\mathbf{r}_{ji}) = \sum_{L_1}\sum_{L_2} \beta^N_{L,L_1}(\mathbf{r}_{jJ})\alpha^N_{L_1,L_2}(\mathbf{r}_{JI})\beta^N_{L_2,L'}(\mathbf{r}_{Ii}) \qquad (5.108)$$

This is the core equation of the 3D LF-MLFMA for very low frequency problems.

The frequency-dependent terms in $\overline{\alpha}$ and $\overline{\beta}$ are the spherical Bessel-type functions. We define the normalized spherical Bessel-type functions as

$$j_l^N(kr) = \frac{1}{t^l} j_l(kr) \tag{5.109}$$

$$h_l^{(1)N}(kr) = t^{l+1} h_l^{(1)}(kr) \tag{5.110}$$

which are all moderate values. The relative normalized scalar wave functions can be defined as

$$\Psi_L^N(k, \mathbf{r}) = h_l^{(1)N}(kr) Y_l^m(\theta, \phi) \tag{5.111}$$

and

$$\Re g \Psi_L^N(k, \mathbf{r}) = j_l^N(kr) Y_l^m(\theta, \phi) \tag{5.112}$$

Then, from (5.105) and (5.107), we have

$$\alpha_{L',L}^N(\mathbf{r}) = \sum_{L''} 4\pi i^{(l'+l''-l)} t^{l+l'-l''} \Psi_{L''}^N(k, \mathbf{r}) A_{L,L',L''} \tag{5.113}$$

$$\beta_{L',L}^N(\mathbf{r}) = \sum_{L''} 4\pi i^{(l'+l''-l)} t^{l^<-l^>+l''} \Re g \Psi_{L''}^N(k, \mathbf{r}) A_{L,L',L''} \tag{5.114}$$

With the normalized core equation of 3D LF-MLFMA (5.108), the normalized multipole amplitudes of the outgoing spherical wave at every level are

$$\left[\mathbf{b}^N\right]_L = \frac{1}{t^l} \left[\mathbf{b}\right]_L \tag{5.115}$$

and the normalized multipole amplitudes of the incoming spherical waves at every level are

$$\left[\mathbf{S}^N\right]_L = t^l \left[\mathbf{S}\right]_L \tag{5.116}$$

where vector **S** has the same definition as it is in (5.8) for 2D cases. Following the same proof for the 2D case, we can prove that all of the normalized amplitudes of the outgoing and incoming multipole spherical waves during computation are of moderate value.

From (5.107), we have

$$\left|\beta_{L^i,L^j}^N(\mathbf{r})\right| \approx t^{l^<-l^>+|l^<-l^>|} \leq 1 \tag{5.117}$$

Only floating-point underflows may appear during computation, and they are easily controlled as these underflow terms are exponential functions. Also, because the normalized multipole amplitudes at any level are of moderate value, $\left[\overline{\beta}^N\right]_L$ with very small value plays a trivial role. Ignoring elements which may cause a floating-point underflow will not decrease the precision of the final results. Therefore, the

3D LF-MLFMA is precise and the floating-point overflow during computing can be easily controlled.

In the above, in order to describe the algorithm simply and clearly, we choose a uniform normalization coefficient for all aggregations, translations, and disaggregations. For problems that have significantly different properties of characteristics on geometry or material properties at different subareas, adaptive meshing of these subareas yields a range of mesh sizes. Then, we need to choose different normalization coefficients at different subareas. In this case, we modify the uniformly normalized 3D LF-MLFMA to the nonuniformly normalized form of the 3D LF-MLFMA as

$$\alpha^N_{L,L'}(\mathbf{r}_{ji}, t_{ji}) = \sum_{L_1} \sum_{L_2} \left(\frac{t_{ji}}{t_{jJ}}\right)^l \beta^N_{L,L_1}(\mathbf{r}_{jJ}, t_{jJ}) \left(\frac{t_{jJ}}{t_{JI}}\right)^{l_1}$$
$$\cdot \frac{t_{ji}}{t_{JI}} \alpha^N_{L_1,L_2}(\mathbf{r}_{JI}, t_{JI}) \left(\frac{t_{Ii}}{t_{JI}}\right)^{l_2} \beta^N_{L_2,L'}(\mathbf{r}_{Ii}, t_{Ii}) \left(\frac{t_{ji}}{t_{Ii}}\right)^{l'} \quad (5.118)$$

where t_{ji}, t_{JI}, t_{Ii}, and t_{jJ}, respectively, are normalization coefficients for $\alpha^N_{L,L'}(\mathbf{r}_{ji}, t_{ji})$, $\alpha^N_{L_1,L_2}(\mathbf{r}_{JI}, t_{JI})$, $\beta^N_{L,L_1}(\mathbf{r}_{jJ}, t_{jJ})$, and $\beta^N_{L_2,L'}(\mathbf{r}_{Ii}, t_{Ii})$ and are indicated in brackets of these translators.

The 3D LF-MLFMA can be applied to solve a matrix equation

$$\overline{\mathbf{Z}} \cdot \mathbf{I} = \mathbf{V} \quad (5.119)$$

derived by integral equation and MOM for any kind of 3D surface or volume problem at very low frequencies. To apply the LF-MLFMA to solve (5.119), the unknowns need to be grouped first. At very low frequencies or static cases, the number of the multipole spherical harmonics can be chosen the same at different levels. The nonuniform partition method is applied here. The entire structure is first enclosed in a large cube, which is partitioned into eight smaller subcubes. If a subcube contains more unknowns than a given number which is used to determine the smallest cube size, it is recursively subdivided into smaller cubes. Otherwise, we stop the partition of this subcube. We continue this procedure until no subcube needs to be partitioned. The matrix-vector multiply is then written as

$$\overline{\mathbf{Z}} \cdot \mathbf{I} = \overline{\mathbf{Z}}_n \cdot \mathbf{I} + \overline{\mathbf{Z}}_f \cdot \mathbf{I} \quad (5.120)$$

Here, the first term on the right-hand side of (5.120) is the interaction from the nearby groups and is calculated directly. The second term is the interactions from the well-separated groups. For a large system, the second term is the most time-consuming part and is expedited by the LF-MLFMA.

To test the relative errors of 3D LF-MLFMA, we choose a very narrow conducting strip, which is 5 mm wide and 1m long. It is discretized into two sections in the narrow lateral and 400 sections along the strip. There are 399 loop basis functions and 1,599 tree basis functions for simulating the current distribution. The maximum number of unknowns in the smallest cube is chosen as 20, and six levels of tree structures for LF-MLFMA and static MLFMA are created. The harmonic expansions are truncated at $l = 5$. The interactions from separated groups $\overline{\mathbf{Z}}_f \cdot \mathbf{I}$ are calculated by the 3D LF-MLFMA and compared with the direct results with the assumption that all elements in $\mathbf{I}$ are 1. Figure 5.10 shows relative errors of $\overline{\mathbf{Z}}_f \cdot \mathbf{I}$ by applying LF-MLFMA compared with direct results at 10 MHz and 1 Hz. It can be seen that relative errors are very small and almost the same at different low frequencies. The error is even smaller if the near term is included in $\overline{\mathbf{Z}}$.

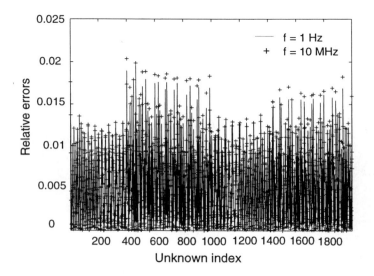

Figure 5.10 The relative errors of $\overline{\mathbf{Z}}_f \cdot \mathbf{I}$ by applying 3D LF-MLFMA compared with direct results at 10 MHz and 1 Hz. Here, all elements in $\mathbf{I}$ are assumed to be 1, and six-level LF-MLFMA is applied.

5.3.4 Computational Complexity of 3D LF-MLFMA

As with the 2D case, we can easily prove that the memory requirement, setup time, and workload per iteration all scale as $O(N)$, where N is the number of unknowns.

At very low frequencies, the matrix equation from the RWG surface basis is singular or nearly singular, and thus the RWG basis is no longer useful. The loop-tree or loop-star decomposition of the currents can yield at a near Helmholtz decomposition of the unknown currents at very low frequencies [34], (see Section 5.3.7). Therefore, the MOM matrix equation from loop-tree or loop star basis can be solved accurately and efficiently by iterative solvers. To further improve the convergence property, a basis rearrangement approach for tree basis is presented [35]. The memory requirement and the floating-point operations for applying the basis rearrangement all scale as $O(N)$, where N is the number of unknowns. Here, we apply the loop-tree basis in testing the computational complexity of the 3D LF-MLFMA.

Figures 5.11–5.13, respectively, show the setup time, the CPU time per iteration, and the memory requirements as functions of the number of unknowns for the 3D LF-MLFMA. The test object is a perfectly conducting sphere and the code runs on a DEC PC. As discussed above, the computational complexity and memory requirements of the 3D LF-MLFMA all scale as $O(N)$.

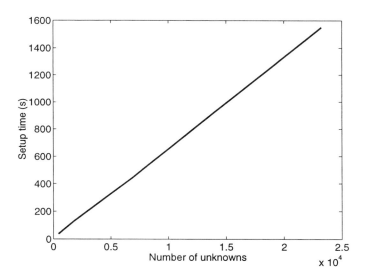

Figure 5.11 Setup time as a function of the number of unknowns for LF-MLFMA. The test object is a perfectly conducting sphere.

5.3.5 Core Equation for 3D Static MLFMA

When frequency tends to zero ($k \to 0$), (5.108) becomes the core equation of the 3D static MLFMA. By substituting (5.85) into (5.105), choosing the wavenumber k as the normalization factor, and replacing the Hankel function with its small argument

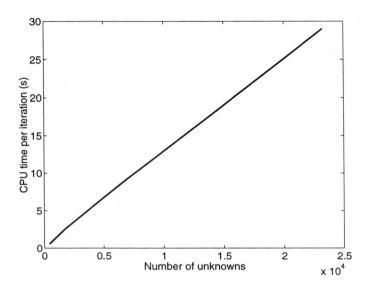

Figure 5.12 CPU time per iteration as a function of the number of unknowns for LF-MLFMA. The test object is a perfectly conducting sphere.

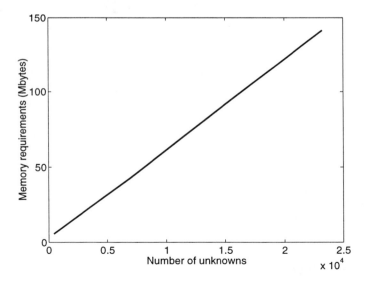

Figure 5.13 Memory requirements as a function of the number of unknowns for LF-MLFMA. The test object is a perfectly conducting sphere.

approximation, we have

$$\alpha_{L',L}^N(\mathbf{r}) = \sum_{l''=|l-l'|}^{l+l'} 4\pi i^{l'+l''-l} k^{l+l'+1} \left\{ -\frac{i(2l''-1)!!}{k^{l''+1} r^{l''+1}} \right\}$$
$$\sum_{m''=-l''}^{l''} Y_{L''}(\Omega) A_{L,L',L''} \tag{5.121}$$

Letting $k = 0$ in (5.121), we have the formulation for the static case as

$$\alpha_{L',L}^0(\mathbf{r}) = -(-1)^{l'} 4\pi i(2(l+l')-1)!! \frac{1}{r^{l'+l+1}}$$
$$\sum_{m''=-(l'+l)}^{l'+l} Y_{(l'+l),m''}(\Omega) A_{L,L',(l+l',m'')} \tag{5.122}$$
$$= -(-1)^{l'} 4\pi i(2(l+l')-1)!! \frac{1}{r^{l'+l+1}}$$
$$Y_{(l'+l),(m-m')}(\Omega) A_{L,L',(l+l',m-m')}$$

Here, the superscript 0 denotes the static case. Similarly, by substituting (5.86) into (5.107), choosing the wavenumber k as the normalization factor, and replacing the spherical Bessel function with its small argument approximation, we have

$$\beta_{L',L}^N(\mathbf{r}) = k^{l<-l>} \sum_{l''=|l-l'|}^{l+l'} 4\pi i^{l'+l''-l} \frac{k^{l''}}{(2l''+1)!!} r^{l''}$$
$$\sum_{m''=-l''}^{l''} Y_{l''m''}(\Omega) A_{L,L',L''} \tag{5.123}$$

Letting $k = 0$ in (5.123) gives

$$\beta_{L',L}^0(\mathbf{r}) = \begin{cases} 4\pi i^{(l'-l)+(l>-l<)} \dfrac{r^{l>-l<}}{(2(l>-l<)+1)!!} Y_{(l>-l<)(m-m')}(\Omega) \\ \qquad A_{L,L',(l>-l<,m-m')}, \qquad |m-m'| \leq l>-l< \\ 0, \qquad \text{otherwise} \end{cases} \tag{5.124}$$

Therefore, the core equation for the 3D static MLFMA is

$$\alpha_{L,L'}^0(\mathbf{r}_{ji}) = \sum_{L_1} \sum_{L_2} \beta_{L,L_1}^0(\mathbf{r}_{jJ}) \alpha_{L_1,L_2}^0(\mathbf{r}_{JI}) \beta_{L_2,L'}^0(\mathbf{r}_{Ii}) \tag{5.125}$$

The derivation is similar to that in [31]. Equation (5.125) is essentially the same as that in [31, 36]. But the new representation of (5.125) for the 3D static MLFMA is more explicit and succinct than those in [29] and is easier to implement.

Figure 5.14 shows relative errors of the separated interactions $\overline{\mathbf{Z}}_f \cdot \mathbf{I}$ by applying the static MLFMA compared with the direct results. The testing object and its discretization are the same as in Figure 5.10. It has 1,600 pulse basis functions to simulate the charge densities. The maximum number of unknowns in the smallest cube is chosen as 20, and six levels of tree structures for static MLFMA are created. The harmonic expansions are truncated at $l = 5$. The relative errors for the statics case are always smaller than for the low-frequency cases. This stems from the need to use Bessel functions in LF-MLFMA, which can be programmed to a finite accuracy, whereas the static MLFMA requires no special function.

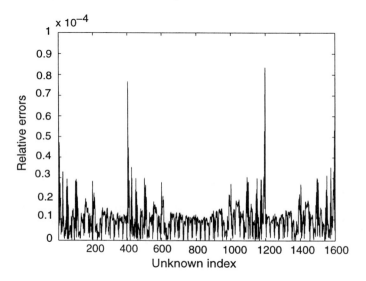

Figure 5.14 The relative errors of $\overline{\mathbf{Z}}_f \cdot \mathbf{I}$ by applying the static MLFMA compared with direct results for static case. Here, all elements in $\mathbf{I}$ are assumed to be 1, and six-level static MLFMA is applied.

5.3.6 Rotation of the Translation Matrices for 3D LF-MLFMA and 3D Static MLFMA

The normalized translation matrices $\overline{\alpha}^N$ and $\overline{\beta}^N$ are in general represented by full matrices. The coordinate system rotation technique [29, 37] can also be used here directly to save storage. The coordinate system is rotated to make the translation direction point in the z direction in the new coordinate system and then is rotated back to the original coordinate system. With these procedures, the workloads are of the same order, but the storage is reduced [29].

An arbitrary rotation of a coordinate axis can be achieved by three rotations performed successively by performing a rotation α about the z axis, a rotation β about the y axis of the resulting coordinate frame, and a rotation γ about the z axis of the latter coordinate frame. The three angles α, β, and γ are known as the Euler angles. If the matrix representation of an arbitrary rotation of spherical harmonics is denoted by $\overline{\mathbf{D}}(\alpha, \beta, \gamma)$, we can write [29, 30, 37, 38]

$$\overline{\mathbf{D}}(\alpha, \beta, \gamma) = \overline{\mathbf{D}}(0, 0, \gamma) \cdot \overline{\mathbf{D}}(0, \beta, 0) \cdot \overline{\mathbf{D}}(\alpha, 0, 0) \quad (5.126)$$

For our problems, $\overline{\mathbf{D}}(0, 0, \gamma)$ and $\overline{\mathbf{D}}(\alpha, 0, 0)$ are diagonal matrices with diagonal elements being $e^{im\gamma}$ and $e^{im\alpha}$, respectively. The desired rotations that bring the z axis to the translation direction can be achieved by choosing $\alpha = \phi$, $\beta = \theta$, and $\gamma = 0$, where θ and ϕ are the spherical angles corresponding to the translation direction.

$\overline{\mathbf{D}}(0, \beta, 0)$ is usually a dense matrix. It can be simplified by expressing it in terms of a rotation about the z axis as shown in [29, 30, 37, 38]:

$$\overline{\mathbf{D}}(0, \beta, 0) = \overline{\mathbf{D}}(-\frac{\pi}{2}, 0, 0) \cdot \overline{\mathbf{D}}(0, -\frac{\pi}{2}, 0) \\ \cdot \overline{\mathbf{D}}(\beta, 0, 0) \cdot \overline{\mathbf{D}}(0, \frac{\pi}{2}, 0) \cdot \overline{\mathbf{D}}(\frac{\pi}{2}, 0, 0) \quad (5.127)$$

This reduces the problem of computing the matrix representation of any rotation to that of computing the one matrix $\overline{\mathbf{S}}(\frac{\pi}{2}) = \overline{\mathbf{D}}(0, \frac{\pi}{2}, 0) \cdot \overline{\mathbf{D}}(\frac{\pi}{2}, 0, 0)$, which is block diagonal. The use of $\overline{\mathbf{S}}$ matrices does not decrease the number of operations required for multiplication, but it does reduce the memory requirements since only a single $\overline{\mathbf{S}}$ matrix is required for all rotations.

With the definitions above, a normalized translation $\overline{\tau}(\mathbf{r})$ $\left(\overline{\alpha}^N(\mathbf{r}) \text{ or } \overline{\beta}^N(\mathbf{r})\right)$ with arbitrary direction $\mathbf{r}$ can be written as

$$\overline{\tau}(\mathbf{r}(r, \theta, \phi)) = \overline{\mathbf{D}}(-\phi, 0, 0) \cdot \overline{\mathbf{D}}(0, -\theta, 0) \cdot \overline{\tau}(r\hat{z}) \cdot \overline{\mathbf{D}}(0, \theta, 0) \cdot \overline{\mathbf{D}}(\phi, 0, 0) \quad (5.128)$$

where $\overline{\tau}(r\hat{z})$ is the matrix representation of the z axis translation with many of zero elements.

When $\mathbf{r}$ points to the z axis, the harmonics functions in (5.113) and (5.114) are nonzero only for $L'' = (l'', 0)$. From (5.95), the second 3-j symbol is nonzero only when $m'' = m - m'$. Therefore, the translation matrices $\overline{\alpha}^N(r\hat{z})$ and $\overline{\beta}^N(r\hat{z})$ are nonzero only when $m = m'$. From (5.113) and (5.114), we obtain that the matrix elements of $\overline{\alpha}^N$ and $\overline{\beta}^N$ for a translation along the z axis are

$$\alpha^N_{L',L}(r\hat{z}) = \begin{cases} \sum_{l''} 4\pi i^{(l'+l''-l)} t^{l+l'-l''} \Psi^N_{(l'',0)}(k, r\hat{z}) A_{L,L',(l'',0)}, \\ \qquad\qquad\qquad\qquad\qquad\qquad\qquad\qquad m' = m \\ 0, \qquad\qquad\qquad\qquad\qquad\qquad\qquad\qquad m' \neq m \end{cases} \quad (5.129)$$

$$\beta_{L',L}^N(r\hat{z}) = \begin{cases} \sum_{l''} 4\pi i^{(l'+l''-l)} t^{l^<-l^>+l''} \Re g \Psi_{(l'',0)}^N(k,r\hat{z}) A_{L,L',(l'',0)}, & m'=m \\ 0, & m' \neq m \end{cases} \tag{5.130}$$

Many of the elements of $\overline{\alpha}^N(r\hat{z})$ and $\overline{\beta}^N(r\hat{z})$ are zero. For the nonzero elements, by applying the identities of Clebsch-Gordon coefficients [33]

$$(j_1 j_2 m_1 m_2 | j_1 j_2 jm) = (-1)^{j_1+j_2-j} (j_1 j_2 - m_1 - m_2 | j_1 j_2 j - m) \tag{5.131}$$

and

$$(j_1 j_2 m_1 m_2 | j_1 j_2 jm) = (-1)^{j_1+j_2-j} (j_2 j_1 m_1 m_2 | j_2 j_1 jm) \tag{5.132}$$

we obtain the symmetrical properties of the nonzero elements of matrices $\overline{\alpha}^N(r\hat{z})$ and $\overline{\beta}^N(r\hat{z})$,

$$\alpha_{(l',m)(l,m)}^N(r\hat{z}) = \alpha_{(l',-m)(l,-m)}^N(r\hat{z}) \tag{5.133}$$

$$\alpha_{(l',m)(l,m)}^N(r\hat{z}) = (-1)^{l-l'} \alpha_{(l,m)(l',m)}^N(r\hat{z}) \tag{5.134}$$

and

$$\frac{1}{t^{l_L^<-l_L^>}} \beta_{(l',m)(l,m)}^N(r\hat{z}) = \frac{1}{t^{l_R^<-l_R^>}} \beta_{(l',-m)(l,-m)}^N(r\hat{z}) \tag{5.135}$$

$$\frac{1}{t^{l_L^<-l_L^>}} \beta_{(l',m)(l,m)}^N(r\hat{z}) = (-1)^{l-l'} \frac{1}{t^{l_R^<-l_R^>}} \beta_{(l,m)(l',m)}^N(r\hat{z}) \tag{5.136}$$

where the subscripts L and R represent the terms related to the translation matrix elements in the left-hand side and the right-hand side of (5.135) or (5.136), respectively.

The translation matrices along the z axis have many zero elements. The nonzero elements are subject to the symmetrical properties (5.133)–(5.136). Therefore, only a very small portion of the elements need to be stored. In addition, in (5.128), only one translation matrix needs to be generated for all translations with the same distance but different directions. Therefore, applying matrix rotations decreases the storage significantly.

Using the notations in [38], the nonzero elements of $\overline{\mathbf{D}}(0, \beta, 0)$ are

$$
\begin{aligned}
d^{(l)}_{m'm}(\beta) &= D(0,\beta,0)_{(l,m')(l,m)} \\
&= \sum_{m''=-l}^{l} e^{-im'\frac{\pi}{2}} d^{(l)}_{m''m'}\left(\frac{\pi}{2}\right) e^{im''\beta} d^{(l)}_{m''m}\left(\frac{\pi}{2}\right) e^{im\frac{\pi}{2}} \\
&= \sum_{m''=-l}^{l} e^{i(m-m')\frac{\pi}{2}} d^{(l)}_{m''m'}\left(\frac{\pi}{2}\right) d^{(l)}_{m''m}\left(\frac{\pi}{2}\right) e^{im''\beta} \\
&= \begin{cases}
D_0 + 2\sum_{m''=1}^{l} d^{(l)}_{m''m'}\left(\frac{\pi}{2}\right) d^{(l)}_{m''m}\left(\frac{\pi}{2}\right) \cos(m''\beta), & M(m',m) = 0 \\
-2\sum_{m''=1}^{l} d^{(l)}_{m''m'}\left(\frac{\pi}{2}\right) d^{(l)}_{m''m}\left(\frac{\pi}{2}\right) \sin(m''\beta), & M(m',m) = 1 \\
-D_0 - 2\sum_{m''=1}^{l} d^{(l)}_{m''m'}\left(\frac{\pi}{2}\right) d^{(l)}_{m''m}\left(\frac{\pi}{2}\right) \cos(m''\beta), & M(m',m) = 2 \\
2\sum_{m''=1}^{l} d^{(l)}_{m''m'}\left(\frac{\pi}{2}\right) d^{(l)}_{m''m}\left(\frac{\pi}{2}\right) \sin(m''\beta), & M(m',m) = 3
\end{cases}
\end{aligned}
$$
(5.137)

where

$$
D_0 = \begin{cases} d^{(l)}_{0m'}\left(\frac{\pi}{2}\right) d^{(l)}_{0m}\left(\frac{\pi}{2}\right), & \mathrm{mod}(l,2) = \mathrm{mod}(m,2) \\ 0, & \mathrm{mod}(l,2) \neq \mathrm{mod}(m,2) \end{cases}
$$
(5.138)

$$M(m',m) = \mathrm{mod}(4 + \mathrm{mod}(m-m',4),4) \tag{5.139}$$

There are typographical errors in [38], and they are corrected in (5.137). The value of $d^{(l)}_{m''m}\left(\frac{\pi}{2}\right)$ ($m \neq l$) can be calculated recursively. The recursion relation for the $d^{(l)}_{m''m}(\beta)$ is [38]

$$
d^{(l)}_{m''m}(\beta) = \left(\frac{l-m''}{l-m}\right)^{\frac{1}{2}} d^{(l-\frac{1}{2})}_{(m'+\frac{1}{2})(m+\frac{1}{2})}(\beta) \cdot \cos\frac{\beta}{2} \\
+ \left(\frac{l+m''}{l-m}\right)^{\frac{1}{2}} d^{(l-\frac{1}{2})}_{(m''-\frac{1}{2})(m+\frac{1}{2})}(\beta) \cdot \sin\frac{\beta}{2}, \qquad m \neq l
$$
(5.140)

The sign error in [38] before the second term in the right-hand side of (5.140) is already corrected here. When $m = l$,

$$d_{m''l}^{(l)}(\beta) = (-1)^{l-m''} \sqrt{\frac{(2l)!}{(l+m'')!(l-m'')!}} \left(\cos\frac{\beta}{2}\right)^{l+m''} \left(\sin\frac{\beta}{2}\right)^{l-m''} \tag{5.141}$$

Some lower order values of $d_{m''m}^{(l)}\left(\frac{\pi}{2}\right)$ are tabulated in [38].

By substituting (5.137) into (5.128), we have

$$\tau_{L',L}(\mathbf{r}(r,\theta,\phi))$$

$$= \sum_{m''=-min(l',l)}^{min(l',l)} e^{-im'\phi} d_{m'm''}^{(l')}(-\theta) \tau_{(l',m'')(l,m'')}(r\hat{z}) d_{m''m}^{(l)}(\theta) e^{im\phi}$$

$$= e^{i(m-m')\phi} \sum_{m''=-min(l',l)}^{min(l',l)} d_{m''m'}^{(l')}(\theta) \tau_{(l',m'')(l,m'')}(r\hat{z}) d_{m''m}^{(l)}(\theta) \tag{5.142}$$

$$= e^{i(m-m')\phi} \Big\{ d_{0m'}^{(l')}(\theta) \tau_{(l',0)(l,0)}(r\hat{z}) d_{0m}^{(l)}(\theta)$$

$$+ \sum_{m''=1}^{min(l',l)} \left[d_{m''m'}^{(l')}(\theta) d_{m''m}^{(l)}(\theta) + d_{-m''m'}^{(l')}(\theta) d_{-m''m}^{(l)}(\theta) \right]$$

$$\cdot \tau_{(l',m'')(l,m'')}(r\hat{z}) \Big\}$$

From (5.122) and (5.124), we obtain the translation matrices along the z direction for the static case as

$$\alpha_{L',L}^0(r\hat{z}) \begin{cases} -(-1)^{l'} 4\pi i(2(l+l')-1)!! \frac{1}{r^{l'+l+1}} Y_{(l'+l),0}(\Omega) \\ \quad \cdot A_{L,L',(l+l',0)}, & m=m' \\ 0, & m \neq m' \end{cases} \tag{5.143}$$

$$\beta_{L',L}^0(r\hat{z}) = \begin{cases} 4\pi i^{(l'-l)+(l^>-l^<)} \frac{r^{l^>-l^<}}{(2(l^>-l^<)+1)!!} Y_{(l^>-l^<),0}(\Omega) \\ \quad \cdot A_{L,L',(l^>-l^<,0)}, & m=m' \text{ and } l^> \geq l^< \\ 0, & \text{otherwise} \end{cases} \tag{5.144}$$

The symmetrical properties (5.133) and (5.134) are also applicable for the static case; however, (5.135) and (5.136) are not. For the static case, the symmetrical properties become

$$\alpha_{(l',m)(l,m)}^0(r\hat{z}) = \alpha_{(l',-m)(l,-m)}^0(r\hat{z}) \tag{5.145}$$

$$\alpha^0_{(l',m)(l,m)}(r\hat{z}) = (-1)^{l-l'} \alpha^0_{(l,m)(l',m)}(r\hat{z}) \tag{5.146}$$

and

$$\beta^0_{(l',m)(l,m)}(r\hat{z}) = \beta^0_{(l',-m)(l,-m)}(r\hat{z}), \qquad l^<_L - l^>_L = l^<_R - l^>_R \tag{5.147}$$

The saving of storage by using a matrix rotation technique has been thoroughly studied in [29, 37]. The application of symmetrical properties of translations will further decrease the memory requirements. Here we only test the errors introduced by the matrix rotation. We choose a very narrow conducting strip, which is 5 mm wide and 1m long and is discretized into two sections in the narrow lateral and 400 sections along the strip. There are 399 loop basis functions and 1,599 tree basis functions for simulating the current distribution at very low frequencies and 1,600 pulse basis functions to simulate the charge densities for the static case. Under the assumption that all elements in $\mathbf{I}$ are 1, the interactions between separated terms $\overline{\mathbf{Z}}_f \cdot \mathbf{I}$ are calculated by direct method and the 3D LF-MLFMA or static MLFMA with and without matrix rotation. Here, the six-level MLFMA is applied and harmonic expansions are truncated at $l = 5$ for both low-frequency and static cases. Figure 5.15 shows relative errors of $\overline{\mathbf{Z}}_f \cdot \mathbf{I}$ by using the 3D LF-MLFMA with matrix rotation compared with direct results at 10 MHz and 1 Hz. Figure 5.16 shows relative errors of results from the 3D LF-MLFMA with matrix rotation compared with results from the plain 3D LF-MLFMA at 10 MHz and 1 Hz. It can be seen that relative errors are almost the same at different low frequencies, and the order of numerical errors from the matrix rotation is always much smaller than the order of errors from the plain LF-MLFMA for fixed harmonic expansion truncation. Figure 5.17 shows relative errors of $\overline{\mathbf{Z}}_f \cdot \mathbf{I}$ by using the static MLFMA with matrix rotation compared with direct results. Figure 5.18 shows relative errors of results from the static MLFMA with matrix rotation compared with the results from the plain static MLFMA. As with the low-frequency case, numerical errors from the matrix rotation are always smaller than the order of errors from the plain static MLFMA for fixed harmonic expansion truncation. Therefore, the implementation of matrix rotation techniques does not increase the error order for both low-frequency and static cases.

5.3.7 3D LF-MLFMA Based on RWG Basis

By using the loop-tree basis designed for low-frequency problems

$$\mathbf{J}(\mathbf{r'}) = \mathbf{J}^t_L(\mathbf{r'}) \cdot \mathbf{I}_L + \mathbf{J}^t_T(\mathbf{r'}) \cdot \mathbf{I}_T \tag{5.148}$$

and the Galerkin testing procedure and applying $\nabla \cdot \mathbf{J}_L(\mathbf{r}) = 0$, we have the matrix equation:

$$\begin{bmatrix} \overline{\mathbf{Z}}_{LL} & \overline{\mathbf{Z}}_{LT} \\ \overline{\mathbf{Z}}_{TL} & \overline{\mathbf{Z}}_{TT} \end{bmatrix} \cdot \begin{bmatrix} \mathbf{I}_L \\ \mathbf{I}_T \end{bmatrix} = \begin{bmatrix} \mathbf{V}_L \\ \mathbf{V}_T \end{bmatrix} \tag{5.149}$$

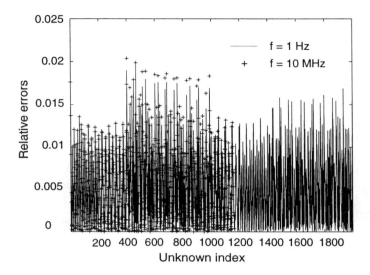

Figure 5.15 The relative errors of $\overline{\mathbf{Z}}_f \cdot \mathbf{I}$ by using LF-MLFMA with matrix rotation compared to direct results at low frequencies of 10 MHz and 1 Hz. Here, all elements in $\mathbf{I}$ are assumed to be 1. The six-level LF-MLFMA is applied.

Equation (5.149) is the matrix equation based on the loop-tree basis for solving very low-frequency problems. The application of loop-tree basis remedies the low-frequency breakdown for RWG-type basis. Here, we use RWG-type basis to mean the RWG basis, wire basis, and wire-surface basis [39]. To further improve the spectral property of the matrix equation, a basis rearrangement approach is developed to improve the spectral property of the impedance matrix of loop-tree basis [35].

For plane-wave excitation, when the frequency is very low, the elements of the impedance matrix, the excitation vector, and the results scale with respect to the frequency ω as following

$$\begin{bmatrix} \overline{\mathbf{Z}}_{LL}(O(\omega)) & \overline{\mathbf{Z}}_{LT}(O(\omega)) \\ \overline{\mathbf{Z}}_{TL}(O(\omega)) & \overline{\mathbf{Z}}_{TT}(O(\frac{1}{\omega})) \end{bmatrix} \begin{bmatrix} \mathbf{I}_L(O(1)) \\ \mathbf{I}_T(O(\omega)) \end{bmatrix} = \begin{bmatrix} \mathbf{V}_L(O(\omega)) \\ \mathbf{V}_T(O(1)) \end{bmatrix} \quad (5.150)$$

where the expressions in the brackets are the scaling properties of the matrix elements at very low frequencies or when $\omega \to 0$. The matrix equation in (5.150) is unbalanced when $\omega \to 0$. This imbalance stems from the fact that we use an electric field integral equation, but the inductive part produces predominantly a magnetic field, with a subdominant electric field. Hence, the upper left hand block of the matrix becomes subdominant in an electric field equation. To remedy this, (5.150) can be frequency

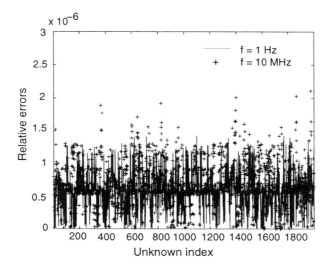

Figure 5.16 The relative errors of $\overline{\mathbf{Z}}_f \cdot \mathbf{I}$ by applying LF-MLFMA with matrix rotation compared with results of plain LF-MLFMA at low frequencies of 10 MHz and 1 Hz. Here, all elements in $\mathbf{I}$ are assumed to be 1. The six-level LF-MLFMA is applied.

normalized in a balanced manner as

$$\begin{bmatrix} \frac{1}{i\omega}\overline{\mathbf{Z}}_{LL}(O(1)) & \overline{\mathbf{Z}}_{LT}(O(\omega)) \\ \overline{\mathbf{Z}}_{TL}(O(\omega)) & i\omega\overline{\mathbf{Z}}_{TT}(O(1)) \end{bmatrix} \begin{bmatrix} \mathbf{I}_L(O(1)) \\ \frac{1}{i\omega}\mathbf{I}_T(O(1)) \end{bmatrix} = \begin{bmatrix} \frac{1}{i\omega}\mathbf{V}_L(O(1)) \\ \mathbf{V}_T(O(1)) \end{bmatrix} \quad (5.151)$$

After the normalization, $\frac{1}{\omega}\overline{\mathbf{Z}}_{LL}$, $\omega\overline{\mathbf{Z}}_{TT}$, $\frac{1}{\omega}\mathbf{V}_L$, and $\mathbf{V}_T$ remain finite, $\overline{\mathbf{Z}}_{LT} = \overline{\mathbf{Z}}_{TL}^t \to 0$ when $\omega \to 0$.

When using iterative solvers to solve (5.151), to further improve the convergence property, we can apply diagonal or block diagonal preconditioner on it:

$$\begin{bmatrix} \overline{\mathbf{P}}_{LL} & \mathbf{0} \\ \mathbf{0} & \overline{\mathbf{P}}_{TT} \end{bmatrix} \begin{bmatrix} \frac{1}{i\omega}\overline{\mathbf{Z}}_{LL} & \overline{\mathbf{Z}}_{LT} \\ \overline{\mathbf{Z}}_{TL} & i\omega\overline{\mathbf{Z}}_{TT} \end{bmatrix} \begin{bmatrix} \mathbf{I}_L \\ \frac{1}{i\omega}\mathbf{I}_T \end{bmatrix} = \begin{bmatrix} \frac{1}{i\omega}\overline{\mathbf{P}}_{LL} \cdot V_L \\ \overline{\mathbf{P}}_{TT} \cdot V_T \end{bmatrix} \quad (5.152)$$

where $\overline{\mathbf{P}}_{LL}$ and $\overline{\mathbf{P}}_{TT}$ are the diagonal or block diagonal preconditioner for $\frac{1}{i\omega}\overline{\mathbf{Z}}_{LL}$ and $i\omega\overline{\mathbf{Z}}_{TT}$, respectively.

For a large linear system, the traditional matrix equation solver cannot be used directly due to limited computer resources. LF-MLFMA is an efficient method to solve large linear systems without compromising precision [28]. But for complex structures, some global loop basis functions are involved, such as the global loop basis function around a hole, the global wire-surface loop basis function for wire-surface loop, the global loop basis function for a strip band whose two ends are all connected

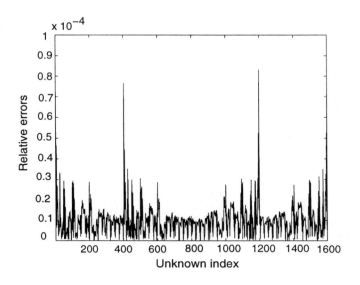

Figure 5.17 The relative errors of $\overline{\mathbf{Z}}_f \cdot \mathbf{I}$ by applying static MLFMA with matrix rotation compared to direct results. Here, all elements in $\mathbf{I}$ are assumed to be 1. The six-level static MLFMA is applied.

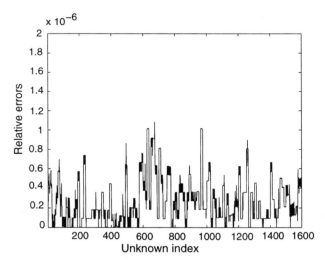

Figure 5.18 The relative errors of $\overline{\mathbf{Z}}_f \cdot \mathbf{I}$ by the static MLFMA with matrix rotation compared with results by the plain static MLFMA. Here, all elements in $\mathbf{I}$ are assumed to be 1. The six-level static MLFMA is applied.

directly or indirectly to a patch surface or wire, the global loop basis function for closed surface with toruses, and the global wire loop basis function. If we apply LF-MLFMA directly based on the loop-tree basis, the group size at the finest level is limited by the size of global loop basis functions. This may make LF-MLFMA inefficient and inapplicable. If there are only a few global loop basis functions, we can extract the matrix-vector multiplication that is associated with these global loop basis functions and evaluate it directly. The rest of the matrix-vector multiplication is expedited by LF-MLFMA. But for integrated circuits problems, a preponderance of global loop basis functions may be involved and some of the global loop basis functions will be very large. Hence, the method of extracting the global loop basis functions is inefficient and therefore cannot be applied. Here, we apply LF-MLFMA based on small local basis transformation instead of using loop-tree basis directly. We can apply LF-MLFMA based on integration points [40], patch/segment, or RWG-type basis. Because all the loop-tree basis functions are superpositions of RWG-type basis functions, we apply LF-MLFMA based on RWG-type basis functions here [39].

Global loop basis functions, local loop basis functions, and tree basis functions all can be represented as the superposition of RWG-type basis functions:

$$\mathbf{J}_L(\mathbf{r}) = \overline{\mathbf{F}}_{LR} \cdot \mathbf{J}_{RWG}(\mathbf{r}) \tag{5.153}$$

$$\mathbf{J}_T(\mathbf{r}) = \overline{\mathbf{F}}_{TR} \cdot \mathbf{J}_{RWG}(\mathbf{r}) \tag{5.154}$$

where $\overline{\mathbf{F}}_{LR}$ and $\overline{\mathbf{F}}_{TR}$ are connection matrices between loop and tree basis functions with RWG-type basis functions. Here, the subscript RWG denotes terms that are associated with RWG-type basis functions. Equations (5.153) and (5.154) can be combined as

$$\begin{bmatrix} \mathbf{J}_L(\mathbf{r}) \\ \mathbf{J}_T(\mathbf{r}) \end{bmatrix} = \overline{\mathbf{F}}_{LTR} \cdot \mathbf{J}_{RWG}(\mathbf{r}) \tag{5.155}$$

where

$$\overline{\mathbf{F}}_{LTR} = \begin{bmatrix} \overline{\mathbf{F}}_{LR} \\ \overline{\mathbf{F}}_{TR} \end{bmatrix} \tag{5.156}$$

By applying (5.153) in matrix elements in (5.149), we have

$$\begin{aligned} \overline{\mathbf{Z}}_{LL} &= i\omega\mu\langle \mathbf{J}_L(\mathbf{r}), g(\mathbf{r},\mathbf{r}'), \mathbf{J}_L^t(\mathbf{r}')\rangle \\ &= i\omega\mu\langle \overline{\mathbf{F}}_{LR} \cdot \mathbf{J}_{RWG}(\mathbf{r}), g(\mathbf{r},\mathbf{r}'), \mathbf{J}_{RWG}^t(\mathbf{r}') \cdot \overline{\mathbf{F}}_{LR}^t \rangle \\ &= \overline{\mathbf{F}}_{LR} \cdot \overline{\mathbf{Z}}_{RWG}^V \cdot \overline{\mathbf{F}}_{LR}^t \end{aligned} \tag{5.157}$$

where

$$\overline{\mathbf{Z}}_{RWG}^V = i\omega\mu\langle \mathbf{J}_{RWG}(\mathbf{r}), g(\mathbf{r},\mathbf{r}'), \mathbf{J}_{RWG}^t(\mathbf{r}')\rangle \tag{5.158}$$

and

$$\langle \mathbf{A}(\mathbf{r}), g(\mathbf{r},\mathbf{r}'), \mathbf{B}(\mathbf{r}')\rangle = \int d\mathbf{r}\,\mathbf{A}(\mathbf{r}) \cdot \int d\mathbf{r}'\, g(\mathbf{r},\mathbf{r}')\mathbf{B}(\mathbf{r}') \tag{5.159}$$

Similarly, we have

$$\overline{\mathbf{Z}}_{LT} = \overline{\mathbf{F}}_{LR} \cdot \overline{\mathbf{Z}}_{RWG}^V \cdot \overline{\mathbf{F}}_{TR}^t \quad (5.160)$$

$$\overline{\mathbf{Z}}_{TL} = \overline{\mathbf{F}}_{TR} \cdot \overline{\mathbf{Z}}_{RWG}^V \cdot \overline{\mathbf{F}}_{LR}^t \quad (5.161)$$

$$\overline{\mathbf{Z}}_{TT} = \overline{\mathbf{F}}_{TT} \cdot \left(\overline{\mathbf{Z}}_{RWG}^V + \overline{\mathbf{Z}}_{RWG}^S \right) \cdot \overline{\mathbf{F}}_{TT}^t \quad (5.162)$$

where

$$\overline{\mathbf{Z}}_{RWG}^S = -\frac{i}{\omega\epsilon} \langle \nabla \cdot \mathbf{J}_{RWG}(\mathbf{r}), g(\mathbf{r},\mathbf{r}'), \nabla' \cdot \mathbf{J}_{RWG}^t(\mathbf{r}') \rangle \quad (5.163)$$

From (5.157), (5.160)–(5.162), we have

$$\begin{bmatrix} \frac{1}{i\omega}\overline{\mathbf{Z}}_{LL} & \overline{\mathbf{Z}}_{LT} \\ \overline{\mathbf{Z}}_{TL} & i\omega\overline{\mathbf{Z}}_{TT} \end{bmatrix}$$
$$= \begin{bmatrix} \frac{1}{i\omega}\overline{\mathbf{F}}_{LR} \cdot \overline{\mathbf{Z}}_{RWG}^V \cdot \overline{\mathbf{F}}_{LR}^t & \overline{\mathbf{F}}_{LR} \cdot \overline{\mathbf{Z}}_{RWG}^V \cdot \overline{\mathbf{F}}_{TR}^t \\ \overline{\mathbf{F}}_{TR} \cdot \overline{\mathbf{Z}}_{RWG}^V \cdot \overline{\mathbf{F}}_{LR}^t & i\omega\overline{\mathbf{F}}_{TR} \cdot \left(\overline{\mathbf{Z}}_{RWG}^V + \overline{\mathbf{Z}}_{RWG}^S\right) \cdot \overline{\mathbf{F}}_{TR}^t \end{bmatrix} \quad (5.164)$$

Define

$$\overline{\mathbf{F}}'_{LR} = \frac{1}{\sqrt{i\omega}} \overline{\mathbf{F}}_{LR} \quad (5.165)$$

$$\overline{\mathbf{F}}'_{TR} = \sqrt{i\omega} \overline{\mathbf{F}}_{TR} \quad (5.166)$$

$$\overline{\mathbf{F}}'_{LTR} = \begin{bmatrix} \overline{\mathbf{F}}'_{LR} \\ \overline{\mathbf{F}}'_{TR} \end{bmatrix} \quad (5.167)$$

Substituting (5.165)–(5.167) in (5.164) leads to

$$\begin{bmatrix} \frac{1}{i\omega}\overline{\mathbf{Z}}_{LL} & \overline{\mathbf{Z}}_{LT} \\ \overline{\mathbf{Z}}_{TL} & i\omega\overline{\mathbf{Z}}_{TT} \end{bmatrix} = \overline{\mathbf{F}}'_{LTR} \cdot \overline{\mathbf{Z}}_{RWG}^V \cdot \overline{\mathbf{F}}'^t_{LTR} + \begin{bmatrix} \overline{\mathbf{0}} & \overline{\mathbf{0}} \\ \overline{\mathbf{0}} & \overline{\mathbf{F}}'_{TR} \cdot \overline{\mathbf{Z}}_{RWG}^S \cdot \overline{\mathbf{F}}'^t_{TR} \end{bmatrix} \quad (5.168)$$

By applying (5.168) in (5.152), we have

$$\begin{bmatrix} \overline{\mathbf{P}}_{LL} & \overline{\mathbf{0}} \\ \overline{\mathbf{0}} & \overline{\mathbf{P}}_{TT} \end{bmatrix} \left\{ \overline{\mathbf{F}}'_{LTR} \cdot \overline{\mathbf{Z}}_{RWG}^V \cdot \overline{\mathbf{F}}'^t_{LTR} + \begin{bmatrix} \overline{\mathbf{0}} & \overline{\mathbf{0}} \\ \overline{\mathbf{0}} & \overline{\mathbf{F}}'_{TR} \cdot \overline{\mathbf{Z}}_{RWG}^S \cdot \overline{\mathbf{F}}'^t_{TR} \end{bmatrix} \right\}$$
$$\begin{bmatrix} \mathbf{I}_L \\ \frac{1}{i\omega}\mathbf{I}_T \end{bmatrix} = \begin{bmatrix} \frac{1}{i\omega}\overline{\mathbf{P}}_{LL} \cdot V_L \\ \overline{\mathbf{P}}_{TT} \cdot V_T \end{bmatrix} \quad (5.169)$$

Denote

$$\overline{\mathbf{Z}}' = \begin{bmatrix} \overline{\mathbf{P}}_{LL} & \overline{\mathbf{0}} \\ \overline{\mathbf{0}} & \overline{\mathbf{P}}_{TT} \end{bmatrix}$$
$$\left\{ \overline{\mathbf{F}}'_{LTR} \cdot \overline{\mathbf{Z}}_{RWG}^V \cdot \overline{\mathbf{F}}'^t_{LTR} + \begin{bmatrix} \overline{\mathbf{0}} & \overline{\mathbf{0}} \\ \overline{\mathbf{0}} & \overline{\mathbf{F}}'_{TR} \cdot \overline{\mathbf{Z}}_{RWG}^S \cdot \overline{\mathbf{F}}'^t_{TR} \end{bmatrix} \right\} \quad (5.170)$$

The transpose of matrix $\overline{\mathbf{Z}}'$ is

$$\overline{\mathbf{Z}}'^t = \left\{ \overline{\mathbf{F}}'_{LTR} \cdot \overline{\mathbf{Z}}^V_{RWG} \cdot \overline{\mathbf{F}}'^t_{LTR} + \begin{bmatrix} \overline{\mathbf{0}} & \overline{\mathbf{0}} \\ \overline{\mathbf{0}} & \overline{\mathbf{F}}'_{TR} \cdot \overline{\mathbf{Z}}^S_{RWG} \cdot \overline{\mathbf{F}}'^t_{TR} \end{bmatrix} \right\} \begin{bmatrix} \overline{\mathbf{P}}^t_{LL} & \overline{\mathbf{0}} \\ \overline{\mathbf{0}} & \overline{\mathbf{P}}^t_{TT} \end{bmatrix} \quad (5.171)$$

By using the notation of (5.170), (5.169) becomes

$$\overline{\mathbf{Z}}' \cdot \begin{bmatrix} \mathbf{I}_L \\ \frac{1}{i\omega}\mathbf{I}_T \end{bmatrix} = \begin{bmatrix} \frac{1}{i\omega}\overline{\mathbf{P}}_{LL} \cdot V_L \\ \overline{\mathbf{P}}_{TT} \cdot V_T \end{bmatrix} \quad (5.172)$$

When an iterative solver is used to solve the matrix equation (5.171), in each iterative step, we need to multiply the matrix $\overline{\mathbf{Z}}'$ or its conjugate transpose by a vector, the trial solution. The most time-consuming parts are the multiplication of dense matrix $\overline{\mathbf{Z}}^V_{RWG}$ and $\overline{\mathbf{Z}}^S_{RWG}$ with vectors. This multiplication can be expedited by LF-MLFMA. Here, the basis is the RWG-type basis. No large global loop basis function is involved now. Therefore, LF-MLFMA can be performed efficiently.

By applying the above techniques, the inductance and input susceptance of a 12-turn spiral inductor are calculated. The radius of the wire is 2.76 μm. The track spacing is 24 μm. The wire spiral inductor is 250 μm above and connected to a ground plate. The ground plate is 800×800 μm^2. The structure of the spiral inductor is shown in Figure 5.19. The input susceptance of the spiral as a function of frequency is shown in Figure 5.20. The rate of change of the susceptance with frequency is always negative. This characteristic meets the Foster's reactance/susceptance theorem. The inductance as a function of frequency is shown in Figure 5.21. At very low frequencies, the capacitance is very small, and the spiral inductor can be considered as a pure inductor. But when the frequency becomes higher, the effect of the parasitic capacitance becomes larger. The structure is resonant at high frequencies due to the presence of parasitic capacitance in the inductor loop.

By using the loop-tree basis for low-frequency cases and RWG-type basis for high-frequency cases, we can solve electromagnetic problems from zero frequency to microwave frequencies. Here, the input impedance of a Hertzian dipole is calculated. The so-called Hertzian dipole comprises two conducting spheres connected by a length of wire, and fed at the center of the wire. We first use the loop-tree basis at very low frequencies and then switch over to the RWG basis when frequency is high. At very low frequencies, the Hertzian dipole resembles a series LC network. The input impedance is

$$Z_{in} = -i\omega L - \frac{1}{i\omega C} \quad (5.173)$$

where L is the inductance, and C is the mutual capacitance between the two spheres. Because the inductance of the wire is very small, at very low frequencies, the input

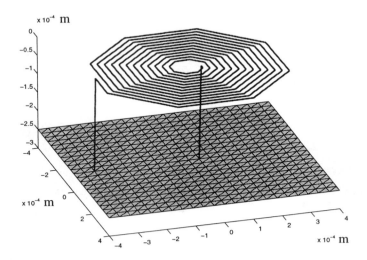

Figure 5.19 Structure of the wire spiral inductor connected to a ground plate.

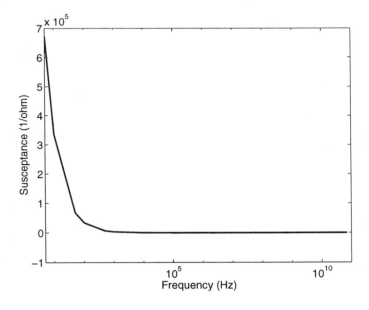

Figure 5.20 Input susceptance of the wire spiral inductor versus frequency.

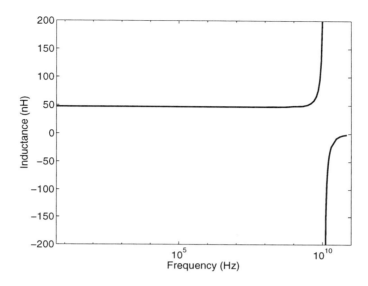

Figure 5.21 Inductance of the wire spiral inductor versus frequency.

impedance is approximated by

$$Z_{in} \approx -\frac{1}{i\omega C} \quad (5.174)$$

Figure 5.22 shows the comparison of the reactance from plain CG, LF-MLFMA, and the capacitor approximation from 1 Hz to 10 MHz. The loop-tree basis is used in the LF-MLFMA and the plain CG method. Figure 5.23 shows the input resistance and reactance from 10 to 100 MHz. We use loop-tree basis at frequencies near 10 MHz and then switch to the RWG basis. Only plain CG method is used in this frequency range. We observe that the electromagnetic physics is correctly captured by the solution method. At low frequencies, the Hertzian dipole correctly behaves like a capacitor, while at higher frequencies, it behaves like an LC tank circuit.

5.4 CONCLUSIONS

In this chapter, the LF-MLFMA in 2D and 3D with a computational complexity of $O(N)$ for very low-frequency problems is developed. The LF-MLFMA can be used not only independently for very low-frequency cases, but also to solve large-scale structures with rapidly varying areas when merged with a dynamic algorithm described in Chapters 2 and 3. A more explicit and succinct representation of the

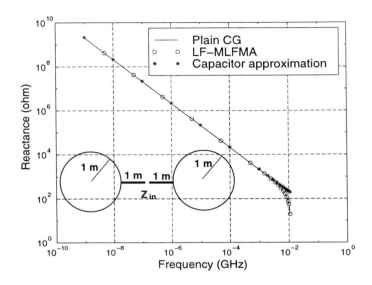

Figure 5.22 Input reactance of a Hertzian dipole at very low frequencies. Loop-tree basis is used.

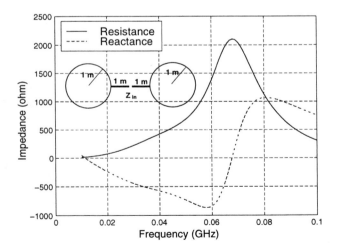

Figure 5.23 Input resistance and reactance from the plain CG method. Loop-tree and RWG bases are separately used at different frequencies.

3D static MLFMA is also derived. The matrix rotation techniques is applied to the 3D LF-MLFMA to save storage. Some symmetries of the translation matrices along the z direction for the very low-frequency case and the static case are also derived to further reduce the memory requirements. A loop-tree based method that is also efficient for iterative solvers is described.

REFERENCES

1. R. F. Harrington, *Field Computation by Moment Methods*, Marlabar, FL: Krieger, 1982.

2. R. L. Wagner and W. C. Chew, "A study of wavelets for the solution of electromagnetic integral equations," *IEEE Trans. Antennas Propagat.*, vol. 43, pp. 802–810, Aug. 1995.

3. F. X. Canning, "Solution of IML form of moment method problems in 5 iterations," *Radio Sci.*, vol. 30, no. 5, pp. 1371–1384, Sept.–Oct. 1995.

4. E. Michielssen and A. Boag, "Multilevel evaluation of electromagnetic fields for the rapid solution of scattering problems," *Microwave Opt. Tech. Lett.*, vol. 7, no. 17, pp. 790–795, Dec. 5, 1994.

5. A. Boag and R. Mittra, "Complex multipole beam approach to electromagnetic scattering problems," *IEEE Trans. Antennas Propagat.*, vol. 42, pp. 366–372, March 1994.

6. B. Hu, W. C. Chew, E. Michielssen, and J. Zhao, "Fast inhomogenous plane wave algorithm for the fast analysis of two-dimensional scattering problems," *Radio Science*, vol. 34, no. 4, pp. 759–772, July–Aug. 1999.

7. B. Hu and W. C. Chew "Fast inhomogeneous plane wave algorithm for multilayered medium problems," *IEEE Geosci. Remote Sensing*, accepted for publication, 2001.

8. J. Carrier, L. Greengard, and V. Rokhlin, "A fast adaptive multipole algorithm for particle simulations," *SIAM J. Sci. Stat. Comput.*, vol. 9, no. 4, pp. 669–686, July 1988.

9. V. Rokhlin, "Rapid solution of integral equations of scattering theory in two dimensions," *J. Comput. Phys.*, vol. 35, no. 2, pp. 414–439, 1990.

10. R. Coifman, V. Rokhlin, and S. Wandzura, "The fast multipole method for the wave equation: A pedestrian prescription," *IEEE Ant. Propag. Mag.*, vol. 35, no. 3, pp. 7–12, 1993.

11. V. Rokhlin, "Diagonal forms of translation operators for the Helmholtz equation in three dimensions," *Appl. Comp. Harmon. Anal.*, vol. 1, pp. 82–93, 1993.

12. C. C. Lu and W. C. Chew, "A fast algorithm for solving hybrid integral equation," *IEE Proc.-H*, vol. 140, no. 6, pp. 455–460, Dec. 1993.

13. R. Coifman, V. Rokhlin, and S. Wandzura, "Faster single-stage multipole method for the wave equation," *10th Annual Review of Progress in Applied Computational Electromagnetics*, pp. 19–24, 1994.

14. R. L. Wagner and W. C. Chew, "A ray-propagation fast multipole algorithm," *Microwave Opt. Tech. Lett.*, vol. 7, no. 10, pp. 435–438, 1994.

15. J. M. Song and W. C. Chew, "Fast multipole method solution using parametric geometry," *Microwave Opt. Tech. Lett.*, vol. 7, no. 16, pp. 760–765, 1994.

16. K. Nobors, S. Kim, and J. White, "Fast capacitance extraction of general three-dimensional structures," *IEEE Trans. Microwave Theory Tech.*, vol. 40, no. 7, pp. 1496–1506, July 1992.

17. M. Epton and B. Dembart, "Multipole translation theory for the three-dimensional Laplace and Helmholtz equations," *SIAM J. Sci. Comput.*, vol. 16, no. 4, pp. 865–897, 1995.

18. B. Dembart and E. Yip, "A 3D fast multipole method for electromagnetics with multiple levels," *11th Annual Review of Progress in Applied Computational Electromagnetics*, p. 621, 1995.

19. A. Brandt, "Multilevel computations of integral transforms and particle interactions with oscillatory kernels," *Compt. Phys. Commun.*, vol. 65, pp. 24–38, 1991.

20. C. C. Lu, and W. C. Chew, "A multilevel algorithm for solving a boundary integral equation of wave scattering," *Micro. Opt. Tech. Lett.*, vol. 7, no. 10, pp. 456–470, July 1994.

21. J. M. Song and W. C. Chew, "Multilevel fast-multipole algorithm for solving combined field integral equations of electromagnetic scattering," *Microwave Opt. Tech. Lett.*, vol. 10, no. 1, pp. 14–19, 1995.

22. W. C. Chew, S. Koc, J. M. Song, C. C. Lu, and E. Michielssen, "A succinct way to diagonalize the translation matrix in three dimensions," *Microwave Opt. Tech. Lett.*, vol. 15, no. 3, pp. 144–147, 1997.

23. J. M. Song, C. C. Lu, and W. C. Chew, "Multilevel fast multipole algorithm for electromagnetic scattering by large complex objects," *IEEE Trans. Antennas Propagat.*, vol. 45, no. 10, pp. 1488–1493, Oct. 1997.

24. Z. Wang, Y. Yuan, and Q. Wu, "A parallel multipole accelerated 3-D capacitance simulator based on an improved model," *IEEE Trans. Computer-Aided Design*, vol. 15, no. 12, pp. 1441–1450, Dec. 1996.

25. G. Hoyler and R. Unbehauen, "The fast multipole method for EMC problem," *Electrical Engineering* (Germany), vol. 80, no. 6, pp. 403–411, Dec. 1997.

26. J. S. Zhao and W. C. Chew, "MLFMA for solving boundary equations of 2D electromagnetic scattering from static to electrodynamic," *Microwave Opt. Tech. Lett.*, vol. 20, no. 5, pp. 306–311, March 1999.

27. W. C. Chew, *Waves and Fields in Inhomogeneous Media*, New York: Van Nostrand Reinhold, 1990; reprinted by IEEE Press, 1995.

28. J. S. Zhao and W. C. Chew, "Three dimensional multilevel fast multipole algorithm from static to electrodynamic," *Microwave Opt. Tech. Lett.*, vol. 26, no. 1, pp. 43–48, 2000.

29. L. Greengard and V. Rokhlin, "A new version of the fast multipole method for the Laplace equation in three dimensions," research report YALE, EUDCSRR-1115, September 1996.

30. J. S. Zhao and W. C. Chew, "Applying matrix rotation to the three-dimensional low-frequency multilevel fast multipole algorithm," *Microwave Opt. Tech. Lett.*, vol. 26, no. 2, pp. 105–110, 2000.

31. W. C. Chew, J. A. Friedrich, and Robert Geiger, "A multiple scattering solution for the effective permittivity of a sphere mixture," *IEEE Geosci. Remote Sensing*, vol. 28, no. 2, pp. 207–214, March 1990.

32. G. A. Watson, *A Treatise on the Theory of Bessel Functions*, Second Edition, Cambridge, U.K.: Cambridge University Press, 1958.

33. M. Abramowitz and I. A. Stegun, *Handbook of Mathematical Functions*, New York: Dover Publications, Inc., 1972.

34. D. R. Wilton and A. W. Glisson, "On improving the electric field integral equation at low frequencies," *URSI Radio Science Meeting Digest*, p. 24, Los Angeles, CA, June 1981.

35. J. S. Zhao and W. C. Chew, "Integral equation solution of Maxwell's equations from zero frequency to microwave frequencies," *IEEE Trans. Antennas Propagat.*, James R. Wait Memorial Special Issue, vol. 48, no. 10, pp. 1635–1645, Oct. 2000.

36. J. Dull, K. Gallivan, J. M. Song, and W. C. Chew, "Parallel fast multipole capacitance solver," *IEEE Antennas and Propagation Society International Symposium*, Atlanta, GA, vol. 3, pp. 1766–1769, 1998.

37. S. Koc and W. C. Chew, "Calculation of acoustical scattering from a cluster of scatterers," *J. Acoust. Soc. Am.*, vol. 103, no. 2, pp. 721–734, Feb. 1998.

38. A. R. Edmonds, *Angular Momentum in Quantum Mechanics*, Princeton, NJ: Princeton University Press, 1957.

39. J. S. Zhao and W. C. Chew, "Applying LF-MLFMA to solve complex PEC structures," *Microwave Opt. Tech. Lett.*, vol. 28, no. 3, pp. 155-160, 2001.

40. K. C. Donepudi, J. M. Jin, S. Velamparambil, J. M. Song, and W. C. Chew, "A higher-order parallelized multilevel fast multipole algorithm for 3D scattering," research report CCEM-36-99, Nov. 2, 1999.

6

Error Analysis of Surface Integral Equation Methods

Karl F. Warnick and Weng Cho Chew

6.1 INTRODUCTION

Numerical methods based on surface integral equations have enjoyed widespread use in computational electromagnetics for many years, since the introduction of the method of moments in the early days of the field of computational electromagnetics [1]. The recent development of techniques for fast evaluation of interactions has greatly extended the frontier of problems that can be analyzed using integral equation solvers [2]. In this chapter, we survey results on the error analysis of this approach to numerical simulation of electromagnetics problems.

An understanding of the convergence behavior of a numerical method requires an estimate or bound on the final solution error in terms of parameters of the algorithm and physical properties of the problem to be solved. If the solution is obtained using an iterative method, then a complete understanding of the numerical method requires iteration count estimates as well. We obtain solution error and iteration count estimates for the method of moments applied to several canonical two- and three-dimensional scattering problems. These results illustrate the relationship between electromagnetic effects and numerical performance and provide insights into the behavior of numerical methods as commonly observed in computational electromagnetics.

6.1.1 Surface Integral Equations and the Method of Moments

For conducting or dielectric bodies, Maxwell's equations and the boundary conditions on the electromagnetic fields at the surface of the scatterer can be cast into an equivalent system of surface integral equations. These equations are based on integral operators that relate an unknown, equivalent current on the surface of the body to the fields scattered in response to a given incident field. In this chapter, we consider only the special case of the electric field integral equation (EFIE) applied to a perfectly electrically conducting object, although the approach used here could be modified to apply to other integral equation formulations and types of scatterers.

Let S denote the surface of a perfectly electrically conducting (PEC) body, and $\mathbf{J}$ the surface current on S induced by an incident field $\mathbf{E}^{\text{inc}}$. The incident field $\mathbf{E}^{\text{inc}}$ is the total field which would exist if the scatterer were not present. The surface current $\mathbf{J}$ in the absence of the scatterer would radiate the scattered field $\mathbf{E}^{\text{sc}}$ such that $\mathbf{E} = \mathbf{E}^{\text{inc}} + \mathbf{E}^{\text{sc}}$. The EFIE is of the form

$$\hat{n} \times \mathcal{T}\mathbf{J} = \hat{n} \times \mathbf{E}^{\text{inc}} \tag{6.1}$$

where the integral operator is [3, 4]

$$\mathcal{T}\mathbf{J} = -ik_0\eta \int_S ds'\, g(\mathbf{r},\mathbf{r}')\mathbf{J}(\mathbf{r}') - \frac{i\eta}{k_0}\nabla \int_S ds'\, g(\mathbf{r},\mathbf{r}')\nabla' \cdot \mathbf{J}(\mathbf{r}') \tag{6.2}$$

This definition differs from that of [4] in that it does not include the cross product with the surface normal vector $\hat{n}$. In this expression, $\eta = \sqrt{\mu/\epsilon}$ is the characteristic impedance of the medium surrounding the scatterer, ϵ and μ are the permittivity and permeability of the medium, k_0 is the wavenumber of the time harmonic incident field, and the singular kernel is the free space Green's function

$$g(\mathbf{r},\mathbf{r}') = \frac{e^{ik_0 R}}{4\pi R} \tag{6.3}$$

where $R = |\mathbf{r} - \mathbf{r}'|$.

For a translationally invariant body, the three-dimensional scattering problem reduces to a pair of two-dimensional Helmholtz boundary value problems. If the incident electric field is in the invariant direction, the problem is transverse magnetic (TM), and the EFIE reduces to a scalar integral equation of the form

$$\mathcal{L}J = E^{\text{inc}} \tag{6.4}$$

where $J(\rho)$ is the longitudinal component of the surface current vector and the integral operator is

$$\mathcal{L}J = \frac{k_0 \eta}{4} \int_D ds' \, H_0^{(1)}(k_0|\rho - \rho'|) J(\rho') \tag{6.5}$$

For 2D problems, D denotes an arc in the plane, or a longitudinal cross-section of the 3D scatterer surface S.

For the transverse electric (TE) polarization, the integral operator becomes

$$\begin{aligned}\mathcal{N}J &= \frac{k_0\eta}{4}\hat{t}\cdot\int_D ds'\,H_0^{(1)}(k_0|\rho-\rho'|)\hat{t}'J(\rho')\\ &\quad +\frac{1}{4k_0\eta}\hat{t}\cdot\nabla\int_D ds'\,H_0^{(1)}(k_0|\rho-\rho'|)\nabla'\cdot[\hat{t}'J(\rho')]\end{aligned} \tag{6.6}$$

where $J(\rho)$ is the tangential component of the surface current vector, and $\hat{t}$ is a unit vector tangential to D and transverse to the invariant direction.

The surface integral equations of electromagnetics can be solved approximately by projecting the operator onto a finite-dimensional subspace. This yields a linear system of equations for the weights corresponding to the unknown surface current on the scatterer. This approach is based on an approximate represention of the current as a linear combination of a finite set of basis functions. For the scalar or 2D case, this is

$$J(x) \simeq \hat{J}(x) = \sum_{n=1}^{N} I_n f_n(x) \tag{6.7}$$

where x is a parameter for the curve D, the $f_n(x)$, $n = 1, \ldots, N$ are expansion functions, and the I_n are weights for the approximate current solution. The span of the f_n is the trial subspace in which the numerical solution lies. The basis functions are generally defined on a discrete geometrical representation, or mesh, for the scatterer surface. The average length of the elements of the mesh is the discretization length h. The dimensionless discretization density or unknowns per wavelength $n_\lambda = \lambda/h$ can also be used to characterize the fineness of the mesh.

Common types of basis functions are the delta function (collocation), piecewise constant (pulse), and piecewise linear (triangle):

$$f(x) = \delta(x) \tag{6.8}$$

$$f(x) = \begin{cases} 1 & -h/2 \leq x \leq h/2 \\ 0 & |x| > h/2 \end{cases} \tag{6.9}$$

$$f(x) = \begin{cases} 1 - (x/h)^2 & -h \leq x \leq h \\ 0 & |x| > h \end{cases} \tag{6.10}$$

If the mesh is regular, so that each element is of length h, then the basis functions are of the form $f_n(x) = f(x - x_n)$, where x_n is the nth node of the mesh. In the 3D case, f_n becomes a vector function defined on surface patches [5].

Substituting (6.7) into (6.4) and testing the integral equation with another set of functions $t_n(x)$ produces the linear system

$$\sum_n Z_{mn} I_n = V_m \qquad (6.11)$$

where

$$Z_{mn} = h^{-1} \int dx\, t_m(x) \mathcal{L} f_n(x) \qquad (6.12)$$

$$V_n = h^{-1} \int dx\, t_m(x) E^{\text{inc}}(x) \qquad (6.13)$$

By solving (6.11), an approximate solution $\hat{J}$ for the current on the scatterer can be obtained. This can be done by direct factorization of the moment matrix $\overline{Z}$ or by making use of an iterative linear system solution algorithm.

This type of discretization is known as the method of moments (MOM) or the Galerkin-Petrov method [6]. If the trial and testing subspaces are identical, then MOM reduces to Galerkin's method. In other communities, the method of moments for surface integral equations with basis and testing functions defined on mesh elements is known as the boundary element method (BEM).

Throughout this chapter, the hat on a quantity such as $\hat{J}$ denotes an approximate value obtained from the method of moments. The hat is also used for unit vectors, but the meaning of the notation should be clear from context.

6.1.2 Error Measures

For smooth scatterers, if no sources lie on the scatterer surface, then the surface current is bounded and hence square integrable. In this case, the L^2 error of the current $\hat{\mathbf{J}}$ with respect to the exact current $\mathbf{J}$ is in principle a meaningful error measure, which could be used to assess the accuracy of a numerical method. The relative L^2 error is

$$\begin{aligned} \text{Err}_{L^2} &= \frac{\|\hat{\mathbf{J}} - \mathbf{J}\|}{\|\mathbf{J}\|} \\ &= \left[\int_S ds\, \left|\hat{\mathbf{J}}(\mathbf{r}) - \mathbf{J}(\mathbf{r})\right|^2 \right]^{1/2} \left[\int_S ds\, |\mathbf{J}(\mathbf{r})|^2 \right]^{-1/2} \end{aligned} \qquad (6.14)$$

The relative root mean square (RMS) error is also of interest:

$$\text{Err}_{\text{RMS}} = \left[\sum_{n=1}^N \left|\hat{\mathbf{J}}_n - \mathbf{J}_n\right|^2 \right]^{1/2} \left[\sum_{n=1}^N |\mathbf{J}_n|^2 \right]^{-1/2} \qquad (6.15)$$

where $\mathbf{J}_n$ is the surface current density vector evaluated at the nth node point of the surface mesh.

One drawback of the current error is that the L^2 norm breaks down in some cases. If the scatterer has an edge or corner, then the current is in general singular, and the L^2 norm may not exist. In this case, the current error for singular regions of the scatterer must be studied using a Sobolev norm, as discussed in Section 6.1.3.1.

Another approach is to consider the error in a scattering amplitude or scattering cross-section. These scalar quantities are often the final desired result of a numerical simulation. Let the incident electric field be a plane wave of the form

$$\mathbf{E}^{\text{inc}}(\mathbf{r}) = \mathbf{E}_0 e^{i\mathbf{k}^{\text{inc}} \cdot \mathbf{r}} \tag{6.16}$$

where the constant vector $\mathbf{E}_0$ gives the polarization and magnitude of the field, and $\mathbf{k}^{\text{inc}}$ is the wavevector. The scattering amplitude S is then defined by [7]

$$E^{\text{sc}}(\hat{k}^{\text{sc}}, \hat{k}^{\text{inc}}) \to \frac{e^{ikr}}{kr} S(\hat{k}^{\text{sc}}, \hat{k}^{\text{inc}}) E^{\text{inc}}, \quad r \to \infty \tag{6.17}$$

and gives the amplitude of the outgoing plane wave propagating in the direction $\hat{k}^{\text{sc}}$. The scattering cross-section or radar cross-section (RCS) is $\sigma(\hat{k}^{\text{sc}}, \hat{k}^{\text{inc}}) = (4\pi/k_0^2)|S|^2$.

For 2D problems, the scattering amplitude satisfies the definition

$$E^{\text{sc}}(\phi^{\text{sc}}, \phi^{\text{inc}}) \to \sqrt{\frac{-2i}{\pi k_0 \rho}} e^{ik_0\rho} S(\phi^{\text{sc}}, \phi^{\text{inc}}), \quad \rho \to \infty \tag{6.18}$$

where ϕ^{inc} and ϕ^{sc} are the angles of the propagation directions of the incident and scattered fields. By making use of the far field expansion of the Green's function in (6.5), it can be shown that the scattering amplitude is related to the surface current on the scatterer by

$$S(\phi^{\text{inc}}, \phi^{\text{sc}}) = -\frac{k_0 \eta}{4} \langle E^{\text{sc}}, J \rangle \tag{6.19}$$

where $\langle \cdot, \cdot \rangle$ denotes the L^2 inner product.

The relative scattering amplitude error for given incident and scattered directions is

$$\text{Err}_S = \frac{|\hat{S} - S|}{|S|} \tag{6.20}$$

where $\hat{S}$ is the computed scattering amplitude obtained from the approximate current solution. Error measures which take into account more than one scattered direction are also commonly used in computational electromagnetics. These include the maximum relative RCS error

$$\text{Err}_{\max} = \max_{1 \leq m \leq M} |\hat{\sigma}(\theta_m)/\sigma(\theta_m) - 1| \qquad (6.21)$$

and the relative RMS RCS error

$$\text{Err}_{\text{RMS}} = \left\{ \frac{1}{M} \sum_{m=1}^{M} |10 \log_{10}[\hat{\sigma}(\theta_m)/\sigma(\theta_m)]|^2 \right\}^{1/2} \qquad (6.22)$$

where the θ_m represent M selected scattering directions. A drawback of these error measures is that the scattering amplitude or RCS errors may be small even when the surface current error is large.

6.1.3 Approaches to Error Analysis

There are two main approaches to the analysis of numerical methods in electromagnetics: theoretical and empirical. The first has been pursued almost exclusively by the numerical analysis community, through the application of fundamental theorems of analysis and operator theory, and has resulted in the asymptotic solution error estimates described in Section 6.1.3.1. The second area involves comparison of results with benchmarks for which an analytical solution or measured data is available, and the use of insight into the physical behavior of fields to predict or improve the accuracy and efficiency of a numerical method. Empirical approaches have been highly succesful over the past two decades in driving research in the field of computational electromagnetics.

6.1.3.1 Asymptotic Error Estimates

The asymptotic error estimates developed by the numerical analysis community for electromagnetics problems [8–16] have their roots in the study of Laplace's equation and Sobolev space theory. The goals of this work are to prove that numerical solutions converge as the mesh is refined, determine the asymptotic convergence rate, and obtain preconditioners that reduce the computational cost to $O(1)$ computations per degree of freedom as the mesh density increases. This requires solution error estimates and condition number estimates for the moment matrix.

The key parameter is h, the mesh or discretization length. Error and condition number estimates results obtained using the Sobolev space approach are asymptotic as $h \to 0$. The underlying theorem is that the convergence behavior of MOM for a dynamic problem ($k_0 > 0$) for small h is close in a certain sense to that of the static or Laplace problem ($k_0 = 0$). Estimates can then be obtained by studying the boundary integral operator for the Laplace problem. From the operator point of view, this amounts to making the decomposition

$$\mathcal{L} = -(i/k_0)\mathcal{A} + \mathcal{K} \tag{6.23}$$

where the static part $\mathcal{A}$ is self-adjoint and positive definite, and $\mathcal{K}$ is compact, and neglecting the influence of $\mathcal{K}$ on the convergence behavior of the method of moments.

The static approximation leads to the theorem of quasioptimality, that the MOM solution is close to the best possible solution in the trial subspace, where "best possible" is defined in terms of a Sobolev norm. As a consequence, the actual error is close to the approximation error, or the error of the best possible solution with respect to the exact solution. The error analysis of MOM is thus reduced to a problem in approximation theory.

Physical properties of the scattering problem determine the smoothness of the surface current. The ability of the basis functions to represent functions of the given smoothness class then determines a solution error estimate of the form

$$\|\hat{J} - J\|_{H^s} \leq ch^\alpha \tag{6.24}$$

for 2D problems. The parameter α is the order of convergence of the numerical method and gives the log-log slope of the error as a function of the number of unknowns. Convergence theorems guarantee that estimates of this type holds for h smaller than some unknown, problem-dependent constant H.

The norm in (6.24) is the Sobolev norm associated with the space of functions of the smoothness class to which the surface current J belongs. The Sobolev norm is an alternative to the scattering amplitude in obtaining a scalar quantity which is finite for singular currents and can be used to quantify numerical error. On the real line, the Sobolev norm can be interpreted in the Fourier domain as a weighted norm, so that [17]

$$\|u\|_{H^s} = \int_{-\infty}^{\infty} dk \, (1 + |k|)^{2s} |U(k)|^2 \tag{6.25}$$

where $U(k)$ is the Fourier transform of $u(x)$ and $s = \pm 1/2$. The space of functions for which this norm is finite is denoted by H^s. For $s = 1/2$, the space $H^{1/2}$ is smaller than the square integrable function space L^2, and consists of functions smooth enough that the decay of $U(k)$ for large $|k|$ offsets the growth of the weight function. For $s = -1/2$, the weight function acts as a smoothing term, so that functions which are less smooth than L^2 functions belong to the space $H^{-1/2}$. In particular, this Sobolev space contains functions with singularities of the form $x^{-1/2}$, $x \to 0$.

The connection of (6.25) to electromagnetics arises because the weight function has the same asymptotic behavior for large k as the one-dimensional Fourier representations of the kernels of (6.5) and (6.6). This can be seen from results given in (6.75) for the TM polarization, and (6.100) for the TE polarization. Surface currents due to TM polarized incident fields belong to $H^{-1/2}$, and $H^{1/2}$ for the TE case. For the TM polarization, it is well known that currents can have $x^{-1/2}$ singularities at edges [18]. The L^2 norm of such a current is infinite and cannot be used to define a meaningful current error, whereas the $H^{-1/2}$ norm remains finite, and so (6.24) can be used as an error measure.

The order of convergence α in (6.24) can be evaluated for smooth scatterers and scatterers with edges. For smooth scatterers, the current is continuous and differentiable everywhere, as long as there are no sources on the scatterer itself. In this case, for localized, low-order basis functions, the exponent is $\alpha = 2$. In the nonsmooth case, if the current near an edge behaves as $x^{-1/2}$, existing convergence theories predict $\alpha = 1/2$ [10]. As shown in Section 6.2.5, scattering amplitudes are accurate to a higher order than would be expected from the bound (6.24), due to the variational nature of the scattering amplitude as computed by the method of moments [19].

A posteriori residual-based error bounds [17, 20, 21] have also been developed using Sobolev norms and have been applied to adaptive grid algorithms [12]. These bounds give the solution error in terms of the residual error, or error in the scattered field, and require accurate computation of residuals in order to bound the solution error.

In principle, the asymptotic error bound (6.24) holds for any scattering problem. For electrically small problems, the error analysis of the method of moments is essentially a closed problem. Most results apply to the scalar case, but the 3D vector case has been considered as well [8, 9].

For the electrically large problems of interest in computational electromagnetics, however, asymptotic bounds can fail for practical reasons. The influence of the global geometrical shape and of wave phenomena such as resonance are contained in the constant H. As the electrical size grows, the effects of long-range interactions become stronger, and the asymptotic regime may be pushed to smaller values of H. At the same time, due to finite computational resources, the minimum value of h that can be attained in practice increases with frequency. Due to the Nyquist criterion, the number of unknowns required to model an oscillatory surface current is proportional to the electrical size of the scatterer. The order of the total number of unknowns is

$$N \simeq \left(\frac{k_0 d}{h}\right)^{\dim-1} \quad (6.26)$$

where dim is the dimensionality of the scattering problem and d is a linear measure of the scatterer. For large electrical lengths $k_0 d$, the number of unknowns can be so great for the coarsest possible mesh that it is impractical to refine the mesh by reducing h. As a result, for electrically large problems it may not be possible to reach the asymptotic regime for which the bound (6.24) holds. In order to obtain quantitative information about accuracy and convergence, absolute error estimates are required.

6.1.3.2 Empirical Methods

In practice, numerical methods used in computational electromagnetics are validated by checking canonical test cases for which exact or tabulated solutions are available. This approach is used almost exclusively in the field of computational electromagnetics. Although this type of verification can provide reasonable confidence of accuracy over classes of similar scatterer geometries, it yields only a limited understanding of the underlying causes of solution error and condition number growth.

6.1.4 Spectral Convergence Theory

Both the theoretical and empirical approaches to error analysis have led to important contributions in the field of computational electromagnetics. The intent of this work is to address the drawbacks of these approaches and provide an understanding of error and convergence behavior for the electrically large, complex problems of current interest.

The spectral convergence theory developed in this chapter extends asymptotic convergence theories by providing absolute, nonasymptotic estimates that are valid for large scattering problems. These estimates also provide a theoretical understanding of accuracy for benchmark problems, and a sound basis for extrapolating empirical observations to more general problems of practical interest.

The fundamental results on which this approach is based are spectral estimates for the surface integral operators of scattering. These estimates in turn provide solution error and condition number estimates for the method of moments, in terms of the properties of the scatterer, the mesh used to obtain a discrete representation of the scatterer, and the numerical method.

The influence of the scatterering problem on the solution accuracy results from the combined effects of:

- Smooth regions;
- Edge, corner, and point singularities;
- Resonance;
- Low-frequency breakdown;
- Incident electromagnetic field.

Error estimates also depend on the numerical method through the choices of:

- Expansion and testing functions;
- Quadrature rule used to evaluate moment matrix elements;
- Linear system solution algorithm.

These properties of the scattering problem and numerical method determine the accuracy and convergence rate of the solution.

In this chapter, we consider each of these factors separately through a detailed study of several canonical examples. To leading order, these error contributions can in principle be superimposed to estimate the error for more complex scatterers. Although it may not be possible to take into account higher order interactions between edges, smooth regions, and resonant structures, the order-of-magnitude error is all that is needed for practical purposes, and the present theory should suffice for many scattering problems encountered in computational electromagnetics.

The circular cylinder is considered as an example of a smooth scatterer. The errors on the interiors of the flat strip and flat plate, away from edge singularities, provide additional examples of error on smooth regions. We then study the error due to the singular current at the edges of a flat strip and combine the edge error and the error on the smooth interior of the strip in order to obtain a total error estimate for the strip.

Internal and real resonances also affect the solution convergence rate. The cylinder exhibits internal resonances, which are nonphysical but still influence numerical accuracy. The rectangular cavity is an example of a scatterer with real, physical resonances. Spectral estimates for these scatterers allow the error due to resonance to be quantified.

The cylinder, strip, and cavity are 2D problems. We give spectral estimates for the flat plate in order to extend the error analysis to the more general 3D case. We consider also a method for regulating the kernel of the EFIE to reduce the error in scattering amplitude computations.

For each of the scatterers, and for common types of basis functions and quadrature rules, the goal is to estimate the solution error and moment matrix condition number in terms of the discretization density of the mesh. Both current and scattering amplitude errors are to be considered. As expressed in general form in (6.24), the exponent of the dependence on h is the order of convergence of the numerical method, and the constant gives the absolute accuracy of the result. The error also depends on the incident field in the EFIE (6.1). For a smooth scatterer, this results from an approximation error, which is analogous to the dispersion error of finite difference methods. Condition number estimates quantify the difficulty of solving a linear system for the unknown current in terms of the mesh density, the electrical size of the scatterer, resonance, and polynomial order of the basis functions.

Some of the text and figures included in this chapter have appeared previously in expanded form in [22–27].

6.2 SPECTRAL CONVERGENCE THEORY—2D

In general, the integral operators $\mathcal{L}$, $\mathcal{N}$, and $\mathcal{T}$ are non-self-adjoint, so that the usual theory of self-adjoint operators on Hilbert spaces cannot be applied to study the behavior of numerical solutions. One approach to overcoming this is the use of the static decomposition (6.23). For high frequencies or scatterers with large electrical size, the static limit fails to capture all the relevant physics. In this work, we employ a decomposition of the form

$$\mathcal{L} = \mathcal{H} + \mathcal{R} \tag{6.27}$$

where $\mathcal{H}$ is normal and $\mathcal{R}$ is a nonnormal perturbation. A normal operator satisfies $\mathcal{H}^\dagger \mathcal{H} = \mathcal{H}\mathcal{H}^\dagger$ and has a spectral decomposition [28]. In this work, the operator $\mathcal{H}$ corresponds to a high-frequency limit of $\mathcal{L}$. If $\mathcal{R}$ is in some sense small, the eigenvalues of $\mathcal{H}$ provide spectral estimates for $\mathcal{L}$ and its matrix discretization, which arises from the method of moments. These estimates lead to the notion of spectral error, from which the current solution and scattering amplitude errors can be obtained.

In the following sections, we apply this method to the circular cylinder, the flat strip, and the rectangular cavity. To complete the analysis of the flat strip, we also consider the error due to edge singularities of the strip, using a nonspectral approach.

6.2.1 Circular Cylinder—TM

Because an analytical solution for the plane wave scattering problem for the circular cylinder exists, this scatterer is often used as a benchmark for numerical methods. In this section, we provide a theoretical understanding of the error behavior of the method of moments for the circular cylinder in order to better understand how benchmark results generalize to more complex problems.

For the cylinder, $\mathcal{L}$ is normal, and an exact spectral decomposition exists [14], so that the operator $\mathcal{R}$ in (6.27) vanishes. Here, we obtain the spectrum of the discretized operator, and thereby determine the solution error in terms of the choice of basis functions, the discretization density, and the quadrature rule used to evaluate moment matrix elements. The dependence of the conditioning of the moment matrix on the inner product used to discretize the EFIE has been studied for the cylinder using a similar approach [6].

We now expand the kernel of the EFIE in terms of eigenfunctions of the Helmholtz operator $\nabla^2 + k_0^2$ in order to obtain a spectral decomposition for the moment matrix. The cylindrical mode expansion of the kernel is

$$H_0^{(1)}(k_0|\boldsymbol{\rho} - \boldsymbol{\rho}'|) = \sum_{l=-\infty}^{\infty} J_l(k_0 a) H_l^{(1)}(k_0 a) e^{il(\phi - \phi')} \tag{6.28}$$

where ρ and ρ' lie on a circle of radius a and ϕ and ϕ' are the corresponding angles in the cylindrical coordinate system. The expansion and testing functions are taken to be of the form $f_n(\phi) = f(\phi - \phi_n)$ and $t_n(\phi) = t(\phi - \phi_n)$. The nodes are evenly spaced, so that $\phi_n = (n - 1/2)\theta_0$, where $\theta_0 = 2\pi/N$. In terms of cylindrical modes, the moment matrix becomes

$$Z_{mn} = \frac{\eta \pi k_0 a}{2N} \sum_{l=-\infty}^{\infty} J_l(k_0 a) H_l^{(1)}(k_0 a) \tilde{t}_{-l} \tilde{f}_l e^{il(\phi_m - \phi_n)} \quad (6.29)$$

where $\tilde{t}_l$ is the Fourier transform of $t(\phi)$,

$$\tilde{t}_l = \frac{1}{\theta_0} \int d\phi \, t(\phi) e^{-il\phi} \quad (6.30)$$

evaluated at l and normalized by $1/\theta_0$, and $\tilde{f}_l$ is defined similarly.

From (6.29), the eigenvectors of the moment matrix are of the form $e^{ir\phi_n}$, where r is an integer and n indexes components of the eigenvector. The corresponding eigenvalues can be determined from

$$\sum_{n=1}^{N} Z_{mn} e^{ir\phi_n} = \frac{\eta \pi k_0 a}{2N} \sum_{l=-\infty}^{\infty} J_l(k_0 a) H_l^{(1)}(k_0 a) \tilde{t}_{-l} \tilde{f}_l e^{il\phi_m} \sum_{n=1}^{N} e^{i(r-l)\phi_n} \quad (6.31)$$

The sum over n can be evaluated using

$$\sum_{n=1}^{N} e^{ir\phi_n} = (-1)^r \frac{\sin(\pi r)}{\sin(\pi r/N)} \quad (6.32)$$

The right-hand side of (6.32) is equal to $(-1)^s N$ if $r = sN$, where s is an integer, and vanishes otherwise. Equation (6.31) then becomes

$$\sum_{n=1}^{N} Z_{mn} e^{ir\phi_n} = \left[\frac{\eta \pi k_0 a}{2} \sum_{q=-\infty}^{\infty} J_{r+qN}(k_0 a) H_{r+qN}^{(1)}(k_0 a) \tilde{t}_{-r-qN} \tilde{f}_{r+qN} \right] e^{ir\phi_m} \quad (6.33)$$

From this expression, we can identify

$$\hat{\Lambda}_r = \frac{\eta \pi k_0 a}{2} \sum_{q=-\infty}^{\infty} J_{r+qN}(k_0 a) H_{r+qN}^{(1)}(k_0 a) \tilde{t}_{-r-qN} \tilde{f}_{r+qN} \quad (6.34)$$

as the eigenvalue of $e^{ir\phi_n}$.

In the limit as $N \to \infty$, $\hat{\Lambda}_r$ is equal to the rth eigenvalue Λ_r of the continuous operator $\mathcal{L}$. In this limit, all terms in the sum in (6.34) vanish except the $q = 0$ term, and $\tilde{t}_{-r} \tilde{f}_r \to 1$, so that we recover the exact eigenvalue [14]:

$$\Lambda_r = (\eta \pi k_0 a/2) J_r(k_0 a) H_r^{(1)}(k_0 a) \quad (6.35)$$

Error Analysis of Surface Integral Equation Methods

The eigenvalues of the moment matrix can be written in the form

$$\hat{\Lambda}_r = \Lambda_r + \Delta_r \tag{6.36}$$

where

$$\Delta_r = \frac{\eta \pi k_0 a}{2} \sum_{q \neq 0} J_{r+qN}(k_0 a) H^{(1)}_{r+qN}(k_0 a) \tilde{t}_{-r-qN} \tilde{f}_{r+qN} + \Lambda_r (\tilde{t}_{-r} \tilde{f}_r - 1) \tag{6.37}$$

is the spectral error caused by discretization. Since $\overline{\mathbf{Z}}$ has N eigenvalues, the index r lies in the range $-N/2+1 \leq r \leq N/2$ if N is even or $-(N-1)/2 \leq r \leq (N-1)/2$ if N is odd. We refer to the eigenvalues of $\mathcal{L}$ with order $|r| > N/2$ as unmodeled eigenvalues, since the spatial frequency of the corresponding eigenfunctions is greater than the Nyquist frequency of the sample points x_n, and they cannot be represented in the trial subspace.

6.2.1.1 Spectral Error

In this section, we study the spectral error introduced by the discretization of the EFIE, as defined in (6.36). Using the large order expansion $J_\nu(x) H^{(1)}_\nu(x) \sim -i(\pi|\nu|)^{-1}$, the relative spectral error $E_r = \Delta_r / \Lambda_r$ can be approximated as

$$E_r \simeq -\frac{i\eta}{2n_\lambda \Lambda_r} \sum_{q \neq 0} \frac{\tilde{t}_{-r-qN} \tilde{f}_{r+qN}}{|q + r/N|} + \tilde{t}_{-r} \tilde{f}_r - 1 \tag{6.38}$$

The first term of this expression is determined by the asymptotic behavior of the spectral representation of the kernel, which is associated with the singularity of the Green's function $g(\rho, \rho')$ at $\rho = \rho'$. This contribution to E_r might be called sampling error, since it arises from aliasing of high spatial frequency ($|r| > N/2$) components of the kernel. The second term, $\tilde{t}_{-r} \tilde{f}_r - 1$, is the approximation error or smoothing error caused by inaccurate representation of low spatial frequency ($|r| \leq N/2$) eigenfunctions of the kernel.

We now consider the spectral error for piecewise polynomial expansion and testing functions. The particular basis functions studied here are generated by convolutions of the pulse function, and are splines of the type studied in [6]. For large p, these piecewise polynomials become Gaussian-like, with support that grows in width with p. The functions could be rescaled so that the width of the support is close to h, but we do not consider this possibility here, since in practice, $p \leq 1$ for this class of basis functions. Greater accuracy can be obtained by p-refinement, for which the basis consists of a set of polynomials complete up to order p, but we defer the treatment of this type of discretization to Section 6.2.7.

For the piecewise polynomial bases, the product of the Fourier transforms of the testing and expansion functions $\tilde{t}_{-r}\tilde{f}_r$ is s_r^b, where

$$s_r = \frac{\sin(\pi r/N)}{\pi r/N} \qquad (6.39)$$

The exponent is

$$b = p + p' + 2 \qquad (6.40)$$

where p and p' are the polynomial orders of the testing and expansion functions. The pulse function (piecewise constant basis) is of order 0, and the triangle function (piecewise linear basis) is order 1. The delta function (point matching) can also be considered in this scheme and has order -1.

By making use of (6.38), the spectral error for these types of testing and expansion functions is

$$E_{r,b} \simeq -\frac{i\eta}{2n_\lambda \Lambda_r} \sum_{q \neq 0} \frac{\operatorname{sgn}(q) \sin^b \pi(q + \beta_r/n_\lambda)}{\pi^b (q + \beta_r/n_\lambda)^{b+1}} + \left[\frac{\sin(\pi \beta_r/n_\lambda)}{\pi \beta_r/n_\lambda}\right]^b - 1 \qquad (6.41)$$

where $\beta_r = r/(k_0 a)$ is the normalized spatial frequency of the rth mode.

For the $b = 0$ discretization, the leading order term of the sampling error can be made to vanish by a proper choice of the locations of the delta functions on each element of the discretization. Because of the singularity of the kernel, the testing and expansion points cannot coincide. In order to obtain finite moment matrix elements, their relative locations must be shifted. We take the testing function to be $\delta(\phi)$ and the expansion function to be $\delta(\phi + \alpha\theta_0/2)/2 + \delta(\phi - \alpha\theta_0/2)/2$, where the parameter α specifies the relative shift of the testing and expansion points. In this case, $\tilde{t}_{-r}\tilde{f}_r = \cos(\alpha \pi r/N)$. For small β_r/n_λ, the spectral error evaluates to

$$E_{r,0} \simeq \frac{i\eta \ln[2\sin(\alpha\pi/2)]}{n_\lambda \Lambda_r} \qquad (6.42)$$
$$-\frac{i\eta}{4n_\lambda^3 \Lambda_r} \left\{ g_3(\alpha) + 2\beta_r^2 \left[\frac{\alpha^2 \pi^2}{4} g_1(\alpha) + \alpha\pi g_2(\alpha) + g_3(\alpha)\right] \right\} - \frac{\pi^2 \alpha^2 \beta_r^2}{2n_\lambda^2}$$

where $g_n(x) = i^{n+1}[\operatorname{Li}_n(e^{i\pi x}) - (-1)^n \operatorname{Li}_n(e^{-i\pi x})]$ and $\operatorname{Li}_n(x)$ is the polylogarithm function of order n. In obtaining this result, we have included the third order term of the expansion $J_\nu(x) H_\nu^{(1)}(x) \sim -i/(\pi|\nu|) - ix^2/(2\pi|\nu|^3)$ in (6.34). The leading order sampling error term of (6.42) vanishes for a shift of $\alpha = 1/3$. In this case, the spectral error becomes

$$E_{r,0} \simeq -\frac{i\eta(0.2 + 1.5\beta_r^2)}{n_\lambda^3 \Lambda_r} - \frac{\pi^2 \beta_r^2}{18n_\lambda^2} \qquad (6.43)$$

The first term represents sampling error, and the second term is smoothing error.

For pulse expansion functions and point testing ($b = 1$), the spectral error for small β_r/n_λ is

$$E_{r,1} \simeq -\frac{1.8\,i\eta\beta_r^2}{n_\lambda^3 \Lambda_r} - \frac{\pi^2 \beta_r^2}{6n_\lambda^2} \tag{6.44}$$

where the constant is $3\zeta(3)/2 \simeq 1.8$ and $\zeta(x)$ is the Riemann zeta function. For point testing and triangle expansion functions or pulse testing and expansion functions ($b = 2$), the spectral error is

$$E_{r,2} \simeq -\frac{1.2\,i\eta\beta_r^2}{n_\lambda^3 \Lambda_r} - \frac{\pi^2 \beta_r^2}{3n_\lambda^2} \tag{6.45}$$

where the constant is $\zeta(3) \simeq 1.2$. For all three discretizations, the sampling error is third order in n_λ^{-1} and the smoothing error is second order. For $b > 2$, the sampling error term becomes higher order in n_λ^{-1} as b increases. The smoothing error remains second order, but its magnitude grows with b.

6.2.1.2 Quadrature Error

The expressions for the spectral error obtained in the previous section were based on the assumption of exact integration of the moment matrix elements Z_{mn}. In practice, numerical quadrature rules are employed to evaluate the matrix elements. The effect of approximate integration can be taken into account in (6.38) by replacing the continuous integral in the Fourier transform of the expansion function $f(\phi)$ with the quadrature rule, so that $\tilde{f}_r$ becomes

$$\tilde{f}_{r,M} = \frac{1}{\theta_0} \sum_{n=1}^{M} w_n f(\xi_n) e^{-ir\xi_n} \tag{6.46}$$

where M is the order of the quadrature rule and w_n are the weights corresponding to the abscissas ξ_n. The Fourier transform $\tilde{t}_r$ of the testing function $t(\phi)$ is modified similarly.

For the M-point first order Riemann integration rule, the weights are $w_n = \delta = \theta_0/M$ and the abscissas are $\xi_n = (n - 1/2)\delta - \theta_0/2$. In the case of piecewise constant expansion functions ($b = 1$), $\tilde{f}_{r,M}$ becomes the periodic sinc function

$$\tilde{f}_{r,M} = \frac{\sin\frac{\pi r}{N}}{M \sin\frac{\pi r}{MN}} \tag{6.47}$$

The maxima of $\tilde{f}_{p,M}$ lie at $p = MNs$, where $s = 0, \pm 1, \pm 2, \ldots$. Thus, for small r, the terms of the summation over q in (6.38) for which $q = Ms$ yield the leading contribution to the sampling error. Since $\tilde{f}_{r+qN,M} \simeq (-1)^{s(M+1)}$ for these terms, we find that

$$E_{r,M}^{(1)} \simeq -\frac{i\eta}{2n_\lambda \Lambda_r} \sum_{s \neq 0} \frac{(-1)^{s(M+1)}}{|Ms|} \qquad (6.48)$$

If M is even, the sum over s is finite and can be evaluated in closed form, and the sampling error becomes

$$E_{r,M}^{(1)} \simeq \frac{i\eta \ln 2}{M n_\lambda \Lambda_r} \qquad (6.49)$$

for small β_r. The quadrature rule also has a small effect on the smoothing error term of the spectral error, but the additional contribution is of the same order in n_λ as the smoothing error for exact integration.

The quadrature error contribution (6.49) is first order in n_λ^{-1}, and so can dominate the higher order spectral error terms in (6.38) for small values of M. In order for the quadrature error to be as small as the sampling error in (6.44) for the $b = 1$ discretization, M must increase as n_λ^2. Commonly, matrix elements near the singularity of the kernel are evaluated using a more accurate method such as analytical integration of the singularity of the kernel [29]. For a single point integration rule ($M = 1$), analytical integration of the diagonal matrix elements reduces the quadrature error roughly by a factor of three [26].

The use of a specialized integration rule such as Gaussian quadrature does not necessarily improve the order of the spectral error in $M n_\lambda$. Because of the singularity of the kernel, moment matrix integrals are in general only first order accurate, regardless of the type of integration rule. In order to significantly reduce the error using a Gaussian quadrature rule, the weights and abscissas must be computed by taking into account the specific form of the singularity of the kernel [29].

6.2.1.3 Current Error

The surface current solution error for the method of moments can be determined from the spectral error due to discretization. If the incident field is a plane wave traveling along the x axis, then the weights J_n, $n = 1, 2, \ldots, N$, of the expansion functions for the approximate current are obtained by solving a linear system with a vector on the right-hand side having components

$$E_n^{\text{inc}} = \int d\phi \, t_n(\phi) e^{ik_0 a \cos \phi} \qquad (6.50)$$

Using the cylindrical mode expansion of a plane wave, E_n^{inc} can be written as

$$E_n^{\text{inc}} = \sum_{r=-\infty}^{\infty} i^r J_r(k_0 a) \tilde{t}_{-r} e^{ir\phi_n} \qquad (6.51)$$

Applying $\overline{Z}^{-1}$ to this expression gives

$$\hat{J}_n = \sum_{r=-\infty}^{\infty} \frac{i^r J_r(k_0 a)\tilde{t}_{-r}}{\hat{\Lambda}_r} e^{ir\phi_n} \qquad (6.52)$$

since each term of the sum in (6.51) is an eigenvector of $\overline{Z}$. The exact current evaluated at the angle ϕ_n on the cylinder is

$$J_n = \sum_{r=-\infty}^{\infty} \frac{i^r J_r(k_0 a)}{\Lambda_r} e^{ir\phi_n} \qquad (6.53)$$

The current error $\Delta J_n = J_n - \hat{J}_n$ is then

$$\Delta J_n = \frac{2}{\eta \pi k_0 a} \sum_{r=-\infty}^{\infty} \frac{i^r (1 + E_r - \tilde{t}_{-r})}{H_r^{(1)}(k_0 a)(1 + E_r)} e^{ir\phi_n} \qquad (6.54)$$

at the node with angle ϕ_n.

Using this result for the error at each node, the RMS current error becomes

$$\|\Delta \mathbf{J}\| \simeq \frac{2}{\eta \pi k_0 a} \left[\sum_{r=-\infty}^{\infty} \left| \frac{E_r^{(1)} + \tilde{t}_{-r}(\tilde{f}_r - 1)}{H_r^{(1)}(k_0 a)(1 + E_r)} \right|^2 \right]^{1/2} \qquad (6.55)$$

where $E_r^{(1)}$ is the sampling error term of (6.38). To leading order, the approximate current solution is not affected by the smoothing error introduced by the testing functions. This occurs because the left- and right-hand sides of the EFIE in (6.4) are tested using the same set of functions, and the effect of the testing functions cancels for each mode.

A comparison of this result with numerical experiments is shown in Figure 6.1. The approximate expression (6.44) is employed for the sampling error. The number of integration points is sufficiently large that the quadrature error (6.49) is negligible for this range of n_λ. For larger n_λ, beyond the axis limit of Figure 6.1, the curves would eventually level off at a slope of -1 due to quadrature error.

Due to the smoothness of the current solution for a plane incident wave, the current error is determined by the spectral error for modes of low spatial frequency ($|r|$ near $k_0 a$ or smaller). A less smooth incident field leads to larger amplitudes for higher-order modes. If the incident field is produced by a line source located at a distance d from the cylinder, for example, then it can be shown that the current error depends on the spectral error E_r for roughly $|r| \leq 2a/d$, for small d, which exceeds $k_0 a$ when the line source is closer than λ/π to the cylinder.

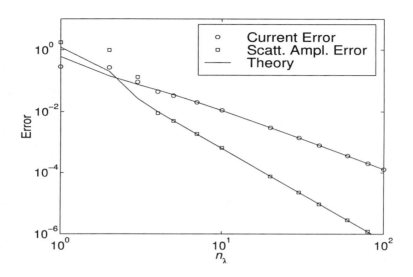

Figure 6.1 Relative RMS surface current error and backscattering amplitude error for a cylinder, TM polarization, $k_0 a = \pi$, and piecewise constant expansion functions with point testing. The current error is second order, whereas the scattering amplitude error is third order.

6.2.1.4 Scattering Amplitude Error

The bistatic scattering amplitude can be obtained from the approximate current $\hat{J}_n$ on the scatterer by discretizing the inner product in (6.19) to obtain

$$\hat{S}(\phi^{\text{inc}}, \phi^{\text{sc}}) = -\frac{k_0 \eta}{4} h \sum_{n=1}^{N} E_n^{\text{sc}*} \hat{J}_n \qquad (6.56)$$

The scattered plane wave E_n^{sc} is discretized on the surface of the conductor using the expansion functions $f_n(\phi)$ in a manner similar to (6.50) for the incident field.

By making use of (6.52), the numerical bistatic scattering amplitude for the cylinder becomes

$$\hat{S}(\phi^{\text{inc}}, \phi^{\text{sc}}) = -\sum_{r=-\infty}^{\infty} \frac{J_r(k_0 a) \tilde{f}_r \tilde{t}_{-r}}{H_r^{(1)}(k_0 a)(1 + E_r)} e^{ir\phi} \qquad (6.57)$$

where $\phi = \phi^{\text{sc}} - \phi^{\text{inc}}$. Subtracting the exact value of the scattering amplitude gives

$$\Delta S(\phi^{\text{inc}}, \phi^{\text{sc}}) = -\sum_{r=-\infty}^{\infty} \frac{J_r(k_0 a)}{H_r^{(1)}(k_0 a)} \frac{E_r^{(1)}}{1 + E_r} e^{ir\phi} \qquad (6.58)$$

for the error, where $E_r^{(1)}$ is the sampling error component of the spectral error E_r. This result shows that to leading order, the smoothing error term $\tilde{t}_{-r}\tilde{f}_r - 1$ of the spectral error E_r does not contribute to the scattering amplitude error. The increased accuracy of the scattering amplitude relative to the current is a consequence of the scattering amplitude being stationary with respect to perturbations of the solution [19]. Physically, each mode is scaled by a smoothing factor due to testing and expansion of the incident and scattered fields, but discretization of the EFIE introduces an identical scaling factor into the mode interaction as computed by the moment matrix, and the two factors cancel in the computed scattering amplitude. Numerical results for the relative error for the backscattering amplitude $S(\phi^{\text{inc}}, \phi^{\text{inc}})$ are shown in Figure 6.1.

6.2.1.5 Internal Resonance

If $k_0 a$ is such that the cylinder has an internal resonance, or a nontrivial solution to the interior Dirichlet problem, then the EFIE breaks down from a numerical point of view. A common solution is to employ a different integral equation formulation which remains stable at internal resonances. We study here the behavior of the EFIE near an internal resonance, and show that it is possible to obtain accurate scattering amplitudes using the EFIE at an internal resonance.

In the continuous case, both the resonant eigenvalue and the amplitude of the corresponding mode in the incident field vanish together as $k_0 a$ approaches a resonance, leading to a finite limit for the amplitude of that mode in the exact surface current. The effect of discretization is to change the resonant eigenvalue by an amount Δ_r, so that the location of the resonance is shifted as a function of $k_0 a$. If χ is a zero of $J_r(x)$, then for $k_0 a$ near χ, the eigenvalue of the EFIE of order r can be expanded as

$$\Lambda_r \simeq i\frac{\eta\pi\chi}{2} J_r'(\chi) Y_r(\chi)(k_0 a - \chi) \qquad (6.59)$$

The numerical internal resonance is the value of $k_0 a$ for which $\hat{\Lambda}_r = \Lambda_r + \Delta_r$ vanishes. Since the discretization error is nearly pure imaginary, discretization does not lead to complex resonances.

At the exact resonance, the amplitude of the resonant mode in the incident field is negligible, whereas the eigenvalue is equal to Δ_r. As a consequence, the approximate current solution is missing the resonant mode. Unless rounding error is significant due to the large condition number of the moment matrix, the amplitudes of other modes remains accurate.

At the numerical internal resonance, the moment matrix is singular, and direct linear system solution methods can fail due to rounding error. In this case, the conjugate gradient (CG) iterative solver can be applied to the normal form $\overline{\mathbf{Z}}^\dagger \cdot \overline{\mathbf{Z}} \cdot \mathbf{J} = \overline{\mathbf{Z}}^\dagger \cdot \mathbf{E}$ of the linear system. The amplitude of the resonant mode in the new right-hand side is zero, since the corresponding eigenvalue is zero. Because this eigenvector is not present in the right-hand side, the behavior of the CG algorithm is governed by an effective matrix operator that no longer has the vanishing eigenvalue, and the iteration converges at a rate similar to nonresonant frequencies.

At both the numerical and exact resonances, even if a convergent matrix iteration is employed, the current error is large, since the amplitude of the resonant mode in the computed current is incorrect. Since an internally resonant mode does not radiate outside the scatterer, scattering amplitudes and cross-sections can still be accurate.

6.2.2 Circular Cylinder—TE

The TE polarization can be treated in a similar manner as the TM case above. The main difference is the stronger singularity of the kernel for the TE case. Using the expansion (6.28), the moment matrix elements arising from the discretization of the operator $\mathcal{N}$ in (6.6) can be expressed as

$$Z_{mn} = \frac{\eta k_0 a}{4\theta_0} \int\int d\phi\, d\phi'\, t_m(\phi) \sum_l J_l(k_0 a) H_l^{(1)}(k_0 a)$$
$$\times \left[\cos(\phi - \phi') f_n(\phi') + \frac{il}{(k_0 a)^2} \frac{\partial f_n(\phi')}{\partial \phi'}\right] e^{il(\phi - \phi')} \quad (6.60)$$

where the parameters and basis functions are defined as before. Expanding $\cos(\phi - \phi')$ into exponentials, integrating the second term by parts, and making use of the recursion relations for the derivatives of the Bessel and Hankel functions yields

$$Z_{mn} = \frac{\eta \pi k_0 a}{2} \sum_l J'_l(k_0 a) H_l^{(1)\prime}(k_0 a) \tilde{t}_{-l} \tilde{f}_l e^{il(\phi_m - \phi_n)} \quad (6.61)$$

By proceeding as in the previous section, the eigenvalues of $\overline{Z}$ are found to be

$$\hat{\Lambda}_r = \frac{\eta \pi k_0 a}{2} \sum_{q=-\infty}^{\infty} J'_{r+qN}(k_0 a) H_{r+qN}^{(1)\prime}(k_0 a) \tilde{t}_{-r-qN} \tilde{f}_{r+qN} \quad (6.62)$$

for the TE polarization. This expression is identical to (6.34), except that the Bessel and Hankel functions are replaced with their first derivatives. The eigenvalues of the continuous EFIE are $\Lambda_r = (\eta \pi k_0 a/2) J'_r(k_0 a) H_r^{(1)\prime}(k_0 a)$.

6.2.2.1 Spectral Error

From (6.62), the relative spectral error is

$$E_r \simeq \frac{i\eta n_\lambda}{2\Lambda_r} \sum_{q \neq 0} |q + r/N| \tilde{t}_{-r-qN} \tilde{f}_{r+qN} + \tilde{t}_{-r} \tilde{f}_r - 1 \quad (6.63)$$

where we have used the asymptotic expansion $J'_\nu(x) H_\nu^{(1)\prime}(x) \sim i|\nu|/(\pi x^2)$, $\nu \to \infty$. The current and scattering amplitude errors have the same forms as (6.55) and (6.58) for the TM polarization, but the Bessel and Hankel functions are replaced with their first derivatives. Since $H_r^{(1)\prime}(k_0 a)$ is smaller than $H_r^{(1)}(k_0 a)$ as $|r|$ approaches $|k_0 a|$ from below, surface wave modes with spatial frequencies near $k_0 a$ contribute more strongly to the current for the TE polarization.

If the pulse expansion ($b = 1$) described in Section 6.2.1.1 is used to discretize the TE EFIE, the leading-order term of the spectral error as given by (6.63) vanishes. To obtain the spectral error for small β_r/n_λ, we must employ an additional term of the expansion $J'_\nu(x) H_\nu^{(1)\prime}(x) \sim i|\nu|/(\pi x^2) - ix^2/(2\pi|\nu|)$ in (6.62). This leads to

$$E_{r,1} \simeq \frac{0.9\, i\eta \beta_r^2}{n_\lambda^3 \Lambda_r} - \frac{\pi^2 \beta_r^2}{6 n_\lambda^2} \tag{6.64}$$

where the constant is $3\zeta(3)/4 \simeq 0.9$.

For the $b = 2$ discretization, the diagonal moment matrix element Z_{nn} diverges if the testing point is located at the apex of the triangle expansion function. The testing functions must be shifted in order for the matrix elements to be finite. Thus, we employ the shifted, symmetric testing function $\delta(\phi + \alpha\theta_0/2)/2 + \delta(\phi - \alpha\theta_0/2)/2$. In this case, a factor of $\ln[2\sin(\alpha\pi/2)]$ appears in the leading-order term of the sampling error. This term vanishes for a shift of $\alpha = 1/3$, and the total spectral error becomes

$$E_{r,2} \simeq \frac{i\eta(-0.2\,\beta_r^2 + 1.5\,\beta_r^4)}{n_\lambda^3 \Lambda_r} - \frac{\pi^2 \beta_r^2}{3 n_\lambda^2} \tag{6.65}$$

for small β_r/n_λ. The constants are $g_4(1/3)/4 \simeq 0.2$ and $\pi g_2(1/3)/6 + g_3(1/3)/2 \simeq 1.5$. The error for this scheme is similar to that of the $b = 0$ discretization of Section 6.2.1.1 for the TM polarization.

For the $b = 3$ discretization, the spectral error is

$$E_{r,3} \simeq \frac{1.8\, i\eta \beta_r^4}{n_\lambda^3 \Lambda_r} - \frac{\pi^2 \beta_r^2}{2 n_\lambda^2} \tag{6.66}$$

for small β_r/n_λ. The constant is $3\zeta(3)/2 \simeq 1.8$. Numerical results for the $b = 3$ discretization (pulse testing and triangle expansion functions, or point testing with piecewise quadratic expansion functions) are shown in Figure 6.2.

In all cases, the sampling error is third order in n_λ^{-1}, and the smoothing error is second order. The order of the sampling error depends on the locations of the testing functions relative to the expansion functions, so that the error is strongly sensitive to irregular testing.

The spectral error for the $b = 0$ discretization does not decrease as n_λ becomes large, so that this scheme is not viable for the TE polarization.

6.2.2.2 Quadrature Error

Since the kernel of the EFIE for the TE polarization has a stronger singularity than in the TM case, the hypersingular term of (6.6) is often integrated by parts to reduce

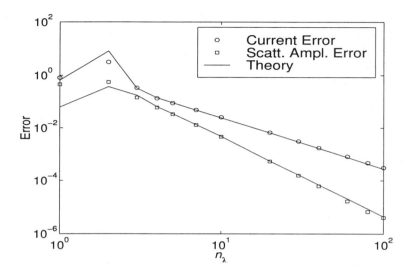

Figure 6.2 Relative RMS surface current error and backscattering amplitude error for a cylinder, TE polarization, $k_0 a = \pi$, and the $b = 3$ discretization scheme (pulse/triangle or point/quadratic testing and expansion functions).

the singularity before application of a numerical quadrature rule. We consider here the spectral error introduced by the quadrature rule with and without integration by parts, for the case of pulse testing and triangle expansion functions ($b = 3$) with the M-point integration rule described in Section 6.2.1.2.

If the EFIE is discretized directly as in (6.6), without integration by parts, the sampling error arising from the hypersingular term is given by

$$E_{r,M}^{(1)} = \frac{i\eta\pi}{2k_0 a \Lambda_r} \sum_{q \neq 0}(r + qN) J_{r+qN}(k_0 a) H_{r+qN}(k_0 a) \tilde{t}_{-r-qN,M} \tilde{f}'_{r+qN,M}$$

(6.67)

In this expression, $\tilde{t}_{r,M}$ is equal to the periodic sinc function of (6.47). The term $\tilde{f}'_{r,M}$ is the Fourier transform of the derivative of the triangle function, sampled using the M-point quadrature rule, and is

$$\tilde{f}'_{r,M} = e^{i\alpha\pi r/N} \frac{2i \sin\frac{\pi r}{N}}{\theta_0 \sin\frac{\pi r}{MN}}$$

(6.68)

where α specifies the relative shift of the testing and expansion functions. Employing a large order expansion of $J_\nu(x) H_\nu(x)$ and evaluating the summation over q for small

β_r yields

$$E_{r,M}^{(1)} = \frac{\eta \beta_r}{2\Lambda_r} \tan(\alpha \pi M/2) \qquad (6.69)$$

For $\alpha = 0$, the leading term of the error given by (6.69) vanishes, and the error becomes higher order in n_λ^{-1}. If the locations of the testing functions are shifted away from $\alpha = 0$, then the method is nonconvergent, since the spectral error does not vanish as $n_\lambda \to \infty$.

With integration by parts, the quadrature error due to the hypersingular term of the EFIE is

$$E_{r,M}^{(1)} = -\frac{\eta \pi}{2k_0 a \Lambda_r} \sum_{q \neq 0} J_{r+qN}(k_0 a) H_{r+qN}(k_0 a) \tilde{t}'_{-r-qN,M} \tilde{f}'_{r+qN,M} \qquad (6.70)$$

where $\tilde{t}'_{r,M}$ is the Fourier transform of the derivative of the pulse function, so that

$$\tilde{t}'_{r,M} = 2i\theta_0^{-1} \sin\frac{\pi r}{N} \qquad (6.71)$$

Expanding the Bessel and Hankel functions in (6.70) and evaluating the sum over q leads to

$$E_{r,M}^{(1)} = -\frac{i\eta \beta_r^2}{n_\lambda M \Lambda_r} \ln[2\cos(\alpha \pi M/2)] \qquad (6.72)$$

for small β_r, which is of the same order as (6.49) for the weakly singular kernel. The total sampling error for the TE polarization can be approximated by the sum of (6.49) and (6.72).

Although the integrand of the hypersingular term has the same spectral content with or without integration by parts, the two formulations differ by a vanishing integral of a total derivative. When a numerical quadrature rule is employed, this integral no longer vanishes. This accounts for the difference between (6.69) and (6.72).

6.2.3 Flat Strip—TM

The next scatterer that we consider is a 2D cross-section of an infinite, perfectly conducting strip. For this scatterer, the EFIE operator is more difficult to analyze than for the cylinder. It can be shown by making use of the Fourier representation employed below that the EFIE is nonnormal. In this case, the eigenfunctions of the operator do not form an orthogonal basis in L^2, and an exact modal expansion of the type used in Sections 6.2.1 and 6.2.2 is not available. (It has been shown for the scalar, weakly singular integral equation in three dimensions, that the eigenfunctions together with a finite number of adjoint functions for each eigenvalue are complete in L^2 [30].) For the strip, we must employ a decomposition of the form (6.27) to analyze the operator $\mathcal{L}$, with a nonvanishing perturbation $\mathcal{R}$. The normal part $\mathcal{H}$ will have a diagonal Fourier representation, and is related to the physical optics or infinite plane approximation for scattering from the strip.

This treatment provides error estimates for the solution on the interior of the strip, away from the edges. The error at the edges, where the solution is singular, must be estimated separately, as is done in Section 6.2.5.

The operator $\mathcal{L}$ has an infinite-dimensional Fourier representation, with matrix elements given by

$$L_{rs} = d^{-1}\langle e^{i\beta_r k_0 x}, \mathcal{L} e^{i\beta_s k_0 x}\rangle \tag{6.73}$$

where d is the width of the strip and $\langle \cdot, \cdot \rangle$ is the L^2 inner product. The normalized spatial frequency β_r is defined by $\beta_r = r/D$, $r = 0, \pm 1, \pm 2, \ldots$, where $D = d/\lambda$. Using the spectral representation

$$g(x, x') = \frac{i}{4\pi} \int_{-\infty}^{\infty} \frac{dk_x}{\sqrt{k_0^2 - k_x^2}} e^{ik_x(x-x')} \tag{6.74}$$

for the kernel of (6.4), L_{rs} becomes

$$L_{rs} = \frac{\eta}{2\pi^2 D} \int_{-\infty}^{\infty} \frac{dk}{\sqrt{1-k^2}} \frac{\sin[\pi D(k-\beta_r)]}{k - \beta_r} \frac{\sin[\pi D(k-\beta_s)]}{k - \beta_s} \tag{6.75}$$

where $k = k_x/k_0$ is a normalized spatial frequency.

In [26], the asymptotic expansion

$$L_{rr} \sim \frac{\eta}{2\sqrt{1-\beta_r^2}} - \frac{\eta}{2\pi^2(1-\beta_r^2)}\left[i + \frac{\beta_r \ln\left(i\beta_r + \sqrt{1-\beta_r^2}\right)}{\sqrt{1-\beta_r^2}}\right] D^{-1}$$

$$+ O(D^{-3/2}), \quad D \to \infty \tag{6.76}$$

is obtained for $|\beta_r| \neq 1$. The first term is the limiting value for an infinite plane, and higher order terms represent the effect of diffraction by the edges of the strip. Equation (6.76) breaks down if $|\beta_r| = 1$. These values of β_r correspond to surface wave current modes with spatial frequency k_0. The radiated fields for these modes travel along the strip. For $|\beta_r| = 1$, the asymptotic expansion becomes

$$L_{rr} \sim \frac{\sqrt{2}\eta}{3}(1-i)D^{1/2} + \frac{\eta}{8\sqrt{2\pi}}(1+i)D^{-1/2}$$

$$- \frac{\eta}{6\pi^2}\left[i + \frac{7\sqrt{2}}{4}\right]D^{-1} + O(D^{-3/2}), \quad D \to \infty \tag{6.77}$$

as shown in [26].

The Fourier representation of the operator $\mathcal{H}$ can now be given as the diagonal operator with elements L_{rr}. Since $\mathcal{H}$ is diagonal, it is a normal operator. The nonnormal part of the EFIE is $\mathcal{R} = \mathcal{L} - \mathcal{H}$. We seek to estimate the eigenvalues of $\mathcal{L}$ using the eigenvalues of $\mathcal{H}$, which are the diagonal elements L_{rr}. As a consequence of the Bauer-Fike theorem [31], we have the relative error bound

$$\min_r \frac{|L_{rr} - \Lambda_s|}{|L_{rr}|} \leq \|\mathcal{H}^{-1/2}\mathcal{R}\mathcal{H}^{-1/2}\| \tag{6.78}$$

where Λ_s is an eigenvalue of $\mathcal{L}$. It can be shown [26] that the norm on the right of (6.78) is asymptotically $O(1)$ as $D \to \infty$, so that the diagonal elements L_{rr} provide estimates of the eigenvalues of $\mathcal{L}$, with bounded error as D becomes large.

6.2.3.1 Discretized Operator

We now study the method of moments for the flat strip, using the spectral estimates obtained above. For a regular mesh, the expansion and testing functions are of the form $f_n(x) = f(x - x_n)$ and $t_n(x) = t(x - x_n)$, where $x_n = (n - 1/2)h - d/2$. The index n ranges from 1 to N, where the total number of degrees of freedom is $N = d/h$.

The Fourier representation of the discretized EFIE is

$$\hat{L}_{rs} = \frac{1}{N} \sum_{m,n=1}^{N} e^{-ik_0(\beta_r x'_m - \beta_s x'_n)} Z_{mn} \tag{6.79}$$

This can be viewed as a change of basis to surface currents of the form

$$u_r(x) = \sum_{m=1}^{N} e^{i\beta_r x} t_m(x) \tag{6.80}$$

evaluated at the node points x_n. By making use of (6.74), the moment matrix elements can be written as

$$Z_{mn} = \frac{\eta}{2n_\lambda} \int_{-\infty}^{\infty} \frac{dk}{\sqrt{1-k^2}} e^{i2\pi k(m-n)/n_\lambda} \tilde{t}(-k)\tilde{f}(k) \tag{6.81}$$

where $\tilde{t}(k)$ and $\tilde{f}(k)$ are the Fourier transforms of the basis functions $t(x)$ and $f(x)$, normalized by $1/h$. The Fourier representation can then be expressed as

$$\hat{L}_{rs} = \frac{\eta}{2n_\lambda^2 D} \int_{-\infty}^{\infty} \frac{dk}{\sqrt{1-k^2}} F_r(k) F_s(k) \tilde{t}(-k) \tilde{f}(k) \tag{6.82}$$

where

$$F_r(k) = \frac{\sin[\pi D(k - \beta_r)]}{\sin[\pi(k - \beta_r)/n_\lambda]} \tag{6.83}$$

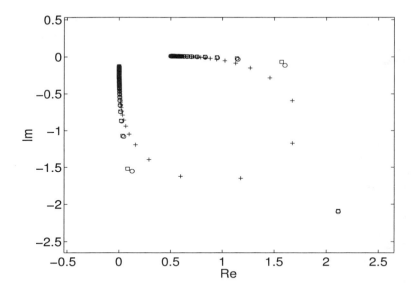

Figure 6.3 Spectrum of $\overline{Z}$ normalized to $\eta = 1$ for strip, $D = 20$, $n_\lambda = 10$. Pluses: computed. Circles: first order theoretical approximation, (6.76) and (6.77). Squares: numerical evaluation of L_{rr}. The zero spatial frequency eigenvalue is on the real axis, the surface wave eigenvalues are farthest from the origin, and the eigenvalues of high order, evanescent modes approach the negative imaginary axis.

is a periodic sinc function scaled by a factor of N.

Since $F_r(k)$ becomes a series of delta functions as D becomes large, the integral in (6.82) can be evaluated approximately by expanding the integrand about each of the maxima of $F_r(k)$ at $k_q = \beta_r + qn_\lambda$, $q = 0, \pm 1, \pm 2, \ldots$. Retaining the leading order contribution for each q leads to

$$\hat{L}_{rr} \simeq \sum_q \tilde{t}(k_q)\tilde{f}(k_q) L_{r+qN, r+qN}, \quad D \to \infty \quad (6.84)$$

The N eigenvalues of the moment matrix $\overline{Z}$ can be estimated by $\hat{\Lambda}_r \simeq \hat{L}_{rr}$, $-N/2 + 1 \le r \le N/2$, or $-(N-1)/2 \le r \le (N-1)/2$ if N is odd. As discussed for the cylinder in Section 6.2.1, the eigenvalues Λ_r of $\mathcal{L}$ with order $|r| > N/2$ are termed unmodeled eigenvalues.

Figure 6.3 compares the numerically computed spectrum of the moment matrix for a strip of width 20λ to the first order approximation $\hat{\Lambda}_r \simeq \hat{L}_{rr}$. For an infinite plane, modes of the form $e^{i\beta k_0 x}$ are eigenfunctions of $\mathcal{L}$. For high spatial frequencies ($|\beta| > 1$), the eigenvalues are equal to $-i(\eta/2)/\sqrt{\beta^2 - 1}$ and lie on the negative imaginary axis. The low-frequency modes ($|\beta| < 1$) have eigenvalues $(\eta/2)/\sqrt{1-\beta^2}$ on the real axis. If the width of the strip is finite, the spectrum becomes discrete, and edge diffraction couples the low-frequency and high-frequency modes with large eigenvalues ($|\beta_r| \simeq 1$), so that the two groups of eigenvalues join to form a loop in the complex plane. The approximate eigenvalues $\hat{L}_{rr}$ are degenerate for $\pm r$, whereas the eigenvalues $\hat{\Lambda}_r$ of $\overline{Z}$ are distinct, since the nonnormal perturbation $\mathcal{R}$ removes the degeneracy of the even and odd modes $\cos(k_0 \beta_r x)$ and $\sin(k_0 \beta_r x)$.

6.2.3.2 Spectral Error

The shift $\Delta L_{rr} = \hat{L}_{rr} - L_{rr}$ provides an estimate of the spectral error $\Delta \Lambda_{rr} = \hat{\Lambda}_{rr} - \Lambda_{rr}$ introduced by discretization. From (6.84), the approximate relative spectral error $E_r \simeq \Delta L_{rr}/L_{rr}$ is

$$E_r(n_\lambda, \alpha) \simeq \sqrt{1-\beta_r^2} \sum_{q \neq 0} \frac{\tilde{t}(k_q)\tilde{f}(k_q)}{\sqrt{1-k_q^2}} + \tilde{t}(-\beta_r)\tilde{f}(\beta_r) - 1 \quad (6.85)$$

where $k_q = \beta_r + qn_\lambda$. In deriving this expression, we have retained only the leading term in the asymptotic expansion of L_{rr}. Following the terminology introduced in Section 6.2.1.1, the first term is sampling error due to the unmodeled eigenvalues of $\mathcal{L}$. These eigenvalues influence the eigenvalues of the moment matrix, since the unmodeled eigenfunctions are aliased by discretization to lower order, modeled modes. The second term, $\tilde{t}(-\beta_r)\tilde{f}(\beta_r) - 1$, is smoothing error due to inaccurate representation of the modeled eigenfunctions ($|\beta_r| \leq n_\lambda/2$) by the expansion and testing functions.

We now specialize the treatment to the same piecewise polynomial basis functions used to discretize the EFIE for the cylinder in Section 6.2.1.1. In this case, the window function $\tilde{t}(-k)\tilde{f}(k)$ becomes $s^b(k)$, where

$$s(k) = \frac{\sin(\pi k/n_\lambda)}{\pi k/n_\lambda} \quad (6.86)$$

which is equivalent to (6.39). The exponent b is defined in (6.40).

For the piecewise polynomial basis functions, the spectral error becomes

$$E_{r,b} \simeq \frac{-i\sqrt{1-\beta_r^2}}{n_\lambda} \frac{\sin^b(\pi\beta_r/n_\lambda)}{\pi^b} \sum_{q \neq 0} \frac{(-1)^{bq} e^{i\pi\alpha k_q/n_\lambda}}{(k_q/n_\lambda)^b \sqrt{(k_q/n_\lambda)^2 - 1/n_\lambda^2}}$$
$$+ \frac{\sin^b(\pi\beta_r/n_\lambda)}{(\pi\beta_r/n_\lambda)^b} e^{i\pi\alpha\beta_r/n_\lambda} - 1 \quad (6.87)$$

For the $b = 0$ discretization, the testing function is taken to be symmetric about $x = 0$, so that $t(x) = \delta(x + \alpha h/2)/2 + \delta(x - \alpha h/2)/2$, and the expansion function is $f(x) = \delta(x)$. In this case, the spectral error reduces to

$$E_{r,0} \simeq \frac{2i}{n_\lambda} \ln[2\sin(\pi\alpha/2)] + \frac{2i}{5n_\lambda^3} - \frac{\pi^2\alpha^2\beta_r^2}{2n_\lambda^2} \tag{6.88}$$

for small β_r/n_λ. The first two terms represent the sampling error and the third term is the smoothing error. The sampling error is singular at $\alpha = 0$, due to the singularity of the Green's function at $x = x'$. If $\alpha = 1/3$, the leading term of $E_{r,0}$ vanishes, and the error becomes second order in n_λ^{-1}.

For $1 \leq b \leq 3$, the spectral error is smallest if $\alpha = 0$. In this case, the smoothing error is dominant for small β_r/n_λ, so that

$$E_{r,b} \simeq -\frac{b\pi^2\beta_r^2}{6n_\lambda^2} \tag{6.89}$$

The sampling error term can be approximated by expressions similar to those obtained for the cylinder in Section 6.2.1.1.

6.2.3.3 Quadrature Error

The use of approximate numerical quadrature to evaluate the integrals in (6.12) leads to an additional sampling error component. Following the treatment in Section 6.2.1.2, we employ the M-point integration rule

$$\int_{-h/2}^{h/2} dx\, f(x) \simeq \sum_{n=1}^{M} f(\xi_n) w_n \tag{6.90}$$

The function $\tilde{f}$ appearing in (6.82) is replaced by

$$\tilde{f}_M(k_x) = \frac{1}{h}\sum_{n=1}^{M} w_n f(\xi_n) e^{-ik_x\xi_n} \tag{6.91}$$

where $k_x = k_0 k$ and $f(x)$ is the basis function used to discretize the EFIE. The Fourier transform $\tilde{t}$ is modified similarly. The M-point first order integration rule is given by $w_n = \delta = h/M$ and $\xi_n = (n - 1/2)\delta - h/2$. For the $b = 1$ discretization, $\tilde{f}_M(k)$ becomes the periodic sinc function

$$\tilde{f}_M(k) = \frac{\sin\frac{\pi k}{n_\lambda}}{M\sin\frac{\pi k}{Mn_\lambda}} \tag{6.92}$$

Integrating (6.82) by including the leading contributions from the maxima of $\tilde{f}_M(k)$ at $k = qMn_\lambda$ for $q = \pm 1, \pm 2, \ldots$ yields the relative spectral error for small β_r,

$$E_{r,1,M} \simeq \frac{i2\ln 2}{Mn_\lambda} \tag{6.93}$$

Since $\Lambda_r \simeq \eta/2$ for small β_r, this result is equivalent to (6.49). The order of the quadrature error is the same for other expansion and testing functions.

6.2.3.4 Current Error

In general, the current solution $J(x)$ is singular at the edges $x = \pm d/2$ of the strip. For the TM polarization, the singularity is of the form $x^{-1/2}$, so that the current is not in the space of square integrable functions and its L^2 norm is infinite. As discussed above in Section 6.1.3.1, J belongs to a larger Hilbert space, the fractional-order Sobolev space $H^{-1/2}$. The current error including the edges of the strip can be studied in the norm associated with that Sobolev space [9–16]. In Section 6.2.5, we study the contribution of edge error to the forwared scattering amplitude. Here, we restrict attention to the current error on the interior region of the strip.

We define the interior $\tilde{S}$ of the strip by removing regions near the edges on the order of a wavelength in size. The current is finite on $\tilde{S}$, and so its L^2 norm exists and can be used to quantify the solution error. The interior current error can be estimated from the spectral error studied above.

The interior current error can be related to the spectral error by expanding the inverse of $\mathcal{L} = \mathcal{H} + \mathcal{R}$ as $\mathcal{L}^{-1} = \mathcal{H}^{-1} + \mathcal{R}\mathcal{H}^{-1}\mathcal{R} + \cdots$. The first term of this expansion, $\mathcal{H}^{-1}$, can be viewed as a first order scattering approximation, since it corresponds to the physical optics approximation, or the leading term of (6.76), together with a first order correction arising from higher order terms of (6.76). If the incident field is a plane wave, with an angle of incidence such that $\cos\phi^{\text{inc}} = \beta_r$ for some r, then the Fourier representation of the discretized incident field has components $\tilde{t}(\beta_r)\delta_{rs}$. In the first order scattering approximation, the approximate current solution has the Fourier representation $\hat{J}_s \simeq \hat{L}_{rr}^{-1}\tilde{t}(\beta_r)\delta_{rs}$.

The relative interior current error can then be estimated as

$$\frac{\|\hat{J} - J\|_{\text{RMS}(\tilde{S})}}{\|J\|_{\text{RMS}(\tilde{S})}} \simeq \left|\frac{\hat{L}_{rr} - L_{rr}\tilde{t}(\beta_r)}{\hat{L}_{rr}}\right| \tag{6.94}$$

where $\text{RMS}(\tilde{S})$ denotes the norm used in (6.15), but with a change of limits in the summation to $n_\lambda + 1 \leq n \leq N - n_\lambda - 1$. This eliminates the edge singularity from the error computation. By making use of the definition of $E_{r,b}$, and assuming that the spectral error is small, the relative RMS error becomes

$$\text{Err}_{\text{RMS}(\tilde{S})} \simeq |E_{r,b}(\beta_r) + 1 - \tilde{t}(\beta_r)| \tag{6.95}$$

We refer to the quantity on the right as the modified spectral error. The additional term $1 - \tilde{t}(\beta_r)$ eliminates the smoothing error due to the testing functions and reduces the smoothing error term of (6.85) to $\tilde{f}(\beta_r) - 1$. If the testing functions are delta functions located at the node points x_n, then $\tilde{t}(\beta_r) = 1$ and the right-hand side reduces to the relative spectral error $|E_{r,b}^{(1)}(\beta_r)|$.

Figure 6.4 shows the relative RMS current error on the interior of a strip of width $D = 10$ for a plane wave incident field as a function of the angle of incidence. The discretization density is fixed at $n_\lambda = 10$, and the reference solution is obtained using

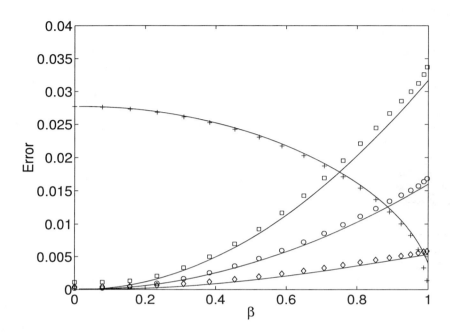

Figure 6.4 Relative current error norm for a plane wave incident at an angle of $\phi^{\text{inc}} = \cos^{-1}\beta$, $D = 10$, $n_\lambda = 10$. Error computed over 8λ interior region of the strip. Diamonds: $b = 0$, $\alpha = 1/3$. Pluses: $b = 1$ (point matching, pulse basis, single-point integration rule). Circles: $b = 1$ (point matching, pulse basis, high order integration rule). Squares: $b = 2$ (point matching, piecewise linear basis). Solid lines: theoretical approximations, (6.95) and (6.87), and the single point integration rule estimate from [26].

a discretization density of $n_\lambda = 100$ with geometrical grid refinement near the edges of the strip.

Figure 6.5 shows the relative RMS current error on the interior of the strip for a fixed incidence angle as a function of discretization density. The reference solution is obtained using a discretization density of $n_\lambda = 140$ with geometrical grid refinement at the edges of the strip. The error is asymptotically first order in n_λ^{-1} for the $b = 1$ case with a single-point integration rule, and is second order for the other discretization schemes.

6.2.3.5 Scattering Amplitude Error

The spectral error can also be related to the error in the forward scattering amplitude $S(\phi^{\text{inc}})$. The scattered field E^{sc} is discretized using the expansion functions f_n, to

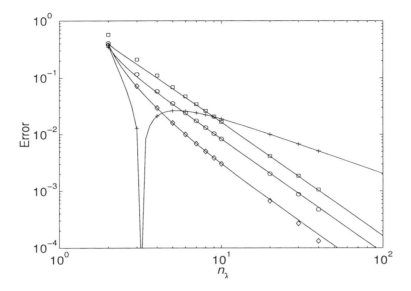

Figure 6.5 Relative current error as a function of n_λ for a plane wave incident at an angle of $\phi^{\text{inc}} = \pi/4$, $D = 10$, $n_\lambda = 10$, for 8λ interior region of the strip. Diamonds: $b = 0$, $\alpha = 1/3$. Pluses: $b = 1$ (point matching, pulse basis, single-point integration rule). Circles: $b = 1$ (point matching, pulse basis, high order integration rule). Squares: $b = 2$ (point matching, piecewise linear basis). Solid lines: theoretical approximations, (6.95) and (6.87), and the single point integration rule estimate from [26].

yield a vector with components

$$E_n^{\text{sc}} = \int dx\, f_n(x) E^{\text{sc}}(x) \tag{6.96}$$

If the scattered field is a plane wave traveling away from the scatterer at an angle $\phi^{\text{sc}} = \pi - \phi^{\text{inc}}$, the Fourier representation of the discretized scattered field is $\tilde{f}(\beta_r)\delta_{rs}$, where $\beta_r = \cos\phi^{\text{inc}}$, and we assume for simplicity that ϕ^{inc} is such that r is an integer. In the first order scattering approximation, the forward scattering amplitude is

$$\hat{S}(\phi^{\text{inc}}) \simeq \frac{\tilde{t}(\beta_r)\tilde{f}(\beta_r)}{\hat{\Lambda}_r} \tag{6.97}$$

From this approximation, we obtain the relative error

$$\frac{|S(\phi^{\text{inc}}) - \hat{S}(\phi^{\text{inc}})|}{|S(\phi^{\text{inc}})|} \simeq |E_r^{(1)}(\beta_r)| \tag{6.98}$$

where $E_r^{(1)}$ is the first term (sampling error) of (6.85) and $S(\phi^{\text{inc}})$ is the exact scattering amplitude. To leading order, the smoothing error does not contribute to the error in the forward scattering amplitude. Numerical results for the scattering amplitude are shown in Section 6.2.5.

6.2.4 Flat Strip—TE

The domain of $\mathcal{N}$ is the fractional-order Sobolev space $H^{1/2}$, which contains the square integrable functions vanishing at the edges of the strip [15]. Accordingly, we define the Fourier representation to be

$$N_{rs} = d^{-1} c_r c_s \langle \cos(\beta_r k_0 x), \mathcal{N} \cos(\beta_s k_0 x) \rangle \qquad (6.99)$$

for $r = 0, 2, 4, \ldots$ and $s = 0, 2, 4, \ldots$. Here, c_r is a normalization constant such that the transformation to the Fourier representation is unitary, and we also redefine β_r to be $r/(2D)$. For $r = 1, 3, 5, \ldots$ or $s = 1, 3, 5, \ldots$, the corresponding cosine function is replaced by a sine function in (6.99).

As in the TM case, $\mathcal{N}$ can be divided into a normal part and a nonnormal perturbation, such that the diagonal elements N_{rr} provide a first order approximation to the eigenvalues of $\mathcal{N}$. For the even modes, the diagonal elements become

$$N_{rr} = \frac{\eta}{2\pi^2 D} \int_{-\infty}^{\infty} dk \, \sin^2(\pi Dk) \sqrt{1 - k^2} \left[\frac{1}{(k - \beta_r)^2} - \frac{1}{(k - \beta_r)(k + \beta_r)} \right] \qquad (6.100)$$

For the odd modes, $\sin^2(\pi Dk)$ is replaced by $\cos^2(\pi Dk)$. In [26], the expansion

$$N_{rr} \sim \frac{\eta \sqrt{1 - \beta_r^2}}{2} - \frac{\eta}{2\pi^2} \left[i - \frac{\ln\left(i\beta_r + \sqrt{1 - \beta_r^2}\right)}{\beta_r \sqrt{1 - \beta_r^2}} \right] D^{-1}$$
$$+ O(D^{-2}), \quad D \to \infty \qquad (6.101)$$

is derived for $\beta_r \neq 1$. This result is valid for both odd and even modes. For $\beta_r = 1$,

$$N_{rr} \sim \frac{\eta \sqrt{2}(1 + i)}{2\pi} D^{-1/2} - \frac{i\eta}{\pi^2} D^{-1} + O(D^{-3/2}), \quad D \to \infty \qquad (6.102)$$

which is the self-coupling of the surface wave mode. From (6.101) and (6.102), it can be seen that up to a factor of $\eta/4$, the spectrum of the EFIE for the TE polarization is approximately the inverse of the spectrum of the TM EFIE. This is to be expected, since the product of the two operators is a compact perturbation of the identity [32].

6.2.4.1 Discretized Operator

Discretizing the integral operator $\mathcal{N}$ in the same manner as for the TM case leads to moment matrix elements

$$Z_{mn} = \frac{\eta}{2n_\lambda} \int_{-\infty}^{\infty} dk \sqrt{1-k^2} e^{i\pi k\alpha/n_\lambda} e^{i2\pi k(m-n)/n_\lambda} \tilde{t}(-k)\tilde{f}(k) \qquad (6.103)$$

where $\tilde{t}$ and $\tilde{f}$ are defined as before. The Fourier representation of the approximate operator $\hat{\mathcal{N}}$ becomes

$$\hat{N}_{rs} = \frac{\eta}{2n_\lambda^2 D} \int_{-\infty}^{\infty} dk \sqrt{1-k^2} e^{i\pi k\alpha/n_\lambda} F_r(k) F_s(k) \tilde{t}(-k)\tilde{f}(k) \qquad (6.104)$$

The function $F_r(k)$ is

$$F_r(k) = c_r' \left[\frac{\sin\left[\pi D(k-\beta_r)\right]}{\sin\left[\pi(k-\beta_r)/n_\lambda\right]} - (-1)^r \frac{\sin\left[\pi D(k+\beta_r)\right]}{\sin\left[\pi(k+\beta_r)/n_\lambda\right]} \right] \qquad (6.105)$$

where c_r' is a normalization constant.

Figure 6.6 shows the first order theoretical approximation to the spectrum of $\mathcal{N}$, together with the computed spectrum for a strip of width 10λ. The EFIE is discretized using pulse expansion functions and point matching with a density of $n_\lambda = 10$.

6.2.4.2 Spectral Error

Following the derivation of (6.87) for the TM polarization, the relative spectral error for piecewise polynomial basis functions is

$$E_{r,b} \simeq \frac{i}{n_\lambda \sqrt{1-\beta_r^2}} \frac{\sin^b (\pi\beta_r/n_\lambda)}{\pi^b} \sum_{q \neq 0} \frac{(-1)^{bq} e^{i\pi\alpha k_q/n_\lambda} \sqrt{(k_q/n_\lambda)^2 - 1/n_\lambda^2}}{(k_q/n_\lambda)^b}$$

$$+ \frac{\sin^b (\pi\beta_r/n_\lambda)}{(\pi\beta_r/n_\lambda)^b} e^{i\pi\alpha\beta_r/n_\lambda} - 1 \qquad (6.106)$$

where $k_q = \beta_r + qn_\lambda$. For small β_r/n_λ, the summation over q can be evaluated in closed form, which leads to

$$E_{r,0} \simeq -\frac{in_\lambda}{2} \csc^2(\pi\alpha/2) + \frac{i\alpha\pi\beta_r}{n_\lambda} \qquad (6.107a)$$

$$E_{r,1} \simeq \beta_r \tan(\pi\alpha/2) - \frac{i\pi\alpha\beta_r}{n_\lambda} - \frac{\pi^2\beta_r^2}{6n_\lambda^2} \qquad (6.107b)$$

$$E_{r,2} \simeq -\frac{2i\beta_r^2}{n_\lambda} \ln\left[2\sin(\pi\alpha/2)\right] - (2/3 + \alpha^2)\frac{\pi^2\beta_r^2}{2n_\lambda^2} \qquad (6.107c)$$

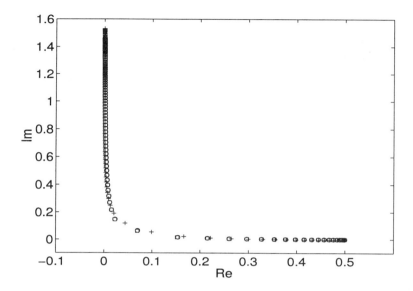

Figure 6.6 Spectrum of TE EFIE normalized to $\eta = 1$ for strip, $D = 10$, $b = 1$ discretization, $n_\lambda = 10$. Pluses: computed. Circles: first order theoretical approximation, (6.101), (6.102), and (6.107b). Squares: numerical computation of $\hat{N}_{rr}$. The zero spatial frequency eigenvalue is on the real axis, the surface wave eigenvalues are closest to the origin, and the eigenvalues of high order, evanescent modes approach the positive imaginary axis.

As for the cylinder, $E_{r,0}$ becomes large as the discretization density increases. The error $E_{r,1}$ for the $b = 1$ discretization is infinite at $\alpha = 1$, which corresponds to point matching at the edges of the pulse expansion functions, where the radiated electric field is singular. For $b = 2$, $E_{r,2}$ is infinite at $\alpha = 0$, due to the discontinuity of the triangle function at the apex, and the leading term vanishes at $\alpha = 1/3$. For the $b = 2$ case, we employ the same symmetric testing function $t(x) = \delta(x+\alpha h/2)/2 + \delta(x - \alpha h/2)/2$ used in Section 6.2.2.1.

6.2.4.3 Current Error

For the TE polarization, the L^2 norm of the error exists over the entire strip, since the current is finite as long as there are no sources on the strip itself. The current is singular, but the singularity is weaker than for the TM polarization and has the form $x^{1/2}$ near the edges of the strip. As in the TM case, we eliminate the edge regions and consider the interior current error.

An additional complication of the error analysis for the TE polarization is the strong surface wave components of the surface current. From (6.102), it can be seen

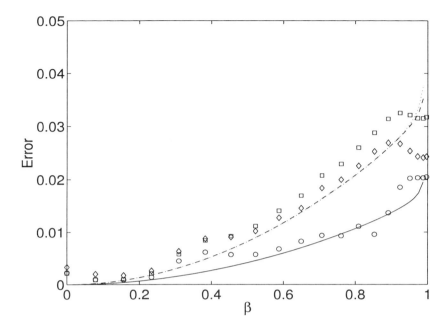

Figure 6.7 Relative error in the mode with spatial frequency $k_0\beta$ for a TE-polarized plane wave incident at an angle of $\phi^{\text{inc}} = \cos^{-1}\beta$, $D = 10$, $n_\lambda = 10$. Error computed over 8λ interior region of the strip. Circles: $b = 1$ (point matching, pulse basis). Squares: $b = 2$ (symmetric point matching, piecewise linear basis, $\alpha = 1/3$). Diamonds: $b = 3$ (pulse testing, piecewise linear basis). Theoretical approximations, (6.106); solid line: $b = 1$, dashed line: $b = 2$, dotted line: $b = 3$.

that the eigenvalue of the surface wave mode becomes smaller as the width of the strip increases. Because of this, the amplitudes of modes with spatial frequencies near k_0 are larger relative to the dominant physical optics mode than is the case for the TM polarization, and they cannot be ignored in the error analysis. To overcome this, for the TE polarization we compute the error in the amplitude of the dominant physical optics mode alone, rather than the total current error.

Figure 6.7 shows the relative RMS error for the dominant mode of the current on the interior of a strip of width $D = 10$ for a plane wave incident field as a function of the angle of incidence. The discretization density is fixed at $n_\lambda = 10$. Figure 6.8 shows the relative dominant-mode current error using the same norm as in Figure 6.7 for a fixed incidence angle as a function of discretization density. The reference solution is obtained using a discretization density of $n_\lambda = 140$ with geometrical grid refinement.

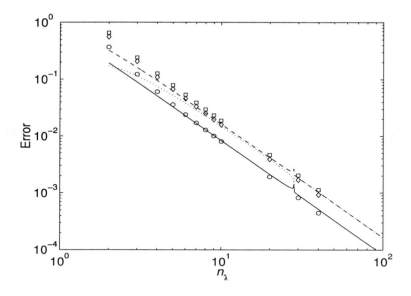

Figure 6.8 Relative error in the mode with spatial frequency $k_0\beta$ as a function of n_λ for a plane wave incident at an angle of $\phi^{\text{inc}} = \pi/4$, $D = 10$, $n_\lambda = 10$, for 8λ interior region of the strip. Circles: $b = 1$ (point matching, pulse basis). Squares: $b = 2$ (symmetric point matching, piecewise linear basis, $\alpha = 1/3$). Diamonds: $b = 3$ (pulse testing, piecewise linear basis). Theoretical approximations, (6.106); solid line: $b = 1$, dashed line: $b = 2$, dotted line: $b = 3$.

6.2.4.4 Quadrature Error

For the $b = 0$ and $b = 1$ discretizations, the contribution to the moment matrix elements of the hypersingular term in (6.6) can be evaluated analytically. The remaining term which must be integrated numerically has the same kernel as for the TM polarization. Since the spectra for the two polarizations are approximately equal for small β_r, the spectral error due to quadrature error for the TE polarization is the same as that given by (6.93) for $b \leq 1$.

For $b > 1$, the hypersingular term must be integrated numerically. As was shown in Section 6.2.2.2, the spectral error can be reduced if the singularity of the integrand is weakened by integrating by parts. If this is done, it can be shown that the moment matrix elements are given by

$$Z_{mn} = \frac{\eta}{2n_\lambda} \int_{-\infty}^{\infty} \frac{dk}{\sqrt{1-k^2}} e^{i\pi k\alpha/n_\lambda} e^{i2\pi k(m-n)/n_\lambda} \left[\tilde{t}(-k)\tilde{f}(k) - \frac{\tilde{t}'(-k)\tilde{f}'(k)}{k_0^2} \right]$$

(6.108)

where $\tilde{t}'(k)$ and $\tilde{f}'(k)$ are the Fourier transforms of $t'(x)$ and $f'(x)$, scaled by a factor of $1/h$. The Fourier representation of $\overline{Z}$ is then

$$N_{rs} = \frac{\eta}{2n_\lambda^2 D} \int_{-\infty}^{\infty} \frac{dk}{\sqrt{1-k^2}} e^{i\pi k\alpha/n_\lambda} F_r(k) F_s(k) \left[\tilde{t}(-k)\tilde{f}(k) - \frac{\tilde{t}'(-k)\tilde{f}'(k)}{k_0^2} \right] \quad (6.109)$$

For $b = 3$ and the first order, M-point integration rule with weights $w_n = \delta = h/M$ and abscissas $\xi_n = (n - 1/2)\delta - h/2$, we have

$$\tilde{t}_M(k) = \frac{\sin\frac{\pi k}{n_\lambda}}{M \sin\frac{\pi k}{Mn_\lambda}} \quad (6.110a)$$

$$\tilde{f}_M(k) = \cos\frac{\pi k}{Mn_\lambda} \left(\frac{\sin\frac{\pi k}{n_\lambda}}{M \sin\frac{\pi k}{Mn_\lambda}} \right)^2 \quad (6.110b)$$

$$\tilde{t}'_M(k) = 2ih^{-1} \sin\frac{\pi k}{n_\lambda} \quad (6.110c)$$

$$\tilde{f}'_M(k) = \frac{2i \sin^2\frac{\pi k}{n_\lambda}}{hM \sin\frac{\pi k}{Mn_\lambda}} \quad (6.110d)$$

The spectral error introduced by the quadrature rule can be approximated by expanding the integrand of (6.109) around the maxima at $k = qMn_\lambda$ for $q = \pm 1, \pm 2, \ldots$ and retaining the leading term for each q. This leads to an error of

$$E_{r,3,M} = -\frac{i}{Mn_\lambda \sqrt{1-\beta_r^2}} \sum_{q \neq 0} \frac{(-1)^q \left[1 - \beta_r^2 (-1)^{qM}\right]}{\sqrt{\left(q + \frac{\beta_r}{Mn_\lambda}\right)^2 - \left(\frac{1}{Mn_\lambda}\right)^2}} \quad (6.111)$$

For small β_r, the resulting error is identical to (6.93).

6.2.5 Flat Strip—Edge Error

In the preceding sections, we analyzed the solution and scattering amplitude error on the interior or smooth part of the flat strip. We now consider the error at the edges of the strip, due to inaccurate representation of the singularity of the current. We treat the case of the TM polarization, since the singularity is strongest and the edge effect has the greatest impact on the scattered fields. As discussed earlier, the current is not square integrable near the edges, so the appropriate error measure here is the scattering amplitude error.

A direct analysis of the scattering amplitude error using the principle of quasioptimality described in Section 6.1.3.1 leads to an error estimate of order $h^{1/2}$. The actual error is much smaller than this estimate. The scattering amplitude (6.19) is stationary with respect to perturbations of the solution [19], so that the scattering amplitude error is second order relative to the current error. Here, we employ this property of the scattering amplitude to estimate the error due to the edge regions of the strip. This result is then combined with the interior current error estimate obtained in Section 6.2.3, yielding an error estimate for the entire strip.

By introducing the adjoint equation $\mathcal{L}^\dagger J^a = E^{sc}$, we can write the scattering amplitude in variational form as [1, 9]

$$S = -\frac{k_0 \eta}{4} \frac{\langle E^{sc}, J \rangle \langle J^a, E^{inc} \rangle}{\langle \mathcal{L}J, J^a \rangle} \quad (6.112)$$

By substituting the approximate currents $\hat{J} = J + \Delta J$ and $\hat{J}^a = J^a + \Delta J^a$, the leading scattering amplitude error is found to be

$$\Delta S = (k_0 \eta / 4) \left(\langle \mathcal{L}\Delta J, \Delta J^a \rangle - S^{-1} \langle \Delta J, E^{sc} \rangle \langle \Delta J^a, E^{inc} \rangle \right) \quad (6.113)$$

which is second order in the solution errors ΔJ and ΔJ^a.

The first order term $\langle \Delta J, E^{sc} \rangle$ can be estimated by assuming the optimality of $\hat{J}$. For a piecewise constant expansion, $\hat{J}(x) = J(x_n) f(x - x_n)$, where $f(x)$ is the pulse function given by (6.9), and here x is the distance from the edge of the strip. If the amplitude of the incident field is unity, then by the results of Section 6.2.3, the amplitude of the current away from the edge is Λ_r^{-1}, where $r = D \cos \phi^{inc}$. The current at the edge is then on the order of $J(x) \simeq \Lambda_r^{-1}(x/\lambda)^{-1/2}$. By making use of this approximation, the first order edge error can be estimated as

$$\langle \Delta J, E^{sc} \rangle \simeq \Lambda_r^{-1} h \sum_{n=1}^{n_\lambda} (x_n/\lambda)^{-1/2} - \Lambda_r^{-1} \int_0^\lambda dx \, (x/\lambda)^{-1/2} \quad (6.114)$$

where $r = D \cos \phi^{inc}$ and $x_n = (n - 1/2)h$. It can be shown that $N^{-1/2} \sum_{n=1}^{N} (n - 1/2)^{-1/2} \sim 2 - cN^{-1/2}$, $N \to \infty$, where $c = \zeta(2)(1 - \sqrt{2}) \simeq 0.6$. The integral on the right evaluates to 2λ. The first order error is thus

$$\langle \Delta J, E^{sc} \rangle \simeq 0.6 \, \Lambda_r^{-1} \lambda n_\lambda^{-1/2} \quad (6.115)$$

which is of order $h^{1/2}$.

We now assume that the first term on the right of (6.113) is on the order of the second term or smaller, and that the solution error for the adjoint equation is similar to that of the original integral equation. This leads to the estimate

$$\Delta S \simeq (k_0 \eta / 4) 0.2 \Lambda_r^{-1} \lambda n_\lambda^{-1} \quad (6.116)$$

for one of the edges. This result indicates that the current error at the strip edges produces a first order scattering amplitude error.

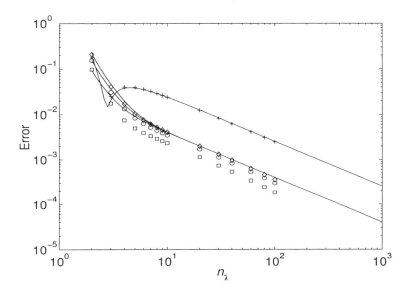

Figure 6.9 Relative scattering amplitude error for a strip, TM polarization, $D = 10$, $\phi^{\text{inc}} = -\phi^{\text{sc}} = \pi/3$. Diamonds: $b = 0$, $\alpha = 1/3$. Pluses: $b = 1$ (point matching, pulse basis, single-point integration rule). Circles: $b = 1$ (point matching, pulse basis, high order integration rule). Squares: $b = 2$ (point matching, piecewise linear basis). Solid lines: theoretical approximation, (6.117).

Normalizing this result by the approximate total scattering amplitude, which is $S \simeq (k_0\eta/4)\Lambda_r^{-1}(d + 4\lambda)$, assuming that d is large, and combining this with the interior error given by (6.98) leads to the total strip error estimate

$$\text{Err}_S = \frac{|S - \hat{S}|}{|S|} \simeq \left| E_r^{(1)}(\beta_r) + 0.4\, D^{-1} n_\lambda^{-1} \right| \quad (6.117)$$

For the pulse expansion ($b = 1$), the error estimate becomes

$$\text{Err}_S \simeq \left| -1.8\, i\eta \beta_r^2 \Lambda_r^{-1} n_\lambda^{-3} + 0.4\, D^{-1} n_\lambda^{-1} \right| \quad (6.118)$$

where $\Lambda_r \simeq (\eta/2)/\sqrt{1 - \beta_r^2}$ and $\beta_r = \cos\phi^{\text{inc}}$.

Figure 6.9 compares the estimate (6.117) to the computed error for a flat strip of length 10λ. The incident and scattered angles are $\phi^{\text{inc}} = -\phi^{\text{sc}} = \pi/3$. The reference solution is obtained using a discretization density of $n_\lambda = 200$.

6.2.6 Rectangular Cavity

For the circular cylinder, the moment matrix becomes ill-conditioned near internal resonances. Internal resonances are nonphysical, since the internally resonant mode does not couple to the external fields. If a small cut is made in the cylinder, the resonant mode radiates outside the scatterer, and the resonance becomes physical. For open, cavity-like scatterers in general, physical resonances occur at resonant frequencies, which are near the internal resonances of a similarly shaped closed scatterer. At these resonances, the moment matrix also becomes ill-conditioned.

A scatterer of this type is the open parallel strip or rectangular cavity, consisting of two perfectly conducting strips of width d, separated by a distance w. We consider also the half-open cavity, for which one of the ends is closed by a side wall, so that the cavity is open only on one end.

In order to analyze the spectrum of the EFIE for this scatterer, we employ the same Fourier series representation L_{rs} used for the flat strip above. Since the resonant modes of a closed rectangular cavity vanish at the endpoints of each cavity wall, we consider only the modes which vanish at the ends of the strip. For the TM polarization,

$$L_{rs} = c_r c_s \frac{k_0 \eta}{4} \int_{-d/2}^{d/2} \int_{-d/2}^{d/2} dx\, dx'\, H_0^{(1)}(k_0 |x - x'|) \sin(\beta_r k_0 x) \sin(\beta_s k_0 x') \tag{6.119}$$

where $\beta_r = r/(2D)$ is the normalized spatial frequency of the mode of order r and c_r is a normalization constant. If r or s is even, the corresponding sine function is replaced by the cosine function. For these modes, (6.75) becomes

$$L_{rs} = \frac{\eta}{2\pi^2 D} \int_{-\infty}^{\infty} \frac{dk}{\sqrt{1 - k^2}} F_r(-k) F_s(k) \tag{6.120}$$

where F_r is given by (6.105). For the TE polarization, the expression for L_{rs} has the same form as (6.120), but the factor of $\sqrt{1 - k^2}$ appears in the numerator of the integrand.

The cross-coupling between modes on parallel strips separated by a distance W in wavelengths can be found by modifying (6.75), so that

$$L_{rs}^c = \frac{\eta}{2\pi^2 D} \int_{-\infty}^{\infty} \frac{dk}{\sqrt{1 - k^2}} F_r(-k) F_s(k) e^{i 2\pi W \sqrt{1 - k^2}} \tag{6.121}$$

The presence of the second strip splits the spectrum of the EFIE into sums and differences of the self-coupling and cross-coupling. The sum and difference eigenvalues correspond to the family of cavity modes supported by this scatterer, and can be approximated by sums and differences of the diagonal elements of $\overline{\mathbf{L}}$ and $\overline{\mathbf{L}}^c$.

In this treatment, we consider the resonances as a function of the cavity length and width. Scattering resonances are often studied as a function of k_0, rather than the dimensionless lengths D and W employed here. Changing k_0 is equivalent to scaling the size of the cavity with the aspect ratio D/W fixed.

6.2.6.1 Resonant Case

If the cavity dimensions are such that one of the modes is at resonance, the eigenvalue corresponding to the resonant mode is purely real. The imaginary part of each cavity mode eigenvalue is proportional to the difference between the stored electric and magnetic energies associated with the mode. Poynting's theorem is

$$-\int_V dv\, \mathbf{E} \cdot \mathbf{J}^* = \oint_A d\hat{s} \cdot (\mathbf{E} \times \mathbf{H}^*) + i\omega \int_V dv\, (\mathbf{E} \cdot \mathbf{D}^* - \mathbf{B} \cdot \mathbf{H}^*) \qquad (6.122)$$

where V is a large region containing the scatterer, with boundary A, and the current $\mathbf{J}$ radiates the fields $\mathbf{E}$ and $\mathbf{H}$. This can be used to relate the eigenvalue of a mode to the stored and radiated energy. If the $\hat{z}$ component J of the surface current is an eigenfunction of $\mathcal{L}$, such that $\mathcal{L}J = \Lambda J$, then the left-hand side of (6.122) becomes

$$\Lambda \int_D ds\, |J|^2 \qquad (6.123)$$

The change in sign arises because the definition (6.4) implies that $\mathcal{L}J$ is equal to the negative of the field radiated by the current in free space at the surface of the scatterer. We assume that the eigenfunction J is normalized so that the integral in this expression evaluates to unity. Identifying the terms on the right of (6.122) as radiated power and stored power shows that the eigenvalues of the EFIE are of the form

$$\Lambda = P_{\text{rad}} + i2\omega(U_E - U_H) \qquad (6.124)$$

where P_{rad} is the radiation loss for the mode and U_E and U_H are the stored electric and magnetic energies, for a current with unit L^2 norm. At resonance, the energy storage is balanced, and the imaginary part vanishes. The radiation loss depends on the enclosed area and the size of the cavity opening. For an internal resonance of a closed scatterer, $P_{\text{rad}} = 0$.

By making use of (6.75) and (6.121), the difference eigenvalue for the mode of order r can be approximated as

$$\Lambda_r \simeq \frac{\eta}{2\pi^2 D} \int_{-\infty}^{\infty} \frac{dk}{\sqrt{1-k^2}} F_r(-k) F_r(k) \left[1 - (-1)^n e^{i2\pi W\sqrt{1-k^2}}\right] \qquad (6.125)$$

where n is an integer to be specified below. Since $F_r(k)$ is strongly peaked at $k = \beta_r$, Λ_r is small when the phase term in square brackets vanishes at β_r. This leads to the resonance condition

$$W\sqrt{1-\beta_r^2} = \frac{n}{2} \qquad (6.126)$$

The resonances of the open cavity are associated with the homogeneous solutions of the interior Helmholtz problem for a rectangular domain. By making use of the definition of β_r, it can be seen that (6.126) is equivalent to

$$\sqrt{[r/(2D)]^2 + [n/(2W)]^2} = 1 \qquad (6.127)$$

which is the resonance condition for the TM$_{rn}$ mode of a closed rectangular cavity with side lengths of D and W in wavelengths.

By solving (6.126) for β_r, we find that the normalized spatial frequency of the resonant mode is given by $\beta_r = \sqrt{1 - [n/(2W)]^2}$. Since β_r is real, the condition $n < 2W$ must hold. As n increases, the angle of propagation of the field radiated by the corresponding mode becomes closer to normal to the cavity walls, so that the field is more strongly confined to the cavity for larger n, and less power is radiated by the mode. Since the radiated power of each mode is given by the real part of the eigenvalue, the real part decreases as n increases. In order to find the index of the eigenvalue with smallest real part, we choose the largest possible value of n, which is

$$n = \lfloor 2W \rfloor \qquad (6.128)$$

where $\lfloor x \rfloor$ is the integer part of x. The corresponding value of β_r is approximately $\sqrt{\alpha/W}$, where $\alpha = 2W - \lfloor 2W \rfloor$ is the fractional part of $2W$. For a given surface current mode with order r, the corresponding difference eigenvalue will be smallest if there exists a closed cavity mode TM$_{rn}$ such that n is given by (6.128) and both mode numbers satisfy the resonance condition (6.127).

The real part of the eigenvalue corresponding to the resonant cavity mode can be estimated by taking the real part of (6.125). After combining the integrand for $k > 0$ with $k < 0$, we obtain

$$\text{Re}\{\Lambda_r\} \simeq \frac{4\eta\beta_r^2}{\pi^2 D} \int_0^1 \frac{dk}{\sqrt{1-k^2}} \frac{\sin^2[\pi D(k-\beta_r)]}{(k^2 - \beta_r^2)^2} \sin^2\left(\pi W \sqrt{1-k^2}\right) \qquad (6.129)$$

The integral over k can be estimated as [23]

$$\text{Re}\{\Lambda_r\} \simeq \frac{\eta\sqrt{W}}{2D}\left(\alpha + \frac{\pi^2\alpha^2}{18}\right) \qquad (6.130)$$

This analysis applies to the TE polarization as well if β_r is small.

Since the fields radiated by the cavity modes propagate predominantly in a direction near normal to the strips, the eigenvalue of the resonant mode for the half-open cavity (one end closed by a conducting wall) behaves similarly to that of the open cavity. The end wall decreases the radiation loss associated with the resonant mode, so that in general the real part of the resonant eigenvalue for the half-open cavity is smaller than that of the open cavity.

6.2.6.2 Near-Resonant Case

If the resonance condition (6.126) is not satisfied exactly, then the imaginary part of the smallest cavity mode eigenvalue is nonzero. For a given set of cavity dimensions D and W, let r be the positive integer for which (6.126) is closest to equality, and let W_r be the nearest cavity width for which the resonance condition holds, so that

$$W_r \sqrt{1 - \beta_r^2} = n/2 \qquad (6.131)$$

We then define the shift ν_r away from resonance by

$$\nu_r = W_r - W \qquad (6.132)$$

The difference eigenvalue as given by (6.125) can be estimated as [23]

$$\text{Im}\{\Lambda_r\} \simeq i\eta\pi\nu_r \qquad (6.133)$$

for the imaginary part of the smallest cavity mode eigenvalue.

6.2.6.3 Spectral Error

Smoothing error enters into the spectrum multiplicatively, so that its relative effect is unchanged for the small eigenvalues of resonant modes. Sampling error is additive, so that the relative error becomes large as the magnitude of the eigenvalue decreases. The sampling error is negligible for the cross-coupling between the two strips, since the kernel of the EFIE is smooth when the source and observation points are well separated, but sampling error does affect the value of the resonant eigenvalue through the self-coupling of the mode. We assume the use of an M-point integration rule for moment matrix elements, and employ the spectral error estimate (6.93) in the analysis given below.

Since the spectral shift due to quadrature error is imaginary, the effect of the error is to shift the locations of the resonances of the cavity. The resonances occur approximately at cavity dimensions such that $\nu_r = 0$. Including the discretization error, the imaginary part of the smallest eigenvalue becomes

$$\text{Im}\{\Lambda_r\} \simeq i\eta\pi\nu_r + \frac{i\eta \ln 2}{Mn_\lambda} \qquad (6.134)$$

The resulting shift in the location of the resonance, considered as a function of cavity width W, is approximately $\Delta W \simeq \ln 2/(\pi M n_\lambda)$. The shift in the location of the resonant frequency relative to k_0 is

$$\frac{\Delta k}{k_0} \simeq \frac{\ln 2}{\pi M n_\lambda W} \qquad (6.135)$$

which is small for typical values of $n_\lambda = 10$ and $M = 1$, as long as $W \geq 1$.

Because low-order resonances with small values of r are closely spaced, a more stringent error criterion is the shift in the location of the resonance relative to the spacing between adjacent resonances. The distance between resonances is $W_{r+1} - W_r \simeq rW/(4D^2)$, so that the shift in the resonance relative to the spacing between resonances is $4D^2 \ln 2/(\pi r M n_\lambda W)$. If we require the relative shift to be less than 10%, we obtain the condition

$$n_\lambda > \frac{9D^2}{rMW} \qquad (6.136)$$

on the discretization density. This restriction is equivalent to requiring that the discretization error ΔL_{rr} be much smaller than the distance between neighboring cavity mode eigenvalues. This result shows the increased sensitivity to discretization error of resonance structures, which has been observed in numerical experiments [33].

6.2.6.4 Numerical Results

In this section, we validate the theoretical spectral estimates for the cavity by comparison with numerical results. Eigenvalues are estimated by numerical diagonalization of the moment matrix. All computed results are given for the TM polarization. For numerical results on the effects of discretization error, pulse basis functions are employed with a single point integration rule for the off-diagonal elements of the moment matrix and a first order analytical approximation for the diagonal elements. The spectral error for this discretization scheme is obtained in [26] as

$$\Delta L_{rr} \simeq \frac{i\eta}{n_\lambda} (\ln \pi - 1) \qquad (6.137)$$

for small β_r. This low-order method is convenient for illustrating the effect of discretization, since the spectral error is significant for relatively small cavity dimensions. By comparison with (6.93), the spectral error for this discretization scheme is equivalent to that of an M-point integration rule with $M \simeq 5$.

Figure 6.10 shows the real and imaginary parts of the smallest cavity mode eigenvalue of the EFIE, as a function of the cavity depth D, for a width of $W = 3$. The computed and theoretical results differ for the real part of the eigenvalue. The analysis given above neglects the radiation loss due to coupling of the resonant mode to other modes. Because of this, the theoretical estimates underestimate the real part of the eigenvalue.

Figure 6.11 compares the imaginary part of the resonant eigenvalue of the moment matrix for both open and half-open cavities for $W = 2$ and $n_\lambda = 10$ to the theoretical estimates obtained above, in the presence of discretization error. For $W = 2$, the lowest order cavity mode is nearest to resonance. This width is slightly below the resonant value W_1 as defined by (6.131), but approaches resonance as D increases,

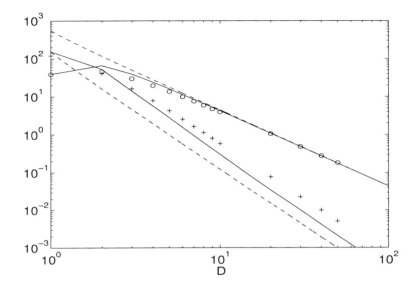

Figure 6.10 Smallest cavity mode eigenvalue as a function of depth, open cavity, $W = 3$. Circles: Imaginary part, computed value. Pluses: real part, computed value. Dashed lines: theoretical estimates, (6.130) and (6.133). Solid lines: numerical integration of (6.125).

since $W_1 \to 2$ as $D \to \infty$, so that the imaginary part of the exact eigenvalue decreases as D^{-2}. Since the location of the resonance is shifted by the discretization error, the imaginary part of the eigenvalue of the moment matrix does not fall off for large D, but approaches the discretization error (6.137) as D increases.

6.2.7 Higher-Order Basis Functions

The piecewise polynomial discretizations studied in the previous sections employ a single degree of freedom for each mesh element. Accuracy can be improved by using higher-order polynomials and allowing more than one degree of freedom per element. As before, we discretize the operator $\mathcal{L}$ for the flat strip on the regular mesh with midpoints $x_n = (n - 1/2)h - d/2$, $n = 1, \ldots, N$. On each element, let the finite-dimensional trial subspace V^p consist of the $p + 1$ basis functions $f_{na} = f_a(x - x_n)$, $a = 0, \ldots, p$. For interpolatory polynomials, we also choose $p + 1$ nodes $x_{na} = x_n + y_a$ on each element, so that $f_a(y_b) = \delta_{ab}$, and assume that the nodes y_a on the reference element are evenly spaced. Continuity is not enforced at the element boundaries. The density of the degrees of freedom (unknowns) is $(p + 1)n_\lambda$.

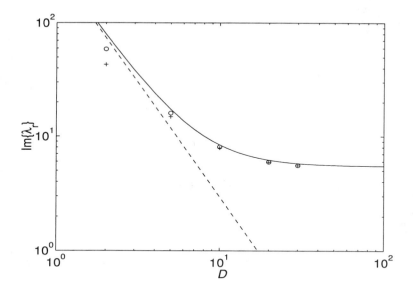

Figure 6.11 Imaginary part of minimal eigenvalue for open cavity as a function of depth, $W = 2$, $n_\lambda = 10$ discretization. Pluses: computed value, open cavity. Circles: computed value, half-open cavity. Solid line: theoretical result, including discretization error (6.137). Dashed line: theoretical result, without discretization error.

The moment method discretization of $\mathcal{L}$ using the basis defined above has matrix elements

$$\hat{L}_{mn,ab} = h^{-1} \langle f_{ma}, \mathcal{L} f_{nb} \rangle \tag{6.138}$$

By making use of (6.74), the elements of the moment matrix can be expressed as

$$\hat{L}_{mn,ab} = \frac{\eta}{2n_\lambda} \int \frac{dk}{\sqrt{1-k^2}} e^{i2\pi k(m-n)/n_\lambda} F_a(-k) F_b(k) \tag{6.139}$$

where $k = k_x/k_0$ and $F_a(k)$ is the Fourier transform of $f_a(x)$, normalized by h^{-1}.

6.2.7.1 Orthonormal Polynomials

For the case of an orthonormal basis, the approximate operator can be represented in the trial space as

$$\hat{\mathcal{L}} J = h^{-1} \sum_{m,n=1}^{N} \sum_{a,b=0}^{p} f_{ma} \hat{L}_{mn,ab} \langle f_{nb}, J \rangle \tag{6.140}$$

where the angle brackets denote the L^2 inner product. As in Section 6.2.3, we obtain a spectral estimate for this operator using the normal part of $\hat{\mathcal{L}}$. This leads to

$$\hat{\Lambda}_r \simeq \frac{\eta}{2n_\lambda^2 D} \int_{-\infty}^{\infty} \frac{dk}{\sqrt{1-k^2}} \frac{\sin^2[\pi D(k-\beta_r)]}{\sin^2[\pi(k-\beta_r)/n_\lambda]} F^{(p)^2}(\beta_r, k) \qquad (6.141)$$

In this expression, we have defined

$$F^{(p)}(k_1, k_2) = \sum_{a=0}^{p} F_a(k_1) F_a(-k_2) \qquad (6.142)$$

The function $F^{(p)}(k_1, k_2)$ is the Fourier representation of the projection operator from the space of L^2 functions onto the trial subspace spanned by the basis polynomials. Where $F^{(p)}(k, k)$ is close to unity, the Fourier mode with normalized spatial frequency k can be accurately represented by the basis functions.

Expanding the integrand of (6.141) about the local maxima at $k_q = \beta_r + qn_\lambda$, $q = 0, \pm 1, \pm 2$, and retaining the leading order term for large D at each maximum yields

$$\hat{\Lambda}_r \simeq \frac{\eta}{2} \sum_{q=-\infty}^{\infty} \frac{F^{(p)^2}(\beta_r, k_q)}{\sqrt{1-k_q^2}}, \quad D \to \infty \qquad (6.143)$$

From (6.76) and (6.143), the relative spectral error can be estimated as

$$E_r^{(p)}(n_\lambda) \simeq \frac{\eta}{2\Lambda_r} \sum_{q \neq 0} \frac{F^{(p)^2}(\beta_r, k_q)}{\sqrt{1-k_q^2}} + F^{(p)^2}(\beta_r) - 1 \qquad (6.144)$$

where $F^{(p)}(k) \equiv F^{(p)}(k, k)$. As before, we decompose the spectral error into two contributions,

$$E_r^{(p)} = E_r^{(p),1} + E_r^{(p),2} \qquad (6.145)$$

where the sampling error $E_r^{(p),1}$ is the first term on the left-hand side of (6.144) and the smoothing error is $E_r^{(p),2}(n_\lambda) = F^{(p)^2}(\beta_r) - 1$.

The Legendre polynomials are an orthonormal set on the interval $[-1, 1]$. We set $f_{na} = c_a P_a[2(x - x_n)/h]$, where $P_a(x)$ is the Legendre polynomial of order a, and $c_a = \sqrt{2a+1}$ is a normalization constant. Since the Fourier transform of a Lagrange polynomial is proportional to a spherical Bessel function, we have the explicit formula:

$$F^{(p)}(k_1, k_2) = \sum_{a=0}^{p} (2a+1) j_a(\pi k_1/n_\lambda) j_a(\pi k_2/n_\lambda) \qquad (6.146)$$

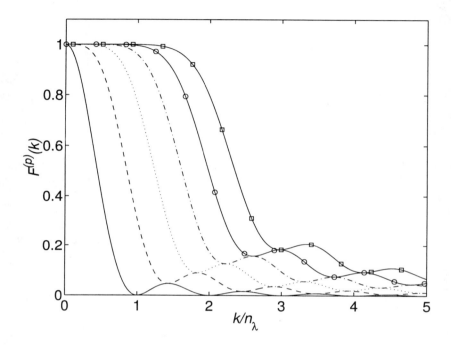

Figure 6.12 $F^{(p)}(k)$ for Legendre expansion. Solid line: $p = 0$. Dashed: $p = 1$. Dotted: $p = 2$. Dash-dot: $p = 3$. Solid/circles: $p = 4$. Solid/squares: $p = 5$.

This result is shown in Figure 6.12. The improvement in the approximation power as p increases can be seen by expanding $F^{(p)}(k)$ for small k:

$$\begin{aligned}
F^{(0)}(k) &\simeq 1 - \frac{\pi^2 k^2}{3n_\lambda^2} \\
F^{(1)}(k) &\simeq 1 - \frac{\pi^4 k^4}{45 n_\lambda^4} \\
F^{(2)}(k) &\simeq 1 - \frac{\pi^6 k^6}{1575 n_\lambda^6}
\end{aligned} \qquad (6.147)$$

The exponential convergence obtained with p-refinement is evident in the decrease of the constant with p.

6.2.7.2 Interpolatory Polynomials

For an interpolatory basis, we estimate the eigenvalues of $\hat{\mathcal{L}}$ as an operator in the discrete space of values at the nodes x_{ma}, so that

$$\hat{\Lambda}_r \simeq N^{-1} \left\langle e^{i\beta_r k_0 x}, \hat{\mathcal{L}} e^{i\beta_r k_0 x} \right\rangle_{x_{ma}} \quad (6.148)$$

where the subscript x_{ma} indicates that the inner product is discrete. This implies that the resulting error estimates are RMS error at the nodes x_{ma}, rather than L^2 error. Evaluating the inner product yields

$$\hat{\Lambda}_r \simeq N^{-1} \sum_{m,n=1}^{N} \sum_{a,b=0}^{p} e^{-i\beta_r k_0 x_{ma}} \hat{L}_{mn,ab} e^{i\beta_r k_0 x_{nb}} \quad (6.149)$$

Using (6.139) for $\hat{L}_{mn,ab}$ shows that the forms of (6.141) and (6.144) remain the same for an interpolatory basis, but with the new definition

$$F^{(p)}(k_1, k_2) = \sum_{a=0}^{p} e^{ik_1 t_a} F_a(k_2) \quad (6.150)$$

where $t_a = k_0 y_a$. This is the Fourier representation of the truncated completeness relation for the interpolatory basis, composed with the projection of the trial subspace onto the discrete space of values at the nodes y_a.

The Lagrange polynomials are an example of an interpolatory basis. In this case, we set $f_{na} = L_{p,a}(x - x_n)$, where $L_{p,a}(x)$ is the ath canonical polynomial of order p with respect to the $p+1$ nodes y_a. If the nodes y_a are equally spaced on $[-h/2, h/2]$, then by expanding (6.150) we obtain

$$\begin{aligned}
F^{(0)}(k) &\simeq 1 - \frac{\pi^2 k^2}{6 n_\lambda^2} \\
F^{(1)}(k) &\simeq 1 - \frac{\pi^2 k^2}{3 n_\lambda^2} \\
F^{(2)}(k) &\simeq 1 - \frac{\pi^4 k^4}{60 n_\lambda^4} \\
F^{(3)}(k) &\simeq 1 - \frac{\pi^4 k^4}{405 n_\lambda^4}
\end{aligned} \quad (6.151)$$

Figure 6.13 shows $F^{(p)}(k)$ for this basis.

6.2.7.3 Current Error

As in Sections 6.2.3 and 6.2.4, we study the error of the solution on the interior region $\tilde{S}$ of the strip, away from the edges, where the solution is singular. For a plane

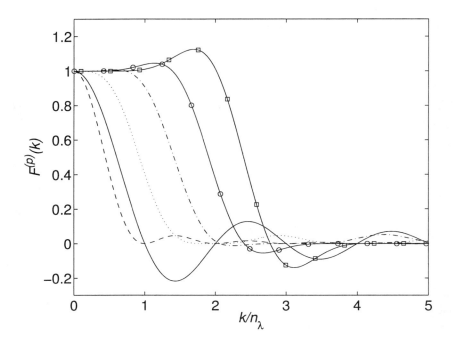

Figure 6.13 $F^{(p)}(k)$ for Lagrange interpolatory polynomial expansion. Solid line: $p = 0$. Dashed: $p = 1$. Dotted: $p = 2$. Dash-dot: $p = 3$. Solid/circles: $p = 4$. Solid/squares: $p = 5$.

incident wave, in the normal approximation the leading order solution is

$$u(x) \simeq \Lambda_r^{-1} e^{ik_0 x \cos \phi} \tag{6.152}$$

where $r = D \cos \phi$ and we neglect the fractional part of $D \cos \phi$ for large D. The approximate solution obtained using the method of moments is

$$\hat{u}(x) \simeq \hat{\Lambda}_r^{-1} F^{(p)}(\beta_r) e^{ik_0 x \cos \phi} \tag{6.153}$$

which is equivalent to the expression used in Section 6.2.3.4 for low-order basis functions.

The interior error of $\hat{J}$ relative to the exact solution can be estimated as

$$\frac{\|J - \hat{J}\|_{\tilde{S}}}{\|J\|_{\tilde{S}}} \simeq \left| \frac{\hat{\Lambda}_r - \Lambda_r F^{(p)}(\beta_r)}{\hat{\Lambda}_r} \right| \tag{6.154}$$

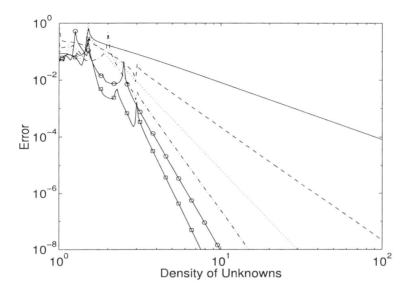

Figure 6.14 Relative error estimate (6.155) for Legendre expansion, $\beta_r = 1/2$. The independent variable is the total density of the degrees of freedom per wavelength, $(p+1)n_\lambda$. Solid line: $p = 0$. Dashed: $p = 1$. Dotted: $p = 2$. Dash-dot: $p = 3$. Solid/circles: $p = 4$. Solid/squares: $p = 5$.

where the L^2 norm is used for orthogonal polynomials and RMS error at the nodes for interpolatory polynomials. Assuming that the error is small relative to Λ_r, this becomes

$$\text{Err} \simeq \left| E^{(p),1} + F^{(p)}(\beta_r)[F^{(p)}(\beta_r) - 1] \right| \quad (6.155)$$

This error estimate is shown for the Legendre and Lagrange expansions in Figures 6.14 and 6.15 as a function of the density of degrees of freedom per wavelength.

The asymptotic order of the current solution error in the grid size $h = \lambda/n_\lambda$ for the bases of Examples 1 and 2 can be obtained from the smoothing error term $F^{(p)}(\beta_r) - 1$ of (6.155). For the Legendre expansion, from (6.147), the order of convergence is $2p + 2$. For the Lagrange expansion, from (6.151), the order is $p + 2$ for p even, and $p + 1$ for p odd.

Figure 6.16 shows the error estimate (6.155) as a function of order. The exponential convergence of the interior solution error is evident in the figure, since the order is shown on a linear scale and the logarithmic error falls off roughly linearly.

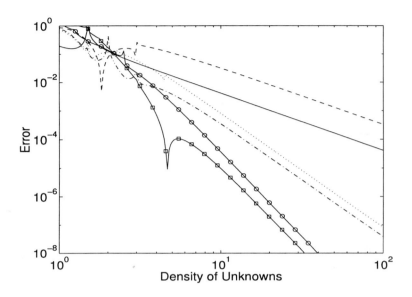

Figure 6.15 Relative error estimate (6.155) for Lagrange expansion, $\beta_r = 1/2$. The independent variable is the total density of the degrees of freedom per wavelength, $(p+1)n_\lambda$. Solid line: $p = 0$. Dashed: $p = 1$. Dotted: $p = 2$. Dash-dot: $p = 3$. Solid/circles: $p = 4$. Solid/squares: $p = 5$.

6.2.7.4 Scattering Amplitude Error

If the approximate forward scattering amplitude is computed by projecting the scattered field into the trial subspace before forming the inner product with the current solution, then the relative error is

$$\frac{|S(\phi) - \hat{S}(\phi)|}{|S(\phi)|} \simeq E_1^{(p)}(\cos\phi, n_\lambda) \tag{6.156}$$

As before, this estimate neglects the approximation error near the edges of the strip.

For the Legendre and Lagrange expansions, the terms of the summation over q in (6.144) fall off as $1/q$ or faster, and the order of the sum in $h = \lambda/n_\lambda$ is the same as that of the $q = 1$ term, so that

$$E_1^{(p)} \simeq \frac{\eta}{2\Lambda_r n_\lambda} F^{(p)2}(\beta_r, \beta_r + n_\lambda) \tag{6.157}$$

By examination of $F^{(p)}$, we find that the sampling error, as well as the error in the forward scattering amplitude, is of order $2p + 3$ for both sets of basis functions.

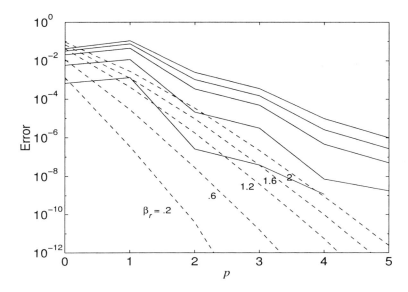

Figure 6.16 Relative error estimate (6.155) as a function of order, for several values of β_r, $n_\lambda = 10$. Solid lines: Lagrange polynomials. Dashed lines: Legendre polynomials.

6.2.8 Summary

In this section, we have defined the concept of spectral error and applied this to obtain solution error estimates for the cylinder, strip, and cavity. The solution error is qualitatively different for these scatterers, due to resonance, edge effect, and sensitivity to incidence angle. This has been an obstacle to the practical error analysis of the method of moments. The spectral error, on the other hand, is similar for all the scatterers considered. Equations (6.38) and (6.85) show that the absolute sampling error is the same for the cylinder and strip to leading order in N, and the relative smoothing error is identical for the two scatterers. This arises because the spectral error is locally determined by the singularity of the kernel and the smoothness of the eigenfunctions. The absolute sampling error and the relative smoothing error are relatively insensitive to the scatterer geometry.

The strong dependence of solution error on the scatterer geometry is caused by three properties: the edge effect, the modes excited by the incident field, and the change in the magnitudes of the eigenvalues with the scatterer shape. The edge effect leads to large solution error at scatterer corners and edges, as studied in Section 6.2.5. The modes excited by the incident field determine the weighting of the spectral error in the solution error estimate. The third effect arises when the geometry is such that a low-order, propagating mode is near resonance and has a small eigenvalue. The small eigenvalue leads to large relative sampling error. The canonical scatterer illustrating this effect is the cavity studied in Section 6.2.6.

6.3 SPECTRAL CONVERGENCE THEORY—3D

Three-dimensional electromagnetic scattering problems represent a significant computational challenge, since vector integral operators are involved, and many degrees of freedom are generally required. In this section, we study the properties of the EFIE for a flat plate and obtain spectral error estimates for the method of moments for the vector EFIE. We also discuss a method for reducing the spectral error by regulating the singular kernel of the integral equation.

6.3.1 Flat Plate

In Section 6.2.3, we observed that spectral estimates can be obtained for an infinite strip by approximating the eigenfunctions of the two-dimensional EFIE as modes of the form $e^{ik_0\beta x}$. Here, we apply this approach to the three-dimensional case.

Spectral estimates for the operator $\mathcal{T}$ can be obtained by approximating the eigenfunctions as $\hat{t}e^{ik_0\boldsymbol{\beta}\cdot\mathbf{r}}$, where $\hat{t}$ is a constant unit vector tangential to the scatterer S. Applying $\mathcal{T}$ and making use of the Fourier representation of the kernel $g(\mathbf{r}, \mathbf{r}')$ yields

$$\mathcal{T}\hat{t}e^{ik_0\boldsymbol{\beta}\cdot\mathbf{r}} = \frac{k_0\eta d^2}{8\pi^2}\int dk_x\,dk_y\,\frac{e^{i\mathbf{k}\cdot\mathbf{r}}}{k_z}\frac{\sin\left[(k_x - k_0\beta_x)d/2\right]}{(k_x - k_0\beta_x)d/2}\frac{\sin\left[(k_y - k_0\beta_y)d/2\right]}{(k_y - k_0\beta_y)d/2}$$
$$\times \left[\hat{t} - \frac{\mathbf{k}(\boldsymbol{\beta}\cdot\hat{t})}{k_0}\right] \qquad (6.158)$$

where $k_z = \sqrt{k_0^2 - k_x^2 - k_y^2}$. By expanding the integrand about the maxima of the sinc functions at $\mathbf{k} = k_0\boldsymbol{\beta}$, (6.158) can be approximated as

$$\mathcal{T}\hat{t}e^{ik_0\boldsymbol{\beta}\cdot\mathbf{r}} \sim \frac{\eta d^2}{8\pi^2}\frac{e^{ik_0\boldsymbol{\beta}\cdot\mathbf{r}}}{\sqrt{1-\beta^2}}\left[\hat{t} - \boldsymbol{\beta}(\boldsymbol{\beta}\cdot\hat{t})\right]\int dk_x\,dk_y \qquad (6.159)$$
$$\times \frac{\sin\left[(k_x - k_0\beta_x)d/2\right]}{(k_x - k_0\beta_x)d/2}\frac{\sin\left[(k_y - k_0\beta_y)d/2\right]}{(k_y - k_0\beta_y)d/2}, \quad d \to \infty$$

Evaluating the integrals yields

$$\mathcal{T}\hat{t}e^{ik_0\boldsymbol{\beta}\cdot\mathbf{r}} \sim \frac{\eta}{2\sqrt{1-\beta^2}}\left[\hat{t} - \boldsymbol{\beta}(\boldsymbol{\beta}\cdot\hat{t})\right]e^{ik_0\boldsymbol{\beta}\cdot\mathbf{r}}, \quad d \to \infty \qquad (6.160)$$

If $\hat{t}$ is parallel to $\boldsymbol{\beta}$, then the vector in square brackets is a scalar multiple of $\hat{t}$, so that $\hat{t}e^{ik_0\boldsymbol{\beta}\cdot\mathbf{r}}$ is an approximate eigenfunction of $\mathcal{T}$, with the eigenvalue

$$\Lambda^{\text{TE}} \simeq \frac{\eta}{2}\sqrt{1-\beta^2} \qquad (6.161)$$

This result corresponds to the curl-free modes on the scatterer, which are related to the eigenfunctions of the 2D operator $\mathcal{N}$ for the TE polarization. For the divergence-free modes, $\hat{t}$ is perpendicular to β, and (6.160) yields the approximate eigenvalue

$$\Lambda^{\text{TM}} \simeq \frac{\eta}{2} \frac{1}{\sqrt{1-\beta^2}} \qquad (6.162)$$

These modes are similar to eigenfunctions of the 2D operator $\mathcal{L}$ for the TM polarization. The estimates (6.161) and (6.162) break down as β approaches the singularities at $|\beta| = 1$.

The continuous operator $\mathcal{T}$ has an infinite number of discrete eigenvalues. The spectrum has accumulation points at the origin and at $i\infty$, due to the curl-free and divergence-free modes for large β. As $|\beta| \to \infty$, the estimate (6.161) approaches $i\infty$, whereas (6.162) goes to zero.

If $\mathcal{T}$ is discretized using the method of moments, a matrix operator is obtained. This matrix has a finite number of eigenvalues. The eigenvectors of the moment matrix correspond approximately to the eigenfunctions of $\mathcal{T}$ with normalized spatial frequency β in the range $-k_{\max}/k_0 \leq \beta \leq k_{\max}/k_0$, where $k_{\max}$ is the maximum spatial frequency represented in the trial space of basis functions, or the Nyquist frequency of the mesh. If the mesh length or average element size is h, then the maximum frequency is $k_{\max} = \pi/h$.

From (6.161), the eigenvalue of the moment matrix with largest magnitude corresponds to $\beta = k_{\max}/k_0$ and can be estimated as

$$\Lambda^{\text{TE}}_{\max} \simeq \frac{i\eta}{2} \frac{\pi}{k_0 h} \qquad (6.163)$$

This can also be written as $i\eta n_\lambda/4$, where $n_\lambda = \lambda/h$ is the linear discretization density or number of unknowns per wavelength used in the previous section. This estimate is valid for low-order basis functions. For higher-order bases (p-refinement), more precise estimates similar to those obtained in Section 6.2.7 for the 2D case are required

If $k_0 d \ll 1$, the smallest curl-free eigenvalue corresponds to a mode with the lowest possible nonzero spatial frequency, which is $\beta = \pi/(k_0 d)$. Equation (6.161) yields the estimate

$$\Lambda^{\text{TE}}_{\min} \simeq \frac{i\eta}{2} \frac{\pi}{k_0 d} \qquad (6.164)$$

for this eigenvalue. The curl-free part of the spectrum of $\mathcal{T}$ lies between the two extremal eigenvalues given by (6.163) and (6.164).

The spectrum of $\mathcal{T}$ also has an accumulation point at the origin, corresponding to divergence-free modes with large spatial frequency. The eigenvalue of the discretized operator with the least magnitude arises from the divergence-free mode with maximal spatial frequency, $\beta = \pi/(k_0 h)$. Equation (6.162) yields the estimate

$$\Lambda_{\min}^{\text{TM}} \simeq -\frac{i\eta}{2}\frac{k_0 h}{\pi} \tag{6.165}$$

for the smallest eigenvalue.

For small $k_0 h$, the largest divergence-free eigenvalue corresponds to the mode with lowest spatial frequency ($\beta \simeq 0$). If $k_0 d \ll 1$, the approximation used in obtaining (6.159) breaks down for $\beta = 0$, since k_z is rapidly varying near the maxima of the sinc functions at $k_x = 0$, $k_y = 0$. A more accurate evaluation of the integral in (6.158) is obtained by expanding the sinc functions, retaining terms up to second order, and integrating up to the first zeros of the sinc functions. This leads to the estimate

$$\Lambda_{\max}^{\text{TM}} \simeq \frac{k_0\eta d^2}{2\pi^2} \int_0^{2\pi/d} \int_0^{2\pi/d} \frac{dk_x\, dk_y}{k_z}\left[1 - \frac{k_x^2 d^2}{24} - \frac{k_y^2 d^2}{24}\right] \tag{6.166}$$

Evaluating the integral in the limit as $k_0 \to 0$ yields

$$\Lambda_{\max}^{\text{TM}} \simeq -\frac{i\eta k_0 d}{2\pi}\left[(4 - \pi^2/9)\log(1 + \sqrt{2}) - \pi^2\sqrt{2}/9\right] \simeq -\frac{i\eta}{2}\frac{k_0 d}{\pi} \tag{6.167}$$

At low frequencies, the divergence-free part of the spectrum lies between the values given by (6.165) and (6.167).

Figure 6.17 shows computed spectra of the discretized EFIE for a plate of side $d = 1\text{m}$ at various values of k_0. Rao-Wilton-Glisson (RWG) vector basis functions [5] are employed on a regular triangular mesh. The discretization length h is taken to be the distance between nodes, 0.1m. Qualitatively, the spectrum of $\mathcal{T}$ is a combination of the spectra of $\mathcal{L}$ and $\mathcal{N}$ in Figures 6.3 and 6.6. For large $k_0 d$, the spectrum lies on a curve in the complex plane, from the positive imaginary axis, through $\eta/2$ on the real axis, where the curl-free and divergence-free parts meet, and ending near the origin along the negative imaginary axis, as can be seen in Figure 6.17(a). As k_0 decreases, the two parts of the spectrum separate, moving towards the accumulation points of the spectrum of $\mathcal{T}$ at 0 and $+i\infty$. This is the phenomenon of low-frequency breakdown, which will be considered in Section 6.4.2.9.

6.3.2 Rooftop Basis Functions

In the previous section, we obtained spectral estimates for the continuous operator $\mathcal{T}$ on the flat plate. Here, we study the discretization error for a simple type of vector basis function, the rooftop. We define two sets of node points, $\mathbf{r}_{mn1} = mh\hat{x} + (n - 1/2)h\hat{y}$, $m = 1, \ldots, M - 1$, $n = 1, \ldots, M$, and $\mathbf{r}_{mn2} = (m - 1/2)h\hat{x} + nh\hat{y}$, $m = 1, \ldots, M$, $n = 1, \ldots, M - 1$. For convenience, we assume that $d = Mh$. The

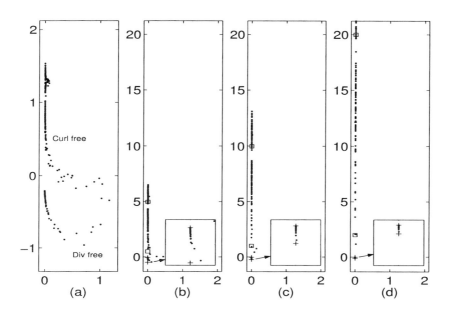

Figure 6.17 Moment matrix spectra for 1m × 1m plate at various wavenumbers ($\eta = 1$): (a) $k_0 = 4\pi$; (b) $k_0 = \pi$; (c) $k_0 = \pi/2$; (d) $k_0 = \pi/4$. Squares: theoretical extremal curl-free eigenvalue estimates, (6.163) and (6.164). Pluses: theoretical extremal divergence-free eigenvalue estimates, (6.165) and (6.167). As the frequency becomes small, the two parts of the spectrum separate; the curl-free eigenvalues moving to $+i\infty$, and the divergence-free eigenvalues to the origin.

rooftop functions are $\mathbf{f}_{mnj}(\mathbf{r}) = \mathbf{f}_j(\mathbf{r} - \mathbf{r}_{mnj})$, $j = 1, 2$, where

$$\mathbf{f}_1(\mathbf{r}) = \begin{cases} \hat{x}(1 - |x|/h) & -h \le x \le h, \ -h/2 \le y \le h/2 \\ 0 & \text{otherwise} \end{cases} \quad (6.168)$$

$$\mathbf{f}_2(\mathbf{r}) = \begin{cases} \hat{y}(1 - |y|/h) & -h/2 \le x \le h/2, \ -h \le y \le h \\ 0 & \text{otherwise} \end{cases} \quad (6.169)$$

These basis functions are a combination of the triangle function and the pulse function studied in Section 6.2.

For the rooftop basis, the moment matrix elements are

$$Z_{mnj,m'n'j'} = -ik_0\eta h^{-2} \int\int d\mathbf{r}\, d\mathbf{r}'\, g(\mathbf{r},\mathbf{r}') \Big[\mathbf{f}_{mnj}(\mathbf{r}) \cdot \mathbf{f}_{m'n'j'}(\mathbf{r}') \\ + k_0^{-2} \nabla \cdot \mathbf{f}_{mnj}(\mathbf{r}) \nabla' \cdot \mathbf{f}_{m'n'j'}(\mathbf{r}') \Big] \quad (6.170)$$

We obtain spectral estimates using the approximate eigenfunction $\hat{t}e^{ik_0\boldsymbol{\beta}\cdot\mathbf{r}}$ evaluated at the node points $\mathbf{r}_{mnj}$. This leads to the expression

$$\hat{\Lambda} \simeq \frac{1}{M^2} \sum_{mnj,m'n'j'} t_j e^{ik_0\boldsymbol{\beta}\cdot(\mathbf{r}_{mnj}-\mathbf{r}_{m'n'j'})} Z_{mnj,m'n'j'} \qquad (6.171)$$

where t_j, $j = 1, 2$ denotes the x and y components of $\hat{t}$. By applying the derivation leading to (6.158), we obtain

$$\hat{\Lambda} \simeq \frac{k_0 \eta h^2}{8M^2\pi^2} \int \frac{d\mathbf{k}}{k_z} \sum_{j,j'} t_j t_{j'} R^2(\mathbf{k} - k_0\boldsymbol{\beta})$$

$$\times \left[\mathbf{F}_j(-\mathbf{k}) \cdot \mathbf{F}_{j'}(\mathbf{k}) - \frac{1}{k_0^2} \mathbf{k} \cdot \mathbf{F}_j(-\mathbf{k})\, \mathbf{k} \cdot \mathbf{F}_{j'}(\mathbf{k}) \right] \qquad (6.172)$$

where

$$R(\mathbf{k}) = \frac{\sin(k_x d/2)\sin(k_y d/2)}{\sin(k_x h/2)\sin(k_y h/2)} \qquad (6.173)$$

$$\mathbf{F}_1(\mathbf{k}) = \hat{x} \left[\frac{\sin(k_x h/2)}{k_x h/2}\right]^2 \frac{\sin(k_y h/2)}{k_y h/2} \qquad (6.174)$$

$$\mathbf{F}_2(\mathbf{k}) = \hat{y} \frac{\sin(k_x h/2)}{k_x h/2} \left[\frac{\sin(k_y h/2)}{k_y h/2}\right]^2 \qquad (6.175)$$

By expanding the integrand about each of the maxima of $R^2(\mathbf{k} - k_0\boldsymbol{\beta})$ as in the derivation of (6.84), we arrive at the estimate

$$\hat{\Lambda} \simeq \frac{k_0\eta}{2} \sum_{p,q=-\infty}^{\infty} \sum_{j,j'=1}^{2} \frac{1}{k_{pqz}} t_j t_{j'} F_{jj}(\mathbf{k}_{pq}) F_{j'j'}(\mathbf{k}_{pq}) (\delta_{jj'} - k_{pqj} k_{pqj'}/k_0^2)$$

$$(6.176)$$

where $\mathbf{k}_{pq} = \hat{x}(k_0\beta_x + 2\pi p/h) + \hat{y}(k_0\beta_y + 2\pi q/h)$, the second subscript on F_{jj} denotes the x and y components of $\mathbf{F}_j$, and the subscript j on k_{pqj} denotes the x and y components k_{pqx} and k_{pqy}.

We can determine the order of the spectral error in h by inspection of (6.176). The smoothing error arises from the $p = q = 0$ term and is of order h^2, which can be seen by expanding $R(\mathbf{k})$ for small $\mathbf{k}$. For the sampling error, the term $1/k_{pqz}$ contributes a factor of h, the term $F_{jj}F_{j'j'}$ contributes a factor of at most h^4, and $k_{pqj}k_{pqj'}$ contributes a factor of h^{-2}, for an overall order of h^3. From this result, we see that the spectral convergence rate for the rooftop vector basis functions is identical to that of the low-order bases studied in Sections 6.2.1.1, 6.2.2.1, and 6.2.3.2 for the 2D case.

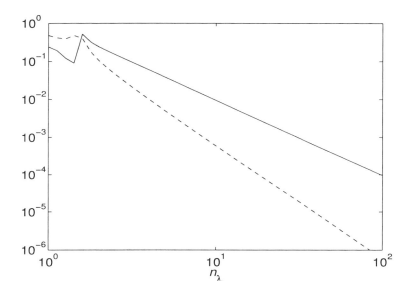

Figure 6.18 Spectral error for rooftop basis as a function of discretization density, divergence-free mode, normalized spatial frequency $\beta = 1/2$. Solid line: total error. Dashed line: sampling error.

Figure 6.18 shows the spectral error as a function of the discretization density n_λ, for $\hat{t} = \hat{x}$ and $\beta = \hat{y}/2$. Since the direction and phase vectors are perpendicular, the mode is divergence-free. Figure 6.19 shows the spectral error for a fixed discretization density as the vectors $\hat{t}$ and β are rotated. The spectral error is larger when $\hat{t}$ and β are parallel. In this case, the mode has the largest divergence, and the error contribution from the hypersingular term of the EFIE is greatest. The curves in Figure 6.19 are analogous to dispersion error plots for finite difference and finite element schemes.

6.4 ITERATIVE SOLUTION METHODS

For large linear systems, direct linear system solution methods are impractical, due to the high computational cost of matrix factorization. For electrically large scattering problems, the number of unknowns can be so large that filling the moment matrix is not feasible, and fast methods [2] must be used to compute matrix-vector multiplications with the moment matrix indirectly. For these reasons, iterative linear system solution algorithms are widely used in computational electromagnetics.

262 Fast and Efficient Algorithms in CEM

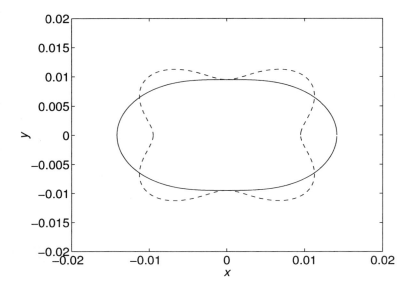

Figure 6.19 Spectral error as a function of angle, $n_\lambda = 10$. Solid line: $\hat{t}$ rotated; β fixed at $\hat{x}/2$. Dashed line: both $\hat{t}$ and β rotated.

There are two main classes of iterative algorithms. The first consists of stationary iterations based on matrix splittings [34, 35]. The second class is the nonstationary iterations, the most important of which are methods of the conjugate gradient (CG) type, or Krylov subspace iterations [36]. Both stationary and nonstationary iterations are used in computational electromagnetics [37]. Stationary iterations converge most rapidly for some problems, but in general are less robust than the Krylov subspace iterations. In this section, we restrict attention to methods of the latter type.

6.4.1 Iteration Count Estimates

The difficulty of solving a linear system by a Krylov subspace iteration is determined in large part by the matrix condition number. The condition number $\kappa(\overline{\mathbf{Z}})$ in the L^2 norm is the ratio of the largest and smallest singular values of the matrix. In this section, we review the classical estimates for the number of iterations required to solve a linear system to a given error tolerance.

The simplest Krylov subspace iteration is the conjugate gradient method. This algorithm applies only to Hermitian matrices. Since the moment matrix is non-

Hermitian, CG must be applied to the normal form

$$\overline{Z}^\dagger \cdot \overline{Z} \cdot x = \overline{Z}^\dagger \cdot b \qquad (6.177)$$

of the linear system (6.11) arising from the method of moments. This approach is termed CGNE. The natural error measure is the relative residual norm $r_n = |\overline{Z}^\dagger \cdot \overline{Z} \cdot x_n - \overline{Z}^\dagger \cdot b|/r_0$, where $r_0 = |\overline{Z}^\dagger \cdot \overline{Z} \cdot x_0 - \overline{Z}^\dagger \cdot b|$ and x_0 is the initial guess. Asymptotically, as the iteration count n increases, the residual error decays as $r_n \simeq \rho^n$, where ρ is the asymptotic convergence factor of the iteration. Approximating the spectrum of $\overline{Z}^\dagger \cdot \overline{Z}$ by an interval leads to the estimate [38]

$$\rho = \frac{\sqrt{\kappa(\overline{Z}^\dagger \cdot \overline{Z})} - 1}{\sqrt{\kappa(\overline{Z}^\dagger \cdot \overline{Z})} + 1} \qquad (6.178)$$

The condition number of $\overline{Z}^\dagger \cdot \overline{Z}$ is approximately $\kappa^2(\overline{Z})$, so that for large κ the number of iterations required to obtain an error of $r_n \leq \epsilon$ is

$$n_{\text{ITER}} \simeq \kappa(\overline{Z}) \frac{|\ln \epsilon|}{2} \qquad (6.179)$$

The iteration count for CGNE is thus asymptotically proportional to the condition number of the moment matrix.

For other iterative methods such as the biconjugate gradient method (BCG) or generalized minimum residual (GMRES), which can be applied to indefinite, non-self-adjoint matrices, the optimal convergence factor depends on the distribution of the spectrum of $\overline{Z}$ in the complex plane. If the spectrum is approximated by a disk not containing the origin, the convergence factor becomes [39]

$$\rho = \frac{\kappa(\overline{Z}) - 1}{\kappa(\overline{Z}) + 1} \qquad (6.180)$$

which leads to the same iteration count estimate (6.179) as CGNE.

In practice, actual convergence rates can deviate significantly from the estimate (6.179). If the right-hand side belongs to an invariant subspace, then the effective condition number is reduced to that of the operator restricted to the subspace. For a smooth scatterer, the current solution is close to the physical optics current, which is proportional to the incident field. In this case, the right-hand side is approximately in a small invariant subspace, and the iteration converges rapidly. For nonsmooth scatterers or singular source excitations, higher-order scattering couples this small subspace to the full solution space, leading to slower convergence.

6.4.2 Condition Number Estimates

The eigenvalue estimates of Sections 6.2 and 6.3.1 can be used to obtain moment matrix condition number estimates for the scatterer geometries considered. Together with the iteration count estimates of the previous section, these results determine the computational cost of obtaining the unknown current, if an indirect linear system solver is employed.

6.4.2.1 Circular Cylinder—TM

For the circular cylinder, (6.29) shows that for a regular discretization, the discretized EFIE has a complete system of eigenvectors and is therefore a normal matrix. The singular values of a normal matrix are equal to the magnitudes of the eigenvalues, so that we can obtain the condition number from the extremal eigenvalues.

The largest eigenvalue of the moment matrix arises from maximizing (6.35) over r. As a function of r, $|J_r(k_0 a) H_r^{(1)}(k_0 a)|$ is oscillatory and increasing for $|r| < k_0 a$, and decays monotonically for $|r| > k_0 a$. The maximum value occurs at $|r| \simeq k_0 a$. Using expansions of the Bessel and Hankel functions ([40], Equations 8.441 #3, 8.443, and 8.454), we arrive at the asymptotic expression

$$J_\nu(\nu) H_\nu^{(1)}(\nu) \simeq \frac{6^{2/3}(1 - i\sqrt{3})}{9\,\Gamma^2(2/3)} \nu^{-2/3} \quad \nu \to \infty \qquad (6.181)$$

From this result, it can be seen that the largest eigenvalue of $\overline{\mathbf{Z}}$ is

$$\lambda_{\max} \simeq \frac{\eta 2\pi(1 - i\sqrt{3})}{6^{4/3}\,\Gamma^2(2/3)} (k_0 a)^{1/3} \qquad (6.182)$$

where we have neglected the small shift due to discretization error. This eigenvalue corresponds to a surface wave mode with spatial frequency k_0 on the cylinder. For a large cylinder, the corresponding current mode radiates fields that travel in a direction tangential to the surface of the cylinder. The magnitude of the eigenvalue grows with the $1/3$ power of the electrical size of the cylinder. In Section 6.4.2.3, we will see that the corresponding growth rate exponent for the strip is $1/2$, since the surface wave mode is more strongly self-coupled for a flat scatterer.

The smallest eigenvalue of the moment matrix is more difficult to determine, due to the internal resonances associated with closed conducting bodies. It is well known that the EFIE becomes numerically unstable if $k_0 a$ is such that the interior boundary value problem with the Dirichlet condition on the surface of the cylinder has a nontrivial solution. As shown in Section 6.2.1.5, the corresponding eigenvalue of the moment matrix is then determined by the discretization error for the resonant mode, and will in general be small. If the frequency or the size of the cylinder is perturbed slightly, so that the resonant eigenvalue is canceled by the discretization error, the moment matrix becomes singular and has an infinite condition number. In general, near an internal resonance, the condition number is determined by the small eigenvalue of the resonant mode.

If no modes are near resonances, then the smallest eigenvalues of the moment matrix correspond to modes with rapid spatial oscillation. The spectrum of $\mathcal{L}$ has an

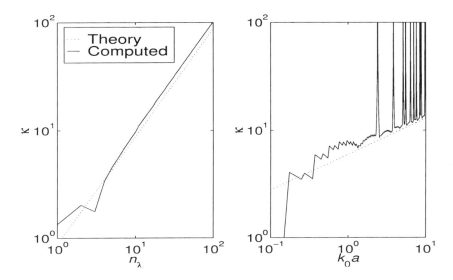

Figure 6.20 Condition number of the moment matrix, TM polarization, with point testing and pulse basis functions, as a function of discretization density n_λ for a cylinder of radius $k_0 a = \pi$, and as a function of $k_0 a$ for $n_\lambda = 10$. The dotted lines are the theoretical result given by (6.184).

accumulation point at the origin, due to the vanishing eigenvalues of eigenfunctions of increasingly large order. Employing a finite basis to discretize the EFIE leads to a cutoff of the spectrum near this accumulation point at the maximum spatial frequency representable in the discrete basis, so that the spectrum of $\overline{Z}$ corresponds to the N lowest order eigenvalues of $\mathcal{L}$. The smallest high-order eigenvalue arises from the discrete mode with largest spatial frequency, which corresponds to $|r| = N/2$. Applying the large-order expansion $J_\nu(x) H_\nu^{(1)}(x) \sim -i(\pi|\nu|)^{-1} + O(\nu^{-3})$, $\nu \to \infty$ ([40], Equation 8.452) to $\Lambda_{\min} = (\eta \pi k_0 a/2) J_{N/2}(k_0 a) H_{N/2}^{(1)}(k_0 a)$ leads to the result

$$\Lambda_{\min} \simeq -\frac{i\eta}{n_\lambda} \quad (6.183)$$

for the highest-order eigenvalue of the moment matrix. As long as n_λ is large enough that this eigenvalue is smaller in magnitude than the eigenvalues of low-order modes which may be near to internal resonance, the condition number can be approximated

as
$$\kappa(\overline{\mathbf{Z}}) \simeq 0.6\, n_\lambda (k_0 a)^{1/3} \qquad (6.184)$$

by making use of (6.182) and (6.183). Since internal resonances cause the magnitude of the smallest eigenvalue of the moment matrix to decrease, this estimate is in general a lower bound for the condition number. This condition number estimate is compared to computed values using the singular value decomposition of the moment matrix in Figure 6.20.

The condition number estimate (6.184) neglects the dependence of $\hat{\Lambda}_{\min}$ on the choices of expansion and testing functions. In order to include this dependence, we must take into account the spectral error in the estimate for $\hat{\Lambda}_{\min}$. If there is no relative shift between the basis and testing functions ($\alpha = 0$), then the smoothing error contribution is dominant, and the estimate for the smallest eigenvalue becomes $\Lambda_r \tilde{t}_r \tilde{f}_r$, where $r = N/2$. For the piecewise polynomial bases described in Section 6.2, from (6.39) $\tilde{t}_r \tilde{f}_r$ is equal to the bth power of the sinc function evaluated at $\pi/2$, where b is the order of the discretization scheme. This contributes a factor of $(2/\pi)^b$ to the denominator of the condition number, and we obtain the estimate

$$\kappa(\overline{\mathbf{Z}}) \simeq 0.6\, n_\lambda (k_0 a)^{1/3} \left(\frac{\pi}{2}\right)^b \qquad (6.185)$$

which takes into account the smoothing error. The exponential dependence on b matches the numerical observations reported by Dallas et al. [6]. If the relative shift α of the expansion and testing functions is nonzero, then the condition number may grow more rapidly than linearly with n_λ due to the sampling error, which is neglected in the derivation of (6.185). Certain values of the relative shift also cause the moment matrix to become singular, if the sampling error exactly cancels one of the eigenvalues of $\mathcal{L}$.

6.4.2.2 Circular Cylinder—TE

Since $|J'_\nu(x) H^{(1)\prime}_\nu(x)|$ grows as $|\nu|$ becomes large, in contrast to the TM case, the modes with high spatial frequencies have the largest eigenvalues for the TE polarization. This arises because the operator $\mathcal{N}$ has an accumulation point at $+i\infty$. The maximum eigenvalue of $\overline{\mathbf{Z}}$ corresponds to $q = N/2$ and is

$$\Lambda_{\max} \simeq \frac{i\eta n_\lambda}{4} \qquad (6.186)$$

where we have made use of $J'_\nu(x) H^{(1)\prime}_\nu(x) \sim i|\nu|/(\pi x^2)$, $\nu \to \infty$. Near internal resonances lead to small eigenvalues if $k_0 a$ is such that $J'_q(k_0 a)$ is small for some q. These small eigenvalues dominate the condition number for large $k_0 a$. For values of $k_0 a$ such that internal resonances are not dominant, the smallest eigenvalues correspond to the modes with $|q| \simeq k_0 a$, which are the surface wave modes for the cylinder. In this case, since $J'_\nu(\nu)H^{(1)'}_\nu(\nu) \simeq 0.2\,(1+i\sqrt{3})\nu^{-4/3}$ for large $|\nu|$, the smallest eigenvalue is

$$\Lambda_{\min} \simeq 0.3\,\eta(1+i\sqrt{3})(k_0 a)^{-1/3} \tag{6.187}$$

The resulting condition number estimate is

$$\kappa(\overline{\mathbf{Z}}) \simeq 0.4\,n_\lambda(k_0 a)^{1/3} \tag{6.188}$$

which is of the same order as the TM result.

6.4.2.3 Flat Strip—TM

As for the cylinder, employing a finite basis to discretize the EFIE leads to a cutoff of the spectrum near this accumulation point, so that the spectrum of $\overline{\mathbf{Z}}$ corresponds to the N lowest-order eigenvalues of $\mathcal{L}$. From (6.84), the eigenvalue of the moment matrix with the smallest magnitude corresponds to $\beta_r = n_\lambda/2$. For the moment, we retain only the $q = 0$ term of the summation in (6.84), and assume that $\tilde{t}(-n_\lambda/2)\tilde{f}(n_\lambda/2) \simeq 1$, and thereby obtain

$$\hat{\Lambda}_{\min} \simeq -\frac{i\eta}{n_\lambda} \tag{6.189}$$

in the infinite plane approximation. This eigenvalue is proportional to the discretization length $h = \lambda/n_\lambda$, as expected [11, 41, 42]. Since the corresponding eigenfunction has the highest spatial frequency, its self-coupling is the most strongly localized.

The largest eigenvalue of $\overline{\mathbf{Z}}$ arises from $|\beta_r| = 1$, which corresponds to the surface wave mode with spatial frequency k_0, and from (6.77) is

$$\Lambda_{\max} \simeq \frac{\eta\sqrt{2}}{3}(1-i)D^{1/2} \tag{6.190}$$

The field radiated by this current mode travels parallel to the surface, so that the surface wave has the strongest long range coupling. The surface wave mode can be considered to be antiresonant, since its approximate eigenvalue (which represents the energy stored by the mode for a unit surface current) grows with the width of the strip. If the strip lies in a lossy medium, so that k_0 has a nonzero imaginary part, then the singularities in the integrand of (6.82) at $|k| = 1$ are eliminated, and for large D the maximum eigenvalue becomes independent of the scatterer size.

In the normal approximation, the condition number of the moment matrix is $|\hat{\Lambda}_{\max}|/|\hat{\Lambda}_{\min}|$. By making use of the extremal eigenvalue estimates obtained above, we have

$$\kappa(\overline{\mathbf{Z}}) \simeq \frac{2}{3}n_\lambda D^{1/2} \tag{6.191}$$

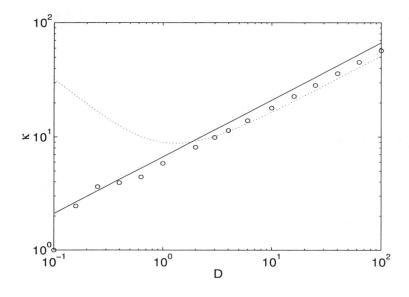

Figure 6.21 Condition number of moment matrix for strip, TM polarization, $n_\lambda = 10$, $b = 1$ discretization with single point integration rule. Circles: computed using SVD. Solid line: theoretical approximation, (6.191). Dotted line: theoretical approximation, (6.76) and (6.77), including discretization error.

Since $\hat{\Lambda}_{min}$ is determined by localized interactions, for an irregular discretization of a smooth scatterer with slowly varying discretization lengths, the condition number is determined by the smallest discretization length. As for the cylinder, the condition number depends on the choices of testing and expansion functions, but we neglect that dependence in (6.191) since it is the same as that of (6.185). Figure 6.21 shows the computed condition number for $n_\lambda = 10$ as a function of D and the theoretical estimate with and without discretization error and higher-order terms in the asymptotic expansion of L_{rr}.

6.4.2.4 Flat Strip—TE

For the TE polarization, the largest eigenvalue of the moment matrix arises from the discrete mode with largest spatial frequency. This corresponds to the estimate (6.101) evaluated at $r = N/2$, which leads to

$$\Lambda_{max} \simeq \frac{i\eta n_\lambda}{4} \qquad (6.192)$$

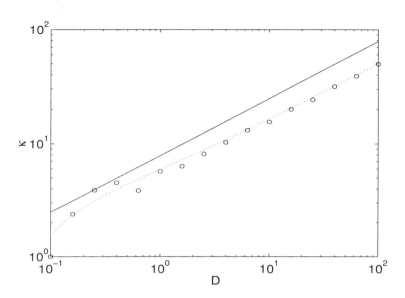

Figure 6.22 Condition number of moment matrix for strip, TE polarization, $n_\lambda = 10$, $b = 1$ discretization. Circles: computed using SVD. Solid line: theoretical approximation, (6.193). Dotted line: theoretical approximation, (6.101) and (6.102), including discretization error, (6.106).

where n_λ is the discretization density or number of nodes per wavelength. The smallest eigenvalue arises from the surface wave mode and is approximated by (6.102). The condition number of $\overline{Z}$ for the TE polarization is then

$$\kappa \simeq \frac{\pi}{4} n_\lambda D^{1/2} \qquad (6.193)$$

We neglect the dependence on the choices of testing and expansion functions in (6.193). Figure 6.22 shows the condition number as a function of D for a discretization density of $n_\lambda = 10$.

6.4.2.5 Rectangular Cavity

The small eigenvalues corresponding to resonant modes of the cavity were estimated in Section 6.2.6. In order to obtain condition number estimates, we also require estimates of the maximal eigenvalues of the EFIE. For the TM polarization, the maximal eigenvalue is given by (6.190). This eigenvalue corresponds to the surface wave mode on a single strip. Since the fields radiated by this mode travel primarily in a direction tangential to the strip, it is not significantly affected by the presence of a second, parallel strip, so that (6.190) also provides an estimate of the largest eigenvalue of the EFIE for the cavity. The maximal eigenvalue for the TE polarization for a single strip is given by (6.192). Since the fields radiated by the high-frequency modes decay exponentially away from the source, this eigenvalue is locally determined, and (6.192) is valid for the cavity as well.

Since the spacing of the resonances as a function of the cavity dimensions decreases as D becomes large, the imaginary part of the smallest cavity mode eigenvalue as estimated by (6.133) is small, and (6.130) can be employed as an estimate of the magnitude of Λ_r. For the TM polarization, this leads to a condition number estimate of

$$\kappa^{TM} \simeq \frac{4}{3} \frac{D^{3/2}}{\alpha \sqrt{W}} \qquad (6.194)$$

where we have retained only the leading order term of (6.130), and the maximal eigenvalue is obtained from (6.190). For the TE polarization,

$$\kappa^{TE} \simeq \frac{n_\lambda D}{2\alpha \sqrt{W}} \qquad (6.195)$$

The growth rate for the TE polarization with D is not as large as that of the TM case, since the largest eigenvalue of $\mathcal{N}$ does not depend on electrical size, whereas the largest eigenvalue of $\mathcal{L}$ increases in magnitude with electrical size.

From (6.194) and (6.195), it can be seen that the condition number of the moment matrix is maximal at the smallest possible value of α. This corresponds to the resonance of the TM_{1n} mode, for which the normalized spatial frequency is $\beta_1 = 1/(2D)$. By expanding (6.126) for large D, we see that $\alpha \simeq Wr^2/(4D^2)$, so that the maximum condition number is

$$\kappa_{max}^{TM} \simeq \frac{16}{3} \frac{D^{7/2}}{W^{3/2}} \qquad (6.196)$$

for the TM polarization. For the TE polarization,

$$\kappa_{max}^{TE} \simeq \frac{2n_\lambda D^3}{W^{3/2}} \qquad (6.197)$$

As shown by these estimates, the condition number of the moment matrix grows rapidly with the cavity depth D. Since the parameter α increases with the mode number r, the most extreme ill-conditioning is confined to narrow bands near the lowest-order resonances of the cavity corresponding to small values of r.

These theoretical condition number estimates for the cavity can be validated by comparison with numerical results. Condition numbers are obtained from the ratio of extremal singular values, using a numerical singular value decomposition. All computed results are given for the TM polarization.

Error Analysis of Surface Integral Equation Methods 271

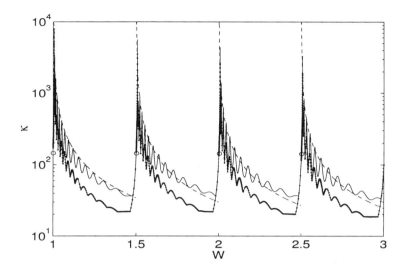

Figure 6.23 Condition number of moment matrix for cavity of depth $D = 10$, as a function of width, $n_\lambda = 10$ discretization. Solid line/dots: computed value, open cavity. Solid line: computed value, half-open cavity. Dashed line: theoretical estimate, (6.194). Circles: (6.130) and (6.133), including discretization error (6.137), at exact resonances of the lowest-order mode ($r = 1$).

Figure 6.23 shows the condition number of the moment matrix for a cavity of length $D = 10$ as a function of width. The more precise theoretical estimate obtained by taking into account both (6.130) and (6.133), as well as the spectral error (6.137), is shown at the locations of the resonances of the TM_{1n} mode. As W increases, the cavity mode eigenvalues move past the origin in the complex plane, and the condition number is largest when one of the modes is at resonance and its eigenvalue has a vanishing imaginary part. Away from resonance, the condition number for the half-open cavity is approximately twice that of the open cavity.

Figure 6.24 is similar to Figure 6.23, but the range of widths is smaller. The theoretical locations of the resonances of the EFIE are marked by circles. The shifting of the peaks of the computed condition numbers away from these locations is caused by discretization error.

6.4.2.6 Higher-Order Basis Functions

It is known that the condition number of the moment matrix increases rapidly with the polynomial order p of the basis, at least as fast as p^2 [12]. The spectral estimates obtained in Section 6.2.3 are valid for $|r| \leq N/2$. For $p > 0$, the order of the smallest

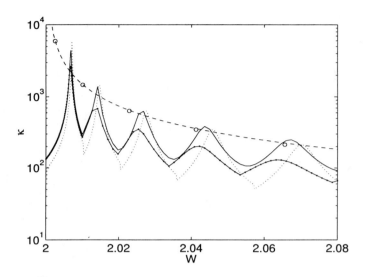

Figure 6.24 Condition number of moment matrix for the same cavity as in Figure 6.23, over a smaller range of the width W. Solid line/dots: computed value, open cavity. Solid line: computed value, half-open cavity. Dashed line: theoretical estimate, (6.194). Dotted line: (6.130) and (6.133), including discretization error (6.137). Circles: (6.130) and (6.133), without discretization error, at exact resonances.

eigenvalue is greater than $N/2$, so that a more refined treatment is required in order to obtain an explicit condition number estimate.

Let U be the unitary transformation with matrix elements

$$U_{nr} = N^{-1/2} e^{i\beta_r k_0 x_n} \qquad (6.198)$$

where β_r and x_n are defined as before. Transforming each block of the matrix representation of $\hat{\mathcal{L}}$ leads to

$$\hat{A}_{rs,ab} = \sum_{m,n=1}^{N} U^*_{mr} \hat{L}_{mn,ab} U_{nr} \qquad (6.199)$$

By making use of (6.139), this matrix element can be written as

$$\frac{\eta}{2n_\lambda^2 D} \int_{-\infty}^{\infty} \frac{dk}{\sqrt{1-k^2}} \frac{\sin[\pi D(k-\beta_r)]}{\sin[\pi(k-\beta_r)/n_\lambda]} \frac{\sin[\pi D(k-\beta_s)]}{\sin[\pi(k-\beta_s)/n_\lambda]} F_a(-k) F_b(k) \qquad (6.200)$$

Approximating this operator by neglecting off-diagonal blocks, for which $r \neq s$, leads to

$$\hat{A}_{rr,ab} \simeq \frac{\eta}{2} \sum_{q=-\infty}^{\infty} \frac{F_a(-k_q)F_b(k_q)}{\sqrt{1-k_q^2}} \qquad (6.201)$$

where $k_q = \beta_r + qn_\lambda$ as before, and $|\beta_r| \neq 1$.

The minimal eigenvalue of $\hat{\mathcal{L}}$ can be estimated from the minimal eigenvalue of $\hat{A}_{rr,ab}$ for the largest value of r, which is $r = N/2$. For high-order modes, the operator $-i\hat{\mathcal{L}}$ behaves like the positive definite operator arising from the static limit, so that $-i\hat{A}_{rr,ab}$ for large, fixed r is approximately Hermitian. We employ the variational bound

$$\left|\hat{\Lambda}_{\min}\right| \leq \left|\frac{\mathbf{v}^\dagger \hat{A}_{N/2\ N/2} \mathbf{v}}{\mathbf{v}^\dagger \mathbf{v}}\right| \qquad (6.202)$$

to estimate the smallest eigenvalue. Choosing the trial vector $v_a = F_a(k)$ leads to

$$\left|\hat{\Lambda}_{\min}\right| \leq \left|\frac{\eta}{2F^{(p)}(k,k)} \sum_q \frac{F^{(p)^2}(k,k_q)}{\sqrt{1-k_q^2}}\right| \qquad (6.203)$$

By the variational expression (6.202), minimizing this expression over k leads to an upper bound for the smallest eigenvalue of $\hat{\mathcal{L}}$.

We now apply this eigenvalue bound to the Legendre expansion. Using the asymptotic approximation $j_a(x) \sim \cos[x + (a+1/2)^2/(2x) - \pi(a-1)/2]/x$ for the spherical Bessel function, the minimum of the right-hand side of (6.203) occurs for $k = n_\lambda/2 + n_\lambda q'$, where q' is a large integer. For k of this form, $F^{(p)}(k,k) \simeq (p+1)(p+2)/(2k^2)$, and

$$F^{(p)}(k,k_q) \simeq (kk_q)^{-1} \sum_{a=0}^{p/2} (4a+1) \cos\left[\frac{(2a+1/2)^2}{2\pi q}\right] \qquad (6.204)$$

This leads to the estimate

$$\left|\hat{\Lambda}_{\min}\right| \simeq \frac{\eta}{n_\lambda(p+1)(p+2)} \sum_{q=p/2}^{\infty} (2/\pi^2)q^{-3} \left\{\sum_{a=0}^{p/2} (4a+1) \cos\left[\frac{(2a+1/2)^2}{2\pi q}\right]\right\}^2 \qquad (6.205)$$

for p even, with a similar result for p odd. The summation over q can be evaluated numerically and is nearly independent of p, with an approximate value of 3.1. The resulting condition number estimate is

$$\kappa \simeq 0.2\, n_\lambda(p+1)(p+2) D^{1/2} \qquad (6.206)$$

For the Legendre expansion applied to a flat strip of length $D = 1$, this estimate is shown in Figure 6.25.

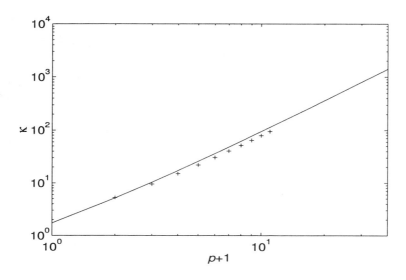

Figure 6.25 Moment matrix condition number as a function of polynomial order for Legendre polynomial expansion applied to a flat strip, $D = 1$, $n_\lambda = 4$. Solid line: theoretical estimate, (6.206). Pluses: computed matrix condition number.

6.4.2.7 Flat Plate

The extremal eigenvalues of the discretized operator correspond to the largest curl-free eigenvalue (6.163) and the smallest divergence-free eigenvalue (6.165). These estimates lead to

$$\kappa \simeq \frac{\pi^2}{k_0^2 h^2} = \frac{n_\lambda^2}{4} \tag{6.207}$$

for the condition number. This result breaks down as the plate size d becomes large relative to the wavelength $\lambda = 2\pi/k_0$, since the eigenvalues of the surface wave modes ($\beta = k_0$) depend on $k_0 d$ [22, 26] and eventually will surpass the eigenvalues of the high-frequency modes ($\beta = \pi/h$).

Since the high-frequency modes that determine (6.207) radiate evanescent fields, the eigenvalues depend only on local properties of the scatterer. Thus, unless resonance leads to eigenvalues that are smaller than (6.165), the estimate (6.207) holds for arbitrary scatterers that are smooth on the scale of a wavelength.

Figure 6.26 compares the theoretical condition number estimate to numerical results for a square plate. For the computed values, RWG basis functions are employed on a regular triangular mesh. The plate size is $d = 1\lambda$ for the smaller values of n_λ, and $d = 0.1\lambda$ for the largest.

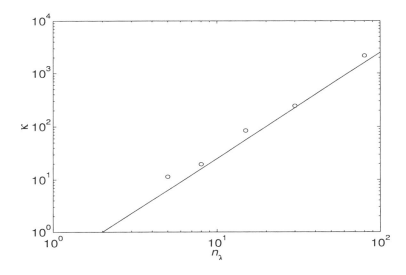

Figure 6.26 Condition number of moment matrix for square plate. Solid line: theoretical estimate, (6.207). Circles: computed.

6.4.2.8 Preconditioners

Near-neighbor preconditioners are often employed to reduce the condition number of the moment matrix. For 2D problems, this type of preconditioner corresponds to a diagonal band of $\overline{Z}$. For 3D problems, the near-neighbor preconditioner is a block sparse matrix. By solving a linear system with the preconditioning matrix at each step of an iterative algorithm, the effective matrix which governs the convergence of the iteration is the inverse of the preconditioner multiplied by the original moment matrix.

The near-neighbor preconditioner overcomes the growth of the condition number with the discretization density n_λ, because the preconditioner accurately models the localized self-interaction of evanescent modes with high spatial frequency. Since the preconditioner does not model the long-range interactions that determine the surface wave eigenvalues, the condition number can still increase with electrical size of the scatterer. Near-neighbor preconditioners also do not overcome ill-conditioning due to resonant effects, since resonance is produced by global interactions.

6.4.2.9 Low-Frequency Breakdown

For a fixed mesh, the estimate (6.207) shows that the condition number of the moment matrix for the vector EFIE grows as $1/k_0^2$ as the frequency decreases. This is observed in practice as the low-frequency breakdown of the method of moments. If the basis used to discretize the EFIE is such that the divergence-free modes can be separated from the curl-free modes, then the two subspaces can be rescaled to improve the conditioning of the discretized operator at low frequencies. This can be accomplished using the loop-star or loop-tree techniques [43–46]. In this section, we analyze this approach to overcoming low-frequency breakdown.

We first assume that the Helmholtz decomposition is exact, so that the curl-free and divergence-free modes can be explicitly separated. If the divergence-free block is scaled by $1/(k_0 h)$, and the curl-free block by $k_0 h$, then the condition number becomes

$$\kappa \simeq \frac{d^2}{\pi^2 h^2} \tag{6.208}$$

which is independent of frequency. The optimal scaling factors are $k_0 d/\pi$ and $-\pi/(k_0 h)$, in which case the spectrum of the scaled operator is a single interval in the complex plane with extrema at i and id/h, and the condition number is

$$\kappa \simeq d/h \tag{6.209}$$

which is the same as that of the discretized static problem ($k_0 \to 0$).

For the loop-star and loop-tree bases, the loop space is divergence free, and the restriction of the discretized operator to this space is well conditioned, with $\kappa \simeq d/h$. The restriction to the star or tree spaces, however, has been observed empirically to be poorly conditioned [47, 48] so that the estimates (6.208) and (6.209) break down in practice.

This ill-conditioning of the tree matrix is due to the existence of modes in the discrete tree space with eigenvalues that are much smaller than eigenvalues associated with a continuous curl-free space. Figure 6.27(a) shows the eigenfunction corresponding to the eigenvalue with smallest magnitude of the tree space matrix for a square plate. The charge accumulation for this mode is small, since the current flows back and forth between the vertical cuts that define the tree space.

Figure 6.27(b) shows the approximate curl-free mode with smallest eigenvalue for the RWG discretization of the same problem, which corresponds to the eigenvalue with smallest positive imaginary part in Figure 6.17(d). Strong charge accumulations exist at the corners of the plate. For this example, the ratio of the smallest curl-free eigenvalue to the smallest tree space eigenvalue is approximately 30. The condition number of the tree space matrix is roughly 30 times larger than that of the curl-free restriction of the full moment matrix, due to the presence of low-charge or nearly divergence-free modes of the type shown in Figure 6.27(a) in the tree space. This problem can be overcome by preconditioning the tree space matrix with the inverse of a discretization of the divergence operator [47, 48] Since the low-charge modes have small divergence, multiplication by this inverse operator increases the eigenvalues of these modes.

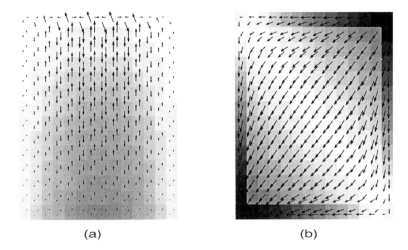

Figure 6.27 (a) Mode corresponding to smallest eigenvalue of the tree space interaction matrix, for a square plate of side 0.025λ. (b) Mode corresponding to smallest curl-free eigenvalue. Superimposed on the vector fields are the charge densities associated with each mode. The curl-free mode is associated with a relatively strong charge density or divergence, whereas the tree space mode is nearly divergence-free.

For electrically large scatterers, the eigenvalues corresponding to the curl-free and divergence-free modes join in the complex plane at the zero spatial frequency eigenvalue $\Lambda \simeq \eta/2$. Scaling of eigenvalues near $\eta/2$ by $k_0 h$ and $1/(k_0 h)$ causes the condition number of the preconditioned matrix to become $1/(k_0 h)^2$, which is similar to the unpreconditioned value given by (6.207). Thus, this remedy for low-frequency breakdown does not work if the electrical size of the scatterer is on the order of a wavelength or larger.

6.5 CONCLUSION

In this chapter, we have studied the solution error and the condition number of the method of moments for several canonical 2D and 3D scatterers. The convergence behavior is determined by physical effects such as resonance, edge singularities, and low-frequency breakdown, and by properties of the numerical method, including mesh density, the polynomial order of basis functions, and quadrature rules used to integrate moment matrix elements. The notion of spectral error as defined in this chapter allows the contributions of these factors to be separated and understood individually.

In the case of the flat strip, we are able to combine the error contributions due to the choice of basis functions and the mesh density on the smooth, interior region of the strip, and the error due to the singularity of the current at the edges of the strip, to obtain an error estimate for the forward scattering amplitude of the strip. This example shows how the results of this work on spectral convergence can be used to analyze the numerical behavior of the method of moments for complex scatterers with multiple physical effects contributing to the solution error.

From a computational point of view, the solution of three-dimensional vector scattering problems is more difficult than 2D problems, but the spectral convergence theory shows that the close physical relationship between 2D and 3D scattering problems leads to similarities in the numerical behaviors of 2D and 3D solution methods. The spectrum of the EFIE for a flat plate, for example, is roughly the union of the spectra for the 2D operators of the TE and TM polarizations. The spectral convergence rate with discretization density of the vector rooftop basis functions is also the same as that of the 2D problems when scalar pulse or triangle basis functions are employed.

In addition to the solution error analysis of the method of moments, we also study the condition number of the moment matrix, which is an indicator of the difficulty of numerically solving the discretized integral equation for the unknown surface current. Theoretical condition number estimates show the dependence of matrix conditioning on the discretization density, choice of basis functions, type of scatterer, and electrical size of the problem.

REFERENCES

1. R. F. Harrington, *Field Computation by Moment Method*, Marlabar, FL: Krieger, 1982.

2. C. C. Lu and W. C. Chew, "A multilevel algorithm for solving a boundary integral equation of scattering," *Micro. Opt. Tech. Lett.*, vol. 7, pp. 466–470, July 1994.

3. W. C. Chew, *Waves and Fields in Inhomogeneous Media*, New York: IEEE Press, 1995.

4. G. C. Hsiao and R. E. Kleinman, "Mathematical foundations for error estimation in numerical solutions of integral equation in electromagnetics," *IEEE Trans. Antennas Propagat.*, vol. 45, pp. 316–328, March 1997.

5. S. M. Rao, D. R. Wilton, and A. W. Glisson, "Electromagnetic scattering by surfaces of arbitary shape," *IEEE Trans. Antennas Propagat.*, vol. 30, pp. 409–418, May 1982.

6. A. G. Dallas, G. C. Hsiao, and R. E. Kleinman, "Observations on the numerical stability of the Galerkin method," *Adv. Comput. Math.*, vol. 9, pp. 37–67, 1998.

7. J. J. Bowman, T. B. A. Senior, and P. L. E. Uslenghi, *Electromagnetic and Acoustic Scattering by Simple Shapes*, New York: Hemisphere, 1987.

8. L. Demkowicz, "Asymptotic convergence in finite and boundary element methods: Part 1: Theoretical results," *Computers Math. Applic.*, vol. 27, no. 12, pp. 69–84, 1994.

9. H. Holm and E. P. Stephan, "A boundary element method for electromagnetic transmission problems," *Appl. Anal.*, vol. 56, pp. 213–226, 1995.

10. H. Holm, M. Maischak, and E. P. Stephan, "The hp-version of the boundary element method for Helmholtz screen problems," *Computing*, vol. 57, pp. 105–134, 1996.

11. M. Ainsworth, W. McLean, and T. Tran, "The conditioning of boundary element equations on locally refined meshes and preconditioning by diagonal scaling," *SIAM J. Numer. Anal.*, vol. 36, no. 6, pp. 1901–1932, 1999.

12. M. Maischak, P. Mund, and E. P. Stephan, "Adaptive multilevel BEM for acoustic scattering," *Comp. Meth. Appl. Mech. Eng.*, vol. 150, pp. 351–367, 1997.

13. T. Tran, E. P. Stephan, and P. Mund, "Hierarchical basis preconditioners for first kind integral equations," *Appl. Anal.*, vol. 65, pp. 353–372, 1997.

14. S. Amini and N. D. Maines, "Preconditioned Krylov subspace methods for boundary element solution of the Helmholtz equation," *Int. J. Num. Meth. Engr.*, vol. 41, pp. 875–898, 1998.

15. E. P. Stephan and T. Tran, "Domain decomposition algorithms for indefinite hypersingular integral equations: The h and p versions," *SIAM J. Sci. Comput.*, vol. 19, pp. 1139–1153, July 1998.

16. E. P. Stephan and T. Tran, "Domain decomposition algorithms for indefinite weakly singular integral equations: The h and p versions," Tech. Rep. AMR95/48, Univ. New South Wales, Sidney, Dec. 1995.

17. E. F. Kuester, "Computable error bounds for variational functionals of solutions of a convolution integral equations of the first kind," *Wave Motion*, vol. 22, pp. 171–185, 1995.

18. J. Van Bladel, *Singular Electromagnetic Fields and Sources*, New York: IEEE Press, 1995.

19. D. S. Jones, "A critique of the variational method in scattering problems," *IRE Trans. Antennas Propagat.*, vol. AP-4, no. 3, pp. 297–301, 1956.

20. M. Feistauer, G. C. Hsiao, and R. E. Kleinman, "Asymptotic and a posteriori error estimates for boundary element solutions of hypersingular integral equations," *SIAM J. Numer. Anal.*, vol. 33, pp. 666–685, April 1996.

21. G. C. Hsiao and R. E. Kleinman, "Feasible error estimates in boundary element methods," in *Boundary Element Technology VII* (C. A. Brebbia and M. S. Ingber, eds.), pp. 875–886, Southampton: Computational Mechanics Publ., 1991.

22. K. F. Warnick and W. C. Chew, "Accuracy and conditioning of the method of moments for the 2D EFIE," in *15th Annual Review of Progress in Applied Computational Electromagnetics*, (Monterey, CA), pp. 198–204, Naval Postgraduate School, March 15–20, 1999.

23. K. F. Warnick and W. C. Chew, "Convergence of moment method solutions of the electric field integral equation for a 2D open cavity," *Micr. Opt. Tech. Lett.*, vol. 23, pp. 212–218, Nov. 20, 1999.

24. K. F. Warnick, G. Kang, and W. C. Chew, "Regulated kernel for the electric field integral equation," in *IEEE Antennas and Propagation Society Int. Symp.*, Salt Lake City, UT, July 16–22, 2000.

25. K. F. Warnick and W. C. Chew, "Accuracy of the method of moments for scattering by a cylinder," *IEEE Trans. Micr. Th. Tech.*, vol. 48, pp. 1652–1660, Oct. 2000.

26. K. F. Warnick and W. C. Chew, "On the spectrum of the electric field integral equation and the convergence of the moment method," *Int. J. Numer. Meth. Engr.*, vol. 51, pp. 31–56, 2001.

27. K. F. Warnick and W. C. Chew, "Spectral viewpoint on low frequency breakdown of the EFIE," *IEEE Trans. Antennas Propagat.*, in review.

28. A. W. Naylor and G. R. Sell, *Linear Operator Theory in Engineering and Science*, New York: Springer-Verlag, 1982.

29. S. Wandzura, "Accuracy in computation of matrix elements of singular kernels," in *11th Annual Review of Progress in Applied Computational Electromagnetics*, vol. II, (Monterey, CA), pp. 1170–1176, Naval Postgraduate School, March 20–25, 1995.

30. A. G. Ramm, "Eigenfunction expansion of a discrete spectrum in diffraction problems," *Radiotek. i Elektron.*, vol. 18, pp. 364–369, 1973.

31. S. C. Eisenstat and I. C. F. Ipsen, "Three absolute perturbation bounds for matrix eigenvalues imply relative bounds," *SIAM J. Matrix Anal. Appl.*, vol. 20, no. 1, pp. 149–158, 1998.

32. S. Amini and S. M. Kirkup, "Solution of Helmholtz equation in the exterior domain by elementary boundary integral methods," *J. Comp. Phys.*, vol. 118, pp. 208–221, 1995.

33. M. D. Pocock and S. P. Walker, "The complex bi-conjugate gradient solver applied to large electromagnetic scattering problems; computational costs, and cost scalings," *IEEE Trans. Antennas Propagat.*, vol. 45, no. 1, pp. 140–146, 1997.

34. L. Tsang, C. H. Chan, and H. Sangani, "Application of a banded matrix iterative approach to Monte Carlo simulations of scattering of waves by random rough surface: TM case," *Microw. Opt. Tech. Lett.*, vol. 6, pp. 148–151, Feb. 1993.

35. D. A. Kapp and G. S. Brown, "A new numerical method for rough-surface scattering calculations," *IEEE Trans. Antennas Propagat.*, vol. 44, pp. 711–721, May 1996.

36. G. H. Golub and H. A. van der Vorst, "Closer to the solution: Iterative linear solvers," in I. S. Duff and G. A. Watson, Eds., *The State of the Art in Numerical Analysis*, pp. 63–92, Oxford, U.K.: Clarendon Press, 1997.

37. J. C. West and J. M. Sturm, "On iterative approaches for electromagnetic rough-surface scattering problems," *IEEE Trans. Antennas Propagat.*, vol. 47, pp. 1281–1288, Aug. 1999.

38. G. H. Golub and C. F. V. Loan, *Matrix Computations*, Second Edition, Baltimore: Johns Hopkins University Press, 1993.

39. T. A. Driscoll, K.-C. Toh, and L. N. Trefethen, "From potential theory to matrix iterations in six steps," *SIAM Review*, vol. 40, pp. 547–578, 1998.

40. I. S. Gradshteyn and I. M. Ryzhik, *Table of Integrals, Series, and Products*, Fifth Edition, San Diego, CA: Academic Press, 1994.

41. G. C. Hsiao and W. L. Wendland, "The Aubin-Nitsche lemma for integral equations," *J. Integral Eqs.*, vol. 3, pp. 299–315, 1981.

42. F. X. Canning and J. F. Scholl, "Diagonal preconditioners for the EFIE using a wavelet basis," *IEEE Trans. Antennas Propagat.*, vol. 44, pp. 1239–1246, Sept. 1996.

43. J. R. Mautz and R. F. Harrington, "An E-field solution for a conducting surface small or comparable to the wavelength," *IEEE Trans. Antennas Propagat.*, vol. 32, pp. 330–339, April 1984.

44. E. Arvas, R. F. Harrington, and J. R. Mautz, "Radiation and scattering from electrically small conducting bodies of arbitrary shape," *IEEE Trans. Antennas Propagat.*, vol. 34, pp. 66–77, Jan. 1986.

45. M. Burton and S. Kashyap, "A study of a recent, moment-method algorithm that is accurate to very low frequencies," *Appl. Comp. Electromag. Soc. J.*, vol. 10, pp. 58–68, Nov. 1995.

46. W. Wu, A. W. Glisson, and D. Kajfez, "A study of two numerical solution procedures for the electric field integral equation at low frequency," *Appl. Comp. Electromag. Soc. J.*, vol. 10, pp. 69–80, Nov. 1995.

47. W. C. Chew, J. S. Zhao, and J. M. Song, "Solving Maxwell's equations from zero to microwave frequencies," in *30th Plasmadynamics and Lasers Conference*, Paper 99–3729, (Norfolk, VA), American Institute of Aeronautics and Astronautics, June 28–July 1, 1999.

48. J. S. Zhao and W. C. Chew, "Integral equation solution of Maxwell's equations from zero frequency to microwave frequencies," *IEEE Trans. Antennas Propagat.*, vol. 48, pp. 1635–1645, 2000.

7

Advances in the Theory of Perfectly Matched Layers

Fernando L. Teixeira and Weng C. Chew

7.1 INTRODUCTION

The finite-difference time-domain (FDTD) method [1–4] is a very popular numerical method for full-wave simulation of electromagnetic fields in complex environments. It is an efficient, second-order accurate (both in space and time) scheme that combines a leapfrog update in time with a staggered central differencing in space to simulate Maxwell's equations. Because the FDTD method is a partial differential equation (PDE)-based algorithm, there is no need to obtain a Green's function (i.e., to invert a differential operator), and, as a result, arbitrary geometries and media (including dispersive and nonlinear) can be easily studied. FDTD has also been used in other instances where hyperbolic partial differential equations (wave propagation) occur, such as in elastodynamics or acoustics. Some of the concepts discussed in this chapter have direct application in those areas as well.

Among the issues facing the implementation of PDE-based algorithms such as FDTD or the finite element method (FEM) is the proper truncation of the computational domain. In many situations, the problem to be simulated corresponds to an open-region problem. The finite grid requirement of any practical numerical implementation requires a proper treatment of the grid boundaries. Ideally, one must ensure that spurious reflections from the grid boundaries are small enough so that the solution is not contaminated. Usually this is accomplished through the use of an absorbing boundary condition (ABC) on or near the grid termination.

The perfectly matched layer (PML) is a very efficient ABC introduced in the literature by Berenger for Cartesian coordinates in 1994 [5] and fervently studied since then [6–24]. The PML has since then been shown to outperform previously proposed ABCs by orders of magnitude in terms of reduced reflection coefficients. The PML essentially achieves a reflectionless absorption of electromagnetic waves in the continuum limit as the mesh discretization size goes to zero. The absorption inside the PML operates through conductive losses, so that an exponential decay for the fields inside the PML is obtained. Therefore, when the computational domain is surrounded by a PML region, the spurious reflection from the grid boundaries can be made exponentially smaller. Moreover, being a material ABC, the PML retains the nearest-neighbor interaction characteristic of the FDTD method, and therefore it is particularly suited for implementation on parallel computers. Also because of this property, the PML retains the (low) computational complexity of the FDTD, which is $O(N)$ per time step.

In this chapter, we discuss the theoretical foundations and various implementations of the PML concept from the complex stretching viewpoint [7, 14–17]. Sections of this chapter are based on the authors' previously published research [28–34]. It is not our intention here to present a comprehensive review on the PML. Among the issues not addressed here are implementation details for PDE-based methods other than FDTD (such as the FEM or the transmission-line method), PML profile and parameters optimization, and applications of the PML to areas other than electromagnetics (such as elastodynamics [25, 26], acoustics [17], and quantum mechanics [27]). As a result, many references on the subject have not been included and, therefore, the reference list at the end of this chapter is by no means complete. Again, this simply reflects the thematic perspective we have chosen here. Other reviews on the PML topic which have appeared recently can be found in [23, 24].

7.2 PML VIA COMPLEX SPACE COORDINATES

7.2.1 Frequency Domain Analysis

The PML was originally derived through the introduction of matched artificial electric and magnetic conductivities, and through a particular splitting of the electromagnetic field components into subcomponents [5]. An alternative derivation was later given in [7], where it was shown that the PML can be related to a complex stretching of the Cartesian coordinates in the frequency domain. Through this complex stretching, source-free Maxwell's equations in the PML are modified to ($e^{-i\omega t}$ convention)

$$\nabla_s \times \mathbf{H} = -i\omega\epsilon\mathbf{E} \qquad (7.1)$$

$$\nabla_s \times \mathbf{E} = i\omega\mu\mathbf{H} \tag{7.2}$$

$$\nabla_s \cdot \epsilon\mathbf{E} = 0 \tag{7.3}$$

$$\nabla_s \cdot \mu\mathbf{H} = 0 \tag{7.4}$$

where

$$\nabla_s = \hat{x}\frac{1}{s_x}\frac{\partial}{\partial x} + \hat{y}\frac{1}{s_y}\frac{\partial}{\partial y} + \hat{z}\frac{1}{s_z}\frac{\partial}{\partial z} \tag{7.5}$$

in Cartesian coordinates, and the s_ζ, $\zeta = x, y, z$, are the so-called *complex stretching variables*, given by

$$s_\zeta(\zeta, \omega) = a_\zeta(\zeta) + i\frac{\Omega_\zeta(\zeta)}{\omega} \tag{7.6}$$

with $a_\zeta \geq 1$ and $\Omega_\zeta \geq 0$ (profile functions). The first inequality ensures that evanescent waves will have an exponential decay faster than in the non-PML region, and the second inequality ensures that propagating waves will also decay exponentially. The ordinary Maxwell's equations are a special case of the above equations when $s_\zeta = 1$. Therefore, the complex stretching variables can be seen as added degrees of freedom to Maxwell's equations.

The reflectionless property of a PML interface can be easily verified by writing down the TE and TM reflection coefficients, R^{TE} and R^{TM}, for a planar interface and verifying that they are zero [7]. Note that, for a single interface (half space problem), only one single complex stretching variable is different from unity (the normal coordinate to the interface). In the corner regions of the computational domain, two or three coordinates need to be stretched simultaneously [7].

Perhaps a more direct and elegant way to verify the reflectionless characteristic is to observe that the coordinate stretching is just a particular mapping of the coordinate space to a *complex* coordinate space (i.e., an analytic continuation of the spatial variables) [14]. This mapping is defined through

$$\zeta \to \tilde{\zeta} = \int_0^\zeta s_\zeta(\zeta')d\zeta' = \int_0^\zeta \left(a_\zeta(\zeta') + i\frac{\Omega_\zeta(\zeta')}{\omega}\right)d\zeta' = b_\zeta(\zeta) + i\frac{\Delta_\zeta(\zeta)}{\omega} \tag{7.7}$$

so that

$$\frac{1}{s_\zeta}\frac{\partial}{\partial \zeta} = \frac{\partial}{\partial \tilde{\zeta}} \tag{7.8}$$

Therefore, the modified source-free Maxwell's equations (7.1)–(7.4) can be written as Maxwell's equations in a *complex space*:

$$\tilde{\nabla} \times \mathbf{H} = i\omega\epsilon\mathbf{E} \tag{7.9}$$

$$\tilde{\nabla} \times \mathbf{E} = i\omega\mu\mathbf{H} \tag{7.10}$$

$$\tilde{\nabla} \cdot \epsilon\mathbf{E} = 0 \tag{7.11}$$

$$\tilde{\nabla} \cdot \mu\mathbf{H} = 0 \qquad (7.12)$$

with

$$\tilde{\nabla} = \hat{x}\frac{\partial}{\partial \tilde{x}} + \hat{y}\frac{\partial}{\partial \tilde{y}} + \hat{z}\frac{\partial}{\partial \tilde{z}} = \nabla_s \qquad (7.13)$$

Because (7.9)–(7.12) are formally the same as usual Maxwell's equations except for the change of coordinates, their solutions will be the usual solutions to Maxwell's equations but with a change of variables according to (7.7). In particular, closed-form solutions for the fields inside the PML can be written by inspection from the knowledge of the ordinary closed-form solutions of Maxwell's equations. Furthermore, because the complex variables $\tilde{\zeta}(\zeta)$ are always continuous functions of the real variables ζ for bounded s_ζ according to (7.7) (regardless of the continuity of s_ζ), the resultant fields inside the PML will be continuous everywhere if the original fields are continuous (the composition of two continuous functions is a continuous function). In particular, boundary conditions are preserved by this transformation.

By writing down the Green's function inside a metallic box (which can be thought as an FDTD or FEM computational domain), and doing the transformation (7.7), one can easily verify that the Green's function converges to the free-space Green's function as Ω_ζ is increased [14]. Moreover, any propagating eigenfunction (mode) satisfying Sommerfeld's radiation condition is continuously mapped, according to (7.7), into an exponentially decaying function; that is,

$$e^{ik\zeta} \to e^{-\frac{\Delta_\zeta}{c}} \cdot e^{ikb_\zeta}$$

which characterizes the reflectionless absorption. Note also that, similarly, a purely evanescent mode is mapped into a mode having both evanescent and propagating factors. This is one of the reasons why the performance of the PML deteriorates for evanescent modes and why it is important to choose a profile function $a_\zeta(\zeta) > 1$ whenever evanescent modes are present.

The above observations are related to the behavior of the PML in the continuum space. In the continuum, the complex stretching variables s_ζ do not need to be continuous to arrive at the reflectionless absorption. However, in numerical implementations, the space is discretized and material discontinuities will cause spurious reflections in the grid. In order to minimize such spurious reflections due to discretization, the complex stretching coordinates are chosen to be smooth, so that, when passing from the continuum to the discrete case, material discontinuities are minimized. This amounts to choosing the profile functions $a_\zeta(\zeta)$ and $\Omega_\zeta(\zeta)$ in (7.7) to be smooth functions of ζ.

In the discrete space, spurious reflection from the PML has two main sources. The first cause is the reflection due to the discontinuity of the material parameters discussed in the last paragraph. This reflection is dominated by the discontinuity at the physical domain to PML interface [11]. The second cause is the reflection from the grid termination itself. This reflection is present because, in all practical instances, the

PML layer thickness is finite and a residual reflection from the termination will always be present. This fact establishes the basic trade-off on the practical performance of the PML. On one hand, it would be better to have the PML loss as large as possible to reduce the second cause of spurious reflections. On the other hand, larger losses would mean larger variations on the material parameters, which will increase the contribution of the first cause of spurious reflections.

7.2.2 Time Domain Analysis

Because the complex coordinates involve a frequency dependence, they would be represented as convolutional operators in the time domain. However, convolutions can be avoided by splitting the electromagnetic fields in the same manner as suggested by Berenger [5]. If we rewrite the x component of (7.1)

$$-i\omega\epsilon E_x = \frac{1}{s_y}\frac{\partial}{\partial y}H_z - \frac{1}{s_z}\frac{\partial}{\partial z}H_y \tag{7.14}$$

and split the fields as $E_x = E_{xy} + E_{xz}$, such that

$$-i\omega\epsilon E_{xy} = \frac{1}{s_y}\frac{\partial}{\partial y}(H_{zy} + H_{zx}) \tag{7.15}$$

$$i\omega\epsilon E_{xz} = \frac{1}{s_z}\frac{\partial}{\partial z}(H_{yz} + H_{yx}) \tag{7.16}$$

then the corresponding time domain equations become

$$a_x \frac{\partial}{\partial t} E_{xy} + \Omega_x E_{xy} = \frac{\partial}{\partial y}(H_{zy} + H_{zx}) \tag{7.17}$$

$$a_x \frac{\partial}{\partial t} E_{xz} + \Omega_x E_{xz} = \frac{\partial}{\partial z}(H_{yz} + H_{yx}) \tag{7.18}$$

where the same procedure applies for the other electric field components in (7.1) and for the magnetic field components in (7.2). Such splitting may be done only inside the PML domain (to save memory) or everywhere (i.e., both inside the PML and in the physical domain to facilitate parallelization of the code). The above time-domain equations have the same form as in the original Berenger formulation, with the added generality of a_x to address evanescent waves. Their implementation in an FDTD algorithm is straightforward [5, 10].

Apart from its conceptual simplicity, the main advantage of the complex stretching approach to PML is that it provides a route to generalize the PML concept for curvilinear geometries and for media with more complex constitutive properties (e.g., dispersive and anisotropic). In the next sections, we will treat those generalizations.

7.3 PML-FDTD FOR DISPERSIVE MEDIA WITH CONDUCTIVE LOSS

7.3.1 Time Domain Analysis

Linear time-dispersive media are often encountered in nature. In such media, broadband electromagnetic waves will propagate and attenuate in a frequency-dependent manner. Therefore, to have a realistic model of propagation of electromagnetic waves in such media, it is prudent to include the effects of dispersion.

To achieve perfect matching in dispersive media with conductive loss, we assume the *same* frequency-dependent parameters *everywhere* [28]. The PML region is then set up as the region where the analytic continuation of the spatial variables is enforced.

We start by writing the modified Maxwell's equations (7.1)–(7.4) in terms of **D** and with a medium conductivity σ explicitly included:

$$i\omega a_x \mathbf{B}_{sx} - \Omega_x \mathbf{B}_{sx} = \frac{\partial}{\partial x}(\hat{x} \times \mathbf{E}) \tag{7.19}$$

$$-i\omega a_x \mathbf{D}_{sx} + \Omega_x \mathbf{D}_{sx} + a_x \sigma \mathbf{E}_{sx} + i\frac{\Omega_x}{\omega}\sigma \mathbf{E}_{sx} = \frac{\partial}{\partial x}(\hat{x} \times \mathbf{H}) \tag{7.20}$$

Transforming the above back into the time domain, we obtain

$$-a_x \partial_t \mathbf{B}_{sx} - \Omega_x \mathbf{B}_{sx} = \frac{\partial}{\partial x}(\hat{x} \times \mathbf{E}) \tag{7.21}$$

$$a_x \partial_t \mathbf{D}_{sx} + \Omega_x \mathbf{D}_{sx} + a_x \sigma \mathbf{E}_{sx} + \Omega_x \sigma \int_0^t \mathbf{E}_{sx}(\tau)\, d\tau = \frac{\partial}{\partial x}(\hat{x} \times \mathbf{H}) \tag{7.22}$$

In the above, for a dispersive medium, we let $\mathbf{B}_{sx} = \mu \mathbf{H}_{sx}$ while

$$\mathbf{D}_{sx}(t) = \epsilon(t) * \mathbf{E}_{sx}(t) \tag{7.23}$$

7.3.2 Dispersive Medium Models

For the dispersive medium models, we will assume either a Lorentz relaxation model or a Debye relaxation model. Both are causal models, so that the Kramers-Kronig relations are automatically satisfied. Because of this, the permittivity value will be complex having both frequency-dependent real and imaginary parts. The imaginary part can be thought of as frequency-dependent loss or conductivity. Furthermore, normally encountered dispersive response can be modeled as a sum of Lorentz and/or Debye terms. This can be done, for example, by first evaluating the dielectric permittivity and the effective conductivity for various frequencies and then curve fitting the result by a meromorphic function expanded as a partial fraction expansion with (possibly) single poles (Debye terms), complex conjugate poles (Lorentz terms), and a pole at $\omega_p = 0$ (static conductivity term), plus a constant term standing for the permittivity at infinite frequency.

An N-species Lorentzian dispersive medium is characterized by a frequency-dependent relative permittivity function given by

$$\epsilon(\omega) = \epsilon_0 \left[\epsilon_\infty + \chi(\omega)\right] = \epsilon_0 \epsilon_\infty + \epsilon_0 (\epsilon_s - \epsilon_\infty) \sum_{p=1}^{N} \frac{G_p \omega_p^2}{\omega_p^2 - i 2 \omega \alpha_p - \omega^2} \quad (7.24)$$

where $\chi(\omega)$ is the medium susceptibility, ω_p is the resonant frequency for the pth species, α_p is the corresponding damping factor, and ϵ_0, ϵ_∞ are the static and infinite frequency permittivities, respectively. In the time domain, a corresponding complex susceptibility function can be defined:

$$\hat{\chi}(t) = \sum_{p=1}^{P} i \gamma_p e^{-(\alpha_p + i \beta_p) t} u(t) \quad (7.25)$$

where

$$\beta_p = \sqrt{\omega_p^2 - \alpha_p^2} \quad (7.26)$$

$$\gamma_p = \frac{\omega_p^2 G_p (\epsilon_s - \epsilon_\infty)}{\sqrt{\omega_p^2 - \alpha_p^2}} \quad (7.27)$$

and

$$\sum_{p=1}^{P} G_p = 1 \quad (7.28)$$

so that the time-domain susceptibility function is $\mathcal{F}^{-1}[\chi(\omega)] = \chi(t) = \Re[\hat{\chi}(t)]$ and $\epsilon(t) = \epsilon_0 [\epsilon_\infty \delta(t) + \chi(t)]$, where $\mathcal{F}$ denotes Fourier transform. For a Debye model, the frequency-dependent permittivity function is written as

$$\epsilon(\omega) = \epsilon_0 [\epsilon_\infty + \chi(\omega)] = \epsilon_0 \epsilon_\infty + \epsilon_0 \sum_{p=1}^{N} \frac{A_p}{1 - i \omega \tau_p} \quad (7.29)$$

In this formula, A_p is the pole amplitude and τ_p is the relaxation time for the pth species. Note that the complex susceptibility function for the Debye relaxation model can be considered as a special case of (7.25) when $\alpha_p > \omega_p$ and $\Im(\beta_p) < 0$. Hence, the expression (7.25) applies to *both* models through a proper choice of parameters.

The electric flux is related to the electric field via

$$\mathbf{D}(t) = \epsilon_0 \epsilon_\infty \mathbf{E}(t) + \epsilon_0 \chi(t) * \mathbf{E}(t) \quad (7.30)$$

Using (7.25) in (7.30), we have, for both models,

$$\mathbf{D}(t) = \epsilon_0 \epsilon_\infty \mathbf{E}(t) + \epsilon_0 \sum_{p=1}^{P} \Re\left[\hat{\chi}(t) * \mathbf{E}(t)\right] \quad (7.31)$$

We need to incorporate this convolution in the FDTD scheme at a minimal computational cost. In principle, a convolution would require the storage of all time history of the fields. However, because we are dealing with susceptibilities (convolutional kernels) that have an exponential dependence, it is possible to calculate this convolution recursively. We start by writing $t = l\Delta_t$, corresponding to the time discretization of our FDTD scheme and using a piecewise linear approximation [35] for the electric field in the time-discretization scheme such that

$$\mathbf{E}(t) = \mathbf{E}^l + \frac{(t - l\Delta_t)}{\Delta_t}\left(\mathbf{E}^{l+1} - \mathbf{E}^l\right) \quad (7.32)$$

Equation (7.31) then becomes

$$\mathbf{D}^l = \epsilon_0 \epsilon_\infty \mathbf{E}^l + \epsilon_0 \sum_{p=1}^{P} \Re e\left[\mathbf{Q}_p^l\right] \quad (7.33)$$

where

$$\mathbf{Q}_p^l = \sum_{m=0}^{l-1}\left[\left(\hat{\chi}_p^0 - \hat{\zeta}_p^0\right)\mathbf{E}^{l-m} + \hat{\zeta}_p^0 \mathbf{E}^{l-m-1}\right] e^{-(\alpha_p + i\beta_p)m\Delta_t} \quad (7.34)$$

In the above, $\hat{\chi}_p^0$ and $\hat{\zeta}_p^0$ are constants, which depend on the parameters of the particular model, given by

$$\hat{\chi}_p^0 = \int_0^{\Delta_t} \hat{\chi}_p(t)dt = \frac{i\gamma_p}{\alpha_p + i\beta_p}\left[1 - e^{-(\alpha_p + i\beta_p)\Delta_t}\right] \quad (7.35)$$

$$\hat{\zeta}_p^0 = \int_0^{\Delta_t} t\hat{\chi}_p(t)dt = \frac{i\gamma_p}{\Delta_t(\alpha_p + i\beta_p)^2}\left\{1 - \left[(\alpha_p + i\beta_p)\Delta_t + 1\right]e^{-(\alpha_p + i\beta_p)\Delta_t}\right\} \quad (7.36)$$

Furthermore, $\mathbf{Q}_p^l$ can be calculated recursively through $\mathbf{Q}_p^0 = 0$ and

$$\mathbf{Q}_p^l = \nu_p^0 \mathbf{E}^l + \hat{\zeta}_p^0 \mathbf{E}^{l-1} + \mathbf{Q}_p^{l-1} e^{-i\tilde{\omega}_p \Delta_t} \quad (7.37)$$

for $l > 0$, where $\tilde{\omega}_p = \beta_p - i\alpha_p$, and $\hat{\nu}_p^0 = \hat{\chi}_p^0 - \hat{\zeta}_p^0$.

The above equations allow the computation of $\mathbf{D}^l$ given $\mathbf{E}^l$ as the input. However, one would like to compute $\mathbf{E}^l$ given $\mathbf{D}^l$ as the input in an FDTD scheme as we shall see later. To this end, we substitute (7.35) and (7.36) into (7.37) to obtain

$$\mathbf{D}^l = \epsilon_0\left(\epsilon_\infty + \sum_{p=1}^{P}\Re e\left[\hat{\nu}_p^0\right]\right)\mathbf{E}^l + \epsilon_0 \sum_{p=1}^{P}\Re e\left[\hat{\zeta}_p^0 \mathbf{E}^{l-1} + \mathbf{Q}_p^{l-1} e^{-i\tilde{\omega}_p \Delta_t}\right] \quad (7.38)$$

or that

$$\mathbf{D}^l = \epsilon_0\left(\lambda_0 \mathbf{E}^l + \lambda_1 \mathbf{E}^{l-1} + \mathbf{P}^{l-1}\right) \quad (7.39)$$

where

$$\lambda_0 = \epsilon_\infty + \sum_{p=1}^{P} \Re[\hat{\nu}_p^0] \qquad (7.40)$$

$$\lambda_1 = \sum_{p=1}^{P} \Re[\hat{\zeta}_p^0] \qquad (7.41)$$

$$\mathbf{P}^{l-1} = \sum_{p=1}^{P} \Re\left(\mathbf{Q}_p^{l-1} e^{-i\tilde{\omega}_p \Delta_t}\right) \qquad (7.42)$$

depends only on $\mathbf{Q}_p^{l-1}$.

7.3.3 Incorporation into FDTD Update

To incorporate the time-domain equations of the previous section into an FDTD scheme, we need to devise a time-stepping scheme. The space discretization is a simple issue because those equations involve the same spatial curl operators of the usual Maxwell's equations. The time discretization for them is as follows:

$$\frac{-a_x\left(\mathbf{B}_{sx}^{l+\frac{1}{2}} - \mathbf{B}_{sx}^{l-\frac{1}{2}}\right)}{\Delta_t} - \Omega_x \mathbf{B}_{sx}^{l+\frac{1}{2}} = \partial_x \hat{x} \times \mathbf{E}^l \qquad (7.43)$$

$$\frac{a_x\left(\mathbf{D}_{sx}^{l+1} - \mathbf{D}_{sx}^l\right)}{\Delta_t} + \Omega_x \mathbf{D}_{sx}^{l+1} + a_x \sigma \mathbf{E}_{sx}^{l+1} + \sigma \Omega_x \mathbf{F}_{sx}^l = \partial_x \hat{x} \times \mathbf{H}^{l+\frac{1}{2}} \qquad (7.44)$$

where $\mathbf{F}(t) = \int_0^t \mathbf{E}(\tau) \, d\tau$. Equation (7.43) can be easily rearranged for time stepping

$$\mathbf{B}_{sx}^{l+\frac{1}{2}} = -\frac{1}{h_x}\left[\Delta_t\left(\partial_x \hat{x} \times \mathbf{E}^l\right) - a_x \mathbf{B}_{sx}^{l-\frac{1}{2}}\right] \qquad (7.45)$$

$$h_x \mathbf{D}_{sx}^{l+1} + a_x \sigma \Delta_t \mathbf{E}_{sx}^{l+1} = \Delta_t\left(\partial_x \hat{x} \times \mathbf{H}^{l+\frac{1}{2}}\right) + a_x \mathbf{D}_{sx}^l - \sigma \Omega_x \Delta_t \mathbf{F}_{sx}^l \qquad (7.46)$$

with $h_x = a_x + \Omega_x \Delta_t$. However, the left-hand side of (7.46) depends on both $\mathbf{D}_{sx}^{l+1}$ and $\mathbf{E}_{sx}^{l+1}$ making it unsuitable for time stepping. To remove this problem, we substitute (7.39) into the left-hand side of (7.46) so that we have

$$q_x \mathbf{E}_{sx}^{l+1} = \Delta_t\left(\partial_x \hat{x} \times \mathbf{H}^{l+\frac{1}{2}}\right) + a_x \mathbf{D}_{sx}^l - \sigma \Omega_x \Delta_t \mathbf{F}_{sx}^l - h_x \epsilon_0\left(\lambda_1 \mathbf{E}_{sx}^l + \mathbf{P}_{sx}^l\right) \qquad (7.47)$$

with $q_x = (a_x + \Omega_x \Delta_t) \lambda_0 \epsilon_0 + a_x \sigma \Delta_t$ The above equation is now suitable for time stepping and updating $\mathbf{E}_{sx}^{l+1}$. After $\mathbf{B}_{sx}^{l+\frac{1}{2}}$ is updated (and hence $\mathbf{H}_{sx}^{l+\frac{1}{2}}$ is updated), it is used in (7.47) to update $\mathbf{E}_{sx}^{l+1}$. On the right-hand side of (7.47), the other pertinent quantities may be updated as follows:

$$\mathbf{D}_{sx}^l = \epsilon_0\left(\lambda_0 \mathbf{E}_{sx}^l + \lambda_1 \mathbf{E}_{sx}^{l-1} + \mathbf{P}_{sx}^{l-1}\right) \qquad (7.48)$$

$$\mathbf{F}_{sx}^{l} = \mathbf{F}_{sx}^{l-1} + \frac{1}{2}\left(\mathbf{E}_{sx}^{l} + \mathbf{E}_{sx}^{l-1}\right)\Delta_{t} \tag{7.49}$$

$$\mathbf{Q}_{p,sx}^{l} = \hat{\nu}_{p}^{0}\mathbf{E}_{sx}^{l} + \hat{\zeta}_{p}^{0}\mathbf{E}_{sx}^{l-1} + \mathbf{Q}_{p,sx}^{l-1}e^{-i\bar{\omega}_{p}\Delta_{t}} \tag{7.50}$$

$$\mathbf{P}_{sx}^{l} = \sum_{p=1}^{P} \Re e\left[\mathbf{Q}_{p,sx}^{l}e^{-i\bar{\omega}_{p}\Delta_{t}}\right] \tag{7.51}$$

The above scheme is repeated for x replaced with y and z. Hence, (7.45), (7.47)–(7.51) constitute the complete updating scheme for the electromagnetic fields in dispersive media including the PML. Storage is required for $\mathbf{H}_{s\xi}$, $\mathbf{E}_{s\xi}$, $\mathbf{F}_{s\xi}$, $\mathbf{Q}_{p,s\xi}$, $p = 1, \cdots, P$, $\xi = x, y, z$, and since each $\mathbf{A}_{s\xi}$ has two vector components, we need to store $(18 + 6P)N$ values where N is the number of simulation nodes, and P is the number of species in the relaxation model. The added storage cost of simulating a PML dispersive medium is $6PN$ while the added cost of a PML conductive medium is to store $\mathbf{F}_{s\xi}$ which is $6N$. A nondispersive, nonconductive PML medium will require $12N$ storage as opposed to the $6N$ needed in the plain Yee's FDTD scheme [1]. Although (7.47)–(7.51) seem to suggest the need to store the electric field at two successive time steps, this can be avoided in the numerical algorithm by storing, at each time step, the electric field at the previous time step in a temporary variable.

To illustrate the accuracy of the PML-FDTD in dispersive media, results from the FDTD simulation for a homogeneous dispersive half-space problem with conductive loss are compared against a pseudo-analytical solution. The pseudo-analytical solution is obtained by integrating the frequency-domain Sommerfeld integrals of the dispersive half-space problem for many excitation frequencies. The result is then multiplied by the spectrum of the source pulse, and subsequently inverse-Fourier transformed to yield the time-domain solution. Figure 7.1 compares the results for the FDTD simulation using both a PML-RC (piecewise-constant electric field) approach and a PML-PLRC (piecewise-linear electric field) in a dispersive half-space against the pseudo-analytical solution. The half-space dispersion parameters are obtained by fitting a two-species ($P = 2$) Debye model to the experimental data reported by Hipp for the Puerto Rico type of claim loams [28]. The PML for this example is set up with 10 layers and a quadratic taper. The source pulse is the first derivative of of the Blackmann-Harris pulse [28], at central frequency $f_c = 200$ MHz. The half-space occupies 60% of the vertical height of the cubic simulation region. The simulation is done with a $N_x \times N_y \times N_z = 50 \times 50 \times 50$ grid with a space discretization size $\Delta_s = 5.86$ cm and a time step $\Delta_t = 90.2$ ps. Assuming that the origin is at a corner of the cube, then the source is a vertical electric dipole (y-directed) located at $(x, y, z) = (25, 35, 25)\Delta_s$ and the x component of the electric field is sampled at $(x, y, z) = (15, 25, 25)\Delta_s$. The field is deliberately sampled inside the half-space so that it is more sensitive to its dispersive properties. The results of Figure 7.1 show a good agreement between the formulations. Also, no noticeable reflection due to the grid termination is present.

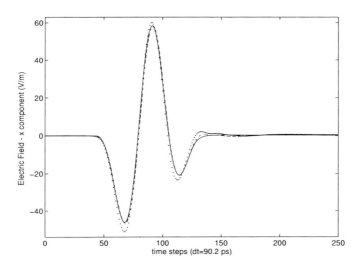

Figure 7.1 Sommerfeld solution (solid line) versus PML-FDTD solution using RC (dotted line) and PLRC (dashed line) for an infinitesimal vertical electric dipole radiating on top of dispersive half-space modeled by a two-species Debye model. The PLRC approach presents a better agreement against the pseudo-analytical Sommerfeld solution. (*Source:* [28], ©1998, IEEE. Reprinted with permission.)

7.4 MAXWELLIAN PML

In the previous sections, we have been considering the PML as an analytic continuation of Maxwell's equations and deriving the time-domain equation by splitting the electromagnetic fields into subcomponents. Because the resultant fields do not satisfy Maxwell's equations, this is sometimes called a non-Maxwellian PML.

There is another (dual) formulation of the PML that is quite attractive also. Through field transformations, it is possible to cast the modified Maxwell's equations in the PML, (7.1)–(7.4), into the familiar Maxwell's equations but for a modified medium, with complex anisotropic permittivity and permeability tensors [8,10]. This results in what is usually called the *Maxwellian* PML. To show that, we write the x component of the Faraday equation in complex space:

$$i\omega\mu H_x^c = \frac{\partial E_z^c}{\partial \tilde{y}} - \frac{\partial E_y^c}{\partial \tilde{z}} = \frac{1}{s_y}\frac{\partial E_z^c}{\partial y} - \frac{1}{s_z}\frac{\partial E_y^c}{\partial z} \quad (7.52)$$

The fields in (7.52) do not satisfy Maxwell's equations when $s_\zeta \neq 1$ (i.e., inside the PML), and to make this fact more explicit, the superscript c is added onto the field variables. However, if we multiply (7.52) by $s_y s_z$ and using the fact that s_ζ and $\partial/\partial \zeta'$ commute when $\zeta \neq \zeta'$, we arrive at

$$i\omega\mu \frac{s_y s_z}{s_x}(s_x H_x^c) = \frac{\partial}{\partial y}(s_z E_z^c) - \frac{\partial}{\partial z}(s_y E_y^c) \qquad (7.53)$$

If we then repeat the same procedure for the other components of the curl equations and introduce a new set of fields defined as $E_\zeta^a = s_\zeta E_\zeta^c$ and $H_\zeta^a = s_\zeta H_\zeta^c$, then this new set of fields obeys the usual Maxwell's equations but on an anisotropic medium of constitutive parameters $\overline{\mu} = \mu \overline{\overline{\Lambda}}$ and $\overline{\epsilon} = \epsilon \overline{\overline{\Lambda}}$, with

$$\overline{\overline{\Lambda}} = \hat{x}\hat{x}\left(\frac{s_y s_z}{s_x}\right) + \hat{y}\hat{y}\left(\frac{s_z s_x}{s_y}\right) + \hat{z}\hat{z}\left(\frac{s_x s_y}{s_z}\right) \qquad (7.54)$$

Note that the Maxwellian fields E_ζ^a, H_ζ^a coincide with the complex-space fields E_ζ^c, H_ζ^c when $s_\zeta = 1$ (i.e., in the physical domain). This is necessary since both formulations are expected to recover the original Maxwellian fields at those points. Equation (7.54) is the most general form for the constitutive tensors on the Cartesian, anisotropic PML formulation (corresponding to the PML medium at corner interfaces). In a single planar interface case, only the stretching coordinate normal to the interface has $s_\zeta \neq 1$ and, as a result, the medium is uniaxial.

Both Maxwellian and non-Maxwellian PML formulations satisfy the same boundary conditions on the continuity of *tangential* components of **E** and **H** across the PML interface, another requirement of consistency. This is because their tangential components differ by factors (tangential stretching variables) that are continuous across PML interfaces. However, the *normal* components of **E** and **H** satisfy *in general* different boundary conditions, since the normal stretching is not necessarily continuous (although in the practical numerical implementation, this is usually imposed to minimize spurious reflections due to discretization) across the interfaces.

The Maxwellian PML not only provides an interesting setting to study the PML concept but also provides a theoretical basis (blueprint) for the development of engineered absorbers [12]. Furthermore, it is more easily implemented on methods based in variational formulations such as the FEM.

7.5 EXTENSION TO (BI)ANISOTROPIC MEDIA

Using the analytic continuation of Maxwell's equations to a complex space, it is very simple to extend the perfect matching condition in (bi)anisotropic media. All that is needed is to assume the *same* constitutive tensors $\overline{\epsilon}$ and $\overline{\mu}$ everywhere. The PML region is then set up as the region where the analytic continuation of the spatial variables is enforced [29].

7.5.1 Non-Maxwellian Formulation

In a general anisotropic medium, $\overline{\epsilon}, \overline{\mu}$, the fields obey

$$-i\omega\overline{\epsilon} \cdot \mathbf{E} = \nabla \times \mathbf{H} \tag{7.55}$$

$$i\omega\overline{\mu} \cdot \mathbf{H} = \nabla \times \mathbf{E} \tag{7.56}$$

Inside the PML, the fields simply obey

$$-i\omega\overline{\epsilon} \cdot \mathbf{E} = \tilde{\nabla} \times \mathbf{H} \tag{7.57}$$

$$i\omega\overline{\mu} \cdot \mathbf{H} = \tilde{\nabla} \times \mathbf{E} \tag{7.58}$$

The major difference here is that, in the anisotropic case, each field component needs to be split in general into *three* subcomponents, $E_x = E_{xx} + E_{xy} + E_{xz}$, as opposed to only two, $E_x = E_{xy} + E_{xz}$, in the isotropic case. This is because the temporal derivative of the electric (magnetic) field components now depends on all three spatial derivatives of the magnetic (electric) field, as opposed to only the transverse ones in the isotropic case. This is related to the fact that such media may support longitudinal waves, which also need to be absorbed. For example, by letting $\overline{a} = \overline{\epsilon}^{-1}$, we have the following update equations for the E_x field:

$$-(i\omega s_x)E_{xx} = a_{xz}\frac{\partial}{\partial x}H_y - a_{xy}\frac{\partial}{\partial x}H_z \tag{7.59}$$

$$-(i\omega s_y)E_{xy} = a_{xx}\frac{\partial}{\partial y}H_z - a_{xz}\frac{\partial}{\partial y}H_x \tag{7.60}$$

$$-(i\omega s_z)E_{xz} = a_{xy}\frac{\partial}{\partial z}H_x - a_{xx}\frac{\partial}{\partial z}H_y \tag{7.61}$$

and analogous equations for the other components. Alternatively, one can use the **D** and **B** fields (with component splitting into two subcomponents) in the update equations and apply the constitutive relations after each time step. However, this two-step approach is not strictly necessary. For simplicity, we have considered up to this moment only the anisotropic case. The PML for (bi)anisotropic media follows along similar lines. In the next section, where we treat the Maxwellian PML case, this added degree of generality will be included.

7.5.2 Maxwellian Formulation

It is also possible to develop a Maxwellian PML formulation to match an interior media that is (bi)anisotropic. Such PML media will be represented by constitutive tensors $\overline{\epsilon}_{PML}(\overline{\epsilon}), \overline{\mu}_{PML}(\overline{\mu}), \overline{\xi}_{PML}(\overline{\xi})$, and $\overline{\zeta}_{PML}(\overline{\zeta})$, which depend on the particular tensors $\overline{\epsilon}, \overline{\mu}, \overline{\xi}$, and $\overline{\zeta}$ of the interior (physical) media. The isotropic, anisotropic, and bi-isotropic cases can be seen as special cases of the general formulation described below.

We start by rewriting the analytic continuation given by (7.7) by means of a dyadic function $\overline{\Gamma}$ such that

$$\mathbf{r} \to \tilde{\mathbf{r}} = \overline{\Gamma} \cdot \mathbf{r} \tag{7.62}$$

$$\overline{\Gamma} = \hat{x}\hat{x}\left(\frac{\tilde{x}}{x}\right) + \hat{y}\hat{y}\left(\frac{\tilde{y}}{y}\right) + \hat{z}\hat{z}\left(\frac{\tilde{z}}{z}\right) \tag{7.63}$$

Furthermore, we can write the modified nabla operator of (7.5) more compactly as

$$\tilde{\nabla} = \overline{\mathbf{S}} \cdot \nabla \tag{7.64}$$

with

$$\overline{\mathbf{S}}(\omega) = \hat{x}\hat{x}\left(\frac{1}{s_x}\right) + \hat{y}\hat{y}\left(\frac{1}{s_y}\right) + \hat{z}\hat{z}\left(\frac{1}{s_z}\right) \tag{7.65}$$

Noting that $s_\zeta(\zeta)$ and $\partial/\partial\zeta'$ commute for $\zeta \neq \zeta'$, and that $\overline{\mathbf{S}}$ is a diagonal tensor, the following identity can be verified for any vector function $\mathbf{a}(\mathbf{r})$ in Cartesian coordinates:

$$\nabla \times \left(\overline{\mathbf{S}}^{-1} \cdot \mathbf{a}\right) = (\det \overline{\mathbf{S}})^{-1} \overline{\mathbf{S}} \cdot (\overline{\mathbf{S}} \cdot \nabla) \times \mathbf{a} \tag{7.66}$$

where $\det \overline{\mathbf{S}} = (s_x s_y s_z)^{-1}$. The dyadic $\overline{\Gamma}$ also satisfies a similar equation.

In a bianisotropic and dispersive media, the Maxwell's equations are

$$\nabla \times \mathbf{E}(\mathbf{r}) = i\omega \mathbf{B}(\mathbf{r}) \tag{7.67}$$

$$\nabla \times \mathbf{H}(\mathbf{r}) = -i\omega \mathbf{D}(\mathbf{r}) \tag{7.68}$$

with

$$\mathbf{D}(\mathbf{r}) = \overline{\epsilon} \cdot \mathbf{E}(\mathbf{r}) + \overline{\xi} \cdot \mathbf{H}(\mathbf{r}) \tag{7.69}$$

$$\mathbf{B}(\mathbf{r}) = \overline{\zeta} \cdot \mathbf{E}(\mathbf{r}) + \overline{\mu} \cdot \mathbf{H}(\mathbf{r}) \tag{7.70}$$

where it is understood that both the fields and the constitutive tensors are functions of frequency.

The PML in complex space for such a medium is obtained by just keeping the *same* constitutive parameters *everywhere* and enforcing the complex stretching on the PML region. Inside the complex-space PML, the modified Maxwell's equations then simply read

$$\tilde{\nabla} \times \mathbf{E}^c(\tilde{\mathbf{r}}) = i\omega \mathbf{B}^c(\tilde{\mathbf{r}}) \tag{7.71}$$

$$\tilde{\nabla} \times \mathbf{H}^c(\tilde{\mathbf{r}}) = -i\omega \mathbf{D}^c(\tilde{\mathbf{r}}) \tag{7.72}$$

with

$$\mathbf{D}^c(\tilde{\mathbf{r}}) = \overline{\epsilon} \cdot \mathbf{E}^c(\tilde{\mathbf{r}}) + \overline{\xi} \cdot \mathbf{H}^c(\tilde{\mathbf{r}}) \tag{7.73}$$

$$\mathbf{B}^c(\tilde{\mathbf{r}}) = \overline{\zeta} \cdot \mathbf{E}^c(\tilde{\mathbf{r}}) + \overline{\mu} \cdot \mathbf{H}^c(\tilde{\mathbf{r}}) \tag{7.74}$$

Again, the subscript c indicates that the fields in (7.73) and (7.74) are not Maxwellian because they are in complex space. Using (7.62) and (7.64) we can recast (7.73) and (7.74) in the real space domain:

$$(\overline{\mathbf{S}} \cdot \nabla) \times \mathbf{E}^c(\overline{\boldsymbol{\Gamma}} \cdot \mathbf{r}) = i\omega \mathbf{B}^c(\overline{\boldsymbol{\Gamma}} \cdot \mathbf{r}) \tag{7.75}$$

$$(\overline{\mathbf{S}} \cdot \nabla) \times \mathbf{H}^c(\overline{\boldsymbol{\Gamma}} \cdot \mathbf{r}) = -i\omega \mathbf{D}^c(\overline{\boldsymbol{\Gamma}} \cdot \mathbf{r}) \tag{7.76}$$

Using (7.66), we write (7.75) and (7.76) as

$$\nabla \times \left[\overline{\mathbf{S}}^{-1} \cdot \mathbf{E}^c(\overline{\boldsymbol{\Gamma}} \cdot \mathbf{r}) \right] = i\omega (\det \overline{\mathbf{S}})^{-1} \overline{\mathbf{S}} \cdot \mathbf{B}^c(\overline{\boldsymbol{\Gamma}} \cdot \mathbf{r}) \tag{7.77}$$

$$\nabla \times \left[\overline{\mathbf{S}}^{-1} \cdot \mathbf{H}^c(\overline{\boldsymbol{\Gamma}} \cdot \mathbf{r}) \right] = -i\omega (\det \overline{\mathbf{S}})^{-1} \overline{\mathbf{S}} \cdot \mathbf{D}^c(\overline{\boldsymbol{\Gamma}} \cdot \mathbf{r}) \tag{7.78}$$

Introducing a new set of fields defined as

$$\mathbf{E}^a(\mathbf{r}) = \overline{\mathbf{S}}^{-1} \cdot \mathbf{E}^c(\overline{\boldsymbol{\Gamma}} \cdot \mathbf{r}) \tag{7.79}$$

$$\mathbf{H}^a(\mathbf{r}) = \overline{\mathbf{S}}^{-1} \cdot \mathbf{H}^c(\overline{\boldsymbol{\Gamma}} \cdot \mathbf{r}) \tag{7.80}$$

$$\mathbf{D}^a(\mathbf{r}) = (\det \overline{\mathbf{S}})^{-1} \overline{\mathbf{S}} \cdot \mathbf{D}^c(\overline{\boldsymbol{\Gamma}} \cdot \mathbf{r}) \tag{7.81}$$

$$\mathbf{B}^a(\mathbf{r}) = (\det \overline{\mathbf{S}})^{-1} \overline{\mathbf{S}} \cdot \mathbf{B}^c(\overline{\boldsymbol{\Gamma}} \cdot \mathbf{r}) \tag{7.82}$$

and substituting back in (7.77) and (7.78), we have

$$\nabla \times \mathbf{E}^a(\mathbf{r}) = i\omega \mathbf{B}^a(\mathbf{r}) \tag{7.83}$$

$$\nabla \times \mathbf{H}^a(\mathbf{r}) = -i\omega \mathbf{D}^a(\mathbf{r}) \tag{7.84}$$

with

$$\mathbf{D}^a(\mathbf{r}) = \left[(\det \overline{\mathbf{S}})^{-1} \left(\overline{\mathbf{S}} \cdot \overline{\boldsymbol{\epsilon}} \cdot \overline{\mathbf{S}} \right) \right] \cdot \mathbf{E}^a(\mathbf{r}) + \left[(\det \overline{\mathbf{S}})^{-1} \left(\overline{\mathbf{S}} \cdot \overline{\boldsymbol{\xi}} \cdot \overline{\mathbf{S}} \right) \right] \cdot \mathbf{H}^a(\mathbf{r}) \tag{7.85}$$

$$\mathbf{B}^a(\mathbf{r}) = \left[(\det \overline{\mathbf{S}})^{-1} \left(\overline{\mathbf{S}} \cdot \overline{\boldsymbol{\zeta}} \cdot \overline{\mathbf{S}} \right) \right] \cdot \mathbf{E}^a(\mathbf{r}) + \left[(\det \overline{\mathbf{S}})^{-1} \left(\overline{\mathbf{S}} \cdot \overline{\boldsymbol{\mu}} \cdot \overline{\mathbf{S}} \right) \right] \cdot \mathbf{H}^a(\mathbf{r}) \tag{7.86}$$

Therefore, the fields $\mathbf{E}^a, \mathbf{H}^a, \mathbf{D}^a, \mathbf{B}^a$ obey the Maxwell's equations. They also coincide with the original fields inside the computational domain (non-PML region), since $\overline{\boldsymbol{\Gamma}} = \overline{\mathbf{S}} = \overline{\mathbf{I}}$ there. Furthermore, from (7.79)–(7.82), it is seen that they present the same attenuative behavior of $\mathbf{E}^c, \mathbf{H}^c, \mathbf{D}^c, \mathbf{B}^c$ inside the PML, while preserving perfectly matching conditions. As a result, the Maxwellian PML bianisotropic constitutive parameters are given as

$$\overline{\boldsymbol{\epsilon}}_{PML}(\omega) = (\det \overline{\mathbf{S}})^{-1} \left(\overline{\mathbf{S}} \cdot \overline{\boldsymbol{\epsilon}} \cdot \overline{\mathbf{S}} \right) \tag{7.87}$$

$$\overline{\boldsymbol{\xi}}_{PML}(\omega) = (\det \overline{\mathbf{S}})^{-1} \left(\overline{\mathbf{S}} \cdot \overline{\boldsymbol{\xi}} \cdot \overline{\mathbf{S}} \right) \tag{7.88}$$

$$\overline{\boldsymbol{\zeta}}_{PML}(\omega) = (\det \overline{\mathbf{S}})^{-1} \left(\overline{\mathbf{S}} \cdot \overline{\boldsymbol{\zeta}} \cdot \overline{\mathbf{S}} \right) \tag{7.89}$$

$$\overline{\mu}_{PML}(\omega) = (\det \overline{\mathbf{S}})^{-1} \left(\overline{\mathbf{S}} \cdot \overline{\mu} \cdot \overline{\mathbf{S}} \right) \quad (7.90)$$

These formulas give directly the bianisotropic constitutive parameters that have to be present both in single interface problems and 2D or 3D corner interfaces. Maxwellian PML tensors to match isotropic, anisotropic, or biisotropic media problems can be taken as special cases of the above formulas. Note that (7.62) bears a formal resemblance to the expression of an affine transformation. In this respect, we should note that $\overline{\Gamma}(\mathbf{r})$ is already a function of position, and therefore, (7.62) defines a *nonlinear* transformation on **r**. Moreover, it always preserves the orthogonality of the metric, since $\overline{\Gamma}$ is diagonal. In the Fourier domain, the PML may be viewed simply as a *complex* mapping (stretching) of the metric.

The fact that a change on the constitutive parameters can mimic the mapping of the metric (allowing for a Maxwellian PML formulation) is best appreciated using the language of differential forms for the electromagnetic fields [36]. Using this language, the constitutive parameters are not simply tensors, but Hodge operators relating 1-forms (H, E) to 2-forms (B, D), and carrying all the information about the metric of the space. The Maxwellian PML formulation is then just a consequence of the metric independence of Maxwell's equations. This will be discussed in Section 7.10.

7.6 PML FOR INHOMOGENEOUS MEDIA

By recognizing the PML as a mapping of the coordinate space to a complex space (and therefore transparent to the constitutive properties of the medium), the implementation of the PML in inhomogeneous media follows exactly the same lines of the homogeneous case. Because the PML is a *local* boundary condition, its implementation depends only locally on the properties of the medium at the grid termination surface. The basic difference in this case is that the PML for inhomogeneous media will also be inhomogeneous at the interface, with its constitutive parameters mirroring those at the interface plane. In other words, the PML *inherits* the constitutive parameters of the interior domain at each point of the interface.

In inhomogeneous media, Maxwell's equations are

$$\nabla \times \mathbf{H} = -i\omega\epsilon(\mathbf{r})\mathbf{E} \quad (7.91)$$

$$\nabla \times \mathbf{E} = i\omega\mu(\mathbf{r})\mathbf{H} \quad (7.92)$$

$$\nabla \cdot \epsilon(\mathbf{r})\mathbf{E} = 0 \quad (7.93)$$

$$\nabla \cdot \mu(\mathbf{r})\mathbf{H} = 0 \quad (7.94)$$

The modified Maxwell's equations in the PML will then just read

$$\nabla_s \times \mathbf{H} = -i\omega\epsilon(\mathbf{r})\mathbf{E} \quad (7.95)$$

$$\nabla_s \times \mathbf{E} = i\omega\mu(\mathbf{r})\mathbf{H} \tag{7.96}$$

$$\nabla_s \cdot \epsilon(\mathbf{r})\mathbf{E} = 0 \tag{7.97}$$

$$\nabla_s \cdot \mu(\mathbf{r})\mathbf{H} = 0 \tag{7.98}$$

where $\epsilon(\mathbf{r})$ and $\mu(\mathbf{r})$ at the PML interface inherit the values from the interior equations so that $\epsilon(\mathbf{r})$ and $\mu(\mathbf{r})$ are kept continuous at the physical domain–PML interface. *Inside* the PML, it is more natural to choose $\epsilon(\mathbf{r})$ and $\mu(\mathbf{r})$ to be constant along the normal direction to the interface to minimize any reflection.

For the Maxwellian PML case, the inhomogeneity is incorporated directly into the PML constitutive tensors, as they explicitly depend on the ϵ and μ of the media. Again, because the perfect matching is a *local* condition, the generalization is achieved by just replacing ϵ and μ by $\epsilon(\mathbf{r})$ and $\mu(\mathbf{r})$ in the expressions for the PML constitutive tensors (i.e., $\overline{\epsilon} = \epsilon(\mathbf{r})\overline{\Lambda}$ and $\overline{\mu} = \mu(\mathbf{r})\overline{\Lambda}$).

7.7 CURVILINEAR PML

In a variety of problems, it is of interest to employ the FDTD method in non-Cartesian systems. This is true especially in problems with some kind of cylindrical or spherical symmetry or to avoid the staircase approximation of curved boundaries [2].

However, the original PML concept applied only to Cartesian coordinates. To extend its range of applicability, the PML concept was also applied to nonorthogonal FDTD grids [37, 38] with very good results. However, an only approximate impedance matching condition was obtained, since the perfect matching condition was derived based on the assumption that the metric coefficients were independent of the spatial coordinates.

In this section, we derive the PML media for curvilinear orthogonal coordinate systems with an exact formulation in the sense that it provides a reflectionless termination in the continuum limit. The formulation is based on the complex coordinate stretching approach.

7.7.1 Cylindrical PML-FDTD

The PML formulation for a cylindrical coordinate system proceeds by writing the Maxwell's equations on a complex cylindrical coordinate [15]. For brevity, only the TM_z case will be addressed here, since the TE_z case follows by duality. Also, since in the z direction the PML formulation does not change, only the 2D problem is treated here. To achieve the reflectionless absorption of the outward traveling waves, the radial coordinate is mapped through

$$\rho \to \tilde{\rho} = \int_0^\rho s_\rho(\rho')d\rho' = b_\rho(\rho) + i\frac{\Delta_\rho(\rho)}{\omega} \tag{7.99}$$

with
$$s_\rho(\rho) = a_\rho(\rho) + i\frac{\Omega_\rho(\rho)}{\omega} \quad (7.100)$$

The variables a_ρ and Ω_ρ are added degrees of freedom. In these equations, Ω_ρ controls the absorption for propagating waves, and a_ρ increases the decay rate of evanescent waves, if they exist, along the ρ direction. In the physical region, we must set $\Omega_\rho = 0$ and $a_\rho = 1$ in order to reduce the modified equations to the original Maxwell's equations. In the 3D case, the above transformation is combined with a similar transformation on the z variable, $z \to \tilde{z}$ as in (7.7), at the bottom and top regions of the cylindrical domain. Because ρ and z are orthogonal everywhere, one transformation is carried independently of the other.

To understand the origin of the reflectionless property of this transformation, we can employ essentially the same arguments used in the Cartesian case and consider the continuity properties of the transformed variables and fields. As in the Cartesian case, closed-form solutions for the fields inside the PML can be written by inspection from the knowledge of the ordinary closed-form solutions of Maxwell's equations in free space, by a simple change of variables, $\rho \to \tilde{\rho}$ and $z \to \tilde{z}$. Furthermore, the complex variables $\tilde{\rho}(\rho)$ and $\tilde{z}(z)$ are always continuous functions of their arguments (real variables). This is because they are defined in terms of integrals: (7.7) and (7.99) of bounded functions $s_\rho(\rho)$ and $s_z(z)$ (note that $s_\rho(\rho)$ and $s_z(z)$ themselves need not be continuous for the continuity of $\tilde{\rho}(\rho)$ and $\tilde{z}(z)$ to hold). As a result, the fields inside the PML will be continuous everywhere if the original fields are continuous (because the composition of two continuous functions is a continuous function). In particular, all boundary conditions are preserved by this transformation, and, hence, a perfectly matched interface is obtained.

To avoid *numerical* reflections due to the discretization process, both a_ρ and Ω_ρ can be chosen to increase gradually in the PML region (similarly to the Cartesian case). The Maxwell's equations in the cylindrical complex space then read

$$\frac{1}{\tilde{\rho}}\frac{\partial}{\partial\tilde{\rho}}(\tilde{\rho}H_\phi) - \frac{1}{\tilde{\rho}}\frac{\partial H_\rho}{\partial\phi} = -i\omega\epsilon E_z \quad (7.101)$$

$$\frac{1}{\tilde{\rho}}\frac{\partial E_z}{\partial\phi} = i\omega\mu H_\rho \quad (7.102)$$

$$\frac{\partial E_z}{\partial\tilde{\rho}} = -i\omega\mu H_\phi \quad (7.103)$$

By splitting the z component of the electric field $E_z = E_{z\rho} + E_{z\phi}$, the above can be rearranged in a form suitable for time stepping as follows:

$$(i\omega s_\rho)\epsilon \tilde{E}_{z\rho} = -\frac{\partial}{\partial\rho}(\tilde{\rho}H_\phi) \quad (7.104)$$

$$E_{z\rho} = \tilde{\rho}^{-1}\tilde{E}_{z\rho} \quad (7.105)$$

$$(i\omega\tilde{\rho})\epsilon E_{z\phi} = \frac{\partial H_\rho}{\partial \phi} \tag{7.106}$$

$$(i\omega\tilde{\rho})\epsilon E_{z\phi} = \frac{\partial H_\rho}{\partial \phi} \tag{7.107}$$

$$E_z = E_{z\rho} + E_{z\phi} \tag{7.108}$$

$$(i\omega s_\rho)\mu H_\phi = -\frac{\partial E_z}{\partial \rho} \tag{7.109}$$

$$(i\omega\tilde{\rho})\mu H_\rho = \frac{\partial E_z}{\partial \phi} \tag{7.110}$$

We note here that the splitting is necessary only in the E_z component (H_z in the TE_z case). This is because the transversal problem has stretching only in the ρ direction. This is in contrast to the Cartesian case where all field components need to be split in the split-field formulation. However, an additional field $\tilde{E}_{z\rho}$ (convolution of the original field with the stretched radius) is needed, as well as auxiliary fields for the convolutions present in (7.104) and (7.105). Substituting (7.99) and (7.100) into (7.104)–(7.110) and inverse-Fourier transforming, the following time-domain equations are obtained:

$$\left(a_\rho \frac{\partial}{\partial t} + \Omega_\rho\right) \epsilon \tilde{E}_{z\rho}(t) = \frac{\partial}{\partial \rho}\left(\tilde{\rho}(t) * H_\phi(t)\right) \tag{7.111}$$

$$E_{z\rho}(t) = \tilde{\rho}^{-1}(t) * \tilde{E}_{z\rho}(t) \tag{7.112}$$

$$\left(b_\rho \frac{\partial}{\partial t} + \Delta_\rho\right) \epsilon E_{z\phi}(t) = -\frac{\partial H_\rho(t)}{\partial \phi} \tag{7.113}$$

$$E_z(t) = E_{z\rho}(t) + E_{z\phi}(t) \tag{7.114}$$

$$\left(a_\rho \frac{\partial}{\partial t} + \Omega_\rho\right) \mu H_\phi(t) = \frac{\partial E_z(t)}{\partial \rho} \tag{7.115}$$

$$\left(b_\rho \frac{\partial}{\partial t} + \Delta_\rho\right) \mu H_\rho(t) = -\frac{\partial E_z(t)}{\partial \phi} \tag{7.116}$$

The above equations can be easily implemented by a cylindrical staggered-grid FDTD algorithm [39]. The convolutions in (7.111) and (7.112) are explicitly given by

$$\tilde{\rho}(t) * H_\phi(t) = b_\rho H_\phi(t) + \Delta_\rho \int_0^t H_\phi(t')dt' \tag{7.117}$$

$$\tilde{\rho}^{-1}(t) * \tilde{E}_{z\rho}(t) = \frac{1}{b_\rho}\left(\tilde{E}_{z\rho}(t) - \frac{1}{\tau_0}\int_0^t \tilde{E}_{z\rho}(t-t')e^{-\frac{t'}{\tau_0}}dt'\right) \tag{7.118}$$

where $\tau_0 = b_\rho/\Delta_\rho$. Equations (7.117) and (7.118) can be incorporated in the time discretization scheme at a minimal computational cost and memory requirement. For

(7.117) this is trivial. For (7.118) this is done by recursive convolution [17], reducing (7.118) to a recursive relation that is updated iteratively. With $t = l\Delta t$, we have

$$E_{z\rho}^l = \frac{1}{b_\rho}\left[\tilde{E}_{z\rho}^l - \left(1 - e^{-\frac{\Delta t}{\tau_0}}\right)Q^l\right] \quad (7.119)$$

where $Q^l = \tilde{E}_{z\rho}^l + Q^{l-1}e^{-\frac{\Delta t}{\tau_0}}$ for $l > 0$, and $Q^0 = 0$. This assumes that the electric field is constant over each time interval Δt. A similar formula can be derived assuming piecewise linear functional dependence over Δt [17].

It is important to point out here that the time-domain equations (7.111)–(7.116) are not unique. Many other time-domain equations can be derived from the complex-space frequency domain equations. The scheme of (7.111)–(7.116) involves convolutions and it is used to illustrate that, because of the (Debye-like) frequency dependence of the complex variables $s_\rho(\omega)$ and $\tilde{\rho}(\omega)$, such convolutions can be reduced to a recursive update and therefore have a very small associated computational cost. However, alternative algorithms can also be derived that avoid any convolutions at all. One such algorithm is presented in [15]. For instance, this could be achieved by expanding the radial derivative in (7.101) and setting

$$(i\omega\tilde{\rho})\epsilon E_{z\rho 1} = -H_\phi \quad (7.120)$$

$$(i\omega s_\rho)\epsilon E_{z\rho 2} = -\frac{\partial H_\phi}{\partial \rho} \quad (7.121)$$

$$E_{z\rho} = E_{z\rho 1} + E_{z\rho 2} \quad (7.122)$$

where, in a staggered-grid spatial discretization scheme, the H_ϕ in (7.120) should be taken as an average over half-grid points to preserve second-order accuracy in space.

The effect of the transformation in (7.99) can be appreciated by considering the Green's function for a point source at (ρ', ϕ') inside a perfect electrically conducting (PEC) cylinder of radius a:

$$g(\boldsymbol{\rho}, \boldsymbol{\rho}') = g(\rho, \phi, \rho', \phi')$$
$$= \frac{i}{4}H_0^{(1)}(k|\boldsymbol{\rho} - \boldsymbol{\rho}'|) - \frac{i}{4}\sum_{n=-\infty}^{+\infty} J_n(k\rho)\frac{J_n(k\rho')H_n^{(1)}(ka)}{J_n(ka)}e^{in(\phi-\phi')}$$
(7.123)

When the PEC boundary is backed by a concave cylindrical PML, the above maps to

$$g(\tilde{\boldsymbol{\rho}}, \boldsymbol{\rho}') = g(\tilde{\rho}, \phi, \rho', \phi')$$
$$= \frac{i}{4}H_0^{(1)}(k|\tilde{\boldsymbol{\rho}} - \boldsymbol{\rho}'|) - \frac{i}{4}\sum_{n=-\infty}^{+\infty} J_n(k\tilde{\rho})\frac{J_n(k\rho')H_n^{(1)}(k\tilde{a})}{J_n(k\tilde{a})}e^{in(\phi-\phi')}$$
(7.124)

where $\tilde{\rho}$ is defined in (7.99). Note that $\tilde{a}$ is also complex-valued. The effect of the cylindrical PML, therefore, is to make the reflected field (summation term in (7.124)) exponentially small. Furthermore, this analytic continuation preserves the analyticity on the upper half complex ω plane [30]. This is because the singularities of the Green's function in (7.123) are due to branch points (located on the real axis) of the $H_n^{(1)}(\cdot)$, and zeros of the denominator functions $J_n(\cdot)$ (all on the real axis) [30]. When $\rho \to \tilde{\rho}$ with $\Delta_\rho > 0$, all singularities are translated down to the lower half ω-plane, so that the upper half plane is kept free of any singularity. This is important to ensure that causality is preserved and that the resultant solutions are dynamically stable [30]. This will be discussed in more detail in Section 7.8.

To illustrate the absorption of the spurious reflection from the grid termination by the cylindrical PML, we compare the PML-FDTD solution against an analytic solution obtained by solving a free-space problem in the frequency domain, multiplying by the source pulse spectrum, and inverse Fourier-transforming.

Figure 7.2 shows the normalized E_z field computed using the analytic formulation and the 2D FDTD algorithm for a line source in a cylindrical grid. The excitation is the derivative of a Blackmann-Harris pulse centered at $f_c = 300$ MHz. The line source is at $(r, \phi) = (7.5\lambda_c, 0^0)$ and the field is sampled at $(r, \phi) = (6.5\lambda_c, 0^0)$, where $\lambda_c = c/f_c$. The grid has a hard termination at $r = 9\lambda_c$. The FDTD algorithm includes an eight-layer cylindrical PML region before the grid ends, with $a_\rho = 1$ and a quadratic taper on σ_ρ. The PML thickness is $0.5\lambda_c$, for a cell size $\Delta\rho = \lambda_c/16$ in the radial direction. The curves are in excellent agreement and no reflection is visible. To illustrate the high absorption achieved, the inset shows the simulation of the same problem without the PML. In the PML-FDTD simulation, the maximum residual amplitude of the normalized E_z field over the time-window of the reflected pulse is less than $5. \times 10^{-3}$, which is less than 0.8% of the maximum amplitude present in the simulation without PML. This residual field can be attributed not only to spurious reflections but also to numerical dispersion effects.

To extract out numerical dispersion effects so as to accurately quantify the amount of spurious reflection caused by the discrete cylindrical PML, we follow a methodology discussed in [3]. Two cylindrical FDTD grids are used, a test domain Ω_T and a benchmark domain Ω_B. The test domain includes the cylindrical PML, and the benchmark domain has untreated boundaries but a much larger size. The outer boundaries of Ω_B are defined so that its spurious reflections can be causally isolated during the time-stepping when comparing simulations on the two grids. Therefore, the grid Ω_B effectively simulates an infinite grid. Any discrepancies between the computed field values in the two grids are produced by spurious reflection from the PML in the Ω_T domain. Figure 7.3 presents the local reflection errors obtained from an eight-layer cylindrical PML. Three results are presented, corresponding to three different angular grid resolutions along the ϕ coordinate. The simulations employ a (hard) line source excited by a Blackmann-Harris pulse with central frequency $f_c = 300$ MHz. The local reflection error is obtained by field sampling at two grid

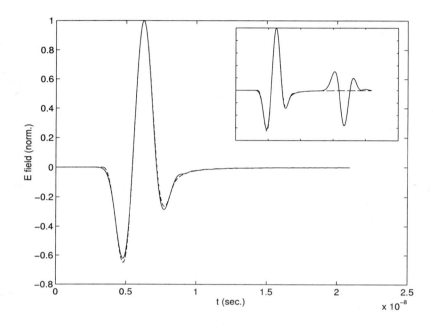

Figure 7.2 Analytic solution for a line source on free space (dashed line) versus 2D cylindrical-grid FDTD solution with eight-layer cylindrical PML (solid line). The inset illustrates the result of the simulation without the PML. (*Source:* [15], ©1997, IEEE. Reprinted with permission.)

cells away from the free-space/PML interface of the Ω_T domain. The line source location is eight cells away from the free-space/PML interface. The radial resolution in all cases is fixed at $\Delta\rho = \lambda_c/16$. The free-space/PML interface in Ω_T is situated at $\rho_0 = 4\lambda_c$. The eight-layer PML employs a fourth-order tapered profile on the imaginary part of the stretching variables and no stretching on the real part. The benchmark domain is truncated at $\rho = 8\lambda_c$. The local error data is normalized to the peak value of the incident pulse at the sampling location. The three angular resolutions considered correspond to values of $\Delta\phi$ such that $\rho_0\Delta\phi = \lambda/8, \lambda/12, \lambda/16$, where ρ_0 is value of the ρ coordinate corresponding to the free-space/PML interface in Ω_T. The results of Figure 7.3 exemplify the very low reflection levels incurred by the cylindrical PML. A reflection level better than -70 dB is obtained with the eight-layer cylindrical PML.

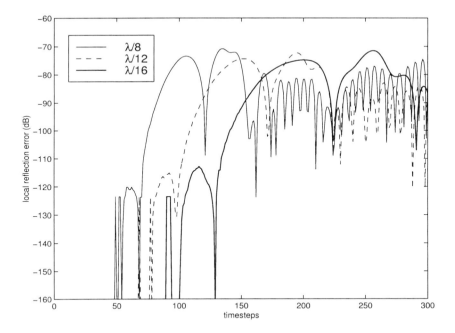

Figure 7.3 Local reflection error within test grid observed over the first 300 time steps for an eight-layer cylindrical PML. Three different *angular* resolutions for the cylindrical grid are considered. (*Source:* [34], ©2000, IEEE. Reprinted with permission.)

7.7.2 Spherical PML-FDTD

In spherical coordinates, the reflectionless absorption of outward traveling waves is achieved through the analytic continuation on the radial variable:

$$r \to \tilde{r} = \int_0^r s_r(r')dr' = b_r(r) + i\frac{\Delta_r(r)}{\omega} \qquad (7.125)$$

with

$$s_r(r) = a_r(r) + i\frac{\Omega_r(r)}{\omega} \qquad (7.126)$$

from which the generalized Faraday's law in a form suitable for time-stepping reads

$$(i\omega\tilde{r})\mu H_r = \frac{1}{\sin\theta}\left[\frac{\partial}{\partial\theta}(\sin\theta E_\phi) - \frac{\partial E_\theta}{\partial\phi}\right] \qquad (7.127)$$

$$(i\omega s_r)\mu\tilde{H}_\theta = \frac{1}{\sin\theta}\frac{\partial}{\partial\phi}(s_r E_r) - \frac{\partial \tilde{E}_\phi}{\partial r} \qquad (7.128)$$

$$(i\omega s_r)\mu \tilde{H}_\phi = \frac{\partial \tilde{E}_\theta}{\partial r} - \frac{\partial}{\partial \theta}(s_r E_r) \tag{7.129}$$

$$H_\theta = \tilde{r}^{-1} \tilde{H}_\theta \tag{7.130}$$

$$H_\phi = \tilde{r}^{-1} \tilde{H}_\phi \tag{7.131}$$

The time-domain version of the above is

$$\left(b_r \frac{\partial}{\partial t} + \Delta_r\right)\mu H_r(t) = -\frac{1}{\sin\theta}\left[\frac{\partial}{\partial \theta}(\sin\theta E_\phi(t)) - \frac{\partial E_\theta(t)}{\partial \phi}\right] \tag{7.132}$$

$$\left(a_r \frac{\partial}{\partial t} + \Omega_r\right)\mu \tilde{H}_\theta(t) = -\frac{1}{\sin\theta}\frac{\partial}{\partial \phi}[s_r(t) * E_r(t)] + \frac{\partial \tilde{E}_\phi(t)}{\partial r} \tag{7.133}$$

$$\left(a_r \frac{\partial}{\partial t} + \Omega_r\right)\mu \tilde{H}_\phi(t) = -\frac{\partial \tilde{E}_\theta(t)}{\partial r} + \frac{\partial}{\partial \theta}(s_r(t) * E_r(t)) \tag{7.134}$$

$$H_\theta(t) = \tilde{r}^{-1}(t) * \tilde{H}_\theta(t) \tag{7.135}$$

$$H_\phi(t) = \tilde{r}^{-1}(t) * \tilde{H}_\phi(t) \tag{7.136}$$

Similar generalization is made on Ampere's law:

$$(i\omega\tilde{r})\epsilon E_r = -\frac{1}{\sin\theta}\left[\frac{\partial}{\partial \theta}(\sin\theta H_\phi) - \frac{\partial H_\theta}{\partial \phi}\right] \tag{7.137}$$

$$(i\omega s_r)\epsilon \tilde{E}_\theta = -\frac{1}{\sin\theta}\frac{\partial}{\partial \phi}(s_r H_r) + \frac{\partial \tilde{H}_\phi}{\partial r} \tag{7.138}$$

$$(i\omega s_r)\epsilon \tilde{E}_\phi = -\frac{\partial \tilde{H}_\theta}{\partial r} + \frac{\partial}{\partial \theta}(s_r H_r) \tag{7.139}$$

$$E_\theta = \tilde{r}^{-1}\tilde{E}_\theta \tag{7.140}$$

$$E_\phi = \tilde{r}^{-1}\tilde{E}_\phi \tag{7.141}$$

resulting in following time-domain equations suitable for time stepping:

$$\left(b_r \frac{\partial}{\partial t} + \Delta_r\right)\epsilon E_r(t) = \frac{1}{\sin\theta}\left[\frac{\partial}{\partial \theta}(\sin\theta H_\phi(t)) - \frac{\partial H_\theta(t)}{\partial \phi}\right] \tag{7.142}$$

$$\left(a_r \frac{\partial}{\partial t} + \Omega_r\right)\epsilon \tilde{E}_\theta(t) = \frac{1}{\sin\theta}\frac{\partial}{\partial \phi}[s_r(t) * H_r(t)] - \frac{\partial \tilde{H}_\phi(t)}{\partial r} \tag{7.143}$$

$$\left(a_r \frac{\partial}{\partial t} + \Omega_r\right)\epsilon \tilde{E}_\phi(t) = \frac{\partial \tilde{H}_\theta(t)}{\partial r} - \frac{\partial}{\partial \theta}(s_r(t) * H_r(t)) \tag{7.144}$$

$$E_\theta(t) = \tilde{r}^{-1}(t) * \tilde{E}_\theta(t) \tag{7.145}$$

$$E_\phi(t) = \tilde{r}^{-1}(t) * \tilde{E}_\phi(t) \qquad (7.146)$$

From the above equations, it is seen that, in spherical coordinates, there is no need to split the fields at all. This is because the PML can be achieved through complex stretching in the radial variable r only. This is in contrast to the Cartesian case, where in the 3D situation there are 12 field components after field splitting and six boundary surfaces to treat. However, additional fields components are needed inside the spherical PML: $s_r E_r$, $\tilde{E}_\theta$, $\tilde{E}_\phi$, $s_r H_r$, $\tilde{H}_\theta$, and $\tilde{H}_\phi$ (as well as one auxiliary field for the convolutions). Analogously to the cylindrical case, those convolutions in can be calculated recursively as in the cylindrical case. Moreover, alternative time-domain PML algorithms in spherical coordinates which avoid convolutions altogether can also be easily derived, as described in [15].

To illustrate the absorption of the spurious reflection from the grid termination by the spherical PML, we again compare the PML-FDTD solution against an analytic solution in free space. Figure 7.4 shows the normalized E_θ field computed with the analytic formulation and the 3D FDTD algorithm for a point source in a spherical grid with the same excitation pulse. The θ-polarized dipole is at $(r, \theta, \phi) = (2.5\lambda_c, 90^0, 0^0)$. The field is sampled at $(r, \theta, \phi) = (3\lambda_c, 90^0, 0^0)$. The grid is terminated at $r = 4\lambda_c$. The FDTD algorithm includes an eight-layer spherical PML region before the grid ends, with $a_r = 1$ and a quadratic taper on σ_r. The PML thickness is $0.8\lambda_c$, for a cell size $\Delta r = \lambda_c/10$ in the radial direction. Again, no reflection is visible. The small oscillation after the passage of the incident pulse is also present in the simulation without PML and can be attributed to the discretization errors and numerical dispersion effects due to the high grid curvature in the simulation region. For illustration, the inset shows the simulation of the same problem without the PML. In the PML-FDTD simulation, the maximum residual amplitude of the normalized E_z field for $t > 9$ ns is less than 1×10^{-2}, which is less than 3% of the maximum amplitude present in the simulation without PML. Again, the residual field includes not only the spurious reflection but also numerical dispersion effects.

7.7.3 Maxwellian PML in Cylindrical and Spherical Coordinates

In previous sections, we have shown that the Cartesian PML admits two distinct formulations in the frequency domain. In the first formulation (non-Maxwellian), the PML fields are the analytic continuation of the original physical fields. In the second formulation, the PML fields obey Maxwell's equation in a medium with complex permittivity and permeability constitutive tensors. This is also true for dispersive and/or (bi)anisotropic interior media.

By analogy to the Cartesian case, it is also possible to derive corresponding Maxwellian PMLs in cylindrical and spherical coordinates through suitable transformations in the analytically continued fields [31].

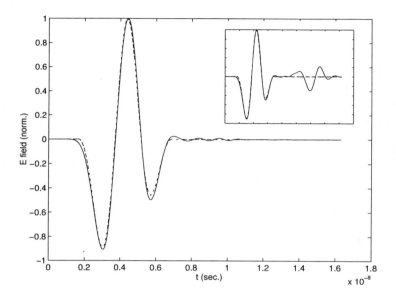

Figure 7.4 Analytic solution for an infinitesimal electric dipole on free-space (dashed line) versus 3D spherical-grid FDTD solution with eight-layer spherical PML (solid line). The inset illustrates the result of the simulation without the PML. (*Source:* [15], ©1997, IEEE. Reprinted with permission.)

In the cylindrical system, we write Faraday's law in the complex space as

$$i\omega\mu H_\rho^c = \frac{1}{\tilde{\rho}}\frac{\partial E_z^c}{\partial \phi} - \frac{\partial E_\phi^c}{\partial \tilde{z}} \tag{7.147}$$

$$i\omega\mu H_\phi^c = \frac{\partial E_\rho^c}{\partial \tilde{z}} - \frac{\partial E_z^c}{\partial \tilde{\rho}} \tag{7.148}$$

$$i\omega\mu H_z^c = \frac{1}{\tilde{\rho}}\frac{\partial}{\partial \tilde{\rho}}(\tilde{\rho}E_\phi^c) - \frac{1}{\tilde{\rho}}\frac{\partial E_\rho^c}{\partial \phi} \tag{7.149}$$

From the definition of the complex variables $\tilde{\rho}$ and $\tilde{z}$ in (7.99) and (7.7), respectively, we have $\partial/\partial\tilde{\rho} = (1/s_\rho)\partial/\partial\rho$ and $\partial/\partial\tilde{z} = (1/s_z)\partial/\partial z$. If we substitute these last identities in (7.147)–(7.149), multiply (7.147) by $s_z(\tilde{\rho}/\rho)$, (7.148) by $s_z s_\rho$, and (7.149) by $s_\rho(\tilde{\rho}/\rho)$, then (7.147)–(7.149) can be recast into the following form:

$$i\omega\mu\left[\left(\frac{\tilde{\rho}}{\rho}\right)\frac{s_z}{s_\rho}\right](s_\rho H_\rho^c) = \frac{1}{\rho}\frac{\partial}{\partial \phi}(s_z E_z^c) - \frac{\partial}{\partial z}\left(\frac{\tilde{\rho}E_\phi^c}{\rho}\right) \tag{7.150}$$

$$i\omega\mu\left[\left(\frac{\rho}{\tilde{\rho}}\right)s_zs_\rho\right]\left(\frac{\tilde{\rho}H_\phi^c}{\rho}\right) = \frac{\partial}{\partial z}(s_\rho E_\rho^c) - \frac{\partial}{\partial \rho}(s_z E_z^c) \quad (7.151)$$

$$i\omega\mu\left[\left(\frac{\tilde{\rho}}{\rho}\right)\frac{s_\rho}{s_z}\right](s_z H_z^c) = \frac{1}{\rho}\frac{\partial}{\partial\rho}\left[\rho\left(\frac{\tilde{\rho}E_\phi^c}{\rho}\right)\right] - \frac{1}{\rho}\frac{\partial}{\partial\phi}(s_\rho E_\rho^c) \quad (7.152)$$

From (7.150)–(7.152) and their duals (Ampere's law), we see that a new set of fields defined by $E_\rho^a = s_\rho E_\rho^c$, $E_\phi^a = (\tilde{\rho}/\rho)E_\phi^c$, $E_z^a = s_z E_z^c$ (similarly for the **H** field), obeys Maxwell's equations on an anisotropic medium of constitutive parameters $\overline{\mu} = \mu\overline{\Lambda}$ and $\overline{\epsilon} = \epsilon\overline{\Lambda}$, with

$$\overline{\Lambda} = \hat{\rho}\hat{\rho}\left(\frac{\tilde{\rho}}{\rho}\right)\left(\frac{s_z}{s_\rho}\right) + \hat{\phi}\hat{\phi}\left(\frac{\rho}{\tilde{\rho}}\right)(s_zs_\rho) + \hat{z}\hat{z}\left(\frac{\tilde{\rho}}{\rho}\right)\left(\frac{s_\rho}{s_z}\right) \quad (7.153)$$

An alternative derivation of the 2D version of (7.153) (with $s_z = 1$) through a graphical approach is presented in [40]. By *defining* $s_\phi = (\tilde{\rho}/\rho)$, the above tensor and the field mapping equations have the same *formal* appearance as in the Cartesian case. Both Maxwellian and non-Maxwellian formulations satisfy the same boundary conditions on the continuity of tangential fields across the PML interface. This is because the corresponding tangential fields differ by factors that are continuous across PML interfaces $z = z_0$ and $\rho = \rho_0$ (since s_ρ is a function of ρ only, s_z is a function of z only, and $\tilde{\rho}/\rho$ is continuous everywhere as implied by their definition through an integral). Because of this, the perfect matching condition for one of the formulations follows automatically from the other, a requirement of consistency.

In spherical coordinates, we write Faraday's law on complex space as

$$i\omega\mu H_r^c = \frac{1}{\tilde{r}\sin\theta}\left[\frac{\partial}{\partial\theta}(\sin\theta E_\phi^c) - \frac{\partial E_\theta^c}{\partial\phi}\right] \quad (7.154)$$

$$i\omega\mu H_\theta^c = \frac{1}{\tilde{r}\sin\theta}\frac{\partial E_r^c}{\partial\phi} - \frac{1}{\tilde{r}}\frac{\partial}{\partial\tilde{r}}(\tilde{r}E_\phi^c) \quad (7.155)$$

$$i\omega\mu H_\phi^c = \frac{1}{\tilde{r}}\frac{\partial}{\partial\tilde{r}}(\tilde{r}E_\theta^c) - \frac{1}{\tilde{r}}\frac{\partial E_r^c}{\partial\theta} \quad (7.156)$$

with $\tilde{r}$ being defined as in (7.125). If we substitute $\partial/\partial\tilde{r} = (1/s_r)\partial/\partial r$ in (7.155) and (7.156), multiply (7.154) by $(\tilde{r}/r)^2$, and (7.155) and (7.156) by $s_r(\tilde{r}/r)$, then (7.154)–(7.156) can be recast as

$$i\omega\mu\left[\left(\frac{\tilde{r}}{r}\right)^2\frac{1}{s_r}\right](s_r H_r^c) = \frac{1}{r\sin\theta}\left\{\frac{\partial}{\partial\theta}\left[\sin\theta\left(\frac{\tilde{r}E_\phi^c}{r}\right)\right] - \frac{\partial}{\partial\phi}\left(\frac{\tilde{r}E_\theta^c}{r}\right)\right\} \quad (7.157)$$

$$i\omega\mu s_r\left(\frac{\tilde{r}H_\theta^c}{r}\right) = \frac{1}{r\sin\theta}\frac{\partial}{\partial\phi}(s_r E_r^c) - \frac{1}{r}\frac{\partial}{\partial r}\left[r\left(\frac{\tilde{r}E_\phi^c}{r}\right)\right] \quad (7.158)$$

$$i\omega\mu s_r\left(\frac{\tilde{r}H_\phi^c}{r}\right) = \frac{1}{r}\frac{\partial}{\partial r}\left[r\left(\frac{\tilde{r}E_\theta^c}{r}\right)\right] - \frac{1}{r}\frac{\partial}{\partial\theta}(s_r E_r^c) \quad (7.159)$$

and similarly for Ampere's law. A set of fields defined by $E_r^a = s_r E_\rho^c$, $E_\phi^a = (\tilde{r}/r) E_\phi^c$, $E_\theta^a = (\tilde{r}/r) E_\theta^c$ (similarly for the **H** field), obeys Maxwell's equations in an anisotropic medium of constitutive parameters $\overline{\mu} = \mu \overline{\Lambda}$ and $\overline{\epsilon} = \epsilon \overline{\Lambda}$, with

$$\overline{\Lambda} = \hat{r}\hat{r} \left(\frac{\tilde{r}}{r}\right)^2 \left(\frac{1}{s_r}\right) + \left(\overline{\mathbf{I}} - \hat{r}\hat{r}\right) s_r \qquad (7.160)$$

By defining $s_\theta = s_\phi = (\tilde{r}/r)$, then this tensor and the field mapping equations have the same formal appearance as the previous ones. The continuity of $\tilde{r}/r$ everywhere implies that both formulations will satisfy the same boundary conditions on the continuity of the tangential fields across a PML interface at $r = r_0$. The perfectly matched condition for one of the formulations then follows automatically from the other.

7.7.4 Conformal (Doubly Curved) PML

As with any ABC, it is of interest to investigate the possibility of further extending the PML concept to a conformal PML. A conformal ABC has the advantage of promoting, when used in combination with conformal computational grids, a further reduction of the amount of buffer space in the computational domain around the scatterer.

In this section, we present an analytic derivation of a 3D conformal PML on a general orthogonal curvilinear coordinate system [32]. The conformal PML also admits a Maxwellian formulation where it can be expressed in terms of an anisotropic constitutive tensor depending on the local radii of curvature of the termination surface. The derivation is achieved through a complex stretching of the *normal* coordinate along the PML (or termination surface). The previously derived PMLs in Cartesian, cylindrical, and spherical coordinates are shown to be special cases of this conformal PML.

We start by introducing a *convex* (when viewed from the outside), closed-surface S around the scatterer(s), representing the interface between free space and the PML region. For a concave scatterer or for a group of scatterers, such a surface can always be chosen by considering its convex hull. Note that a convex surface S defines a *concave* surface PML as seen from inside the computational domain. The restriction to a concave PML is an important one, as will be discussed later on.

At any given point P on S, a *local*, right-handed reference frame can be defined through the orthonormal vectors $\hat{t}_1$, $\hat{t}_2$, $\hat{n}$. The unit vectors $\hat{t}_1$ and $\hat{t}_2$ are tangent to S at P along the principal lines of curvature, and $\hat{n} = \hat{t}_1 \times \hat{t}_2$ is the unit vector that is outwardly normal to S at this point. In terms of local coordinates ξ_1, ξ_2, ξ_3, we write $\hat{t}_i = (\partial \mathbf{r}/\partial \xi_i) / |\partial \mathbf{r}/\partial \xi_i|$, $i = 1, 2$, and $\hat{n} = (\partial \mathbf{r}/\partial \xi_3) / |\partial \mathbf{r}/\partial \xi_3|$, where **r** is the position vector. This reference frame is called a Darboux-Dupin frame, and the first fundamental form induced by the coordinates ξ_1 and ξ_2 on the surface S is diagonal [41]. As a consequence, the curvilinear coordinate system

ξ_1, ξ_2, ξ_3 is *orthogonal*, having a diagonal metric. Any point P' in this local reference frame is uniquely denoted by the local coordinates ξ_1, ξ_2, ξ_3. The equation $\xi_3 = 0$ represents the surface S. The points of constant ξ_3 correspond to *parallel surfaces* in distance ξ_3 to S, and the unit vectors are functions of ξ_1 and ξ_2, only: $\hat{t}_1 = \hat{t}_1(\xi_1, \xi_2)$, $\hat{t}_2 = \hat{t}_2(\xi_1, \xi_2)$, $\hat{n} = \hat{n}(\xi_1, \xi_2)$. If the principal radii of curvature at the point P in S are given by $r_{01}(\xi_1, \xi_2)$ and $r_{02}(\xi_1, \xi_2)$ (both positive for S convex), then at a point P', they will be given by $r_1(\xi_1, \xi_2, \xi_3) = r_{01}(\xi_1, \xi_2) + \xi_3$ and $r_2(\xi_1, \xi_2, \xi_3) = r_{02}(\xi_1, \xi_2) + \xi_3$.

It is well known that in any orthogonal system of curvilinear coordinates ξ_1, ξ_2, ξ_3 defined by the diagonal metric $g_{ij} = g_{ii}\delta_{ij}$ with $g_{ii} = h_i^2$, $i = 1, 2, 3$, the Maxwell's equations for an isotropic medium are written as [42]

$$\frac{1}{h_2 h_3}\left[\frac{\partial}{\partial \xi_2}(h_3 E_3) - \frac{\partial}{\partial \xi_3}(h_2 E_2)\right] - i\omega\mu H_1 = 0 \quad (7.161)$$

$$\frac{1}{h_3 h_1}\left[\frac{\partial}{\partial \xi_3}(h_1 E_1) - \frac{\partial}{\partial \xi_1}(h_3 E_3)\right] - i\omega\mu H_2 = 0 \quad (7.162)$$

$$\frac{1}{h_1 h_2}\left[\frac{\partial}{\partial \xi_1}(h_2 E_2) - \frac{\partial}{\partial \xi_2}(h_1 E_1)\right] - i\omega\mu H_3 = 0 \quad (7.163)$$

$$\frac{\partial}{\partial \xi_1}(h_2 h_3 \epsilon E_1) + \frac{\partial}{\partial \xi_2}(h_3 h_1 \epsilon E_2) + \frac{\partial}{\partial \xi_3}(h_1 h_2 \epsilon E_3) = h_1 h_2 h_3 \rho \quad (7.164)$$

$$\frac{1}{h_2 h_3}\left[\frac{\partial}{\partial \xi_2}(h_3 H_3) - \frac{\partial}{\partial \xi_3}(h_2 H_2)\right] + i\omega\epsilon E_1 = J_1 \quad (7.165)$$

$$\frac{1}{h_3 h_1}\left[\frac{\partial}{\partial \xi_3}(h_1 H_1) - \frac{\partial}{\partial \xi_1}(h_3 H_3)\right] + i\omega\epsilon E_2 = J_2 \quad (7.166)$$

$$\frac{1}{h_1 h_2}\left[\frac{\partial}{\partial \xi_1}(h_2 H_2) - \frac{\partial}{\partial \xi_2}(h_1 H_1)\right] + i\omega\epsilon E_3 = J_3 \quad (7.167)$$

$$\frac{\partial}{\partial \xi_1}(h_2 h_3 \mu H_1) + \frac{\partial}{\partial \xi_2}(h_3 h_1 \mu H_2) + \frac{\partial}{\partial \xi_3}(h_1 h_2 \mu H_3) = 0 \quad (7.168)$$

Using the local coordinate system defined above, we have $h_1 = r_1/r_{01}$, $h_2 = r_2/r_{02}$, and $h_3 = 1$.

It has been observed previously that the modified Maxwell's equations for PML media in Cartesian, cylindrical, and spherical coordinates reduce to the ordinary Maxwell's equations on a complex space, where the x, y, z (Cartesian), ρ, z (cylindrical), and r (spherical) coordinates are analytically continued to a complex space to achieve a reflectionless absorption on the corresponding directions.

In strict analogy to the Cartesian, cylindrical, and spherical PMLs, the conformal PML can be obtained through complex stretching (analytic continuation to the upper-half complex plane) on the *normal* coordinate ξ_3:

$$\xi_3 \rightarrow \tilde{\xi}_3 = \int_0^{\xi_3} s(\zeta) d\zeta = \int_0^{\xi_3} \left(a(\zeta) + i\frac{\Omega(\zeta)}{\omega} \right) d\zeta = b(\xi_3) + i\frac{\Delta(\xi_3)}{\omega} \quad (7.169)$$

where $a \geq 1$ and $\Omega \geq 0$

The effect of this stretching on a vector propagating wave can be seen, for example, by locally expanding the wave in terms of a generalized Wilcox expansion [43, 44] in terms of the coordinates ξ_1, ξ_2, ξ_3:

$$\mathbf{E}(\xi_1, \xi_2, \xi_3) = \frac{e^{ik_0 \xi_3}}{4\pi (r_1 r_2)^{1/2}} \sum_{n=0}^{+\infty} \frac{\mathbf{E}_n(\xi_1, \xi_2)}{(r_1 r_2)^{n/2}} \quad (7.170)$$

where $k_0 = \omega/c$. Note that the lowest-order term in (7.170) corresponds to the geometrical optics spreading factor for a doubly curved wavefront. By applying the mapping of (7.169) in (7.170), we arrive at

$$\mathbf{E}(\xi_1, \xi_2, \tilde{\xi}_3) = \frac{e^{-c^{-1}\Delta(\xi_3)} e^{ik_0 b(\xi_3)}}{4\pi (\tilde{r}_1 \tilde{r}_2)^{1/2}} \sum_{n=0}^{+\infty} \frac{\mathbf{E}_n(\xi_1, \xi_2)}{(\tilde{r}_1 \tilde{r}_2)^{n/2}} \quad (7.171)$$

where $\tilde{r}_1 = r_{01} + \tilde{\xi}_3$, $\tilde{r}_2 = r_{02} + \tilde{\xi}_3$, and the induced exponential decay along the normal coordinate for $\Omega \geq 0$ is evident. Also, if $a \geq 1$, additional attenuation can be achieved for evanescent waves, if they exist. This is analogous to the Cartesian PML. Note also that the complex stretching on the normal coordinate preserves the transverse field distributions. Furthermore, if the original field $\mathbf{E}(\xi_1, \xi_2, \xi_3)$ is continuous everywhere, then the analytically continued field $\mathbf{E}(\xi_1, \xi_2, \tilde{\xi}_3)$ will also be continuous everywhere, from the continuity of $\tilde{\xi}(\xi)$. Boundary conditions over the PML interface are automatically preserved. However, the field in (7.171) does not obey Maxwell's equations. Instead, the substitution of (7.169) in (7.161)–(7.168) leads to the following set of equations *inside* the conformal PML:

$$\frac{1}{\tilde{h}_2} \left[\frac{\partial}{\partial \xi_2} (E_3) - \frac{1}{s} \frac{\partial}{\partial \xi_3} (\tilde{h}_2 E_2) \right] - i\omega \mu H_1 = 0 \quad (7.172)$$

$$\frac{1}{\tilde{h}_1} \left[\frac{1}{s} \frac{\partial}{\partial \xi_3} (\tilde{h}_1 E_1) - \frac{\partial}{\partial \xi_1} (E_3) \right] - i\omega \mu H_2 = 0 \quad (7.173)$$

$$\frac{1}{\tilde{h}_1 \tilde{h}_2} \left[\frac{\partial}{\partial \xi_1} (\tilde{h}_2 E_2) - \frac{\partial}{\partial \xi_2} (\tilde{h}_1 E_1) \right] - i\omega \mu H_3 = 0 \quad (7.174)$$

$$\frac{\partial}{\partial \xi_1} (\tilde{h}_2 \epsilon E_1) + \frac{\partial}{\partial \xi_2} (\tilde{h}_1 \epsilon E_2) + \frac{1}{s} \frac{\partial}{\partial \xi_3} (\tilde{h}_1 \tilde{h}_2 \epsilon E_3) = 0 \quad (7.175)$$

$$\frac{1}{\tilde{h}_2}\left[\frac{\partial}{\partial \xi_2}(H_3) - \frac{1}{s}\frac{\partial}{\partial \xi_3}\left(\tilde{h}_2 H_2\right)\right] + i\omega\epsilon E_1 = 0 \quad (7.176)$$

$$\frac{1}{\tilde{h}_1}\left[\frac{1}{s}\frac{\partial}{\partial \xi_3}\left(\tilde{h}_1 H_1\right) - \frac{\partial}{\partial \xi_1}(H_3)\right] + i\omega\epsilon E_2 = 0 \quad (7.177)$$

$$\frac{1}{\tilde{h}_1\tilde{h}_2}\left[\frac{\partial}{\partial \xi_1}\left(\tilde{h}_2 H_2\right) - \frac{\partial}{\partial \xi_2}\left(\tilde{h}_1 H_1\right)\right] + i\omega\epsilon E_3 = 0 \quad (7.178)$$

$$\frac{\partial}{\partial \xi_1}\left(\tilde{h}_2\mu H_1\right) + \frac{\partial}{\partial \xi_2}\left(\tilde{h}_1\mu H_2\right) + \frac{1}{s}\frac{\partial}{\partial \xi_3}\left(\tilde{h}_1\tilde{h}_2\mu H_3\right) = 0 \quad (7.179)$$

where we have used $\tilde{h}_1 = \tilde{r}_1/r_{01}$, $\tilde{h}_2 = \tilde{r}_2/r_{02}$ (since these metric coefficients are functions of ξ_3 and must be changed accordingly), $h_3 = 1$, and $\partial/\partial\tilde{\xi}_3 = (1/s)(\partial/\partial\xi_3)$. No sources are assumed inside the PML.

A system of differential equations first-order in time can be derived from (7.172)–(7.175) and (7.176)–(7.178) with the aid of auxiliary fields, in a manner very similar to that for cylindrical and spherical coordinates [15]. A time-stepping scheme can then be easily implemented.

We can summarize some basic properties of this new system of partial differential equations, in the case of a concave PML, as follows: (1) In the physical region (i.e., where $s = 1$), it reduces to the usual Maxwell's equations. (2) Any closed-form field solution of the Maxwell's equations in this general orthogonal curvilinear system can be mapped to solutions of this new system through a simple analytic continuation on the normal variable $\xi_3 \to \tilde{\xi}_3$, as is done in the passage from (7.170) to (7.171). No reflected field is induced due to this analytic continuation (in the continuum). (3) This analytic continuation preserves the analyticity of the solutions on the upper half complex ω plane, as long as we limit ourselves to positive radii of curvature ($r_{01} > 0$ and $r_{02} > 0$) (concave or planar PML). This means that, in this case, the resultant frequency-domain solutions are still causal in terms of a real-axis Fourier inversion contour, or, equivalently, that the solutions are dynamically stable. Otherwise (nonconcave, nonplanar PML), the solutions will contain singularities in the upper half plane implying time-domain solutions that may grow unbounded, as will be discussed later.

The fact that this system of equations is not the original Maxwell's equations is a drawback for some applications. To derive a Maxwellian conformal PML absorber, we introduce a new set of fields $\tilde{E}_i$, $\tilde{H}_i$ obtained by using the following transformations on the E_i, H_i fields of (7.172)–(7.179): $\tilde{E}_1 = (\tilde{h}_1/h_1)E_1$, $\tilde{E}_2 = (\tilde{h}_2/h_2)E_2$, $\tilde{E}_3 = sE_3$, $\tilde{H}_1 = (\tilde{h}_1/h_1)H_1$, $\tilde{H}_2 = (\tilde{h}_2/h_2)H_2$, $\tilde{H}_3 = sH_3$.

We note that, since the factors $(\tilde{h}_i/h_i)$ are continuous along the free-space/PML interface (both r_i and $\tilde{\xi}_3$ are continuous), the original tangential fields E_1, H_1, E_2, H_2, and the transformed tangential fields $\tilde{E}_1$, $\tilde{H}_1$, $\tilde{E}_2$, $\tilde{H}_2$ obey the same set of boundary conditions at the free-space/PML interface. Therefore, since E_1, H_1, E_2, H_2 are perfectly matched at this interface, $\tilde{E}_1$, $\tilde{H}_1$, $\tilde{E}_2$, $\tilde{H}_2$ are also perfectly matched.

Furthermore, by substituting the transformed fields into (7.172)–(7.179), we arrive at the following equations:

$$\frac{1}{h_2}\left[\frac{\partial}{\partial \xi_2}\tilde{E}_3 - \frac{\partial}{\partial \xi_3}\left(h_2\tilde{E}_2\right)\right] - i\omega\mu\left(\frac{s h_1 \tilde{h}_2}{\tilde{h}_1 h_2}\right)\tilde{H}_1 = 0 \quad (7.180)$$

$$\frac{1}{h_1}\left[\frac{\partial}{\partial \xi_3}\left(h_1\tilde{E}_1\right) - \frac{\partial}{\partial \xi_1}\tilde{E}_3\right] - i\omega\mu\left(\frac{s\tilde{h}_1 h_2}{h_1 \tilde{h}_2}\right)\tilde{H}_2 = 0 \quad (7.181)$$

$$\frac{1}{h_1 h_2}\left[\frac{\partial}{\partial \xi_1}\left(h_2\tilde{E}_2\right) - \frac{\partial}{\partial \xi_2}\left(h_1\tilde{E}_1\right)\right] - i\omega\mu\left(\frac{\tilde{h}_1 \tilde{h}_2}{s h_1 h_2}\right)\tilde{H}_3 = 0 \quad (7.182)$$

$$\frac{\partial}{\partial \xi_1}\left[h_2\epsilon\left(\frac{s h_1 \tilde{h}_2}{\tilde{h}_1 h_2}\right)\tilde{E}_1\right] + \frac{\partial}{\partial \xi_2}\left[h_1\epsilon\left(\frac{s\tilde{h}_1 h_2}{h_1 \tilde{h}_2}\right)\tilde{E}_2\right] + \frac{\partial}{\partial \xi_3}\left[h_1 h_2\epsilon\left(\frac{\tilde{h}_1 \tilde{h}_2}{s h_1 h_2}\right)\tilde{E}_3\right] = 0 \quad (7.183)$$

$$\frac{1}{h_2}\left[\frac{\partial}{\partial \xi_2}\tilde{H}_3 - \frac{\partial}{\partial \xi_3}\left(h_2\tilde{H}_2\right)\right] + i\omega\epsilon\left(\frac{s h_1 \tilde{h}_2}{\tilde{h}_1 h_2}\right)\tilde{E}_1 = 0 \quad (7.184)$$

$$\frac{1}{h_1}\left[\frac{\partial}{\partial \xi_3}\left(h_1\tilde{H}_1\right) - \frac{\partial}{\partial \xi_1}\tilde{H}_3\right] + i\omega\epsilon\left(\frac{s\tilde{h}_1 h_2}{h_1 \tilde{h}_2}\right)\tilde{E}_2 = 0 \quad (7.185)$$

$$\frac{1}{h_1 h_2}\left[\frac{\partial}{\partial \xi_1}\left(h_2\tilde{H}_2\right) - \frac{\partial}{\partial \xi_2}\left(h_1\tilde{H}_1\right)\right] + i\omega\epsilon\left(\frac{\tilde{h}_1 \tilde{h}_2}{s h_1 h_2}\right)\tilde{E}_3 = 0 \quad (7.186)$$

$$\frac{\partial}{\partial \xi_1}\left[h_2\mu\left(\frac{s h_1 \tilde{h}_2}{\tilde{h}_1 h_2}\right)\tilde{H}_1\right] + \frac{\partial}{\partial \xi_2}\left[h_1\mu\left(\frac{s\tilde{h}_1 h_2}{h_1 \tilde{h}_2}\right)\tilde{H}_2\right] + \frac{\partial}{\partial \xi_3}\left[h_1 h_2\mu\left(\frac{\tilde{h}_1 \tilde{h}_2}{s h_1 h_2}\right)\tilde{H}_3\right] = 0 \quad (7.187)$$

A careful look at (7.180)–(7.187) reveals that they are just the Maxwell's equations on the original orthogonal curvilinear coordinates system of (7.161)–(7.168) characterized by the (real) metric $g_{ij} = g_{ii}\delta_{ij}$ with $g_{ii} = h_i^2$, $i = 1, 2, 3$, and $h_3 = 1$, but now for an *anisotropic* medium, whose constitutive parameters are given by $\overline{\mu} = \mu\overline{\Lambda}$ and $\overline{\epsilon} = \epsilon\overline{\Lambda}$, with

$$\overline{\Lambda} = \hat{t}_1\hat{t}_1\left(\frac{s h_1 \tilde{h}_2}{\tilde{h}_1 h_2}\right) + \hat{t}_2\hat{t}_2\left(\frac{s\tilde{h}_1 h_2}{h_1 \tilde{h}_2}\right) + \hat{n}\hat{n}\left(\frac{\tilde{h}_1 \tilde{h}_2}{s h_1 h_2}\right) \quad (7.188)$$

The significance of this result is that it is possible to achieve reflectionless absorption of electromagnetic waves incident on a smooth, concave surface having anisotropic constitutive tensors given by (7.188), depending on the local principal radii of curvatures. Note that since (7.188) is a formula of a *constitutive* parameter, it is independent of any coordinate system (even though we have used a particular one to *derive* it). This simply means that, given a concave surface termination (as viewed from inside the computational domain), we can represent this anisotropic conformal PML *in any coordinate system*, by expressing the local radii of curvature and the normal stretching as functions of the new coordinates.

An interesting point to observe is the local interplay between the physics of the medium and the geometry, as the (local) *constitutive* parameters depend on the (local) *geometry* of the termination.

The previously derived Cartesian, cylindrical, and spherical Maxwellian PMLs are just special cases of the above. The Cartesian PML is obtained by setting $r_{01} = r_{02} = \infty$, so that $\tilde{h}_1 = \tilde{h}_2 = 1$. Furthermore, if $\hat{t}_1 = \hat{x}$ and $\hat{t}_2 = \hat{y}$, then $\hat{n} = \hat{z}$, and $s = s_z(z)$ for attenuation in the z direction. Since $\overline{\Lambda}_z(z)$ is a function of z only, we can combine it with subsequent stretchings in the x and y directions (for corner regions).

The cylindrical PML is obtained by setting $r_{01} = \infty$, $r_{02} = \rho$, so that $\tilde{h}_1/h_1 = 1$, and $\tilde{h}_2/h_2 = \tilde{\rho}/\rho$. Furthermore, $\hat{t}_1 = \hat{\phi}$, $\hat{t}_2 = \hat{z}$, $\hat{n} = \hat{\rho}$, and $s = s_\rho(\rho)$ for attenuation in the ρ direction. Since $\overline{\Lambda}_\rho(\rho)$ is a function of ρ only, we can combine it with a subsequent stretching (Cartesian) on the z direction (orthogonal to ρ everywhere) also.

The spherical PML is obtained by setting $r_{01} = r_{02} = r$, so that $\tilde{h}_1/h_1 = \tilde{h}_2/h_2 = \tilde{r}/r$. Furthermore, $\hat{t}_1 = \hat{\theta}$, $\hat{t}_2 = \hat{\phi}$, $\hat{n} = \hat{r}$, and $s = s_r(r)$ for attenuation in the r direction.

In the following of this section, we review results of the FDTD simulations on conformal (body-fitted) grids employing the conformal PML on the outer grid termination [45].

Figure 7.5 depicts three examples of grids where the application of a conformal PML is advantageous. These grids are employed to reduce the FDTD staircasing error around curved objects. The first grid is for a circular cylinder, the second one for an elliptical cylinder with eccentricity $e = 4$, and the third one for a NACA 4415 airfoil.

In these examples, a two-layered hyperbolic grid generation technique is used to achieve good orthogonality properties in the PML region (the conformal PML assumes exact orthogonality of the coordinate system) [45].

Figure 7.6 shows the reflection error from these three grid systems when employing a conformal PML on the five outermost cells of the grids. The errors are measured along a contour where the transverse coordinate is constant so that the local variation in the reflection error can be observed. The results show a very low reflection error, below -40 dB over most of the angular span for a five-cell thick PML. As expected, the circular PML does not produce angular variations on its reflection level (reduces to the cylindrical PML discussed before). The angular variation of the reflection error

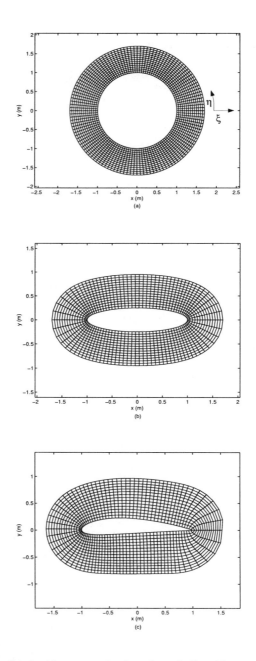

Figure 7.5 Body-fitted grids generated using a hyperbolic grid generation technique. The conformal PML comprises the 5 outermost cells: (a) circular cylinder; (b) elliptical cylinder; (c) NACA 4415 airfoil. (*Source:* [45], ©1999, IEEE. Reprinted with permission.)

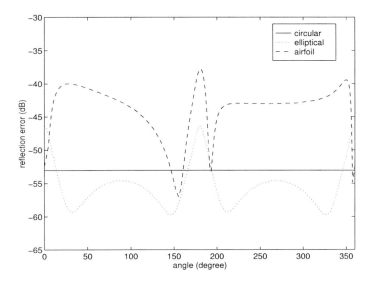

Figure 7.6 Reflection errors in logarithmic scale for the three body-fitted grids of Figure 7.5. (*Source:* [45], ©1999, IEEE. Reprinted with permission.)

of the PML for the other cases can be attributed to two major sources. The first is the deviation from a true orthogonal coordinate system caused by numerical limitations on the grid generation algorithm and the second is the error in the (numerical) computation of the local radius of curvature. These limitations are not present in the Cartesian, cylindrical, and spherical PMLs (global orthogonal coordinate systems).

7.8 STABILITY ISSUES

One of the desirable properties of the artificial PML media in Cartesian or curvilinear coordinates is that they should retain the causality conditions observed by the original Maxwell's equations. This naturally leads to the study of the analytic properties on the complex ω plane of either the PML constitutive tensor parameters (in the case of the Maxwellian PML), or the complex-space Green's functions (in the case of the complex-space PML).

In this section, we study these analytic properties for the various coordinate system PMLs: Cartesian, cylindrical, and spherical [30]. The discussion is centered on the *spectral* properties of these PMLs. Therefore, the conclusions do not depend on the peculiarities of the time-domain implementation (e.g., whether or not it utilizes field-splitting), but have consequences on any time-domain implementation. In particular, we point out conditions under which causality is violated in the sense of a real-axis inverse Fourier contour and the consequences of this fact for the dynamic stability of the PML-FDTD simulations. The conclusions have an impact on the design of PMLs for FDTD simulations and on the use of PMLs as a physical basis for engineered artificial absorbers.

We first examine the connection between violation of causality in the sense of a real-axis Fourier inversion contour and the dynamic stability of the FDTD algorithm. The discussion is centered on the analytical properties of the $\bar{\epsilon}(\omega)$ tensor. For brevity, we restrict our attention to the electric field constitutive relation. By invoking the duality principle [4], all conclusions below also apply to the magnetic case.

In the frequency domain, the electric field constitutive relation is written as

$$\mathbf{D}(\omega) = \bar{\epsilon}(\omega) \cdot \mathbf{E}(\omega) \tag{7.189}$$

where $\bar{\epsilon}$ is the 3×3 permittivity tensor. In the time domain, (7.189) becomes

$$\mathbf{D}(t) = \int_{-\infty}^{\infty} \bar{\epsilon}(\tau) \cdot \mathbf{E}(t - \tau) d\tau \tag{7.190}$$

Invoking causality, we have $\bar{\epsilon}(t) = 0$ for $t < 0$. By writing

$$\bar{\epsilon}(t) = \frac{1}{2\pi} \int_{-\infty}^{\infty} [\bar{\epsilon}(\omega) - \bar{\epsilon}(\infty)] e^{-i\omega t} d\omega + \delta(t)\bar{\epsilon}(\infty) \tag{7.191}$$

where $\bar{\epsilon}(\infty)$ is a real-valued constant tensor and the integration is along the real axis, it is possible to show [46–48] that the condition $\bar{\epsilon}(t) = 0$ for $t < 0$ implies that the integrand in (7.191) must be analytic (holomorphic) in the upper half-plane. Zeros are also forbidden on the upper half-plane (for a detailed discussion of the connection between causality and the analytic properties of the $\bar{\epsilon}(\infty)$ tensor, see [46]). We note here that when we refer to causality in the sense of a real-axis Fourier inversion contour, we are restricting ourselves to a *primitive* causality condition, as defined in [48]. The term causality also appears in other contexts. For instance, it is sometimes used to refer to a *relativistic* causality condition.

When $\bar{\epsilon}(\omega)$ is not analytic in the upper half-plane, causality can still be preserved, provided that the Fourier inversion contour is taken *above* any singularities [4]. In this case, the medium will behave as an *active* medium and its response will not be dynamically stable anymore. The definition of a dynamically stable system adopted here is such that all its eigenfunctions approach zero as $t \to \infty$ (asymptotically stable) or remain bounded as $t \to \infty$ (marginally stable). In FDTD simulations, the fields assume a causal behavior by the very nature of the method (explicit time-stepping scheme), regardless of the analytic properties on the frequency domain. This means that the Fourier inversion contour should always be taken above any singularities [4]. Therefore, following the discussion of the previous paragraph, violation of causality in the sense of real ω axis Fourier inversion contour implies a dynamically unstable FDTD method.

Note that the dynamic stability criterion is distinct from two other stability criteria:
(1) The stability criterion incurred by a particular numerical discretization scheme. For a dynamically unstable system, no convergent numerical discretization scheme prevents the solutions from growing unbounded. For a further discussion of this aspect, see [49]. Note, however, that the definition of dynamic stability in [49] is slightly different from that adopted here, and the anisotropic PML constitutive parameters considered there are frequency independent.

(2) The stability criterion incurred by a particular field-splitting of Maxwell's equations in the PML. It is known that the field-splitting induces a weak well-posedness on the resulting system of partial differential equations [50]. Although such weak well-posedness is also a characteristic of the equations at the continuum level, we should stress that it is a distinct property from the dynamic stability issue discussed here. The discussion here is centered on *spectral* characteristics of the PML equations and do not depend on the particular field-splitting scheme (if any) employed in the time-domain implementation of the PML.

When $\overline{\epsilon}(\omega)$ is analytic on the upper half ω plane, it can also be *defined* via the general time-domain relationship between $\mathbf{E}(t)$ and $\mathbf{D}(t)$:

$$\mathbf{D}(t) = \overline{\epsilon}(\infty) \cdot \mathbf{E}(t) + \int_0^\infty \overline{\boldsymbol{\alpha}}(\tau) \cdot \mathbf{E}(t-\tau) d\tau \qquad (7.192)$$

which is a particular case of (7.190), through

$$\overline{\epsilon}(\omega) = \overline{\epsilon}(\infty) + \int_0^\infty \overline{\boldsymbol{\alpha}}(\tau) e^{i\omega\tau} d\tau \qquad (7.193)$$

We immediately see that, if the function $\overline{\boldsymbol{\alpha}}(t)$ (generalized time-domain complex susceptibility kernel) is finite throughout the range of integration, the above integral converges in the upper half-plane (real axis excluded); therefore, $\overline{\epsilon}(\omega)$ is properly defined. In some instances, the integral in (7.193) diverges for ω on the real axis or in the lower half-plane, and the definition (7.193) is invalid at these points. However, the function $\overline{\epsilon}(\omega)$ can still be defined as the analytic continuation of (7.193) from the upper half-plane. It will, in general, contain singularities in the lower half-plane. This is the case, for instance, for an isotropic medium with static conductivity, where the complex dielectric constant is usually written for real frequencies as the analytic continuation of (7.193) containing a pole at the origin

$$\epsilon(\mathbf{r},\omega) = \epsilon_0 \left[\epsilon_D(\mathbf{r}) + i \frac{\sigma(\mathbf{r})}{\omega \epsilon_0} \right] \qquad (7.194)$$

A useful test for the violation of causality for the Maxwellian PML formulation is to check if the resultant frequency-domain constitutive tensors $\overline{\epsilon}(\mathbf{r},\omega) = \epsilon \overline{\Lambda}(\mathbf{r},\omega)$ and $\overline{\mu}(\mathbf{r},\omega) = \mu \overline{\Lambda}(\mathbf{r},\omega)$ satisfy the Kramers-Kronig relations along with the crossing relation $\overline{\Lambda}^*(\mathbf{r},\omega) = \overline{\Lambda}(\mathbf{r},-\omega^*)$ (the star denotes complex conjugation). This should also be true for their inverses $\overline{\epsilon}(\mathbf{r},\omega)^{-1}$ and $\overline{\mu}(\mathbf{r},\omega)^{-1}$ [33].

7.8.1 Cartesian PML Analysis

The Cartesian PML constitutive tensor $\overline{\Lambda}(\mathbf{r}, \omega)$ is given by

$$\overline{\Lambda}(\mathbf{r},\omega) = \hat{x}\hat{x}\left(\frac{s_y s_z}{s_x}\right) + \hat{y}\hat{y}\left(\frac{s_z s_x}{s_y}\right) + \hat{z}\hat{z}\left(\frac{s_x s_y}{s_z}\right) \tag{7.195}$$

where

$$s_\zeta(\zeta,\omega) = a_\zeta(\zeta) + i\frac{\Omega_\zeta(\zeta)}{\omega} \tag{7.196}$$

and $a_\zeta(\zeta) \geq 1$, $\Omega_\zeta(\zeta) \geq 0$, $\zeta = x, y, z$. Note that (7.196) resembles the expression of a complex dielectric constant of a conductive medium. The tensor in (7.195) is just the product of three simpler, 3×3 uniaxial tensors:

$$\overline{\Lambda}(\mathbf{r},\omega) = \overline{\Lambda}_x(x,\omega) \cdot \overline{\Lambda}_y(y,\omega) \cdot \overline{\Lambda}_z(z,\omega) \tag{7.197}$$

with

$$\overline{\Lambda}_x = \hat{x}\hat{x}\left(\frac{1}{s_x}\right) + \left(\overline{\mathbf{I}} - \hat{x}\hat{x}\right)s_x \tag{7.198}$$

and analogously for $\overline{\Lambda}_y(y,\omega)$ and $\overline{\Lambda}_z(z,\omega)$. In order for $\overline{\Lambda}(\mathbf{r},\omega)$ to satisfy causality, each of the $\overline{\Lambda}_\zeta(\zeta,\omega)$ tensors must satisfy it individually (there are no pole-zero cancellations). Since $\overline{\Lambda}_\zeta(\zeta,\omega)$ and $\overline{\Lambda}_\zeta(\zeta,\omega)^{-1}$ are directly dependent on $1/s_\zeta$ and s_ζ, Kramers-Kronig relations must be satisfied by these functions individually. Kramers-Kronig relations are a consequence of the application of the Cauchy's theorem to the function $\overline{\epsilon}(\omega) - \overline{\epsilon}(\infty)$ in the upper half-plane. When presented in their usual form [4], the analyticity of $\overline{\epsilon}(\omega) - \overline{\epsilon}(\infty)$ over the entire real axis is implicitly assumed. However, this is not the case, for instance, with conductive media, in which the complex dielectric constant has a term $i\sigma/\omega\epsilon_0$ and therefore has a pole singularity present at $\omega = 0$. The Kramers-Kronig relations for conductive-like media must be properly modified to account for the deformation of the closed contour in the Cauchy's theorem to avoid the pole at the origin by means of an infinitesimal semicircle above it. In this more general form, the Kramers-Kronig relations read as [51]

$$\overline{\epsilon}_r(\omega) - \overline{\epsilon}(\infty) = \frac{1}{\pi}PV\int_{-\infty}^{+\infty}\frac{\overline{\epsilon}_i(\omega')}{\omega' - \omega}d\omega' \tag{7.199}$$

$$\overline{\epsilon}_i(\omega) = -\frac{1}{\pi}PV\int_{-\infty}^{+\infty}\frac{\overline{\epsilon}_r(\omega') - \overline{\epsilon}(\infty)}{\omega' - \omega}d\omega' + \frac{\sigma}{\omega} \tag{7.200}$$

where $\overline{\epsilon}_r(\omega)$ and $\overline{\epsilon}_i(\omega)$ are the real and imaginary parts of $\overline{\epsilon}(\omega)$, PV denotes the Cauchy principal value, and the integrals are carried out along the real axis. The $\overline{\epsilon}_r(\omega)$ and $\overline{\epsilon}_i(\omega)$ are to be interpreted as the limiting values as ω approaches the real axis from above. The last term in (7.200) is the pole contribution at the origin and represents the modification from the usual Kramers-Kronig relations. Whenever the stretching variables s_ζ are defined as having a pole at $\omega = 0$ as in (7.196), the Kramers-Kronig relations to be used are as in (7.199) and (7.200). To apply relations (7.199) and (7.200) to the expression (7.196), the following immediate identifications are made:

$$\bar{\epsilon}_r(\omega) \to a_\zeta \qquad (7.201)$$

$$\bar{\epsilon}(\infty) \to a_\zeta \qquad (7.202)$$

$$\bar{\epsilon}_i(\omega) \to \Omega_\zeta/\omega \qquad (7.203)$$

so that (7.199) and (7.200) become

$$0 = \frac{1}{\pi} PV \int_{-\infty}^{+\infty} \frac{\Omega_\zeta}{\omega'(\omega' - \omega)} d\omega' \qquad (7.204)$$

$$\Omega_\zeta/\omega = 0 + \Omega_\zeta/\omega \qquad (7.205)$$

We see immediately that (7.205) verifies (7.204). Equation (7.205) also verifies (7.200) since [42]

$$\frac{1}{\pi} PV \int_{-\infty}^{+\infty} \frac{\Omega_\zeta}{\omega'(\omega' - \omega)} d\omega' = \pi \Omega_\zeta \delta(\omega) \qquad (7.206)$$

and thus (7.204) is true in the upper half-plane. As a result, expression (7.196) obeys the Kramers-Kronig relations, being causal in the sense previously stated. The fact that in the real axis we get a Dirac delta function in (7.206) is related to the divergence of the integral in (7.193) for real ω. Equation (7.196) does not include the Dirac delta function since it must be treated as an analytic continuation from the upper half-plane expression, in a similar fashion as in (7.194). Such conclusions should be expected from the resemblance of (7.196) to the complex dielectric constant of a conductive medium. Indeed, given a general field relationship in the frequency domain,

$$\psi(\omega) = s_\zeta(\omega)\chi(\omega) = \left(a_\zeta + i\frac{\Omega_\zeta}{\omega}\right)\chi(\omega) \qquad (7.207)$$

means having, in the time domain,

$$\psi(t) = a_\zeta \chi(t) + \Omega_\zeta \int_{-\infty}^{t} \chi(\tau) d\tau \qquad (7.208)$$

where causality and (marginal) stability are evident.

It is a simple exercise to verify that, if $a_\zeta \geq 1$ and $\Omega_\zeta \geq 0$, the function $1/s_\zeta$ also satisfies the Kramers-Kronig relations. In this case, since there are no singularities present on the real axis, the usual form of Kramers-Kronig relations can be used.

In summary, the Cartesian anisotropic PML medium with frequency dependence given by (7.196) is causal in the sense of real-axis Fourier inversion contour, and no *dynamic* instability should be expected.

For the complex-space PML formulation, the most direct route to investigate the dynamic stability of the Cartesian PML is to simply write down the frequency-domain closed-form field solutions (or the Green's functions) for the modified Maxwell's equations inside the PML and study its analytical behavior in the complex ω plane. This is because the solutions of the modified Maxwell's equations inside the PML are just the analytic continuation to complex space of the closed-form solutions of the ordinary Maxwell's equations. We shall illustrate this with a simple example.

We take the 2D Green's function $g(\rho, \rho')$ for a cavity enclosed by perfectly electrically conducting (PEC) walls at $x = \pm d_x, y = \pm d_y$ (which can be thought of as the FDTD computational domain when using hard boundary conditions), which is written as

$$g(\rho, \rho') = g(x, y, x', y') = \sum_{m=0}^{+\infty} \sum_{n=0}^{+\infty} \frac{Z_{mx}(x') Z_{ny}(y') Z_{mx}(x) Z_{ny}(y)}{\omega^2 \mu \epsilon - \left(\frac{m\pi}{2d_x}\right)^2 - \left(\frac{n\pi}{2d_y}\right)^2} \quad (7.209)$$

for $-d_x < x < d_x, -d_y < y < d_y$, where

$$Z_{p\zeta}(\zeta) = \frac{1}{\sqrt{d_\zeta}} \sin\left[k_{p\zeta}(\zeta + d_\zeta)\right] \quad (7.210)$$

$k_{p\zeta} = p\pi/(2d_\zeta)$, $p = m, n$, $\zeta = x, y$. By inserting a Cartesian PML before the box walls, $g(\rho, \rho')$ will be analytically continued to the complex plane through the following mapping of the coordinates:

$$\zeta \to \tilde{\zeta} = \zeta_0 + \int_{\zeta_0}^{\zeta} s_\zeta(\zeta') d\zeta' = \zeta_0 + \int_{\zeta_0}^{\zeta} \left(a_\zeta + i\frac{\Omega_\zeta}{\omega}\right) = b_\zeta(\zeta) + i\frac{\Delta_\zeta(\zeta)}{\omega} \quad (7.211)$$

where ζ_0 is on the physical (real) space and $a_\zeta = 1$, $\Omega_\zeta = 0$ in the physical space. From (7.211), we see that the boundaries of the box are complex-valued inside a PML layer, $d_x \to \tilde{d}_x$, $d_y \to \tilde{d}_y$ and, because of this, the reflection coefficient can be made very small. The function in (7.209) has an infinite number of poles located along the real ω axis (natural frequencies of the cavity). With the mapping given by (7.211), these poles are translated to the roots of the equation:

$$\omega^2 \mu \epsilon = \left(\frac{m\pi/2}{b_x(d_x) + i\frac{\Delta_x(d_x)}{\omega}}\right)^2 + \left(\frac{n\pi/2}{b_y(d_y) + i\frac{\Delta_y(d_y)}{\omega}}\right)^2 \quad (7.212)$$

which are always located on the real axis or lower half-plane for $a_\zeta \geq 1$ and $\Omega_\zeta \geq 0$.

In addition to this, it should be noted that $\sin z$ is an entire function (no poles) and that the branch points induced by the $1/\sqrt{d_\zeta}$ factors in (7.210) (which give rise to simple poles in (7.212) for this 2D case) are also on the lower half-plane. Consequently, the resultant PML Green's function is also analytic over the entire upper half ω plane, and causality in the sense of real-axis Fourier inversion contour is preserved. This is true for any observation point (x, y) inside the physical or PML domains. A similar situation occurs in the 3D case. Consequently, and similarly to the Maxwellian PML, the complex-space PML in Cartesian coordinates is dynamically stable.

7.8.2 Cylindrical PML Analysis

For brevity, the discussion for the cylindrical PML stability will focus only on those aspects differing from the previous analysis of the Cartesian case.

In cylindrical coordinates, the constitutive tensor $\overline{\Lambda}(\rho, z; \omega)$ for the anisotropic PML formulation is given by

$$\overline{\Lambda}(\rho, z; \omega) = \hat{\rho}\hat{\rho}\left(\frac{s_\phi s_z}{s_\rho}\right) + \hat{\phi}\hat{\phi}\left(\frac{s_z s_\rho}{s_\phi}\right) + \hat{z}\hat{z}\left(\frac{s_\rho s_\phi}{s_z}\right) \quad (7.213)$$

In the above, s_ρ and s_z are the stretching parameters in the ρ and z directions. The variable s_ϕ on the other hand is defined as

$$s_\phi(\rho, \omega) = \frac{\tilde{\rho}}{\rho} = \frac{1}{\rho}\left(\rho_0 + \int_{\rho_0}^{\rho} s_\rho(\rho')d\rho'\right) = \frac{1}{\rho}\left(b_\rho(\rho) + i\frac{\Delta_\rho(\rho)}{\omega}\right) \quad (7.214)$$

It can be seen as a "pseudo-stretching" parameter in the ϕ coordinate that accounts for the modification in the metric coefficient after the stretching in the ρ direction. Moreover, ρ_0 is a reference radius in the physical (real) space and $a_\rho = 1, \Omega_\rho = 0$ in the physical space. The tensor in (7.213) can also be written as the product of three simpler, 3×3 uniaxial tensors:

$$\overline{\Lambda}(\rho, z; \omega) = \overline{\Lambda}_\rho(\rho, \omega) \cdot \overline{\Lambda}_\phi(\rho, \omega) \cdot \overline{\Lambda}_z(z, \omega) \quad (7.215)$$

with

$$\overline{\Lambda}_\rho = \hat{\rho}\hat{\rho}\left(\frac{1}{s_\rho}\right) + \left(\overline{\mathbf{I}} - \hat{\rho}\hat{\rho}\right)s_\rho \quad (7.216)$$

and analogously for $\overline{\Lambda}_\phi(\rho, \omega)$ and $\overline{\Lambda}_z(z, \omega)$. In order for $\overline{\Lambda}(\rho, z; \omega)$ to satisfy causality, each of the tensors in (7.216) must satisfy it individually. Since the tensors $\overline{\Lambda}_\rho$ and $\overline{\Lambda}_z$ have the same analytical properties of the Cartesian PML tensors studied before, we shall limit ourselves to study the analytical properties of the $\overline{\Lambda}_\phi(\rho, \omega)$ tensor. The frequency dependence of $\overline{\Lambda}_\phi(\rho, \omega)$ (and $\overline{\Lambda}_\phi(\rho, \omega)^{-1}$) is determined by s_ϕ and $1/s_\phi$, and therefore we must focus our attention on these factors (dependence on the "pseudo-stretching") and the amount by which they differ from the Cartesian case. Again, to ensure causality in the sense of a real-axis Fourier inversion contour for $\overline{\Lambda}_\phi(\rho, \omega)$, we must ensure (necessary condition) that there are no poles due to s_ϕ or $1/s_\phi$ above the real axis. For s_ϕ, this is evident, as the only pole is at $\omega = 0$, and it can be shown that this function satisfies the Kramers-Kronig relations. However, a major difference arises in the angular factor $1/s_\phi$. It is due to the fact that the factor Δ_ρ in the imaginary part of s_ϕ may, at certain instances, be *negative*. This is in contrast to the corresponding σ factors in the imaginary part of s_x, s_y, s_z, or s_ρ, which are also chosen to be positive to achieve absorption. Due to the fact that Ω_ρ is positive, when used in (7.214) for a concave cylindrical PML (at the outer boundary), we will still have $\Delta_\rho > 0$. But for a *convex* cylindrical PML (inner boundary), we have $\Delta_\rho < 0$, as the integral in (7.214) is carried over *decreasing* values of ρ. The net effect of this is that the factor $1/s_\phi$ will then have poles in the upper half-plane, and the resultant $\overline{\Lambda}(\rho, z; \omega)$ will be noncausal in the sense of real-axis Fourier inversion contour. Consequently, we should expect dynamic instability on FDTD simulations when employing a convex PML on cylindrical coordinates. Note that we may enforce $\Delta_\rho > 0$ on convex surfaces if we choose $\Omega_\rho < 0$ at the convex PML. However, in this case, it would be the factor $1/s_\rho$ which would give rise to singularities in the upper half-plane. This is also related to the fact that, in order to have absorption in the *radial* direction, Ω_ρ has to be positive, irrespective of the concave or convex geometry.

Such dynamically unstable behavior can also be expected from a direct analysis of the properties of the formal complex-space Green's functions of the cylindrical PML.

For example, the 2D Green's function for a point source at (ρ', ϕ') inside a PEC cylinder of radius a is given by

$$g(\boldsymbol{\rho}, \boldsymbol{\rho}') = g(\rho, \phi, \rho', \phi')$$

$$= \frac{i}{4} H_0^{(1)}(k|\boldsymbol{\rho} - \boldsymbol{\rho}'|) - \frac{i}{4} \sum_{n=-\infty}^{+\infty} J_n(k\rho) \frac{J_n(k\rho') H_n^{(1)}(ka)}{J_n(ka)} e^{in(\phi - \phi')}$$

(7.217)

so that, when backed by a concave cylindrical PML, the field is analytically continued to

$$g(\tilde{\boldsymbol{\rho}}, \boldsymbol{\rho}') = g(\tilde{\rho}, \phi; \rho', \phi')$$

$$= \frac{i}{4} H_0^{(1)}(k|\tilde{\boldsymbol{\rho}} - \boldsymbol{\rho}'|) - \frac{i}{4} \sum_{n=-\infty}^{+\infty} J_n(k\tilde{\rho}) \frac{J_n(k\rho') H_n^{(1)}(k\tilde{a})}{J_n(k\tilde{a})} e^{in(\phi - \phi')}$$

(7.218)

where $\tilde{\rho}$. Note that $\tilde{a}$ is also complex-valued. As discussed in Section 7.7, the effect of the cylindrical PML is to make the reflected field (summation term) in the expression above exponentially small. Furthermore, this analytic continuation preserves the analyticity on the upper half-plane of the resultant field in (7.218). This is because the singularities in (7.217) are due to branch points (located on the real axis) of the $H_n^{(1)}(\cdot)$, and zeros of the denominator functions $J_n(\cdot)$ (all on the real axis). When $\rho \to \tilde{\rho}$ have a positive imaginary part $\Delta_\rho > 0$, all singularities are translated down to the lower half ω plane, so that the upper half-plane is kept free of any singularity in the resultant expression (7.218).

In contrast, the 2D Green's function for a point source at (ρ', ϕ') located *outside* a perfectly conducting cylinder of radius a is given by

$$g(\boldsymbol{\rho}, \boldsymbol{\rho}') = g(\rho, \phi, \rho', \phi')$$
$$= \frac{i}{4} H_0^{(1)}(k|\boldsymbol{\rho} - \boldsymbol{\rho}'|) - \frac{i}{4} \sum_{n=-\infty}^{+\infty} H_n^{(1)}(k\rho) \frac{H_n^{(1)}(k\rho') J_n(ka)}{H_n^{(1)}(ka)} e^{-in(\phi-\phi')}$$
(7.219)

so that the solution when the perfectly conducting cylinder is "coated" by a convex cylindrical PML is given by the analytic continued expression

$$g(\boldsymbol{\rho}, \boldsymbol{\rho}') = g(\tilde{\rho}, \phi, \rho', \phi')$$
$$= \frac{i}{4} H_0^{(1)}(k|\tilde{\boldsymbol{\rho}} - \boldsymbol{\rho}'|) - \frac{i}{4} \sum_{n=-\infty}^{+\infty} H_n^{(1)}(k\tilde{\rho}) \frac{H_n^{(1)}(k\rho') J_n(k\tilde{a})}{H_n^{(1)}(k\tilde{a})} e^{-in(\phi-\phi')}$$
(7.220)

This analytic continuation does not preserve analyticity in the upper half-plane as the variable $\tilde{\rho}$ now has $\Delta_\rho < 0$. The zeros of the denominator functions $H_n^{(1)}(\cdot)$ located in the lower half-plane (see illustration in Figure 7.7) are translated *upwards* in the complex ω plane and will eventually appear as poles on the upper half-plane of reflected field terms in (7.220) [30]. Branch points of the $H_n^{(1)}(\cdot)$ functions on the real axis will also eventually be translated to the upper half-plane. The lack of analyticity on the upper half-plane makes these solutions grow unbounded in the time domain (dynamically unstable behavior).

For generality, we have been focusing on the spectral analysis, but from an analysis of the resulting PML equations in the time domain, it is observed that the unstable behavior is usually associated with update equations involving the angular ϕ derivatives. This can be seen as a consequence of the way the stretched differential arc lengths behave under stretching for the convex and concave situations. Although the stretched differential arc lengths $d\tilde{z} = s_z dz$ and $d\tilde{\rho} = s_\rho d\rho$ have positive imaginary parts for both the convex and concave cases, resulting in attenuation for z and ρ directions in both cases, the stretched angular arclength $\tilde{\rho} d\phi$ has a *positive* imaginary part for the concave PML and a *negative* imaginary part for the convex part. The resultant effect, in the concave PML case, is a complex "stretching" on the angular coordinate, but translates, in the convex PML case, to a "squeezing" on the angular coordinate.

To illustrate the previous discussion, a numerical result of cylindrical grid FDTD simulations with convex and concave cylindrical PML is presented, corresponding to Configurations A and B of Figure 7.8.

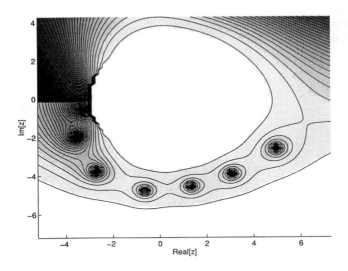

Figure 7.7 Contour map of the magnitude of the function $H_7^{(1)}(z)$ on the complex plane. Darker regions represent smaller values. The zeros on the lower half-plane and the branch cut on the negative real axis are clearly visible. (*Source:* [30], ©1999, IEEE. Reprinted with permission.)

Figure 7.9 shows the normalized E_z field from a line source radiating in the presence of a perfectly conducting circular cylinder and computed using a cylindrical grid 2D PML-FDTD algorithm. The line source is located at $(r, \phi) = (4.25\lambda_c, 0^0)$ and the field is sampled at $(r, \phi) = (3.75\lambda_c, 0^0)$, where λ_c is the (free-space) wavelength corresponding to the central frequency of the excitation pulse. The interior conducting cylinder is centered at the origin $\rho = 0$ and has a radius $a = \lambda_c$. The excitation pulse is the first derivative of the Blackman-Harris pulse with central frequency $f_c = 300$ MHz.

The cylindrical grid is terminated at $r = 5\lambda_c$, where a hard boundary condition is set ($E_z = 0$). In all simulations, the FDTD algorithm includes an eight-layer concave cylindrical PML region before the grid termination (at the outer boundary). The solid line in Figure 7.9 shows the result when using only a concave cylindrical PML before the outer boundary. This corresponds to the Configuration A depicted in Figure 7.8. Both the direct pulse and the nonspurious reflected pulse due to the inner perfect conducting cylinder are visible. No spurious reflections due to the outer boundary are visible. The dotted line is the result of the simulation when an eight-layer convex cylindrical PML is placed around the inner cylinder. This corresponds

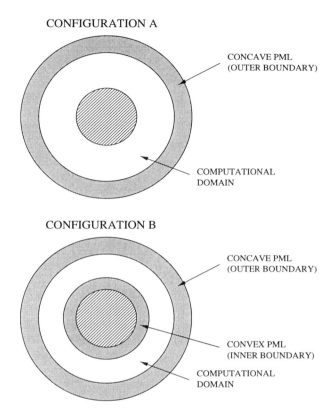

Figure 7.8 Configurations for the stability test of the PML-FDTD algorithm in cylindrical and spherical coordinates. The computational domain surrounds a cylindrical or spherical PEC scatterer, coated or not by a convex PML. (*Source:* [30], ©1999, IEEE. Reprinted with permission.)

to Configuration B depicted in Figure 7.8. The same value of $\sigma_{\rho,max}$ is used for the concave and convex PML in this case. A reflected wave from the (PML-coated) inner cylinder is nonetheless present and, more importantly, the instability of the resultant FDTD algorithm in this case is dramatic and occurs soon after the wave reaches the convex PML (early-time effect).

7.8.3 Spherical PML Analysis

In this section, we examine the causality and dynamic stability of the spherical PML. The focus is on those aspects that differ from the previous cases.

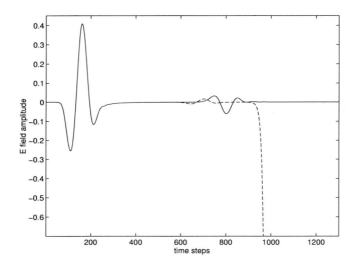

Figure 7.9 2D cylindrical FDTD solution with eight-layer cylindrical PML. The solid line represents the case with the concave PML only. Both the direct and the reflected pulses due to the inner PEC cylinder are visible. No reflected fields due to the grid termination are visible. The dashed line represents the FDTD simulation with the convex cylindrical PML coating the inner PEC cylinder. A dramatic early-time instability is evident. (*Source:* [30], ©1999, IEEE. Reprinted with permission.)

In the spherical system, the constitutive tensor $\overline{\Lambda}(r;\omega)$ for the Maxwellian PML formulation is given by

$$\overline{\Lambda}(r;\omega) = \hat{r}\hat{r}\left(\frac{s_\theta s_\phi}{s_r}\right) + \hat{\theta}\hat{\theta}\left(\frac{s_\phi s_r}{s_\theta}\right) + \hat{\phi}\hat{\phi}\left(\frac{s_r s_\theta}{s_\phi}\right) \qquad (7.221)$$

In the above, s_r is the stretching parameter in the r direction, $s_r(r,\omega) = a_r(r) + i\frac{\Omega_r(r)}{\omega}$, with $a_r \geq 1$ and $\Omega_r \geq 0$. The variables s_θ and s_ϕ are "pseudo-stretching" parameters in the θ and ϕ angular coordinates that account for the modification in the metric coefficient after the stretching in the r direction

$$s_\theta(r,\omega) = s_\phi(r,\omega) = \frac{\tilde{r}}{r} = \frac{1}{r}\left(r_0 + \int_{r_0}^{r} s_r(r')dr'\right) = \frac{1}{r}\left(b_r(r) + i\frac{\Delta_r(r)}{\omega}\right) \qquad (7.222)$$

where r_0 is in the physical (real) space and $a_r = 1$, $\Omega_r = 0$ in the physical space.

Since $s_\theta = s_\phi$, the constitutive tensor in (7.221) can be simplified to

$$\overline{\Lambda}(r;\omega) = \hat{r}\hat{r}\left(\frac{\tilde{r}}{r}\right)^2\left(\frac{1}{s_r}\right) + (\overline{\mathbf{I}} - \hat{r}\hat{r})\,s_r \qquad (7.223)$$

Following the same reasoning used in the Cartesian PML and cylindrical PML cases, it can be shown that $\overline{\Lambda}(r;\omega)$ has no poles in the upper half-plane for a *concave* spherical PML, and therefore the resultant FDTD scheme will be dynamically stable in this case.

In the case of a *convex* spherical PML, we note that, due to the pole cancellation in the angular terms $\hat{\theta}\hat{\theta}$ and $\hat{\phi}\hat{\phi}$, the tensor $\overline{\Lambda}(r;\omega)$ has no poles in the upper half-plane. However, its inverse $\overline{\Lambda}(r;\omega)^{-1}$ has poles in the upper half-plane due to the factor $\tilde{r}/r$ in the radial term $\hat{r}\hat{r}$ of (7.223), and therefore dynamic instability should be expected on FDTD simulation using a convex PML on spherical coordinates.

This is also predicted by a direct analysis of the Green's function $g(\mathbf{r},\mathbf{r}')$ in the frequency domain. Analogously to the Cartesian and cylindrical cases, the solutions of the modified Maxwell's equations inside the spherical PML are just the analytic continuation of the solutions of the usual Maxwell's equations in spherical coordinates. When using a convex spherical PML, the analyticity of the solutions in the upper half-plane is again not preserved. The basic distinction from the cylindrical case is that the Hankel $H_n^{(1)}(\cdot)$ and Bessel $J_n(\cdot)$ functions are replaced by its spherical counterparts $h_n^{(1)}(\cdot)$ and $j_n(\cdot)$. Although not having branch points on the real axis as $H_n^{(1)}(\cdot)$, the *denominator* functions $h_n^{(1)}(\cdot)$ in the spherical case still have zeros on the lower half-plane [30]. These zeros will eventually appear as poles on the upper half-plane of the field solutions under complex stretching.

In what follows, numerical results using the spherical PML are presented to illustrate the previous discussion. Figure 7.10 depicts the normalized E_θ field from a Hertzian-dipole radiating in the presence of a perfect conducting sphere. The field is computed using a spherical-grid 3D PML-FDTD algorithm. The θ-polarized Hertzian dipole is located at $(r,\theta,\phi) = (1.9\lambda_c, 90^0, 0^0)$ and the resultant field is sampled at $(r,\theta,\phi) = (2.5\lambda_c, 90^0, 0^0)$, where λ_c is the (free-space) wavelength corresponding to the central frequency of the excitation pulse. The perfectly conducting sphere is centered at the origin $r = 0$ and has a radius $a = 0.5\lambda_c$. The spherical grid is terminated at $r = 3.5\lambda_c$, where a hard boundary condition is set (zero tangential fields). The FDTD algorithm includes an eight-layer concave spherical PML region before the grid ends. The excitation pulse is the first derivative of the Blackman-Harris pulse with central frequency $f_c = 300$ MHz. The solid line in the Figure 7.10 shows the result when using only a concave spherical PML before the outward boundary. This corresponds to the Configuration A depicted in Figure 7.8. Both the direct pulse and the nonspurious reflected pulse due to the inner perfect conducting sphere are visible. No spurious reflection due to the outer boundary is visible. The small oscillations after the passage of the direct pulse are due to the numerical dispersion effects, caused by the coarse grid density adopted. The dotted line is the

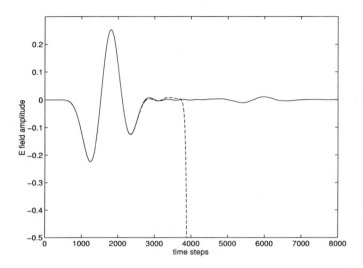

Figure 7.10 3D spherical FDTD solution with eight-layer spherical PML. The solid line represents the case with the concave PML only. Both the direct and the reflected pulse due to the inner PEC sphere are visible. No reflected fields due to the grid ends are visible. The dashed line represents the FDTD simulation with the convex spherical PML coating the inner PEC sphere. A dramatic early-time instability is evident. (*Source:* [30], ©1999, IEEE. Reprinted with permission.)

result of the simulation when an eight-layer convex spherical PML is placed around the inner sphere. This corresponds to the Configuration B depicted in Figure 7.8. The same value of $\sigma_{r,max}$ is used for the concave and convex PML in this case. As in the cylindrical case, we observe the dramatic instability of the resultant FDTD algorithm, with an exponential growth soon after the wave reaches the convex spherical PML.

7.8.4 Imposing Stability a Posteriori: The Quasi-PML

The above conclusions have an impact on the design of electromagnetic PML absorbers as tools for numerical simulations and on their use (anisotropic medium formulation) as a physical basis for engineered artificial absorbers.

An alternative to avoid the singularities in the upper half-plane for the convex case would be to *impose* $\Delta_{\rho,r} \geq 0$ at inner boundaries (irrespective of $\Omega_{\rho,r} \geq 0$). In this manner, a dynamically stable scheme can be obtained. However, the resultant cylindrical interface is not perfectly matched anymore (since (7.214) is not true anymore). It should be more appropriately called a quasi-PML [19]. This approximation behaves as a true PML only in the limit $\rho, r \to \infty$, when the cylindrical and spherical PML reduces to the Cartesian PML. From our preliminary experience, the two important features exhibited by a quasi-PML when compared to a true PML (when the latter is applicable) are (1) a nonzero reflection coefficient in the continuum limit and (2) the need for a more finely tuned profile to achieve the best results.

The inner versus outer domain differentiation in separable geometries is a major asymmetry present on the cylindrical or spherical PML and has no direct analogy in the Cartesian PML. The spatial domain enclosed by an inner boundary on the cylindrical or spherical is of finite extent, in contrast to the Cartesian case and to the outer boundaries case. A wave propagating into an inner cylindrical or spherical domain eventually emerges back on the outer domain,

Still, a quasi-PML may have practical applications both for numerical simulations and as a physical basis for engineered artificial absorbers when the radius of curvature ρ is such that $\rho \gg \lambda$.

7.9 GENERALIZED PML-FDTD SCHEMES

The extensions of the PML treated in the previous section can be easily combined. For instance, one may have a medium which is (bi)anisotropic, inhomogeneous, and dispersive simultaneously, and, in this case, the formulation described previously for (bi)anisotropic interior media can automatically incorporate the fact that now the physical constitutive tensors are themselves functions of frequency; that is,

$$\overline{\epsilon}_{PML}(\mathbf{r}; \omega) = (\det \overline{\mathbf{S}})^{-1} \left(\overline{\mathbf{S}} \cdot \overline{\epsilon}(\mathbf{r}; \omega) \cdot \overline{\mathbf{S}} \right) \quad (7.224)$$

$$\overline{\xi}_{PML}(\mathbf{r}; \omega) = (\det \overline{\mathbf{S}})^{-1} \left(\overline{\mathbf{S}} \cdot \overline{\xi}(\mathbf{r}; \omega) \cdot \overline{\mathbf{S}} \right) \quad (7.225)$$

$$\overline{\zeta}_{PML}(\mathbf{r}; \omega) = (\det \overline{\mathbf{S}})^{-1} \left(\overline{\mathbf{S}} \cdot \overline{\zeta}(\mathbf{r}; \omega) \cdot \overline{\mathbf{S}} \right) \quad (7.226)$$

$$\overline{\mu}_{PML}(\mathbf{r}; \omega) = (\det \overline{\mathbf{S}})^{-1} \left(\overline{\mathbf{S}} \cdot \overline{\mu}(\mathbf{r}; \omega) \cdot \overline{\mathbf{S}} \right) \quad (7.227)$$

Again, this is possible because the analytic continuation is done in the frequency domain and the perfect matching condition is a *local* condition on the fields. In the following, we will illustrate this by describing in detail the derivation of two PML-FDTD algorithms in 3D cylindrical coordinates and dispersive media (with conductive loss) [34]. The first is a split-field algorithm akin to the original Berenger's Cartesian formulation. The second is an unsplit algorithm, where the PML is represented as an anisotropic medium (Maxwellian PML).

7.9.1 Cylindrical PML-PLRC-FDTD: Split-Field Formulation

The complex-space nabla operator in cylindrical coordinates is written as

$$\tilde{\nabla} = \hat{\rho}\frac{\partial}{\partial\tilde{\rho}} + \hat{\phi}\frac{1}{\tilde{\rho}}\frac{\partial}{\partial\phi} + \hat{z}\frac{\partial}{\partial\tilde{z}} = \hat{\rho}\frac{1}{s_\rho}\frac{\partial}{\partial\rho} + \hat{\phi}\frac{1}{\tilde{\rho}}\frac{\partial}{\partial\phi} + \hat{z}\frac{1}{s_z}\frac{\partial}{\partial z} \quad (7.228)$$

Due to the frequency dependence, the complex-space nabla operator in the time-domain will also be a convolutional operator in time. Three distinct frequency-dependent terms are identified: s_ρ, $\tilde{\rho}$, and s_z, each one associated with a given direction. To facilitate the solution in the time domain without convolutions, Maxwell's equations with complex cylindrical coordinates are split as follows:

$$i\omega \mathbf{B}_{s\rho} = \left(\hat{\rho}\frac{1}{s_\rho}\frac{\partial}{\partial\rho}\right) \times \mathbf{E} \quad (7.229)$$

$$i\omega \mathbf{B}_{s\phi} = \left(\hat{\phi}\frac{1}{\tilde{\rho}}\frac{\partial}{\partial\phi}\right) \times \mathbf{E} \quad (7.230)$$

$$i\omega \mathbf{B}_{sz} = \left(\hat{z}\frac{1}{s_z}\frac{\partial}{\partial z}\right) \times \mathbf{E} \quad (7.231)$$

$$-i\omega \mathbf{D}_{s\rho} + \sigma \mathbf{E}_{s\rho} = \left(\hat{\rho}\frac{1}{s_\rho}\frac{\partial}{\partial\rho}\right) \times \mathbf{H} \quad (7.232)$$

$$-i\omega \mathbf{D}_{s\phi} + \sigma \mathbf{E}_{s\phi} = \left(\hat{\phi}\frac{1}{\tilde{\rho}}\frac{\partial}{\partial\phi}\right) \times \mathbf{H} \quad (7.233)$$

$$-i\omega \mathbf{D}_{sz} + \sigma \mathbf{E}_{sz} = \left(\hat{z}\frac{1}{s_z}\frac{\partial}{\partial z}\right) \times \mathbf{H} \quad (7.234)$$

By doing so, each frequency-dependent coordinate stretching term s_ρ, $\tilde{\rho}$, s_z acquires a one-to-one relationship with a given split field; therefore, each split field will have its time-domain evolution affected by only one associated stretching coordinate.

In a dispersive and inhomogeneous medium,

$$\mathbf{B} = \mathbf{B}_{s\rho} + \mathbf{B}_{s\phi} + \mathbf{B}_{sz} = \mu\mathbf{H} = \mu\left(\mathbf{H}_{s\rho} + \mathbf{H}_{s\phi} + \mathbf{H}_{sz}\right) \quad (7.235)$$

$$\mathbf{D} = \mathbf{D}_{s\rho} + \mathbf{D}_{s\phi} + \mathbf{D}_{sz} = \epsilon(\mathbf{r};\omega)\mathbf{E} = \epsilon(\mathbf{r};\omega)\left(\mathbf{E}_{s\rho} + \mathbf{E}_{s\phi} + \mathbf{E}_{sz}\right) \quad (7.236)$$

Transforming back to the time domain, (7.229)–(7.234) become

$$a_\rho \frac{\partial}{\partial t}\mathbf{B}_{s\rho} + \Omega_\rho \mathbf{B}_{s\rho} = -\frac{\partial}{\partial\rho}\left(\hat{\rho}\times\mathbf{E}\right) \quad (7.237)$$

$$b_\rho \frac{\partial}{\partial t}\mathbf{B}_{s\phi} + \Delta_\rho \mathbf{B}_{s\phi} = -\hat{\phi}\times\frac{\partial}{\partial\phi}\mathbf{E} \quad (7.238)$$

Advances in the Theory of PMLs 333

$$a_z \frac{\partial}{\partial t}\mathbf{B}_{sz} + \Omega_z \mathbf{B}_{sz} = -\frac{\partial}{\partial z}(\hat{z} \times \mathbf{E}) \qquad (7.239)$$

$$a_\rho \frac{\partial}{\partial t}\mathbf{D}_{s\rho} + \Omega_\rho \mathbf{D}_{s\rho} + a_\rho \sigma \mathbf{E}_{s\rho} + \Omega_\rho \sigma \int_0^t \mathbf{E}_{s\rho}(\tau)\,d\tau = \frac{\partial}{\partial \rho}(\hat{\rho} \times \mathbf{H}) \qquad (7.240)$$

$$b_\rho \frac{\partial}{\partial t}\mathbf{D}_{s\phi} + \Delta_\rho \mathbf{D}_{s\phi} + b_\rho \sigma \mathbf{E}_{s\phi} + \Delta_\rho \sigma \int_0^t \mathbf{E}_{s\phi}(\tau)\,d\tau = \hat{\phi} \times \frac{\partial}{\partial \phi}\mathbf{H} \qquad (7.241)$$

$$a_z \frac{\partial}{\partial t}\mathbf{D}_{sz} + \Omega_z \mathbf{D}_{sz} + a_z \sigma \mathbf{E}_{sz} + \Omega_z \sigma \int_0^t \mathbf{E}_{sz}(\tau)\,d\tau = \frac{\partial}{\partial z}(\hat{z} \times \mathbf{H}) \qquad (7.242)$$

where σ may also be a function of position, $\sigma = \sigma(\mathbf{r})$. In terms of components, the right-hand side of (7.237)–(7.239) becomes

$$\frac{\partial}{\partial \rho}(\hat{\rho} \times \mathbf{E}) = -\hat{\phi}\frac{\partial E_z}{\partial \rho} + \hat{z}\frac{\partial E_\phi}{\partial \rho} \qquad (7.243)$$

$$\hat{\phi} \times \frac{\partial}{\partial \phi}\mathbf{E} = \hat{z}\left(E_\phi - \frac{\partial E_\rho}{\partial \phi}\right) + \hat{\rho}\frac{\partial E_z}{\partial \phi} \qquad (7.244)$$

$$\frac{\partial}{\partial z}(\hat{z} \times \mathbf{E}) = \hat{\phi}\frac{\partial E_\rho}{\partial z} - \hat{\rho}\frac{\partial E_\phi}{\partial z} \qquad (7.245)$$

and similarly for (7.240)–(7.242). From (7.237)–(7.242), we see that each split-field $\mathbf{E}_{s\nu}, \mathbf{D}_{s\nu}, \mathbf{B}_{s\nu}, \mathbf{H}_{s\nu}, \nu = \rho, \phi, z$, has two components, being everywhere perpendicular to the ν direction.

The same update scheme may be used everywhere, since the interior domain (non-PML region) is a special case of a PML with $a_\rho = a_z = 1$ and $\Omega_\rho = \Omega_z = 0$. This allows for an easier parallelization of the resulting FDTD code.

An N-species Lorentzian dispersive medium is characterized by a frequency-dependent permittivity function given by (7.24). For an N-species Debye model, the frequency-dependent permittivity function is given by (7.29).

The dispersion model is included in the FDTD time-stepping scheme using the piecewise linear recursive convolution (PLRC) algorithm. The intermediate steps in the derivation of the combined PML-PLRC-FDTD time-stepping scheme in cylindrical coordinates are analogous to the Cartesian case discussed before, and will not be discussed here. The final equations for the PML-PLRC-FDTD time-stepping scheme for the electromagnetic fields in cylindrical coordinates are

$$\mathbf{H}_{s\rho}^{l+\frac{1}{2}} = -(a_\rho + \Omega_\rho \Delta_t)^{-1}\left[\frac{1}{\mu}\Delta_t \frac{\partial}{\partial \rho}(\hat{\rho} \times \mathbf{E}^l) - a_\rho \mathbf{H}_{s\rho}^{l-\frac{1}{2}}\right] \qquad (7.246)$$

$$\mathbf{H}_{s\phi}^{l+\frac{1}{2}} = -(b_\rho + \Delta_\rho \Delta_t)^{-1}\left[\frac{1}{\mu}\Delta_t (\hat{\phi} \times \frac{\partial}{\partial \phi}\mathbf{E}^l) - b_\rho \mathbf{H}_{s\phi}^{l-\frac{1}{2}}\right] \qquad (7.247)$$

$$\mathbf{H}_{sz}^{l+\frac{1}{2}} = -(a_z + \Omega_z \Delta_t)^{-1}\left[\frac{1}{\mu}\Delta_t \frac{\partial}{\partial z}(\hat{z} \times \mathbf{E}^l) - a_z \mathbf{H}_{sz}^{l-\frac{1}{2}}\right] \qquad (7.248)$$

$$[(a_\rho + \Omega_\rho \Delta_t) \lambda_0 \epsilon_0 + a_\rho \sigma \Delta_t] \mathbf{E}_{s\rho}^{l+1} =$$
$$\Delta_t \frac{\partial}{\partial \rho} \left(\hat{\rho} \times \mathbf{H}^{l+\frac{1}{2}} \right) + a_\rho \mathbf{D}_{s\rho}^l - \sigma \Omega_\rho \Delta_t \mathbf{F}_{s\rho}^l - (a_\rho + \Omega_\rho \Delta_t)\epsilon_0 \left(\lambda_1 \mathbf{E}_{s\rho}^l + \mathbf{P}_{s\rho}^l \right)$$
(7.249)

$$[(b_\rho + \Delta_\rho \Delta_t) \lambda_0 \epsilon_0 + b_\rho \sigma \Delta_t] \mathbf{E}_{s\phi}^{l+1} =$$
$$\Delta_t \left(\hat{\phi} \times \frac{\partial}{\partial \phi} \mathbf{H}^{l+\frac{1}{2}} \right) + b_\rho \mathbf{D}_{s\rho}^l - \sigma \Delta_\rho \Delta_t \mathbf{F}_{s\phi}^l - (b_\rho + \Delta_\rho \Delta_t)\epsilon_0 \left(\lambda_1 \mathbf{E}_{s\phi}^l + \mathbf{P}_{s\phi}^l \right)$$
(7.250)

$$[(a_z + \Omega_z \Delta_t) \lambda_0 \epsilon_0 + a_z \sigma \Delta_t] \mathbf{E}_{sz}^{l+1} =$$
$$\Delta_t \frac{\partial}{\partial z} \left(\hat{z} \times \mathbf{H}^{l+\frac{1}{2}} \right) + a_z \mathbf{D}_{sz}^l - \sigma \Omega_z \Delta_t \mathbf{F}_{sz}^l - (a_z + \Omega_z \Delta_t)\epsilon_0 \left(\lambda_1 \mathbf{E}_{sz}^l + \mathbf{P}_{sz}^l \right)$$
(7.251)

On the right-hand side of (7.249)–(7.251), the other pertinent quantities are updated as in (7.48)–(7.51). These equations constitute the complete split-field PML-PLRC-FDTD formulation in cylindrical coordinates. Note the parallelism between this scheme and the scheme for dispersive media in Cartesian coordinates treated before. The complex parameters $\hat{\zeta}_p^a$, $\hat{\zeta}_p^b$ are the zeroth- and first-order momenta of the (time-domain) complex susceptibility functions taken over a time-discretization interval, λ_0 and λ_1 are real constants that give instantaneous corrections for the dispersive dielectric constant, and $\tilde{\omega}_p = \beta_p - i\alpha_p$ are the complex frequency poles of the model. These constants depend on the particular choice of parameters for the Debye or Lorentz dispersive model chosen. For the pth species, they depend on the $\alpha_p, \beta_p, \gamma_p$ parameters as given by (7.35)–(7.36) and (7.40)–(7.42). Furthermore, for an inhomogeneous media, these parameters are functions of position.

7.9.2 Cylindrical PML-PLRC-FDTD: Maxwellian Formulation

In case of 3D cylindrical coordinates, the PML constitutive tensors matched to a homogeneous nondispersive medium characterized by constitutive parameters ϵ, μ are written as

$$\overline{\epsilon}_{PML} = \epsilon \overline{\Lambda}_{[\rho,\phi,z]}(\rho, z; \omega) \quad (7.252)$$

$$\overline{\mu}_{PML} = \mu \overline{\Lambda}_{[\rho,\phi,z]}(\rho, z; \omega) \quad (7.253)$$

with

$$\overline{\Lambda}_{[\rho,\phi,z]}(\rho, z; \omega) = \hat{\rho}\hat{\rho} \frac{\tilde{\rho} s_z}{\rho s_\rho} + \hat{\phi}\hat{\phi} \frac{\rho s_z s_\rho}{\tilde{\rho}} + \hat{z}\hat{z} \frac{\tilde{\rho} s_\rho}{\rho s_z} \quad (7.254)$$

and $\tilde{\rho}, s_\rho, s_z$ defined as before. For inhomogeneous and dispersive interior media $\epsilon(\mathbf{r}, \omega)$, we have that $\overline{\epsilon}_{PML} = \epsilon(\mathbf{r}, \omega)\overline{\Lambda}_{[\rho,\phi,z]}(\rho, z; \omega)$ and $\overline{\mu}_{PML} = \mu \overline{\Lambda}_{[\rho,\phi,z]}(\rho, z; \omega)$. In the unsplit field formulation, we write for Maxwell's equations

$$i\omega \mu \overline{\Lambda}_{[\rho,\phi,z]} \cdot \mathbf{H} = \nabla \times \mathbf{E} \quad (7.255)$$

$$-i\omega \left(1 + \frac{i\sigma(\mathbf{r})}{\omega\epsilon(\mathbf{r},\omega)}\right) \epsilon(\mathbf{r},\omega)\overline{\Lambda}_{[\rho,\phi,z]} \cdot \mathbf{E} = \nabla \times \mathbf{H} \qquad (7.256)$$

The above may be written in terms of auxiliary fields $\mathbf{B}_a$, $\mathbf{D}_a$, $\mathbf{E}_a$ as

$$i\omega\mu\mathbf{H}_a = \nabla \times \mathbf{E} \qquad (7.257)$$

$$-i\omega\mathbf{D}_a + \sigma\mathbf{E}_a = \nabla \times \mathbf{H} \qquad (7.258)$$

where, by definition, $\mathbf{H}_a = \overline{\Lambda}_{[\rho,\phi,z]} \cdot \mathbf{H}$, $\mathbf{D}_a = \overline{\Lambda}_{[\rho,\phi,z]} \cdot \mathbf{D}$, and $\mathbf{E}_a = \overline{\Lambda}_{[\rho,\phi,z]} \cdot \mathbf{E}$.

The auxiliary fields are introduced as a computational convenience. This is because these equations look like the ordinary Maxwell's equations except that the field quantities on the left side are auxiliary fields. Therefore, the update of those auxiliary fields in terms of the ordinary fields is just the unsplit version of the update equations (7.246)–(7.251), with $s_\rho = s_z = 1$ (i.e., an update of ordinary Maxwell's equations in a dispersive media):

$$\mathbf{H}_a^{l+\frac{1}{2}} = \mathbf{H}_a^{l-\frac{1}{2}} - \frac{1}{\mu}\Delta_t \nabla \times \mathbf{E}^l \qquad (7.259)$$

$$(\lambda_0\epsilon_0 + \sigma(\mathbf{r})\Delta_t)\mathbf{E}_a^{l+1} = \Delta_t \nabla \times \mathbf{H}^{l+\frac{1}{2}} + \mathbf{D}_a^l - \epsilon_0\left(\lambda_1 \mathbf{E}_a^l + \mathbf{P}_a^l\right) \qquad (7.260)$$

and with the other pertinent auxiliary quantities (functions of position) on the right-hand side of (7.260) being updated analogously as in (7.48)–(7.51).

The overall update scheme is incomplete without specifying how one updates the original fields from the auxiliary ones. This is done using the definitions of the auxiliary fields themselves that is,

$$\mathbf{E} = \overline{\Lambda}_{[\rho,\phi,z]}^{-1} \cdot \mathbf{E}_a \qquad (7.261)$$

$$\mathbf{H} = \overline{\Lambda}_{[\rho,\phi,z]}^{-1} \cdot \mathbf{H}_a \qquad (7.262)$$

$$\mathbf{D} = \overline{\Lambda}_{[\rho,\phi,z]}^{-1} \cdot \mathbf{D}_a \qquad (7.263)$$

or, in terms of components,

$$E_\rho = \frac{s_\rho}{s_\phi s_z} E_{a,\rho} \qquad (7.264)$$

$$H_\rho = \frac{s_\rho}{s_\phi s_z} H_{a,\rho} \qquad (7.265)$$

$$D_\rho = \frac{s_\rho}{s_\phi s_z} D_{a,\rho} \qquad (7.266)$$

and analogously for the other components by cyclic permutation of indices.

A simple system of first-order differential equations can be derived from the above equations through the use of another set of auxiliary fields $\mathbf{E}_e$, $\mathbf{H}_e$, $\mathbf{D}_e$, so that (7.264) is replaced by

$$(i\omega s_z)E_{e,\rho} = (i\omega s_\rho)E_{a,\rho} \qquad (7.267)$$

$$(i\omega s_\phi) E_\rho = i\omega E_{e,\rho} \tag{7.268}$$

and analogously for (7.265) and (7.266). In the time domain, the above become

$$\left(a_z \frac{\partial}{\partial t} + \Omega_z\right) E_{e,\rho} = \left(a_\rho \frac{\partial}{\partial t} + \Omega_\rho\right) E_{a,\rho} \tag{7.269}$$

$$\left(b_\rho \frac{\partial}{\partial t} + \Delta_\rho\right) E_\rho = \frac{\partial}{\partial t} E_{e,\rho} \tag{7.270}$$

The time update equations for the original field in terms of the auxiliary field then become

$$(a_z + \Omega_z \Delta_t) E_{e,\rho}^l = a_z E_{e,\rho}^{l-1} + (a_\rho + \Omega_\rho \Delta_t) E_{a,\rho}^l - a_\rho E_{e,\rho}^{l-1} \tag{7.271}$$

$$(b_\rho + \Delta_\rho \Delta_t) E_\rho^l = b_\rho E_\rho^{l-1} + E_{e,\rho}^l - E_{e,\rho}^{l-1} \tag{7.272}$$

where we have used a backward Euler scheme for simplicity. A central differencing scheme is also possible. In this case, a linear interpolation can be used for the constant terms to maintain second-order accuracy in time. Analogous update equations apply for other components and fields. In the case of the **H** field components, however, one may incorporate one of the complex stretching variables of (7.265) already into a modified (lossy) update for (7.257) instead. In this case, one fewer auxiliary field is needed. Together with (7.259) and (7.260), they constitute the complete update scheme for the electromagnetic fields in the unsplit PML-PLRC-FDTD formulation for cylindrical coordinates and dispersive media.

7.10 UNIFIED THEORY: BRIEF DISCUSSION

7.10.1 PML as a Change on the Metric of Space

Under the analytic continuation of the spatial coordinates, the elementary arclength is transformed to

$$ds^2 \to (d\tilde{s})^2 = (d\tilde{x})^2 + (d\tilde{y})^2 + (d\tilde{z})^2 \tag{7.273}$$

or, using (7.7),

$$(d\tilde{s})^2 = s_x^2 (dx)^2 + s_y^2 (dy)^2 + s_z^2 (dz)^2 \tag{7.274}$$

where we used the fact that s_ζ is a function of ζ only. From (7.274), we recognize that this analytic continuation on the Cartesian coordinates is equivalent to a change in the metric of space. From the Euclidean metric tensor $[g_{ij}] = [\delta_{ij}]$, or

$$[g_{ij}] = \begin{bmatrix} 1 & 0 & 0 \\ 0 & 1 & 0 \\ 0 & 0 & 1 \end{bmatrix} \tag{7.275}$$

we are led to a *complex* metric tensor given by (covariant components)

$$[\tilde{g}_{ij}(x,y,z)] = [S_{ij}(x,y,z)] \cdot [\delta_{ij}] \cdot [S_{ij}(x,y,z)] \quad (7.276)$$

where

$$[S_{ij}(x,y,z)] = \begin{bmatrix} s_x(x) & 0 & 0 \\ 0 & s_y(y) & 0 \\ 0 & 0 & s_z(z) \end{bmatrix} \quad (7.277)$$

Therefore, the PML can be interpreted as a complexification of the metric tensor of space in the Fourier domain [33].

In a general orthogonal curvilinear coordinate system (u, v, w), if we choose w to be analytically continued as

$$w \to \tilde{w} = \int_0^w s_w(w') dw' \quad (7.278)$$

then, the original metric tensor given in terms of the metric coefficients h_i

$$[g_{ij}(u,v,w)] = [h_i^2(u,v,w)\delta_{ij}] \quad (7.279)$$

$$[g_{ij}] = \begin{bmatrix} (h_u)^2 & 0 & 0 \\ 0 & (h_v)^2 & 0 \\ 0 & 0 & (h_w)^2 \end{bmatrix} \quad (7.280)$$

is mapped to

$$[\tilde{g}_{ij}] = \begin{bmatrix} (\tilde{h}_u)^2 & 0 & 0 \\ 0 & (\tilde{h}_v)^2 & 0 \\ 0 & 0 & (\tilde{h}_w)^2 \end{bmatrix} \quad (7.281)$$

where $\tilde{h}_u = h_u(u,v,\tilde{w})$, $\tilde{h}_v = h_v(u,v,\tilde{w})$, and $\tilde{h}_w = s_w(w)h_w(u,v,\tilde{w})$. The new metric can be recast as

$$[\tilde{g}_{ij}(u,v,w)] = [S_{ij}(u,v,w)] \cdot [g_{ij}(u,v,w)] \cdot [S_{ij}(u,v,w)] \quad (7.282)$$

with

$$[S_{ij}(u,v,w)] = \begin{bmatrix} (\tilde{h}_u/h_u) & 0 & 0 \\ 0 & (\tilde{h}_v/h_v) & 0 \\ 0 & 0 & (\tilde{h}_w/h_w) \end{bmatrix} \quad (7.283)$$

The PML on doubly curved mesh terminations (conformal PML) [32] is obtained by attaching an orthogonal coordinate system to the termination surface (with local coordinates along the principal curvatures and along the normal to the surface) and enforcing the analytic continuation on the normal coordinate. If w is the normal coordinate to the mesh termination (given by $w = 0$), and if we set $h_3 = 1$, the conformal PML is built over *parallel surfaces* to the mesh termination (i.e., $w = c$, $c \geq 0$), as illustrated in Figure 7.11.

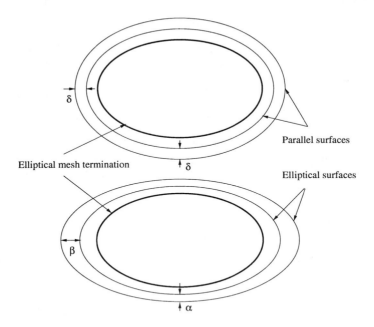

Figure 7.11 (Top) PML in curvilinear coordinates. A *local* coordinate system is attached to the termination surface (an ellipse in the figure), with local metric coefficients. The resulting PML is defined over layers which comprise *parallel surfaces*. (Bottom) A system of elliptical surfaces, which are not parallel anymore. (*Source:* Teixeira and Chew, *International Journal of Numerical Modelling*, 2000, pp. 441–455, ©2000, John Wiley. Reprinted with permission.)

The transverse metric coefficients h_u, h_v are given as $h_u = (r_u + w)/r_u$, $h_v = (r_v + w)/r_v$, where r_u, r_v are the local radii of curvature of the termination surface and u, v are the principal directions on the termination surface. In this case, the original metric tensor of such curvilinear coordinate system

$$[g_{ij}(u,v,w)] = \begin{bmatrix} (h_u)^2 & 0 & 0 \\ 0 & (h_v)^2 & 0 \\ 0 & 0 & 1 \end{bmatrix} \quad (7.284)$$

is mapped to

$$[\tilde{g}_{ij}(u,v,w)] = [S_{ij}(u,v,w)] \cdot [g_{ij}(u,v,w)] \cdot [S_{ij}(u,v,w)] \quad (7.285)$$

inside the PML, with

$$[S_{ij}(u,v,w)] = \begin{bmatrix} (\tilde{h}_u/h_u) & 0 & 0 \\ 0 & (\tilde{h}_v/h_v) & 0 \\ 0 & 0 & s_w \end{bmatrix} \quad (7.286)$$

where $\tilde{h}_u = (r_u + \tilde{w})/r_u$, $\tilde{h}_v = (r_v + \tilde{w})/r_v$. Note that, in order to achieve an homotopy between the ordinary fields and the resultant fields after the analytic continuation (so that the fields inside the PML can be written as the original fields after a change of variables) it is important to have the metric coefficients properly modified accordingly, since they are, in general, functions of the coordinates themselves. This peculiarity is not encountered in the Cartesian PML case because, in that case, the metric coefficients are independent of the spatial coordinates. In early developments of the PML concept, extensions of the Cartesian PML to curvilinear coordinates were suggested in which the metric coefficients were left unchanged, leading to only an approximate PML (*quasi-PML*). This is discussed in more detail in [52].

The Cartesian, cylindrical, and spherical PMLs are special cases of the general orthogonal curvilinear case, followed (possibly) by a successive application of the analytic continuation in orthogonal directions, if needed to achieve absorption in corner regions. For example, in cylindrical coordinates, if we choose $u = \rho$, $v = \phi$, $w = z$, the metric tensor is given by $h_\rho = 1$, $h_\phi = \rho$, and $h_z = 1$, or

$$[g_{ij}(\rho)] = \begin{bmatrix} 1 & 0 & 0 \\ 0 & \rho^2 & 0 \\ 0 & 0 & 1 \end{bmatrix} \quad (7.287)$$

If we enforce an analytic continuation along ρ (i.e., $\rho \to \tilde{\rho}$), the end result is equivalent to changing the metric tensor to

$$[\tilde{g}_{ij}(\rho, z)] = [S_{ij}(\rho, z)] \cdot [g_{ij}(\rho)] \cdot [S_{ij}(\rho, z)] \quad (7.288)$$

with

$$[S_{ij}(\rho, z)] = \begin{bmatrix} s_\rho & 0 & 0 \\ 0 & \tilde{\rho}/\rho & 0 \\ 0 & 0 & 1 \end{bmatrix} \quad (7.289)$$

In cylindrical coordinates, since the z coordinate is everywhere orthogonal to ρ, we may also simultaneously enforce an analytic continuation on z, $z \to \tilde{z}$, so that the above matrix $[S_{ij}(\rho, z)]$ is changed to

$$[S_{ij}(\rho, z)] = \begin{bmatrix} s_\rho & 0 & 0 \\ 0 & \tilde{\rho}/\rho & 0 \\ 0 & 0 & s_z \end{bmatrix} \quad (7.290)$$

7.10.2 Metric and Topological Structure of Maxwell's Equations

The fact that it is possible to derive a Maxwellian PML for different coordinate systems as well as for arbitrary (dispersive and bianisotropic) linear media, is too remarkable to be considered a mere coincidence. Instead, it should be considered an indication of some inherent symmetry. The fact that the PML can be viewed as a geometric concept (a change on the metric of space) is a strong indication that a possible hidden geometric property of Maxwell's equations is at work. Such a property is briefly discussed next.

Maxwell's equations, like most equations of physics, are surprisingly rich in symmetries. Some of these symmetries are evident (easily traced) in the vector calculus language. Others, however, are not well appreciated within this language. Our symmetry of interest here is a geometric one: the metric invariance of Maxwell's equations, first discovered by Weyl and Cartan, and rediscovered a number of times [36, 53]. In the vector calculus language, such invariance is not obvious because the metric structure of Maxwell's equations is intertwined with their topological structure. To uncover such invariance, a mathematical language where the topological structure of Maxwell's equations is factored out from their metric structure is needed. Such a language already exists, but is not commonly used (due to the lack of exposure in the engineering literature) by the electromagnetics community: it is the language of differential forms (exterior calculus).

The use of differential forms as a concise and elegant mathematical framework to study electromagnetics has been pioneered, among others, by Deschamps [36]. Over the years, it has been proved useful to obtain a number of results involving, for example, Green's functions in complex media, boundary conditions [54, 55], or the construction of consistent basis functions in the finite-element method [56–58].

In this language, the source-free Maxwell's equations are written as

$$dE = i\omega B \tag{7.291}$$

$$dH = -i\omega D \tag{7.292}$$

$$dD = 0 \tag{7.293}$$

$$dB = 0 \tag{7.294}$$

where E and H are electric and magnetic field intensity 1-forms, D and B are electric and magnetic flux density 2-forms. The operator d is the usual exterior derivative, which plays the role of the curl and div operators of vector calculus. The exterior derivative is an operator applicable to any differentiable manifold, even without a defined metric. In other words, such an operator is independent of the concept of distance. The Maxwell's equations in the above form are *topological* equations (i.e., manifestly invariant under diffeomorphisms). This is in marked contrast to the vector calculus operators, which depend on metric factors. Moreover, the constitutive parameters of a given medium relate the 1-forms E, H to the 2-forms D, B and are given in terms of the so-called *Hodge operators* $\star_e$ and $\star_h$ as [55, 59]:

$$D = \star_e E \tag{7.295}$$

$$B = \star_h H \tag{7.296}$$

In three dimensions, the Hodge operators establish a 1:1 map (isomorphism) between the space of 1-forms as E and H and the space of 2-forms as D and B. Hodge operators depend on a metric so that all the information about the metric of space is contained in the constitutive relations.

In the differential forms context, the PML is obtained through a modification on the Hodge operators:

$$\star_e \to \tilde{\star}_e \tag{7.297}$$

$$\star_h \to \tilde{\star}_h \tag{7.298}$$

induced by the complexification of the metric. The resultant forms inside the PML, $\tilde{E}, \tilde{D}, \tilde{H}, \tilde{B}$ therefore obey the modified Hodge relations

$$\tilde{D} = \tilde{\star}_e \tilde{E} \tag{7.299}$$

$$\tilde{B} = \tilde{\star}_h \tilde{H} \tag{7.300}$$

but still obey the same equations (7.291)–(7.294)

Because of such metric-topological factorization, the distinction between the Maxwellian PML and the complex-space PML is not present in the differential forms language.

7.10.3 Hybrid PMLs

Different PML formulations in the vector calculus language arises from how the (unique) differential form fields can be mapped into corresponding vector fields. This mapping is an isomorphism governed by the metric of space [59]. If the original, real metric is chosen to govern such isomorphism, then the Maxwellian PML formulation in the vector language is recovered. On the other hand, if the modified, complex metric is chosen (a canonical isomorphism since this metric also defines the Hodge operators for the PML in (7.299) and (7.300)), then the complex-space PML formulation in vector language is recovered. Because of this, the resulting vector fields in each case differ only through metric factors. The general relationship between the resultant electric vector fields is given as [33]

$$E_i^m = \frac{\tilde{h}_i}{h_i} E_i^c \tag{7.301}$$

where E_i^m stands for the Maxwellian PML vector field components and E_i^c for the complex-space PML vector field components. A similar relationship holds for the magnetic fields H_i as well.

The differential forms viewpoint also reveals that if other metrics are chosen to govern the form-vector isomorphism (e.g., hybridizations of the previous ones), other, infinitely many frequency-domain PML formulations are possible (albeit more cumbersome) [33]. In such context, the complex space PML and Maxwellian PML are particular cases of these choices.

REFERENCES

1. K. S. Yee, "Numerical solution of initial boundary value problems involving Maxwell's equation in isotropic media," *IEEE Trans. Antennas Propagat.*, vol. 14, pp. 302–307, 1966.

2. T. Weiland, "Time domain electromagnetic field computations with finite difference methods," *Int. J. Num. Model*, vol. 9, pp. 295–319, 1996.

3. A. Taflove and S. Hagness, Eds., *Advances in Computational Electrodynamics: The Finite-Difference Time-Domain Method*, second edition, Norwood, MA: Artech House, 2000.

4. W. C. Chew, *Waves and Fields in Inhomogeneous Media*, Piscataway, NJ: IEEE Press, 1995.

5. J. P. Berenger, "A perfectly matched layer for the absorption of electromagnetic waves," *J. Comput. Phys.*, vol. 114, no. 2, pp. 185–200, 1994.

6. D. S. Katz, E. T. Thiele, and A. Taflove, "Validation and extension to three dimensions of the Berenger PML absorbing boundary condition," *IEEE Microwave Guided Wave Lett.*, vol. 4, no. 8, pp. 268–270, 1994.

7. W. C. Chew and W. Weedon, "A 3D perfectly matched medium from modified Maxwell's equations with stretched coordinates," *Microwave Opt. Tech. Lett.*, vol. 7, no. 13, pp. 599–604, 1994.

8. Z. S. Sacks, D. M. Kingsland, R. Lee, and J.-F. Lee, "A perfectly matched anisotropic absorber for use as an absorbing boundary condition," *IEEE Trans. Antennas Propagat.*, vol. 43, no. 12, pp. 1460–1463, 1995.

9. C. M. Rappaport, "Interpreting and improving the PML absorbing boundary condition using anisotropic lossy mapping of space," *IEEE Trans. Magn.*, vol. 32, no. 3, pp. 968–974, 1996.

10. S. D. Gedney, "An anisotropic PML absorbing media for the FDTD simulation of fields in lossy and dispersive media," *Electromagn.*, vol. 16, pp. 399–415, 1996.

11. W. C. Chew and J. M. Jin, "Perfectly matched layers in the discretized space: An analysis and optimization," *Electromagn.*, vol. 16, pp. 325–340, 1996.

12. R. W. Ziolkoswki, "The design of Maxwellian absorbers for numerical boundary conditions and for practical applications using artificial engineered materials," *IEEE Trans. Antennas Propagat.*, vol. 45, pp. 656–671, 1997.

13. J.-Y. Wu, D. M. Kingsland, J.-F. Lee, and R. Lee, "A comparison of anisotropic PML to Berenger's PML and its application to the finite-element method for EM scattering," *IEEE Trans. Antennas Propagat.*, vol. 45, no. 1, pp. 40–50, 1997.

14. W. C. Chew, J. M. Jin, , and E. Michielssen, "Complex coordinate stretching as a generalized absorbing boundary condition," *Microwave Opt. Technol. Lett.*, vol. 15, no. 6, pp. 363–369, 1997.

15. F. L. Teixeira and W. C. Chew, "PML-FDTD in cylindrical and spherical grids," *IEEE Microwave Guided Wave Lett.*, vol. 7, no. 9, pp. 285–287, 1997.

16. F. Collino and P. Monk, "The perfectly matched layer in curvilinear coordinates," *SIAM J. Sci. Computing*, vol. 19, pp. 2061–2090, 1998.

17. F. L. Teixeira and W. C. Chew, "Extension of the PML absorbing boundary condition to 3D spherical coordinates: scalar case," *IEEE Trans. Magn.*, vol. 34, no. 5, pp. 2680–2683, 1998.

18. R. W. Ziolkoswki, "Maxwellian material based absorbing boundary conditions," *Comp. Meth. Appl. Mech. Eng.* vol. 169, no. 3–4, pp. 237–262, 1999.

19. J.-Q. He and Q. H. Liu, "A nonuniform cylindrical FDTD algorithm with improved PML and quasi-PML absorbing boundary conditions," *IEEE Trans. Geosci. Remote Sens.*, vol. 37, no. 2, pp. 1066–1072, 1999.

20. N. V. Kantartzis, P. G. Petropoulos, and T. D. Tsiboukis, "Performance evaluation and absorption enhancement of the Grote-Keller and unsplit PML boundary conditions for the 3-D FDTD method in spherical coordinates," *IEEE Trans. Magn.*, vol. 35, no. 3, pp. 1418–1421, 1999.

21. J.-Y. Wu, J. Nehrbass, and R. Lee, "A comparison of PML for TVFEM and FDTD," *Int. J. Num. Model.*, vol. 13, no. 2–3, pp. 233–244, 2000.

22. P. G. Petropoulos, "Reflectionless sponge layers as absorbing boundary conditions for the numerical solution of Maxwell equations in rectangular, cylindrical, and spherical coordinates," *SIAM J. Appl. Math.*, vol. 60, no. 3, pp. 1037–1058, 2000.

23. S. D. Gedney, "The perfectly matched layer absorbing medium," in *Advances in Computational Electrodynamics: The Finite-Difference Time-Domain Method*, A. Taflove, Ed., Chapter 5, pp. 263–344, Norwood, MA: Artech House, 1998.

24. M. Kuzuoglu and R. Mittra, "A systematic study of perfectly matched absorbers," in *Frontiers in Electromagnetics*, D. H. Werner and R. Mittra, Eds., Chapter 14, pp. 609–643, Piscataway, NJ: IEEE Press, 2000.

25. F. D. Hastings, J. B. Schneider, and J. N. Chun, "Application of the perfectly matched layer (PML) absorbing boundary condition to elastic wave propagation," *J. Acoust. Soc. Am.*, vol. 100, no. 5, pp. 3061–3069, 1996.

26. W. C. Chew and Q. H. Liu, "Perfectly matched layers for elastodynamics: a new absorbing boundary condition," *J. Comp. Acoust.*, vol. 4, no. 4, pp. 341–360, 1997.

27. A. Ahland, D. Schulz, and E. Voges, "Accurate mesh truncation for Schroedinger equations by a perfectly matched layer absorber: Applications to the calculation of optical spectra," *Phys. Rev. B*, vol. 60, no. 8, pp. R1509–R5112, 1999.

28. F. L. Teixeira, W. C. Chew, M. Straka, M. L. Oristaglio, and T. Wang, "Finite-difference time-domain simulation of ground penetrating radar on dispersive, inhomogeneous, and conductive media," *IEEE Trans. Geosci. Remote Sensing*, vol. 36, no. 6, pp. 1928–1937, 1998.

29. F. L. Teixeira and W. C. Chew, "General closed-form PML constitutive tensors to match arbitrary bianisotropic and dispersive linear media," *IEEE Microwave Guided Wave Lett.*, vol. 8, no. 6, pp. 223–225, 1998.

30. F. L. Teixeira and W. C. Chew, "On causality and dynamic stability of perfectly matched layers for FDTD simulations," *IEEE Trans. Microwave Theory Tech.*, vol. 47, no. 6, pp. 775–785, 1999.

31. F. L. Teixeira and W. C. Chew, "Systematic derivation of anisotropic PML absorbing media in cylindrical and spherical coordinates," *IEEE Microwave Guided Wave Lett.*, vol. 7, no. 11, pp. 371–373, 1997.

32. F. L. Teixeira and W. C. Chew, "Analytical derivation of a conformal perfectly matched absorber for electromagnetic waves," *Microwave Opt. Technol. Lett.*, vol. 17, no. 4, pp. 231–236, 1998.

33. F. L. Teixeira and W. C. Chew, "Differential forms, metrics, and the reflectionless absorption of electromagnetic waves," *J. Electromagn. Waves Appl.*, vol. 13, no. 5, pp. 665–686, 1999.

34. F. L. Teixeira and W. C. Chew, "Finite-difference simulation of transient electromagnetic fields for cylindrical geometries in complex media," *IEEE Trans. Geosci. Remote Sensing*, vol. 38, no. 4, pp. 1530–1543, 2000.

35. D. F. Kelley and R. J. Luebbers, "Piecewise linear recursive convolution for dispersive media using FDTD," *IEEE Trans. Antennas Propagat.*, vol. 44, no. 6, pp. 792–797, 1996.

36. G. A. Deschamps, "Electromagnetics and differential forms," *Proc. IEEE*, vol. 69, no. 6, pp. 676–696, 1981.

37. E. A. Navarro, C. Wu, P. Y. Chung, and J. Litva, "Application of PML super-absorbing boundary condition to nonorthogonal FDTD method," *Electron. Lett.*, vol. 30, no. 20, pp. 1654–1655, 1994.

38. J. A. Roden and S. D. Gedney, "Efficient implementation of the uniaxial-based PML media in three-dimensional nonorthogonal coordinates with the use of the FDTD technique," *Microwave Opt. Technol. Lett.*, vol. 14, no. 2, pp. 71–75, 1997.

39. M. Fusco, "FDTD algorithm in curvilinear coordinates," *IEEE Trans. Ant. Propagat.*, vol. 38, no. 1, pp. 76–89, 1990.

40. J. Maloney, M. Kesler, and G. Smith, "Generalization of PML to cylindrical geometries," in *Proc. of 13th Annual Rev. of Prog. Appl. Comp. Electromag.*, March 17-21, 1997, Monterey, CA, 1997, vol. 2, pp. 900–908.

41. H. W. Guggenheimer, *Differential Geometry*, New York: Dover, 1977.

42. J. A. Stratton, *Electromagnetic Theory*, New York: McGraw Hill, 1941.

43. A. Chatterjee and J. L. Volakis, "Conformal absorbing boundary conditions for the vector wave equation," *Microw. Opt. Technol. Lett.*, vol. 6, no. 16, pp. 886–889, 1993.

44. C. H. Wilcox, "An expansion theorem for electromagnetic fields," *Comm. Pure Appl. Math.*, vol. 9, pp. 115–134, 1956.

45. K.-P. Hwang and J. M. Jin, "Application of a hyperbolic grid generation technique to a conformal PML implementation," *IEEE Microwave Guided Wave Lett.*, vol. 9, no. 4, pp. 137–139, 1999.

46. E. M. Lifshitz and L. P. Pitaevskii, *Statistical Physics: Part I*, Oxford, U.K.: Pergamon Press, 1980.

47. L. D. Landau, E. M. Lifshitz, and L. P. Pitaevskii, *Electrodynamics of Continuous Media*, Oxford, U.K.: Pergamon Press, 1984.

48. H. M. Nussenzveig, *Causality and Dispersion Relations*, New York: Academic Press, 1972.

49. J. W. Nehrbass, J.-F. Lee, and R. Lee, "Stability analysis for perfectly matched layered absorbers," *Electromagn.*, vol. 16, pp. 385–397, 1996.

50. S. Abarbanel and D. Gottlieb, "A mathematical analysis of the PML method," *J. Comput. Phys.*, vol. 134, pp. 357–363, 1997.

51. J. A. Kong, *Electromagnetic Wave Theory*, Cambridge, MA: EMW Publishing, 1999.

52. F. Collino and P. Monk, "Optimizing the perfectly matched layer," *Comp. Meth. Appl. Mech. Eng.*, vol. 164, no. 1–2, pp. 157–171, 1998.

53. D. Van Dantzig, "The fundamental equations of electromagnetism, independent of metrical geometry," *Proc. Cambridge Phil. Soc.*, vol. 37, pp. 421–427, 1934.

54. K. F. Warnick, R. H. Selfridge, and D. V. Arnold, "Electromagnetic boundary conditions and differential forms," *IEE Proc., Pt. H*, vol. 142, pp. 326–332, 1995.

55. K. F. Warnick and D. V. Arnold, "Green forms for anisotropic, inhomogeneous media," *J. Electromagn. Waves Appl.*, vol. 11, pp. 1145–1164, 1997.

56. A. Bossavit, "Whitney forms: a class of finite elements for three-dimensional computations in electromagnetism," *IEE Proc., Pt. A*, vol. 135, pp. 493–500, 1988.

57. A. Bossavit, "Differential forms and the computation of fields and forces in electromagnetism," *Eur. J. Mech. B*, vol. 10, pp. 474–488, 1991.

58. R. Hiptmair, "Canonical construction of finite elements," *Math. Comp.*, vol. 68, no. 228, pp. 1325–1346, 1999.

59. F. L. Teixeira and W. C. Chew, "Lattice electromagnetic theory from a topological viewpoint," *J. Math. Phys.*, vol. 40, no. 1, pp. 169–187, 1999.

8

Fast Forward and Inverse Methods for Buried Objects

Tie Jun Cui and Weng Cho Chew

8.1 INTRODUCTION

Electromagnetic (EM) scattering and inverse scattering by objects in free space or homogeneous background have been well studied since the 1980s. Although these studies have proven useful in many applications where the effect of the environment can be neglected, they exclude many problems of practical interest where the effects of the ground, ocean, and atmosphere must be accounted for. In fact, in many cases the effect of these environments can be simply represented by a model consisting of one or more planar dielectric layers. So EM scattering and inverse scattering by objects that reside in layered media become very important and have wide applications in radar remote sensing, land mine detection, geophysical probing, medical imaging, nondestructive testing, and target identification.

In the forward scattering of buried objects, many methods have been proposed to analyze wire structures [1–5], two-dimensional (2D) conducting strips [6–8], 2D dielectric and conducting cylinders [9–18], and three-dimensional (3D) dielectric and conducting objects [19–25] using the Sommerfeld-integral representations. However, these approaches are directly based on the conventional method of moments (MOM), which handles small and moderately sized scatterers. When the buried object becomes very large, the evaluation of Sommerfeld integrals, the storage of impedance matrix, and the matrix inversion will be impossible for a small computer if the above methods are used. Although some efficient methods have been proposed for the fast evaluation of Sommerfeld integrals [18, 25], the memory requirement and matrix inversion are still the main bottlenecks for large-scale problems. On the other hand, nonlinear inverse scattering methods require repeated solutions of forward scattering problems. Therefore, developing fast forward scattering algorithms for buried objects is very important.

In the first part of this chapter, three fast algorithms are presented for the forward scattering by buried 2D dielectric cylinders, 3D conducting plates, and 3D dielectric objects of arbitrary shape using the conjugate gradient (CG) method and fast Fourier transform (FFT). The CG-FFT method is one of the most efficient techniques to analyze large-scale problems; it has been widely used and investigated for EM scattering in a homogeneous space and in microstrip antennas [26–36]. Compared with objects in the homogeneous space, the main difference in the buried object problem is that the integral equation contains a reflected-field term from the ground, which is expressed by the Sommerfeld integrals, besides the primary-field term in homogeneous space. In this chapter, the Galerkin's method is utilized to discretize the electric field integral equations (EFIE). After discretization, the primary-field term yields a cyclic convolution like that in a homogeneous space, while the reflected-field term yields a cyclic correlation, both of which can be rapidly evaluated by FFT. Due to the use of FFT to handle the cyclic convolution and correlation, the Sommerfeld integrals' evaluation has been reduced to a minimum. In the meantime, the memory required in these algorithms is only of order N, and the computational complexity is of order $N \log N$, where N is the total unknown number. Therefore, it is possible to solve large buried object problems on a small computer by using these algorithms.

The fast forward solver of buried objects provides a simple way to detect the buried targets. In the second part of this chapter we introduce a very early time electromagnetic (VETEM) system, which is a pulsed time-domain system, as are most ground-penetrating radar (GPR). The VETEM system consists of a transmitting loop antenna and a receiving loop antenna, which run over a lossy ground to detect buried objects. Using the fast forward solvers and wire-antenna modeling, we can numerically simulate the VETEM system. From the spatial-time domain scattered magnetic field received at the receiving loop, the approximate target location and buried depth can be estimated. However, the shape and dielectric property of the buried target cannot be determined. In order to reconstruct the buried objects accurately, we have to use inverse scattering methods.

In the EM area, inverse scattering is a difficult and complex topic. Solution methods have been proposed for problems involving one-dimensional (1D) unknowns [37–39], higher-dimensional unknowns in a homogeneous space with linear scattering approximations [40–43], and higher-dimensional unknowns in a homogeneous space considering multiple scattering mechanisms [44–53]. In geophysical exploration, well-logging is one of the more important techniques, where low-frequency measurement methods are presented by using an axisymmetric multilayered model [54, 55]. For inverse scattering that involves the sensing of buried targets and land mine detection, simple models have been proposed by using the GPR technique without accounting for the air-earth interface [56, 57]. To develop more accurate models, a half space problem must be considered to represent the air-earth interface. Many inversion methods have been investigated to reconstruct the objects buried in a half space or multilayered media [58–62]. Among these methods, the modified gradient approach [58–60] and the diffraction tomographic scheme [62] are efficient algorithms. However, either only single profile (permittivity or conductivity) can be reconstructed or the background is assumed lossless in these methods.

In the third part of this chapter, several fast and efficient inverse scattering algorithms are presented to detect 2D and 3D buried objects using the distorted Born iterative method (DBIM) and diffraction tomographic (DT) scheme. In these algorithms, the air-earth interface has been taken into account and the earth can be either lossy or lossless. Both the permittivity and conductivity profiles of buried objects are reconstructed.

The organization of this chapter is as follows. After the introduction, Green's functions of the buried problem are first discussed in Section 8.2. Then, fast forward scattering methods are introduced in Section 8.3, which includes 2D buried dielectric cylinders, 3D buried conducting plates, and 3D dielectric objects. In Section 8.4, we briefly introduce the detection of buried targets using the VETEM system and fast forward methods. Finally, the fast inverse scattering methods are discussed in Section 8.5, where we develop a single-frequency DBIM algorithm, a multiple-frequency DBIM algorithm, a new DT algorithm for 2D buried objects, and a 3D general DT algorithm. Many numerical simulations and reconstruction results are presented in these sections.

8.2 GREEN'S FUNCTIONS

The Green's function of an electromagnetic problem is the solution of the wave equation when the source is a point source. When the Green's function is known, the solution due to a general source can be obtained by the principle of linear superposition. Hence, the Green's function is the fundamental solution of an EM problem.

For a buried object problem shown in Figure 8.1, several kinds of Green's functions exist due to multiple spatial regions and different kinds of sources. Now, we discuss different Green's functions involved in the buried object problem. In this chapter, the time dependence of $\exp(-i\omega t)$ is assumed and suppressed.

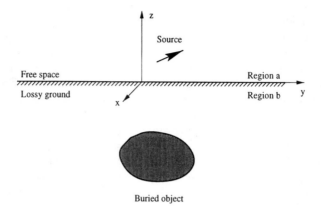

Figure 8.1 Detection of buried objects. The source can be either electric type or magnetic type, and the buried objects can be either dielectrics or perfect conductors.

8.2.1 Green's Function in Integral Equation

In practical problems, most buried objects have electrical properties, for example, the perfectly conducting objects and dielectric objects. Such objects will interact with the incident electric field, and then the electric current is induced on the surface of conducting objects or inside dielectric objects. Hence, we first consider the electric field Green's function produced by an electric dipole when the source point and field point are both in Region b, the lossy ground.

Under the Cartesian coordinate system shown in Figure 8.1, the electric-field dyadic Green's function produced by a 3D electric dipole that is oriented in the $\hat{\alpha}'$ direction can be formulated as follows when the source point and field point are both in Region b [25, 63]:

$$\hat{\alpha} \cdot \bar{\mathbf{G}}_{ee}^{bbP}(\mathbf{r}, \mathbf{r}', k) \cdot \hat{\alpha}' = \frac{i\omega\mu_0}{4\pi} \left(\hat{\alpha} \cdot \hat{\alpha}' + \frac{1}{k_b^2} \hat{\alpha} \cdot \nabla\nabla \cdot \hat{\alpha}' \right) g^P \quad (8.1)$$

$$\hat{\alpha} \cdot \bar{\mathbf{G}}_{ee}^{bbR}(\mathbf{r}, \mathbf{r}', k) \cdot \hat{\alpha}' = \frac{i\omega\mu_0}{4\pi} \left(\hat{\alpha}_s \cdot \hat{\alpha}'_s g_{\text{TE}}^R + \alpha_z \alpha'_z g_{\text{TM}}^R \right)$$

$$+ \frac{i\omega\mu_0}{4\pi k_b^2} \left(\hat{\alpha} \cdot \nabla\nabla \cdot \hat{\alpha}'' g_{\text{TM}}^R + 2\hat{\alpha}_s \cdot \nabla_s \nabla_s \cdot \hat{\alpha}'_s g_{\text{EM}}^R \right) \quad (8.2)$$

in which the subscript ee represents the electric field produced by the electric dipole, bb represents both the field point and source point in Region b, and P and R denote the primary and reflected fields, respectively. Here,

$$g^P(\mathbf{r} - \mathbf{r}') = \frac{e^{ik_b|\mathbf{r}-\mathbf{r}'|}}{|\mathbf{r} - \mathbf{r}'|} \tag{8.3}$$

is the scalar Green's function in a homogeneous space, and

$$g^R_{\text{TE,TM,EM}}(\mathbf{r}, \mathbf{r}') = \frac{i}{2\pi} \int_{-\infty}^{+\infty} \int_{-\infty}^{+\infty} dk_x dk_y \frac{R^{\text{TE,TM,EM}}}{k_{bz}} e^{-ik_{bz}(z+z')}$$
$$\cdot e^{i[k_x(x-x')+k_y(y-y')]} \tag{8.4}$$

are Sommerfeld integrals, where R^{TE} and R^{TM} are reflection coefficients of the TE wave and TM wave from Region b to Region a. The mixed reflection coefficient R^{EM} is defined as

$$R^{\text{EM}} = \frac{k_b^2}{2k_s^2} \left(R^{\text{TM}} + R^{\text{TE}} \right)$$

In the above expressions, $\hat{\alpha}$ is the polarization direction of the electric field; $k_{az,bz} = \sqrt{k_{a,b}^2 - k_s^2}$; $k_s^2 = k_x^2 + k_y^2$; $k_{a,b}$ are wave numbers in Regions a and b: $k_{a,b}^2 = k^2 \epsilon_{a,b} + ik\eta_0 \sigma_{a,b}$, in which $\epsilon_{a,b}$ and $\sigma_{a,b}$ are the relative permittivity and conductivity in Regions a and b, respectively; k and η_0 are the wave number and wave impedance in free space; $\nabla_s = \hat{x}\partial_x + \hat{y}\partial_y$; $\nabla = \nabla_s + \hat{z}\partial_z$; $\mathbf{r} = \hat{x}x + \hat{y}y + \hat{z}z$; $\hat{\alpha} = \hat{\alpha}_s + \hat{z}\alpha_z$; $\hat{\alpha}' = \hat{\alpha}'_s + \hat{z}\alpha'_z$; and $\hat{\alpha}'' = -\hat{\alpha}'_s + \hat{z}\alpha'_z$.

In (8.1) and (8.2), we retain the nabla operators explicitly in the Green's functions to later transfer them to the basis and testing functions so as to simplify the problem.

8.2.2 Green's Function for Incident Field

As stated above, the buried objects will interact with the incident electric field. Hence, we only consider the electric-field Green's function to formulate the incident field.

When the source is a 3D electric dipole which resides in Region a, the electric-field dyadic Green's function in Region b can be expressed as

$$\overline{\mathbf{G}}_{ee}^{ba}(\mathbf{r}, \mathbf{r}', k) = \frac{1}{4\pi^2} \int_{-\infty}^{\infty} \int_{-\infty}^{\infty} dk_x dk_y \widetilde{\mathbf{G}}_{ee}^{ba} e^{i(k_{az}z' - k_{bz}z)}$$
$$\cdot e^{i[k_x(x-x')+k_y(y-y')]} \tag{8.5}$$

in which the subscript ba represents the field point in Region b and source point in Region a. The "transmitted" subscript T has been omitted because the field from a to b is always a transmitted field. The spectral dyadic Green's function is given by

$$\widetilde{\mathbf{G}}_{ee}^{ba} = \frac{g_0}{k^2} \begin{bmatrix} k_{az}k_{bz} + k_y^2 & -k_x k_y & k_x k_{bz} \\ -k_x k_y & k_{az}k_{bz} + k_x^2 & k_y k_{bz} \\ k_x k_{az} & k_y k_{az} & k_x^2 + k_y^2 \end{bmatrix} \tag{8.6}$$

in which
$$g_0 = 1/(\tilde{\epsilon}_a k_{bz} + \tilde{\epsilon}_b k_{az})$$

When the source is a 3D magnetic dipole which resides in Region a, the electric-field dyadic Green's function in Region b is written as

$$\overline{\mathbf{G}}_{em}^{ba}(\mathbf{r}, \mathbf{r}', k) = \frac{1}{4\pi^2} \int_{-\infty}^{\infty} \int_{-\infty}^{\infty} dk_x dk_y \tilde{\overline{\mathbf{G}}}_{em}^{ba} e^{i(k_{az}z' - k_{bz}z)}$$
$$\cdot e^{i[k_x(x-x') + k_y(y-y')]} \quad (8.7)$$

in which the subscript em denotes the electric field produced by a magnetic dipole, and the spectral dyadic Green's function $\tilde{\overline{\mathbf{G}}}_{em}^{ba}$ is defined as

$$\tilde{\overline{\mathbf{G}}}_{em}^{ba} = \begin{bmatrix} k_x k_y g_4 & k_y^2 g_4 - \tilde{\epsilon}_a k_{bz} g_0 & -k_y f_0 \\ -k_x^2 g_4 + \tilde{\epsilon}_a k_{bz} g_0 & -k_x k_y g_4 & k_x f_0 \\ \tilde{\epsilon}_a k_y g_0 & -\tilde{\epsilon}_a k_x g_0 & 0 \end{bmatrix} \quad (8.8)$$

in which
$$f_0 = 1/(k_{bz} + k_{az}), \quad g_4 = (\tilde{\epsilon}_b - \tilde{\epsilon}_a) f_0 g_0$$

8.2.3 Green's Function for Scattered Field

The scattered field, which can be either an electric or magnetic field, is measured in Region a. If the receiving antenna is an electric type like dipoles and horns, the electric field is measured. If the receiving antenna is a magnetic type like loops, the magnetic field is measured. Because electric currents induced on (or inside) the buried objects will be obtained by solving the integral equations, the electric dipole is the only source in the Green's function for the scattered field.

When the scattered electric field is measured, the electric-field dyadic Green's function produced by a 3D electric dipole in Region b can be written as

$$\overline{\mathbf{G}}_{ee}^{ab}(\mathbf{r}, \mathbf{r}', k) = \frac{1}{4\pi^2} \int_{-\infty}^{\infty} \int_{-\infty}^{\infty} dk_x dk_y \tilde{\overline{\mathbf{G}}}_{ee}^{ab} e^{i(k_{az}z - k_{bz}z')}$$
$$\cdot e^{i[k_x(x-x') + k_y(y-y')]} \quad (8.9)$$

in which the subscript ab represents the field point in Region a and source point in Region b. The spectral dyadic Green's function $\tilde{\overline{\mathbf{G}}}_{ee}^{ab}$ has a simple relation with $\tilde{\overline{\mathbf{G}}}_{ee}^{ba}$

$$\tilde{\overline{\mathbf{G}}}_{ee}^{ab}(k_x, k_y, k) = \left[\tilde{\overline{\mathbf{G}}}_{ee}^{ba}(-k_x, -k_y, k)\right]^t \quad (8.10)$$

where t represents a transposition.

When the scattered magnetic field is measured, the magnetic-field dyadic Green's function produced by a 3D electric dipole in Region b is expressed as

$$\overline{\mathbf{G}}_{me}^{ab}(\mathbf{r},\mathbf{r}',k) = \frac{1}{4\pi^2}\int_{-\infty}^{\infty}\int_{-\infty}^{\infty} dk_x dk_y \widetilde{\overline{\mathbf{G}}}_{me}^{ab} e^{i(k_{az}z - k_{bz}z')}$$

$$\cdot e^{i[k_x(x-x') + k_y(y-y')]} \quad (8.11)$$

in which the subscript me represents the magnetic field produced by the electric dipole. The spectral dyadic Green's function $\widetilde{\overline{\mathbf{G}}}_{me}^{ab}$ is defined as

$$\widetilde{\overline{\mathbf{G}}}_{me}^{ab} = \frac{1}{k^2}\begin{bmatrix} k_x k_y g_1 & k_y^2 g_1 + k^2 g_3 & -\tilde{\epsilon}_a k^2 k_y g_0 \\ -k_x^2 g_1 - k^2 g_3 & -k_x k_y g_1 & \tilde{\epsilon}_a k^2 k_x g_0 \\ k^2 k_y f_0 & -k^2 k_x f_0 & 0 \end{bmatrix} \quad (8.12)$$

in which

$$g_1 = (k_{bz} - k_{az})/(\tilde{\epsilon}_a k_{bz} + \tilde{\epsilon}_b k_{az}), \quad g_3 = k_{az}/(k_{bz} + k_{az})$$

8.2.4 Reduction to 2D Case

In some circumstances, 2D Green's functions are required, for example, in the detection of long pipes. In this chapter, we will only consider the TM case where the source is an infinite electric current filament directed to $\hat{y}$ (2D point source). Then, the dyadic Green's functions discussed above reduce to scalar Green's functions.

When the source point and field point are both in Region b, the electric field Green's function is written as

$$g_{ee}^{bb}(\boldsymbol{\rho},\boldsymbol{\rho}') = \frac{i}{4\pi}\int_{-\infty}^{\infty} dk_x k_{bz}^{-1}\left[e^{ik_{bz}|z-z'|} + R^{\text{TM}} e^{-ik_{bz}(z+z')}\right] e^{ik_x(x-x')} \quad (8.13)$$

which will be used in the electric field integral equation. Here, $\boldsymbol{\rho} = \hat{x}x + \hat{z}z$ and R^{TM} is again the reflection coefficient of the TM wave from Region b to Region a. In the 2D case, $k_{az,bz}$ are defined as $k_{az,bz} = \sqrt{k_{a,b}^2 - k_x^2}$.

For the incident electric field, the source point is in Region a and the field point is in Region b. Then the electric field Green's function is given by

$$g_{ee}^{ba}(\boldsymbol{\rho},\boldsymbol{\rho}') = \frac{i}{4\pi}\int_{-\infty}^{\infty} dk_x k_{az}^{-1} T_{ab}^{\text{TM}} e^{i(k_{az}z' - k_{bz}z)} e^{ik_x(x-x')} \quad (8.14)$$

in which T_{ab}^{TM} is the transmission coefficient of the TM wave from Region a to Region b.

Similarly, the electric-field Green's function for the scattered field in which the source point is in Region b and the field point is in Region a is expressed as

$$g_{ee}^{ab}(\boldsymbol{\rho},\boldsymbol{\rho}') = \frac{i}{4\pi}\int_{-\infty}^{\infty} dk_x k_{bz}^{-1} T_{ba}^{\text{TM}} e^{i(k_{az}z - k_{bz}z')} e^{ik_x(x-x')} \quad (8.15)$$

where T_{ba}^{TM} is the transmission coefficient of the TM wave from Region b to Region a. After some derivation, the above Green's functions can be simplified to

$$g_{ee}^{bb}(\rho,\rho') = \frac{i}{4}\left[H_0^{(1)}(k_b|\rho-\rho'|) - H_0^{(1)}(k_b\rho_I)\right] + \frac{i}{2\pi}I_1(\rho,\rho') \tag{8.16}$$

$$g_{ee}^{ab}(\rho,\rho') = g_{ee}^{ba}(\rho',\rho) = \frac{i}{2\pi}I_2(\rho,\rho') \tag{8.17}$$

where $H_0^{(1)}(\cdot)$ is the first-kind zeroth-ordered Hankel function, and

$$I_1(\rho,\rho') = \int_{-\infty}^{\infty} dk_x\, f_0(k_x) e^{-ik_{bz}(z+z')} e^{ik_x(x-x')} \tag{8.18}$$

$$I_2(\rho_0,\rho) = \int_{-\infty}^{\infty} dk_x\, f_0(k_x) e^{i(k_{az}z - k_{bz}z')} e^{ik_x(x-x')} \tag{8.19}$$

are Sommerfeld-like integrals, in which f_0 has the same definition as that in Section 8.2.2, and ρ_I is the distance between the field point and the image of source point.

8.3 FAST FORWARD SCATTERING METHODS

In this section, we will introduce the CG-FFT algorithms for buried 2D dielectric cylinders, 3D conducting plates, and 3D dielectric objects. These algorithms are also the basis for inverse scattering.

8.3.1 2D Buried Dielectric Cylinders

As illustrated in Figure 8.2, a dielectric cylinder $\mathcal{D}$ with relative permittivity $\epsilon_r(\rho)$ and conductivity $\sigma_r(\rho)$ is buried in Region b, the lossy ground. For easy implementation of the problem by the CG-FFT, a different coordinate system is used.

8.3.1.1 Integral Equation

From the Green's function discussed above, we easily obtain the electric-field integral equation for the internal electric field inside $\mathcal{D}$:

$$E_b(\rho) = E_b^{\text{inc}}(\rho) + k^2 \int_{\mathcal{D}} d\rho'\, g_{ee}^{bb}(\rho,\rho')\left[\tilde{\epsilon}_r(\rho') - \tilde{\epsilon}_b\right] E_b(\rho') \tag{8.20}$$

in which $\tilde{\epsilon}_r$ and $\tilde{\epsilon}_b$ are complex relative permittivity of the buried object and background

$$\tilde{\epsilon}_r(\rho) = \epsilon_r(\rho) + i\frac{\eta_0}{k}\sigma_r(\rho), \quad \tilde{\epsilon}_b = \epsilon_b + i\frac{\eta_0}{k}\sigma_b \tag{8.21}$$

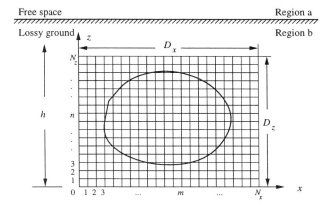

Figure 8.2 The 2D dielectric cylinder is inscribed in a rectangle of size $D_x \times D_z$.

and $E_b^{\text{inc}}(\rho)$ is the incident electric field of the buried object. If the source is a 2D point source located at ρ_t in Region a, the incident field is given by the Green's function

$$E_b^{\text{inc}}(\rho) = ik\eta_0 I g_{ee}^{ba}(\rho, \rho_t) \tag{8.22}$$

where I is the source current value. Similarly, the scattered electric field at the receiving position ρ_r is expressed as

$$E_a^s(\rho_r) = k^2 \int_D d\rho g_{ab}(\rho_r, \rho) \left[\tilde{\epsilon}_r(\rho) - \tilde{\epsilon}_b\right] E_b(\rho) \tag{8.23}$$

It is convenient to use the CG-FFT method to solve the forward problem rapidly. However, the integral equation (8.20) is not the appropriate form to be directly solved by CG-FFT because the integral in (8.20) cannot be written as a convolution or correlation of the Green's function and the internal electric field. If we let

$$\alpha(\rho) = \tilde{\epsilon}_r(\rho) - \tilde{\epsilon}_b$$
$$J(\rho) = \alpha(\rho) E_b(\rho)$$
$$J_b^{\text{inc}}(\rho) = \alpha(\rho) E_b^{\text{inc}}(\rho) \tag{8.24}$$

then the integral equation (8.20) can be written by the following operator form:

$$\mathcal{L}J(\rho) = J_b^{\text{inc}}(\rho) \tag{8.25}$$

where the operator $\mathcal{L}$ is defined as

$$\mathcal{L} = 1 - k^2 \alpha(\rho) \int_{\mathcal{D}} d\rho' g_{ee}^{bb}(\rho, \rho') \tag{8.26}$$

which can be easily implemented by using the CG-FFT method.

8.3.1.2 Galerkin's Method

In this section, we use Galerkin's method to discretize the above integral equation. Suppose that the buried dielectric cylinder is inscribed in a rectangle $D_x \times D_z$. This rectangle is divided by $N_x \times N_z$ rectangular cells of area $\Delta x \Delta z$. Because there is no derivative operation in the integral equation, we choose a pulse function as both the basis and testing functions:

$$f_{mn}(x, z) = w_{mn}(x, z)$$
$$= P_0[x - (m - 0.5)\Delta x] P_0[z - (n - 0.5)\Delta z] \tag{8.27}$$

where

$$P_0(\xi) = \begin{cases} 1, & |\xi| \leq \tfrac{1}{2}\Delta \xi \\ 0, & \text{else} \end{cases}$$

Expanding $J(x, z)$ by the basis function yields

$$J(x, z) = \sum_{m=1}^{N_x} \sum_{n=1}^{N_y} J^D(m, n) f_{mn}(x, z) \tag{8.28}$$

Substituting (8.28) into (8.25) and testing with the weighting function $w_{mn}(x, z)$, one obtains

$$\sum_{m'=1}^{N_x} \sum_{n'=1}^{N_z} G(m, n, m', n') J^D(m', n') = J_b^{\text{inc}D}(m, n) \tag{8.29}$$

where

$$G(m, n, m', n') = \langle \mathcal{L} f_{m'n'}, w_{mn} \rangle, \quad J_b^{\text{inc}D}(m, n) = \langle J_b^{\text{inc}}, w_{mn} \rangle \tag{8.30}$$

in which $\langle f, g \rangle$ is the inner product defined as

$$\langle f, g \rangle = \int f(x, z) g(x, z) \, dx \, dz \tag{8.31}$$

Considering the definition of operator $\mathcal{L}$ and Green's function g_{ee}^{bb} in (8.16), the impedance matrix $G(m, n, m', n')$ can be split into three parts:

$$G(m, n, m', n') = G^0(m - m', n - n') + G^P(m - m', n - n')$$
$$+ G^R(m - m', n + n') \tag{8.32}$$

in which $G^0(m-m', n-n')$ is the contribution from the constant term in the operator (8.26)

$$G^0(m - m', n - n') = \Delta x \Delta z \delta_{(m-m'),(n-n')} \quad (8.33)$$

$G^P(m-m', n-n')$ is the contributions from the primary term in the Green's function

$$G^P(m - m', n - n')$$
$$= -\frac{ik^2}{4} \int_{(m-1)\Delta x}^{m\Delta x} \int_{(n-1)\Delta z}^{n\Delta z} \int_{(m'-1)\Delta x}^{m'\Delta x} \int_{(n'-1)\Delta z}^{n'\Delta z} \alpha(\rho)$$
$$\cdot H_0^{(1)}(k_b|\rho - \rho'|) dx' dz' dx dz \quad (8.34)$$

and $G^R(m - m', n + n')$ is the contribution from the reflected term in the Green's function

$$G^R(m - m', n + n')$$
$$= -\frac{ik^2}{4} \int_{(m-1)\Delta x}^{m\Delta x} \int_{(n-1)\Delta z}^{n\Delta z} \int_{(m'-1)\Delta x}^{m'\Delta x} \int_{(n'-1)\Delta z}^{n'\Delta z} \alpha(\rho)$$
$$\cdot \left[-H_0^{(1)}(k_b \rho_I) + \frac{2}{\pi} I_1(\rho, \rho') \right] dx' dz' dx dz \quad (8.35)$$

Here, $I_1(\rho, \rho')$ is the integral defined in (8.18). However, z and z' in (8.18) must be replaced by $z - h$ and $z' - h$ because two different coordinate systems are used in Figures 8.1 and 8.2. Therefore, (8.29) will be written as

$$\sum_{m'=1}^{N_x} \sum_{n'=1}^{N_z} \left[G^P(m - m', n - n') + G^R(m - m', n + n') \right] J^D(m', n')$$
$$+ \Delta x \Delta z J^D(m, n) = J_b^{incD}(m, n) \quad (8.36)$$

8.3.1.3 Cyclic Convolution and Correlation

From the theory of Fourier transform and discrete Fourier transform (DFT), the cyclic convolution of discrete signals $f(m)$ and $g(m)$ is defined as

$$h(m) = f(m) \otimes g(m) = \sum_{n=0}^{N-1} f(m - n) g(n) \quad (8.37)$$

which can be rapidly computed using FFT:

$$h(m) = \mathcal{F}^{-1} \{ F(j) G(j) \} \quad (8.38)$$

in which $F(j)$ and $G(j)$ are the DFT of $f(m)$ and $g(m)$. Similarly, from the continuous correlation, we can define a cyclic correlation of discrete signals $f(m)$ and $g(m)$

$$t(m) = f(m) \star g(m) = \sum_{n=0}^{N-1} f(m+n)g(n) \qquad (8.39)$$

which can be easily shown to satisfy

$$t(m) = \mathcal{F}^{-1}\left\{F(j)G(-j)\right\} \qquad (8.40)$$

Note that in (8.37)–(8.40), both the discrete signals and their DFT have a cyclic property:

$$f(-m) = f(N-m), \quad f(n) = f(n-N), \quad F(-j) = F(N-j)$$

Using the above definition and property, we can rapidly calculate the summations in (8.36) by FFT because they resemble a 2D cyclic convolution in the primary term and a 1D cyclic convolution in x and 1D correlation in z in the reflected term. However, the computational domain of these discrete functions must be extended to $2N_x \times 2N_z$ from $N_x \times N_z$ since

$$G^P(-m,-n) = G^P(m,n), \quad G^R(-m,n) = G^R(m,n)$$

and

$$G^R(m,n) \neq G^R(m, n - N_z) \text{ for } n > N_y$$

which do not satisfy the cyclic properties. Hence, we define new discrete Green's functions and current distribution in the extended domain:

$$G^{Pe}(m,n) = G^P(m_0, n_0), \quad G^{Re}(m,n) = G^R(m_0, n) \qquad (8.41)$$

$$J^{De}(m,n) = \begin{cases} J^D(m,n), & 1 \leq m \leq N_x, 1 \leq n \leq N_z \\ 0, & \text{else} \end{cases} \qquad (8.42)$$

in which $1 \leq m \leq 2N_x, 1 \leq n \leq 2N_z$; and

$$m_0 = \begin{cases} m, & 1 \leq m \leq N_x \\ 2N_x - m, & \text{else} \end{cases} \quad n_0 = \begin{cases} n, & 1 \leq n \leq N_z \\ 2N_z - n, & \text{else} \end{cases}$$

With new definitions, the terms in (8.36) can be rapidly and exactly computed by using FFT:

$$\Delta x \Delta z J^D(m,n) + \mathcal{F}^{-1}\left\{\tilde{G}^{Pe}(i,j)\tilde{J}^{De}(i,j) + \tilde{G}^{Re}(i,j)\tilde{J}^{De}(i,-j)\right\}$$

$$= J_b^{incD}(m,n) \qquad (8.43)$$

where $\tilde{G}^{Pe}(i,j)$, $\tilde{G}^{Re}(i,j)$, and $\tilde{J}^{De}(i,j)$ are the DFTs of $G^{Pe}(m,n)$, $G^{Re}(m,n)$, and $J^{De}(m,n)$, respectively.

8.3.1.4 CG-FFT Algorithm

The CG algorithm is an efficient method to solve linear system equations. In this algorithm, an adjoint operation $\mathcal{L}^a f$ defined by

$$\langle \mathcal{L}^a f, g \rangle = \langle f, \mathcal{L}g \rangle$$

is required. From the above definition and properties of convolution and correlation, we will obtain

$$\langle \mathcal{L}^a J, w_{mn} \rangle = \Delta x \Delta z J^D(m,n)$$
$$+ \mathcal{F}^{-1} \left\{ \tilde{G}^{Pe*}(i,j) \tilde{J}^{De}(i,j) + \tilde{G}^{Re*}(i,-j) \tilde{J}^{De}(i,-j) \right\} \tag{8.44}$$

Then, the CG-FFT algorithm can be performed as follows with given initial guess of J^D, which is usually set to zero:

$$r_0 = J_b^{\mathrm{inc}D} - \Delta x \Delta z J_0^D$$
$$- \mathcal{F}^{-1} \left\{ \tilde{G}^{Pe}(i,j) \tilde{J}_0^{De}(i,j) + \tilde{G}^{Re}(i,j) \tilde{J}_0^{De}(i,-j) \right\} \tag{8.45}$$

$$s_0 = \Delta x \Delta z r_0$$
$$+ \mathcal{F}^{-1} \left\{ \tilde{G}^{Pe*}(i,j) \tilde{r}_0(i,j) + \tilde{G}^{Re*}(i,-j) \tilde{r}_0(i,-j) \right\} \tag{8.46}$$

$$p_0 = b_0 s_0, \quad b_0 = \|s_0\|_2^{-2} \tag{8.47}$$

For $\nu = 0, 1, 2, \cdots$,

$$t_\nu = \Delta x \Delta z p_\nu$$
$$+ \mathcal{F}^{-1} \left\{ \tilde{G}^{Pe}(i,j) \tilde{p}_\nu(i,j) + \tilde{G}^{Re}(i,j) \tilde{p}_\nu(i,-j) \right\} \tag{8.48}$$

$$a_\nu = \|t_\nu\|_2^{-2} \tag{8.49}$$

$$J_{\nu+1}^D = J_\nu^D + a_\nu p_\nu, \quad r_{\nu+1} = r_\nu - a_\nu t_\nu \tag{8.50}$$

$$s_{\nu+1} = \Delta x \Delta z r_{\nu+1}$$
$$+ \mathcal{F}^{-1} \left\{ \tilde{G}^{Pe*}(i,j) \tilde{r}_{\nu+1}(i,j) + \tilde{G}^{Re*}(i,-j) \tilde{r}_{\nu+1}(i,-j) \right\} \tag{8.51}$$

$$p_{\nu+1} = p_\nu + b_{\nu+1} s_{\nu+1}, \quad b_{\nu+1} = \|s_{\nu+1}\|_2^{-2} \tag{8.52}$$

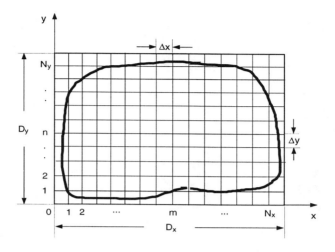

Figure 8.3 The 3D conducting plate is inscribed in a rectangle of size $D_x \times D_y$. The rectangle is parallel to the air-earth interface.

Here, $\tilde{p}_\nu$ and $\tilde{r}_\nu$ are the DFTs of p_ν and r_ν, respectively. The error of this algorithm can be controlled by

$$\mathcal{E} = \frac{\|r_\nu\|_2}{\|J_b^{\text{inc}D}\|_2} < \text{tolerance} \tag{8.53}$$

where the norm is defined as $\|f\|_2^2 = \langle f, f \rangle$.

From the above procedure, we clearly see that the storage requirement of the CG-FFT algorithm is only of order N, and the computational complexity is of order $N \log N$, where $N = N_x N_z$.

8.3.2 3D Buried Conducting Plates

A 3D conducting plate buried in lossy earth is shown in Figure 8.3. In this figure, we have supposed that the conducting plate is parallel to the air-earth interface and the buried depth is h. Although a 2D mesh is used to discretize the conducting plate, the integral equation and corresponding CG-FFT algorithm [64] will be quite different from those in the 2D case.

8.3.2.1 Integral Equation

From the dyadic Green's functions (8.1) and (8.2), the scattered electric field in Region b by the buried conducting plate can be formulated as

$$\mathbf{E}_b^s(\mathbf{r}) = \int_S \left[\bar{\mathbf{G}}_{ee}^{bbP}(\mathbf{r}, \mathbf{r}') + \bar{\mathbf{G}}_{ee}^{bbR}(\mathbf{r}, \mathbf{r}') \right] \cdot \mathbf{J}(\mathbf{r}') d\mathbf{r}' \tag{8.54}$$

where $\mathbf{J}(\mathbf{r})$ is the induced electric current on the plate. Because the plate is parallel to the air-earth interface, the above equation will be greatly reduced since $J_z = 0$:

$$\mathbf{E}_b^s(\mathbf{r}_s) = \frac{i\omega\mu_0}{4\pi} \int_S g_{TE}^{PR}(\mathbf{r}_s - \mathbf{r}_s') \mathbf{J}_s(\mathbf{r}_s') d\mathbf{r}_s'$$
$$+ \frac{i\omega\mu_0}{4\pi k_b^2} \nabla_s \int_S g_{EM}^{PR}(\mathbf{r}_s - \mathbf{r}_s') \nabla_s' \cdot \mathbf{J}_s(\mathbf{r}_s') d\mathbf{r}_s' \tag{8.55}$$

in which

$$g_{TE}^{PR} = g^P + g_{TE}^R, \quad g_{EM}^{PR} = g^P - g_{TM}^R + 2g_{EM}^R$$

as shown in (8.3) and (8.4). Using the boundary condition on the surface of the conducting plate, we easily obtain the EFIE in scalar form:

$$k_b^2 \int_S g_{TE}^{PR}(\mathbf{r}_s - \mathbf{r}_s') J_\xi(\mathbf{r}_s') d\mathbf{r}_s' + \frac{\partial}{\partial \xi} \int_S g_{EM}^{PR}(\mathbf{r}_s - \mathbf{r}_s') \left(\frac{\partial J_x}{\partial x'} + \frac{\partial J_y}{\partial y'} \right) d\mathbf{r}_s'$$
$$= -\hat{E}_{b\xi}^i(\mathbf{r}_s) \tag{8.56}$$

where, $\xi = x$ and y, and

$$\hat{\mathbf{E}}_b^i(\mathbf{r}_s) = \frac{4\pi k_b^2}{i\omega\mu_0} \mathbf{E}_b^i(\mathbf{r}_s)$$

in which $\mathbf{E}_b^i$ is the incident electric field in Region b, which can be obtained by the Green's function (8.5) or (8.7) if the source is an electric dipole or a magnetic dipole.

8.3.2.2 Galerkin's Method

We still use the Galerkin's method to discretize the above integral equation. Suppose that the arbitrarily shaped plate is inscribed in a rectangle $D_x \times D_y$. This rectangle is divided by $(N_x + 1) \times (N_y + 1)$ instead of $N_x \times N_y$ rectangular cells because there exist derivative operations in the integral equation, as shown in Figure 8.3.

From (8.56), both the surface currents J_ξ and their derivatives $\partial J_\xi / \partial \xi'$ are included in the EFIE. To ensure the existence of the derivatives, the basis function of J_ξ must be continuous in the ξ-direction. A simple but efficient basis function is a triangular function in the ξ-direction and a pulse function in the other direction. Hence, we choose the rooftop function, which has this property as both the basis and testing functions:

$$f_{mn}^x = w_{mn}^x = T_0(x - m\Delta x) P_0[y - (n - 0.5)\Delta y] \tag{8.57}$$

$$f^y_{mn} = w^y_{mn} = P_0[x - (m - 0.5)\Delta x]T_0(y - n\Delta y) \tag{8.58}$$

where $P_0(\xi)$ is pulse function defined before, and $T_0(\xi)$ is triangle function:

$$T_0(\xi) = \begin{cases} 1 - \frac{|\xi|}{\Delta \xi}, & |\xi| \leq \Delta \xi \\ 0, & \text{else} \end{cases}$$

To simplify the expressions in this chapter, we define the following numbers:

$$i_{\xi\zeta} = \begin{cases} 0, & \xi = \zeta \\ 1, & \xi \neq \zeta \end{cases} \quad \text{and} \quad n_{\xi\zeta} = \begin{cases} 1, & \xi = \zeta \\ 0, & \xi \neq \zeta \end{cases}$$

Then the currents J_x and J_y can be expanded as

$$J_\xi(x,y) = \sum_{m=1}^{N_x + i_{\xi x}} \sum_{n=1}^{N_y + i_{\xi y}} J^D_\xi(m,n) f^\xi_{mn}(x,y) \tag{8.59}$$

Notice that the discrete functions $J^D_\xi(m,n)$ should be zero when the basis function is located outside the actual plate. From (8.57)–(8.59), one easily obtains the derivatives of J_ξ:

$$\frac{\partial J_\xi}{\partial \xi} = \frac{1}{\Delta \xi} \sum_{m=1}^{N_x+1} \sum_{n=1}^{N_y+1} \left[J^D_\xi(m,n) - J^D_\xi(m - n_{\xi x}, n - n_{\xi y}) \right] P_{mn}(x,y) \tag{8.60}$$

which consist of two adjacent 2D pulses with opposite amplitudes, representing two opposite electric charges. Here, the 2D pulse function P_{mn} is defined as

$$P_{mn}(x,y) = P_0[x - (m - 0.5)\Delta x] P_0[y - (n - 0.5)\Delta y]$$

Similarly, the derivatives of the testing functions are also two adjacent 2D pulses:

$$\frac{\partial w^\xi_{mn}}{\partial \xi} = \frac{1}{\Delta \xi} \left[P_{mn} - P_{(m+n_{\xi x})(n+n_{\xi y})} \right] \tag{8.61}$$

Substituting (8.59) and (8.60) into the integral equation (8.56) and testing with the weighting function w^ξ_{mn} yields

$$\langle \mathcal{L}_\xi J_\xi, w^\xi_{mn} \rangle + \langle \mathcal{L}_{\xi x} J_x, w^\xi_{mn} \rangle + \langle \mathcal{L}_{\xi y} J_y, w^\xi_{mn} \rangle = -\hat{E}^{iD}_{b\xi}(m,n) \tag{8.62}$$

in which $\mathcal{L}_\xi$, $\mathcal{L}_{\xi x}$, and $\mathcal{L}_{\xi y}$ are operator expressions of the first, second, and third terms in the integral equations, and

$$\hat{E}^{iD}_{b\xi}(m,n) = \langle \hat{E}^i_{b\xi}, w^\xi_{mn} \rangle \tag{8.63}$$

Using the expression (8.61) and integrating by parts, the inner products in (8.61) are expressed as

$$\langle \mathcal{L}_\xi J_\xi, w_{mn}^\xi \rangle = k_b^2 \sum_{m'=1}^{N_x+i_{\xi x}} \sum_{n'=1}^{N_y+i_{\xi y}} G_{\text{TE}\xi}^{PR}(m-m', n-n') J_\xi^D(m', n') \quad (8.64)$$

$$\langle \mathcal{L}_{\xi\zeta} J_\zeta, w_{mn}^\xi \rangle$$
$$= -\frac{1}{\Delta\xi\Delta\zeta} \sum_{m'=1}^{N_x+1} \sum_{n'=1}^{N_y+1} [J_\zeta^D(m', n') - J_\zeta^D(m' - n_{\zeta x}, n' - n_{\zeta y})]$$
$$\cdot [G_{\text{EM}}^{PR}(m-m', n-n') - G_{\text{EM}}^{PR}(m + n_{\xi x} - m', n + n_{\xi y} - n')] \quad (8.65)$$

in which

$$G_{\text{TE}\xi}^{PR}(m-m', n-n')$$
$$= \int_{(m-1)\Delta x}^{(m+n_{\xi x})\Delta x} \int_{(n-1)\Delta y}^{(n+n_{\xi y})\Delta y} \int_{(m'-1)\Delta x}^{(m'+n_{\xi x})\Delta x} \int_{(n'-1)\Delta y}^{(n'+n_{\xi y})\Delta y} g_{\text{TE}}^{PR}$$
$$\cdot T_0(\xi - p\Delta\xi) T_0(\xi' - p'\Delta\xi) dx' dy' dx dy \quad (8.66)$$

$$G_{\text{EM}}^{PR}(m-m', n-n')$$
$$= \int_{(m-1)\Delta x}^{m\Delta x} \int_{(n-1)\Delta y}^{n\Delta y} \int_{(m'-1)\Delta x}^{m'\Delta x} \int_{(n'-1)\Delta y}^{n'\Delta y} g_{\text{EM}}^{PR} dx' dy' dx dy \quad (8.67)$$

where $p = m$ and n when $\xi = x$ and y, respectively.

8.3.2.3 CG-FFT Algorithm

Similar to the 2D case, the discrete Green's functions $G_{\text{TE}\xi}^{PR}$ and G_{EM}^{PR}, and the discrete electric current J_ξ^D should be extended to $G_{\text{TE}\xi}^{PRe}$, G_{EM}^{PRe}, and J_ξ^{De}, which are defined in the computational domain $2(N_x + 1) \times 2(N_y + 1)$, through the relations similar to (8.41) and (8.42). Then, the summations in (8.64) and (8.65) can be rapidly computed by using FFT. After some derivations, (8.62) is written as

$$\mathcal{F}^{-1}\left\{\tilde{G}_{\xi x}(i,j) \tilde{J}_x^{De}(i,j) + \tilde{G}_{\xi y}(i,j) \tilde{J}_y^{De}(i,j)\right\} = -\hat{E}_{b\xi}^{iD}(m,n) \quad (8.68)$$

where

$$\tilde{G}_{\xi\xi}(i,j) = k_b^2 \tilde{G}_{\text{TE}\xi}^{PRe}(i,j) + \frac{F_{\xi\xi}}{\Delta\xi^2}\tilde{G}_{\text{EM}}^{PRe}(i,j) \tag{8.69}$$

$$\tilde{G}_{\xi\zeta}(i,j) = \frac{F_{\xi\zeta}}{\Delta\xi\Delta\zeta}\tilde{G}_{\text{EM}}^{PRe}(i,j), \quad (\xi \neq \zeta) \tag{8.70}$$

in which $\xi, \zeta = x, y$; $\tilde{G}_{\text{TE}\xi}^{PRe}$, $\tilde{G}_{\text{EM}}^{PRe}$, and $\tilde{J}_\xi^{De}$ are the DFTs of $G_{\text{TE}\xi}^{PRe}$, G_{EM}^{PRe}, and J_ξ^{De}, respectively; and

$$F_{\xi\zeta} = (f_\xi^* - 1)(1 - f_\zeta)$$

where $f_x = f_x(i)$ and $f_y = f_y(j)$ are given by

$$f_\xi(t) = \exp\left(-\frac{it\pi}{N_\xi + 1}\right)$$

Similarly, the adjoint operations corresponding to (8.69) and (8.70), which are used in the CG algorithm, are expressed as

$$\tilde{G}_{\xi\xi}^a(i,j) = k_b^{2*} \tilde{G}_{\text{TE}\xi}^{PRe*}(i,j) + \frac{F_{\xi\xi}}{\Delta\xi^2}\tilde{G}_{\text{EM}}^{PRe*}(i,j) \tag{8.71}$$

$$\tilde{G}_{\xi\zeta}^a(i,j) = \frac{F_{\xi\zeta}}{\Delta\xi\Delta\zeta}\tilde{G}_{\text{EM}}^{PRe*}(i,j), \quad (\xi \neq \zeta) \tag{8.72}$$

Considering the fact that $F_{\xi\xi}^* = F_{\xi\xi}$ and $F_{\xi\zeta}^* = F_{\zeta\xi}$, we easily obtain

$$\tilde{G}_{\xi\zeta}^a = \tilde{G}_{\zeta\xi}^* \tag{8.73}$$

which can greatly reduce the memory requirement. Substituting (8.68)–(8.72) into the iteration procedure similar to (8.45)–(8.53), the CG-FFT algorithm for 3D conducting plates can be easily performed.

8.3.3 3D Buried Dielectric Objects

Consider a 3D dielectric object of arbitrary shape that is buried in Region b. The dielectric object with a complex permittivity $\tilde{\epsilon}_r(\mathbf{r})$ is assumed to be inscribed in a cuboid $D_x \times D_y \times D_z$ that is parallel to the air-earth interface. The bottom of the cuboid is separated from the interface by h, as shown in Figure 8.4.

8.3.3.1 Integral Equation

The general expression of the scattered electric field in Region b by the buried dielectric object is the same as that by conducting plates shown in (8.54). However,

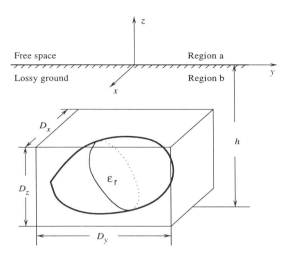

Figure 8.4 The 3D dielectric object is inscribed in a cuboid of size $D_x \times D_y \times D_z$.

the induced electric current $\mathbf{J}(\mathbf{r})$ that is inside the buried object has all x, y, and z components. Hence, the scattered electric field has a more general form [65]:

$$\begin{aligned}
\mathbf{E}_b^s(\mathbf{r}) =\ & \frac{i\omega\mu_0}{4\pi} \int_V g^P(\mathbf{r} - \mathbf{r}')\mathbf{J}(\mathbf{r}')d\mathbf{r}' \\
& + \frac{i\omega\mu_0}{4\pi k_b^2} \nabla \int_V g^P(\mathbf{r} - \mathbf{r}')\nabla' \cdot \mathbf{J}(\mathbf{r}')d\mathbf{r}' \\
& + \frac{i\omega\mu_0}{4\pi} \int_V g_{\text{TE}}^R(\mathbf{r}_s - \mathbf{r}'_s, z + z')\mathbf{J}_s(\mathbf{r}')d\mathbf{r}' \\
& + \frac{i\omega\mu_0}{4\pi} \int_V g_{\text{TM}}^R(\mathbf{r}_s - \mathbf{r}'_s, z + z')\mathbf{J}_z(\mathbf{r}')d\mathbf{r}' \\
& - \frac{i\omega\mu_0}{4\pi k_b^2} \nabla \int_V g_{\text{TM}}^R(\mathbf{r}_s - \mathbf{r}'_s, z + z')\nabla' \cdot \mathbf{J}(\mathbf{r}')d\mathbf{r}' \\
& + \frac{i\omega\mu_0}{2\pi k_b^2} \nabla_s \int_V g_{\text{EM}}^R(\mathbf{r}_s - \mathbf{r}'_s, z + z')\nabla'_s \cdot \mathbf{J}(\mathbf{r}')d\mathbf{r}' \quad (8.74)
\end{aligned}$$

where $\mathbf{J} = \mathbf{J}_s + \hat{z}J_z$ is related to the total electric field $\mathbf{E}_b$ inside the dielectric object by

$$\mathbf{J}(\mathbf{r}) = -i\omega\epsilon_0 \left[\tilde{\epsilon}_r(\mathbf{r}) - \tilde{\epsilon}_b\right] \mathbf{E}_b(\mathbf{r})$$

Considering the relationship of incident, scattered and total fields inside the dielectric object: $\mathbf{E}_b^i + \mathbf{E}_b^s = \mathbf{E}_b$, one easily obtains the electric field integral equation for the induced current

$$\gamma(\mathbf{r})J_\xi(\mathbf{r}) - k_b^2 \int_V g_{TE}^{PR}(\mathbf{r},\mathbf{r}')J_\xi(\mathbf{r}')d\mathbf{r}'$$
$$- \frac{\partial}{\partial \xi} \int_V g_{EM}^{PR}(\mathbf{r},\mathbf{r}') \left(\frac{\partial J_x}{\partial x'} + \frac{\partial J_y}{\partial y'} \right) d\mathbf{r}'$$
$$- \frac{\partial}{\partial \xi} \int_V g_{TM}^{PR1}(\mathbf{r},\mathbf{r}') \frac{\partial J_z}{\partial z'} d\mathbf{r}' = \hat{E}_{b\xi}^i(\mathbf{r}), \quad (\xi = x, y) \qquad (8.75)$$

$$\gamma(\mathbf{r})J_z(\mathbf{r}) - k_b^2 \int_V g_{TM}^{PR}(\mathbf{r},\mathbf{r}')J_z(\mathbf{r}')d\mathbf{r}'$$
$$- \frac{\partial}{\partial z} \int_V g_{TM}^{PR1}(\mathbf{r},\mathbf{r}') \left(\frac{\partial J_x}{\partial x'} + \frac{\partial J_y}{\partial y'} + \frac{\partial J_z}{\partial z'} \right) d\mathbf{r}'$$
$$= \hat{E}_{bz}^i(\mathbf{r}) \qquad (8.76)$$

Here, g_{TE}^{PR} and g_{EM}^{PR} have been defined before, and

$$g_{TM}^{PR} = g^P + g_{TM}^R, \quad g_{TM}^{PR1} = g^P - g_{TM}^R \qquad (8.77)$$

as well as

$$\gamma(\mathbf{r}) = \frac{4\pi\epsilon_b}{\epsilon_r(\mathbf{r}) - \epsilon_b}, \quad \hat{E}_{b\xi}^i(\mathbf{r}) = \frac{4\pi k_b^2}{i\omega\mu_0} E_{b\xi}^i(\mathbf{r}), \quad (\xi = x, y, z)$$

in which $E_{b\xi}^i(\mathbf{r})$ is the incident electric field in Region b that can be obtained by the Green's function (8.5) or (8.7) if the source is electric dipole or magnetic dipole.

8.3.3.2 Galerkin's Method

The discretization of the 3D buried dielectric object is shown in Figure 8.5, where we divide the bounded box $D_x D_y D_z$ into $(N_x + 1) \times (N_y + 1) \times (N_z + 1)$ cuboidal cells of volume $\Delta V = \Delta x \Delta y \Delta z$.

From (8.75) and (8.76), both the volumetric currents J_ξ and their derivatives $\partial J_\xi / \partial \xi'$ are included in the EFIE. To ensure the existence of the derivatives, the basis function of J_ξ must be continuous in the ξ-direction. Thus, we choose the basis and testing functions to be a triangle in the ξ-direction and pulses in the other two

Fast Forward and Inverse Methods for Buried Objects 367

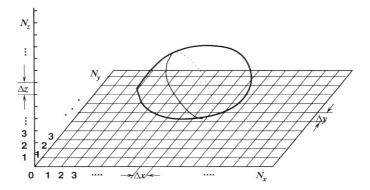

Figure 8.5 The 3D mesh of the buried dielectric object.

directions:

$$f^x_{mnk} = w^x_{mnk}$$
$$= T_0(x - m\Delta x)P_0[y - (n - 0.5)\Delta y]P_0[z - (k - 0.5)\Delta z] \quad (8.78)$$

$$f^y_{mnk} = w^y_{mnk}$$
$$= P_0[x - (m - 0.5)\Delta x]T_0(y - n\Delta y)P_0[z - (k - 0.5)\Delta z] \quad (8.79)$$

$$f^z_{mnk} = w^z_{mnk}$$
$$= P_0[x - (m - 0.5)\Delta x]P_0[y - (n - 0.5)\Delta y]T_0(z - k\Delta z) \quad (8.80)$$

Using the basis functions and the predefined numbers $i_{\xi\zeta}$ and $n_{\xi\zeta}$, the electric currents J_ξ can be expressed as

$$J_\xi(x, y, z) = \sum_{m=1}^{N_x+i_{\xi x}} \sum_{n=1}^{N_y+i_{\xi y}} \sum_{k=1}^{N_z+i_{\xi z}} J_\xi^D(m, n, k) f^\xi_{mnk}(x, y, z) \quad (8.81)$$

Notice that the discrete functions $J_\xi^D(m, n, k)$ should be zero when the basis function is located outside the actual dielectric object. From (8.78)–(8.80), one easily obtains the derivatives of J_ξ:

$$\frac{\partial J_\xi}{\partial \xi} = \frac{1}{\Delta \xi} \sum_{m=1}^{N_x+1} \sum_{n=1}^{N_y+1} \sum_{k=1}^{N_z+1} \left[J_\xi^D(m, n, k) \right.$$
$$\left. - J_\xi^D(m - n_{\xi x}, n - n_{\xi y}, k - n_{\xi z}) \right] P_{mnk}(x, y, z) \quad (8.82)$$

which consists of two adjacent 3D pulses with opposite amplitude, representing two opposite electric charges. Here, the 3D pulse function P_{mnk} is defined as

$$P_{mnk}(x,y,z) = P_{mn}(x,y)P_0[z - (k - 0.5)\Delta z]$$

Similarly, the derivatives of the testing functions are also two adjacent 3D pulses:

$$\frac{\partial w_{mnk}^{\xi}}{\partial \xi} = \frac{1}{\Delta \xi}\left[P_{mnk} - P_{(m+n_{\xi x})(n+n_{\xi y})(k+n_{\xi z})}\right] \quad (8.83)$$

Substituting (8.81) and (8.82) into the EFIE and testing with the weighting functions w_{mnk}^{ξ} yields

$$\langle \gamma J_\xi, w_{mnk}^{\xi}\rangle - \langle \mathcal{L}_\xi J_\xi, w_{mnk}^{\xi}\rangle - \sum_{\zeta=x,y,z}\langle \mathcal{L}_{\xi\zeta} J_\zeta, w_{mnk}^{\xi}\rangle = \hat{E}_{b\xi}^{iD}(m,n,k) \quad (8.84)$$

in which $\mathcal{L}_\xi$ and $\mathcal{L}_{\xi\zeta}$ are operator expressions of the second to fifth terms in the integral equations (8.75) and (8.76), and

$$\hat{E}_{b\xi}^{iD}(m,n,k) = \langle \hat{E}_{b\xi}^{i}, w_{mnk}^{\xi}\rangle \quad (8.85)$$

Using the expression (8.83) and integrating by parts, the inner products in (8.84) are expressed as

$$\langle \gamma J_\xi, w_{mnk}^{\xi}\rangle = \Delta x \Delta y \Delta z \gamma_{mnk}^{\xi} J_\xi^D(m,n,k) \quad (8.86)$$

$$\langle \mathcal{L}_\xi J_\xi, w_{mnk}^{\xi}\rangle = k_b^2 \sum_{m'=1}^{N_x+i_{\xi x}} \sum_{n'=1}^{N_y+i_{\xi y}} \sum_{k'=1}^{N_z+i_{\xi z}} G_\xi^{PR}(m-m', n-n', k, k')$$
$$\cdot J_\xi^D(m',n',k') \quad (8.87)$$

$$\langle \mathcal{L}_{\xi\zeta} J_\zeta, w_{mnk}^{\xi}\rangle = -\frac{1}{\Delta\xi\Delta\zeta} \sum_{m'=1}^{N_x+1} \sum_{n'=1}^{N_y+1} \sum_{k'=1}^{N_z+1}\left[J_\zeta^D(m',n',k')\right.$$
$$\left. - J_\zeta^D(m'-n_{\zeta x}, n'-n_{\zeta y}, k'-n_{\zeta z})\right]$$
$$\cdot \left[G_{\xi\zeta}^{PR}(m-m', n-n', k, k')\right.$$
$$\left. - G_{\xi\zeta}^{PR}(m+n_{\xi x}-m', n+n_{\xi y}-n', k+n_{\xi z}, k')\right] \quad (8.88)$$

in which $G_\xi^{PR} = G_{TE}^{PR}$ and $G_{\xi\zeta}^{PR} = G_{EM}^{PR}$ when $\xi, \zeta = x, y$, $G_z^{PR} = G_{TM}^{PR}$ and $G_{\xi z}^{PR} = G_{z\xi}^{PR} = G_{TM}^{PR1}$. Here, the superscript PR indicates the summation of the primary-field part and the reflected-field part:

$$G_{TE,TM,EM}^{PR} = G^P(m-m', n-n', k-k')$$
$$+ G_{TE,TM,EM}^R(m-m', n-n', k+k')$$

$$G_{TM}^{PR1} = G^P(m-m', n-n', k-k')$$
$$- G_{TM}^R(m-m', n-n', k+k')$$

The discrete Green's functions G^P and $G^R_{\text{TE,TM,EM}}$ have similar definitions as (8.66) and (8.67) [65].

8.3.3.3 CG-FFT Algorithm

In order to use FFT to compute the inner products (8.87) and (8.88), the discrete Green's functions G^P, $G^R_{\text{TE,TM,EM}}$, and discrete electric current J^D_ξ must be extended to $2(N_x+1) \times 2(N_y+1) \times 2(N_z+1)$ from $(N_x+1) \times (N_y+1) \times (N_z+1)$, through the relations similar to (8.41) and (8.42). After some complicated derivations, (8.84) can be written as

$$-\mathcal{F}^{-1}\left\{\sum_{\zeta=x,y,z}\left[\tilde{G}^P_{\xi\zeta}(i,j,l)\tilde{J}^{De}_\zeta(i,j,l) + \tilde{G}^R_{\xi\zeta}(i,j,l)\tilde{J}^{De}_\zeta(i,j,-l)\right]\right\}$$
$$+ \Delta V \gamma^\xi_{mnk} J^D_\xi(m,n,k) = \hat{E}^{iD}_{b\xi}(m,n,k) \tag{8.89}$$

where

$$\tilde{G}^P_{\xi\xi}(i,j,l) = \left(k_b^2 + \frac{F_{\xi\xi}}{\Delta\xi^2}\right)\tilde{G}^{Pe}(i,j,l) \tag{8.90}$$

$$\tilde{G}^P_{\xi\zeta}(i,j,l) = \frac{F_{\xi\zeta}}{\Delta\xi\Delta\zeta}\tilde{G}^{Pe}(i,j,l), \quad (\xi \neq \zeta) \tag{8.91}$$

$$\tilde{G}^R_{\xi\xi}(i,j,l) = k_b^2\tilde{G}^{Re}_{\text{TE}}(i,j,l) + \frac{F_{\xi\xi}}{\Delta\xi^2}\tilde{G}^{Re}_{\text{EM}}(i,j,l), \quad (\xi = x,y) \tag{8.92}$$

$$\tilde{G}^R_{zz}(i,j,l) = \left(k_b^2 - \frac{F_{zz2}}{\Delta z^2}\right)\tilde{G}^{Re}_{\text{TM}}(i,j,l) \tag{8.93}$$

$$\tilde{G}^R_{\xi\zeta}(i,j,l) = -\frac{F_{\xi\zeta}}{\Delta\xi\Delta\zeta}\tilde{G}^{Re}_{\text{EM}}(i,j,l), \quad (\xi\zeta = xy, yx) \tag{8.94}$$

$$\tilde{G}^R_{\xi\zeta}(i,j,l) = -\frac{F_{\xi\zeta 2}}{\Delta\xi\Delta\zeta}\tilde{G}^{Re}_{\text{TM}}(i,j,l), \quad (\xi\zeta = xz, yz, zx, zy) \tag{8.95}$$

in which $\tilde{G}^{Pe,Re}(i,j,l)$ and $\tilde{J}^{De}_\zeta(i,j,l)$ are the DFTs of $G^{Pe,Re}(m,n,k)$ and $J^{De}_\xi(m,n,k)$, respectively; $F_{\xi\zeta}$ has been defined before; and

$$F_{\xi z2} = F_{z\xi 2} = (1 - f^*_\xi)(f^*_z - 1)$$

Similarly, the adjoint operator of (8.89) can be obtained, which is expressed as

$$\mathcal{L}^a J = \Delta V \gamma^{\xi *}_{mnk} J^D_\xi(m,n,k)$$

$$-\mathcal{F}^{-1}\left\{\sum_{\zeta=x,y,z}\left[\tilde{G}^{Pa}_{\xi\zeta}(i,j,l)\tilde{J}^{De}_\zeta(i,j,l) + \tilde{G}^{Ra}_{\xi\zeta}(i,j,l)\tilde{J}^{De}_\zeta(i,j,-l)\right]\right\} \tag{8.96}$$

where

$$\tilde{G}^{Pa}_{\xi\xi}(i,j,l) = \left(k_b^{2*} + \frac{F_{\xi\xi}}{\Delta\xi^2}\right)\tilde{G}^{Pe*}(i,j,l) \tag{8.97}$$

$$\tilde{G}^{Pa}_{\xi\zeta}(i,j,l) = \frac{F_{\xi\zeta}}{\Delta\xi\Delta\zeta}\tilde{G}^{Pe*}(i,j,l), \quad (\xi \neq \zeta) \tag{8.98}$$

$$\tilde{G}^{Ra}_{\xi\xi}(i,j,l) = k_b^{2*}\tilde{G}^{Re*}_{\text{TE}}(i,j,-l) + \frac{F_{\xi\xi}}{\Delta\xi^2}\tilde{G}^{Re*}_{\text{EM}}(i,j,-l),$$
$$(\xi = x, y) \tag{8.99}$$

$$\tilde{G}^{Ra}_{zz}(i,j,l) = \left(k_b^{2*} - \frac{F_{zz2}}{\Delta z^2}\right)\tilde{G}^{Re*}_{\text{TM}}(i,j,-l) \tag{8.100}$$

$$\tilde{G}^{Ra}_{\xi\zeta}(i,j,l) = \frac{F_{\xi\zeta}}{\Delta\xi\Delta\zeta}\tilde{G}^{Re*}_{\text{EM}}(i,j,-l), \quad (\xi\zeta = xy, yx) \tag{8.101}$$

$$\tilde{G}^{Ra}_{\xi z}(i,j,l) = \frac{F^*_{\xi z}}{\Delta\xi\Delta z}\tilde{G}^{Re*}_{\text{TM}}(i,j,-l), \quad (\xi = x, y) \tag{8.102}$$

$$\tilde{G}^{Ra}_{z\xi}(i,j,l) = \frac{F_{z\xi}}{\Delta\xi\Delta z}\tilde{G}^{Re*}_{\text{TM}}(i,j,-l), \quad (\xi = x, y) \tag{8.103}$$

Substituting (8.89) and (8.96) into the CG iteration procedure similar to (8.45)–(8.53), the CG-FFT algorithm for 3D dielectric objects can be easily performed.

8.4 DETECTION OF BURIED OBJECTS USING FORWARD METHOD

In this section, we investigate the possibility of detecting buried objects using the fast forward methods and a very early time electromagnetic (VETEM) system.

8.4.1 VETEM System

The very early time electromagnetic system is a product of the VETEM program, whose broad goal is to enhance the state of the art in electromagnetic methods as applied to environmental problems at sites in the Department of Energy (DOE) complex [66–68]. The VETEM system is a pulsed, time-domain system, as are most ground penetrating radar (GPR) systems. Its potential environmental applications include delineation of the boundaries of waste burial pits and trenches, characterization of the contents of those pits and trenches, measurement of the thickness and assessment of the integrity of clay caps, mapping liquid spill plumes, grout injection monitoring, and so on.

The motivation for developing the VETEM system is to make deeper penetration into electrically conductive ground than GPR, improve resolution in the first 5m

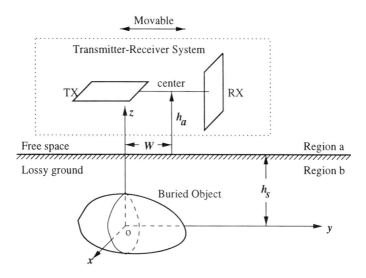

Figure 8.6 A typical VETEM system. The transmitter and receiver are usually orthogonal to avoid receiving direct EM waves.

below the surface, and provide response to dielectric permittivity in an attempt to fill a gap between GPR and available EM instruments. GPR is an excellent tool because it provides high resolution, is fast, and produces data that are often partially interpretable by inspection. However, GPR does not penetrate far through material with high electrical conductivity. High electrical conductivity is, unfortunately, common in waste pits that are often covered with earth materials that have high clay mineral content. Another class of tools, the time-domain EM (TEM) tools, on the other hand, were not designed with very shallow investigations in mind. The VETEM system fills the gap between the TEM and GPR instruments.

Usually, the VETEM system consists of a transmitting loop antenna, TX, and a receiving loop antenna, RX, which run over a lossy ground to detect buried objects. A typical configuration of the VETEM system is shown in Figure 8.6. The source of the system is an input electric current driven by the transmitter, while the measured signal is the output current on the shorted-turn receiving loop at the receiver location.

Numerical modeling of the VETEM system is very important because it provides physical insight, improves interpretation of VETEM field data, and predicts the outcome of a measurement. Next, we introduce two numerical models to simulate the VETEM system.

8.4.2 Numerical Modeling: Loop-Antenna Model

Because the current distributions along transmitting and receiving loops are unknown, the modeling of wire antennas above lossy ground is first studied in this approach to obtain the current distributions.

Many methods have been proposed on the numerical modeling of wire antennas above or inside the lossy ground [4]. In these methods, however, there are limitations. First, point matching is usually used. Second, in most methods the current is assumed to flow on the axis of wire and testing is performed on the surface or vice versa. Third, a delta gap is popularly adopted as the source model, and finally, the input admittance of the wire is simply defined as the ratio of current to voltage at the driving point, which is not variational.

To overcome the above disadvantages, we proposed an accurate model to analyze wire antennas above or inside ground by applying Galerkin's method [5]. In this model, we assumed that:

- The current flows along the surface of the wire;
- The testing points are also on the surface of the wire;
- The current distribution is uniform around the circumference of the wire;
- The current is flowing in the longitudinal direction of the wire;
- The wire radius and the gap at the driving point are electrically small.

In addition, a new source model has been used to replace the delta-gap model. Compared with experimental data, this method gives more accurate results than the old methods [5].

Because the frequency spectrum of the VETEM system can be very low, the accurate model [5] will break down when the working frequency is very low, like most EFIE solvers for EM problems. This is the consequence of the decoupling of electric and magnetic fields at zero frequency. Recently, we have improved the accurate model by introducing the loop-tree basis functions. These bases separate contributions from the vector potential and the scalar potential in the impedance matrix. Then, the contribution from the vector potential will not be swamped by that from the scalar potential after an appropriate frequency normalization. Numerical simulations show that the updated model can provide accurate results in both low and high frequencies.

After the current distribution along the transmitting loop is determined using the wire-antenna model, we can easily obtain the incident electric field of the buried objects from the dyadic Green's function (8.5):

$$\mathbf{E}_b^{\text{inc}}(\mathbf{r}, \mathbf{r}_t, k) = -k\eta_0 \int_C dl' I(l') \overline{\mathbf{G}}_{ee}^{ba}(\mathbf{r}, \mathbf{r}(l'), k) \cdot \hat{\alpha}(l') \qquad (8.104)$$

in which $I(l')$ is the current distribution along the transmitting loop C, and $\hat{\alpha}(l')$ is the current direction. Applying the fast algorithms introduced in the above section, we will get the induced electric current on the conducting plate or inside the dielectric object.

The output electric current measured at the receiving loop antenna is proportional to the total magnetic field perpendicular to the loop. Generally, the total magnetic field is composed of primary field $\mathbf{H}_a^P$ from the transmitter, reflected field $\mathbf{H}_a^R$ from the lossy ground, and scattered field $\mathbf{H}_a^S$ from the buried objects:

$$\mathbf{H}_a(\mathbf{r}_r, k) = \mathbf{H}_a^P(\mathbf{r}_r, k) + \mathbf{H}_a^R(\mathbf{r}_r, k) + \mathbf{H}_a^S(\mathbf{r}_r, k) \qquad (8.105)$$

In the antenna setup shown in Figure 8.6, the transmitter is orthogonal to the receiver. Thus, the contribution from the primary field is zero. However, it is difficult to make RX and TX exactly orthogonal and a slight misalignment can cause a nonnegligible contribution. Hence, we will keep the primary field. For the reflected field, it can be split into two parts:

$$\mathbf{H}_a^R(\mathbf{r}_r, k) = -\mathbf{H}_a^I(\mathbf{r}_r, k) + \mathbf{H}_a^T(\mathbf{r}_r, k) \qquad (8.106)$$

where $\mathbf{H}_a^I$ is imaging field and T represents the transmitted coefficient. From the half-space Green's functions, we have

$$\mathbf{H}_a^{P,I}(\mathbf{r}_r, k) = -\frac{1}{4\pi} \int_C dl' I(l') \overline{\mathbf{G}}_{me}^{P,I}(\mathbf{r}_r, \mathbf{r}(l'), k) \cdot \hat{\alpha}(l') \qquad (8.107)$$

$$\mathbf{H}_a^T(\mathbf{r}_r, k) = \int_C dl' I(l') \overline{\mathbf{G}}_{me}^{aa}(\mathbf{r}_r, \mathbf{r}(l'), k) \cdot \hat{\alpha}(l') \qquad (8.108)$$

$$\mathbf{H}_a^S(\mathbf{r}_r, k) = \int_S d\mathbf{r}' \overline{\mathbf{G}}_{me}^{ab}(\mathbf{r}_r, \mathbf{r}', k) \cdot \mathbf{J}(\mathbf{r}') \qquad (8.109)$$

where $\mathbf{J}(\mathbf{r}')$ is the induced current on the buried scatterer S, $\overline{\mathbf{G}}_{me}^{ab}$ is the dyadic Green's function defined in (8.11), and $\overline{\mathbf{G}}_{me}^{P,I}$ are dyadic Green's functions in Region a:

$$\overline{\mathbf{G}}_{me}^{P}(\mathbf{r}, \mathbf{r}', k) = \frac{1 - ik_a R}{R^3} e^{ik_a R} \begin{bmatrix} 0 & z' - z & y - y' \\ z - z' & 0 & x' - x \\ y' - y & x - x' & 0 \end{bmatrix} \qquad (8.110)$$

$$\overline{\mathbf{G}}_{me}^{I}(\mathbf{r}, \mathbf{r}', k) = \frac{1 - ik_a R_I}{R_I^3} e^{ik_a R_I} \begin{bmatrix} 0 & -z' - z & y - y' \\ z + z' & 0 & x' - x \\ y' - y & x - x' & 0 \end{bmatrix} \qquad (8.111)$$

in which R is the distance from $\mathbf{r}$ to $\mathbf{r}'$, and R_I is the distance from $\mathbf{r}$ to the image of $\mathbf{r}'$ with respect to the air-earth interface. The other dyadic Green's function $\overline{\mathbf{G}}_{me}^{aa}$

represents the magnetic field in Region a produced by electric dipole in Region a

$$\overline{\mathbf{G}}_{me}^{aa}(\mathbf{r},\mathbf{r}',k) = \frac{1}{4\pi^2} \int_{-\infty}^{\infty} \int_{-\infty}^{\infty} dk_x dk_y \widetilde{\overline{\mathbf{G}}}_{me}^{aa} e^{ik_{az}|z+z'|}$$
$$\cdot e^{i[k_x(x-x')+k_y(y-y')]} \tag{8.112}$$

where

$$\widetilde{\overline{\mathbf{G}}}_{me}^{aa} = \frac{1}{k^2} \begin{bmatrix} -k_x k_y g_1 & -k_y^2 g_1 + k^2 g_2 & -\tilde{\epsilon}_b k^2 k_y g_0 \\ k_x^2 g_1 - k^2 g_2 & k_x k_y g_1 & \tilde{\epsilon}_b k^2 k_x g_0 \\ k^2 k_y f_0 & -k^2 k_x f_0 & 0 \end{bmatrix} \tag{8.113}$$

in which $g_1 = (k_{bz} - k_{az})/(\tilde{\epsilon}_a k_{bz} + \tilde{\epsilon}_b k_{az})$ and $g_2 = k_{bz}/(k_{bz} + k_{az})$.

8.4.3 Numerical Modeling: Magnetic-Dipole Model

Since the working frequency of the VETEM system is usually low (from several 10 kHz to several 10 MHz), we can simplify the numerical modeling by replacing the transmitting and receiving loops with magnetic dipoles. For example, in the antenna setup configuration shown in Figure 8.6, the transmitter is a vertical magnetic dipole while the receiver is a horizontal magnetic dipole. The magnetic current on an equivalent dipole has a simple relationship to the loop current I and loop area A: $M\Delta l = i\omega\mu_0 I A$. Hence, there is no need to analyze the transmitting and receiving antennas in this simplified numerical model. However, all formulations for the incident electric field in Region b and magnetic fields in Region a are changed because of the use of magnetic dipole.

From the dyadic Green's function (8.7), the incident electric field in Region b produced by the magnetic dipole is expressed as

$$\mathbf{E}_b^{\text{inc}}(\mathbf{r},\mathbf{r}_t,k) = -i\omega\mu_0 I A \overline{\mathbf{G}}_{em}^{ba}(\mathbf{r},\mathbf{r}_t,k) \cdot \hat{\alpha}_t \tag{8.114}$$

in which $\hat{\alpha}_t$ is the direction of the transmitting magnetic dipole (i.e., the normal direction of the transmitting loop). From the incident electric field and fast algorithms introduced in the above section, the induced electric current on the buried scatterer will be obtained. Therefore, the total magnetic field in Region a is written as

$$\mathbf{H}_a(\mathbf{r}_r,k) = \mathbf{H}_a^P(\mathbf{r}_r,k) - \mathbf{H}_a^I(\mathbf{r}_r,k) + \mathbf{H}_a^T(\mathbf{r}_r,k) + \mathbf{H}_a^S(\mathbf{r}_r,k) \tag{8.115}$$

in which the formulation of $\mathbf{H}_a^S$ is the same as that in (8.109), but $\mathbf{H}_a^P$, $\mathbf{H}_a^I$, and $\mathbf{H}_a^T$ will be different. From the half-space Green's functions, we have

$$\mathbf{H}_a^{P,I}(\mathbf{r}_r,k) = \frac{IA}{4\pi} \overline{\mathbf{G}}_{mm}^{P,I}(\mathbf{r}_r,\mathbf{r}_t,k) \cdot \hat{\alpha}_t \tag{8.116}$$

$$\mathbf{H}_a^T(\mathbf{r}_r,k) = -iIA \overline{\mathbf{G}}_{mm}^{aa}(\mathbf{r}_r,\mathbf{r}_t,k) \cdot \hat{\alpha}_t \tag{8.117}$$

where $\overline{\mathbf{G}}_{mm}^{P,I}$ are the magnetic-field dyadic Green's functions of a magnetic dipole in Region a:

$$\overline{\mathbf{G}}_{mm}^{P}(\mathbf{r},\mathbf{r}',k) = \frac{e^{ik_a R}}{R^5}$$

$$\cdot \begin{bmatrix} (x-x')^2 f_1 - R^2 f_2 & (x-x')(y-y')f_1 & (x-x')(z-z')f_1 \\ (x-x')(y-y')f_1 & (y-y')^2 f_1 - R^2 f_2 & (y-y')(z-z')f_1 \\ (x-x')(z-z')f_1 & (y-y')(z-z')f_1 & (z-z')^2 f_1 - R^2 f_2 \end{bmatrix} \quad (8.118)$$

in which $f_1 = k_a^2 R^2 + i3k_a R - 3$ and $f_2 = k_a^2 R^2 + ik_a R - 1$. Again, R is the distance from $\mathbf{r}$ to $\mathbf{r}'$. A similar expression for $\overline{\mathbf{G}}_{mm}^{I}$ can be obtained by replacing R with R_I and $z - z'$ with $z + z'$. The other dyadic Green's function $\overline{\mathbf{G}}_{mm}^{aa}$ is given by

$$\overline{\mathbf{G}}_{mm}^{aa}(\mathbf{r},\mathbf{r}',k) = \frac{1}{4\pi^2} \int_{-\infty}^{\infty}\int_{-\infty}^{\infty} dk_x dk_y \tilde{\overline{\mathbf{G}}}_{mm}^{aa} e^{ik_{az}|z+z'|}$$

$$\cdot e^{i[k_x(x-x')+k_y(y-y')]} \quad (8.119)$$

which represents the magnetic field in Region a produced by a magnetic dipole in Region a, and

$$\tilde{\overline{\mathbf{G}}}_{mm}^{aa} = f_0 \begin{bmatrix} k_{az}k_{bz} + k_y^2 g_5 & -k_x k_y g_5 & k_x k_{bz} \\ -k_x k_y g_5 & k_{az}k_{bz} + k_x^2 g_5 & k_y k_{bz} \\ k_x k_{az} & k_y k_{az} & k_x^2 + k_y^2 \end{bmatrix} \quad (8.120)$$

in which $g_5 = (\tilde{\epsilon}_b k_{bz} + \tilde{\epsilon}_a k_{az})/(\tilde{\epsilon}_a k_{bz} + \tilde{\epsilon}_b k_{az})$.

Comparing these two numerical models, the first is more complicated, but it is general and valid for all frequencies. The second model is simple while it is only valid for low frequency or small loops. Although the formulations in these two models are completely different, they should give similar simulation results in the low-frequency range.

8.4.4 Simulation Results

The VETEM system used in the simulation is shown in Figure 8.6, where the sizes of transmitting and receiving loops are the same: 0.762m (30 inches) in side length. The distance between central points of transmitting and receiving loops can be varied, but is 2m in this study. The center of each antenna is 0.5334m (21 inches) above the ground. The transmitting loop is horizontally placed, while the receiver is vertical to avoid receiving the direct wave from the transmitter. Hence, the contribution of the primary magnetic field to the received signal is zero. A conducting plate is horizontally buried in the lossy ground.

We set up the Cartesian coordinate system shown in Figure 8.6. The transmitter-receiver system can move from left to right along the y axis, keeping centers of the conducting plate and loop antennas to have the same x coordinate, where W denotes the horizontal distance between the center of the transmitter-receiver system and the center of the buried plate, as illustrated in Figure 8.6. For free space, $\epsilon_a = 1$ and $\sigma_a = 0$. The relative permittivity of lossy ground is assumed to be 16, while the conductivity will change for different cases.

The input signal used in this paper is a ramped pulse. The simulated waveform and its frequency spectrum are displayed in Figure 8.7. From Figure 8.7(b), we clearly see that the bandwidth of the input signal is 5 MHz. In the following examples, all computations are performed in the frequency domain and FFT is used to obtain the time-domain responses.

As we mentioned before, the numerical solution of Maxwell's equations at very low frequency is usually plagued with numerous problems. In the CG-FFT algorithms, the convergence is extremely slow when the working frequency is very low. As an example of a perfectly conducting plate (2m × 2m) in the lossy earth, which is partitioned by 32 × 32 meshes and excited by a vertical magnetic dipole above the earth, it requires 8,188 iterations to make the relative error $\mathcal{E} = 0.001$ when $f = 0.02$ MHz. At the frequency of 5 MHz, it requires 600 iterations to reach the same relative error. If the working frequency is higher, however, the convergence of the CG-FFT algorithm can be very fast. For example, it needs only 35 iterations when $f = 100$ MHz for the same relative error. Hence, it will consume much computer time to obtain an accurate time-domain scattered magnetic field, because the CG-FFT code has to be run in many frequency points. To accelerate the simulation, we propose a frequency-hopping method.

In order to obtain the time-domain scattered magnetic field, the computation is first performed in the frequency domain. We start the CG-FFT simulation at the higher-frequency edge $f_{\max}$ in the frequency spectrum, because it needs fewer iterations for convergence. Using the final current distribution at this frequency as the initial value of the CG-FFT algorithm at the next frequency $f = f_{\max} - \Delta f$, the CG-FFT algorithm will converge rapidly because $\Delta f = f_{\max}/N$ is very small. Here, N is the number of frequency points used to perform the FFT. We continue this procedure until the simulation at the lowest frequency $f = \Delta f$ has been done. Therefore, it requires fewer iterations in the CG-FFT algorithm for all frequencies except the highest one, because very good initial values are provided.

Now, we consider some simulation results. Under the excitation of the input current shown in Figure 8.7, the reflected, scattered, and total magnetic fields at the receiver computed by the two numerical models are illustrated in Figure 8.8 when the ground conductivity $\sigma_b = 0.1$ S/m and the size of conducting plate is 2m × 2m. Here, the center of the transmitter-receiver system is directly above the conducting-plate center ($W = 0$) and the buried depth $h_s = 1$m. From Figure 8.8, we clearly see that the simulation results from the different models are similar, and have less than 5% difference around the peak of waveforms. Because two completely different sets of formulations are used for the primary and reflected fields in the two models, the close agreement of the results is evidence that these models are valid.

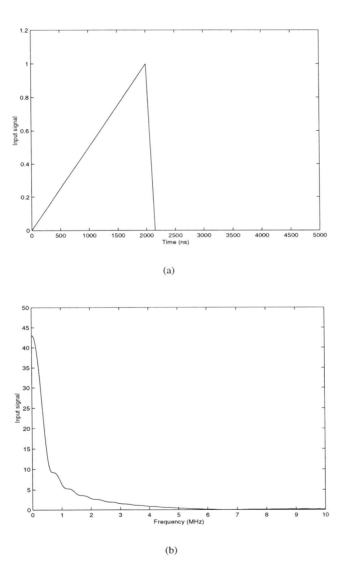

Figure 8.7 Simulated input signal and its frequency spectrum of a typical VETEM system: (a) time-domain wave form; (b) frequency spectrum.

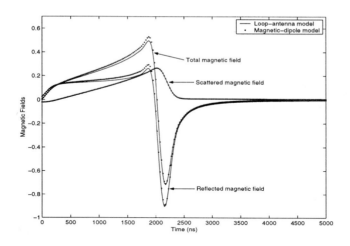

Figure 8.8 Time-domain reflected, scattered, and total magnetic fields at the receiver computed by the loop-antenna model and magnetic-dipole model.

To study the effects of ground conductivity on the received signal, we have computed the reflected field for different conductivity using the magnetic-dipole model, as shown in Figure 8.9. From above numerical results and other simulations, we notice that the reflected response from the earth has the following properties:

- When the ground conductivity increases, the amplitude of earth response increases.

- When the ground conductivity increases, the slope of the ramp in the earth response increases.

- When the ground conductivity is less than 10 S/m, the earth response has a significant tail. When σ_b is larger than 10 S/m, the tail becomes smaller and smaller and disappears at around 50 S/m. After σ_b is larger than 60 S/m, the earth response behaves like that of a perfectly conducting ground.

Therefore, the reflected response is closely related to the ground conductivity, and we can estimate the conductivity from the response.

To test the correctness of the scattered field, we have computed the time-domain responses of different conducting plate sizes when $\sigma_b = 0.1$ S/m, $h_s = 1$m, and $W = 0$, as shown in Figure 8.10. As a reference, we also give the scattered field of a perfectly conducting infinite plane in this figure, which is computed by a closed form in a 1D model. From Figure 8.10, we clearly see that the scattered field becomes

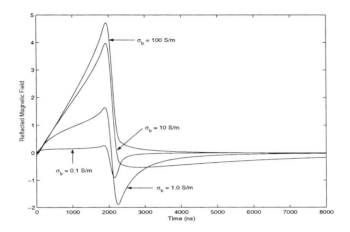

Figure 8.9 Time-domain reflected magnetic fields at the receiver for different earth conductivity.

larger when the plate size increases. As the plate size reaches to 8m × 8m, the scattered field is very close to that of an infinite plane. When the plate size is 16m × 16m, the scattered field is nearly the same as that of an infinite plane. This is because the electric current on the conducting plate induced by the magnetic dipole is concentrated near the dipole. When the conducting plate is large enough, the electric current approaches zero in the region far away from the dipole. Hence, the fitness of scattered fields for a large conducting plate, which is computed by the CG-FFT algorithm, and for infinite conducting plane, which is computed in closed form, shows the validity of the numerical models and CG-FFT algorithm.

Similar to Figure 8.9, Figure 8.11(a) depicts the scattered magnetic field of a 2m × 2m conducting plate buried in different conductivity when $h_s = 1$m and $W = 0$. From this figure, we notice that the scattered field decreases when the earth becomes more lossy. However, the decrease is not significant because the working frequency is low. Hence, it is possible to use the VETEM system to detect buried objects in a very lossy earth. Another important feature of the scattered field response is that the peak of waveform moves to a later time as the ground conductivity increases. As a reference, we also give the scattered field of an infinite conducting plane buried in the same conductivity which is simulated by the closed solution, as shown in Figure 8.11(b). Clearly, a very similar phenomenon occurs for the infinite conducting plane.

Another factor affecting the scattered field is the buried depth. Figure 8.12 displays the time-domain response of the 2m × 2m conducting plate buried in different depth when $\sigma_b = 0.2$ S/m and $W = 0$. From Figure 8.12, we notice that the scattered

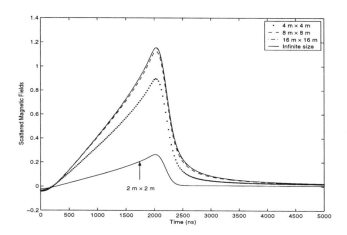

Figure 8.10 Time-domain scattered magnetic fields at the receiver for different conducting plate sizes.

field is very sensitive to the buried depth. Also, when the buried plate is deeper, the waveform peak moves to a later time.

Next, we investigate the spatial-time domain property of the scattered field. When the VETEM system runs over the 2m × 2m conducting plate, the scattered-field responses are illustrated in Figures 8.13 and 8.14 when the earth conductivity is 0.1 S/m and 0.5 S/m, respectively. In these figures, the antenna-location axis represents the distance between the plate center and antenna-system center W. When the antenna center is at the left of the plate center, W is negative, or vice versa. In these figures, (a) and (b) represent the profiles when the buried depth is 0.5m and 2.0m, respectively.

From Figures 8.13 and 8.14, we clearly see that the strongest scattered value occurs around $W = 0$ in either case, where the center of the buried plate is located. Using this property, one can easily estimate the position of the buried target from the spatial-time domain response. The other important property is that the strongest peak value has a positive time shift and the waveform becomes broader as h_s and σ_b increase, from which the buried depth and earth conductivity can be estimated.

8.5 FAST INVERSE SCATTERING METHODS

As discussed in the previous section, fast forward methods can be used to approximate the location of buried objects. However, their shape and dielectric property cannot

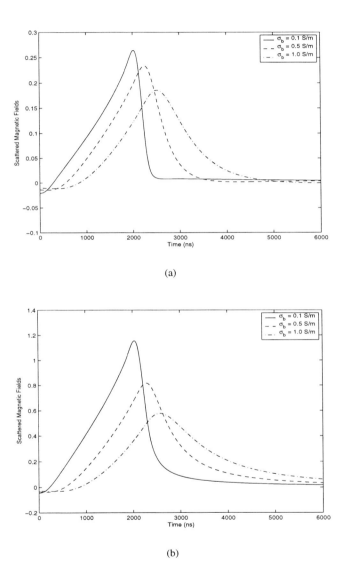

Figure 8.11 Time-domain scattered magnetic fields at the receiver for different earth conductivity: (a) 2m × 2m conducting plate; (b) infinite conducting plane.

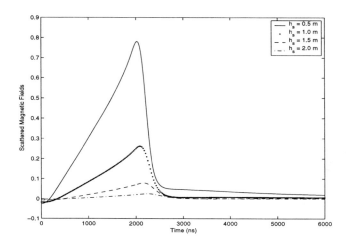

Figure 8.12 Time-domain scattered magnetic fields at the receiver for different buried depths.

be determined. To accurately detect the buried objects, inverse scattering methods have to be used. In this section, we will introduce some inversion algorithms for 2D dielectric cylinders and 3D dielectric objects.

8.5.1 Single-Frequency DBIM Algorithm for 2D Objects

Consider a half-space problem shown in Figure 8.15. Here, 2D dielectric objects are buried in lossy earth. Multiple transmitters TX and receivers RX are located above or on the air-earth interface. When the TX and RX are above the interface, we place them in the same height ($z_t = z_r = z_0$) to reduce the number of Sommerfeld-like integrals.

In inverse scattering, both dielectric properties and shapes of the buried objects are unknown. Hence, we must choose a reconstruction domain $\mathcal{D}$. The domain $\mathcal{D}$ should be large enough so that $\mathcal{D}_i$, the supports of buried objects to be detected, are in $\mathcal{D}$. Hence, the relative permittivity $\epsilon_r(\rho)$ and conductivity $\sigma_r(\rho)$ in the reconstruction domain $\mathcal{D}$ are defined as

$$\epsilon_r(\rho) = \begin{cases} \epsilon_{ri}(\rho), & \rho \in \mathcal{D}_i \\ \epsilon_b, & \text{otherwise,} \end{cases} \qquad \sigma_r(\rho) = \begin{cases} \sigma_{ri}(\rho), & \rho \in \mathcal{D}_i \\ \sigma_b, & \text{otherwise} \end{cases} \qquad (8.121)$$

Conveniently, the shape of $\mathcal{D}$ is chosen as a rectangle to easily implement the forward solver using the CG-FFT.

Fast Forward and Inverse Methods for Buried Objects 383

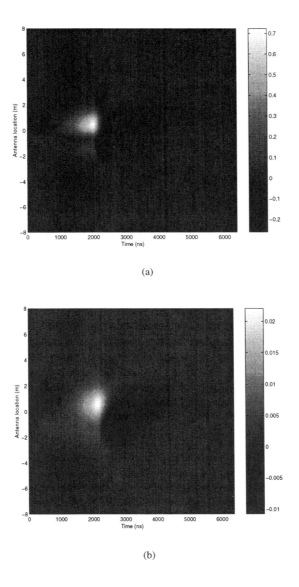

(a)

(b)

Figure 8.13 Spatial-time domain scattered fields of the 2m × 2m conducting plate when $\sigma_b = 0.1$ S/m. The VETEM system moves from left to right in the measurement domain. (a) $h_s = 0.5$m; (b) $h_s = 2.0$m.

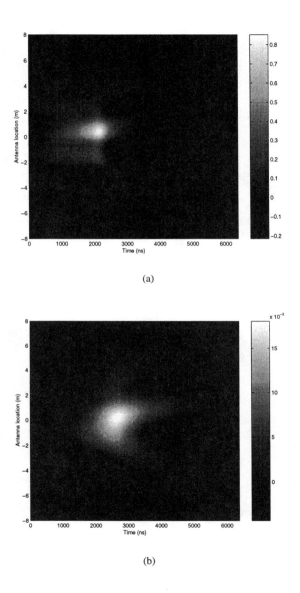

Figure 8.14 Spatial-time domain scattered fields of the 2m × 2m conducting plate when $\sigma_b = 0.5$ S/m. The VETEM system moves from left to right in the measurement domain. (a) $h_s = 0.5$m; (b) $h_s = 2.0$m.

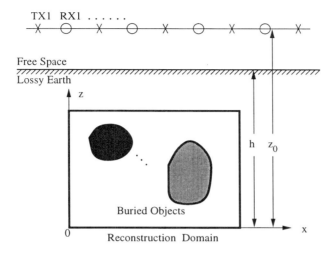

Figure 8.15 Setup configuration of the single-frequency DBIM algorithm. Multiple transmitters and multiple receivers are used.

When the electric current source is a line source, the scattered electric field by buried targets at the receiver location is expressed as

$$E_a^s(\boldsymbol{\rho}_r, \boldsymbol{\rho}_t, \tilde{\epsilon}_r) = E_a^s(\boldsymbol{\rho}_r, \boldsymbol{\rho}_t, \tilde{\epsilon}_{r0})$$
$$+ k^2 \int_{\mathcal{D}} d\boldsymbol{\rho} g_a(\boldsymbol{\rho}_r, \boldsymbol{\rho}, \tilde{\epsilon}_{r0}) E_b(\boldsymbol{\rho}, \boldsymbol{\rho}_t, \tilde{\epsilon}_r) \left[\tilde{\epsilon}_r(\boldsymbol{\rho}) - \tilde{\epsilon}_{r0}(\boldsymbol{\rho}) \right] \quad (8.122)$$

where $E_a^s(\boldsymbol{\rho}_r, \boldsymbol{\rho}_t, \tilde{\epsilon}_r)$ is the scattered electric field at the receivers when the reconstruction domain contains $\tilde{\epsilon}_r$, which can be determined from the measurement data (the total electric field) by subtracting the primary electric field of the transmitter and the reflected electric field from the ground; $E_a^s(\boldsymbol{\rho}_r, \boldsymbol{\rho}_t, \tilde{\epsilon}_{r0})$ is the scattered electric field at the receivers when the reconstruction domain contains the background medium $\tilde{\epsilon}_{r0}$, which can be computed; $E_b(\boldsymbol{\rho}, \boldsymbol{\rho}_t, \tilde{\epsilon}_r)$ is the internal electric field inside the reconstruction domain when it contains $\tilde{\epsilon}_r$, which is generated by the transmitters; $g_a(\boldsymbol{\rho}_r, \boldsymbol{\rho}, \tilde{\epsilon}_{r0})$ is the inhomogeneous medium Green's function when the reconstruction domain contains $\tilde{\epsilon}_{r0}$; and $\tilde{\epsilon}_r(\boldsymbol{\rho})$ and $\tilde{\epsilon}_{r0}(\boldsymbol{\rho})$ are complex permittivity of the reconstructed profile and background profile, which are given as

$$\tilde{\epsilon}_r(\boldsymbol{\rho}) = \epsilon_r(\boldsymbol{\rho}) + i \frac{\eta_0}{k} \sigma_r(\boldsymbol{\rho}), \quad \tilde{\epsilon}_{r0}(\boldsymbol{\rho}) = \epsilon_{r0}(\boldsymbol{\rho}) + i \frac{\eta_0}{k} \sigma_{r0}(\boldsymbol{\rho}) \quad (8.123)$$

The inhomogeneous medium Green's function $g_a(\boldsymbol{\rho}_r, \boldsymbol{\rho}, \tilde{\epsilon}_{r0})$ has a simple relationship with the internal electric field. From the differential equations governing the

Green's function and electric field, we easily obtain

$$g_a(\boldsymbol{\rho}_r, \boldsymbol{\rho}, \tilde{\epsilon}_{r0}) = \frac{1}{ik\eta_0} E_a(\boldsymbol{\rho}_r, \boldsymbol{\rho}, \tilde{\epsilon}_{r0}) \qquad (8.124)$$

where $E_a(\boldsymbol{\rho}_r, \boldsymbol{\rho}, \tilde{\epsilon}_{r0})$ is the total electric field at the receiver location generated by a line source in the reconstruction domain filled by the background medium. Using the reciprocity theorem, the above equation directly gives

$$g_a(\boldsymbol{\rho}_r, \boldsymbol{\rho}, \tilde{\epsilon}_{r0}) = \frac{1}{ik\eta_0} E_b(\boldsymbol{\rho}, \boldsymbol{\rho}_r, \tilde{\epsilon}_{r0}) \qquad (8.125)$$

where $E_b(\boldsymbol{\rho}, \boldsymbol{\rho}_r, \tilde{\epsilon}_{r0})$ is the total (or internal) electric field inside the reconstruction domain filled by the background medium, which is generated by a line source at the receiver location.

Because $E_b(\boldsymbol{\rho}, \boldsymbol{\rho}_t, \tilde{\epsilon}_r)$ is also a function of $\tilde{\epsilon}_r$, (8.122) gives a nonlinear relation between the scattered field and the reconstructed parameters. This equation can be solved using the distorted Born iterative method [45, 69]. With the distorted Born approximation

$$E_b(\boldsymbol{\rho}, \boldsymbol{\rho}_t, \tilde{\epsilon}_r) \simeq E_b(\boldsymbol{\rho}, \boldsymbol{\rho}_t, \tilde{\epsilon}_{r0}) \qquad (8.126)$$

(8.122) can be linearized as

$$e(\boldsymbol{\rho}_r, \boldsymbol{\rho}_t, \tilde{\epsilon}_{r0}) \simeq \frac{k}{i\eta_0} \int_{\mathcal{D}} d\boldsymbol{\rho} E_b(\boldsymbol{\rho}, \boldsymbol{\rho}_r, \tilde{\epsilon}_{r0}) E_b(\boldsymbol{\rho}, \boldsymbol{\rho}_t, \tilde{\epsilon}_{r0})$$
$$\cdot \left[O_\epsilon(\boldsymbol{\rho}) + i\frac{\eta_0}{k} O_\sigma(\boldsymbol{\rho}) \right] \qquad (8.127)$$

where

$$e(\boldsymbol{\rho}_r, \boldsymbol{\rho}_t, \tilde{\epsilon}_{r0}) = E_a^s(\boldsymbol{\rho}_r, \boldsymbol{\rho}_t, \tilde{\epsilon}_r) - E_a^s(\boldsymbol{\rho}_r, \boldsymbol{\rho}_t, \tilde{\epsilon}_{r0}) \qquad (8.128)$$

which can be determined by the measurement data, and

$$O_\epsilon(\boldsymbol{\rho}) = \epsilon_r(\boldsymbol{\rho}) - \epsilon_{r0}(\boldsymbol{\rho}), \quad O_\sigma(\boldsymbol{\rho}) = \sigma_r(\boldsymbol{\rho}) - \sigma_{r0}(\boldsymbol{\rho}) \qquad (8.129)$$

are the object functions, which are real.

Conveniently, the DBIM algorithm starts at $\epsilon_{r0} = \epsilon_b$ and $\sigma_{r0} = \sigma_b$. Numerical examples show that this starting point can make a fast convergence of the DBIM scheme. Actually, the starting point can be different, for example, the free-space parameters: $\epsilon_{r0} = 1$ and $\sigma_{r0} = 0$. However, more iterations are needed in this case for convergence of the DBIM algorithm. Solving the linear integral equation (8.127), one obtains ϵ_r and σ_r. Then, one updates ϵ_{r0} and σ_{r0} by the solved ϵ_r and σ_r, and repeats the procedure until

$$\frac{\|e(\boldsymbol{\rho}_r, \boldsymbol{\rho}_t, \tilde{\epsilon}_{r0})\|^2}{\|E_a^s(\boldsymbol{\rho}_r, \boldsymbol{\rho}_t, \tilde{\epsilon}_r)\|^2} < \text{tolerance} \qquad (8.130)$$

Therefore, the key step in the DBIM algorithm is in solving the linearized equation (8.127). In this section, we will use single-frequency information. At single frequency, the real object functions $O_\epsilon(\rho)$ and $O_\sigma(\rho)$ can be combined into a complex object function

$$O(\rho) = O_\epsilon(\rho) + i\frac{\eta_0}{k}O_\sigma(\rho) \tag{8.131}$$

The integral equation (8.127) is defined in an infinite-dimensional space. To solve the problem using a computer, the problem is approximated within a finite-dimensional space. Suppose that the reconstruction domain $\mathcal{D}$ is approximated by $N = N_x N_z$ pixels, and the transmitter domain $\mathcal{D}_t$ and receiver domain $\mathcal{D}_r$ are approximated by N_t transmitter locations and N_r receiver locations; then (8.127) can be rewritten as a matrix form

$$\overline{\mathbf{M}} \cdot \mathbf{O} = \mathbf{e} \tag{8.132}$$

where $\mathbf{e}$ is a vector indexed by the transmitter and receiver locations, $\mathbf{O}$ is a vector indexed by the pixels required to model the reconstruction domain, and $\overline{\mathbf{M}}$ is a Fréchet derivative operator that maps $\mathbf{O}$ to $\mathbf{e}$, because the error in $\mathbf{e}$ is of higher order than that in $\mathbf{O}$. Here, the Fréchet derivative operator $\overline{\mathbf{M}}$ is not one-to-one, but many-to-one. Hence, $\overline{\mathbf{M}}$ does not have a unique inverse and (8.132) is ill posed. In order to find an adequate solution to (8.132), the regularization procedure is employed to circumvent the instability of the problem. Therefore, a minimum norm solution to (8.132) is sought by an optimization scheme: Find an $\mathbf{O}$ that will minimize the function

$$\mathcal{E} = \|\mathbf{e} - \overline{\mathbf{M}} \cdot \mathbf{O}\|^2 + \gamma \|\mathbf{O}\|^2 \tag{8.133}$$

in which an L_2 norm has been assumed and the second term on the right-hand side of (8.133) is to ensure that the norm of $\mathbf{O}$ is not too large with the regularization parameter γ. After minimizing (8.133), one obtains the following matrix equation:

$$(\overline{\mathbf{M}}^\dagger \cdot \overline{\mathbf{M}} + \gamma \overline{\mathbf{I}}) \cdot \mathbf{O} = \overline{\mathbf{M}}^\dagger \cdot \mathbf{e} \tag{8.134}$$

where $\overline{\mathbf{I}}$ is the unit vector and $\overline{\mathbf{M}}^\dagger$ is the conjugate transpose of $\overline{\mathbf{M}}$. The above equation can be solved by directly inverting the matrix $\overline{\mathbf{M}}^\dagger \cdot \overline{\mathbf{M}} + \gamma \overline{\mathbf{I}}$. In this direct solver, the choice of the regularization parameter γ is very important. It must be properly balanced so that it is large enough to filter out unstable components to obtain a stable solution, but not too large to avoid filtering out too many useful-frequency components in the solution.

However, the direct solver is too expensive and needs more CPU time and storage requirement when the pixel number N is large. Therefore, we use the CG method to solve (8.134). Generally, a CG method seeks the minimum norm solution even if the matrix is ill-conditioned, if the initial guess of the solution is not polluted by a vector in the null space of the ill-conditioned matrix. Hence, the regularization parameter

γ can, in principle, be chosen as zero in the CG procedure, provided the null space of the matrix is investigated at each iteration. In this paper, we choose $\gamma = 10^{-10}$ when we use the CG method to solve (8.134).

As stated above, the problem is nonunique, and a unique solution is obtained by defining a minimum norm solution. The resulting matrix equation (8.134) can now be ill-conditioned, which means that an iterative solver can stop far from the solution because of the presence of large flat zones in the quadratic functional equation (8.133) to be minimized. Therefore, a stable solution can be obtained by appropriately stopping the iterations.

When the CG method is used to solve (8.134), the most time-consuming part at each CG iteration is the matrix-vector multiplication $\overline{\mathbf{M}} \cdot \mathbf{X}$ and $\overline{\mathbf{M}}^\dagger \cdot \mathbf{Y}$. From the integral expression (8.127) and the definition of adjoint operator, one can easily obtain

$$\left(\overline{\mathbf{M}} \cdot \mathbf{X}\right)_m = \frac{k}{i\eta_0} \Delta x \Delta y \sum_{n=1}^{N_x N_y} E_b(\rho_n, \rho_{rm}, \tilde{\epsilon}_{r0})$$
$$\cdot E_b(\rho_n, \rho_{tm}, \tilde{\epsilon}_{r0}) X(\rho_n) \qquad (8.135)$$

$$\left(\overline{\mathbf{M}}^\dagger \cdot \mathbf{Y}\right)_n = k^2 \Delta x_t \Delta x_r \sum_{m=1}^{N_t N_r} g_a^*(\rho_n, \rho_{rm}, \tilde{\epsilon}_{r0}) E_b^*(\rho_n, \rho_{tm}, \tilde{\epsilon}_{r0})$$
$$\cdot Y(\rho_{rm}, \rho_{tm}, \tilde{\epsilon}_{r0}) \qquad (8.136)$$

where, $\left(\overline{\mathbf{M}} \cdot \mathbf{X}\right)_m$ is the mth element of the $N_t N_r$-ordered vector $\overline{\mathbf{M}} \cdot \mathbf{X}$, and $\left(\overline{\mathbf{M}}^\dagger \cdot \mathbf{Y}\right)_n$ is the nth element of the N-ordered vector $\overline{\mathbf{M}}^\dagger \cdot \mathbf{Y}$. In the adjoint operation (8.136), * denotes the complex conjugate, and $g_a(\rho_n, \rho_{rm}, \tilde{\epsilon}_{r0})$ represents the back-propagated field in the reconstruction domain generated by the point source at receiver location, which satisfies

$$g_a(\rho_n, \rho_{rm}, \tilde{\epsilon}_{r0}) = \frac{1}{ik\eta_0} E_b(\rho_n, \rho_{rm}, \tilde{\epsilon}_{r0}) \qquad (8.137)$$

Because the problem (8.132) is an approximation of the continuous problem (8.127), a Riemann sum has been used in the computation of $\overline{\mathbf{M}} \cdot \mathbf{X}$ over the "pixellated" domain $\mathcal{D}$ and $\overline{\mathbf{M}}^\dagger \cdot \mathbf{Y}$ over the transmitter-receiver locations which are in $\mathcal{D}_t$ and $\mathcal{D}_r$. The transmitter-receiver locations should be sufficiently dense over the continuous domains $\mathcal{D}_t$ and $\mathcal{D}_r$. Generally, a large measurement domain with dense transmitter-receiver grid can improve the spatial resolution of the reconstruction, but it is expensive and computationally slow. Hence, we usually choose a truncated measurement domain, which is about four or five times of the reconstruction domain, and we use a relatively loose transmitter-receiver location grid to reduce the cost and numerical complexity.

Once the complex object function is solved by the CG algorithm, both the permittivity and conductivity in the reconstruction domain can be reconstructed:

$$\epsilon_r(\rho) = \epsilon_{r0}(\rho) + \Re e\left[O(\rho)\right], \quad \sigma_r(\rho) = \sigma_{r0}(\rho) + \frac{k}{\eta_0}\Im m\left[O(\rho)\right] \quad (8.138)$$

From (8.135) and (8.136), we clearly see that only one call to the forward solver is needed at each transmitter location and another call is needed at each receiver location in the whole CG procedure, to compute the internal electric fields $E_b(\rho, \rho_t, \tilde{\epsilon}_{r0})$ and $E_b(\rho, \rho_r, \tilde{\epsilon}_{r0})$. If the transmitter and receiver share the same location, the total calling number reduces to one. Hence, the computational complexity to solve the inverse problem is greatly reduced. The other merit of using the CG algorithm to solve (8.134) is that we do not need to store the Fréchet derivative matrix, but only $E_b(\rho, \rho_t, \tilde{\epsilon}_{r0})$ and $E_b(\rho, \rho_r, \tilde{\epsilon}_{r0})$, which are much smaller than $\overline{\mathbf{M}}$.

Now we see some reconstruction examples using this algorithm where the lower region is a wet soil ($\epsilon_b = 4$ and $\sigma_b = 0.005$ S/m). Unless otherwise specified, the reconstruction domain $\mathcal{D}$ is a 2×2 m² square region, whose bottom side is 2.25m below the air-soil interface. This reconstruction domain contains $32 \times 32 = 1{,}024$ pixels. The measurement domain is 8m on the air-soil interface centered about $\mathcal{D}$, and 16 transmitter locations and 16 receiver locations are used. The scattered data are simulated using the CG-FFT algorithm by adding some random error.

First, we see the effect of changing the working frequency on the image resolution. The original object profiles to be reconstructed are shown in the left side of Figure 8.16, where two off-set square dielectric objects ($\epsilon_r = 10$ and $\sigma_r = 0.01$ S/m) are buried in the soil. When the working frequency is 10 MHz, the inversion results after five DBIM iterations from noiseless data are illustrated in the right side of Figure 8.16, which shows that the two objects can be clearly detected and the contrast of permittivity and conductivity can be well evaluated. However, the image resolution of the lower object is poor.

When the working frequency increases to 50 MHz, the reconstructed profiles after the first and fifth iterations from noiseless data are displayed in Figure 8.17. From Figure 8.17, the contrast of reconstructed profiles after the first iteration (Born's approximation) is not accurate. In the permittivity profile, the image of the lower object has a long tail. In the conductivity profile, however, the image of the lower object is not clear. After five DBIM iterations, the image quality of both permittivity and conductivity profiles has been notably improved: the long tail in the permittivity image has disappeared, and the lower object is much clearer in the conductivity image. Also, the contrast of reconstructed profiles is much more accurate than that in the Born approximation.

Comparing Figure 8.17 with Figure 8.16, we clearly see that the image resolution of both objects at higher frequencies is much better than that at lower frequencies, especially for the conductivity profile. However, when the working frequency is much higher, the reconstruction domain will be smaller and the EM wave cannot

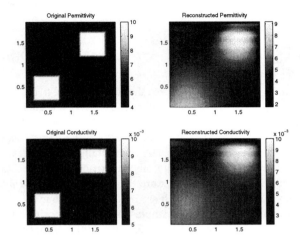

Figure 8.16 The original and reconstructed profiles of two offset dielectric objects when the working frequency is 10 MHz. A single-frequency DBIM algorithm with 16 transmitters and 16 receivers is used. Left side: original profiles. Right side: reconstructed profiles.

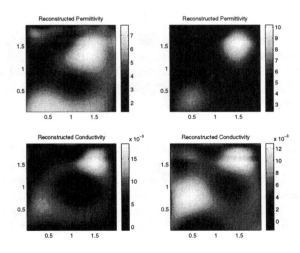

Figure 8.17 Reconstructed profiles of two offset dielectric objects when the working frequency is 50 MHz. A single-frequency DBIM algorithm with 16 transmitters and 16 receivers is used. Left side: the first iteration. Right side: the fifth iteration.

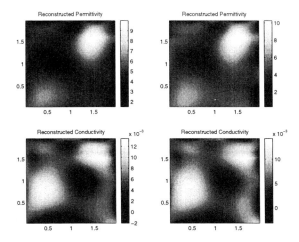

Figure 8.18 Reconstructed profiles of two offset dielectric objects from noisy scattered data. A single-frequency DBIM algorithm with 16 transmitters and 16 receivers is used. Left side: 5% random error. Right side: 10% random error.

penetrate deeper. In the following examples, the working frequencies are all set to 50 MHz.

To show the stability of the proposed DBIM algorithm, we next consider a reconstruction example from noisy data. When 5% and 10% random error have been added to the scattered field collected on the air-earth interface, the reconstructed results are illustrated in Figure 8.18 after five DBIM iterations. Comparing Figure 8.18 with Figure 8.17, we find that the noisy data can still provide a very good resolution of the image, although the contrast of permittivity and conductivity becomes less accurate and there is a stronger ghost image at the right-lower corner as the noise increases. Hence, this algorithm is very tolerant of the measurement error.

Finally, we consider a larger reconstruction domain: a 4×4 m^2 region, whose bottom side is 4.25m below the air-soil interface. This reconstruction domain is divided into $64 \times 64 = 4,096$ pixels. The measurement domain is 16m on the air-earth interface centered about the reconstruction domain, and 32 transmitter locations and 32 receiver locations are used. The original profiles to be reconstructed are shown in the left side of Figure 8.19. When the scattered fields contain 5% random error, the inversion results after 5 DBIM iterations are displayed in the right side of Figure 8.19. From both permittivity and conductivity profiles, it is evident that the two buried objects can be clearly detected although more pixels have been used and the objects are not at the center of the reconstruction domain.

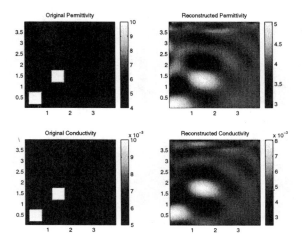

Figure 8.19 The original and reconstructed profiles of two offset dielectric objects in a large reconstruction domain. A single-frequency DBIM algorithm with 32 transmitters and 32 receivers is used. Left side: original profiles. Right side: reconstructed profiles.

8.5.2 Multifrequency DBIM Algorithm for 2D Objects

The single-frequency DBIM algorithm provides an efficient way to accurately detect buried objects. However, this algorithm requires an array of multiple transmitters and multiple receivers, which is expensive to build. In this subsection, multifrequency information is used. Thus, we need only a single transmitter and a receiver which have a fixed offset d, as shown in Figure 8.20. Hence, $z_r = z_t = z_0$ and $x_r = x_t + d$. The transmitter-receiver system is moveable in the measurement plane $x_t \in \mathcal{D}_t$, to make multiple transmitter-receiver locations. The working frequency of the system ranges from $f_{\min}$ to $f_{\max}$.

Using similar formulations and procedures to those in the single-frequency algorithm, we can set up the multifrequency DBIM algorithm by solving a different linearized equation. In the multifrequency case, the linearized equation (8.127) can be written in the following operator form [70]:

$$\overline{\mathbf{M}}_\epsilon \cdot \mathbf{O}_\epsilon + \overline{\mathbf{M}}_\sigma \cdot \mathbf{O}_\sigma = \mathbf{e} \tag{8.139}$$

where $\overline{\mathbf{M}}_\epsilon$ and $\overline{\mathbf{M}}_\sigma$ are the Fréchet derivative operators and $\overline{\mathbf{M}}_\sigma = i\overline{\mathbf{M}}_\epsilon/k$. To solve (8.139), we consider the following regularization equation:

$$\mathcal{E} = \|\mathbf{e} - \overline{\mathbf{M}}_\epsilon \cdot \mathbf{O}_\epsilon - \overline{\mathbf{M}}_\sigma \cdot \mathbf{O}_\sigma\|_A^2 + \delta_\epsilon \|\mathbf{O}_\epsilon\|_{A_0}^2 + \delta_\sigma \|\mathbf{O}_\sigma\|_{A_0}^2 \tag{8.140}$$

Fast Forward and Inverse Methods for Buried Objects   393

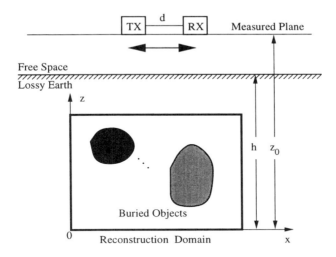

Figure 8.20 Setup configuration of the multifrequency DBIM algorithm. The transmitter and receiver have a fixed offset.

where $\|\mathbf{a}\|_A$ is A-norm, which is defined as $\|\mathbf{a}\|_A^2 = \mathbf{a}^\dagger \cdot \overline{\mathbf{A}} \cdot \mathbf{a}$.

To minimize (8.140), we set the derivatives of $\mathcal{E}$ with respect to $\mathbf{O}_\epsilon$ and $\mathbf{O}_\sigma$ to be zero. Then one obtains the following operator equation:

$$\begin{bmatrix} \overline{\mathbf{M}}_\epsilon^\dagger \cdot \overline{\mathbf{A}} \cdot \overline{\mathbf{M}}_\epsilon + \delta_\epsilon \overline{\mathbf{A}}_0 & \overline{\mathbf{M}}_\epsilon^\dagger \cdot \overline{\mathbf{A}} \cdot \overline{\mathbf{M}}_\sigma \\ \overline{\mathbf{M}}_\sigma^\dagger \cdot \overline{\mathbf{A}} \cdot \overline{\mathbf{M}}_\epsilon & \overline{\mathbf{M}}_\sigma^\dagger \cdot \overline{\mathbf{A}} \cdot \overline{\mathbf{M}}_\sigma + \delta_\sigma \overline{\mathbf{A}}_0 \end{bmatrix} \cdot \begin{bmatrix} \mathbf{O}_\epsilon \\ \mathbf{O}_\sigma \end{bmatrix}$$

$$= \begin{bmatrix} \overline{\mathbf{M}}_\epsilon^\dagger \cdot \overline{\mathbf{A}} \cdot \mathbf{e} \\ \overline{\mathbf{M}}_\sigma^\dagger \cdot \overline{\mathbf{A}} \cdot \mathbf{e} \end{bmatrix} \qquad (8.141)$$

If $\overline{\mathbf{A}}$ and $\overline{\mathbf{A}}_0$ are Hermitian matrices, then the operator

$$\begin{bmatrix} \overline{\mathbf{M}}_\epsilon^\dagger \cdot \overline{\mathbf{A}} \cdot \overline{\mathbf{M}}_\epsilon + \delta_\epsilon \overline{\mathbf{A}}_0 & \overline{\mathbf{M}}_\epsilon^\dagger \cdot \overline{\mathbf{A}} \cdot \overline{\mathbf{M}}_\sigma \\ \overline{\mathbf{M}}_\sigma^\dagger \cdot \overline{\mathbf{A}} \cdot \overline{\mathbf{M}}_\epsilon & \overline{\mathbf{M}}_\sigma^\dagger \cdot \overline{\mathbf{A}} \cdot \overline{\mathbf{M}}_\sigma + \delta_\sigma \overline{\mathbf{A}}_0 \end{bmatrix}$$

is easily shown to be Hermitian.

When a CG method is used to solve the operator equation (8.141), three kinds of operations are involved in the CG algorithm: $\overline{\mathbf{M}}_\epsilon \cdot \mathbf{X}$, $\overline{\mathbf{M}}_\epsilon^\dagger \cdot \overline{\mathbf{A}} \cdot \mathbf{Y}$, and $\overline{\mathbf{M}}_\sigma^\dagger \cdot \overline{\mathbf{A}} \cdot \mathbf{Y}$.

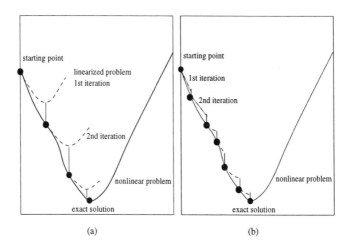

Figure 8.21 Implementations of the multifrequency DBIM algorithm. (a) Accurate solution of the linearized equation. Fewer DBIM iterations are needed. (b) Approximate solution of the linearized equation. More DBIM iterations are needed.

The evaluation of $\overline{\mathbf{M}}_\epsilon \cdot \mathbf{X}$ is quite similar to (8.135), and

$$\overline{\mathbf{M}}_\epsilon^\dagger \cdot \overline{\mathbf{A}} \cdot \mathbf{Y} = \frac{i}{\eta_0} \int_{k_{\min}}^{k_{\max}} dk \cdot k \int_{x_{t\min}}^{x_{t\max}} dx_t E_b^*(\boldsymbol{\rho}, \boldsymbol{\rho}_r, k, \tilde{\epsilon}_{r0})$$
$$\cdot E_b^*(\boldsymbol{\rho}, \boldsymbol{\rho}_t, k, \tilde{\epsilon}_{r0}) Y(\boldsymbol{\rho}_t, k, \tilde{\epsilon}_{r0}) A(\boldsymbol{\rho}_t, k) \qquad (8.142)$$

$$\overline{\mathbf{M}}_\sigma^\dagger \cdot \overline{\mathbf{A}} \cdot \mathbf{Y} = \frac{1}{\eta_0} \int_{k_{\min}}^{k_{\max}} dk \int_{x_{t\min}}^{x_{t\max}} dx_t E_b^*(\boldsymbol{\rho}, \boldsymbol{\rho}_r, k, \tilde{\epsilon}_{r0})$$
$$\cdot E_b^*(\boldsymbol{\rho}, \boldsymbol{\rho}_t, k, \tilde{\epsilon}_{r0}) Y(\boldsymbol{\rho}_t, k, \tilde{\epsilon}_{r0}) A(\boldsymbol{\rho}_t, k) \qquad (8.143)$$

In trading off between the computer time and storage requirement, different approaches can be used to implement the DBIM algorithm. Two typical methods are introduced below:

Implementation A: The schematic outline of the first implementation method is shown in Figure 8.21(a). Starting from the initial guess of the permittivity and conductivity, the linearized operator equation (8.141) is accurately solved using the CG algorithm. Then, the accurate CG solution acts as the starting point of the next iteration and the procedure is repeated.

When the operator equation is accurately solved, the integrals (8.135), (8.142), and (8.143) need to be repeatedly evaluated. Hence, the internal fields $E_b(\rho, \rho_t, k, \tilde{\epsilon}_{r0})$ and $E_b(\rho, \rho_r, k, \tilde{\epsilon}_{r0})$ have to be stored. Then, only one call to the forward solver is needed at each transmitter location and another call is needed at each receiver location for each frequency in the whole CG procedure. That is to say, totally $2N_f N_t$ calls to the forward solver are required to perform the CG algorithm, where N_f is the number of working frequency, and N_t is the number of transmitter-receiver locations.

The merit of this implementation is that only a few DBIM iterations are needed and small computational complexity is required in each DBIM iteration. However, we have to store the internal fields for each transmitter location and frequency.

Implementation B: For a large problem, it is impossible to store the internal fields for all transmitter locations and frequencies. In this case, the implementation method shown in Figure 8.21(b) should be used: we do not solve the linearized operator equation accurately and use only two or three CG iterations. Then, the approximate CG solution acts as the starting point of the next DBIM iteration and the procedure is repeated. Clearly, the DBIM iteration number of this implementation will be larger than that of Implementation A.

Because only a few CG iterations are used, it is not necessary to store the internal fields for each transmitter-receiver location and frequency. But two calls to the forward solver are needed to perform the operation (8.135), and another two calls are needed to perform the adjoint operation (8.142) or (8.143) at each transmitter-receiver location and each frequency for each CG iteration. Therefore, totally $4N_{\text{CG}}^{\text{iter}} N_f N_t$ calls to the forward solver are required to perform the CG algorithm, where $N_{\text{CG}}^{\text{iter}}$ is the CG iteration number.

Compared with the single-frequency DBIM algorithm, both the computational complexity and storage requirement in this algorithm have increased due to the use of multiple frequencies. However, this algorithm is of more practical interest because the data collection is easier and cheaper.

Now, we consider some reconstruction examples using this algorithm. In these examples, the lower region consists of wet soil ($\epsilon_b = 4$ and $\sigma_b = 0.005$ S/m) and the reconstruction domain $\mathcal{D}$ is a 2×2 m^2 region, whose bottom side is 2.25m below the air-soil interface. This reconstruction domain is divided into $32 \times 32 = 1{,}024$ pixels. The measurement plane is 0.5m above the air-soil interface, and 15 transmitter-receiver locations are used. The distance between the transmitter and receiver is fixed to 2m. The measurement length is 10m centered about the reconstruction domain. The working frequency ranges from 0.5 to 50.5 MHz and 11 frequency points are taken. For the parameters contained in the inversion algorithm, we have chosen $\overline{\mathbf{A}}_0 = \overline{\mathbf{I}}$, $A(\rho_t, k) = 1$ and $\delta_\epsilon = \delta_\sigma = 0.001$. The measurement data are simulated from the forward CG-FFT solver by adding 10% random error.

When the first implementation is used to perform the DBIM algorithm, the reconstructed profile of two offset objects is illustrated in Figure 8.22 after five DBIM

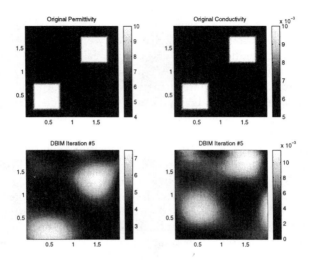

Figure 8.22 The original and reconstructed profiles of two offset dielectric objects. The first implementation of the multifrequency DBIM algorithm where 15 TX-RX locations and 11 frequency points are used.

iterations. Here, 50 iterations have been used to solve the linearized equation by the CG method. For easy comparison, the original profile is also provided in this figure. From Figure 8.22, we clearly see that both the permittivity and conductivity images give good approximations to the real objects in the object locations and object contrast, and two adjacent objects can be clearly identified. Therefore, this inversion algorithm can provide good resolution image even when 10% error is introduced in the measurement data.

To compare the two implementation methods shown in Figure 8.21, we next give reconstruction results of two offset objects using the second implementation. After five and 20 DBIM iterations, the reconstructed profiles are shown in Figure 8.23. Here, only two iterations are used to solve the linearized equation by the CG method. Comparing Figures 8.23 and 8.22, the reconstruction results from the second implementation are not good at the fifth DBIM iteration. However, when the iteration number increases to 20, the reconstruction results become much better.

8.5.3 DT Algorithm for 2D Objects

In the above two sections, two nonlinear inverse scattering algorithms are proposed using the distorted Born iterative method, which can efficiently detect the buried

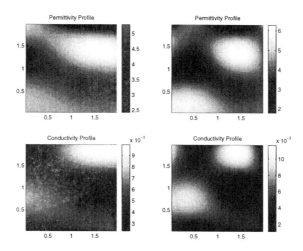

Figure 8.23 The reconstructed profiles of two offset dielectric objects. The second implementation of the multifrequency DBIM algorithm where 15 TX-RX locations and 11 frequency points are used. Left side: 5 DBIM iterations. Right side: 20 DBIM iterations.

objects by accurately giving their locations and dielectric properties. However, these algorithms are slow because many calls to the forward solvers are needed. Although fast algorithms have been used to solve the forward problem, it still takes much CPU time to obtain the buried-object image in a large reconstruction domain. Usually, some practical systems like GPR and VETEM require that the location of buried objects be determined in a few minutes using a portable computer at the site where the data are collected. For this purpose, a linearized inversion algorithm based on diffraction tomography is a good choice for on-site processing.

Consider a half-space problem shown in Figure 8.24. Multiple transmitters and multiple receivers are located in the free space, in a plane that is parallel to the air-earth interface: $z_t = z_r = z_0$. Under the coordinate system shown in this figure, the region $z \leq 0$ is the domain of interest. Suppose that all buried dielectric cylinders are contained in $\mathcal{D}$, the reconstruction domain. Unlike that in Figures 8.15 and 8.20, the top edge of $\mathcal{D}$ is just the air-earth interface. The relative permittivity $\epsilon_r(\rho)$ and conductivity $\sigma_r(\rho)$ inside $\mathcal{D}$ are defined by (8.121).

Using the Green's function (8.15), the scattered electric field from 2D buried cylinders at the receiver location is expressed as

$$E_a^s(\boldsymbol{\rho}_r, \boldsymbol{\rho}_t, k) = k^2 \int_{\mathcal{D}} d\boldsymbol{\rho} \, g_{ee}^{ab}(\boldsymbol{\rho}_r, \boldsymbol{\rho}, k) E_b(\boldsymbol{\rho}, \boldsymbol{\rho}_t, k) O(\boldsymbol{\rho}, k) \tag{8.144}$$

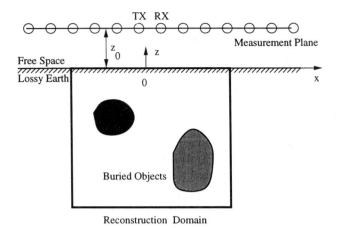

Figure 8.24 Setup configuration of the 2D diffraction tomographic algorithm. Multiple transmitter locations, multiple receiver locations, and multiple frequencies are used.

This is a reduced form of (8.122) when $\epsilon_{r0} = \epsilon_b$ and $\sigma_{r0} = \sigma_b$. Here $E_b(\rho, \rho_t, k)$ is the internal electric field inside the reconstruction domain, and $O(\rho, k)$ is the complex object function defined as

$$O(\rho, k) = \tilde{\epsilon}_r(\rho) - \tilde{\epsilon}_b = O_\epsilon(\rho) + i\frac{\eta_0}{k} O_\sigma(\rho) \qquad (8.145)$$

in which

$$O_\epsilon(\rho) = \epsilon_r(\rho) - \epsilon_b, \qquad O_\sigma(\rho) = \sigma_r(\rho) - \sigma_b \qquad (8.146)$$

are the permittivity-object and conductivity-object functions, which are slightly different from (8.129).

Under the Born approximation, the internal electric field E_b can be approximated by the incident electric field when the buried scatterers have a low contrast:

$$E_b(\rho, \rho_t, k) \cong E_b^{\text{inc}}(\rho, \rho_t, k) = ik\eta_0 I(k) g_{ee}^{ba}(\rho, \rho_t, k) \qquad (8.147)$$

where $g_{ee}^{ba}(\rho, \rho_t, k)$ is the Green's function defined in (8.14), and $I(k)$ is the electric current of the transmitter. Thus, the scattered electric field is further written as

$$E_a^s(\rho_r, \rho_t, k) = ik^3 \eta_0 I(k) \int_D d\rho g_{ee}^{ab}(\rho_r, \rho, k) g_{ee}^{ba}(\rho, \rho_t, k) O(\rho, k) \qquad (8.148)$$

Substituting (8.17) and (8.19) into the above equation and considering that $z_r = z_t = z_0 =$ constant, one easily obtains

$$E_a^s(x_r, x_t, k)$$
$$= -\frac{ik^3\eta_0 I(k)}{4\pi^2} \int_{-\infty}^{\infty} dx \int_{-\infty}^{0} dz O(x, z, k) \int_{-\infty}^{\infty} dk_x \int_{-\infty}^{\infty} dk'_x f_0(k_x)$$
$$\cdot f_0(k'_x) e^{i[(k_{az}+k'_{az})z_0 - (k_{bz}+k'_{bz})z - (k_x+k'_x)x + k_x x_r + k'_x x_t]} \quad (8.149)$$

Let $\tilde{E}_a^s(k_x, k'_x, k)$ be the 2D spatial Fourier transform of $E_a^s(x_r, x_t, k)$; then we have

$$\tilde{E}_a^s(k_x, k'_x, k) = \int_{-\infty}^{\infty} dx_r \int_{-\infty}^{\infty} dx_t E_a^s(x_r, x_t, k) e^{-i(k_x x_r + k'_x x_t)} \quad (8.150)$$

$$E_a^s(x_r, x_t, k) = \frac{1}{4\pi^2} \int_{-\infty}^{\infty} dk_x \int_{-\infty}^{\infty} dk'_x \tilde{E}_a^s(k_x, k'_x, k) e^{i(k_x x_r + k'_x x_t)} \quad (8.151)$$

Comparing (8.149) and (8.151), we get the following relation:

$$\tilde{E}_a^s(k_x, k'_x, k) = -ik^3\eta_0 I(k) f_0(k_x) f_0(k'_x) e^{i(k_{az}+k'_{az})z_0} \int_{-\infty}^{\infty} dx \int_{-\infty}^{0} dz$$
$$\cdot O(x, z, k) e^{-i[(k_x+k'_x)x + (k_{bz}+k'_{bz})z]} \quad (8.152)$$

Let the spatial Fourier transform of the object function be $\tilde{O}(k_x, k_y, k)$; then we have

$$\tilde{O}(k_x, k_z, k) = \int_{-\infty}^{\infty} dx \int_{-\infty}^{0} dz O(x, z, k) e^{-i(k_x x + k_z z)} \quad (8.153)$$

$$O(x, z, k) = \frac{1}{4\pi^2} \int_{-\infty}^{\infty} dk_x \int_{-\infty}^{\infty} dk_z \tilde{O}(k_x, k_z, k) e^{i(k_x x + k_z z)} \quad (8.154)$$

Substituting (8.153) into (8.152), one obtains a simple relationship between the Fourier transform of the scattered field and object function:

$$\tilde{E}_a^s(k_x, k'_x, k) = -\frac{ik^3\eta_0 I(k) e^{i(k_{az}+k'_{az})z_0}}{(k_{az}+k_{bz})(k'_{az}+k'_{bz})} \tilde{O}(k_x + k'_x, k_{bz} + k'_{bz}, k) \quad (8.155)$$

From the above derivation, we clearly see that the Sommerfeld-like integrals have been exactly accounted for in (8.155). The cost is, however, the need to collect measurement data at different transmitter and receiver locations. In the Fourier transform of the scattered field, when we let $k_x = k'_x$, (8.155) gives

$$\tilde{O}(2k_x, 2k_{bz}, k) = \frac{i(k_{az}+k_{bz})^2}{k^3\eta_0 I(k)} e^{-i2k_{az}z_0} \tilde{E}_a^s(k_x, k_x, k) \quad (8.156)$$

The above equation illustrates the object-function spectrum in a specific complex plane. In fact, one can obtain different object spectra by taking different parts of the scattered-field spectra. For example, when we let $k'_x = 0$, (8.155) gives

$$\tilde{O}(k_x, k_{bz} + k_b, k) = \frac{i(k_a + k_b)(k_{az} + k_{bz})}{k^3 \eta_0 I(k)} e^{-i(k_a + k_{az})z_0} \tilde{E}^s_a(k_x, 0, k) \quad (8.157)$$

Comparing (8.156) with (8.157), the first has a simpler form and hence is used in this algorithm.

Just as stated in [62], it is impossible to simultaneously reconstruct the permittivity and conductivity profiles using the proposed DT algorithm without additional information. If $\sigma_r = \sigma_b$ or the contribution of permittivity is much larger than that of conductivity (the displacement current is much larger than the conduction current), the object function (8.145) equals or approaches the permittivity object function. In this case, (8.156) will reduce to

$$\tilde{O}_\epsilon(2k_x, 2k_{bz}) = \frac{i(k_{az} + k_{bz})^2}{k^3 \eta_0 I(k)} e^{-i2k_{az}z_0} \tilde{E}^s_a(k_x, k_x, k) \quad (8.158)$$

In contrast, if $\epsilon_r = \epsilon_b$ or the contribution of conductivity is much larger than that of permittivity (the conduction current is much larger than the displacement current), the object function equals or approaches the conductivity-object function multiplied by $i\eta_0/k$. In this case, we obtain from (8.156)

$$\tilde{O}_\sigma(2k_x, 2k_{bz}) = \frac{(k_{az} + k_{bz})^2}{k^2 \eta_0^2 I(k)} e^{-i2k_{az}z_0} \tilde{E}^s_a(k_x, k_x, k) \quad (8.159)$$

Here, $\tilde{O}_\epsilon(k_x, k_z)$ and $\tilde{O}_\sigma(k_x, k_z)$ are Fourier transforms of the permittivity-object function $O_\epsilon(x, z)$ and conductivity-object function $O_\sigma(x, z)$, respectively.

To reconstruct the object functions $O_\epsilon(x, z)$ and $O_\sigma(x, z)$ using (8.158) and (8.159), we reconsider the Fourier transform pair (8.153) and (8.154). Physically speaking, the above Fourier transform pair correspond to a lossless background, where both of k_x and k_z are real. If the background is lossy, the spectra of object functions are given along a contour in the complex-k_z plane, as shown in (8.158) and (8.159). Therefore, the Fourier transform should be extended to a complex plane in the lossy case. A well-known extension of the Fourier transform is the Laplace transform:

$$L(s) = \int_0^\infty dt f(t) e^{-st} \quad (8.160)$$

$$f(t) = \frac{1}{2\pi i} \int_{\sigma - i\infty}^{\sigma + i\infty} ds L(s) e^{st} \quad (8.161)$$

where $s = \sigma + ik_z$, $\sigma > 0$, and the real part σ is kept as a constant, i.e., the integration path in (8.161) is a straight line parallel to the imaginary axis. If the integration path

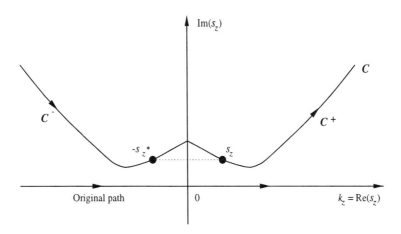

Figure 8.25 Integration contour of the inverse Fourier transform in the upper Riemann sheet of the complex s_z plane.

in the inverse transform is required to be a contour as shown in Figure 8.25, which is just the contour of k_{bz} in the lossy case when k varies from zero to infinity, however, (8.161) will be divergent because the radiation condition is not satisfied. Hence, the Laplace transform cannot be used for this purpose.

To derive closed-form reconstruction formulas in the lossy background, we introduce an approximate generalized Fourier transform pair [72]

$$F(s_z) = \int_{-\infty}^{0} dz f(z) e^{-is_z z} \tag{8.162}$$

$$f(z) \approx \frac{1}{2\pi} \int_{C} ds_z F(s_z) e^{is_z^* z} \tag{8.163}$$

in which s_z is defined in the upper Riemann sheet: $\Im m(s_z) \geq 0$, $*$ denotes the complex conjugation, and $C = C^+ + C^-$ is a contour as shown in Figure 8.25. Clearly, the radiation condition is satisfied in both the forward and inverse transforms. When s_z is real, (8.162) and (8.163) will reduce to the original definition of the Fourier transform.

Although the inversion formula (8.163) is not exact, it is convergent for all integration paths in the upper half-plane, and it approximates the behavior of the original function $f(z)$ very well. Figure 8.26 illustrates the comparisons of original functions and reconstructed profiles from (8.163) using the spectrum along a contour $s_z = \sqrt{k^2 \epsilon_b + ik\eta_0 \sigma_b}$, in which $\epsilon_b = 4$ and $\sigma_b = 0.1$ mS/m. It is clear that (8.163)

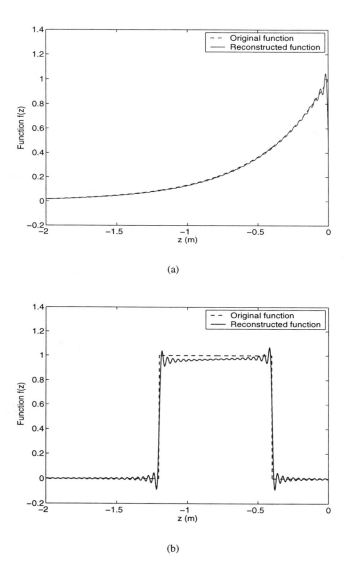

Figure 8.26 Comparisons of original functions and reconstructed profiles using the approximate generalized Fourier transform: (a) exponential function; (b) pulse function.

provides a good approximation of the original function $f(z)$, no matter if it is a continuous function or a pulse function.

Now we consider the extension of the 2D Fourier transform pair (8.153) and (8.154). If we keep k_x in the real axis and extend k_z from real axis to the complex plane s_z, we have the following approximate generalized Fourier transform pair

$$\tilde{O}_\alpha(k_x, s_z) = \int_{-\infty}^{\infty} dx \int_{-\infty}^{0} dz O_\alpha(x, z) e^{-i(k_x x + s_z z)} \tag{8.164}$$

$$O_\alpha(x, z) = \frac{1}{4\pi^2} \int_{-\infty}^{\infty} dk_x \int_{C} ds_z \tilde{O}_\alpha(k_x, s_z) e^{i(k_x x + s_z^* z)} \tag{8.165}$$

where $\Im m(s_z) \geq 0$ and $\alpha = \epsilon$ or σ. Considering the symmetrical property of C^+ and C^-, (8.165) can be rewritten as

$$O_\alpha(x, z) = \frac{1}{4\pi^2} \int_{-\infty}^{\infty} dk_x \int_{C^+} \left[ds_z \tilde{O}_\alpha(k_x, s_z) e^{i(k_x x + s_z^* z)} \right.$$

$$\left. + ds_z^* \tilde{O}_\alpha(-k_x, -s_z^*) e^{-i(k_x x + s_z z)} \right] \tag{8.166}$$

Since $O_\alpha(x, z)$ is a real function, it is easily shown from (8.164) that

$$\tilde{O}_\alpha^*(k_x, s_z) = \tilde{O}_\alpha(-k_x, -s_z^*)$$

Hence, the second integral in (8.166) is just the complex conjugation of the first integral, and then (8.166) is furthermore reduced to

$$O_\alpha(x, z) = \frac{1}{2\pi^2} \Re e \left\{ \int_{-\infty}^{\infty} dk_x e^{ik_x x} \int_{C^+} ds_z \tilde{O}_\alpha(k_x, s_z) e^{is_z^* z} \right\} \tag{8.167}$$

where $\Re e(s_z) \geq 0$ and $\Im m(s_z) \geq 0$. Because the branch cut of k_{bz} also satisfies $\Re e(k_{bz}) \geq 0$ and $\Im m(k_{bz}) \geq 0$, we can directly rewrite (8.167) by letting $s_z = 2k_{bz}$ as

$$O_\alpha(x, z) = \frac{2}{\pi^2} \Re e \left\{ \int_{-\infty}^{\infty} dk_x e^{i2k_x x} \int_{C^+} dk_{bz} \tilde{O}_\alpha(2k_x, 2k_{bz}) e^{i2k_{bz}^* z} \right\} \tag{8.168}$$

Actually, the contour C can be arbitrarily chosen in the upper Riemann sheet if there are no poles between the real axis and C. But we choose C^+ as a specified one, as shown in Figure 8.25, so that the contour integral can be converted into an integral along k through $k_{bz} = \sqrt{k^2 \tilde{\epsilon}_b - k_x^2}$. Hence, the spatial object function is finally written as

$$O_\alpha(x,z) = \frac{2}{\pi^2}\Re e\left\{\int_{-\infty}^{\infty} dk_x e^{i2k_x x} \int_0^\infty dk \frac{2\epsilon_b k + i\eta_0 \sigma_b}{2k_{bz}} \tilde{O}_\alpha(2k_x, 2k_{bz})e^{i2k_{bz}^* z}\right\}$$
(8.169)

Therefore, the formula to reconstruct the permittivity profile of a buried pure-dielectric cylinder is

$$\epsilon_r(x,z) = \epsilon_b + \frac{2}{\pi^2}\Re e\left\{\int_{-\infty}^{\infty} dk_x e^{i2k_x x} \int_0^\infty dk \frac{2\epsilon_b k + i\eta_0 \sigma_b}{2k_{bz}}\right.$$
$$\left. \cdot \frac{i(k_{az}+k_{bz})^2}{k^3 \eta_0 I(k)} e^{-i2k_{az}z_0} \tilde{E}_a^s(k_x, k_x, k) e^{i2k_{bz}^* z}\right\} \quad (8.170)$$

Similarly, the formula to reconstruct the conductivity profile of a buried pure-conductive cylinder becomes

$$\sigma_r(x,z) = \sigma_b + \frac{2}{\pi^2}\Re e\left\{\int_{-\infty}^{\infty} dk_x e^{i2k_x x} \int_0^\infty dk \frac{2\epsilon_b k + i\eta_0 \sigma_b}{2k_{bz}}\right.$$
$$\left. \cdot \frac{(k_{az}+k_{bz})^2}{k^2 \eta_0^2 I(k)} e^{-i2k_{az}z_0} \tilde{E}_a^s(k_x, k_x, k) e^{i2k_{bz}^* z}\right\} \quad (8.171)$$

Although the inversion formulas (8.170) and (8.171) are theoretically derived from pure-dielectric and pure-conductive cylinders, they can heuristically be used to reconstruct simultaneously the permittivity and conductivity profiles of "general" buried targets. Numerical results show that such reconstruction can give good resolution of the buried objects. However, the contrast amplitude of the conductivity is not accurate.

Now, we see some reconstruction examples using the novel DT algorithm where the relative permittivity and conductivity of the lossy earth are $\epsilon_b = 4$ and $\sigma_b = 0.1$ mS/m, respectively. The reconstruction domain $\mathcal{D}$ is a 10×10 m² square region, which is divided into $64 \times 64 = 4{,}096$ pixels. The buried objects are three circular dielectric pipes, which are diagonally located in the reconstruction domain. The diameter of each pipe is 2.5m. The reason we choose three offset pipes is to examine if the inversion algorithm can separate multiple scatterers and reconstruct well the deeper object.

The transmitters and receivers are located 0.5m above the air-earth interface. Here, 64 transmitter locations and 64 receiver locations are used on the measurement domain, which is 80m long centered about the reconstruction domain. Therefore, the distance between TX and TX (or RX and RX) is 1.25m. If the working frequency ranges from 0.5 to 60.5 MHz, then the corresponding wavelength will vary from 300m to 2.48m in the earth, which is suitable for reconstructing the dielectric cylinders with a diameter of 2.5m. In the following examples, 60 frequency points are taken between 0.5 and 60.5 MHz. The measurement data are again simulated from the CG-FFT algorithm by adding a 10% random error.

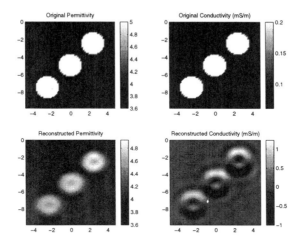

Figure 8.27 Original and reconstructed profiles from noisy data when $\epsilon_r = 5$ and $\sigma_r = 0.2$ mS/m. 2D DT algorithm where 64 transmitter locations, 64 receiver locations, and 60 frequency points are used.

As the first example, we consider the inversion of a weak buried object ($\epsilon_r = 5$ and $\sigma_r = 0.2$ mS/m). Using (8.170) and (8.171), both permittivity and conductivity profiles can be reconstructed simultaneously, as shown in Figure 8.27. From Figure 8.27, the permittivity image has very good resolution giving accurate locations, shapes, and contrasts of buried pipes. In the conductivity image, the upper portion of each pipe is quite clear and a semicircular shape can be clearly seen, but the lower portion has bad resolution and only a small part is visible. Right below each pipe, there is a "ghost image," which gives nonphysical conductivity: negative conductivity. Because of the ghost image, the image of the pipe has a slight shift towards the air-earth interface. The center between the pipe image and ghost image is just the correct location of the pipe.

The above example illustrates the reconstruction of weak buried scatterers for which the Born approximation is valid. When the contrast of buried targets becomes higher, (8.170) and (8.171) can still be used to approximate the image. Figure 8.28 shows the original and reconstructed profiles of stronger buried targets when $\epsilon_r = 10$ and $\sigma_r = 1$ mS/m. Clearly, the shapes of images in both permittivity and conductivity images have a distortion, but the permittivity images have a good spatial resolution and provide an accurate estimate of the contrast. In the conductivity image, the three buried pipes can also be clearly observed.

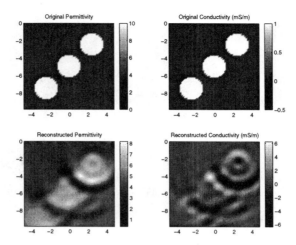

Figure 8.28 Original and reconstructed profiles from noisy data when $\epsilon_r = 10$ and $\sigma_r = 1.0$ mS/m. 2D DT algorithm where 64 transmitter locations, 64 receiver locations, and 60 frequency points are used.

From the above examples, we notice that high-resolution imaging results for weak scatterers are obtained from the proposed inversion algorithm, although only the reflection setup is used (i.e., transmitters and receivers are only placed above the air-earth interface). Mathematically speaking, this is because nearly all spectral information of the object function can be achieved from the reflection-setup scattered data, as shown in (8.156). For example, the spectrum of single-frequency scattered data gives the object-function spectrum along a circle with radius $k\sqrt{\epsilon_b}$ centered at $(k_x, k_z) = (0, 0)$ when the earth is lossless. As the frequency changes from 0 to ∞, we obtain all the object-function spectrum in the whole k_x-k_z domain, and hence the object function can be exactly reconstructed using the inverse Fourier transform. In the above reconstruction examples, only a limited frequency band (0.5 ~ 60.5 MHz) is used. However, the spectrum of the triple-pipe object (with diameter of 2.5m) is nearly zero when the frequency is larger than 60 MHz, and thus we still obtain high-resolution imaging results.

Comparing the reconstruction results from the novel DT algorithm with those from the DBIM algorithms, we notice that the DT algorithm can provide better spatial resolution of image, because more scattered data have been used and the Sommerfeld-like integral has been exactly incorporated. Also, it is very fast to handle large reconstruction domain. However, the DT algorithm cannot give correct conductivity contrast of the buried object.

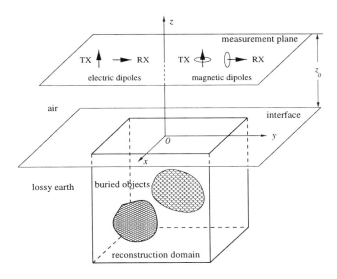

Figure 8.29 Setup configuration of the 3D diffraction tomographic algorithm. The transmitter and receiver have a fixed offset.

8.5.4 DT Algorithm for 3D Objects

In this section, we consider a general DT algorithm for 3D buried objects. As illustrated in Figure 8.29, several 3D dielectric objects $\mathcal{D}_i$ are buried in the lossy earth. The transmitter TX and receiver RX, which can be electric dipoles or magnetic dipoles (i.e., small loop antennas), are located in a plane $z_t = z_r = z_0$ that is parallel to the air-earth interface. Under the coordinate system shown in Figure 8.29, the region $z \leq 0$ is the domain of interest. Suppose that all the buried dielectric objects to be reconstructed are contained in $\mathcal{D} = D_x \times D_y \times D_z$, the reconstruction domain. Then, the relative permittivity $\epsilon_r(\mathbf{r})$ and conductivity $\sigma_r(\mathbf{r})$ inside $\mathcal{D}$ can be defined in a similar way to (8.121).

8.5.4.1 Electric-Dipole Transmitter and Receiver

When the TX-RX are electric dipoles, the electric field is measured at the receiver location. From the half-space Green's function (8.5), the incident electric field of $\mathcal{D}$ is expressed as

$$\mathbf{E}_b^{\text{inc}}(\mathbf{r}, \mathbf{r}_t, k) = -\omega \mu_0 \Delta l_t I_t^e(k) \overline{\mathbf{G}}_{ee}^{ba}(\mathbf{r}, \mathbf{r}_t, k) \cdot \hat{\alpha}_t^e \qquad (8.172)$$

in which $\mathbf{r} \in \mathcal{D}$, $\mathbf{r}_t$ is the transmitter location in the air, and $I_t^e(k)$ and $\hat{\alpha}_t^e$ are the electric current and direction of the transmitting electric dipole with length Δl_t. Under the illumination of the above incident electric field, the scattered electric field by a buried object at the receiver location is written as

$$\mathbf{E}_a^{\text{sca}}(\mathbf{r}_r, \mathbf{r}_t, k) = ik^2 \int_{\mathcal{D}} d\mathbf{r} \overline{\mathbf{G}}_{ee}^{ab}(\mathbf{r}_r, \mathbf{r}, k) \cdot \mathbf{E}_b(\mathbf{r}, \mathbf{r}_t, k) O(\mathbf{r}, k) \qquad (8.173)$$

in which $\mathbf{E}_b(\mathbf{r}, \mathbf{r}_t, k)$ is the internal electric field inside the reconstruction domain, $O(\mathbf{r}, k)$ is the complex object function defined similarly to (8.145), and $\overline{\mathbf{G}}_{ee}^{ab}$ is the electric-field dyadic Green's function given in (8.9).

Generally, the internal electric field $\mathbf{E}_b(\mathbf{r}, \mathbf{r}_t, k)$ can be accurately obtained by solving the EFIE. However, it will take a huge amount of CPU time to evaluate $\mathbf{E}_b(\mathbf{r}, \mathbf{r}_t, k)$ at each frequency and each transmitter location. When the object function has a low contrast, the Born approximation can be used: $\mathbf{E}_b(\mathbf{r}, \mathbf{r}_t, k) \cong \mathbf{E}_b^{\text{inc}}(\mathbf{r}, \mathbf{r}_t, k)$. Hence, the scattered electric field in $\hat{\alpha}_r^e$ polarization is written as

$$\hat{\alpha}_r^e \cdot \mathbf{E}_a^{\text{sca}}(\mathbf{r}_r, \mathbf{r}_t, k)$$
$$= -\frac{ik^3 \eta_0}{16\pi^4} I_t^e(k) \Delta l_t \int_{-\infty}^{\infty} dk_x \int_{-\infty}^{\infty} dk_y \int_{-\infty}^{\infty} dk_x' \int_{-\infty}^{\infty} dk_y' \hat{\alpha}_r^e$$
$$\cdot \widetilde{\mathbf{G}}_{ee}^{ab}(k_x, k_y, k) \cdot \widetilde{\mathbf{G}}_{ee}^{ba}(-k_x', -k_y', k) \cdot \hat{\alpha}_t^e e^{i(k_x x_r + k_y y_r + k_x' x_t + k_y' y_t)}$$
$$\cdot e^{i(k_{az} + k_{az}')z_0} \int_{-\infty}^{\infty} dx \int_{-\infty}^{\infty} dy \int_{-\infty}^{0} dz O(x, y, z, k)$$
$$\cdot e^{-i[(k_{bz} + k_{bz}')z + (k_x + k_x')x + (k_y + k_y')y]} \qquad (8.174)$$

in which we assume that $z_r = z_t = z_0$.

8.5.4.2 Magnetic-Dipole Transmitter and Receiver

When the TX-RX are magnetic dipoles (i.e., small loop antennas), the magnetic field is measured at the receiver location. However, the buried objects are usually dielectric instead of magnetic, and the incident electric field of $\mathcal{D}$ has to be evaluated. From the half-space Green's function (8.7), one obtains

$$\mathbf{E}_b^{\text{inc}}(\mathbf{r}, \mathbf{r}_t, k) = -i\omega\mu_0 A_t I_t^m(k) \overline{\mathbf{G}}_{em}^{ba}(\mathbf{r}, \mathbf{r}_t, k) \cdot \hat{\alpha}_t^m \qquad (8.175)$$

in which $I_t^m(k)$ and A_t are the electric current and area of the transmitting loop, and $\hat{\alpha}_t^m$ is the normal direction of the loop or direction of the magnetic dipole.

Under the illumination of the above incident electric field, the internal electric field $\mathbf{E}_b(\mathbf{r}, \mathbf{r}_t, k)$ inside the reconstruction domain can be obtained, and then the scattered magnetic field at the receiver location can be given by

$$\mathbf{H}_a^{\text{sca}}(\mathbf{r}_r, \mathbf{r}_t, k) = -i\omega\epsilon_0 \int_{\mathcal{D}} d\mathbf{r} \overline{\mathbf{G}}_{me}^{ab}(\mathbf{r}_r, \mathbf{r}, k) \cdot \mathbf{E}_b(\mathbf{r}, \mathbf{r}_t, k) O(\mathbf{r}, k) \qquad (8.176)$$

in which the magnetic-field dyadic Green's function $\overline{\mathbf{G}}_{me}^{ab}$ is shown in (8.11). After using the Born approximation, we obtain the scattered magnetic field in $\hat{\alpha}_r^m$ polarization:

$$\hat{\alpha}_r^m \cdot \mathbf{H}_a^{\text{sca}}(\mathbf{r}_r, \mathbf{r}_t, k)$$
$$= -\frac{k^2}{16\pi^4} I_t^m(k) A_t \int_{-\infty}^{\infty} dk_x \int_{-\infty}^{\infty} dk_y \int_{-\infty}^{\infty} dk'_x \int_{-\infty}^{\infty} dk'_y \hat{\alpha}_r^m$$
$$\cdot \tilde{\mathbf{G}}_{me}^{ab}(k_x, k_y, k) \cdot \tilde{\mathbf{G}}_{em}^{ba}(-k'_x, -k'_y, k) \cdot \hat{\alpha}_t^m e^{i(k_x x_r + k_y y_r + k'_x x_t + k'_y y_t)}$$
$$\cdot e^{i(k_{az} + k'_{az})z_0} \int_{-\infty}^{\infty} dx \int_{-\infty}^{\infty} dy \int_{-\infty}^{0} dz O(x, y, z, k)$$
$$\cdot e^{-i[(k_{bz} + k'_{bz})z + (k_x + k'_x)x + (k_y + k'_y)y]} \quad (8.177)$$

From (8.174) and (8.177), both scattered electric field and magnetic field contain the information about the buried objects.

8.5.4.3 Inversion Algorithms

Comparing (8.174) with (8.177), the scattered electric field has a structure similar to that of the magnetic field. Hence, one can rewrite (8.174) and (8.177) in a compact form:

$$\hat{\alpha}_r^p \cdot \mathbf{P}_a^{\text{sca}}(\mathbf{r}_r, \mathbf{r}_t, k)$$
$$= \frac{1}{16\pi^4} Q_t^p(k) \int_{-\infty}^{\infty} dk_x \int_{-\infty}^{\infty} dk_y \int_{-\infty}^{\infty} dk'_x \int_{-\infty}^{\infty} dk'_y \hat{\alpha}_r^p$$
$$\cdot \tilde{\mathbf{G}}_{pe}^{ab}(k_x, k_y, k) \cdot \tilde{\mathbf{G}}_{ep}^{ba}(-k'_x, -k'_y, k) \cdot \hat{\alpha}_t^p e^{i(k_x x_r + k_y y_r + k'_x x_t + k'_y y_t)}$$
$$\cdot e^{i(k_{az} + k'_{az})z_0} \int_{-\infty}^{\infty} dx \int_{-\infty}^{\infty} dy \int_{-\infty}^{0} dz O(x, y, z, k)$$
$$\cdot e^{-i[(k_{bz} + k'_{bz})z + (k_x + k'_x)x + (k_y + k'_y)y]} \quad (8.178)$$

in which $p = e, m$, $\mathbf{P} = \mathbf{E}, \mathbf{H}$, and

$$Q_t^e(k) = -ik^3 \eta_0 I_t^e(k) \Delta l, \quad Q_t^m(k) = -k^2 I_t^m(k) A_t \quad (8.179)$$

Since $z_r = z_t = z_0 =$ constant, the scattered field $\mathbf{P}_a^{\text{sca}}(\mathbf{r}_r, \mathbf{r}_t, k)$ is only a function of x_r, y_r, x_t, y_t, and k. If we know all the scattered field information at different transmitter locations (x_t, y_t) and receiver locations (x_r, y_r) and assume that the spatial four-dimensional (4D) Fourier transform of $\hat{\alpha}_r^p \cdot \mathbf{P}_a^{\text{sca}}(x_r, y_r, x_t, y_t, k)$ is

$\hat{\alpha}_r^p \cdot \tilde{\mathbf{P}}_a^{\text{sca}}(k_x, k_y, k_x', k_y', k)$; that is,

$$\hat{\alpha}_r^p \cdot \mathbf{P}_a^{\text{sca}}(x_r, y_r, x_t, y_t, k)$$
$$= \frac{1}{(2\pi)^4} \int_{-\infty}^{\infty} dk_x \int_{-\infty}^{\infty} dk_y \int_{-\infty}^{\infty} dk_x' \int_{-\infty}^{\infty} dk_y' \hat{\alpha}_r$$
$$\cdot \tilde{\mathbf{P}}_a^{\text{sca}}(k_x, k_y, k_x', k_y', k) \exp[i(k_x x_r + k_y y_r + k_x' x_t + k_y' y_t)] \quad (8.180)$$

then we easily obtain the following relation from (8.178) and (8.180):

$$\hat{\alpha}_r^p \cdot \tilde{\mathbf{P}}_a^{\text{sca}}(k_x, k_y, k_x', k_y', k)$$
$$= Q_t^p(k) \hat{\alpha}_r^p \cdot \widetilde{\mathbf{G}}_{pe}^{ab}(k_x, k_y, k) \cdot \widetilde{\mathbf{G}}_{ep}^{ba}(-k_x', -k_y', k) \cdot \hat{\alpha}_t^p e^{i(k_{az} + k_{az}')z_0}$$
$$\cdot \int_{-\infty}^{\infty} dx \int_{-\infty}^{\infty} dy \int_{-\infty}^{0} dz\, O(x, y, z, k)$$
$$\cdot e^{-i[(k_x + k_x')x + (k_y + k_y')y + (k_{bz} + k_{bz}')z]} \quad (8.181)$$

Let the spatial Fourier transform of the object function $O(x, y, z, k)$ be $\tilde{O}(k_x, k_y, k_z, k)$; that is,

$$\tilde{O}(k_x, k_y, k_z, k) = \int_{-\infty}^{\infty} dx \int_{-\infty}^{\infty} dy \int_{-\infty}^{0} dz\, O(x, y, z, k)$$
$$\cdot e^{-i(k_x x + k_y y + k_z z)} \quad (8.182)$$

then one obtains from (8.181) and (8.182)

$$\hat{\alpha}_r^p \cdot \tilde{\mathbf{P}}_a^{\text{sca}}(k_x, k_y, k_x', k_y', k)$$
$$= Q_t^p(k) \hat{\alpha}_r^p \cdot \widetilde{\mathbf{G}}_{pe}^{ab}(k_x, k_y, k) \cdot \widetilde{\mathbf{G}}_{ep}^{ba}(-k_x', -k_y', k) \cdot \hat{\alpha}_t^p$$
$$\cdot \tilde{O}(k_x + k_x', k_y + k_y', k_{bz} + k_{bz}', k) e^{i(k_{az} + k_{az}')z_0} \quad (8.183)$$

Equation (8.183) is a general formulation relating the full-information scattered field data and the object function in the spectral domain, where the Sommerfeld integrals have been exactly incorporated. Therefore, this equation is valid for all frequencies and can be applied to both high-frequency and low-frequency methods.

In the Fourier domain of the scattered data, when we let $k_x = k_x'$ and $k_y = k_y'$, (8.183) will be simplified to

$$\hat{\alpha}_r^p \cdot \tilde{\mathbf{P}}_a^{\text{sca}}(k_x, k_y, k_x, k_y, k)$$
$$= Q_t^p(k) \hat{\alpha}_r^p \cdot \widetilde{\mathbf{G}}_{pe}^{ab}(k_x, k_y, k) \cdot \widetilde{\mathbf{G}}_{ep}^{ba}(-k_x, -k_y, k) \cdot \hat{\alpha}_t^p$$
$$\cdot \tilde{O}(2k_x, 2k_y, 2k_{bz}, k) e^{i2k_{az}z_0} \quad (8.184)$$

which gives the object-function spectrum in a specific complex plane. In fact, one can obtain different spectra by taking different parts of scattered data in the Fourier domain.

Equations (8.183) and (8.184) provide an accurate way to retrieve the object-function spectrum, which are valid for all frequencies. Using the above equations, two inversion algorithms are easily generated when the TX-RX are electric dipoles and magnetic dipoles, respectively. However, these algorithms are too expensive because a large amount of scattered data at different transmitter and receiver locations must be collected, making them impractical.

In a practical setup of measurement system, the TX and RX are usually placed on a cart and the cart moves on the measurement plane. In this case, the TX and RX have a fixed offset, $x_t = x_r + \Delta x$ and $y_t = y_r + \Delta y$, and hence the scattered field $\mathbf{P}_a^{\text{sca}}(\mathbf{r}_r, \mathbf{r}_t, k)$ in (8.178) is only a function of x_r, y_r and k. Assume that the spatial 2D Fourier transform of $\hat{\alpha}_r^p \cdot \mathbf{P}_a^{\text{sca}}(x_r, y_r, k)$ is $\hat{\alpha}_r^p \cdot \tilde{\mathbf{P}}_a^{\text{sca}}(k_x, k_y, k)$; that is,

$$\hat{\alpha}_r^p \cdot \mathbf{P}_a^{\text{sca}}(x_r, y_r, k) = \frac{1}{(2\pi)^2} \int_{-\infty}^{\infty} dk_x \int_{-\infty}^{\infty} dk_y \hat{\alpha}_r \cdot \tilde{\mathbf{P}}_a^{\text{sca}}(k_x, k_y, k)$$
$$\cdot \exp[i(k_x x_r + k_y y_r)] \tag{8.185}$$

then one easily obtains from (8.178) and (8.185) after some derivation

$$\hat{\alpha}_r^p \cdot \tilde{\mathbf{P}}_a^{\text{sca}}(k_x, k_y, k)$$
$$= \frac{1}{4\pi^2} Q_t^p(k) \int_{-\infty}^{\infty} dx \int_{-\infty}^{\infty} dy \int_{-\infty}^{0} dz O(x, y, z, k) e^{-i(k_x x + k_y y)}$$
$$\cdot \int_{-\infty}^{\infty} dk_x' \int_{-\infty}^{\infty} dk_y' \hat{\alpha}_r^p \cdot \tilde{\mathbf{G}}_{pe}^{ab}(k_x - k_x', k_y - k_y', k) \cdot \tilde{\mathbf{G}}_{ep}^{ba}(-k_x', -k_y', k) \cdot \hat{\alpha}_t^p$$
$$\cdot e^{i(k_x' \Delta x + k_y' \Delta y)} e^{i[\sqrt{k_a^2 - (k_x - k_x')^2 - (k_y - k_y')^2} + \sqrt{k_a^2 - k_x'^2 - k_y'^2}]z_0}$$
$$\cdot e^{-i[\sqrt{k_b^2 - (k_x - k_x')^2 - (k_y - k_y')^2} + \sqrt{k_b^2 - k_x'^2 - k_y'^2}]z} \tag{8.186}$$

Clearly, a Sommerfeld integral is contained in the above equation:

$$I_S(k_x, k_y, k) = \int_{-\infty}^{\infty} dk_x' \int_{-\infty}^{\infty} dk_y' f(k_x', k_y', k)$$
$$\cdot e^{-i[\sqrt{k_b^2 - (k_x - k_x')^2 - (k_y - k_y')^2} + \sqrt{k_b^2 - k_x'^2 - k_y'^2}]z}$$
$$\cdot e^{i[\sqrt{k_a^2 - (k_x - k_x')^2 - (k_y - k_y')^2} + \sqrt{k_a^2 - k_x'^2 - k_y'^2}]z_0} \tag{8.187}$$

where the integrand

$$f(k_x', k_y', k) = \hat{\alpha}_r^p \cdot \tilde{\mathbf{G}}_{pe}^{ab}(k_x - k_x', k_y - k_y', k) \cdot \tilde{\mathbf{G}}_{ep}^{ba}(-k_x', -k_y', k) \cdot \hat{\alpha}_t^p$$
$$\cdot e^{i(k_x' \Delta x + k_y' \Delta y)} \tag{8.188}$$

is a slowly varying function and $|z|$ or z_0 can be a large parameter.

In the DT algorithm proposed in [62], a similar Sommerfeld integral to (8.187) was encountered. However, a very small z_0 was assumed in [62] and the exponential term

$$e^{i[\sqrt{k_a^2-(k_x-k'_x)^2-(k_y-k'_y)^2}+\sqrt{k_a^2-k'^2_x-k'^2_y}]z_0}$$

has been put into $f(k'_x, k'_y, k)$ as the slowly varying function. However, z_0 is not usually very small in the actual measurement system, for example, in the case when TX and RX are placed in a cart. Hence, the above term should also be considered as a fast-varying function, as illustrated in (8.187).

To evaluate the Sommerfeld integral (8.187) using the asymptotic expansion method, the stationary point should be first calculated. From the exponential terms in (8.187), we can easily show that

$$k'_x = \frac{1}{2}k_x, \quad k'_y = \frac{1}{2}k_y \tag{8.189}$$

is the stationary point. After some derivation, we obtain the asymptotic form of (8.187), which is expressed as

$$I_S(k_x, k_y, k) \cong -\frac{i\pi \gamma_a \gamma_b}{\sqrt{(\gamma_b z_0 - \gamma_a z)\left(\frac{k_a^2}{\gamma_a^2}\gamma_b z_0 - \frac{k_b^2}{\gamma_b^2}\gamma_a z\right)}} f\left(\frac{1}{2}k_x, \frac{1}{2}k_y\right)$$

$$\cdot e^{i2(\gamma_a z_0 - \gamma_b z)} \tag{8.190}$$

The above equation is equivalent to

$$I_S(k_x, k_y, k) \cong \frac{i\pi}{k_b z}\gamma_b^2 \left[\left(1 - \frac{\gamma_b}{\gamma_a}\frac{z_0}{z}\right)\left(1 - \frac{k_a^2 \gamma_b^3}{k_b^2 \gamma_a^3}\frac{z_0}{z}\right)\right]^{-\frac{1}{2}}$$

$$\cdot f\left(\frac{1}{2}k_x, \frac{1}{2}k_y\right) e^{i2(\gamma_a z_0 - \gamma_b z)} \tag{8.191}$$

where $z \leq 0$ and

$$\gamma_{a,b} = \sqrt{k_{a,b}^2 - \frac{1}{4}k_x^2 - \frac{1}{4}k_y^2} \tag{8.192}$$

When $|z| \gg z_0$, the last term on the right-hand side of (8.191) approximates to one, and then (8.191) reduces to

$$I_S(k_x, k_y, k) \cong \frac{i\pi}{k_b z}\gamma_b^2 f\left(\frac{1}{2}k_x, \frac{1}{2}k_y\right) e^{i2(\gamma_a z_0 - \gamma_b z)} \tag{8.193}$$

which is just the asymptotic formula in [62]. However, (8.190) is more general, which is even valid when $z = 0$ and z_0 is large.

Substituting (8.190) into (8.186), one obtains a general formula relating the Fourier-domain scattered field and the object function

$$\hat{\alpha}_r^p \cdot \tilde{\mathbf{P}}_a^{\text{sca}}(k_x, k_y, k)$$
$$= -\frac{i\gamma_a \gamma_b}{4\pi} Q_t^p(k) f\left(\frac{1}{2}k_x, \frac{1}{2}k_y\right) e^{i2\gamma_a z_0} \int_{-\infty}^{\infty} dx \int_{-\infty}^{\infty} dy \int_{-\infty}^{0} dz$$
$$\cdot e^{-i(k_x x + k_y y + 2\gamma_b z)} \frac{O(x,y,z,k)}{\sqrt{(\gamma_b z_0 - \gamma_a z)\left(\frac{k_a^2}{\gamma_a^2}\gamma_b z_0 - \frac{k_b^2}{\gamma_b^2}\gamma_a z\right)}} \quad (8.194)$$

Using the scattered information $\hat{\alpha}_r^p \cdot \tilde{\mathbf{P}}_a^{\text{sca}}(k_x, k_y, k)$, the object function can be reconstructed by solving the integral equation (8.194). However, it requires much numerical work. For the purpose of a fast reconstruction of the object function, we make the following approximations. When $-z \gg z_0$, (8.194) gives

$$\hat{\alpha}_r^p \cdot \tilde{\mathbf{P}}_a^{\text{sca}}(k_x, k_y, k) = \frac{i\gamma_b^2}{4\pi k_b} Q_t^p(k) f\left(\frac{1}{2}k_x, \frac{1}{2}k_y\right) e^{i2\gamma_a z_0}$$
$$\cdot \tilde{\hat{O}}(k_x, k_y, 2\gamma_b, k) \quad (8.195)$$

where $\tilde{\hat{O}}(k_x, k_y, k_z, k)$ is the Fourier transform of

$$\hat{O}(x, y, z, k) = O(x, y, z, k)/z$$

When $-z \ll z_0$, (8.194) gives

$$\hat{\alpha}_r^p \cdot \tilde{\mathbf{P}}_a^{\text{sca}}(k_x, k_y, k) = -\frac{i\gamma_a^2}{4\pi k_a z_0} Q_t^p(k) f\left(\frac{1}{2}k_x, \frac{1}{2}k_y\right) e^{i2\gamma_a z_0}$$
$$\cdot \tilde{O}(k_x, k_y, 2\gamma_b, k) \quad (8.196)$$

in which $\tilde{O}(k_x, k_y, k_z, k)$ is just the Fourier transform of $O(x, y, z, k)$ given by (8.182). In the intermediate region where $-z$ is comparable to z_0, we still use (8.195) as the approximation.

Equations (8.195) and (8.196) provide an efficient way to retrieve the object-function spectrum, which are valid for high frequencies due to the asymptotic expansion. Using these equations, two inversion algorithms are easily generated when the TX-RX are electric dipoles and magnetic dipoles, respectively.

As stated in the above section, it is impossible to simultaneously reconstruct the permittivity and conductivity profiles of buried objects using the DT algorithms without additional information. When $\sigma_r = \sigma_b$ or the contribution of permittivity is much larger than that of conductivity, we obtain

$$\tilde{O}_\epsilon(2k_x, 2k_y, 2k_{bz}) = \frac{\hat{\alpha}_r^p \cdot \tilde{\mathbf{P}}_a^{\text{sca}}(k_x, k_y, k_x, k_y, k) e^{-i2k_{az} z_0}}{Q_t^p(k) \hat{\alpha}_r^p \cdot \tilde{\overline{\mathbf{G}}}_{pe}^{ab}(k_x, k_y, k) \cdot \tilde{\overline{\mathbf{G}}}_{ep}^{ba}(-k_x, -k_y, k) \cdot \hat{\alpha}_t^p}$$
$$(8.197)$$

which is suitable for the full-information algorithms. For the partial-information algorithms, we obtain from (8.195) and (8.196)

$$\tilde{O}_\epsilon(k_x, k_y, 2\gamma_b) = -\frac{i4\pi k_b}{\gamma_b^2} \frac{\hat{\alpha}_r^p \cdot \tilde{\mathbf{P}}_a^{\text{sca}}(k_x, k_y, k)e^{-i2k_{az}z_0}}{Q_t^p(k)f\left(\frac{1}{2}k_x, \frac{1}{2}k_y\right)} \quad \text{for } -z \gg z_0 \quad (8.198)$$

$$\tilde{O}_\epsilon(k_x, k_y, 2\gamma_b) = \frac{i4\pi k_a z_0}{\gamma_a^2} \frac{\hat{\alpha}_r^p \cdot \tilde{\mathbf{P}}_a^{\text{sca}}(k_x, k_y, k)e^{-i2k_{az}z_0}}{Q_t^p(k)f\left(\frac{1}{2}k_x, \frac{1}{2}k_y\right)} \quad \text{for } -z \ll z_0 \quad (8.199)$$

In contrast, when $\epsilon_r = \epsilon_b$ or the contribution of conductivity is much larger than that of permittivity, one obtains

$$\tilde{O}_\sigma(2k_x, 2k_y, 2k_{bz}) = -\frac{ik}{\eta_0} \frac{\hat{\alpha}_r^p \cdot \tilde{\mathbf{P}}_a^{\text{sca}}(k_x, k_y, k_x, k_y, k)e^{-i2k_{az}z_0}}{Q_t^p(k)\hat{\alpha}_r^p \cdot \tilde{\mathbf{G}}_{pe}^{ab}(k_x, k_y, k) \cdot \tilde{\mathbf{G}}_{ep}^{ba}(-k_x, -k_y, k) \cdot \hat{\alpha}_t^p} \tag{8.200}$$

which is suitable for the full-information algorithms, and

$$\tilde{O}_\sigma(k_x, k_y, 2\gamma_b) = -\frac{4\pi k k_b}{\eta_0 \gamma_b^2} \frac{\hat{\alpha}_r^p \cdot \tilde{\mathbf{P}}_a^{\text{sca}}(k_x, k_y, k)e^{-i2k_{az}z_0}}{Q_t^p(k)f\left(\frac{1}{2}k_x, \frac{1}{2}k_y\right)} \quad \text{for } -z \gg z_0 \quad (8.201)$$

$$\tilde{O}_\sigma(k_x, k_y, 2\gamma_b) = \frac{4\pi k k_a z_0}{\eta_0 \gamma_a^2} \frac{\hat{\alpha}_r^p \cdot \tilde{\mathbf{P}}_a^{\text{sca}}(k_x, k_y, k)e^{-i2k_{az}z_0}}{Q_t^p(k)f\left(\frac{1}{2}k_x, \frac{1}{2}k_y\right)} \quad \text{for } -z \ll z_0 \quad (8.202)$$

which are suitable for the partial-information algorithms. In the above equations, $\tilde{O}_\epsilon(k_x, k_y, k_z)$ and $\tilde{O}_\sigma(k_x, k_y, k_z)$ are Fourier transforms of the permittivity-object function $O_\epsilon(x, y, z)$ and conductivity-object function $O_\sigma(x, y, z)$, respectively; and $\tilde{\hat{O}}_\epsilon(k_x, k_y, k_z)$ as well as $\tilde{\hat{O}}_\sigma(k_x, k_y, k_z)$ are Fourier transforms of $\hat{O}_\epsilon(x, y, z)$ and $\hat{O}_\sigma(x, y, z)$, which satisfy $\hat{O}_{\epsilon,\sigma}(x, y, z) = O_{\epsilon,\sigma}(x, y, z)/z$.

The inversion of object functions from their Fourier spectra $\tilde{O}_{\epsilon,\sigma}(k_x, k_y, k_z)$ or $\tilde{\hat{O}}_{\epsilon,\sigma}(k_x, k_y, k_z)$ can be performed in a similar way to that for 2D objects.

8.5.4.4 Inversion Results

In the following examples, the relative permittivity and conductivity of the lossy earth are chosen as $\epsilon_b = 4$ and $\sigma_b = 1$ mS/m. The reconstruction domain $\mathcal{D}$ and measurement domain $\mathcal{S}$ vary with working frequencies. However, $64 \times 64 \times 64 = 262{,}144$ pixels are always used to divide the reconstruction domain. The real buried target is a dielectric cube centered in $\mathcal{D}$, whose sidelength is one-fourth of that of $\mathcal{D}$, and dielectric parameters are set to $\epsilon_r = 6$ and $\sigma_r = 2$ mS/m if it is not specified. To view the inversion results, two slices of the reconstruction domain will be shown in these examples.

To reduce the experimental cost, partial-information algorithms are used in these examples, where the transmitter and receiver have a fixed offset: $\Delta x = 0$ and

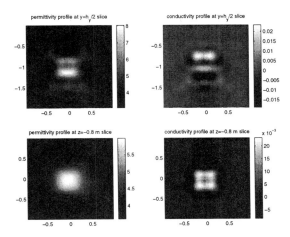

Figure 8.30 Reconstruction results from the electric-dipole algorithm when the working frequency ranges from 1.5 to 300 MHz. 3D DT algorithm where 4,096 TX-RX locations and 50 frequency points are used. Left side: permittivity profiles. Right side: conductivity profiles.

$\Delta y = 0.05$m. In each example, the experimental data are collected at $64 \times 64 = 4,096$ transmitter-receiver locations, which are simulated from the CG-FFT solver by adding a 5% random error. In most of the examples, the transmitter is a vertical electric (or magnetic) dipole while the receiver is a horizontal electric (or magnetic) dipole except it is specified. The reason why we use an orthogonal arrangement for TX and RX is to avoid receiving a direct wave from the transmitter so that scattered fields can be easily picked up at the receiver.

First, we consider the reconstruction from relatively high-frequency data that are measured from 1.5 to 300 MHz by using 50 sampling points. In this example, the reconstruction domain $\mathcal{D}$ is $2 \times 2 \times 2$ m^3 and the measurement domain $\mathcal{S}$ is 8×8 m^2. Using the electric-dipole algorithm, the reconstructed profiles at two slices $y = h_y/2$ and $z = -0.8$m are shown in Figure 8.30. From this figure, we notice that high-resolution inversion results are obtained where the corners of dielectric cube are clearly visible and the reconstructed permittivity gives a good approximation to the original value. However, the reconstructed conductivity is not accurate because (8.201) and (8.202) are derived for pure-conductive objects. Similarly, the magnetic-dipole algorithm can also give a high-resolution inversion result.

When the upper frequency increases, for example, to 600 MHz, the image resolution can be further improved. However, the reconstruction domain cannot be

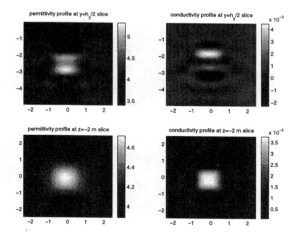

Figure 8.31 Reconstruction results from the electric-dipole algorithm when the working frequency ranges from 1 to 100 MHz. 3D DT algorithm where 4,096 TX-RX locations and 50 frequency points are used. Left side: permittivity profiles. Right side: conductivity profiles.

large at high frequencies due to limited computer memory. Hence, we should use lower-frequency data to reconstruct the larger domain.

When the working frequency ranges from 1 to 100 MHz with 50 sampling points, the reconstructed profiles from the electric-dipole algorithm at two slices $y = h_y/2$ and $z = -2$m are illustrated in Figure 8.31. Here, the reconstruction domain is a $5 \times 5 \times 5$ m^3 cubic region and the measurement domain is 20×20 m^2. From Figure 8.31, high-resolution images are still obtained at the given frequency range, in which the shape of a dielectric cube is quite clear. Because a lower frequency range is used, a larger object (1.25m in sidelength) buried in a deeper region can be detected.

To investigate the impact of polarization on the reconstruction quality, we repeat the above arrangement by using a vertical receiver instead of a horizontal receiver, where z-polarized electric fields are measured. In this case, the spatial resolution of reconstruction profiles in the horizontal direction becomes worse for this vertical-receiver arrangement. However, the reconstructed permittivity is more accurate than that in Figure 8.31 [72].

Next, we come back to the horizontal-receiver arrangement and see what happens if the working frequency is even lower. When the working frequency ranges from 1 to 50 MHz with 50 sampling points, the reconstructed profiles from the electric-dipole algorithm shows that good-resolution images can still be obtained, although the shape of the dielectric cube becomes ambiguous [72].

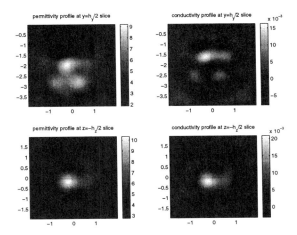

Figure 8.32 Reconstruction results from the electric-dipole algorithm for high-contrast object when $\epsilon_r = 30$ and $\sigma_r = 500$ S/m. 3D DT algorithm where 4,096 TX-RX locations and 30 frequency points are used. Left side: permittivity profiles. Right side: conductivity profiles.

Actually, the proposed partial-information algorithms are high-frequency methods because asymptotic expansions are used to evaluate the Sommerfeld integrals. However, the above examples illustrate that they may also be used for lower frequencies and good-resolution images may be obtained. Our test shows that even when the upper frequency goes down to 30 MHz, both electric-dipole and magnetic-dipole algorithms can give good-resolution results. When the upper frequency decreases to 10 MHz, the electric-dipole algorithm can still be used to detect the targets, but the magnetic-dipole algorithm cannot. The physical reason why the magnetic-dipole algorithm breaks down is that a magnetic dipole primarily radiates magnetic field. However, the buried targets are usually dielectric, requiring electric field for interaction. At low frequencies, however, the magnetic field and electric field are weakly coupled.

In the above examples, the buried dielectric cube has a low contrast compared to the background, which is just the case the proposed algorithms are valid for. However, in most practical problems, the buried targets have usually a high contrast. Hence, the proposed methods will be very useful if they can detect high-contrast objects.

Next, we consider the reconstruction of a high-contrast dielectric object. Figure 8.32 displays the inversion result from the electric-dipole algorithm when the dielectric parameters are $\epsilon_r = 30$ and $\sigma_r = 500$ mS/m. Here, the working frequency ranges from 1 to 100 MHz, and 30 sampling points are taken. From Figure 8.32, the high-contrast object can be clearly detected using the proposed inversion algorithm, where the buried depth is accurately given and the size of the cube is well approximated. Similar reconstruction results can be obtained from the magnetic-dipole algorithm.

REFERENCES

1. P. Parhami and R. Mittra, "Wire antenna over lossy half space," *IEEE Trans. Antennas Propagat.*, vol. 28, pp. 397–403, 1980.

2. G. J. Burke, E. K. Miller, J. N. Brittingham, D. L. Lager, R. J. Lytle, and J. T. Okada, "Computer modeling of antennas near the ground," *Electromagnetics*, vol. 1, pp. 29–49, 1981.

3. G. J. Burke and E. K. Miller, "Modeling antennas near to and penetrating a lossy interface," *IEEE Trans. Antennas Propagat.*, vol. 32, pp. 1040–1049, 1984.

4. T. J. Cui and W. C. Chew, "Modeling of arbitrary wire antennas above ground," *IEEE Trans. Geosci. Remote Sensing*, vol. 38, pp. 357–365, no.1, Part 2, Jan. 2000.

5. T. J. Cui and W. C. Chew, "Accurate model of arbitrary wire antennas in free space, above or inside ground," *IEEE Trans. Antennas Propagat.*, vol. 48, pp. 482–493, no. 4, Apr. 2000.

6. C. M. Butler, "Current induced on a conducting strip which resides on the planar interface between two semi-infinite half spaces," *IEEE Trans. Antennas Propagat.*, vol. 32, pp. 226–231, March 1984.

7. C. M. Butler, "TM scattering by a narrow conducting strip located at the interface between two semi-infinite half-spaces," *Radio Sci.*, vol. 22, pp. 983–986, 1987.

8. B. A. Baertlein, J. R. Wait, and D. G. Dudley, "Scattering by a conducting strip over a lossy half-space," *Radio Sci.*, vol. 24, pp. 485–497, 1989.

9. S. F. Mahmoud, S. M. Ali, and J. R. Wait, "Electromagnetic scattering from a buried cylindrical inhomogeneity inside a lossy earth," *Radio Sci.*, vol. 16, pp. 1285–1298, Nov.-Dec. 1981.

10. C. M. Butler, X. B. Xu, and A. W. Glisson, "Current induced on a conducting cylinder located near the planar interface between two semi-infinite half-space," *IEEE Trans. Antennas Propagat.*, vol. 33, pp. 616–624, June 1985.

11. X. B. Xu and C. M. Butler, "Current induced by TE excitation on a conducting cylinder located near the planar interface between two semi-infinite half-spaces," *IEEE Trans. Antennas Propagat.*, vol. 34, pp. 880–890, July 1986.

12. X. B. Xu and C. M. Butler, "Scattering of TM excitation by coupled and partially buried cylinders at the interface between two media," *IEEE Trans. Antennas Propagat.*, vol. 35, pp. 529–538, May 1987.

13. P. G. Cottis and J. D. Kanellopoulos, "Scattering from a conducting cylinder above a lossy medium," *Int. J. Electron.*, vol. 65, pp. 1031–1038, 1988.

14. X. B. Xu and C. M. Butler, "Current induced by TE excitation on coupled and partially buried cylinders at the interface between two media," *IEEE Trans. Antennas Propagat.*, vol. 38, pp. 1823–1828, Nov. 1990.

15. Y. Leviatan and Y. Meyouhas, "Analysis of electromagnetic scattering from buried cylinders using a multi-filament current model," *Radio Sci.*, vol. 25, pp. 1231–1244, 1990.

16. T. J. Cui and W. Wiesbeck, "TE wave scattering by multiple two-dimensional scatterers buried under one-dimensional multi-layered media," *Proc. IEEE International Geoscience and Remote Sensing Symposium*, pp. 766–768, vol. I, Lincoln, Nebraska, May 27–31, 1996.

17. T. J. Cui and W. Wiesbeck, "TM wave scattering by multiple two-dimensional scatterers buried under one-dimensional multi-layered media," *Proc. IEEE International Geoscience and Remote Sensing Symposium*, pp. 763–765, vol. I, Lincoln, Nebraska, May 27–31, 1996.

18. T. J. Cui and W. C. Chew, "Efficient evaluation of Sommerfeld integrals for TM wave scattering by buried objects," *J. Electromagnetic Waves Appl.*, vol. 12, pp. 607–657, 1998.

19. P. E. Wannamaker, G. W. Hohmann, and W. A. SanFilipo, "Electromagnetic modeling of three-dimensional bodies in layered earths using integral equations," *Geophys.*, vol. 49, pp. 60–74, Jan. 1984.

20. K. A. Michalski and D. Zheng, "Electromagnetic scattering and radiation by surfaces of arbitrary shape in layered media, Part I: Theory," *IEEE Trans. Antennas Propagat.*, vol. 38, pp. 335–344, March 1990.

21. K. A. Michalski and D. Zheng, "Electromagnetic scattering and radiation by surfaces of arbitrary shape in layered media, Part II: Implementation and results for contiguous half-spaces," *IEEE Trans. Antennas Propagat.*, vol. 38, pp. 345–352, March 1990.

22. T. J. Cui, W. Wiesbeck, and A. Herschlein, "Electromagnetic scattering by multiple dielectric and conducting objects buried under multi-layered media, Part II: Numerical implementation and results," *IEEE Trans. Geosci. Remote Sensing*, vol. 36, pp. 526–546, March 1998.

23. T. J. Cui, W. Wiesbeck, and A. Herschlein, "Electromagnetic scattering by multiple dielectric and conducting objects buried under multi-layered media, Part I: Theory," *IEEE Trans. Geosci. Remote Sensing*, vol. 36, pp. 526–546, March 1998.

24. T. J. Cui and W. C. Chew, "Efficient method for the near-field scattering by buried dielectric and conducting objects," *Electromagnetics*, vol. 18, pp. 555–573, 1998.

25. T. J. Cui and W. C. Chew, "Fast evaluation of Sommerfeld integrals for EM scattering and radiation by three-dimensional buried objects," *IEEE Trans. Geosci. Remote Sensing*, vol. 37, pp. 887–900, March 1999.

26. T. K. Sarkar, E. Arvas, and S. M. Rao, "Application of FFT and the conjugate method for the solution of electromagnetic radiation from electrically large and small conducting bodies," *IEEE Trans. Antennas Propagat.*, vol. 34, pp. 635–640, May 1986.

27. D. T. Borup, D. M. Sullivan, and O. P. Gandhi, "Comparison of the FFT conjugate gradient method and the finite-difference time-domain method for the 2-D absorption problem," *IEEE Trans. Microwave Theory Tech.*, vol. 35, pp. 383–395, April 1987.

28. T. J. Peters and J. L. Volakis, "Application of a conjugate gradient FFT method to scattering from thin planar material plates," *IEEE Trans. Antennas Propagat.*, vol. 36, pp. 518–526, April 1988.

29. M. F. Catedra, J. G. Cuevas, and L. Nuno, "A scheme to analyze conducting plates of resonant size using the conjugate-gradient method and the fast Fourier transform," *IEEE Trans. Antennas Propagat.*, vol. 36, pp. 1744–1752, Dec. 1988.

30. M. F. Catedra, E. Gago, and L. Nuno, "A numerical scheme to obtain the RCS of three-dimensional bodies of resonant size using the conjugate gradient method and the fast Fourier transform," *IEEE Trans. Antennas Propagat.*, vol. 37, pp. 528–537, May 1989.

31. A. P. M. Zwamborn and P. M. van dan Berg, "The weak form of the conjugate gradient method for plate problems," *IEEE Trans. Antennas Propagat.*, vol. 39, pp. 224–228, Feb. 1991.

32. J. M. Jin and J. L. Volakis, "A biconjugate gradient FFT solution for scattering by planar plates," *Electromagnetics*, vol. 12, pp. 105–109, Jan.-March 1992.

33. T. V. Tran and A. McCowen, "An improved pulse-basis conjugate gradient FFT method for the thin conducting plate problem," *IEEE Trans. Antennas Propagat.*, vol. 41, pp. 185–190, Feb. 1993.

34. C. C. Su, "The three-dimensional algorithm of solving the electric field integral equation using face-centered node points, conjugate gradient method, and FFT," *IEEE Trans. Microwave Theory Tech.*, vol. 41, pp. 510–515, March 1993.

35. H. Gan and W. C. Chew, "A discrete BCG-FFT algorithm for solving 3D inhomogeneous scatterer problems," *J. Electromagnetic Waves Appl.*, vol. 9, pp. 1339–1357, 1995.

36. Y. Zhuang, K.-L. Wu, C. Wu, and J. Litva, "A combined full-wave CG-FFT method for rigorous analysis of large microstrip antenna arrays," *IEEE Trans. Antennas Propagat.*, vol. 44, pp. 102–109, Jan. 1996.

37. W. M. Boerner, A. K. Jordan, and I. W. Kay, (eds.), "Special Issue on Inverse Methods in Electromagnetics," *IEEE Trans. Antennas Propagat.*, vol. 29, no. 2, March 1981.

38. K. P. Bube and R. Burridge, "The one-dimensional inverse problem of reflection seismology," *SIAM Rev.*, vol. 25, no. 4, pp. 497–559, 1983.

39. T. J. Cui and C. H. Liang, "Inverse scattering method for one-dimensional inhomogeneous lossy medium by using a microwave networking technique," *IEEE Trans. Microwave Theory Tech.*, vol. 43, no. 8, pp. 1773–1781, Aug. 1995.

40. R. K. Mueller, M. Kaveh, and G. Wade, "Reconstructive tomography and applications to ultrasonics," *Proc. IEEE*, vol. 67, pp. 567–587, 1979.

41. D. Colton, "The inverse electromagnetic scattering problem for a perfectly conducting cylinder," *IEEE Trans. Antennas Propag.*, vol. 29, pp. 364–368, 1981.

42. A. J. Devaney, "A computer simulation study of diffraction tomography," *IEEE Trans. Biomed. Eng.*, vol. 30, pp. 377–386, 1983.

43. M. Slaney, A. C. Kak, and L. E. Larsen, "Limitations of imaging with first-order diffraction tomography," *IEEE Trans. Microwave Theory Tech.*, vol. 32, pp. 860–874, 1984.

44. D. K. Ghodgonkar, O. P. Gandhi, and M. J. Hagmann, "Estimation of complex permittivities of three-dimensional inhomogeneous biological bodies," *IEEE Trans. Microwave Theory Tech.*, vol. 31, pp. 442–446, 1983.

45. W. C. Chew and Y. M. Wang, "Reconstruction of two-dimensional permittivity distribution using the distorted Born iterative method," *IEEE Trans. Medical Imaging*, vol. 9, pp. 218–225, 1990.

46. R. Kleinman and P. M. van den Berg, "An extended modified gradient technique for profile inversion," *Radio Sci.*, vol. 28, pp. 887–894, 1993.

47. S. Barkeshli and R. G. Lautzenheiser, "An iterative method for inverse scattering problems based on an exact gradient research," *Radio Sci.*, vol. 29, pp. 1119–1130, 1994.

48. W. C. Chew, G. P. Otto, W. H. Weedon, J. H. Lin, C. C. Lu, Y. M. Wang, and M. Moghaddam, "Nonlinear diffraction tomography: the use of inverse scattering for imaging," *Int. J. Imag. Sys. Tech.*, vol. 7, pp. 16–24, 1996.

49. J. H. Lin and W. C. Chew, "Solution of the three-dimensional electromagnetic inverse problem by the local shape function and the conjugate gradient fast Fourier transform methods," *J. Opt. Soc. Am. A*, vol. 14, no. 11, pp. 3037–3045, Nov. 1997.

50. W. C. Chew, "Imaging and inverse problems in electromagnetics," in *Advances in Computational Electrodynamics*, A. Taflove, (Ed.), Boston: Artech House, pp. 653–702, 1998.

51. T. Isernia, V. Pascazio, and R. Pierri, "A nonlinear estimation method in tomographic imaging," *IEEE Trans. Geosci. Remote Sensing*, vol. 35, pp. 910–923, 1997.

52. A. Franchois and C. Pichot, "Microwave imaging — Complex permittivity reconstruction with a Levenberg-Marquardt method," *IEEE Trans. Antennas Propag.*, vol. 45, pp. 203–215, Feb. 1997.

53. R. Pierri and G. Leone, "Inverse scattering of dielectric cylinders by a second-order Born approximation," *IEEE Trans. Geosci. Remote Sensing*, vol. 37, pp. 374–382, Jan. 1999.

54. Q. H. Liu, "Reconstruction of two-dimensional axisymmetric inhomogeneous media," *IEEE Trans. Geosci. Remote Sensing*, vol. 31, pp. 587–594, 1993.

55. W. C. Chew and Q. H. Liu, "Inversion of induction tool measurements using the distorted Born iterative method and CG-FFHT," *IEEE Trans. Geosci. Remote Sensing*, vol. 32, pp. 878–884, 1994.

56. A. J. Witten, J. E. Molyneux, and J. E. Nyquist, "Ground penetrating radar tomography: algorithm and case studies," *IEEE Trans. Geosci. Remote Sensing*, vol. 32, pp. 461–467, 1994.

57. R. Deming and A. J. Devaney, "Diffraction tomography for multi-monostatic ground penetrating radar imaging," *Inverse Probl.*, vol. 13, pp. 29–45, 1997.

58. P. Chaturvedi and R. G. Plumb, "Electromagnetic imaging of underground targets using constrained optimization," *IEEE Trans. Geosci. Remote Sensing*, vol. 33, pp. 551–561, 1995.

59. L. Souriau, B. Duchene, D. Lesselier, and R. Kleinman, "Modified gradient approach to inverse scattering for binary objects in stratified media," *Inverse Problem*, vol. 12, pp. 463–481, 1996.

60. M. Lambert, D. Lesselier, and B. J. Kooij, "The retrieval of a buried cylindrical obstacle by a constrained modified gradient method in the H-polarization case and for Maxwellian materials," *Inverse Problem*, vol. 14, pp. 1265–1283, 1998.

61. O. M. Bucci, L. Crocco, T. Isernia, V. Pascazio, and R. Pierri, "A bilinear approach for subsurface sensing: numerical results," *Proc. International Geoscience and Remote Sensing Symposium*, July 1998, Seattle, WA.

62. T. B. Hansen and P. M. Johansen, "Inversion scheme for monostatic ground penetrating radar that takes into account the planar air-soil interface," *IEEE Trans. Geosci. Remote Sensing*, vol. 38, pp. 496–506, Jan. 2000.

63. J. S. Zhao, W. C. Chew, C. C. Lu, E. Michielssen, and J. M. Song, "Thin-stratified medium fast-multipole algorithm for solving microstrip structures," *IEEE Trans. Microwave Theory Tech.*, vol. 46, pp. 395–403, April 1998.

64. T. J. Cui and W. C. Chew, "Fast algorithm for electromagnetic scattering by buried conducting plates of large size," *IEEE Trans. Antennas Propagat.*, vol. 47, pp. 1116–1118, June 1999.

65. T. J. Cui and W. C. Chew, "Fast algorithm for electromagnetic scattering by buried 3D dielectric objects of large size," *IEEE Trans. Geosci. Remote Sensing*, vol. 37, pp. 2597–2608, Sept. 1999.

66. D. L. Wright, T. P. Grover, V. F. Labson, and L. Pellerin, "The very early time electromagnetic (VETEM) system: first field test results," *Proceeding of the 9th Annual Symposium on the Applications of Geophysics to Environmental and Engineering Problem (SAGEEP)*, pp. 81–90, Keystone, CO, April 28–May 2, 1996.

67. D. L. Wright, D. V. Smith, J. D. Abraham, R. T Smith, T. J. Cui, and W. C. Chew, "New field and modeling results from a simulated waste pit using the enhanced very early time electromagnetic (VETEM) prototype system," *Proceeding of the 12th Annual Symposium on the Applications of Geophysics to Environmental and Engineering Problem (SAGEEP)*, Oakland, CA, March 14–18, 1999.

68. T. J. Cui, W. C. Chew, A. A. Aydiner, D. L. Wright, D. V. Smith, and J. D. Abraham, "Numerical modeling of an enhanced very early time electromagnetic (VETEM) prototype system," *IEEE Antennas and Propagation Magazine*, vol. 42, no. 2, pp. 17–27, April 2000.

69. T. J. Cui, W. C. Chew, A. A. Aydiner, and S. Y. Chen, "Inverse scattering of 2D dielectric objects buried in a lossy earth using the distorted Born iterative method," *IEEE Trans. Geosci. Remote Sensing*, vol. 39, pp. 339–346, Feb. 2001.

70. T. J. Cui and W. C. Chew, "Frequency-spatial domain inverse scattering of two-dimensional dielectric objects buried under a lossy earth," *Research Report*, Electromagnetics Laboratory, University of Illinois at Urbana-Champaign, no. CCEM-19-99, July 1999.

71. T. J. Cui and W. C. Chew, "Novel diffraction tomographic algorithm for imaging two-dimensional targets buried under a lossy earth," *IEEE Trans. Geosci. Remote Sensing*, vol. 38, pp. 2033–2041, July 2000.

72. T. J. Cui and W. C. Chew, "Diffraction tomographic algorithms for the detection of three-dimensional objects buried in a lossy half space," *Research Report*, Electromagnetics Laboratory, University of Illinois at Urbana-Champaign, no. CCEM-6-00, Feb. 2000; to be published in *IEEE Trans. Antennas Propagat.*, vol. 49, no. 12, Dec. 2001.

9
Low-Frequency Scattering from Penetrable Bodies

Siyuan Chen and Weng Cho Chew

9.1 INTRODUCTION

Computational electromagnetics has recently become an important analysis tool due to the recent advent of fast algorithms that allow integral equation solutions involving millions of unknowns [1, 2], (see Chapter 3). Computational electromagnetics has also played a very important role in geoscience and remote sensing, for instance, in the design of electrical logging tools [3], interpretation of measurement data [3, 4], as well as in understanding rough-surface scattering [5, 6], and remotely sensed data [7, 8].

However, computational electromagnetics in an electrical prospection setting often has a different set of requirements. First, the need for large penetration depth dictates that the frequency of operation is low, and that the wavelength is long. Therefore, the measurement apparatus or the features being probed is often much smaller than the wavelength. This regime of operation is often known as the quasi-static regime. However, it is well known that many computational electromagnetics methods, such as the method of moments (MOM) and the finite element method (FEM) developed for electrodynamics, break down when the frequency is very low [9–15]. Hence, it is often simpler to solve a problem right at static with Laplace's or Poisson's equations, but not at very low frequency. But it is well known that many electrical logging tools operate at very low frequencies, which is not right at static. Examples are the induction logging tool and the laterolog, where low-frequency physics is sometimes important.

In this work, we will study the solution of integral equations using the method of moments such that the solution is stable all the way from static to some moderately low frequencies. The low-frequency breakdown problem can be described in terms of the natural Helmholtz decomposition of Maxwell's equations. As the frequency decreases, the electric and magnetic fields are decoupled. The unknown current consists of two components, a curl-free part and a divergence-free part. At zero frequency, the divergence-free part produces only the magnetic field. The electric field is produced by the charge, which is related to the curl-free current through the current continuity equation

$$\nabla \cdot \epsilon \mathbf{E} = \rho = \lim_{\omega \to 0} \nabla \cdot \mathbf{J}/i\omega \qquad (9.1)$$

Hence, to maintain a physically finite charge, it is necessary that the divergence of the curl-free part scales as $O(\omega)$ as the frequency approaches zero. No such frequency scaling is needed for the divergence-free part [15].

The low-frequency breakdown can be demonstrated using the following example. The simulated object is a dielectric sphere with radius $r = 0.5$m and relative permittivity $\epsilon_r = 5.0$. The sphere is placed in free space and illuminated by a plane wave; 72 curvilinear patches are used to discretize the sphere, corresponding to 216 unknowns. We compute the bistatic radar cross-section (RCS) (dBsm) at different frequencies. The results are shown in Figure 9.1, from which we can see that the results gradually become less accurate as the frequency decreases.

Many researchers have investigated this problem. One way to overcome this difficulty is to use loop-tree or loop-star basis functions, which separate the contribution from the divergence-free current and the curl-free current [9–12, 15].

So far, most of the works are based on solving the EFIE to determine scattering from a perfect electric conducting (PEC) body. In geophysics and optics applications, penetrable scatterers are frequently encountered. The use of loop-tree formulations to study scattering from a penetrable scatterer is not a trivial extension of the existing methods. Another integral operator that is related to the magnetic current is involved. The frequency scaling property of each operator must be analyzed for the proper frequency normalization.

Many applications involve more than one body. Many researchers have studied this problem; some of the pioneering work can be found in [16–21]. Most of the research work has been done in the high-frequency regime, where the traditional Rao-Wilton-Glisson (RWG) basis function is used. The commonly used approach is to match the boundaries on the interface that separates two materials, and the resulting integral equations are tested to obtain the MOM equations. Since the RWG basis function is "localized," the implementation is straightforward.

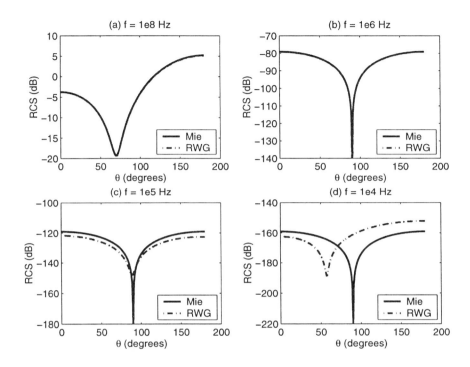

Figure 9.1 Bistatic RCS computed using RWG basis at different frequencies. Seventy-two curvilinear patches are used. The RWG basis results gradually lose accuracy.

In the low-frequency regime, which is important for geoscience, the loop-tree basis function is used to overcome the low-frequency breakdown problem. Combined with a proper frequency normalization scheme, the penetrable single body problem has been solved [22, 23].

For multibody problems, if the bodies do not touch each other, the procedure developed for the single-body case can be applied to each body. However, once the bodies touch each other, it is difficult to match the boundaries on the interface which separates two materials, since the loop-tree basis is not "localized." To overcome this difficulty, we view the touched bodies as the limiting case of the separated bodies. However, as the bodies touch each other, the unknowns originally defined on each body become redundant at the common surfaces. A number-of-unknowns reduction scheme (which we shall subsequently call "NOUR" for brevity) is needed for this case. In the high-frequency regime, where the RWG basis function is used, the construction of the NOUR scheme is straightforward, since the support of the

RWG basis function is "localized." For the loop-tree basis function used in the low-frequency regime, the NOUR scheme is more complicated. This will be the case not only because the support of the loop-tree basis function is "nonlocalized" but also because frequency normalization is needed.

This chapter is organized as follows. In Section 9.2, we discuss low-frequency scattering from a single penetrable body. We start with the analysis of frequency scaling property of the matrix element, which is obtained from the PMCHWT formulation using a loop-tree basis function. A careful analysis of the frequency scaling property of each operator allows us to introduce a frequency normalization scheme to reduce the condition number of the MOM matrix. In Section 9.3, we discuss the low-frequency scattering from many bodies. By studying some generic problems, we propose a systematic way to solve the multibody problem and generalize the rules to construct the NOUR matrix. Numerical simulations are used to validate our approach and to demonstrate some applications in the RCS computation and induction well-logging modeling.

9.2 LOW-FREQUENCY SCATTERING FROM A SINGLE PENETRABLE BODY

In this section, we first briefly review the concepts of loop-tree basis functions. After obtaining the MOM matrix using loop-tree basis, we analyze the frequency scaling property of each operator and give the corresponding frequency normalization scheme. Numerical results and a physical interpretation are used to show how and why the method works.

9.2.1 A Brief Review of Basis Functions

The definition of RWG basis function associated with the nth interior edge is given by [24, 25]

$$\Lambda_n(\mathbf{r}) = \begin{cases} \frac{1}{2S_n^+} \rho_n^+, \mathbf{r} \in S_n^+ \\ \frac{1}{2S_n^-} \rho_n^-, \mathbf{r} \in S_n^- \end{cases} \tag{9.2}$$

where $S_n^\pm$ is the area of the two triangular patches, respectively, and $\rho_n^\pm$ is shown in Figure 9.2. This definition is slightly different from the original in that it is normalized by the interior edge length.

The nth loop basis function is defined on all the triangular patches attached to the nth interior node as shown in Figure 9.3(a). The definition of the loop basis function is [9–11]

$$\mathbf{O}_n(\mathbf{r}) = \sum_{i=1}^{N_n} \frac{\mathbf{L}_i}{S_i} f_i(\mathbf{r}) \tag{9.3}$$

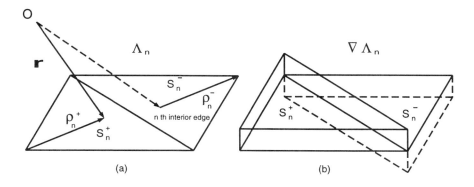

Figure 9.2 RWG basis function: (a) one RWG basis function defined on a pair of patches; (b) divergence of the RWG basis function.

where N_n is the total number of patches attached to the nth interior node and

$$f_i(\mathbf{r}) = \begin{cases} 1, & \mathbf{r} \in S_i \\ 0, & \text{otherwise} \end{cases} \quad (9.4)$$

The vector $\mathbf{L}_i$ is parallel to the edge opposite the nth interior node in the ith triangular patch and its length is equal to the length of the edge. It is easy to prove that the loop basis function is a linear combination of a set of RWG basis functions as shown in Figure 9.3(b). The loop basis function describes the current flowing around the interior node.

A loop basis function is divergence-free. This property allows us to rewrite the loop basis as [11, 15]

$$\mathbf{O}(\mathbf{r}) = \sum_i \hat{n}_i \times \nabla_S \Psi_i(\mathbf{r}) \quad (9.5)$$

where

$$\hat{n}_i = \frac{\mathbf{r}_{be} \times \mathbf{r}_{db}}{|\mathbf{r}_{be} \times \mathbf{r}_{db}|} \quad (9.6)$$

$$\Psi_i = 1 + \frac{1}{2S_i} \boldsymbol{\rho} \cdot (\mathbf{r}_{be} \times \hat{n}_i) \quad (9.7)$$

The definitions of vectors $\mathbf{r}_{be}$ and $\mathbf{r}_{db}$ are given in Figure 9.3(c) and the value of the scalar function Ψ_i has a pyramidal shape as shown in Figure 9.3(d).

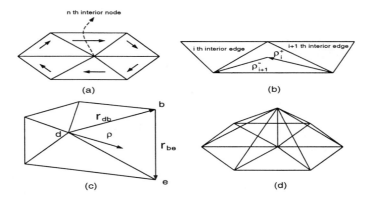

Figure 9.3 Loop basis function: (a) a loop basis function; (b) a loop basis function can be viewed as a linear combination of several RWG basis functions; (c) definitions of vectors $\mathbf{r}_{db}$ and $\mathbf{r}_{be}$; (d) value of the scalar function Ψ_i on the support of the loop basis function.

The tree basis function is complementary to the loop basis function. Combined, they form a complete set in the RWG space, which is spanned by the RWG basis functions. Tree basis functions are a subset of conventional RWG basis functions: this subset is formed by excluding all the RWG basis functions associated with the branches of a tree structure. Two tree structures are shown in Figures 9.4(a) and 9.4(b).

It should be noted that tree basis functions, as opposed to the loop basis, are neither curl-free nor divergence-free. The dimension of the MOM matrix obtained by using loop-tree functions is the same as that obtained by using conventional RWG basis functions. Even though the tree structure for one object is not unique, a carefully chosen tree structure is essential for the whole scheme to work efficiently.

Another complementary basis function to the loop basis function is the star basis function. The nth star basis function is defined on the nth triangular patch and all patches attached to it. It is constructed by placing the RWG basis functions associated with the edges of the nth patch in such a way that the current flows out of the nth patch as shown in Figure 9.5. The expression of the star basis is [9, 12]

$$\star_n(\mathbf{r}) = \sum_{i=1}^{3} g_{ni} \mathbf{\Lambda}_{\mathbf{ni}}(\mathbf{r})$$

where the value of g_{ni} is chosen from $\{-1, 0, 1\}$ to guarantee the current to flow out of the nth patch or to eliminate the contribution from the boundary edge.

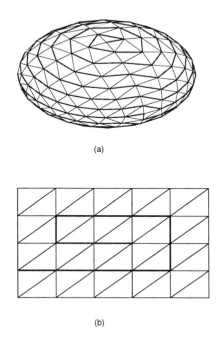

Figure 9.4 Denoted by bold solid lines: (a) a tree structure on a sphere; (b) a tree structure on a plate.

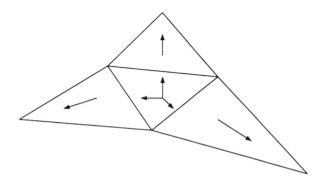

Figure 9.5 A star basis function.

It should be noted that neither the tree nor the star basis function is curl-free or divergence-free. The loop basis functions can be combined with either of them to form a complete set in the RWG space. The dimension of the MOM matrix obtained by using loop-tree or loop-star basis functions is the same as that obtained by using conventional RWG basis functions.

One tree function is defined on two patches, but one star function is usually defined on four patches. It is therefore easier to perform the basis function rearrangement using loop-tree basis functions [15]. Thus, the following discussion is based on loop-tree basis functions only.

9.2.2 General Equations and Frequency Normalization

Consider a penetrable scatterer with its parameters and outward normal vector $\hat{n}$ as shown in Figure 9.6. After applying the equivalence principle and boundary conditions at the surface S, we obtain a coupled set of integral equations for the unknown electric current $\mathbf{J}(\mathbf{r})$ and magnetic current $\mathbf{M}(\mathbf{r})$ as [19]

$$-\mathbf{E}^{\text{inc}}(\mathbf{r})|_{\tan} = (\mathcal{L}_1 + \mathcal{L}_2)\mathbf{J}(\mathbf{r})|_{\tan} - (\mathcal{K}_1 + \mathcal{K}_2)\mathbf{M}(\mathbf{r})|_{\tan} \quad (9.8)$$

$$-\eta_1 \mathbf{H}^{\text{inc}}(\mathbf{r})|_{\tan} = \eta_1(\mathcal{K}_1 + \mathcal{K}_2)\mathbf{J}(\mathbf{r})|_{\tan} - \left(\mathcal{L}_1 + \frac{\eta_1^2}{\eta_2^2}\mathcal{L}_2\right)\mathbf{M}(\mathbf{r})|_{\tan} \quad (9.9)$$

where $\mathbf{E}^{\text{inc}}(\mathbf{r})$ and $\mathbf{H}^{\text{inc}}(\mathbf{r})$ are incident electric and magnetic fields, respectively; η_i ($i = 1, 2$) is the wave impedance of medium i; and ϵ_{ri} and μ_{ri} ($i = 1, 2$) are the relative permittivity and permeability, respectively. The two operators $\mathcal{L}_i$ and $\mathcal{K}_i$ ($i = 1, 2$) are defined as [19]

$$\mathcal{L}_i \mathbf{X}(\mathbf{r}) = \int_S \left[i\omega\mu_i \mathbf{X}(\mathbf{r}') - \frac{i}{\omega\epsilon_i}\nabla\nabla' \cdot \mathbf{X}(\mathbf{r}')\right] g_i(\mathbf{r} - \mathbf{r}')d\mathbf{r}' \quad (9.10)$$

$$\mathcal{K}_i \mathbf{X}(\mathbf{r}') = \int_S \mathbf{X}(\mathbf{r}') \times \nabla g_i(\mathbf{r} - \mathbf{r}')d\mathbf{r}' \quad (9.11)$$

where the integrals are taken as Cauchy principal values. The vector function $\mathbf{X}(\mathbf{r})$ is defined on the boundary S; $\epsilon_i = \epsilon_0 \epsilon_{ri}$, $\mu_i = \mu_0 \mu_{ri}$, and ϵ_0 and μ_0 are the permittivity and permeability of free space, respectively. The Green's function is defined as

$$g_i(\mathbf{r} - \mathbf{r}') = \frac{\exp(ik_i|\mathbf{r} - \mathbf{r}'|)}{4\pi|\mathbf{r} - \mathbf{r}'|} \quad (9.12)$$

where

$$k_i = \omega\sqrt{\mu_i \epsilon_i} \quad (9.13)$$

The Galerkin procedure is used to obtain the impedance matrix. The loop-tree basis function is used to expand the unknown electric and magnetic currents as

$$\mathbf{J}(\mathbf{r}') = \sum_{n=1}^{N_O} J_n^O \mathbf{O}_n(\mathbf{r}') + \sum_{n=1}^{N_T} J_n^T \mathbf{T}_n(\mathbf{r}') \quad (9.14)$$

Low-Frequency Scattering from Penetrable Bodies 433

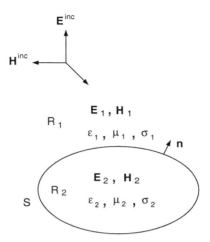

Figure 9.6 A generic penetrable 3-D scatter.

$$\mathbf{M}(\mathbf{r}') = \sum_{n=1}^{N_O} M_n^O \mathbf{O}_n(\mathbf{r}') + \sum_{n=1}^{N_T} M_n^T \mathbf{T}_n(\mathbf{r}') \qquad (9.15)$$

where $\mathbf{O}_n(\mathbf{r}')$ and $\mathbf{T}_n(\mathbf{r}')$ are the loop and tree basis functions, respectively. Using matrix notation, we can rewrite (9.14) and (9.15) as

$$\mathbf{J}(\mathbf{r}') = \mathbf{O}(\mathbf{r})^t \cdot \mathbf{J}_O + \mathbf{T}(\mathbf{r})^t \cdot \mathbf{J}_T \qquad (9.16)$$

$$\mathbf{M}(\mathbf{r}') = \mathbf{O}(\mathbf{r})^t \cdot \mathbf{M}_O + \mathbf{T}(\mathbf{r})^t \cdot \mathbf{M}_T \qquad (9.17)$$

where $\mathbf{O}(\mathbf{r})$, $\mathbf{T}(\mathbf{r})$, $\mathbf{J}_O$, $\mathbf{J}_T$, $\mathbf{M}_O$ and $\mathbf{M}_T$ are the column vectors.

The first terms in (9.16) and (9.17) are divergence-free. Since the second terms are not curl-free, (9.16) and (9.17) do not represent complete Helmholtz decompositions. However, a complete Helmholtz decomposition is not mandatory as long as the tree function always has a component in the curl-free space [15]. The basis function rearrangement will be introduced to represent the complete Helmholtz decomposition as the frequency approaches zero. Substituting (9.16) and (9.17) into (9.8) and (9.9), and testing them with the loop-tree basis functions $\mathbf{O}(\mathbf{r}')$ and $\mathbf{T}(\mathbf{r}')$, we obtain the matrix equation and corresponding frequency dependence as

$$\begin{bmatrix} \overline{\mathbf{A}}_{OO}(\omega) & \overline{\mathbf{A}}_{OT}(\omega) & -\eta_1\overline{\mathbf{B}}_{OO}(\omega^2) & -\eta_1\overline{\mathbf{B}}_{OT}(1) \\ \overline{\mathbf{A}}_{TO}(\omega) & \overline{\mathbf{A}}_{TT}(\omega^{-1}) & -\eta_1\overline{\mathbf{B}}_{TO}(1) & -\eta_1\overline{\mathbf{B}}_{TT}(1) \\ \eta_1\overline{\mathbf{C}}_{OO}(\omega^2) & \eta_1\overline{\mathbf{C}}_{OT}(1) & \overline{\mathbf{D}}_{OO}(\omega) & \overline{\mathbf{D}}_{OT}(\omega) \\ \eta_1\overline{\mathbf{C}}_{TO}(1) & \eta_1\overline{\mathbf{C}}_{TT}(1) & \overline{\mathbf{D}}_{TO}(\omega) & \overline{\mathbf{D}}_{TT}(\omega^{-1}) \end{bmatrix} \begin{bmatrix} \mathbf{J}_O(1) \\ \mathbf{J}_T(\omega) \\ \eta_1^{-1}\mathbf{M}_O(1) \\ \eta_1^{-1}\mathbf{M}_T(\omega) \end{bmatrix}$$

$$= \begin{bmatrix} \mathbf{V}_O^E(\omega) \\ \mathbf{V}_T^E(1) \\ \eta_1 \mathbf{V}_O^H(\omega) \\ \eta_1 \mathbf{V}_T^H(1) \end{bmatrix} \qquad (9.18)$$

where the quantities in the parentheses denote the scaling properties of the relevant terms. The frequency scaling is quite easy to derive except for the elements in $\overline{\mathbf{B}}_{OO}$, $\overline{\mathbf{C}}_{OO}$, and excitation vectors $\mathbf{V}_O^E$ and $\mathbf{V}_O^H$. The following derivation shows that they do have the frequency dependence shown in (9.18).

The elements of $\overline{\mathbf{B}}_{OO}$ are given by

$$B_{OO}^{mn} = \int_S d\mathbf{r} \mathbf{O}_m(\mathbf{r}) \cdot \int_{S'} d\mathbf{r}' \nabla g(\mathbf{r}, \mathbf{r}') \times \mathbf{O}_n(\mathbf{r}') \qquad (9.19)$$

where S and S' are the support of expanding and testing functions. At low frequencies, we can expand the gradient of the Green's function as

$$4\pi \nabla g(\mathbf{r}, \mathbf{r}') = \nabla \frac{1}{R} + \frac{\mathbf{R}}{R^3} \left[-\frac{(kR)^2}{2} - \frac{i(kR)^3}{3} + \cdots \right] \qquad (9.20)$$

where $\mathbf{R} = \mathbf{r} - \mathbf{r}'$. The first term in (9.20) will not contribute to the integral in (9.19); that is,

$$\int_S d\mathbf{r} \mathbf{O}_m(\mathbf{r}) \cdot \int_{S'} d\mathbf{r}' \nabla \frac{1}{R} \times \mathbf{O}_n(\mathbf{r}') = 0 \qquad (9.21)$$

since the static field which is produced by a loop source is curl-free outside the source region. Therefore, the elements of $\overline{\mathbf{B}}_{OO}$ scale as $O(\omega^2)$. The same procedure can be used to show that the elements of $\overline{\mathbf{C}}_{OO}$ also scale as $O(\omega^2)$.

We now analyze the frequency scaling properties of excitation vectors $\mathbf{V}_O^E$ and $\mathbf{V}_O^H$. It can be shown that the mth element of excitation vector $\mathbf{V}_O^E$ can be written as

$$\mathbf{V}_{Om}^E = i\omega\mu \sum_i \int \Psi_i(\mathbf{r}) \hat{n}_i \cdot \mathbf{H}^{\text{inc}}(\mathbf{r}) dS \qquad (9.22)$$

where scalar function Ψ_i is given in (9.7). If the incident wave is a plane wave $\mathbf{E}^{\text{inc}} = \mathbf{E}_0 \exp(i\mathbf{k} \cdot \mathbf{r})$, we have

$$\mathbf{V}_{Om}^E = ik \sum_i \int \Psi_i(\mathbf{r}) (\hat{n}_i \times \hat{k}) \cdot \mathbf{E}^{\text{inc}}(\mathbf{r}) dS \qquad (9.23)$$

It is obvious that the elements of the excitation vector $\mathbf{V}_O^E$ scale as $O(\omega)$. The frequency-dependence property is the same for elements of excitation vector $\mathbf{V}_O^H$.

The above frequency-dependence analysis not only provides the information for the frequency normalization, but also makes the numerical scheme stable at very low frequency. In conclusion, we should not use the first term in the RHS of (9.20) in computations. Theoretically, the $\nabla \frac{1}{R}$ term's contribution cancels out. However, this cancellation is never complete due to numerical error. This error sets a lower frequency bound for the frequency normalization scheme. Worse still, for the near term, $\nabla \frac{1}{R}$ gives rise to a second-order singularity that is difficult to handle even by existing singularity extraction techniques [26]. Consequently, the scheme will not be stable.

Frequency normalization is necessary for balancing the matrix elements and excitation vectors. To do so, we normalize (9.18) as

$$\overline{Z}' \cdot X' = Y' \tag{9.24}$$

where
$$\overline{Z}' = \overline{D}_1 \cdot \overline{Z} \cdot \overline{D}_2, \qquad X' = \overline{D}_2^{-1} \cdot X, \qquad Y' = \overline{D}_1 \cdot Y \tag{9.25}$$

$$\overline{D}_1 = \text{diag}\left\{k_1^{-1}\overline{I}_{N_1}, \overline{I}_{N_2}, k_1^{-1}\overline{I}_{N_1}, \overline{I}_{N_2}\right\} \tag{9.26}$$

$$\overline{D}_2 = \text{diag}\left\{\overline{I}_{N_1}, ik_1\overline{I}_{N_2}, \overline{I}_{N_1}, ik_1\overline{I}_{N_2}\right\} \tag{9.27}$$

and $\overline{Z}$, X, and Y are the impedance matrix, unknown current vector, and excitation vector given by (9.18), respectively; $\overline{I}_{N_1}$ is a $N_1 \times N_1$ identity matrix and N_1 is the number of loop basis functions; $\overline{I}_{N_2}$ is a $N_2 \times N_2$ identity matrix and N_2 is the number of tree basis functions.

The normalized matrix elements, unknowns, and excitation vectors of (9.24) now scale as

$$\begin{bmatrix} O(1) & O(\omega) & O(\omega) & O(1) \\ O(\omega) & O(1) & O(1) & O(\omega) \\ O(\omega) & O(1) & O(1) & O(\omega) \\ O(1) & O(\omega) & O(\omega) & O(1) \end{bmatrix} \begin{bmatrix} O(1) \\ O(1) \\ O(1) \\ O(1) \end{bmatrix} = \begin{bmatrix} O(1) \\ O(1) \\ O(1) \\ O(1) \end{bmatrix} \tag{9.28}$$

After the normalization, matrix equation (9.27) is well behaved even at very low frequencies. The contribution from the vector potential is preserved because it has been separated from the contribution from the scalar potential.

Using the loop-tree basis function and corresponding frequency normalization, we recompute the example in Section 9.1. In Figure 9.7, we show that the results are accurate even at very low frequency.

In geophysical applications, dipole sources are used and lossy media are encountered. Two examples are used to show the performance of low-frequency MOM in both lossless and lossy media. We use an electric dipole pointing in the x direction and operating at 20 kHz as a source. The media and geometry parameters used in the two examples are given in Figure 9.8. The electric fields are recorded at $\phi = 45°$, $R = 0.75$m, and $\theta = 0$ to $180°$. The results are shown in Figure 9.9(a–d) and Figure 9.10(a–d). The results agree well with Mie series solutions.

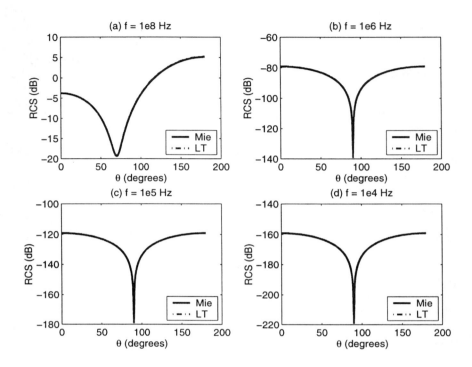

Figure 9.7 Bistatic RCS computed using loop-tree basis at different frequencies. Seventy-two curvilinear patches are used.

9.2.3 Interpretations

It is helpful to understand the physics that happens in the whole procedure. At zero frequency, a dynamic problem decouples into electrostatic and magnetostatic parts, which can be solved independently from each other. This decoupling is represented in (9.28), in which the first and fourth row equations represent the magnetostatic problem while the second and third row equations represent the electrostatic problem. This will be the case if the field is excited by a magnetic dipole or an electric dipole at low frequency as shown below.

For a magnetic dipole pointing in the $\hat{a}$ direction ($IA \Rightarrow 1$), the excitation vectors can be written as

$$\mathbf{V'}_O^E = -\frac{i\eta}{4\pi} \left\langle \hat{r} \times \hat{a} \left(\frac{ik}{r} - \frac{1}{r^2} \right) e^{ikr}, \mathbf{O} \right\rangle \qquad (9.29)$$

Low-Frequency Scattering from Penetrable Bodies 437

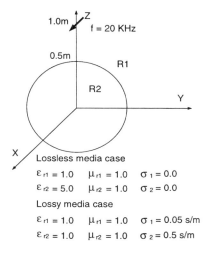

Figure 9.8 The medium and geometry parameters used for computing the scattered field from the lossless and lossy spheres.

$$\mathbf{V'}_T^E = -\frac{ik\eta}{4\pi}\left\langle \hat{r}\times\hat{a}\left(\frac{ik}{r}-\frac{1}{r^2}\right)e^{ikr}, \mathbf{T}\right\rangle \tag{9.30}$$

$$\mathbf{V'}_O^H = -\frac{k\eta}{4\pi}\left\langle \hat{a}\frac{e^{ikr}}{r}, \mathbf{O}\right\rangle \tag{9.31}$$

$$\mathbf{V'}_T^H = -\frac{\eta}{4\pi}\left\langle \left[\left(\frac{k^2}{r}+\frac{ik}{r^2}-\frac{1}{r^3}\right)\hat{a}\right.\right.$$
$$\left.\left.+\hat{r}(\hat{r}\cdot\hat{a})\left(-\frac{k^2}{r}-\frac{3ik}{r^2}+\frac{3}{r^3}\right)\right]e^{ikr}, \mathbf{T}\right\rangle \tag{9.32}$$

As the frequency approaches zero, the excitation vectors given by (9.30) and (9.31) are zero. This corresponds to the magnetostatic case. For an electric dipole, we choose

$$IL \Rightarrow i\omega\sqrt{\mu\epsilon} \tag{9.33}$$

The excitation vectors are

$$\mathbf{V'}_O^E = -\frac{k\eta}{4\pi}\left\langle \hat{a}\frac{e^{ikr}}{r}, \mathbf{O}\right\rangle \tag{9.34}$$

$$\mathbf{V'}_T^E = -\frac{\eta}{4\pi}\left\langle \left[\left(\frac{k^2}{r}+\frac{ik}{r^2}-\frac{1}{r^3}\right)\hat{a}\right.\right.$$

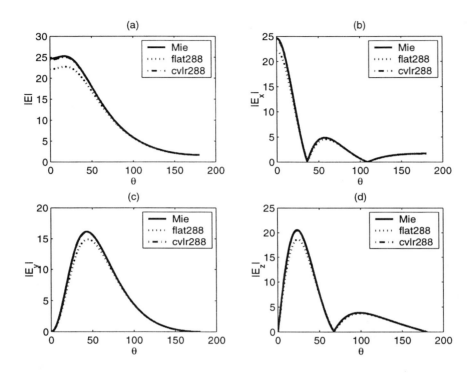

Figure 9.9 Scattered field from a lossless sphere. The field is excited by an electric dipole operating at $f = 20$ kHz. The parameters used are shown in Figure 9.8.

$$+ \hat{r}(\hat{r} \cdot \hat{a}) \left(-\frac{k^2}{r} - \frac{3ik}{r^2} + \frac{3}{r^3} \right) \Big] e^{ikr}, \mathbf{T} \Big\rangle \tag{9.35}$$

$$\mathbf{V}'^H_O = \frac{i\eta}{4\pi} \Big\langle \hat{r} \times \hat{a} \left(\frac{ik}{r} - \frac{1}{r^2} \right) e^{ikr}, \mathbf{O} \Big\rangle \tag{9.36}$$

$$\mathbf{V}'^H_T = \frac{ik\eta}{4\pi} \Big\langle \hat{r} \times \hat{a} \left(\frac{ik}{R} - \frac{1}{r^2} \right) e^{ikr}, \mathbf{T} \Big\rangle \tag{9.37}$$

As the frequency approaches zero, the excitation vectors given by (9.34) and (9.37) are zero. This corresponds to the electrostatic case.

If a penetrable body is illuminated by the static electric field, (9.24) implies the object can be replaced with an electric charge $\nabla \cdot \mathbf{J}_T/i\omega$ and a magnetic current $\mathbf{M}_O$ on the bounding surface. As we know from the extinction theorem for scalar waves, the field can be terminated using electric charge and a dipole layer instead of

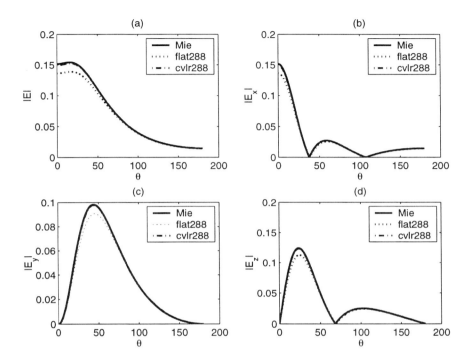

Figure 9.10 Scattered field from a lossy sphere buried in the lossy medium. The field is excited by an electric dipole operating at $f = 20$ kHz. The parameters used are also shown in Figure 9.8.

a magnetic current. As shown below, the dipole layer and magnetic current actually produce the same electric field.

The electric field produced by the magnetic current is

$$\mathbf{E}(\mathbf{r}) = -\int_S dS' \nabla \times \left[g(\mathbf{r},\mathbf{r}')\mathbf{M}(\mathbf{r}')\right] \qquad (9.38)$$

For the static case, (9.38) can be written in terms of the scalar potential $\phi(\mathbf{r}')$ as

$$\mathbf{E}(\mathbf{r}) = -\int_S dS' \nabla \times \left\{\mathbf{n}' \times \nabla'[\phi(\mathbf{r}')g(\mathbf{r},\mathbf{r}')] - \phi(\mathbf{r}')\mathbf{n}' \times \nabla' g(\mathbf{r},\mathbf{r}')\right\} \qquad (9.39)$$

By Stokes' theorem, the integration of the first term on the RHS of (9.39) is zero since the surface is closed. Using the vector identity, we have

$$\mathbf{E}(\mathbf{r}) = \int_S dS' \left\{ \phi(\mathbf{r}')\mathbf{n}'\nabla \cdot \nabla' g(\mathbf{r},\mathbf{r}') - [\phi(\mathbf{r}')\mathbf{n}' \cdot \nabla]\nabla' g(\mathbf{r},\mathbf{r}') \right\} \quad (9.40)$$

The first term on the RHS of (9.40) vanishes because the integral is taken as the Cauchy principal value. Applying the vector identity to the second term again, we have

$$\mathbf{E}(\mathbf{r}) = -\nabla \int_S dS' \phi(\mathbf{r}')\mathbf{n}' \cdot \nabla' g(\mathbf{r},\mathbf{r}') \quad (9.41)$$

which is exactly the same as the electric field produced by the dipole layer.

The above analysis demonstrates the important fact that if proper basis functions are chosen, the equations governing the dynamic problem should naturally reduce to those for the static problem after a frequency normalization process. Consequently, the solution will be accurate and stable in the low-frequency regime.

9.3 SCATTERING FROM A MULTIBODY

In this section, we will discuss the scattering from a multibody. We first modify the PMCHWT formulation for the multibody and solve the problem using the RWG basis function in the high-frequency regime. The NOUR scheme for RWG and its meaning will be discussed. For the low-frequency problem, a similar idea will be used to obtain the NOUR scheme. To guarantee the frequency normalization, several rules will be generalized for the construction of the NOUR scheme for loop-tree basis functions.

9.3.1 PMCHWT Formulation for Multibody Problem

For a multibody problem, in which both penetrable and PEC bodies are involved, the PMCHWT formulation needs to be modified. To this end, we now consider a multibody problem as shown Figure 9.11. It involves three closed surfaces S_1, S_2, and S_3, which separate four regions R_1, R_2, R_3, and R_4. We define the unknown electric and magnetic currents $\mathbf{J}_1$, $\mathbf{M}_1$, $\mathbf{J}_2$, $\mathbf{M}_2$, $\mathbf{J}_3$, and $\mathbf{M}_3$ on S_1, S_2, and S_3, respectively. The fields outside and inside of S_1 can be written as

$$\mathbf{E}^{\text{out}} = \mathbf{E}^{\text{inc}} + \mathcal{L}_1 \mathbf{J}_1 - \mathcal{K}_1 \mathbf{M}_1 + \mathcal{L}_1 \mathbf{J}_3 - \mathcal{K}_1 \mathbf{M}_3 \quad (9.42)$$

$$\mathbf{E}^{\text{in}} = -\mathcal{L}_2 \mathbf{J}_1 + \mathcal{K}_2 \mathbf{M}_1 + \mathcal{L}_2 \mathbf{J}_2 - \mathcal{K}_2 \mathbf{M}_2 \quad (9.43)$$

$$\mathbf{H}^{\text{out}} = \mathbf{H}^{\text{inc}} + \mathcal{K}_1 \mathbf{J}_1 + \frac{1}{\eta_1^2} \mathcal{L}_1 \mathbf{M}_1 + \mathcal{K}_1 \mathbf{J}_3 + \frac{1}{\eta_1^2} \mathcal{L}_1 \mathbf{M}_3 \quad (9.44)$$

Low-Frequency Scattering from Penetrable Bodies

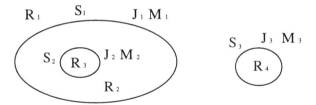

Figure 9.11 A generic multibody problem.

$$\mathbf{H}^{\text{in}} = -\mathcal{K}_2 \mathbf{J}_2 - \frac{1}{\eta_2^2}\mathcal{L}_2 \mathbf{M}_1 + \mathcal{K}_2 \mathbf{J}_1 + \frac{1}{\eta_2^2}\mathcal{L}_2 \mathbf{M}_2 \tag{9.45}$$

Applying the boundary conditions for both **E** and **H** fields on S_1, we have

$$-\mathbf{E}^{\text{inc}} = (\mathcal{L}_1 + \mathcal{L}_2)\mathbf{J}_1 - \eta_0(\mathcal{K}_1 + \mathcal{K}_2)\frac{\mathbf{M}_1}{\eta_0}$$

$$- \mathcal{L}_2 \mathbf{J}_2 + \eta_0 \mathcal{K}_2 \frac{\mathbf{M}_1}{\eta_0} + \mathcal{L}_1 \mathbf{J}_3 - \eta_0 \mathcal{K}_1 \frac{\mathbf{M}_3}{\eta_0}\bigg|_{\text{tan}} \tag{9.46}$$

$$-\eta_0 \mathbf{H}^{\text{inc}} = \eta_0(\mathcal{K}_1 + \mathcal{K}_2)\mathbf{J}_1 + \left(\frac{\eta_0^2}{\eta_1^2}\mathcal{L}_1 + \frac{\eta_0^2}{\eta_2^2}\mathcal{L}_2\right)\frac{\mathbf{M}_1}{\eta_0}$$

$$- \eta_0 \mathcal{K}_2 \mathbf{J}_2 - \frac{\eta_0^2}{\eta_2^2}\mathcal{L}_2 \frac{\mathbf{M}_2}{\eta_0} + \eta_0 \mathcal{K}_1 \mathbf{J}_3 + \frac{\eta_0^2}{\eta_1^2}\mathcal{L}_1 \frac{\mathbf{M}_3}{\eta_0}\bigg|_{\text{tan}} \tag{9.47}$$

We have six current unknowns and two equations, and the rest of the equations can be obtained by applying the same procedure to surfaces S_2 and S_3. It will be more convenient to write the equations in a matrix form as

$$\begin{bmatrix} \overline{\mathbf{A}} & \overline{\mathbf{B}} & \overline{\mathbf{C}} \\ \overline{\mathbf{D}} & \overline{\mathbf{E}} & \overline{\mathbf{F}} \\ \overline{\mathbf{G}} & \overline{\mathbf{H}} & \overline{\mathbf{N}} \end{bmatrix} \begin{bmatrix} \mathbf{X}_1 \\ \mathbf{X}_2 \\ \mathbf{X}_3 \end{bmatrix} = \begin{bmatrix} \mathbf{V}_1 \\ \mathbf{V}_2 \\ \mathbf{V}_3 \end{bmatrix} \tag{9.48}$$

where

$$\overline{\mathbf{A}} = \begin{bmatrix} (\mathcal{L}_1 + \mathcal{L}_2) & -\eta_0(\mathcal{K}_1 + \mathcal{K}_2) \\ \eta_0(\mathcal{K}_1 + \mathcal{K}_2) & \left(\frac{\eta_0^2}{\eta_1^2}\mathcal{L}_1 + \frac{\eta_0^2}{\eta_2^2}\mathcal{L}_2\right) \end{bmatrix} \tag{9.49}$$

$$\overline{\mathbf{B}} = \overline{\mathbf{D}} = \begin{bmatrix} -\mathcal{L}_2 & \eta_0 \mathcal{K}_2 \\ -\eta_0 \mathcal{K}_2 & -\frac{\eta_0^2}{\eta_2^2}\mathcal{L}_2 \end{bmatrix} \tag{9.50}$$

$$\overline{\mathbf{C}} = \overline{\mathbf{G}} = \begin{bmatrix} \mathcal{L}_1 & -\eta_0 \mathcal{K}_1 \\ \eta_0 \mathcal{K}_1 & \frac{\eta_0^2}{\eta_1^2}\mathcal{L}_1 \end{bmatrix} \tag{9.51}$$

$$\overline{\mathbf{E}} = \begin{bmatrix} (\mathcal{L}_2 + \mathcal{L}_3) & -\eta_0(\mathcal{K}_2 + \mathcal{K}_3) \\ \eta_0(\mathcal{K}_2 + \mathcal{K}_3) & \left(\frac{n_0^2}{n_2^2}\mathcal{L}_2 + \frac{n_0^2}{n_3^2}\mathcal{L}_3\right) \end{bmatrix} \quad (9.52)$$

$$\overline{\mathbf{F}} = \overline{\mathbf{H}} = \begin{bmatrix} 0 & 0 \\ 0 & 0 \end{bmatrix} \quad (9.53)$$

$$\overline{\mathbf{N}} = \begin{bmatrix} (\mathcal{L}_1 + \mathcal{L}_4) & -\eta_0(\mathcal{K}_1 + \mathcal{K}_4) \\ \eta_0(\mathcal{K}_1 + \mathcal{K}_4) & \left(\frac{n_0^2}{n_1^2}\mathcal{L}_1 + \frac{n_0^2}{n_4^2}\mathcal{L}_4\right) \end{bmatrix} \quad (9.54)$$

$$\mathbf{X}_1 = \begin{bmatrix} \mathbf{J}_1 \\ \eta_0^{-1}\mathbf{M}_1 \end{bmatrix}, \quad \mathbf{X}_2 = \begin{bmatrix} \mathbf{J}_2 \\ \eta_0^{-1}\mathbf{M}_2 \end{bmatrix}, \quad \mathbf{X}_3 = \begin{bmatrix} \mathbf{J}_3 \\ \eta_0^{-1}\mathbf{M}_3 \end{bmatrix} \quad (9.55)$$

$$\mathbf{V}_1 = \begin{bmatrix} -\mathbf{E}^{\text{inc}} \\ -\eta_0\mathbf{H}^{\text{inc}} \end{bmatrix}, \quad \mathbf{V}_2 = \begin{bmatrix} 0 \\ 0 \end{bmatrix}, \quad \mathbf{V}_3 = \begin{bmatrix} -\mathbf{E}^{\text{inc}} \\ -\eta_0\mathbf{H}^{\text{inc}} \end{bmatrix} \quad (9.56)$$

To interpret (9.48), we first define the relation between two surfaces. For two closed surfaces, there are three kinds of relations we are interested in. For the problem shown in Figure 9.11, we define the relation between S_1 and S_2 as "contained." Surfaces S_1 and S_2 can "see" each other through region R_2, which is characterized by operators $\mathcal{L}_2$ and $\mathcal{K}_2$. We define the relation between S_1 and S_3 as "parallel." Surfaces S_1 and S_3 can "see" each other through region R_1, which is characterized by operators $\mathcal{L}_1$ and $\mathcal{K}_1$. The relation between surfaces S_2 and S_3 is defined as "isolated" and they cannot "see" each other, which implies that there is no direct coupling between these two surfaces.

The first two equations, which are contained in blocks $\overline{\mathbf{A}}$, $\overline{\mathbf{B}}$, $\overline{\mathbf{C}}$, $\mathbf{X}_1$, and $\mathbf{V}_1$, are the $\mathbf{E}$ and $\mathbf{H}$ field equations obtained by applying boundary conditions on surface S_1. Block $\overline{\mathbf{A}}$ is the contribution from $\mathbf{J}_1$ and $\mathbf{M}_1$, which is on S_1; therefore, we call it the "self-block." Blocks $\overline{\mathbf{B}}$ and $\overline{\mathbf{C}}$ are the contributions from the currents on surfaces S_2 and S_3, respectively; therefore, we call them the "cross-blocks." We notice that the signs in blocks $\overline{\mathbf{B}}$ and $\overline{\mathbf{C}}$ are opposite to each other. The reason is that S_2 is contained in S_1 and hence currents $\mathbf{J}_2$ and $\mathbf{M}_2$ contribute to $\mathbf{E}^{\text{in}}$ using operators $\mathcal{L}_2$ and $\mathcal{K}_2$; meanwhile, surface S_3 is "parallel" to S_1, and hence $\mathbf{J}_3$ and $\mathbf{M}_3$ contribute to $\mathbf{E}^{\text{out}}$ using operators $\mathcal{L}_1$ and $\mathcal{K}_1$.

The rest of the blocks represent the equations obtained from surfaces S_2 and S_3, and they have similar interpretations. We notice that blocks $\overline{\mathbf{F}}$ and $\overline{\mathbf{H}}$ are zero. This is because surfaces S_2 and S_3 are "isolated" and hence they do not contribute to each other. We also notice that the self-blocks always have a form

$$\begin{bmatrix} (\mathcal{L}_i + \mathcal{L}_j) & -\eta_0(\mathcal{K}_i + \mathcal{K}_j) \\ \eta_0(\mathcal{K}_i + \mathcal{K}_j) & \left(\frac{n_0^2}{n_i^2}\mathcal{L}_i + \frac{n_0^2}{n_j^2}\mathcal{L}_j\right) \end{bmatrix} \quad (9.57)$$

where i and j are the region index outside and inside the surface. If the surface is a PEC, which implies that the magnetic current is zero, (9.57) reduces to

$$[\mathcal{L}_i] \quad (9.58)$$

Low-Frequency Scattering from Penetrable Bodies

The cross-blocks always have the form

$$\begin{bmatrix} \pm \mathcal{L}_i & \mp \eta_0 \mathcal{K}_i \\ \pm \eta_0 \mathcal{K}_i & \pm \frac{\eta_0^2}{\eta_i^2} \mathcal{L}_i \end{bmatrix} \tag{9.59}$$

The upper signs are used for the "parallel" relation and the lower signs are for the "contained" relation. Index i denotes the region between the two surfaces. If the relation between the two surfaces is "isolated," (9.59) reduces to (9.53). If one of the surfaces is a PEC, (9.59) reduces to either

$$\begin{bmatrix} \pm \mathcal{L}_i \\ \pm \eta_0 \mathcal{K}_i \end{bmatrix} \tag{9.60}$$

or

$$\begin{bmatrix} \pm \mathcal{L}_i & \mp \eta_0 \mathcal{K}_i \end{bmatrix} \tag{9.61}$$

depending on whether the PEC surface is used as a source point or a field point. If both surfaces are PEC, (9.59) reduces to

$$[\mathcal{L}_i] \tag{9.62}$$

and the relations between the two surfaces can only be "parallel" in this case. As we have discussed before, (9.48) can be discretized using either the RWG or loop-tree basis. After the testing procedure, we obtain a set of linear equations, which can be solved for the unknown electric and magnetic currents.

9.3.2 Number-of-Unknowns Reduction Scheme for RWG Basis

So far, we have only considered the situation in which none of the surfaces touch another. The technique we have developed can be applied to each block in (9.48) [22, 23]. However, in many applications, the bodies touch each other. Two touched surfaces can be thought as a limiting case of two "parallel" surfaces as shown in Figure 9.12, so (9.48) is still valid. However, as two surfaces S_2 and S_3 merge together, some important physics happens. From the definition of impressed current, the unknowns defined on S_2 and S_3 in Figure 9.12(a) are independent of each other. But when surfaces S_2 and S_3 touch each other, we know from the boundary conditions that the currents on surfaces S_2 and S_3 should be the same. Therefore, originally defined current unknowns have redundancy, which makes the MOM solution unstable. In this section, we will discuss the removal of the redundancy and study what the NOUR scheme implies when the RWG basis is used. Even though our approach does not provide a computational advantage over the traditional way when the RWG basis function is used, it does provide a more general way to solve multibody problems and it is necessary for solving low-frequency problems using the loop-tree basis function.

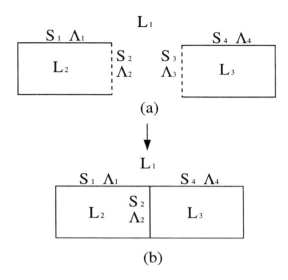

Figure 9.12 Two bodies merged together. The unknowns defined on surfaces S_2 and S_4 become redundant.

To simplify the formulation, we introduce the following short-hand notation:

$$\mathcal{L}_i = \begin{bmatrix} \mathcal{L}_i & -\eta_0 \mathcal{K}_i \\ \eta_0 \mathcal{K}_i & \frac{\eta_0^2}{\eta_i^2} \mathcal{L}_i \end{bmatrix}, \quad \mathbf{J}_i = \begin{bmatrix} \mathbf{J}_i \\ \eta_0^{-1} \mathbf{M}_i \end{bmatrix}, \quad \mathbf{E}^{\text{inc}} = \begin{bmatrix} \mathbf{E}^{\text{inc}} \\ \eta_0 \mathbf{H}^{\text{inc}} \end{bmatrix} \quad (9.63)$$

For the problem shown in Figure 9.12(a), we define four basis functions Λ_1, Λ_2, Λ_3, and Λ_4, on surfaces S_1, S_2, S_3, and S_4, respectively. Surfaces S_2 and S_3, which are the supports for Λ_2 and Λ_3, respectively, will finally merge together as shown in Figure 9.12(b). The currents on the two bodies can be expressed in terms of basis functions as

$$\mathbf{J}_1 = a_1 \Lambda_1 + a_2 \Lambda_2 \quad (9.64)$$

$$\mathbf{J}_2 = a_3 \Lambda_3 + a_4 \Lambda_4 \quad (9.65)$$

where a_1, a_2, a_3, and a_4 are the unknown current coefficients we are solving for. From the discussion in Section 9.2, we obtain the equations by applying boundary conditions. After testing these equations, we obtain the MOM equation

$$\begin{bmatrix} \langle \Lambda_1, (\mathcal{L}_1+\mathcal{L}_2)\Lambda_1 \rangle & \langle \Lambda_1, (\mathcal{L}_1+\mathcal{L}_2)\Lambda_2 \rangle & \langle \Lambda_1, \mathcal{L}_1 \Lambda_3 \rangle & \langle \Lambda_1, \mathcal{L}_1 \Lambda_4 \rangle \\ \langle \Lambda_2, (\mathcal{L}_1+\mathcal{L}_2)\Lambda_1 \rangle & \langle \Lambda_2, (\mathcal{L}_1+\mathcal{L}_2)\Lambda_2 \rangle & \langle \Lambda_2, \mathcal{L}_1 \Lambda_3 \rangle & \langle \Lambda_2, \mathcal{L}_1 \Lambda_4 \rangle \\ \langle \Lambda_3, \mathcal{L}_1 \Lambda_1 \rangle & \langle \Lambda_3, \mathcal{L}_1 \Lambda_2 \rangle & \langle \Lambda_3, (\mathcal{L}_1+\mathcal{L}_3)\Lambda_3 \rangle & \langle \Lambda_3, (\mathcal{L}_1+\mathcal{L}_3)\Lambda_4 \rangle \\ \langle \Lambda_4, \mathcal{L}_1 \Lambda_1 \rangle & \langle \Lambda_4, \mathcal{L}_1 \Lambda_2 \rangle & \langle \Lambda_4, (\mathcal{L}_1+\mathcal{L}_3)\Lambda_3 \rangle & \langle \Lambda_4, (\mathcal{L}_1+\mathcal{L}_3)\Lambda_4 \rangle \end{bmatrix}$$

Low-Frequency Scattering from Penetrable Bodies 445

$$\times \begin{bmatrix} a_1 \\ a_2 \\ a_3 \\ a_4 \end{bmatrix} = \begin{bmatrix} -\langle \mathbf{\Lambda}_1, \mathbf{E}^{\text{inc}} \rangle \\ -\langle \mathbf{\Lambda}_2, \mathbf{E}^{\text{inc}} \rangle \\ -\langle \mathbf{\Lambda}_3, \mathbf{E}^{\text{inc}} \rangle \\ -\langle \mathbf{\Lambda}_4, \mathbf{E}^{\text{inc}} \rangle \end{bmatrix} \qquad (9.66)$$

which can be used to solve for the unknown current coefficients a_1 to a_4. When surfaces S_2 and S_3 merge together as shown in Figure 9.12(b), we have

$$a_2 = a_3 \qquad (9.67)$$

if we define

$$\mathbf{\Lambda}_2 = -\mathbf{\Lambda}_3 \qquad (9.68)$$

As mentioned before, (9.66) is still valid for this case. However, if we solve (9.66) without enforcing (9.67), then the solution is unstable. Therefore, we need to reduce the unknowns as follows:

$$\overline{\mathbf{A}} \cdot \overline{\mathbf{Z}} \cdot \overline{\mathbf{A}}^t \cdot \mathbf{a}' = \overline{\mathbf{A}} \cdot \mathbf{V} \qquad (9.69)$$

where $\overline{\mathbf{Z}}$ is the impedance matrix given in (9.66); $\mathbf{V}$ is the excitation vector; and $\mathbf{a}'$ is the unknown vector without redundancy and it is related to the original unknown vector $\mathbf{a}$ through the NOUR matrix $\overline{\mathbf{A}}^t$

$$\begin{bmatrix} a_1 \\ a_2 \\ a_3 \\ a_4 \end{bmatrix} = \begin{bmatrix} 1 & 0 & 0 \\ 0 & 1 & 0 \\ 0 & 1 & 0 \\ 0 & 0 & 1 \end{bmatrix} \begin{bmatrix} a_1 \\ a_2 \\ a_4 \end{bmatrix} \qquad (9.70)$$

Equation (9.69) is well conditioned and could be used to solve for $\mathbf{a}'$ and $\mathbf{a}$. It will be insightful to study the meaning of the NOUR scheme. When $\overline{\mathbf{A}}^t$ operates on the impedance matrix $\overline{\mathbf{Z}}$, it is nothing but a column combination operator, which combines columns 2 and 3 of $\overline{\mathbf{Z}}$. Similarly, operator $\overline{\mathbf{A}}$ combines rows 2 and 3 of $\overline{\mathbf{Z}}$. After the row and column combinations, (9.66) becomes

$$\begin{bmatrix} \langle \mathbf{\Lambda}_1, (\mathcal{L}_1 + \mathcal{L}_2)\mathbf{\Lambda}_1 \rangle & \langle \mathbf{\Lambda}_1, \mathcal{L}_2\mathbf{\Lambda}_2 \rangle & \langle \mathbf{\Lambda}_1, \mathcal{L}_1\mathbf{\Lambda}_4 \rangle \\ \langle \mathbf{\Lambda}_2, \mathcal{L}_2\mathbf{\Lambda}_1 \rangle & \langle \mathbf{\Lambda}_2, (\mathcal{L}_2 + \mathcal{L}_3)\mathbf{\Lambda}_2 \rangle & -\langle \mathbf{\Lambda}_2, \mathcal{L}_3\mathbf{\Lambda}_4 \rangle \\ \langle \mathbf{\Lambda}_4, \mathcal{L}_1\mathbf{\Lambda}_1 \rangle & -\langle \mathbf{\Lambda}_4, \mathcal{L}_3\mathbf{\Lambda}_2 \rangle & \langle \mathbf{\Lambda}_4, (\mathcal{L}_1 + \mathcal{L}_3)\mathbf{\Lambda}_4 \rangle \end{bmatrix}$$

$$\times \begin{bmatrix} a_1 \\ a_2 \\ a_4 \end{bmatrix} = \begin{bmatrix} -\langle \mathbf{\Lambda}_1, \mathbf{E}^{\text{inc}} \rangle \\ 0 \\ -\langle \mathbf{\Lambda}_4, \mathbf{E}^{\text{inc}} \rangle \end{bmatrix} \qquad (9.71)$$

If the problem shown in Figure 9.12(b) is solved using the traditional approach, which applies the boundary conditions on open surfaces S_1, S_2, and S_4 instead of the two bodies, the MOM equation obtained will be the same as (9.71). Due to the definitions of the basis functions, no junction is considered so far. In fact, we need to define

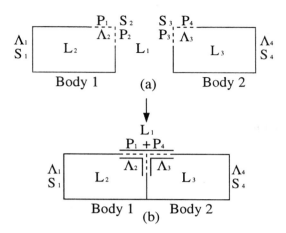

Figure 9.13 Two bodies merged together. The unknowns defined on surfaces S_2 and S_4 become redundant and form a junction.

the basis across surfaces S_1, S_2 and surfaces S_3, S_4 as shown in Figure 9.13, which form a junction. We show next how the combination scheme handles the junction problem implicitly. The currents on the surfaces of the two bodies in Figure 9.13(a) can be written as

$$\mathbf{J}_1 = a_1 \mathbf{\Lambda}_1 + a_2 \mathbf{\Lambda}_2 \tag{9.72}$$

$$\mathbf{J}_2 = a_3 \mathbf{\Lambda}_3 + a_4 \mathbf{\Lambda}_4 \tag{9.73}$$

where

$$\mathbf{\Lambda}_2 = \mathbf{P}_1 + \mathbf{P}_2 \tag{9.74}$$

$$\mathbf{\Lambda}_3 = \mathbf{P}_3 + \mathbf{P}_4 \tag{9.75}$$

This problem is different from the one shown in Figure 9.12 in the sense that only part of $\mathbf{\Lambda}_2$ and $\mathbf{\Lambda}_3$ will merge together. Since we are only interested in the junction part, after testing with $\mathbf{\Lambda}_2$ and $\mathbf{\Lambda}_3$, we have

$$\begin{bmatrix} \langle \mathbf{\Lambda}_2, (\mathcal{L}_1+\mathcal{L}_2)\mathbf{\Lambda}_1 \rangle & \langle \mathbf{\Lambda}_2, (\mathcal{L}_1+\mathcal{L}_2)\mathbf{\Lambda}_2 \rangle & \langle \mathbf{\Lambda}_2, \mathcal{L}_1\mathbf{\Lambda}_3 \rangle & \langle \mathbf{\Lambda}_2, \mathcal{L}_1\mathbf{\Lambda}_4 \rangle \\ \langle \mathbf{\Lambda}_3, \mathcal{L}_1\mathbf{\Lambda}_1 \rangle & \langle \mathbf{\Lambda}_3, \mathcal{L}_1\mathbf{\Lambda}_2 \rangle & \langle \mathbf{\Lambda}_3, (\mathcal{L}_1+\mathcal{L}_3)\mathbf{\Lambda}_3 \rangle & \langle \mathbf{\Lambda}_3, (\mathcal{L}_1+\mathcal{L}_3)\mathbf{\Lambda}_4 \rangle \end{bmatrix}$$

$$\times \begin{bmatrix} a_1 \\ a_2 \\ a_3 \\ a_4 \end{bmatrix} = \begin{bmatrix} -\langle \mathbf{\Lambda}_2, \mathbf{E}^{\text{inc}} \rangle \\ -\langle \mathbf{\Lambda}_3, \mathbf{E}^{\text{inc}} \rangle \end{bmatrix} \tag{9.76}$$

When the two bodies touch each other, we define

$$\mathbf{P}_2 = -\mathbf{P}_3 \tag{9.77}$$

so we have

$$a_2 = a_3 \tag{9.78}$$

Similar to the case shown in Figure 9.12(b), the combination scheme implies that rows 1 and 2 and columns 2 and 3 of (9.76) should be combined. Making use of (9.77) and (9.78), we have

$$\langle \mathbf{\Lambda}_2, \mathcal{L}_2(a_1\mathbf{\Lambda}_1 + a_2\mathbf{\Lambda}_2)\rangle + \langle \mathbf{\Lambda}_3, \mathcal{L}_3(a_2\mathbf{\Lambda}_3 + a_4\mathbf{\Lambda}_4)\rangle$$
$$+\langle (\mathbf{P}_1 + \mathbf{P}_4), \mathcal{L}_1(a_1\mathbf{\Lambda}_1 + a_2(\mathbf{P}_1 + \mathbf{P}_4) + a_4\mathbf{\Lambda}_4)\rangle = \langle -(\mathbf{P}_1 + \mathbf{P}_4), \mathbf{E}^{\text{inc}}\rangle \tag{9.79}$$

which is the junction condition and has an interpretation shown in Figure 9.13(b). The first term $\langle \mathbf{\Lambda}_2, \mathcal{L}_2(a_1\mathbf{\Lambda}_1 + a_2\mathbf{\Lambda}_2)\rangle$ is the result of testing the field using $\mathbf{\Lambda}_2$, and the field is generated by the current flowing inside body 1. The second term $\langle \mathbf{\Lambda}_3, \mathcal{L}_3(a_2\mathbf{\Lambda}_3 + a_4\mathbf{\Lambda}_4)\rangle$ is the result of testing the field using $\mathbf{\Lambda}_3$, and the field is generated by the current flowing inside body 2. The third term $\langle (\mathbf{P}_1 + \mathbf{P}_4), \mathcal{L}_1(a_1\mathbf{\Lambda}_1 + a_2(\mathbf{P}_1 + \mathbf{P}_4) + a_4\mathbf{\Lambda}_4)\rangle$ is the result of testing the field by $(\mathbf{P}_1 + \mathbf{P}_4)$, which is a basis implicitly formed by the combination scheme, and the field is generated by the current flowing outside body 1 and body 2.

Equation (9.79) is the same as the junction condition which can be obtained using the traditional approach [17]. Understanding the meaning of the NOUR scheme also provides some computational advantages. We notice that some terms, such as $\langle \mathbf{P}_2, \mathcal{L}_1\mathbf{\Lambda}_1\rangle$ and $\langle \mathbf{P}_3, \mathcal{L}_1\mathbf{\Lambda}_1\rangle$, finally cancel out, and therefore, there is no need to compute them.

As an example, we compute the bistatic RCS of a coated PEC sphere with radius $a = 0.2\lambda$, $b = 0.18\lambda$ and relative permittivity $\epsilon_r = 1.7 + 0.3i$. In this case, no touched surfaces are involved, and therefore, no NOUR scheme is applied. The results agree well with the Mie series solutions shown in Figure 9.14.

In Figure 9.15, we show the monostatic RCS of an inhomogeneous cylinder with $a = 10.16$ cm, $b = 5.08$ cm, $d = 7.62$ cm, and relative permittivity $\epsilon_{r1} = 3.4$, $\epsilon_{r2} = 2.6$, respectively. The frequency is 3.5 GHz. In this case, part of the surfaces of the two bodies touch each other and the unknowns defined on the common surface become redundant. The NOUR scheme is applied to obtain accurate results.

In Figure 9.16, we show the monostatic RCS of an inhomogeneous PEC-dielectric cylinder. The setup is the same except that half of the cylinder is PEC and the frequency is 3.0 GHz. The results shown in Figures 9.15 and 9.16 agree well with those published in [17].

The NOUR scheme for RWG basis has been implemented successfully to compute the RCS in the high-frequency regime. It is straightforward to construct the NOUR scheme for the RWG basis, since the unknown coefficients have a simple "one-to-one" relation given by (9.67) and (9.78). The ideas developed in this section will be used to obtain the NOUR scheme for the loop-tree basis.

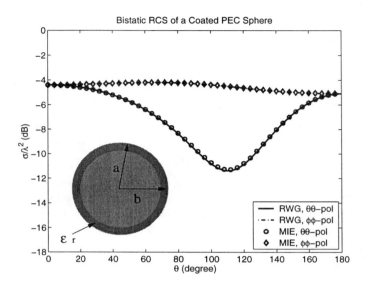

Figure 9.14 Bistatic RCS for a coated PEC sphere; radius $a = 0.2\lambda$, $b = 0.18\lambda$; relative permittivity of the coating material $\epsilon_r = 1.7 + 0.3i$.

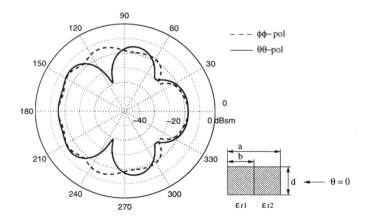

Figure 9.15 Monostatic RCS for an inhomogeneous cylinder; $a = 10.16$ cm, $b = 5.08$ cm, $d = 7.62$ cm; relative permittivity $\epsilon_{r1} = 3.4$, $\epsilon_{r2} = 2.6$; frequency $f = 3.5$ GHz.

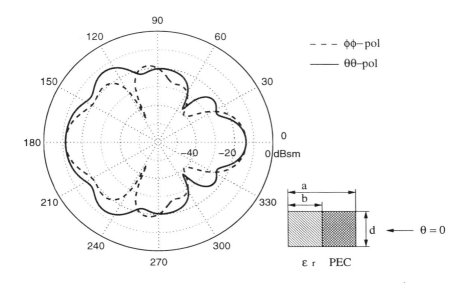

Figure 9.16 Monostatic RCS of an inhomogeneous PEC-dielectric cylinder; $a = 10.16$ cm, $b = 5.08$ cm, $d = 7.62$ cm; relative permittivity $\epsilon_r = 2.6$; frequency $f = 3.0$ GHz.

9.3.3 Number-of-Unknowns Reduction Scheme for Loop-Tree Basis

The loop-tree basis is formed by recombining the RWG basis. When combined with the frequency normalization scheme, the loop-tree basis can be used in the low-frequency regime. Compared with the RWG basis function, the support of the loop-tree basis is not "localized," which implies the simple "one-to-one" relation does not exist anymore. Since the purpose of using the loop-tree basis is to apply the frequency normalization scheme, a valid NOUR scheme should preserve the frequency scaling properties of the matrix elements. Nevertheless, the idea used for the RWG basis is still valid. In this section, we propose a systematic approach to construct the NOUR scheme for the loop-tree basis. The discussion will focus on (1) finding the relation of redundant loop-tree unknown coefficients, and (2) eliminating the redundant unknown coefficients without affecting the frequency normalization scheme.

Consider a simple two-body problem; we define edges e_1 to e_{12} and nodes n_1 to n_5 as shown in Figure 9.17. Furthermore, we define loops and trees for body 1 and body 2 as follows:

	Body 1:			Body 2:		
RWG :	$\mathbf{R}_1$	e_1	$n_4 \to n_3$	$\mathbf{R}_7$	e_7	$n_3 \to n_5$
	$\mathbf{R}_2$	e_2	$n_2 \to n_4$	$\mathbf{R}_8$	e_8	$n_5 \to n_2$
	$\mathbf{R}_3$	e_3	$n_4 \to n_1$	$\mathbf{R}_9$	e_9	$n_1 \to n_5$
	$\mathbf{R}_4$	e_4	$n_1 \to n_3$	$\mathbf{R}_{10}$	e_{10}	$n_3 \to n_1$
	$\mathbf{R}_5$	e_5	$n_2 \to n_3$	$\mathbf{R}_{11}$	e_{11}	$n_3 \to n_2$
	$\mathbf{R}_6$	e_6	$n_1 \to n_2$	$\mathbf{R}_{12}$	e_{12}	$n_1 \to n_2$
LOOP :	$\mathbf{O}_1$	$\mathbf{R}_1$	$+1$	$\mathbf{O}_4$	$\mathbf{R}_7$	-1
		$\mathbf{R}_2$	$+1$		$\mathbf{R}_8$	-1
		$\mathbf{R}_5$	-1		$\mathbf{R}_{11}$	$+1$
	$\mathbf{O}_2$	$\mathbf{R}_1$	$+1$	$\mathbf{O}_5$	$\mathbf{R}_8$	$+1$
		$\mathbf{R}_3$	-1		$\mathbf{R}_9$	$+1$
		$\mathbf{R}_4$	-1		$\mathbf{R}_{12}$	-1
	$\mathbf{O}_3$	$\mathbf{R}_4$	-1	$\mathbf{O}_6$	$\mathbf{R}_{10}$	$+1$
		$\mathbf{R}_5$	$+1$		$\mathbf{R}_{11}$	-1
		$\mathbf{R}_6$	$+1$		$\mathbf{R}_{12}$	$+1$
TREE :	$\mathbf{T}_1$	$\mathbf{R}_3$		$\mathbf{T}_4$	$\mathbf{R}_7$	
	$\mathbf{T}_2$	$\mathbf{R}_4$		$\mathbf{T}_5$	$\mathbf{R}_9$	
	$\mathbf{T}_3$	$\mathbf{R}_5$		$\mathbf{T}_6$	$\mathbf{R}_{12}$	

We define RWG $\mathbf{R}_1$ to $\mathbf{R}_6$ on body 1 and RWG $\mathbf{R}_7$ to $\mathbf{R}_{12}$ on body 2. For example, $\mathbf{R}_1$ is defined on edge e_1 and the current flowing direction is from n_4 to n_3. On body 1, we define three loops $\mathbf{O}_1$, $\mathbf{O}_2$, and $\mathbf{O}_3$. For example, loop $\mathbf{O}_1$ consists of three RWGs, $\mathbf{R}_1$, $\mathbf{R}_2$, and $\mathbf{R}_5$. The signs associated with the three RWGs are $+1$, $+1$, and -1. In other words, we have

$$\mathbf{O}_1 = \mathbf{R}_1 + \mathbf{R}_2 - \mathbf{R}_5 \tag{9.80}$$

We also define three trees on body 1, and they are R_3, R_4, and R_5. The rest of the RWGs are cuts. The definitions of loops and trees on body 2 are similar. Expanding the currents on body 1 and body 2 in terms of the loop-tree basis, we have

$$\mathbf{J}_1 = a_1\mathbf{O}_1 + a_2\mathbf{O}_2 + a_3\mathbf{O}_3 + b_1\mathbf{T}_1 + b_2\mathbf{T}_2 + b_3\mathbf{T}_3 \tag{9.81}$$

and

$$\mathbf{J}_2 = a_4\mathbf{O}_4 + a_5\mathbf{O}_5 + a_6\mathbf{O}_6 + b_4\mathbf{T}_4 + b_5\mathbf{T}_5 + b_6\mathbf{T}_6 \tag{9.82}$$

where a_1 to a_6 and b_1 to b_6 are the unknown coefficients we are solving for. As before, we omitted the magnetic current part to simplify the notation. Obviously, we

Low-Frequency Scattering from Penetrable Bodies 451

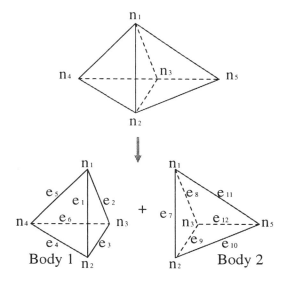

Figure 9.17 A simple two-body problem used to demonstrate how to construct the NOUR scheme for the loop-tree basis.

only need nine unknowns if the RWG basis is used. Therefore, we need to know which loop-tree unknown is redundant. If this problem is solved using the RWG basis, the coefficients associated with RWG $\mathbf{R}_1$ and $\mathbf{R}_7$ must be the same. This is also true for RWG $\mathbf{R}_2$, $\mathbf{R}_8$ and $\mathbf{R}_3$, $\mathbf{R}_9$, since they share the same edges. To find the relation among the unknown coefficients, we rewrite (9.81) and (9.82) in terms of the RWG basis as

$$\begin{aligned}\mathbf{J}_1 =& (a_1 + a_2)\mathbf{R}_1 + a_1\mathbf{R}_2 + (-a_2 + b_1)\mathbf{R}_3 \\ &+ (-a_2 + a_3 + b_2)\mathbf{R}_4 + (-a_1 + a_3 + b_3)\mathbf{R}_5 + a_3\mathbf{R}_6\end{aligned} \quad (9.83)$$

and

$$\begin{aligned}\mathbf{J}_2 =& (-a_4 + b_4)\mathbf{R}_7 + (-a_4 + a_5)\mathbf{R}_8 + (a_5 + b_5)\mathbf{R}_9 \\ &+ a_6\mathbf{R}_{10} + (a_4 - a_6)\mathbf{R}_{11} + (-a_5 + a_6 + b_6)\mathbf{R}_{12}\end{aligned} \quad (9.84)$$

Comparing the coefficients in front of $(\mathbf{R}_1, \mathbf{R}_7)$, $(\mathbf{R}_2, \mathbf{R}_8)$, and $(\mathbf{R}_3, \mathbf{R}_9)$, we have

$$\begin{aligned} a_1 + a_2 &= -a_4 + b_4 \\ a_1 &= -a_4 + a_5 \\ -a_2 + b_1 &= a_5 + b_5 \end{aligned} \quad (9.85)$$

Rewriting (9.85) in a matrix form, we have

$$\begin{bmatrix} 1 & 1 & 0 & 1 & 0 & -1 & 0 \\ 1 & 0 & 0 & 1 & -1 & 0 & 0 \\ 0 & -1 & 1 & 0 & -1 & 0 & -1 \end{bmatrix} \begin{bmatrix} a_1 \\ a_2 \\ b_1 \\ a_4 \\ a_5 \\ b_4 \\ b_5 \end{bmatrix} = \begin{bmatrix} 0 \\ 0 \\ 0 \end{bmatrix} \quad (9.86)$$

which can be used to eliminate three redundant unknowns. After applying Gaussian elimination, we have

$$\begin{bmatrix} 1 & 0 & 0 & 1 & -1 & 0 & 0 \\ 0 & -1 & 0 & 0 & -1 & 1 & 0 \\ 0 & 0 & 1 & 0 & 0 & -1 & -1 \end{bmatrix} \begin{bmatrix} a_1 \\ a_2 \\ b_1 \\ a_4 \\ a_5 \\ b_4 \\ b_5 \end{bmatrix} = \begin{bmatrix} 0 \\ 0 \\ 0 \end{bmatrix} \quad (9.87)$$

which is equivalent to

$$\begin{bmatrix} a_1 \\ a_2 \\ b_1 \end{bmatrix} = \begin{bmatrix} -1 & 1 & 0 & 0 \\ 0 & -1 & 1 & 0 \\ 0 & 0 & 1 & 1 \end{bmatrix} \begin{bmatrix} a_4 \\ a_5 \\ b_4 \\ b_5 \end{bmatrix} \quad (9.88)$$

Now, we can eliminate redundant unknowns a_1, a_2, and b_1. The unknown vectors with and without redundancy are related through

$$\begin{bmatrix} a_1 \\ a_2 \\ a_3 \\ b_1 \\ b_2 \\ b_3 \\ a_4 \\ a_5 \\ a_6 \\ b_4 \\ b_5 \\ b_6 \end{bmatrix} = \begin{bmatrix} 0 & 0 & 0 & -1 & 1 & 0 & 0 & 0 & 0 \\ 0 & 0 & 0 & 0 & -1 & 0 & 1 & 0 & 0 \\ 1 & 0 & 0 & 0 & 0 & 0 & 0 & 0 & 0 \\ 0 & 0 & 0 & 0 & 0 & 0 & 1 & 1 & 0 \\ 0 & 1 & 0 & 0 & 0 & 0 & 0 & 0 & 0 \\ 0 & 0 & 1 & 0 & 0 & 0 & 0 & 0 & 0 \\ 0 & 0 & 0 & 1 & 0 & 0 & 0 & 0 & 0 \\ 0 & 0 & 0 & 0 & 1 & 0 & 0 & 0 & 0 \\ 0 & 0 & 0 & 0 & 0 & 1 & 0 & 0 & 0 \\ 0 & 0 & 0 & 0 & 0 & 0 & 1 & 0 & 0 \\ 0 & 0 & 0 & 0 & 0 & 0 & 0 & 1 & 0 \\ 0 & 0 & 0 & 0 & 0 & 0 & 0 & 0 & 1 \end{bmatrix} \begin{bmatrix} a_3 \\ b_2 \\ b_3 \\ a_4 \\ a_5 \\ a_6 \\ b_4 \\ b_5 \\ b_6 \end{bmatrix} \quad (9.89)$$

To generalize this approach for complicated problems, we need to know (1) if this elimination process can always be performed, (2) if any unknown can be chosen as

Low-Frequency Scattering from Penetrable Bodies

a redundant unknown, and (3) if the choice is not unique, how it will affect the final results. Since the equations are linearly independent, the elimination can always be performed. However, not any combination of three unknowns can be chosen as a redundant unknown set. Obviously, a set which contains b_1, b_4, and b_5 is not a valid choice, since the rank of the matrix, which is formed by the corresponding columns of b_1, b_4, and b_5 in (9.86), is less than 3.

Obviously, the choice is not unique either. Another choice is

$$\begin{bmatrix} a_1 \\ b_4 \\ b_1 \end{bmatrix} = \begin{bmatrix} -1 & 1 & 0 & 0 \\ 0 & 1 & 1 & 0 \\ 0 & 1 & 1 & 1 \end{bmatrix} \begin{bmatrix} a_4 \\ a_5 \\ a_2 \\ b_5 \end{bmatrix} \qquad (9.90)$$

which is equivalent to

$$\begin{bmatrix} a_1 \\ a_2 \\ a_3 \\ b_1 \\ b_2 \\ b_3 \\ a_4 \\ a_5 \\ a_6 \\ b_4 \\ b_5 \\ b_6 \end{bmatrix} = \begin{bmatrix} 0 & 0 & 0 & 0 & -1 & 1 & 0 & 0 & 0 \\ 1 & 0 & 0 & 0 & 0 & 0 & 0 & 0 & 0 \\ 0 & 1 & 0 & 0 & 0 & 0 & 0 & 0 & 0 \\ 1 & 0 & 0 & 0 & 0 & 1 & 0 & 1 & 0 \\ 0 & 0 & 1 & 0 & 0 & 0 & 0 & 0 & 0 \\ 0 & 0 & 0 & 1 & 0 & 0 & 0 & 0 & 0 \\ 0 & 0 & 0 & 0 & 1 & 0 & 0 & 0 & 0 \\ 0 & 0 & 0 & 0 & 0 & 1 & 0 & 0 & 0 \\ 0 & 0 & 0 & 0 & 0 & 0 & 1 & 0 & 0 \\ 1 & 0 & 0 & 0 & 0 & 1 & 0 & 0 & 0 \\ 0 & 0 & 0 & 0 & 0 & 0 & 0 & 1 & 0 \\ 0 & 0 & 0 & 0 & 0 & 0 & 0 & 0 & 1 \end{bmatrix} \begin{bmatrix} a_2 \\ a_3 \\ b_2 \\ b_3 \\ a_4 \\ a_5 \\ a_6 \\ b_5 \\ b_6 \end{bmatrix} \qquad (9.91)$$

The unknown elimination schemes given by (9.89) and (9.91) are quite different. The first scheme eliminates two loop unknowns, while the second eliminates only one loop unknown. If no frequency normalization is required, both schemes are valid. However, in the low-frequency limit, where frequency normalization is necessary, only the first scheme can be used. To explain this, we need to consider the impedance matrix obtained by using the loop-tree basis:

$$\overline{\mathbf{Z}}^{LT} = \begin{bmatrix} OO & OT & OO & OT \\ TO & TT & TO & TT \\ OO & OT & OO & OT \\ TO & TT & TO & TT \end{bmatrix} \qquad (9.92)$$

where

$$\begin{aligned} OO &\sim \langle \mathbf{O}, g, \mathbf{O} \rangle \quad \text{or} \quad \langle \mathbf{O}, \nabla g \times \mathbf{O} \rangle \\ OT &\sim \langle \mathbf{O}, g, \mathbf{T} \rangle \quad \text{or} \quad \langle \mathbf{O}, \nabla g \times \mathbf{T} \rangle \\ TO &\sim \langle \mathbf{T}, g, \mathbf{O} \rangle \quad \text{or} \quad \langle \mathbf{T}, \nabla g \times \mathbf{O} \rangle \\ TT &\sim \langle \mathbf{T}, g, \mathbf{T} \rangle \quad \text{or} \quad \langle \mathbf{T}, \nabla g \times \mathbf{T} \rangle \end{aligned} \qquad (9.93)$$

As mentioned before, operation $\overline{\mathbf{Z}}^{LT} \cdot \overline{\mathbf{A}}^t$ is equivalent to performing the column combination for $\overline{\mathbf{Z}}^{LT}$. The column combinations of (9.92) corresponding to multiplying (9.92) by columns 1 and 6 in (9.93) are not allowed, since trees $\mathbf{T}_1$ and $\mathbf{T}_4$ merge into loops $\mathbf{O}_2$, and trees $\mathbf{T}_1$, $\mathbf{T}_4$, and loop $\mathbf{O}_1$ merge into $\mathbf{O}_5$. In other words, since

$$\begin{aligned}\nabla \cdot (\mathbf{T}_1 + \mathbf{T}_4 + \mathbf{O}_2) \neq 0 \\ \nabla \cdot (\mathbf{T}_1 + \mathbf{T}_4 + \mathbf{O}_1 + \mathbf{O}_5) \neq 0\end{aligned} \qquad (9.94)$$

the divergence-free property of the loop basis, which is essential for the frequency normalization scheme, vanishes, and therefore the frequency normalization scheme breaks down.

In general, the rules for constructing the NOUR scheme are: (1) *Tree merges into tree* is allowed. (2) *Tree merges into loop* is not allowed. (3) *Loop merges into loop* is allowed. (4) *Loop merges into tree* is allowed. A valid NOUR scheme should take all these rules into consideration. It can be shown that such a NOUR scheme always exists.

So far we have only considered the penetrable bodies. If a PEC body is involved, the NOUR scheme for the magnetic current is a little different. Assuming that body 1 is a PEC, (9.86) becomes

$$\begin{bmatrix} 0 & 0 & 0 & 1 & -1 & 0 & 0 \\ 0 & 0 & 0 & 0 & -1 & 1 & 0 \\ 0 & 0 & 0 & 0 & 0 & -1 & -1 \end{bmatrix} \begin{bmatrix} a_1 \\ a_2 \\ b_1 \\ a_4 \\ a_5 \\ b_4 \\ b_5 \end{bmatrix} = \begin{bmatrix} 0 \\ 0 \\ 0 \end{bmatrix} \qquad (9.95)$$

which is equivalent to

$$\begin{bmatrix} a_4 \\ a_5 \\ b_4 \end{bmatrix} = \begin{bmatrix} -1 \\ -1 \\ -1 \end{bmatrix} [b_5] \qquad (9.96)$$

Equation (9.96) is the NOUR scheme for the magnetic current and a_4, a_5, b_4, and b_5 are the unknown coefficients for magnetic current.

To verify our approach, we design one example shown in Figure 9.18(a). This example consists of two surfaces, which do not touch each other. Therefore, no NOUR scheme is needed and the results will be used as a standard. Without changing the physics, the problem can also be treated in such a way as shown in Figures 9.18(c) and 9.18(d). In the second treatment, three touched surfaces are involved and the unknowns defined on the common surfaces are redundant. Therefore, the NOUR scheme is needed. If the media of the two outer bodies are the same, then it is equivalent to the one shown in Figure 9.18(a). The relative permittivities of the

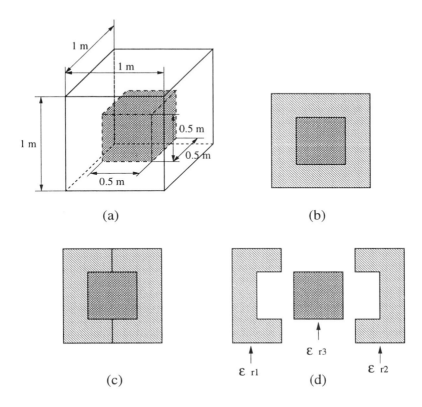

Figure 9.18 Example used to verify the NOUR scheme for loop-tree basis.

two outer bodies are $\epsilon_{r1} = 2.6$ and $\epsilon_{r2} = 2.6$. In the first example, the relative permittivity of the inner body is $\epsilon_{r3} = 3.4$ and in the second example, the inner body is a PEC. The results of the two examples are given in Figures 9.19 and 9.20. Good agreement is observed among them.

As we mentioned before, our method can also be used for induction well-logging modeling. As an example, we compute the apparent conductivity for the structures shown in Figures 9.21(a) and 9.22(a), a vertical case and a dipping case. The generic logging tool consists of one transmitter and one receiver with 20 inches spacing as shown in Figures 9.21(b) and 9.22(b). The tool operates at 20 kHz, and hence, the loop-tree basis function is used. The meshes used are given in Figures 9.21(c) and 9.22(c). For the structure shown in Figure 20(a), we can also use the numerical mode matching (NMM) method. The results given in Figure 9.23 show good agreement.

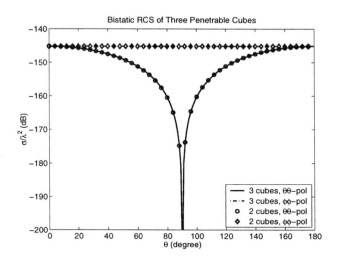

Figure 9.19 Bistatic RCS for the example shown in Figure 9.18; relative permittivity $\epsilon_{r1} = 2.6$, $\epsilon_{r2} = 2.6$, $\epsilon_{r2} = 3.4$; frequency $f = 20$ kHz.

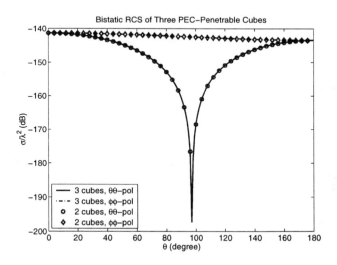

Figure 9.20 Bistatic RCS for the example shown in Figure 9.18; relative permittivity $\epsilon_{r1} = 2.6$, $\epsilon_{r2} = 2.6$; inner cube is PEC; frequency $f = 20$ kHz.

Low-Frequency Scattering from Penetrable Bodies 457

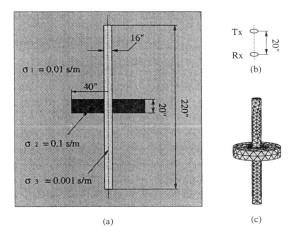

Figure 9.21 (a) Geometry, (b) generic induction tool, and (c) mesh used to compute apparent conductivity for vertical case; frequency $f = 20$ kHz.

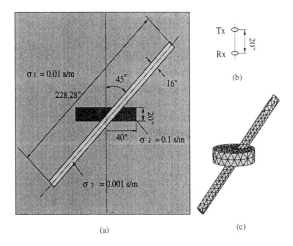

Figure 9.22 (a) Geometry, (b) generic induction tool, and (c) mesh used to compute apparent conductivity for vertical case; frequency $f = 20$ kHz.

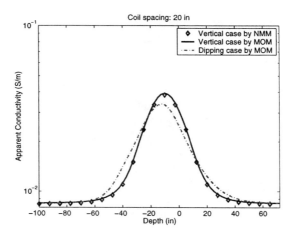

Figure 9.23 Apparent conductivity results for the structures shown in Figures 9.21(a) and 9.22(a).

REFERENCES

1. J. M. Song, C. C. Lu, W. C. Chew, and S. W. Lee, "Fast Illinois Solver Code (FISC)," *IEEE Ant. Propag. Mag.*, vol. 40, no. 3, pp. 27–34, June 1998.

2. J. M. Song and W. C. Chew, "The Fast Illinois Solver Code: Requirements and scaling properties," *IEEE Computational Science and Engineering*, vol. 5, no. 3, pp. 19–23, July–Sept. 1998.

3. B. Anderson and W. C. Chew, "New high speed technique for calculating synthetic induction and DPT logs," *Transactions of the 25th SPWLA Annual Logging Symposium*, vol. 1, paper HH, Houston, TX, 1985.

4. V. Druskin and L. Knizhnerman, "Spectral approach to solving three-dimensional Maxwell's diffusion equations in the time and frequency domains," *Radio Science*, vol. 29, no. 4, July–Aug. 1994.

5. L. Tsang, K. Ding, S. Shih, and J. A. Kong, "Scattering of electromagnetic waves from dense distributions of spheroidal particles based on Monte Carlo simulations," *Journal of the Optical Society of America A-Optics and Image Science*, vol. 15, no. 10, pp. 2660–2669, Oct. 1998.

6. R. J. Adams and G. S. Brown, "Iterative solution of one-dimensional rough surface scattering problems based on a factorization of the Helmholtz operator," *IEEE Trans. Antennas Propagat.*, vol. 47, pp. 765–767, April 1999.

7. K. Sarabandi, E. Li, and A. Nashashibi, "Modeling and measurements of scattering from road surfaces at millimeter-wave frequencies," *IEEE Trans. Antennas Propagat.*, vol. 45, pp. 1679–1688, Nov. 1997.

8. J. T. Johnson, L. Tsang, R. T. Shin, K. Pak, C. H. Chan, A. Ishimaru, and Y. Kuga, "Backscattering enhancement of electromagnetic waves from two-dimensional perfectly conducting random rough surfaces: A comparison of Monte Carlo simulations with experimental data," *IEEE Trans. Antennas Propagat.*, vol. 44, pp. 748–756, May 1996.

9. W. Wu, A. W. Glisson, and D. Kajfez, "A comparison of two low-frequency formulations for the electric field integral equation," *Proceeding of the 10th Annual Review of Progress in Applied Computational Electromagnetics,* vol. 2, pp. 484–491, Monterey, CA, March 1994.

10. D. R. Wilton and A. W. Glisson, "On improving the electric field integral equation at low frequencies," *1981 Spring URSI Radio Science Meeting Digest,* p. 24, Los Angeles, CA, June 1981.

11. J. R. Mautz and R. F. Harrington, "An E-field solution for a conducting surface small or comparable to the wavelength," *IEEE Trans. Antennas Propagat.*, vol. 32, pp. 330–339, April 1984.

12. M. Burton and S. Kashyap, "A study of a recent, moment-method algorithm that is accurate to very low frequencies," *Applied Computational Electromagnetic Society Journal*, vol. 10, no. 3, pp. 58–68, Nov. 1995.

13. J. M. Song and W. C. Chew, "Moment method solutions using parametric geometry," *Journal of Electromagnetic Waves and Applications*, vol. 9, no. 1/2, pp. 71–83, 1995.

14. M. I. Sancer, R. L. McClary, and K. J. Glover, "Electromagnetic computation using parametric geometry," *Electromagnetics*, vol. 10, no. 1–2, pp. 85–103, 1990.

15. J. S. Zhao and W. C. Chew, "Integral equation solution of Maxwell's equations from zero frequency to microwave frequencies," *IEEE Trans. Antennas Propagat.*, vol. 48, no. 10, pp. 1635–1645, Oct. 2000.

16. A. A. Kishk and L. Shafai, "Different formulations for numerical solution of single or multibodies of revolution with mixed boundary conditions," *IEEE Trans. Antennas Propagat.*, vol. 34, no. 5, pp. 666–673, May 1984.

17. L. N. Medgyesi-Mitschang and J. M. Putnam, "Electromagnetic scattering from axially inhomogeneous bodies of revolution," *IEEE Trans. Antennas Propagat.*, vol. 32, no. 8, pp. 797–806, Aug. 1984.

18. S. Ström and W. Zhang, "The null field approach to electromagnetic scatter from composite objects" *IEEE Trans. Antennas Propagat.*, vol. 36, no. 3, pp. 376–382, March 1988.

19. L. N. Medgyesi-Mitschang, J. M. Putnam, and M. B. Gedera, "Generalized method of moments for three-dimensional penetrable scatterers," *J. Opt. Soc. Am. A*, vol. 11, no. 4, pp. 1383–1398, April. 1994.

20. P. M. Goggans, A. A. Kishk, and A. W. Glisson, "Electromagnetic scattering from objects composed of multiple homogeneous regions using a region-by-region solution," *IEEE Trans. Antennas Propagat.*, vol. 42, no. 6, pp. 865–871, June 1994.

21. E. Yip and B. Dembart, "Matrix assembly in FMM-MOM codes," *ISS Tech97-001*, Jan. 1997.

22. S. Chen, W. C. Chew, J. M. Song, and J. S. Zhao, "Low frequency MOM for penetrable scatterers," *Proc. IEEE/APS-2000*, vol. 3, pp. 1838–1841, 2000.

23. S. Chen, W. C. Chew, J. M. Song, and J. S. Zhao, "Analysis of low frequency scattering from penetrable scatterers," *IEEE Trans. on Geoscience and Remote Sensing,* vol. 39, no. 4, April. 2001.

24. S. M. Rao, D. R. Wilton, and A. W. Glisson, "Electromagnetic scattering by surfaces of arbitrary shape," *IEEE Trans. Antennas Propagat.,* vol. 32, pp. 878–884, May 1982.

25. W. A. Johnson, D. R. Wilton, and R. M. Sharpe, "Modeling scattering from and radiation by arbitrary shaped objects with the electric integral equation triangular surface patch code," *Electromagnetics*, vol. 10, no. 1–2, pp. 41–63, 1990.

26. G. D. Roberto, "On the numerical integration of the linear shape function times the 3-D Green's function or its gradient on a plane triangle," *IEEE Trans. Antennas Propagat.*, vol. 41, no. 10, pp. 1448–1455, Oct. 1993.

10
Efficient Analysis of Waveguiding Structures

Kaladhar Radhakrishnan and Weng Cho Chew

10.1 INTRODUCTION

Since their discovery in the last part of the nineteenth century, waveguides have become an indispensable part of several applications ranging from radio frequencies to optical frequencies. Consequently, numerical characterization and modeling of complex waveguide structures has become an important research topic. Because analytic techniques exist only for the simplest waveguide structures, we have to resort to numerical techniques to analyze complicated waveguide structures.

In this chapter, numerical schemes have been developed to analyze waveguiding structures. Section 10.2 describes a two-dimensional finite difference formulation that constitutes the foundation on which the numerical scheme is developed. The inhomogeneous transverse vector wave equation is used as the governing equation. The spatial derivatives along the transverse direction are approximated using finite differences. The use of the inhomogeneous wave equation allows us to model waveguiding structures with arbitrary permittivity profiles without matching boundary conditions explicitly at all the dielectric interfaces. The formulation is also generalized to handle anisotropic substrates. The source is assumed to be time harmonic, which allows us to operate in the frequency domain. The finite difference formulation results in an asymmetric sparse matrix equation. Section 10.3 describes a scheme to solve the sparse matrix equation efficiently. In order to exploit the sparsity of the matrix, a Krylov subspace scheme (bi-Lanczos algorithm) is used to solve the matrix equation.

In Section 10.4, the numerical scheme is modified to solve for the fields in the presence of waveguide discontinuities. This is accomplished by using an implicit mode matching scheme. The field is propagated analytically along the longitudinal direction. By matching the transverse electric and magnetic fields at the boundary, we can solve for the unknown field components. The formulation is generalized to handle n junctions. Numerical examples involving typical waveguide structures are shown in Section 10.5. The results agree well with previously published results.

10.2 FINITE DIFFERENCE FORMULATION

The transverse field components are used to formulate the waveguide problem. In this formulation, the electric field is defined as a fore vector and the magnetic field is defined as a back vector [1, 2]. The x, y, and z components of any given fore vector $\widetilde{\mathbf{f}}_\mathbf{m}$ and any back vector $\widehat{\mathbf{g}}_{\mathbf{m}+\frac{1}{2}}$ are defined at different locations as

$$\widetilde{\mathbf{f}}_\mathbf{m} = \hat{x} f^x_{m+(1/2),n} + \hat{y} f^y_{m,n+(1/2)} + \hat{z} f^z_{m,n} \qquad (10.1)$$

$$\widehat{\mathbf{g}}_{\mathbf{m}+\frac{1}{2}} = \hat{x} g^x_{m,n+(1/2)} + \hat{y} g^y_{m+(1/2),n} + \hat{z} g^z_{m+(1/2),n+(1/2)} \qquad (10.2)$$

Strictly speaking, these fields should be defined in a three-dimensional space like $f^x_{m+(1/2),n,p}$. However, since the discretization is along the xy plane, the third parameter p associated with the z axis is suppressed. Equations (10.1) and (10.2) imply that the electric field and the magnetic field components are defined on a staggered grid just like in the Yee algorithm [1].

The discretized vector wave equation for the electric field in an inhomogeneous, anisotropic source free medium is given by

$$\widehat{\nabla} \times [\widehat{\overline{\mu}}_{\mathbf{m}+\frac{1}{2}}]^{-1} \cdot \widetilde{\nabla} \times \widetilde{\mathbf{E}}_\mathbf{m} - \omega^2 \widetilde{\overline{\epsilon}}_\mathbf{m} \cdot \widetilde{\mathbf{E}}_\mathbf{m} = 0 \qquad (10.3)$$

In the above equation, $\widetilde{\nabla} \times$ is the curl operator using forward difference to approximate the derivatives, whereas $\widehat{\nabla} \times$ represents the curl operator using backward difference to approximate the derivatives. The tensors $\widehat{\overline{\mu}}_{\mathbf{m}+\frac{1}{2}}$ and $\widetilde{\overline{\epsilon}}_\mathbf{m}$ represent the permittivity and permeability tensors for the anisotropic medium and are functions of x and y. A tensor represented with a tilde (e.g., $\widetilde{\overline{\epsilon}}_\mathbf{m}$) is a tensor that operates on a fore vector and outputs a fore vector. Similarly, a tensor represented with a hat (e.g., $\widehat{\overline{\mu}}_{\mathbf{m}+\frac{1}{2}}$) is a tensor that operates on a back vector and outputs a back vector. Any waveguide which has a reflection symmetry about the z axis has the following structure for $\widehat{\overline{\mu}}_{\mathbf{m}+\frac{1}{2}}$ and $\widetilde{\overline{\epsilon}}_\mathbf{m}$ [3].

$$\widetilde{\epsilon}_{\mathbf{m}} = \begin{bmatrix} \epsilon^{xx}_{m+\frac{1}{2},n} & \epsilon^{xy}_{m,n+\frac{1}{2}} & 0 \\ \epsilon^{yx}_{m+\frac{1}{2},n} & \epsilon^{yy}_{m,n+\frac{1}{2}} & 0 \\ 0 & 0 & \epsilon^{zz}_{m,n} \end{bmatrix} \quad (10.4)$$

$$\widehat{\mu}_{\mathbf{m}+\frac{1}{2}} = \begin{bmatrix} \mu^{xx}_{m,n+\frac{1}{2}} & \mu^{xy}_{m+\frac{1}{2},n} & 0 \\ \mu^{yx}_{m,n+\frac{1}{2}} & \mu^{yy}_{m+\frac{1}{2},n} & 0 \\ 0 & 0 & \mu^{zz}_{m+\frac{1}{2},n+\frac{1}{2}} \end{bmatrix} \quad (10.5)$$

Equations (10.4) and (10.5) also give the location at which each component of the tensors is defined. If a tensor of the form shown above operates on a vector, both the x and y components are needed to evaluate the x or the y component of the output vector. For example, the x component of the electric flux density vector is defined as

$$D^x_{m+\frac{1}{2},n} = \epsilon^{xx}_{m+\frac{1}{2},n} E^x_{m+\frac{1}{2},n} + \epsilon^{xy}_{m,n+\frac{1}{2}} E^y_{m,n+\frac{1}{2}} \quad (10.6)$$

However, the second term on the right-hand side is defined at a location different from that of the left-hand side. This situation is similar to what occurs in the discretization of a nonorthogonal grid in an isotropic medium. Holland [4] introduced a second-order spatial interpolation scheme that averages the four neighboring components as shown below:

$$D^x_{m+\frac{1}{2},n} = \epsilon^{xx}_{m+\frac{1}{2},n} E^x_{m+\frac{1}{2},n} + \frac{1}{4}\left(\epsilon^{xy}_{m,n+\frac{1}{2}} E^y_{m,n+\frac{1}{2}} + \epsilon^{xy}_{m,n-\frac{1}{2}} E^y_{m,n-\frac{1}{2}}\right.$$
$$\left. + \epsilon^{xy}_{m+1,n+\frac{1}{2}} E^y_{m+1,n+\frac{1}{2}} + \epsilon^{xy}_{m+1,n-\frac{1}{2}} E^y_{m+1,n-\frac{1}{2}}\right) \quad (10.7)$$

A less accurate, but still second-order scheme can be achieved by just averaging two components as shown below:

$$D^x_{m+\frac{1}{2},n} = \epsilon^{xx}_{m+\frac{1}{2},n} E^x_{m+\frac{1}{2},n} + \frac{1}{2}\left(\epsilon^{xy}_{m,n+\frac{1}{2}} E^y_{m,n+\frac{1}{2}} + \epsilon^{xy}_{m+1,n-\frac{1}{2}} E^y_{m+1,n-\frac{1}{2}}\right) \quad (10.8)$$

Since the error introduced in the finite differencing scheme is second order, the two-term interpolation given in (10.8) will not introduce any lower-order error terms.

Extracting the transverse components of the vector wave equation (10.3), we get

$$\widehat{\nabla}_s \times [\widehat{\mu}^{zz}_{\mathbf{m}+\frac{1}{2}}]^{-1} \widetilde{\nabla}_s \times \widetilde{\mathbf{E}}^s_{\mathbf{m}} + ik_z \hat{z} \times [\widehat{\mu}^s_{\mathbf{m}+\frac{1}{2}}]^{-1} \cdot \widetilde{\nabla}_s \times \widetilde{\mathbf{E}}^s_{\mathbf{m}}$$
$$+ ik_z \hat{z} \times [\widehat{\mu}^s_{\mathbf{m}+\frac{1}{2}}]^{-1} \cdot ik_z \hat{z} \times \widetilde{\mathbf{E}}^s_{\mathbf{m}} - \omega^2 \widetilde{\epsilon}^s_{\mathbf{m}} \cdot \widetilde{\mathbf{E}}^s_{\mathbf{m}} = 0 \quad (10.9)$$

The term $\widetilde{\mathbf{E}}^s_{\mathbf{m}}$ represents the transverse components of the electric field. Similarly, $\widetilde{\epsilon}^s_{\mathbf{m}}$ and $\widehat{\mu}^s_{\mathbf{m}+\frac{1}{2}}$, are 2×2 tensors corresponding to the transverse components. Pre-multiplying (10.9) with $\hat{z} \times \widehat{\overline{\mu}}^{s}_{m+\frac{1}{2}} \cdot \hat{z}\times$, we get

$$\hat{z} \times \widehat{\overline{\mu}}^{s}_{m+\frac{1}{2}} \cdot \hat{z} \times \widetilde{\nabla}_s \times [\widehat{\mu}^{zz}_{m+\frac{1}{2}}]^{-1} \widetilde{\nabla}_s \times \widetilde{\mathbf{E}}^{s}_{\mathbf{m}} - \hat{z} \times \widetilde{\nabla}_s \times ik_z \widetilde{\mathbf{E}}^{z}_{\mathbf{m}}$$
$$- \omega^2 \hat{z} \times \widehat{\overline{\mu}}^{s}_{m+\frac{1}{2}} \cdot \hat{z} \times \widetilde{\overline{\epsilon}}^{s}_{\mathbf{m}} \cdot \widetilde{\mathbf{E}}^{s}_{\mathbf{m}} = k_z^2 \widetilde{\mathbf{E}}^{s}_{\mathbf{m}} \quad (10.10)$$

Using some standard vector identities and the divergence-free condition to eliminate the z component, we arrive at the transverse vector wave equation:

$$\hat{z} \times \widehat{\overline{\mu}}^{s}_{m+\frac{1}{2}} \cdot \hat{z} \times \widehat{\nabla}_s \times [\widehat{\mu}^{zz}_{m+\frac{1}{2}}]^{-1} \widetilde{\nabla}_s \times \widetilde{\mathbf{E}}^{s}_{\mathbf{m}} + \widetilde{\nabla}_s [\widetilde{\epsilon}^{zz}_{\mathbf{m}}]^{-1} \widehat{\nabla}_s \cdot \widetilde{\overline{\epsilon}}^{s}_{\mathbf{m}} \cdot \widetilde{\mathbf{E}}^{s}_{\mathbf{m}}$$
$$- \omega^2 \hat{z} \times \widehat{\overline{\mu}}^{s}_{m+\frac{1}{2}} \cdot \hat{z} \times \widetilde{\overline{\epsilon}}^{s}_{\mathbf{m}} \cdot \widetilde{\mathbf{E}}^{s}_{\mathbf{m}} = k_z^2 \widetilde{\mathbf{E}}^{s}_{\mathbf{m}} \quad (10.11)$$

By invoking the duality principle, we arrive at the transverse vector wave equation for the magnetic field:

$$\hat{z} \times \widetilde{\overline{\epsilon}}^{s}_{\mathbf{m}} \cdot \hat{z} \times \widetilde{\nabla}_s \times [\widetilde{\epsilon}^{zz}_{\mathbf{m}}]^{-1} \widehat{\nabla}_s \times \widehat{\mathbf{H}}^{s}_{m+\frac{1}{2}} + \widehat{\nabla}_s [\widehat{\mu}^{zz}_{m+\frac{1}{2}}]^{-1} \widetilde{\nabla}_s \cdot \widehat{\overline{\mu}}^{s}_{m+\frac{1}{2}} \cdot \widehat{\mathbf{H}}^{s}_{m+\frac{1}{2}}$$
$$- \omega^2 \hat{z} \times \widetilde{\overline{\epsilon}}^{s}_{\mathbf{m}} \cdot \hat{z} \times \widehat{\overline{\mu}}^{s}_{m+\frac{1}{2}} \cdot \widehat{\mathbf{H}}^{s}_{m+\frac{1}{2}} = k_z^2 \widehat{\mathbf{H}}^{s}_{m+\frac{1}{2}} \quad (10.12)$$

Equations (10.11) and (10.12) can be written in the form of an eigenvalue problem:

$$\overline{\mathcal{L}}_e \cdot \widetilde{\mathbf{E}}^{s}_{\mathbf{m}} = k_z^2 \widetilde{\mathbf{E}}^{s}_{\mathbf{m}} \quad (10.13)$$

$$\overline{\mathcal{L}}_h \cdot \widehat{\mathbf{H}}^{s}_{m+\frac{1}{2}} = k_z^2 \widehat{\mathbf{H}}^{s}_{m+\frac{1}{2}} \quad (10.14)$$

Both matrices $\overline{\mathcal{L}}_e$ and $\overline{\mathcal{L}}_h$ are asymmetric and extremely sparse and share the same eigenvalues. Each eigenvector of these matrices corresponds to the transverse components of the electric field and the magnetic field for a particular mode. Iterative solutions are desirable to solve such eigenvalue problems since they exploit the sparsity of the matrix while limiting memory requirements and computational complexity. Explicit storage of the matrix is avoided, and it is accessed in the form of a matrix vector multiply. Each time the matrix $\overline{\mathcal{L}}_e$ operates on a vector $\widetilde{\mathbf{E}}^{s}_{\mathbf{m}}$, the values of the field components are updated based on the values of their neighboring field components.

In the presence of a time harmonic electric current source localized in the $z = z'$ plane, the eigenvalue equations shown in (10.13) and (10.14) become sparse matrix equations of the form

$$\overline{\mathcal{L}}_e \cdot \widetilde{\mathbf{E}}^{s}_{\mathbf{m}} + \frac{\partial^2}{\partial z^2} \widetilde{\mathbf{E}}^{s}_{\mathbf{m}} = \overline{\mathcal{S}}_e \cdot \widetilde{\mathbf{J}}^{s}_{\mathbf{m}} \delta(z - z') \quad (10.15)$$

$$\overline{\mathcal{L}}_h \cdot \widehat{\mathbf{H}}^{s}_{m+\frac{1}{2}} + \frac{\partial^2}{\partial z^2} \widetilde{\mathbf{H}}^{s}_{\mathbf{m}} = \left(\hat{z} \times \widetilde{\mathbf{J}}^{s}_{\mathbf{m}}\right) \delta(z - z') \quad (10.16)$$

The z dependence of the field has been suppressed in (10.15) and (10.16). The operator $\overline{\mathcal{S}}_e$ in (10.16) is defined as

$$\overline{\mathcal{S}}_e \cdot \widetilde{\mathbf{J}}^{s}_{\mathbf{m}} = i\omega \hat{z} \times \widehat{\overline{\mu}}^{s}_{m+\frac{1}{2}} \cdot \hat{z} \times \widetilde{\mathbf{J}}^{s}_{\mathbf{m}} + \widetilde{\nabla}_s [i\omega \widetilde{\epsilon}^{zz}_{\mathbf{m}}]^{-1} \widehat{\nabla}_s \cdot \widetilde{\mathbf{J}}^{s}_{\mathbf{m}} \quad (10.17)$$

Efficient Analysis of Waveguiding Structures

The generalized formal solutions to the sparse matrix equations (10.15) and (10.16), in terms of matrix functions, are

$$\widetilde{\mathbf{E}}_{\mathbf{m}}^s = \frac{1}{2i}\overline{\mathcal{L}}_e^{-\frac{1}{2}} \cdot e^{i\overline{\mathcal{L}}_e^{\frac{1}{2}}|z-z'|} \overline{\mathcal{S}}_e \cdot \widetilde{\mathbf{J}}_{\mathbf{m}}^s \tag{10.18}$$

$$\widehat{\mathbf{H}}_{\mathbf{m}+\frac{1}{2}}^s = \pm\frac{1}{2}e^{i\overline{\mathcal{L}}_h^{\frac{1}{2}}|z-z'|} \cdot (\hat{z} \times \widetilde{\mathbf{J}}_{\mathbf{m}}^s) \tag{10.19}$$

The positive sign in (10.19) is used for values of $z > z'$, and the negative sign is used for $z < z'$. The bi-Lanczos algorithm is used to solve iteratively for the field components. Each iteration in the bi-Lanczos algorithm requires two matrix-vector multiplies. One computes $\overline{\mathcal{L}}_e \cdot \mathbf{v}$ while the other evaluates $\overline{\mathcal{L}}_e^T \cdot \mathbf{w}$. It is straightforward to compute the former since we know the exact definition for the operator $\overline{\mathcal{L}}_e$. In order to compute the latter without writing out the entire matrix, we seek a physical meaning to $\overline{\mathcal{L}}_e^T$. The fact that $\overline{\mathcal{L}}_e$ and $\overline{\mathcal{L}}_h$ share the same eigenvalues suggests that the latter might be related in some form to $\overline{\mathcal{L}}_e^T$. It can be shown along the lines of [5], that the transpose operator of $\overline{\mathcal{L}}_e$ is nothing but

$$\overline{\mathcal{L}}_e^T(\overline{\mu}, \overline{\epsilon}) = -\hat{z} \times \overline{\mathcal{L}}_h(\overline{\mu}^T, \overline{\epsilon}^T) \cdot \hat{z} \times \tag{10.20}$$

This implies that there exists another dual waveguide with the medium $(\overline{\mu}^T, \overline{\epsilon}^T)$ that shares the same eigenvalues as the original waveguide $(\overline{\mu}, \overline{\epsilon})$. In the case of a reciprocal medium where $\overline{\mu}$ and $\overline{\epsilon}$ are symmetric tensors, the dual waveguide is identical to the original waveguide. Equation (10.20) suggests that the electric field modes in the waveguide $(\overline{\mu}, \overline{\epsilon})$ are orthogonal to the magnetic field modes in the dual waveguide.

10.2.1 Boundary Conditions

The formulation described in the previous section is valid for any waveguiding structure. In order to numerically model these waveguiding structures, we need to use boundary conditions at the edges of the computational domain. In metallic waveguides, the walls are approximated as perfect electric conductors to truncate the computational domain. In open waveguiding structures like a dielectric waveguide or a microstrip line, we need to introduce a metallic wall at a sufficient distance to artificially truncate the computational domain.

The boundary conditions satisfied by the transverse electric field components at the metallic walls of a rectangular waveguide of width a and height b are

$$\left.\begin{array}{r}E_x = 0 \\ \dfrac{\partial D_y}{\partial y} = 0\end{array}\right\} x = 0, x = a \tag{10.21}$$

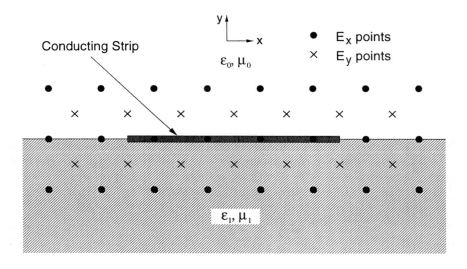

Figure 10.1 Location of the E_x and E_y points around the conducting strip.

$$\left.\begin{array}{r}\dfrac{\partial D_x}{\partial x} = 0 \\ E_y = 0\end{array}\right\} y = 0, y = b \qquad (10.22)$$

In the case of a microstrip or a stripline structure, we need to model the conducting strips as well as the outer walls. Because the formulation is valid for an inhomogeneous substrate, it is possible to model the conducting strip as a part of the substrate with a high conductivity. However, this is not desirable because it leads to an ill-conditioned matrix with slow convergence in the iterative solver. The more efficient way is to model the conducting strip using boundary conditions. Figure 10.1 shows the location of the transverse field components around the conducting strip. In this case, the conducting strip is assumed to be infinitesimally thin and is placed on the plane where the E_x field components are defined. It is important to ensure that the conducting strip does not end on an E_x point. Instead, the conducting strip should end at the midpoint between two adjacent E_x points as shown in the figure. This point corresponds to the location where the E_z component is defined. It has been shown [6] that the electric field component parallel to a conducting edge is always regular. By avoiding the computation of the E_x component at the edge of the conducting strip, we can keep the field finite throughout the computational domain.

In order to model the conducting strip using boundary conditions, we set the tangential electric field to zero on the surface of the conducting strip. This implies that all E_x components that fall within the conducting strip will be set to zero. To impose the boundary conditions on the field points immediately surrounding the conducting strip, we make use of the condition that E_z should be zero on the conducting strip. Because the electric field satisfies the divergence free condition on the area just outside the conducting strip, the E_z component can be expressed in terms of the transverse field components as

$$ik_z \widetilde{\mathbf{E}}_{\mathbf{m}}^z = [\widetilde{\epsilon}_{\mathbf{m}}^{zz}]^{-1} \widehat{\nabla}_s \cdot \widetilde{\epsilon}_{\mathbf{m}}^s \cdot \mathbf{E}_{\mathbf{m}}^s \qquad (10.23)$$

While evaluating the field at the points immediately surrounding the conducting strip, the E_z term makes its presence felt through the second term in (10.11). The second term can be written in terms of E_z as

$$\widetilde{\nabla}_s [\widetilde{\epsilon}_{\mathbf{m}}^{zz}]^{-1} \widehat{\nabla}_s \cdot \widetilde{\epsilon}_{\mathbf{m}}^s \cdot \widetilde{\mathbf{E}}_{\mathbf{m}}^s = ik_z \left[\hat{x} \left(\frac{E_{m+1,n}^z - E_{m,n}^z}{\Delta x} \right) \right. \\ \left. + \hat{y} \left(\frac{E_{m,n+1}^z - E_{m,n}^z}{\Delta y} \right) \right] \qquad (10.24)$$

By setting the E_z components defined on the surface of the conducting strip to zero, it is possible to simulate the presence of the conducting strip. It is straightforward to extend the boundary conditions to model a conducting strip that is not infinitesimally thin. In that case, one needs to ensure that all four edges of the conducting strip end on E_z points.

10.3 SOLUTION TO THE SPARSE MATRIX EQUATION

To solve for the fields using (10.18) and (10.19), one needs to evaluate matrix functions of the form $f(\overline{\mathcal{L}}_e) \cdot \widetilde{\mathbf{s}}_e$. A Krylov-subspace-based method is the ideal choice for solving such sparse matrix functions [7,8]. We use the bi-Lanczos algorithm [9,10] to solve the asymmetric matrix equation. The bi-Lanczos algorithm approximates the original matrix $\overline{\mathcal{L}}_e$ of size $N \times N$ with a smaller tridiagonal matrix of size $M \times M$. Also generated during the bi-Lanczos algorithm are two sets of iteration vectors, $\overline{\mathbf{V}}$ and $\overline{\mathbf{W}}$, each of size $M \times N$. The relation between the iteration vectors and the matrices can be summarized as

$$\overline{\mathcal{L}}_e \cdot \overline{\mathbf{V}} = \overline{\mathbf{V}} \cdot \overline{\mathbf{T}} \qquad (10.25)$$

$$\overline{\mathcal{L}}_e^T \cdot \overline{\mathbf{W}} = \overline{\mathbf{W}} \cdot \overline{\mathbf{T}}^T \qquad (10.26)$$

The iteration vectors $\overline{\mathbf{V}}$ and $\overline{\mathbf{W}}$ in exact arithmetic are biorthogonal to each other. In practice, however they lose their orthogonality after a few iterations. Another alternative is to use the more robust Arnoldi method. The drawback with the Arnoldi method is that it becomes prohibitively expensive for large matrices because it performs explicit reorthogonalization at each iteration.

A matrix function of the form of (10.18) can be solved using the SLDM technique. It can be shown that

$$f(\overline{\mathcal{L}_e}) \cdot \mathbf{v}_1 = \overline{\mathbf{V}} \cdot \overline{\mathbf{Q}} \cdot f(\overline{\mathbf{\Lambda}}) \cdot \overline{\mathbf{Q}}^{-1} \cdot \mathbf{e}_1 \quad (10.27)$$

where $\overline{\mathbf{Q}}$ and $\overline{\mathbf{\Lambda}}$ are the eigenvectors and eigenvalues of the tridiagonal matrix $\overline{\mathbf{T}}$.

Thus, by carrying out the bi-Lanczos algorithm and the spectral decomposition of the resulting tridiagonal matrix $\overline{\mathbf{T}}$, we can use (10.27) to solve the sparse matrix equation. The solution to (10.18) gives the field components at any given z plane for an arbitrary localized current source at $z = z'$. As a by-product of evaluating the current source response, it is also possible to solve for the propagating modes. The eigenvalues of the matrix $\overline{\mathbf{T}}$ are the same as the eigenvalues of the matrix $\overline{\mathcal{L}_e}$. Since these eigenvalues represent k_z^2 for that particular mode, it is possible to isolate the eigenpairs $(\lambda_i, \mathbf{q}_i)$ of the tridiagonal matrix $\overline{\mathbf{T}}$, which correspond to the propagating modes. From these eigenpairs, it is possible to obtain the eigenpairs of the $\overline{\mathcal{L}_e}$ matrix, $(\lambda_i, \overline{\mathbf{V}} \cdot \mathbf{q}_i)$. These eigenpairs contain all information that is required about the propagating modes.

10.3.1 Complexity and Storage Issues

The two main steps in the algorithm are the bi-Lanczos iterations and the spectral decomposition of the tridiagonal matrix. Each bi-Lanczos iteration has two matrix-vector multiplies. Since the matrices are sparse, the matrix-vector multiply is an $O(N)$ operation. The number of bi-Lanczos iterations needed to propagate information transversely across the grid in the simulation domain is observed empirically to scale as $\sqrt{N}$. It should be noted that apart from the number of unknowns, the relative permittivities of the various substrates also play a role in determining the number of iterations required for convergence. This is because of the multiple scattering that ensues, which directly affects the condition number of the matrix. Another factor that might influence the convergence is the number of conducting strips and their location. Because each conducting strip has a singularity associated with its edges, the greater the number of conducting strips, the higher the condition number of the matrix. Conducting strips placed close to each other can also increase the condition number. The overall complexity of the bi-Lanczos algorithm is $O(N^{1.5})$. The complexity of the spectral decomposition procedure is once again $O(N^{1.5})$ because the size of the matrix $\overline{\mathbf{T}}$ scales as $\sqrt{N}$. If it is desired to solve for just the propagating modes, it is possible to solve for just those eigenpairs instead of carrying out the whole spectral decomposition. This will make the algorithm faster, but will not affect the overall complexity.

The iteration vectors $\overline{\mathbf{V}}$ and $\overline{\mathbf{W}}$ are the primary bottlenecks in limiting the storage requirement. Because the iteration vectors $\overline{\mathbf{W}}$ are not needed to compute the field, they can be discarded as they are generated. On the other hand, one needs to store $\overline{\mathbf{V}}$ in order to compute the current response or to compute the field distribution for the propagating modes. Hence, the storage requirements scale as $O(N^{1.5})$, which is not desirable for solving large problems. Fortunately, it is possible to circumvent the storage bottleneck as discussed below.

In order to evaluate $f(\overline{\mathcal{L}}_e) \cdot \mathbf{v}_1$ or the eigenvector corresponding to the propagating modes, we need to evaluate a matrix equation of the form $\overline{\mathbf{V}} \cdot \mathbf{q}$, where $\mathbf{q}$ is a vector of size M. During the bi-Lanczos iterations, the vectors $\overline{\mathbf{V}}$ are evaluated using the recursive relation,

$$\mathbf{v}_{i+1} = [\overline{\mathcal{L}}_e \cdot \mathbf{v}_i - \alpha(i)\mathbf{v}_i - \gamma(i-1)\mathbf{v}_{i-1}]/\beta(i) \quad (10.28)$$

In the above expression α, β, and γ are the diagonal and subdiagonal elements of the matrix $\overline{\mathbf{T}}$. If we store the starting vector $\mathbf{v}_1$, it is possible to recursively obtain the subsequent iteration vectors using (10.28). As each vector $\mathbf{v}_i$ is computed, its contribution to the matrix-vector product $\overline{\mathbf{V}} \cdot \mathbf{q}$ is evaluated, and then the vector is discarded. This process of regenerating the iteration vectors takes less than half the time required for the bi-Lanczos iterations because this process needs only one matrix-vector multiply and the terms α, β, and γ need not be recomputed. This increases the overall computational time of the algorithm by roughly 10%, but reduces the storage to a small fraction of the original requirements. For small problems it is feasible to store the iteration vectors, but for larger problems it is imperative to discard and regenerate the iteration vectors.

10.4 WAVEGUIDE DISCONTINUITIES

Waveguide discontinuity problems are generally solved using the mode-matching method. In this method, the fields are expanded in terms of the modes of the waveguide. Then by matching the transverse electric field and the transverse magnetic field at the boundary, we can solve for the reflected and transmitted fields. This usually involves expressing the electric field in terms of the magnetic field or vice versa. However, in Section 10.2, the electric field was defined as a fore vector and the magnetic field was defined as a back vector. From (10.1) and (10.2), we can see that these vectors are defined at different locations. In order to overcome this problem, we define a new operator $\widetilde{\mathcal{L}}_h$ as

$$\widetilde{\mathcal{L}}_h = -\hat{z} \times \overline{\mathcal{L}}_h \cdot \hat{z} \times \quad (10.29)$$

We can now rewrite (10.16) as

$$\widetilde{\mathcal{L}}_h \cdot (\hat{z} \times \widehat{\mathbf{H}}_s) + \frac{\partial^2}{\partial z^2}(\hat{z} \times \widehat{\mathbf{H}}_s) = \frac{\partial}{\partial z}\widetilde{\mathbf{J}}_s \delta(z-z') \quad (10.30)$$

The new operator $\widetilde{\mathcal{L}}_h$ operates on the vector $(\hat{z} \times \widehat{\mathbf{H}}_s)$ whereas $\overline{\mathcal{L}}_h$ operates on the vector $\widehat{\mathbf{H}}_s$. In many ways, it is more convenient to deal with $(\hat{z} \times \widehat{\mathbf{H}}_s)$ than just $\widehat{\mathbf{H}}_s$.

While $\widehat{\mathbf{H}}_s$ is defined as a back vector, $(\hat{z} \times \widehat{\mathbf{H}}_s)$ has the properties of a fore vector. For example,

$$\widehat{\mathbf{H}}^s_{m+\frac{1}{2}} = \hat{x} H^x_{m,n+\frac{1}{2}} + \hat{y} H^y_{m+\frac{1}{2},n} \tag{10.31}$$

$$(\hat{z} \times \widehat{\mathbf{H}}^s_{m+\frac{1}{2}}) = -\hat{x} H^y_{m+\frac{1}{2},n} + \hat{y} H^x_{m,n+\frac{1}{2}} \tag{10.32}$$

The electric field, $\widetilde{\mathbf{E}}^s_m$, which is a fore vector, is defined as

$$\widetilde{\mathbf{E}}^s_m = \hat{x} E^x_{m+\frac{1}{2},n} + \hat{y} E^y_{m,n+\frac{1}{2}} \tag{10.33}$$

By comparing (10.32) and (10.33), we can see that the x and y components of $(\hat{z} \times \widehat{\mathbf{H}}^s_{m+\frac{1}{2}})$ are defined at the same location as those of a fore vector like $\widetilde{\mathbf{E}}^s_m$. Thus, representing the magnetic field as $(\hat{z} \times \widehat{\mathbf{H}}_s)$ allows us to derive equations relating the magnetic field and the electric field.

The solution to (10.30) in terms of matrix functions is

$$(\hat{z} \times \widehat{\mathbf{H}}_s) = \pm \frac{1}{2} e^{i\widetilde{\overline{\mathcal{L}}}_h^{\frac{1}{2}} |z-z'|} \cdot \widetilde{\mathbf{J}}_s \tag{10.34}$$

where the positive sign is used if $z > z'$ and the negative sign is used otherwise.

From Maxwell's equations, the transverse component of the magnetic field can be derived from the transverse component of the electric field as

$$\frac{\partial}{\partial z}(\hat{z} \times \widehat{\mathbf{H}}_s) = -i\omega\epsilon\widetilde{\mathbf{E}}_s + \frac{i}{\omega}\widehat{\nabla}_s \times \mu^{-1}\widetilde{\nabla}_s \times \widetilde{\mathbf{E}}_s \tag{10.35}$$

Using the finite difference method to discretize the operators, we can write the expression in the form of a matrix equation:

$$\frac{\partial}{\partial z}(\hat{z} \times \widehat{\mathbf{H}}_s) = i\overline{\mathbf{A}} \cdot \widetilde{\mathbf{E}}_s \tag{10.36}$$

Since the operator $\overline{\mathbf{A}}$ is independent of z, (10.36) can be written as

$$(\hat{z} \times \widehat{\mathbf{H}}_s) = i\overline{\mathbf{A}} \cdot \left(\frac{\partial}{\partial z}\right)^{-1} \widetilde{\mathbf{E}}_s \tag{10.37}$$

Using the duality theorem on (10.35) yields another expression relating the magnetic field to the electric field:

$$\frac{\partial}{\partial z}(\hat{z} \times \widetilde{\mathbf{E}}_s) = i\omega\mu\widehat{\mathbf{H}}_s - \frac{i}{\omega}\widetilde{\nabla}_s \times \epsilon^{-1}\widehat{\nabla}_s \times \widehat{\mathbf{H}}_s \tag{10.38}$$

Using finite differences to approximate the spatial derivatives, we can express the above equation in the form of a matrix function:

$$\frac{\partial}{\partial z}(\hat{z} \times \widetilde{\mathbf{E}}_s) = i\overline{\mathbf{B}} \cdot \widehat{\mathbf{H}}_s \tag{10.39}$$

Efficient Analysis of Waveguiding Structures

Since we want to deal with fore-vectors, we define a new operator $\widetilde{\mathbf{B}}$ as

$$\widetilde{\mathbf{B}} = -\hat{z} \times \mathbf{B} \cdot \hat{z} \times \tag{10.40}$$

Using the new operator $\widetilde{\mathbf{B}}$, (10.38) can be expressed in terms of $(\hat{z} \times \widehat{\mathbf{H}}_s)$ and $\widetilde{\mathbf{E}}_s$:

$$\frac{\partial}{\partial z}\widetilde{\mathbf{E}}_s = -i\widetilde{\mathbf{B}} \cdot (\hat{z} \times \widehat{\mathbf{H}}_s) \tag{10.41}$$

By substituting the expressions for the fields given in (10.18) and (10.34), we can derive the following equalities:

$$\overline{\mathcal{L}}_e = \widetilde{\mathbf{B}} \cdot \overline{\mathbf{A}} \tag{10.42}$$

$$\widetilde{\mathcal{L}}_h = \overline{\mathbf{A}} \cdot \widetilde{\mathbf{B}} \tag{10.43}$$

10.4.1 The Single Junction Problem

Figure 10.2 shows a typical waveguide junction with a current source in the first waveguide section. The incident field in the presence of a current source $\widetilde{\mathbf{J}}_s$ at $z = 0$ is

$$\widetilde{\mathbf{E}}_s = \frac{1}{2i}\overline{\mathcal{L}}_{1e}^{-\frac{1}{2}} \cdot e^{i\overline{\mathcal{L}}_{1e}^{\frac{1}{2}}|z|} \cdot \overline{\mathcal{S}}_{1e} \cdot \widetilde{\mathbf{J}}_s \tag{10.44}$$

In the presence of a waveguide junction at $z = d$, the total field in waveguide 1 is

$$\widetilde{\mathbf{E}}_{1s} = \frac{1}{2i}\left[\overline{\mathcal{L}}_{1e}^{-\frac{1}{2}} \cdot e^{i\overline{\mathcal{L}}_{1e}^{\frac{1}{2}}|z|} + e^{-i\overline{\mathcal{L}}_{1e}^{\frac{1}{2}}(z-d)} \cdot \overline{\mathbf{R}}_{12} \cdot \overline{\mathcal{L}}_{1e}^{-\frac{1}{2}} \cdot e^{i\overline{\mathcal{L}}_{1e}^{\frac{1}{2}}d}\right] \cdot \overline{\mathcal{S}}_{1e} \cdot \widetilde{\mathbf{J}}_s \tag{10.45}$$

where $\overline{\mathbf{R}}_{12}$ is a matrix that transforms the incident field to the reflected field at the junction and is yet to be determined. Similarly, we can write down the expression for the field in waveguide 2 as

$$\widetilde{\mathbf{E}}_{2s} = \frac{1}{2i}e^{i\overline{\mathcal{L}}_{2e}^{\frac{1}{2}}(z-d)} \cdot \overline{\mathbf{T}}_{12} \cdot \overline{\mathcal{L}}_{1e}^{-\frac{1}{2}} \cdot e^{i\overline{\mathcal{L}}_{1e}^{\frac{1}{2}}d} \cdot \overline{\mathcal{S}}_{1e} \cdot \widetilde{\mathbf{J}}_s \tag{10.46}$$

where $\overline{\mathbf{T}}_{12}$ is the transmission matrix yet to be determined.

At the waveguide junction, the transverse electric field components must be continuous. However, waveguides 1 and 2 may be different sizes. In which case, the electric field component is matched at the common interface denoted by S_a in Figure 10.2. The electric field is forced to be zero at all other regions, denoted by S_b in Figure 10.2. Thus, in order to match the electric field at the interface, we use the rectangular transformation matrix $\overline{\mathbf{O}}$. This matrix pads the input vector with some additional zeros at the locations, which correspond to the region S_b. There are two possible cases, depending on which waveguide has the higher number of unknowns.

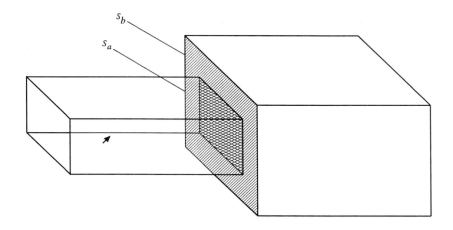

Figure 10.2 Waveguide junction with a current source in waveguide 1.

10.4.1.1 Case 1 : $N_2 > N_1$

In this case, the second waveguide has more unknowns than the first. The junction shown in Figure 10.2 is an example of this case. In such cases, the electric field satisfies the following boundary condition at the interface:

$$\overline{O}_{12} \cdot [\mathbf{E}_{inc} + \mathbf{E}_{ref}] = \mathbf{E}_{trn} \tag{10.47}$$

In the above equation, $\mathbf{E}_{inc}$, $\mathbf{E}_{ref}$, and $\mathbf{E}_{trn}$ represent the incident, reflected, and transmitted field at the interface and are defined as

$$\begin{aligned}
\mathbf{E}_{inc} &= \frac{1}{2i} \overline{\mathcal{L}}_{1e}^{-\frac{1}{2}} \cdot e^{i\overline{\mathcal{L}}_{1e}^{\frac{1}{2}} d} \cdot \overline{S}_{1e} \cdot \widetilde{\mathbf{J}}_s \\
\mathbf{E}_{ref} &= \frac{1}{2i} \overline{\mathbf{R}}_{12} \cdot \overline{\mathcal{L}}_{1e}^{-\frac{1}{2}} \cdot e^{i\overline{\mathcal{L}}_{1e}^{\frac{1}{2}} d} \cdot \overline{S}_{1e} \cdot \widetilde{\mathbf{J}}_s \\
\mathbf{E}_{trn} &= \frac{1}{2i} \overline{\mathbf{T}}_{12} \cdot \overline{\mathcal{L}}_{1e}^{-\frac{1}{2}} \cdot e^{i\overline{\mathcal{L}}_{1e}^{\frac{1}{2}} d} \cdot \overline{S}_{1e} \cdot \widetilde{\mathbf{J}}_s
\end{aligned} \tag{10.48}$$

The transverse magnetic field components are continuous across the waveguide junction. The magnetic field is matched at the common interface S_a between the two waveguides. Unlike the electric field, the magnetic field is not forced to be zero outside the common interface, which corresponds to the region S_b. To satisfy these conditions, the transverse of the transformation matrix is used as shown below:

$$\mathbf{H}_{inc} + \mathbf{H}_{ref} = \overline{O}_{12}^T \cdot \mathbf{H}_{trn} \tag{10.49}$$

Equations (10.37) and (10.48) can be used to express the incident, reflected, and transmitted magnetic fields at the interface in terms of the electric field:

$$\mathbf{H}_{inc} = \overline{\mathbf{A}}_1 \cdot \overline{\mathcal{L}}_{1e}^{-\frac{1}{2}} \cdot \mathbf{E}_{inc}$$
$$\mathbf{H}_{ref} = -\overline{\mathbf{A}}_1 \cdot \overline{\mathcal{L}}_{1e}^{-\frac{1}{2}} \cdot \mathbf{E}_{ref} \qquad (10.50)$$
$$\mathbf{H}_{trn} = \overline{\mathbf{A}}_2 \cdot \overline{\mathcal{L}}_{2e}^{-\frac{1}{2}} \cdot \mathbf{E}_{trn}$$

Using (10.49) in (10.50), we get

$$\overline{\mathbf{A}}_1 \cdot \overline{\mathcal{L}}_{1e}^{-\frac{1}{2}} \cdot [\mathbf{E}_{inc} - \mathbf{E}_{ref}] = \overline{\mathbf{O}}^T \cdot \overline{\mathbf{A}}_2 \cdot \overline{\mathcal{L}}_{2e}^{-\frac{1}{2}} \cdot \mathbf{E}_{trn} \qquad (10.51)$$

Equation (10.51) can be rewritten as

$$\mathbf{E}_{inc} - \mathbf{E}_{ref} = \overline{\mathbf{M}}_{21} \cdot \mathbf{E}_{trn} \qquad (10.52)$$

where the matrix $\overline{\mathbf{M}}_{ij}$ is defined as

$$\overline{\mathbf{M}}_{ij} \cdot \mathbf{x} = \overline{\mathcal{L}}_{je}^{\frac{1}{2}} \cdot \overline{\mathbf{A}}_j^{-1} \cdot \overline{\mathbf{O}}_{ji}^T \cdot \overline{\mathbf{A}}_i \cdot \overline{\mathcal{L}}_{ie}^{-\frac{1}{2}} \cdot \mathbf{x} \qquad (10.53)$$

Equations (10.47) and (10.51) can be used to solve for the reflected and transmitted fields in terms of the incident field:

$$\mathbf{E}_{ref} = \left(\overline{\mathbf{I}} - \overline{\mathbf{M}}_{21} \cdot \overline{\mathbf{O}}_{12}\right) \cdot \left(\overline{\mathbf{I}} + \overline{\mathbf{M}}_{21} \cdot \overline{\mathbf{O}}_{21}\right)^{-1} \cdot \mathbf{E}_{inc} \qquad (10.54)$$

$$\mathbf{E}_{trn} = 2 \left(\overline{\mathbf{I}} + \overline{\mathbf{O}}_{12} \cdot \overline{\mathbf{M}}_{12}\right)^{-1} \cdot \overline{\mathbf{O}}_{12} \cdot \mathbf{E}_{inc} \qquad (10.55)$$

The matrix $\overline{\mathbf{M}}_{21}$ can be written as $(\overline{\mathbf{A}}_1 \cdot \overline{\mathcal{L}}_{1e}^{-\frac{1}{2}})^{-1} \cdot \overline{\mathbf{O}}_{12}^T \cdot \overline{\mathbf{A}}_2 \cdot \overline{\mathcal{L}}_{2e}^{-\frac{1}{2}}$. Since the structure of the matrix $\overline{\mathbf{A}}_1 \cdot \overline{\mathcal{L}}_{1e}^{\frac{1}{2}}$ is very similar to that of the matrix $\overline{\mathbf{A}}_2 \cdot \overline{\mathcal{L}}_{2e}^{\frac{1}{2}}$, the matrix $\overline{\mathbf{M}}_{21}$ is well conditioned with its eigenvalues clustered around 1. This causes (10.54) and (10.55) to converge rapidly.

10.4.1.2 Case 2 : $N_1 > N_2$

In this case, the first waveguide has more unknowns than the first. The microstrip step junction shown in Figure 10.3 is an example of this case. In such cases, the electric and magnetic fields satisfy the following boundary conditions at the interface:

$$\mathbf{E}_{inc} + \mathbf{E}_{ref} = \overline{\mathbf{O}}_{21} \cdot \mathbf{E}_{trn} \qquad (10.56)$$

$$\overline{\mathbf{O}}_{21}^T \cdot [\mathbf{H}_{inc} + \mathbf{H}_{ref}] = \mathbf{H}_{trn} \qquad (10.57)$$

Using (10.50) and (10.52), we can rewrite (10.57) in terms of the electric field components as

$$\overline{\mathbf{M}}_{12} \cdot [\mathbf{E}_{inc} - \mathbf{E}_{ref}] = \mathbf{E}_{trn} \qquad (10.58)$$

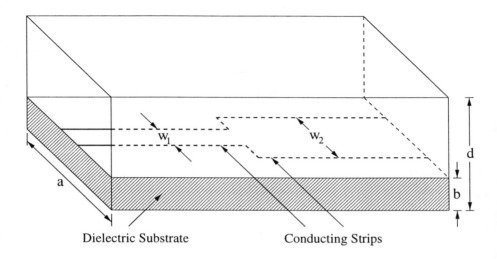

Figure 10.3 Shielded microstrip line with a step discontinuity.

Equations (10.56) and (10.58) can now be solved for the reflected and transmitted fields at the junction:

$$\mathbf{E}_{ref} = \left(\overline{\mathbf{O}}_{21} \cdot \overline{\mathbf{M}}_{12} - \overline{\mathbf{I}}\right) \cdot \left(\overline{\mathbf{O}}_{21} \cdot \overline{\mathbf{M}}_{12} + \overline{\mathbf{I}}\right)^{-1} \cdot \mathbf{E}_{inc} \qquad (10.59)$$

$$\mathbf{E}_{trn} = 2\left(\overline{\mathbf{M}}_{12} \cdot \overline{\mathbf{O}}_{21} + \overline{\mathbf{I}}\right)^{-1} \cdot \overline{\mathbf{M}}_{12} \cdot \mathbf{E}_{inc} \qquad (10.60)$$

Just as in the previous case, (10.59) and (10.60) converge rapidly.

Each iteration involves one matrix vector multiply of the form $\overline{\mathbf{M}}_{ij} \cdot x$. We can avoid inverting the matrix $\overline{\mathbf{A}}_j$ if we redefine $\overline{\mathbf{M}}_{ij}$ as

$$\overline{\mathbf{M}}_{ij} \cdot \mathbf{x} = \overline{\mathcal{L}}_{je}^{-\frac{1}{2}} \cdot \overline{\mathbf{B}}_j \cdot \overline{\mathbf{O}}_{ij}^T \cdot \overline{\mathbf{A}}_i \cdot \overline{\mathcal{L}}_{ie}^{-\frac{1}{2}} \cdot \mathbf{x} \qquad (10.61)$$

We have made use of (10.42) in (10.53) to derive the above expression. Thus, each matrix vector multiply of the form $\overline{\mathbf{M}}_{ij} \cdot \mathbf{x}$ involves two SLDM operations to evaluate $\overline{\mathcal{L}}_{ie}^{-\frac{1}{2}}$ and $\overline{\mathcal{L}}_{je}^{-\frac{1}{2}}$. Hence, if we need n_{it} iterations, we will need to carry out $2n_{it}$ SLDM operations to solve the junction problem.

10.4.2 The n-Junction Problem

In this section, we generalize the single junction formulation to solve for the fields in the presence of multiple junctions. Let us consider a waveguiding structure with n junctions, as shown in Figure 10.4. We can divide the waveguiding structure into

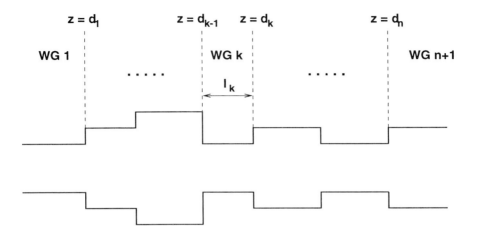

Figure 10.4 An arbitrary n-junction microstrip structure.

$n + 1$ waveguide sections along the z direction. The waveguide cross-section in each section is constant along the z direction. The notation used for the n-junction problem is the same as that used for the two-junction problem. The incident and reflected fields are defined at the first junction ($z = d_1$). The transmitted field is defined at the nth junction. Waveguide section k represents the region in the waveguide between the junctions $z = d_{k-1}$ and $z = d_k$. The waveguide section k has a forward propagating wave component $\mathbf{E}_k^+$ and a backward propagating wave component $\mathbf{E}_k^-$. The forward wave $\mathbf{E}_k^+$ is defined at $z = d_{k-1}$ while the backward wave is defined at $z = d_k$. Thus, there are $2(n-1)$ unknown field components in the interior waveguide sections. These, along with the reflected and transmitted field components, yield a total of $2n$ unknowns. By matching the electric and magnetic field components at the n junctions, we get $2n$ equations to solve for the $2n$ unknowns.

At the first waveguide junction, there are two possibilities. If the number of unknowns in waveguide 1 (N_1) is greater than that for waveguide 2 (N_2), then the boundary conditions for the electric and magnetic fields are

$$\mathbf{E}_{inc} + \mathbf{E}_{ref} = \overline{\mathbf{O}}_{21} \cdot \left[\mathbf{E}_2^+ + e^{i\overline{\mathcal{L}}_{2e}^{\frac{1}{2}} l_2} \cdot \mathbf{E}_2^- \right] \tag{10.62}$$

$$\overline{\mathbf{M}}_{12} \cdot [\mathbf{E}_{inc} - \mathbf{E}_{ref}] = \mathbf{E}_2^+ - e^{i\overline{\mathcal{L}}_{2e}^{\frac{1}{2}} l_2} \cdot \mathbf{E}_2^- \tag{10.63}$$

where l_i is $(d_i - d_{i-1})$ and is the length of waveguide section i. The rectangular matrix $\overline{\mathbf{O}}_{ij}$ is the transformation matrix from waveguide i to waveguide j and the matrix $\overline{\mathbf{M}}_{ij}$ is defined in (10.53). By eliminating the reflected field, we can combine

(10.62) and (10.63) into a single equation:

$$(\overline{\mathbf{M}}_{12} \cdot \overline{\mathbf{O}}_{21} + \overline{\mathbf{I}}) \cdot \mathbf{E}_2^+ + (\overline{\mathbf{M}}_{12} \cdot \overline{\mathbf{O}}_{21} - \overline{\mathbf{I}}) \cdot e^{i\overline{\mathcal{L}}_{2e}^{\frac{1}{2}} l_2} \cdot \mathbf{E}_2^- = 2\overline{\mathbf{M}}_{12} \cdot \mathbf{E}_{inc} \quad (10.64)$$

If the number of unknowns in the first waveguide N_1 is smaller than that in the second waveguide N_2, then the boundary conditions are written as

$$\overline{\mathbf{O}}_{12} \cdot [\mathbf{E}_{inc} + \mathbf{E}_{ref}] = \mathbf{E}_2^+ + e^{i\overline{\mathcal{L}}_{2e}^{\frac{1}{2}} l_2} \cdot \mathbf{E}_2^- \quad (10.65)$$

$$\mathbf{E}_{inc} - \mathbf{E}_{ref} = \overline{\mathbf{M}}_{21} \cdot \left[\mathbf{E}_2^+ - e^{i\overline{\mathcal{L}}_{2e}^{\frac{1}{2}} l_2} \cdot \mathbf{E}_2^- \right] \quad (10.66)$$

These equations can once again be combined into a single equation as

$$(\overline{\mathbf{I}} + \overline{\mathbf{O}}_{12} \cdot \overline{\mathbf{M}}_{21}) \cdot \mathbf{E}_2^+ + (\overline{\mathbf{I}} - \overline{\mathbf{O}}_{12} \cdot \overline{\mathbf{M}}_{21}) \cdot e^{i\overline{\mathcal{L}}_{2e}^{\frac{1}{2}} l_2} \cdot \mathbf{E}_2^- = 2\overline{\mathbf{O}}_{12} \cdot \mathbf{E}_{inc} \quad (10.67)$$

Next, we consider the boundary conditions at one of the interior junctions, $z = d_k$. If $N_k > N_{k+1}$, the boundary conditions satisfied by the electric and magnetic fields are

$$e^{i\overline{\mathcal{L}}_{ke}^{\frac{1}{2}} l_k} \cdot \mathbf{E}_k^+ + \mathbf{E}_k^- = \overline{\mathbf{O}}_{k+1,k} \cdot \left[\mathbf{E}_{k+1}^+ + e^{i\overline{\mathcal{L}}_{k+1,e}^{\frac{1}{2}} l_{k+1}} \cdot \mathbf{E}_{k+1}^- \right] \quad (10.68)$$

$$\overline{\mathbf{M}}_{k,k+1} \cdot \left[e^{i\overline{\mathcal{L}}_{ke}^{\frac{1}{2}} l_k} \cdot \mathbf{E}_k^+ - \mathbf{E}_k^- \right] = \mathbf{E}_{k+1}^+ - e^{i\overline{\mathcal{L}}_{k+1,e}^{\frac{1}{2}} l_{k+1}} \cdot \mathbf{E}_{k+1}^- \quad (10.69)$$

If the number of unknowns in waveguide k (N_k) is less than the number of unknowns in waveguide $k+1$ (N_{k+1}), then the boundary conditions are written as

$$\overline{\mathbf{O}}_{k,k+1} \cdot \left[e^{i\overline{\mathcal{L}}_{ke}^{\frac{1}{2}} l_k} \cdot \mathbf{E}_k^+ + \mathbf{E}_k^- \right] = \mathbf{E}_{k+1}^+ + e^{i\overline{\mathcal{L}}_{k+1,e}^{\frac{1}{2}} l_{k+1}} \cdot \mathbf{E}_{k+1}^- \quad (10.70)$$

$$e^{i\overline{\mathcal{L}}_{ke}^{\frac{1}{2}} l_k} \cdot \mathbf{E}_k^+ - \mathbf{E}_k^- = \overline{\mathbf{M}}_{k+1,k} \cdot \left[\mathbf{E}_{k+1}^+ - e^{i\overline{\mathcal{L}}_{k+1,e}^{\frac{1}{2}} l_{k+1}} \cdot \mathbf{E}_{k+1}^- \right] \quad (10.71)$$

Similarly, there are two cases for the boundary condition at the final junction. If $N_n > N_{n+1}$, then

$$e^{i\overline{\mathcal{L}}_{ne}^{\frac{1}{2}} l_n} \cdot \mathbf{E}_n^+ + \mathbf{E}_n^- = \overline{\mathbf{O}}_{n+1,n} \cdot \mathbf{E}_{trn} \quad (10.72)$$

$$\overline{\mathbf{M}}_{n,n+1} \cdot \left[e^{i\overline{\mathcal{L}}_{ne}^{\frac{1}{2}} l_n} \cdot \mathbf{E}_n^+ - \mathbf{E}_n^- \right] = \mathbf{E}_{trn} \quad (10.73)$$

By eliminating the transmitted field, we can combine (10.72) and (10.73) as

$$(\overline{\mathbf{I}} - \overline{\mathbf{O}}_{n+1,n} \cdot \overline{\mathbf{M}}_{n,n+1}) \cdot e^{i\overline{\mathcal{L}}_{ne}^{\frac{1}{2}} l_n} \cdot \mathbf{E}_n^+ + (\overline{\mathbf{I}} + \overline{\mathbf{O}}_{n+1,n} \cdot \overline{\mathbf{M}}_{n,n+1}) \cdot \mathbf{E}_n^- = 0 \quad (10.74)$$

On the other hand, if $N_n < N_{n+1}$, then the boundary conditions are written as

$$\overline{\mathbf{O}}_{n,n+1} \cdot \left[e^{i\overline{\mathcal{L}}_{n e}^{\frac{1}{2}} l_n} \cdot \mathbf{E}_n^+ + \mathbf{E}_n^- \right] = \mathbf{E}_{trn} \tag{10.75}$$

$$e^{i\overline{\mathcal{L}}_{n e}^{\frac{1}{2}} l_n} \cdot \mathbf{E}_n^+ - \mathbf{E}_n^- = \overline{\mathbf{M}}_{n+1,n} \cdot \mathbf{E}_{trn} \tag{10.76}$$

Eliminating the transmitted field from these equations yields

$$(\overline{\mathbf{M}}_{n+1,n} \cdot \overline{\mathbf{O}}_{n,n+1} - \overline{\mathbf{I}}) \cdot e^{i\overline{\mathcal{L}}_{n e}^{\frac{1}{2}} l_n} \cdot \mathbf{E}_n^+ + (\overline{\mathbf{M}}_{n+1,n} \cdot \overline{\mathbf{O}}_{n,n+1} + \overline{\mathbf{I}}) \cdot \mathbf{E}_n^- = 0 \tag{10.77}$$

After eliminating the reflected and transmitted fields, we now have $2n - 2$ unknowns and an equal number of equations to solve for them. These equations can be written in the form of a block matrix structure. For example, in the case of a four-junction problem, the block matrix will have the following structure:

$$\begin{bmatrix} \overline{\mathbf{A}}_{11} & \overline{\mathbf{A}}_{12} & \overline{\emptyset} & \overline{\emptyset} & \overline{\emptyset} & \overline{\emptyset} \\ \overline{\mathbf{A}}_{21} & \overline{\mathbf{A}}_{22} & \overline{\mathbf{A}}_{23} & \overline{\mathbf{A}}_{24} & \overline{\emptyset} & \overline{\emptyset} \\ \overline{\mathbf{A}}_{31} & \overline{\mathbf{A}}_{32} & \overline{\mathbf{A}}_{33} & \overline{\mathbf{A}}_{34} & \overline{\emptyset} & \overline{\emptyset} \\ \overline{\emptyset} & \overline{\emptyset} & \overline{\mathbf{A}}_{43} & \overline{\mathbf{A}}_{44} & \overline{\mathbf{A}}_{45} & \overline{\mathbf{A}}_{46} \\ \overline{\emptyset} & \overline{\emptyset} & \overline{\mathbf{A}}_{53} & \overline{\mathbf{A}}_{54} & \overline{\mathbf{A}}_{55} & \overline{\mathbf{A}}_{56} \\ \overline{\emptyset} & \overline{\emptyset} & \overline{\emptyset} & \overline{\emptyset} & \overline{\mathbf{A}}_{65} & \overline{\mathbf{A}}_{66} \end{bmatrix} \cdot \begin{bmatrix} \mathbf{E}_2^+ \\ \mathbf{E}_2^- \\ \mathbf{E}_3^+ \\ \mathbf{E}_3^- \\ \mathbf{E}_4^+ \\ \mathbf{E}_4^- \end{bmatrix} = \begin{bmatrix} \mathbf{x}_1 \\ \emptyset \\ \emptyset \\ \emptyset \\ \emptyset \\ \emptyset \end{bmatrix} \tag{10.78}$$

In the above equations, $\mathbf{x}_1$, $\overline{\mathbf{A}}_{11}$, and $\overline{\mathbf{A}}_{12}$ correspond to the incident field and the matrices associated with the first junction and can be determined using (10.64) or (10.67), depending on the relative sizes of waveguides 1 and 2. Similarly, matrices $\overline{\mathbf{A}}_{65}$ and $\overline{\mathbf{A}}_{66}$ correspond to the matrices associated with the last junction and are determined using either (10.74) or (10.77), depending on the sizes of the last two waveguide sections. All the other matrices in (10.78) are determined either by using (10.68) and (10.69) or by using (10.70) and (10.71), depending on the size of the waveguide sections associated with that junction. Each block matrix-vector multiply for an n-junction problem involves n matrix-vector multiplies of the form $\overline{\mathbf{M}}_{ij} \cdot \mathbf{x}$. As the $\overline{\mathbf{M}}_{ij}$ matrices approach the identity matrix, the structure of the block matrix becomes very simple, and it is possible to write down the inverse of the block matrix by inspection. Since the eigenvalues of the $\overline{\mathbf{M}}_{ij}$ matrices are clustered around 1, the block matrix tends to be well conditioned and converges quickly when solved using an iterative solver.

10.5 NUMERICAL EXAMPLES

In Section 10.2, it was shown that the each eigenvector of the matrix $\overline{\mathcal{L}}_e$ corresponds to the transverse electric field component for a given mode. Figure 10.5 plots the

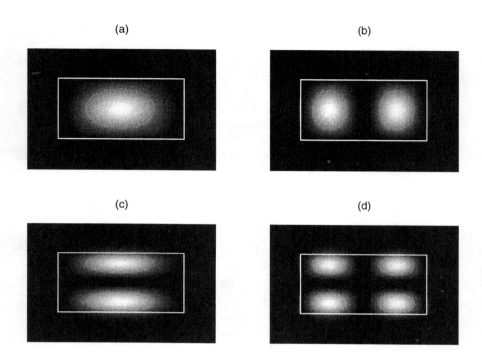

Figure 10.5 Field intensity plots for the modes (a) E_{11}^x, (b) E_{21}^x, (c) E_{12}^x, and (d) E_{22}^x in a rectangular dielectric waveguide. The white line indicates the interface between the core and the cladding.

field intensity plots for the first four propagating modes in a rectangular dielectric waveguide with an aspect ratio of 2 and a dielectric constant of 2.25. In order to obtain the dispersion curves for a waveguiding structure, one needs to solve for the eigenvalues of the matrix $\overline{\mathcal{L}}_e$. In the next example, we consider a dual plane triple microstrip structure on an anisotropic substrate as shown in Figure 10.6(a). Dispersion curves for the first three modes are shown in Figure 10.6(b). Very good agreement is seen between these curves and the ones obtained using a vector finite element based method [11].

The first junction problem to be analyzed is the microstrip step discontinuity shown in Figure 10.3. The structure shown in the figure was originally analyzed by Koster and Jansen [12] using the spectral domain approach. The same structure was later analyzed by Meyer [13] using a hybrid mode matching method of lines (M3OL) technique. The substrate has a relative permittivity of 2.32 and a thickness of $b = 2.32$ mm. The microstrip width w_1 is 2.286 mm and $w_2 = 3w_1$. The height of the cover plane d is 5 mm while the width of the lateral shielding a is 11.276 mm.

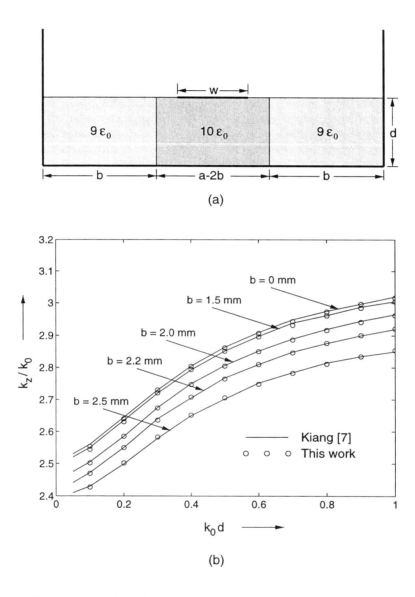

Figure 10.6 (a) Dual plane triple microstrip structure, $\epsilon_{x1} = \epsilon_{x2} = 9.4\epsilon_0$, $\epsilon_{y1} = \epsilon_{y2} = 11.6\epsilon_0$, $\epsilon_{z1} = \epsilon_{z2} = 9.4\epsilon_0$, $d = w = 1.0$ mm, $h = 4.0$ mm, $a = 10.0$ mm, $s = 2.0$ mm. (b) Dispersion curves for the structure shown in (a).

The implicit mode-matching scheme described in the previous section is used to solve for the incident, reflected, and transmitted fields at the step discontinuity. We can then use the spectral Lanczos decomposition technique to solve for the fields at any given z. The incident field is just the fundamental quasi-TEM mode with very little contribution from the higher-order modes. The reflected mode has significant contributions from the higher-order modes, which decay as the reflected wave is propagated away from the boundary. Figure 10.7 plots the total field components in the first and second waveguides as functions of the longitudinal distance from the interface. The former is just the sum of the reflected and incident fields and we can see the higher-order modes from the reflected field decaying as the wave moves away from the boundary. The total field component in the second waveguide is just the transmitted field which is composed primarily of the fundamental mode. The transmitted field does have some higher order modes which decay with increasing z. As stipulated by the boundary conditions, we can see that the total field at the interface is identical in both waveguides. Figure 10.8 plots the scattering parameters for the reflected and transmitted field as a function of frequency. The results seem to agree well with those published by Koster and Jansen [12] and Meyer [13].

The final structure to be analyzed is a microstrip taper discontinuity. We analyze this structure by using a staircasing approximation to model the microstrip taper. The dimensions of the microstrip taper are $w = 2.4$ mm, $l = 2.0$ mm and $h = 1.6$ mm. The taper is modeled using three different step sizes. For the coarsest step size, the taper is approximated as a single-junction problem. The same problem is also solved using finer step sizes, which correspond to a two-junction or a four-junction problem. Figure 10.9 shows the S-parameter plots for the microstrip taper solved using different step sizes. Figure 10.10 shows the step size for the three different cases. At low frequencies, the simple step discontinuity does an adequate job of modeling the taper since the electrical length of the taper is relatively small. The two-junction model does a better job of modeling the microstrip taper over a slightly larger frequency range. However, even the two-junction model starts to deviate from the more accurate four-junction model at high frequencies.

10.6 CONCLUSIONS

In this chapter, we have presented an efficient scheme to perform full wave analysis of waveguiding structures. Using the finite difference method to discretize the waveguide cross-section allows us to model arbitrary structures with inhomogeneous substrates. Since the discretization is only along two dimensions, the size of the problem is much smaller than conventional three-dimensional finite difference schemes. Despite discretizing only the waveguide cross-section, this method can analyze three-dimensional discontinuities by propagating the field analytically along the longitudinal direction. Further improvements in complexity and storage

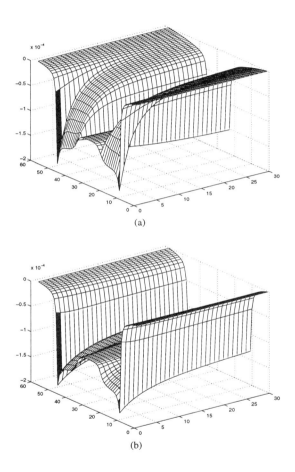

Figure 10.7 E_y component of the total field just under the microstrip in the (a) first and (b) second waveguides as a function of z away from the interface.

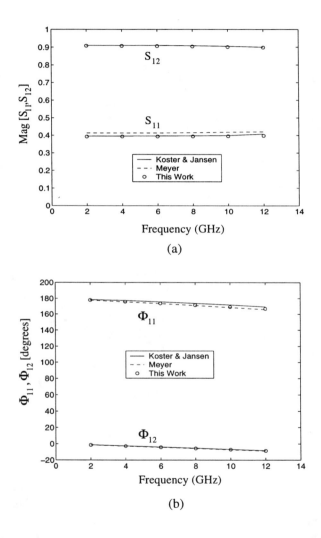

Figure 10.8 Scattering parameters for the step discontinuity: (a) magnitude; (b) phase.

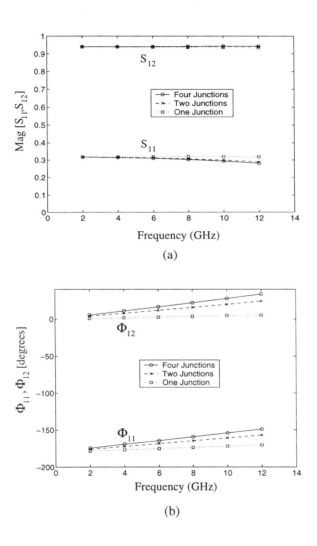

Figure 10.9 Scattering parameters for the microstrip taper: (a) magnitude; (b) phase.

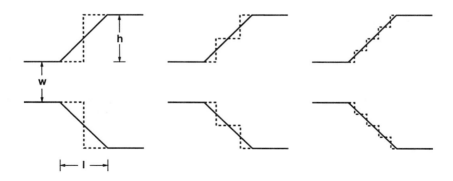

Figure 10.10 A microstrip taper modeled using the staircasing approximation with three different step sizes.

are gained by using model-order reduction techniques. The formulation is extended to the analysis of microstrip lines by modeling the conductive strip inside the waveguide using boundary conditions. Since a large number of microstrip structures are etched on an anisotropic substrate, the finite difference formulation was generalized to include anisotropic substrates.

Microstrip discontinuities are an inherent part of any microstrip circuit. As a result, in order to analyze these microstrip structures, one needs to be able to analyze these microstrip discontinuities. In Section 10.4, an implicit mode-matching scheme is described to solve the single junction problem. By using the Krylov subspace-based reduction techniques (bi-Lanczos), we can avoid explicit evaluation of the modes and solve for the fields with a computational complexity of $O(N^{1.5})$. The storage requirements can also be made to scale as only $O(N)$. Since most practical structures have multiple discontinuities, the single junction formulation was generalized to solve for the fields in the presence of n junctions.

REFERENCES

1. K. S. Yee, "Numerical solution of initial boundary value problems involving Maxwell's equations in isotropic media," *IEEE Trans. Antennas Propagat.*, vol. 14, pp. 302–307, 1966.

2. W. C. Chew, "Electromagnetic theory on a lattice," *J. Appl. Phys.*, vol. 75, pp. 4843–4850, 1994.

3. R. E. Collin, *Field Theory of Guided Waves*, New York: McGraw-Hill, 1960.

4. R. Holland, "Finite-difference solution of Maxwell's equations in generalized nonorthogonal coordinates," *IEEE Trans. Nuclear Science*, vol. NS-30, pp. 4589–4591, 1983.

5. W. C. Chew and M. Nasir, "A variational analysis of anisotropic, inhomogeneous dielectric waveguides," *IEEE Trans. Micro. Theory Tech.*, vol. 37, pp. 661–668, 1989.

6. J. Meixner, "The behavior of electromagnetic fields at edges," *IEEE Trans. Antennas Propagat.*, vol. 20, pp. 442–446, 1992.

7. V. Druskin and L. Knizhnerman, "Krylov subspace approximation of eigenpairs and matrix functions in exact and computer arithmetic," *Numer. Linear Algebra Appl.*, vol. 2, pp. 205–217, 1995.

8. V. Druskin and L. Knizhnerman, "Two polynomial methods of calculating functions of symmetric matrices," *U.S.S.R. Comput. Maths. Math. Phys.*, vol. 29, pp. 112–121, 1989.

9. C. Lanczos, "An iteration method for the solution of the eigenvalue problem of linear differential and integral operators," *J. Res. Nat. Bur. Stand.*, vol. 45, pp. 255–282, 1950.

10. G. H. Golub and C. F. Loan, *Matrix Computations*, Baltimore, MD: Johns Hopkins University Press, 1989.

11. M. S. Alam, M. Koshiba, K. Hirayama, and Y. Hayashi, "Hybrid-mode analysis of multilayered and multiconductor transmission lines," *IEEE Trans. Microwave Theory Tech.*, vol. 45, pp. 205–211, 1997.

12. N. H. L. Koster and R. H. Jansen, "The microstrip step discontinuity: A revised description," *IEEE Trans. Microwave Theory Tech.*, vol. 34, no. 2, pp. 213–223, 1986.

13. P. Meyer, "Solving microstrip discontinuities with a combined mode-matching and method-of-lines procedure," *Microwave Opt. Tech. Lett.*, vol. 8, no. 1, pp. 4–8, 1995.

11
Volume-Surface Integral Equation

Cai-Cheng Lu

11.1 INTRODUCTION

The solution to electromagnetic wave interaction with material coated objects has applications in radar cross-section prediction for coated targets, printed circuit and microstrip antenna analysis [1–3]. This problem can be studied using either differential equation or integral equation solvers. The hybrid boundary integral equation method and finite element method have also been applied to solve this type of problem [4]. The finite-difference time-domain (FDTD) method is a popular differential equation solver, as it is suitable for wide bandwidth modeling. However, implementation of the absorbing boundary condition in the FDTD requires additional meshing, thus resulting in extra memory and computational cost.

On the other hand, the frequency domain integral equation approach does not require extra mesh and usually gives more accurate results. A typical method in solving coated object scattering problems is to use the surface integral equation, in which the unknown functions (electric current and magnetic current) are assumed on the interfaces between different materials and between material and conductors. Then the surface integral equation is formulated using the free space Green's function. This method, however, is suitable for homogeneous material coating [5–7].

In printed circuit and microstrip antenna simulation problems, the surface integral equation (SIE) is formulated using the multilayer medium Green's function [3]. The advantage of this SIE is the reduction of the unknown domain (because the mesh and the basis functions are only assigned for the conducting patch). This reduction of unknown domain is at the cost of evaluating the Green's function, which usually is given in terms of Sommerfeld integrals. Another disadvantage of this SIE formulation is the assumption that the layered medium is of infinite extent. The model fidelity is compromised for a finite size substrate.

Another approach to solving this problem is to use the finite-element boundary-integral equation approach. However, poor conditioning of the matrix system may occur. In this chapter, we introduce a hybrid integral equation approach that combines the volume integral equation (VIE) and the surface integral equation to model the mixed dielectric and conducting structures [8, 9]. The volume integral equation is applied to the material region and the surface integral equation is enforced over the conducting surface. This results in a very general model as all the volume and surface regions are modeled properly. The advantage of this approach is that in the coated object scattering problem, the coating material can be inhomogeneous, and in the printed circuit and microstrip antenna simulation problems the substrate can be of finite size. Another advantage of this approach is the simplicity of the Green's function in both the VIE and SIE; this has an important impact on the implementation of fast solvers. However, the additional cost here is the increase of the number of unknowns since the volume that is occupied by the dielectric material is meshed. This results in larger memory requirement and longer solution time in solving the corresponding matrix equation. But this deficiency can be overcome by applying fast integral equation solvers such as the multilevel fast multipole algorithm.

In this chapter, we will discuss the formulation and the discretization of the hybrid volume-surface integral equation (VSIE). Numerical examples are provided at the end of the chapter to demonstrate the validity and the application of the algorithm developed.

11.2 THE FORMULATION OF THE INTEGRAL EQUATIONS

Numerical treatment of the volume integral equation can be traced to as early as three decades ago [10, 11]. Since then many researchers have used the VIE for various applications [12–19] and a number of methods have been developed for solving VIE. The application of the surface integral equations also started a long time ago, especially when Harrington introduced the method of moments (MOM) to solve integral equations in the 1960s [20]. In this section, we will introduce a hybrid integral equation approach that combines the volume integral equation with the surface integral equation to solve problems that involve conductors as well as dielectrics. Since the surface integral equation techniques have been discussed in

Volume-Surface Integral Equation

detail in other chapters in this book, in the following, we will first introduce the formulation of the volume integral equation, followed by a discussion of the hybrid integral equations.

11.2.1 Volume Integral Equation (VIE)

It is known that a current source $\mathbf{J}_i$ in a linear, isotropic, and homogeneous medium k_b generates an electromagnetic field $\mathbf{E}^i, \mathbf{H}^i$ that satisfies Ampere's equation:

$$\nabla \times \mathbf{H}^i = -i\omega\epsilon_b \mathbf{E}^i + \mathbf{J}_i \tag{11.1}$$

The solution for the electric field is written as an integral of the source current,

$$\mathbf{E}^i(\mathbf{r}) = i\omega\mu_b \int_{\text{source region}} \overline{\mathbf{G}}(\mathbf{r}, \mathbf{r}'; k_b) \cdot \mathbf{J}_i(\mathbf{r}') \, d\mathbf{r}' \tag{11.2}$$

where

$$\overline{\mathbf{G}}(\mathbf{r}, \mathbf{r}'; k_b) = \left(\overline{\mathbf{I}} + \frac{1}{k_b^2}\nabla\nabla\right) \frac{e^{ik_b|\mathbf{r}-\mathbf{r}'|}}{4\pi|\mathbf{r}-\mathbf{r}'|} \tag{11.3}$$

is the 3D dyadic Green's function for homogeneous medium k_b. When a scattering dielectric object with $k(\mathbf{r})$ is introduced in the background medium, then the wavenumber becomes

$$k(\mathbf{r}) = \begin{cases} k(\mathbf{r}), & \text{object region V} \\ k_b, & \text{background} \end{cases} \tag{11.4}$$

and the new field distribution $\mathbf{E}(\mathbf{r})$ satisfies

$$\nabla \times \mathbf{H} = -i\omega\epsilon(\mathbf{r})\mathbf{E} + \mathbf{J}_i \tag{11.5}$$

By writing $-\epsilon(\mathbf{r}) = -\epsilon_b + [\epsilon_b - \epsilon(\mathbf{r})]$, we have

$$\nabla \times \mathbf{H} = -i\omega\epsilon_b \mathbf{E} + i\omega[\epsilon_b - \epsilon(\mathbf{r})]\mathbf{E} + \mathbf{J}_i \tag{11.6}$$

Comparing (11.6) with (11.1), it is recognized that the new field is generated by two sources: $\mathbf{J}_i$ and $\mathbf{J}_V = i\omega[\epsilon_b - \epsilon(\mathbf{r})]\mathbf{E}$. The second source is called the induced source and exists only inside the scattering object region because $\epsilon(\mathbf{r}) = \epsilon_b$ when $\mathbf{r}$ is outside the dielectrics. The induced current $\mathbf{J}_V$ is a function of space since $\mathbf{E}(\mathbf{r})$ is an unknown vector. By the superposition theorem, the total electric field $\mathbf{E}(\mathbf{r})$ inside the dielectric object region is the sum of the primary field $\mathbf{E}^i(\mathbf{r})$, generated by the original source $\mathbf{J}_i$, and the scattered field $\mathbf{E}^s(\mathbf{r})$, generated by $\mathbf{J}_V$. Since $\mathbf{J}_V$ radiates in the homogeneous background medium k_b, the scattered field can be written in the same form as (11.2):

$$\mathbf{E}^s(\mathbf{r}) = i\omega\mu_b \int_V \overline{\mathbf{G}}(\mathbf{r}, \mathbf{r}'; k_b) \cdot \mathbf{J}_V(\mathbf{r}') \, d\mathbf{r}' \tag{11.7}$$

As a result, the total field in the object region (V) is given by

$$\mathbf{E}(\mathbf{r}) = \mathbf{E}^i(\mathbf{r}) + i\omega\mu_b \int_V \overline{\mathbf{G}}(\mathbf{r},\mathbf{r}';k_b) \cdot \mathbf{J}_V(\mathbf{r}') \, d\mathbf{r}', \qquad \mathbf{r} \in V \qquad (11.8)$$

Equation (11.8) is called the electric field volume integral equation. It is also a Fredholm integral equation of the second kind because the unknown is both inside and outside the integral operator. Equation (11.8) is derived for the three-dimensional case. For two-dimensional applications, (11.8) is modified so that the volume integral is replaced by an integration over the cross-section of the object. More over, in the two-dimensional TM (transverse magnetic) case, the electric field has only an axial component, and the vector volume integral equation is then reduced to a scalar integral equation.

Since $\mathbf{J}_V$ and $\mathbf{E}$ are related by

$$\mathbf{J}_V = i\omega \left[\epsilon_b - \epsilon(\mathbf{r})\right] \mathbf{E}(\mathbf{r}) \qquad (11.9)$$

either $\mathbf{J}_V$ or $\mathbf{E}$ can be used as unknown distribution functions. In the next section, there will be further discussion about which vector quantities to choose as the unknown distribution function.

Substituting (11.9) into (11.8), we obtain an equation for the total electric field

$$\mathbf{E}(\mathbf{r}) = \mathbf{E}^i(\mathbf{r}) + \omega^2 \mu_b \epsilon_b \int_V \overline{\mathbf{G}}(\mathbf{r},\mathbf{r}';k_b) \cdot [\epsilon(\mathbf{r})/\epsilon_b - 1] \mathbf{E}(\mathbf{r}') \, d\mathbf{r}', \quad \mathbf{r} \in V \quad (11.10)$$

For small dielectric contrast (i.e., $\epsilon(\mathbf{r}) \sim \epsilon_b$), the second term on the right-hand side of (11.10) is much smaller than the first. As a first order approximation, one can use $\mathbf{E}^i$ to replace $\mathbf{E}$ in the integral to get

$$\mathbf{E}(\mathbf{r}) = \mathbf{E}^i(\mathbf{r}) + \omega^2 \mu_b \epsilon_b \int_V \overline{\mathbf{G}}(\mathbf{r},\mathbf{r}';k_b) \cdot [\epsilon(\mathbf{r})/\epsilon_b - 1] \mathbf{E}^i(\mathbf{r}') \, d\mathbf{r}', \quad \mathbf{r} \in V \quad (11.11)$$

The approximation to reduce (11.10) to (11.11) is called the first-order Born approximation that has been widely used in inverse scattering applications [21,22]. The approximate solution $\mathbf{E}(\mathbf{r})$ from (11.11) can be used again in the integral in (11.10) to derive a new solution to the total electric field. This process can be continued so that (11.10) is satisfied. The convergence rate of the above iteration depends on the dielectric contrast $\epsilon(\mathbf{r})/\epsilon_b$. For high-contrast dielectrics, the convergence is very slow, and numerical solution techniques are necessary to solve (11.10). This will be discussed in the next section.

It should be pointed out that the solution $\mathbf{E}$ from the integral equation (11.8) or (11.10) satisfies (11.6). Hence the induced volume electric current $\mathbf{J}_V$ radiates in the background medium. In this sense, we state that the dielectric object is removed and an induced current distribution is placed within the object region.

In the above derivation, it has been assumed that the dielectric is nonmagnetic (i.e., $\mu = \mu_b$). If the material has permeability μ that is different from the background medium, then a magnetic current $\mathbf{M}_V$ is induced in addition to the electric current $\mathbf{J}_V$. In this case, there are two unknown functions to be determined. Hence two equations are needed for this purpose. One is the electric field volume integral equation, and the other is the magnetic field volume integral equation. The two equations are formulated by the linear superposition principle (i.e., the total field is the sum of the primary and the scattered field). Now we have two induced sources, and the scattered electric field contains two contributions, one from $\mathbf{J}_V$, and one from $\mathbf{M}_V$. The contribution from $\mathbf{J}_V$ is given by (11.7), and the contribution by $\mathbf{M}_V$ can be shown to be

$$\mathbf{E}(\mathbf{r}) = -\nabla \times \int_V \overline{\mathbf{G}}(\mathbf{r},\mathbf{r}';k_b) \cdot \mathbf{M}_V(\mathbf{r}')\, d\mathbf{r}' \tag{11.12}$$

Hence the electric field volume integral equation is

$$\begin{aligned}\mathbf{E}(\mathbf{r}) =& \mathbf{E}^i(\mathbf{r}) + i\omega\mu_b \int_V \overline{\mathbf{G}}(\mathbf{r},\mathbf{r}';k_b) \cdot \mathbf{J}_V(\mathbf{r}')\, d\mathbf{r}' \\ & - \nabla \times \int_V \overline{\mathbf{G}}(\mathbf{r},\mathbf{r}';k_b) \cdot \mathbf{M}_V(\mathbf{r}')\, d\mathbf{r}', \quad \mathbf{r} \in V\end{aligned} \tag{11.13}$$

The magnetic field volume integral equation can be formulated in the same way as

$$\begin{aligned}\mathbf{H}(\mathbf{r}) =& \mathbf{H}^i(\mathbf{r}) + \nabla \times \int_V \overline{\mathbf{G}}(\mathbf{r},\mathbf{r}';k_b) \cdot \mathbf{J}_V(\mathbf{r}')\, d\mathbf{r}' \\ & + i\omega\epsilon_b \int_V \overline{\mathbf{G}}(\mathbf{r},\mathbf{r}';k_b) \cdot \mathbf{M}_V(\mathbf{r}')\, d\mathbf{r}', \quad \mathbf{r} \in V\end{aligned} \tag{11.14}$$

11.2.2 Hybrid Volume-Surface Integral Equation (VSIE)

If the scattering target consists of perfect conductors as well as dielectrics, the scattered field can also be solved using the integral equation techniques. In this case, a surface integral equation is formed for the conducting surfaces, and a volume integral equation is formed for the dielectrics. We first concentrate on nonmagnetic material case so that $\mu = \mu_b$ for all regions. By the equivalence principle, the total electric field in space consists of the primary field $\mathbf{E}^i$ and the scattered field. The latter is made up of the radiation from the induced surface current $\mathbf{J}_S$ on the conducting surfaces, and the radiation from the induced volume current $\mathbf{J}_V$ in the dielectric region. Again, these two currents radiate in the background medium which is assumed to be homogeneous and of infinite extent. Hence the scattered field can be formally written as

$$\mathbf{E}^s(\mathbf{r}) = i\omega\mu_b \int_S \overline{\mathbf{G}}(\mathbf{r},\mathbf{r}';k_b) \cdot \mathbf{J}_S(\mathbf{r}')\, d\mathbf{r}' + i\omega\mu_b \int_V \overline{\mathbf{G}}(\mathbf{r},\mathbf{r}';k_b) \cdot \mathbf{J}_V(\mathbf{r}')\, d\mathbf{r}' \tag{11.15}$$

The surface integral equation is formed by imposing the zero-tangential electric component on the conducting surfaces, which is given by

$$\left[\mathbf{E}^i(\mathbf{r}) + i\omega\mu_b \int_S \overline{\mathbf{G}}(\mathbf{r},\mathbf{r}';k_b) \cdot \mathbf{J}_S(\mathbf{r}') \, d\mathbf{r}' \right. \\ \left. + i\omega\mu_b \int_V \overline{\mathbf{G}}(\mathbf{r},\mathbf{r}';k_b) \cdot \mathbf{J}_V(\mathbf{r}') \, d\mathbf{r}' \right]_{\tan} = 0, \quad \mathbf{r} \in S \quad (11.16)$$

The subscript "tan" stands for taking the tangent component of the vector before it. The volume integral equation is formed by writing the total electric field as the summation of the primary field and the scattered fields, that is,

$$\mathbf{E}(\mathbf{r}) = \mathbf{E}^i(\mathbf{r}) + i\omega\mu_b \int_S \overline{\mathbf{G}}(\mathbf{r},\mathbf{r}';k_b) \cdot \mathbf{J}_S(\mathbf{r}') \, d\mathbf{r}' \\ + i\omega\mu_b \int_V \overline{\mathbf{G}}(\mathbf{r},\mathbf{r}';k_b) \cdot \mathbf{J}_V(\mathbf{r}') \, d\mathbf{r}', \quad \mathbf{r} \in V \quad (11.17)$$

Equations (11.16) and (11.17) are called the hybrid volume-surface integral equation (or combined volume-surface integral equations). They must be solved simultaneously in order to determine the surface current $\mathbf{J}_S$ and the volume current $\mathbf{J}_V$. Since the dielectric volume region is considered part of the solution domain, the volume can assume arbitrary shape, and hence this hybrid integral equation approach can be applied to materials of arbitrary shape as well as of inhomogeneous dielectric permittivity. This is in contrast to the surface integral equation methods, in which the material has to be homogeneous.

11.3 NUMERICAL SOLUTION OF THE HYBRID VSIE

In many practical applications, problems usually involve complicated object geometries, and hence numerical methods are needed to solve the integral equations. We will follow the general steps of the method of moments [20] to discretize the integral equations. In general this consists of three steps: (1) selecting a set of basis functions, and expanding the unknown function in terms of the basis function, (2) testing the resultant integral equation by a testing procedure to convert the integral equation into a matrix equation, and (3) solving the matrix equation to find the numerical values of the expansion coefficient. Our interest is in solving the integral equation formed over complex scattering objects; hence we will use the subdomain basis function to expand the unknown currents.

11.3.1 Mesh Generating

The solution domains are first subdivided into small patches (for the conducting surface) and small volumes (for the dielectrics). The specific shapes of the cells used depend on applications as well as on the availability of the CAD models. In general, the following shapes for the surface patch are considered:

(1) Flat triangular patch;
(2) Flat rectangular patch;
(3) Curvilinear triangle;
(4) Curvilinear quadrangle.

For volume cells, we also have a number of options. Typical shapes are

(1) Flat-faced tetrahedron;
(2) Flat-faced hexahedron;
(3) Curvilinear-faced tetrahedron;
(4) Curvilinear-faced hexahedron.

In theory, the surface and the volume can be discretized independently. However, for the convenience of singular integral processing as well as for accurate geometrical representation, we choose to impose a connection condition of the surface patch and the volume cells, as shown in Figure 11.1. To satisfy this condition, an external face of a volume cell and a surface patch are either fully overlapped or not overlapped at all (in other words, partial overlap is not allowed under the connection condition). While this will burden the mesh generation process, it removes the integration difficulty when a conducting patch cuts through the dielectric due to mesh errors. As a result, the above listed shapes must be selected in pairs to satisfy the connection condition. The possible pairs are listed below:

- Flat triangle for surface and flat-faced tetrahedron for volume;

- Flat quadrangle for surface and flat-faced hexahedron for volume;

- Curvilinear-faced tetrahedron for surface and the same order curvilinear-faced tetrahedron for volume;

- Curvilinear-faced hexahedron for surface and the same order curvilinear-faced hexahedron for volume.

Though there are essentially many choices of mesh type, practical selection is usually restricted by a number of factors such as availability of original data or limited computer resources. This is because a different mesh generally results in a different number of unknowns. For example, the difference between a tetrahedron mesh and a hexahedron mesh of the same edge length can be more than five times.[1]

---

[1]The volume for an equilateral tetrahedron is $a^3/6\sqrt{2}$ if the edge length is a, and the volume for an equilateral hexahedron is a^3. The number of unknowns for a tetrahedron mesh is approximately proportional to $2N_1$ and that for hexahedron mesh is about $3N_2$. Here, N_1 and N_2 are the number of volume cells for tetrahedron and hexahedron shapes. Since $N_1/N_2 = 6\sqrt{2}$, we have $2N_1/3N_2 = 4\sqrt{2} \sim 5.6$.

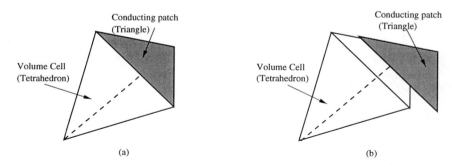

Figure 11.1 (a) The correct selection of surface and volume cells: the conducting (PEC) patch lies right on top of a tetrahedron face. (b) Incorrect selection of surface and volume cells: the conducting patch partially overlaps with a face of the tetrahedron.

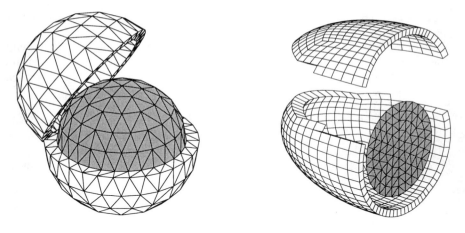

Figure 11.2 The triangle-tetrahedron mesh of a coated sphere (left), and the triangle-hexahedron mesh of an antenna-radome structure (right).

However, the hexahedron mesh is more difficult to generate and is not as flexible for representing some shapes, such as the tip region of a sharp cone. Figure 11.2 shows the mesh examples of a coated sphere and an antenna radome with plate antenna. It can be seen that arbitrarily oriented hexahedron cells can model curved dielectric layers accurately, and yet the number of unknowns is much smaller than using the tetrahedron shapes of equal sizes.

11.3.2 Discretization of the Hybrid Integral Equation

We will follow the general steps of the method of moments to discretize the hybrid volume-surface integral equation. To illustrate the discretization process, we choose to use the quadrangle-hexahedron pair for the surface and volume mesh. We will use the four-point quadrangle and eight-point hexahedron to derive the discretized equation. It should be seen that by increasing the order of the polynomials in the following equations, one can easily extend the method to curvilinear quadrangle-hexahedron cases.

In the surface discretization, each quadrangle cell is determined by four vertices, as shown in Figure 11.3, and are denoted by $\mathbf{r}_i$, $i = 1, 2, 3, 4$. Then, any point inside the quadrangle can be written in terms of two parameters u and v:

$$\mathbf{r} = (1-u)(1-v)\mathbf{r}_1 + u(1-v)\mathbf{r}_2 + uv\mathbf{r}_3 + (1-u)v\mathbf{r}_4, \quad u, v \in [0, 1] \quad (11.18)$$

In the above, we have mapped $\mathbf{r}_1, \mathbf{r}_2, \mathbf{r}_3, \mathbf{r}_4$ to $(u, v)=(0,0),(1,0),(1,1)$, and $(0,1)$, respectively. For this type of mesh, the basis function used most often is the roof-top basis function, which is defined on two adjacent patches that share a common edge. If we denote $u = 1$ as the common edge for both patches, then the basis function can be defined as

$$\mathbf{f}_j^S(\mathbf{r}) = \frac{\pm l_j}{\sqrt{g}} u \frac{\partial \mathbf{r}}{\partial u}, \quad \mathbf{r} \in \Omega_j^\pm, j = 1, 2, \cdots, N_S \quad (11.19)$$

where g is the surface Jacobian for the mapping in (11.18) and is given by $g = \det(g_{ij})$, and $\Omega_j^\pm$ stands for the first and the second patch associated with the jth surface basis function, as shown in Figure 11.3. The metric length variable g_{ij} is defined as

$$g_{ij} = \frac{\partial \mathbf{r}}{\partial u^i} \cdot \frac{\partial \mathbf{r}}{\partial u^j}, \quad i, j = 1, 2, \quad u^1 = u, u^2 = v \quad (11.20)$$

The surface current is then expressed as the superposition of the basis functions by

$$\mathbf{J}_S(\mathbf{r}) = \sum_{j=1}^{N_S} a_j^S \mathbf{f}_j^S(\mathbf{r}) \quad (11.21)$$

where N_S is the total number of surface basis functions.

In the volume integral discretization, each hexahedron cell is determined by its eight vertices. By allowing the hexahedron cells to assume arbitrary orientation, a collection of these cells can easily represent any complex shape. Moreover, the number of unknowns corresponding to this cell selection is much smaller than for the tetrahedron cells. For this type of cell shape, the most frequently used basis functions include the pulse basis function and the roof-top basis function. The pulse basis function has the advantage of accepting not-well-connected meshes. We will use the roof-top basis function as it gives much higher solution accuracy than the pulse basis

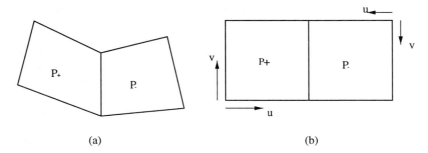

Figure 11.3 (a) Two adjacent quadrangles in the (x, y, z) system, and (b) their maps in (u, v)-space. Note that the common edge is indicated by $u = 1$ for both patches.

function for the same grid density. Since a roof-top basis function is defined on two adjacent cells that share a common face, the mesh for this selection must be well connected.

Since $\mathbf{J}_V$ and the total electric field $\mathbf{E}$ is related by (11.9), in theory either vector can be used as the unknown function in the integral equation. However, it is preferable to choose a vector that is proportional to $\mathbf{D}(\mathbf{r}) = \epsilon(\mathbf{r})\mathbf{E}(\mathbf{r})$ as a distribution function. Here, we have used $i\omega\epsilon(\mathbf{r})\mathbf{E}(\mathbf{r})$ as the distribution function, which is approximated by the following expansion:

$$i\omega\epsilon(\mathbf{r})\mathbf{E}(\mathbf{r}) = \sum_{j=1}^{N_S} a_j^S \mathbf{f}_j^S(\mathbf{r}), \quad \mathbf{r} \in V \quad (11.22)$$

As a result, the volume electric current is given by

$$\mathbf{J}_V(\mathbf{r}) = \left[\frac{\epsilon_b}{\epsilon(\mathbf{r})} - 1\right] i\omega\epsilon(\mathbf{r})\mathbf{E}(\mathbf{r}) = \chi(\mathbf{r}) \sum_{j=1}^{N_S} a_j^S \mathbf{f}_j^S(\mathbf{r}), \quad \mathbf{r} \in V \quad (11.23)$$

Assume that the eight vertices for a cell are denoted $\mathbf{r}_i$, $i = 1, 2, \cdots, 8$ and are ordered as shown in Table 11.1. Then any point inside the cell is given by

$$\mathbf{r} = \sum_{l=1}^{8} P_l(u, v, w)\mathbf{r}_l, \quad u, v, w \in [0, 1] \quad (11.24)$$

where

$P_1 = (1-u)(1-v)(1-w) \quad P_2 = u(1-v)(1-w)$
$P_3 = (1-u)v(1-w) \quad P_4 = uv(1-w) \quad P_5 = (1-u)(1-v)w$
$P_6 = u(1-v)w \quad P_7 = (1-u)vw \quad P_8 = uvw$

Table 11.1 Map of Hexahedron Vertices in (u, v, w) Space.

	$\mathbf{r}_1$	$\mathbf{r}_2$	$\mathbf{r}_3$	$\mathbf{r}_4$	$\mathbf{r}_5$	$\mathbf{r}_6$	$\mathbf{r}_7$	$\mathbf{r}_8$
u	0	1	0	1	0	1	0	1
v	0	0	1	1	0	0	1	1
w	0	0	0	0	1	1	1	1

The jth basis function $\mathbf{f}_j^V$, defined on two adjacent hexahedrons that share a common face, can be written as

$$\mathbf{f}_j^V(\mathbf{r}) = \frac{\pm S_j}{\sqrt{g^{\pm}}} u \frac{\partial \mathbf{r}}{\partial u}, \quad \mathbf{r} \in \Omega_j^{\pm}, \quad j = 1, 2, \cdots, N_V \tag{11.25}$$

In (11.25), S_j is the center Jacobian of the face that is shared by the two hexahedrons. This face is indicated by $u = 1$, and $\sqrt{g}$ is the volume Jacobian defined as $g = \det(g_{ij})$, where

$$g_{ij} = \frac{\partial \mathbf{r}}{\partial u^i} \cdot \frac{\partial \mathbf{r}}{\partial u^j}, \quad i, j = 1, 2, 3, \quad u^1 = u, u^2 = v, u^3 = w \tag{11.26}$$

Comparing the hybrid integral (11.16) and (11.17), we found that both equations have the scattered electric field, which consists of the radiation from surface current $\mathbf{J}_S$ and the volume current $\mathbf{J}_V$. Using the mixed potential notation, they are

$$\mathbf{E}(\mathbf{J}_S) = i\omega\mu_b \int_S G\mathbf{J}_S dS' - \frac{\nabla}{i\omega\epsilon_b} \int_S G\nabla' \cdot \mathbf{J}_S dS' = i\omega\mathbf{A}^S - \nabla\Phi^S \tag{11.27}$$

$$\mathbf{E}(\mathbf{J}_V) = i\omega\mu_b \int_V G\mathbf{J}_V dV' - \frac{\nabla}{i\omega\epsilon_b} \int_V G\nabla' \cdot \mathbf{J}_V dV' = i\omega\mathbf{A}^V - \nabla\Phi^V \tag{11.28}$$

In the above, $G = \exp(ikR)/4\pi R$ is the scalar Green's function, and R is the distance $|\mathbf{r}-\mathbf{r}'|$ from source point to field point. Using the mixed potential expression in (11.16) and (11.17), the hybrid volume-surface integral equation can be rewritten as

$$\left[i\omega\mathbf{A}^S - \nabla\Phi^S + i\omega\mathbf{A}^V - \nabla\Phi^V\right]_{\tan} = -E_{\tan}^{\text{inc}}, \quad \mathbf{r} \in S \tag{11.29}$$

$$-\mathbf{E} + \left[i\omega\mathbf{A}^S - \nabla\Phi^S + i\omega\mathbf{A}^V - \nabla\Phi^V\right] = -E^{\text{inc}}, \quad \mathbf{r} \in V \tag{11.30}$$

Using the approximate expression for $\mathbf{J}_S$ from (11.21) and $\mathbf{J}_V$ from (11.23), the potentials can be written as the weighted summation of the elemental potentials (the contribution from single basis function):

$$\mathbf{A}^{\Omega}(\mathbf{r}) = \mu_b \sum_j a_j^{\Omega} \int_{\Omega_j} G\chi^{\Omega} \mathbf{f}_j^{\Omega}(\mathbf{r}') d\Omega' = \sum_j a_j^{\Omega} \mathbf{A}_j^{\Omega}(\mathbf{r}) \tag{11.31}$$

$$\Phi^{\Omega}(\mathbf{r}) = \sum_j \frac{a_j^{\Omega}}{i\omega\epsilon_b} \int_{\Omega_j} G\nabla' \cdot [\chi^{\Omega} \mathbf{f}_j^{\Omega}(\mathbf{r}')] \, d\Omega' = \sum_j a_j^{\Omega} \Phi_j^{\Omega}(\mathbf{r}) \qquad (11.32)$$

In (11.31) and (11.32), $\Omega = S$ or V, and the range of the indices is not shown because the summation here is a formal notation. The actual range will depend on which domain the summation is running. In order to use a uniform basis index, we arrange the basis function such that the surface basis functions are ordered from 1 to N_S, and the volume basis functions from $N_S + 1$ to $N_S + N_V$, where N_S and N_V are the total number of surface basis functions and the volume basis functions, respectively. Now we test (11.29) using the surface basis function and test (11.30) using the volume basis function. This is called the extended Galerkin's method. After testing, the hybrid integral equations are converted into a matrix equation that is formally written as

$$\begin{bmatrix} Z^{SS} & Z^{SV} \\ Z^{VS} & Z^{VV} \end{bmatrix} \cdot \begin{bmatrix} a^S \\ a^V \end{bmatrix} = \begin{bmatrix} U^S \\ U^V \end{bmatrix} \qquad (11.33)$$

where a^S and a^V stand for the vectors of the expansion coefficients for the surface current and volume current, respectively. The excitation vectors are given by

$$U_i^{\Omega} = -\int_{\Omega_i} \mathbf{f}_i^{\Omega}(\mathbf{r}) \cdot \mathbf{E}^{\text{inc}}(\mathbf{r}) \, d\Omega, \quad i = 1, 2, \cdots, N_{\Omega}, \Omega \in \{S, V\} \qquad (11.34)$$

The elements of the block matrices are given by

$$Z_{ij}^{S\Omega} = \int_{S_i} \mathbf{f}_i^S(\mathbf{r}) \cdot [i\omega \mathbf{A}_j^{\Omega}(\mathbf{r}) - \nabla \Phi_j^{\Omega}(\mathbf{r})] \, dS, \quad i = 1, 2, \cdots, N_S \qquad (11.35)$$

$$Z_{ij}^{V\Omega} = -\delta_{V\Omega} \int_{V_i} \frac{\mathbf{f}_i^V \cdot \mathbf{f}_j^{\Omega}}{i\omega\epsilon(\mathbf{r})} dV + \int_{V_i} \mathbf{f}_i^V(\mathbf{r}) \cdot [i\omega \mathbf{A}_j^{\Omega}(\mathbf{r}) - \nabla \Phi_j^{\Omega}(\mathbf{r})] \, dV$$
$$i = N_S + 1, N_S + 2, \cdots, N_S + N_V \qquad (11.36)$$

In (11.36), $\delta_{V\Omega} = 1$ when $\Omega = V$, and $\delta_{V\Omega} = 0$ when $\Omega = S$. One advantage of using the mixed potential formulation is that the gradient operation on the scalar potential can be transferred to the testing function. This is done by applying the identity

$$\mathbf{f}_i \nabla \Phi_j = \nabla \cdot (\Phi_j \mathbf{f}_i) - \Phi_j \nabla \cdot \mathbf{f}_i \qquad (11.37)$$

to both equations, and then using the divergence theorem to convert the integral of the divergence term into a contour integral (for surface testing function) or surface integral (for volume testing function). This results in alternative forms that have reduced order of singularity for the matrix elements:

$$Z_{ij}^{S\Omega} = \int_{S_i} [i\omega \mathbf{f}_i^S \mathbf{A}_j^{\Omega} + \Phi_j^{\Omega} \nabla \cdot \mathbf{f}_i^S] \, dS - \int_{\partial S_i} \Phi_j^{\Omega} \mathbf{f}_i^S \cdot d\mathbf{l}, i = 1, 2, \cdots, N_S \qquad (11.38)$$

Volume-Surface Integral Equation

$$Z_{ij}^{V\Omega} = -\delta_{V\Omega} \int_{V_i} \frac{\mathbf{f}_i^V \cdot \mathbf{f}_j^\Omega}{i\omega\epsilon(\mathbf{r})} dV + \int_{V_i} \left[i\omega \mathbf{f}_i^V \mathbf{A}_j^\Omega + \Phi_j^\Omega \nabla \cdot \mathbf{f}_i^V\right] dV$$
$$- \int_{\partial V_i} \Phi_j^\Omega \mathbf{f}_i^V \cdot d\mathbf{S}, \qquad i = N_S + 1, N_S + 2, \cdots, N_S + N_V$$
(11.39)

In (11.38) and (11.39), ∂S_i stands for the boundary of S_i that is a closed contour, and ∂V_i stands for the boundary of V_i that is a closed surface enclosing the combined volume region of V_i^+ and V_i^-. From (11.38) and (11.39) we can see that the calculation of the matrix elements is straightforward. Both the vector potentials $\mathbf{A}_j^\Omega$ and the scalar potentials Φ_j^Ω are generally smooth in the nonzero region of the test function (which are usually about 0.1λ). As a result, the integrals in (11.38) and (11.39) can be easily evaluated by numerical integration schemes. For quadrangle and hexahedron cell shapes, the cell can be transformed into a unit cell (unit square in u-v space for quadrangle, and unit cubic in u-v-w space for hexahedron). One of the best numerical integration schemes is by the Gaussian quadrature. For triangle and tetrahedron shapes, the integraton rules can be found in [23, 24]. Singular integrals occurred in the calculation of the vector or scalar potentials will be discussed in the next section.

The matrix equation (11.33) can be solved by many numerical matrix equation solvers such as the Gauss elimination method, the lower-upper triangle decomposition method, and conjugate gradient method (which solves for the unknown vector iteratively with a finite number of iterations). Once the equation is solved, the expansion coefficients a^S and a^V can be used to calculate the scattered field and radar cross-sections. In circuit modeling, the coefficients can also be used to extract various circuit parameters such as the S-parameters. Since the dielectric substrate is considered part of the solution domain, it can be of arbitrary shape and thickness, and finite (in the layered media model of printed circuit, the dielectric has to be flat and of infinite extent). In antenna analysis problems, the coefficients can be used to retrieve the antenna's input impedance and calculate the antenna's radiation pattern. For example, when a radome is applied to an antenna, the combined system consists of both dielectrics and conductors; hence the hybrid surface-volume integral equation is ideal for this problem. In Section 11.8, some numerical examples will be presented to demonstrate the applications of the hybrid surface-volume integral equation approaches.

11.3.3 Mesh Termination

Now we want to point out that special attention should be given to the integral on volume cells. Recall that in the surface integral equation processing, each surface basis function is associated with two surface cells. In the volume integral equation, when we select $i\omega\epsilon\mathbf{E}$ as the unknown vector and expand it by volume basis functions, we realize that this vector is generally nonzero outside the scattering object. Theoretically speaking, in order to represent this vector, we must extend the mesh to cover all background media (except for the interior of the conductors where the vector is zero). Since we are only interested in its value in the dielectric, it is necessary to truncate the mesh. To this end, we divide all the volume cells into two categories: background cells and dielectric cells. Then only basis functions that contain at least one dielectric cell are retained. Thus some volume basis functions are defined on two dielectric cells, and the others are defined on one dielectric cell and one background cell. The background cell is called the auxiliary cell, as shown in Figure 11.4.

It is clear that the auxiliary cells are introduced to truncate the mesh. The existence of auxiliary cells introduces several difficulties: (1) it extends the source integration in (11.27), and the field integration in (11.28) to outside the true dielectric object, (2) it slightly increases the computation time because of the extra integral regions, and (3) it needs extra mesh generating tasks, which may not always be possible. As a result, in practical implementation of the algorithm, the auxiliary cells are eliminated by, for example, taking the limit of $h \to 0$. This is possible because there is no limit to the size of the auxiliary cell as long as it shares a face with a cell in the dielectric region. As a result, a boundary basis function that is originally associated with a dielectric cell and an auxiliary cell now is defined on the dielectric cell only. This basis function is called "half" basis as it is associated with one cell.

Another point that is worth mentioning is the numerical process of the integrand in (11.32). Here, we assume that the dielectric permittivity $\epsilon(\mathbf{r})$ is constant within each small volume cell. Hence if the permittivities in the two adjacent cells are different in values, and hence $\chi(\mathbf{r})$ are different, then there will be a discontinuity in the normal component of the induced volume current $\mathbf{J}_V$ since $\mathbf{J}_V = \chi(\mathbf{r})i\omega\epsilon(\mathbf{r})\mathbf{E}$ and the u-dependence of $i\omega\epsilon(\mathbf{r})\mathbf{E}$ is continuous across the dielectric interface. The continuity of $i\omega\epsilon(\mathbf{r})\mathbf{E}$ is ensured by the roof-top basis function for rectangular cells and the RWG basis function [25] for triangular cells. This is explained by the fact

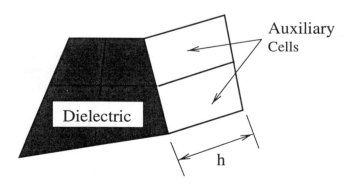

Figure 11.4 Cells in dielectrics (shaded) and auxiliary cells introduced to terminate the mesh.

that an electric charge distribution exists on the interface. In fact, if we denote the u-dependence of $\chi(\mathbf{r})$ by $\chi(u) = \chi_j[U(u) - U(u-1)], u \in [0,1]$, then

$$\nabla' \cdot [\chi(u')\mathbf{f}_j] = \nabla'\chi(u') \cdot \mathbf{f}_j(\mathbf{r}') + \chi(u')\nabla' \cdot \mathbf{f}_j$$
$$= [\delta(u') - \delta(u'-1)]\chi_j \hat{u}' \cdot \mathbf{f}_j(\mathbf{r}') + \chi(u')\nabla' \cdot \mathbf{f}_j \quad (11.40)$$

From (11.19), $\mathbf{f}_j(\mathbf{r}') = 0$ when $u' = 0$. Hence the term containing $\delta(u')$ on the right-hand side of the above vanishes. For the term with $\delta(u'-1)$, we can show that

$$\int_{V_j^\pm} G\delta(u'-1)\chi_j^\pm \hat{u}' \cdot \mathbf{f}_j dV' = \int_0^1 \int_0^1 \int_0^1 G\delta(u'-1)\chi_j^\pm \hat{u}' \cdot \mathbf{f}_j \sqrt{g}\, du' dv' dw'$$
$$= \int_0^1 \int_0^1 G\chi_j^\pm \hat{u}' \cdot \mathbf{f}_j(\mathbf{r}') \sqrt{g}|_{u'=1}\, dv' dw'$$
$$= \int_0^1 \int_0^1 G\chi_j^\pm (\pm S_j)\, dv' dw'$$
$$(11.41)$$

It can be seen that the volume integral for this term is converted to an integral over the face. Combining the two volume integrals in V_j^+ and V_j^-, we have

$$\int_{V_j^+ + V_j^-} G\delta(u'-1)\chi_j \hat{u}' \cdot \mathbf{f}_j(\mathbf{r}') dV' = \int_0^1 \int_0^1 (\chi_j^+ - \chi_j^-) S_j G\, dv' dw' \quad (11.42)$$

The result in (11.42) indicates that the induced surface charge density is proportional to $\chi_j^+ - \chi_j^-$, the difference in the permittivities of the two volume cells. As is expected, if the two adjacent cells have the same permittivity, leading to the same χ value for both cells, then there is no net surface charge. The above result is also obtained by Schaubert [14] for the tetrahedron cells.

11.3.4 Enforcing the Continuity Condition

Material mesh termination can be made simpler if the material is terminated by conducting surfaces. This is true for the material external faces that are in exact contact with a perfectly conducting surface. The exact contact is necessary because this termination applies the continuity condition: $\nabla \cdot \mathbf{J}_S - i\omega\rho_s = 0$. Here ρ_s is the surface charge density on the conducting patch. Since $\rho_s = \hat{n} \cdot \mathbf{D}$ on a conducting patch that is on dielectric, we have $\hat{n} \cdot \mathbf{D} = \nabla \cdot \mathbf{J}_S/i\omega$. This equation is used to calculate $\mathbf{D}$ on conducting patches which lie on dielectric. To this end, assuming that patch p is an exterior face of a tetrahedron, then the volume basis associated with this face is $\mathbf{f}_n^V(\mathbf{r})$, and hence, according to (11.22), the normal component of electric flux density vector $\mathbf{D}$ over this surface is given by $\hat{n} \cdot \mathbf{D}(\mathbf{r}) = a_n^V \hat{n} \cdot \mathbf{f}_n^V(\mathbf{r})/i\omega$. On the other hand, if this face S_p is overlaid with a conducting patch S_q (in fact, S_q and

S_p indicate exactly the same geometrical surface region), then the surface current on patch S_q is given by

$$\mathbf{J}_S(\mathbf{r}) = \sum_{m \in M_q} a_m^S \mathbf{f}_m^S(\mathbf{r}), \qquad \mathbf{r} \in S_q$$

where M_q is a set of indices for the basis functions that are associated with patch S_q. As a result, applying the continuity condition implies

$$a_n^V = \frac{1}{\hat{n} \cdot \mathbf{f}_n^V} \sum_{m \in M_q} a_m^S \nabla \cdot \mathbf{f}_m^S$$

Note that though $\mathbf{f}_n^V$ and $\mathbf{f}_n^S$ are functions of space, $\hat{n} \cdot \mathbf{f}_n^V$ and $\nabla \cdot \mathbf{f}_m^S$ are constants [25, 26].

The advantage of using the continuity condition is that the number of unknowns can be reduced. The number of basis functions that can be removed is equal to the number of basis functions on the conducting surface that is in direct contact with material. For thin material coating when the coating is full, this saving can be significant. On the other hand, this causes implementation complexity, especially if the integral equation is to be solved by fast solvers. It should be pointed out that it is not necessary to enforce the continuity condition explicitly as shown above. The condition is automatically satisfied if the shared basis function is not removed. Hence for the purpose of easier implementation of the fast solvers, one usually chooses not to apply the continuity condition explicitly.

11.3.5 Other Cell Shapes

It can be seen that the derivation from (11.29) through (11.42) does not depend on the choice of mesh cell shapes and the choice of basis functions. Hence it applies to any allowable mesh shapes and basis functions. In the following, we will briefly discuss some other mesh choices: the triangle-tetrahedron pair, and the curvilinear-quadrangle and curvilinear-hexahedron pair.

11.3.5.1 Triangle Patch and Tetrahedron Volume Cell

In this case, the triangle is specified by three points $\mathbf{r}_1$, $\mathbf{r}_2$, and $\mathbf{r}_3$. The position vector $\mathbf{r}$ inside the triangle is given in terms of two parameters u and v such that

$$\mathbf{r} = (1 - u - v)\mathbf{r}_1 + u\mathbf{r}_2 + v\mathbf{r}_3, \quad u, v \in [0, 1], \quad u + v \leq 1 \qquad (11.43)$$

It can be seen that the three vertices $\mathbf{r}_1$, $\mathbf{r}_2$, and $\mathbf{r}_3$ are mapped into the u-v space as $(0, 0)$, $(1, 0)$, and $(0, 1)$, respectively. In general, there is no restriction on the order of the three vertices for solving the electric field (surface) integral equation. However, if the magnetic field integral equation (MFIE) or the combined field integral equation

Volume-Surface Integral Equation

(CFIE) is to be solved, then the normal of each triangle cell must be defined. Usually the normal directions of the triangles are pointing to the external direction. To enforce this condition, it is necessary to order the three vertices in each triangle such that the vector $(\mathbf{r}_2 - \mathbf{r}_1) \times (\mathbf{r}_3 - \mathbf{r}_1)$ points toward outward normal direction. For this type of surface patch, the most frequently used basis function is the RWG basis function [25] that is defined on two triangles that share a common edge. The RWG basis function is defined as

$$\mathbf{f}_j^S(\mathbf{r}) = \frac{\pm l_j}{\sqrt{g^{\pm}}} (\mathbf{r} - \mathbf{r}_{3,4}), \quad \mathbf{r} \in S_j^{\pm} \tag{11.44}$$

In the above, l_j is the length of the common edge, $\sqrt{g^{\pm}}$ is the Jacobian of the transformation given in (11.43), and S_j^+ and S_j^- stand for two triangles specified by $\mathbf{r}_1, \mathbf{r}_2, \mathbf{r}_3$, and $\mathbf{r}_1, \mathbf{r}_4, \mathbf{r}_2$, respectively, as shown in Figure 11.5. Since the transformation given in (11.43) is a first-order polynomial of variables u and v, the Jacobian is a constant. It can be shown that $\sqrt{g^{\pm}} = 2A_j^{\pm}$, where $A_j^{\pm}$ is the area of $S_j^{\pm}$.

A tetrahedron is specified by four vertices $\mathbf{r}_1, \mathbf{r}_2, \mathbf{r}_3$, and $\mathbf{r}_4$. The position vector $\mathbf{r}$ inside the tetrahedron is given in terms of three parameters u, v, and w such that

$$\mathbf{r} = (1 - u - v - w)\mathbf{r}_1 + u\mathbf{r}_2 + v\mathbf{r}_3 + w\mathbf{r}_4, \quad u, v, w \in [0, 1], \text{ and } u + v + w \leq 1 \tag{11.45}$$

It can be seen that the four vertices $\mathbf{r}_1, \mathbf{r}_2, \mathbf{r}_3$, and $\mathbf{r}_4$ are mapped into u-v-w space as $(0, 0, 0)$, $(1, 0, 0)$, $(0, 1, 0)$, and $(0, 0, 1)$, respectively. There are four faces in a tetrahedron. We use face-i to indicate the face that is opposite to the vertex-i. For numerical implementation purposes, the four vertices in a tetrahedron are ordered such that the normal direction of the face 2-3-4 calculated using the right-hand rule is directed outward.

The 3D version of the RWG basis vector basis function is defined over two adjacent tetrahedrons that share a common face. If V_j^+ and V_j^- are two such tetrahedrons,

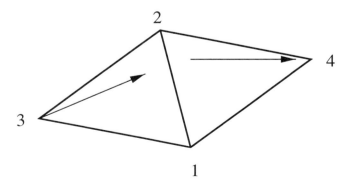

Figure 11.5 Two adjacent triangles that form the domain of an RWG basis function.

then the corresponding basis function is defined as

$$\mathbf{f}_j^V(\mathbf{r}) = \frac{\pm S_j}{\sqrt{g^\pm}}(\mathbf{r} - \mathbf{r}_{1,5}), \quad \mathbf{r} \in V_j^\pm \qquad (11.46)$$

where S_j is the area of the common face, $\sqrt{g^\pm}$ is the volume Jacobian for tetrahedron $V_j^\pm$, and $\mathbf{r}_1$ and $\mathbf{r}_5$ are the two vertices that are not shared by the two tetrahedrons (called free nodes in a basis function), as shown in Figure 11.6. Since the transformation given in (11.45) is a first-order polynomial of variables u, v, and w, the Jacobian is a constant. Again, one can show that the Jacobian is related to the volume of the tetrahedron by $\sqrt{g^\pm} = 3V_j^\pm$.

11.3.5.2 Curvilinear Quadrangle and Curvilinear Hexahedron

Using curvilinear mesh to solve surface integral equation and volume integral equation separately has been studied previously [26–28]. The curvilinear mesh is most useful for higher-order basis functions where the basis function domain is much larger than 0.1λ. However, for some cases where small and rapidly varying geometries occur, a curvilinear mesh is also very useful. Here we use a curvilinear quadrangle to explain the formulation.

An mth order curvilinear quadrangle is specified by $(m+1)^2$ points. For example, the second-order curvilinear quadrangle is specified by nine-points, as shown in Figure 11.7.

We will build the mapping of the u-v space to the points on the surface using the interpolatory polynomial. To this end, consider the second-order quadrangle first. The three independent Lagrange interpolatory polynomials of order 2 over the nodes of $u = 0, 0.5$, and 1, are

$$L_1^2(u) = (1 - 2u)(1 - u), \quad L_2^2(u) = 4u(1 - u), \quad L_3^2(u) = u(2u - 1) \quad (11.47)$$

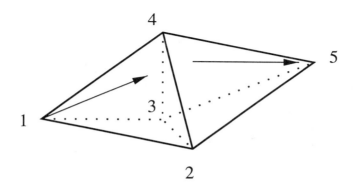

Figure 11.6 Two adjacent tetrahedrons that form the domain of a 3D RWG basis function.

where the superscript "2" in L_j^2 indicates that the order is 2. Using these polynomials, the position vector on the second-order curvilinear quadrangle can be written as

$$\mathbf{r}(u, v) = \sum_{j=1}^{3} \sum_{i=1}^{3} L_i^2(u) L_j^2(v) \mathbf{r}_{i+3(j-1)}, \quad u, v \in [0, 1] \qquad (11.48)$$

In general, for mth order curvilinear quadrangle, the position vector is

$$\mathbf{r}(u, v) = \sum_{j=1}^{m+1} \sum_{i=1}^{m+1} L_i^m(u) L_j^m(v) \mathbf{r}_{i+(m+1)(j-1)}, \quad u, v \in [0, 1] \qquad (11.49)$$

where the interpolatory polynomials L_j^m are given by

$$L_j^m(u) = \prod_{\substack{k=1 \\ k+j \neq 0}}^{m+1} \frac{u - u_k}{u_j - u_k}, \quad j = 1, 2, \cdots, m+1, \quad u_k = (k-1)/m \qquad (11.50)$$

In the same manner, the mth order curvilinear hexahedron is specified by $(m+1)^3$ points, $\mathbf{r}_i$, $i = 1, 2, \cdots, (m+1)^3$, and the position vector $\mathbf{r}$ inside the hexahedron is

$$\mathbf{r}(u, v, w) = \sum_{k,j,i=1}^{m+1} L_i^m(u) L_j^m(v) L_k^m(w) \mathbf{r}_{i+(m+1)[(j-1)+(m+1)(k-1)]}, \qquad (11.51)$$

$$u, v, w \in [0, 1]$$

For the curvilinear surface patches and volume cells, the basis functions are defined in the same way as in the lower-order mesh shape given by (11.19) and (11.25), respectively. And hence the difference in the implementation between using lower-order

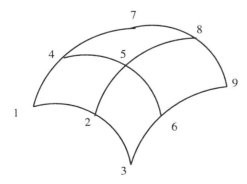

Figure 11.7 A second-order curvilinear quadrilateral defined by nine-points.

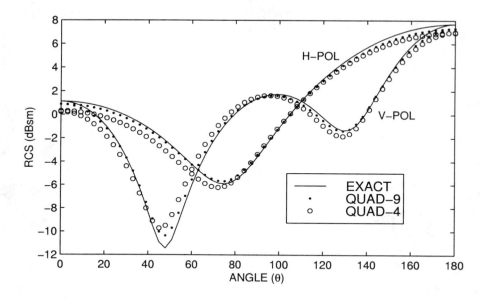

Figure 11.8 The RCS of a coated sphere calculated using first order mesh (QUAD-4, circles) and second-order mesh (QOAD-9, dots). The radius of the sphere has radius 0.3m, coating thickness is 0.1m, and the dielectric permittivity is $\epsilon_r = 2.5 - 0.9i$. The exact solution using Mie series is also shown as reference (solid line).

mesh and high-order mesh is in the position vector calculation. Using curvilinear mesh allows more accurate representation of the object geometry, and in general will give rise to a more accurate solution, especially for objects that contain small but curved faces, such as small spheres and thin rods. Figure 11.8 shows the comparison of RCSs calculated for a coated small sphere ($R = 0.3$m) using first-order and second-order meshes (the number of patches is 96 for both meshes). It can be seen that the second-order mesh gives more accurate solution. Here the solution accuracy difference is mostly attributed to error in representing the sphere's surface.

11.4 COMBINED FIELD INTEGRAL EQUATION

The combined field integral equation (CFIE) [29] can be applied to the closed conducting surfaces to increase the rate of convergence if the iterative method is used to solve the matrix equation. CFIE is also useful for cases with large grid desparities (such as large aspect ratio in a triangle patch) and very small grid sizes (corresponding to low-frequency break down). The CFIE for the hybrid surface-volume integral equations is formulated similarly as in the surface integral equation case, namely,

$$\text{CFIE} = \alpha \cdot \text{EFIE} + (1-\alpha) \cdot \text{MFIE} \tag{11.52}$$

where α is the combination parameter and its value is usually between 0 and 1. It should be pointed out that in general the target may consists of dielectric and open or conducting surfaces. The CFIE applies to the closed conducting surfaces only. The magnetic field integral equation (MFIE) is formed using the condition

$$\mathbf{J}^S(\mathbf{r}) = \hat{n} \times \mathbf{H}(\mathbf{r}), \quad \mathbf{r} \in S \tag{11.53}$$

where $\mathbf{H}(\mathbf{r})$ is the total magnetic field on the conducting surface; it is the sum of the incident field and the scattered field. When the object consists of conductors and dielectrics, the scattered magnetic field is contributed by the surface current and the volume current,

$$\mathbf{H}(\mathbf{r}) = \nabla \times \int_S G \mathbf{J}^S(\mathbf{r}')dS' + \nabla \times \int_V G \mathbf{J}^V(\mathbf{r}')dV' \tag{11.54}$$

Hence the magnetic field integral equation in terms of the induced electric currents $\mathbf{J}^S$ and $\mathbf{J}^V$ is

$$-\eta_b \mathbf{J}^S + \eta_b \hat{n} \times \nabla \times \int_S G \mathbf{J}^S(\mathbf{r}')dS' + \eta_b \hat{n} \times \nabla \times \int_V G \mathbf{J}^V(\mathbf{r}')dV' \\ = -\eta_b \hat{n} \times \mathbf{H}^{\text{inc}}, \quad \mathbf{r} \in S \tag{11.55}$$

The wave impedance η_b is multiplied on both sides of the equation to balance the units between the EFIE and MFIE. In numerical implementation, the singular surface integral in (11.55) is usually processed analytically so that the surface integral is written as a sum of the singular integral and a primary integral. The singular integral can be evaluated into closed form for continuous surface to $(\mathbf{J}^S/2)$ [30]; hence, (11.55) is usually rewritten in the following form:

$$-\frac{\eta_b}{2}\mathbf{J}^S + \eta_b \hat{n} \times \nabla \times \left[\oint_S G \mathbf{J}^S(\mathbf{r}')dS' + \int_V G \mathbf{J}^V(\mathbf{r}')dV' \right] = -\eta_b \hat{n} \times \mathbf{H}^{\text{inc}}, \mathbf{r} \in S \tag{11.56}$$

When this integral equation is discretized using the Galerkin's method, the matrix elements are given by

$$Z_{ij}^{\text{MFIE}} = -\frac{\eta_b}{2}\sum_{\alpha,\beta} \delta_{\alpha\beta} \int_\alpha \mathbf{f}_i^\alpha \cdot \mathbf{f}_j^\beta dS + \eta_b \int_\alpha \mathbf{f}_i^\alpha \cdot \hat{n} \times \oint_\beta \nabla G \times \mathbf{f}_j^S dS' dS,$$

$$j = 1, 2, \cdots, N_S$$

$$\tag{11.57}$$

$$Z_{ij}^{\mathrm{MFIE}} = \eta_b \int_\alpha \mathbf{f}_i^\alpha \cdot \hat{n} \times \oint_\beta \nabla G \times \mathbf{f}_j^S dV' dS, \tag{11.58}$$

$$j = N_S + 1, \cdots, N_S + N_V$$

In (11.57), $\alpha = S_i^\pm$, $\beta = S_j^\pm$. The Dirac function $\delta_{\alpha\beta}$ is multiplied to the leading term so that only integration on self cells is accounted for.

11.5 SINGULAR INTEGRAL TREATMENTS

Because of the $1/R$ terms in the integrand, the integrals in (11.31) and (11.32) are singular if a testing point $\mathbf{r}$ is inside a source cell. It can be shown that the $1/R$ terms are integrable. However, one cannot use the regular numerical scheme to obtain accurate results. In other words, the singular integrals must be processed by special methods. There are many methods that can be used to evaluate the singular integrals that occurred in the method of moment matrix elements calculation, such as the singularity extraction method [31–33] and the Duffy's transform method [34].

For example, the singularity extraction method is very useful for the singular integration on a flat surface cell or over a flat-faced volume cell. In this method, the asymptotic singular function is first identified by taking the limit of $R \to 0$ to the integral, and then the singular integral is rewritten as

$$\int_\Omega \frac{F(R, \mathbf{r}, \mathbf{r}')}{R\sqrt{g}} d\mathbf{r}' = \int_\Omega \frac{F(R, \mathbf{r}, \mathbf{r}') - F(0, \mathbf{r}, \mathbf{r}')}{R\sqrt{g}} d\mathbf{r}' + \int_\Omega \frac{F(0, \mathbf{r}, \mathbf{r}')}{R\sqrt{g}} d\mathbf{r}' \tag{11.59}$$

where Ω stands for either a surface domain S or a volume domain V, and $F(R, \mathbf{r}, \mathbf{r}')$ is a regular function of $\mathbf{r}'$. In the above, the first term to the right of the equation is a regular integral and can be evaluated by a proper numerical integral scheme. The second integral can be evaluated into closed form if the domain is a flat polygon or flat-faced volume cell (in these cells, the Jacobian are constants). Interested readers can refer to [22] and [23] for details.

A more general and useful singular integral processing is the Duffy's transformation method [35]. The advantage of this method is that it applies to singular integrals on general curvilinear faces and/or volume cells with curvilinear faces. In the following, we will consider the integration of

$$I_S^{\mathrm{EFIE}}(\mathbf{r}) = \int_S \frac{F(R, \mathbf{r}, \mathbf{r}')}{R\sqrt{g}} dS', \quad \mathbf{r} \in S \tag{11.60}$$

The superscript "EFIE" is used to distinguish the singular integral that occurred in the magnetic field integral equation (MFIE) that will be discussed shortly. If the integral domain S in (11.60) is a quadrangle (or curvilinear quadrangle), then (11.18) or (11.48) can be used to map the quadrangle into a square in u-v space. Since

$dS' = \sqrt{g}dv'du'$, the above integral can be written as

$$I_S^{EFIE}(\mathbf{r}) = \int_0^1 \int_0^1 \frac{F(R,\mathbf{r},\mathbf{r}')}{R}dv'du', \quad \mathbf{r} \in S \quad (11.61)$$

From (11.49), the position vector $\mathbf{r}$ in the above is written in terms of variables u', v' as

$$\mathbf{r}' = \sum_{j=1}^{m+1}\sum_{i=1}^{m+1} L_i^m(u')L_j^m(v')\mathbf{r}_{i+(m+1)(j-1)} \quad (11.62)$$

Since $\mathbf{r} \in S$, we can find $u_0, v_0 \in [0,1]$ such that

$$\mathbf{r} = \sum_{j=1}^{m+1}\sum_{i=1}^{m+1} L_i^m(u_0)L_j^m(v_0)\mathbf{r}_{i+(m+1)(j-1)} \quad (11.63)$$

Then the source to field vector $\mathbf{R}$ is

$$\mathbf{R} = \mathbf{r} - \mathbf{r}' = \sum_{j=1}^{m+1}\sum_{i=1}^{m+1} \left[L_i^m(u_0)L_j^m(v_0) - L_i^m(u')L_j^m(v')\right]\mathbf{r}_{i+(m+1)(j-1)} \quad (11.64)$$

Since $L_i^m(u')$ and $L_j^m(v')$ are both polynomials of order m, $\mathbf{R}$ can be expressed in the polynomial form of variables u' and v':

$$\mathbf{R} = \sum_{\substack{l,k=0 \\ l+k \neq 0}}^{m} (u'-u_0)^l(v'-v_0)^k \mathbf{P}_{lk} \quad (11.65)$$

where the vector coefficients $\mathbf{P}_{lk}$ for $l,k = 0, 1, \cdots, m$, are given by

$$\mathbf{P}_{lk} = \sum_{j=1}^{m+1}\sum_{i=1}^{m+1} \frac{\partial^l L_i^m(u_0)}{l!\partial u_0^l} \frac{\partial^k L_j^m(v_0)}{k!\partial v_0^k} \mathbf{r}_{i+(m+1)(j-1)} \quad (11.66)$$

Note that in (11.65) we have applied the condition $\mathbf{P}_{00} = 0$ because $\mathbf{R} = 0$ when $u' = u_0, v' = v_0$. Now we divide the integral domain in u-v space (a unit rectangle) into four triangles that share the same vertex (u_0, v_0), as shown in Figure 11.9.

As a result, the integral in (11.61) can be expressed as the sum of four integrals:

$$I_S^{EFIE} = I_1^{EFIE} + I_2^{EFIE} + I_3^{EFIE} + I_4^{EFIE}$$

Let us consider one of the four integrals:

$$I_j^{EFIE} = \int_{S_j} \frac{F(R,\mathbf{r},\mathbf{r}')}{R}dv'du' \quad (11.67)$$

510 Fast and Efficient Algorithms in CEM

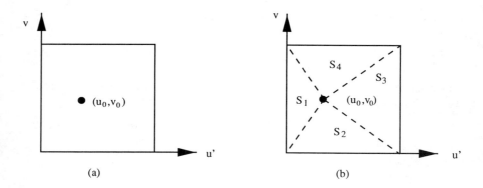

Figure 11.9 (a) The square integral domain in the u-v space. (b) The partition of the square domain into four triangular domains: S_1, S_2, S_3, and S_4.

Table 11.2 Table of u_a, v_a, u_b, v_b Values for the Four Triangles in Figure 11.9

	S_1	S_2	S_3	S_4
u_a	1	0	1	1
v_b	0	0	0	1
u_b	0	1	1	0
v_b	0	0	1	1

The domain is shown in Figure 11.10 in which the values of u_a, v_a, u_b, and v_b for the four triangles are given in Table 11.2.

By a transformation of $(u', v') \to (u_1, v_1)$ such that

$$u' = u_0 + (u_a - u_0)u_1 + (u_b - u_a)v_1 = u_0 + a_{11}u_1 + a_{12}v_1 \qquad (11.68)$$

$$v' = v_0 + (v_a - v_0)u_1 + (v_b - v_a)v_1 = v_0 + a_{21}u_1 + a_{22}v_1 \qquad (11.69)$$

the integral in (11.67) is converted to

$$I_j^{\mathrm{EFIE}} = \int_{S_j} \frac{F(R,\mathbf{r},\mathbf{r}')}{R} dv' du' = \int_0^1 \int_0^{u_1} \frac{F(R,\mathbf{r},\mathbf{r}')}{R} \sqrt{g_j} dv_1 du_1$$

Since the transformation from the u'-v' system to the u_1-v_1 system is linear, the Jacobian is a constant. In fact, it is proportional to the area of S_j in the u'-v' system:

$$\sqrt{g_j} = \begin{vmatrix} u_a - u_0 & u_b - u_a \\ v_a - v_0 & v_b - v_a \end{vmatrix} = (u_a - u_0)(v_b - v_a) - (u_b - u_a)(v_a - v_0)$$

Volume-Surface Integral Equation

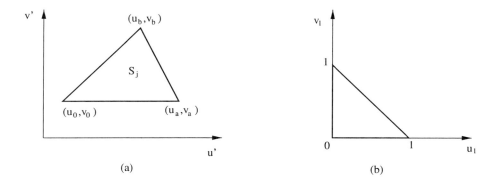

Figure 11.10 The mapping from the $u' - v'$ system in (a) to the $u_1 - v_1$ system in (b)

Using the transformation in (11.68) and (11.69), the source-to-field vector $\mathbf{R}$ given in (11.65) is expressed in the u_1-v_1 system as

$$\mathbf{R} = \sum_{\substack{l,k=0 \\ l+k \neq 0}}^{m} (u' - u_0)^l (v' - v_0)^k \mathbf{P}_{lk}$$

$$= \sum_{\substack{l,k=0 \\ l+k \neq 0}}^{m} (a_{11}u_1 + a_{12}v_1)^l (a_{21}u_1 + a_{22}v_1)^k \mathbf{P}_{lk} \quad (11.70)$$

The last step we need is to introduce a new variable t such that $v_1 = tu_1$ and $dv_1 = tdu_1$. This leads to $a_{11}u_1 + a_{12}v_1 = (a_{11} + a_{12}t)u_1$, $a_{21}u_1 + a_{22}v_1 = (a_{21} + a_{22}t)u_1$. Hence,

$$\mathbf{R} = \sum_{\substack{l,k=0 \\ l+k \neq 0}}^{m} (a_{11} + a_{12}t)^l (a_{21} + a_{22}t)^k u_1^{l+k} \mathbf{P}_{lk}, = u_1 \mathbf{R}_0 \quad (11.71)$$

where

$$\mathbf{R}_0 = \sum_{\substack{l,k=0 \\ l+k \neq 0}}^{m} (a_{11} + a_{12}t)^l (a_{21} + a_{22}t)^k u_1^{l+k-1} \mathbf{P}_{lk} \quad (11.72)$$

Note that $\mathbf{R}_0 \neq 0$ for all $t \in [0, 1]$, $u_1 \in [0, 1]$ since $l + k \neq 0$. From the transformation $v_1 = tu_1$, when v_1 varies from 0 to u_1, t varies from 0 to 1. Hence the integral in (11.67) is finally transformed to a form that can be evaluated using a numerical method:

$$I_j^{\text{EFIE}} = \int_0^1 \int_0^1 \frac{F(R,\mathbf{r},\mathbf{r}')}{u_1 R_0} \sqrt{g_j} u_1 dt du_1 = \sqrt{g_j} \int_0^1 \int_0^1 \frac{F(u_1 R_0,\mathbf{r},\mathbf{r}')}{R_0} dt du_1 \quad (11.73)$$

For a singular integral over a curvilinear volume cell, the process is similar to the above. The idea is to partition the domain in the u'-v'-w' system (a unit cubic) into six tetrahedrons that share a common vertex (u_0, v_0, w_0), where $\mathbf{r} = \mathbf{r}'(u_0, v_0, w_0)$. Then use the two steps of transformations to transfer the integration in a tetrahedron to a regular integration that can be calculated with a numerical method. The transformation needed is

$$\begin{bmatrix} u' - u_0 \\ v' - v_0 \\ w' - w_0 \end{bmatrix} = \begin{bmatrix} u_a - u_0 & u_b - u_a & u_c - u_b \\ v_a - v_0 & v_b - v_a & v_c - v_b \\ w_a - w_0 & w_b - w_a & w_c - w_b \end{bmatrix} \begin{bmatrix} u_1 \\ v_1 \\ w_1 \end{bmatrix} \quad (11.74)$$

and

$$u_1 = u_1, \quad v_1 = tu_1, \quad w_1 = su_1 \quad (11.75)$$

where the parameters u_a, v_a, etc., are listed in Table 11.3. After applying (11.74) and (11.75), we get

$$\begin{aligned} I_j^{\text{EFIE}} &= \int_0^1 \int_0^1 \int_0^1 \frac{F(R,\mathbf{r},\mathbf{r}')}{R} dw' dv' du' \\ &= \sqrt{g_j} \int_0^1 \int_0^1 \int_0^1 \frac{F(u_1 R_0,\mathbf{r},\mathbf{r}')}{R_0} u_1 ds dt du_1 \end{aligned} \quad (11.76)$$

where the Jacobian $\sqrt{g_j}$ is a constant and it is the determinant of the transformation matrix in (11.74), and R_0 is given by

$$\sum_{\substack{\alpha,\beta,\gamma = 0 \\ \alpha+\beta+\gamma \neq 0}} Q_1(\alpha,t,s) Q_2(\beta,t,s) Q_3(\gamma,t,s) u_1^{\alpha+\beta+\gamma-1} \mathbf{P}_{\alpha\beta\gamma}$$

where

$$\begin{aligned} Q_1(\alpha,t,s) &= (a_{11} + a_{12}t + a_{13}s)^\alpha \\ Q_2(\beta,t,s) &= (a_{21} + a_{22}t + a_{23}s)^\beta \\ Q_3(\gamma,t,s) &= (a_{31} + a_{32}t + a_{33}s)^\gamma \end{aligned}$$

$$\mathbf{P}_{\alpha\beta\gamma} = \sum_{i,j,k=1}^{m+1} \frac{\partial^\alpha L_i^m(u_0)}{\alpha! \partial u_0^\alpha} \frac{\partial^\beta L_i^m(v_0)}{\beta! \partial v_0^\beta} \frac{\partial^\gamma L_i^m(w_0)}{\gamma! \partial w_0^\gamma} \mathbf{r}_{i+(m+1)[(j-1)+(m+1)(k-1)]}$$

Volume-Surface Integral Equation   513

Table 11.3 The Parameters in Matrix $\overline{\mathbf{A}}$ of (11.74)

	u_a	v_a	w_a	u_b	v_b	w_b	u_c	v_c	w_c
V_1	0	1	0	0	0	0	0	0	1
V_2	1	0	0	1	1	0	1	1	1
V_3	0	0	0	1	0	0	1	0	1
V_4	1	1	0	0	1	0	0	1	1
V_5	0	0	0	0	1	0	1	1	0
V_6	0	0	1	1	0	1	1	1	1

11.6 SOLUTION OF VSIE BY FAST MULTIPOLE METHOD

The fast multipole method (FMM) for solving the SIE has been discussed in previous chapters. Here, we will show that the FMM can also be applied to solve the VIE. Recall that in FMM, the matrix elements are calculated differently depending on the distance between a testing function and a basis function. For near-neighbor matrix elements, the calculation remains unchanged. The well-separated elements is calculated by

$$Z_{ij} = i\omega\mu_b \int_{\Omega_i}\int_{\Omega_j} \mathbf{f}_i^{\Omega_i}(\mathbf{r}) \cdot \left(\overline{\mathbf{I}} + \frac{\nabla\nabla}{k_b^2}\right) G(\mathbf{r},\mathbf{r}')\chi(\mathbf{r}')\mathbf{f}_j^{\Omega_j}\, d\mathbf{r}'d\mathbf{r} \qquad (11.77)$$

If the above equation is compared to that for the SIE case, one finds that the only difference is the factor $\chi(\mathbf{r}')$, which is equal to unit for SIE and is defined by (11.23) for VIE. Hence (11.77) can be expanded as

$$Z_{ij} = \eta_b \left(\frac{ik_b}{4\pi}\right)^2 \int d^2\hat{k}\, \mathbf{V}_{fmi}(\hat{k}) \cdot T_{mn}(\hat{k}) \mathbf{V}_{snj}(\hat{k}) \qquad (11.78)$$

where the integration over $\hat{k}$ is on a unit sphere, and

$$\mathbf{V}_{fmi}(\hat{k}) = \int_{\Omega_i}\left(\overline{\mathbf{I}} - \hat{k}\hat{k}\right) \cdot \mathbf{f}_i^{\Omega_i}(\mathbf{r}) e^{i\mathbf{k}\cdot(\mathbf{r}-\mathbf{r}_m)}\, d\mathbf{r} \qquad (11.79)$$

$$\mathbf{T}_{mn}(\hat{k}) = \sum_{l=0}^{L} i^l (2l+1) h_l^{(1)}(kr_{mn}) P_l(\hat{k}\cdot\hat{r}_{mn}) \qquad (11.80)$$

$$\mathbf{V}_{snj}(\hat{k}) = \int_{\Omega_j}\left(\overline{\mathbf{I}} - \hat{k}\hat{k}\right) \cdot \mathbf{f}_j^{\Omega_j}(\mathbf{r}')\chi(\mathbf{r}') e^{-i\mathbf{k}\cdot(\mathbf{r}'-\mathbf{r}_n)}\, d\mathbf{r}' \qquad (11.81)$$

In the surface integral equation case, $\mathbf{V}_{fmi}$ and $\mathbf{V}_{snj}$ are complex conjugates to each other if the Galerkin's test is applied. This allows one to store only one of $\mathbf{V}_{fmi}$ or $\mathbf{V}_{snj}$. However with material presenting, these two vectors are no longer related

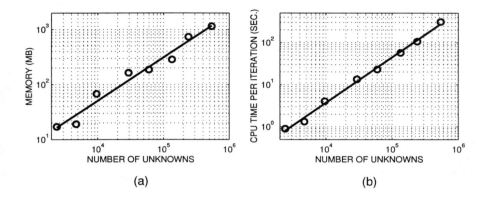

Figure 11.11 (a) The memory (in MB) requirement versus N. (b) The CPU time (in seconds) for one CG iteration versus N. Here N is the number of unknowns N.

and storage spaces must be allocated for both of them. Other implementations of MLFMA for VIE will be almost identical to that for SIE. The CPU time and memory requirements as functions for the number of unknowns for solving the VSIE for one-layer coated sphere is shown in Figure 11.11. It can be found that the slopes of the two curves are very close to unit.

To demonstrate the performance of the MLFMA for the hybrid surface-volume integral equation, we calculate the RCS of a conducting plate with rim coating on its four sides, as shown in Figure 11.12 and Figure 11.13. The conducting plate is modeled by 3,862 quadrangles, and the coating material is modeled by 1,062 hexahedrons. The total number of unknowns is 11,799. The calculation used five-level MLFMA to perform the matrix-vector multiplication. The RCS is calculated for every $0.5°$ angle in ϕ on the $\theta = 90°$ plane, and the average number of iterations needed to reduce the relative residue error of 0.002 is six. This small number of iteration is due to the use of CFIE and a right-hand-side updating technique that uses the phase-corrected previous solution as the initial guess for the next iteration. If the initial guess technique is not used, or if angles are picked randomly, the average iteration number would be more than 50. Without using CFIE, the iteration would not converge in 500 iterations.

11.7 NUMERICAL EXAMPLES

In this section, we will present a number of numerical examples to demonstrate the validity and the application of the volume integral equation and the hybrid surface-volume integral equation.

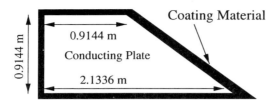

Figure 11.12 The geometry of the EMCC dart object: the coating material has permittivity $\epsilon_r = 4.5 + 9i$. The dimensions are for the conducting plate. Other dimensions are: thickness = 0.0508m and width of material coating = 0.0508m.

11.7.1 Results by Volume Integral Equation

In this section, we present calculated results for (a) the RCS of a homogeneous dielectric sphere, (b) the near-field of dielectric spherical shell, and (c) the RCS of dielectric plates. For the spherical symmetric objects, the Mie series solutions are used for comparison. For other homogeneous dielectric objects, the surface integral equation is used as comparison.

First we show the RCS at 300 MHz for a dielectric sphere of radius 0.5m and $\epsilon_r = 1.63$. The sphere is subdivided into 2,592 first-order hexahedrons (a first-order hexahedron is specified by eight points at the vertices). The bistatic RCSs for both vertical (V) and horizontal (H) polarization incidences are plotted in Figure 11.14.

To model objects with edges and corners, it is necessary to use a denser mesh than that used for smoothing regions or to use a nonuniform mesh. This is to ensure that the rapid current variations are accurately represented. It is found in our numerical simulation that for comparable solution accuracy, the volume integral equation technique needs coarser grids compared to the surface integral equation approach (using the PMCHW formulation). In the next example, we show the mesh convergence tests using the two methods for a rectangular dielectric box of size $0.72\lambda_0 \times 0.36\lambda_0 \times 0.24\lambda_0$ at 300 MHz. Four meshes of different grid size are constructed for this model. They are $\Delta = 0.027$m, 0.04m, 0.06m, and 0.08m, respectively. The calculated RCS using the SIE and the VIE are shown in Figure 11.15. The solutions from the VIE for the four models are very close, and hence only one curve is drawn for the VIE. However, the SIE solutions converge much slower than that of the VIE.

To further demonstrate the grid convergence property for the SIE and VIE, another dielectric plate of size $3.5\lambda_0 \times 2.0\lambda_0 \times 0.25\lambda_0$ is investigated. For this plate, three meshes are constructed and the RCSs for the plate are shown in Figure 11.16. (There are only two SIE results due to the limited computer memory and the finest mesh has not been applied to the SIE.) The faster grid convergence for VIE solution is probably

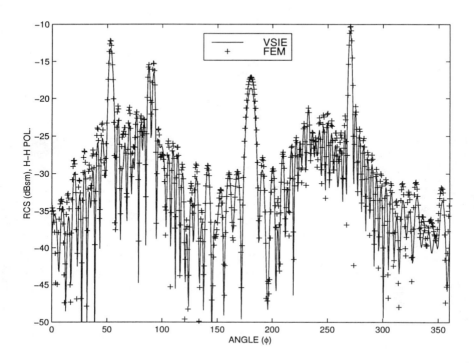

Figure 11.13 The radar cross-section of an EMCC dart object for $\theta = 90°$ observation plane calculated by the hybrid volume-surface integral equation (VSIE) and the hybrid FEM and boundary integral equation technique (FEM).

due to the fact that the volume current distribution is generally smoother than that of the surface current.

Using the tetrahedron mesh or the hexahedron mesh also has an advantage over the structured cubic mesh. A structured grid is the one with grid line parallel to the coordinate lines. Hence, using a structured grid to model curved boundaries the staircase cannot be avoided. In an unstructured grid, however, the grid line can be of arbitrary orientation. Hence it is very flexible for modeling curved boundaries. In the inverse scattering algorithms, the forward scattering is most often solved by the CGFFT method in which pulse or roof-top basis functions are used. However, because of the staircase geometry error in modeling curved objects, accurate solution often requires much denser grid points than with the unstructured tetrahedron/hexahedron meshes. This is especially true when strong dielectric contrast presents in the object. As an example, Figure 11.17 shows the RCSs of a dielectric sphere with structured cubic mesh and tetrahedron mesh. It can be seen that very fine meshes are needed

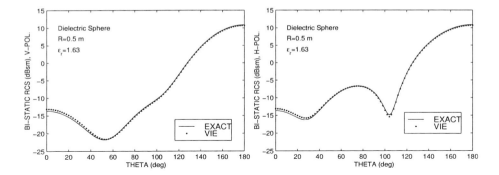

Figure 11.14 The RCS of V-polarized incidence (left) and H-polarized incidence (right), of a homogeneous dielectric sphere calculated by the volume integral equation (dotted line) and the series solution (solid line).

for the structured cubic mesh to reach an accuracy similar to that of the tetrahedron mesh.

Similarly, unstructured hexahedron meshes represent a curved object more accurately than the structured cubic mesh for the same grid size. Figure 11.18 shows the RCSs for a spherical dielectric shell when the dielectric constant varies from $\epsilon_r = 1.1$ to 8.8. It can be seen that for low dielectric contrasts, the structured cubic mesh gives solution accuracy comparable to that of the unstructured hexahedron meshes. However, for high dielectric contrast ($\epsilon_r = 8.8$) the solution from even the refined cubic mesh does not agree with the exact solution.

The volume integral equation has also been applied to the two-layer problem. In this problem, we have two layers of sphereical shells. The inner radius of the inner shell is 1.2m, and each layer is 0.1m thick. The dielectric constants are $\epsilon_r = 2.0 + 0.5i$ for the inner layer and $1.5 + 0.0i$ for the outer layer. The incident wave is from the $\theta = 0$ direction, and the frequency is 300 MHz. The results are shown in Figure 11.19.

11.7.2 Results by Hybrid Volume-Surface Integral Equation

In this section, we present calculated results for (1) the RCS of a coated sphere, and (2) the RCS of a coated ogive. Again, for the spherically symmetric objects, the Mie series solutions are used for comparison. For other homogeneous coated objects, the surface integral equation is used as comparison.

Figure 11.20 shows the RCS of a coated sphere at $f = 300$ MHz. The conducting core has a radius of 0.3m, and the coating is uniform around the sphere with a thickness of 0.05m. The dielectric permittivity of the coating material is $\epsilon_r = 4.0 + 0.1i$. The conducting surface is modeled by 600 quadrilaterals, and the coating dielectric is modeled by 600 hexahedrons. The calculated RCS is compared with the Mie-series solution, and the results are shown in Figure 11.20. The results of another coated sphere model that has larger radius is shown in Figure 11.21.

Next we show the RCS of a coated ogive object in Figure 11.22. The object is formed by combining two identical cones in a base-to-base manner. The cone base radius is 0.24m, and the height is 1m. The coating thickness is 0.07m. The profile of the cone is a parabolic line. The results from the SIE are also shown in the figure for comparison.

One application of the volume integral equation is to simulate the indoor radio wave propagation where the near field is needed. In the next section we will have an example for simulating the radio wave propagation in site-specific environments. Here, we present an example to show the near-field calculation for a dielectric shell of spherical shape. For this structure, there is analytical solution by Mie series. Hence in Figure 11.23, we show the calculated near field using the VIE and Mie series. The excitation is a plane wave travelling in the $\theta = 0$ direction. The cross view of the shell geometry and the observation points are shown in Figure 11.23(a). The near-field distribution is shown in Figure 11.23(b).

11.8 OTHER APPLICATIONS

In addition to the calculation of radar cross-section of coated targets, the hybrid volume-surface integral equation technique has a number of other applications such as the radar antenna radome analysis, printed circuit modeling, radio wave simulation in complex environments, and electromagnetic wave interaction with biomedical materials. In this section we will present selected numerical examples to demonstrate the application.

11.8.1 Indoor Radio Wave Propagation Simulation

A big advantage of using the volume integral equation for analyzing the site-specific wave propagation is that it is able to accurately consider the small structural features that are comparable to or smaller than a wavelength. The first example considers a dipole radiation in the presence of a dielectric shell structure made of four piece of dielectric walls [Figure 11.24(a)]. The radiated total field is calculated and is shown in Figure 11.24(b) (also shown in the figure are the results by the surface integral equation). It can be seen that the two results agree very well. Next we consider small dielectric poles within the rooms in a three-room structure (this example is designed to demonstrate the capability of the VSIE technique for modeling large sized structures with small features). The rooms are 3m×3m in floor size with a height of 2.4m (no ceiling and floor is considered here). A dipole is put in one room and the total electric field distributions are calculated for all the rooms in the $h = 1.5$m plane. To compare the results from the dielectric pole, the calculation is done for two cases: (1) without the pole, and (2) with pole. The results are shown in Figures 11.25 and 11.26. It can be seen that the poles have a significant effect on the field distributions for x- and z-directed antennas. More simulation results can be found in [35].

11.8.2 Microwave Thermal Effect Simulation

When an electromagnetic wave is applied to lossy dielectrics, the EM energy will be converted to thermal energy, which will be absorbed by the dielectrics. Figure 11.27 is an example configuration of realizing the EM to thermal energy conversion. The absorption of EM energy causes the temperature of the dielectrics to rise. During a short time period in which the heat conduction can be ignored, the rate of temperature rise is approximately proportional to $\sigma |\mathbf{E}|^2/2$. Here σ is the effective conductivity of the dielectrics, and $\mathbf{E}$ is the electric field vector. If the application time of the EM field is long and the heat conduction effect cannot be ignored, then the temperature rise in the dielectric must be determined through a combined electromagnetic and heat transfer analysis [36]. To illustrate the combined analysis process, consider a dielectric sample in a microwave cavity. The process starts with the initial temperature of the tissue. The electromagnetic field distribution inside the sample is determined first by solving hybrid surface-volume integral equations. This solution provides a thermal source term for the heat transfer equation:

$$\rho C \frac{\partial T}{\partial t} = \nabla \cdot (k \nabla T) + \frac{1}{2}\sigma |\mathbf{E}|^2$$

where k is the specific heat coefficient, ρ is the mass of the sample, and C is the heat conduction coefficient. Then a finite difference scheme is applied to solve the heat transfer equation, which determines the temperature distribution inside the tissue for the next time step. Since the tissue's electrical characteristics (ϵ and σ) are functions of temperature, their values are then updated based on the new temperature distribution. The iteration continues until a termination condition is satisfied. This combined iterative solution of wave equation and heat transfer equation allows us to model the complex rewarming process. Figure 11.28 shows the temperature distribution in a dielectric sample in a circular microwave cavity (detailed configuration parameters can be found in [36]).

11.8.3 Antenna Radome Modeling

The presence of radomes will generally affect the radiation characteristics of antennas. Techniques for analyzing the antenna pattern deformation by radomes include high-frequency methods such as GTD and PO, and numerical methods such as the method of moments. The numerical methods provide accurate solutions to the problem, as demonstrated in [37]. However, because the number of unknowns increases rapidly with the electrical size of the antenna-radome structure, the numerical methods were either applied to two-dimensional problems or to 3D problems with electrically small sized radomes. In practical characterization of electrically larger sized radomes, the ray tracing methods based on high-frequency techniques are necessary. An important assumption to high-frequency methods is that the structures have smooth surfaces and electrically large radii of curvature. For most portions of a realistic radome, this assumption is valid. However, for some radome types that have sharp tips, such as the ogive or cone type, the radius of curvature may be comparable to or smaller than a wavelength; the high-frequency approximation is not valid in that region, and hence the solution accuracy by high-frequency methods is reduced.

Now we present the calculation of radome effects by the hybrid volume-surface integral equation techniques. The hybrid VSIE can be applied to efficiently and accurately analyze dielectric radomes of sizes much larger than that can be analyzed by a numerical method, and yet VSIE provides much higher solution accuracy than a high-frequency method in general.

Figure 11.29 shows the radiation pattern of a dipole array with the presence of differently shaped radomes. Three types of radomes are considered: (1) tangent ogive with height $6.67\lambda_0$ and base diameter $5.33\lambda_0$ (modeled by 76,914 tetrahedrons), (2) straight cone with the same dimension as the ogive (modeled by 55,308 tetrahedrons), (3) hemisphere of base diameter of $5.33\lambda_0$ (modeled by 45,000 tetrahedrons). The thickness and the permittivity of all three radomes are the same. They are $t = 0.008\lambda_0$ and $\epsilon_r = 2+1i$. The dipole array is on the z-axis and is made of three dipole elements of equal amplitude and phase. The locations of the elements are $-1\lambda_0$, 0, and $1\lambda_0$. The calculated results for $\phi = 0$ and θ from $0°$ to $180°$ are shown in Figure 11.29. As expected, the interferences by the ogive- and cone-shaped radomes are much more severe than those by the hemisphere.

It should be pointed out that for the same grid size (which results in the same order of accuracy), using a hexahedron mesh can significantly reduce the number of unknowns and hence can increase the electrical size of the radome that can be analyzed by MLFMA. In the last example, if the hexahedron model is used, the number of cells is only 6,912, and the memory is about 80 MB (compared to the memory for the tetrahedron model, which is about 690 MB). To demonstrate the efficiency of using the hexahedron mesh, Figure 11.30 shows the radiations of a dipole array in the presence of hemisphere radomes of one, two, and three layers. The three-layer radome problem is modeled by 72,900 hexahedrons and the memory needed for generating the result for the three-layer radome in Figure 11.30 is 1,088 MB.

It is known that radome will generally affect the performance of the antenna covered by it. Knowing the radome's influence in advance is important to make appropriate compensations in antenna design. The algorithm introduced in this chapter can be used to simulate antenna beam's pointing error due to dielectric radomes. As the last example, consider three dielectric radomes: a tangent ogive radome, a cone radome, and a hemisphere radome. The three radomes are of the same size, same thickness, and made of same dielectrics ($\epsilon_r = 2 + 1i$). The antenna is a dipole array of eight dipoles with triangular amplitude distribution and varying progressive phase. The dipoles are equally spaced on the z-axis in $z = -1.75\lambda_0$ to $z = 1.75\lambda_0$. Figure 11.31 shows the calculated beam pointing error (the reference beam is the beam of the same array in free-space).

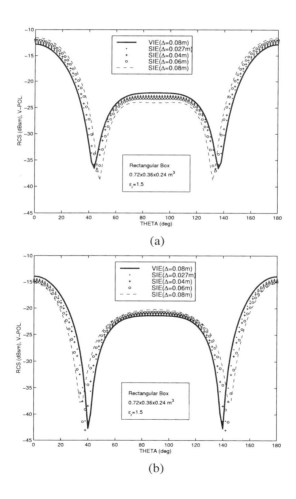

Figure 11.15 The backscattered RCS of a rectangular dielectric box for the $\phi = 0$ cutting plane: (a) V-polarized incidence, (b) H-polarized incidence.

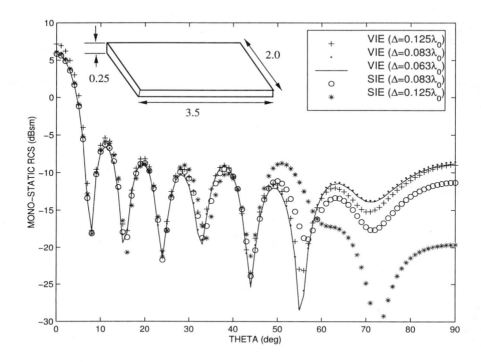

Figure 11.16 The RCS (vertical polarization incident case) of a dielectric plate of dimension $3.5\lambda_0 \times 2.0\lambda_0 \times 0.25\lambda_0$ at frequency of 1 GHz. The dielectric constant is $\epsilon_r = 3.0 + 0.09i$.

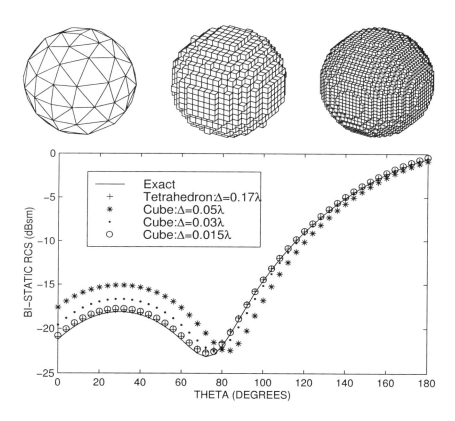

Figure 11.17 The volume meshes of a dielectric sphere and the RCS solved using different meshes: tetrahedron mesh (top left), structured cubic mesh ($\Delta = 0.05\lambda$) (top middle), structured cubic mesh (dense) ($\Delta = 0.03\lambda$) (top right), and the RCS comparison for different meshes (bottom) (the cubic mesh for $\Delta = 0.015\lambda$ is not shown).

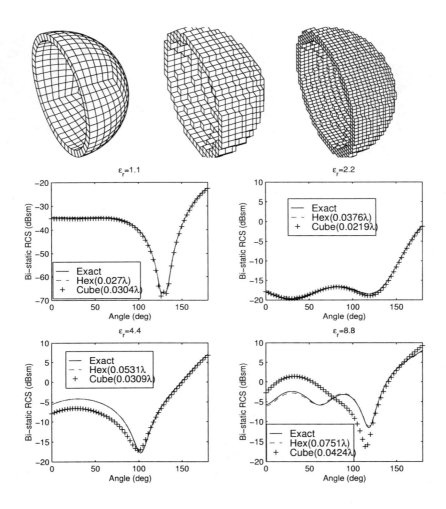

Figure 11.18 The volume meshes of a dielectric spherical shell and the RCS solved using different meshes: hexahedron mesh (top left), structured cubic mesh 1 (top middle), a denser structured cubic mesh 2 (top right), and the RCS comparison for different meshes and different dielectric constants (bottom).

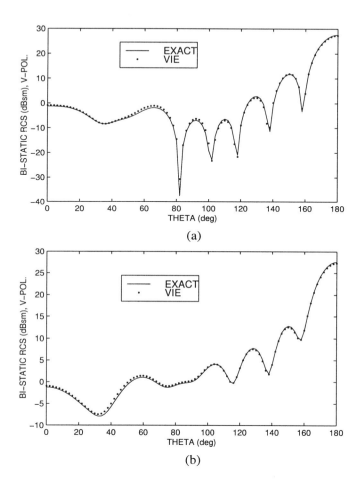

Figure 11.19 The RCS for two layers of spherical dielectric shells: (a) V-polarized incidence; (b) H-polarized incidence.

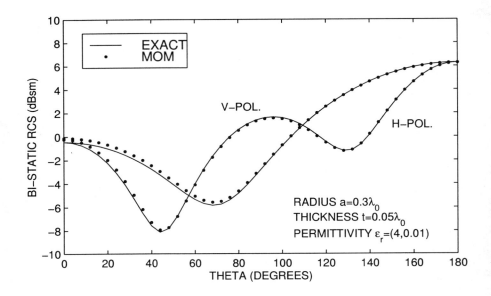

Figure 11.20 The RCS of a coated sphere calculated by hybrid integral equation (dotted line), surface integral equation (dash-line), and the series solution (solid line).

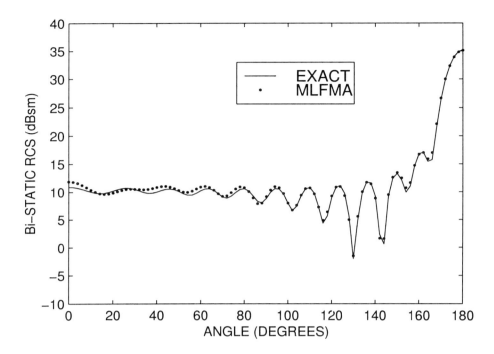

Figure 11.21 The RCS (V-polarized) of a coated conducting sphere calculated by analytic series (solid line), and by the multilevel fast multipole algorithm (MLFMA) for the hybrid surface-volume integral equation. The radius of the sphere is $R = 2$m, and the thickness of the coating material is $t = 0.05$m. The dielectric material of the coating is $\epsilon_r = 4.0 + 9i$.

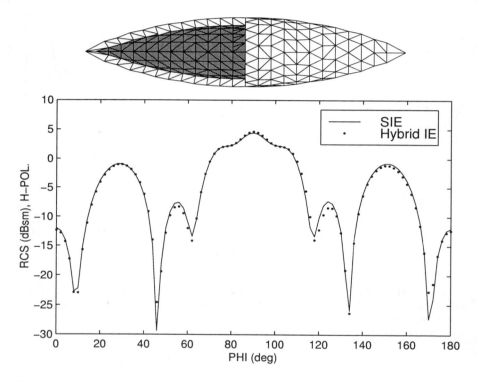

Figure 11.22 The tetrahedron mesh of a coated double cone (top), and the monostatic RCS (bottom) of a coated double cone calculated using surface integral equation (solid line), and hybrid volume-surface integral equation (dotted line).

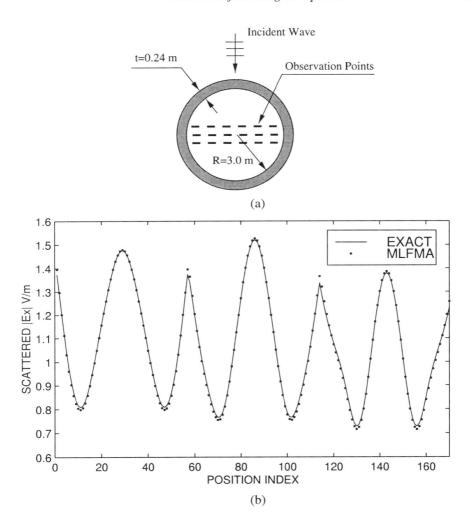

Figure 11.23 (a) The cross-sectional view of the dielectric shell (shaded area) and the observation points (dotted line). (b) The near-field comparison of VIE and Mie series along the observation line.

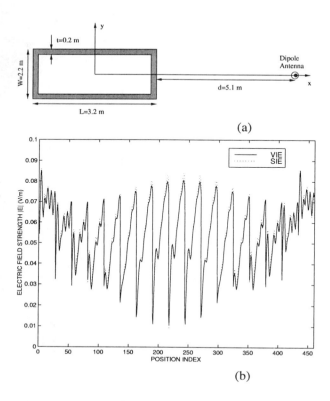

Figure 11.24 (a) The mesh of a three-dimensional dielectric object formed by four pieces of dielectric walls. The inner dimensions are 1.5m, 1.0m, and 0.5m, respectively. The thickness is 0.25m. (b) The electric fields inside the object calculated by the volume integral equation approach (solid), and by the surface integral equation approach (dotted). The fields are calculated over a planar region specified by $-1.3 \leq x \leq 1.3$, $-0.8 \leq y \leq 0.8$, and $z = 0$ (all units here are in meters). The step sizes in the x- and y-directions are both 0.1m.

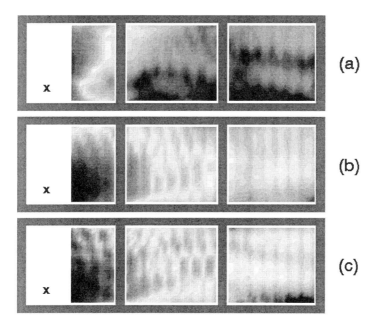

Figure 11.25 The field distribution in three rooms without poles (the symbol "x" indicates the location of the dipole antenna): (a) x-directed dipole, (b) y-directed dipole, (c) z-directed dipole.

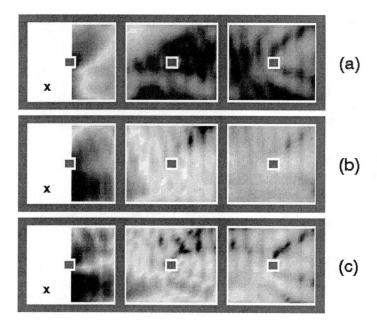

Figure 11.26 The field distribution in three rooms with poles (the symbol "x" indicates the location of the dipole antenna): (a) x-directed dipole, (b) y-directed dipole, (c) z-directed dipole.

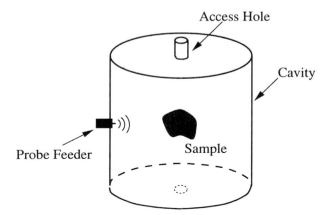

Figure 11.27 The geometry of a circular cylindrical cavity and dielectric sample inside the cavity to be processed.

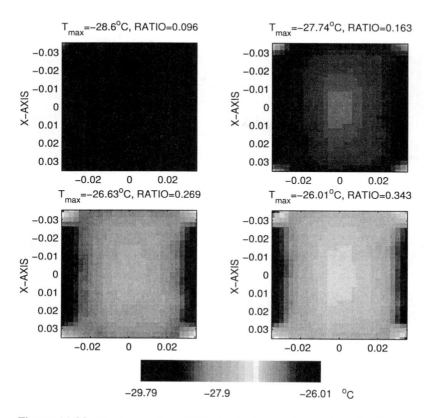

Figure 11.28 The temperature distribution in the sample at various time instances.

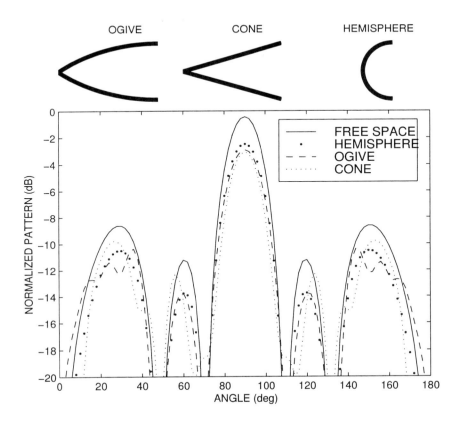

Figure 11.29 The comparison of the normalized radiation of a dipole array in the presence of three types of radome shapes: (a) ogive (dashed line), (b) straight cone (dotted line), and (c) hemisphere (dots). The radiation of the same dipole array in free space is also plotted for reference. (The cross-sectional view of the radome generating curves are shown on top of the plot.) The array axis is perpendicular to the radome axis and is located on the base plane of the radome. The radiation pattern is calculated in the plane containing the radome axis and the array axis and the zero angle refers to the tip direction of the radome.

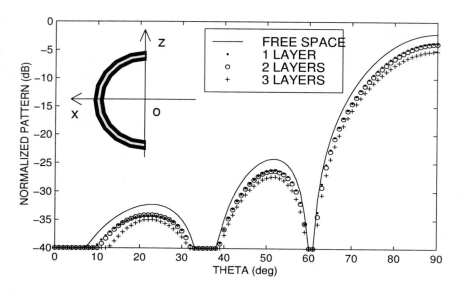

Figure 11.30 The radiation of a dipole array in the presence of three hemisphere radomes. The base diameter is $10\lambda_0$ and the permittivities are $2 + 1i$ for layer 1 and 3, $3 + 0.1i$ for the middle layer. The dipole array consists of 16 z-axis oriented dipoles with equal amplitude and equal phase. The dipole elements are equally spaced with inter-element space of $0.5\lambda_0$. The frequency of operation is 7.49475 GHz.

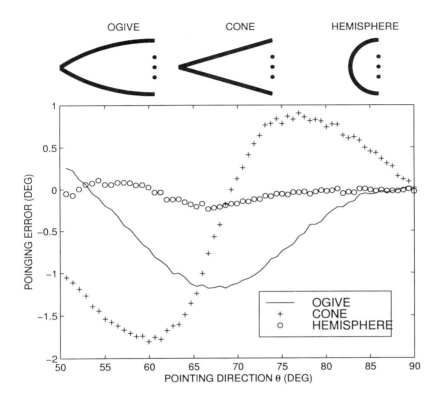

Figure 11.31 The beam pointing error or a dipole array in three types of radomes: (a) a tangent ogive radome (the solid line), (b) a cone radome (cross), and (c) a hemisphere radome (circles). The horizontal axis is the beam direction θ of the array in free-space, with $\theta = 0$ being the array axis. The dots in the above figures indicate the dipole elements.

REFERENCES

1. T. Vaupel and V. Hansen, "Electrodynamic analysis of combined microstrip and coplanar/slotline structure with 3-D components based on a surface/volume integral equation approach," *IEEE Trans. Micro. Theory Tech.*, vol. 47, no. 9, Sept. 1999.

2. B. M. Kolundzuja, "Electromagnetic modeling of composite metallic and dielectric structures," *IEEE Trans. Micro. Theory Tech.*, vol. 47, no. 7, pp. 1021–1032, July 1999.

3. C. F. Wang and J. M. Jin, "A fast full-wave analysis of scattering and radiation from large finite arrays of microstrip antennas," *IEEE Trans. Antennas Propag.*, vol. 46, no. 10, pp. 1467–1474, Oct. 1998.

4. X. Sheng, J. M. Jin, J. Song, C. C. Lu, and W.C. Chew, "On the formulation of hybrid finite-element and boundary-integral methods for 3D scattering," *IEEE Trans. Antennas Propag.*, vol. 46, no. 3, pp. 303–311, March 1998.

5. S. M. Rao, C. C. Cha, R. L. Cravey, and D. L. Wilkes, "Electromagnetic scattering from arbitrary shaped conducting bodies coated with lossy materials of arbitrary thickness," *IEEE Trans. Antennas Propag.*, vol. 39, no. 5, 627–631, May 1991.

6. R. F. Harrington, "Boundary integral formulations for homogeneous material bodies," *J. Electromag. Waves Appl.*, vol. 3, no. 1, pp. 1–15, 1989.

7. L. N. Medgyesi-Mitschang, J. M. Putnam, and M. B. Gedera, "Generalized method of moments for three-dimensional penetrable scatterers," *J. Opt. Soc. Am.*, A. vol. 11, pp. 1383–98, April 1994.

8. T. K. Shark, S. M. Rao, and A. R. Djordievic, "Electromagneitc scattering and radiation from finite microstrip structures," *IEEE Trans. Micro. Theory Tech.*, vol. 38, no. 11, pp. 1568–1575, Nov. 1990.

9. C. C. Lu and W. C. Chew, "A coupled surface-volume integral equation approach for the calculation of electromagnetic scattering from composite metallic and material targets," *IEEE Trans. Antennas Propag.*, vol. 48, no. 12, pp. 1866–1868, Dec. 2000.

10. J. H. Richmond, "Scattering by a dielectric cylinder of arbitrary cross section shape," *IEEE Trans. Antennas Propag.*, vol. 13, no. 5, pp. 334–341, May 1965.

11. J. H. Richmond, "TM scattering by a dielectric cylinder of arbitrary cross section shape," *IEEE Trans. Antennas Propag.*, vol. 14, no. 7, pp. 460–464, July 1966.

12. T. K. Wu and L. L. Tsai, "Scattering from arbitrarily-shaped lossy dielectric bodies of revolution," *Radio Science*, vol. 12, pp. 709–718, 1977.

13. D. E. Livesay and K. M. Chen, "Electromagnetic fields induced inside arbitrary shaped biological bodies," *IEEE Trans. Micro. Theory Tech.*, vol. 22, no. 12, pp. 1273–1280, Dec. 1974.

14. D. H. Schaubert, D. R. Wilton, and A. W. Glisson, "Tetrahedral modeling method for electromagnetic scattering by arbitrary shaped inhomogeneous dielectric bodies," *IEEE Trans. Antennas Propag.*, vol. 32, no. 1, pp. 77–85, Jan. 1984.

15. M. F. Catedra, E. Gago, and L. Nulo, "A numerical scheme to obtain the RCS of three dimensional bodies of resonant size using the conjugate gradient method

and the fast Fourier transform," *IEEE Trans. Antennas Propag.*, vol. 37, no. 5, pp. 528–537, May 1989.

16. H. Gan and W. C. Chew, "A discrete BCG-FFT algorithm for solving 3D inhomogeneous scatterer problems," *J. Electromag. Waves Appl.*, vol. 9, no. 10, pp. 1339–1357, 1995.

17. R. D. Graglia, "The use of parametric elements in the moment method solution of static and dynamic volume integral equations," *IEEE Trans. Antennas Propag.*, vol. 36, no. 5, pp. 636–646, May 1996.

18. X. Q. Sheng, J. M. Jin, J. M. Song, W. C. Chew, and C. C. Lu, "Solution of combined-field integral equation using multi-level fast multipole algorithm for scattering by homogeneous bodies," *IEEE Trans. Antennas Propag.*, vol. 46, no. 11, pp. 1718–1726, Nov. 1998.

19. K. Umashankar, A. Taflove, and S. M. Rao, "Electromagnetic scattering by arbitrary shaped three-dimensional homogeneous lossy dielectric objects," *IEEE Trans. Antennas Propag.*, vol. 34, no. 6, pp. 758–766, June 1986.

20. R. F. Harrington, *Field Computation by Moment Methods*, New York: The MacMillan Company, 1968.

21. J. H. Lin, "A Study of Iterative Methods on Forward and Inverse Scattering Problems", Ph.D. dissertation, University of Illinois at Urbana-Champaign, 1995.

22. Y. M. Wang, "Theory and Applications of Scattering and Inverse Scattering Problems", Ph.D. Dissertation, University of Illinois at Urbana-Champaign, 1991.

23. A. H. Stroud, *Approximate Calculation of Multiple Integrals*, Englewood Cliffs, NJ: Prentice-Hall, 1971.

24. M. Abramowitz and I. A. Stegun, *Handbook of Mathematical Functions with Formulas, Graphs, and Mathematical Tables*, New York: Dover, 1972.

25. S. M. Rao, D. R. Wilton, and A. W. Glisson, "Electromagnetic scattering by surfaces of arbitrary shape," *IEEE Trans. Antennas Propag.*, vol. 30, no. 3, pp. 409–418, May 1982.

26. J. M. Song and W. C. Chew, "Moment method solution with parametric geometry," *J. Electromagnetic Waves and Appl.* vol. 9, no. 1/2, pp. 71–83, 1995.

27. M. I. Sancer, R. L. McClary, and K. J. Glover, "Electromagnetic computation using parametric geometry," *Electromagn.*, vol. 10, no. 1, pp. 85–103, Jan. 1990.

28. D. L. Wilkes, and C.-C. Cha, "Method of moments solution with parametric curved triangular patches," *IEEE APS Int. Symp.*, pp. 1512–1515, London, Ontario, Canada, 1991.

29. J. M. Song and W. C. Chew, "Fast multipole method solution of the combined field integral equation," in *11th Annual ACES Conference*, May 1995.

30. W. C. Chew, *Waves and Fields in Inhomogeneous Media,* New York: Van Nostrand Reinhold, 1990.

31. R. D. Graglia, "On the numerical integration of the linear shape functions times the 3-D Green's function or its gradient on a plane triangle," *IEEE Trans. Antennas Propag.*, vol. 41, no. 10, pp. 1448–1455, Oct. 1993.

32. S. W. Lee, J. Boersma, C. L. Law, and G. A. Deschamps, "Singularity in Green's function and its numerical evaluations," *IEEE Trans. Antennas Propag.*, vol. 28, no. 3, May 1980.

33. D. R. Wilton, S. M. Rao, A. W. Glisson, D. H. Schaubert, O. M. Al-Bundak, and C. M. Butler, "Potential integrals for uniform and linear source distributions on polygonal and polyhedral domains," *IEEE Trans. Antennas Propag.*, vol. 32, no. 3, pp. 276–281, March 1984.

34. M. G. Duffy, "Quadrature over a pyramid or cube of integrands with a singularity at a vertex," *SIAM J. Numer. Anal.*, vol. 19, no. 6, pp. 1260–1262, Dec. 1982.

35. C. C. Lu, "Indoor radio wave propagation modeling by multilevel fast multipole algorithm," to be published by *Micro. Opt. Tech. Lett.*, in the issue of May 5, 2001.

36. C. C. Lu, H. Z. Li, and D. Gao, "A Combined heat transfer and electromagnetic analysis of rapid rewarming of biological tissues," *IEEE Trans. Micro. Theory Tech.*, vol. 48, no. 11, pp. 1083–1086, Nov. 2000.

37. E. Arvas, A. Rahhalarabi, U. Pekel, and E. Gundogan, "Electromagnetic transmission through a small radome of arbitrary shape," *IEE Proc. H, Micro. Antennas Propag.*, vol. 137, no. 6, pp. 401–405, Dec. 1990.

12
Finite Element Analysis of Complex Axisymmetric Problems

Andrew D. Greenwood and Jian-Ming Jin

12.1 INTRODUCTION

Axisymmetric objects, also known as bodies of revolution (BOR), are important both as radar targets and as radiating structures. Thus, their scattering and radiation properties are the subject of extensive studies. The rotational symmetry present in both problems allows a solution with a two-dimensional (2D) computational technique. The primary method is the method of moments (MOM) based on integral equation formulations [1–8]. This method works well for problems consisting of perfectly conducting or homogeneous materials, but the computational complexity increases rapidly when inhomogeneous materials are present. The computational complexity increases because a volume formulation, rather than a surface formulation, is required for inhomogeneous materials and an integral equation generates a dense system of equations. Another popular method for BOR problems is the finite difference method (FDM) [9]. This method results in a sparse matrix and hence is computationally efficient. However, it relies on rectangular, cylindrical, or spherical computational grids, making it difficult to model arbitrarily shaped geometries.

The finite element method (FEM) is an effective method both for BOR scattering and radiation problems because the FEM can remove all of the difficulties associated with the MOM and the FDM [10–14]. For example, the FEM is much more versatile than the MOM because the same formulation is used for conducting, homogeneous, and inhomogeneous geometries. Also, the FEM always generates a sparse system of equations, which can be solved using much less computer memory and CPU time than a dense MOM matrix. Furthermore, the FEM conveniently and accurately models arbitrary shapes using triangular elements and has a great material handling capability [15–18]. As a result, the FEM has significant computational advantages for the electromagnetic problems involving an arbitrary, inhomogeneous BOR.

In both scattering and radiation problems, the main issues involving the FEM arise from two sources. First, the FEM formulation must be accurate and free of spurious modes. Second, because both problems involve an open-region domain, it is necessary to truncate the FEM mesh with an appropriate absorbing boundary condition (ABC). The mixed-basis formulation that uses vector (edge) basis functions for the transverse field components and scalar (nodal) basis functions for the angular component is accurate and free of spurious modes [13, 14] and is described in Section 12.2 of this chapter. The use of cylindrical perfectly matched layers (PML) as an effective mesh truncation is the subject of Section 12.3. Numerical results to validate the method are given in Section 12.4. Section 12.5 describes an approximate method to deal with a BOR with small appendages, and finally a conclusion is given in Section 12.6.

12.2 FORMULATION

Past FEM formulations for axisymmetric problems employ either the coupled azimuth potential (CAP) formulation [9–11, 15, 16] or the three-component, node-based formulation [12]. Both formulations make use of a cylindrical (ρ, ϕ, z) coordinate system. In the CAP formulation, the problem is formulated in terms of the angular electric field (E_ϕ) and the angular magnetic field (H_ϕ). All other components are found in terms of E_ϕ, H_ϕ, and their derivatives. The main advantage of this formulation is that it requires approximately one-third fewer unknowns than other formulations. Further, because E_ϕ and H_ϕ are everywhere continuous in a rotationally symmetric problem, they can be represented by traditional node-based basis functions without the problem of spurious modes. However, an imperfect cancelation problem limits the accuracy when computing the transverse field components from E_ϕ and H_ϕ [19]. In the three-component, node-based formulation, the problem is formulated in terms of the entire electric field (**E**) or the entire magnetic field (**H**). Each component of **E** or **H** is expanded in terms of node-based, scalar basis functions. The main advantage of this formulation is that it yields all components of either the electric field or the magnetic field directly and therefore avoids the imperfect cancelation problem associated with the CAP formulation. However, the expansion scheme makes it difficult to enforce the proper boundary conditions at material discontinuities and sharp conductor edges. This limits the kinds of material that can be conveniently considered. Further, the problem of spurious modes can be overcome only by the use of a penalty term [12], and it is difficult to choose the proper penalty factor.

The use of vector finite elements, also known as edge elements, eliminates many of the disadvantages of the three-component, node-based formulation while retaining the advantage of computing directly either the electric or the magnetic field. In the 2D FEM for axisymmetric problems, a mixed edge-node formulation is developed by expanding the transverse $(\hat{\rho}, \hat{z})$ field components using a 2D edge (vector) basis, and expanding the angular $(\hat{\phi})$ field component using a 2D nodal (scalar) basis. At a material discontinuity, this expansion scheme automatically constrains the tangential field components to be continuous without similarly constraining the normal field component. Thus, the proper, physical boundary condition is automatically satisfied at both electric and magnetic material interfaces, allowing both material types in the same problem. Further, this expansion scheme is free of spurious modes without requiring a penalty factor.

12.2.1 Problem Definition

The computational domain for a typical problem is shown in Figure 12.1. The figure shows a slice of a rotationally symmetric geometry. For an electromagnetic scattering problem, the geometry is illuminated by an incident plane wave, and for a radiation problem, it contains nonrotationally symmetric current sources that are modeled as a Fourier series. Before discussing the derivation of the FEM equations, the form of the constitutive parameters of the media is considered. As shown in Figure 12.1, the outer boundary of the mesh is lined with an absorbing PML. The PML can be conveniently interpreted as an anisotropic medium [20,21]. Using this interpretation, the constitutive parameters take the form

$$\mu = \mu_0 \mu_r \overline{\mathbf{L}} \qquad \epsilon = \epsilon_0 \epsilon_r \overline{\mathbf{L}} \qquad (12.1)$$

where $\overline{\mathbf{L}}$ is a diagonal tensor which can be written as

$$\overline{\mathbf{L}} = \hat{\rho}\hat{\rho} L_\rho + \hat{\phi}\hat{\phi} L_\phi + \hat{z}\hat{z} L_z \qquad (12.2)$$

The tensor $\overline{\mathbf{L}}$ contains information about the PML, and except in the PML itself, $\overline{\mathbf{L}} = \overline{\mathbf{I}}$ where $\overline{\mathbf{I}}$ denotes the unit dyad. Note also that in the PML, $\mu_r = \epsilon_r = 1$.

12.2.2 Variational Formulation

The vector wave equation, which follows from the constitutive parameters in (12.1), is given by

$$\nabla \times \frac{1}{\mu_r}\overline{\mathbf{L}}^{-1} \cdot \nabla \times \mathbf{E} - k_0^2 \epsilon_r \overline{\mathbf{L}} \cdot \mathbf{E} = -jk_0\eta_0 \mathbf{J} \qquad (12.3)$$

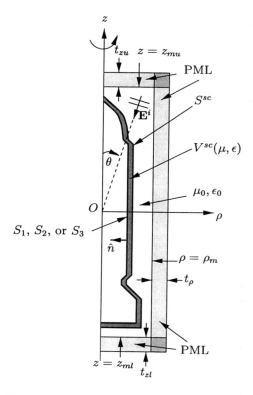

Figure 12.1 Slice of a typical geometry with the PML enclosure.

where $k_0 = \omega\sqrt{\mu_0\epsilon_0}$ is the free-space wave number, $\eta_0 = \sqrt{\mu_0/\epsilon_0}$ denotes the free-space wave impedance, and **J** represents a source current. The boundary conditions are [22]

$$\hat{n} \times \mathbf{E} = 0 \quad \text{on } S_1 \tag{12.4}$$

$$\hat{n} \times (\nabla \times \mathbf{E}) = 0 \quad \text{on } S_2 \tag{12.5}$$

$$\frac{1}{\mu_r}\hat{n} \times \overline{\mathbf{L}}^{-1} \cdot (\nabla \times \mathbf{E}) + \gamma_e \hat{n} \times \hat{n} \times \mathbf{E} = 0 \quad \text{on } S_3 \tag{12.6}$$

where S_1 denotes a perfect electric conductor (PEC) surface, S_2 denotes a perfect magnetic conductor (PMC) surface, S_3 denotes a surface with an impedance boundary condition, and $\hat{n}$ denotes the unit normal to the surface. Note that $S_1 + S_2 + S_3$ makes up the surface of an impenetrable region as shown in Figure 12.1.

According to the generalized variational principle [22], the functional for this problem is written as

$$F(\mathbf{E}) = \frac{1}{2}\iiint_V \left[\frac{1}{\mu_r}(\nabla \times \mathbf{E}) \cdot \overline{\mathbf{L}}^{-1} \cdot (\nabla \times \mathbf{E}) - k_0^2 \epsilon_r \mathbf{E} \cdot \overline{\mathbf{L}} \cdot \mathbf{E}\right] dV$$

$$+ \frac{1}{2}\iint_{S_3} \gamma_e \left[\mathbf{E} \cdot \mathbf{E} - (\hat{n} \cdot \mathbf{E})(\hat{n} \cdot \mathbf{E})\right] dS$$

$$+ jk_0\eta_0 \iiint_V \mathbf{E} \cdot \mathbf{J} dV \tag{12.7}$$

For scattering problems, the source current $\mathbf{J}$ is typically zero, and the excitation is a known incident plane wave. To model an incident plane wave, the known incident field (denoted $\mathbf{E}^i$) is separated from the remainder of the electric field, which is denoted $\mathbf{E}^s$. For radiation problems, $\mathbf{E}^i = 0$ and $\mathbf{E}^s$ is used to denote the radiated field. Thus, $\mathbf{E} = \mathbf{E}^i + \mathbf{E}^s$ is substituted into (12.7), and terms that do not depend on $\mathbf{E}^s$ are dropped. This yields

$$F(\mathbf{E}^s) = \frac{1}{2}\iiint_V \left[\frac{1}{\mu_r}(\nabla \times \mathbf{E}^s) \cdot \overline{\mathbf{L}}^{-1} \cdot (\nabla \times \mathbf{E}^s) - k_0^2 \epsilon_r \mathbf{E}^s \cdot \overline{\mathbf{L}} \cdot \mathbf{E}^s\right] dV$$

$$+ \iiint_{V^{sc}} \left[\frac{1}{\mu_r}(\nabla \times \mathbf{E}^s) \cdot \overline{\mathbf{L}}^{-1} \cdot (\nabla \times \mathbf{E}^i) - k_0^2 \epsilon_r \mathbf{E}^s \cdot \overline{\mathbf{L}} \cdot \mathbf{E}^i\right] dV$$

$$+ \frac{1}{2}\iint_{S_3} \gamma_e \left[\mathbf{E}^s \cdot \mathbf{E}^s - (\hat{n} \cdot \mathbf{E}^s)(\hat{n} \cdot \mathbf{E}^s)\right] dS$$

$$+ \iint_{S_3} \gamma_e \left[\mathbf{E}^s \cdot \mathbf{E}^i - (\hat{n} \cdot \mathbf{E}^s)(\hat{n} \cdot \mathbf{E}^i)\right] dS$$

$$- \iint_{S^{sc}} \mathbf{E}^s \cdot (\hat{n} \times \nabla \times \mathbf{E}^i) dS$$

$$+ jk_0\eta_0 \iiint_V \mathbf{E}^s \cdot \mathbf{J} dV \tag{12.8}$$

where V^{sc} is a penetrable region with $\mu_r \neq 1$ and/or $\epsilon_r \neq 1$, and S^{sc} is its boundary, as shown in Figure 12.1. Note that $\hat{n}$ on S^{sc} points inside, and in the derivation of (12.8), the fact that $\nabla \times \nabla \times \mathbf{E}^i - k_0^2 \mathbf{E}^i = 0$ is used. To take advantage of the rotational symmetry of the problem, the fields and source current densities are expanded in Fourier modes as

$$\mathbf{E}^{i,s} = \sum_{m=-\infty}^{\infty} \left[\mathbf{E}_{t,m}^{i,s}(\rho,z) + \hat{\phi} E_{\phi,m}^{i,s}(\rho,z)\right] e^{jm\phi}$$

$$\mathbf{J} = \sum_{m=-\infty}^{\infty} \left[\mathbf{J}_{t,m}(\rho,z) + \hat{\phi} J_{\phi,m}(\rho,z)\right] e^{jm\phi} \tag{12.9}$$

The expansions in (12.9) are substituted into (12.8), and the integrations with respect to ϕ are then performed, yielding

$$F(\mathbf{E}^s) = 2\pi \sum_{m=-\infty}^{\infty} F_m(\mathbf{E}^s) \qquad (12.10)$$

where

$$\begin{aligned}
F_m(\mathbf{E}^s) = \frac{1}{2} \iint_\Omega &\left\{ \frac{\rho}{\mu_r} \left[\frac{1}{L_\phi} (\nabla_t \times \mathbf{E}^s_{t,-m}) \cdot (\nabla_t \times \mathbf{E}^s_{t,m}) \right.\right. \\
&+ \frac{1}{\rho^2} \left[\nabla_t(\rho E^s_{\phi,-m}) + jm\mathbf{E}^s_{t,-m} \right] \cdot \tilde{\overline{\mathbf{L}}}_t^{-1} \cdot \left[\nabla_t(\rho E^s_{\phi,m}) - jm\mathbf{E}^s_{t,m} \right] \right] \\
&\left. - k_0^2 \epsilon_r \rho \left[\mathbf{E}^s_{t,-m} \cdot \overline{\mathbf{L}}_t \cdot \mathbf{E}^s_{t,m} + L_\phi E^s_{\phi,-m} E^s_{\phi,m} \right] \right\} d\Omega \\
+ \iint_{\Omega^{sc}} &\left\{ \frac{\rho}{\mu_r} \left[(\nabla_t \times \mathbf{E}^s_{t,-m}) - \frac{\hat{\phi}}{\rho} \times \left[\nabla_t(\rho E^s_{\phi,-m}) + jm\mathbf{E}^s_{t,-m} \right] \right] \right. \\
&\left. \cdot \overline{\mathbf{L}}^{-1} \cdot (\nabla \times \mathbf{E}^i_m) - k_0^2 \epsilon_r \rho \left[\mathbf{E}^s_{t,-m} \cdot \overline{\mathbf{L}}_t \cdot \mathbf{E}^i_{t,m} + L_\phi E^s_{\phi,-m} E^i_{\phi,m} \right] \right\} d\Omega \\
+ \frac{1}{2} \int_{C_3} &\gamma_e \rho \left[\mathbf{E}^s_{t,-m} \cdot \mathbf{E}^s_{t,m} - (\hat{n} \cdot \mathbf{E}^s_{t,-m})(\hat{n} \cdot \mathbf{E}^s_{t,m}) + E^s_{\phi,-m} E^s_{\phi,m} \right] d\ell \\
+ \int_{C_3} &\gamma_e \rho \left[\mathbf{E}^s_{t,-m} \cdot \mathbf{E}^i_{t,m} - (\hat{n} \cdot \mathbf{E}^s_{t,-m})(\hat{n} \cdot \mathbf{E}^i_{t,m}) + E^s_{\phi,-m} E^i_{\phi,m} \right] d\ell \\
- \int_{C^{sc}} &\rho \left[\mathbf{E}^s_{t,-m} + \hat{\phi} E^s_{\phi,-m} \right] \cdot \left[\hat{n} \times (\nabla \times \mathbf{E}^i_m) \right] d\ell \\
+ jk_0\eta_0 \iint_\Omega &\rho \left[\mathbf{E}^s_{t,-m} \cdot \mathbf{J}_{t,m} + E^s_{\phi,-m} J_{\phi,m} \right] d\Omega \qquad (12.11)
\end{aligned}$$

in which Ω is the 2D slice of V, C_3 is the 2D slice of S_3, and so forth, and

$$\nabla_t = \hat{\rho} \frac{\partial}{\partial \rho} + \hat{z} \frac{\partial}{\partial z} \qquad (12.12)$$

$$\overline{\mathbf{L}}_t = \hat{\rho}\hat{\rho} L_\rho + \hat{z}\hat{z} L_z \qquad (12.13)$$

$$\tilde{\overline{\mathbf{L}}}_t = \hat{\rho}\hat{\rho} L_z + \hat{z}\hat{z} L_\rho \qquad (12.14)$$

The problem defined by (12.3)–(12.6) is now solved by seeking the stationary point of (12.10) subject to (12.4). To accomplish this task, an FEM expansion is substituted into (12.10). An appropriate FEM expansion must consider the conditions the field must satisfy at points along the z-axis because an improper expansion results in nonintegrable singularities in the functional. All field values must be continuous when the z-axis is approached along any $\phi =$ constant line. The conditions along the z-axis that ensure this are

$$E_{\rho,0} = E_{\phi,0} = (\nabla \times \mathbf{E})_{\rho,0} = (\nabla \times \mathbf{E})_{\phi,0} = 0 \quad (12.15)$$

for $m = 0$,

$$\begin{aligned} E_{\rho,\pm 1} &= \mp j E_{\phi,\pm 1} \\ (\nabla \times \mathbf{E})_{\rho,\pm 1} &= \mp j (\nabla \times \mathbf{E})_{\phi,\pm 1} \\ E_{z,\pm 1} &= (\nabla \times \mathbf{E})_{z,\pm 1} = 0 \end{aligned} \quad (12.16)$$

for $m = \pm 1$, and

$$E_{\rho,m} = E_{\phi,m} = E_{z,m} = (\nabla \times \mathbf{E})_{\rho,m} = (\nabla \times \mathbf{E})_{\phi,m} = (\nabla \times \mathbf{E})_{z,m} = 0 \quad (12.17)$$

for $|m| > 1$. In each element of the FEM mesh, expansions which, together with a homogeneous Dirichlet condition on E_ϕ for $m \neq \pm 1$, satisfy these z-axis conditions are [23]

$$E^s_{\phi,0} = \sum_{i=1}^{3} e^e_{\phi,i} N^e_i$$

$$\mathbf{E}^s_{t,0} = \sum_{i=1}^{3} e^e_{t,i} \mathbf{N}^e_i \quad (12.18)$$

for $m = 0$,

$$E^s_{\phi,\pm 1} = \sum_{i=1}^{3} e^e_{\phi,i} N^e_i$$

$$\mathbf{E}^s_{t,\pm 1} = \sum_{i=1}^{3} \left[\mp j \hat{\rho} e^e_{\phi,i} N^e_i + e^e_{t,i} \rho \mathbf{N}^e_i \right] \quad (12.19)$$

for $m = \pm 1$, and

$$E^s_{\phi,m} = \sum_{i=1}^{3} e^e_{\phi,i} N^e_i$$

$$\mathbf{E}^s_{t,m} = \sum_{i=1}^{3} e^e_{t,i} \rho \mathbf{N}^e_i \quad (12.20)$$

for $|m| > 1$, where N^e_i is a standard 2D nodal-element basis function, and $\mathbf{N}^e_i$ is a standard 2D edge-element basis function [22]. When the FEM expansions are substituted into (12.10), a system of the form

$$F(\mathbf{E}^s) = \sum_{m=-\infty}^{\infty} \left(\frac{1}{2} \begin{Bmatrix} e_t^{-m} \\ e_\phi^{-m} \end{Bmatrix}^T \begin{bmatrix} A_{tt}^m & A_{t\phi}^m \\ A_{\phi t}^m & A_{\phi\phi}^m \end{bmatrix} \begin{Bmatrix} e_t^m \\ e_\phi^m \end{Bmatrix} - \begin{Bmatrix} e_t^{-m} \\ e_\phi^{-m} \end{Bmatrix}^T \begin{Bmatrix} B_t^m \\ B_\phi^m \end{Bmatrix} \right)$$
(12.21)

results, where $[A_{tt}^m]$, $[A_{t\phi}^m]$, $[A_{\phi t}^m]$, $[A_{\phi\phi}^m]$, $\{B_t^m\}$, and $\{B_\phi^m\}$ are all assembled from the elemental matrices and vectors, which are obtained directly from the substitution of (12.18)–(12.20) into (12.10). Note that $[A_{tt}^m]$ and $[A_{\phi\phi}^m]$ are symmetric and $[A_{t\phi}^m] = [A_{\phi t}^m]^T$ so that the entire matrix in (12.21) is symmetric in addition to being sparse. The stationary point of the functional is found by differentiating (12.21) with respect to

$$\begin{Bmatrix} e_t^{-m} \\ e_\phi^{-m} \end{Bmatrix}^T$$

and setting the result to zero, giving systems of the form

$$\begin{bmatrix} A_{tt}^m & A_{t\phi}^m \\ A_{\phi t}^m & A_{\phi\phi}^m \end{bmatrix} \begin{Bmatrix} e_t^m \\ e_\phi^m \end{Bmatrix} = \begin{Bmatrix} B_t^m \\ B_\phi^m \end{Bmatrix}; \qquad m = 0, \pm 1, \pm 2, \ldots \qquad (12.22)$$

For plane wave incidence, it can be shown that

$$\mathbf{E}_{t,m}^i = \begin{cases} \mathbf{E}_{t,-m}^i & \text{V-pol incidence} \\ -\mathbf{E}_{t,-m}^i & \text{H-pol incidence} \end{cases}$$

$$E_{\phi,m}^i = \begin{cases} -E_{\phi,-m}^i & \text{V-pol incidence} \\ E_{\phi,-m}^i & \text{H-pol incidence} \end{cases}$$

$$(\nabla \times \mathbf{E}^i)_{t,m} = \begin{cases} -(\nabla \times \mathbf{E}^i)_{t,-m} & \text{V-pol incidence} \\ (\nabla \times \mathbf{E}^i)_{t,-m} & \text{H-pol incidence} \end{cases}$$

$$(\nabla \times \mathbf{E}^i)_{\phi,m} = \begin{cases} (\nabla \times \mathbf{E}^i)_{\phi,-m} & \text{V-pol incidence} \\ -(\nabla \times \mathbf{E}^i)_{\phi,-m} & \text{H-pol incidence} \end{cases} \qquad (12.23)$$

where V-pol and H-pol represent vertically polarized and horizontally polarized, respectively. Using (12.23) and symmetry properties of the FEM equations, it follows that

$$\{e_t^m\} = \begin{cases} \{e_t^{-m}\} & \text{V-pol incidence} \\ -\{e_t^{-m}\} & \text{H-pol incidence} \end{cases}$$

$$\{e_\phi^m\} = \begin{cases} -\{e_\phi^{-m}\} & \text{V-pol incidence} \\ \{e_\phi^{-m}\} & \text{H-pol incidence} \end{cases} \qquad (12.24)$$

Finite Element Analysis of Complex Axisymmetric Problems

for all $m \neq 0$. When source currents are present in the problem, the Fourier series can be separated into even and odd functions of m such that a relation similar to (12.24) can be derived. Thus, the solution of the FEM equations is needed for positive numbered modes only.

12.2.3 Solution of the Equations

The solution of a scattering or radiation problem by the FEM involves the assembly and solution of the system of equations of the form of (12.22). As mentioned in Section 12.2.2, the matrix of (12.22) is sparse and symmetric. Also, in (12.22), the FEM matrix $[A^m]$ is independent of the excitation of the system. Thus, if the solution is computed by matrix decomposition techniques, the decomposition of the matrix is computed only once, even if the solution corresponding to multiple excitations (for example, multiple incident angles or polarizations in a scattering problem) is required. Finally, when the unknowns are ordered according to the reverse Cuthill-McKee (RCM) ordering, the FEM matrix is highly banded. The RCM algorithm is discussed in [24], and a typical result for a case with 5,000 unknowns is shown in Figure 12.2.

The first step in solving the FEM equations is to use the RCM ordering to assemble the FEM matrix for a given mode number (m). In each row of the matrix, the first nonzero element through the diagonal element are stored in an array. This storage scheme facilitates the second step of the solution, the efficient decomposition of the matrix by a band solver. The band solver employed computes the LDL^T decomposition of the matrix using Crout decomposition techniques of $O(NB^2)$ computational complexity, where N is the number of unknowns and B is the matrix half-bandwidth [22]. After factoring the matrix, the third solution step is to assemble an excitation vector; then, the fourth step is the solution of triangular systems using forward and back substitutions of $O(NB)$ computational complexity. The contribution of mode m to the far-field result is then computed, and the next step is to repeat the assembly of an excitation vector, the forward and back substitutions, and the far-field calculation for each electromagnetic excitation. The final step is the repetition of the whole process for each mode number required. For scattering problems, a rule of thumb for the number of modes required is $M_{max} = k_0 \rho_{max} \sin\theta + 6$ [1], where ρ_{max} is the maximum radius of the scatterer. This rule of thumb is valid for $k_0 \rho_{max} \sin\theta > 3$. For radiation problems, the number of modes is set by the accuracy with which the Fourier series is required to represent the original source current density.

It is noteworthy that a minimum degree ordering of the unknowns, which seeks to minimize the number of nonzeros in the factored matrix, may offer improved efficiency in the matrix solution. For example, some preliminary experiments show that the matrix storage can be reduced to about half of the current banded storage. Further, the incorporation of partial pivoting into the matrix decomposition step seriously disrupts the computation by spoiling the symmetry of the matrix and increasing

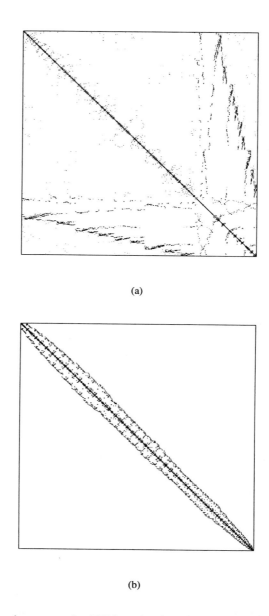

Figure 12.2 Sparsity pattern of an FEM matrix with 5,000 unknowns: (a) original ordering and (b) RCM ordering [13].

the bandwidth of nonzeros. Although these FEM matrices do not satisfy the criteria for guaranteed numerical stability, experience with a large number of problems has yet to reveal a case of significant error. Should such a case be encountered, partial pivoting or an alternative matrix solution should be used.

12.2.4 Far-Field Calculations

The solution of the FEM equations yields the near-zone electric field in the solution domain. For scattering problems, the far-field radar cross-section (RCS) of the target is often of interest. Similarly, for radiation problems, the far-field radiation pattern is usually of interest. Thus, the electric field far from the geometry is computed from near-field values. Consider an infinitesimal dipole, located at the observation point, oriented in the $\hat{u}$ direction, and excited such that it produces a plane wave of unit amplitude at the coordinate origin. Denoting the far-zone electric field as $\mathbf{E}^s(\mathbf{r})$ and the field radiated by the dipole as $\mathbf{E}^d$, by reciprocity

$$\mathbf{E}^s(\mathbf{r}) \cdot \hat{u} = \frac{e^{-jk_0 r}}{4\pi r} \iint_{S'} \left[\mathbf{E}^d \cdot (\hat{n} \times \nabla \times \mathbf{E}^s) + (\nabla \times \mathbf{E}^d) \cdot (\hat{n} \times \mathbf{E}^s) \right] dS' \tag{12.25}$$

where S' is any rotationally symmetric surface that encloses the entire non-free-space region of the geometry. The fields in (12.25) are expanded in Fourier modes as in (12.9), and the ϕ integration is performed, giving

$$\mathbf{E}^s(\mathbf{r}) \cdot \hat{u} = \frac{e^{-jk_0 r}}{2r} \sum_{m=-\infty}^{\infty} j^m e^{jm\phi^d} \int_{C'} \rho' \left[\mathbf{E}^d_{-m} \cdot (\hat{n} \times \nabla \times \mathbf{E}^s_m) \right. $$
$$\left. + (\nabla \times \mathbf{E}^d_{-m}) \cdot (\hat{n} \times \mathbf{E}^s_m) \right] d\ell' \tag{12.26}$$

where θ^d and ϕ^d denote the observation direction. The components of $\mathbf{E}^s(\mathbf{r})$ that are of interest are the vertically polarized component $E^s_\theta(\mathbf{r})$ and the horizontally polarized component $E^s_\phi(\mathbf{r})$. For $\hat{u} = \hat{\theta}$ and $\hat{u} = \hat{\phi}$, expressions for $\mathbf{E}^d_{-m}$ and $(\nabla \times \mathbf{E}^d_{-m})$ are found and substituted into (12.26), giving

$$E^s_\theta(\mathbf{r}) = \frac{e^{-jk_0 r}}{2r} \sum_{m=-\infty}^{\infty} j^m e^{jm\phi^d} \int_{C'} \rho' e^{jk_0 z' \cos\theta^d} \left\{ (\hat{n} \times \nabla \times \mathbf{E}^s_m) \right.$$
$$\cdot \left[-\hat{\rho} \cos\theta^d j J'_m(k_0\rho' \sin\theta^d) - \hat{\phi} \cos\theta^d \frac{m J_m(k_0\rho' \sin\theta^d)}{k_0\rho' \sin\theta^d} \right.$$
$$\left. - \hat{z} \sin\theta^d J_m(k_0\rho' \sin\theta^d) \right] + k_0(\hat{n} \times \mathbf{E}^s_m) \cdot \left[\hat{\rho} j \frac{m J_m(k_0\rho' \sin\theta^d)}{k_0\rho' \sin\theta^d} \right.$$
$$\left. \left. + \hat{\phi} J'_m(k_0\rho' \sin\theta^d) \right] \right\} d\ell' \tag{12.27}$$

and

$$E_\phi^s(\mathbf{r}) = \frac{e^{-jk_0 r}}{2r} \sum_{m=-\infty}^{\infty} j^m e^{jm\phi^d} \int_{C'} \rho' e^{jk_0 z' \cos\theta^d} \left\{ (\hat{n} \times \nabla \times \mathbf{E}_m^s) \right.$$
$$\cdot \left[\hat{\rho} \frac{m J_m(k_0 \rho' \sin\theta^d)}{k_0 \rho' \sin\theta^d} - \hat{\phi} j J_m'(k_0 \rho' \sin\theta^d) \right] + k_0 (\hat{n} \times \mathbf{E}_m^s)$$
$$\cdot \left[-\hat{\rho} \cos\theta^d J_m'(k_0 \rho' \sin\theta^d) + \hat{\phi} \cos\theta^d j \frac{m J_m(k_0 \rho' \sin\theta^d)}{k_0 \rho' \sin\theta^d} \right.$$
$$\left. \left. + \hat{z} \sin\theta^d j J_m(k_0 \rho' \sin\theta^d) \right] \right\} d\ell' \quad (12.28)$$

where $J_m(x)$ denotes the Bessel function of order m and $J_m'(x)$ denotes the derivative of $J_m(x)$ with respect to its argument. For scattering problems, the RCS is found from the definition

$$\sigma = \lim_{r \to \infty} 4\pi r^2 \frac{|\mathbf{E}^s(\mathbf{r})|^2}{|\mathbf{E}^i(\mathbf{r})|^2} \quad (12.29)$$

The dimension of the RCS is inverse length squared, and it is commonly expressed in decibels per square meter, abbreviated dBsm.

12.3 CYLINDRICAL PML

Both exact and approximate FEM mesh truncations for axisymmetric problems are found in the literature. The unimoment method uses the CAP formulation with a spherical harmonic expansion for mesh truncation [11]. Although the spherical harmonic expansion is an exact mesh truncation, it requires a spherical mesh boundary. Hence, it is not efficient for arbitrarily shaped geometries. ABCs derived from the Wilcox expansion theorem are also used [9, 12]. These ABCs approximate the radiation condition, requiring the outer mesh boundary to be placed far from the geometry of interest. While some of these ABCs also require a spherical mesh boundary, others can be applied at a cylindrical boundary. However, the ABCs that can be applied at a cylindrical boundary require an increase in the FEM matrix bandwidth for accurate implementation [9].

A more recent method of mesh truncation is to line the inside of the FEM mesh with a lossy PML. A PML is an artificial medium designed such that, first, waves propagating through the PML are attenuated and, second, the reflection coefficient at the air-to-PML interface is zero for all incidence angles and all frequencies. A PML that satisfies both of these conditions is available in cylindrical coordinates [20, 21, 25]. Hence, the mesh is truncated with a cylindrical boundary, which is much more efficient for arbitrarily shaped geometries than is a spherical boundary.

Further, because any reflection from the outer mesh boundary is attenuated by the PML, any convenient boundary condition can be applied there. There is no need for an increase in the bandwidth of the FEM matrix, and the air-to-PML interface can be placed very close to the scatterer or the antenna. Although PML is an approximate mesh truncation, it can be made very accurate because reflection errors can be systematically controlled. For a more accurate mesh truncation, one needs only increase the thickness of the PML region.

12.3.1 Parameter Definitions

The diagonal tensor $\overline{L}$ contains the properties of the PML. In order to effectively eliminate artificial reflections, the air-to-PML interface must be reflectionless in cylindrical coordinates, and waves that propagate through the PML must be attenuated. These conditions are satisfied when the elements of $\overline{L}$ are given by [20, 21, 25]

$$L_\rho = \frac{S_z \tilde{\rho}}{S_\rho \rho} \qquad L_\phi = \frac{S_z S_\rho \rho}{\tilde{\rho}} \qquad L_z = \frac{S_\rho \tilde{\rho}}{S_z \rho} \qquad (12.30)$$

in which

$$S_\rho = S_\rho(\rho) = \begin{cases} 1 & 0 \le \rho \le \rho_m \\ 1 - j\alpha \left(\dfrac{\rho - \rho_m}{t_\rho} \right)^2 & \rho > \rho_m \end{cases} \qquad (12.31)$$

$$S_z = S_z(z) = \begin{cases} 1 - j\alpha \left(\dfrac{z_{ml} - z}{t_{zl}} \right)^2 & z < z_{ml} \\ 1 & z_{ml} \le z \le z_{mu} \\ 1 - j\alpha \left(\dfrac{z - z_{mu}}{t_{zu}} \right)^2 & z > z_{mu} \end{cases} \qquad (12.32)$$

$$\tilde{\rho} = \begin{cases} \rho & 0 \le \rho \le \rho_m \\ \rho - j\alpha \dfrac{(\rho - \rho_m)^3}{3 t_\rho^2} & \rho > \rho_m \end{cases} \qquad (12.33)$$

where t_ρ, t_{zl}, and t_{zu} are the PML thicknesses; $\rho = \rho_m$, $z = z_{ml}$, and $z = z_{mu}$ are the locations of the air-to-PML interfaces (see Figure 12.1), and α is a real parameter to be selected. While the air-to-PML interface is reflectionless for all incidence angles and all frequencies in continuous space, some spurious reflection results from the discretization of the FEM mesh [26]. Smaller values of α lower the contrast at the air-to-PML interface, thus reducing this spurious reflection, while larger values of α increase the attenuation of waves propagating through the PML. Thus, there is a trade-off in the selection of the PML loss parameter α.

Table 12.1 Memory and CPU Requirements (the values listed are those required to compute monostatic scattering for two polarizations and 181 incidence angles on a 44-Mflop DEC Alpha workstation).

Target	Number of unknowns	Matrix half-bandwidth	Memory (Mbytes)	CPU time (min)
Ogive	18,429	197	13	25.5
Double ogive	15,348	201	11	21.3
Conesphere with gap	42,848	379	33	135.1

12.3.2 Systematic Error Reduction

Reflection errors from the PML are systematically reduced by keeping the thickness of the PML fixed and finding the value of α that produces the minimum reflection error. If this reflection error is too large for a given application, the thickness of the PML is increased and the process is repeated.

As an example, the selection of α is investigated by exciting a current loop in a free-space FEM mesh and examining the relative error in the fields computed by the FEM code, and by computing the far field bistatic RCS of a conducting sphere of radius 2λ and comparing to the exact Mie series result. In both cases, it is found that for a mesh length of $\lambda/20$ and a PML thickness of 0.25λ, the optimum value of α is around $5.5 \sim 6.0$.

12.4 NUMERICAL RESULTS

A number of numerical results are presented to show the validity and capability of the mixed-element FEM technique. Results from scattering problems are presented first, and then results from radiation problems are considered. Unless otherwise stated, mesh length for each example is $\lambda/20$, and the air-to-PML interface is placed 0.25λ from the geometry.

12.4.1 Scattering

A number of Electromagnetic Code Consortium (EMCC) benchmark targets are considered, and the FEM results are compared to measurements first published in [27], where detailed descriptions of the targets are also found. To establish a reference on the capability of the code, information is presented in Table 12.1 about the memory and CPU time required to generate the monostatic scattering results for two polarizations and 181 incidence angles on a DEC Alpha workstation with an average throughput of 44 million floating point operations per second (44 Mflop). For the

EMCC benchmark targets, the mesh is truncated with a PML 0.5λ thick with $\alpha = 10$.

The first EMCC benchmark target is a metallic ogive. The ogive has a length of 25.4 cm, maximum diameter of 5.08 cm, and half-angle of 22.62° at each tip. The monostatic RCS of the ogive at 9 GHz is computed and compared to measurements in Figure 12.3. The agreement is generally good; however, some error is observed around 15° elevation for both polarizations. This error must be caused by errors in measurement as the target is symmetric about the 0° elevation plane.

The second EMCC benchmark target is a metallic double ogive. The double ogive is formed by joining two different half-ogives. The top piece has a half-length of 12.7 cm, maximum diameter of 5.08 cm, and half-angle of 22.62° at the tip; the bottom has a half-length of 6.35 cm, maximum diameter of 5.08 cm, and half-angle of 46.4° at the tip. The computed monostatic RCS of the double ogive at 9 GHz is compared to measurements in Figure 12.4, and except for some discrepancy near the bottom of the dynamic range, the agreement is generally good.

The third EMCC benchmark is a metallic conesphere. The sphere has a radius of 7.485 cm; the cone tip has a half-angle of 7°; the cone is 60.505 cm long, and the cone is tangent to the sphere at the junction. There is a 0.635-cm-wide × 0.635-cm-deep gap located at the junction between the cone and the sphere. The monostatic RCS at 9 GHz is computed, and a comparison to measured values is shown in Figure 12.5. In computing the result for this target, portions of the mesh between the conesphere and the PML interface are coarsened to a mesh length of $\lambda/10$. This reduces the number of unknowns in the problem, which becomes large because the target is over 20λ long.

As a final example, consider the electromagnetic scattering from a spherical dielectric Luneburg lens [28], which is characterized by a permittivity profile of

$$\epsilon_r(r) = 2 - \left(\frac{r}{a}\right)^2 \qquad (0 \leq r \leq a) \tag{12.34}$$

where a is the radius of the lens. This lens is commonly analyzed using Fermat's principle from geometrical optics (GO). According to the GO, when a beam of parallel rays is incident upon this lens, it is focused to a point on the other side of the lens. By reciprocity, when a point source is placed on the surface of the lens, the lens transforms the rays into parallel rays. Although the GO ray picture gives some understanding of the lens, field pictures from the FEM can improve the understanding, especially in the vicinity of the focal point, where GO fails to provide an accurate picture [29]. The bistatic scattering patterns for Luneburg lenses with diameters of 4λ, 7λ, and 10λ where λ denotes the free-space electromagnetic wavelength are shown in Figure 12.6. The scattering patterns are normalized by πa^2 where a is the radius of the lens, and because the patterns are symmetric, one half of the E-plane and one half of the H-plane pattern are shown on the same polar plot. Note the low level of the backscatter and the similarity between the E-plane and the H-plane patterns when the size of the lens increases, indicating the diminishing role of the

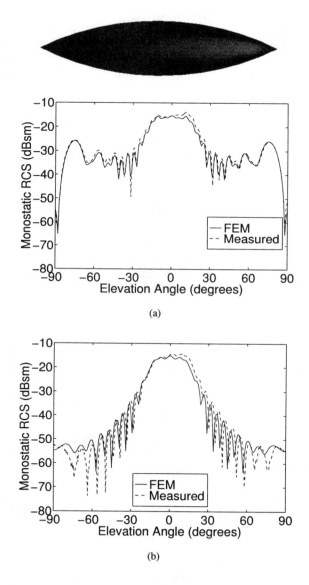

Figure 12.3 RCS of a metallic ogive at 9 GHz: (a) VV-pol; (b) HH-pol [13]. The ogiv has a height of 25.4 cm (7.63λ), maximum diameter of 5.08 cm (1.53λ), and half-angle c 22.62° at each tip.

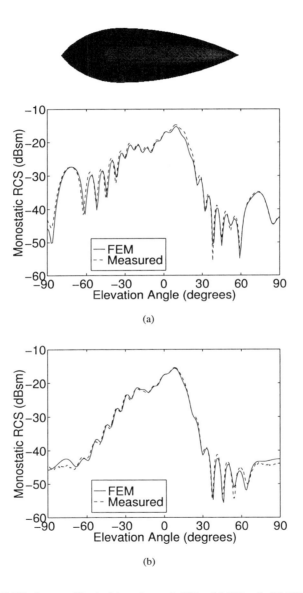

Figure 12.4 RCS of a metallic double ogive at 9 GHz: (a) VV-pol; (b) HH-pol [13]. The top piece of the double ogive has a half-height of 12.7 cm (3.81λ), maximum diameter of 5.08 cm (1.53λ), and half-angle of 22.62° at the tip; the bottom has a half-height of 6.35 cm (1.91λ), maximum diameter of 5.08 cm (1.53λ), and half-angle of 46.4° at the tip.

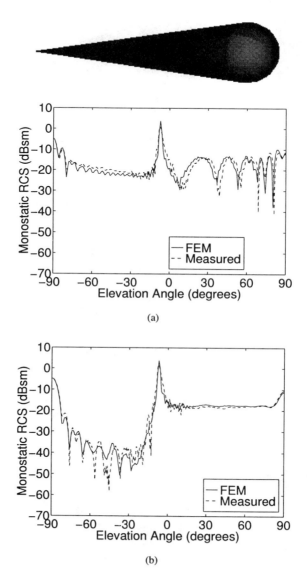

Figure 12.5 RCS of a metallic conesphere with a gap at 9 GHz: (a) VV-pol; (b) HH-pol [13]. The sphere has a radius of 7.485 cm (2.247λ), and the cone has a half-angle of $7°$ at the tip and a length of 60.505 cm (18.164λ). At the junction between the cone and the sphere, there is a 0.635-cm-wide × 0.635-cm-deep ($0.19\lambda \times 0.19\lambda$) gap.

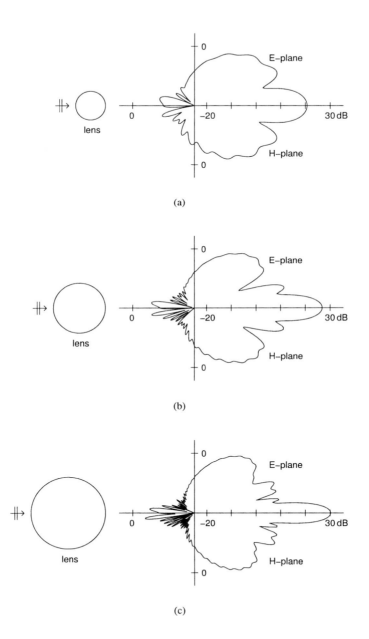

Figure 12.6 Normalized bistatic scattering from a Luneburg lens: (a) 4λ diameter; (b) 7λ diameter; (c) 10λ diameter [29].

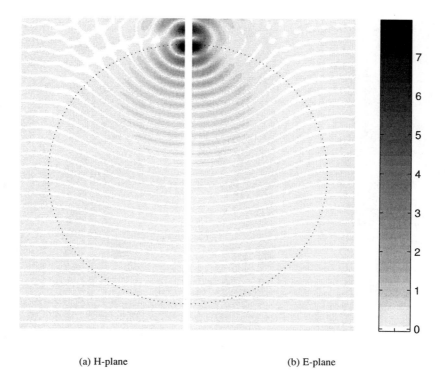

(a) H-plane (b) E-plane

Figure 12.7 Snapshot of the electric field near a 10λ-diameter Luneburg lens excited by an incident plane wave from −90° elevation [29].

field polarization. The field distribution for the 10λ case is shown in Figure 12.7, where the absolute value of the real part of the electric field is shown.

12.4.2 Radiation

To show the validity and capability of the method for radiation problems, radiation patterns from a corrugated horn antenna are computed. Waveguides and horn antennas with corrugated boundaries are extensively investigated in [30] as antennas that can radiate circularly polarized waves over a wide beamwidth and wide bandwidth. As part of the investigation, measurements on several corrugated horn antennas are presented. A diagram showing one such antenna is shown in Figure 12.8. The antenna is constructed by bolting together metal washers with thicknesses alternating between 0.07938 and 0.3175 cm. In the section of the waveguide in front of the

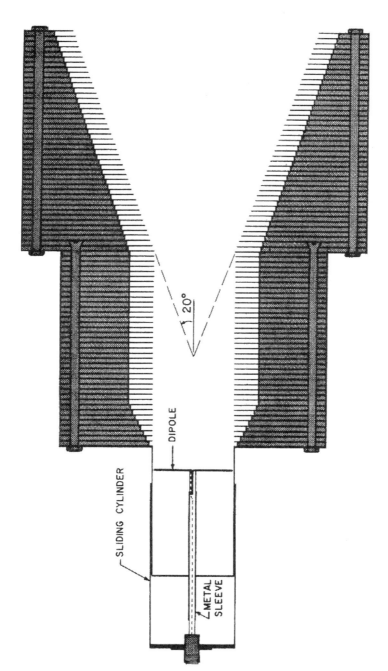

Figure 12.8 Diagram of the corrugated horn antenna. Reproduced from [30].

Figure 12.9 Photo of the corrugated horn antenna. Reproduced from [30].

20° flare, the washers have inner radii of 3 and 4.8 cm, respectively. A photograph of the antenna is shown in Figure 12.9. Measurements of the radiation pattern of this antenna are given in [30] at 4.5, 5.0, 5.2, 5.5, 6.0, and 6.5 GHz. The radiation pattern is computed by the FEM at these same frequencies, and some of the FEM and measured results are shown in Figures 12.10 through 12.12. The agreement between the FEM results and the measured results is excellent. At power levels above -30 dB, note the agreement between the FEM and measurement in predicting slight offsets in the E-plane and H-plane patterns with respect to each other. The similar

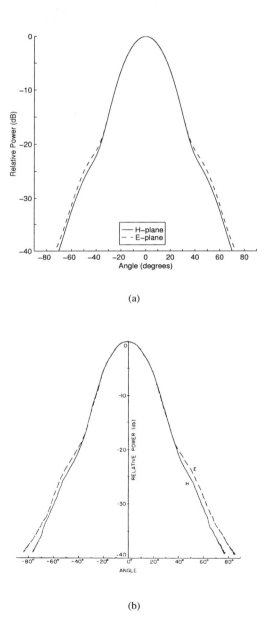

Figure 12.10 Radiation pattern of the corrugated horn antenna at 4.5 GHz: (a) FEM [14]; (b) measured results reproduced from [30].

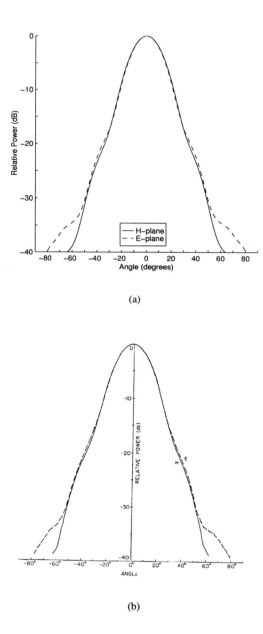

Figure 12.11 Radiation pattern of the corrugated horn antenna at 5.2 GHz: (a) FEM [14]; (b) measured results reproduced from [30].

Finite Element Analysis of Complex Axisymmetric Problems 565

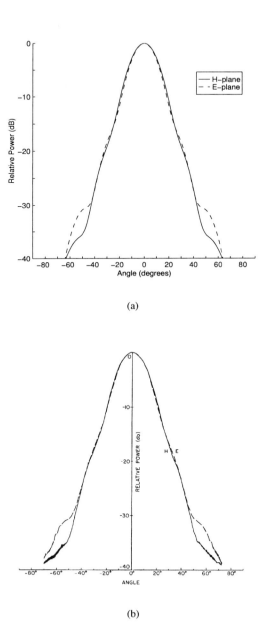

Figure 12.12 Radiation pattern of the corrugated horn antenna at 6.0 GHz: (a) FEM [14]; (b) measured results reproduced from [30].

E-plane and H-plane patterns indicate that, when properly fed, this antenna can be used to radiate circular polarization.

The Luneburg lens, discussed in connection with the scattering examples above, exhibits interesting and useful properties as a directional radiator. It has the property that the transmitting beam can be steered by moving the antenna feed. The radiation patterns of a Hertzian dipole placed on the surface of 4λ-, 7λ-, and 10λ-diameter Luneburg lenses are shown in Figure 12.13. Clearly, the radiation pattern becomes more directive as the size of the lens increases. Note that the side lobe level is at -15 dB. The field distribution in the 10λ-diameter lens is displayed in Figure 12.14. The figure shows how a spherical wave emitted by the dipole is converted into a wave with a locally planar phase front. However, the plane wave generated by the Luneburg lens is imperfect. This is due to the split field focus observed in the Luneburg lens scattering example.

12.5 BOR WITH APPENDAGES

In BOR scattering problems, the computational efficiency is achieved by exploiting the rotational symmetry of the problem. The rotational symmetry is the property that allows the problem to be solved using a 2D computational method. There is, however, a class of problems in which the geometry consists of a large BOR with one or more small appendages. An example of such a problem is shown in Figure 12.15. In the geometry shown in the figure, the rotational symmetry of the problem is broken by the presence of two small fins. Thus, a 3D computational method is required to rigorously compute the electromagnetic scattering from this target. A 3D computational method is much more expensive than the BOR methods discussed in the preceding sections. Because of the increased computational complexity of a 3D method, a hybrid method is developed which allows the scattering from small appendages to be approximately combined with the scattering from a large BOR [31–34]. The rotational symmetry can then be exploited in the computation of the scattering from the large BOR.

The basis of the hybrid method is the integral equation given by

$$\mathbf{E}(\mathbf{r}) = \mathbf{E}^i_{\text{BOR}}(\mathbf{r}) - jk_0\eta_0 \iint_{S_{\text{App}}} \overline{\mathbf{G}}_{\text{BOR}}(\mathbf{r}, \mathbf{r}') \cdot \mathbf{J}(\mathbf{r}')dS' \tag{12.35}$$

where $\mathbf{E}^i_{\text{BOR}}$ represents the incident field on the appendages in the presence of the BOR, S_{App} represents the surface of the appendages, $\mathbf{J}$ denotes the induced surface current density, and $\overline{\mathbf{G}}_{\text{BOR}}$ represents the dyadic Green's function in the presence of the BOR. The solution of (12.35) requires the discretization of S_{App} rather than the surface of the entire object, and thus is much less computationally intensive. The incident field $\mathbf{E}^i_{\text{BOR}}$ is calculated using the FEM. The Green's function $\overline{\mathbf{G}}_{\text{BOR}}$ is, in general, unknown, and is here approximated by the GO [35]. The scattered field is

Finite Element Analysis of Complex Axisymmetric Problems 567

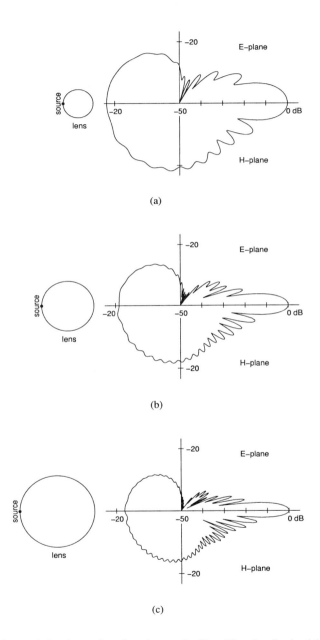

Figure 12.13 Radiation from a Luneburg lens excited by a Hertzian dipole: (a) 4λ diameter; (b) 7λ diameter; (c) 10λ diameter [29].

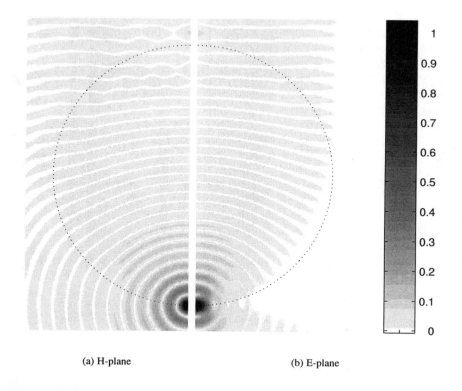

(a) H-plane (b) E-plane

Figure 12.14 Snapshot of the electric field near a 10λ-diameter Luneburg lens excited by a Hertzian dipole [29].

Figure 12.15 Example of a large BOR with small appendages.

computed using the reciprocity theorem, which gives

$$\mathbf{E}^s(\mathbf{r}) \cdot \hat{u} = -jk_0\eta_0 \frac{e^{-jk_0 r}}{4\pi r} \iint_{S_{\text{App}}} \mathbf{E}^r_{\text{BOR}}(\mathbf{r}') \cdot \mathbf{J}(\mathbf{r}')dS' \qquad (12.36)$$

where $\mathbf{E}_{\text{BOR}}^r$ is the field radiated by the dipole in the presence of the large BOR. Recall that this field can be computed by the FEM. In fact, when backscattering is being computed, $\mathbf{E}_{\text{BOR}}^r$ is the same as $\mathbf{E}_{\text{BOR}}^i$ that is used in (12.35).

The validity of the hybrid technique is tested by computing the scattering from a metallic cylinder with four wings. The scattering is compared with computations from the Fast Illinois Solver Code (FISC) [36], which is an MOM program that uses a multilevel fast multipole algorithm to speed up the matrix solution. The cylinder considered has a radius of 1.25λ and height of 5λ, where λ is the free-space electromagnetic wavelength. The attached wings are $1\lambda \times 0.5\lambda \times 0.025\lambda$. The monostatic RCS is shown in Figure 12.16, where the agreement between FISC and the hybrid method is rather good.

12.6 CONCLUSION

The FEM with the mixed edge-node formulation and cylindrical PML is efficient, accurate, and flexible in solving both BOR scattering problems and axisymmetric radiation problems. Rotational symmetry allows these 3D problems to be solved using a 2D numerical method. The mixed edge-node formulation expands the transverse field component using an edge-based vector basis and the angular field component using a node-based scalar basis. This expansion scheme allows easy treatment of boundary conditions on both material interfaces and conductor edges, and makes the method free of spurious solutions. In addition, the FEM has very flexible material handling capabilities, and an FEM mesh of triangular elements easily conforms to arbitrary geometries. The use of PML for FEM mesh truncation also permits an efficient computational domain for almost any problem geometry. Further, PML can be made very accurate so it can be placed near the problem geometry, and by changing the PML thickness and loss parameter, reflection errors can be systematically reduced.

This chapter shows the efficiency and accuracy of the mixed edge-node FEM with cylindrical PML for BOR scattering and axisymmetric radiation problems. This is done by first developing the FEM equations and showing that the resulting system of equations is sparse, symmetric, and efficiently solved using banded matrix techniques. It is further seen that PML mesh truncation requires no special handling during the matrix solution process. After developing the FEM equations, numerical examples are presented which show that the FEM agrees very well with benchmark measured results. Additional numerical results show that the method can handle realistic-size radar targets and antennas. The method can be hybridized with an MOM to approximately deal with BORs with small appendages.

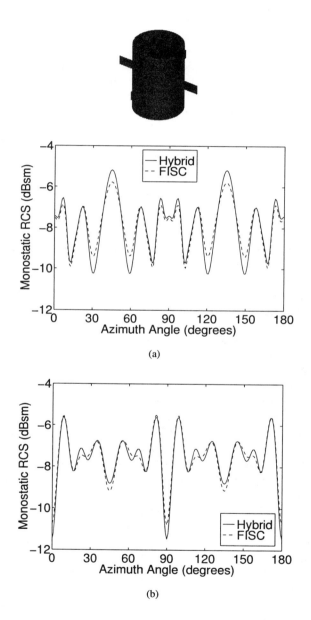

Figure 12.16 RCS of a metallic cylinder with four wings: (a) VV-pol; (b) HH-pol. The cylinder has a radius of 1.25λ and height of 5λ, and the wings are $1\lambda \times 0.5\lambda \times 0.025\lambda$.

REFERENCES

1. M. G. Andreasen, "Scattering from bodies of revolution," *IEEE Trans. Antennas Propagat.*, vol. 13, pp. 303–310, 1965.

2. J. R. Mautz and R. F. Harrington, "Electromagnetic scattering from a homogeneous material body of revolution," *Arch. Elektron. Uebertragungstech.*, vol. 33, pp. 71–80, 1979.

3. L. N. Medgyesi-Mitschang and J. M. Putnam, "Electromagnetic scattering from axially inhomogeneous bodies of revolution," *IEEE Trans. Antennas Propagat.*, vol. 32, pp. 797–806, 1984.

4. D. E. Baker, "Pattern prediction of broadband monopole antennas on finite ground planes using the BOR moment method," *Microw. J.*, vol. 31, pp. 153–164, 1988.

5. A. Berthon and R. P. Bills, "Integral equation analysis of radiating structures of revolution," *IEEE Trans. Antennas Propagat.*, vol. 37, pp. 159–170, 1989.

6. P. Steyn and D. B. Davidson, "A moment method formulation for electromagnetic radiation and scattering from composite bodies of revolution," in *9th Ann. Rev. Progress Appl. Computat. Electromag.*, Monterey, CA, pp. 64–71, 1993.

7. J. Liu, J. Wang, and Y. Gao, "Computation of E-field distribution of low gain antenna on conducting body of revolution," in *11th Ann. Rev. Progress Appl. Computat. Electromag.*, Monterey, CA, pp. 687–694, 1995.

8. F. L. Teixeira and J. R. Bergman, "B-spline basis functions for moment-method analysis of axisymmetric reflector antennas," *Microwave Opt. Tech. Lett.*, vol. 14, pp. 188–191, 1997.

9. R. Gordon and R. Mittra, "PDE techniques for solving the problem of radar scattering by a body of revolution," *Electromagn.*, vol. 10, pp. 163–174, 1990.

10. M. A. Morgan, S. K. Chang, and K. K. Mei, "Coupled azimuth potentials for electromagnetic field problems in inhomogeneous axially symmetric media," *IEEE Trans. Antennas Propagat.*, vol. 25, pp. 413–417, 1977.

11. M. A. Morgan and K. K. Mei, "Finite-element computation of scattering by inhomogeneous penetrable bodies of revolution," *IEEE Trans. Antennas Propagat.*, vol. 27, pp. 202–214, 1979.

12. A. Khebir, J. D'Angelo, and J. Joseph, "A new finite element formulation for RF scattering by complex bodies of revolution," *IEEE Trans. Antennas Propagat.*, vol. 41, pp. 534–541, 1993.

13. A. D. Greenwood and J. M. Jin, "A novel, efficient algorithm for scattering from a complex BOR using mixed finite elements and cylindrical PML," *IEEE Trans. Antennas Propagat.*, vol. 47, pp. 620–629, 1999.

14. A. D. Greenwood and J. M. Jin, "Finite element analysis of complex axisymmetric radiating structures," *IEEE Trans. Antennas Propagat.*, vol. 47, pp. 1260–1266, 1999.

15. K. K. Mei, "Unimoment method for solving antenna and scattering problems," *IEEE Trans. Antennas Propagat.*, vol. 22, pp. 760–766, 1974.

16. R. K. Gordon and R. Mittra, "Finite element analysis of axisymmetric radomes," *IEEE Trans. Antennas Propagat.*, vol. 41, pp. 975–981, 1993.

17. G. C. Chinn, L. W. Epp, and D. J. Hoppe, "A hybrid finite-element method for axisymmetric waveguide feed horns," *IEEE Trans. Antennas Propagat.*, vol. 44, pp. 280–285, 1996.

18. E. Richalot, M. F. Wong, V. Fouad-Hanna, and H. Baudrand, "Antenna analysis using edge elements and spherical mode expansion," in *27th European Microwave 97 Conference and Exhibition*, pp. 854–859, 1997.

19. A. D. Greenwood and J. M. Jin, "Computation of the RCS of a complex BOR using FEM with coupled azimuth potentials and PML," *Electromagn.*, vol. 19, pp. 147–170, 1999.

20. F. L. Teixeira and W. C. Chew, "Systematic derivation of anisotropic PML absorbing media in cylindrical and spherical coordinates," *IEEE Microwave Guided Wave Lett.*, vol. 7, pp. 371–373, 1997.

21. J. Maloney, M. Kesler, and G. Smith, "Generalization of PML to cylindrical geometries," in *13th Ann. Rev. Progress Appl. Computat. Electromag.*, Monterey, CA, pp. 900–908, 1997.

22. J. M. Jin, *The Finite Element Method in Electromagnetics*, New York: John Wiley & Sons, 1993.

23. M. F. Wong, M. Prak, and V. Fouad-Hanna, "Axisymmetric edge-based finite element formulation for bodies of revolution: Application to dielectric resonators," *IEEE MTT-S Digest*, pp. 285–288, 1995.

24. A. George and J. W. Liu, *Computer Solution of Large Sparse Positive Definite Systems*, Englewood Cliffs, NJ: Prentice-Hall, 1981.

25. W. C. Chew, J. M. Jin, and E. Michielssen, "Complex coordinate stretching as a generalized absorbing boundary condition," *Microwave Opt. Tech. Lett.*, vol. 15, pp. 363–369, 1997.

26. W. C. Chew and J. M. Jin, "Perfectly matched layers in the discretized space: an analysis and optimization," *Electromagn.*, vol. 16, pp. 325–340, 1996.

27. A. C. Woo, H. T. G. Wang, M. J. Schuh, and M. L. Sanders, "Benchmark radar targets for the validation of computational electromagnetics programs," *IEEE Antennas Propagat. Mag.*, vol. 35, pp. 84–89, 1993.

28. R. K. Luneburg, *The Mathematical Theory of Optics*, Providence, RI: Brown University Press, 1944.

29. A. D. Greenwood and J. M. Jin, "A field picture of wave propagation in inhomogeneous dielectric lenses," *IEEE Antennas Propagat. Mag.*, vol. 41, pp. 9–18, 1999.

30. M. J. Al-Hakkak and Y. T. Lo, "Circular waveguides and horns with anisotropic and corrugated boundaries," Tech. Rep. 73-3, Antenna Laboratory, Department of Electrical Engineering, University of Illinois, Urbana, IL, 1973.

31. J. M. Jin, S. S. Ni, and S. W. Lee, "Hybridization of SBR and FEM for scattering by large bodies with cracks and cavities," *IEEE Trans. Antennas Propagat.*, vol. 43, pp. 1130–1139, 1995.

32. A. D. Greenwood, S. S. Ni, and J. M. Jin, "Hybrid FEM/SBR method to compute the radiation pattern from a microstrip patch antenna in a complex geometry," *Microwave Opt. Tech. Lett.*, vol. 13, pp. 84–87, 1996.

33. J. M. Jin, F. Ling, S. T. Carolan, J. M. Song, W. C. Gibson, and W. C. Chew, "A hybrid SBR/MoM technique for analysis of scattering from small protrusions on a large conducting body," *IEEE Trans. Antennas Propagat.*, vol. 43, pp. 1130–1139, 1995.

34. A. D. Greenwood and J. M. Jin, "Hybrid MoM/SBR method to compute the scattering from a slot array antenna in a complex geometry," *Appl. Comput. Electromagn. Soc. J.*, vol. 13, pp. 43–51, 1998.

35. A. D. Greenwood and J. M. Jin, "A hybrid MoM/FEM technique for scattering from a complex BOR with appendages," *Appl. Comput. Electromagn. Soc. J.*, vol. 15, pp. 13–19, 2000.

36. J. M. Song, C. C. Lu, and W. C. Chew, "Multilevel fast multipole algorithm for electromagnetic scattering by large complex objects," *IEEE Trans. Antennas Propagat.*, vol. 45, pp. 1488–1493, 1997.

13
Hybridization in Computational Electromagnetics

Jian-Ming Jin and Jian Liu

13.1 INTRODUCTION

The word "hybridization" in computational electromagnetics refers to the procedure of combining two or more different numerical or asymptotic methods to expand their capabilities and enhance their efficiency in handling large-scale, complex electromagnetic problems. The resulting numerical tools are often referred to as *hybrid techniques*.

The need for hybridization arises from practical problems that are usually so large and so complex that no single numerical or asymptotic method can provide an accurate and efficient solution. Consider the problem of electromagnetic scattering from a realistic airplane as an example (Figure 13.1). The airplane is not only electrically very large, but also contains many small features that contribute significantly to its overall backscatter radar cross-section (RCS). For instance, it has a radome, made of a frequency selective surface, at its nose, and behind the radome there is a slot array antenna. It has one or two engine inlets, which are often deep and terminated with rotating engine blades. The airplane's surface has many small gaps/grooves formed by imperfect connections of panels, and there are many antennas used for a variety of different purposes. To reduce the RCS, some portions of the surface are coated with

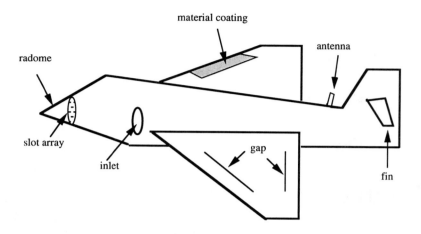

Figure 13.1 Illustration of a typical airplane.

radar absorbing materials (RAMs). Because of these complex structures, the ray-based high-frequency asymptotic methods cannot provide accurate results for the RCS prediction, and because of its electrically large size, numerical methods are simply too time-consuming and memory-intensive for any existing powerful supercomputers. However, if all the small features on the airplane can be removed (with the radome and RAMs replaced by perfectly conducting surfaces, the inlets and gaps filled with perfect conductors, and the antennas removed), the airplane would become a simple large object with conducting surfaces. It is well known that the RCS of such an object can be calculated very efficiently and accurately using a ray-based high-frequency asymptotic method. Furthermore, if the scattering of each individual small feature can be characterized using a suitable numerical method and then added to the scattering of the large object with their major interactions included, the RCS of the airplane can be predicted efficiently with reasonable accuracy. This example clearly demonstrates the importance of hybridization for large-scale, complex problems.

There are two classes of hybrid techniques. The first combines different numerical methods to form a more powerful hybrid numerical technique. For example, the finite element method (FEM) [1–3] is known for its excellent capability to model arbitrary geometries and ability to deal with complex inhomogeneous materials. Its system matrix is sparse, which leads to a very low computational complexity. However, the method is not suited for open-region problems because it cannot incorporate the radiation condition efficiently. To use the FEM for open-region problem, one has to discretize a large region truncated using an approximate boundary condition. On the other hand, the method of moments (MOM) [4–7] based on integral equations is very well suited for dealing with open-region problems because integral equations are formulated using appropriate Green's functions that incorporate the radiation conditions. As a result, the discretization is limited to either the surface or the volume of the object. However, because of the use of Green's functions, the MOM produces a fully populated, dense matrix, which is very expensive to store and solve. This is especially true for an inhomogeneous object that requires a volume discretization. Therefore, given an inhomogeneous volumetric object, an ideal solution would be to use the FEM to formulate the solution inside the object and to use the MOM to formulate the solution outside the object. Such a combination would give an accurate and efficient solution to the problem.

The second class of hybrid techniques combines a numerical method (such as the FEM and MOM) with an asymptotic method to deal with very large objects with small features. The example of Figure 13.1 has clearly illustrated its importance. The asymptotic method can range from simple geometrical optics (GO) and physical optics (PO) to more complicated geometrical theory of diffraction (GTD), uniform theory of diffraction (UTD), physical theory of diffraction (PTD), and uniform asymptotic theory (UAT) [8–13]. One very practical asymptotic method is the shooting-and-bouncing-ray (SBR) method [14–17], which combines the GO, PO, and some useful features of the GTD and UTD. Two practically very important problems that can be handled by this class of hybrid techniques are (1) scattering by large bodies having small features such as cracks, cavities, and conformal antennas on their surfaces, and (2) scattering by large bodies with small protruding objects.

In this chapter, we describe several hybrid techniques to demonstrate the basic principles, formulations, and power of hybrid techniques. The first combines the FEM and an absorbing boundary condition (ABC) [18–24] for an approximate analysis of scattering by an arbitrary object. The second combines the FEM and a boundary integral equation (BIE) for a more accurate analysis of scattering by a complex volumetric object [25–40]. The third is a recently developed novel hybridization of the FEM and BIE, which can be considered as an improvement of the hybrid FEM/ABC and FEM/BIE techniques [41]. It is based on a concept called adaptive absorbing boundary condition (AABC) derived iteratively from BIE [42–45]. The fourth is the hybrid FEM/SBR technique for scattering by a large object with small indentations such as cracks, gaps, and cavities [46, 47]. The fifth is the hybrid MOM/SBR technique for scattering by a large object with small protrusions [48–50].

13.2 HYBRID FEM/ABC TECHNIQUE

This section describes an approximate analysis of scattering problems using the FEM. Its main purpose is to describe the basic formulation of the FEM to facilitate the presentation of the hybrid FEM/BIE, FEM/AABC, and FEM/SBR techniques. Of course, the approximate method itself is also useful for problems that do not require an exact solution.

13.2.1 Problem Statement

The problem to be considered is that of electromagnetic scattering by a free-standing arbitrarily shaped object in free space. The permittivity and permeability of the object are denoted by ϵ and μ; they revert to their free-space values ϵ_0 and μ_0 outside the object. Since in this case the solution domain is the entire infinite space and the FEM is only applicable to a finite domain, we must first truncate the solution domain into a finite domain. This can be done by introducing an artificial surface S to enclose the scatterer (Figure 13.2). The electric field inside S satisfies Maxwell's equations, which lead to the second-order vector wave equation

$$\nabla \times \left(\frac{1}{\mu_r} \nabla \times \mathbf{E} \right) - k_0^2 \epsilon_r \mathbf{E} = 0 \quad \text{in } V \tag{13.1}$$

where V denotes the volume enclosed by S. In the above, $\epsilon_r = \epsilon/\epsilon_0$, $\mu_r = \mu/\mu_0$, and k_0 is the free-space wave number. To uniquely define the boundary-value problem, a boundary condition has to be prescribed on S. Assuming that S is placed sufficiently far away from the scatterer, the scattered field $\mathbf{E}^{\text{scat}}$ satisfies approximately the Sommerfeld radiation condition

$$\hat{n} \times (\nabla \times \mathbf{E}^{\text{scat}}) + jk_0 \hat{n} \times (\hat{n} \times \mathbf{E}^{\text{scat}}) = 0 \quad \text{on } S \tag{13.2}$$

where $\hat{n}$ denotes the outward normal unit vector. Since $\mathbf{E}^{\text{scat}} = \mathbf{E} - \mathbf{E}^{\text{inc}}$ where $\mathbf{E}^{\text{inc}}$ denotes the incident field, (13.2) can be written as

$$\hat{n} \times (\nabla \times \mathbf{E}) + jk_0 \hat{n} \times (\hat{n} \times \mathbf{E}) = \mathbf{U} \quad \text{on } S \tag{13.3}$$

where $\mathbf{U} = \hat{n} \times (\nabla \times \mathbf{E}^{\text{inc}}) + jk_0 \hat{n} \times (\hat{n} \times \mathbf{E}^{\text{inc}})$.

13.2.2 Finite Element Analysis

One of the approaches to formulate FEM is to start from the variational expression. In accordance with the variational principle, the solution to the boundary-value problem defined by (13.1) and (13.3) can be obtained by solving the equivalent variational problem

$$\delta F(\mathbf{E}) = 0 \tag{13.4}$$

where F denotes the functional given by

$$F(\mathbf{E}) = \frac{1}{2} \iiint_V \left[\frac{1}{\mu_r} (\nabla \times \mathbf{E}) \cdot (\nabla \times \mathbf{E}) - k_0^2 \epsilon_r \mathbf{E} \cdot \mathbf{E} \right] dV$$
$$+ \iint_S \left[\frac{jk_0}{2} (\hat{n} \times \mathbf{E}) \cdot (\hat{n} \times \mathbf{E}) + \mathbf{E} \cdot \mathbf{U} \right] dS \tag{13.5}$$

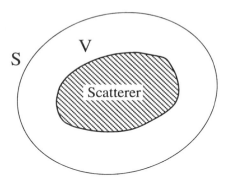

Figure 13.2 A free-standing scatterer enclosed by an artificial surface S.

This variational expression can be derived from (13.1) and (13.3) using the generalized variational principle described in [2]. It can also be proven by taking the first variation of the functional F. Here, we take the second approach to better understand the nature of the variational solution to (13.4).

Taking the first variation of (13.5), one obtains

$$\delta F(\mathbf{E}) = \iiint_V \left[\frac{1}{\mu_r} (\nabla \times \mathbf{E}) \cdot (\nabla \times \delta \mathbf{E}) - k_0^2 \epsilon_r \mathbf{E} \cdot \delta \mathbf{E} \right] dV$$
$$+ \iint_S [jk_0(\hat{n} \times \mathbf{E}) \cdot (\hat{n} \times \delta \mathbf{E}) + \delta \mathbf{E} \cdot \mathbf{U}] dS \qquad (13.6)$$

Application of the vector identity

$$\frac{1}{\mu_r}(\nabla \times \mathbf{E}) \cdot (\nabla \times \delta \mathbf{E}) = \nabla \cdot \left[\delta \mathbf{E} \times \frac{1}{\mu_r}(\nabla \times \mathbf{E}) \right] + \delta \mathbf{E} \cdot \nabla \times \left(\frac{1}{\mu_r} \nabla \times \mathbf{E} \right)$$
$$(13.7)$$

yields

$$\delta F(\mathbf{E}) = \iiint_V \left[\nabla \times \left(\frac{1}{\mu_r} \nabla \times \mathbf{E} \right) - k_0^2 \epsilon_r \mathbf{E} \right] \cdot \delta \mathbf{E} \, dV$$
$$+ \iiint_V \nabla \cdot \left[\delta \mathbf{E} \times \frac{1}{\mu_r} (\nabla \times \mathbf{E}) \right] dV$$
$$+ \iint_S [jk_0(\hat{n} \times \mathbf{E}) \cdot (\hat{n} \times \delta \mathbf{E}) + \delta \mathbf{E} \cdot \mathbf{U}] dS \qquad (13.8)$$

Using the divergence theorem, one has

$$\iiint_V \nabla \cdot \left[\delta \mathbf{E} \times \frac{1}{\mu_r}(\nabla \times \mathbf{E}) \right] dV = \iint_S [\delta \mathbf{E} \times (\nabla \times \mathbf{E})] \cdot \hat{n} \, dS$$
$$= -\iint_S [\hat{n} \times (\nabla \times \mathbf{E})] \cdot \delta \mathbf{E} \, dS \quad (13.9)$$

where $1/\mu_r$ is dropped from the right-hand side since $\mu_r = 1$ on S. Substituting this into (13.8), one obtains

$$\delta F(\mathbf{E}) = \iiint_V \left[\nabla \times \left(\frac{1}{\mu_r} \nabla \times \mathbf{E} \right) - k_0^2 \epsilon_r \mathbf{E} \right] \cdot \delta \mathbf{E} \, dV$$
$$- \iint_S [\hat{n} \times (\nabla \times \mathbf{E}) + jk_0 \hat{n} \times (\hat{n} \times \mathbf{E}) - \mathbf{U}] \cdot \delta \mathbf{E} \, dS \quad (13.10)$$

Since $\delta \mathbf{E}$ is an arbitrary variation, the necessary condition for $\delta F(\mathbf{E}) = 0$ is that

$$\nabla \times \left(\frac{1}{\mu_r} \nabla \times \mathbf{E} \right) - k_0^2 \epsilon_r \mathbf{E} = 0 \quad \text{in } V \quad (13.11)$$
$$\hat{n} \times (\nabla \times \mathbf{E}) + jk_0 \hat{n} \times (\hat{n} \times \mathbf{E}) = \mathbf{U} \quad \text{on } S \quad (13.12)$$

which are recognized as the original vector wave equation and the boundary condition that define the boundary-value problem.

Having established the equivalence between the original boundary-value problem and the variational problem, we now consider the solution of the variational problem using the FEM. The first step of FEM is to subdivide the solution domain into small elements such as tetrahedral elements. We then construct basis functions for each element to represent the electric field $\mathbf{E}$ within the element. Many types of basis functions have been developed in the past. It is found that for vector electric and magnetic fields, the best choice is to use the vector basis functions that can guarantee the continuity of the tangential component of the represented field [51–56]. Once the basis functions are constructed, the field in the entire volume V can be expressed as

$$\mathbf{E} = \sum_{i=1}^{N} E_i \mathbf{N}_i \quad (13.13)$$

where $\mathbf{N}_i$ denotes the basis function associated with the unknown expansion coefficient E_i and N denotes the total number of such basis functions.

Substituting (13.13) into (13.5), one obtains

$$F = \frac{1}{2} \{E\}^T [K] \{E\} - \{E\}^T \{b\} \quad (13.14)$$

where $\{E\} = \{E_1, E_2, \ldots, E_N\}^T$ denotes a column vector, $[K]$ is a symmetric square matrix with the elements given by

$$K_{ij} = \iiint_V \left[\frac{1}{\mu_r}(\nabla \times \mathbf{N}_i) \cdot (\nabla \times \mathbf{N}_j) - k_0^2 \epsilon_r \mathbf{N}_i \cdot \mathbf{N}_j\right] dV$$
$$+ jk_0 \iint_S (\hat{n} \times \mathbf{N}_i) \cdot (\hat{n} \times \mathbf{N}_j) dS \tag{13.15}$$

and finally, $\{b\}$ is a column vector with the elements given by

$$b_i = -\iint_S \mathbf{N}_i \cdot \mathbf{U} \, dS \tag{13.16}$$

Clearly, the functional F is now a quadratic function of E_i. It is well known that the necessary condition for $\delta F = 0$ is that

$$\frac{\partial F}{\partial E_i} = 0 \quad (i = 1, 2, \ldots, N) \tag{13.17}$$

When this is applied to (13.14), one obtains the matrix equation

$$[K]\{E\} = \{b\} \tag{13.18}$$

The solution of this matrix equation yields the numerical values of the expansion coefficients, from which the electric field everywhere can be calculated from (13.13).

The FEM described above has several important advantages. First, it can model arbitrarily shaped geometries. Theoretically, there is no limitation on the complexity of geometries to be modeled provided that the solution domain can be subdivided or meshed into small elements. Of course, in reality most mesh generators have a limited capability, which is often the bottleneck for the application of the FEM. Second, the FEM can easily handle inhomogeneous and anisotropic materials, which is important because of the widespread use of materials in engineering applications. Third, the method can be implemented easily on computers, and general-purpose computer codes can be developed for different electromagnetics problems. Fourth, and probably most important, the FEM matrix is symmetric and sparse; hence, it can be generated, stored, and solved very efficiently. The symmetry of the matrix $[K]$ can be seen easily from (13.15). The sparsity of the matrix can be recognized from the local nature of $\mathbf{N}_i$ and $\mathbf{N}_j$. Since $\mathbf{N}_i$ and $\mathbf{N}_j$ are formed by the basis functions defined over each element, they overlap with each other (thus, leading to a nonzero K_{ij}) only when they belong to the same element or the neighboring elements that share a common edge. As a result, K_{ij} is always zero when $\mathbf{N}_i$ and $\mathbf{N}_j$ do not belong to the same element or the neighboring elements that share a common edge. No matter how large the solution domain or what the size of $[K]$, each row and column of $[K]$ contains only a few nonzero elements. Hence, the memory required to store $[K]$ is $O(N)$ and the computational cost to generate $[K]$ and perform a matrix product of $[K]$ with a vector is also $O(N)$, where N denotes the dimension of $[K]$. This property allows the simulation of a very large electromagnetic problem that may require millions of unknowns from its numerical discretization.

It is evident that the FEM is most suited for a bounded problem with a finite solution domain. In that kind of problem, the differential equation and boundary conditions are often defined accurately and the only error in the FEM solution would be the numerical error introduced by the FEM discretization. However, for scattering problems, the solution obtained using an ABC is approximate, and to reduce its error, the truncation surface must be placed at some distance away from the scatterer and must have a convex shape so that it does not intercept multiple bounces of the scattered field. These two constraints on the truncation surface result in a large computational domain, which not only produces more unknowns, but also requires a finer mesh to suppress the dispersion error of the FEM solution [57]. Even when the truncation surface is placed sufficiently far away from the scatterer, the solution accuracy is still unpredictable because of its dependence on many factors such as the incidence angle and polarization of the excitation and the shape and material composition of the scatterer. Although in general the solution accuracy can be improved using a higher-order ABC [21–24], a method has yet to appear that can quantify and subsequently control the accuracy of the final solution.

13.2.3 Numerical Results

We consider the scattering by a finite circular conducting cylinder having a length of 10λ and a diameter of 2λ, where λ denotes the free-space wavelength. The truncation surface is placed 0.4λ away from the scatterer's surface. The backscatter RCS calculated by the FEM/ABC method is compared to the more accurate MOM calculation in Figure 13.3. The comparison shows that the FEM/ABC result agrees generally with the MOM solution; however, there is about 5-dB error in the low RCS region.

13.3 HYBRID FEM/BIE TECHNIQUE

As stated in the preceding section, the ABC for scattering analysis is approximate, which produces a solution with an unpredictable accuracy, and to use the ABC the truncation surface must be placed sufficiently far away from the scatterer. In this section, we describe a method that can alleviate these two disadvantages. This method replaces the ABC with the exact BIE representation for the field outside the truncation surface. Since the BIE is exact, the truncation surface can be placed as close to the scatterer as possible and it can take an arbitrary shape.

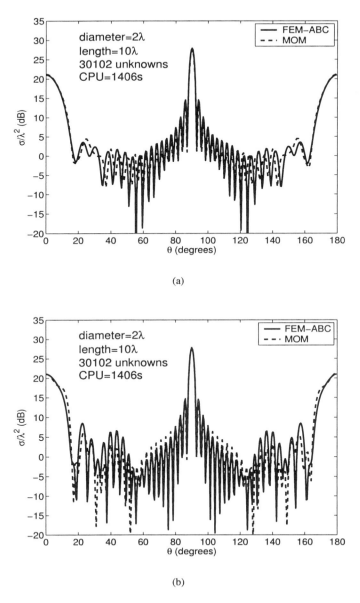

Figure 13.3 Monostatic RCS of a conducting cylinder having a length of 10λ and a diameter of 2λ: (a) HH-polarization; (b) VV-polarization.

13.3.1 Formulation

Consider the problem of electromagnetic wave scattering by an arbitrarily shaped, inhomogeneous body characterized by relative permittivity and permeability (ϵ_r, μ_r). To solve this problem using the FEM/BIE method, we first introduce an artificial surface S (which can be the surface of the body) to enclose the body and divide the problem into an interior and an exterior one. The field inside S satisfies the vector wave equation (13.1). On the truncation surface S, we assume a boundary condition

$$\hat{n} \times \left(\frac{1}{\mu_r} \nabla \times \mathbf{E} \right) = -jk_0 \hat{n} \times \bar{\mathbf{H}} \tag{13.19}$$

which comes from one of Maxwell's equations, where $\bar{\mathbf{H}} = \eta_0 \mathbf{H}$ and η_0 is the free-space wave impedance. The boundary-value problem defined by (13.1) and (13.19) can be formulated into an equivalent variational problem with the functional

$$F(\mathbf{E}) = \frac{1}{2} \iiint_V \left[\frac{1}{\mu_r} (\nabla \times \mathbf{E}) \cdot (\nabla \times \mathbf{E}) - k_0^2 \epsilon_r \mathbf{E} \cdot \mathbf{E} \right] dV$$
$$+ jk_0 \iint_S (\mathbf{E} \times \bar{\mathbf{H}}) \cdot \hat{n} \, dS \tag{13.20}$$

This functional can be discretized using the FEM described in the preceding section. Since in this case, $\bar{\mathbf{H}}$ is unknown, we expand it using the vector basis functions defined for the surface elements on S:

$$\bar{\mathbf{H}} = \sum_{i=1}^{N_S} \bar{H}_i \mathbf{N}_i^S \tag{13.21}$$

where $\mathbf{N}_i^S$ denotes the surface vector basis function associated with the unknown expansion coefficient $\bar{H}_i$ and N_S denotes the total number of such basis functions. Following the standard procedure, we obtain the matrix equation

$$\begin{bmatrix} K_{VV} & K_{VS} & 0 \\ K_{SV} & K_{SS} & B_{SS} \end{bmatrix} \begin{Bmatrix} E_V \\ E_S \\ \bar{H}_S \end{Bmatrix} = \begin{Bmatrix} 0 \\ 0 \end{Bmatrix} \tag{13.22}$$

where $\{E_V\}$ is a vector containing the discrete electric fields inside V, $\{E_S\}$ and $\{\bar{H}_S\}$ are the vectors containing the discrete electric and magnetic fields on S, respectively. Furthermore, $[K_{VV}]$, $[K_{VS}]$, $[K_{SV}]$, and $[K_{SS}]$, and $[B_{SS}]$ are sparse matrices and, in particular, $[K_{VV}]$ and $[K_{SS}]$ are symmetric and $[K_{VS}] = [K_{SV}]^T$. For the convenience of description, we assumed that for the electric field, the interior unknowns are numbered first and then the unknowns on the surface S.

Equation (13.22) cannot be solved unless a relation between $\{E_S\}$ and $\{\bar{H}_S\}$ is established. Such a relation is provided by BIE for the exterior field. Outside S, the field **E** satisfies the vector wave equation

$$\nabla \times \nabla \times \mathbf{E}(\mathbf{r}) - k_0^2 \mathbf{E}(\mathbf{r}) = -jk_0\eta_0 \mathbf{J}_i(\mathbf{r}) \qquad \mathbf{r} \in V_\infty \quad (13.23)$$

where $\mathbf{J}_i$ denotes the electric current density of the source that produces the incident field and V_∞ denotes the exterior region.

To derive an integral equation for **E** and $\bar{\mathbf{H}}$, we introduce the three-dimensional, free-space dyadic Green's function $\overline{\mathbf{G}}_{e0}$ that satisfies the vector equation

$$\nabla \times [\nabla \times \overline{\mathbf{G}}_{e0}(\mathbf{r}, \mathbf{r}')] - k_0^2 \overline{\mathbf{G}}_{e0}(\mathbf{r}, \mathbf{r}') = \bar{\mathbf{I}}\delta(\mathbf{r} - \mathbf{r}') \quad (13.24)$$

and the Sommerfeld radiation condition. This function has a well-known solution [58]

$$\overline{\mathbf{G}}_{e0}(\mathbf{r}, \mathbf{r}') = \left(\bar{\mathbf{I}} - \frac{1}{k_0^2}\nabla\nabla'\right) g(\mathbf{r}, \mathbf{r}') \quad (13.25)$$

with

$$g(\mathbf{r}, \mathbf{r}') = \frac{e^{-jk_0|\mathbf{r}-\mathbf{r}'|}}{4\pi|\mathbf{r}-\mathbf{r}'|} \quad (13.26)$$

Next, we make use of the vector-dyadic Green's second identity [59]

$$\iiint_{V_o} [(\nabla \times \nabla \times \mathbf{A}) \cdot \overline{\mathbf{D}} - \mathbf{A} \cdot (\nabla \times \nabla \times \overline{\mathbf{D}})] dV$$
$$= \iint_{S_o} [(\hat{n} \times \mathbf{A}) \cdot (\nabla \times \overline{\mathbf{D}}) + (\hat{n} \times \nabla \times \mathbf{A}) \cdot \overline{\mathbf{D}}] dS \quad (13.27)$$

where V_o is an arbitrary volume enclosed by S_o. By letting $V_o = V_\infty$, $\mathbf{A} = \mathbf{E}$, and $\overline{\mathbf{D}} = \overline{\mathbf{G}}_{e0}$ in (13.27), we find

$$\mathbf{E}(\mathbf{r}) = \mathbf{E}^{\mathrm{inc}}(\mathbf{r})$$
$$+ \iint_S \left\{ \overline{\mathbf{G}}_{m0}(\mathbf{r}, \mathbf{r}') \cdot [\hat{n}' \times \mathbf{E}(\mathbf{r}')] - jk_0 \overline{\mathbf{G}}_{e0}(\mathbf{r}, \mathbf{r}') \cdot [\hat{n}' \times \bar{\mathbf{H}}(\mathbf{r}')] \right\} dS'$$
$$\mathbf{r} \in V_\infty \quad (13.28)$$

where

$$\overline{\mathbf{G}}_{m0}(\mathbf{r}, \mathbf{r}') = \nabla \times \overline{\mathbf{G}}_{e0}(\mathbf{r}, \mathbf{r}') = \nabla g(\mathbf{r}, \mathbf{r}') \times \bar{\mathbf{I}} \quad (13.29)$$

and

$$\mathbf{E}^{\mathrm{inc}}(\mathbf{r}) = -jk_0\eta_0 \iiint_{V_s} \overline{\mathbf{G}}_{e0}(\mathbf{r}, \mathbf{r}') \cdot \mathbf{J}_i(\mathbf{r}') dV' \quad (13.30)$$

which represents the electric field produced by $\mathbf{J}_i$ in the infinite space without the object, where V_s denotes the volume occupied by $\mathbf{J}_i$. Similarly, we obtain

$$\bar{\mathbf{H}}(\mathbf{r}) = \bar{\mathbf{H}}^{\text{inc}}(\mathbf{r})$$
$$+ \iint_S \{\overline{\mathbf{G}}_{m0}(\mathbf{r},\mathbf{r}') \cdot [\hat{n}' \times \bar{\mathbf{H}}(\mathbf{r}')] + jk_0 \overline{\mathbf{G}}_{e0}(\mathbf{r},\mathbf{r}') \cdot [\hat{n}' \times \mathbf{E}(\mathbf{r}')]\} dS'$$
$$\mathbf{r} \in V_\infty \quad (13.31)$$

where

$$\bar{\mathbf{H}}^{\text{inc}}(\mathbf{r}) = \eta_0 \iiint_{V_s} \overline{\mathbf{G}}_{m0}(\mathbf{r},\mathbf{r}') \cdot \mathbf{J}_i(\mathbf{r}') dV' \quad (13.32)$$

which represents the magnetic field produced by $\mathbf{J}_i$ in the infinite space without the object.

To write (13.28) and (13.31) in compact form and in terms of more familiar scalar Green's function, we define the operators

$$\mathbf{L}(\mathbf{X}) = jk_0 \iint_S \left[\mathbf{X}(\mathbf{r}')g(\mathbf{r},\mathbf{r}') + \frac{1}{k_0^2} \nabla' \cdot \mathbf{X}(\mathbf{r}') \nabla g(\mathbf{r},\mathbf{r}') \right] dS' \quad (13.33)$$

$$\mathbf{K}(\mathbf{X}) = \iint_S \mathbf{X}(\mathbf{r}') \times \nabla g(\mathbf{r},\mathbf{r}') \, dS' \quad (13.34)$$

and introduce the equivalent surface currents

$$\bar{\mathbf{J}}_s = \hat{n} \times \bar{\mathbf{H}} = \eta_0 \, \hat{n} \times \mathbf{H} \qquad \mathbf{M}_s = \mathbf{E} \times \hat{n}. \quad (13.35)$$

As a result, (13.28) and (13.31) can be written as

$$\mathbf{E}(\mathbf{r}) = \mathbf{E}^{\text{inc}}(\mathbf{r}) - \mathbf{L}(\bar{\mathbf{J}}_s) + \mathbf{K}(\mathbf{M}_s) \qquad \mathbf{r} \in V_\infty \quad (13.36)$$

and

$$\bar{\mathbf{H}}(\mathbf{r}) = \bar{\mathbf{H}}^{\text{inc}}(\mathbf{r}) - \mathbf{K}(\bar{\mathbf{J}}_s) - \mathbf{L}(\mathbf{M}_s) \qquad \mathbf{r} \in V_\infty \quad (13.37)$$

These two equations provide the foundation to derive integral equations for $\bar{\mathbf{J}}_s$ and $\mathbf{M}_s$. Taking the cross product of these with $\hat{n}$ and letting $\mathbf{r}$ approach S, we have

$$-\frac{1}{2}\mathbf{M}_s(\mathbf{r}) + \hat{n} \times \mathbf{L}(\bar{\mathbf{J}}_s) - \hat{n} \times \tilde{\mathbf{K}}(\mathbf{M}_s) = \hat{n} \times \mathbf{E}^{\text{inc}}(\mathbf{r}) \qquad \mathbf{r} \in S \quad (13.38)$$

$$\frac{1}{2}\bar{\mathbf{J}}_s(\mathbf{r}) + \hat{n} \times \tilde{\mathbf{K}}(\bar{\mathbf{J}}_s) + \hat{n} \times \mathbf{L}(\mathbf{M}_s) = \hat{n} \times \bar{\mathbf{H}}^{\text{inc}}(\mathbf{r}) \qquad \mathbf{r} \in S \quad (13.39)$$

where $\tilde{\mathbf{K}}$ is the same integral as in (13.34), except that the singular point $\mathbf{r} = \mathbf{r}'$ is now removed. Equation (13.38) is known as the electric field integral equation (EFIE) and (13.39) is called the magnetic field integral equation (MFIE).

Any of the EFIE and MFIE can be used to obtain a relation between $\bar{\mathbf{J}}_s$ and $\mathbf{M}_s$, which can be discretized to give a relation between $\{E_S\}$ and $\{\bar{H}_S\}$. However, for a given S, $\mathbf{L}$ and $\mathbf{K}$ can be singular at certain frequencies when the exterior medium is lossless. Consequently, both EFIE and MFIE may give an erroneous solution at these frequencies. This is known as the problem of interior resonance [7] and the singular frequencies correspond to the resonant frequencies of a cavity formed by filling the interior of S with the exterior medium. To eliminate this problem, we can combine (13.38) and (13.39) to find

$$\alpha\,\hat{n}\times\left[\hat{n}\times\mathbf{L}(\bar{\mathbf{J}}_s)-\frac{1}{2}\mathbf{M}_s-\hat{n}\times\tilde{\mathbf{K}}(\mathbf{M}_s)\right]$$
$$+(1-\alpha)\left[\frac{1}{2}\bar{\mathbf{J}}_s+\hat{n}\times\tilde{\mathbf{K}}(\bar{\mathbf{J}}_s)+\hat{n}\times\mathbf{L}(\mathbf{M}_s)\right]$$
$$=\alpha\,\hat{n}\times\hat{n}\times\mathbf{E}^{\text{inc}}(\mathbf{r})+(1-\alpha)\,\hat{n}\times\bar{\mathbf{H}}^{\text{inc}}(\mathbf{r})\quad\mathbf{r}\in S\quad(13.40)$$

which is known as the combined field integral equation (CFIE) [60, 61]. This combination results in an integral operator corresponding to that for a cavity with a resistive wall whose resonant frequencies are complex. As a result, it cannot be singular for a real frequency. The combination parameter α is usually chosen anywhere between 0.2 and 0.8.

To discretize (13.40), we can expand $\bar{\mathbf{J}}_s$ and $\mathbf{M}_s$ as

$$\bar{\mathbf{J}}_s = \sum_{i=1}^{N_S} \bar{H}_i \mathbf{g}_i \quad (13.41)$$

$$\mathbf{M}_s = -\sum_{i=1}^{N_S} E_i \mathbf{g}_i \quad (13.42)$$

where $\mathbf{g}_i$ denote the vector basis functions that are related to $\mathbf{N}_i^S$ in (13.21) by

$$\mathbf{g}_i = \hat{n}\times\mathbf{N}_i^S \quad (13.43)$$

Substituting (13.41) and (13.42) into (13.40), we obtain

$$\sum_{j=1}^{N_S}\bar{H}_j\left\{\alpha\,\hat{n}\times[\hat{n}\times\mathbf{L}(\mathbf{g}_j)]+(1-\alpha)\left[\frac{1}{2}\mathbf{g}_j+\hat{n}\times\tilde{\mathbf{K}}(\mathbf{g}_j)\right]\right\}$$
$$+\sum_{j=1}^{N_S}E_j\left\{\alpha\,\hat{n}\times\left[\frac{1}{2}\mathbf{g}_j+\hat{n}\times\tilde{\mathbf{K}}(\mathbf{g}_j)\right]-(1-\alpha)\hat{n}\times\mathbf{L}(\mathbf{g}_j)\right\}$$
$$=\alpha\,\hat{n}\times\hat{n}\times\mathbf{E}^{\text{inc}}(\mathbf{r})+(1-\alpha)\hat{n}\times\bar{\mathbf{H}}^{\text{inc}}(\mathbf{r})\quad\mathbf{r}\in S\quad(13.44)$$

To convert this into a matrix equation, we choose a set of testing functions $\mathbf{t}_i$ to obtain

$$[P_{SS}]\{E_S\}+[Q_{SS}]\{\bar{H}_S\}=\{b_S\} \quad (13.45)$$

where $[P_{SS}]$ and $[Q_{SS}]$ are square matrices with the elements given by

$$[P_{SS}]_{ij} = \iint_S \mathbf{t}_i \cdot \left\{ \alpha\,\hat{n} \times \left[\frac{1}{2}\mathbf{g}_j + \hat{n} \times \tilde{\mathbf{K}}(\mathbf{g}_j)\right] - (1-\alpha)\hat{n} \times \mathbf{L}(\mathbf{g}_j) \right\} dS \tag{13.46}$$

$$[Q_{SS}]_{ij} = \iint_S \mathbf{t}_i \cdot \left\{ \alpha\,\hat{n} \times [\hat{n} \times \mathbf{L}(\mathbf{g}_j)] + (1-\alpha)\left[\frac{1}{2}\mathbf{g}_j + \hat{n} \times \tilde{\mathbf{K}}(\mathbf{g}_j)\right] \right\} dS \tag{13.47}$$

and $\{b_S\}$ is a column vector with the elements given by

$$\{b_S\}_i = \iint_S \mathbf{t}_i \cdot \left\{ \alpha\,\hat{n} \times [\hat{n} \times \mathbf{E}^{\text{inc}}] + (1-\alpha)\hat{n} \times \bar{\mathbf{H}}^{\text{inc}} \right\} dS \tag{13.48}$$

From the expressions of $\mathbf{L}$ and $\mathbf{K}$, it is clear that $\mathbf{g}_i$ is a good testing function for $\mathbf{L}(\mathbf{g}_j)$ and a poor testing function for $\mathbf{K}(\mathbf{g}_j)$. On the other hand, $\hat{n} \times \mathbf{g}_i$ is a good testing function for $\mathbf{K}(\mathbf{g}_j)$ and a poor testing function for $\mathbf{L}(\mathbf{g}_j)$. Hence, using $\mathbf{g}_i$ for $\mathbf{t}_i$ would result in a well conditioned $[Q_{SS}]$ and an ill conditioned $[P_{SS}]$, and using $\hat{n} \times \mathbf{g}_i$ for $\mathbf{t}_i$ would result in a well conditioned $[P_{SS}]$ and an ill conditioned $[Q_{SS}]$. Therefore, to test both $\mathbf{L}$ and $\mathbf{K}$ well and make both $[P_{SS}]$ and $[Q_{SS}]$ well conditioned, we can use the combination of $\mathbf{g}_i$ and $\hat{n} \times \mathbf{g}_i$ for testing [40, 62], or in other words, we let

$$\mathbf{t}_i = \mathbf{g}_i + \hat{n} \times \mathbf{g}_i \tag{13.49}$$

Combining (13.22) and (13.45), we obtain the complete system

$$\begin{bmatrix} K_{VV} & K_{VS} & 0 \\ K_{SV} & K_{SS} & B_{SS} \\ 0 & P_{SS} & Q_{SS} \end{bmatrix} \begin{Bmatrix} E_I \\ E_S \\ \bar{H}_S \end{Bmatrix} = \begin{Bmatrix} 0 \\ 0 \\ b_S \end{Bmatrix} \tag{13.50}$$

which can be solved for the fields inside V and on S. Equation (13.50) is a partly sparse, partly full matrix equation and is no longer symmetric. For relatively small problems, one can first solve (13.22) to find

$$\{E_S\} = -[K_{SS} - K_{SV} K_{VV}^{-1} K_{VS}]^{-1} [B_{SS}]\{\bar{H}_S\} \tag{13.51}$$

and then substitute this into (13.45) to obtain

$$([Q_{SS}] - [P_{SS}][K_{SS} - K_{SV} K_{VV}^{-1} K_{VS}]^{-1}[B_{SS}])\{\bar{H}_S\} = \{b_S\} \tag{13.52}$$

Note that the calculation of (13.51) is independent of the excitation vector $\{b_S\}$ and can make full use of the symmetry and sparsity of the FEM matrices. For large problems, iterative solvers applied directly to (13.50) are usually the method of choice.

13.3.2 Application of MLFMA

The FEM/BIE method described above has a bottleneck, which is the dense matrices generated by the BIE. This bottleneck severely limits the capability of the FEM/BIE method in dealing with large objects since the dense matrices $[P_{SS}]$ and $[Q_{SS}]$ have a memory requirement of $O(N_S^2)$ and a computational complexity of $O(N_S^2)$ to compute a matrix-vector product.

One solution to the problem discussed above is to compute the matrix-vector products using FMM [63–65]. The basic idea of FMM is first to divide the surface subscatterers into groups. The addition theorem is then used to translate the scattered field of different scattering centers within a group into a single center, and this process is called aggregation. Doing this, the number of scattering centers is reduced significantly. Similarly, for each group, the field scattered by all the other group centers can be first received by the group center, and then redistributed to the subscatterers belonging to the group. This process is called disaggregation. It has been shown that FMM can reduce the memory requirement and computational complexity to $O(N_S^{1.5})$.

The memory requirement and computational complexity can be further reduced to $O(N_S \log N_S)$ using MLFMA [66–68]. To implement MLFMA, the entire object is first enclosed in a large cube, which is divided into eight smaller cubes. Each subcube is then recursively subdivided into smaller cubes until the edge length of the finest cube is about half a wavelength. For two points in the same or nearby finest cubes, their interaction is calculated in a direct manner. However, when the two points reside in different nonnearby cubes, their interaction is calculated by FMM, as described above. The level of cubes on which FMM is applied depends on the distance between the two points. The detailed description of MLFMA is given in [66] and is not repeated here although the equations to be treated are different.

The basic formulas, derived with the addition theorem, to calculate the matrix elements for nonnearby groups are given by

$$[P_{SS}]_{ij} = \left(\frac{k_0}{4\pi}\right)^2 \oint \mathbf{V}_{im}^P T_{mm'}(\hat{k}\cdot\hat{r}_{mm'}) \cdot \mathbf{V}_{jm'} d^2\hat{k} \qquad (13.53)$$

$$[Q_{SS}]_{ij} = \left(\frac{k_0}{4\pi}\right)^2 \oint \mathbf{V}_{im}^Q T_{mm'}(\hat{k}\cdot\hat{r}_{mm'}) \cdot \mathbf{V}_{jm'} d^2\hat{k} \qquad (13.54)$$

where

$$\begin{aligned}\mathbf{V}_{im}^P = \iint_S & e^{-j\mathbf{k}_0\cdot\mathbf{r}_{im}}[\alpha\hat{k}\times(\mathbf{g}_i+\hat{n}\times\mathbf{g}_i) \\ & -(1-\alpha)(\bar{\mathbf{I}}-\hat{k}\hat{k})\cdot(\mathbf{g}_i-\hat{n}\times\mathbf{g}_i)]dS\end{aligned} \qquad (13.55)$$

$$\begin{aligned}\mathbf{V}_{im}^Q = \iint_S & e^{-j\mathbf{k}_0\cdot\mathbf{r}_{im}}[-\alpha(\bar{\mathbf{I}}-\hat{k}\hat{k})\cdot(\mathbf{g}_i+\hat{n}\times\mathbf{g}_i) \\ & -(1-\alpha)\hat{k}\times(\mathbf{g}_i-\hat{n}\times\mathbf{g}_i)]dS\end{aligned} \qquad (13.56)$$

$$\mathbf{V}_{jm'} = \iint_S e^{j\mathbf{k}_0 \cdot \mathbf{r}_{jm'}} \mathbf{g}_j dS \tag{13.57}$$

and

$$T_{mm'}(\hat{k} \cdot \hat{r}_{mm'}) = \sum_{l=0}^{L}(-j)^l(2l+1)h_l^{(2)}(k_0 r_{mm'})P_l(\hat{r}_{mm'} \cdot \hat{k}) \tag{13.58}$$

In the above, the integrals in (13.53) and (13.54) are over the unit spherical surface, $\mathbf{g}_i$ resides in a group G_m centered at $\mathbf{r}_m$, $\mathbf{g}_j$ resides in a group $G_{m'}$ centered at $\mathbf{r}_{m'}$, $\mathbf{r}_{im} = \mathbf{r}_i - \mathbf{r}_m$, $\mathbf{r}_{jm'} = \mathbf{r}_j - \mathbf{r}_{m'}$, and $\mathbf{r}_{mm'} = \mathbf{r}_m - \mathbf{r}_{m'}$. Also in (13.58), $h_l^{(2)}$ denotes the spherical Hankel function of the second kind, P_l denotes the Legendre polynomial of degree l, and L denotes the number of multipole expansion terms, whose choice is discussed in [66].

As described earlier, MLFMA converts the direct interaction component $[P_{SS}]_{ij}$ or $[Q_{SS}]_{ij}$ between two "far-away" points i and j into three indirect components: the radiation component from the point j to the group center m', which is represented by $\mathbf{V}_{jm'}$; the translation component from the group center m' to another group center m, represented by $T_{mm'}$; and the receiving component from the group center m to the point i, which is represented by $\mathbf{V}_{im}$. Among these three components, only the receiving component is different for different formulations; the other two components, the translation and the radiation components, are the same.

13.3.3 Numerical Results

Figure 13.4 shows the bistatic RCS of a conducting sphere having a diameter d and coated with a lossy dielectric layer having a thickness t. The results are compared to those obtained using the Mie series and good agreement is observed. The memory requirement and the total CPU time on one processor of an SGI Power Challenge (R8000) are given in Table 13.1. These results are obtained without using a preconditioner. Figure 13.5 shows the result for a sphere coated with two dielectric layers.

13.4 HYBRID FEM/AABC TECHNIQUE

In the preceding two sections, we described the hybrid FEM/ABC and FEM/BIE methods. Both methods have their advantages and disadvantages. The major advantage of the FEM/ABC method is that it produces a purely sparse matrix that can be stored and solved efficiently using a variety of well-developed sparse matrix solvers. However, the solution obtained using an ABC is approximate, and to reduce its error, the truncation surface must be placed at some distance away from the scatterer and must have a convex shape. These two constraints on the truncation surface result in a

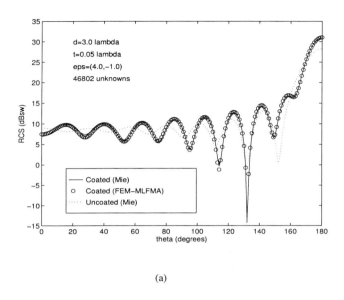

(a)

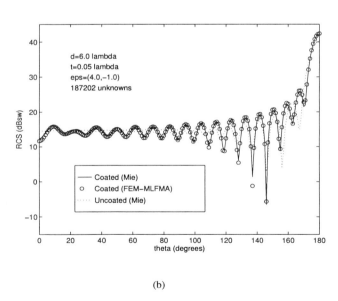

(b)

Figure 13.4 Bistatic RCS of a coated sphere: (a) $d = 3\lambda$; (b) $d = 6\lambda$ [40]. The conducting sphere has a diameter d and the coating has a thickness $t = 0.05\lambda$, a relative permittivity $\epsilon_r = 4.0 - j1.0$, and a relative permeability $\mu_r = 1$.

Table 13.1 Memory Requirement and CPU Time for the FEM/MLFMA Solution of Scattering from a Coated Sphere [40]

Sphere diameter	Number of unknowns	Level of MLFMA	Memory requirement	CPU time per iteration	Total CPU time
0.75λ	3,330	2	14.7 MB	0.424 s	683 s
1.5λ	11,704	3	37.6 MB	3.343 s	3,750 s
3λ	46,802	4	132.9 MB	18.46 s	20,728 s
6λ	187,202	5	522.5 MB	87.84 s	130,501 s

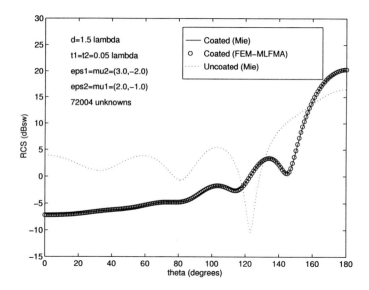

Figure 13.5 Bistatic RCS of a sphere coated with two dielectric layers [40]. The conducting sphere has a diameter $d = 1.5\lambda$, the inner layer has a thickness $t_1 = 0.05\lambda$, a relative permittivity $\epsilon_{r1} = 3.0 - j2.0$, and a relative permeability $\mu_{r1} = 2.0 - j1.0$, and the outer layer has a thickness $t_2 = 0.05\lambda$, a relative permittivity $\epsilon_{r2} = 2.0 - j1.0$, and a relative permeability $\mu_{r2} = 3.0 - j2.0$.

large computational domain. Even when the truncation surface is placed sufficiently far away from the scatterer, the solution accuracy is still unpredictable because of its dependence on many factors such as the incidence angle and polarization of the excitation and the shape and material composition of the scatterer. It is evident that an ideal FEM/ABC method would be the one that (1) retains the advantage of the

ABC, (2) allows an arbitrarily shaped truncation surface that can be placed very close to the scatterer, and (3) has a mechanism to systematically reduce its truncation error.

On the other hand, the FEM/BIE method permits the use of an arbitrarily shaped truncation surface that can be placed very close to the scatterer. However, the method is relatively inefficient. Since a BIE produces a full matrix equation, the final system of equations becomes of a partly full, partly sparse matrix. To date, there is no efficient algorithm developed specifically for solving this type of matrix equation. The most efficient approach to deal with a partly full, partly sparse matrix is, perhaps, to use iterative solvers. Such an approach requires the evaluation of the matrix-vector product in each iteration. The computational complexity of this evaluation consists of two parts: the part associated with FEM is $O(N)$ and the part related to BIE is $O(N_S^2)$, where N denotes the total number of unknowns and N_S denotes the number of unknowns on the truncation surface. Although the second part can be reduced to $O(N_S \log N_S)$ by using the MLFMA, it still consumes more computing time (usually over 10 times) than the first part because of a large constant associated with its computational complexity. Furthermore, because the FEM matrices are less well conditioned, the condition of the final combined system is compromised and it often takes hundreds and thousands of iterations (thus MLFMA calculations) to reach convergence. In addition, special care must be exercised to eliminate the problem of interior resonance. From the above discussions, it is clear that an ideal FEM/BIE method would be the one that (1) is free of interior resonance, (2) produces a purely sparse FEM matrix, and (3) requires only a few MLFMA evaluations.

This section describes a novel method [41] that meets the requirements for both the ideal FEM/ABC and FEM/BIE methods described above. It can be regarded as the improvement of the FEM/BIE method using the ABC concept and the improvement of the FEM/ABC method using the BIE concept. It retains all the advantages of both methods and discards all of their disadvantages. In this method, an artificial, arbitrarily shaped artificial surface is first introduced, as usual, to truncate the computational domain. A boundary condition, in the form of ABC, is then derived from the BIE. This boundary condition, together with the governing partial differential equation, is solved by FEM. The resulting solution is then used to calculate a more accurate ABC on the truncation surface. This process is continued until the ABC is converged and adapted to the specific problem and the solution is converged to the true solution. The method is hence named the FEM/adaptive absorbing boundary condition (FEM/AABC) method.

13.4.1 Formulation

Consider a free-standing inhomogeneous dielectric object that may contain metallic structures. In what follows, the object is assumed to be recursively embedded in two domains V_1 and V_2 with bounding surfaces S_1 and S_2, respectively (Figure 13.6). In addition, position-dependent unit vectors normal to S_i ($i = 1, 2$) are denoted by $\hat{n}_i$,

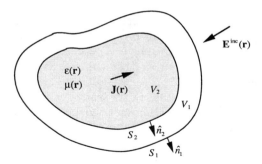

Figure 13.6 A free-standing object recursively embedded in two domains V_1 and V_2 with bounding surfaces S_1 and S_2.

which are assumed outward pointing. The permittivity and permeability of the object and its surroundings are denoted by $\epsilon(\mathbf{r})$ and $\mu(\mathbf{r})$; they revert to their free-space values ϵ_0 and μ_0 outside S_2.

In V_2, $\mathbf{E}(\mathbf{r})$ obeys

$$\nabla \times \left[\frac{1}{\mu_r}\nabla \times \mathbf{E}(\mathbf{r})\right] - k_0^2 \epsilon_r \mathbf{E}(\mathbf{r}) = 0 \tag{13.59}$$

subject to the boundary condition

$$\hat{n} \times \mathbf{E}(\mathbf{r}) = 0 \tag{13.60}$$

on the metallic surfaces. In V_1, $\bar{\mathbf{H}}(\mathbf{r})$ satisfies

$$\nabla \times [\nabla \times \bar{\mathbf{H}}(\mathbf{r})] - k_0^2 \bar{\mathbf{H}}(\mathbf{r}) = 0 \tag{13.61}$$

subject to a postulated boundary condition

$$\hat{n}_1 \times [\nabla \times \bar{\mathbf{H}}(\mathbf{r})] + jk_0 \hat{n}_1 \times [\hat{n}_1 \times \bar{\mathbf{H}}(\mathbf{r})] = \mathbf{V}(\mathbf{r}) \qquad \mathbf{r} \in S_1 \tag{13.62}$$

The left-hand side of (13.62) is simply a linear combination of the tangential electric and magnetic fields on S_1. The right-hand side expresses the same quantity in terms of the tangential electric and magnetic fields on S_2. To be more specific, $\mathbf{V}(\mathbf{r})$ is given by (13.62) with

$$E(r) = E^{inc}(r)$$
$$+ \iint_{S_2} \{\overline{G}_{m0}(r,r') \cdot [\hat{n}'_2 \times E(r')] - jk_0\overline{G}_{e0}(r,r') \cdot [\hat{n}'_2 \times \bar{H}(r')]\} dS' \qquad (13.63)$$

$$\bar{H}(r) = \bar{H}^{inc}(r)$$
$$+ \iint_{S_2} \{\overline{G}_{m0}(r,r') \cdot [\hat{n}'_2 \times \bar{H}(r')] + jk_0\overline{G}_{e0}(r,r') \cdot [\hat{n}'_2 \times E(r')]\} dS' \qquad (13.64)$$

where $\overline{G}_{e0}$ and $\overline{G}_{m0}$ denote the electric- and magnetic-type free-space dyadic Green's functions defined in (13.25) and (13.29), respectively. It is apparent that both tangential electric and magnetic fields on S_2 are needed to evaluate the fields on S_1. This is why the electric field formulation is used in V_2 and the magnetic field formulation is used in V_1; this will become more obvious later.

The problem statement described above lays the foundation for a new hybridization scheme to couple the FEM and BIE. This hybridization scheme does not suffer the problem of interior resonance associated with the EFIE and MFIE because the postulated boundary condition in (13.62) is actually an impedance boundary condition. As a result, the cavity formed by such a surface cannot support any real resonance. To a certain extent, the proposed BI formulation resembles that of the CFIE. We also note that if the BI terms on the right-hand side of (13.63) and (13.64) are neglected, (13.62) reduces to the standard first-order ABC. Such an ABC is reflectionless only for the normally incident waves. Clearly, the BI terms provide necessary corrections to the first-order ABC so that (13.62) becomes an exact ABC.

The boundary-value problem defined above can be discretized using the FEM. In accordance with the variational principle, the field in V_2 can be obtained by seeking the weak solution of (13.59), which minimizes the functional

$$F_2(E) = \frac{1}{2} \iiint_{V_2} \left[\frac{1}{\mu_r}(\nabla \times E) \cdot (\nabla \times E) - k_0^2\epsilon_r E \cdot E\right] dV$$
$$+ jk_0 \iint_{S_2} \hat{n}_2 \cdot (E \times \bar{H}) dS \qquad (13.65)$$

Subdividing V_2 into finite elements and using vector basis functions to expand both E and $\bar{H}$, we obtain the matrix equation

$$\begin{bmatrix} K_{V_2V_2} & K_{V_2S_2} & 0 \\ K_{S_2V_2} & K_{S_2S_2} & B_{S_2S_2} \end{bmatrix} \begin{Bmatrix} E_{V_2} \\ E_{S_2} \\ \bar{H}_{S_2} \end{Bmatrix} = \begin{Bmatrix} 0 \\ 0 \end{Bmatrix} \qquad (13.66)$$

where $\{E_{V_2}\}$ is a vector containing the discrete electric field inside V_2, $\{E_{S_2}\}$ and $\{\bar{H}_{S_2}\}$ are vectors containing the discrete electric and magnetic fields on S_2, respectively, and finally $[K_{V_2V_2}]$, $[K_{V_2S_2}]$, $[K_{S_2V_2}]$, $[K_{S_2S_2}]$, and $[B_{S_2S_2}]$ denote the corresponding FEM matrices, which are completely sparse.

Similarly, the field in V_1 can be obtained from the weak solution of (13.61) and (13.62), which minimizes the functional

$$F_1(\bar{\mathbf{H}}) = \frac{1}{2} \iiint_{V_1} [(\nabla \times \bar{\mathbf{H}}) \cdot (\nabla \times \bar{\mathbf{H}}) - k_0^2 \bar{\mathbf{H}} \cdot \bar{\mathbf{H}}] dV + jk_0 \iint_{S_2} \hat{n}_2 \cdot (\mathbf{E} \times \bar{\mathbf{H}}) dS$$
$$+ \frac{jk_0}{2} \iint_{S_1} (\hat{n}_1 \times \bar{\mathbf{H}}) \cdot (\hat{n}_1 \times \bar{\mathbf{H}}) dS + jk_0 \iint_{S_1} \bar{\mathbf{H}} \cdot (\mathbf{P} + \mathbf{R}) dS \quad (13.67)$$

where $\mathbf{P}$ is given by

$$\mathbf{P}(\mathbf{r}) = \hat{n}_1 \times \mathbf{E}^{\text{inc}}(\mathbf{r}) + \hat{n}_1 \times [\hat{n}_1 \times \bar{\mathbf{H}}^{\text{inc}}(\mathbf{r})] \quad (13.68)$$

and $\mathbf{R}$ is given by

$$\mathbf{R}(\mathbf{r}) = \hat{n}_1 \times \iint_{S_2} \{\overline{\mathbf{G}}_{m0}(\mathbf{r}, \mathbf{r}') \cdot [\hat{n}_2' \times \mathbf{E}(\mathbf{r}')] - jk_0 \overline{\mathbf{G}}_{e0}(\mathbf{r}, \mathbf{r}') \cdot [\hat{n}_2' \times \bar{\mathbf{H}}(\mathbf{r}')]\} dS'$$
$$+ \hat{n}_1 \times \left(\hat{n}_1 \times \iint_{S_2} \{\overline{\mathbf{G}}_{m0}(\mathbf{r}, \mathbf{r}') \cdot [\hat{n}_2' \times \bar{\mathbf{H}}(\mathbf{r}')] \right.$$
$$\left. + jk_0 \overline{\mathbf{G}}_{e0}(\mathbf{r}, \mathbf{r}') \cdot [\hat{n}_2' \times \mathbf{E}(\mathbf{r}')]\} dS' \right) . \quad (13.69)$$

Subdividing V_1 into finite elements and using vector basis functions to expand both $\mathbf{E}$ and $\bar{\mathbf{H}}$, we obtain the matrix equation

$$\begin{bmatrix} B_{S_2S_2}^T & L_{S_2S_2} & L_{S_2V_1} & 0 \\ 0 & L_{V_1S_2} & L_{V_1V_1} & L_{V_1S_1} \\ 0 & 0 & L_{S_1V_1} & L_{S_1S_1} \end{bmatrix} \begin{Bmatrix} E_{S_2} \\ \bar{H}_{S_2} \\ \bar{H}_{V_1} \\ \bar{H}_{S_1} \end{Bmatrix} = \begin{Bmatrix} 0 \\ 0 \\ b_{S_1} \end{Bmatrix} \quad (13.70)$$

where $\{\bar{H}_{V_1}\}$ is a vector containing the discrete magnetic field inside V_1, $\{\bar{H}_{S_1}\}$ is a vector containing the discrete magnetic field on S_1, and $[L_{S_2S_2}]$, $[L_{S_2V_1}]$, $[L_{V_1S_2}]$, $[L_{V_1V_1}]$, $[L_{V_1S_1}]$, $[L_{S_1V_1}]$, and $[L_{S_1S_1}]$ are FEM matrices, which are purely sparse. Also, $\{b_{S_1}\}$ is the right-hand side vector, given by

$$\{b_{S_1}\}_i = -jk_0 \iint_{S_1} \mathbf{N}_i^{S_1} \cdot (\mathbf{P} + \mathbf{R}) dS \quad (13.71)$$

where $\mathbf{N}_i^{S_1}$ denotes the vector basis function on S_1. The part associated with $\mathbf{R}$ can be written as

$$\iint_{S_1} \mathbf{N}_i^{S_1} \cdot \mathbf{R}\, dS = \iint_{S_1} (\hat{n}_1 \times \mathbf{N}_i^{S_1})$$
$$\cdot \iint_{S_2} \{jk_0 \overline{\mathbf{G}}_{e0}(\mathbf{r},\mathbf{r}') \cdot [\hat{n}_2' \times \bar{\mathbf{H}}(\mathbf{r}')] - \overline{\mathbf{G}}_{m0}(\mathbf{r},\mathbf{r}') \cdot [\hat{n}_2' \times \mathbf{E}(\mathbf{r}')]\}\, dS'\, dS$$
$$+ \iint_{S_1} [\hat{n}_1 \times (\hat{n}_1 \times \mathbf{N}_i^{S_1})]$$
$$\cdot \iint_{S_2} \{\overline{\mathbf{G}}_{m0}(\mathbf{r},\mathbf{r}') \cdot [\hat{n}_2' \times \bar{\mathbf{H}}(\mathbf{r}')] + jk_0 \overline{\mathbf{G}}_{e0}(\mathbf{r},\mathbf{r}') \cdot [\hat{n}_2' \times \mathbf{E}(\mathbf{r}')]\}\, dS'\, dS$$
(13.72)

for $i = 1, 2, \ldots, N_{S_1}$, where N_{S_1} denotes the number of unknowns on S_1. Equation (13.72) is actually a measure of the field radiated by the equivalent sources $\hat{n}_2 \times \mathbf{E}$ and $\hat{n}_2 \times \bar{\mathbf{H}}$ on S_2 and received by $\mathbf{N}_i^{S_1}$ on S_1. By denoting

$$\{R_{S_1}\}_i = \iint_{S_1} \mathbf{N}_i^{S_1} \cdot \mathbf{R}\, dS \qquad (i = 1, 2, \ldots, N_{S_1}) \tag{13.73}$$

(13.72) can further be written in matrix form as

$$\{R_{S_1}\} = [Y_{S_1 S_2}]\{E_{S_2}\} + [Z_{S_1 S_2}]\{\bar{H}_{S_2}\} \tag{13.74}$$

where

$$[Y_{S_1 S_2}]_{ij} = jk_0 \iint_{S_1} [\hat{n}_1 \times (\hat{n}_1 \times \mathbf{N}_i^{S_1})] \cdot \iint_{S_2} \overline{\mathbf{G}}_{e0}(\mathbf{r},\mathbf{r}') \cdot (\hat{n}_2' \times \mathbf{N}_j^{S_2})\, dS'\, dS$$
$$- \iint_{S_1} (\hat{n}_1 \times \mathbf{N}_i^{S_1}) \cdot \iint_{S_2} \overline{\mathbf{G}}_{m0}(\mathbf{r},\mathbf{r}') \cdot (\hat{n}_2' \times \mathbf{N}_j^{S_2})\, dS'\, dS \qquad (13.75)$$
$$[Z_{S_1 S_2}]_{ij} = jk_0 \iint_{S_1} (\hat{n}_1 \times \mathbf{N}_i^{S_1}) \cdot \iint_{S_2} \overline{\mathbf{G}}_{e0}(\mathbf{r},\mathbf{r}') \cdot (\hat{n}_2' \times \mathbf{N}_j^{S_2})\, dS'\, dS$$
$$+ \iint_{S_1} [\hat{n}_1 \times (\hat{n}_1 \times \mathbf{N}_i^{S_1})] \cdot \iint_{S_2} \overline{\mathbf{G}}_{m0}(\mathbf{r},\mathbf{r}') \cdot (\hat{n}_2' \times \mathbf{N}_j^{S_2})\, dS'\, dS$$
(13.76)

for $i = 1, 2, \ldots, N_{S_1}$ and $j = 1, 2, \ldots, N_{S_2}$, where $\mathbf{N}_j^{S_2}$ denotes the vector basis functions on S_2 and N_{S_2} is the number of such basis functions. The evaluation of $\{R_{S_1}\}$ can be accelerated by using MLFMA [66–68], which reduces the computational complexity to $O(\sqrt{N_{S_1} N_{S_2}} \log(N_{S_1} N_{S_2}))$.

Equations (13.66) and (13.70) can be combined to form a complete linear system

$$\begin{bmatrix} K_{V_2V_2} & K_{V_2S_2} & 0 & 0 & 0 \\ K_{S_2V_2} & K_{S_2S_2} & B_{S_2S_2} & 0 & 0 \\ 0 & B_{S_2S_2}^T & L_{S_2S_2} & L_{S_2V_1} & 0 \\ 0 & 0 & L_{V_1S_2} & L_{V_1V_1} & L_{V_1S_1} \\ 0 & 0 & 0 & L_{S_1V_1} & L_{S_1S_1} \end{bmatrix} \begin{Bmatrix} E_{V_2} \\ E_{S_2} \\ \bar{H}_{S_2} \\ \bar{H}_{V_1} \\ \bar{H}_{S_1} \end{Bmatrix} = \begin{Bmatrix} 0 \\ 0 \\ 0 \\ 0 \\ b_{S_1} \end{Bmatrix} \quad (13.77)$$

Although this equation looks complicated, it is actually a simple FEM system of equations, whose coefficient matrix is purely sparse and symmetric. More importantly, this system is always lossy even if the FEM region is lossless. Therefore, the matrix is always nonsingular and relatively well conditioned.

However, the matrix equation (13.77) cannot be solved because $\{b_{S_1}\}$ has to be evaluated from the tangential electric and magnetic fields on S_2, which are unknown. This problem is solved through iterations, as described in the following three steps:

1. Let $\mathbf{R} = 0$, calculate $\{b_{S_1}\}$, and solve (13.77) for $\{E_{S_2}\}$ and $\{\bar{H}_{S_2}\}$.

2. Use $\{E_{S_2}\}$ and $\{\bar{H}_{S_2}\}$ to calculate $\mathbf{R}$ and then $\{b_{S_1}\}$.

3. Solve (13.77) for $\{E_{S_2}\}$ and $\{\bar{H}_{S_2}\}$ and check their convergence. If the solution is not converged, go back to step 2; otherwise, terminate iteration.

Note that in the above iterative process, the matrix in (13.77) remains the same; hence, there is no need to generate a new matrix. More importantly, if a direct sparse solver is used to solve (13.77), one has to factorize the matrix only once. This is also true even when the excitation is changed. A powerful direct solver developed recently is the multifrontal algorithm [69–71], which is the extension of the well-known frontal algorithm [72].

13.4.2 Numerical Results

A rigorous mathematical analysis [41] shows that the iterative process described above is guaranteed to converge exponentially. To demonstrate this convergence, we consider a conducting cube having a side length of 1λ, where λ denotes the free-space wavelength [Figure 13.7(a)]. The truncation surface S_1 is placed 0.05λ away from the surface of the cube, which is used as the integration surface S_2. The normalized residual error in the solution is plotted in Figure 13.8(a) as a function of the iteration number. Clearly, the iterative process converges very quickly. An exception to this is a cavity geometry, which can support many multiple bounces. In this exceptional case, it is better off to use a convex S_1 and let the FEM deal with the cavity. To show this, we consider the conducting cube again. In one case, we cut off one of its corners [Figure 13.7(b)] and in the other case, we cut a cavity ($0.5\lambda \times 0.5\lambda \times 0.5\lambda$) on one of its faces [Figure 13.7(c)]. In both cases, S_1 is placed 0.05λ away from the

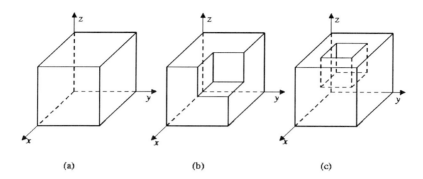

Figure 13.7 Three geometries to investigate the convergence of the adaptive absorbing boundary condition approach: (a) a complete cube; (b) a cube with an indented corner; (c) a cube with a cavity [41].

conducting surface. As shown in Figure 13.8(a), the convergence for the first case is still very fast even though the truncation surface is concave, but the convergence for the cavity case is much slower, as shown in Figure 13.8(b). However, when a convex S_1 is used and the cavity is modeled by the FEM, the convergence is significantly improved.

The second example is a metallic sphere having a diameter of 6λ. The sphere is coated with two 0.025λ-thick lossy dielectric layers. The relative permittivity is $\epsilon_r = 4.0 - j2.0$ for the inner layer and $\epsilon_r = 4.0 - j1.0$ for the the outer layer. The bistatic RCS in the E-plane is shown in Figure 13.9 and compared to the Mie solution. The RMS error is 0.08 dB. The third example is a metallic sphere having a diameter of 10λ. The sphere is coated with a 0.05λ-thick lossy dielectric layer having a relative permittivity of $\epsilon_r = 4.0 - j1.0$. The bistatic RCS in the E-plane is shown in Figure 13.10 and the RMS error is 0.12 dB. This kind of RMS error is considered quite small since the RCS has many deep nulls, whose values are more difficult to compute accurately.

The last example is a benchmark target recently designed by the Electromagnetic Code Consortium (EMCC). It is basically a 1-in thick trapezoidal metallic plate with its edges coated a 2-in-wide lossy dielectric of the same thickness. The detailed information about the geometry is shown in Figure 13.11. This target is designed to test the accuracy and capability of CEM codes for calculating the RCS from metallic and dielectric composites. The multiple interactions between the plate corners and the multiple bounces inside the dielectric coating make accurate calculation a very challenging problem. The calculated monostatic RCS at 1.0 GHz in the x-y plane for two polarizations is given in Figure 13.12.

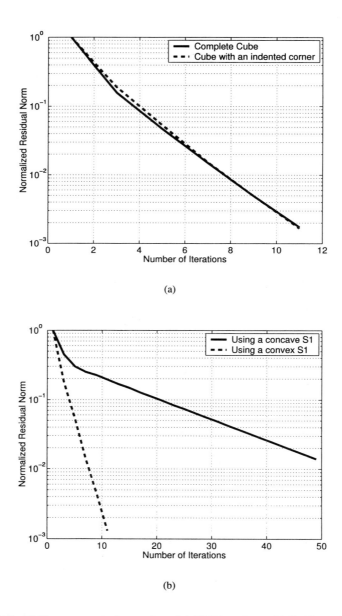

Figure 13.8 RMS error versus the number of AABC iterations: (a) for the complete cube and the cube with an indented corner; (b) for the cube with a cavity [41].

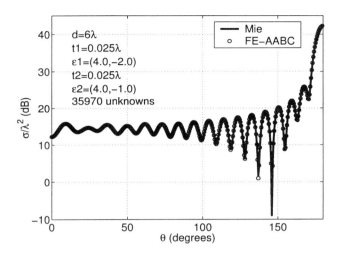

Figure 13.9 Comparison between the exact and FEM/AABC computations of the VV-polarized bistatic RCS of a metallic sphere of 6λ in diameter coated with two dielectric layers [41].

Table 13.2 Computational Information about the FEM/AABC Calculations [41]

Problem	Unknowns	Iterations	Memory	Setup time	Solution time
Fig. 13.9	35,970	32	420 MB	4,756 s	171 s
Fig. 13.10	46,660	61	618 MB	7,628 s	580 s
Fig. 13.12	56,598	45	800 MB	7,001 s	220 s

The computational information (the number of unknowns, the number of AABC iterations, memory used, CPU times) for each example is given in Table 13.2. The AABC iteration is terminated when the normalized residual norm reaches 0.001, which is a rather conservative number. The total computing time consists of two parts. The first part is for setup, which includes the times to set up MLFMA and the multifrontal algorithm. This setup is performed only once for each problem. Of the total setup time, the MLFMA consumes over 90%. The second part is the solution time performed for each right-hand side. Again, the MLFMA consumes a significant portion of this time. In the case of monostatic calculations, the number of AABC iterations and the solution time are averaged for each incident angle. All the computations are carried out on a DEC personal workstation (500-MHz Alpha 21164 processor).

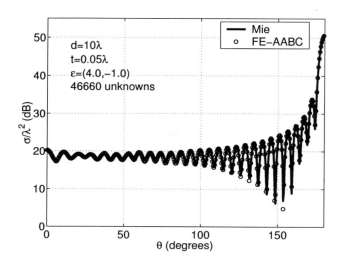

Figure 13.10 Comparison between the exact and FEM/AABC computations of the VV-polarized bistatic RCS of a metallic sphere of 10λ in diameter coated with one dielectric layer [41].

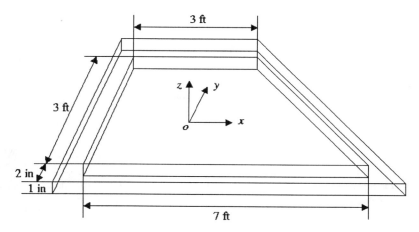

Figure 13.11 A trapezoidal plate with dielectric coated edges [41].

13.5 HYBRID FEM/SBR TECHNIQUE

Electromagnetic scattering by large bodies with cracks and cavities on their surfaces is a very important problem for practical applications. Because of complex wave

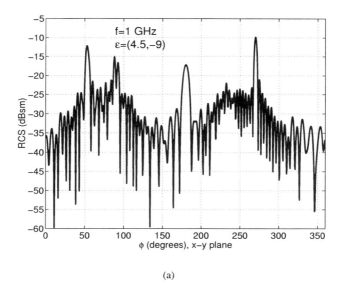

(a)

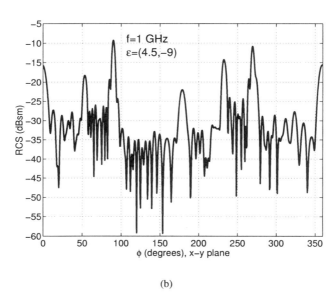

(b)

Figure 13.12 Monostatic RCS of the coated trapezoidal plate: (a) HH-polarized RCS in the x-y plane; (b) VV-polarized RCS in the x-y plane [41].

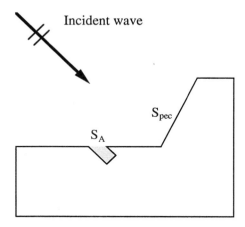

Figure 13.13 A large body with a crack.

phenomena, these cracks and cavities can make very significant contributions to the backscatter RCS. In this section, we describe a robust hybrid technique for this type of problem. This technique employs the SBR method to compute the scattering from the large bodies with the cracks and cavities filled with perfect conductors and the FEM to characterize the cracks and cavities whose openings are covered with perfect conductors. A coupling scheme is then developed to combine these two methods in such a manner that it includes all significant interactions. This results in an efficient and accurate computation of the scattering by large bodies with cracks and cavities.

13.5.1 Formulation

The generic problem under consideration is the scattering of electromagnetic waves by large perfectly electrical conducting (PEC) bodies having cracks and small cavities on their surfaces, which may be filled with inhomogeneous materials. For the sake of a clear description, we consider a single body with a single crack, whose cross section is illustrated in Figure 13.13. The dimension of the large body is on the order of tens or hundreds of wavelengths, and the crack is usually several wavelengths long and several tenths of a wavelength wide.

The field outside the scatterer satisfies the vector wave equation given in (13.23). To formulate an integral equation for this field, we introduce a dyadic Green's function $\overline{\mathbf{G}}_e$ that satisfies the differential equation (13.24) and the Sommerfeld radiation condition. By following the same procedure as outlined in Section 13.3, we obtain

$$\mathbf{E}(\mathbf{r}) = \mathbf{E}^{\text{inc}}(\mathbf{r}) - \iint_S \left[\overline{\mathbf{G}}_m(\mathbf{r}, \mathbf{r}') \cdot \mathbf{M}_s(\mathbf{r}') + jk_0 \overline{\mathbf{G}}_e(\mathbf{r}, \mathbf{r}') \cdot \bar{\mathbf{J}}_s(\mathbf{r}') \right] dS'$$
$$\mathbf{r} \in V_\infty \quad (13.78)$$

where $\overline{\mathbf{G}}_m(\mathbf{r}, \mathbf{r}') = \nabla \times \overline{\mathbf{G}}_e(\mathbf{r}, \mathbf{r}')$, $\bar{\mathbf{J}}_s$ and $\mathbf{M}_s$ are defined by (13.35), and

$$\mathbf{E}^{\text{inc}}(\mathbf{r}) = -jk_0 \eta_0 \iiint_{V_s} \overline{\mathbf{G}}_e(\mathbf{r}, \mathbf{r}') \cdot \mathbf{J}_i(\mathbf{r}') dV' \quad (13.79)$$

In this case, S is the surface of the scatterer, which consists of the surface of the PEC body, denoted by S_{pec}, and the aperture of the crack, denoted by S_A. Since on S_{pec}, $\mathbf{M}_s(\mathbf{r}') = \mathbf{E}(\mathbf{r}') \times \hat{n}' = 0$, (13.36) is reduced to

$$\mathbf{E}(\mathbf{r}) = \mathbf{E}^{\text{inc}}(\mathbf{r}) - \iint_{S_A} \overline{\mathbf{G}}_m(\mathbf{r}, \mathbf{r}') \cdot \mathbf{M}_s(\mathbf{r}') dS' - jk_0 \iint_S \overline{\mathbf{G}}_e(\mathbf{r}, \mathbf{r}') \cdot \bar{\mathbf{J}}_s(\mathbf{r}') dS'$$
$$\mathbf{r} \in V_\infty \quad (13.80)$$

Since S is very large, (13.80) cannot be used for an efficient numerical solution. However, if we can find a dyadic Green's function $\overline{\mathbf{G}}_e(\mathbf{r}, \mathbf{r}')$ such that

$$\hat{n}' \times \overline{\mathbf{G}}_e(\mathbf{r}, \mathbf{r}') = 0 \quad \mathbf{r}' \in S \quad (13.81)$$

(13.80) can be further reduced to

$$\mathbf{E}(\mathbf{r}) = \mathbf{E}^{\text{inc}}(\mathbf{r}) - \iint_{S_A} \overline{\mathbf{G}}_m(\mathbf{r}, \mathbf{r}') \cdot \mathbf{M}_s(\mathbf{r}') dS' \quad \mathbf{r} \in V_\infty \quad (13.82)$$

The corresponding magnetic field is given by

$$\bar{\mathbf{H}}(\mathbf{r}) = \bar{\mathbf{H}}^{\text{inc}}(\mathbf{r}) - jk_0 \iint_{S_A} \overline{\mathbf{G}}_e(\mathbf{r}, \mathbf{r}') \cdot \mathbf{M}_s(\mathbf{r}') dS' \quad \mathbf{r} \in V_\infty \quad (13.83)$$

where

$$\bar{\mathbf{H}}^{\text{inc}}(\mathbf{r}) = \eta_0 \iiint_{V_s} \overline{\mathbf{G}}_m(\mathbf{r}, \mathbf{r}') \cdot \mathbf{J}_i(\mathbf{r}') dV' \quad (13.84)$$

Apparently, (13.82) and (13.83) are much more efficient for a numerical solution since the unknown current $\mathbf{M}_s$ is now distributed only over the aperture of the crack S_A. Unfortunately, given an arbitrary S, it is impossible to find a dyadic Green's function $\overline{\mathbf{G}}_e$ that satisfies (13.81) in addition to (13.24) and the Sommerfeld radiation condition. This leads to two difficulties for a numerical solution of (13.82) and (13.83). The first is how to evaluate $\mathbf{E}^{\text{inc}}$ and $\bar{\mathbf{H}}^{\text{inc}}$. The second is how to evaluate the second term on the right-hand side of (13.82) and (13.83).

The first difficulty can be avoided by using a different method to evaluate $\mathbf{E}^{inc}$ and $\bar{\mathbf{H}}^{inc}$. Since $\mathbf{E}^{inc}$ and $\bar{\mathbf{H}}^{inc}$ represent the fields produced by $\mathbf{J}_i$ in the presence of a PEC S, which is the surface of the large body with the crack filled with a perfect conductor, it can be calculated very efficiently and reasonably accurately using a high-frequency asymptotic method, such as the SBR.

The SBR method is a high-frequency method for computing scattering of electromagnetic waves by electrically large bodies [14–17]. In this method, a very dense grid of rays, typically 10–20 rays per wavelength, representing the incident field, is launched toward the target. Each ray is traced as it bounces around within the target region and is governed by the GO. At the last hit point or at each and every hit point, a PO-type integration is performed to determine the ray contribution to the scattered field. The final result is a summation of the contributions from all the rays. The method can incorporate the ray-divergence factor and multilayered materials. Through the use of the PTD, the first-order edge diffraction can also be included. As a result, the method is efficient and accurate for large bodies and, thus, is ideal for computing $\mathbf{E}^{inc}$ and $\bar{\mathbf{H}}^{inc}$ in (13.82) and (13.83). To calculate $\bar{\mathbf{H}}^{inc}$ over the aperture of the crack whose typical width is only a small fraction of a wavelength, we have to realize that, because of the discrete nature of the rays, the hit-points of the rays on the surface of the target are also discrete. As a result, it is difficult to obtain the accurate value of $\bar{\mathbf{H}}^{inc}$ over the aperture of the crack. This difficulty can be alleviated by simply increasing the density of rays. However, since one does not know beforehand which group of the rays will hit the crack, it is necessary to increase the density of the entire grid of rays. This will increase dramatically the SBR computation time and, thus, reduce its efficiency. This problem is avoided in this work by first dividing the aperture S_A into small cells, and then introducing a fictitious boundary, slightly larger than S_A, around the aperture. If a ray is hit within this boundary, its ray tube area is subdivided into several small areas. Each of these small subareas is then assigned to the corresponding cells, depending on its location, with a phase correction. With this approach, $\bar{\mathbf{H}}^{inc}$ can be computed efficiently and accurately with a relative error of less than one percent.

The second difficulty mentioned above is how to evaluate the second term on the right-hand side of (13.82) and (13.83), which represents the field radiated by $\mathbf{M}_s$ in the presence of the large body with the crack filled with a perfect conductor. To deal with this difficulty, let us understand first how this term is used in the numerical solution. To use (13.83) for a numerical solution, we apply it to the aperture of the crack to find

$$\hat{n}_a \times \bar{\mathbf{H}}(\mathbf{r}) = \hat{n}_a \times \bar{\mathbf{H}}^{inc}(\mathbf{r}) - jk_0 \hat{n}_a \times \iint_{S_A} \overline{\mathbf{G}}_e(\mathbf{r},\mathbf{r}') \cdot \mathbf{M}_s(\mathbf{r}') dS'$$
$$\mathbf{r} \in S_A \quad (13.85)$$

where $\hat{n}_a$ denotes the unit vector normal to S_A and pointing away from the crack. By subdividing S_A into small surface elements and using vector basis functions to expand $\bar{\mathbf{H}}$ and $\mathbf{M}_s$, we obtain

$$[B_{AA}]\{\bar{H}_A\} + [P_{AA}]\{E_A\} = \{b_A\} \tag{13.86}$$

where

$$[B_{AA}]_{ij} = jk_0 \iint_{S_A} (\mathbf{g}_i \times \mathbf{g}_j) \cdot \hat{n}_a \, dS \tag{13.87}$$

$$[P_{AA}]_{ij} = k_0^2 \iint_{S_A} \left[\mathbf{g}_i \cdot \iint_{S_A} \overline{\mathbf{G}}_e(\mathbf{r}, \mathbf{r}') \cdot \mathbf{g}_j \, dS' \right] dS \tag{13.88}$$

$$\{b_A\}_i = jk_0 \iint_{S_A} \mathbf{g}_i \cdot \bar{\mathbf{H}}^{\text{inc}}(\mathbf{r}) dS \tag{13.89}$$

It is obvious that $\overline{\mathbf{G}}_e(\mathbf{r}, \mathbf{r}')$ is used only in the evaluation of $[P_{AA}]_{ij}$. Since in this evaluation, both $\mathbf{r}$ and $\mathbf{r}'$ are on S_A, $\overline{\mathbf{G}}_e(\mathbf{r}, \mathbf{r}')$ represents the interaction between two points on S_A. In such a case, if S_A is located on a locally planar surface, $\overline{\mathbf{G}}_e(\mathbf{r}, \mathbf{r}')$ can be written as

$$\overline{\mathbf{G}}_e(\mathbf{r}, \mathbf{r}') = \overline{\mathbf{G}}_{\text{half}}(\mathbf{r}, \mathbf{r}') + \overline{\mathbf{G}}_{\text{diff}}(\mathbf{r}, \mathbf{r}') \tag{13.90}$$

where $\overline{\mathbf{G}}_{\text{half}}(\mathbf{r}, \mathbf{r}')$ denotes the half-space Green's function [58] given by

$$\overline{\mathbf{G}}_{\text{half}}(\mathbf{r}, \mathbf{r}') = \overline{\mathbf{G}}_{e0}(\mathbf{r}, \mathbf{r}') - \overline{\mathbf{G}}_{e0}(\mathbf{r}, \mathbf{r}'_i) + 2\hat{n}_a \hat{n}_a g(\mathbf{r}, \mathbf{r}'_i) \tag{13.91}$$

where $\overline{\mathbf{G}}_{e0}$ denotes the free-space dyadic Green's function and $\mathbf{r}'_i$ denotes the image point of $\mathbf{r}'$. The second term in (13.90), $\overline{\mathbf{G}}_{\text{diff}}(\mathbf{r}, \mathbf{r}')$, denotes the difference between $\overline{\mathbf{G}}_e(\mathbf{r}, \mathbf{r}')$ and $\overline{\mathbf{G}}_{\text{half}}(\mathbf{r}, \mathbf{r}')$. This difference is caused by the reflection and diffraction of the large body, and the expression for $\overline{\mathbf{G}}_{\text{diff}}(\mathbf{r}, \mathbf{r}')$ is often difficult, if not impossible, to obtain. To avoid this difficulty, we neglect $\overline{\mathbf{G}}_{\text{diff}}(\mathbf{r}, \mathbf{r}')$ in the evaluation of $[P_{AA}]_{ij}$. Doing so, we neglect the field scattered by the crack, diffracted and/or reflected back to the crack by the large body, and scattered by the crack again. In most cases, this field is unimportant. Note that with the neglect of $\overline{\mathbf{G}}_{\text{diff}}(\mathbf{r}, \mathbf{r}')$, (13.88) can be written as

$$[P_{AA}]_{ij} = k_0^2 \iint_{S_A} \left[\mathbf{g}_i \cdot \iint_{S_A} g(\mathbf{r}, \mathbf{r}') \mathbf{g}_j(\mathbf{r}') dS' \right] dS$$

$$- \iint_{S_A} (\nabla \cdot \mathbf{g}_i) \left[\iint_{S_A} g(\mathbf{r}, \mathbf{r}') \nabla' \cdot \mathbf{g}_j \, dS' \right] dS \tag{13.92}$$

With (13.86), it remains to find another relation between $\{E_A\}$ and $\{\bar{H}_A\}$ from the interior of the crack. Because of the possible complexity of the interior of the crack, the FEM is used to formulate the field inside the crack. The functional for the interior field is

$$F(\mathbf{E}) = \frac{1}{2} \iiint_V \left[\frac{1}{\mu_r} (\nabla \times \mathbf{E}) \cdot (\nabla \times \mathbf{E}) - k_0^2 \epsilon_r \mathbf{E} \cdot \mathbf{E} \right] dV$$
$$+ jk_0 \iint_{S_A} (\mathbf{E} \times \bar{\mathbf{H}}) \cdot \hat{n}_a \, dS \quad (13.93)$$

where V denotes the volume of the cavity. Application of FEM yields

$$\begin{bmatrix} K_{VV} & K_{VA} & 0 \\ K_{AV} & K_{AA} & B_{AA} \end{bmatrix} \begin{Bmatrix} E_V \\ E_A \\ \bar{H}_A \end{Bmatrix} = \begin{Bmatrix} 0 \\ 0 \end{Bmatrix} \quad (13.94)$$

which can be solved together with (13.86). Note that the combined final system can also be written as

$$\begin{bmatrix} K_{VV} & K_{VA} \\ K_{AV} & K_{AA} - P_{AA} \end{bmatrix} \begin{Bmatrix} E_V \\ E_A \end{Bmatrix} = \begin{Bmatrix} 0 \\ b_A \end{Bmatrix} \quad (13.95)$$

whose coefficient matrix remains symmetric.

13.5.2 Scattered Field Calculation

Once the electric field, or the equivalent magnetic current, over the crack's aperture is obtained, the field scattered by the crack can be calculated by

$$\bar{\mathbf{H}}^{\text{scat}}(\mathbf{r}) = -jk_0 \iint_{S_A} \overline{\mathbf{G}}_e(\mathbf{r}, \mathbf{r}') \cdot \mathbf{M}_s(\mathbf{r}') dS' \quad \mathbf{r} \in V_\infty \quad (13.96)$$

However, as mentioned earlier, in many cases we do not have the expression for $\overline{\mathbf{G}}_{\text{diff}}(\mathbf{r}, \mathbf{r}')$, and neglecting this term here would neglect the field radiated by $\mathbf{M}_s$ and reflected by the large body, resulting in a significant error in the scattered field computation. To alleviate this difficulty, we employ the reciprocity theorem [73, 74]. In accordance with the reciprocity theorem [73], the field $\mathbf{H}^{\text{scat}}$ radiated by $\mathbf{M}_s$ in the presence of the large body is related to $\mathbf{M}_s$ by

$$\iiint_{V'} \mathbf{H}^{\text{scat}} \cdot \mathbf{M}' \, dV' = \iint_{S_A} \mathbf{H}' \cdot \mathbf{M}_s \, dS \quad (13.97)$$

where $\mathbf{M}'$ denotes an arbitrary magnetic current and $\mathbf{H}'$ is the field radiated by $\mathbf{M}'$ in the presence of the large body. By choosing an infinitesimal magnetic current element as $\mathbf{M}'$ and placing it at the observation point, we obtain the far field radiated by $\mathbf{M}_s$ in the presence of the large body as

$$E_{\theta,\phi}^{\text{scat}}(\mathbf{r}) = jk_0 \eta_0 \frac{e^{-jk_0 r}}{4\pi r} \iint_{S_A} \mathbf{M}_s \cdot \mathbf{H}'_{v,h} dS \quad (13.98)$$

where $\mathbf{H}'_v$ denotes the magnetic field due to the vertically polarized incident field (whose electric field has a unit amplitude and θ-component only) in the presence of the large body with the crack filled with a perfect conductor. Likewise, $\mathbf{H}'_h$ denotes the magnetic field due to the horizontally polarized incident field (whose electric field has a unit amplitude and ϕ-component only) under the same condition. Both $\mathbf{H}'_v$ and $\mathbf{H}'_h$ can be computed efficiently and accurately using the SBR method. In the backscatter case, they are the same as the incident field used in (13.83).

13.5.3 Analysis of the Hybrid Technique

Rigorously speaking, the formulation of the hybrid technique described above is an exact formulation because other than the errors inherited from the FEM and SBR method, there is no approximation introduced in the formulation. However, as we mentioned above, it is often more convenient and efficient to omit the second part of (13.90) in the calculation of (13.88). It is then instructive to examine the effect of this omission on the accuracy of the scattered-field computation.

First of all, because of the use of the SBR method, which includes multiple bounces (and edge diffraction, if necessary), for computing $\bar{\mathbf{H}}^{\text{inc}}$ in (13.89), the incident field reflected (and diffracted) by the large body is included in the hybrid technique. This inclusion is important since the magnitude of the reflected incident field is often comparable to that of the field incident directly on the crack, as shown in Figure 13.14(a). Second, because of the use of the reciprocity theorem for calculating the scattered field and the use of the SBR method for computing $\mathbf{H}'_{v,h}$ in (13.98), the scattered field reflected (and diffracted) by the large body is also included in the scattered-field calculation. This inclusion is also very important since the magnitude of the reflected scattered field is comparable to the field scattered directly into the observation direction. Because of these two features, the crack can contribute to the scattered field even if it is in the shadow region, which is illustrated clearly in Figure 13.14(b).

Now, let us examine the role of the second part of (13.90) in the calculation of (13.88). First, we recognize that $\overline{\mathbf{G}}_e$ in (13.88) represents the interaction between the two points on the crack's aperture, or more specifically, the field at $\mathbf{r}$ produced by the source at $\mathbf{r}'$. This field consists of two parts. The first part is the direct field from $\mathbf{r}'$ to $\mathbf{r}$ in the presence of the conductor beneath the current, which is represented by $\overline{\mathbf{G}}_{\text{half}}$. The second part is the field produced by the source at $\mathbf{r}'$, then diffracted or reflected by the large body to the point at $\mathbf{r}$, and is represented by $\overline{\mathbf{G}}_{\text{diff}}$. Therefore, neglecting the second part (13.90) means that we neglect the field scattered by the crack, diffracted or reflected back into the crack, and scattered by the crack again, as illustrated in Figure 13.15. If the crack is several wavelengths away from the edges and reflecting surfaces, $\overline{\mathbf{G}}_{\text{diff}}$ is much smaller than $\overline{\mathbf{G}}_{\text{half}}$, and neglecting it in (13.90) will not introduce a noticeable error in the scattered-field computation. However, if the crack is very close to the edges and reflecting surfaces, $\overline{\mathbf{G}}_{\text{diff}}$ can be comparable

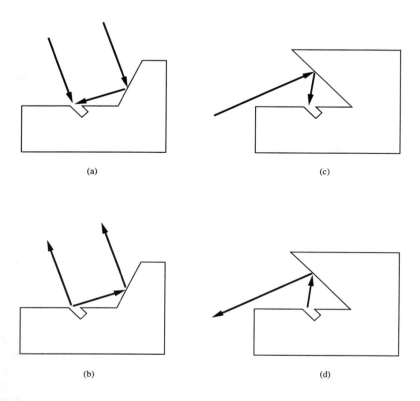

Figure 13.14 Effects included in the formulation: (a, b) indirect incident fields; (c, d) indirect scattered fields.

to $\overline{\mathbf{G}}_{\text{half}}$, and in this case we must include the effect of $\overline{\mathbf{G}}_{\text{diff}}$ in a manner similar to that described in [75–77].

13.5.4 Numerical Results

In this section, we present several results obtained with the hybrid FEM/SBR technique to demonstrate its accuracy, efficiency, and capability.

The first problem is a $5\lambda \times 5\lambda \times 5\lambda$ conducting cube having a crack on its upper surface. The crack is 5λ long, 0.2λ wide, and 0.25λ deep. It is well known that the solution to the scattering by this geometry can be obtained from a two-dimensional solution if the incident plane wave is perpendicular to the axis of the crack. The backscatter and bistatic RCS are shown in Figure 13.16. The results of the hybrid technique are compared with the scaled 2D solution obtained using the MOM. As

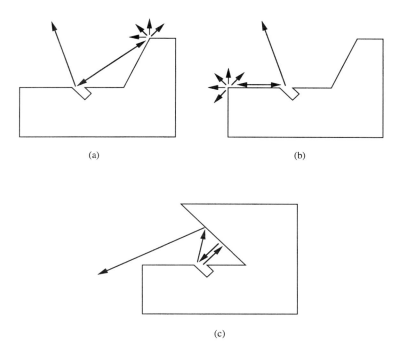

Figure 13.15 Effects not included when $\overline{\mathbf{G}}_{\text{diff}}(\mathbf{r}, \mathbf{r}')$ is omitted in the calculation of $[P_{AA}]_{ij}$: (a, b) field scattered by the crack, then diffracted back to the crack, and scattered by the crack again; (c) field scattered by the crack, then reflected back to the crack, and scattered by the crack again.

can be seen, the agreement between the two solutions is very good, demonstrating the validity of the hybrid technique.

The scatterer for the second problem is illustrated in Figure 13.17, where a 5λ long, 0.5λ wide, and 0.25λ deep crack is situated on an L-shaped conducting object. This problem differs from the first in that both incident and scattered fields can have multiple bounces. The solution to this problem can also be obtained using the two-dimensional MOM. Again, the agreement between the hybrid solution and the scaled 2D MOM result for the backscatter RCS is very good for both VV- and HH-polarizations.

The third problem is plane wave scattering by a 40λ long, 15.5λ wide, and 7.2λ thick conducting almond. The top of the almond is flattened into an ogive-shaped planar surface carved with a 16.256λ long, 0.2λ wide, and 0.85λ deep rectangular crack (see [78] for a detailed description of this geometry). The RCS computed using

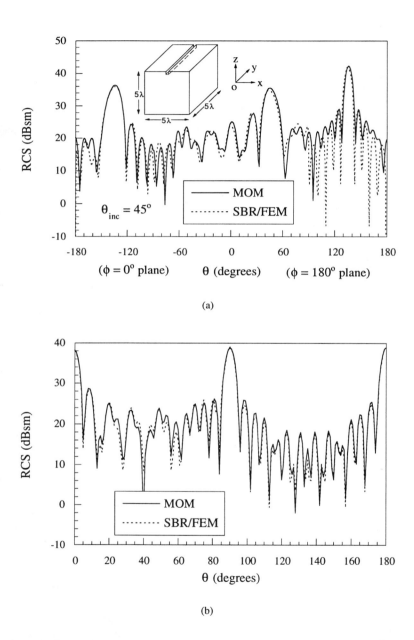

Figure 13.16 Comparison of the results computed using the hybrid FEM/SBR and the MOM for VV-polarization: (a) bistatic RCS; (b) monostatic RCS [46].

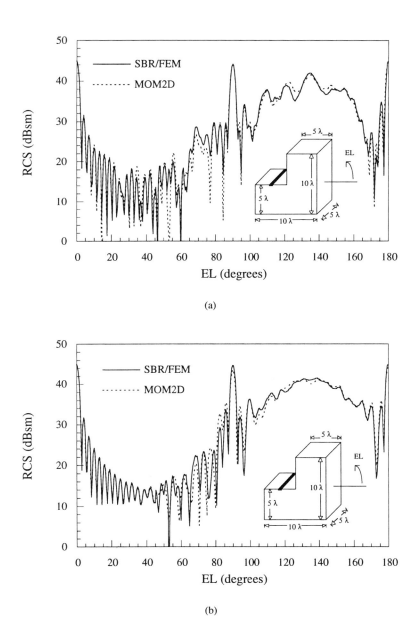

Figure 13.17 Comparison of the monostatic RCS computed using the hybrid FEM/SBR and the MOM: (a) VV-polarization; (b) HH-polarization [46].

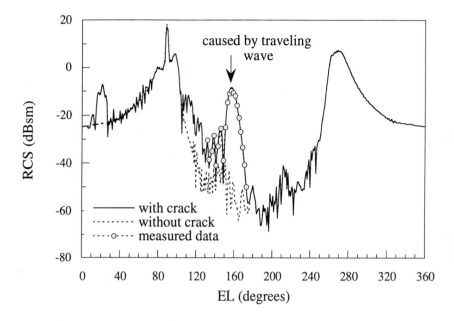

Figure 13.18 Monostatic RCS of an almond with a crack on its upper surface for HH-polarization. The crack is 16.256λ long, 0.2λ wide, and 0.85λ deep, and the almond is 40λ long, 15.5λ wide, and 7.2λ thick.

our hybrid technique is given in Figure 13.18, which shows good agreement with the measured data [78]. The peaks around the angles of 20 and 160 degrees are caused by the traveling waves supported by the crack.

13.6 HYBRID MOM/SBR TECHNIQUE

Another problem of practical importance is the scattering by a large body with small protruding structures. A promising approach for this problem is again to hybridize a numerical method with an asymptotic method. There are two extremes for this type of hybridization. One is simply to superimpose solutions from asymptotic and numerical methods. While this approach is most widely used in practical applications, it neglects the interactions between the two solutions, which can be significant in many problems. The other extreme is to combine an asymptotic and a numerical method in an exact manner, such as the classical work of combining the MOM with the GTD by Thiele et al. [75–77]. In this approach, the effect of a large body is included by incorporating its diffraction into the Green's function in the integral equation for the small structures, which accounts for all interactions. The approach is particularly attractive for analyzing the radiation of an antenna placed on a large body, and it has recently been extended to scattering by finned convex objects [79]. While this approach is accurate, it is difficult to implement in a general-purpose computer code because of its complex nature.

A more practical approach is to develop a technique that can include all significant interactions and neglect all trivial interactions, similar to the one described in the preceding section. In this section, we employ the same philosophy to develop a technique that combines the SBR method and the MOM to solve for the scattering by large conducting bodies with small structures mounted on their surfaces.

13.6.1 Formulation

Consider the problem of wave scattering by a large, perfectly conducting body with a small protruding structure, illustrated in Figure 13.19(a). By using the dyadic Green's function $\overline{\mathbf{G}}_e$ that satisfies (13.24) and the Sommerfeld radiation condition, we can derive the integral equation for the electric field as

$$\mathbf{E}(\mathbf{r}) = \mathbf{E}^{\text{inc}}(\mathbf{r}) - jk_0 \iint_S \overline{\mathbf{G}}_e(\mathbf{r},\mathbf{r}') \cdot \bar{\mathbf{J}}_s(\mathbf{r}')dS' \tag{13.99}$$

where S denotes the surface of the entire scatterer, $\mathbf{J}_s$ is the surface current density induced on S, and $\mathbf{E}^{\text{inc}}$ represents the incident electric field given by

$$\mathbf{E}^{\text{inc}}(\mathbf{r}) = -jk_0\eta_0 \iiint_{V_s} \overline{\mathbf{G}}_e(\mathbf{r},\mathbf{r}') \cdot \mathbf{J}_i(\mathbf{r}')dV' \tag{13.100}$$

in which V_s denotes the volume occupied by $\mathbf{J}_i$. If $\overline{\mathbf{G}}_e$ is the free-space dyadic Green's function, $\mathbf{E}^{\text{inc}}$ is then the electric field produced by $\mathbf{J}_i$ in the free space without the scatterer.

Equation (13.99) with the free-space dyadic Green's function provides the necessary integral equation for a MOM solution of $\mathbf{J}_s$ and then $\mathbf{E}$. However, since S includes the entire surface of the scatterer, a MOM solution of (13.99) requires the discretization of the entire surface, resulting in a very large number of unknowns for a large scatterer. As a result, the size of a scatterer to be handled by MOM is very limited because of the limitation of computer memory and time. To alleviate this limitation for the special scatterers considered here, we decompose the scatterer in Figure 13.19(a) into two parts: one is the large body, whose surface is denoted as S_b, and the other is the small protrusion, whose surface is represented by S_p, as illustrated in Figure 13.19(b). Assuming that $\overline{\mathbf{G}}_e$ is a dyadic Green's function that satisfies the boundary condition

$$\hat{n}' \times \overline{\mathbf{G}}_e(\mathbf{r},\mathbf{r}') = 0 \qquad \mathbf{r}' \in S_b \tag{13.101}$$

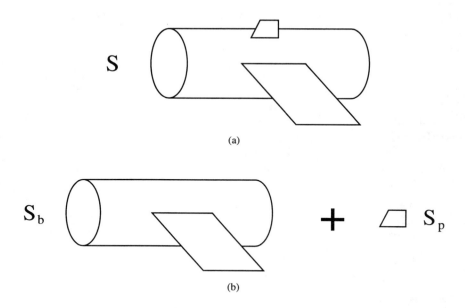

Figure 13.19 (a) Original problem and (b) decomposed problem.

(13.99) then becomes

$$\mathbf{E}(\mathbf{r}) = \mathbf{E}^{\text{inc}}(\mathbf{r}) - jk_0 \iint_{S_p} \overline{\mathbf{G}}_e(\mathbf{r},\mathbf{r}') \cdot \bar{\mathbf{J}}_s(\mathbf{r}')dS' \tag{13.102}$$

where $\mathbf{E}^{\text{inc}}$ now becomes the electric field produced by $\mathbf{J}_i$ in the presence of the large body without the protrusion. Since S_p is small, (13.102) can be solved by MOM efficiently.

The formulation described above can be readily applied to problems where the large body has one of the canonical shapes for which the dyadic Green's function is available. However, for a general shape of S_b, the explicit expression of $\overline{\mathbf{G}}_e$ is usually unknown. As a result, the MOM solution of (13.102) has two difficulties: (1) the required excitation $\mathbf{E}^{\text{inc}}$ cannot be calculated, and (2) the elements of the MOM impedance matrix cannot be computed. In the following, we discuss the approaches to alleviate these difficulties.

Like in the preceding section, the first difficulty can be alleviated by using a different method to calculate $\mathbf{E}^{\text{inc}}$. Since $\mathbf{E}^{\text{inc}}$ is the incident electric field in the presence of the large body without the protrusion, it can be calculated efficiently and accurately using the SBR method with proper care [50, 80].

To alleviate the second difficulty, we first consider the role of $\overline{\mathbf{G}}_e$ in the MOM solution. To solve (13.102) by MOM, we apply it on S_p and take the cross product with the normal of S_p, yielding

$$jk_0\hat{n}_p \times \iint_{S_p} \overline{\mathbf{G}}_e(\mathbf{r},\mathbf{r}') \cdot \bar{\mathbf{J}}_s(\mathbf{r}')dS' = \hat{n}_p \times \mathbf{E}^{\text{inc}}(\mathbf{r}) \qquad \mathbf{r} \in S_p \qquad (13.103)$$

Since both $\mathbf{r}$ and $\mathbf{r}'$ are on S_p, the $\overline{\mathbf{G}}_e(\mathbf{r},\mathbf{r}')$ in this equation represents the interaction between the two points on the protrusion. This interaction includes three contributions. The first is the direct interaction between the two points (the field produced by the point source at $\mathbf{r}'$ reaches directly at $\mathbf{r}$) and this interaction can be represented by the free-space dyadic Green's function. The second is the interaction through the base on which the protrusion is placed (the field produced by the point source at $\mathbf{r}'$ is reflected by the base to the point $\mathbf{r}$) and this interaction can be represented by the free-space dyadic Green's function with the $\mathbf{r}'$ replaced by the image point $\mathbf{r}'_i$. The combination of these two interactions can be represented by the half-space dyadic Green's function denoted as $\overline{\mathbf{G}}_{\text{half}}$. The third contribution is the interaction through the other parts of the large body such as the edges (the field produced by the point source at $\mathbf{r}'$ is diffracted by the edges to the point $\mathbf{r}$). We denote this interaction as $\overline{\mathbf{G}}_{\text{diff}}$, which is the difference between $\overline{\mathbf{G}}_e$ and $\overline{\mathbf{G}}_{\text{half}}$. Therefore, we have

$$\overline{\mathbf{G}}_e(\mathbf{r},\mathbf{r}') = \overline{\mathbf{G}}_{\text{half}}(\mathbf{r},\mathbf{r}') + \overline{\mathbf{G}}_{\text{diff}}(\mathbf{r},\mathbf{r}') \qquad (13.104)$$

where $\overline{\mathbf{G}}_{\text{half}}$ is given in (13.91). Whereas the expression for $\overline{\mathbf{G}}_{\text{half}}$ is available, the expression for $\overline{\mathbf{G}}_{\text{diff}}$ is often difficult, if not impossible, to obtain. Although the effect of $\overline{\mathbf{G}}_{\text{diff}}$ can still be included in the MOM solution, as was done in [76–79], its numerical implementation is complicated and dependent on the geometry of the large object. To simplify the MOM solution and effectively decouple the MOM and SBR computations, we neglect $\overline{\mathbf{G}}_{\text{diff}}$ in the MOM solution of (13.103), and, doing so, we neglect the field scattered by the protrusion, diffracted and/or reflected back to the protrusion by the large body, and scattered by the protrusion again. In most cases, this field is unimportant. However, when necessary it can be recovered by using an iterative approach discussed in the next section.

The MOM solution of (13.103) is straightforward. First, S_p is subdivided into small triangular elements and the current on S_p is expanded as

$$\bar{\mathbf{J}}_s = \sum_{n=1}^{N} \bar{I}_n \mathbf{f}_n(\mathbf{r}) \qquad (13.105)$$

where N is the number of unknowns and $\mathbf{f}_n(\mathbf{r})$ denotes the vector basis functions, which can be the same as $\mathbf{g}_i$ in the preceding sections. One popular choice for $\mathbf{f}_n(\mathbf{r})$ is the Rao-Wilton-Glisson (RWG) basis functions [81]. In the usual MOM analysis, N includes only the interior edges; however, here N must include the boundary edges connecting the protrusion to the large body to allow the continuous flow of the current from the protrusion to the large body or vice versa. Applying Galerkin's method to (13.103) results in a matrix equation

$$[Z]\{\bar{I}\} = \{V\} \qquad (13.106)$$

whose elements are given by

$$Z_{mn} = jk_0 \iint_{S_p} \iint_{S_p} \left\{ \left[\mathbf{f}_m(\mathbf{r}) \cdot \mathbf{f}_n(\mathbf{r}') - \frac{1}{k_0^2} \nabla \cdot \mathbf{f}_m(\mathbf{r}) \nabla' \cdot \mathbf{f}_n(\mathbf{r}') \right] \right.$$
$$\left. \cdot [g(\mathbf{r},\mathbf{r}') - g(\mathbf{r},\mathbf{r}'_i)] + 2\hat{n}_b \cdot \mathbf{f}_m(\mathbf{r})\hat{n}_b \cdot \mathbf{f}_n(\mathbf{r}')g(\mathbf{r},\mathbf{r}'_i) \right\} dS' dS \quad (13.107)$$

$$V_m = \iint_{S_p} \mathbf{E}^{\text{inc}}(\mathbf{r}) \cdot \mathbf{f}_m(\mathbf{r}) dS \qquad (13.108)$$

where $\hat{n}_b$ denotes the unit vector normal to the base of the protrusion.

13.6.2 Scattered Field Calculation

Once the surface current on the protrusion is obtained, the scattered field radiated by this current in the far zone can be evaluated using

$$\mathbf{E}^{\text{scat}}(\mathbf{r}) = -jk_0 \iint_{S_p} \overline{\mathbf{G}}_e(\mathbf{r},\mathbf{r}') \cdot \bar{\mathbf{J}}_s(\mathbf{r}') dS' \qquad (13.109)$$

Since $\mathbf{r}$ in this equation represents the observation point in the far zone, one cannot replace $\overline{\mathbf{G}}_e$ with $\overline{\mathbf{G}}_{\text{half}}$ because such a replacement would neglect the field scattered by the protrusion and diffracted and/or reflected to the observation point, resulting in an error whose magnitude is comparable to the field scattered directly to the observation point. This problem can be alleviated by using the approach based on the reciprocity theorem, as done in the FEM/SBR method. In this approach, we place an infinitesimal electric current element at the observation point, either vertically polarized or horizontally polarized. We then compute the electric field $\mathbf{E}'_{v,h}$ produced by this current element on S_p in the presence of the large body without the protrusion using the SBR method. In the backscatter case, $\mathbf{E}'_{v,h}$ is the same as $\mathbf{E}^{\text{inc}}$. From the reciprocity theorem, the scattered field can be obtained as

$$E^{\text{scat}}_{\theta,\phi}(\mathbf{r}) = -jk_0 \frac{e^{-jk_0 r}}{4\pi r} \iint_{S_p} \bar{\mathbf{J}}_s \cdot \mathbf{E}'_{v,h} dS' \qquad (13.110)$$

The total scattered field from the entire scatterer is the superposition of this field and the field scattered by the large body without the protrusion, which can be calculated efficiently and accurately using the SBR method.

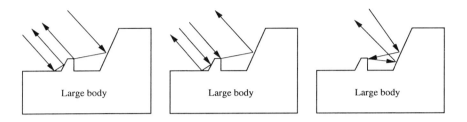

Figure 13.20 Effects included in the hybrid technique: direct and indirect incident fields and direct and indirect scattered fields.

13.6.3 Iterative Improvement

Because of the use of the SBR method, $\mathbf{E}^{\text{inc}}$ in (13.103) includes not only the direct incident field, but also the fields multiply reflected by the large body. Generally speaking, the magnitude of the indirect incident field is comparable to that of the direct field, so neglecting either of them will result in a significant error in the calculation of $\mathbf{E}^{\text{inc}}$. Similarly, since $\mathbf{E}'_{v,h}$ in (13.110) is calculated using the SBR method, the reflection and multiple bounces are also included in the scattered-field calculation. Therefore, all major interactions, as illustrated in Figure 13.20, between the SBR and MOM have been included in the hybrid technique.

The only approximation in the hybrid technique is introduced by the approximate Green's function, formed by neglecting the second term in (13.104). As pointed out earlier, this neglects the field scattered by the protrusion, reflected and/or diffracted back to the protrusion by the large object, and scattered by the protrusion again, as illustrated in Figure 13.21. In most problems, this contribution is insignificant. However, when the protrusion is very close to edges and reflecting surfaces, this contribution can become significant and its omission can cause a substantial error in the solution. Here, we describe an iterative approach, similar to those in [82–84], to reduce the error systematically.

In this iterative approach, we use the current on the protrusion obtained from (13.106) as the initial value and then calculate the field produced by this current in the presence of the large body. This field can be considered as the secondary incident field, which, when superimposed to the $\mathbf{E}^{\text{inc}}$, yields a new incident field on the protrusion. Using this as the incident field in (13.108), we obtain a new, improved current on the protrusion. This process is repeated several times until a stable value for the current is reached. The iterative process can be expressed as

$$\{I\}_i = [Z]^{-1}\{V[\mathbf{E}^{\text{inc}} + \mathbf{E}(\{I\}_{i-1})]\} \qquad (13.111)$$

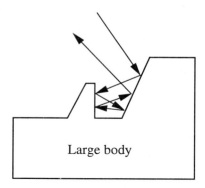

Figure 13.21 Effects not included in the hybrid technique, but can be recovered by an iterative approach: field scattered by the protrusion, diffracted and/or reflected back to the protrusion by the large object, and scattered by the protrusion again.

where i denotes the number of iterations, and $\mathbf{E}(\{I\}_{i-1})$ denotes the field on the protrusion produced by the current $\{I\}_{i-1}$ on the protrusion, which can be calculated using either PO or the SBR method.

A similar approach can be employed for large bodies with multiple protrusions. When each protrusion is characterized using the MOM, the interaction between them is neglected. To recover this interaction, we can first analyze protrusions separately and obtain the current on each protrusion. We then choose the current on one of the protrusions as the excitation to obtain the secondary incident fields on other protrusions, which then yield new currents. This process can be repeated until the convergence is reached. For the case with two protrusions, the process can be expressed as

$$\{I_1\}_i = [Z_1]^{-1} \{V_1[\mathbf{E}_1^{\text{inc}} + \mathbf{E}_1(\{I_2\}_{i-1})]\} \tag{13.112}$$

$$\{I_2\}_i = [Z_2]^{-1} \{V_2[\mathbf{E}_2^{\text{inc}} + \mathbf{E}_2(\{I_1\}_{i-1})]\} \tag{13.113}$$

where $\mathbf{E}_j(\{I_k\})$ denotes the field on the jth protrusion produced by the current on the kth protrusion, which can again be calculated using either PO or the SBR method.

13.6.4 Numerical Results

This section gives some examples to demonstrate the accuracy and capability of the hybrid MOM/SBR technique.

The first example is a scatterer consisting of two plates of different sizes ($2\lambda \times 2\lambda$ and $1\lambda \times 1\lambda$, respectively) placed 1λ apart on an $8\lambda \times 8\lambda$ large plate having a thickness of 1λ. The hybrid solution is shown in Figure 13.22 and is compared to the result obtained using FISC [85]. The agreement between the two solutions is good. In calculating the hybrid solution, we apply a single MOM to the two small plates. As a result, the multiple interactions between the two small plates are included in the MOM solution.

To show the effectiveness of the iterative approach, we reconsider the previous example. We first apply the MOM to each of the two plates independently; hence, no interaction between them is included. The hybrid solution so obtained is represented by the dotted line in Figure 13.23, which shows a significant disagreement with the FISC result, especially in the region between 130 and 160 degrees, because of the strong multiple interactions. This error is, however, reduced significantly when the iterative approach is applied with only two iterations. The result is shown by the dashed line, which agrees very well with the FISC solution.

To further demonstrate the effectiveness of the iterative approach, we consider an L-shaped conducting body with a protrusion on the surface. This problem differs from the previous one in that both the incident and scattered fields can have multiple bounces. First, we neglect the contribution of $\overline{G}_{diff}$, and the results obtained are given in Figure 13.24 where a noticeable error is observed. This is expected because the protrusion is close to the reflecting surface. The results obtained using the iterative approach with 10 iterations are shown in Figure 13.25, which demonstrates clearly again that the iterative approach is an effective method to improve the accuracy for such a problem.

In the last example, the protrusion is a geometrically tapered circular disk supported by a small circular cylinder having a radius of 0.25m and a length of 0.7m. This object is placed on the VFY218 airplane, as shown in Figure 13.26. The RCS of this target is given in Figure 13.27. The hybrid solution is compared with the result obtained by XPATCH, which applies the SBR directly to the entire target. The two solutions agree remarkably well. Also shown is the RCS of the target without the protrusion calculated using XPATCH. The results show that the protrusion has a significant effect on the total RCS in the low-observable directions.

13.7 SUMMARY

In this chapter, we discussed an important issue in computational electromagnetics—the development of hybrid techniques for large-scale, complex electromagnetics problems. We demonstrated clearly that properly formulated hybrid techniques can be more efficient than the first-principle numerical methods and more accurate than the high-frequency asymptotic methods. We also illustrated clearly that the development of hybrid techniques requires good understanding of each of the numerical and asymptotic methods: their detailed formulation, implementation, capabilities, and limitations.

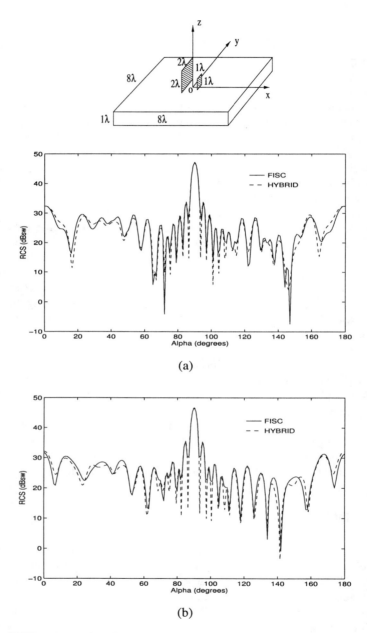

Figure 13.22 Monostatic RCS of two small plates placed on a large thick plate in the xz-plane (multiple interactions are included in the MOM solution): (a) VV-polarization; (b) HH-polarization [50].

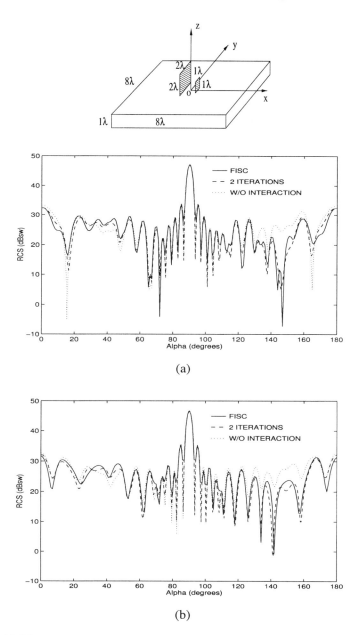

Figure 13.23 Monostatic RCS of two small plates placed on a large thick plate in the xz-plane (multiple interactions are not included in the MOM solution, but recovered by the iterative approach): (a) VV-polarization; (b) HH-polarization [50].

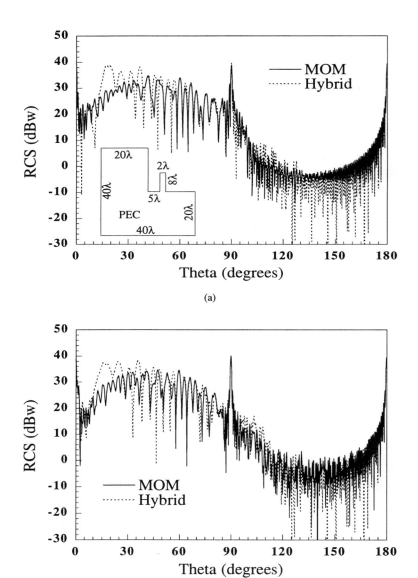

Figure 13.24 Comparison of the monostatic echo-width calculated by the hybrid MOM/SBR and the MOM for an L-shaped body with a protrusion: (a) TM polarization; (b) TE polarization [48].

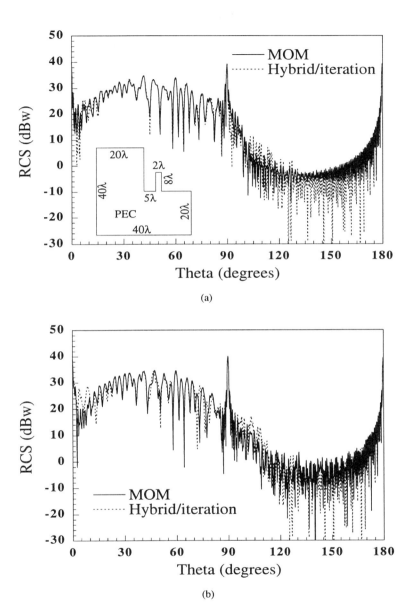

Figure 13.25 Comparison of the monostatic echo-width calculated by the hybrid MOM/SBR in conjunction with the iterative approach and the MOM for an L-shaped body with a protrusion: (a) TM polarization; (b) TE polarization [48].

Figure 13.26 The VFY218 airplane (15.5m long, 9m wide, and 4.1m thick) with a mushroom-shaped protrusion.

We considered two classes of hybrid techniques. The first class combines two numerical methods to form a more powerful hybrid numerical technique. An example given in this chapter was the one that combines the FEM and BIE. The resulting FEM/BIE method is especially suited to dealing with open-region scattering and radiation problems that involve complex, inhomogeneous materials. Two hybridization schemes were presented to combine FEM and BIE. The traditional one employs a single artificial surface to separate the interior and exterior regions and then applies the FEM to the interior region and the BIE to the exterior region. In this scheme, care must be exercised to (1) accurately evaluate the singular integrals, (2) effectively remove the problem of interior resonance, and (3) properly test all integral operators. The resulting system matrix is partly sparse and partly full. When an iterative solver is used, the efficiency of this scheme is mainly limited by the need to evaluate the boundary integrals in each iteration because the number of iterations required to reach convergence can be quite high. These problems are eliminated in a recently developed hybridization scheme that employs two artificial surfaces so that the FEM and BIE regions overlap with each other. In this scheme, an inhomogeneous boundary condition is postulated on the FEM truncation surface. This boundary condition

Hybridization in Computational Electromagnetics 627

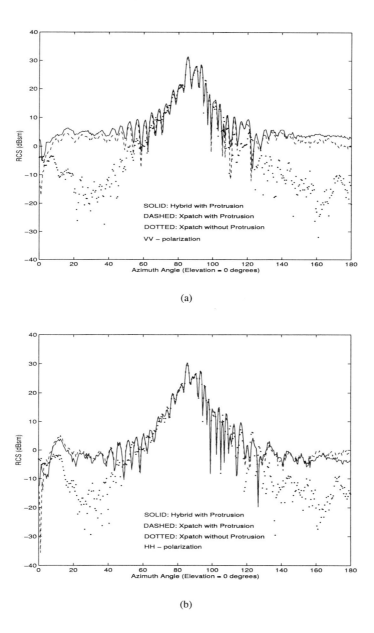

Figure 13.27 Monostatic RCS of the VFY218 airplane with a mushroom-shaped protrusion at $f = 300$ MHz: (a) VV-polarization; (b) HH-polarization [50].

is updated through an iterative procedure and it eventually converges to the exact boundary condition. This scheme is free of interior resonance, avoids the evaluation of singular integrals, and produces a purely sparse system matrix, which can be solved very efficiently. Also, the number of boundary-integral evaluations is greatly reduced. In both schemes, the boundary integrals are evaluated using the MLFMA.

The second class of hybrid techniques combines a numerical and an asymptotic method. We described two such hybrid techniques to outline the basic hybridization procedure. One combines the FEM and SBR to deal with large conducting scatterers with small indentations. The other hybridizes the MOM and SBR to simulate scattering by large conducting bodies with small protrusions. Both techniques use an approximate Green's function in the discretization of the integral equation over the aperture of an indentation or the surface of a protrusion. Both employ the SBR method to compute the incident field in the presence of the large scatterer and the scattered field in conjunction with the reciprocity theorem. This type of hybridization incorporates the major interactions between the large scatterer and the small structures and yet greatly simplifies its implementation on computers. For most problems, this type of hybridization can yield a satisfactory solution. For problems where the accuracy is not sufficient, an iterative approach can be employed to systematically reduce the error in the hybrid solution. Our particular implementation assumes that the indentations and protrusions are located on a locally flat surface because of the use of the half-space Green's function. If this assumption is not valid, the FEM/ABC method can be used to characterize the effect of the indentations or protrusions without a need to use a Green's function. One such example is given in [86].

Finally, we note that for the sake of clarity, we omitted details in the formulations and numerical implementations of the presented hybrid techniques (such as how to decouple the FEM and SBR calculations in the FEM/SBR method and the MOM and SBR calculations in the MOM/SBR method, how to improve the SBR for a more accurate calculation of the near field, the convergence analysis of the FEM/AABC method, and the use of the multifrontal algorithm). The interested reader is referred to the related journal articles and technical reports for such details and more numerical examples. Although we considered only scattering problems, the hybrid techniques can also be applied effectively to antenna radiation problems.

REFERENCES

1. O. C. Zienkiewicz and R. L. Taylor, *The Finite Element Method, Vol. 1: Basic Formulation and Linear Problems,* Fourth Edition, New York: McGraw-Hill, 1989.

2. J. M. Jin, *The Finite Element Method in Electromagnetics*, New York: Wiley, 1993.

3. P. P. Silvester and G. Pelosi, Eds., *Finite Elements for Wave Electromagnetics: Methods and Techniques,* New York: IEEE Press, 1994.

4. R. F. Harrington, *Field Computation by Moment Methods,* New York: Macmillan, 1968.

5. E. K. Miller, L. Medgyesi-Mitschang, and E. H. Newman, Eds., *Computational Electromagnetics: Frequency-Domain Method of Moments,* New York: IEEE Press, 1992.

6. J. J. H. Wang, *Generalized Moment Methods in Electromagnetics,* New York: Wiley, 1991.

7. A. F. Peterson, S. L. Ray, and R. Mittra, *Computational Methods for Electromagnetics,* New York: IEEE Press, 1997.

8. J. B. Keller, "Geometrical theory of diffraction," *J. Opt. Soc. Am.*, vol. 52, pp. 116–130, 1962.

9. P. Ya. Ufimtsev, "Method of edge waves in the physical theory of diffraction," *Izd-Vo Sov. Radio*, pp. 1–243, 1962.

10. S. W. Lee, "Comparison of uniform asymptotic theory and Ufimtsev's theory of EM edge diffraction," *IEEE Trans. Antennas Propagat.*, vol. 25, pp. 162–170, March 1977.

11. P. H. Pathak, "Techniques for high frequency problems," in *Antenna Handbook–Theory, Applications, and Design,* Y. T. Lo and S. W. Lee, Eds., New York: Van Nostrand Reinhold, 1988, ch. 4.

12. R. G. Kouyoumjian and P. H. Pathak, "A uniform geometrical theory of diffraction for an edge in a perfectly conducting surfaces," *Proc. IEEE*, vol. 62, pp. 1448–1461, Nov. 1974.

13. S. W. Lee and G. A. Deschamps, "A uniform asymptotic theory of electromagnetic diffraction by a curved wedge," *IEEE Trans. Antennas Propagat.*, vol. 24, pp. 25–34, Jan. 1976.

14. H. Ling, R. C. Chou, and S. W. Lee, "Rays versus modes: Pictorial display of energy flow in an open-ended waveguide," *IEEE Trans. Antennas Propagat.*, vol. 35, pp. 605–607, May 1987.

15. H. Ling, R. C. Chou, and S. W. Lee, "Shooting and bouncing rays: Calculating the RCS of an arbitrary shaped cavity," *IEEE Trans. Antennas Propagat.*, vol. 37, pp. 194–205, Feb. 1989.

16. J. Baldauf, S. W. Lee, L. Lin, S. K. Jeng, S. M. Scarborough, and C. L. Yu, "High frequency scattering from trihedral corner reflectors and other benchmark targets: SBR vs. experiments," *IEEE Trans. Antennas Propagat.*, vol. 39, pp. 1345–1351, Sept. 1991.

17. D. J. Andersh, M. Hazlett, S. W. Lee, D. D. Reeves, D. P. Sullivan, and Y. Chu, "XPATCH: A high-frequency electromagnetic-scattering prediction code and environment for complex three-dimensional objects," *IEEE Antennas Propagat. Mag.*, vol. 36, pp. 65–69, Feb. 1994.

18. B. Engquist and A. Majda, "Absorbing boundary conditions for the numerical simulation of waves," *Math. Comput.*, vol. 31, pp. 421–435, 1977.

19. G. Mur, "Absorbing boundary conditions for the finite-difference approximation of time-domain electromagnetic field equations," *IEEE Trans. Electromagn. Compat.*, vol. 23, pp. 377–382, Nov. 1981.

20. A. Bayliss, M. Gunzburger, and E. Turkel, "Boundary conditions for the numerical solution of elliptic equations in exterior regions," *SIAM J. Appl. Math.*, vol. 42, pp. 430–451, April 1982.

21. A. F. Peterson, "Absorbing boundary conditions for the vector wave equation," *Microwave Opt. Tech. Lett.*, vol. 1, pp. 62–64, April 1988.

22. J. P. Webb and V. N. Kanellopoulos, "Absorbing boundary conditions for the finite element solution of the vector wave equation," *Microwave Opt. Tech. Lett.*, vol. 2, pp. 370–372, Oct. 1989.

23. R. Mittra and O. M. Ramahi, "Absorbing boundary conditions for the direct solution of partial differential equations arising in electromagnetic scattering problems," in *PIER 2: Finite Element and Finite Difference Methods in Electromagnetic Scattering,* M. A. Morgan, Ed., New York: Elsevier, 1990.

24. B. Stupfel, "Absorbing boundary conditions for the scalar and vector wave equations," *IEEE Trans. Antennas Propagat.*, vol. 42, pp. 773–780, 1994.

25. S. P. Marin, "Computing scattering amplitudes for arbitrary cylinders under incident plane waves," *IEEE Trans. Antennas Propagat.*, vol. 30, pp. 1045–1049, Nov. 1982.

26. J. M. Jin and V. V. Liepa, "Application of hybrid finite element method to electromagnetic scattering from coated cylinders," *IEEE Trans. Antennas Propagat.*, vol. 36, pp. 50–54, Jan. 1988.

27. J. M. Jin and V. V. Liepa, "A note on hybrid finite element method for solving scattering problems," *IEEE Trans. Antennas Propagat.*, vol. 36, pp. 1486–1490, Oct. 1988.

28. Z. Gong and A. W. Glisson, "A hybrid equation approach for the solution of electromagnetic scattering problems involving two-dimensional inhomogeneous dielectric cylinders," *IEEE Trans. Antennas Propagat.*, vol. 38, pp. 60–68, Jan. 1990.

29. K. L. Wu, G. Y. Delisle, D. G. Fang, and M. Lecours, "Coupled finite element and boundary element methods in electromagnetic scattering," in *PIERS 2: Finite Element and Finite Difference Methods in Electromagnetic Scattering*, M. A. Morgan, Ed., New York: Elsevier, 1990.

30. X. Yuan, D. R. Lynch, and J. W. Strohbehn, "Coupling of finite element and moment methods for electromagnetic scattering from inhomogeneous objects," *IEEE Trans. Antennas Propagat.*, vol. 38, pp. 386–393, March 1990.

31. X. Yuan, "Three-dimensional electromagnetic scattering from inhomogeneous objects by the hybrid moment and finite element method," *IEEE Trans. Microwave Theory Tech.*, vol. 38, pp. 1053–1058, Aug. 1990.

32. J. M. Jin and J. L. Volakis, "A finite element–boundary integral formulation for scattering by three-dimensional cavity-backed apertures," *IEEE Trans. Antennas Propagat.*, vol. 39, pp. 97–104, Jan. 1991.

33. J. M. Jin and J. L. Volakis, "A hybrid finite element method for scattering and radiation by microstrip patch antennas and arrays residing in a cavity," *IEEE Trans. Antennas Propagat.*, vol. 39, pp. 1598–1064, Nov. 1991.

34. J.-J. Angelini, C. Soize, and P. Soudais, "Hybrid numerical methods for harmonic 3-D Maxwell equations: Scattering by a mixed conducting and inhomogeneous anisotropic dielectric medium," *IEEE Trans. Antennas Propagat.*, vol. 41, pp. 66–76, May 1993.

35. T. Eibert and V. Hansen, "Calculation of unbounded field problems in free space by a 3-D FEM/BEM-hybrid apporach," *J. Electromagn. Waves Appl.*, vol. 11, no. 4, pp. 1445–1457, April 1994.

36. W. E. Boyes and A. A. Seidl, "A hybrid finite element method for 3-D scattering using nodal and edge elements," *IEEE Trans. Antennas Propagat.*, vol. 42, pp. 1436–1442, Oct. 1994.

37. G. E. Antilla and N. G. Alexopoulos, "Scattering from complex three-dimensional geometries by a curvilinear hybrid finite-element integral equation apporach," *J. Opt. Soc. Am. A*, vol. 10, no. 1, pp. 61–77, 1996.

38. T. Cwik, C. Zuffada, and V. Jamnejad, "Modeling three-dimensional scatterers using a coupled finite element–integral equation formulation," *IEEE Trans. Antennas Propagat.*, vol. 44, pp. 453–459, April 1996.

39. N. Lu and J. M. Jin, "Application of fast multipole method to finite-element boundary-integral solution of scattering problems," *IEEE Trans. Antennas Propagat.*, vol. 44, pp. 781–786, June 1996.

40. X. Q. Sheng, J. M. Jin, J. M. Song, C. C. Lu, and W. C. Chew, "On the formulation of hybrid finite-element and boundary-integral methods for 3-D scattering," *IEEE Trans. Antennas Propagat.*, vol. 46, pp. 303–311, March 1998.

41. J. Liu and J. M. Jin, "A novel hybridization of higher order finite element and boundary integral methods for electromagnetic scattering and radiation problems," *IEEE Trans. Antennas Propagat.*, vol. 49, no. 11, Nov. 2001.

42. Y. Li and Z. J. Cendes, "High-accuracy absorbing boundary conditions for scattering," *IEEE Trans. Magn.*, vol. 31, pp. 1524–1529, March 1995.

43. J. M. Jin and N. Lu, "Application of adaptive absorbing boundary condition to finite element solution of three-dimensional scattering," *IEE Proceeding-Microw. Antennas Propag.*, vol. 143, no. 1, pp. 57–61, Feb. 1996.

44. S. Alfonzetti, G. Borzi, and N. Salerno, "Iteratively-improved Robin boundary conditions for the finite element solution of scattering problems in unbounded domains," *Int. J. Num. Meth. Eng.*, vol. 42, pp. 601–629, 1998.

45. S. Alfonzetti and G. Borzi, "Accuracy of the Robin boundary condition iteration method for the finite element solution of scattering problems," *Int. J. Num. Model.*, vol. 13, no. 2/3, pp. 217–231, March–June 2000.

46. J. M. Jin, S. S. Ni, and S. W. Lee, "Hybridization of SBR and FEM for scattering by large bodies with cracks and cavities," *IEEE Trans. Antennas Propagat.*, vol. 43, pp. 1130–1139, Oct. 1995.

47. A. D. Greenwood, S. S. Ni, J. M. Jin, and S. W. Lee, "Hybrid FEM/SBR method to compute the radiation pattern from a microstrip patch antenna in a complex geometry," *Microwave Opt. Tech. Lett.*, vol. 13, pp. 84–87, Oct. 1996.

48. F. Ling and J. M. Jin, "Hybridization of SBR and MoM for scattering by large bodies with inhomogeneous protrusions," *Progress in Electromagnetics Research*, PIER 17, pp. 25–43, 1997.

49. A. D. Greenwood and J. M. Jin, "Hybrid MoM/SBR method to compute scattering from a slot array in a complex geometry," *Appl. Comput. Electromagn. Soc. J.*, vol. 13, no. 1, pp. 43–51, March 1998.

50. J. M. Jin, F. Ling, S. Carolan, J. M. Song, W. C. Gibson, W. C. Chew, C. C. Lu, and R. Kipp, "A hybrid SBR/MoM for analysis of scattering by small protrusion on a large conducting body," *IEEE Trans. Antennas Propagat.*, vol. 46, no. 9, pp. 1349–1357, Sept. 1998.

51. J. C. Nedelec, "Mixed finite elements in R^3," *Num. Meth.*, vol. 35, pp. 315–341, 1980.

52. M. L. Barton and Z. J. Cendes, "New vector finite elements for three-dimensional magnetic field computation," *J. Appl. Phys.*, vol. 61, no. 8, pp. 3919–3921, April 1987.

53. A. Bossavit and I. Mayergoyz, "Edge-elements for scattering problems," *IEEE Trans. Magnetics*, vol. 25, pp. 2816–2821, July 1989.

54. Z. J. Cendes, "Vector finite elements for electromagnetic field," *IEEE Trans. Magn.*, vol. 27, pp. 3958–3966, Sept. 1991.

55. J. S. Savage and A. F. Peterson, "Higher-order vector finite elements for tetrahedral cells," *IEEE Trans. Microwave Theory Tech.*, vol. 44, pp. 874–879, June 1996.

56. R. D. Graglia, D. R. Wilton, and A. F. Peterson, "Higher order interpolatory vector bases for computational electromagnetics," *IEEE Trans. Antennas Propagat.*, vol. 45, pp. 329–342, March 1997.

57. R. Lee and A. C. Cangellaris, "A study of discretization error in the finite-element approximation of wave solutions," *IEEE Trans. Antennas Propagat.*, vol. 40, pp. 542–549, May 1992.

58. C. T. Tai, *Dyadic Green Functions in Electromagnetic Theory*, Second Edition, New York: IEEE Press, 1994.

59. C. T. Tai, *Generalized Vector and Dyadic Analysis*, Second Edition, New York: IEEE Press, 1997.

60. J. R. Mautz and R. F. Harrington, "H-field, E-field, and combined-field solutions for conducting body of revolution," *AEU*, vol. 32, pp. 157–164, April 1978.

61. S. M. Rao and D. R. Wilton, "E-field, H-field, and combined field solution for arbitrarily shaped three-dimensional dielectric bodies," *Electromagnetics*, vol. 10, no. 4, pp. 407–421, 1990.

62. X. Q. Sheng, J. M. Jin, J. M. Song, W. C. Chew, and C. C. Lu, "Solution of combined-field integral equation using multi-level fast multipole method for scattering by homogeneous bodies," *IEEE Trans. Antennas Propagat.*, vol. 46, pp. 1718–1726, Nov. 1998.

63. V. Rokhlin, "Rapid solution of integral equations of scattering theory in two dimensions," *J. Comput. Phys.*, vol. 86, pp. 414–439, Feb. 1990.

64. R. Coifman, V. Rokhlin, and S. Wandzura, "The fast multipole method for the wave equation: A pedestrian prescription," *IEEE Antennas Propagat. Mag.*, vol. 35, pp. 7–12, June 1993.

65. C. C. Lu and W. C. Chew, "Fast algorithm for solving hybrid integral equations," *IEE Proceedings-H*, vol. 140, pp. 455–460, Dec. 1993.

66. J. M. Song and W. C. Chew, "Multilevel fast-multipole algorithm for solving combined field integral equations of electromagnetic scattering," *Microwave Opt. Tech. Lett.*, vol. 10, no. 1, pp. 14–19, Sept. 1995.

67. B. Dembart and E. Yip, "A 3D fast multipole method for electromagnetics with multiple levels," *11th Ann. Rev. Progress Appl. Computat. Electromag.*, pp. 621–628, March 1995.

68. J. M. Song, C. C. Lu, and W. C. Chew, "MLFMA for electromagnetic scattering by large complex objects," *IEEE Trans. Antennas Propagat.*, vol. 45, pp. 1488–1493, Oct. 1997.

69. P. R. Amestoy and I. S. Duff, "Vectorization of a multiprocessor multifrontal code," *Int. J. Supercomput. Appl.*, vol. 3, pp. 41–59, 1989.

70. T. A. Davis and I. S. Duff, "An unsymmetric-pattern multifrontal methods for parallel sparse LU factorization," *SIAM J. Matrix Anal. Appl.*, vol. 18, no. 1, pp. 140–158, 1997.

71. J. W. H. Liu, "The multifrontal method for sparse matrix solution: Theory and practice," *SIAM Rev.*, vol. 34, no. 1, pp. 82–109, March 1992.

72. B. M. Irons, "A frontal method solution program for finite element analysis," *Int. J. Numer. Meth. Eng.*, vol. 2, pp. 5–32, 1970.

73. R. F. Harrington, *Time-Harmonic Electromagnetic Fields,* New York: McGraw-Hill, 1961.

74. P. H. Pathak and R. J. Burkholder, "A reciprocity formulation for the EM scattering by an obstacle within a large open cavity," *IEEE Trans. Microwave Theory Tech.*, vol. 41, pp. 702–707, April 1993.

75. G. A. Thiele and T. H. Newhouse, "A hybrid technique for combining moment methods with the geometrical theory of diffraction," *IEEE Trans. Antennas Propagat.*, vol. 23, pp. 62–69, Jan. 1975.

76. E. P. Ekelman and G. A. Thiele, "A hybrid technique for combining moment method treatment of wire antennas with the GTD for the curved surfaces," *IEEE Trans. Antennas Propagat.*, vol. 28, pp. 813–839, Nov. 1980.

77. L. W. Henderson and G. A. Thiele, "A hybrid MM-GTD technique for treatment of wire antennas near a curved surface," *Radio Sci.*, vol. 16, pp. 1125–1130, 1981. Also, related paper in *IEEE Trans. Antennas Propagat.*, vol. 30, pp. 1257–1261, April 1982.

78. A. K. Dominek, H. T. Shamansky, and N. Wang, "Scattering from three-dimensional cracks," *IEEE Trans. Antennas Propagat.*, vol. 37, pp. 586–591, May 1989.

79. M. Hsu and P. H. Pathak, "Hybrid analysis (MM-UTD) of EM scattering from finned convex objects," *1995 IEEE AP-S Int. Symp. Dig.*, vol. 3, pp. 1456–1459, June 1995.

80. S. T. Carolan and J. M. Jin, "Hybridization of the method of moments and the shooting-and-bouncing-ray method for scattering from large geometries with small protrusions," Univ. of Illinois, Research Report: CCEM-16-97, July 1997.

81. S. M. Rao, D. R. Wilton, and A. W. Glisson, "Electromagnetic scattering by surface of arbitrary shape," *IEEE Trans. Antennas Propagat.*, vol. 30, pp. 409–418, May 1982.

82. D. S. Wang, "Current-based hybrid analysis for surface-wave effects on large scatterers," *IEEE Trans. Antennas Propagat.*, vol. 39, pp. 839–849, June 1991.

83. U. Jakobus and F. M. Landstorfer, "Improved PO-MM hybrid formulation for scattering from three-dimensional perfectly conducting bodies of arbitrary shape," *IEEE Trans. Antennas Propagat.*, vol. 43, pp. 162–169, Feb. 1995.

84. F. Obelleiro-Basteiro, J. L. Rodriguez, and R. J. Burkholder, "An iterative physical optics approach for analyzing the electromagnetic scattering by large open-ended cavities," *IEEE Trans. Antennas Propagat.*, vol. 43, pp. 356–362, April 1995.

85. J. M. Song, C. C. Lu, W. C. Chew, and S. W. Lee, "Fast Illinois solver code (FISC)," *IEEE Antennas Propag. Mag.*, vol. 40, no. 3, pp. 27–34, June 1998.

86. X. Sheng and J. M. Jin, "Hybrid FEM/SBR method to compute scattering by large bodies with small protruding scatterers," *Microwave Opt. Tech. Lett.*, vol. 15, no. 2, pp. 78–84, June 1997.

14

High-Order Methods in Computational Electromagnetics

Jian-Ming Jin, Kalyan C. Donepudi, Jian Liu, Gang Kang, Jiming Song, and Weng Cho Chew

14.1 INTRODUCTION

The research on computational electromagnetics has a history of more than 40 years. Early efforts in this area concentrated on the development of numerical methods for solving a variety of electromagnetics problems. These efforts have resulted in a number of numerical methods, such as the method of moments (MOM), the finite element method (FEM), and the finite-difference time-domain (FDTD) method. Within each method, many formulations and implementations have been developed. It is fair to say that today numerical methods are available to solve most electromagnetics problems, although there are always new problems whose numerical solutions have not yet been formulated. Therefore, the present focus of computational electromagnetics has been on the enhancement of the efficiency and accuracy of the existing numerical methods. One approach to achieving this goal is through the development of fast algorithms, such as the FFT-based methods and fast multipole methods. Another approach is the use of higher-order basis functions, which lead to higher-order methods.

A higher-order method has two aspects: the higher-order geometrical modeling and the higher-order representation of the unknown quantities to be solved for. The unknown quantities are either the electric and/or magnetic fields for the FEM and FDTD or the electric and/or magnetic currents for the MOM, although the fields and the currents are actually closely related. Whereas the higher-order geometrical modeling can be easily achieved using a higher-order transformation or mapping, the higher-order representation of the fields or currents is more involved because the fields and currents, both of which are vectors, have some special properties. Traditional higher-order scalar polynomials have been proven inadequate to be used as basis functions for representing the vector fields and currents. In this chapter, we discuss the formulation of higher-order vector basis functions and their application in the MOM and FEM, together with some numerical examples to demonstrate the performance of the resulting higher-order methods.

14.2 HIGHER-ORDER MOM AND MLFMA

The MOM is a popular discretization method in electromagnetics to solve integral equations of scattering [1]. In this method, the scatterer's surface is first divided into a number of connected triangular or quadrilateral patches. The surface electric current is then approximately represented by roof-top functions for quadrilateral patches and the Rao-Wilton-Glisson (RWG) functions for triangular patches [2]. Since it is easy to model an arbitrary surface using triangular patches and straightforward to refine big triangles to smaller ones, the RWG functions have gained widespread use.

The RWG functions represent the current passing through the edges of a triangular patch as a constant and hence are complete to the zeroth order. As a result, small patches (typically, over 100 patches per square wavelength) have to be used to yield a sufficiently accurate representation. This leads to a large number of unknowns for a large scatterer. In addition, they exhibit a low-order convergence rate—the solution accuracy increases slowly with the number of unknowns. Therefore, it is very costly to obtain high accuracy using low-order basis functions. A remedy is to employ higher-order basis functions.

The development of higher-order basis functions for modeling electromagnetic fields has received intense attention recently because of their faster convergence, permitting more accurate results with less effort than the low-order basis functions. In the context of MOM, Wandzura extended the RWG basis functions to a higher order, where most unknown parameters are related to interior edges [3]. Hamilton et al. extended the RWG basis functions to a set of basis functions consisting of edge-based functions and two kinds of patch-based functions [4, 5]. In the context of FEM, higher-order edge-based basis functions have also been developed [6–12]. Based on Nedelec's approach, Graglia et al. developed general, unified expressions for higher-order interpolatory vector basis functions [11], which are adopted in this work.

14.2.1 Formulation

The higher-order interpolatory vector basis functions proposed by Graglia et al. [11] are based on the RWG basis functions. Consider a curvilinear triangular patch in the xyz-space, as illustrated in Figure 14.1(a). Using a parametric transformation or mapping, this patch can be transformed into a planar right-angle triangle in the ξ-space, as shown in Figure 14.1(b). Take the second-order transformation as an example; a curvilinear triangular patch described by six nodes can be transformed into a planar right-angle triangle using the transformation

$$\mathbf{r} = \sum_{j=1}^{6} \varphi_j(\xi_1, \xi_2, \xi_3)\mathbf{r}_j \tag{14.1}$$

where the shape functions φ_j are defined in terms of the parametric coordinates ξ_1, ξ_2, ξ_3 as

$$\begin{aligned}
\varphi_1 &= \xi_1(2\xi_1 - 1) & \varphi_2 &= \xi_2(2\xi_2 - 1) \\
\varphi_3 &= \xi_3(2\xi_3 - 1) & \varphi_4 &= 4\xi_1\xi_2 \\
\varphi_5 &= 4\xi_2\xi_3 & \varphi_6 &= 4\xi_3\xi_1
\end{aligned} \tag{14.2}$$

It can be seen easily that $\xi_1 + \xi_2 + \xi_3 = 1$. Such a parametric transformation is rather standard in the FEM [13] and can be accomplished for any orders.

As in Figure 14.1, the three nodes of a triangular patch are labeled by $(1, 2, 3)$, and their opposite edges can be labeled accordingly. The RWG basis functions for the three edges are defined as

$$\begin{aligned}
\mathbf{\Lambda}_1(\mathbf{r}) &= \frac{1}{\mathcal{J}}(\xi_2\boldsymbol{\ell}_3 - \xi_3\boldsymbol{\ell}_2) \\
\mathbf{\Lambda}_2(\mathbf{r}) &= \frac{1}{\mathcal{J}}(\xi_3\boldsymbol{\ell}_1 - \xi_1\boldsymbol{\ell}_3) \\
\mathbf{\Lambda}_3(\mathbf{r}) &= \frac{1}{\mathcal{J}}(\xi_1\boldsymbol{\ell}_2 - \xi_2\boldsymbol{\ell}_1)
\end{aligned} \tag{14.3}$$

where $\mathcal{J}$ denotes the *Jacobian* given by $\mathcal{J} = \boldsymbol{\ell}^1 \cdot (\boldsymbol{\ell}^2 \times \boldsymbol{\ell}^3)$ with the unitary basis vectors $\boldsymbol{\ell}^i$ defined as

$$\boldsymbol{\ell}^i = \frac{\partial \mathbf{r}}{\partial \xi_i} \qquad i = 1, 2, 3 \tag{14.4}$$

Also, $\boldsymbol{\ell}_1, \boldsymbol{\ell}_2$, and $\boldsymbol{\ell}_3$ represent the edge vectors, which are related to the unitary basis vectors by

$$\boldsymbol{\ell}_1 = -\boldsymbol{\ell}^2 \qquad \boldsymbol{\ell}_2 = \boldsymbol{\ell}^1 \qquad \boldsymbol{\ell}_3 = \boldsymbol{\ell}^2 - \boldsymbol{\ell}^1 \tag{14.5}$$

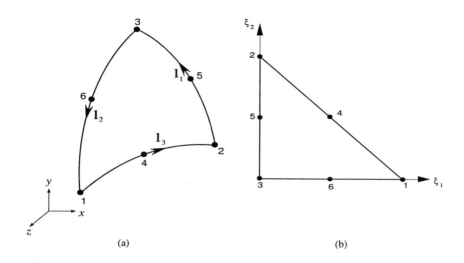

Figure 14.1 (a) A curvilinear triangular patch with six nodes in the xyz-space and (b) its parametric representation in the ξ-space.

A unique feature of the RWG functions so defined is that they have a normal component only on the associated edges and this normal component is a constant, which is often normalized to 1. As a result, these functions are well suited to representing the current density because they guarantee the normal continuity. Note that the three expressions in (14.3) can be written uniformly as

$$\Lambda_\beta(\mathbf{r}) = \frac{1}{\mathcal{J}}(\xi_s \boldsymbol{\ell}_t - \xi_t \boldsymbol{\ell}_s) \qquad (14.6)$$

where (β, s, t) are cyclic permutations of $(1, 2, 3)$.

The higher-order vector basis functions are formed by multiplying the RWG functions Λ_β ($\beta = 1, 2, 3$) by a set of interpolatory polynomial functions, which are complete to the specified order. To construct a set of interpolatory polynomials for a specified order, say order p, we can choose a set of interpolation nodes arranged in a similar form to that of the Lagrange bases of the same order. However, to stay away from the vertices of the triangle, the interpolation nodes are shifted away from the two edges where the normal components of Λ_β vanish (Figure 14.2). Multiplying Λ_β by the Lagrange interpolation polynomials generates a set of higher-order basis functions associated with edge β. Since there are three edges, one has three sets of interpolation nodes and must use three sets of labels.

To avoid the use of the three sets of labels for the interpolation nodes and use one set instead, we can first form a set of nodes arranged in a similar form to that

High-Order Methods in Computational Electromagnetics

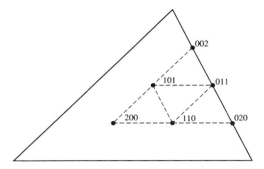

Figure 14.2 Interpolation nodes for higher-order ($p = 2$) basis functions associated with Λ_1.

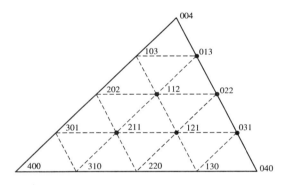

Figure 14.3 A subset of nodes taken to be interpolation nodes for higher-order ($p = 2$) basis functions associated with Λ_1.

of the Lagrange bases of order $p + 2$ and then choose a subset of the nodes to be interpolation nodes. For the vector basis functions associated with Λ_β, this subset is formed by the interior nodes and the nodes residing on edge β except for the two at the ends of the edge (Figure 14.3). By labeling each interpolation node with the indices (i, j, k), which satisfy $i + j + k = p + 2$, the unified expression for the vector basis functions so constructed are given by [11]

$$\Lambda^1_{ijk}(\mathbf{r}) = N^1_{ijk} R_i(p+2, \xi_1) \hat{R}_j(p+2, \xi_2) \hat{R}_k(p+2, \xi_3) \Lambda_1(\mathbf{r})$$
$$i = 0, 1, \ldots, p; j, k = 1, 2, \ldots, p+1 \qquad (14.7)$$

$$\Lambda^2_{ijk}(\mathbf{r}) = N^2_{ijk} \hat{R}_i(p+2, \xi_1) R_j(p+2, \xi_2) \hat{R}_k(p+2, \xi_3) \Lambda_2(\mathbf{r})$$
$$j = 0, 1, \ldots, p; i, k = 1, 2, \ldots, p+1 \qquad (14.8)$$

$$\Lambda^3_{ijk}(\mathbf{r}) = N^3_{ijk} \hat{R}_i(p+2, \xi_1) \hat{R}_j(p+2, \xi_2) R_k(p+2, \xi_3) \Lambda_3(\mathbf{r})$$
$$k = 0, 1, \ldots, p; i, j = 1, 2, \ldots, p+1 \qquad (14.9)$$

Here, N^β_{ijk} ($\beta = 1, 2, 3$) are normalization factors, R and $\hat{R}$ denote the Silvester-Lagrange and shifted Silvester polynomials defined by

$$R_i(p, \xi) = \begin{cases} \dfrac{1}{i!} \prod_{k=0}^{i-1} (p\xi - k), & 1 \leq i \leq p \\ 1, & i = 0 \end{cases} \quad (14.10)$$

$$\hat{R}_i(p, \xi) = \begin{cases} \dfrac{1}{(i-1)!} \prod_{k=1}^{i-1} (p\xi - k), & 2 \leq i \leq p+1 \\ 1, & i = 1 \end{cases} \quad (14.11)$$

Equations (14.7)–(14.9) can be written uniformly as

$$\boldsymbol{\Lambda}^\beta_{ijk}(\mathbf{r}) = N^\beta_{ijk} R_{i_\beta}(p+2, \xi_\beta) \hat{R}_{i_s}(p+2, \xi_s) \hat{R}_{i_t}(p+2, \xi_t) \boldsymbol{\Lambda}_\beta(\mathbf{r})$$
$$i_\beta = 0, 1, \ldots, p; i_s, i_t = 1, 2, \ldots, p+1 \quad (14.12)$$

where i_β is taken to be i, j, or k for $\beta = 1, 2,$ or 3, and similarly for i_s and i_t.

Equation (14.12) provides one basis function for an edge node and three basis functions for an interior node. Since a surface vector function has only two degrees of freedom, one of the three basis functions is dependent and has to be discarded. This leads to $3(p+1)$ degrees on the boundary and $p(p+1)$ in the interior, resulting in three unknowns for the zeroth-order, eight unknowns for the first-order, 15 unknowns for the second-order patch, and so on. Since the unknowns on edges are shared by two triangles, on the average there are 1.5, 5, and 10.5 unknowns per patch for the zeroth-, first-, and second-order basis functions, respectively.

With the basis functions given above, we can now apply them to the MOM solution of scattering. Consider an arbitrarily shaped three-dimensional (3D) conducting object illuminated by an incident field ($\mathbf{E}^{\text{inc}}$, $\mathbf{H}^{\text{inc}}$). The electric field integral equation (EFIE) and magnetic field integral equation (MFIE) are given by

$$\hat{n} \times \mathbf{L}(\mathbf{J}) = \hat{n} \times \mathbf{E}^{\text{inc}}(\mathbf{r}) \qquad \mathbf{r} \in S \quad (14.13)$$

$$\frac{1}{2}\mathbf{J}(\mathbf{r}) + \hat{n} \times \tilde{\mathbf{K}}(\mathbf{J}) = \hat{n} \times \mathbf{H}^{\text{inc}}(\mathbf{r}) \qquad \mathbf{r} \in S \quad (14.14)$$

respectively, where $\mathbf{J}(\mathbf{r})$ denotes the unknown surface current density and the integral operators $\mathbf{L}$ and $\mathbf{K}$ are defined by

$$\mathbf{L}(\mathbf{J}) = jk_0 \eta_0 \iint_S \left[\mathbf{J}(\mathbf{r}')g(\mathbf{r},\mathbf{r}') + \frac{1}{k^2} \nabla' \cdot \mathbf{J}(\mathbf{r}') \nabla g(\mathbf{r},\mathbf{r}') \right] dS' \quad (14.15)$$

$$\tilde{\mathbf{K}}(\mathbf{J}) = \iint_S \mathbf{J}(\mathbf{r}') \times \nabla g(\mathbf{r},\mathbf{r}') dS' \quad (14.16)$$

in which S denotes the conducting surface of the object, k_0 is the free-space wavenumber, η_0 is the free-space wave impedance, $\hat{n}$ is an outwardly directed normal, and

$g(\mathbf{r}, \mathbf{r}')$ is the well-known free-space Green's function given by

$$g(\mathbf{r}, \mathbf{r}') = \frac{e^{-jk_0|\mathbf{r}-\mathbf{r}'|}}{4\pi|\mathbf{r}-\mathbf{r}'|}. \tag{14.17}$$

When $\mathbf{r} = \mathbf{r}'$, the integral in (14.16) is interpreted in the principal value sense.

Either EFIE or MFIE can be used to solve for $\mathbf{J}$. However, for a given closed S, $\mathbf{L}$ can be singular at certain frequencies when the exterior medium is lossless. Consequently, (14.13) may give an erroneous solution at these frequencies. This is known as the problem of interior resonance and the singular frequencies correspond to the resonant frequencies of a cavity formed by filling the interior of S with the exterior medium. A similar problem occurs in (14.14) as well. To eliminate this problem, one can combine (14.13) and (14.14) to find

$$\alpha\,\hat{n} \times \hat{n} \times \mathbf{L}(\mathbf{J}) + (1-\alpha)\eta_0 \left[\frac{1}{2}\mathbf{J}(\mathbf{r}) + \hat{n} \times \tilde{\mathbf{K}}(\mathbf{J})\right]$$
$$= \alpha\,\hat{n} \times \hat{n} \times \mathbf{E}^{\text{inc}}(\mathbf{r}) + (1-\alpha)\eta_0 \hat{n} \times \mathbf{H}^{\text{inc}}(\mathbf{r}) \quad \mathbf{r} \in S \tag{14.18}$$

which is known as the CFIE [14]. This combination results in an integral operator whose singular frequencies correspond to the resonant frequencies of a cavity with a resistive wall, which are complex. As a result, the CFIE operator cannot be singular for a real frequency. The combination parameter α is usually chosen anywhere between 0.2 and 0.8.

The EFIE, MFIE, and CFIE can be solved by MOM. Since both EFIE and MFIE can be considered the special case of CFIE, we consider the MOM solution of CFIE here. For this, the conducting surface S is subdivided into small triangular patches and the current on the surface is expanded as

$$\mathbf{J}(\mathbf{r}') = \sum_{n=1}^{N} I_n \mathbf{\Lambda}_n(\mathbf{r}') \tag{14.19}$$

where N is the number of unknowns and $\mathbf{\Lambda}_n(\mathbf{r}')$ denote the vector basis functions. Applying Galerkin's method results in a matrix equation

$$\sum_{n=1}^{N} Z_{mn} I_n = V_m \quad m = 1, 2, \ldots, N \tag{14.20}$$

in which

$$Z_{mn} = \alpha \iint_S \mathbf{\Lambda}_m(\mathbf{r}) \cdot \mathbf{L}(\mathbf{\Lambda}_n) dS$$
$$+ (1-\alpha)\eta_0 \iint_S \mathbf{\Lambda}_m(\mathbf{r}) \cdot \left[\frac{1}{2}\mathbf{\Lambda}_n(\mathbf{r}) + \hat{n} \times \tilde{\mathbf{K}}(\mathbf{\Lambda}_n)\right] dS \tag{14.21}$$

$$V_m = \iint_S [\alpha \mathbf{E}^{\text{inc}}(\mathbf{r}) + (1-\alpha)\eta_0\,\hat{n} \times \mathbf{H}^{\text{inc}}(\mathbf{r})] \cdot \mathbf{\Lambda}_m(\mathbf{r}) dS \tag{14.22}$$

It remains to evaluate Z_{mn}, which contains double integrals: one over primed variables and the other over unprimed variables. This evaluation can be carried out accurately using Gaussian quadrature rules if the supports of the basis functions Λ_m and Λ_n are far apart. However, when these supports coincide with or are close to each other, Gaussian quadrature rules will not be sufficient because of the singularity associated with Green's function in the integrand. To circumvent this problem, the singularity of the EFIE term is first converted into a weaker form by using Gauss' theorem, and Duffy's method [15] is then employed to generate a set of quadrature rules around the point of singularity, as described below.

First, the integral over primed variables in Z_{mn} is still evaluated using the standard Gaussian quadrature rules. The integration point then becomes the singular point for the remaining integral over unprimed variables. This integral can be written in the following form:

$$I = \iint_\Delta \frac{f(x,y,z)}{\rho} dS = \int_0^1 d\xi_1 \int_0^{1-\xi_1} \frac{f(\xi_1, \xi_2)}{\rho} \mathcal{J}(\xi_1, \xi_2) \, d\xi_2 \qquad (14.23)$$

where Δ denotes a curvilinear triangular patch, $f(x,y,z)$ is either a continuous vector or scalar function, $\mathcal{J}(\xi_1, \xi_2)$ is the Jacobian, and ρ represents the distance between the field and source points. In Duffy's method, the integral I over the source patch is first written in terms of subtriangles defined by the primed integration point and the three vertices of the triangle (Figure 14.4). As a result, I can be written as

$$I = \sum_{e=1}^{3} \int_0^1 d\xi_1^e \int_0^{1-\xi_1^e} \frac{f(\xi_1^e, \xi_2^e)}{\rho} \mathcal{J}(\xi_1^e, \xi_2^e) \, d\xi_2^e \qquad (14.24)$$

where ξ_1^e and ξ_2^e are the parametric coordinates of the eth subtriangle. The integral over each subtriangle can be transformed into an integration over a square by introducing the transformation $\xi_2^e = (1 - \xi_1^e)u$. By doing this, one can rewrite (14.24) as

$$I = \sum_{e=1}^{3} \int_0^1 d\xi_1^e \int_0^1 (1 - \xi_1^e) \frac{f\{\xi_1^e, (1 - \xi_1^e)u\}}{\rho} \mathcal{J}\{\xi_1^e, (1 - \xi_1^e)u\} \, du \qquad (14.25)$$

with

$$\rho = (1 - \xi_1^e)\sqrt{q(\xi_1^e, u, \mathbf{r}_1, \mathbf{r}_2, \ldots, \mathbf{r}_6)} \qquad (14.26)$$

where the function q is defined in terms of the coordinate ξ_1^e, the transformation variable u, and $\mathbf{r}_1, \mathbf{r}_2, \ldots, \mathbf{r}_6$, which are the position vectors of the six nodes of the patch. The numerator term $1 - \xi_1^e$ cancels the singular nature of ρ. Consequently, each of the above integrals can be accurately evaluated by using one-dimensional Gauss-Legendre quadrature rules.

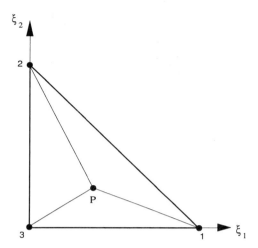

Figure 14.4 A triangle is divided into three subtriangles.

With the accurate evaluation of Z_{mn}, the linear system of equations in (14.20) can now be solved using MLFMA [16–18]. Note that in MLFMA, Z_{mn} is evaluated only when Λ_m and Λ_n are inside the same or nearby groups. The MLFMA kernel used here is ScaleME [19–21].

14.2.2 Numerical Examples

In this section, we consider several examples to demonstrate the performance of the higher-order method. In all the examples, the solution is obtained using either the biconjugate gradient stabilized (BICGSTAB) [22] or the generalized minimal residual (GMRES) [23] with a block diagonal preconditioner [24].

We first consider the convergence of the higher-order formulation, using a conducting sphere having a diameter of 9λ. The parameter used in the study is the root-mean-square (RMS) error defined as

$$\text{RMS} = \sqrt{\frac{1}{N_s} \sum_{i=1}^{N_s} |\sigma_{\text{ref}} - \sigma_{\text{cal}}|^2} \qquad (14.27)$$

where σ_{cal} denotes the calculated radar cross-section (RCS) and σ_{ref} denotes the reference Mie solution, both measured in dB, and N_s is the number of sampling points, which are the angles of observation here. Figure 14.5 displays the RMS error in the bistatic RCS as a function of the unknown density for both E- and H-planes (the difference between the two planes is that there are many deep nulls in the E-plane bistatic RCS, which tend to exaggerate the RMS error). A dramatic improvement is observed when using higher-order basis functions. The improvement is more pronounced when a higher accuracy is desired. To obtain a solution with the RMS error less than 0.2 dB in the E-plane, the zeroth-order basis functions use about 43,200 unknowns whereas the first- and second-order basis functions require only about 9,600 and 7,000 unknowns, respectively. If the desired RMS error is set to be 0.1 dB, the zeroth-order basis functions would require about 80,688 unknowns whereas the first- and second-order basis functions would require only about 14,440 and 8,400 unknowns, respectively. The computation is carried out on a 16-node Linux cluster called *Orion* [25]. Each node of the cluster consists of a 100-MHz motherboard and a 350-MHz AMD K6-2 processor. Each board is equipped with a 256 SDRAM card leading to a total of 4 GB of dynamic random access memory. The corresponding computation times and memory requirements are summarized in Table 14.1 for a set of computations.

Next, we present a few examples to demonstrate the self consistency in the higher-order solutions and the capability of the higher-order method. The first example is a $2.5\lambda \times 2.5\lambda \times 3.75\lambda$ open cavity. The cavity opening is in the direction of $\theta = 0°$. The monostatic RCS patterns for the $\theta\theta$ and $\phi\phi$ polarizations are computed using the zeroth-, first-, and second-order basis functions. The results are shown in Figure 14.6. The second- and first-order basis functions require about only 1,179 and 1,640 unknowns to attain the same accuracy as those obtained by using the zeroth-order basis functions with 2,182 unknowns.

The second example is a conducting cube, whose sides are 15λ long. The bistatic RCS patterns in both E- and H-planes are given in Figure 14.7 (the angle of incidence is $\theta^{\text{inc}} = 0°$, which is normal to one of the faces of the cube). The zeroth-order basis functions use about 271,872 unknowns whereas the first- and second-order basis functions require only about 73,200 and 43,200 unknowns, respectively. The computations were carried out on the SGI Origin 2000 (250-MHz processors) and the corresponding computation times and memory requirements are summarized in Table 14.2.

The third example is the 9.936-in NASA almond, which is an EMCC benchmark geometry [26]. The almond is placed horizontally with its longer axis aligned with the x axis and its sharp tip pointing in the $\hat{x}$ direction. Its monostatic RCS is computed

Table 14.1 Computation Times and Memory Requirements on a Linux Cluster *Orion* [24]

Basis	Unknowns	Processors	Setup time	Solution time	Memory
0th order	43,200	16	2,305 s	134 s	601 MB
1st order	9,000	8	1,029 s	94 s	207 MB
2nd order	6,804	8	1,403 s	79 s	263 MB

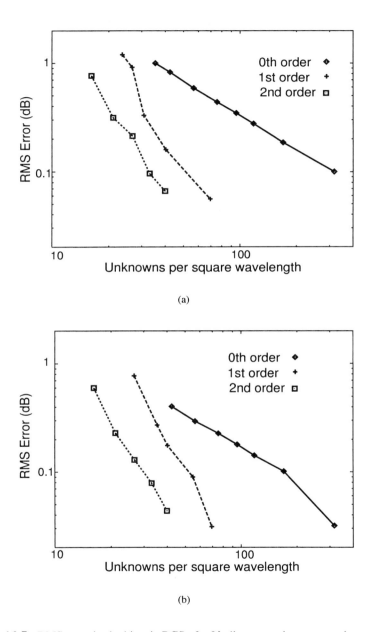

Figure 14.5 RMS error in the bistatic RCS of a 9λ diameter sphere versus the number of unknowns per square wavelength: (a) E-plane; (b) H-plane [24].

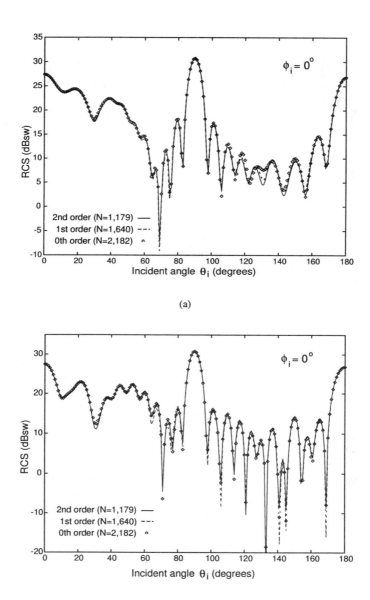

Figure 14.6 Monostatic RCS of a $2.5\lambda \times 2.5\lambda \times 3.75\lambda$ open cavity: (a) $\theta\theta$-polarization; (b) $\phi\phi$-polarization [24].

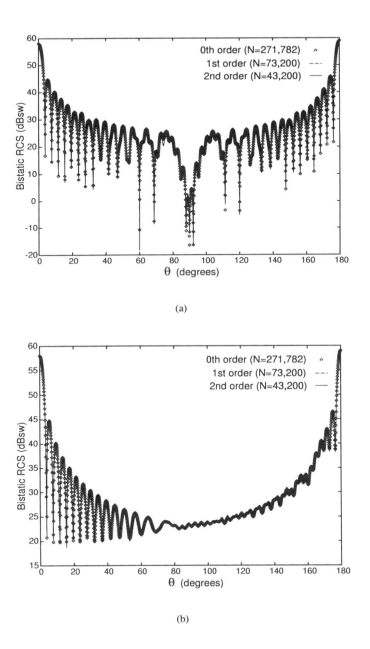

Figure 14.7 Bistatic RCS patterns of a conducting cube having side length of 15λ: (a) E-plane; (b) H-plane [24].

Table 14.2 Computation Times and Memory Requirements on SGI Origin 2000 [24]

Problem	Basis	Unknowns	Processors	Setup	Solution	Memory
Fig. 14.7	0th order	271,872	8	1,390 s	478 s	1.6 GB
Fig. 14.7	1st order	73,200	8	926 s	233 s	750 MB
Fig. 14.7	2nd order	43,200	8	1,270 s	202 s	970 MB
Fig. 14.8	1st order	17,480	8	442 s	47.5 s	674 MB
Fig. 14.9	2nd order	537,600	32	3,700 s	4,860 s	18.8 GB

using the first-order basis functions (3,496 triangular patches) for both the $\theta\theta$ and $\phi\phi$ polarizations in the $\theta = 90°$ plane. This plane, especially in the tip region around $\phi = 0°$, is considered the most difficult for the RCS computation because of diffraction due to the tip of the almond. The results are shown in Figure 14.8 and compared to those obtained by FISC [27] using 53,092 triangular patches, which is based on the zeroth-order basis functions and flat patches. While good agreement is obtained from 30° to 180°, we also observed the disagreement in the tip region from 0° to 30°. We conjectured that the disagreement is caused by the difference between the two meshes used in the calculations. To verify this conjecture, we carried out another calculation using the second-order basis functions and the same mesh (3,496 triangular patches) as used in the first-order calculation. The result agrees with that obtained using the first-order basis functions. We then used another finer mesh (6,708 triangular patches) and the first-order basis functions for calculation and obtained the results shown in the figure, which still exhibit some disagreement with the FISC solution. Obviously, for this geometry the RCS in the tip region is very sensitive to the mesh and a slight difference in the mesh can result in different solutions. The computations were carried out on the SGI Origin 2000 and the corresponding computation times and memory requirements are summarized in Table 14.2 (note that the solution time in this case is averaged for one incident angle).

The final example is a conducting sphere of 72λ in diameter, with a surface area of 16,286 λ^2. The bistatic RCS in the E-plane is computed using the second-order basis functions with 537,600 unknowns and the result is compared with the Mie series solution in Figure 14.9 (the angle of incidence is $\theta^{inc} = 0°$). The RMS error is about 0.22 dB over the entire region. Again, the computations were carried out on the SGI Origin 2000 and the corresponding computation times and memory requirements are summarized in Table 14.2.

All the examples above suggest that the higher-order method can significantly reduce the number of unknowns, especially from the zeroth to the first order, without compromising the accuracy of geometry modeling. Typical mesh density to obtain a 0.1-dB solution accuracy is 330, 55, 33 unknowns per square wavelength for the solution of CFIE using the zeroth-, first-, and second-order basis functions, respectively.

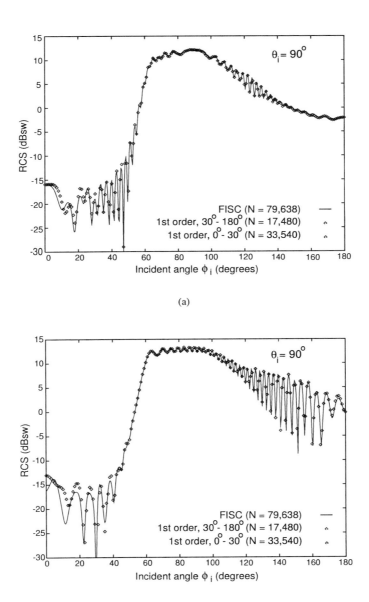

Figure 14.8 Monostatic RCS of a 9.936-in NASA almond at 20 GHz: (a) $\theta\theta$ polarization; (b) $\phi\phi$ polarization [24].

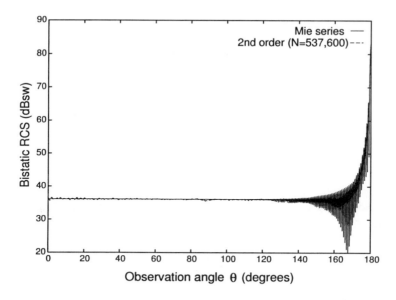

Figure 14.9 E-plane bistatic RCS pattern of a conducting sphere having a diameter of 72λ [24].

14.3 POINT-BASED IMPLEMENTATION OF HIGHER-ORDER MLFMA

The preceding section demonstrated clearly the fast convergence of the higher-order method—the solution error decreased quickly with the increase in the number of unknowns. However, when MLFMA is used to accelerate the solution of the higher-order Galerkin-based MOM, its performance is limited for the following reason. In MLFMA, only the near interactions of the MOM matrix are computed explicitly whereas the far interactions are computed implicitly by using the multipole expansion of current distributions. In this procedure, the scatterer is first enclosed in a large cube, which is then divided into eight equally sized small cubes. Each of the small cubes is further divided into eight smaller cubes recursively until the smallest cube size is about several times bigger than the longest patch edge. Thus, the edge length limits the number of levels used in MLFMA.

In the Galerkin-based MOM using the RWG basis functions, since the edge length is about 0.1λ, the finest cube is about a quarter of a wavelength. However, because of the large patch size used with higher-order basis functions, the number of levels in MLFMA is at least one or two less than that for the RWG basis functions. In MLFMA, the radiation pattern of each basis is calculated and stored. The number of

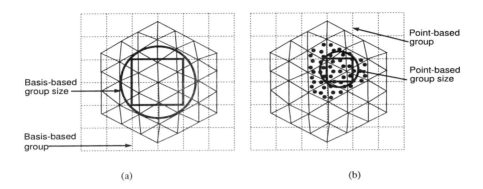

Figure 14.10 Applying MLFMA to (a) basis-based sources and (b) point-based sources.

samplings (K) in the radiation pattern is given by

$$K = 2L^2, \qquad L = k_0 d + \alpha (k_0 d)^{1/3} \tag{14.28}$$

where k_0 is the wavenumber, α depends on the accuracy, and d is the diameter of the group size, which is bigger than the cube as shown in Figure 14.10(a). Each higher-order basis needs many more samplings than the RWG basis. The average number of unknowns required is about 1.5 per patch for RWG basis, 5 and 10.5 for the first- and second-order basis functions, respectively. Thus, the number of unknowns in each group for higher-order basis functions is larger than that for RWG basis. Therefore, more near interactions have to be calculated and stored for each higher-order basis. Consequently, MLFMA cannot work very efficiently with the higher-order Galerkin's method.

To overcome this problem, we proposed to implement MLFMA based on point-to-point interactions, instead of the traditional basis-to-basis interactions [28]. When we calculate the matrix elements for which the testing and source bases are not close to each other, we can apply Gaussian quadrature to evaluate the integrals. This process can be interpreted as replacing a continuous source distribution with discrete sources as shown in Figure 14.10(b). Thus, one matrix-vector multiply is similar to the calculation of the electromagnetic fields for a given distribution of N source bases and then testing them with these bases. In this implementation, we first find Q equivalent point sources from these N source bases, then calculate electromagnetic fields at these Q points, and finally test them with each testing basis. The value of Q depends on the number of patches and the quadrature rule used for each patch. The MLFMA is used to calculate electromagnetic fields at Q points generated by Q point sources. By doing so, the number of levels used is not limited by the size of basis functions, making MLFMA more efficient. Furthermore, the near interaction part of the MOM matrix is redefined as the difference between the original matrix and the interactions calculated by MLFMA. Consequently, the memory requirement can be reduced as well.

14.3.1 Formulation

By testing the EFIE with the basis function $\mathbf{t}_j$, the matrix element Z_{ji} is written as

$$Z_{ji} = \iint \mathbf{t}_j(\mathbf{r}) \cdot \iint \overline{\mathbf{G}}(\mathbf{r},\mathbf{r}') \cdot \mathbf{t}_i(\mathbf{r}')dS'dS \tag{14.29}$$

where the dyadic Green's function is given by

$$\overline{\mathbf{G}}(\mathbf{r},\mathbf{r}') = -\frac{jk_0\eta_0}{4\pi}\left(\overline{\mathbf{I}} - \frac{\nabla\nabla'}{k_0^2}\right)\frac{e^{-jk_0|\mathbf{r}-\mathbf{r}'|}}{|\mathbf{r}-\mathbf{r}'|}$$

and η_0 denotes the impedance. When the testing basis ($\mathbf{t}_j$) is sufficiently far away from the source basis ($\mathbf{t}_i$), the integrals can be evaluated by using Gaussian quadrature to give

$$Z_{ji}^{\text{far}} = \sum_{p=1}^{Q} W_{jp}\mathbf{t}_j(\mathbf{r}_p) \cdot \sum_{q=1}^{Q} \overline{\mathbf{G}}(\mathbf{r}_p,\mathbf{r}_q) \cdot W_{iq}\mathbf{t}_i(\mathbf{r}_q)$$

$$= \sum_{p=1}^{Q} \mathbf{J}_j(\mathbf{r}_p) \cdot \sum_{q=1}^{Q} \overline{\mathbf{G}}(\mathbf{r}_p,\mathbf{r}_q) \cdot \mathbf{J}_i(\mathbf{r}_q)$$

where Q is the total number of quadrature points on all patches, W_{jp} is the quadrature weight at the pth point, which is nonzero only on the patch supporting the basis $\mathbf{t}_j$, and $\mathbf{J}_j(\mathbf{r}_p)$ is defined as $W_{jp}\mathbf{t}_j(\mathbf{r}_p)$. Correspondingly, the matrix-vector multiply for the far interactions is written as

$$\sum_{i=1}^{N} Z_{ji}^{\text{far}} x_i = \sum_{p=1}^{Q} \mathbf{J}_j(\mathbf{r}_p) \cdot \sum_{q=1}^{Q} \overline{\mathbf{G}}(\mathbf{r}_p,\mathbf{r}_q) \cdot \sum_{i=1}^{N} \mathbf{J}_i(\mathbf{r}_q) x_i \tag{14.30}$$

where x_i is the coefficient of the ith basis function. By defining

$$\mathbf{E}_p = \sum_{q=1}^{Q} \overline{\mathbf{G}}(\mathbf{r}_p,\mathbf{r}_q) \cdot \mathbf{J}_q \qquad \mathbf{J}_q = \sum_{i=1}^{N} \mathbf{J}_i(\mathbf{r}_q) x_i \tag{14.31}$$

(14.30) is simplified as

$$\sum_{i=1}^{N} Z_{ji}^{\text{far}} x_i = \sum_{p=1}^{Q} \mathbf{J}_j(\mathbf{r}_p) \cdot \mathbf{E}_p$$

The above formulation means that for a given distribution of N source bases, calculate the electric fields and test them with these bases. The MLFMA can be used to calculate the fields generated by current sources very efficiently. Defining a variable I_{pq} such that

$$I_{pq} = \begin{cases} 1 & p\text{-}q \text{ interaction is calculated by MLFMA} \\ 0 & \text{otherwise} \end{cases}$$

the matrix-vector multiply calculated by MLFMA is modified as

$$\sum_{i=1}^{N} Z_{ji}^{\text{far}} x_i = \sum_{p=1}^{Q} \mathbf{J}_j(\mathbf{r}_p) \cdot \sum_{q=1}^{Q} I_{pq} \overline{\mathbf{G}}(\mathbf{r}_p, \mathbf{r}_q) \cdot \sum_{i=1}^{N} \mathbf{J}_i(\mathbf{r}_q) x_i \qquad (14.32)$$

Now, the near interaction part of the MOM matrix is redefined as the difference between the original matrix and the interactions calculated by MLFMA, so we have

$$Z_{ji}^{\text{near}} = Z_{ji} - Z_{ji}^{\text{far}} \qquad (14.33)$$

where

$$Z_{ji}^{\text{far}} = \sum_{p} \mathbf{J}_j(\mathbf{r}_p) \cdot \sum_{q} I_{pq} \overline{\mathbf{G}}(\mathbf{r}_p, \mathbf{r}_q) \cdot \mathbf{J}_i(\mathbf{r}_q)$$

Hence,

$$\begin{aligned} Z_{ji}^{\text{near}} &= Z_{ji} - Z_{ji}^{\text{far}} \\ &= \iint \mathbf{t}_j(\mathbf{r}) \cdot \iint \overline{\mathbf{G}}(\mathbf{r}, \mathbf{r}') \mathbf{t}_i(\mathbf{r}') dS' dS \\ &\quad - \sum_{p} \mathbf{J}_j(\mathbf{r}_p) \cdot \sum_{q} I_{pq} \overline{\mathbf{G}}(\mathbf{r}_p, \mathbf{r}_q) \cdot \mathbf{J}_i(\mathbf{r}_q) \end{aligned}$$

If the testing basis $\mathbf{t}_j$ and the source basis $\mathbf{t}_i$ belong to the nearby groups but are not very close to each other, the same quadrature rule used for evaluating the far interactions is used to calculate the first term of the above equation. If all point-point interactions are calculated by MLFMA, Z_{ji}^{near} equals to zero. Therefore, Z^{near} is a sparse matrix and has fewer nonzero elements than the matrix for near interactions between bases.

Since the current vector in (14.31) has only two tangential components, we write it in terms of two independent tangential vectors as

$$\mathbf{t}_i(\mathbf{r}_q) = \sum_{l'=1}^{2} \mathbf{\Lambda}_{l'}(\mathbf{r}_q) y_{il'}(\mathbf{r}_q)$$

Therefore, current $\mathbf{J}_q$ at the qth source point $\mathbf{r}_q$ is written as

$$\mathbf{J}_q = \sum_{i=1}^{N} W_{iq} \mathbf{t}_i(\mathbf{r}_q) x_i = \sum_{i=1}^{N} W_{iq} \sum_{l'=1}^{2} \mathbf{\Lambda}_{l'}(\mathbf{r}_q) y_{il'}(\mathbf{r}_q) x_i$$

$$= \sum_{l'=1}^{2} \mathbf{\Lambda}_{l'}(\mathbf{r}_q) Y_{l'}(\mathbf{r}_q) \quad (14.34)$$

where

$$Y_{l'}(\mathbf{r}_q) = \sum_{i=1}^{N} W_{iq} y_{il'}(\mathbf{r}_q) x_i$$

Substituting the above expression into (14.30) yields

$$\sum_{i=1}^{N} Z_{ji}^{\text{far}} x_i = \sum_{p=1}^{Q} \sum_{l=1}^{2} W_{jp} y_{jl} \mathbf{\Lambda}_l(\mathbf{r}_p) \cdot \sum_{q=1}^{Q} \overline{\mathbf{G}}(\mathbf{r}_p, \mathbf{r}_q) \cdot \sum_{l'=1}^{2} \mathbf{\Lambda}_{l'}(\mathbf{r}_q) Y_{l'}(\mathbf{r}_q) \quad (14.35)$$

Applying the addition theorem to the dyadic Green's function [29] yields

$$\overline{\mathbf{G}}(\mathbf{r}_p, \mathbf{r}_q) = -\frac{jk_0\eta_0}{4\pi}\left(\overline{\mathbf{I}} + \frac{\nabla\nabla}{k_0^2}\right)\frac{e^{-jk_0|\mathbf{r}_p-\mathbf{r}_q|}}{|\mathbf{r}_p-\mathbf{r}_q|}$$

$$= \int d^2\hat{k}(\overline{\mathbf{I}} - \hat{k}\hat{k})e^{-j\mathbf{k}\cdot(\mathbf{r}_{pm}-\mathbf{r}_{qm'})}\alpha(\mathbf{k},\mathbf{r}_{mm'}) \quad (14.36)$$

where

$$\alpha(\mathbf{k}, \mathbf{r}_{mm'}) = -\frac{jk_0\eta_0}{4\pi} \sum_{l=0}^{L}(-j)^l(2l+1)h_l^{(2)}(k_0 r_{mm'})P_l(\hat{r}_{mm'}\cdot\hat{k})$$

$$\mathbf{r}_{pq} = \mathbf{r}_p - \mathbf{r}_q = \mathbf{r}_p - \mathbf{r}_m + \mathbf{r}_m - \mathbf{r}_{m'} - \mathbf{r}_q = \mathbf{r}_{pm} + \mathbf{r}_{mm'} - \mathbf{r}_{qm'}$$

and $\mathbf{r}_m$ and $\mathbf{r}_{m'}$ are the group centers for the points p and q, respectively. Consequently, the far interactions of the matrix-vector multiply can be written as

$$\sum_{i=1}^{N} Z_{ji}^{\text{far}} x_i = \sum_{p=1}^{Q} \sum_{l=1}^{2} U_{jpl} \int d^2\hat{k} \mathbf{V}_{mlp}(\mathbf{k})$$

$$\cdot \sum_{m'} \alpha(\mathbf{k}, \mathbf{r}_{mm'}) \sum_{q\in G_{m'}} \sum_{l'=1}^{2} \mathbf{V}^*_{m'ql'}(\mathbf{k}) Y_{l'}(\mathbf{r}_q) \quad (14.37)$$

where $U_{jpl} = W_{jp} y_{jl}$, the summation over m' is carried out for all nonneighbor groups, denoted by $G_{m'}$, and the radiation and receiving patterns are given by

$$\mathbf{V}_{mlp}(\mathbf{k}) = e^{-j\mathbf{k}\cdot\mathbf{r}_{pm}}(\overline{\mathbf{I}} - \hat{k}\hat{k}) \cdot \mathbf{\Lambda}_l(\mathbf{r}_p) \quad (14.38)$$

Since the point sources do not have any spatial extent, we can refine the finest cube in MLFMA as small as a quarter of a wavelength. There are two advantages in applying MLFMA to point-to-point interactions. First of all, the group size is the same as the cube and there is no overlapping in sources as shown in Figure 14.10. Secondly, smaller group size means that many fewer samplings of radiation patterns are needed. As a consequence, the memory requirements for the radiation pattern can be reduced. Further reductions can be obtained if the radiation patterns are calculated on the fly since no numerical integration is needed to calculate the radiation pattern.

14.3.2 Complexity Analysis

The memory requirements in the MLFMA implementation have three parts: one for near interaction elements Z^{near}, another for the radiation and receiving patterns for each basis or point as given by (14.38), and the third for the radiation patterns for all nonempty groups at each level. The first part is unchanged from the traditional basis-based approach to the point-based approach. This, however, is not the case for the second part. When the second-order basis is used, the number of levels in the point-based MLFMA is increased by two compared to that for the basis-based MLFMA. The cubic size in the finest level is then reduced to one-fourth, and by (14.28), the number of samplings of the radiation and receiving patterns for each point is reduced to one-eighth. Since the number of points is about twice of the number of bases, the second part of the memory requirement is reduced to one-fourth. From one level to a coarser level, the samplings in the radiation patterns are increased by a factor of 4, and since the number of nonempty cubes is reduced to one-fourth, the memory requirement is a constant for all nonempty cubes at each level. Consequently, the point-based approach needs more memory than the traditional basis-based approach in the third part, but, this increase is insignificant compared to the reduction achieved in the second part. Hence, if we write the total memory required as $C_1 N \log N + C_2 N$, the proposed method maintains C_1 while it reduces C_2 by a factor of 4.

The number of operations in each matrix-vector multiply consists of three parts similar to the memory requirements: one for near interactions, another for calculating the radiation patterns for each group from its points or bases at the finest level, corresponding to the last two summations in (14.37), and evaluating the fields at each basis, corresponding to the first two summations and the integral in (14.37), and the third for the translation at each level and interpolation/anterpolation from one level to another [18]. Since a large number of groups is needed in the translation, which is the summation over m' in (14.37), and the interpolation and anterpolation are required for both θ and ϕ, the number of operations in the third part is dominant. Hence, the point-based approach needs more operations than the basis-based approach because of the added levels. If we write the total number of operations as $C_3 N \log N + C_4 N$, the proposed method maintains C_3 while it increases C_4.

14.3.3 Numerical Results

The inherent advantage in applying MLFMA to the point-based sources is shown by computing the RCS of a perfect electric conducting (PEC) sphere with different diameters using the second-order basis functions [24] and then comparing the memory needed for applying MLFMA to the basis-based sources. The computation is carried out on a 16-node linux cluster *Orion* [25]. The discretization size is kept constant at 33 unknowns per square wavelength and the diameter of the sphere is increased from 9λ to 36λ. The memory required is shown in Figure 14.11(a), which clearly indicates that the memory requirements are reduced by a factor of 2. The corresponding CPU times recorded for a matrix-vector multiply are plotted in Figure 14.11(b). The CPU time of the point-based approach is initially increased for small problems, but eventually converges to that of the basis-based approach for larger problems. This is due to the increased memory access time in the basis-based approach because of its larger memory requirements.

Next, the RCS of a 50λ-diameter sphere is calculated using 322,896 second-order basis functions. The RMS error is about 0.0877 dB and the corresponding bistatic RCS pattern in the E-plane is plotted in Figure 14.12. Seven levels of the point-based MLFMA with radiation patterns computed on the fly are used, which requires 5.43 GB of memory. The use of the basis-based approach requires 12.69 GB of memory.

Finally, the RCS of a conducting cube having a side length of 15λ is calculated using 45,320 second-order basis functions. The corresponding bistatic RCS patterns in the E-plane are plotted in Figure 14.13. In this case, we use six levels of MLFMA with radiation patterns computed on the fly. The point-based approach requires 474 MB of memory, whereas the basis-based approach requires 988 MB of memory.

The numerical examples above show clearly that the point-based implementation of the higher-order MLFMA allows one to use more levels in MLFMA than applying MLFMA to basis interactions directly, and hence reduces the memory requirements significantly.

14.4 HIGHER-ORDER FEM

The most popular higher-order vector elements used in electromagnetics are those introduced by Nedelec [6]. These elements are distinguished by their reduced degrees of freedom. It has been shown in the preceding chapter that in the finite element analysis of the vector field problems, the curl of the field is as important as the field itself. Therefore, even if the field itself is represented by a polynomial of a given order (say, order p), its curl will be represented by a polynomial of order $p - 1$. The overall accuracy of the solution will then be limited by the lower order. Therefore, the accuracy of the solution will not be affected if the terms (the gradient terms of order p) that do not contribute to the curl representation are removed while keeping

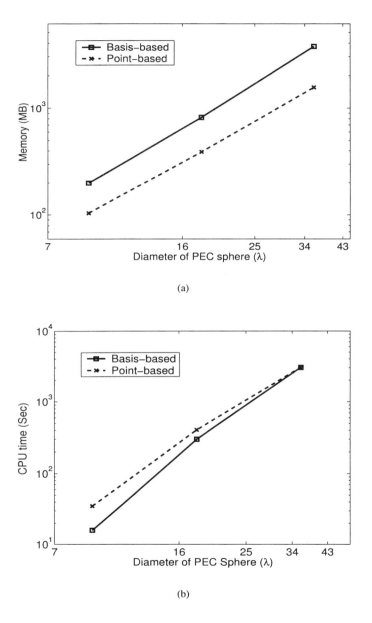

Figure 14.11 (a) Memory requirements and (b) CPU time for a matrix-vector multiply of MLFMA as applied to the basis-based and point-based sources [28].

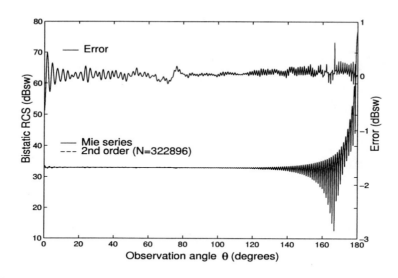

Figure 14.12 Bistatic RCS of a PEC sphere having a diameter of 50λ using the point-based approach (7-level MLFMA, 5.43 GB of memory) [28].

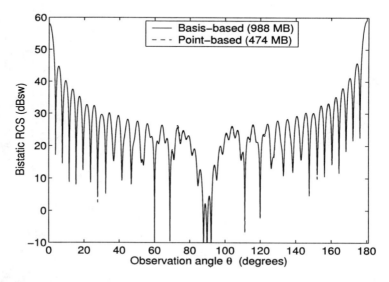

Figure 14.13 Bistatic RCS of a PEC cube having a side length of 15λ calculated by the MLFMA as applied to the basis-based and point-based sources, respectively [28].

the field representation complete to order $p - 1$. This results in fewer degrees of freedom, and thus increases the solution efficiency. The solution obtained will have a better balance in the accuracy of the field and its curl. The removal of the unnecessary gradient terms makes the basis function of mixed degree.

Like scalar elements, there are two types of higher-order vector elements. The first type uses interpolatory basis functions, and the second type uses hierarchal basis functions. Both types span the same vector space and start from the zeroth-order basis functions. Their major difference is in their construction. The interpolatory basis functions are defined on a set of points on the element, such that each basis function vanishes at all the points except for one. This type of basis function has several advantages. It has a good linear independence, thus resulting in a better conditioned matrix system. The coefficients have a physical interpretation as components of the vector field at the interpolation points. Since they interpolate the tangential component, their use makes it easy to enforce boundary conditions. Finally, it has a unified expression, which significantly simplifies the implementation of computer codes for the generation of arbitrary order basis functions. However, interpolatory basis functions of a given order are all different from those of the lower-order ones. Hence, different order basis functions cannot be used together, which makes it impossible to implement p-adaption (iterative increase of the element orders in different regions until the solution is converged to a specified accuracy).

In contrast, the hierarchal basis functions are not defined on a set of points. Higher-order hierarchal basis functions are formed by adding some new functions to the lower-order basis functions. Thus, the first-order basis functions contain those of the zeroth order, the second-order ones contain those of the zeroth and first orders, and so on. The distinct advantage of this type of basis function is that it permits the use of different orders in a problem, and hence can be employed for p-adaption. In this section, we describe higher-order interpolatory vector basis functions; for hierarchal ones the reader is referred to Webb [12].

14.4.1 Higher-Order Tetrahedral Elements

Consider a curvilinear tetrahedral element in the xyz-space. This element can be mapped to a rectilinear element in the ξ-space. For a quadratic element specified by 10 points (Figure 14.14), the mapping is given by

$$\mathbf{r} = \sum_{j=1}^{10} \varphi_j(\xi_1, \xi_2, \xi_3, \xi_4) \mathbf{r}_j \tag{14.39}$$

where the shape functions φ_j are defined in terms of the parametric coordinates $\xi_1, \xi_2, \xi_3, \xi_4$ as

$$\varphi_1 = \xi_1(2\xi_1 - 1) \qquad \varphi_2 = \xi_2(2\xi_2 - 1)$$
$$\varphi_3 = \xi_3(2\xi_3 - 1) \qquad \varphi_4 = \xi_4(2\xi_4 - 1)$$

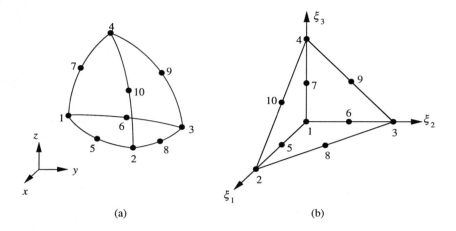

Figure 14.14 (a) A quadratic curvilinear tetrahedral element in the xyz coordinate system and (b) its parametric representation in the ξ space.

$$\varphi_5 = 4\xi_1\xi_2 \qquad \varphi_6 = 4\xi_1\xi_3$$
$$\varphi_7 = 4\xi_1\xi_4 \qquad \varphi_8 = 4\xi_2\xi_3$$
$$\varphi_9 = 4\xi_3\xi_4 \qquad \varphi_6 = 4\xi_2\xi_4 \qquad (14.40)$$

It can be seen easily that $\xi_1 + \xi_2 + \xi_3 + \xi_4 = 1$. Such a parametric transformation is rather standard in the FEM and can be accomplished for any orders.

As in the MOM, the higher-order vector basis functions are based on the zeroth-order basis functions, which are also known as Whitney's elements. Consider a tetrahedral element whose four nodes are labeled by (γ, β, m, n). Its four faces can be labeled by (γ, β, m, n) as well, where face γ is the one opposite to node γ and so on. The vector basis function associated with the edge shared by faces γ and β (which is also the edge connecting nodes m and n) is given by [8]

$$\mathbf{\Omega}_{\gamma\beta}(\mathbf{r}) = \xi_n \nabla \xi_m - \xi_m \nabla \xi_n \qquad (14.41)$$

where ξ_m and ξ_n denote the normalized coordinates such that ξ_m or ξ_n has a value of 1 at node m or n and 0 on face m or n. It can be shown that the basis functions $\mathbf{\Omega}_{\gamma\beta}$ have tangential components only on faces γ and β and they guarantee the tangential continuity of the interpolated field while allowing the normal component to be discontinuous.

The higher-order interpolatory vector basis functions [11] are formed by multiplying the zeroth-order basis functions $\mathbf{\Omega}_{\gamma\beta}$ by a set of interpolatory polynomial functions, which are complete to the specified order. To construct a set of interpolatory polynomials for a specified order, say order p, we can choose a set of interpolation

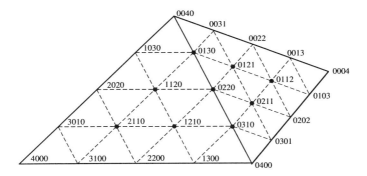

Figure 14.15 A subset of nodes taken to be interpolation nodes for higher-order ($p = 2$) basis functions associated with edge 1 (interior nodes omitted for clarity).

nodes arranged in a similar form to that of the Lagrange bases of the same order. However, to stay away from the vertices of the tetrahedron, the interpolation nodes are shifted away from the two faces where the tangential components of $\Omega_{\gamma\beta}$ vanish. A more systematic choice is, however, to first form a set of nodes arranged in a similar form to that of the Lagrange bases of order $p + 2$ and then choose a subset of the nodes to be interpolation nodes. For the vector basis functions associated $\Omega_{\gamma\beta}$, this subset is formed by the remaining nodes after those residing on faces m and n have been thrown away (Figure 14.15). By labeling each interpolation node with the indexes (i, j, k, l), which satisfy $i + j + k + l = p + 2$, the unified expression for the vector basis function so constructed is given by [11]

$$\Omega_{ijkl}^{\gamma\beta}(\mathbf{r}) = N_{ijkl}^{\gamma\beta} R_{i_\gamma}(p+2, \xi_\gamma) R_{i_\beta}(p+2, \xi_\beta) \hat{R}_{i_m}(p+2, \xi_m)$$
$$\times \hat{R}_{i_n}(p+2, \xi_n) \Omega_{\gamma\beta}(\mathbf{r})$$
$$i_\gamma, i_\beta = 0, 1, \ldots, p; i_m, i_n = 1, 2, \ldots, p+1 \quad (14.42)$$

In the above, $N_{ijkl}^{\gamma\beta}$ is a normalization factor, i_γ is taken to be i, j, k, or l for $\gamma = 1, 2, 3$, or 4, respectively, and similarly for i_β, i_m, and i_n, and R and $\hat{R}$ denote the Silvester-Lagrange and shifted Silvester polynomials defined in (14.10) and (14.11), respectively.

Equation (14.42) provides one basis function for an interpolation node on an edge of the tetrahedron. It provides three basis functions for an interpolation node on a face since a face is associated with three edges. Since a vector function has only two degrees of freedom in its tangential components, one of the basis functions is dependent and has to be discarded. For an interior interpolation node, (14.42) provides six basis functions, among which only three are independent and the other three must be discarded. Hence, for each tetrahedron of order p, there are $(p+1)(p+3)(p+4)/2$ basis functions. Assuming that in a finite element mesh, the number of edges, faces, and tetrahedra are n_{edge}, n_{face}, and n_{tet}, respectively, the total number of unknowns n_{ukwn} is

$$n_{\text{ukwn}} = (p+1) \times n_{\text{edge}} + p(p+1) \times n_{\text{face}} + p(p^2-1)/2 \times n_{\text{tet}} \qquad (14.43)$$

14.4.2 Application to Cavity Scattering

In this section, we demonstrate the performance of the higher-order FEM in the numerical computation of the RCS of a deep open cavity, which is often considered a grand challenge in computational electromagnetics. Recently, an efficient numerical technique was developed for the analysis of this problem [30, 31]. This technique is based on the FEM, which is known for its capability to handle arbitrary geometries and complex material composition. The FEM mesh at the cavity aperture is terminated in an exact manner by the boundary integral (BI) method. The technique exploits the unique features of the FE–BI equations and, more importantly, the unique features of the problem of scattering by a large and deep cavity. It is designed in such a manner that it uses minimal memory, which is proportional to the maximum cross-section of the cavity and independent of the depth of the cavity, and its computation time increases only linearly with the depth of the cavity. Furthermore, it computes the scattered fields for all angles of incidence without requiring significant additional time. The performance of this technique can further be enhanced using higher-order curvilinear elements. The use of these higher-order elements significantly reduces the dispersion error and also permits a more accurate modeling of the problem geometry.

Consider the problem of plane wave scattering by a large, deep, and arbitrarily shaped open cavity (Figure 14.16). In accordance with the FE–BI method [13], the electric field inside the cavity and at the aperture of the cavity can be obtained by seeking the stationary point of the functional

$$\begin{aligned}
F = {} & \frac{1}{2} \iiint_V \left[\frac{1}{\mu_r} (\nabla \times \mathbf{E}) \cdot (\nabla \times \mathbf{E}) - k_0^2 \epsilon_r \mathbf{E} \cdot \mathbf{E} \right] dV \\
& - k_0^2 \iint_S \mathbf{M}(\mathbf{r}) \cdot \left[\iint_S \mathbf{M}(\mathbf{r}') g(\mathbf{r},\mathbf{r}') \, dS' \right] dS \\
& + \iint_S \nabla \cdot \mathbf{M}(\mathbf{r}) \left[\iint_S g(\mathbf{r},\mathbf{r}') \nabla' \cdot \mathbf{M}(\mathbf{r}') dS' \right] dS \\
& - 2jk_0 \eta_0 \iint_S \mathbf{M}(\mathbf{r}) \cdot \mathbf{H}^{\text{inc}}(\mathbf{r}) \, dS
\end{aligned} \qquad (14.44)$$

where V denotes the volume of the cavity and S denotes its aperture, $\mathbf{M} = \mathbf{E} \times \hat{z}$ is the equivalent magnetic current over the aperture, $\mathbf{H}^{\text{inc}}$ denotes the incident magnetic field, k_0 is the free-space wavenumber, η_0 is the free-space wave impedance, and $g(\mathbf{r},\mathbf{r}')$ denotes the free-space Green's function.

High-Order Methods in Computational Electromagnetics

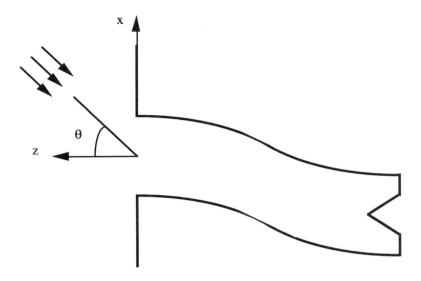

Figure 14.16 Illustration of a large, deep, and arbitrarily shaped open cavity.

This functional can be discretized by first subdividing the volume of the cavity into small tetrahedral elements and then representing the field as

$$\mathbf{E} = \sum_{i=1}^{N} E_i \mathbf{N}_i \qquad (14.45)$$

where $\mathbf{N}_i$ denotes the vector basis function, E_i denotes the expansion coefficient, and N denotes the total number of expansion terms.

Substituting (14.45) into (14.44) and applying the Rayleigh-Ritz procedure, we obtain the matrix equation

$$[A]\{E\} = \{b\} \qquad (14.46)$$

where $[A]$ is a symmetric matrix, $\{E\}$ is a vector storing the discrete unknowns, and $\{b\}$ is a known vector determined from the incident field. The matrix $[A]$ is contributed by the first three integrals in (14.44) and can be decomposed into two parts: the part contributed by the volume integral is sparse and the other part contributed by the two surface integrals is fully populated. The problem is to solve (14.46) for $\{E\}$, from which the electric field can be calculated using (14.45).

Although the formulation of the FEM matrix equation is straightforward, its solution is difficult because of a large number of unknowns for a large and deep cavity. To illustrate this difficulty, consider a circular cavity with a diameter of 5 wavelengths and a depth of 50 wavelengths. If we use 20 zeroth-order elements per wavelength, we would have about 1,000 layers along the depth of the cavity with about 24,000 unknowns per layer, resulting in 24 million unknowns. Clearly, for a problem of this size, the most important factor for its numerical solution is the memory requirement.

The most memory-efficient method to solve (14.46) is an iterative method such as the conjugate gradient (CG), the biconjugate gradient (BCG), and the quasi-minimum residual (QMR) methods. Using such methods, the memory requirement is proportional to $O(N)$. For the cavity considered here, the memory required is estimated to be about 7.68 GB. However, an iterative method for this problem has two major problems. The most serious problem is its slow or even nonconvergence. Since the finite element matrix has a relatively poor condition number, when its dimension is large, the convergence of an iterative solution is very slow and, in most cases that we have tested, the solution even fails to converge. Although the slow convergence can be improved using a preconditioning technique, its implementation requires a large amount of memory, usually exceeding that for the storage of $[A]$. The less serious, but also important, problem is that an iterative method must repeat the entire solution process for a new right-hand side. Since we are interested in the RCS calculation, the number of right-hand sides can be very large.

It is well known that a direct method can find the solution of a matrix equation with a fixed number of operations and, moreover, it can find the solution for many right-hand sides with a negligible amount of extra computing time. Therefore, it seems that a direct method is preferred for this problem. However, the major problem for a direct method is its huge memory requirement. For the cavity considered here, with a proper numbering of the unknowns, the matrix $[A]$ can be stored in a banded matrix whose half bandwidth is about 24,000. This would require 4,608 GB, which far exceeds the capability of most available computing facilities.

From the above analysis, it is clear that one must reduce the memory requirement to render a direct method useful for a large problem. In general, this is impossible; however, the problem considered here possesses several unique features that make this reduction possible. If we number the unknowns starting from the bottom of the cavity, we obtain the matrix $[A]$ with two very unique features (Figure 14.17). First, $[A]$ is a symmetric and banded matrix, which is mostly sparse except for the right-bottom part that is a full matrix generated from the first two surface integrals in (14.44). The half bandwidth is approximately equal to the number of unknowns for one layer. For the cavity considered here, it is about 24,000. Second, $\{b\}$ is a vector whose elements are zeros except for a small part at the bottom, which is contributed by the last surface integral in (14.44). For the cavity considered here, out of 24 million elements in $\{b\}$, only 24,000 of them are nonzeros and they are at the bottom of $\{b\}$. Recognizing these two unique features, we can design a special method to solve (14.46) efficiently.

High-Order Methods in Computational Electromagnetics   667

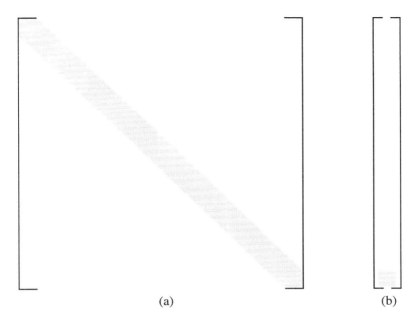

Figure 14.17 (a) The structure of the finite element matrix, whose nonzero elements are contained within a narrow band. (b) The structure of the right-hand side vector, which contains very few nonzero elements at the bottom of the vector.

This special method is based on the band solver. As pointed out above, a band solver for a general problem would require the storage of the banded matrix, which requires about 4,608 GB for the cavity considered here. However, a careful examination of the Gaussian elimination process reveals that for a symmetric banded matrix having a half bandwidth of iwb, the elimination of an equation only involves the previous iwb equations. Other equations preceding these iwb equations are never needed in the elimination process; they are needed only in the back substitution process. Since again $\{b\}$ is nonzero only at its bottom and the calculation of the RCS requires only the electric field at the aperture of the cavity, these equations are actually never needed after the elimination process. Recognizing this fact, we can modify the band algorithm in such a manner that only a matrix of $iwb \times iwb$ is needed for its implementation. For the cavity considered here, the memory required is about 4.6 GB. In this modified band algorithm, we begin with the first equation and generate the equations one by one. Once an equation is generated, we immediately apply Gaussian elimination to this equation and keep the reduced equation in the memory. This process continues until we encounter the $(iwb + 1)$th equation. Since the first equation is not needed in the Gaussian elimination of this equation, we place this equation in the memory occupied by the first equation. This process continues

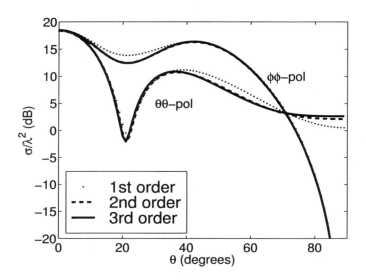

Figure 14.18 Monostatic RCS of a $1.5\lambda \times 1.5\lambda \times 0.6\lambda$ rectangular cavity modeled with tetrahedral elements [31].

until all the equations have been processed. The information remaining in the core memory is all that is needed to find the solution of the unknowns at the aperture of the cavity via the back substitution process.

When higher-order tetrahedral elements are used, for a fixed-size cavity the memory requirement scales as $(p+1)^2 D^4$ and the computing time scales as $(p+1)^2 D^7$, where p denotes the order of the elements and D denotes the mesh density (number of unknowns per wavelength). Although there is an additional coefficient $(p+1)^2$ in both the memory requirement and computing time, the higher-order elements use a much lower mesh density D, which can lead to a significant saving in the memory and computing time.

Next, we present some numerical results to demonstrate the validity and capability of the method. As a first check, we test the self-consistency of the method. The first example is a $1.5\lambda \times 1.5\lambda \times 0.6\lambda$ rectangular cavity. In this case, the surface of the cavity is flat; hence, there is no issue about geometry modeling. The RCS is given in Figure 14.18 as a function of the angle of incidence, obtained by the first-, second-, and third-order elements using the same number of elements (90). The figure demonstrates clearly the convergence with respect to the order of elements. The second example is a circular cavity having a diameter of 0.61λ and depth of 2.1λ. In this case, the surface of the cavity is curved and it is critical to have accurate geometry modeling. The RCS of this cavity is shown in Figure 14.19 as a function of the azimuth angle, calculated by the first-, second-, and third-order elements using

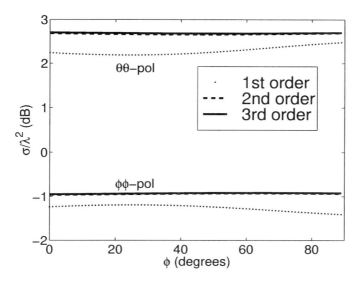

Figure 14.19 Monostatic RCS of a circular cavity having a diameter of 0.61λ and depth of 2.1λ modeled with tetrahedral elements [31].

the same number of elements (427). The exact RCS should be a constant because of the rotational symmetry of the problem. The results show clearly that this is the case when the order of elements increases.

In the third example, we examine the efficiency of higher-order elements. The parameter used in this examination is the RMS error defined in (14.27). The reference solution in this case is obtained using the third-order elements with an overly dense mesh, and it does not change anymore when either the order of elements or the mesh density is increased. Figure 14.20(a) displays the RMS error in the monostatic RCS of a $1.0\lambda \times 1.0\lambda \times 4.0\lambda$ rectangular cavity as a function of the number of unknowns. It is evident that, for the same number of unknowns, the higher-order elements produce more accurate results. For a desired accuracy, the number of unknowns required for the higher-order elements is smaller than that for lower-order elements. Figure 14.20(b) displays the corresponding CPU time. Although the CPU time increases with the order of elements for the same number of unknowns, a careful comparison between Figures 14.20(a,b) indicates that for a given accuracy, the higher-order elements consume less CPU time than the lower-order elements. Hence, the higher-order elements are both more accurate and efficient than the lower-order ones.

The last example considered using tetrahedral elements is a circular cavity having a diameter of 2.0λ and depth of 10.0λ. Its monostatic RCS is shown in Figure 14.21 and the information about the discretization, computer memory, and CPU time is

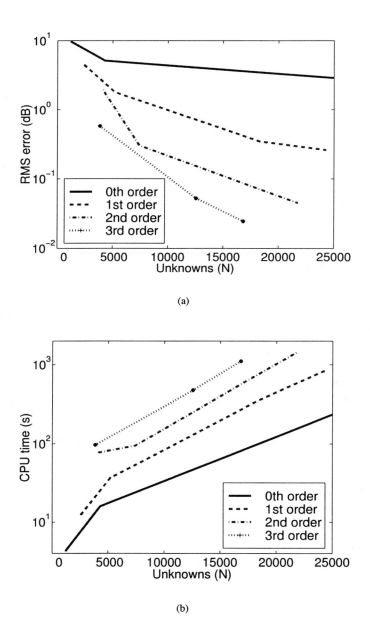

Figure 14.20 (a) RMS error versus the number of unknowns for the monostatic RCS of a $1.0\lambda \times 1.0\lambda \times 4.0\lambda$ rectangular cavity and (b) corresponding CPU time [31].

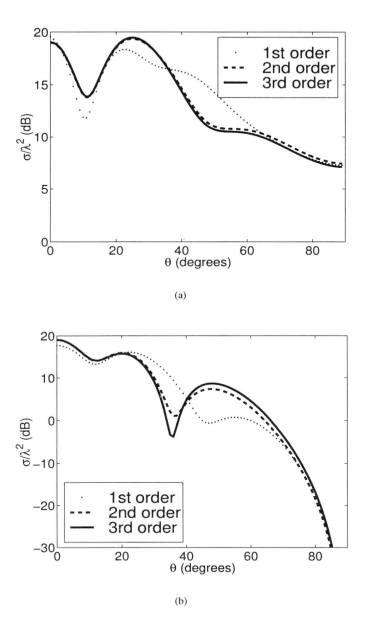

Figure 14.21 Monostatic RCS of a circular cavity having a diameter of 2.0λ and depth of 10.0λ modeled with tetrahedral elements: (a) $\theta\theta$-polarization; (b) $\phi\phi$-polarization [31].

Table 14.3 Information about RCS Calculation of a Circular Cavity [31]

Order of elements	Number of unknowns	Memory (MB)	CPU time (s)
First	14,218	6.8	381
Second	42,555	21.5	9,985
Third	94,844	77.9	89,435

given in Table 14.3. The calculation is carried out on a DEC personal workstation (500-MHz Alpha 21164 processor). Clearly, the solution converges when the order of elements increases.

14.5 MIXED-ORDER PRISM ELEMENTS

Although the use of higher-order tetrahedral elements yields a remarkable improvement in the accuracy and efficiency of the special FEM, the coefficient $(p+1)^2$ reduces the amount of savings one can achieve in the memory and computing time. However, if we can design a special element that has a higher-order interpolation in the transverse plane and a lower-order (say the zeroth-order here) interpolation along the depth of a cavity, the computational complexity indicates that for a fixed-size cavity, the memory requirement scales as D_t^4 and the computing time scales as $D_t^6 D_l$, where D_t denotes the mesh density (number of unknowns per wavelength) in the transverse plane and D_l denotes the mesh density (number of unknowns per wavelength) along the depth of the cavity. Since we use a higher-order interpolation in the transverse plane, D_t can be reduced. A reduction by a factor of 2 can reduce the memory requirement by a factor of 16 and the computing time by a factor of 64, and a reduction of mesh density by a factor of 3 can reduce the memory requirement by a factor of 81 and the computing time by a factor of 729. Although a lower-order interpolation is used along the depth of a cavity, an increase in the mesh density along this direction does not increase the memory requirement and increases the computing time only linearly. Clearly, such an element is optimal for the proposed special method and can be implemented on triangular prisms. Compared to the higher-order tetrahedral elements, the mixed-order prism elements using the same order and same mesh density in the transverse plane and the zeroth order along the depth reduce the memory requirement by a factor of $(p+1)^2$ and the computing time by a factor of $(p+1)^2 D/D_l$.

Similar to the case of the tetrahedral element, a curved triangular prism element is first transformed into one with straight edges (Figure 14.22). The order of transformation can be specified independently. Since we intend to use higher-order basis functions in the transverse plane and lower-order ones in the longitudinal direction,

High-Order Methods in Computational Electromagnetics

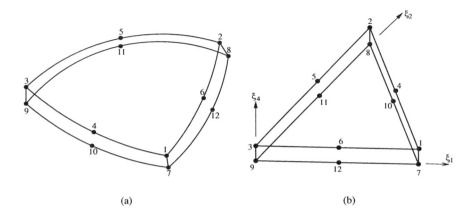

Figure 14.22 A mixed second/zeroth-order triangular prism element: (a) curved element; (b) rectilinear element.

the prism element is usually very thin. As a result, we use higher-order transformation in the transverse plane and lower-order one in the longitudinal direction. For the element illustrated in Figure 14.22, the transformation functions are given by

$$\varphi_1 = \xi_1(2\xi_1 - 1)\xi_4 \qquad \varphi_2 = \xi_2(2\xi_2 - 1)\xi_4$$
$$\varphi_3 = \xi_3(2\xi_3 - 1)\xi_4 \qquad \varphi_4 = 4\xi_1\xi_2\xi_4$$
$$\varphi_5 = 4\xi_2\xi_3\xi_4 \qquad \varphi_6 = 4\xi_1\xi_3\xi_4$$
$$\varphi_7 = \xi_1(2\xi_1 - 1)\xi_5 \qquad \varphi_8 = \xi_2(2\xi_2 - 1)\xi_5$$
$$\varphi_9 = \xi_3(2\xi_3 - 1)\xi_5 \qquad \varphi_{10} = 4\xi_1\xi_2\xi_5$$
$$\varphi_{11} = 4\xi_2\xi_3\xi_5 \qquad \varphi_{12} = 4\xi_1\xi_3\xi_5 \qquad (14.47)$$

Apparently, (ξ_1, ξ_2, ξ_3) describe the transformation in the transverse plane and (ξ_4, ξ_5) describe that in the longitudinal direction. They satisfy the relations $\xi_1 + \xi_2 + \xi_3 = 1$ and $\xi_4 + \xi_5 = 1$.

The zeroth-order vector basis functions for this element to interpolate a transverse field are given by [32]

$$\mathbf{\Omega}_{\beta,m}(\mathbf{r}) = \xi_n(\xi_s \nabla \xi_t - \xi_t \nabla \xi_s) \qquad (14.48)$$

where (β, s, t) are the cyclic permutation of $(1, 2, 3)$ and (m, n) are the cyclic mutation of $(4, 5)$. Those to interpolate a longitudinal field are given by

$$\mathbf{\Omega}_\beta(\mathbf{r}) = \xi_\beta \nabla \xi_4 \qquad \beta = 1, 2, 3 \qquad (14.49)$$

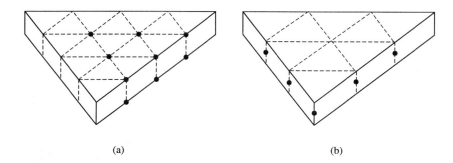

Figure 14.23 Interpolation points for (a) the transverse field and (b) the longitudinal field.

To construct mixed-order basis functions, say the pth-order in the transverse plane and the zeroth-order in the longitudinal direction, we multiply the zeroth-order functions with a polynomial of the pth-order in the transverse plane and the zeroth-order in the longitudinal direction. Following the approach proposed by Graglia et al. [32], we have

$$\Omega_{ijk}^{1,m}(\mathbf{r}) = N_{ijk}^{1,m} R_i(p+2,\xi_1)\hat{R}_j(p+2,\xi_2)\hat{R}_k(p+2,\xi_3)\Omega_{1,m}(\mathbf{r})$$
$$i = 0, 1, \ldots, p; j, k = 1, 2, \ldots, p+1 \quad (14.50)$$

$$\Omega_{ijk}^{2,m}(\mathbf{r}) = N_{ijk}^{2,m} \hat{R}_i(p+2,\xi_1)R_j(p+2,\xi_2)\hat{R}_k(p+2,\xi_3)\Omega_{2,m}(\mathbf{r})$$
$$j = 0, 1, \ldots, p; i, k = 1, 2, \ldots, p+1 \quad (14.51)$$

$$\Omega_{ijk}^{3,m}(\mathbf{r}) = N_{ijk}^{3,m} \hat{R}_i(p+2,\xi_1)\hat{R}_j(p+2,\xi_2)R_k(p+2,\xi_3)\Omega_{3,m}(\mathbf{r})$$
$$k = 0, 1, \ldots, p; i, j = 1, 2, \ldots, p+1 \quad (14.52)$$

to interpolate a transverse field, where (i,j,k) denote the interpolation points, shown in Figure 14.23(a). These can be written uniformly as

$$\Omega_{ijk}^{\beta,m}(\mathbf{r}) = N_{ijk}^{\beta} R_{i_\beta}(p+2,\xi_\beta)\hat{R}_{i_s}(p+2,\xi_s)\hat{R}_{i_t}(p+2,\xi_t)\Omega_\beta(\mathbf{r})$$
$$i_\beta = 0, 1, \ldots, p; i_s, i_t = 1, 2, \ldots, p+1 \quad (14.53)$$

where i_β is taken to be i, j, or k for $\beta = 1, 2$, or 3, and similarly for i_s and i_t. The degrees of freedom provided by (14.53) are $6(p+1) + 2p(p+1) = 2p^2 + 8p + 6$.

Table 14.4 Comparison Between Higher-Order Tetrahedral and Mixed-Order Prism Elements [31]

Type of element	Unknowns	Half-bandwidth	Memory	CPU time
3rd-order tetrahedra	86,600	3,452	63 MB	28,003 s
3rd/0th-order prisms	84,500	421	9 MB	1,473 s

The mixed-order basis functions to interpolate a longitudinal field are given by

$$\Omega^1_{i'j'k'}(\mathbf{r}) = N^1_{i'j'k'}\hat{R}_{i'}(p+1,\xi_1)R_{j'}(p+1,\xi_2)R_{k'}(p+1,\xi_3)\Omega_1(\mathbf{r})$$
$$i' = 1, 2, \ldots, p+1; j', k' = 0, 1, \ldots, p \quad (14.54)$$

$$\Omega^2_{i'j'k'}(\mathbf{r}) = N^2_{i'j'k'}R_{i'}(p+1,\xi_1)\hat{R}_{j'}(p+1,\xi_2)R_{k'}(p+1,\xi_3)\Omega_2(\mathbf{r})$$
$$j' = 1, 2, \ldots, p+1; i', k' = 0, 1, \ldots, p \quad (14.55)$$

$$\Omega^3_{i'j'k'}(\mathbf{r}) = N^3_{i'j'k'}R_{i'}(p+1,\xi_1)R_{j'}(p+1,\xi_2)\hat{R}_{k'}(p+1,\xi_3)\Omega_3(\mathbf{r})$$
$$k' = 1, 2, \ldots, p+1; i', j' = 0, 1, \ldots, p \quad (14.56)$$

where (i', j', k') denote the interpolation points, shown in Figure 14.23(b). These can be written uniformly as

$$\Omega^\beta_{i'j'k'}(\mathbf{r}) = N^\beta_{i'j'k'}\hat{R}_{i_\beta}(p+1,\xi_1)R_{i_s}(p+1,\xi_2)R_{i_t}(p+1,\xi_3)\Omega_\beta(\mathbf{r})$$
$$i_\beta = 1, 2, \ldots, p+1; i_s, i_t = 0, 1, \ldots, p \quad (14.57)$$

where i_β is taken to be i', j', or k' for $\beta = 1$, 2, or 3, and similarly for i_s and i_t. The degrees of freedom provided by (14.57) are $(p+2)(p+3)/2$, resulting in $(5p^2 + 21p + 18)/2$ degrees of freedom for each prism.

As pointed out by Graglia et al. [32], these basis functions provide element-to-element tangential continuity even in the curvilinear case. Note that the extension from the zeroth order in the longitudinal direction to a higher order is very straightforward.

To demonstrate the advantage of this type of element, we computed the RCS of a $2\lambda \times 2\lambda \times 10\lambda$ rectangular cavity using the third-order tetrahedral elements and the mixed-order prism elements with third-order in the transverse plane and zeroth-order in the longitudinal direction. The number of unknowns in each case was chosen to yield a comparable accuracy (the one using the prism elements is slightly more accurate). The RCS is given in Figure 14.24 and the information about the discretization, computer memory, and CPU time is given in Table 14.4. As can be seen, both types of element used about the same number of unknowns. But, the mixed-order prism elements have a much smaller half-bandwidth and hence use a much smaller memory and CPU time.

Next, we examine the efficiency of mixed-order prism elements with respect to the order in the transverse plane while fixing the one in the longitudinal direction to be

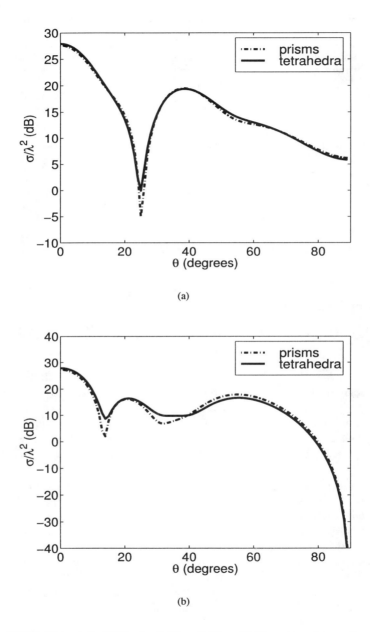

Figure 14.24 Monostatic RCS of a $2.0\lambda \times 2.0\lambda \times 10.0\lambda$ rectangular cavity: (a) $\theta\theta$-polarization; (b) $\phi\phi$-polarization [31].

High-Order Methods in Computational Electromagnetics

Table 14.5 Information about Memory Requirements and CPU Times for RCS Calculations [31]

Problem	Unknowns	Half-bandwidth	Memory	CPU time
Fig. 14.26(a)	1,354,240	2,645	98 MB	161 hr
Fig. 14.26(b)	5,365,353	2,793	109 MB	652 hr

zeroth order. The parameter used in this examination is again the RMS error defined in (14.27). The reference solution is obtained using third-order in the transverse with an overly dense mesh and zeroth-order in the longitudinal direction with a mesh density of 80 points per wavelength. The reference solution so obtained does not change anymore when either the order of elements or the mesh density is increased. For all the calculated results, the mesh density in the longitudinal direction is fixed at 40 points per wavelength so that the accuracy of the results is determined only by the order in the transverse plane. Figure 14.25(a) displays the RMS error in the monostatic RCS of a $1.5\lambda \times 1.5\lambda \times 2.0\lambda$ rectangular cavity as a function of the number of unknowns. It is evident that, for the same number of unknowns, the higher-order elements produce more accurate results. For a desired accuracy, the number of unknowns required for the higher-order elements is much smaller than that for lower-order elements. Figure 14.25(b) displays the corresponding CPU time. Clearly, for a given accuracy, the higher-order elements consume much less CPU time than the lower-order elements. Compared to the case of tetrahedra [Figure 14.20(b)], here the increase in the CPU time when the order is increased is much less significant.

Finally, we present two examples, all illustrated in Figure 14.26, to demonstrate the capability of the proposed method. The first one is a $5\lambda \times 5\lambda \times 10\lambda$ rectangular cavity. The computed RCS results are compared with a modal solution [33] in Figure 14.27 and excellent agreement is observed. The second is an offset bend cavity having a total depth of 32λ at 10 GHz. The computed results are compared in Figure 14.28 with the measured data and a 2D solution based on a hybrid boundary integral method and modal approach (BIM/MODE) [34]. For the $\theta\theta$-polarization, our method agrees better with the measured data, whereas for the $\phi\phi$-polarization, our solution is very similar to that of the BIM/MODE. For the sake of comparison, the results in Figures 14.27 and 14.28 are calculated for the cavities without the ground plane. All the calculations are done using the mixed 3rd/0th-order prism elements. The information about the number of unknowns, half-bandwidth, memory requirements, and CPU times is given in Table 14.5. The CPU times are measured on a DEC personal workstation (500-MHz Alpha 21164 processor).

We should note that in all examples presented here, the computations are done without making any assumption about the variation of the cross-section of the cavity along its depth. In other words, the computing times would remain the same even if the cavity's cross-section varies along its depth. However, if a cavity or some sections

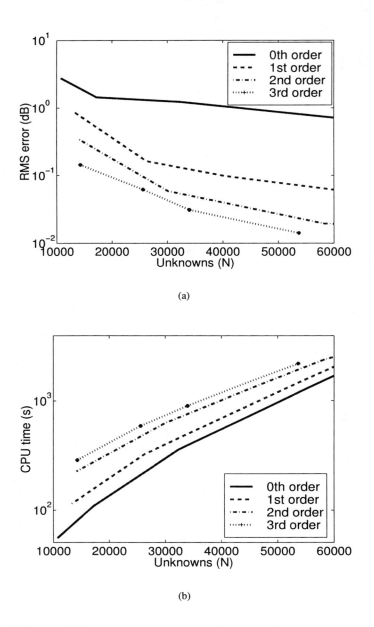

Figure 14.25 (a) RMS error versus the number of unknowns for the monostatic RCS of a 1.5λ × 1.5λ × 2.0λ rectangular cavity modeled with prism elements and (b) corresponding CPU time [31].

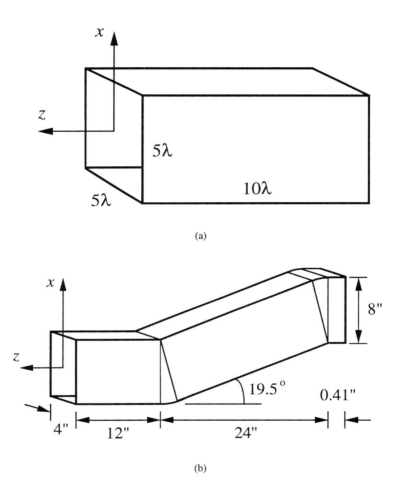

Figure 14.26 (a) A $5\lambda \times 5\lambda \times 10\lambda$ rectangular cavity. (b) An offset bend cavity having a total depth of 32λ at 10 GHz.

of a cavity have a constant cross-section, our special algorithm can be modified such that the corresponding computing time is proportional to $\log L$, instead of L, where L is the length of the cavity or the length of the section with a constant cross-section, at a cost of increasing the memory requirements by a factor of 1.5. If we utilize this property, the CPU times for the calculation of the two examples in Figure 14.26 are reduced to 7.1 and 23.7 hrs, respectively, with the corresponding memory requirements increased to 149 and 166 MB, respectively.

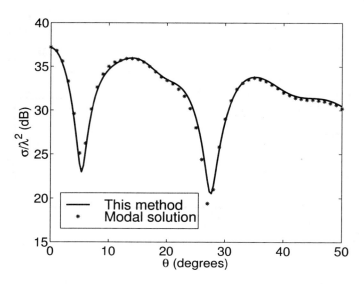

Figure 14.27 Monostatic $\theta\theta$-polarized RCS of the rectangular cavity modeled with prism elements [31].

14.6 POINT-BASED GRID-ROBUST HIGHER-ORDER BASES

Among the present higher-order methods, one class uses higher-order basis functions in projection methods such as the FEM and Galerkin-based MOM described in the preceding sections. The other class uses the Nyström approach with high-order discretization [35].

In the first class of higher-order methods, the basis functions used to expand the unknown surface current have an important property in that their normal components are always continuous across the sides of triangular patches representing the surface of the object to be analyzed. This property precludes the infinite accumulation of electric charges along the patch sides and thus permits the use of Galerkin's method in the solution of the integral equation for the surface current.

In the second class of higher-order methods, the unknown surface currents are represented by their samples at a set of discrete points on the surface of the object [35]. The Nyström approach is then employed to discretize the integral equation to be solved. Additional efforts are often required to deal with the singular integrals by using local correction techniques, which use either the LU decomposition or the singular value decomposition (SVD) to solve certain matrix equations.

Comparing the two classes of higher-order methods discussed above, the first class produces more accurate solutions for the same number of unknowns. However, the evaluation of the integrals in the first class of methods is more time consuming

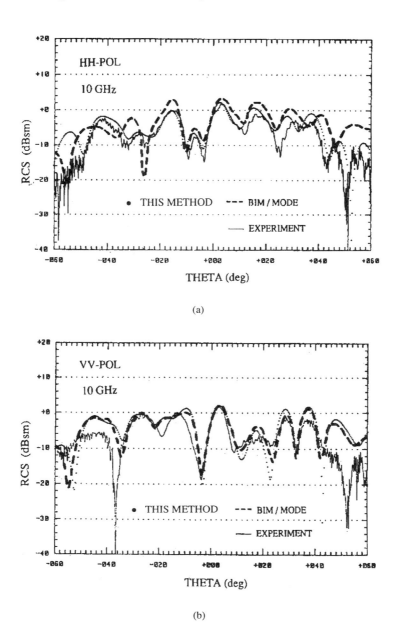

Figure 14.28 Monostatic RCS of the offset bend cavity in the xz plane at 10 GHz modeled with prism elements: (a) $\theta\theta$-polarization; (b) $\phi\phi$-polarization [31].

because there are two-fold integrations to be evaluated numerically. Additionally, the basis functions used in the first class require the mesh of triangular patches to be well connected: a side of a triangular patch has to be entirely shared by another patch and no vertices of a triangular patch can reside in the middle of its neighboring patch's sides. This requirement is also found in the FEM [13]. However, for many practical applications, this requirement is too stringent and it precludes the use of defective meshes.

This section describes a set of novel, grid-robust, higher-order vector basis functions for the MOM solution of integral equations for 3D electromagnetics problems. These basis functions are defined over curvilinear triangular patches and represent the unknown electric current density within each patch using the Lagrange interpolation polynomials. The highlight of these basis functions is that the Lagrange interpolation points are chosen to be the same as the nodes of the well-developed Gaussian quadratures. As a result, the evaluation of the integrals in the MOM is greatly simplified. Additionally, the surface of an object to be analyzed can be easily meshed because the new basis functions do not require the side of a triangular patch to be entirely shared by another triangular patch. These basis functions are implemented with point-matching for the MOM solution of the EFIE, the MFIE, and the CFIE. Numerical examples are presented to demonstrate the higher-order convergence and the grid robustness for defective meshes using the new basis functions.

14.6.1 Vector Basis Functions

Given an object, its surface can be meshed into curvilinear triangular patches. The current on the surface can then be represented as a summation of the currents on each of the patches:

$$\mathbf{J}(\mathbf{r}) = \sum_{p=1}^{P} \mathbf{J}_p(\mathbf{r}) \tag{14.58}$$

where $\mathbf{J}_p(\mathbf{r})$ denotes the current on the pth patch. The $\mathbf{J}_p(\mathbf{r})$ can be interpolated by the Lagrange interpolator:

$$\mathbf{J}_p(\mathbf{r}) = \sum_{i=1}^{I_p} \overline{\mathbf{L}}_{(i,p)}(\mathbf{r}) \cdot \mathbf{J}_p(\mathbf{r}_i) \tag{14.59}$$

where $\mathbf{r}_i$ ($i = 1, 2, \ldots, I_p$) are the interpolation points and $\overline{\mathbf{L}}_{(i,p)}(\mathbf{r})$ is a vector interpolator with the property that

$$\overline{\mathbf{L}}_{(i,p)}(\mathbf{r}_j) = \overline{\mathbf{I}} \delta_{ij} \tag{14.60}$$

where δ_{ij} is the Kronecker delta function and $\overline{\mathbf{I}}$ is the unit dyadic. Equation (14.59) can also be interpreted as the expansion of the current using $\overline{\mathbf{L}}_{(i,p)}(\mathbf{r})$ as the basis functions and $\mathbf{J}_p(\mathbf{r}_i)$ as the unknown expansion coefficients, which are to be determined.

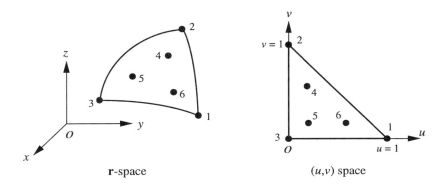

Figure 14.29 A curvilinear triangular patch in the **r** space mapped to a planar right-angle triangle in the (u, v) space.

With the property in (14.60), the evaluation of the surface integrals in the MOM can be greatly simplified. For example, consider a surface integral over a triangular patch

$$\iint_{\triangle p} f(\mathbf{r}, \mathbf{r}')\overline{\mathbf{L}}_{(i,p)}(\mathbf{r}') \cdot \mathbf{J}_p(\mathbf{r}_i) ds' \tag{14.61}$$

where $f(\mathbf{r}, \mathbf{r}')$ is the kernel, which may be singular at $\mathbf{r} = \mathbf{r}'$. When the point $\mathbf{r}$ is not close to the triangular patch p, the Gaussian quadrature yields

$$\sum_{s=1}^{I} w_s f(\mathbf{r}, \mathbf{r}_s)\overline{\mathbf{L}}_{(i,p)}(\mathbf{r}_s) \cdot \mathbf{J}_p(\mathbf{r}_i) \tag{14.62}$$

where I is the number of points in the quadrature rule and w_s denote the weighting coefficients. If the quadrature points are the same as the interpolation points in the Lagrange polynomial, (14.60) can be used to simplify (14.63) to give

$$w_i f(\mathbf{r}, \mathbf{r}_i) \mathbf{J}_p(\mathbf{r}_i) \tag{14.63}$$

which greatly simplifies the result of integration.

To design the basis functions $\overline{\mathbf{L}}_{(i,p)}(\mathbf{r})$, we first map a curvilinear triangular patch in the **r** space into the (u, v) space, as shown in Figure 14.29. Since the surface current has two degrees of freedom, we can write any surface current in the (u, v) space as

$$\mathbf{J}_p = \frac{1}{\mathcal{J}}[J_p^u \mathbf{u} + J_p^v \mathbf{v}] \tag{14.64}$$

where

$$\mathbf{u} = \frac{\partial \mathbf{r}}{\partial u} \qquad \mathbf{v} = \frac{\partial \mathbf{r}}{\partial v} \tag{14.65}$$

$$\mathcal{J} = \sqrt{g_{11}g_{22} - g_{12}g_{21}} \tag{14.66}$$

in which

$$g_{11} = \mathbf{u} \cdot \mathbf{u} \qquad g_{12} = g_{21} = \mathbf{v} \cdot \mathbf{u} \qquad g_{22} = \mathbf{v} \cdot \mathbf{v} \tag{14.67}$$

From (14.64), the current at any point $\mathbf{r}_i$ on the patch can be written as

$$\mathbf{J}_{(i,p)} = \frac{1}{\mathcal{J}_{(i,p)}}[J^u_{(i,p)}\mathbf{u}_{(i,p)} + J^v_{(i,p)}\mathbf{v}_{(i,p)}] \tag{14.68}$$

The current $\mathbf{J}_p$ can then be interpolated by the Lagrange interpolator as

$$\mathbf{J}_p = \sum_{i=1}^{I_p} \frac{1}{\mathcal{J}} L_{(i,p)}(u,v)[J^u_{(i,p)}\mathbf{u} + J^v_{(i,p)}\mathbf{v}] \tag{14.69}$$

where $L_{(i,p)}(u,v)$ is a two-dimensional Lagrange interpolator with the property that

$$L_{(i,p)}(u_j, v_j) = \delta_{ij} \tag{14.70}$$

Equation (14.69) can also be written as

$$\mathbf{J}_p = \sum_{i=1}^{I_p} \frac{1}{\mathcal{J}} L_{(i,p)}(u,v)[\mathbf{u}(g_{22}\mathbf{u} - g_{12}\mathbf{v})_{(i,p)} + \mathbf{v}(g_{11}\mathbf{v} - g_{21}\mathbf{u})_{(i,p)}] \cdot \mathbf{J}_{(i,p)}$$

$$= \sum_{i=1}^{I_p} \overline{\mathbf{L}}_{(i,p)}(u,v) \cdot \mathbf{J}_{(i,p)} \tag{14.71}$$

It then remains to design $\overline{\mathbf{L}}_{(i,p)}(u,v)$ in the $\mathbf{r}$ space to satisfy (14.60) or $L_{(i,p)}(u,v)$ in the (u,v) space to satisfy (14.70) (for simplicity, $L_{(i,p)}(u,v)$ is denoted as $L_i(u,v)$ in the remainder of this section). In addition, we also wish the Lagrange interpolation polynomials $L_i(u,v)$ to have the following properties. First, the interpolation points are distributed inside a triangular patch for the reason to be discussed later. Second, the interpolation points are the same as the nodes of a Gaussian quadrature rule. Third, the chosen interpolation points should give the highest degree of accuracy among the quadrature rules that have the same number of interpolation points.

The three properties above provide a guideline to choose the Gaussian quadratures. Once the quadrature points are chosen, one can proceed to find the explicit expression of $L_i(u,v)$. In the (u,v) space, we can define the space of n-degree polynomials [36] as

$$P_n^2 = \text{span}\{u^i v^j; i,j \geq 0; i+j \leq n\} \tag{14.72}$$

Table 14.6 Information about the Vector Basis Functions and the Chosen Quadrature Rules

Order of basis functions	Number of interpolation points	Accuracy of quadrature rule	Unknowns in each patch
0	1	1	2
1	3	2	6
2	6	4	12

such that the dimension of the space is

$$\dim P_n^2 = C_{n+2}^2 = \frac{(n+2)(n+1)}{2} \qquad (14.73)$$

For $n = 1$, $\dim P_n^2 = 3$, and $P_n^2 = \text{span}\{1, u, v\}$, and in this case, we choose the interpolation points from the three-point quadrature rules. For $n = 2$, $\dim P_n^2 = 6$, and $P_n^2 = \text{span}\{1, u, v, u^2, uv, v^2\}$, and consequently, we choose the interpolation points from the six-point quadrature rules. This can be extended to the higher-order ones in a straightforward manner.

Once the polynomials P_n^2 are chosen, $L_i(u, v)$ can be determined by solving the matrix equation

$$\begin{bmatrix} P_1(u_1, v_1) & P_1(u_2, v_2) & \cdots & P_1(u_m, v_m) \\ P_2(u_1, v_1) & P_2(u_2, v_2) & \cdots & P_2(u_m, v_m) \\ \vdots & \vdots & \ddots & \vdots \\ P_m(u_1, v_1) & P_m(u_2, v_2) & \cdots & P_m(u_m, v_m) \end{bmatrix} \begin{bmatrix} L_1(u, v) \\ L_2(u, v) \\ \vdots \\ L_m(u, v) \end{bmatrix}$$
$$= \begin{bmatrix} P_1(u, v) \\ P_2(u, v) \\ \vdots \\ P_m(u, v) \end{bmatrix} \qquad (14.74)$$

where (u_i, v_i) denote the interpolation points and $m = I_p$ is the total number of such points. If the Vandermonde determinant of the above equation is nontrivial, $L_i(u, v)$ can be obtained uniquely. Therefore, an additional requirement for the interpolation points is that they should guarantee $|VDM| \neq 0$.

Table 14.6 lists the information about the zeroth-, first-, second-, and third-order vector basis functions that are designed based on the procedure described above. The zeroth-order basis function is the same as the well-known pulse function. For the zeroth-, first-, and second-order basis functions, we choose the quadrature rules [37] whose number of points matches the degrees of polynomials, $\dim P_n^2$.

14.6.2 MOM Formulation

The new basis functions can be implemented in the MOM solution of the EFIE, MFIE, and CFIE using the point-matching approach. The EFIE is given by

$$jk_0\eta_0 \hat{t} \cdot \iint_S \overline{\mathbf{G}}(\mathbf{r},\mathbf{r}') \cdot \mathbf{J}(\mathbf{r}')ds' = -\hat{t} \cdot \mathbf{E}^{\text{inc}}(\mathbf{r}) \qquad (14.75)$$

where $\overline{\mathbf{G}}(\mathbf{r},\mathbf{r}')$ is the dyadic Green's function, $\hat{t}$ denotes the tangential vector on the surface S, and $\mathbf{E}^{\text{inc}}$ denotes the incident electric field. Substituting the current expansion into the EFIE and applying the point-matching, we obtain

$$\sum_{p=1}^{P}\sum_{i=1}^{I_p} \begin{bmatrix} A_{uu} & A_{uv} \\ A_{vu} & A_{vv} \end{bmatrix} \begin{bmatrix} J^u_{(i,p)} \\ J^v_{(i,p)} \end{bmatrix} = \begin{bmatrix} U_u \\ U_v \end{bmatrix} \qquad (14.76)$$

where

$$A_{\alpha\beta} = jk_0\eta_0 \iint_{\Delta p} \boldsymbol{\alpha}_{(j,q)} \cdot \overline{\mathbf{G}}(\mathbf{r}_{(j,q)},\mathbf{r}') \cdot \boldsymbol{\beta} L_{(i,p)}(\mathbf{r}')\mathcal{J}^{-1}ds' \qquad (14.77)$$

$$U_\alpha = \boldsymbol{\alpha}_{(j,q)} \cdot \mathbf{E}^{\text{inc}}(\mathbf{r}_{(j,q)}) \qquad (14.78)$$

where Δp denotes the area of the pth patch, and (j,q) denotes that the matched point is at the jth node of the qth patch.

When the pth and qth patches are far apart, $A_{\alpha\beta}$ can be evaluated using Gaussian quadrature to yield

$$A_{\alpha\beta} = jk_0\eta_0 w_{(i,p)}\boldsymbol{\alpha}_{(j,q)} \cdot \overline{\mathbf{G}}(\mathbf{r}_{(j,q)},\mathbf{r}_{(i,p)}) \cdot \boldsymbol{\beta}_{(i,p)}\mathcal{J}^{-1}_{(i,p)} \qquad (14.79)$$

where $w_{(i,p)}$ denotes the weight in the quadrature. When the pth and qth patches are close to each other, we can employ the Gauss' theorem to transfer one del operator away from the Green's function, obtaining

$$\begin{aligned}A_{\alpha\beta} = &\; jk_0\eta_0 \iint_{\Delta p} \boldsymbol{\alpha}_{(j,q)} \cdot \boldsymbol{\beta} g(\mathbf{r}_{(j,q)},\mathbf{r}')L_{(i,p)}(\mathbf{r}')\mathcal{J}^{-1}ds' \\ &+ \frac{j\eta_0}{k_0} \iint_{\Delta p} \boldsymbol{\alpha}_{(j,q)} \cdot \nabla g(\mathbf{r}_{(j,q)},\mathbf{r}') \frac{\partial L_{(i,p)}(\mathbf{r}')}{\partial \beta} \mathcal{J}^{-1}ds' \\ &- \frac{j\eta_0}{k_0} \int_{\partial p} \boldsymbol{\alpha}_{(j,q)} \cdot \nabla g(\mathbf{r}_{(j,q)},\mathbf{r}')L_{(i,p)}(\mathbf{r}')\boldsymbol{\beta}\cdot\hat{n}_e\mathcal{J}^{-1}d\ell'\end{aligned} \qquad (14.80)$$

where $g(\mathbf{r},\mathbf{r}')$ is the scalar Green's function, ∂p denotes the contour of Δp, and $\hat{n}_e$ denotes the unit tangential vector along ∂p. The resulting integrals can be evaluated using a higher-degree Gaussian quadrature. When $p = q$, the first integral in (14.80) has a $1/R$ ($R = |\mathbf{r} - \mathbf{r}'|$) singularity, which can be evaluated using Duffy's transformation method [15]. The third integral is nonsingular if the Gaussian quadrature points do not reside on ∂p. However, the second integral has a $1/R^2$ singularity. To overcome this problem, we apply the singularity extraction method to find

$$\iint_{\Delta p} \boldsymbol{\alpha}_{(j,p)} \cdot \nabla g(\mathbf{r}_{(j,p)}, \mathbf{r}') \frac{\partial L_{(i,p)}(\mathbf{r}')}{\partial \beta} \mathcal{J}^{-1} ds'$$

$$= \iint_{\Delta p} \nabla g(\mathbf{r}_{(j,p)}, \mathbf{r}') \cdot \left\{ \boldsymbol{\alpha}_{(j,p)} \frac{\partial L_{(i,p)}(\mathbf{r}')}{\partial \beta} - \boldsymbol{\alpha} \left[\frac{\partial L_{(i,p)}(\mathbf{r}')}{\partial \beta} \right]_{\mathbf{r}'=\mathbf{r}_{(j,p)}} \right\} \mathcal{J}^{-1} ds'$$

$$+ \iint_{\Delta p} \boldsymbol{\alpha} \cdot \nabla g(\mathbf{r}_{(j,p)}, \mathbf{r}') \left[\frac{\partial L_{(i,p)}(\mathbf{r}')}{\partial \beta} \right]_{\mathbf{r}'=\mathbf{r}_{(j,p)}} \mathcal{J}^{-1} ds'. \quad (14.81)$$

As a result, the first integral on the right-hand side has a $1/R$ singularity at the Gaussian quadrature points and can be evaluated using Duffy's transformation method. The second integral can be transformed into a nonsingular contour integral using Gauss' theorem:

$$\iint_{\Delta p} \boldsymbol{\alpha} \cdot \nabla g(\mathbf{r}_{(j,p)}, \mathbf{r}') \left[\frac{\partial L_{(i,p)}(\mathbf{r}')}{\partial \beta} \right]_{\mathbf{r}'=\mathbf{r}_{(j,p)}} \mathcal{J}^{-1} ds'$$

$$= -\iint_{\Delta p} \nabla' \cdot \left\{ \boldsymbol{\alpha} g(\mathbf{r}_{(j,p)}, \mathbf{r}') \left[\frac{\partial L_{(i,p)}(\mathbf{r}')}{\partial \beta} \right]_{\mathbf{r}'=\mathbf{r}_{(j,p)}} \mathcal{J}^{-1} \right\} ds'$$

$$= -\int_{\partial p} \boldsymbol{\alpha} \cdot \hat{n}_e \, g(\mathbf{r}_{(j,p)}, \mathbf{r}') \left[\frac{\partial L_{(i,p)}(\mathbf{r}')}{\partial \beta} \right]_{\mathbf{r}'=\mathbf{r}_{(j,p)}} \mathcal{J}^{-1} d\ell' \quad (14.82)$$

which can be evaluated using a higher-degree Gaussian quadrature.

A similar formulation can be applied to the MFIE, which is given by

$$\frac{\mathbf{J}(\mathbf{r})}{2} - \hat{n} \times \nabla \times \iint_S g(\mathbf{r}, \mathbf{r}') \mathbf{J}(\mathbf{r}') ds' = \hat{n} \times \mathbf{H}^{\text{inc}}(\mathbf{r}) \quad (14.83)$$

where $\hat{n}$ denotes the normal unit vector on the surface S, and $\mathbf{H}^{\text{inc}}$ denotes the incident magnetic field. Substituting the current expansion into the MFIE and applying the point-matching, we obtain

$$\sum_{p=1}^{P} \sum_{i=1}^{I_p} \begin{bmatrix} B_{uu} & B_{uv} \\ B_{vu} & B_{uu} \end{bmatrix} \begin{bmatrix} J_{(i,p)}^u \\ J_{(i,p)}^v \end{bmatrix} = \begin{bmatrix} V_u \\ V_v \end{bmatrix} \quad (14.84)$$

where

$$B_{\alpha\beta} = \frac{1}{2} \boldsymbol{\alpha}_{(j,q)} \cdot \boldsymbol{\beta} \mathcal{J}_{(i,p)}^{-1} \delta_{ij} \delta_{pq} - \boldsymbol{\alpha}_{(j,q)}$$

$$\cdot \left[\hat{n} \times \iint_{\Delta p} \nabla g(\mathbf{r}_{(j,q)}, \mathbf{r}') \times \boldsymbol{\beta} L_{(i,p)}(\mathbf{r}') \mathcal{J}^{-1} ds' \right] \quad (14.85)$$

$$V_\alpha = \boldsymbol{\alpha}_{(j,q)} \cdot \left[\hat{n} \times \mathbf{H}^{\text{inc}}(\mathbf{r}_{(j,q)}) \right]. \quad (14.86)$$

When the pth and qth patches are far apart, $B_{\alpha\beta}$ can be evaluated using Gaussian quadrature to yield

$$B_{\alpha\beta} = w_{(i,p)}\boldsymbol{\alpha}_{(j,q)} \cdot \left[\hat{n} \times \nabla g(\mathbf{r}_{(j,q)}, \mathbf{r}_{(i,p)}) \times \boldsymbol{\beta}_{(i,p)} \mathcal{J}_{(i,p)}^{-1}\right] \quad (14.87)$$

When the pth and qth patches are close to each other, $B_{\alpha\beta}$ can be evaluated using a higher-degree Gaussian quadrature. When $p = q$, the second term in the right-hand of (14.85) contains

$$\hat{n} \times [g(\mathbf{r}_{(j,q)}, \mathbf{r}') \times \boldsymbol{\beta}] = [(\hat{n} \cdot \hat{R})\boldsymbol{\beta} - (\hat{n} \cdot \boldsymbol{\beta})\hat{R}]\frac{jk_0 R + 1}{R^2}e^{-jk_0 R} \quad (14.88)$$

Since when $R \to 0$, both $\hat{R}$ and $\boldsymbol{\beta}$ are perpendicular to $\hat{n}$ and, hence, $\hat{n} \cdot \boldsymbol{\beta} \to 0$ and $\hat{n} \cdot \hat{R} \to 0$. As a result, (14.88) has a $1/R$ singularity and, therefore, it can be evaluated using Duffy's transformation method.

14.7 NUMERICAL RESULTS

A conducting sphere having a diameter of 1m is used to test the MOM solution using the new higher-order vector basis functions. The parameter used in the study is the RMS error defined in (14.27).

Figures 14.30 and 14.31 show the convergence behavior of the higher-order basis functions obtained at 0.3 GHz. It is clear that when the number of unknowns increases, the results obtained using higher-order basis functions converge more quickly than those of lower-order ones. This is particularly true when a high solution accuracy is desired. We also note that the solution using the MFIE has a much higher accuracy than the one based on the EFIE. This is in contrast to the Galerkin-based MOM using traditional basis functions, where the EFIE result is found more accurate than the MFIE result. The CFIE results are obtained using two different weights on the EFIE and MFIE: one with $\alpha = 0.2$ and the other with $\alpha = 0.5$. As expected, for a small α, the accuracy is close to that of the MFIE, whereas for a large α, the accuracy approaches that of the EFIE. Since the MFIE has a better accuracy and its matrix is better conditioned than the EFIE, a small α such as $\alpha = 0.2$ is a good choice for this method.

Next, the proposed basis functions are validated on defective meshes. Here, a defective mesh refers to one in which an edge of a triangular patch is partially, instead of totally, shared by another triangular patch, or in other words, two neighboring triangular patches partially share one edge. The validation is carried out on a 1λ diameter sphere using the MFIE with the third-order basis functions. By mapping the triangles of an inscribed regular octahedron onto the surface of the sphere, we obtain an exact description of the sphere with well-connected curvilinear triangular patches, as illustrated in Figure 14.32(a). If we rotate the front half of the octahedron

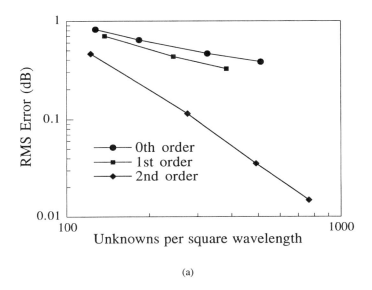

(a)

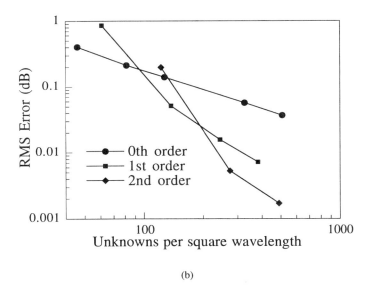

(b)

Figure 14.30 RMS error in the bistatic RCS versus the number of unknowns per square wavelength: (a) EFIE; (b) MFIE [38].

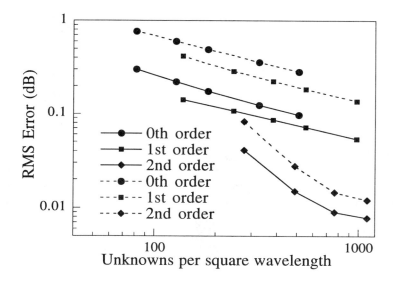

Figure 14.31 RMS error in the bistatic RCS versus the number of unknowns per square wavelength obtained using the CFIE [38]. The solid lines correspond to $\alpha = 0.2$ and the dash lines correspond to $\alpha = 0.5$.

by an angle of $\Delta\phi$ and then map the eight triangles onto the surface of the sphere, we obtain a defective mesh, which also describes the sphere exactly, as shown in Figure 14.32(b). By subdividing each of the eight triangles into four subtriangles, we generate a mesh with 32 triangular patches, which is used for the study here. Figure 14.33 shows the RMS error as a function of the rotating angle $\Delta\phi$. It is seen that the RMS errors from a series of defective meshes are within ±0.004 dB of the RMS error of the well-connected mesh, which is 0.013 dB. This demonstrates clearly the grid-robustness of the proposed new higher-order basis functions.

Finally, we consider the 9.936-in NASA almond, which is an EMCC benchmark geometry [26]. The almond is placed horizontally with its longer axis aligned with the x axis and its sharp tip pointing toward the $\hat{x}$ direction. Its monostatic RCS is computed at 5 GHz using the second-order basis functions (496 triangular patches and 5,952 unknowns) for both the VV and HH polarizations in the horizontal plane. The results are shown in Figure 14.34 and compared to those obtained by a second-order Galerkin-based MOM [24] using the same number of patches and 5,208 unknowns. Excellent agreement is obtained in the VV polarization and only a relatively small error is observed in the HH polarization. It is our experience that in the tip regions

High-Order Methods in Computational Electromagnetics 691

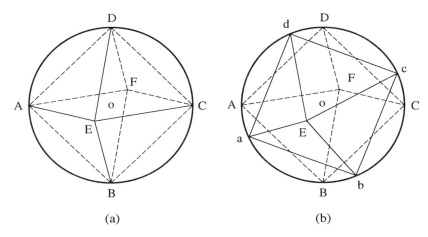

Figure 14.32 (a) An inscribed regular octahedron. (b) The front half of the octahedron is rotated by an angle of $\Delta\phi$.

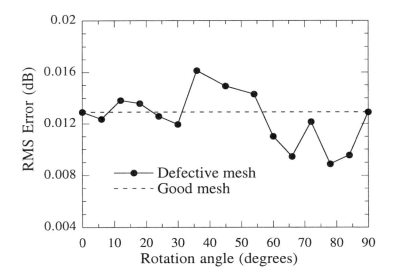

Figure 14.33 RMS error in the bistatic RCS versus the angle of rotation [38].

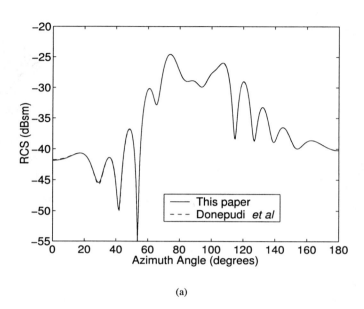

(a)

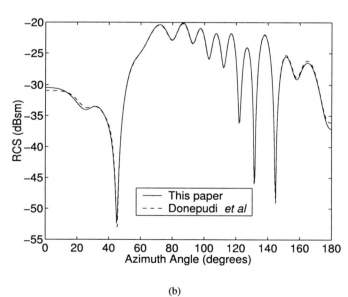

(b)

Figure 14.34 The monostatic RCS of a 9.936-in NASA almond at 5 GHz: (a) VV polarization; (b) HH polarization [38].

around $AZ = 0°$ and $AZ = 180°$, it is very difficult to obtain the HH polarized RCS because of diffraction due to the tip of the almond.

14.8 SUMMARY

This chapter described higher-order methods and their applications in computational electromagnetics. It started with a description of higher-order vector basis functions on a surface for the MOM solution of integral equations of scattering. This method employed higher-order parametric elements to provide accurate modeling of the scatterer's geometry and higher-order interpolatory vector basis functions for an accurate representation of the electric current density on the scatterer's surface. The resultant numerical system of equations was then solved using the MLFMA with a minimum computational complexity. It was found that the higher-order method can significantly reduce the number of unknowns, especially from the zeroth to the first order, without compromising the accuracy of geometry modeling. Typical mesh density to obtain a 0.1-dB solution accuracy is 330, 55, 33 unknowns per square wavelength for the solution of CFIE using the zeroth-, first-, and second-order basis functions, respectively.

However, when the MLFMA was applied directly to solve the system of equations resulting from the higher-order MOM, the number of levels used in MLFMA is determined by the patch size. Because of the larger patch size used with higher-order basis functions, the number of levels is smaller than that in the MLFMA using the zeroth-order basis. This limits the performance of the higher-order MLFMA. This problem was then investigated and overcome using a novel point-based implementation, in which higher-order basis functions are first represented by a set of point sources and the MLFMA is then applied to calculate the point-to-point interactions. Since the point sources have no spatial extent, the number of levels in the MLFMA can be made the same as that in the MLFMA using the zeroth-order basis, which results in a significant reduction in the memory requirements.

We then described higher-order vector basis functions for tetrahedral and triangular prism elements and their application to the RCS calculation of a deep, arbitrarily shaped cavity. A special solver was designed for this particular problem, which exploits the unique features of the FEM equations for cavity scattering. The solver was designed in such a manner that it uses minimal memory, which is proportional to the maximum cross section of the cavity and independent of the depth of the cavity, and its computation time increases only linearly with the depth of the cavity (or even less if the cavity or some of its sections have a constant cross section). Furthermore, it computes the scattered fields for all angles of incidence without requiring significant additional time. The technique was implemented with higher-order tetrahedral and mixed-order prism elements, both having curved sides to allow for accurate modeling of arbitrary geometries. Although both types of element yield a remarkably more accurate and efficient solution for scattering by 3D cavities, the mixed-order prism is optimal for the special solver.

Finally, a set of novel, grid-robust, higher-order vector basis functions was proposed for the MOM solution of integral equations for 3D electromagnetics problems. These basis functions were defined over curvilinear triangular patches and based on the Lagrange interpolation polynomials. The Lagrange interpolation points were chosen to be the same as the nodes of the Gaussian quadratures, which greatly simplified the evaluation of the integrals in the MOM. The new basis functions were also applicable to defective meshes where the side of a triangular patch is only partially shared by another triangular patch. These basis functions were implemented with point-matching for the MOM solution of the EFIE, the MFIE, and the CFIE. Numerical examples were presented to demonstrate the higher-order convergence and the grid robustness for defective meshes.

REFERENCES

1. R. F. Harrington, *Field Computation by Moment Methods,* New York: Macmillan, 1968.

2. S. M. Rao, D. R. Wilton, and A. W. Glisson, "Electromagnetic scattering by surfaces of arbitrary shape," *IEEE Trans. Antennas Propagat.*, vol. 30, pp. 409–418, May 1982.

3. S. Wandzura, "Electric current basis functions for curved surfaces," *Electromagn.*, vol. 12, no. 1, pp. 77–91, Jan. 1992.

4. L. R. Hamilton, P. A. Macdonald, M. A. Stalzer, R. S. Turley, J. L. Visher, and S. M. Wandzura, "Electromagnetic scattering computations using high-order basis functions in the method of moments," *IEEE APS Int. Symp. Dig.*, vol. 3, pp. 2166–2169, June 1994.

5. L. R. Hamilton, J. J. Ottusch, M. A. Stalzer, R. S. Turley, J. L. Visher, and S. M. Wandzura, "Accuracy estimation and high order methods," *11th Ann. Rev. Progress Appl. Computat. Electromag.*, vol. 2, pp. 1177–1184, March 1995.

6. J. C. Nedelec, "Mixed finite elements in R^3," *Numerische Mathematik*, vol. 35, pp. 315–341, 1980.

7. J. C. Nedelec, "A family of mixed finite elements in R^3," *Numerische Mathematik*, vol. 50, pp. 57–81, 1986.

8. A. Bossavit and I. Mayergoyz, "Edge elements for scattering problems," *IEEE Trans. Magn.*, vol. 25, pp. 2816–2821, 1989.

9. J. P. Webb and B. Forghani, "Hierarchal scalar and vector tetrahedra", *IEEE Trans. Magn.*, vol. 29, pp. 1495–1498, March 1993.

10. J. S. Savage and A. F. Peterson, "Higher-order vector finite elements for tetrahedral cells," *IEEE Trans. Microwave Theory Tech.*, vol. 44, pp. 874–879, June 1996.

11. R. D. Graglia, D. R. Wilton, and A. F. Peterson, "Higher order interpolatory vector bases for computational electromagnetics," *IEEE Trans. Antennas Propagat.*, vol. 45, pp. 329–342, March 1997.

12. J. P. Webb, "Hierarchal vector basis functions of arbitrary order for triangular and tetrahedral finite elements," *IEEE Trans. Antennas Propagat.*, vol. 47, pp. 1244–1253, Aug. 1999.

13. J. M. Jin, *The Finite Element Method in Electromagnetics*, New York: Wiley, 1993.

14. J. R. Mautz and R. F. Harrington, "H-field, E-field, and combined-field solutions for conducting body of revolution," *AEU*, vol. 32, pp. 157–164, April 1978.

15. M. G. Duffy, "Quadrature over a pyramid or cube of integrands with a singularity at a vertex," *J. Numer. Anal.*, vol. 19, no. 6, pp. 1260–1262, Dec. 1982.

16. J. M. Song and W. C. Chew, "Multilevel fast-multipole algorithm for solving combined field integral equations of electromagnetic scattering," *Microwave Opt. Tech. Lett.*, vol. 10, no. 1, pp. 14–19, Sept. 1995.

17. B. Dembart and E. Yip, "A 3D fast multipole method for electromagnetics with multiple levels," *11th Ann. Rev. Progress Appl. Computat. Electromag.*, vol. 1, pp. 621–628, March 1995.

18. J. M. Song, C. C. Lu, and W. C. Chew, "MLFMA for electromagnetic scattering by large complex objects," *IEEE Trans. Antennas Propagat.*, vol. 45, pp. 1488–1493, Oct. 1997.

19. S. Velamparambil, J. Song, and W. C. Chew, "A portable parallel multilevel fast multipole solver for scattering from perfectly conducting bodies," *IEEE APS Int. Symp. Dig.*, vol. 1, pp. 648–651, July 1999.

20. S. Velamparambil, J. Song, and W. C. Chew, "ScaleMe: A portable, distributed memory multilevel fast multipole kernel for electromagnetic and acoustic integral equation solvers," Research Report: CCEM-23-99, Univ. of Illinois, Sept. 7, 1999.

21. S. Velamparambil, J. Song, and W. C. Chew, "ScaleMe: Application interface. A programmer's guide and reference," Research Report: CCEM-27-99, Univ. of Illinois, Oct. 8, 1999.

22. H. A. van Der Vorst, "BI-CGSTAB: A fast and smoothly converging variant of BI-CG for the solution of nonsymmetric linear systems," *SIAM J. Sci. Statist. Comput.*, vol. 13, pp. 631–644, 1992.

23. Y. Saad, "GMRES: A generalized minimal residual algorithm for solving nonsymmetric linear systems," *SIAM J. Sci. Statist. Comput.*, vol. 7, pp. 856–869, 1986.

24. K. C. Donepudi, J. M. Jin, S. Velamparambil, J. M. Song, and W. C. Chew, "A higher-order parallelized multilevel fast multipole algorithm for 3D scattering," *IEEE Trans. Antennas Propagat.*, vol. 49, June 2001.

25. S. Velamparambil, J. Schutt-Aine, J. Nickel, J. Song, and W. Chew, "Solving large scale electromagnetic problems using a linux cluster and parallel MLFMA," *IEEE APS Int. Symp. Dig.*, vol. 1, pp. 636–639, July 1999.

26. A. C. Woo, H. T. G. Wang, M. J. Schuh, and M. L. Sanders, "Benchmark radar targets for the validation of computational electromagnetics programs," *IEEE Antennas Propagat. Mag.*, vol. 35, pp. 84–89, Feb. 1993.

27. J. M. Song, C. C. Lu, W. C. Chew, and S. W. Lee, "Fast Illinois solver code (FISC)" *IEEE Antennas Propag. Mag.*, vol. 40, no. 3, pp. 27–34, June 1998.

28. K. C. Donepudi, J. M. Song, J. M. Jin, G. Kang, and W. C. Chew, "A novel implementation of multilevel fast multipole algorithm for higher-order Galerkin's method," *IEEE Trans. Antennas Propagat.*, vol. 48, pp. 1192–1197, Aug. 2000.

29. R. Coifman, V. Rokhlin, and S. Wandzura, "The fast multipole method for the wave equation: A pedestrian prescription," *IEEE Antennas Propagat. Mag.*, vol. 35, no. 3, pp. 7–12, June 1993.

30. J. M. Jin, "Electromagnetic scattering from large, deep, and arbitrarily-shaped open cavities," *Electromagn.*, vol. 18, no. 1, pp. 3–34, Jan.-Feb. 1998.

31. J. Liu and J. M. Jin, "A special higher-order finite element method for scattering by deep cavities," *IEEE Trans. Antennas Propagat.*, vol. 48, pp. 494–703, May 2000.

32. R. D. Graglia, D. R. Wilton, A. F. Peterson, and I. L. Gheorma, "Higher order interpolatory vector bases on prism elements," *IEEE Trans. Antennas Propagat.*, vol. 46, pp. 442–450, March 1998.

33. S. W. Lee and H. Ling, "Data book for cavity RCS," Electromagn. Lab., Univ. Illinois, Urbana, Tech. Rep. SWL89-1, Jan. 1989.

34. H. Ling, "RCS of waveguide cavities: A hybrid integral/modal approach," *IEEE Trans. Antennas Propagat.*, vol. 38, pp. 1413–1419, Sept. 1990.

35. L. S. Canino, J. J. Ottusch, M. A. Stalzer, J. L. Visher, and S. Wandzura, "Numerical solution of the Helmholtz equation in 2D and 3D using a high-order Nyström discretization," *J. Comput. Phys.*, vol. 146, pp. 627–663, 1998.

36. J. S. Hesthaven, "From electrostatics to almost optimal nodal sets for polynomial interpolation in a simplex," *SIAM J. Numer. Anal.*, vol. 35, no. 2, pp. 655–676, April 1998.

37. D. A. Dunavant, "High degree efficient symmetrical Gaussian quadrature rules for the triangle," *Int. J. Numer. Meth. Eng.*, vol. 21, pp. 1129–1148, 1985.

38. G. Kang, J. M. Song, W. C. Chew, K. Donepudi, and J. M. Jin, "A novel grid-robust higher-order vector basis function for the method of moments," *IEEE Trans. Antennas Propagat.*, vol. 49, May 2001.

15
Asymptotic Waveform Evaluation for Broadband Calculations

Dan Jiao and Jian-Ming Jin

15.1 INTRODUCTION

Many electromagnetics applications require the computation of frequency responses over a broad band of frequencies rather than at one or a few isolated frequencies. For example, for radar target recognition, one has to compute the radar cross-section (RCS) of a target over a wide frequency band to generate the range profiles and synthetic aperture radar (SAR) images. For analysis of antennas, especially broadband antennas, one has to calculate the input impedance at many frequencies. Such calculations can be very time consuming when a traditional frequency-domain numerical method is used because a set of algebraic equations must be solved repeatedly at many frequencies. The number of algebraic equations is proportional to the electrical size of the problem and can be large for most applications. Therefore, there is a need to find approximate solution techniques that can efficiently simulate a frequency response over a broad band. Over the past few years, some approaches have been developed to address this issue; these approaches include the methods of model-based parameter estimation (MBPE) [1, 2] and impedance matrix interpolation [3].

A technique similar to the MBPE is the method of asymptotic waveform evaluation (AWE) [4], which was originally developed for high-speed circuit analysis. In

AWE, the transfer function of a circuit is expanded into a series, and the circuit model is then approximated with a lower-order transfer function by moment matching. A detailed description of the method can be found in [5], and a good tutorial article [6] on AWE is also available for electromagnetics researchers. The AWE method has recently been applied to the finite element and finite difference analysis of electromagnetics problems [7–13]. In these applications, the implementation of AWE is straightforward since the resultant matrix equation has a simple dependence on frequency. In the limited applications of AWE to integral equations, the dependence of the scalar Green's function on frequency is often neglected in order to arrive at a simplified final numerical system [14].

For electromagnetic scattering and radiation by conducting and dielectric objects, a very useful solution technique is the method of moments (MOM), which solves a surface integral equation (SIE) for the electric current on the surface of an object. This method is advantageous because (1) it limits the unknown current on the surface of an object and (2) it satisfies the radiation condition via the Green's function. However, the method results in a dense matrix that is computationally expensive to generate and invert. Since this matrix depends on frequency in a complex manner, one has to repeat the calculations at each frequency to obtain the solution over a band of frequencies. Recently, Reddy et al. [15] applied AWE to the MOM solution of the electric field integral equation (EFIE) for the fast computation of the RCS of a perfectly electrically conducting (PEC) object. It was shown that the AWE method requires less CPU time to obtain a frequency response and it can speed up the RCS calculation by a factor of three, compared with the direct calculation over a band of interest.

In this chapter, we describe the application of the AWE method to a variety of electromagnetics problems for a fast frequency-sweep analysis [16–21]. These problems include (1) radiation of antennas made of metallic surfaces and/or wires, (2) scattering by a PEC body, (3) scattering by a dielectric body, which can be dispersive, and (4) scattering and radiation of conformal cavity-backed microstrip patch antennas, whose substrates can also be dispersive. All of these problems are formulated in terms of an SIE or its combination with the finite element method (FEM).

15.2 THE AWE METHOD

Given an electromagnetics problem, its numerical analysis usually results in a matrix equation in the following form:

$$A(k)x(k) = y(k) \qquad (15.1)$$

where A is a square matrix, x is an unknown vector, y is a known vector associated with the source or excitation, and k is the wave number related to frequency. Since the matrix A depends on frequency, it must be generated and solved repeatedly at each individual frequency in order to obtain a solution over a frequency band. This can be time-consuming, especially for problems whose response varies drastically with frequency. This difficulty can be alleviated using the AWE method.

In accordance with the AWE method, to obtain the solution of (15.1) over a wide frequency band, we expand $x(k)$ into a Taylor series

$$x(k) = \sum_{n=0}^{Q} m_n (k - k_0)^n \qquad (15.2)$$

where k_0 is the expansion point, m_n denote the unknown coefficients, and Q denotes the total number of such coefficients. Substituting this into (15.1), expanding the impedance matrix $A(k)$ and the excitation vector $y(k)$ into a Taylor series, and finally matching the coefficients of the equal powers of $k - k_0$ on both sides yield the recursive relation for the moment vectors:

$$m_0 = A^{-1}(k_0) y(k_0) \qquad (15.3)$$

$$m_n = A^{-1}(k_0) \left[\frac{y^{(n)}(k_0)}{n!} - \sum_{i=1}^{n} \frac{A^{(i)}(k_0) m_{n-i}}{i!} \right] \qquad n \geq 1 \qquad (15.4)$$

where A^{-1} denotes the inverse of A, $A^{(i)}$ denotes the ith derivative of A, and likewise $y^{(n)}$ denotes the nth derivative of y.

The Taylor expansion has a limited bandwidth. To obtain a wider bandwidth, we represent $x(k)$ with a better behaved rational Padé function:

$$x(k) = \frac{\sum_{i=0}^{L} a_i (k - k_0)^i}{1 + \sum_{j=1}^{M} b_j (k - k_0)^j} \qquad (15.5)$$

where $L + M = Q$. The unknown coefficients a_i and b_j can be calculated by substituting (15.2) into (15.5), multiplying (15.5) by the denominator of the Padé expansion, and matching the coefficients of the equal powers of $k - k_0$. This leads to the matrix equation

$$\begin{bmatrix} m_L & m_{L-1} & m_{L-2} & \cdots & m_{L-M+1} \\ m_{L+1} & m_L & m_{L-1} & \cdots & m_{L-M+2} \\ m_{L+2} & m_{L+1} & m_L & \cdots & m_{L-M+3} \\ \vdots & \vdots & \vdots & \ddots & \vdots \\ m_{L+M-1} & m_{L+M-2} & m_{L+M-3} & \cdots & m_L \end{bmatrix} \begin{bmatrix} b_1 \\ b_2 \\ b_3 \\ \vdots \\ b_M \end{bmatrix}$$

$$= - \begin{bmatrix} m_{L+1} \\ m_{L+2} \\ m_{L+3} \\ \vdots \\ m_{L+M} \end{bmatrix} \qquad (15.6)$$

which can be solved for b_j. Once b_j are obtained, the unknown coefficients a_i can then be calculated as

$$a_i = \sum_{j=0}^{i} b_j m_{i-j} \qquad 0 \leq i \leq L \qquad (15.7)$$

Clearly, in the procedure described above the impedance matrix $A(k)$ is inverted only once, which is the main reason for the efficiency of the AWE method. With AWE, one obtains a solution that is accurate at a frequency near the point of expansion. The accuracy of the solution decreases when the frequency moves away from the point of expansion. In many practical applications, one is often required to find the solution over a specified frequency band. In such cases, one point of expansion may not be able to yield an accurate solution over the entire band, and multiple points of expansion may become necessary. These points can be selected automatically using the complex frequency hopping (CFH) technique [8], which can be realized with a simple binary search algorithm [12], as described below.

Given a frequency band $[f_1, f_2]$ and an error tolerance ϵ for the desired quantity, denoted by σ, the following steps can be used to select the points of expansion to generate a solution of a desired accuracy within the given frequency band. (1) Let $f_{\min} = f_1$ and $f_{\max} = f_2$. (2) Apply AWE at $f_{\min}$ and $f_{\max}$ and obtain $\sigma_1(f)$ and $\sigma_2(f)$. (3) Choose $f_{\mathrm{mid}} = (f_{\max} + f_{\min})/2$ and calculate $\sigma_1(f_{\mathrm{mid}})$ and $\sigma_2(f_{\mathrm{mid}})$. (4) If $|\sigma_1(f_{\mathrm{mid}}) - \sigma_2(f_{\mathrm{mid}})| < \epsilon$, stop. (5) Otherwise, apply AWE at f_{mid} and repeat the steps above for the subregions $[f_{\min}, f_{\mathrm{mid}}]$ and $[f_{\mathrm{mid}}, f_{\max}]$. This process continues until an accurate solution is obtained over the entire frequency band.

Note that in the binary search algorithm described above, one can also use more points to check the accuracy of the solution. For example, one can choose two points at $f_{\mathrm{comp1}} = f_{\min} + 0.4(f_{\max} - f_{\min})$ and $f_{\mathrm{comp2}} = f_{\min} + 0.6(f_{\max} - f_{\min})$ and check the accuracy at these two points. This can reduce the probability of false termination of the binary search. The implementation described here is one of the two approaches of the CFH technique. The other approach constructs one rational transfer function over the entire frequency band by matching the moments calculated at different frequency points. As pointed out by [8], the approach employed here is generally more efficient for generating frequency responses.

15.3 ANALYSIS OF METALLIC ANTENNAS

Starting from this section, we describe the application of the AWE method to the analysis of a number of electromagnetic scattering and radiation problems. We first consider the analysis of antennas made of metallic surfaces.

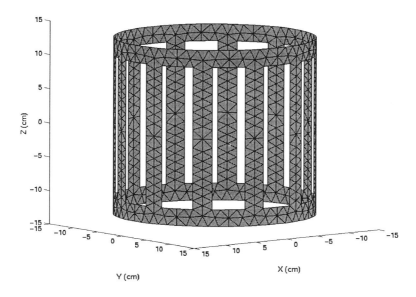

Figure 15.1 Subdivision of a structure into small triangular elements [18].

15.3.1 Formulation

Consider an arbitrarily shaped metallic antenna. A voltage source applied to the antenna will induce an electric current on the antenna's surface. The electric field intensity produced by this current can be expressed as

$$\mathbf{E}(\mathbf{r}) = -jk\eta \iint_S \overline{\mathbf{G}}(\mathbf{r}, \mathbf{r}') \cdot \mathbf{J}(\mathbf{r}') dS' \tag{15.8}$$

where k is the free-space wavenumber, η is the free-space wave impedance, S denotes the conducting surface of the antenna, $\mathbf{J}$ denotes the unknown surface current density on S, and $\overline{\mathbf{G}}(\mathbf{r}, \mathbf{r}')$ is the well-known free-space dyadic Green's function given by

$$\overline{\mathbf{G}}(\mathbf{r}, \mathbf{r}') = \left(\overline{\mathbf{I}} + \frac{\nabla \nabla}{k^2} \right) g(\mathbf{r}, \mathbf{r}') \qquad g(\mathbf{r}, \mathbf{r}') = \frac{e^{-jk|\mathbf{r} - \mathbf{r}'|}}{4\pi |\mathbf{r} - \mathbf{r}'|} \tag{15.9}$$

with $\overline{\mathbf{I}}$ being the unit dyad.

To determine the unknown surface current density $\mathbf{J}$, we first subdivide the conducting surface S into small triangular elements. One example of such a subdivision is shown in Figure 15.1. The surface current density can then be expanded using the

Rao-Wilton-Glisson (RWG) basis function $\mathbf{f}_n(\mathbf{r})$ [22]:

$$\mathbf{J}(\mathbf{r}) = \sum_{n=1}^{N} I_n \mathbf{f}_n(\mathbf{r}) \tag{15.10}$$

where N is the number of unknowns, which is the number of edges shared by two triangular elements, and I_n denotes the unknown expansion coefficient. The RWG basis function $\mathbf{f}_n(\mathbf{r})$, also known as the triangular roof-top function, is defined over two triangular elements joined at a common edge ℓ_n:

$$\mathbf{f}_n(\mathbf{r}) = \begin{cases} \dfrac{\ell_n}{2A_n^+} \boldsymbol{\rho}_n^+, & \mathbf{r} \text{ in } T_n^+ \\ \dfrac{\ell_n}{2A_n^-} \boldsymbol{\rho}_n^-, & \mathbf{r} \text{ in } T_n^- \\ 0, & \text{otherwise} \end{cases} \tag{15.11}$$

where $T_n^\pm$ denote the two triangles associated with the nth edge, $A_n^\pm$ are the areas of triangles $T_n^\pm$, ℓ_n is the length of the nth edge, and $\boldsymbol{\rho}_n^\pm$ are the vectors defined in Figure 15.2(a). The vector plot of $\mathbf{f}_n(\mathbf{r})$ is illustrated in Figure 15.2(b). The most important feature of this basis function is that its normal component on edge ℓ_n is a constant (normalized to 1) whereas the normal components on other edges are zero. This feature guarantees the continuity of current flow over all edges and makes I_n the current density passing through edge ℓ_n. Triangular elements are chosen because of their excellent capability to model arbitrary geometries. Higher-order basis functions can also be employed without any difficulty.

Applying Galerkin's method to (15.8) results in a matrix equation

$$Z(k)I(k) = V(k) \tag{15.12}$$

in which the impedance matrix Z and vector V have the elements given by

$$Z_{mn}(k) = jk\eta \iint_{T_m} \iint_{T_n} \left[\mathbf{f}_m(\mathbf{r}) \cdot \mathbf{f}_n(\mathbf{r}') - \frac{1}{k^2} \nabla \cdot \mathbf{f}_m(\mathbf{r}) \nabla \cdot \mathbf{f}_n(\mathbf{r}') \right] g(\mathbf{r}, \mathbf{r}') dS' dS \tag{15.13}$$

$$V_m(k) = -\iint_{T_m} \mathbf{E}(\mathbf{r}) \cdot \mathbf{f}_m(\mathbf{r}) dS \tag{15.14}$$

where T_m and T_n denote the support of $\mathbf{f}_m$ and $\mathbf{f}_n$, respectively.

If there is neither a capacitor nor a voltage source applied at T_m, then $V_m(k) = 0$ because the boundary condition requires that the tangential electric field must vanish on a PEC surface. If a capacitor C_m is applied at T_m, as illustrated in Figure 15.3(a), it is easy to find that

$$V_m(k) = -\frac{(\ell_m)^2}{j\omega C_m} I_m(k) \tag{15.15}$$

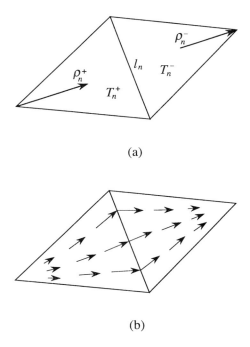

Figure 15.2 (a) Illustration of two joint triangular elements. (b) Vector plot of the RWG function $\mathbf{f}_n$.

where ω denotes the angular frequency. Clearly, this can be moved to the left-hand side of (15.12), which is equivalent to adding $(\ell_m)^2/j\omega C_m$ to Z_{mm}. If a voltage source V_m^{ex} is applied at T_m, as illustrated in Figure 15.3(b), then

$$V_m(k) = \ell_m V_m^{\text{ex}} \tag{15.16}$$

which provides the right-hand side of (15.12) for a solution of $I(k)$.

To apply AWE to (15.12), one has to calculate the derivatives of $Z(k)$. The integrand in Z_{mn} can be written in the form

$$f(k) = \left(\frac{1}{k} - ak\right)\frac{e^{-jkr}}{r} \tag{15.17}$$

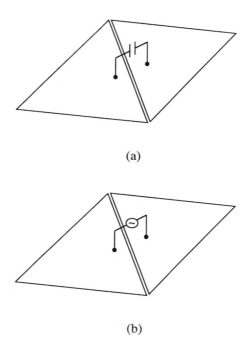

Figure 15.3 (a) A capacitor applied at T_m. (b) A voltage source applied at T_m.

where a is a constant. It can be derived that its nth derivative is given by

$$f^{(n)}(k) = \sum_{i=1}^{n-1} j^{i-1} \frac{(-1)^n n! r^{i-2}}{(i-1)! k^{n-i+2}} e^{-jkr} + n(-j)^{n+1} r^{n-2} e^{-jkr} \left(a + \frac{1}{k^2}\right)$$
$$+ (-j)^n r^{n-1} e^{-jkr} \left(\frac{1}{k} - ak\right) \tag{15.18}$$

A carefully designed recursive formula can allow the calculation of any derivatives.

15.3.2 Numerical Examples

The first example is a radio-frequency (RF) coil, which is an important device in a magnetic resonance imaging (MRI) system. The primary method used for its analysis and design is based on the equivalent lumped-circuit model. In this method, the conductors used to construct an RF coil are modeled as inductors and the RF coil is then modeled as an LC network. The Kirchhoff voltage and current laws

Table 15.1 CPU Timings for Calculations on DEC Personal Workstation

Problem	No. of unknowns	AWE method Freq. pts	AWE method CPU time	Direct method Freq. pts	Direct method CPU time	Speedup
Fig. 15.4	348	25,000	35.6 s	250	1,352 s	38.0
Fig. 15.5	1,043	20,000	719.5 s	400	36,192 s	50.3

are then employed to establish a set of linear equations, whose solution gives the resonant frequencies and the current distributions in the coil. The method is simple and also effective for RF coils operating at a low frequency. However, by modeling a conductor as an inductor, one neglects the current variation in the conductor and also the radiation loss of the conductor. This introduces a significant error in the analysis of RF coils operating at higher frequencies. As a result, the equivalent lumped-circuit method becomes inaccurate and often predicts the current distribution that disagrees with measured data.

A more accurate analysis of RF coils is to employ the method described in this section, which is valid at both low and high frequencies and is also applicable to complex RF coils having structures such as RF shields. Here we apply it to a low-pass birdcage coil whose diameter and length are 26 cm. The coil is made of 2.54-cm-wide conducting strips and has 12 rungs. Its configuration is shown in Figure 15.1. The only things missing there are 12 capacitors, applied at the center of each rung. The capacitors have a value of 1.7 pF to place the dominant mode at 128 MHz and the voltage is applied across one of the capacitors. Figure 15.4 shows the magnitude of the input admittance as a function of frequency. The number of triangular elements used to model the coil is 336 and the number of unknowns is 348. With a frequency increment of 1 MHz, it takes the direct method, which solves (15.12) repeatedly for each frequency, 1,352 s to obtain the solution on a DEC Personal Workstation (500-MHz Alpha 21164 processor). With a sixth-order Taylor expansion ($Q = 6, L = 3, M = 3$), the AWE method produces an accurate solution with 0.01-MHz increments over the entire band in 35.6 s, which is 38 times faster than the direct method.

The above calculations are repeated for the birdcage coil placed in a cylindrical RF shield having a diameter of 32 cm and a length of 30 cm. The capacitors have a value of 2.95 pF to place the dominant mode at 128 MHz. Figure 15.5 gives the magnitude of the input admittance as a function of frequency. The discretization and CPU time information is summarized in Table 15.1, which shows a similar speedup.

The method described in this section can also be applied to compute the input impedances and mutual coupling of wire antennas attached to a conducting body. When a wire antenna is attached to a conducting body, one has to design a model to simulate the current flow at the junction [23]. To demonstrate the usefulness of the AWE method, we consider two antenna configurations, for which the measured

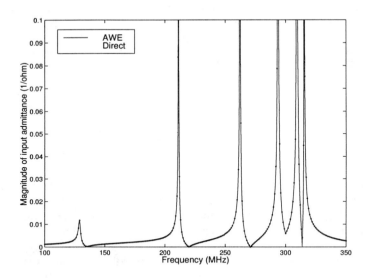

Figure 15.4 Magnitude of the input admittance of a nonshielded low-pass birdcage coil as a function of frequency [18].

data are available for comparison [24, 25]. The first configuration consists of two inverted-L antennas on a finite ground plane and the second consists of two loop antennas on a finite ground plane, both shown in Figure 15.6. As can be seen from Figures 15.7 and 15.8, the agreement between the calculated and measured results is very good.

15.4 ANALYSIS OF METALLIC SCATTERERS

Next, we consider the problem of scattering by a metallic object, which is similar to the antenna problem treated in the preceding section, except for the excitation.

15.4.1 Formulation

The scattering by a PEC body can be formulated in terms of an EFIE or a magnetic field integral equation (MFIE). However, both EFIE and MFIE suffer from the problem of interior resonance, which yields erroneous solutions at certain frequencies. This problem can be overcome by combining EFIE and MFIE to form a combined field integral equation (CFIE) [26].

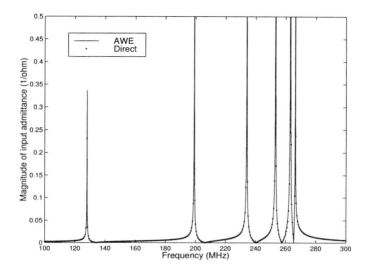

Figure 15.5 Magnitude of the input admittance of a shielded low-pass birdcage coil as a function of frequency [18].

Consider an arbitrarily shaped three-dimensional (3-D) conducting object illuminated by an incident field $\mathbf{E}^{\text{inc}}(\mathbf{r})$. The EFIE is given by

$$\hat{n} \times \iint_S \overline{\mathbf{G}}(\mathbf{r},\mathbf{r}') \cdot \mathbf{J}(\mathbf{r}')dS' = \frac{1}{jk\eta} \hat{n} \times \mathbf{E}^{\text{inc}}(\mathbf{r}) \qquad \text{on } S \qquad (15.19)$$

where S denotes the conducting surface of the object, $\hat{n}$ is an outwardly directed normal, and $\overline{\mathbf{G}}(\mathbf{r},\mathbf{r}')$ is given in (15.9). For a closed conducting object, the MFIE is given by

$$\frac{1}{2}\mathbf{J}(\mathbf{r}) - \hat{n} \times \nabla \times \iint_S g(\mathbf{r},\mathbf{r}')\mathbf{J}(\mathbf{r}')dS' = \hat{n} \times \mathbf{H}^{\text{inc}}(\mathbf{r}) \qquad \text{on } S \qquad (15.20)$$

where $\mathbf{r}$ approaches S from the outside and $g(\mathbf{r},\mathbf{r}')$ is also given in (15.9). The CFIE for a closed conducting object is simply a linear combination of the EFIE [26, 27] and the MFIE and is of the form

$$\alpha \text{ EFIE} + (1-\alpha)\frac{1}{jk}\text{ MFIE} \qquad (15.21)$$

where α is the combination parameter ranging from 0 to 1 and can be chosen to be any value within this range. It is found that $\alpha = 0.8$ is an overall good choice.

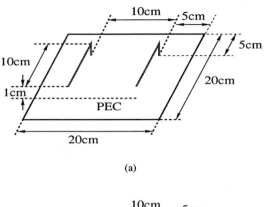

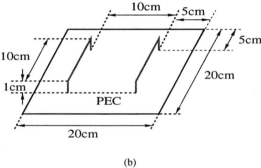

Figure 15.6 (a) Two inverted-L antennas on a finite ground plane. (b) Two loop antennas on a finite ground plane.

Applying the MOM to (15.21) in a procedure similar to that described in the preceding section results in the matrix equation (15.12) with the elements given by

$$Z_{mn}(k) = \alpha \iint_{T_m} \iint_{T_n} \left[\mathbf{f}_m(\mathbf{r}) \cdot \mathbf{f}_n(\mathbf{r}') - \frac{1}{k^2} \nabla \cdot \mathbf{f}_m(\mathbf{r}) \nabla \cdot \mathbf{f}_n(\mathbf{r}') \right] g(\mathbf{r},\mathbf{r}') dS' dS$$
$$+ (1-\alpha) \frac{1}{jk} \left\{ \frac{1}{2} \iint_{T_m} \mathbf{f}_m(\mathbf{r}) \cdot \mathbf{f}_n(\mathbf{r}) dS \right.$$
$$\left. - \iint_{T_m} \mathbf{f}_m(\mathbf{r}) \cdot \hat{n} \times \nabla \times \iint_{T_n} g(\mathbf{r},\mathbf{r}') \mathbf{f}_n(\mathbf{r}') dS' dS \right\} \quad (15.22)$$

$$V_m(k) = \frac{1}{jk\eta} \iint_{T_m} [\alpha \mathbf{E}^{\text{inc}}(\mathbf{r}) + (1-\alpha)\eta\hat{n} \times \mathbf{H}^{\text{inc}}(\mathbf{r})] \cdot \mathbf{f}_m(\mathbf{r}) dS \quad (15.23)$$

Asymptotic Waveform Evaluation for Broadband Calculations

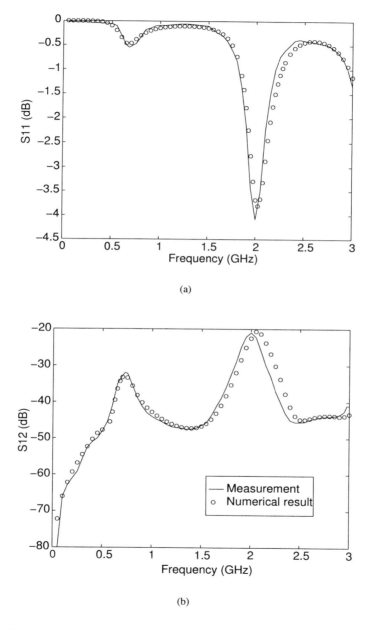

Figure 15.7 S-parameters of two inverted-L antennas on a finite ground plane: (a) S_{11}; (b) S_{12} [16].

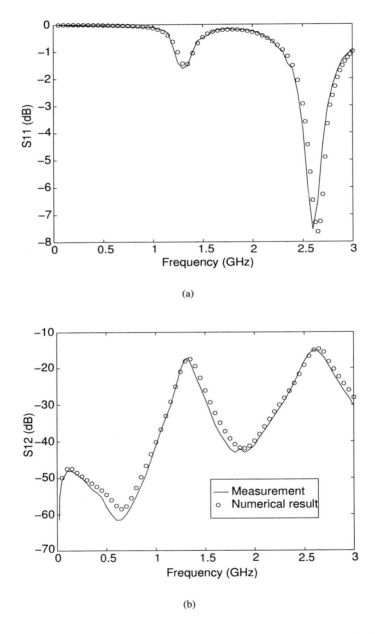

Figure 15.8 S-parameters of two loop antennas on a finite ground plane: (a) S_{11}; (b) S_{12} [16].

Asymptotic Waveform Evaluation for Broadband Calculations

Table 15.2 CPU Timings for RCS Calculations on DEC Personal Workstation

Problem	AWE method		Direct method		Speedup
	Exp. pts	CPU time	Freq. pts	CPU time	
Fig. 15.9	1	10.6 s	61	341.1 s	32.3
Fig. 15.10	7	1,989.3 s	84	23,220 s	11.7

Again, the matrix equation can be solved with the aid of AWE for a broadband calculation. The integrand in the MFIE can be written in the form

$$g(k) = (1 + jkr)\frac{e^{-jkr}}{r^3}$$

whose nth derivative is given by

$$g^{(n)}(k) = je^{-jkr}(-jr)^{n-3}(1 + jkr - n)$$

15.4.2 Numerical Examples

The first example is a PEC sphere with a radius of 0.318 cm. With a frequency step of 1 GHz, it takes the direct method 341.1 s to obtain the solution from 10 to 70 GHz on a DEC Personal Workstation. With one expansion point at 40 GHz and a tenth-order Taylor expansion ($Q = 10, L = 5, M = 5$), the AWE produces an accurate solution with 0.1-GHz increments over the entire band in 10.6 s. A comparison of the CPU time is given in Table 15.2 and the result is shown in Figure 15.9.

The next example is to demonstrate the performance of the binary search algorithm. The scatterer is the 1m long NASA almond [28]. It takes the direct method 23,220 s on a DEC Personal Workstation to calculate the RCS at 84 frequency points from 0 to 1.7 GHz. The number of unknowns varies from 1,560 to 2,148 during the frequency sweep. With the AWE method, it takes only 1,989.3 s to calculate the RCS over the entire band using seven expansion points. The error tolerance ϵ in the RCS was chosen to be very small in order to obtain a very smooth curve (Figure 15.10). This choice results in a relatively small speedup of 11.7. With a slightly larger error tolerance, the number of expansion points can be reduced, leading to a larger speedup. The results presented in Figure 15.10 show the RCS for two different incidence angles and two different polarizations.

15.5 ANALYSIS OF DIELECTRIC SCATTERERS

Now, we proceed to the problem of scattering by a penetrable dielectric object, which may be dispersive. The treatment is correspondingly more complex.

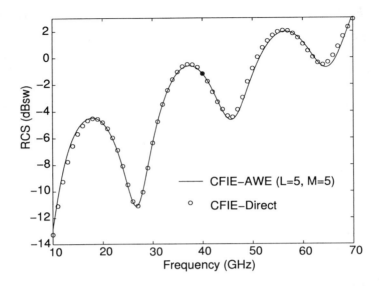

Figure 15.9 RCS frequency response of a sphere with a radius of 0.318 cm from 10 to 70 GHz [16].

15.5.1 Formulation

A formulation that is widely used for scattering by dielectric bodies is the so-called PMCHW [29], named after Poggio, Miller, Chang, Harrington, and Wu, who originally developed the formulation. In this formulation, the EFIE for the field outside the object is combined with the EFIE for the field inside the object to form a combined equation. Similarly, the MFIE for the field outside the object is combined with the MFIE for the field inside the object to form another combined equation. These two equations are then solved by the MOM. This formulation is found to be free of interior resonances and yields accurate and stable solutions. With the AWE method, we can even model dispersive dielectrics.

Consider an arbitrarily shaped homogeneous dielectric scatterer characterized by permittivity ϵ_2 and permeability μ_2 and immersed in an infinite and homogeneous medium having permittivity ϵ_1 and permeability μ_1. Using either the equivalence principle or the vector Green's theorem, one can formulate a set of four integral equations to calculate the electric and magnetic fields **E** and **H** in terms of equivalent electric and magnetic currents **J** and **M** on the surface of the scatterer. The equation to calculate the electric field is known as the EFIE and there are two such equations: one is for the field inside the object (EFIE–I) and the other for the field outside the object (EFIE–O). The equation to calculate the magnetic field is known as the MFIE

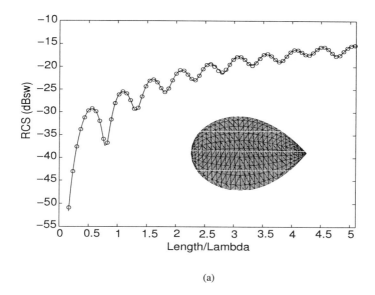

(a)

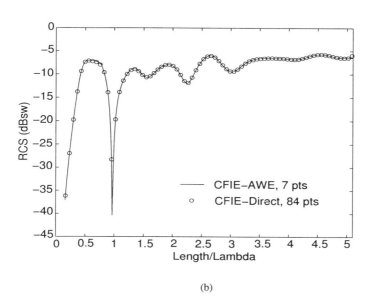

(b)

Figure 15.10 RCS frequency response of the 1m long NASA almond from 0 to 1.7 GHz: (a) VV-polarization with $\theta^{inc} = 90°$ and $\phi^{inc} = 0°$; (b) HH-polarization with $\theta^{inc} = 45°$ and $\phi^{inc} = 45°$ [16].

and there are also two such equations: one is for the field inside the object (MFIE–I) and the other for the field outside the object (MFIE–O). A simple combination of EFIE–I and EFIE–O yields an integral equation

$$[\eta_1 \mathbf{L}_1(\mathbf{J}) + \eta_2 \mathbf{L}_2(\mathbf{J}) - \mathbf{K}_1(\mathbf{M}) - \mathbf{K}_2(\mathbf{M}) = \mathbf{E}^{\text{inc}}]_{\text{tan}} \quad (15.24)$$

and a similar combination of MFIE–I and MFIE–O results in another integral equation

$$[\eta_1 \mathbf{K}_1(\mathbf{J}) + \eta_2 \mathbf{K}_2(\mathbf{J}) + \mathbf{L}_1(\mathbf{M}) + \mathbf{L}_2(\mathbf{M}) = \eta_1 \mathbf{H}^{\text{inc}}]_{\text{tan}} \quad (15.25)$$

where $\eta_i = \sqrt{\mu_i/\epsilon_i}$, and $\mathbf{L}_i$ and $\mathbf{M}_i$ are integral operators defined by

$$\mathbf{L}_i(\mathbf{X}) = jk_i \iint_S \left[\mathbf{X}(\mathbf{r}') + \frac{1}{k_i^2}\nabla\nabla' \cdot \mathbf{X}(\mathbf{r}')\right] g_i(\mathbf{r},\mathbf{r}')dS' \quad (15.26)$$

$$\mathbf{K}_i(\mathbf{X}) = \iint_S \mathbf{X}(\mathbf{r}') \times \nabla g_i(\mathbf{r},\mathbf{r}')dS' \quad (15.27)$$

in which $k_i = \omega\sqrt{\mu_i\epsilon_i}$, S denotes the surface of the scatterer, and $g_i(\mathbf{r},\mathbf{r}')$ is the scalar Green's function given by

$$g_i(\mathbf{r},\mathbf{r}') = \frac{e^{-jk_i|\mathbf{r}-\mathbf{r}'|}}{4\pi|\mathbf{r}-\mathbf{r}'|} \quad (15.28)$$

Equations (15.24) and (15.25) are known as the PMCHW formulation [29]. When $\mathbf{r} = \mathbf{r}'$, the integral in (15.27) is interpreted in the principal value sense.

For a dispersive dielectric object, ϵ_2 and μ_2 can be a function of frequency. For simplicity, we assume that μ_2 is a constant and only ϵ_2 varies with frequency. The complex permittivity of a dielectric can be described by the Debye model [30]

$$\epsilon_2(\omega) = \epsilon_2'(\omega) - j\epsilon_2''(\omega) = \epsilon_{2\infty}' + \frac{\epsilon_{2s}' - \epsilon_{2\infty}'}{1 + j\omega\tau_e} \quad (15.29)$$

where ϵ_{2s}' denotes the static dielectric constant, $\epsilon_{2\infty}'$ is the optical dielectric constant, and τ_e is a relaxation time constant related to the original relaxation time constant τ by

$$\tau_e = \tau \frac{\epsilon_{2s}' + 2\epsilon_0}{\epsilon_{2\infty}' + 2\epsilon_0} \quad (15.30)$$

with ϵ_0 denoting the permittivity of free space.

Equations (15.24) and (15.25) can be solved numerically using the MOM, with the aid of AWE for a broadband calculation.

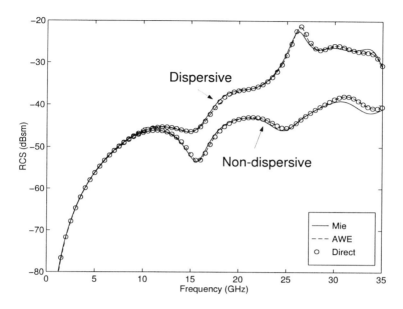

Figure 15.11 Backscatter RCS versus frequency of a dielectric sphere having a radius of 0.5 cm [21].

15.5.2 Numerical Examples

Consider a dielectric sphere of radius 0.5 cm immersed in free space. For the numerical solution, the surface of the sphere is modeled by 648 triangular patches, resulting in 972 edges and thus 1,944 unknowns. The permittivity of the sphere is characterized by $\epsilon'_{rs} = 2.56$, $\epsilon'_{r\infty} = 1.0$, and $\tau = 1.59$ ps/rad. As a result, as frequency varies from 0.5 to 35 GHz, the real part of the relative permittivity varies from 2.56 to 2.22 and the imaginary part varies from -0.012 to -0.65. The backscatter RCS of the sphere is shown in Figure 15.11, where the backscatter RCS of a nondispersive sphere of the same size and having a relative permittivity 2.56 is also given. Three solutions are displayed in the figure. One is the exact Mie series solution, the second is the solution obtained directly at each frequency, and the third is the solution obtained using the AWE method. With a frequency step of 0.5 GHz, it takes the direct method 24,611 s to obtain the solution on a DEC Personal Workstation. In contrast, the AWE method produces the solution with 0.01 GHz increments in 2,206 s on the same computer. Figure 15.12 shows similar results for a 1 cm × 1 cm × 1 cm dielectric cube with normal incidence.

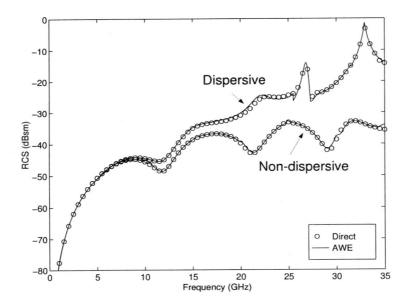

Figure 15.12 Backscatter RCS versus frequency of a dielectric cube having a side length of 1 cm [21].

15.6 ANALYSIS OF MICROSTRIP ANTENNAS

The preceding sections described the application of AWE to the MOM solution. In this section, we demonstrate the use of AWE in the FEM combined with the boundary integral equation (BIE). The detailed formulation of the FEM/BIE method is described in Chapter 13.

15.6.1 Formulation

The FEM/BIE hybrid method is a powerful numerical technique for solving electromagnetic scattering and radiation problems [31]. This techniques divides an open-region problem into interior and exterior problems and employs the FEM to deal with the interior problem. The exterior problem is formulated using the BIE which, when coupled with interior fields, provides an efficient solution to the original problem.

For the problem of scattering and radiation by a cavity-backed microstrip patch antenna recessed in a ground plane (Figure 15.13), it has been shown [31] that the

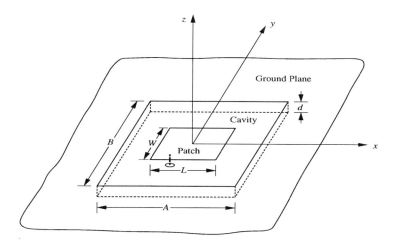

Figure 15.13 Illustration of a cavity-backed microstrip patch antenna.

electric field inside the cavity and over the aperture can be found by seeking the stationary point of the functional

$$F(\mathbf{E}) = \frac{1}{2} \iiint_V \left[\frac{1}{\mu_r} (\nabla \times \mathbf{E}) \cdot (\nabla \times \mathbf{E}) - k^2 \epsilon_r \mathbf{E} \cdot \mathbf{E} \right] dV$$
$$+ \iiint_V \left[jkZ\mathbf{J}^{\text{int}} \cdot \mathbf{E} - \frac{1}{\mu_r} \mathbf{M}^{\text{int}} \cdot (\nabla \times \mathbf{E}) \right] dV$$
$$- k^2 \iint_S \mathbf{M}(\mathbf{r}) \cdot \left[\iint_S \mathbf{M}(\mathbf{r}')g(\mathbf{r},\mathbf{r}') \, dS' \right] dS$$
$$+ \iint_S \nabla \cdot \mathbf{M}(\mathbf{r}) \left[\iint_S g(\mathbf{r},\mathbf{r}')\nabla' \cdot \mathbf{M}(\mathbf{r}')dS' \right] dS$$
$$- 2jk\eta \iint_S \mathbf{M}(\mathbf{r}) \cdot \mathbf{H}^{\text{inc}}(\mathbf{r}) \, dS \qquad (15.31)$$

where V denotes the volume of the cavity and S denotes its aperture, $\mathbf{M} = \mathbf{E} \times \hat{z}$ is the equivalent magnetic current over the aperture, k is the free-space wavenumber, η is the free-space wave impedance, and $g(\mathbf{r},\mathbf{r}')$ denotes the free-space Green's function. For scattering problems, $\mathbf{H}^{\text{inc}}$ denotes the incident magnetic field. For radiation problems, $\mathbf{J}^{\text{int}}$ and $\mathbf{M}^{\text{int}}$ denote the electric and magnetic sources associated with the antenna feeds.

The dielectric substrate is characterized by its relative permittivity ϵ_r and permeability μ_r. For a dispersive substrate, the relative permittivity ϵ_r can be represented by the Debye model described in the preceding section.

The discretization of the above functional has been described in detail in the past [31]. It leads to the matrix equation

$$A(k)E(k) = B(k) \qquad (15.32)$$

where A is a partly sparse and partly full symmetric matrix, E is a vector representing the discretized electric field, and B is the excitation vector related to the incident field for scattering or internal sources for radiation. Equation (15.32) can be solved using the AWE method for a broadband calculation.

15.6.2 Numerical Examples

The first example is scattering by an antenna geometry consisting of a 3.66 cm × 2.60 cm rectangular conducting patch residing on a dielectric substrate having thickness $t = 0.158$ cm, relative permittivity $\epsilon_r = 2.17$, and a loss tangent of 0.001. The substrate is housed in a 7.32 cm × 5.20 cm rectangular cavity recessed in a ground plane. The incident plane wave is polarized along the θ direction and the angle of incidence is $\theta^{\text{inc}} = 60°$ and $\phi^{\text{inc}} = 45°$. The monostatic RCS of the antenna is shown in Figure 15.14 as a function of frequency from 2 to 8 GHz. Clearly, the RCS

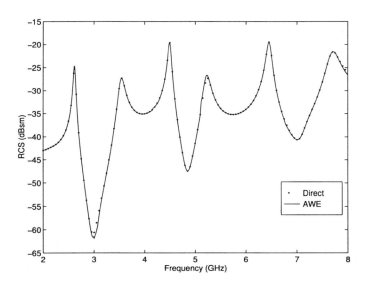

Figure 15.14 Monostatic VV-polarized RCS versus frequency for a cavity-backed patch antenna ($A = 7.32$ cm, $B = 5.20$ cm, $L = 3.66$ cm, $W = 2.60$ cm, $d = 0.158$ cm, $\epsilon_r = 2.17$, $\tan \delta = 0.001$, $\theta^{\text{inc}} = 60°$, $\phi^{\text{inc}} = 45°$) [19].

Table 15.3 CPU Timings for Calculations on DEC Personal Workstation

Problem	No. of unknowns	AWE method Freq. pts	AWE method CPU time	Direct method Freq. pts	Direct method CPU time	Speedup
Fig. 15.14	1,585	600	265.2 s	120	3,254.4 s	12.3
Fig. 15.15	1,741	300	254.2 s	60	4,012.8 s	15.8
Fig. 15.16	1,741	300	254.2 s	60	4,012.8 s	15.8

is characterized by a series of peaks, each corresponding to a resonant mode of the patch. The results compare well with the experimental data for the patch residing on an infinite substrate [32]. The number of unknowns used in the calculation is 1,585. With a frequency increment of 0.05 GHz, it takes the direct method, which solves (15.32) repeatedly for each frequency, 3,254.4 s to obtain the solution on a DEC Personal Workstation. With a sixth-order Taylor expansion ($Q = 6, L = 3, M = 3$), the AWE method produces an accurate solution with 0.01 GHz increments over the entire band in 265.2 s.

The second example is radiation by an antenna geometry consisting of a 5.0 cm × 3.4 cm rectangular conducting patch residing on a dielectric substrate having thickness $t = 0.08770$ cm, relative permittivity $\epsilon_r = 2.17$, and a loss tangent of 0.0015. The substrate is housed in a 7.5 cm × 5.1 cm rectangular cavity recessed in a ground plane. The patch is excited by a current probe applied at $x_f = 1.22$ cm and $y_f = 0.85$ cm. Figure 15.15 shows the input impedance of the antenna as a function of frequency from 1 to 4 GHz. Figure 15.16 shows the input impedance when a 50-Ω impedance load is placed at $x_L = -2.2$ cm and $y_L = -1.5$ cm. The calculated results compare well with the experimental data for the patch residing on an infinite substrate [32]. The number of unknowns used in the calculations is 1,741. With a frequency increment of 0.05 GHz, it takes the direct method 4,012.8 s to obtain the solution. With a sixth-order Taylor expansion ($Q = 6, L = 3, M = 3$), the AWE method produces an accurate solution with 0.01-GHz increments over the entire band in 254.2 s.

The information about the calculations is summarized in Table 15.3 for a clear comparison. In both cases, the speedup is more than an order of magnitude.

To demonstrate the capability of the AWE method to deal with dispersive substrates, we reconsider the second example. We assume that the substrate is dispersive with $\epsilon'_{rs} = 2.17$, $\epsilon'_{r\infty} = 1.0$, and $\tau = 7.96$ ps/rad. As a result, as frequency varies from 1 to 10 GHz, the real part of the relative permittivity varies from 2.16 to 1.79 and the imaginary part varies from -0.081 to -0.55. Figure 15.17(a) shows the input impedance of the antenna from 1 to 4 GHz. Figure 15.17(b) gives the input impedance when a 50-Ω impedance load is placed at $x_L = -2.2$ cm and $y_L = -1.5$ cm. It can be seen clearly that the AWE solution agrees very well with the direct solution.

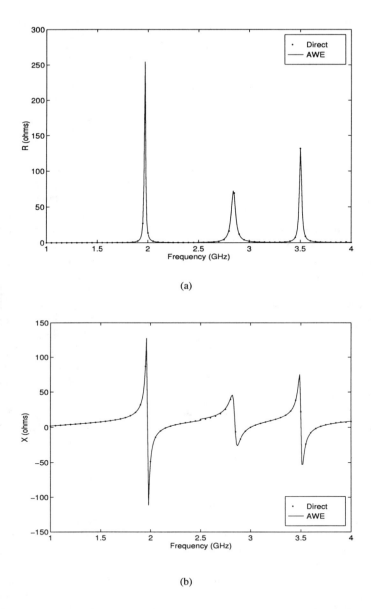

Figure 15.15 Input impedance versus frequency for a cavity-backed patch antenna with a probe feed at $x_f = 1.22$ cm and $y_f = 0.85$ cm ($A = 7.5$ cm, $B = 5.1$ cm, $L = 5.0$ cm, $W = 3.4$ cm, $d = 0.08779$ cm, $\epsilon_r = 2.17$, $\tan\delta = 0.0015$): (a) resistance; (b) reactance [19].

Asymptotic Waveform Evaluation for Broadband Calculations 723

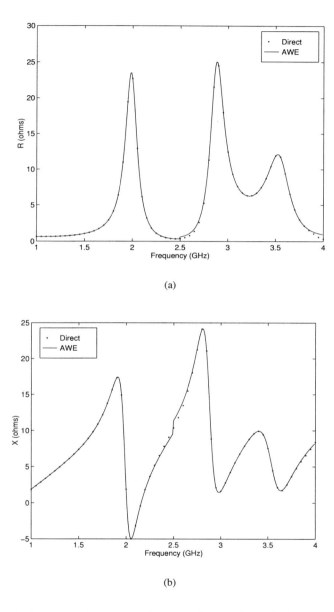

Figure 15.16 Input impedance versus frequency for a cavity-backed patch antenna with a probe feed at $x_f = 1.22$ cm and $y_f = 0.85$ cm and a 50-Ω resistor load at $x_L = -2.2$ cm and $y_L = -1.5$ cm ($A = 7.5$ cm, $B = 5.1$ cm, $L = 5.0$ cm, $W = 3.4$ cm, $d = 0.08779$ cm, $\epsilon_r = 2.17$, $\tan \delta = 0.0015$): (a) resistance; (b) reactance [20].

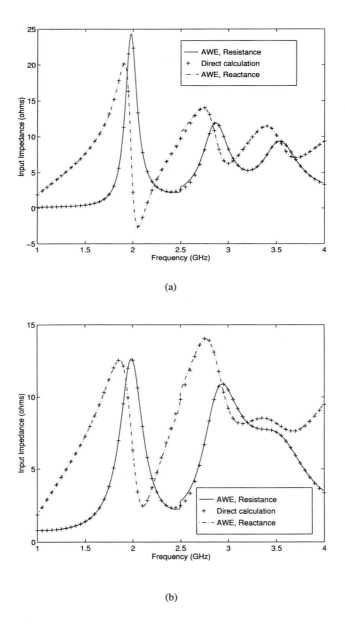

Figure 15.17 Input impedance versus frequency for a cavity-backed patch antenna residing on a dispersive substrate: (a) without an impedance load; (b) with a 50-Ω resistor load.

15.7 SUMMARY

This chapter described the formulation of the AWE method and its application for achieving a fast frequency-sweep analysis with a variety of electromagnetics problems, which include (1) radiation of antennas made of metallic wires and/or surfaces, (2) scattering by a PEC body, (3) scattering by a dispersive dielectric body, and (4) scattering and radiation of conformal cavity-backed microstrip patch antennas. The AWE method has also been applied very successfully to microstrip circuits in a multilayer medium. Some numerical examples of this application are given in Chapter 16. It was shown that the use of AWE can speed up the analysis by more than an order of magnitude.

REFERENCES

1. G. J. Burke, E. K. Miller, S. Chakrabarti, and K. Demarest, "Using model-based parameter estimation to increase the efficiency of computing electromagnetic transfer functions," *IEEE Trans. Magn.*, vol. 25, pp. 2807–2809, July 1989.

2. K. Kottapalli, T. K. Sarkar, Y. Hua, E. K. Miller, and G. J. Burke, "Accurate computation of wide-band response of electromagnetic systems utilizing narrow-band information," *IEEE Trans. Microwave Theory Tech.*, vol. 39, pp. 682–687, April 1991.

3. E. H. Newman, "Generation of wide-band data from the method of moments by interpolating the impedance matrix," *IEEE Trans. Antennas Propagat.*, vol. 36, pp. 1820–1824, Dec. 1988.

4. L. T. Pillage and R. A. Rohrer, "Asymptotic waveform evaluation for timing analysis," *IEEE Trans. Computer-Aided Design*, vol. 9, pp. 352–366, April 1990.

5. E. Chiprout and M. Nakhla, *Asymptotic Waveform Evaluation and Moment Matching for Interconnect Analysis*, Boston, MA: Kluwer Academic Publishers, 1994.

6. W. T. Smith, R. D. Slone, and S. K. Das, "Recent progress in reduced-order modeling of electrical interconnects using asymptotic waveform evaluation and Padé approximation via the Lanczos process," *Appl. Comput. Electromagn. Soc. Newslett.*, vol. 12, pp. 46–71, July 1997.

7. X. Yuan and Z. J. Cendes, "A fast method for computing the spectral response of microwave devices over a broad bandwidth," *1993 URSI Radio Science Meeting*, p. 196, Ann Arbor, MI, 1993.

8. M. A. Kolbehdari, M. Srinivasan, M. S. Nakhla, Q. J. Zhang, and R. Achar, "Simultaneous time and frequency domain solutions of EM problems using finite element and CFH techniques," *IEEE Trans. Microwave Theory Tech.*, vol. 44, pp. 1526–1534, Sept. 1996.

9. M. Li, Q. J. Zhang, and M. S. Nakhla, "Finite difference solution of EM fields by asymptotic waveform techniques," *IEE Proc.-H*, vol. 143, pp. 512–520, Dec. 1996.

10. J. Gong and J. L. Volakis, "AWE implementation for electromagnetic FEM analysis," *Electron. Lett.*, vol. 32, pp. 2216–2217, Nov. 1996.

11. S. V. Polstyanko, R. Dyczij-Edlinger, and J. F. Lee, "Fast frequency sweep technique for the efficient analysis of dielectric waveguides," *IEEE Trans. Microwave Theory Tech.*, vol. 45, pp. 1118–1126, July 1997.

12. J. Zhang and J. M. Jin, "Preliminary study of AWE for FEM analysis of scattering problems," *Microwave Opt. Tech. Lett.*, vol. 17, no. 1, pp. 7–12, 1998.

13. X. M. Zhang and J. F. Lee, "Application of the AWE method with the 3-D TVFEM to model spectral responses of passive microwave components," *IEEE Trans. Microwave Theory Tech.*, vol. 46, pp. 1735–1741, Nov. 1998.

14. J. E. Bracken, D. K. Sun, and Z. J. Cendes, "S-domain methods for simultaneous time and frequency characterization of electromagnetic devices," *IEEE Trans. Microwave Theory Tech.*, vol. 46, pp. 1277–1290, Sept. 1998.

15. C. J. Reddy, M. D. Deshpande, C. R. Cockrell, and F. B. Beck, "Fast RCS computation over a frequency band using method of moments in conjunction with asymptotic waveform evaluation technique," *IEEE Trans. Antennas Propagat.*, vol. 46, pp. 1229–1233, Aug. 1998.

16. D. Jiao, X. Y. Zhu, and J. M. Jin, "Fast and accurate frequency-sweep calculations using asymptotic waveform evaluation and combined-field integral equation," *Radio Science*, vol. 34, pp. 1055–1063, Sept.–Oct. 1999.

17. F. Ling, D. Jiao, and J. M. Jin, "Efficient electromagnetic modeling of microstrip structures in multilayer media," *IEEE Trans. Microwave Theory Tech.*, vol. 47, pp. 1810–1818, Sept. 1999.

18. D. Jiao and J. M. Jin, "Fast frequency-sweep analysis of RF coils for MRI," *IEEE Trans. Biomed. Eng.*, vol. 46, pp. 1387–1390, Nov. 1999.

19. D. Jiao and J. M. Jin, "Fast frequency-sweep analysis of cavity-backed microstrip patch antennas," *Microwave Opt. Tech. Lett.*, vol. 22, no. 6, pp. 389–393, Sept. 1999.

20. D. Jiao and J. M. Jin, "Fast frequency-sweep analysis of microstrip antennas on a dispersive substrate," *Electron. Lett.*, vol. 35, no. 14, pp. 1122–1123, July 1999.

21. D. Jiao and J. M. Jin, "Asymptotic waveform evaluation for scattering by a dispersive dielectric object," *Microwave Opt. Tech. Lett.*, vol. 24, no. 4, pp. 232–234, Feb. 2000.

22. S. M. Rao, D. R. Wilton, and A. W. Glisson, "Electromagnetic scattering by surface of arbitrary shape," *IEEE Trans. Antennas Propagat.*, vol. 30, pp. 409–418, May 1982.

23. H. Y. Chao, "Wire and junction basis function support for the three-dimensional fast multipole method," M.S. thesis, Univ. of Illinois at Urbana-Champaign, 1998.

24. S. V. Georgakopoulos, C. A. Balanis, and C. B. Birtcher, "Coupling between wire antennas: Analytic solution, FDTD, and measurements," *IEEE APS Int. Symp. Dig.*, vol. 1, pp. 504–507, June 1998.

25. S. V. Georgakopoulos, "Coupling between multiple wire antennas on complex structures," M.S. thesis, Arizona State University, May 1998.

26. J. R. Mautz and R. F. Harrington, "H-field, E-field, and combined-field solutions for conducting bodies of revolution," *AEU*, vol. 32, pp. 157–164, April 1978.

27. P. L. Huddleston, L. N. Medgyesi-Mitschang, and J. M. Putnam, "Combined field integral equation formulation for scattering by dielectrically coated conducting bodies," *IEEE Trans. Antennas Propagat.*, vol. 34, pp. 510–520, April 1986.

28. A. C. Woo, H. T. G. Wang, M. J. Schuh, and M. L. Sanders, "Benchmark radar targets for the validation of computational electromagnetics programs," *IEEE Antennas Propagat. Mag.*, vol. 35, no. 1, pp. 84–89, Feb. 1993.

29. J. R. Mautz and R. F. Harrington, "Electromagnetic scattering from a homogeneous material body of revolution," *AEU*, vol. 33, pp. 71–80, Feb. 1979.

30. A. R. von Hippel, *Dielectrics and Waves*, Cambridge, MA: MIT Press, 1954.

31. J. M. Jin, *The Finite Element Method in Electromagnetics*, New York: Wiley, 1993.

32. D. P. Forrai and E. H. Newman, "Radiation and scattering from loaded microstrip antennas over a wide bandwidth," Ohio State Univ. Tech. Rep. 719493-1, Sept. 1988.

16

Full-Wave Analysis of Multilayer Microstrip Problems

Feng Ling and Jian-Ming Jin

16.1 INTRODUCTION

In the past decades, integrated circuits (ICs) have undergone impressive advancements. Planar passive components are extensively used in IC systems. Hence, fast and accurate electromagnetic (EM) analysis of planar components becomes very important for IC design.

To develop a reliable computer-aided design (CAD) tool for this purpose, several aspects have to be considered. First, with the increased complexity of those structures, multilayer media are often employed to allow more versatile designs. This fact necessitates that the modeling be capable of dealing with planar components in multilayer media. Second, as the operating frequency of ICs increases, the empirical formulas and quasi-static models are unable to provide accurate results. Consequently, it is desirable to develop a rigorous full-wave EM modeling. Third, the ability to tackle electrically large problems becomes necessary to perform the integrated simulation.

There have been a variety of full-wave EM methods developed to accomplish the goal mentioned above. These methods can be divided into two broad classes: the differential equation approach and the integral equation approach. The differential equation approach includes the finite-difference time-domain (FDTD) [1–4] and finite element methods (FEM) [5–8]. In the integral equation approach, the method of moments (MOM) is employed to solve the pertinent integral equations [9–24]. The integral equation approach is more attractive for multilayer medium problems since the method allows one to apply Green's theorem to reduce volume integrals to surface integrals, thus reducing the matrix dimension significantly because of the use of surface discretization rather than volume discretization.

In this work, we develop the spatial domain MOM for the analysis of microstrip circuits and antennas in a multilayer medium. Section 16.2 describes the evaluation of Green's functions using the discrete complex image method (DCIM). Special attention is given to the treatment of surface-wave contributions. Section 16.3 discusses the MOM solution using higher-order vector basis functions, which can represent the unknown currents and model geometries more accurately than lower-order basis functions. Section 16.4 employs the asymptotic waveform evaluation (AWE) to rapidly compute the solution over a broad frequency band. This is followed by the presentation of the adaptive integral method (AIM) and the multilevel fast multipole algorithm (MLFMA) for simulating large-scale microstrip problems. Both methods reduce the computational complexity from $O(N^2)$ to $O(N \log N)$, where N denotes the number of unknowns. Numerical examples are given in each section to demonstrate the accuracy and capability of the proposed formulations.

16.2 GREEN'S FUNCTIONS FOR MULTILAYER MEDIA

This section addresses the efficient evaluation of Green's functions, which are expressed in terms of the Sommerfeld integrals (SIs). Several approaches have been developed to efficiently evaluate the Green's functions, such as the fast Hankel transform (FHT) approach [25, 26], the steepest-descent path (SDP) approach [27], and the discrete complex image method [28–36]. In this work, an improved DCIM is employed to efficiently evaluate the Green's functions. The spectral domain Green's functions for multilayer media are first derived from a simple transmission line perspective. Then, the DCIM is employed to efficiently evaluate the SIs, resulting in closed-form spatial domain Green's functions. The issue of surface-wave extraction is discussed. For the far-field Green's functions, it is necessary to extract surface waves to approximate $1/\sqrt{\rho}$ asymptotic behavior. In the previous work of the DCIM, the surface-wave contribution is treated analytically by using residue calculus, which makes it difficult to extend to multilayer cases. Here, the surface-wave contribution is obtained by performing a contour integral recursively in the complex k_ρ plane [37]. Finally, to make the Green's function evaluation more efficient, especially for three-dimensional (3D) structures, an interpolation scheme is employed, which is able to restore the Green's function at the same order of time as that in free-space problems.

Full-Wave Analysis of Multilayer Microstrip Problems

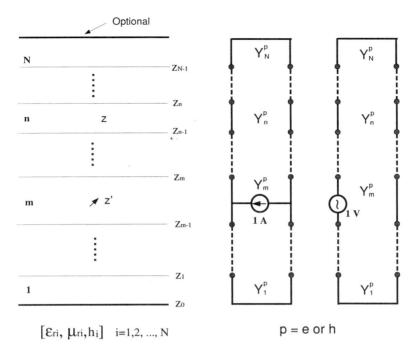

Figure 16.1 A multilayer medium with source and field points in layer m and layer n, respectively, and its transmission line representation.

Consider a current source in a multilayer medium. Each layer is characterized by relative permittivity ϵ_r, relative permeability μ_r, and thickness h, as shown in Figure 16.1. The electric field due to the current can be expressed in a mixed potential form as

$$\mathbf{E} = -j\omega\mu_0 \langle \overline{\mathbf{G}}^A, \mathbf{J} \rangle + \frac{1}{j\omega\epsilon_0} \nabla \langle G^\Phi, \nabla' \cdot \mathbf{J} \rangle \tag{16.1}$$

where $\mathbf{J}$ denotes the electric current density of the source, and $\overline{\mathbf{G}}^A$ and G^Φ are the Green's functions for the vector and scalar potentials, respectively. The detailed discussion in [38] shows that it is preferable to choose $\overline{\mathbf{G}}^A$ as

$$\overline{\mathbf{G}}^A = \begin{bmatrix} G^A_{xx} & 0 & G^A_{xz} \\ 0 & G^A_{xx} & G^A_{yz} \\ G^A_{zx} & G^A_{zy} & G^A_{zz} \end{bmatrix} \tag{16.2}$$

and G^Φ as the scalar potential for a horizontal electric dipole (HED).

In general, the Green's function for a multilayer medium is expressed in terms of an SI, which can be written as

$$G(\rho, z|z') = \frac{1}{2\pi} \int_0^\infty \tilde{G}(k_\rho, z|z') J_0(k_\rho \rho) k_\rho dk_\rho \tag{16.3}$$

where $\tilde{G}$ is the spectral domain counterpart of G and $J_0(\cdot)$ denotes the zeroth-order Bessel function.

To evaluate the SI, we begin with the derivation of the spectral domain Green's functions for an arbitrary dipole in multilayer media, which can be accomplished by constructing equivalent transmission lines [38,39]. Therefore, the original problem to find electric and magnetic fields is converted to the problem of obtaining the voltage and current of the corresponding transmission lines. From the voltages and currents, the Green's functions for the vector potential $\overline{\tilde{G}}^A$ and the scalar potential $\tilde{G}^\Phi$ can be derived as

$$\tilde{G}^A_{xx} = \frac{1}{j\omega\mu_0} V_i^h \tag{16.4}$$

$$\tilde{G}^A_{zt} = \frac{\mu_r k_t}{jk_\rho^2}(I_i^h - I_i^e), \quad t = x, y \tag{16.5}$$

$$\tilde{G}^A_{zz} = \frac{1}{j\omega\epsilon_0}\left[\left(\frac{\mu_r}{\epsilon_r'} - \frac{\mu_r' k_z^2}{\epsilon_r k_\rho^2}\right) I_v^e + \frac{k_0^2 \mu_r \mu_r'}{k_\rho^2} I_v^h\right] \tag{16.6}$$

$$\tilde{G}^\Phi = \frac{j\omega\epsilon_0}{k_\rho^2}(V_i^e - V_i^h) \tag{16.7}$$

where $V_i^{e,h}$ and $I_i^{e,h}$ are the voltages and currents of the equivalent transmission line. The $\tilde{G}^A_{xz}$ and $\tilde{G}^A_{yz}$ are related to $\tilde{G}^A_{zx}$ and $\tilde{G}^A_{zy}$ by the reciprocity principle, which gives

$$\tilde{G}^A_{tz}(z|z') = -\frac{\mu_r'}{\mu_r}\tilde{G}^A_{zt}(z'|z) \tag{16.8}$$

Therefore, the Green's functions required in this method are G^A_{xx}, G^A_{zz}, G^A_{zx}, G^A_{zy}, and G^Φ. Also, G^A_{zx} and G^A_{zy} have the same kernels so that a total of only four Green's functions are required for evaluation.

Once the spectral domain Green's functions are obtained, the DCIM can be applied to rapidly evaluate the SIs. A minor modification of the DCIM is necessary since $\tilde{G}^A_{zx}$ and $\tilde{G}^A_{zy}$ have k_x and k_y dependence, which makes G^A_{zx} and G^A_{zy} in a form of x and y derivative of the same SI. Let us rewrite the spectral domain Green's function in a simple form as

$$\tilde{G} = A\frac{F}{2jk_{zm}} \tag{16.9}$$

where A is a constant. As the first step of the DCIM, the primary field term F_{pr} is extracted from F when the source and observation points are in the same layer. Note that there is no primary field term in $\tilde{G}_{zt}^A$. The static contributions F_{st}, which dominate as $k_\rho \to \infty$, are also extracted, which makes the remaining kernel decay to zero for a sufficiently large k_ρ. This happens only when both source and field points are on the interface of two different layers.

The next step is to extract the guided-mode or surface-wave contributions F_{gm}, which can be written as

$$F_{gm} = 2jk_{zm} \sum_i \frac{2k_{\rho i} Res_i}{k_\rho^2 - k_{\rho i}^2} \qquad (16.10)$$

where Res_i is the residue for the pole $k_{\rho i}$. When transformed to the spatial domain, the guided-mode contributions to the Green's function become

$$G_{gm} = \frac{A}{4\pi} \left[-2\pi j \sum_i Res_i H_0^{(2)}(k_{\rho i}\rho) k_{\rho i} \right] \qquad (16.11)$$

where $H_0^{(2)}$ denotes the zeroth-order Hankel of the second kind.

In the previous work on the DCIM, Res_i is calculated analytically by using residue calculus, which makes it difficult to extend to multilayer cases. Here, we obtain Res_i by evaluating a contour integral numerically in the complex k_ρ-plane recursively. The integration begins with a rectangular contour enclosing the region of interest. If the calculated value is nonzero, then we subdivide the contour into four contours and evaluate the contour integral along each of them. This process is repeated until the location $k_{\rho i}$ and residue Res_i for all the poles are found. These procedures can be illustrated by the following example, which is a five-layer medium with the top layer being free space. The magnitude of $\tilde{G}^\Phi$ in the first and fourth quadrants is plotted in Figure 16.2. The contour integral is repeated until we find the surface-wave poles at $k_\rho = 1.736k_0$ and $2.435k_0$, respectively. Since the top layer is the unbounded free space, there is a branch cut associated with k_{z0}, where the radiation modes lie on [41]. This branch cut is known and can be deliberately avoided. For shielded multilayer media, both of the modes guided by the layers and the ground planes are extracted. The poles on the imaginary axis correspond to the evanescent modes, which are not extracted in this method.

After these extractions, the remainder of F can be approximated as a sum of complex exponentials using the generalized pencil-of-function (GPOF) method [42]. With the aid of the Sommerfeld identity, we can obtain the closed-form Green's functions.

As pointed out in [31], for a multilayer medium with an unbounded top or bottom layer, there is a branch cut. For instance, when the top layer is free space, the branch cut is associated with k_{z0}. When the source is in the bounded region of multilayer

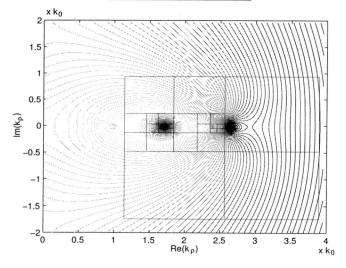

Figure 16.2 Magnitude of $\tilde{G}^\Phi$ in the first and fourth quadrants of the complex k_ρ-plane for a five-layer medium [40]. The frequency is 30 GHz. Unit for thickness is mm.

media, the DCIM needs to be modified. In that case, we rewrite (16.9) as

$$\tilde{G} = A\frac{F'}{2jk_{z0}}, \qquad F' = F\frac{k_{z0}}{k_{zm}}. \tag{16.12}$$

As a consequence, k_{zm} in F_{pr} and F_{gm} is replaced by k_{z0}. The GPOF approximation is performed in terms of k_{z0}. For shielded multilayer media, since there are no branch cuts, such a modification is not necessary.

With the guided-mode extraction, the DCIM works well in the far-field region. However, a problem is encountered for the near-field evaluation. As we know, when $z \neq z'$, the Green's function is not singular at $\rho = 0$; however, the guided-mode term carries the singularity. This phenomenon is shown in Figure 16.3, where a five-layer

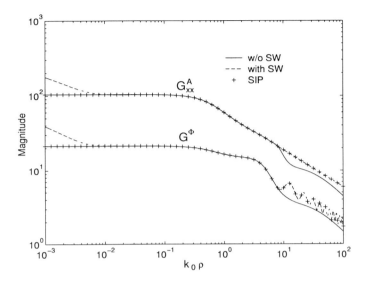

Figure 16.3 Magnitude of Green's functions, G_{xx}^A and G^Φ, for a five-layer medium [40]. $z' = 0.4$ mm, $z = 1.4$ mm.

medium problem is investigated for $z \neq z'$. In the figure, the dashed and solid lines represent those obtained by the DCIM with and without the guided-mode extraction, respectively. The crosses represent those obtained by direct numerical integration along the Sommerfeld integration path (SIP).

To overcome this difficulty, a transition point is introduced, as is done in [28], which divides the near- and far-field regions. Therefore, the DCIM is applied twice: once with and once without the guided-mode extraction. The first yields the Green's function for the near-field region, and the second gives the Green's function for the far-field region. From Figure 16.3, we can see that the two approaches overlap in the middle region and the transition point can be easily picked up, say, at $\log_{10} k_0 \rho = 0$. With this modification, the DCIM works for all the regions. The magnitudes of four Green's functions G_{xx}^A, G_{zx}^A, G_{zz}^A, and G_Φ are shown in Figure 16.4. Compared to the result of SIP, the DCIM with the guided-mode extraction has an average relative error of 0.13% in the far-field region, whereas the error for the DCIM without the guided-mode extraction is over 104%. Note that G_{zx}^A approaches zero as $\rho \to 0$ except in the case that both source and field points are at the different interfaces.

The DCIM provides an efficient way to evaluate the Green's functions; however, issues of computer time still have to be considered since the number of Green's

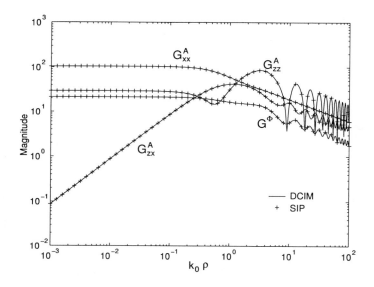

Figure 16.4 Magnitude of Green's functions, G_{xx}^A, G_{zx}^A, G_{zz}^A, and G^Φ, for a five-layer medium [40]. $z' = 0.4$ mm, $z = 1.4$ mm.

functions to be evaluated is proportional to $O(N^2)$ in the MOM analysis, where N denotes the number of unknowns. Since the guided modes are expressed in terms of the Hankel function, their evaluation is expensive compared to the evaluation of the remaining part. Furthermore, for structures supporting vertical currents, the Green's functions for all the different combinations of z and z' are needed. The DCIM has to be performed for every combination, although it is possible for some cases to generate the complex images that are independent of z and z'. To circumvent these problems, an interpolation scheme is usually employed [18, 19, 23, 25, 35]. In this scheme, sections along the z-axis where the structure is located are first determined and then subdivided into N_s sheets, as shown in Figure 16.5, resulting in a total of N_s^2 combinations of z and z'. For each pair of sheets, the DCIM is performed and the Chebyshev interpolation is applied to the variable ρ [43]. For G_{xx}^A, G_{zz}^A, and G^Φ, the reciprocity principle can reduce the number of times to perform the DCIM to $N_s(N_s + 1)/2$. When z and z' are located between those sheets, the Lagrange polynomial interpolation is employed for the variables z and z' [43].

With the interpolation strategy, the computer time to evaluate the Green's functions for multilayer media is further reduced in the MOM analysis. Moreover, it is possible to store the interpolation coefficients as a database instead of performing

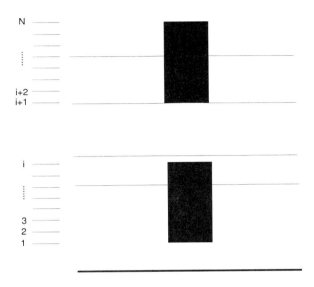

Figure 16.5 Interpolation scheme for a general 3D circuit in multilayer media.

the interpolations on-line. Therefore, the generation of Green's functions can be decoupled from specific circuit geometries.

16.3 THE METHOD-OF-MOMENTS SOLUTION

With the integral equation and the Green's functions available, the MOM can be employed to solve the currents on microstrips. A critical factor for an efficient and accurate MOM analysis is the choice of basis functions. Traditional numerical modeling employs roof-top functions for rectangular discretization or Rao-Wilton-Glisson (RWG) functions [44] for triangular discretization. These functions are complete to the zeroth order. As a result, a very fine discretization is often required to yield an accurate solution. This leads to a large matrix equation, which is expensive to solve. In addition, the numerical solution converges slowly to the exact one when the discretization is made finer. A solution to this problem is to employ higher-order basis functions, which have a better convergence rate and can yield an accurate solution with a rather coarse discretization. The better convergence of higher-order basis functions has been demonstrated in the literature [45–47]. The higher-order interpolatory basis functions developed by Graglia et al. [46] are employed in this work. Also, the curvilinear discretization is used, which provides more flexibility to model arbitrary shapes.

The mixed-potential form of the electric field is given by (16.1). With excitation by an applied field $\mathbf{E}^a$, the induced current on the microstrips can be found by solving the integral equation:

$$\hat{n} \times \left[-j\omega\mu_0 \langle \overline{\mathbf{G}}^A(\mathbf{r},\mathbf{r}'), \mathbf{J}(\mathbf{r}') \rangle + \frac{1}{j\omega\epsilon_0} \nabla \langle G^\Phi(\mathbf{r},\mathbf{r}'), \nabla' \cdot \mathbf{J}(\mathbf{r}') \rangle \right] = -\hat{n} \times \mathbf{E}^a(\mathbf{r}) \tag{16.13}$$

where $\hat{n}$ denotes the unit normal to the surface of the microstrips.

To solve the integral equation (16.13) using the MOM, the surface is first divided into curvilinear triangular patches. As shown in Figure 16.6(a), a quadratic triangular patch described by six nodes can be employed. After the coordinate transformation, shown in Figure 16.6(b), we can easily describe any vector $\mathbf{r}$ on the patch in terms of the quadratic shape function φ

$$\mathbf{r} = \sum_{i=1}^{6} \varphi_i(\xi_1, \xi_2, \xi_3) \mathbf{r}_i \tag{16.14}$$

where the coordinates ξ_1, ξ_2, and ξ_3 have the dependence relation $\xi_1 + \xi_2 + \xi_3 = 1$ and the shape functions are given by

$$\begin{array}{lll} \varphi_1 = \xi_1(2\xi_1 - 1) & \varphi_2 = \xi_2(2\xi_2 - 1) & \varphi_3 = \xi_3(2\xi_3 - 1) \\ \varphi_4 = 4\xi_1\xi_2 & \varphi_5 = 4\xi_2\xi_3 & \varphi_6 = 4\xi_3\xi_1 \end{array} \tag{16.15}$$

The edge vectors can be calculated as

$$\boldsymbol{\ell}_1 = -\frac{\partial \mathbf{r}}{\partial \xi_2}, \quad \boldsymbol{\ell}_2 = -\frac{\partial \mathbf{r}}{\partial \xi_1}, \quad \boldsymbol{\ell}_3 = \frac{\partial \mathbf{r}}{\partial \xi_2} - \frac{\partial \mathbf{r}}{\partial \xi_1} \tag{16.16}$$

The gradient vectors are evaluated by

$$\nabla \xi_1 = \frac{\hat{n} \times \boldsymbol{\ell}_1}{\mathcal{J}}, \quad \nabla \xi_2 = \frac{\hat{n} \times \boldsymbol{\ell}_2}{\mathcal{J}}, \quad \nabla \xi_3 = -\nabla \xi_1 - \nabla \xi_2 \tag{16.17}$$

where $\mathcal{J}$ is the *Jocobian* of the transformation.

The building block of the higher-order interpolatory basis functions is the zeroth-order basis function, which is given by

$$\boldsymbol{\Lambda}_\beta(\mathbf{r}) = \frac{1}{\mathcal{J}} (\xi_{\beta+1} \boldsymbol{\ell}_{\beta-1} - \xi_{\beta-1} \boldsymbol{\ell}_{\beta+1}), \quad \beta = 1, 2, 3 \tag{16.18}$$

The zeroth-order basis function on the flat triangular patch is also known as the RWG basis function [44]. This function has a constant normal and a linear tangential variation along edges.

Full-Wave Analysis of Multilayer Microstrip Problems

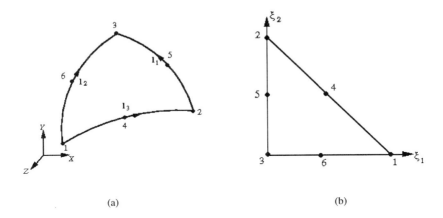

Figure 16.6 A curvilinear patch in (x, y, z) and (ξ_1, ξ_2, ξ_3) coordinates.

The higher-order interpolatory vector basis functions on a given triangular element are constructed by multiplying the zeroth-order basis functions with a set of polynomial functions [46]

$$\Lambda_{ijk}^{\beta}(\mathbf{r}) = N_{\beta} \frac{(p+2)\xi_{\beta}\hat{\alpha}_{ijk}(\boldsymbol{\xi})}{i_{\beta}} \Lambda_{\beta}(\mathbf{r}) \tag{16.19}$$

where β denotes the edge number associated with the zeroth-order basis function; $i, j,$ and k are the indexes for labeling the interpolation points, which satisfy $i + j + k = p + 2$; and i_{β} takes $i, j,$ or k for $\beta = 1, 2,$ or 3, respectively. The normalization coefficients N_{β} are given by

$$N_{\beta} = \frac{p+2}{p+2-i_{\beta}} \ell_{\beta} \tag{16.20}$$

The $\hat{\alpha}_{ijk}(\boldsymbol{\xi})$ is the polynomial function defined in terms of shifted Silvester-Lagrange polynomials $\hat{R}$ as

$$\hat{\alpha}_{ijk}(\boldsymbol{\xi}) = \hat{R}_i(p+2, \xi_1)\hat{R}_j(p+2, \xi_2)\hat{R}_k(p+2, \xi_3) \tag{16.21}$$

where the shifted Silvester-Lagrange polynomial is given by

$$\hat{R}_i(p, \xi) = \begin{cases} \dfrac{1}{(i-1)!} \prod_{k=1}^{i-1}(p\xi - k), & 2 \leq i \leq p+1 \\ 1, & i = 1 \end{cases} \tag{16.22}$$

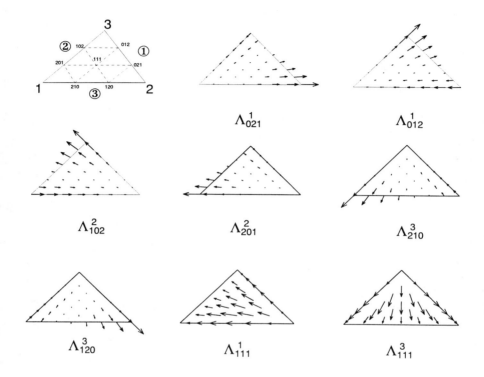

Figure 16.7 The first-order vector basis functions.

The number of degrees of freedom is $N_e = (p+1)(p+3)$ on a triangular element for the basis functions of order p. The first-order basis functions so obtained are shown in Figure 16.7, which shows a linear normal and a quadratic tangential variation at edges.

Therefore, the current density on each patch is expanded as

$$\mathbf{J}(\mathbf{r}) = \sum_{i=1}^{N_e} I_i \mathbf{\Lambda}_i(\mathbf{r}) \qquad (16.23)$$

Substituting (16.23) into (16.13) and applying Galerkin's procedures, we obtain a matrix equation

$$[Z]\{I\} = \{V\} \qquad (16.24)$$

where the global impedance matrix $[Z]$ and the right-hand side (RHS) vector are assembled from the local impedance matrix and local RHS vector, respectively. For example, the local impedance matrix for patch m and patch n has the dimension $N_e \times N_e$, and the local RHS vector has the dimension N_e, whose elements Z_{ij} and V_i are given by

$$Z_{ij}^{mn} = -j\omega\mu_0 \langle \mathbf{\Lambda}_i; \langle \overline{\mathbf{G}}^A; \mathbf{\Lambda}_j \rangle_{S_n} \rangle_{S_m} - \frac{1}{j\omega\epsilon_0} \langle \nabla \cdot \mathbf{\Lambda}_i, \langle G^\Phi, \nabla' \cdot \mathbf{\Lambda} \rangle_{S_n} \rangle_{S_m}$$
(16.25)

$$V_i^m = -\langle \mathbf{\Lambda}_i, \mathbf{E}^a \rangle_{S_m}$$
(16.26)

The double surface integrals are involved in the matrix element, which are evaluated by using Gaussian quadrature when the two elements do not overlap. When they do, the method proposed by Duffy [48] is employed to evaluate the singular integral.

The excitation, or equivalently the RHS of (16.13), is different for different problems. When the scattering properties of microstrip antennas are of interest, the applied field $\mathbf{E}^a$ should be the electric field in the multilayer medium environment without conductors. When we are simulating circuit problems, a voltage delta source is usually applied to the excitation port.

Once the current distribution on the microstrips is obtained, the parameters associated with the current distribution are then extracted. For the scattering/radiation problems, those parameters are far fields, which can be calculated by the standard stationary phase method. A simpler approach is to employ the reciprocity theorem [24]. For the circuit problems, the S-parameters are usually extracted. For N-port circuit problems, it is necessary to extract the scattering parameters. In general, N linearly independent excitations are required for an N-port network. In some practical cases, the properties of symmetry and reciprocity can be utilized to reduce the number of excitations. Two popular extraction schemes can be found from [18] and [49].

The convergence behavior of the higher-order MOM is first analyzed using a microstrip patch antenna as an example. The patch is 40 mm × 40 mm, which resides on a substrate with relative permittivity 2.17 and thickness 1.58 mm. The incident plane wave is assumed to be θ polarized with the incident angle $(\theta^i, \phi^i) = (60°, 45°)$. The frequency is 10 GHz. The θ component of backscatter RCS is calculated. The simulation is carried out at various levels of discretization for basis functions with different orders. The error is plotted in Figure 16.8(a) by computing $|\sigma - \sigma_{\text{ref}}|/|\sigma_{\text{ref}}|$ where σ_{ref} is the reference RCS result obtained using the third-order scheme at a discretization of 400 triangles. The corresponding CPU time for matrix filling is given in Figure 16.8(b). From the figures, we observe that for the same number of unknowns, the higher-order scheme gives more accurate results and the CPU times are comparable for different order schemes. For small problems, the higher-order schemes use more CPU time because the singular and near interaction terms are relatively dominant. When the problem size becomes large, the higher-order schemes become more efficient than the lower-order ones.

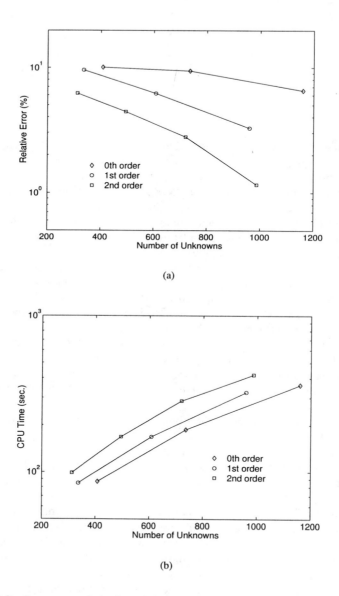

Figure 16.8 Convergence behavior of the high-order basis functions: (a) relative error versus the number of unknowns; (b) CPU time versus the number of unknowns.

Next, we consider several problems to demonstrate the capability of the method. The scattering properties of a microstrip antenna are investigated first. The backscatter RCS $\sigma_{\theta\theta}$ is given in Figure 16.9 as a function of frequency. The zeroth-, first-, and second-order schemes are employed for a coarse discretization with 24 triangles. As can be seen from Figure 16.9, the zeroth-order scheme does not give an accurate result, especially at high frequencies, whereas the first- and second-order schemes converge to the accurate result. The numbers of unknowns for the zeroth-, first-, and second-order approaches are 29, 106, and 231, respectively. The result for the rectangular patch antenna is compared with those from [10]. It is seen that the RCS peaks occur at the frequencies corresponding to the cavity resonant modes. For example, the first two peaks in Figure 16.9 around 2.7 and 3.7 GHz correspond, respectively, to the first two dominant cavity modes (1,0) and (0,1).

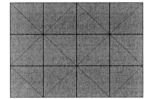

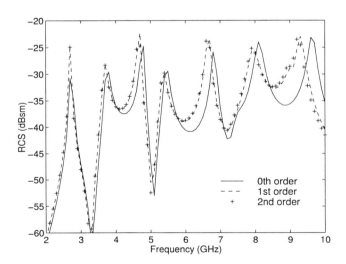

Figure 16.9 Backscatter RCS ($\sigma_{\theta\theta}$) versus frequency for a rectangular microstrip patch antenna [50]. $L = 36.6$ mm, $W = 26$ mm, $h = 1.58$ mm, $\epsilon_r = 2.17$, $\theta^i = 60°$, $\phi^i = 45°$.

The second example is a four-port branch line coupler. This structure has been analyzed using the FDTD method [3], and it is found difficult for regular FDTD grids to match all of the circuit dimensions exactly. In contrast, all the dimensions in our case are precisely modeled. With the mesh shown in Figure 16.10, we employ the second-order scheme, which has 968 unknowns. The S-parameters obtained are shown in Figure 16.10, compared with the measured data from [3]. Good agreement is observed.

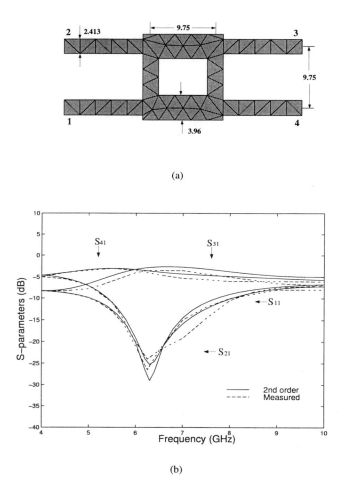

Figure 16.10 (a) Geometry and (b) S-parameters for a branch line coupler (dimensions in mm) [51]. $\epsilon_r = 2.2$, $h = 0.794$ mm.

To show advantages of the curvilinear discretization, the third example is a four-port annular-ring power-divider, which has been analyzed using the approximate planar waveguide models [52]. Here, the circular boundary is precisely modeled by the curvilinear patches, as shown in Figure 16.11. The numbers of unknowns for the

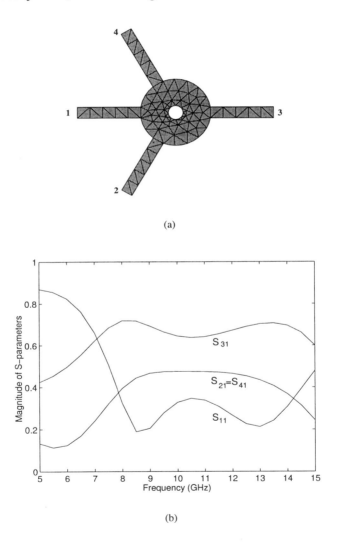

Figure 16.11 (a) Geometry and (b) S-parameters of an annular-ring power-divider [51]. $\epsilon_r = 2.2$, $h = 0.79$ mm. The line width is 2.4 mm. The inner and outer radii of the annular ring are 1.5 mm and 7.2 mm, respectively. The angle between port 1 and port 2 is 60°.

zeroth-, first-, and second-order approaches are 172, 612, and 1,320, respectively. The second-order results are given in Figure 16.11, which agrees reasonably well with those in [52].

All the above examples have no vertical current so that only one component G_{xx}^A in $\overline{\mathbf{G}}^A$ is required to build the impedance matrix. In the next two examples, we show the capability of the method to deal with the 3D structures with both horizontal and vertical currents. Both of them are the spiral inductors, one with a rectangular shape and the other with a curvilinear boundary. The first one has previously been analyzed using the FDTD method [4]. Its discretization is shown in Figure 16.12(a); a top view of the current distribution at $f = 3.5$ GHz is displayed in Figure 16.12(b); and the results obtained using the second-order basis functions with 1,263 unknowns are given in Figure 16.12(c). The results are in good agreement with those in [4]. The detailed information about the geometry of the second example is given in [8]. The only difference here is that the bonding edge of the bridge and the spiral is shifted from the center to the side. The S-parameters calculated using the second-order basis functions with 840 unknowns are given in Figure 16.13.

16.4 FAST FREQUENCY-SWEEP CALCULATION

The MOM analysis in the preceding section is implemented in the frequency domain. To obtain frequency responses over a band of interest, we have to repeat the calculation at each discrete frequency. This can be computationally intensive for devices with complicated frequency responses. For this reason, several different techniques have been proposed to characterize the device by using the reduced-order model, such as the asymptotic waveform evaluation (AWE) [53], the complex frequency hopping (CFH) [54], and the Padé via Lanczos (PVL) [55,56]. All of them were originally developed in the circuit community. The basic idea of these techniques is to approximate the frequency response, or the transfer function, by a low-order rational function, the Padé approximant. In AWE and CFH, the Padé approximant of the frequency response is obtained by the moment matching process. In PVL, the Padé approximant is obtained by the Lanczos process so that the direct calculation of moments is eluded. Recently, these techniques have been extended for electromagnetic analysis [57–60]. A parallel effort ongoing in the electromagnetics community is the development of the model-based parameter estimation (MBPE) [61,62]. Instead of matching moments in AWE, MBPE matches the derivatives of the transfer function to find the Padé approximant.

Although PVL is more stable than AWE, the algorithm requires the submatrices to be frequency independent. This is not the case in MOM since the submatrices are frequency dependent through the Green's functions. Therefore, the AWE is employed in this work to achieve the fast frequency sweep [36]. The unknown current is first expanded as a Taylor series at the expansion point. The Taylor series coefficients, or the moments, are associated with the frequency derivatives of the impedance matrix,

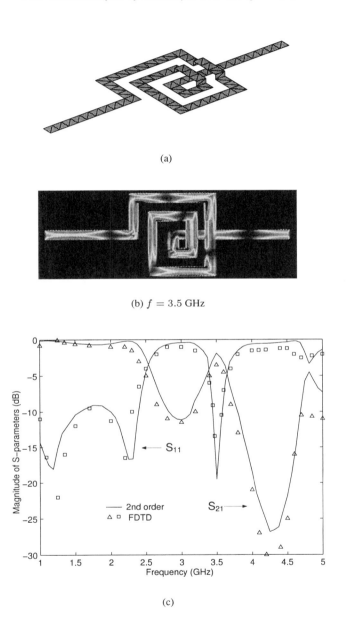

Figure 16.12 (a) Geometry, (b) current distribution, and (c) S-parameters of a spiral inductor [51]. $\epsilon_r = 9.6$, $h = 2.0$ mm. The line widths and spacings are all 2.0 mm. The height and the span of the air bridges are 1.0 mm and 6.0 mm, respectively.

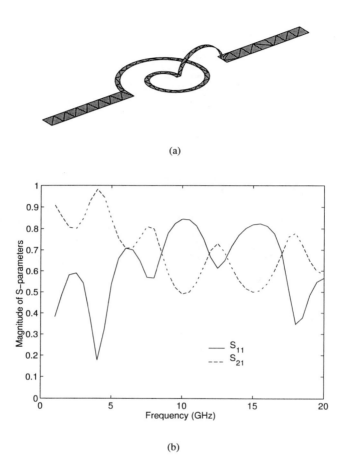

Figure 16.13 (a) Geometry and (b) S-parameters of a spiral inductor [51]. $\epsilon_r = 9.8$, $h = 0.635$ mm (see [8] for detailed information about the geometry).

which can be derived because of the use of the DCIM. The Padé approximant is then obtained by matching the moments. The response over a frequency band near the expansion point can be easily obtained. To obtain the broadband response, more expansion points are required. A simple binary search algorithm, as proposed in [57], is employed to automatically determine the expansion points and obtain the accurate solution over the entire broadband. The details of the AWE method are given in Chapter 15.

Full-Wave Analysis of Multilayer Microstrip Problems

Table 16.1 Comparison of the CPU Time Using the Direct Calculation and AWE [36]

	Direct			AWE			Speedup
	T_{freq} (s)	N_{freq}	T_1 (s)	T_{expn} (s)	N_{expn}	T_2 (s)	
Ex. 1	6.4	60	384	18.1	1	18.1	21.2
Ex. 2	104.0	80	8,320	350.2	3	1,050.6	7.9
Ex. 3	27.0	400	10,800	95.9	9	863.1	12.5

T_{freq}:   CPU time per frequency in the direct calculation;
N_{freq}:   Number of frequencies in the direct calculation;
T_1:   Total CPU time by the direct calculation;
T_{expn}:   CPU time per expansion point with AWE;
N_{expn}:   Number of expansion points with AWE;
T_2:   Total CPU time with AWE.

In the next three examples, AWE is applied to expedite the calculation of frequency response over a specified bandwidth. In all the cases, we use $q = 8$. The CPU time using AWE is compared in Table 16.1 with that of the direct calculation to demonstrate the efficiency. We first consider a microstrip double-stub on a single layer, which has a relative permittivity $\epsilon_r = 9.9$ and a thickness 0.127 mm [17, 18]. The number of unknowns is 205. The direct calculation gives a very good result compared with the measured data [17, 18], as shown in Figure 16.14(a). In this calculation, the CPU time at each frequency is 6.4 s and a total of 60 frequencies are sampled to obtain the accurate results. Therefore, the total CPU time is 384 s. Using AWE and choosing only one expansion point $f = 10.0$ GHz, we can obtain a very good result, as given by Figure 16.14(b). In the calculation with AWE, the CPU time is 18.1 s. Thus, using AWE is approximately 21 times faster. Note that the sampling points are not dense enough to catch the null at about 9.75 GHz for the direct calculation.

We next consider a two-port asymmetric antenna on a two-layer substrate [21]. There are two orthogonally crossed dipoles on the top layer. They have the different lengths so that dual frequency operation is achieved. The longitudinal one has length 11.9 mm and the transversal one has length 10.2 mm. The width of both is 1.7 mm. The two substrate layers have the same relative permittivity $\epsilon_r = 2.17$. The top layer has thickness 1.6 mm and the bottom layer has thickness 0.8 mm. On the bottom layer, the feeding lines are 2.2 mm wide. Fed from port 1, the longitudinal dipole is excited at the resonant frequency of 8.4 GHz. The current distribution is given in Figure 16.15(a). Fed from port 2, the transversal dipole is excited at its resonant frequency 9.6 GHz. At 11.3 GHz, the coupling bend consisting of two perpendicular half-dipoles is resonant. The incident power is essentially transmitted from one port to the other. The magnitude of S-parameters is given in Figure 16.15(b). The

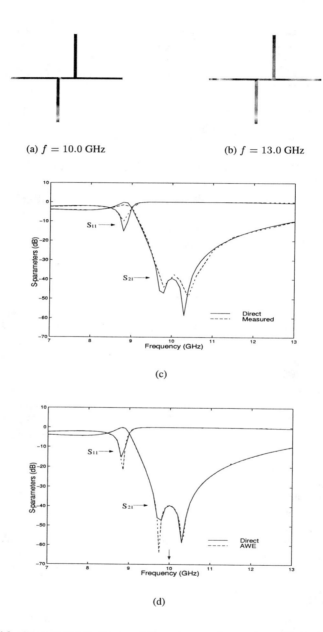

Figure 16.14 Current distributions and S-parameters for a microstrip double-stub [36]. $\epsilon_r = 9.9, h = 0.127$ mm. The line width is 0.122 mm. The stub length is 2.921 mm. The spacing between two stubs is 0.757 mm.

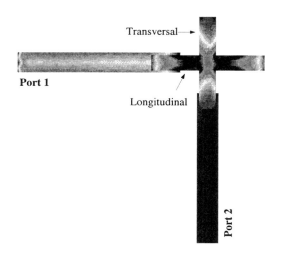

(a) $f = 8.4$ GHz

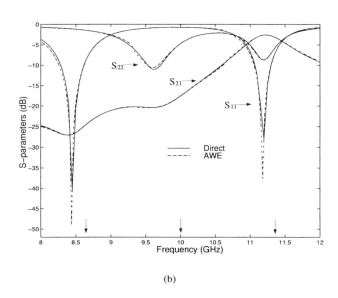

(b)

Figure 16.15 Current distribution and S-parameters for a two-port asymmetric antenna [36].

measured data are available from [21]. Good agreement can be observed. In this example, the number of unknowns is 912. The direct method takes 104 s for each frequency, and a total of 80 sampling frequencies are used to obtain the accurate results. Therefore, the total CPU time is about 139 min. In contrast, with AWE only three expansion frequencies are needed. At each expansion point, the CPU time is 350.2 s. Totally, the CPU time is 17.5 min, making AWE approximately eight times faster.

The last example is an overlap-gap-coupled microstrip filter [23, 63], illustrated in Figure 16.16(a). It is a two-layer geometry. The top layer is alumina with relative permittivity 9.8 and thickness 0.254 mm. The bottom layer is duroid with relative permittivity 2.2 and thickness 0.254 mm. The geometrical parameters are the same as those in [63]. The number of unknowns is 503. In the direct calculation, the CPU time at each frequency is 27 s and a total of 400 sampling frequencies are used to obtain the accurate results. Therefore, the total CPU time is approximately 3 h. The computed S-parameters in comparison with the measured data are shown in Figure 16.16(b). The small discrepancy with the measured data is due to the finite thickness of metalization and the fabrication tolerance of substrates, as discussed in [23] and [63]. In contrast, with AWE only nine expansion frequencies are chosen automatically by the binary search. At each expansion point, the CPU time is 95.9 s. The CPU time totals 863.1 s. Thus, the method with AWE is approximately 12.5 times faster than the direct calculation. The results with AWE agree very well with the direct calculation, as shown in Figure 16.16.

16.5 THE CONJUGATE GRADIENT–FFT METHOD

To simulate large-scale microstrip problems, it is often necessary to employ a large number of unknowns. For the conventional MOM, whether in the spectral or the spatial domain, the memory requirement is always proportional to $O(N^2)$, where N denotes the number of unknowns. This requirement can easily become prohibitive even on the most powerful computers. Even if the memory permits, the computing time can become very excessive because direct matrix inversion solvers, such as Gaussian elimination and LU decomposition methods, require $O(N^3)$ floating-point operations. When an iterative solver such as the conjugate gradient (CG) method is employed for solving the MOM matrix equation, the operation count is $O(N^2)$ per iteration because of the need to evaluate the matrix-vector product. This operation count is still too high for an efficient simulation. To make the iterative method more efficient, it is necessary to speed up the matrix-vector multiplication.

By exploiting the translational invariance of the Green's function, the matrix-vector product can be computed using the FFT. When this is combined with the CG method, the resulting algorithm is called the conjugate gradient–FFT (CGFFT) method [64–72]. Several different schemes of CGFFT have been implemented in

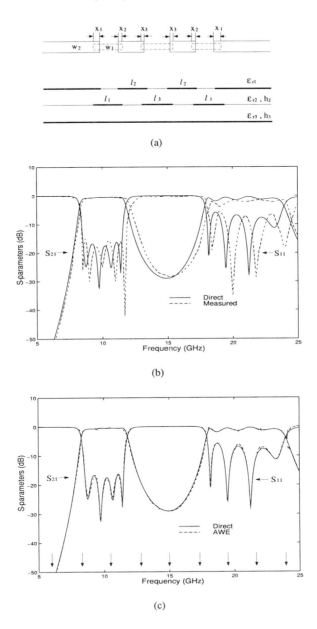

Figure 16.16 S-parameters for an overlap-gap-coupled microstrip filter [36]. $\epsilon_{r1} = 1.0$, $\epsilon_{r2} = 9.8$, $\epsilon_{r3} = 2.2$, $h_2 = h_3 = 0.254$ mm, $w_1 = 0.812$ mm, $w_2 = 0.458$ mm, $l_1 = 6.990$ mm, $l_2 = 6.457$ mm, $l_3 = 7.242$ mm, $x_1 = 1.311$ mm, $x_2 = 0.386$ mm, $x_3 = 0.269$ mm.

the past, which differ primarily in the treatment of the Green's function and the del operators. A review of those works reveals that the spatially discretized Green's functions can eliminate the aliasing and truncation errors, and that the transfer of the del operators to the basis and testing functions can improve the accuracy and efficiency [70–72]. The resulting algorithm is suitable for analysis of large-scale problems because the CPU time per iteration is of $O(N \log N)$ and the memory requirement is of $O(N)$.

To make use of CGFFT, a uniform discretization is required. We first enclose the conducting surface by a rectangle, which is then divided into $M \times N$ small rectangular cells whose side lengths are Δx and Δy along the x and y directions, respectively. Assume that $\mathbf{f}^x_{m,n} = f^x_{m,n}\hat{x}$ and $\mathbf{f}^y_{m,n} = f^y_{m,n}\hat{y}$ are vector basis functions in the x and y directions, respectively, where $f^{x,y}_{m,n}$ represent the roof-top functions. We can expand the surface current distribution $\mathbf{J}$ in a sequence of vector basis functions $\mathbf{f}^x_{m,n}$ and $\mathbf{f}^y_{m,n}$ as

$$\mathbf{J} = \sum_{m,n} I^x_{m,n} \mathbf{f}^x_{m,n} + \sum_{m,n} I^y_{m,n} \mathbf{f}^y_{m,n} \qquad (16.27)$$

Substituting (16.27) into (16.13) and applying Galerkin's procedure, we obtain

$$\begin{bmatrix} \mathbf{G}_{xx} & \mathbf{G}_{xy} \\ \mathbf{G}_{yx} & \mathbf{G}_{yy} \end{bmatrix} \begin{bmatrix} \mathbf{J}_x \\ \mathbf{J}_y \end{bmatrix} = \begin{bmatrix} \mathbf{b}_x \\ \mathbf{b}_y \end{bmatrix} \qquad (16.28)$$

where

$$\mathbf{G}_{xx} = [G_{xx}(m-m', n-n')], \qquad \mathbf{G}_{xy} = [G_{xy}(m-m', n-n')]$$
$$\mathbf{G}_{yx} = [G_{yx}(m-m', n-n')], \qquad \mathbf{G}_{yy} = [G_{yy}(m-m', n-n')]$$
$$\mathbf{b}_x = [\langle \mathbf{f}^x_{m,n}, \mathbf{E}^a(\mathbf{r}) \rangle], \qquad \mathbf{b}_y = [\langle \mathbf{f}^y_{m,n}, \mathbf{E}^a(\mathbf{r}) \rangle]$$

in which

$$G_{xx}(m-m', n-n') = \Gamma^x_a(m-m', n-n') + \Gamma_{xx}(m-m', n-n')$$
$$G_{xy}(m-m', n-n') = \frac{1}{\Delta x \Delta y} \sum_{i=0}^{1} \sum_{k=0}^{-1} (-1)^{i+k} \Gamma_q(m-m'+i, n-n'+k)$$
$$G_{yy}(m-m', n-n') = \Gamma^y_a(m-m', n-n') + \Gamma_{yy}(m-m', n-n')$$
$$G_{yx}(m-m', n-n') = G_{xy}(n'-n, m'-m)$$

and

$$\Gamma_{xx}(m-m', n-n') = \frac{1}{(\Delta x)^2} \sum_{i=0}^{1}\sum_{k=0}^{-1}(-1)^{i+k}\Gamma_q(m-m'+i+k, n-n')$$

$$\Gamma_{yy}(m-m', n-n') = \frac{1}{(\Delta y)^2} \sum_{i=0}^{1}\sum_{k=0}^{-1}(-1)^{i+k}\Gamma_q(m-m', n-n'+i+k)$$

$$\Gamma_a^x(m-m', n-n') = j\omega\mu_0 \iint_S f_{m,n}^x \iint_S G_a f_{m',n'}^x \, ds' ds$$

$$\Gamma_a^y(m-m', n-n') = j\omega\mu_0 \iint_S f_{m,n}^y \iint_S G_a f_{m',n'}^y \, ds' ds$$

$$\Gamma_q(m-m', n-n') = \frac{j}{\omega\epsilon_0} \iint_S \Pi_{m,n} \iint_S G_q \Pi_{m',n'} \, ds' ds$$

In the above, $\Pi_{m,n}$ is the 2D unit pulse function defined over (m, n)-th rectangular cell.

The biconjugate gradient (BCG) algorithm is employed here for the solution of $Ax = b$ in which A is symmetric. The convolution relationship between $\overline{\mathbf{G}}^A$ and $\mathbf{J}$ gives the following equations:

$$\mathbf{G}_{xx}\mathbf{J}_x = DFT^{-1}\{DFT\{G_{xx}(m,n)\} \cdot DFT\{J_x(m,n)\}\} \quad (16.29)$$

$$\mathbf{G}_{yy}\mathbf{J}_y = DFT^{-1}\{DFT\{G_{yy}(m,n)\} \cdot DFT\{J_y(m,n)\}\} \quad (16.30)$$

The convolution relationship between G^Φ and $\nabla \cdot \mathbf{J}$ is also exploited:

$$\mathbf{G}_{xy}\mathbf{J}_y = [D_y(m,n) - D_y(m+1,n)] \quad (16.31)$$

$$\mathbf{G}_{yx}\mathbf{J}_x = [D_x(m,n) - D_x(m,n+1)] \quad (16.32)$$

where

$$D_x(m,n) = \frac{1}{\Delta x \Delta y} DFT^{-1}\{\Gamma_{qx}(m,n)\} \quad (16.33)$$

$$D_y(m,n) = \frac{1}{\Delta x \Delta y} DFT^{-1}\{\Gamma_{qy}(m,n)\} \quad (16.34)$$

and

$$\Gamma_{qx}(m,n) = DFT\{\Gamma_q(m,n)\} \cdot DFT\{J_x(m,n) - J_x(m-1,n)\}$$

$$\Gamma_{qy}(m,n) = DFT\{\Gamma_q(m,n)\} \cdot DFT\{J_y(m,n) - J_y(m,n-1)\}.$$

To demonstrate the accuracy and efficiency of the method, we consider a corporate-fed microstrip array depicted in Figure 16.17. The radiation patterns are given in Figure 16.18. In calculating the radiation patterns, we have increased the number of

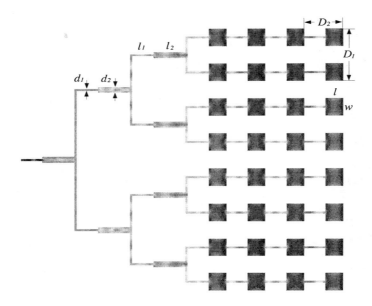

Figure 16.17 Geometry of a corporate-fed microstrip antenna array having 8 × 4 elements [72]. $\epsilon_r = 2.2$, $h = 1.59$ mm, $l = 10.08$ mm, $w = 11.79$ mm, $d_1 = 1.3$ mm, $d_2 = 3.93$ mm, $l_1 = 12.32$ mm, $l_2 = 18.48$ mm, $D_1 = 23.58$ mm, $D_2 = 22.40$ mm, $f = 9.42$ GHz.

microstrip patch elements in each arm from 4 to 16 to show the effect on the radiation pattern. The monostatic RCS of the same structure is given in Figure 16.19. The requirement of computational resources is listed in Table 16.2, where the normalized residual is defined as $Err = \|\mathbf{r}_n\|^2/\|\mathbf{b}\|^2$. All computations are carried out on a DEC alpha workstation.

Table 16.2 CPU Time and Memory on DEC Alpha Workstation

Array	Number of unknowns	CPU time per Iteration	Number of iterations	Computer storage	Tolerance Err
8 × 8	118,073	11.6 s	313	18 MB	10^{-3}
8 × 8	118,073	11.6 s	599	18 MB	10^{-6}
16 × 16	495,044	97.81 s	425	65 MB	10^{-3}
16 × 16	495,044	97.81 s	1,070	65 MB	10^{-6}

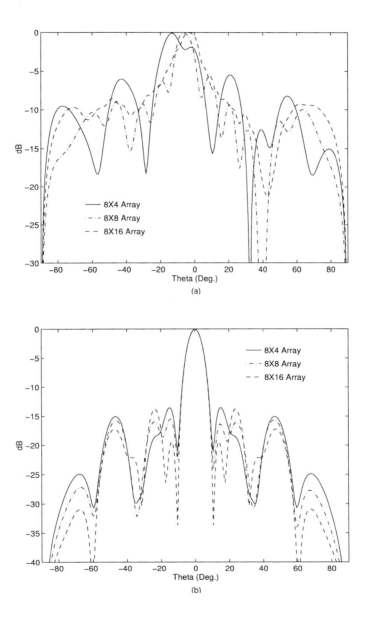

Figure 16.18 Radiation patterns of the $8 \times N$ microstrip corporate-fed microstrip array: (a) E-plane pattern; (b) H-plane pattern [72].

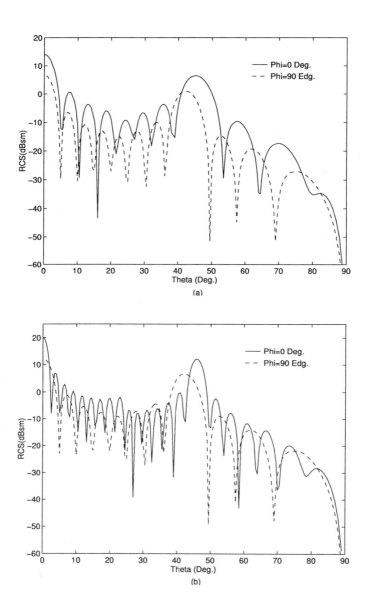

Figure 16.19 Monostatic RCS of a corporate-fed microstrip antenna array: (a) 8 × 8 array; (b) 8 × 16 array [72].

16.6 THE ADAPTIVE INTEGRAL METHOD

The CGFFT method requires a uniform discretization to make use of the translational invariance of the Green's functions. This limits the applicability of the method to complex geometries and results in a staircase approximation for curved boundaries. The AIM is developed to lift this restriction. The basic principle of this method is to translate the triangular basis function onto a regular Cartesian grid and then utilize FFT to carry out the matrix-vector multiplication. This idea was originally proposed by Bleszynski et al. for solving the scattering and radiation problems [73–75]. It was later extended to large-scale microstrip structures [76], where the DCIM is employed to accelerate the computation of the spatial domain Green's functions. This method retains the advantages of the CGFFT method as well as the excellent modeling capability offered by the RWG basis functions. Similar approaches include the sparse-matrix/canonical grid method [77] and the precorrected-FFT method [78].

To employ AIM to accelerate the matrix-vector multiplication, we first enclose the whole structure in a rectangular region and then recursively subdivide it into small rectangular grids, as illustrated in Figure 16.20. Then the original basis functions on the triangular elements are translated to the rectangular grids.

Let us first denote the RWG basis function as $\mathbf{f}_i$ and rewrite the impedance matrix element as

$$Z_{ij} = -j\omega\mu_0 \iint_{T_i} \iint_{T_j} \left[\mathbf{f}_i(\mathbf{r}) \cdot \mathbf{f}_j(\mathbf{r}') G_{xx}^A(\mathbf{r},\mathbf{r}') \right.$$
$$\left. - \frac{1}{k_0^2} \nabla \cdot \mathbf{f}_i(\mathbf{r}) \nabla' \cdot \mathbf{f}_j(\mathbf{r}') G^\Phi(\mathbf{r},\mathbf{r}') \right] ds' ds \quad (16.35)$$

where T_i and T_j denote the support of $\mathbf{f}_i$ and $\mathbf{f}_j$, respectively. If any one of the Cartesian components of $\mathbf{f}_i(\mathbf{r})$ and $\nabla \cdot \mathbf{f}_i(\mathbf{r})$ is denoted as $\psi_i(\mathbf{r})$, the impedance matrix element of (16.35) can be expressed as a linear combination of matrix elements in the form of

$$A_{ij} = \iint_{T_i} \iint_{T_j} \psi_i(\mathbf{r}) g(\mathbf{r},\mathbf{r}') \psi_j(\mathbf{r}') ds' ds \quad (16.36)$$

where g can be either G_{xx}^A or G^Φ. The AIM first approximates $\psi_i(\mathbf{r})$ as a combination of the Dirac delta functions on the rectangular grids:

$$\psi_i(\mathbf{r}) \simeq \hat{\psi}_i(\mathbf{r}) = \sum_{u=1}^{(M+1)^2} \Lambda_{iu} \delta(\mathbf{r} - \mathbf{r}_{iu}) \quad (16.37)$$

where Λ_{iu} is the translation coefficient for the basis function $\psi_i(\mathbf{r})$, M is the order of the translation, and $\mathbf{r}_{iu} = (x_{iu}, y_{iu})$ is the coordinate of the grid.

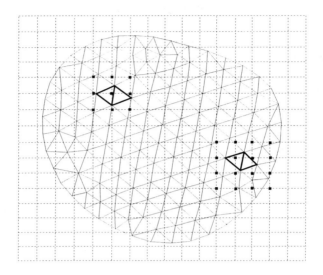

Figure 16.20 Translation of RWG basis functions to rectangular grids [76]. The highlighted triangular basis function on the left is approximated by $(M+1)^2 = 9$ rectangular grids. The one on the right is approximated by $(M+1)^2 = 16$ rectangular grids.

The translation coefficients can be found based on the criterion that the translated basis function produces the same multipole moments as the original basis function

$$\sum_{u=1}^{(M+1)^2} (x_{iu} - x_0)^{q_1} (y_{iu} - y_0)^{q_2} \Lambda_{iu} = \iint_{T_i} \psi_i(\mathbf{r})(x - x_0)^{q_1} (y - y_0)^{q_2} ds$$

$$\text{for } 0 \leq q_1, q_2 \leq M \quad (16.38)$$

where the reference point $\mathbf{r}_0 = (x_0, y_0)$ is chosen as the center of the basis function. The closed-form solution of (16.38) has been given by Bleszynski et al. [74].

Once this translation is found, we can approximate the matrix element of (16.36) as

$$\hat{A}_{ij} = \sum_{u=1}^{(M+1)^2} \sum_{v=1}^{(M+1)^2} \Lambda_{iu} g(\mathbf{r}_u, \mathbf{r}'_v) \Lambda_{jv} \quad (16.39)$$

With the translation formula given above, we can now rewrite the impedance matrix element in the form of (16.35) as

$$\hat{Z}_{ij} = -j\omega\mu_0 \sum_{u=1}^{(M+1)^2} \sum_{v=1}^{(M+1)^2} \left[(\Lambda_{x,iu}\Lambda_{x,jv} + \Lambda_{y,iu}\Lambda_{y,jv}) G_{xx}^A(\mathbf{r}_u, \mathbf{r}'_v) \right.$$
$$\left. - \frac{1}{k_0^2} \Lambda_{d,iu}\Lambda_{d,jv} G^\Phi(\mathbf{r}_u, \mathbf{r}'_v) \right] \quad (16.40)$$

where Λ_x, Λ_y, and Λ_d denote the translation coefficients for the x-component, the y-component, and the divergence of the basis function, respectively.

The $\hat{Z}_{ij}$ in (16.40) offers good accuracy to approximate Z_{ij} in (16.35) when the basis and testing functions are at a sufficiently large distance. The significant error, which occurs when the basis and testing functions are close to each other, is compensated for by introducing a residual matrix $[R]$. Therefore, the impedance matrix is decomposed into

$$[Z] = [\hat{Z}] + [R] \quad (16.41)$$

Note that only when $\mathbf{f}_i$ and $\mathbf{f}_j$ are very close does R_{ij} have an appreciable value, which makes the residual matrix $[R]$ very sparse. In matrix form, $[\hat{Z}]$ can be written as

$$[\hat{Z}] = -j\omega\mu_0 \left[\bar{\Lambda}_x \overline{\mathbf{G}}_{xx}^A \bar{\Lambda}_x^T + \bar{\Lambda}_y \overline{\mathbf{G}}_{xx}^A \bar{\Lambda}_y^T - \frac{1}{k_0^2} \bar{\Lambda}_d \overline{\mathbf{G}}^\Phi \bar{\Lambda}_d^T \right] \quad (16.42)$$

where $\bar{\Lambda}_x, \bar{\Lambda}_y$, and $\bar{\Lambda}_d$ are sparse matrices with each row containing only $(M+1)^2$ nonzero elements. The translational invariance of $\overline{\mathbf{G}}_{xx}^A$ and $\overline{\mathbf{G}}^\Phi$ enables the use of FFT to accelerate the computation of the product of this matrix $[\hat{Z}]$ with a vector. These features make the algorithm less demanding of memory and CPU time. Employing the CG method as the iterative solver, we can write the matrix-vector multiplication as

$$[Z] \cdot \mathbf{I} = [\hat{Z}] \cdot \mathbf{I} + [R] \cdot \mathbf{I} \quad (16.43)$$

where $[\hat{Z}] \cdot \mathbf{I}$ can be evaluated using FFT as

$$[\hat{Z}] \cdot \mathbf{I} = -j\omega\mu_0 \left\{ \bar{\Lambda}_x \mathcal{F}^{-1} \left[\mathcal{F}(\overline{\mathbf{G}}_{xx}^A) \cdot \mathcal{F}(\bar{\Lambda}_x^T \mathbf{I}) \right] + \bar{\Lambda}_y \mathcal{F}^{-1} \left[\mathcal{F}(\overline{\mathbf{G}}_{xx}^A) \cdot \mathcal{F}(\bar{\Lambda}_y^T \mathbf{I}) \right] \right.$$
$$\left. - \frac{1}{k_0^2} \bar{\Lambda}_d \mathcal{F}^{-1} \left[\mathcal{F}(\overline{\mathbf{G}}^\Phi) \cdot \mathcal{F}(\bar{\Lambda}_d^T \mathbf{I}) \right] \right\} \quad (16.44)$$

As we can see from the analysis described above, the memory requirement of AIM is proportional to $O(N)$ due to the sparsity of matrices $[R], \bar{\Lambda}_x$, and $\bar{\Lambda}_y$. The

CPU time per iteration is dominated by the FFT computation of the matrix-by-vector product, which is proportional to $O(N \log N)$. In contrast, for the conventional MOM, the CPU time for the matrix fill is of $O(N^2)$ and the CPU time per iteration is also of $O(N^2)$.

The accuracy of the AIM is first examined. Consider a microstrip line on a substrate with relative permittivity $\epsilon_r = 2.17$ and thickness $h = 1.58$ mm. The frequency is 3.0 GHz. The line is 5 mm wide and 400 mm long, as seen from Figure 16.21(a). Enclose the line by rectangular grids with the grid size being 6.25 mm. Matrix elements calculated by using (16.35) and (16.40) are compared in Figure 16.21(b) where $M = 3$ is chosen. The relative error is also given, which is defined as

$$\Delta_{ij} = \frac{|Z_{ij} - \hat{Z}_{ij}|}{|Z_{ij}|} \qquad (16.45)$$

Figure 16.21(c) gives the relative errors for $M = 1$ and $M = 3$, which shows that the accuracy increases with the increase of M.

Next, the complexity of this algorithm is evaluated. The CPU time per iteration and the memory requirement versus the number of unknowns are plotted by considering a rectangular microstrip patch on the substrate with relative permittivity $\epsilon_r = 2.17$ and thickness $h = 1.58$ mm. The frequency is 3.0 GHz. The patch is discretized with 20 unknowns per wavelength λ_0. It is seen from Figure 16.22 that the CPU time per iteration is scaled as $O(N \log N)$ and the memory requirement is scaled as $O(N)$.

The first example is a parallel-coupled bandpass filter. The dimension of the filter is given by Figure 8.26 in [79]. The substrate has permittivity $\epsilon_r = 10.0$ and thickness $h = 0.635$ mm. The numbers of facets and unknowns are 1086 and 1402, respectively. Again, the current distributions are shown at $f = 9.0$ GHz and $f = 11.0$ GHz. The S-parameters of this bandpass filter are shown in Figure 16.23. Both MOM and AIM results agree very well. The measured data from [79] are also given in Figure 16.23 for comparison. The discrepancy at the low-frequency range is believed to be due to the fabrication error.

The example above demonstrates the accuracy of this algorithm. Because the structure analyzed is relatively small, the saving of CPU time is not expected, although significant memory reduction has been achieved. For large-scale problems, we can predict a substantial reduction of CPU time in both the matrix fill and the matrix solve. To illustrate the efficiency of this method, we now consider some large-scale microstrip structures. First, a structure consisting of five radial stubs is analyzed, as shown in Figure 16.24. The dimension of each radial stub is given in [80]. The spacing between stubs is 7.5 mm. The substrate has the permittivity of 10.0 and the thickness of 0.635 mm. The numbers of facets and unknowns are 3982 and 5580, respectively. The memory requirement is 21 MB and the CPU time per iteration is 1.7 s in the AIM algorithm. In contrast, the conventional MOM requires 250 MB memory and 5.7 s CPU time per iteration. The CPU time for the matrix fill is only

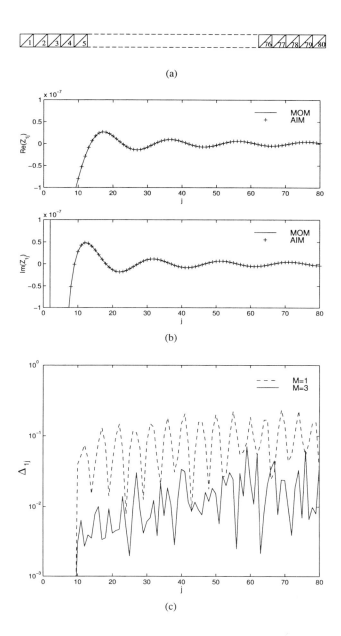

Figure 16.21 Matrix elements Z_{1j} ($j = 1, 2, \ldots, 80$) calculated by the conventional MOM and the AIM [76].

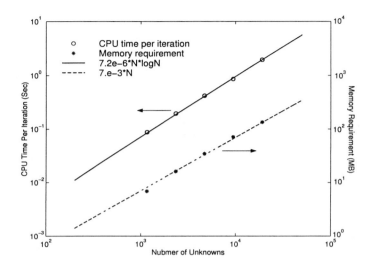

Figure 16.22 Complexity of the AIM [76]. The CPU time per iteration is close to $9 \times 10^{-6} N \log N$, and the memory requirement is close to $7 \times 10^{-3} N$.

40% of that in the conventional MOM. The S-parameters of this structure are shown in Figure 16.24. The current distributions at two frequencies, $f = 8.0$ GHz and $f = 12.0$ GHz, are also given in the figure.

The second problem concerns the scattering properties of a microstrip antenna array. The substrate has relative permittivity $\epsilon_r = 2.17$ and thickness $h = 1.58$ mm. The rectangular element has length 36.6 mm and width 26.0 mm. The element spacing is 55.517 mm in both axes. The incident wave is θ-polarized with incident angles $\theta = 0°$ and $\phi = 45°$. The monostatic RCS (θ-component) of various size arrays is given by Figure 16.25 as a function of frequency, which exhibits good agreement with those data available in [12]. The 31×31 array has 141,267 unknowns. The standard MOM requires about 160 GB memory, which is far beyond the capability of most computers currently available. This method requires only 402 MB memory. The CPU time per iteration is 35.8 s.

16.7 THE MULTILEVEL FAST MULTIPOLE ALGORITHM

Another fast algorithm to speed up the matrix-vector multiplication is the fast multipole method (FMM), which is originally proposed to evaluate particle simulations [81] and later extended to solve electromagnetics problems [82–87]. Some suc-

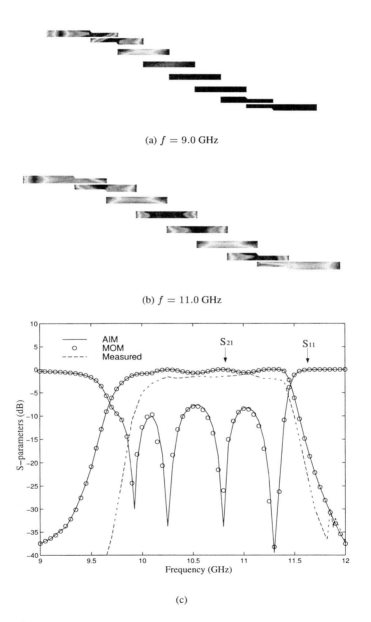

Figure 16.23 (a, b) Current distributions and (c) S-parameters for the parallel-coupled bandpass filter [76]. $\epsilon_r = 10.0$, $h = 0.635$ mm.

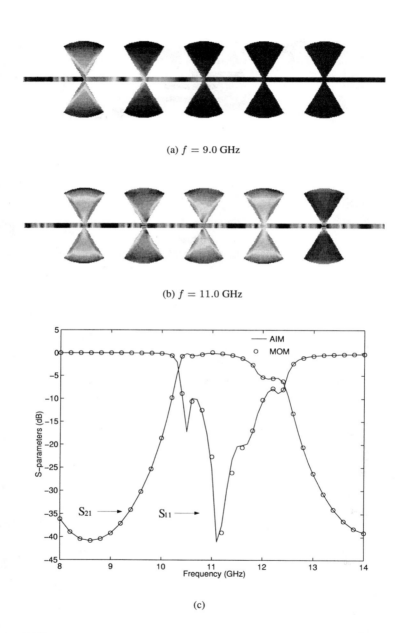

(a) $f = 9.0$ GHz

(b) $f = 11.0$ GHz

(c)

Figure 16.24 (a, b) Current distributions and (c) S-parameters for the cascaded microstrip radial stub [76]. $\epsilon_r = 10.0$, $h = 0.635$ mm. The spacing between stubs is 7.5 mm.

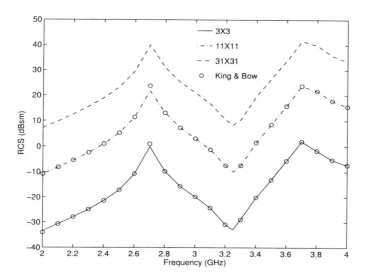

Figure 16.25 Backscatter RCS ($\sigma_{\theta\theta}$) for microstrip antenna arrays with various sizes as a function of frequency [76].

cessful examples using this technique include Fastcap [82], which applies the FMM to extract capacitance for objects in homogeneous media, and FISC (Fast Illinois Solver Code) [87], which uses the multilevel fast multipole algorithm (MLFMA) to deal with free-space scattering problems. After successfully being applied to homogeneous problems, this method has been investigated and extended to microstrip problems. One approach is to combine the FMM with the DCIM [88–90]. In [88] and [89], the equivalent problem is set up by adding N_c images at the corresponding complex coordinates, and therefore represented by $N(N_c + 1)$ basis functions. In the FMM implementation, the translation is different for different images. Also, the static problem and the 2D problem are treated in [88] and [89], respectively. In [90], both the 2D and 3D FMMs are employed because the surface-wave poles are extracted in the DCIM, which makes the implementation complicated. The multilevel algorithm is not implemented in those analyses. The other FMM approach is to express the Green's function in terms of a rapidly converging steepest descent integral, and evaluate the Hankel function arising in the integrand by the FMM [91]. This approach is good for thin-stratified media.

In this section, the MLFMA combined with the DCIM is presented for efficient analysis of microstrip structures [92]. Instead of being treated separately, the image sources are grouped with the original source. By the use of the multilevel algorithm, the complexity is reduced to $O(N \log N)$. The algorithm requires little extra computation compared with that applied to free space problems.

In general, both G_{xx}^A and G^Φ can be expressed in a closed form by using the DCIM

$$G(\mathbf{r},\mathbf{r}') = \sum_{p=0}^{N_c} a_p \frac{e^{-jkr_p}}{4\pi r_p}, \qquad r_p = |\mathbf{r} - (\mathbf{r}' + \hat{z}b_p)| \tag{16.46}$$

where a_p and b_p are the complex coefficients obtained from the DCIM.

To use the FMM, the entire structure is divided into groups denoted by G_m ($m = 1, 2, \ldots, M$). Letting $\mathbf{r}_i$ be the field point in group G_m centered at $\mathbf{r}_m$, and $\mathbf{r}_j$ be the source point in group $G_{m'}$ centered at $\mathbf{r}_{m'}$, we have

$$\begin{aligned}\mathbf{r}_{ij} &= \mathbf{r}_i - (\mathbf{r}_j + \hat{z}b_p) \\ &= (\mathbf{r}_i - \mathbf{r}_m) + (\mathbf{r}_m - \mathbf{r}_{m'}) + (\mathbf{r}_{m'} - \mathbf{r}_j) - \hat{z}b_p \\ &= \mathbf{r}_{im} + \mathbf{r}_{mm'} - \mathbf{r}_{jm'} - \hat{z}b_p\end{aligned} \tag{16.47}$$

Employing the addition theorem and the elementary identity, we can rewrite the Green's function in (16.46) as

$$G(\mathbf{r}_i,\mathbf{r}_j) \approx \frac{k}{j16\pi^2} \oiint \sum_{p=0}^{N_c} a_p e^{j\mathbf{k}\cdot\hat{z}b_p} e^{-j\mathbf{k}\cdot(\mathbf{r}_{im}-\mathbf{r}_{jm'})} T(\hat{r}_{mm'} \cdot \hat{k}) d^2\hat{k} \tag{16.48}$$

where

$$T(\hat{r}_{mm'} \cdot \hat{k}) = \sum_{l=0}^{L}(-j)^l(2l+1)h_l^{(2)}(kr_{mm'})P_l(\hat{r}_{mm'} \cdot \hat{k}) \tag{16.49}$$

Substituting (16.48) into (16.35), we obtain

$$Z_{ij} = \frac{\omega k}{16\pi^2}\Bigg[\oiint S^A(\hat{k})\mathbf{U}_{im}(\hat{k}) \cdot T(\hat{r}_{mm'},\hat{k})\mathbf{U}_{jm'}^*(\hat{k})d^2\hat{k} \\ - \frac{1}{k^2}\oiint S^\Phi(\hat{k})V_{im}(\hat{k})T(\hat{r}_{mm'},\hat{k})V_{jm'}^*(\hat{k})\Bigg]d^2\hat{k} \tag{16.50}$$

where

$$\mathbf{U}_{im}(\hat{k}) = \iint_{T_i} e^{-j\mathbf{k}\cdot\mathbf{r}_{im}}\mathbf{f}_i d\mathbf{r} \tag{16.51}$$

$$V_{im}(\hat{k}) = \iint_{T_i} e^{-j\mathbf{k}\cdot\mathbf{r}_{im}}\nabla \cdot \mathbf{f}_i d\mathbf{r} \tag{16.52}$$

$$S(\hat{k}) = \sum_{p=0}^{N_c} a_p e^{j\mathbf{k}\cdot\hat{z}b_p} \tag{16.53}$$

When we use an iterative method to solve (16.24), the matrix-vector multiplication performed in each iteration can be written as

$$\sum_{j=1}^{N} Z_{ij} I_j = \sum_{m' \in B_m} \sum_{j \in G_{m'}} Z_{ij} I_j + \frac{\omega k}{16\pi^2} \left[\oiint S^A(\hat{k}) \mathbf{U}_{im}(\hat{k}) \right.$$
$$\sum_{m' \notin B_m} T(\hat{r}_{mm'}, \hat{k}) \sum_{j \in G_{m'}} \mathbf{U}^*_{jm'}(\hat{k}) I_j d^2\hat{k}$$
$$\left. - \frac{1}{k^2} \oiint S^\Phi(\hat{k}) V_{im}(\hat{k}) \sum_{m' \notin B_m} T(\hat{r}_{mm'}, \hat{k}) \sum_{j \in G_{m'}} V^*_{jm'}(\hat{k}) I_j d^2\hat{k} \right]$$
(16.54)

for $i \in G_m$, where B_m denotes the neighboring groups of G_m including G_m itself. Therefore, the first term in (16.54) is the contribution from nearby groups and is calculated directly. The second term is the far interaction to be calculated by the FMM.

It has been shown that the operation count for calculating (16.54) is proportional to $O(N^{1.5})$ [85]. This complexity can be reduced to $O(N \log N)$ by using the multilevel algorithm [86].

Before we apply the proposed method to realistic problems, the accuracy of this algorithm is examined. Consider a microstrip line on a substrate with relative permittivity $\epsilon_r = 2.17$ and thickness $h = 1.58$ mm. The frequency is 3.0 GHz. The line is 5 mm wide and 400 mm long. As shown in Figure 16.26, the line is discretized into triangular elements with edge length 5 mm. We plot the values of matrix elements $Z_{1j} (j = 1, 2, \ldots, 80)$ obtained by using two different approaches. One approach is to use the original formulation (16.35). The other approach is to use the MLFMA, where group size d is $0.25\lambda_0$ with λ_0 being the wavelength in free space. Note that $Z_{1j} (j = 1, 2, \ldots, 14)$ is considered as the near interaction so it is calculated directly, and $Z_{1j} (j = 15, 16, \ldots, 80)$ is considered as the far interaction so it is calculated by the MLFMA. As seen from Figure 16.26, these two approaches agree well. In this calculation, the number of modes L is chosen to be $kd + 3 \ln(\pi + kd)$.

Next, the complexity of this algorithm is evaluated. The CPU time per iteration and the memory requirement versus the number of unknowns are plotted by considering a rectangular microstrip patch on the substrate with relative permittivity $\epsilon_r = 2.17$ and thickness $h = 1.58$ mm. The frequency is 3.0 GHz. The patch is discretized with 20 unknowns per wavelength λ_0. It is seen from Figure 16.27 that the CPU time per iteration is scaled as $O(N \log N)$ and the memory requirement is scaled as $O(N)$.

As an example, we recalculate the corporate-fed microstrip antenna array in Figure 16.17, which involves 6,569 facets and 8,668 edges. At frequency $f = 9.42$ GHz, the radiation patterns in the two principal planes $\phi = 0°$ and $\phi = 90°$ are given in Figure 16.28, which shows excellent agreement between this method and the conventional

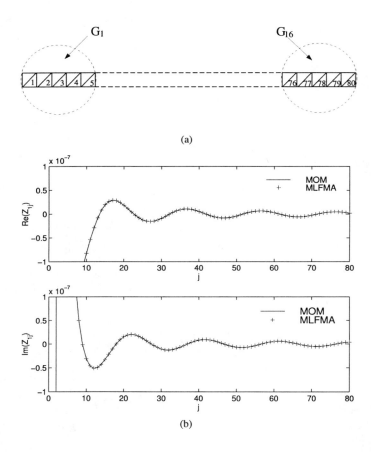

Figure 16.26 Matrix elements Z_{1j} ($j = 1, 2, \ldots, 80$) calculated by the conventional MOM and the MLFMA [92].

MOM. For the conventional MOM, the memory requirement is over 600 MB and the CPU time per iteration is 15.8 s. However, it takes only 36.3 MB and 3.0 s for the 5-level MLFMA. The MLFMA also yields more than a 70% reduction in the CPU time for the matrix fill compared to the conventional MOM.

16.8 SUMMARY

A fast integral equation solver for microstrip structures in multilayer media is presented. In this method, the spectral domain Green's functions are derived from the

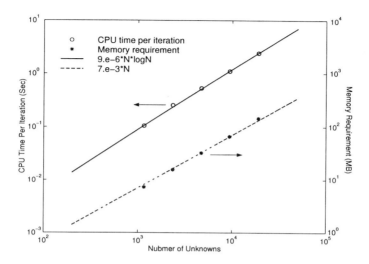

Figure 16.27 Complexity of the MLFMA [92]. The CPU time per iteration is close to $9 \times 10^{-6} N \log N$ and the memory requirement is close to $7 \times 10^{-3} N$.

simple transmission line perspective. The time-consuming numerical integration of the associated SIs is circumvented by applying the DCIM. Furthermore, the interpolation strategy is employed to evaluate the Green's functions, which makes the Green's function calculation in the MOM analysis very efficient. The higher-order interpolatory basis functions defined on the curvilinear triangular elements are employed, which offers a better convergence rate and gives an accurate solution with a rather coarse discretization.

To achieve the fast frequency sweep, a reduced-order model is incorporated into this MOM analysis, in which AWE is employed to obtain the Padé approximant at the expansion point. Then, the frequency response near the expansion point can be easily obtained. When one expansion point is not adequate, a binary search algorithm is employed to automatically determine the expansion points and obtain the accurate solution over the entire broadband. The numerical results show that the method using AWE results in 8 to 22 times faster than the direct calculation.

To simulate large-scale problems, two fast algorithms have been developed for planar structures in multilayer media. One is the FFT-accelerated scheme, AIM. The other is the multipole-accelerated scheme, MLFMA. Both of the algorithms reduce the computational complexity to $O(N \log N)$, which allows us to solve large-scale problems.

As one might see, several techniques have been developed to tackle the multilayer microstrip antennas and circuits, which leaves plenty of room for future studies. For

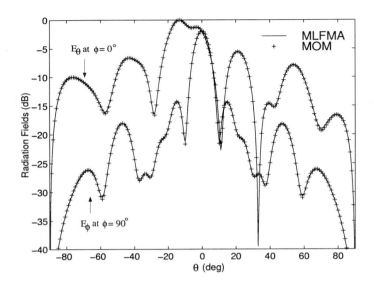

Figure 16.28 Radiation patterns of the corporate-fed microstrip antenna array [92]. $\epsilon_r = 2.2$, $h = 1.59$ mm, $l = 10.08$ mm, $w = 11.79$ mm, $d_1 = 1.3$ mm, $d_2 = 3.93$ mm, $l_1 = 12.32$ mm, $l_2 = 18.48$ mm, $D_1 = 23.58$ mm, $D_2 = 22.40$ mm. $f = 9.42$ GHz.

example, the higher-order basis functions and the AIM have shown their advantages independently. We expect that the combination of both techniques can yield more improvement. Also, the AIM in the work is confined to 2D planar structures. The structures with vertical supporting can also be solved. Because most of the unknowns are on the planar surfaces, the matrix-vector multiplication associated with the vertical currents can be directly performed. The algorithm will maintain the computational complexity as $O(N \log N)$.

REFERENCES

1. K. S. Yee, "Numerical solution of initial boundary value problems involving Maxwell's equations in isotropic media," *IEEE Trans. Antennas Propagat.*, vol. 14, pp. 302–307, May 1966.

2. X. Zhang and K. K. Mei, "Time-domain finite difference approach to the calculation of frequency-dependent characteristics of microstrip discontinuities," *IEEE Trans. Microwave Theory Tech.*, vol. 36, pp. 1775–1787, Dec. 1988.

3. D. M. Sheen, S. M. Ali, M. D. Abouzahra, and J. A. Kong, "Application of the three-dimensional finite-difference time-domain method to the analysis of planar microstrip circuits," *IEEE Trans. Microwave Theory Tech.*, vol. 38, pp. 849–857, July 1990.

4. M. Fujii and W. J. R. Hoefer, "A three-dimensional Harr-wavelet-based multiresolution analysis similar to the FDTD method—Derivation and application," *IEEE Trans. Microwave Theory Tech.*, vol. 46, pp. 2463–2475, Dec. 1998.

5. J. M. Jin, *The Finite Element Method in Electromagnetics*, New York: Wiley, 1993.

6. J. S. Wang and R. Mittra, "Finite element analysis of MMIC structures and electronic packages using absorbing boundary conditions," *IEEE Trans. Microwave Theory Tech.*, vol. 42, pp. 441–449, March 1994.

7. J. G. Yook, N. I. Dib, and L. P. B. Katehi, "Characterization of high frequency interconnects using finite difference time domain and finite element methods," *IEEE Trans. Microwave Theory Tech.*, vol. 42, pp. 1727–1736, Sept. 1994.

8. A. C. Polycarpou, P. A. Tirkas, and C. A. Balanis, "The finite-element method for modeling circuits and interconnects for electronic packaging," *IEEE Trans. Microwave Theory Tech.*, vol. 45, pp. 1868–1874, Oct. 1997.

9. D. M. Pozar, "Input impedance and mutual coupling of rectangular microstrip antennas," *IEEE Trans. Antennas Propagat.*, vol. 30, pp. 1191–1196, Nov. 1982.

10. E. H. Newman and D. Forrai, "Scattering from a microstrip patch," *IEEE Trans. Antennas Propagat.*, vol. 35, pp. 245–251, March 1987.

11. T. S. Horng, N. G. Alexopoulos, S. C. Wu, and H. Y. Yang, "Full-wave spectral-domain analysis for open microstrip discontinuities of arbitrary shape including radiation and surface-wave losses," *Int. J. Microwave Millimeter Wave Comput. Aided Eng.*, vol. 2, pp. 224–240, 1992.

12. A. S. King and W. J. Bow, "Scattering from a finite array of microstrip patches," *IEEE Trans. Antennas Propagat.*, vol. 40, pp. 770–774, July 1992.

13. J. R. Mosig, "Arbitrarily shaped microstrip structures and their analysis with a mixed potential integral equation," *IEEE Trans. Microwave Theory Tech.*, vol. 36, pp. 314–323, Feb. 1988.

14. J. R. Mosig, "Integral equation techniques," in *Numerical Techniques for Microwave and Millimeter-Wave Passive Structures*, T. Itoh, Ed., New York: Wiley, 1988.

15. K. L. Wu, M. Spenuk, J. Litva, and D. G. Fang, "Theoretical and experimental study of feed network effects on the radiation pattern of series-fed microstrip antenna arrays," *IEE Proc.-H*, vol. 138, pp. 238–242, June 1991.

16. K. A. Michalski and D. Zheng, "Analysis of microstrip resonators of arbitrary shape," *IEEE Trans. Microwave Theory Tech.*, vol. 40, pp. 112–119, Jan. 1992.

17. D. I. Wu, D. C. Chang, and B. L. Brim, "Accurate numerical modeling of microstrip junctions and discontinuities," *Int. J. Microwave Millimeter Wave Comput. Aided Eng.*, vol. 1, pp. 48–58, 1991.

18. D. C. Chang and J. X. Zhang, "Electromagnetic modeling of passive circuit elements in MMIC," *IEEE Trans. Microwave Theory Tech.*, vol. 40, pp. 1741–1747, Sept. 1992.

19. F. Alonso-Monferrer, A. A. Kishk, and A. W. Glisson, "Green's functions analysis of planar circuits in a two-layer grounded medium," *IEEE Trans. Antennas Propagat.*, vol. 40, pp. 690–696, June 1992.

20. R. A. Kipp and C. H. Chan, "Triangular-domain basis functions for full-wave analysis of microstrip discontinuities," *IEEE Trans. Microwave Theory Tech.*, vol. 41, pp. 1187–1194, June/July 1993.

21. R. Gillard, J. Corre, M. Drissi, and J. Citerne, "A general treatment of matched terminations using integral equations–modeling and application," *IEEE Trans. Microwave Theory Tech.*, vol. 42, pp. 2545–2553, Dec. 1994.

22. E. K. L. Yeung, J. C. Beal, and Y. M. M. Antar, "Multilayer microstrip structure analysis with matched load simulation," *IEEE Trans. Microwave Theory Tech.*, vol. 43, pp. 143–149, Jan. 1995.

23. M. J. Tsai, F. D. Flaviis, O. Fordham, and N. G. Alexopoulos, "Modeling planar arbitrarily shaped microstrip elements in multilayered media," *IEEE Trans. Microwave Theory Tech.*, vol. 45, pp. 330–337, March 1997.

24. F. Ling and J. M. Jin, "Scattering and radiation analysis of microstrip antennas using discrete complex image method and reciprocity theorem," *Microwave Opt. Tech. Lett.*, vol. 16, pp. 212–216, Nov. 1997.

25. J. Zhao, S. Kapur, D. E. Long, and W. W.-M. Dai, "Efficient three-dimensional extraction based on static and full-wave layered Green's functions," *Proceedings of the 35th Design Automation Conference*, June 1998.

26. R. C. Hsieh and J. T. Kuo, "Fast full-wave analysis of planar microstrip circuit elements in stratified media," *IEEE Trans. Microwave Theory Tech.*, vol. 46, pp. 1291–1297, Sept. 1998.

27. T. J. Cui and W. C. Chew, "Fast evaluation of Sommerfeld integrals for EM scattering and radiation by three-dimensional buried objects," *IEEE Geosci. Remote Sensing*, vol. 37, pp. 887–900, March 1999.

28. D. G. Fang, J. J. Yang, and G. Y. Delisle, "Discrete image theory for horizontal electric dipole in a multilayer medium," *IEE Proc.-H*, vol. 135, pp. 297–303, Oct. 1988.

29. Y. L. Chow, J. J. Yang, D. G. Fang, and G. E. Howard, "A closed-form spatial Green's function for the thick microstrip substrate," *IEEE Trans. Microwave Theory Tech.*, vol. 39, pp. 588–592, March 1991.

30. J. J. Yang, Y. L. Chow, G. E. Howard, and D. G. Fang, "Complex images of an electric dipole in homogeneous and layered dielectrics between two ground planes," *IEEE Trans. Microwave Theory Tech.*, vol. 40, pp. 595–600, March 1992.

31. R. A. Kipp and C. H. Chan, "Complex image method for sources in bounded regions of multilayer structures," *IEEE Trans. Microwave Theory Tech.*, vol. 42, pp. 860–865, May 1994.

32. G. Dural and M. I. Aksun, "Closed-form Green's functions for general sources and stratified media," *IEEE Trans. Microwave Theory Tech.*, vol. 43, pp. 1545–1552, July 1995.

33. M. I. Aksun, "A robust approach for the derivation of closed-form Green's functions," *IEEE Trans. Microwave Theory Tech.*, vol. 44, pp. 651–658, May 1996.

34. C. H. Chan and R. A. Kipp, "Application of the complex image method to multilevel, multiconductor microstrip lines," *Int. J. Microwave Millimeter Wave Comput. Aided Eng.*, vol. 7, pp. 359–367, 1997.

35. C. H. Chan and R. A. Kipp, "Application of the complex image method to characterization of microstrip vias," *Int. J. Microwave Millimeter Wave Comput. Aided Eng.*, vol. 7, pp. 368–379, 1997.

36. F. Ling, D. Jiao, and J. M. Jin, "Efficient electromagnetic modeling of microstrip structures in multilayer media," *IEEE Trans. Microwave Theory Tech.*, vol. 47, pp. 1810–1818, Sept. 1999.

37. B. Hu and W. C. Chew, "Fast inhomogeneous plane wave algorithm for electromagnetic solutions in layered medium structures—2D case," *Radio Science*, vol. 35, no. 1, Jan.-Feb. 2000

38. K. A. Michalski and D. Zheng, "Electromagnetic scattering and radiation by surfaces of arbitrary shape in layered media, Part I: Theory," *IEEE Trans. Antennas Propagat.*, vol. 38, pp. 335–344, March 1990.

39. K. A. Michalski and J. R. Mosig, "Multilayered media Green's functions in integral equation formulations," *IEEE Trans. Antennas Propagat.*, vol. 45, pp. 508–519, March 1997.

40. F. Ling and J. M. Jin, "Discrete complex image method for general multilayer media," *IEEE Microwave Guided Wave Lett.*, vol. 10, pp. 400–402, Oct. 2000.

41. W. C. Chew, *Waves and Fields in Inhomogeneous Media*, Piscataway, NJ: IEEE Press, 1995.

42. Y. Hua and T. K. Sarkar, "Generalized pencil-of-function method for extracting poles of an EM system from its transient response," *IEEE Trans. Antennas Propagat.*, vol. 37, pp. 229–234, Feb. 1989.

43. W. H. Press, S. A. Teukolsky, W. T. Vettering, and B. P. Flannery, *Numerical Recipes in Fortran*, Cambridge: University Press, 1992.

44. S. M. Rao, D. R. Wilton, and A. W. Glisson, "Electromagnetic scattering by surface of arbitrary shape," *IEEE Trans. Antennas Propagat.*, vol. 30, pp. 409–418, May 1982.

45. J. C. Nedelec, "Mixed finite elements in R^3," *Numer. Mathem.*, vol. 35, pp. 315–341, 1980.

46. R. D. Graglia, D. R. Wilton, and A. F. Peterson, "Higher order interpolatory vector bases for computational electromagnetics," *IEEE Trans. Antennas Propagat.*, vol. 45, pp. 329–342, March 1997.

47. J. P. Webb, "Hierarchal vector basis functions of arbitrary order for triangular and tetrahedral finite element," *IEEE Trans. Antennas Propagat.*, vol. 47, pp. 1244–1253, Aug. 1999.

48. M. G. Duffy, "Quadrature over a pyramid or cube of integrands with a singularity at a vertex," *J. Numer. Anal.*, vol. 19, pp. 1260–1262, Dec. 1982.

49. J. C. Rautio, "A de-embedding algorithm for electromagnetics," *Int. J. Microwave Millimeter Wave Comput. Aided Eng.*, vol. 11, pp. 282–287, 1991.

50. F. Ling, K. Donepudi, and J. M. Jin, "Higher-order full-wave analysis of multilayer microstrip structures," *Microwave Opt. Tech. Lett.*, vol. 25, pp. 141–145, April 2000.

51. F. Ling, J. Liu, and J. M. Jin, "Efficient electromagnetic modeling of three-dimensional multilayer microstrip antennas and circuits," *IEEE Trans. Microwave Theory Tech.*, submitted for publication, 2000.

52. F. Tefiku and E. Yamashita, "Improved analysis method for multiport microstrip annular-ring power divider," *IEEE Trans. Microwave Theory Tech.*, vol. 42, pp. 376–381, March 1994.

53. L. T. Pillage and R. A. Rohrer, "Asymptotic waveform evaluation for timing analysis," *IEEE Trans. Computer-Aided Design*, vol. 9, pp. 352–366, April 1990.

54. E. Chiprout and M. S. Nakhla, "Analysis of interconnect networks using complex frequency hopping," *IEEE Trans. Computer-Aided Design*, vol. 14, pp. 186–200, Feb. 1995.

55. P. Feldmann and R. W. Freund, "Efficient linear circuit analysis by Padé approximation via the Lanczos process," *IEEE Trans. Computer-Aided Design*, vol. 14, pp. 639–649, May 1995.

56. M. Celik and A. C. Cangellaris, "Simulation of dispersive multiconductor transmission lines by Padé approximation via the Lanczos process," *IEEE Trans. Microwave Theory Tech.*, vol. 44, pp. 2525–2535, Dec. 1996.

57. M. A. Kolbehdari, M. Srinivasan, M. S. Nakhla, Q. J. Zhang, and R. Achar, "Simultaneous time and frequency domain solutions of EM problems using finite element and CFH techniques," *IEEE Trans. Microwave Theory Tech.*, vol. 44, pp. 1526–1534, Sept. 1996.

58. J. P. Zhang and J. M. Jin, "Preliminary study of AWE method for FEM analysis of scattering problems," *Microwave Opt. Tech. Lett.*, vol. 17, pp. 7–12, Jan. 1998.

59. C. J. Reddy, M. D. Deshpande, C. R. Cockrell, and F. B. Beck, "Fast RCS computation over a frequency band using method of moments in conjunction with asymptotic waveform evaluation technique," *IEEE Trans. Antennas Propagat.*, vol. 46, pp. 1229–1233, Aug. 1998.

60. J. E. Bracken, D. K. Sun, and Z. J. Cendes, "S-domain methods for simultaneous time and frequency characterization of electromagnetic devices," *IEEE Trans. Microwave Theory Tech.*, vol. 46, pp. 1277–1290, Sept. 1998.

61. G. J. Burke, E. K. Miller, S. Chakrabarthi, and K. Demarest, "Using model-based parameter estimation to increase the efficiency of computing electromagnetic transfer functions," *IEEE Trans. Magn.*, vol. 25, pp. 2807–2809, July 1989.

62. J. E. Pekarek and T. Itoh, "Use of frequency derivatives in the three-dimensional full-wave spectral domain technique," *IEEE Trans. Microwave Theory Tech.*, vol. 44, pp. 2466–2472, Dec. 1996.

63. O. Fordham, M. J. Tsai, and N. G. Alexopoulos, "Electromagnetic synthesis of overlap-gap-coupled microstrip filters," *IEEE MTT-S Int. Microwave Symp.*, pp. 1199–1202, 1995.

64. N. N. Bojarski, "k-space formulation of the electromagnetic scattering problem," Air Force Avionics Lab. Technical Report AFAL-TR-71-75, March 1971.

65. T. K. Sarkar, E. Arvas, and S. M. Rao, "Application of FFT and the conjugate gradient method for the solution of electromagnetic radiation from electrically large and small conducting bodies," *IEEE Trans. Antennas Propagat.*, vol. 34, pp. 635–640, May 1986.

66. T. J. Peters and J. L. Volakis, "Application of a conjugate gradient FFT method to scattering from thin planar material plates," *IEEE Trans. Antennas Propagat.*, vol. 36, pp. 518–526, April 1988.

67. M. F. Catedra, J. G. Cuevas, and L. Nuno, "A scheme to analyze conducting plates of resonant size using the conjugate gradient method and the fast Fourier transform," *IEEE Trans. Antennas Propagat.*, vol. 36, pp. 1744–1752, Dec. 1988.

68. M. F. Catedra and E. Gago, "Spectral domain analysis of conducting patches of arbitrary geometry in multilayer media using the CG-FFT method," *IEEE Trans. Antennas Propagat.*, vol. 38, pp. 1530–1536, Oct. 1990.

69. A. P. M. Zwamborn and P. M. van den Berg, "A weak form of the conjugate gradient FFT method for plate problems," *IEEE Trans. Antennas Propagat.*, vol. 39, pp. 224–228, Feb. 1991.

70. J. M. Jin and J. L. Volakis, "A biconjugate gradient solution for scattering by planar plates," *Electromagn.*, vol. 12, pp. 105–119, Jan.-March 1992.

71. Y. Zhuang, K. Wu, C. Wu, and J. Litva, "A combined full-wave CG-FFT method for rigorous analysis of large microstrip antenna array," *IEEE Trans. Antennas Propagat.*, vol. 44, pp. 102–109, Jan. 1996.

72. C. F. Wang, F. Ling, and J. M. Jin, "A fast full-wave analysis of scattering and radiation from large finite arrays of microstrip antennas" *IEEE Trans. Antennas Propagat.*, vol. 46, pp. 1467–1474, Oct. 1998.

73. E. Bleszynski, M. Bleszynski, and T. Jaroszewicz, "A fast integral-equation solver for electromagnetic scattering problem," *IEEE APS Int. Symp. Dig.*, pp. 416–419, 1994.

74. E. Bleszynski, M. Bleszynski, and T. Jaroszewicz, "AIM: Adaptive integral method for solving large-scale electromagnetic scattering and radiation problems," *Radio Science*, vol. 31, pp. 1225–1251, Sept.-Oct. 1996.

75. F. Ling, C. F. Wang, and J. M. Jin, "Application of adaptive integral method to scattering and radiation analysis of arbitrarily shaped planar structures," *J. Electromagn. Waves Appl.*, vol. 12, pp. 1021–1038, Aug. 1998.

76. F. Ling, C. F. Wang, and J. M. Jin, "An efficient algorithm for analyzing large-scale microstrip structures using adaptive integral method combined with discrete complex image method," *IEEE Trans. Microwave Theory Tech.*, vol. 48, pp. 832–839, May 2000.

77. C. H. Chan, C. M. Lin, L. Tsang, and Y. F. Leung, "A sparse-matrix/canonical grid method for analyzing microstrip structures," *IEICE Trans. Electron.*, vol. E80-C, pp. 1354–1359, Nov. 1997.

78. J. R. Phillips and J. K. White, "A precorrected-FFT method for electrostatic analysis of complicated 3-D structures," *IEEE Trans. Computer-Aided Design*, vol. 16, pp. 1059–1072, Oct. 1997.

79. T. Edwards, *Foundations for Microstrip Circuit Design*, Chichester, UK: Wiley, 1991.

80. F. Giannini, R. Sorrentino, and J. Vrba, "Planar circuit analysis of microstrip radial stub," *IEEE Trans. Microwave Theory Tech.*, vol. 32, pp. 1652–1655, Dec. 1984.

81. V. Rokhlin, "Rapid solution of integral equations of scattering theory in two dimensions," *J. Comput. Phys.*, vol. 86, pp. 414–439, Feb. 1990.

82. K. Nabors and J. White, "Fastcap: A multipole-accelerated 3-D capacitance extraction program," *IEEE Trans. Computer-Aided Design*, vol. 10, pp. 1447–1459, Nov. 1991.

83. K. Nabors and J. White, "Multipole-accelerated capacitance extraction for 3-D structures with multilayer dielectrics," *IEEE Trans. Circ. Syst.*, vol. 39, pp. 946–954, Nov. 1992.

84. R. Coifman, V. Rokhlin, and S. Wandzura, "The fast multipole method for the wave equation: A pedestrian prescription," *IEEE Antennas Propagat. Mag.*, vol. 35, pp. 7–12, June 1993.

85. J. M. Song and W. C. Chew, "Fast multipole method solution using parametric geometry," *Microwave Opt. Tech. Lett.*, vol. 7, pp. 760–765, Nov. 1994.

86. J. M. Song, C. C. Lu, and W. C. Chew, "Multilevel fast multipole algorithm for electromagnetic scattering by large complex objects," *IEEE Trans. Antennas Propagat.*, vol. 45, pp. 1488–1493, Oct. 1997.

87. J. M. Song, C. C. Lu, W. C. Chew, and S. W. Lee, "Fast Illinois Solver Code (FISC)," *IEEE Antennas Propagat. Mag.*, vol. 40, pp. 27–34, June 1998.

88. V. Jandhyala, E. Michielssen, and R. Mittra, "Multipole-accelerated capacitance computation for 3-D structures in a stratified dielectric medium using in a closed

form Green's function," *Int. J. Microwave Millimeter Wave Comput. Aided Eng.*, vol. 5, pp. 68–78, May 1995.

89. L. Gurel and M. I. Aksun, "Electromagnetic scattering solution of conducting strips in layered media using the fast multipole method," *IEEE Microwave Guided Wave Lett.*, vol. 6, pp. 277–279, Aug. 1996.

90. P. A. MacDonald and T. Itoh, "Fast simulation of microstrip structures using the fast multipole method," *Int. J. Numerical Modelling: Electron. Networks, Devices and Fields*, vol. 9, pp. 345–357, 1996.

91. J. S. Zhao, W. C. Chew, C. C. Lu, E. Michielssen, and J. M. Song, "Thin-stratified medium fast-multipole algorithm for solving microstrip structures," *IEEE Trans. Microwave Theory Tech.*, vol. 46, pp. 395–403, April 1998.

92. F. Ling, J. M. Song, and J. M. Jin, "Multilevel fast multipole algorithm for analysis of large-scale microstrip structures," *IEEE Microwave Guided Wave Lett.*, vol. 9, pp. 508–510, Dec. 1999.

17

The Steepest-Descent Fast Multipole Method

*Vikram Jandhyala, Eric Michielssen,
Balasubramaniam Shanker, and Weng Cho Chew*

17.1 INTRODUCTION

Quasi-planar structures are those whose transverse dimensions are much larger than their height. Efficient and accurate algorithms for analyzing electromagnetic interactions with such structures are important since they permit the characterization of several real-world electromagnetic analysis problems, including rough surface scattering, radiation from microstrip patch antennas, grating structure design, and the analysis of diffractive optical elements. In the past, algorithms have been developed for the rapid analysis of scattering from quasi-planar structures. Two such algorithms are the sparse-matrix flat-surface iterative algorithm [1] and the fast multipole method (FMM)–fast Fourier transform method [2, 3]. The computational complexity of both these techniques scales as $O(N \log N)$, where N is the number of basis functions used to discretize currents on scatterer surfaces.

In this chapter, an FMM variant with $O(N)$ complexity is developed that permits the rapid full-wave electromagnetic analysis of general quasi-planar structures. In its most general form, this technique, termed the steepest-descent fast multipole method (SDFMM), accelerates the solution of multiregion combined field integral equations

(CFIEs). SDFMM relies on a representation of the 3D Green's function in terms of a steepest-descent integral coupled with a 2D FMM-like multilevel inhomogeneous plane-wave expansion. SDFMM is applied to the analysis of several important electromagnetics problems, including scattering from perfectly conducting and dielectric rough surfaces, optical coupling in quantum-well infrared photodetectors, and radiation from microstrip antennas with finite substrates and ground planes. The material presented in this chapter is drawn from published work [4–8].

17.2 FIELD EVALUATION ON QUASI-PLANAR SURFACES

The solutions of integral equations pertinent to the analysis of electromagnetic scattering problems are accelerated and rendered memory efficient by FMM through fast method-of-moments (MOM) matrix-vector products. The physical meaning of a matrix-vector product in the MOM is the evaluation of fields due to sources of known strength. In this section, SDFMM is developed to rapidly evaluate fields on quasi-planar surfaces, with the intention of accelerating iterative solutions of the associated integral equations.

17.2.1 The Scalar Case

Consider a quasi-planar surface S residing in a homogeneous medium, and characterized by a height profile $z(x, y)$. The lateral dimensions of the surface are assumed to be bounded by $L^{(S)}$, and $H = |max\{z(x,y)\} - min\{z(x,y)\}|$, with $H \ll L^{(S)}$. Thus the entire surface S can be enclosed by a cuboid of volume $L^{(S)} \times L^{(S)} \times H$. Let $j_n(\mathbf{r})$ and $f_n(\mathbf{r})$, with $n = 1, ..., N$ and $\mathbf{r} \in S$, denote sets of N compactly supported scalar source and test functions, respectively. A time variation of $e^{-i\omega t}$ is assumed and suppressed throughout. The objective is to compute the "tested" fields given by

$$\sum_{n=1}^{N} \langle f_m, g * j_n \rangle, \quad m = 1, ..., N \tag{17.1a}$$

where

$$[g * x](\mathbf{r}) = \int_S ds' \, g(\mathbf{r}, \mathbf{r}') x(\mathbf{r}'), \quad \mathbf{r}, \mathbf{r}' \in S \tag{17.1b}$$

Here, $g(\mathbf{r}, \mathbf{r}')$ is

$$g(\mathbf{r}, \mathbf{r}') = \frac{e^{ik|\mathbf{r}-\mathbf{r}'|}}{4\pi|\mathbf{r}-\mathbf{r}'|} \tag{17.1c}$$

and k denotes the wavenumber. Clearly, a direct evaluation of the interactions in (17.1a) would require $O(N^2)$ operations.

A two-level implementation of SDFMM proceeds very much like that of a standard two-level FMM by embedding S into a block of dimensions $L^{(S)} \times L^{(S)} \times H$ and by partitioning this block into smaller blocks of dimensions $l \times l \times H$ (Figure 17.1). Because $H \ll L^{(S)}$, no partitioning is carried out in the z-direction. Also, z-coordinates of block centers are determined by the average heights of the portion of S present in each block. Based on this partitioning, an interaction $\langle f_m, g * j_n \rangle$ is termed a near-field interaction if the source j_n and the observer f_m reside in blocks that are separated by no more than B_{di} blocks in both lateral directions (Figure 17.1). Here, B_{di} is a fixed constant, typically less than three. All other interactions are termed far-field interactions. In SDFMM, near-field interactions are computed in the classical manner, using (17.1b). Far-field interactions, on the other hand, are

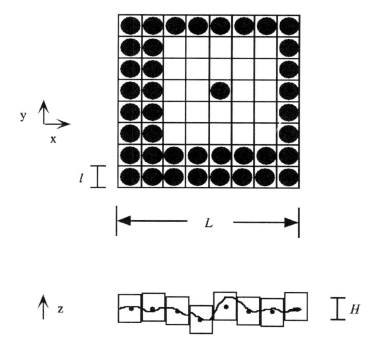

Figure 17.1 Top and side views of a quasi-planar surface S. SDFMM far-field interactions in a two-level implementation are shown. The surface S of lateral dimensions $L^{(S)} \times L^{(S)}$ has been divided into blocks of lateral dimensions $l \times l$. The z-coordinates of their centers (shown as small dark circles) are determined by the mean height of the portion of S they enclose. Circles represent plane-wave expansions. Blocks marked by black circles are in the far field of the central block marked by a lighter circle. In this figure, B_{di} equals two.

computed using a modified expansion of the Green's function. The three-dimensional dynamic scalar Green's function $g(\mathbf{r}, \mathbf{r}')$ can be expressed as a contour integral, using the Sommerfeld identity [9]:

$$g(\mathbf{r}, \mathbf{r}') = \frac{i}{8\pi} \int_{-\infty}^{\infty} dk_z e^{ik_z(z-z')} H_0^{(1)}(k_\rho|\boldsymbol{\rho} - \boldsymbol{\rho}'|) \tag{17.2}$$

where $k_\rho = \sqrt{k^2 - k_z^2}$, $\mathbf{r} = \hat{z}z + \boldsymbol{\rho}$, and $\mathbf{r}' = \hat{z}z' + \boldsymbol{\rho}'$. The Hankel function is factored as

$$H_0^{(1)}(k_\rho|\boldsymbol{\rho} - \boldsymbol{\rho}'|) = \hat{H}_0^{(1)}(k_\rho|\boldsymbol{\rho} - \boldsymbol{\rho}'|) e^{ik_\rho|\boldsymbol{\rho}-\boldsymbol{\rho}'|} \tag{17.3}$$

where $\hat{H}_0^{(1)}(k_\rho|\boldsymbol{\rho} - \boldsymbol{\rho}'|)$ is a slowly varying function for large arguments. Let $\mathbf{r}_s = \hat{z}z_s + \boldsymbol{\rho}_s$, and $\mathbf{r}_t = \hat{z}z_t + \boldsymbol{\rho}_t$ denote the centers of the source and observation blocks, respectively; similarly, let $\mathbf{r}_1 = \hat{z}z_t + \boldsymbol{\rho}$, and $\mathbf{r}_1' = \hat{z}z_s + \boldsymbol{\rho}'$. Using these definitions, (17.2) is rewritten as

$$g(\mathbf{r}, \mathbf{r}') = \frac{i}{8\pi} \int_{-\infty}^{\infty} dk_z e^{ik_z(\delta z - \delta z')} \hat{H}_0^{(1)}(k_\rho|\boldsymbol{\rho} - \boldsymbol{\rho}'|) \tag{17.4}$$
$$e^{ik_\rho|\boldsymbol{\rho}-\boldsymbol{\rho}'|+ik_z(z_t-z_s)}$$

where $\delta z = z - z_t$, and $\delta z' = z' - z_s$. Introducing $k_z = k_0 \cos\alpha$ and $k_\rho = k_0 \sin\alpha$ yields

$$g(\mathbf{r}, \mathbf{r}') = -\frac{ik_0}{8\pi} \int_\Gamma d\alpha \sin\alpha e^{ik_z(\delta z - \delta z')} \hat{H}_0^{(1)}(k_0 \sin\alpha|\boldsymbol{\rho} - \boldsymbol{\rho}'|) \tag{17.5}$$
$$e^{ik_0|\mathbf{r}_1 - \mathbf{r}_1'|\cos(\alpha - \theta_1)}$$

where $\theta_1 = \cot^{-1}[(z_t - z_s)/|\boldsymbol{\rho} - \boldsymbol{\rho}'|]$, with $0 \le \theta_1 \le \pi$. The saddle point is at $\alpha = \theta_1$. In anticipation of the aggregation of fields due to sources residing in a source block and their disaggregation over similar observation blocks, the following steepest-descent path (SDP) is defined:

$$ik_0|\mathbf{r}_1 - \mathbf{r}_1'|\cos(\alpha - \theta_c) = ik_0|\mathbf{r}_1 - \mathbf{r}_1'| - s^2 \tag{17.6}$$

Here $\theta_c = \cot^{-1}[(z_t - z_s)/|\boldsymbol{\rho}_t - \boldsymbol{\rho}_s|]$. Deforming the integrand along the SDP yields

$$g(\mathbf{r}, \mathbf{r}') = -\frac{ik_0}{8\pi} e^{ik_0|\mathbf{r}_1 - \mathbf{r}_1'|} \int_{-\infty}^{\infty} ds \frac{d\alpha}{ds} \sin\alpha \, e^{ik_z(\delta z - \delta z')}$$
$$\hat{H}_0^{(1)}(k_0 \sin\alpha|\boldsymbol{\rho} - \boldsymbol{\rho}'|) e^{ik_0|\mathbf{r}_1 - \mathbf{r}_1'|[\cos(\alpha-\theta_1)-\cos(\alpha-\theta_c)]} e^{-s^2} \tag{17.7}$$

The function $\hat{H}_0^{(1)}(k_0 \sin\alpha|\boldsymbol{\rho} - \boldsymbol{\rho}'|) e^{ik_0|\mathbf{r}_1 - \mathbf{r}_1'|[\cos(\alpha-\theta_1)-\cos(\alpha-\theta_c)]}$ varies slowly for large arguments. Efficient quadrature rules using a small number $n_{(sd)}$ of sampling points can be developed for computing the above integral, as the integrand decays exponentially for $s \to \pm\infty$ and both δz and $\delta z'$ are small for a quasi-planar structure. In other words, the above integral can be efficiently approximated as

$$g(\mathbf{r},\mathbf{r}') = \frac{i}{8\pi} e^{ik_0|\mathbf{r}_1-\mathbf{r}'_1|} \sum_{j=1}^{n_{(sd)}} w_j^{(sd)} k_\rho^{(j)} e^{ik_z^{(j)}(\delta z - \delta z')} \\ \hat{H}_0^{(1)}\left(k_\rho^{(j)}|\boldsymbol{\rho}-\boldsymbol{\rho}'|\right) e^{ik_0|\mathbf{r}_1-\mathbf{r}'_1|[\cos(\alpha_j-\theta_1)-\cos(\alpha_j-\theta_c)]} e^{-s_j^2} \quad (17.8)$$

where $k_\rho^{(j)} = k_0 \sin\alpha_j$, $k_z^{(j)} = k_0 \cos\alpha_j$, and $w_j^{(sd)}$ and s_j are the weights and abscissas of the quadrature rule (the index j labels the integration point).

Combining (17.3), (17.6), and (17.8) yields

$$g(\mathbf{r},\mathbf{r}') = \frac{i}{8\pi} \sum_{j=1}^{n_{(sd)}} w_j^{(sd)} k_\rho^{(j)} H_0^{(1)}\left(k_\rho^{(j)}|\boldsymbol{\rho}-\boldsymbol{\rho}'|\right) e^{ik_z^{(j)}(z-z')} \quad (17.9)$$

Within the context of an MOM scheme, the Hankel function in (17.9) can be computed efficiently through a generalization of the free-space 2D FMM [10, 11] to an inhomogeneous plane-wave basis. Sources lying in each block are represented by plane-wave expansions located at the center of each block, irrespective of the z-location of the source. The Hankel function can then be expressed as

$$H_0^{(1)}\left(k_\rho^{(j)}|\boldsymbol{\rho}-\boldsymbol{\rho}'|\right) = \frac{1}{2\pi} \int_0^{2\pi} d\phi\, e^{ik_\rho^{(j)}(\boldsymbol{\rho}-\boldsymbol{\rho}_t)\cdot\hat{s}} \\ T\left(k_\rho^{(j)},\hat{s},\boldsymbol{\rho}_t-\boldsymbol{\rho}_s\right) e^{ik_\rho^{(j)}(\boldsymbol{\rho}_s-\boldsymbol{\rho}')\cdot\hat{s}} \\ \cong \frac{1}{2\pi} \sum_{j'=1}^{P} w_{j'}^{(fmm)} e^{ik_\rho^{(j)}(\boldsymbol{\rho}-\boldsymbol{\rho}_t)\cdot\hat{s}_{j'}} \\ T\left(k_\rho^{(j)},\hat{s}_{j'},\boldsymbol{\rho}_t-\boldsymbol{\rho}_s\right) e^{ik_\rho^{(j)}(\boldsymbol{\rho}_s-\boldsymbol{\rho}')\cdot\hat{s}_{j'}} \quad (17.10a)$$

$$T\left(k_\rho^{(j)},\hat{s},\boldsymbol{\rho}_t-\boldsymbol{\rho}_s\right) = \sum_{p=-P}^{P} H_p^{(1)}(k_\rho^{(j)}|\boldsymbol{\rho}_t-\boldsymbol{\rho}_s|)\, e^{-ip[\theta-\phi-\pi/2]} \quad (17.10b)$$

where P is the number of integration points required for the azimuthal spectral integration, $w_{j'}^{(fmm)}$ are the quadrature weights defined above, and T is the translation operator, which depends on the complex wavenumber and spectral angle and on the displacement between source and observation blocks. Also, $\hat{s} = \hat{x}\cos\phi + \hat{y}\sin\phi$, and $\cos\theta = \hat{x}\cdot(\boldsymbol{\rho}_t-\boldsymbol{\rho}_s)/|\boldsymbol{\rho}_t-\boldsymbol{\rho}_s|$.

Finally, using the plane-wave basis, the tested fields can be expressed as

$$\langle f_m, g * j_n \rangle = \frac{i}{16\pi^2} \sum_{j=1}^{n_{(sd)}} \sum_{j'=1}^{P} w_j^{(sd)} w_{j'}^{(fmm)} k_\rho^{(j)}$$
$$\int_S d\mathbf{r}\, f_m e^{i\mathbf{k}_0^{(jj')} \cdot (\mathbf{r}-\mathbf{r}_t)} T\left(k_\rho^{(j)}, \hat{s}_{j'}, \boldsymbol{\rho}_t - \boldsymbol{\rho}_s\right) \quad (17.11a)$$
$$\int_S d\mathbf{r'}\, j_n e^{i\mathbf{k}_0^{(jj')} \cdot (\mathbf{r}_s-\mathbf{r'})}$$

where
$$\mathbf{k}_0^{(jj')} = k_\rho^{(j)} \hat{s}_{j'} + k_z^{(j)} \hat{z} \quad (17.11b)$$

The above interactions are depicted schematically in Figure 17.2.

When a matrix-vector product is computed, plane-wave spectra from multiple sources residing in a source block are combined, and this aggregrate spectrum is used to evaluate fields throughout an observer block, thereby decoupling individual sources and observers. Although the technique described above results in substantial CPU time and memory savings, further dramatic savings can be obtained through a multilevel SDFMM. In such a technique, S is recursively enclosed in blocks by hierarchically partitioning a block (the *parent*) at a coarse level into four blocks (the *children*) at a finer level. Plane-wave expansions are shifted to centers of parent blocks, and incoming spectra are shifted to centers of child blocks. Such an operation is termed an FMM traversal. In general, plane-wave spectrum information at different levels is recycled, as in standard FMMs [10, 11]. Complexity estimates of the multilevel SDFMM are discussed later.

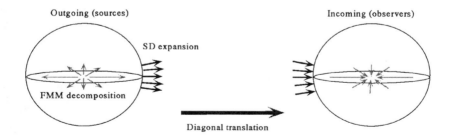

Figure 17.2 Schematic of the SDFMM decomposition. SDP integration is shown by the bold arrows, and azimuthal FMM spectral integration is shown by lighter arrows. The long bold arrow represents the diagonal translation operators. At an observation block, another combined SDFMM integration is used to find field values in the block.

17.2.2 The Vector Case

By building on the results obtained for the scalar case, SDFMM can be used to accelerate Green's functions computations involving vector Green's functions and sources. In what follows, let ϵ and μ represent the permittivity and permeability, respectively, of a homogeneous region. Also, $k = \omega\sqrt{\epsilon\mu}$, and $\eta = \sqrt{\frac{\mu}{\epsilon}}$. Consider $2N$ compactly supported vector sources $\mathbf{j}_n(\mathbf{r})$, and $2N$ similar observers $\mathbf{f}_n(\mathbf{r})$, $n = 1, ..., 2N$, residing on the surfaces bounding the homogeneous region.

The objective is to accelerate the computation of the expressions given by

$$\sum_{n=1}^{N} \langle \mathbf{f}_m, \mathcal{L}\,\mathbf{j}_n - \mathcal{K}\,\mathbf{j}_{n+N} \rangle, \quad m = 1, ..., 2N \tag{17.12a}$$

and

$$\sum_{n=1}^{N} \langle \mathbf{f}_m, \mathcal{K}\,\mathbf{j}_n + \frac{1}{\eta^2}\mathcal{L}\,\mathbf{j}_{n+N} \rangle, \quad m = 1, ..., 2N \tag{17.12b}$$

where the operators $\mathcal{L}$ and $\mathcal{K}$ are defined by

$$\mathcal{L}\,\mathbf{X}(\mathbf{r}) = \int_S ds' \left(-i\omega\mu\,\mathbf{X}(\mathbf{r}') + \frac{i}{\omega\epsilon}\nabla\nabla'\cdot\mathbf{X}(\mathbf{r}') \right) g(\mathbf{r}, \mathbf{r}') \tag{17.12c}$$

$$\mathcal{K}\,\mathbf{X}(\mathbf{r}) = \int_S ds'\,\mathbf{X}(\mathbf{r}') \times \nabla g(\mathbf{r}, \mathbf{r}') \tag{17.12d}$$

and $g(\mathbf{r}, \mathbf{r}')$ is the homogeneous scalar Green's function

$$g(\mathbf{r}, \mathbf{r}') = \frac{e^{ik|\mathbf{r}-\mathbf{r}'|}}{4\pi|\mathbf{r}-\mathbf{r}'|} \tag{17.12e}$$

In this case, SDFMM can be used to represent the action of the operators $\mathcal{L}$ and $\mathcal{K}$ in the following manner:

$$\langle \mathbf{f}_m, \mathcal{L}\,\mathbf{j}_n \rangle = \frac{k\eta}{16\pi^2} \sum_{j=1}^{n_{(sd)}} \sum_{j'=1}^{P} w_j^{(sd)} w_{j'}^{(fmm)} k_\rho^{(j)}$$

$$\int_S d\mathbf{r}\,\mathbf{f}_m e^{i\mathbf{k}^{(jj')}\cdot(\mathbf{r}-\mathbf{r}_t)}$$

$$T\left(k_\rho^{(j)}, \hat{s}_{j'}, \boldsymbol{\rho}_t - \boldsymbol{\rho}_s\right) \left(\bar{\mathbf{I}} - \frac{\mathbf{k}^{(jj')}\mathbf{k}^{(jj')}}{k^2}\right) \tag{17.13a}$$

$$\int_S d\mathbf{r}'\,\mathbf{j}_n e^{i\mathbf{k}^{(jj')}\cdot(\mathbf{r}_s-\mathbf{r}')}$$

and

$$\langle \mathbf{f}_m, \mathcal{K}\, \mathbf{j}_n \rangle = -\frac{1}{16\pi^2} \sum_{j=1}^{n_{(sd)}} \sum_{j'=1}^{P} w_j^{(sd)} w_{j'}^{(fmm)}\, k_\rho^{(j)}$$
$$\int_S d\mathbf{r}\, (\mathbf{f}_m \times \mathbf{k}^{(jj')}) e^{i\mathbf{k}^{(jj')} \cdot (\mathbf{r} - \mathbf{r}_t)}$$
$$T\left(k_\rho^{(j)}, \hat{s}_{j'}, \boldsymbol{\rho}_t - \boldsymbol{\rho}_s\right)$$
$$\int_S d\mathbf{r}'\, \mathbf{j}_n e^{i\mathbf{k}^{(jj')} \cdot (\mathbf{r}_s - \mathbf{r}')} \qquad (17.13\mathrm{b})$$

where

$$\mathbf{k}^{(jj')} = k_\rho^{(j)} \hat{s}_{j'} + k_z^{(j)} \hat{z} \qquad (17.13\mathrm{c})$$

The multilevel SDFMM traversal will proceed exactly in the same manner as for the scalar case, with the distinction that two far-field components of the vector inhomogeneous plane-wave spectra are used in place of scalar spectra.

17.3 COMPUTATIONAL COMPLEXITY ESTIMATES

To analyze the computational complexity and memory requirements of SDFMM, consider a quasi-planar surface S of dimensions $L^{(S)} \times L^{(S)}\, \lambda^2$ with a maximum peak-to-peak height of H, on which a source distribution is modeled using N basis functions. Let the number of SDFMM levels be f. At any intermediate level g, the number of blocks is denoted by $b^{(g)}$, and the size of each block is $l^{(g)} \times l^{(g)}\, \lambda^2$. Furthermore, the number of points used for the steepest descent integration and the number of spectral angles required by SDFMM are $n_{(sd)}^{(g)}$ and $P^{(g)}$, respectively. Finally, (g-independent) constants are denoted by C_i, where i is some integer. The following relations hold:

$$b^{(g)} = (L^{(S)}/l^{(g)})^2 \qquad (17.14\mathrm{a})$$
$$l^{(g-1)} = 2 l^{(g)} \qquad (17.14\mathrm{b})$$

and

$$f = (\log_2(L^{(S)}/l^{(f)})) \qquad (17.14\mathrm{c})$$

Assuming that the basis functions are uniformly distributed over the surface, an upper bound I_d on the total number of near-field interactions is obtained:

$$I_d = (2\, B_{di} + 1)^2\, N^2\, (l^{(f)}/L^{(S)})^2 \qquad (17.15\mathrm{a})$$

Moreover, $[L^{(S)}]^2 = C_1\, N$; therefore,

$$I_d = (2\, B_{di} + 1)^2\, N\, (l^{(f)})^2 / C_1 \qquad (17.15\mathrm{b})$$

The initial step in the computation of the far-field contributions is the projection of the sources onto inhomogeneous plane waves defined with respect to the centers of finest-level blocks. The next step is the recursive generation of all higher-level spectra; this step involves spectrum interpolation and translation of plane-wave reference points from child to parent centers. These operations are carried out while *ascending* the SDFMM tree. Far field expansions will exist at all levels finer than and including a level $g(B_{di})$, determined by the value of B_{di}. At each of these levels, in-level translations using the diagonalized translation operator turn outgoing plane-wave spectra into incoming ones. When *descending* the SDFMM tree, incoming spectra are anterpolated and their reference points shifted from parent to child centers. At the finest level, incoming plane-wave spectra are projected onto observers to obtain actual field values. The overall far-field computation cost is estimated as follows. The total cost I_p of projecting sources onto plane waves and vice versa and is given by

$$I_p = C_2 \, n_{(sd)}^{(f)} \, P^{(f)} \, N \tag{17.16}$$

The cost $I_c^{(g)}$ of all center-to-center translations and interpolation/anterpolation at level g is

$$I_c^{(g)} = C_3 \, n_{(sd)}^{(g)} \, P^{(g)} b^{(g)} \tag{17.17}$$

and the cost of in-level translations $I_t^{(g)}$ is given by

$$I_t^{(g)} = C_4 \, n_{(sd)}^{(g)} \, P^{(g)} \, (2 \, B_{di} + 1)^2 \, b^{(g)} \tag{17.18}$$

A simplistic assumption about $n_{(sd)}^{(g)}$ is that it is independent of the level g and hence equals $n_{(sd)}^{(f)}$. In fact, as explained later, the number of points for SDP integration progressively reduces initially at coarser levels before reaching a saturation value. However, assuming $n_{(sd)}^{(g)} = n_{(sd)}^{(f)}$ provides a simple upper bound on associated costs. The number of spectral angles required depends on the block size and roughly doubles as the block size doubles [12], or $P^{(g)} = 2 \, P^{(g+1)}$. The overall cost of SDFMM is

$$I_{SDFMM} = I_d + I_p + \sum_{g=g(B_{di})}^{f} \left(I_c^{(g)} + I_t^{(g)} \right) \tag{17.19a}$$

$$I_{SDFMM} = (2 \, B_{di} + 1)^2 \, N \, \left(l^{(f)} \right)^2 / C_1 \\ + n_{(sd)}^{(f)} \, P^{(f)} \, N \, \left[C_5 + C_6 \, (2 \, B_{di} + 1)^2 \right] \tag{17.19b}$$

Therefore,

$$I_{SDFMM} = O(N) \tag{17.19c}$$

Not only is the cost of each matrix-vector product of $O(N)$, but the total memory requirements (storing near-field interactions and incoming-outgoing spectra at all levels) are also of $O(N)$. Owing to exponential decay characteristics of the SDP integrand and FMM integration rules, the computational cost for a matrix-vector product is $O(N \ln \epsilon^{-1})$ for an arbitrary and fixed error tolerance of ϵ.

The above complexity calculation assumes a single-region multilevel SDFMM; the extension to the multiregion case is straightforward and essentially involves an added multiplicative factor because distinct SDFMM traversals are performed for each region.

The $O(N)$ cost of SDFMM can be intuitively explained in the following manner. The number of inhomogeneous plane-wave components $n_{(sd)}^{(g)}$ needed to perform the SDP integration at a given level g is proportional to the angle which the observation region subtends as seen from the source region; if the observation region moves closer, more points will be required in the SDP integration rule. If block sizes and minimum block separations are $fixed$ at the finest level, then, clearly, at any coarser level, the angle subtended by an observation region is smaller than that at a finer level. In conclusion, the number of SDP integration points at any level is never larger than that at the finest level. Increasing the size of the rough surface, while keeping surface roughness and finest-level block sizes constant, will not increase the required number of SDP integration points. The cost of SDP integration per block is thus bounded, independently of the number of unknowns. In fact, the angle roughly halves each time one moves up a level; hence, the number of points required for SDP integration roughly decreases initially by a factor of two as one moves to a coarser level, until it saturates to a small constant value. The remaining contribution to the computational complexity of SDFMM arises from the multilevel two-dimensional FMM. The computational labor required per block roughly doubles as the block size doubles [13, 14]. Here, block size refers to the linear dimensions of a block. The increase in computation is brought about by an increase in the number of sampling points $n_{(sd)}^{(g)} P^{(g)}$ required to correctly represent fields due to sources within a block. All four children of an SDFMM parent block are nonempty. Therefore, the total computation per level diminishes by a factor of two at a coarser level, yielding an overall cost proportional to N. The use of a standard three-dimensional FMM in place of SDFMM would result in an $O(N \log N)$ complexity. This is because the total number of harmonics required increases by a factor of four as one goes to a coarser level (proportional to the surface area of the block). Thus, the computation per level remains constant and proportional to N, making the overall cost over $\log N$ levels scale as $O(N \log N)$. The extra $\log N$ term appears because a standard FMM requires that fields be expanded in a plane-wave basis over all 4π steradians as oppposed to a narrow range of angles near the horizontal plane as in SDFMM.

17.4 SCATTERING FROM RANDOM ROUGH SURFACES

The analysis of three-dimensional scattering from two-dimensional rough surfaces is a complex vector problem that requires efficient numerical techniques. For large rough surfaces, solution of the MOM system using direct inversion is practically impossible due to CPU time and memory constraints. In this section, SDFMM is used to rapidly analyze scattering from perfectly electrically conducting (PEC) and penetrable rough surfaces.

17.4.1 Model Development

Random rough surfaces with a Gaussian distribution are generated using a two-step process [2, 15]. First, an uncorrelated Gaussian distribution on a discrete two-dimensional regular grid is assembled. Next, this distribution is filtered in the spectral domain; the filter used also has a Gaussian profile. The parameters associated with this process are the variance σ of the zero-mean Gaussian generator and the correlation length l_c of the filter. The resulting rough surface height $z(x,y)$ follows the statistics

$$\langle z(x,y) \rangle = 0 \qquad (17.20a)$$

$$\langle z(x,y)z(x',y') \rangle = \sigma^2 e^{-\left[(x-x')^2 + (y-y')^2\right]/l_c^2} \qquad (17.20b)$$

where $\langle . \rangle$ denotes an ensemble average.

For a finite-surface sample S to be an accurate representation of an infinite interface, a spatially limited excitation is employed so that edge effects are avoided. Specifically, a Gaussian-weighted superposition of plane waves [2, 15] is used with

$$\mathbf{E}^{inc}(\mathbf{r}) = \frac{2\pi W^2}{[L^{(S)}]^2} \sum_{|\mathbf{K}| \leq k_1} \hat{e}^{inc}(\mathbf{K}, k_z) e^{i\mathbf{K} \cdot \boldsymbol{\rho}} e^{-ik_z z} e^{-|\mathbf{K}-\mathbf{K}_0|^2 W^2/2} \qquad (17.21)$$

and

$$\hat{e}^{inc}(K_x, K_y, k_z) = \frac{1}{\sqrt{K_x^2 + k_z^2}} [k_z, 0, K_x] \qquad (17.22)$$

where $\mathbf{K}_0$ is the average wave vector in the transverse plane. Also, $\mathbf{r} = \hat{z}z + \boldsymbol{\rho}$, and $\mathbf{K} = (K_x, K_y)$, with K_x and K_y multiples of $2\pi/L^{(S)}$. The dispersion relation $K_x^2 + K_y^2 + k_z^2 = k_1^2$ is satisfied, where k_1 is the wavenumber in the homogeneous medium (typically free space) that interfaces the rough surface. Typically, $W = L^{(S)}/4$ [2].

17.4.2 Integral Equation Formulations

For perfectly conducting surfaces, an electric field integral equation (EFIE) is utilized to analyze the field scattered by the surface S when excited by $\mathbf{E}^{inc}(\mathbf{r})$:

$$\mathbf{E}^{scat}(\mathbf{r}, \mathbf{J})\big|_{tan} = -\mathbf{E}^{inc}(\mathbf{r})\big|_{tan} \qquad \mathbf{r}, \mathbf{r}' \in S \qquad (17.23a)$$

where $|_{tan}$ selects the component tangential to S, $\mathbf{E}_{scat}(\mathbf{r}, \mathbf{J})$ denotes the scattered field generated by the surface currents $\mathbf{J}$ and can be expressed as

$$\mathbf{E}^{scat}(\mathbf{r}, \mathbf{J}) = -\mathcal{L}\,\mathbf{J}(\mathbf{r}) \qquad \mathbf{r} \in S \qquad (17.23b)$$

In this expression, the operator $\mathcal{L}$ is as defined in (17.12c). The solution of the EFIE yields the current density $\mathbf{J}(\mathbf{r})$.

To formulate integral equations for a dielectric rough surface, we follow the approach of [16, 17]. As before, a temporal dependency of $e^{-i\omega t}$ is assumed and suppressed. A finite rough surface S, of dimensions $L^{(S)} \times L^{(S)}$, formed at the interface of two dielectric half spaces is assumed. Incident electric and magnetic fields $\mathbf{E}^{inc}(\mathbf{r})$ and $\mathbf{H}^{inc}(\mathbf{r})$ excite S from Region 1, and equivalent electric and magnetic surface currents $\mathbf{J}(\mathbf{r})$ and $\mathbf{M}(\mathbf{r})$ are impressed on the rough surface. In what follows, ϵ_q and μ_q, with $q = 1, 2$, represent the permittivity and permeability, respectively, of the half spaces separated by the surface S. Also, $k_q = \omega\sqrt{\epsilon_q \mu_q}$, and $\eta_q = \sqrt{\frac{\mu_q}{\epsilon_q}}$.

To obtain $\mathbf{J}(\mathbf{r})$ and $\mathbf{M}(\mathbf{r})$, the PMCHWT formulation enforces the continuity of the tangential electric and magnetic field components across S:

$$\mathbf{E}^{inc}(\mathbf{r})\big|_{tan} = (\mathcal{L}_1 + \mathcal{L}_2)\mathbf{J}(\mathbf{r})\big|_{tan} - (\mathcal{K}_1 + \mathcal{K}_2)\mathbf{M}(\mathbf{r})\big|_{tan} \qquad (17.24a)$$

$$\mathbf{H}^{inc}(\mathbf{r})\big|_{tan} = (\mathcal{K}_1 + \mathcal{K}_2)\mathbf{J}(\mathbf{r})\big|_{tan} + \left(\frac{1}{\eta_1^2}\mathcal{L}_1 + \frac{1}{\eta_2^2}\mathcal{L}_2\right)\mathbf{M}(\mathbf{r})\big|_{tan} \qquad (17.24b)$$

where the operators $\mathcal{L}_q$ and $\mathcal{K}_q$ are similar to those defined in (17.12c) and (17.12d), with an additional index $q = 1, 2$ signifying the region:

$$\mathcal{L}_q \mathbf{X}(\mathbf{r}) = \int_S ds' \left(-i\omega\mu_q\,\mathbf{X}(\mathbf{r}') + \frac{i}{\omega\epsilon_q}\nabla\nabla'\cdot\mathbf{X}(\mathbf{r}')\right) g_q(\mathbf{r}, \mathbf{r}') \qquad (17.24c)$$

$$\mathcal{K}_q \mathbf{X}(\mathbf{r}) = \int_S ds'\,\mathbf{X}(\mathbf{r}') \times \nabla g_q(\mathbf{r}, \mathbf{r}') \qquad (17.24d)$$

and $g_q(\mathbf{r}, \mathbf{r}')$ is the scalar Green's function for Region q:

$$g_q(\mathbf{r}, \mathbf{r}') = \frac{e^{ik_q|\mathbf{r}-\mathbf{r}'|}}{4\pi|\mathbf{r}-\mathbf{r}'|} \qquad (17.24e)$$

The solution of the EFIE of (17.23) for perfectly conducting surfaces and of the PMCHWT of (17.24) for dielectric surfaces yields the electric, or electric and magnetic surface current densities $\mathbf{J}(\mathbf{r})$ and $\mathbf{M}(\mathbf{r})$, respectively. These current densities can be used to evaluate the scattered field $\mathbf{E}^{scat}(\mathbf{r})$ required in the computation of the bistatic RCS, which is given by

$$\sigma_{\gamma\delta}(\theta,\phi) = \lim_{r \to \infty} \frac{4\pi r^2 S_\delta(\theta,\phi,\mathbf{r})}{P_\gamma^{inc}} \qquad (17.25)$$

Here, P_γ^{inc} is the incident beam power, $S_\delta(\mathbf{r})$ is the scattered power density, and γ and δ denote the polarization of the incident and the considered scattered electric field, respectively. In a Monte Carlo simulation, the variance of $\mathbf{E}^{scat}(\mathbf{r})$ over an ensemble of rough surfaces possessing the same l_c and σ is used to compute the noncoherent part of the bistatic scattering coefficient [2, 18, 19].

17.4.3 SDFMM-Based Solutions

To solve the EFIE in (17.23), the unknown current density $\mathbf{J}$ is approximated in terms of a linear combination of basis functions $\mathbf{j}_n, n = 1, ..., N$ as

$$\mathbf{J}(\mathbf{r}') \cong \sum_{n=1}^{N} I_n \mathbf{j}_n(\mathbf{r}') \qquad (17.26a)$$

Substituting (17.23b), (17.23c), and (17.26a) into (17.23a), and testing the resulting equation with functions $\mathbf{f}_m, m = 1, ..., N$, results in a system of equations

$$\overline{\mathbf{Z}} \cdot \mathbf{I} = \mathbf{V} \qquad (17.26b)$$

where $\mathbf{I}$ represents the vector of unknown coefficients I_n,

$$Z_{mn} = \langle \mathbf{f}_m, -\mathbf{E}_{scat}(\mathbf{r},\mathbf{j}_n) \rangle \qquad (17.26c)$$

$$V_m = \langle \mathbf{f}_m, \mathbf{E}_{inc}(\mathbf{r}) \rangle \qquad (17.26d)$$

and $\langle,\rangle$ denotes surface integration. The Rao-Wilton-Glisson (RWG) basis [20] is utilized for the functions $\mathbf{j}_n$ and $\mathbf{f}_m$. SDFMM is used to accelerate the iterative solution of the MOM system by the application of (17.13a) in conjunction with the multilevel algorithm described in the previous section.

To solve the PMCHWT in (17.24), the unknown current densities $\mathbf{J}(\mathbf{r})$ and $\mathbf{M}(\mathbf{r})$ are approximated in terms of linear combinations of a set of basis functions $\mathbf{j}_n, n = 1, ..., N$ as

$$\mathbf{J}(\mathbf{r}) \cong \sum_{n=1}^{N} I_n^1 \mathbf{j}_n(\mathbf{r}) \qquad (17.27a)$$

$$\mathbf{M}(\mathbf{r}) \cong \sum_{n=1}^{N} I_n^2 \mathbf{j}_n(\mathbf{r}) \tag{17.27b}$$

Substituting (17.27a) and (17.27b) into (17.24a) and (17.24b), and testing the latter two equations with functions $\mathbf{f}_m$, for $m = 1, \ldots, N$, results in a $2N \times 2N$ system of equations:

$$\bar{\mathbf{Z}} \cdot \mathbf{I} = \mathbf{V} \tag{17.28}$$

Here, the MOM matrix $\bar{\mathbf{Z}}$ has the form

$$\bar{\mathbf{Z}} = \begin{bmatrix} \bar{\mathbf{Z}}^{11} & \bar{\mathbf{Z}}^{12} \\ \bar{\mathbf{Z}}^{21} & \bar{\mathbf{Z}}^{22} \end{bmatrix} \tag{17.29a}$$

with the entries of the four $N \times N$ submatrices given by

$$Z_{mn}^{11} = \langle \mathbf{f}_m, (\mathcal{L}_1 + \mathcal{L}_2)\mathbf{j}_n \rangle \tag{17.29b}$$

$$Z_{mn}^{12} = \langle \mathbf{f}_m, -(\mathcal{K}_1 + \mathcal{K}_2)\mathbf{j}_n \rangle \tag{17.29c}$$

$$Z_{mn}^{21} = \langle \mathbf{f}_m, (\mathcal{K}_1 + \mathcal{K}_2)\mathbf{j}_n \rangle \tag{17.29d}$$

$$Z_{mn}^{22} = \langle \mathbf{f}_m, (\frac{1}{\eta_1^2}\mathcal{L}_1 + \frac{1}{\eta_2^2}\mathcal{L}_2)\mathbf{j}_n \rangle \tag{17.29e}$$

Also, the vector of coefficients $\mathbf{I}$ is

$$\mathbf{I} = \begin{bmatrix} \mathbf{I}^1 \\ \mathbf{I}^2 \end{bmatrix} \tag{17.30}$$

and the excitation vector $\mathbf{V}$ is given by

$$\mathbf{V} = \begin{bmatrix} \mathbf{V}^1 \\ \mathbf{V}^2 \end{bmatrix} \tag{17.31a}$$

where

$$V_m^1 = \langle \mathbf{f}_m, \mathbf{E}^{inc}(\mathbf{r}) \rangle \tag{17.31b}$$

$$V_m^2 = \langle \mathbf{f}_m, \mathbf{H}^{inc}(\mathbf{r}) \rangle \tag{17.31c}$$

The RWG basis [20] is used for both $\mathbf{j}_n$ and $\mathbf{f}_m$.

The iterative solution of the above equations can be accelerated using the SDFMM form similar to that in (17.13a) and (17.13b), with an added index $q = 1, 2$ signifying the region:

$$\langle \mathbf{f}_m, \mathcal{L}_q \mathbf{j}_n \rangle = \frac{k\eta}{16\pi^2} \sum_{j=1}^{n_{(sd),q}} \sum_{j'=1}^{P_q} w_{jq}^{(sd)} w_{j'q}^{(fmm)} k_{\rho q}^{(j)}$$

$$\int_S d\mathbf{r}\, \mathbf{f}_m e^{i\mathbf{k}_q^{(jj')}\cdot(\mathbf{r}-\mathbf{r}_b)}$$

$$T\left(k_{\rho q}^{(j)}, \hat{s}_{j'q}, \boldsymbol{\rho}_b - \boldsymbol{\rho}_a\right) \left(\overline{\mathbf{I}} - \frac{\mathbf{k}_q^{(jj')} \mathbf{k}_q^{(jj')}}{k_q^2}\right) \quad \text{(17.32a)}$$

$$\int_S d\mathbf{r}'\, \mathbf{j}_n e^{i\mathbf{k}_q^{(jj')}\cdot(\mathbf{r}_a-\mathbf{r}')}$$

and

$$\langle \mathbf{f}_m, \mathcal{K}_q \mathbf{j}_n \rangle = -\frac{1}{16\pi^2} \sum_{j=1}^{n_{(sd),q}} \sum_{j'=1}^{P_q} w_{jq}^{(sd)} w_{j'q}^{(fmm)} k_{\rho q}^{(j)}$$

$$\int_S d\mathbf{r}\, (\mathbf{f}_m \times \mathbf{k}_q^{(jj')}) e^{i\mathbf{k}_q^{(jj')}\cdot(\mathbf{r}-\mathbf{r}_b)} \quad \text{(17.32b)}$$

$$T\left(k_{\rho q}^{(j)}, \hat{s}_{j'q}, \boldsymbol{\rho}_b - \boldsymbol{\rho}_a\right)$$

$$\int_S d\mathbf{r}'\, \mathbf{j}_n e^{i\mathbf{k}_q^{(jj')}\cdot(\mathbf{r}_a-\mathbf{r}')}$$

where

$$\mathbf{k}_q^{(jj')} = k_\rho^{(j)} \hat{s}_{j'} + k_z^{(j)} \hat{z} \quad \text{(17.32c)}$$

17.4.4 Simulation Results

SDFMM is used to simulate scattering from individual perfectly conducting and dielectric surfaces, and to characterize their statistical properties through Monte Carlo simulations. The computing platform used is a single R8000 processor on an SGI Power Challenge, with 2 GB of RAM and an average throughput of 60 Mflops.

To validate the SDFMM approach, a comparison is made with results obtained by using the standard MOM technique. The perfectly conducting rough surface used for this purpose has a roughness $\sigma = 0.5\lambda$ and correlation length $l_c = 1.5\lambda$. Its size is $3.9\lambda \times 3.9\lambda$. The density of discretization is 10 nodes per wavelength. Finest-level blocks of size $0.5\lambda \times 0.5\lambda$ are used, and the residual error stopping criterion for the transpose-free quasi-minimal residual (TFQMR) [21] solver used to iteratively reconstruct the surface currents is 10^{-2}. A Gaussian beam lying in the x-z plane and having no amplitude variation in the y- direction is used as the excitation. The beam is incident at an angle of $10°$ from the vertical with a positive x-component (i.e., $\theta = 10°$, $\phi = 180°$). The half-width of the beam is $W = L^{(S)}/4 = 0.975\lambda$, and its electric field vector is polarized in the x-z plane. The current obtained using SDFMM agrees to within 0.5% of that obtained with a classical MOM scheme. Figure 17.3 compares the RCS obtained using SDFMM and MOM. It is evident that the agreement is excellent.

A Monte Carlo simulation is carried out for an ensemble of 50 perfectly conducting rough surfaces of size $5.9\lambda \times 5.9\lambda$ with $\sigma = 0.5\lambda$ and $l_c = 1.5\lambda$. The excitation field is a Gaussian beam similar to the one in the above problem, with $W = L^{(S)}/4 = 1.475\lambda$. The problem involves $N=10{,}325$ unknowns and requires approximately 52 CPU hr for the complete simulation on a 60-Mflop SGI Power Challenge. The noncoherent bistatic scattering coefficients for the cross- and copolarized cases are depicted in Figures 17.4 and 17.5. Backscattering enhancement (at $\theta = 10°$, $\phi = 180°$) is clearly observed for both cases.

The CPU time and memory requirements of SDFMM are examined by applying it to progressively larger penetrable rough surface problems. The rough surfaces utilized for this purpose are characterized by $\sigma = 0.4\lambda$ and $l_c = 1.5\lambda$, and form the

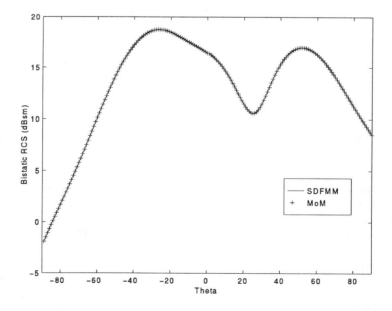

Figure 17.3 Bistatic RCS: SDFMM and MOM results. A Gaussian beam is incident at $\theta = 10°$, $\phi = 180°$, marked as -10.

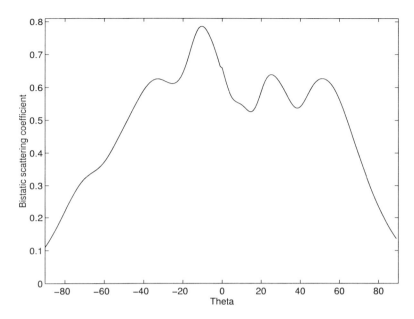

Figure 17.4 Monte Carlo simulation: copolarized bistatic scattering coefficient. A Gaussian beam is incident at $\theta = 10°$, $\phi = 180°$, marked as -10.

interface between half spaces comprised of free space and a dielectric with $\epsilon_2 = 2$. The SDP integration rules utilized in this case are designed for surfaces having a maximum roughness of $\sigma = 1.0\lambda$, and therefore the results for memory and time usage are representative for all surfaces having smaller roughness. Figure 17.6 depicts the CPU time required for a matrix-vector product, while Figure 17.7 shows the memory requirements. The dramatic improvements in speed and memory over standard iterative MOM solution are evident.

An example of scattering from a large surface is shown in Figure 17.8. The area of the surface is $205\lambda^2$. The surface is mildly rough with $\sigma = 0.05\lambda$ and $l_c = 1.0\lambda$. A Gaussian beam is incident at $\theta = 15°$, $\phi = 180°$, and $\epsilon_2 = 2$, making the surface dimensions equal to 410 square wavelengths in the dielectric. This problem requires 191,530 unknowns and can be accommodated in 1.8 GB of memory, as is clear from Figure 17.7. With the classical MOM, and with 2 GB of memory, the largest solvable problem would have been approximately N=16,000 unknowns only. An N=191,530 size problem would have required 325 GB of memory with the classical MOM. Moreover, for an N=191,530 unknown case, a standard iterative MOM solution would entail extremely expensive CPU time requirements of 80 minutes-per-matrix-

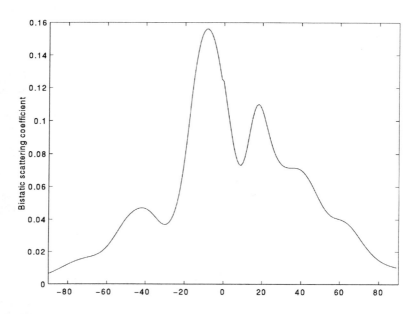

Figure 17.5 Monte Carlo simulation: cross-polarized bistatic scattering coefficient. A Gaussian beam is incident at $\theta = 10°$, $\phi = 180°$, marked as -10.

vector product, and 1,200 hours for matrix fill. The SDFMM correspondingly needs only approximately 14 minutes for a matrix-vector product, and 3 hours for setup.

17.5 QUANTUM WELL GRATING ANALYSIS

17.5.1 Introduction and Formulation

Quantum well infrared photodetectors (QWIPs) with AlGaAs/GaAs quantum wells have shown great potential as sensors in large imaging arrays [22–26]. Owing to quantum-mechanical selection rules, most unstrained n-type QWIPs respond only to the longitudinal component of the optical electrical field (i.e., the field along the growth direction). For such devices to sense normally incident radiation, optical grating couplers are needed to scatter the optical field in directions favorable to intersubband absorption [27–32]. While periodic gratings [27–30] have been extensively studied and modeled using modal expansion methods, aperiodic or quasi-random gratings [31, 32] have not been analyzed in comparable detail. Nonetheless, quasi-random gratings have been experimentally observed to produce better absorption over

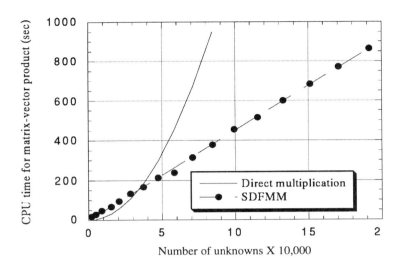

Figure 17.6 Matrix-vector product times: times required for direct multiplication and for SDFMM are shown as a function of problem size.

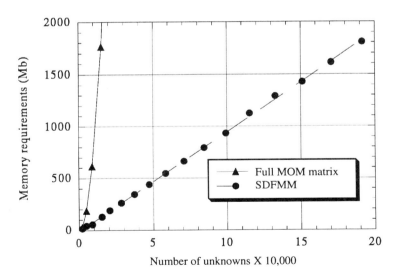

Figure 17.7 Memory requirements for the standard MOM and SDFMM.

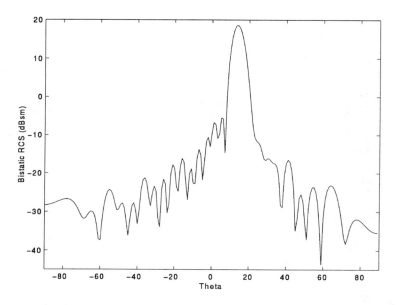

Figure 17.8 Copolarized bistatic RCS for a thin rough surface of area $205\lambda^2$. The roughness parameters are $\sigma = 0.05\lambda$ and $l_c = 1.0\lambda$, where λ is the free-space wavelength. A Gaussian beam is incident at $\theta = 15°$, $\phi = 180°$, marked as -15. The dielectric constant of the surface is $\epsilon_2 = 2$, making the surface dimensions equal to 410 squared wavelengths in the dielectric. The surface is modeled with 191,530 unknowns.

broad spectral ranges than periodic ones. The necessarily large but finite dimensions of realistic gratings place an exacting demand on solution efficiency and accuracy because the computational advantages of small-scale problems or the relative ease of infinite approximations disappear. Unfortunately, the computational burden associated with the solution of a large-scale, three-dimensional, vector electromagnetic problem has restricted the numerical analysis of quasi-random gratings to modal expansions for computationally simpler one-dimensional profiles [32].

In this section, SDFMM is applied to the analysis of scattering from finite two-dimensional quasi-random gratings. This approach permits a fast and memory efficient route to analyzing large and arbitrarily shaped gratings. Furthermore, using SDFMM, we are able to propose new rough surface gratings that display remarkable spectral properties and can potentially improve absorption in QWIPs to a substantial degree.

The back-illuminated QWIP structure to be analyzed is shown in Figure 17.9(a). The metallic grating layer S (modeled as a perfect conductor [29, 32]), which interfaces with GaAs, prevents radiation loss and serves as a contact. An optical wave impinges normally on the grating from the substrate side, and the scattered

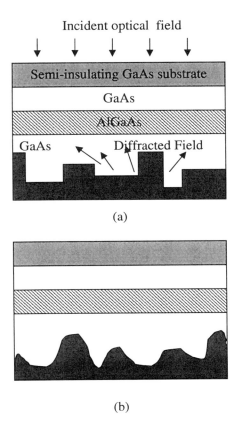

Figure 17.9 Schematic cross-sectional view of a grating coupled QWIP device with (a) a quasi-random grating; (b) a rough surface coupler.

wave excites the quantum wells owing to its nonzero electric field along the growth direction.

The incident field $\mathbf{E}_{inc}(\mathbf{r})$ is assumed to be a tapered Gaussian beam similar to that used in (17.4.1), which suppresses edge scattering effects at the mesa boundary. This assumption is consistent with the observation that a substantial portion of the grating and device near the mesa edge is optically inactive [29]. A surface current $\mathbf{J}(\mathbf{r})$ is generated at the grating due to $\mathbf{E}_{inc}(\mathbf{r})$. The total electric field tangential to S must vanish on S, and therefore the field obeys the same EFIE as in (17.23).

The solution of (17.23) yields $\mathbf{J}(\mathbf{r})$, and the total longitudinal electric field at the device layer can then be obtained via the operator L as in (17.12c). As a figure of merit of the absorption due to a grating, we define an integrated field strength (IFS) as

$$\text{IFS} = \frac{k\eta}{i} \int_D \left| \hat{z} \cdot \int_S \mathcal{L}\, \mathbf{J}(\mathbf{r}')\, dS' \right|^2 dD \qquad (17.33)$$

where D denotes the device layer. The IFS is similar to the measure used earlier [32]. As before, we use the MOM with RWG basis functions to solve the integral equation. For a grating S of size $L^{(S)} \times L^{(S)}$ square wavelengths, the number of unknowns N grows approximately as $200[L^{(S)}]^2$. For the QWIP grating structure under analysis, N can be of the order of 10^4 or 10^5. Hence, direct inversion of $\bar{\mathbf{Z}}$, which entails a computational cost proportional to N^3, is practically impossible. Therefore, recourse is taken to the SDFMM technique as in (17.11), which has CPU time and memory costs that scale as $O(N)$ as demonstrated earlier.

17.5.2 Periodic and Quasi-Random Grating Analysis

Three kinds of periodic and quasi-random gratings have been studied. The P-grating is a doubly *periodic* grating. *Displacing* raised portions of this grating in lateral directions from their original positions on a periodic grid produces a D-grating. *Scaling* raised portions of a P-grating along their diagonal while keeping one corner and the height fixed generates an S-grating. Figure 17.10 shows D- and S-gratings.

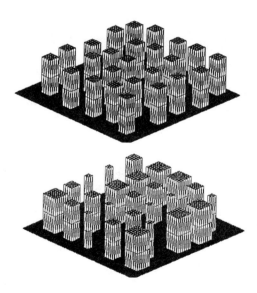

Figure 17.10 Gratings constructed from displacing (top) and scaling (bottom) raised portions of a doubly periodic grating, while keeping heights fixed.

As a precursor to employing the RWG basis in conjunction with SDFMM, the grating surfaces are tessellated into planar triangles with nodes separated by approximately 0.14λ, where λ is the optical wavelength in the GaAs layer. For infrared radiation with a free-space wavelength of 10 μm, $\lambda \cong 3.03$ μm. Current on each grating is modeled in terms of approximately $N=30,000$ basis functions. With SDFMM such problems can be solved in 3–7 hours (depending on the number of iterations required) on a single-processor R8000 SGI Power Challenge. The P-, D-, and S-gratings used here are of dimensions 38×38 μm^2. The periodicity of the P-grating is 4.0 μm along each lateral direction, and the raised portions measure 2.0×2.0 μm^2. The average shift relative to the period size for the specific D-grating considered is 14%, and the maximum is 50%. The average deviation in size (relative compression/elongation) for the S-grating is 32% and the maximum deviation is 50%.

The importance and need for quasi-random gratings is evident when one observes the spectral behavior of the absorption due to the gratings, which is depicted in Figure 17.11. For the P-grating, the absorption diminishes rapidly at larger wavelengths due to the evanescent nature of the higher-order Floquet modes. If such a grating were

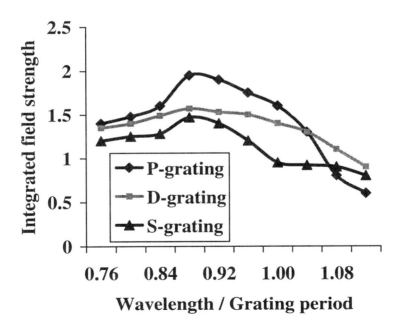

Figure 17.11 IFS in μV^2 versus wavelength/period for the periodic quasi-random gratings. The ratios of maximum to minimum IFS are 3.83, 1.75, and 1.70 for the P-, D-, and S-gratings, respectively.

designed to operate optimally at the nominal wavelength of 10 μm, its performance at the extremal wavelengths (8 μm and 12 μm) would be poor. The D- and S-gratings, on the other hand, exhibit a much smoother spectral behavior over the frequency range of interest, albeit with a slightly reduced peak absorption. Hence there exists a trade-off between peak absorption and smooth spectral behavior.

The results presented here are qualitatively consistent with the findings for one-dimensional gratings reported by Xing and Liu [32]. Hence, we confirm these authors' conjecture that the aforementioned trade-off demonstrated for one-dimensional gratings in their work should extend to two-dimensional gratings.

17.5.3 Random Rough Surface Couplers

The SDFMM-based solution presented herein is sufficiently flexible to permit the rapid analysis of a far larger class of grating and coupling structures than have been analyzed or tested in the past. Specifically, SDFMM can be used to analyze rough surface gratings (shown in Figure 17.9(b)), which are proposed here as suitable candidates for QWIP applications. These rough surfaces are assumed to have correlated Gaussian profiles, and are characterized by two parameters, a correlation length l_c and a mean square height σ. Prior to the analysis of such surfaces, it was expected that their strongly random nature would result in desirable spectral properties without substantial deterioration in peak absorption. This conjecture has been validated on the basis of the following simulation results.

Figure 17.12 shows the spectral dependence of the grating absorption for three rough surface grating realizations with different root mean square heights and a unit wavelength correlation length. All three gratings exhibit an extremely smooth IFS over the entire wavelength band of interest (8–12 μm in free space) without any compromise in the peak IFS. This is in sharp contrast to the behavior exhibited by quasi-random gratings wherein peak absorption is traded for smoothness of spectral behavior. Such behavior as seen in Figure 17.11 suggests that rough surface gratings should be particularly suitable for QWIP applications.

17.6 ANALYSIS OF MICROSTRIP ANTENNA ARRAYS ON FINITE SUBSTRATES

17.6.1 Introduction

Accurate tools for analyzing radiation and scattering from realistic microstrip structures are essential in the design of many modern antenna systems [33, 34]. Unfortunately, the MOM-based assessment of finite-substrate and ground-plane-truncation effects remains a computationally challenging task owing to the large number of basis functions required for the accurate full-wave modeling of these structures. Indeed,

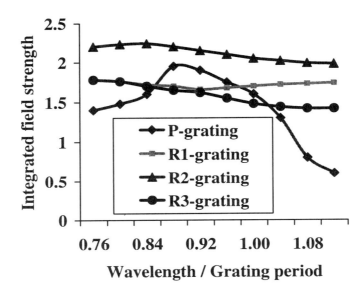

Figure 17.12 IFS in μV^2 versus wavelength/period for rough surface couplers. The three realizations R1, R2, and R3 possess root mean square heights of 0.2λ, 0.4λ, and 0.6λ, respectively. The ratio of maximum to minimum IFS is less than 1.2 for all the three rough surfaces, while the periodic grating has a ratio of 3.83.

such an approach requires one to associate unknown weighting coefficients with basis functions that model currents on microstrip elements, penetrable interfaces, and ground planes, and thereby results in a very large MOM system.

SDFMM permits the analysis of scattering and radiation from large and complex structures within realistic CPU times on laboratory workstations. These structures include large microstrip patch arrays on finite substrates and ground planes modeled using 100,000 or more MOM basis functions.

17.6.2 Integral Equation Formulation and SDFMM Solution

In what follows, it will be assumed that the structure under study is associated with an arbitrary number of regions $q = 1, \cdots, R_T$, with the permittivity and permeability of each region being denoted by ϵ_q and μ_q; apart from the regions comprising the structure, the region exterior to the structure (typically free space) is also included as one of the R_T regions. A simple finite microstrip array consists of PEC patches that lie on a dielectric substrate, and probes that connect the patches to the ground plane.

To formulate integral equations for analyzing scattering and radiation from general finite microstrip structures, the approach outlined in [16] is followed. In general, incident electric and magnetic fields $\mathbf{E}_q^{inc}(\mathbf{r})$ and $\mathbf{H}_q^{inc}(\mathbf{r})$ excite the structure from Region q, and equivalent electric and magnetic surface currents $\mathbf{J}_q^S(\mathbf{r})$ and $\mathbf{M}_q^S(\mathbf{r})$ are impressed on the surfaces bounding Region q. Furthermore, probes are modeled as wires connecting ground planes to conducting patches, and the currents on the wires residing in Region q are denoted by $\mathbf{J}_q^W(\mathbf{r})$. The fields produced by $\mathbf{J}_q = \mathbf{J}_q^S + \mathbf{J}_q^W$ and $\mathbf{M}_q = \mathbf{M}_q^S$ in Region q can be written as

$$\mathbf{E}_q^{scat}(\mathbf{r}) = -\mathcal{L}_q \mathbf{J}_q(\mathbf{r}) + \mathcal{K}_q \mathbf{M}_q(\mathbf{r}) \quad (17.34a)$$

$$\mathbf{H}_q^{scat}(\mathbf{r}) = -\mathcal{K}_q \mathbf{J}_q(\mathbf{r}) - \frac{1}{\eta_q^2} \mathcal{L}_q \mathbf{M}_q(\mathbf{r}) \quad (17.34b)$$

where $k_q = \omega\sqrt{\epsilon_q \mu_q}$, $\eta_q = \sqrt{\mu_q/\epsilon_q}$, and the operators $\mathcal{L}_q$ and $\mathcal{K}_q$ are the same as in (17.24). The PMCHW equations are obtained by enforcing appropriate boundary conditions on the electric and magnetic fields at the interfaces between all regions. Let the total number of surfaces bounding separate volumes be S_T. Also, let q_u and q'_u denote the region separated by surface u ($u = 1, \cdots, S_t$). The following boundary conditions hold true: if surface u is a penetrable interface, then the field constraints are

$$\hat{n} \times \left\{ \mathbf{E}_{q_u}^{inc}(\mathbf{r}) + \mathbf{E}_{q_u}^{scat}(\mathbf{r}) \right\} = \hat{n} \times \left\{ \mathbf{E}_{q'_u}^{inc}(\mathbf{r}) + \mathbf{E}_{q'_u}^{scat}(\mathbf{r}) \right\}, \quad \mathbf{r} \in \text{Surface } u \quad (17.34c)$$

and

$$\hat{n} \times \left\{ \mathbf{H}_{q_u}^{inc}(\mathbf{r}) + \mathbf{H}_{q_u}^{scat}(\mathbf{r}) \right\} = \hat{n} \times \left\{ \mathbf{H}_{q'_u}^{inc}(\mathbf{r}) + \mathbf{H}_{q'_u}^{scat}(\mathbf{r}) \right\}, \quad \mathbf{r} \in \text{Surface } u \quad (17.34d)$$

Here, $\hat{n}$ denotes a position dependent normal to the surface u. If surface u is perfectly conducting, then the constraints are

$$\hat{n} \times \left\{ \mathbf{E}_{q_u}^{inc}(\mathbf{r}) + \mathbf{E}_{q_u}^{scat}(\mathbf{r}) \right\} = 0, \quad \mathbf{r} \in \text{Surface } u \quad (17.34e)$$

and

$$\hat{n} \times \left\{ \mathbf{E}_{q'_u}^{inc}(\mathbf{r}) + \mathbf{E}_{q'_u}^{scat}(\mathbf{r}) \right\} = 0, \quad \mathbf{r} \in \text{Surface } u \quad (17.34f)$$

Finally, along a wire probe residing in Region q

$$\hat{t} \cdot \left\{ \mathbf{E}_q^{inc}(\mathbf{r}) + \mathbf{E}_q^{scat}(\mathbf{r}) \right\} = 0 \quad (17.34g)$$

where $\hat{t}$ denotes a position dependent tangent to the wire in Region q.

The solution of the PMCHW equations yields the electric and magnetic current densities on all interfaces and probes, which can be used to compute antenna radiation patterns, impedances, and radar cross-sections.

17.6.3 MOM Formulation

To solve (17.34), the unknown current densities $\mathbf{J}_q(\mathbf{r})$ and $\mathbf{M}_q(\mathbf{r})$ are approximated in terms of linear combinations of RWG functions, as before. Also, piecewise linear basis functions are used to represent the electric currents on probes, which are modeled as wires of a given radius connecting the ground plane to the microstrip patches. Additional wire-surface junction basis functions model current through terminal wire nodes on wires that are connected to a surface.

Using the above basis functions, the electric and magnetic current densities are approximated as

$$\mathbf{J}_q(\mathbf{r}) \cong \sum_{n=1}^{N_q^S} I_{q,n}^E \mathbf{j}_{q,n}(\mathbf{r}) + \sum_{n=1}^{N_q^W} I_{q,n}^W \mathbf{w}_{q,n}(\mathbf{r}) \qquad (17.35a)$$

and

$$\mathbf{M}_q(\mathbf{r}) \cong \sum_{n=1}^{N_q^S} I_{q,n}^H \mathbf{j}_{q,n}(\mathbf{r}) \qquad (17.35b)$$

where $\mathbf{j}_{q,n}$ denote surface basis functions, and $\mathbf{w}_{q,n}$ represent wire and surface-wire junction basis functions. There are $2N_q^S$ electric and magnetic current surface basis functions and N_q^W wire and surface-wire basis functions associated with Region q. The total number of basis functions N equals $\sum_{q=1}^{R_T}\left(2N_q^S + N_q^W\right)$.

The N basis functions above do not, in general, form an independent set. Indeed, electric and magnetic currents on either side of a penetrable interface flow in opposite directions. Also, dependencies between currents flowing from one surface to another through surface-surface junctions have not been accounted for at this stage. Finally, the fact that the magnetic current is zero on a PEC surface has not yet been exploited. To incorporate the above constraints, the approach outlined in [16, 35] is followed. Let $\mathbf{I} = [I_q, \cdots, I_{R_T}]^T$ denote a vector of N current coefficients, where $I_q = \{[I_{q,n}^E], [I_{q,n}^H], [I_{q,n}^W]\}$, and assume that there are N_{ind} ($N_{ind} < N$) current coefficients that are arranged in a vector $\mathbf{I}_{ind}$. By enforcing the above dependencies, a bipolar matrix $\overline{\mathbf{A}}$ of dimension $N_{ind} \times N$ is constructed such that $\mathbf{I} = \overline{\mathbf{A}}^T \mathbf{I}_{ind}$. Substituting the expansions for the current into the PMCHW equations, and testing the equations with RWG functions $\mathbf{j}_{q,m}$, $m = 1, ..., N_q^S$ and with wire and surface-wire junction basis functions $\mathbf{w}_{q,m}$, $m = 1, ..., N_q^S$, we arrive at the following full-rank matrix equation [16]:

$$\overline{\mathbf{A}}\, \overline{\mathbf{Z}}\, \overline{\mathbf{A}}^T \mathbf{I}_{ind} = \overline{\mathbf{A}}\, \mathbf{V} \qquad (17.36)$$

Here, $\overline{\mathbf{Z}} = \text{Block}[Z_1, \cdots, Z_{R_T}]$ where

$$Z_q = \begin{bmatrix} Z_q^{11} & Z_q^{12} & Z_q^{13} \\ Z_q^{21} & Z_q^{22} & Z_q^{23} \\ Z_q^{31} & Z_q^{32} & Z_q^{33} \end{bmatrix} \qquad (17.37a)$$

$$Z^{11}_{q,mn} = \langle \mathbf{j}_{q,m}, \mathcal{L}_q \mathbf{j}_{q,n}\rangle$$
$$Z^{12}_{q,mn} = \langle \mathbf{j}_{q,m}, -\mathcal{K}_q \mathbf{j}_{q,n}\rangle$$
$$Z^{21}_{q,mn} = \langle \mathbf{j}_{q,m}, \mathcal{K}_q \mathbf{j}_{q,n}\rangle$$
$$Z^{22}_{q,mn} = \langle \mathbf{j}_{q,m}, \mathcal{L}_q \mathbf{j}_{q,n}/\eta_q^2\rangle$$
(17.37b)

The elements $Z^{i3}_{q,mn}$ and $Z^{3i}_{q,mn}$ for $i = 1, 2$ relate to fields radiated by wires being tested by surface basis functions, and vice versa, and $Z^{33}_{q,mn}$ relates to the field radiated by wire basis functions and tested on wires. As a typical microstrip structure has very few probes (and therefore wire basis functions), the subsequent discussions will focus primarily on surface basis functions. It should be noted that each block Z_q depends only on the properties of medium q, which is key to the efficient application of SDFMM. The matrix **V** is the vector of the inner products of incident electric and magnetic fields with testing functions.

In order to rapidly solve (17.36) using an iterative technique, the SDFMM expansion for electromagnetic fields developed earlier is used to efficiently represent terms arising in the product of $\bar{\mathbf{Z}}$ with a trial vector. The fields evaluated by testing functions in a particular block by the operation of L_q and K_q on basis functions in another block can then be represented in the SDFMM representations developed earlier in (17.13a) and (17.13b), and applied to each Region q with the appropriate material parameters.

17.6.4 Simulation Results

Numerical results are presented which demonstrate the efficacy of SDFMM in analyzing scattering and radiation from penetrable and PEC quasi-planar structures. The computing platform used is a single R8000 processor on an SGI Power Challenge, with 2 GB of RAM and an average in-program throughput of 60 Mflops. A TFQMR (transpose-free quasi-minimal residual) [21] iterative solver is used with both SDFMM and MOM.

SDFMM is used to analyze scattering from microstrip arrays, using a large number of MOM unknowns. For this purpose, patches of dimensions $a = 0.27\lambda$ and $b = 0.19\lambda$ are placed on a ground-plane backed dielectric ($\epsilon_r = 2.17$) of thickness $h = 0.05\lambda$. The patch separation in the two lateral directions is given by $L = 0.17\lambda$ and $W = 0.25\lambda$. The dimensions of the substrate $d \times d$ are chosen to be close to the extent of the finite array of patches.

Radiation patterns produced by in-phase excitation are simulated for a large array. This array is modeled using $N = 117,952$ (11×11 array) basis functions, and the pattern is shown in Figure 17.13. In a smaller example considered ($N = 3,720$), the radiation pattern obtained using SDFMM is compared to the standard MOM result (Figure 17.14). This example has $N_{ind} = 53,998$ and a lateral extent of the dielectric substrate and ground plane determined by $d = 6.7\lambda$.

The Steepest-Descent Fast Multipole Method

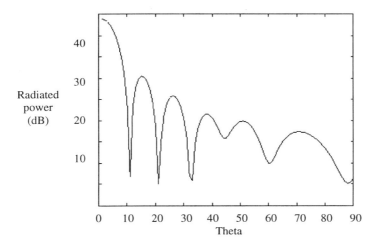

Figure 17.13 Radiation pattern of an 11 × 11 array.

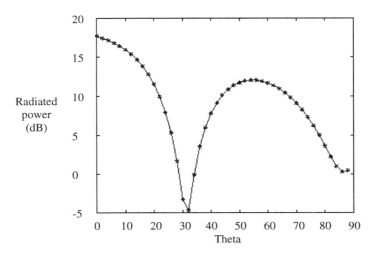

Figure 17.14 Radiation pattern of a 2 × 2 array using SDFMM (solid line), and standard MOM (*).

17.7 CONCLUSION

This chapter presented a review of SDFMM, an FMM-like scheme for rapidly evaluating matrix-vector products that arise in the analysis of scattering and radiation from quasi-planar structures. The CPU time and memory requirements of SDFMM scale as $O(N)$. SDFMM has been applied to the analysis of diverse phenomena including scattering from rough surfaces and quantum-well gratings, and radiation from microstrip antenna arrays, and a similar approach has been devised for multilayered media [36].

REFERENCES

1. J. T. Johnson, L. Tsang, R. T. Shin, K. Pak, C. H. Chan, A. Ishimaru, and Y. Kuga, "Backscattering enhancement of electromagnetic waves from two-dimensional perfectly conducting random rough surfaces—a comparison of Monte Carlo simulations with experimental data," *IEEE Trans. Antennas Propagat.*, vol. 44, pp. 748–756, 1996.

2. R. L. Wagner, J. Song, and W. Chew, "Monte Carlo simulation of electromagnetic scattering from two-dimensional random rough surfaces," *IEEE Trans. Antennas Propagat.*, vol. 45, pp. 235–245, 1997.

3. R. L. Wagner, "Efficient Computational Techniques for Electromagnetic Propagation and Scattering," Ph.D. Dissertation, University of Illinois at Urbana-Champaign, 1996.

4. V. Jandhyala, D. Sengupta, B. Shanker, E. Michielssen, M. Feng, and G. Stillman, "Efficient electromagnetic analysis of two-dimensional finite quasi-random gratings for quantum well infrared photodetectors," *J. Appl. Phys.*, vol. 83, pp. 3360–3363, 1998.

5. V. Jandhyala, B. Shanker, E. Michielssen, and W. Chew, "A combined steepest descent–fast multipole algorithm for the fast analysis of three-dimensional scattering by rough surfaces," *IEEE Trans. Geoscience Rem. Sens.*, vol. 36, pp. 738–748, 1998.

6. V. Jandhyala, B. Shanker, E. Michielssen, and W. Chew, "A fast algorithm for the analysis of scattering by dielectric rough surfaces," *J. Opt. Soc. Am. A*, vol. 15, pp. 1877–1885, 1998.

7. V. Jandhyala, D. Sengupta, B. Shanker, E. Michielssen, M. Feng, and G. Stillman, "Two-dimensional rough surface couplers for broadband quantum-well infrared photodetectors," *Appl. Phys. Lett.*, vol. 73, pp. 3495–3497, 1998.

8. V. Jandhyala, E. Michielssen, B. Shanker, and W. Chew, "A fast algorithm for the analysis of radiation and scattering from microstrip arrays on finite substrates," *Microwave Opt. Tech. Lett.*, vol. 23, pp. 306–310, 1999.

9. W. C. Chew, *Waves and Fields in Inhomogeneous Media*, New York: IEEE Press, 1995.

10. C. C. Lu and W. C. Chew, "Fast algorithm for solving hybrid integral equations," *IEE Proc. H*, vol. 140, pp. 455–460, 1993.

11. J. Rahola, "Efficient solution of linear equations in electromagnetic scattering calculations," Tech. Rep. CSC Research Reports R06/96, Center for Scientific Computing, Espoo, Finland, 1996.

12. O. Bucci, C. Gennarelli, and C. Savarese, "Optimal interpolation of radiated fields over a sphere," *IEEE Trans. Antennas Propagat.*, vol. 39, pp. 1633–1643, 1991.

13. C. C. Lu and W. C. Chew, "A multilevel algorithm for solving a boundary integral equation of wave scattering," *Microwave Opt. Tech. Lett.*, vol. 7, pp. 466–470, 1994.

14. J. M. Song and W. C. Chew, "Multilevel fast-multipole algorithm for solving combined field integral equations of electromagnetic scattering," *Microwave Opt. Tech. Lett.*, vol. 10, pp. 14–19, 1995.

15. P. Tran and A. A. Maradudin, "The scattering of electromagnetic waves from a rough metallic surface," *Opt. Comm.*, vol. 110, pp. 269–273, 1994.

16. L. Medgyesi-Mitschang, J. Putnam, and M. Gedera, "Generalized method of moments for three-dimensional penetrable scatterers," *J. Opt. Soc. Am. A*, vol. 12, pp. 1383–1398, 1994.

17. K. Umashankar, A. Taflove, and S. Rao, "Electromagnetic scattering by arbitrary shaped three-dimensional homogeneous lossy dielectric objects," *IEEE Trans. Antennas Propagat.*, pp. 758–765, June 1986.

18. A. Ishimaru, *Wave Propagation and Scattering in Random Media*, New York: Academic Press, 1978.

19. P. Tran, V. Celli, and A. A. Maradudin, "Electromagnetic scattering from a two dimensional, randomly rough, perfectly conducting surface: Iterative methods," *J. Opt. Soc. Am. A.*, vol. 11, pp. 1686–1689, 1994.

20. S. M. Rao, D. R. Wilton, and A. Glisson, "Electromagnetic scattering by surfaces of arbitrary shape," *IEEE Trans. Antennas Propagat.*, vol. 30, pp. 409–418, 1982.

21. J. Volakis, "EM programmer's notebook," *IEEE Antennas Propagat. Mag.*, vol. 37, pp. 94–100, 1995.

22. B. Levine, "Quantum-well infrared photodetectors," *J. Appl. Phys.*, vol. 74, pp. R1–R81, 1993.

23. C. Bethea, B. Levine, M. Asom, R. Leibenguth, J. Stayt, K. Glogovsky, R. Morgan, J. Blackwell, and W. Parish, "Long wavelength infrared 128×128 $Al_x Ga_{1-x}$ As / GaAs quantum well infrared camera and imaging system," *IEEE Trans. Electron. Devices*, vol. 40, pp. 1957–1963, 1993.

24. L. Kozlowski, G. Williams, G. Sullivan, C. Farley, R. Anderson, J. Chen, D. Cheung, W. Tennant, and R. DeWames, "LWIR 128×128 GaAs/AlGaAs multiple quantum well hybrid focal plane array," *IEEE Trans. Electron. Devices*, vol. 38, pp. 1124–1130, 1991.

25. S. Gunapala, J. Park, G. Sarusi, T. Lin, J. Liu, P. Maker, R. Muller, C. Shott, and T. Hoelter, "15 μ m 128×128 GaAs/$Al_x Ga_{1-x}$As quantum well infrared photodetector focal plane array camera," *IEEE Trans. Electron. Devices*, vol. 38, pp. 1124–1130, 1991.

26. W. Beck, T. Faska, J. Little, J. Albritton, and M. Sensiper, "LWIR imaging performance of 256×256 miniband transport multiple quantum well infrared focal plane arrays," *Proc. Int. Soc. Opt. Eng.*, Oct. 1994.

27. J. Andersson and L. Lundquist, "Near-unity quantum efficiency of AlGaAs/GaAs quantum well infrared detectors using a waveguide with a doubly periodic grating coupler," *Appl. Phys. Lett.*, vol. 59, pp. 857–859, 1991.

28. J. Andersson, L. Lundquist, and Z. Paska, "Quantum efficiency enhancement of AlGaAs/GaAs quantum well infrared detectors using a waveguide with a grating coupler," *Appl. Phys. Lett.*, vol. 58, pp. 2264–2266, 1991.

29. J. Andersson and L. Lundquist, "Grating-coupled quantum-well infrared detectors: theory and performance," *J. Appl. Phys.*, vol. 71, pp. 3600–3610, 1992.

30. L. Lundquist, J. Andersson, Z. Paska, J. Borglind, and D. Haga, "Efficiency of grating-coupled AlGaAs/GaAs quantum well infrared photodetectors," *Appl. Phys. Lett.*, vol. 63, pp. 3361–3363, 1993.

31. G. Sarusi, B. Levine, S. Pearton, K. Bandara, R. Leibenguth, and J. Andersson, "Optimization of two dimensional gratings for very long wavelength quantum well infrared photodetectors," *J. Appl. Phys.*, vol. 76, pp. 4989–4994, 1994.

32. B. Xing and H. Liu, "Simulation of one-dimensional quasi-random gratings for quantum well infrared photodetectors," *J. Appl. Phys.*, vol. 80, pp. 1214–1218, 1996.

33. K. Lee and W. Chen, *Advances in Microstrip and Printed Antennas*, New York: Wiley-Interscience, 1997.

34. D. Pozar and D. Schaubert, *Microstrip Antennas*, Piscataway, NJ: IEEE, 1995.

35. E. Yip and B. Dembart, "Matrix assembly in FMM-MoM codes," Tech. Rep. ISSTECH-97-002, The Boeing Company, Seattle, WA, 1997.

36. J. Zhao, W. Chew, C. Lu, E. Michielssen, and J. Song, "Thin-stratified medium fast-multipole algorithm for solving microstrip structures," *IEEE Trans. Microwave Theory Tech.*, vol. 46, pp. 395–403, 1998.

18

Plane-Wave Time-Domain Algorithms

A. Arif Ergin, Balasubramaniam Shanker, and Eric Michielssen

18.1 INTRODUCTION

Computer simulation of linear transient wave phenomena involving structures that reside in an unbounded medium to generate broadband data is of paramount importance in such disciplines as acoustics, electromagnetics, and geophysics. Numerical techniques for performing such simulations call for the evaluation of retarded-time boundary integrals (RTBIs), which relate known transient source distributions to the fields they radiate. This is due to the fact that RTBIs play a fundamental role in formulation of the integral equation-based numerical techniques such as the marching-on-in-time method and in imposing exact radiation boundary conditions used in differential equation-based methods such as the finite-difference time-domain and finite element time-domain methods. However, evaluation of RTBIs using classical techniques is a computationally expensive procedure limiting the applicability of numerical transient analysis techniques to a small number of problems. This chapter introduces two PWTD algorithms that considerably reduce the cost of evaluating RTBIs by using plane wave bases. Strictly speaking, the introduced PWTD algorithms help evaluate wave fields that are radiated by a cluster of sources and that are tested at a group of observation points that are well separated from the source domain.

The cost of evaluating the field due to an arbitrary surface source distribution can be accelerated by partitioning the whole source distribution into small source clusters and by applying a PWTD algorithm within a two-level or a multilevel framework to calculate the fields due to each cluster. The first PWTD algorithm, introduced in Section 18.3, is based on a Whittaker-type field expansion [1–3] (i.e., a representation of the radiated field in terms of plane waves that propagate in all directions). To some extent, this scheme can be considered the direct time-domain counterpart of the frequency domain fast multipole method (FMM) [4]. Using this PWTD algorithm in two-level and multilevel settings to accelerate MOT schemes reduces the $O(N_t N_s^2)$ computational complexity of a scattering analysis to $O(N_t N_s^{1.5} \log N_s)$ and $O(N_t N_s \log^2 N_s)$, respectively, where N_t and N_s denote the number of temporal and spatial degrees of freedom of the discretized surface equivalent sources. The second PWTD algorithm—introduced in Section 18.5 and henceforth referred to as the *windowed PWTD algorithm*—relies on a finite-cone representation in which only plane waves whose propagation directions fall within a cone encompassing the observation domain are used in representing the radiated field [5]. The computational costs of the PWTD-enhanced MOT algorithms based on the windowed PWTD algorithm can scale as low as $O(N_t N_s^{4/3} \log N_s)$ and $O(N_t N_s \log N_s)$. These algorithms can be considered the time-domain analogs of the windowed fast multipole methods [6–8] and steepest descent path algorithms [9, 10] that have been so remarkably successful in accelerating the solution of frequency domain scattering problems. Since the nature of the wave phenomenon being modeled and the boundary condition being enforced are secondary to the description of the PWTD schemes, the algorithms presented here are for a transient scalar wave solver that imposes a Dirichlet boundary condition on the scatterer. The generalization to different boundary conditions and to a Maxwell equation solver will be outlined in the following chapter.

This chapter is organized as follows. Section 18.2 outlines the classical MOT algorithm for analyzing scalar wave scattering from a scatterer characterized by a Dirichlet boundary condition. The nonwindowed PWTD scheme is presented in Section 18.3. Incorporation of this PWTD algorithm into the MOT scheme of Section 18.2 in two-level and multilevel settings is graphically illustrated in Section 18.4. The resulting algorithms, which have computational complexities of $O(N_t N_s^{1.5} \log N_s)$ and $O(N_t N_s \log^2 N_s)$, are described in sufficient detail to permit the reader to easily retrofit existing MOT codes. Section 18.5 builds upon the ideas presented in Section 18.3 to present the windowed PWTD algorithm. The differences in incorporating the windowed PWTD into MOT schemes is highlighted and the computational complexities of the resulting algorithms are derived in Section 18.6. Finally, a summary of the chapter is given in Section 18.7. The material in this chapter is drawn from [11, 12].

Plane-Wave Time-Domain Algorithms 817

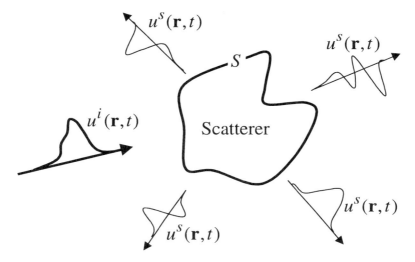

Figure 18.1 The surface scattering problem. The incident field $u^i(\mathbf{r}, t)$ interacts with the scatterer bounded by the surface S and produces the scattered field $u^s(\mathbf{r}, t)$.

18.2 THE MARCHING-ON-IN-TIME METHOD

Consider a field $u^i(\mathbf{r}, t)$ that is incident on a scatterer bounded by a surface S as depicted in Figure 18.1. It is assumed that $u^i(\mathbf{r}, t)$ is temporally bandlimited to ω_{max}. When $u^i(\mathbf{r}, t)$ interacts with S, a scattered field $u^s(\mathbf{r}, t)$ is generated, such that the total field $u(\mathbf{r}, t) = u^i(\mathbf{r}, t) + u^s(\mathbf{r}, t)$ satisfies a given boundary condition on S. In this chapter, the derivations will be carried out assuming the homogeneous Dirichlet condition $u(\mathbf{r}, t) = 0$ holds on S to keep the exposition simple. However, as will be demonstrated in the following chapter, the results of this chapter can be easily extended to cases with different boundary conditions. It is assumed that no interaction takes place before $t = 0$ (i.e., $u^s(\mathbf{r}, t) = 0$ for $t < 0$). Given that the total field satisfies the wave equation

$$\nabla^2 u(\mathbf{r}, t) - \frac{\partial_t^2}{c^2} u(\mathbf{r}, t) = 0 \tag{18.1}$$

everywhere exterior to S and the radiation condition (in addition to the Dirichlet boundary condition on S), it can be shown that $u^s(\mathbf{r}, t)$ can be represented in terms of equivalent surface sources $q(\mathbf{r}, t)$ that reside on S as

$$u^s(\mathbf{r}, t) = \int_S d\mathbf{r}' \frac{\delta(t - R/c)}{4\pi R} * q(\mathbf{r}', t) \tag{18.2}$$

In the above equation, c is the wave speed in the medium, $\delta(\cdot)$ is the Dirac delta, $R = |\mathbf{r} - \mathbf{r}'|$, and $*$ denotes temporal convolution. The unknown source density $q(\mathbf{r}, t)$ in (18.2) can be related to the known incident field by enforcing the Dirichlet boundary condition on S, which yields the integral equation

$$-u^i(\mathbf{r}, t) = \int_S d\mathbf{r}' \frac{\delta(t - R/c)}{4\pi R} * q(\mathbf{r}', t) \qquad \forall \mathbf{r} \in S \qquad (18.3)$$

To solve (18.3) numerically, $q(\mathbf{r}, t)$ is represented in terms of spatial and temporal basis functions $f_n(\mathbf{r})$, $n = 1, \ldots, N_s$, and $T_i(t)$, $i = 0, \ldots, N_t$, as

$$q(\mathbf{r}, t) \cong \sum_{n=1}^{N_s} \sum_{i=0}^{N_t} q_{n,i} f_n(\mathbf{r}) T_i(t) \qquad (18.4)$$

where the $q_{n,i}$ represent unknown expansion coefficients. Choosing the number of basis functions as $N_s \propto S_A (\omega_{max}/c)^2$ and $N_t \propto T\omega_{max}$, where S_A denotes the surface area of the scatterer and T the duration of the analysis, permits accurate representation of $q(\mathbf{r}, t)$. It is assumed that the spatial and temporal basis functions have local supports. Upon substituting (18.4) into (18.3), and testing the resulting equation at time $t = t_j = j\Delta_t$, where Δ_t is the time step size, with the testing functions $\tilde{f}_m(\mathbf{r})$, $m = 1, \ldots, N_s$, the following matrix equation is obtained:

$$\bar{\mathbf{Z}}_0 \mathbf{Q}_j = \mathbf{U}_j^i - \sum_{k=1}^{j-1} \bar{\mathbf{Z}}_k \mathbf{Q}_{j-k} \qquad (18.5)$$

In (18.5), the mth elements of the vectors $\mathbf{Q}_j$ and $\mathbf{U}_j^i$ are given by $q_{m,j}$ and $-\int_S d\mathbf{r}\, \tilde{f}_m(\mathbf{r})\, u^i(\mathbf{r}, t_j)$, respectively, and the elements of the sparse interaction matrix $\bar{\mathbf{Z}}_k$ are given by

$$\bar{Z}_{k,mn} = \int_S d\mathbf{r}\, \tilde{f}_m(\mathbf{r}) \int_S d\mathbf{r}'\, f_n(\mathbf{r}') \left[\frac{\delta(t - R/c)}{4\pi R} * T_{j-k}(t) \right]\bigg|_{t=t_j} \qquad (18.6)$$

Equation (18.5) relates the expansion coefficients at the jth time step to the incident field and the coefficients at prior time steps. Hence, all the expansion coefficients $q_{n,j}$ can be evaluated by starting at the first time step ($j = 0$), forming the right-hand side, and solving (18.5) at each time step. This procedure is known as the MOT scheme [13–15]. The evaluation of the summation appearing on the right-hand side of (18.5) is the most expensive operation in the MOT scheme. In essence, this operation entails the evaluation of the field $u^s(\mathbf{r}, t)$ at $O(N_s)$ points on S due to all past sources, and requires $O(N_s^2)$ operations. Since this operation is repeated for all N_t time steps in the analysis, the computational complexity of the classical MOT scheme scales as $O(N_t N_s^2)$.

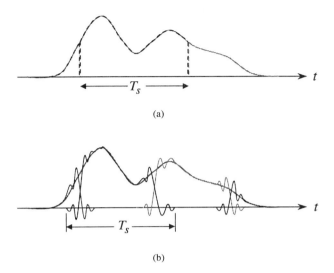

Figure 18.2 Sectioning of a signal into subsignals of duration T_s such that (a) the lth subsignal is zero outside the interval $(l-1)T_s \leq t < lT_s$, and (b) each subsignal is bandlimited to $\omega_s > \omega_{max}$.

18.3 THE PLANE-WAVE TIME-DOMAIN ALGORITHM

Evidently, the cost of the MOT scheme renders infeasible the practical analysis of scattering from large structures. Recent research efforts have focused on reducing this computational burden [16–19]. This section will introduce a PWTD algorithm that achieves this goal by expressing transient fields in terms of plane waves. This algorithm permits the evaluation of transient fields at a set of observers when the sources and observers are clustered in well-separated regions in space. Also, as will be evident in the forthcoming analysis, the algorithm requires all source signals to be of limited temporal extent. However, this limitation does not hinder the applicability of the algorithm since the source signal $q(\mathbf{r}, t)$ can always be broken up into subsignals as

$$q(\mathbf{r}, t) = \sum_{l=1}^{L} q_l(\mathbf{r}, t) \tag{18.7}$$

where each subsignal $q_l(\mathbf{r}, t)$ is identically zero outside of an interval $(l-1)T_s \leq t < lT_s$, as illustrated in Figure 18.2(a). Then, the field radiated by $q_l(\mathbf{r}, t)$, henceforth denoted as $u_l(\mathbf{r}, t)$, can be evaluated using the PWTD algorithm, and the total field can be formed as

$$u(\mathbf{r}, t) = \sum_{l=1}^{L} u_l(\mathbf{r}, t) \tag{18.8}$$

Note that the subsignal $q_l(\mathbf{r}, t)$ and the field $u_l(\mathbf{r}, t)$ are simply related by

$$u_l(\mathbf{r}, t) = \int_S d\mathbf{r}' \frac{\delta(t - R/c)}{4\pi R} * q_l(\mathbf{r}', t) \tag{18.9}$$

18.3.1 Plane Wave Decomposition

Motivated by frequency domain fast multipole schemes [4, 7, 8, 20], which reconstruct the time harmonic field due to a group of sources as a superposition of plane waves, our first task is to postulate a suitable plane wave representation for transient wave fields. Such representations date back to Whittaker [3] and have since been thoroughly studied [1, 2, 5, 21–23]. To arrive at a plane wave representation that facilitates the fast evaluation of transient fields, consider the expression

$$\tilde{u}_l(\mathbf{r}, t) = -\frac{\partial_t}{8\pi^2 c} \int_0^{\theta_{int}} d\theta \sin\theta \int_0^{2\pi} d\phi \\ \int_S d\mathbf{r}' \delta\left[t - \hat{\mathbf{k}} \cdot (\mathbf{r} - \mathbf{r}')/c\right] * q_l(\mathbf{r}', t) \tag{18.10}$$

In (18.10), $\hat{\mathbf{k}} = \hat{\mathbf{x}} \sin\theta \cos\phi + \hat{\mathbf{y}} \sin\theta \sin\phi + \hat{\mathbf{z}} \cos\theta$ is a unit direction vector and the innermost integral over S is a projection of the source distribution $q_l(\mathbf{r}, t)$ onto a plane wave travelling in the $\hat{\mathbf{k}}$ direction. Note that, the directional integration is over a cap of the unit sphere for which $\theta \leq \theta_{int}$, as illustrated in Figure 18.3(a). However, if the upper limit on the θ integral, θ_{int}, is set equal to π, (18.10) expresses $\tilde{u}_l(\mathbf{r}, t)$ as a superposition of plane waves that travel in all directions. Carrying out the θ and ϕ integrations in (18.10) reveals an interesting relation between $\tilde{u}_l(\mathbf{r}, t)$ and $u_l(\mathbf{r}, t)$. To this end, the spherical integral in (18.10) is cast in a new coordinate system $(x', y', z') \equiv (\rho', \theta', \phi')$ in which the z' axis is aligned with $\mathbf{R} = \mathbf{r} - \mathbf{r}'$ as shown in Figure 18.3(a). In this new coordinate system, the upper limit on θ', θ'_{int}, depends on ϕ', $\mathbf{r}$, and $\mathbf{r}'$, and

$$\tilde{u}_l(\mathbf{r}, t) = -\frac{\partial_t}{8\pi^2 c} \int_S d\mathbf{r}' \int_0^{2\pi} d\phi' \\ \int_0^{\theta'_{int}(\phi', \mathbf{r}, \mathbf{r}')} d\theta' \sin\theta' \, \delta\left(t - \hat{\mathbf{k}}' \cdot \mathbf{R}'/c\right) * q_l(\mathbf{r}', t) \tag{18.11}$$

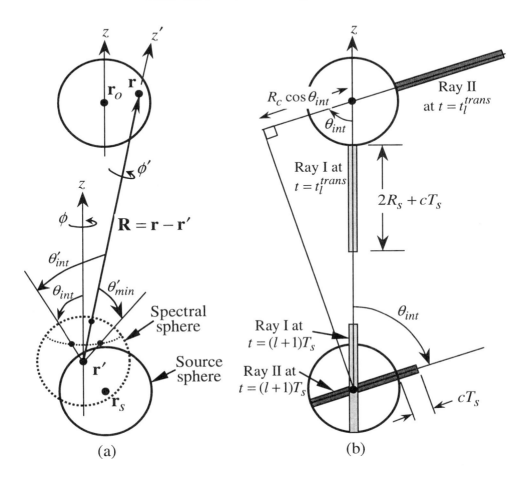

Figure 18.3 (a) Definition of angles in primed coordinate system. (b) Relevant dimensions for ray translation.

where $\hat{\mathbf{k}}' = \hat{\mathbf{x}}' \sin\theta' \cos\phi' + \hat{\mathbf{y}}' \sin\theta' \sin\phi' + \hat{\mathbf{z}}' \cos\theta'$, and $\mathbf{R}' = \hat{\mathbf{z}}' |\mathbf{R}|$. In deriving the limits on the elevation integral in (18.11), it was tacitly assumed that

$$\theta_{int} > \cos^{-1}(\hat{\mathbf{z}} \cdot \mathbf{R}/R) \tag{18.12}$$

Defining $R = |\mathbf{R}|$, using $\hat{\mathbf{k}}' \cdot \mathbf{R}' = R\cos\theta'$, and setting $\tau = R\cos\theta'/c$ in (18.11) yields

$$\begin{aligned}\tilde{u}_l(\mathbf{r},t) &= -\int_S d\mathbf{r}' \int_0^{2\pi} d\phi' \int_{-(R/c)\cos\theta'_{int}(\phi',\mathbf{r},\mathbf{r}')}^{R/c} d\tau \, \frac{\partial_t \delta(t-\tau)}{8\pi^2 R} * q_l(\mathbf{r}',t) \\ &= \int_S d\mathbf{r}' \frac{\delta(t-R/c)}{4\pi R} * q_l(\mathbf{r}',t) \\ &\quad - \int_S d\mathbf{r}' \frac{\delta(t+R\cos\theta'_{int}(\phi',\mathbf{r},\mathbf{r}')/c)}{4\pi R} * q_l(\mathbf{r}',t) \\ &= u_l(\mathbf{r},t) - \int_S d\mathbf{r}' \frac{\delta(t+R\cos\theta'_{int}(\phi',\mathbf{r},\mathbf{r}')/c)}{4\pi R} * q_l(\mathbf{r}',t)\end{aligned} \quad (18.13)$$

The above derivation closely follows that of Heyman [5], who generalizes the results of Tygel and Hubral [2]. For the special case $\theta_{int} = \pi$, it can be shown that the last expression in (18.13) reduces to

$$\tilde{u}_l(\mathbf{r},t) = u_l(\mathbf{r},t) - \int_S d\mathbf{r}' \frac{\delta(t+R/c)}{4\pi R} * q_l(\mathbf{r}',t) \quad (18.14)$$

Were it not for the second term appearing on the right-hand side, $\tilde{u}_l(\mathbf{r},t)$ would equal $u_l(\mathbf{r},t)$! The second term is referred to as the ghost signal and is anticausal (i.e., it appears at the observer before the source signal exists, as shown in Figure 18.4). Clearly, if the signal duration T_s is shorter than R/c, the source signal $q_l(\mathbf{r},t)$ vanishes before any true field $u_l(\mathbf{r},t)$ appears at the observer. This condition also ensures that the true signal and ghost signal never overlap. Thus, choosing $T_s < R/c$ as the subsignal duration permits one to time-gate out the ghost signal. In [2], Tygel and Hubral introduced a similar mechanism to avoid the ghost signal, which they refer to as the "causality trick."

To explore the merits of the field representation provided by (18.10) in the construction of a fast algorithm for evaluating transient fields, consider a source distribution that is confined to a sphere of radius R_s, and a set of observers that are located inside a sphere of equal radius (Figure 18.5). The centers of the source and observer spheres are denoted $\mathbf{r}_s$ and $\mathbf{r}_o$, respectively, and the vector connecting the respective sphere centers is denoted $\mathbf{R}_c = \mathbf{r}_o - \mathbf{r}_s$. It is assumed that $R_c = |\mathbf{R}_c| > 2R_s$ (i.e., the source and observer spheres do not overlap). Next, noting that the vector $\mathbf{r} - \mathbf{r}'$ can be decomposed as $\mathbf{r} - \mathbf{r}' = (\mathbf{r} - \mathbf{r}_o) - \mathbf{R}_c - (\mathbf{r}' - \mathbf{r}_s)$, (18.10) can be rewritten as

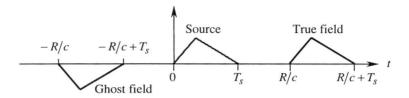

Figure 18.4 Naive application of (18.10) to a source signal results in the true observed field and an anticausal ghost field.

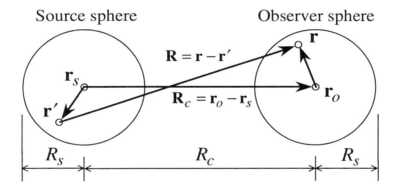

Figure 18.5 Definitions of the vectors used in the three-stage PWTD algorithm.

$$\tilde{u}_l(\mathbf{r},t) = \int d^2\hat{\mathbf{k}}\,\delta\left[t - \hat{\mathbf{k}}\cdot(\mathbf{r}-\mathbf{r}_o)/c\right] * \mathcal{T}(\hat{\mathbf{k}}, \mathbf{R}_c, t)$$
$$* \int_S d\mathbf{r}'\,\delta\left[t + \hat{\mathbf{k}}\cdot(\mathbf{r}'-\mathbf{r}_s)/c\right] * q_l(\mathbf{r}',t) \quad (18.15)$$

where $\int d^2\hat{\mathbf{k}} = \int_0^\pi d\theta \sin\theta \int_0^{2\pi} d\phi$ denotes integration over the unit sphere and the translation function $\mathcal{T}(\hat{\mathbf{k}}, \mathbf{R}_c, t)$ is given by

$$\mathcal{T}(\hat{\mathbf{k}}, \mathbf{R}_c, t) = -\frac{\partial_t}{8\pi^2 c}\delta\left(t - \hat{\mathbf{k}}\cdot\mathbf{R}_c/c\right) \quad (18.16)$$

Next, an algorithm that evaluates $\tilde{u}_l(\mathbf{r},t)$ via a three-stage implementation of (18.15) will be outlined, and the conditions for the reconstruction of $u_l(\mathbf{r},t)$ from $\tilde{u}_l(\mathbf{r},t)$ by proper time gating will be discussed. The scheme proceeds as follows.

1. Perform the rightmost convolution and integration in (18.15); that is, evaluate

$$q_l^{out}(\hat{\mathbf{k}}, t) = \int_S d\mathbf{r}' \delta\left[t + \hat{\mathbf{k}} \cdot (\mathbf{r}' - \mathbf{r}_s)/c\right] * q_l(\mathbf{r}, t) \tag{18.17}$$

which is known as the slant stack transform (SST) of the source distribution $q_l(\mathbf{r}, t)$ [5, 24, 25]. The quantities $q_l^{out}(\hat{\mathbf{k}}, t)$ can be interpreted as outgoing rays (i.e., rays that leave the source sphere in the direction $\hat{\mathbf{k}}$). All $q_l^{out}(\hat{\mathbf{k}}, t)$ together provide the source's transient far field pattern. Note that $q_l^{out}(\hat{\mathbf{k}}, t)$ is completely characterized as soon as $q_l(\mathbf{r}, t)$ ceases to radiate (i.e., at $t = lT_s$).

2. Carry out the center convolution in (18.15); that is, evaluate

$$q_l^{in}(\hat{\mathbf{k}}, t) = \mathcal{T}(\hat{\mathbf{k}}, \mathbf{R}_c, t) * q_l^{out}(\hat{\mathbf{k}}, t) \tag{18.18}$$

The operator $\mathcal{T}(\hat{\mathbf{k}}, \mathbf{R}_c, t) *$ translates each outgoing ray from the source sphere to the observer sphere. The quantities $q_l^{in}(\hat{\mathbf{k}}, t)$ are termed incoming rays (i.e., rays that impinge upon the observation sphere). Since this operation maps an outgoing ray to an incoming ray that travels in the same direction, $\mathcal{T}(\hat{\mathbf{k}}, \mathbf{R}_c, t) *$ can be considered a diagonal translation operator for transient scalar wave fields.

3. Perform the leftmost integration and convolution in (18.15); that is, evaluate

$$\tilde{u}_l(\mathbf{r}, t) = \int d^2\hat{\mathbf{k}} \, \delta\left[t - \hat{\mathbf{k}} \cdot (\mathbf{r} - \mathbf{r}_o)/c\right] * q_l^{in}(\hat{\mathbf{k}}, t) \tag{18.19}$$

This operation superimposes the projections of all the incoming rays onto the observer.

The signals associated with this three-stage algorithm are depicted in Figure 18.6. Note that the observed field depicted in Figure 18.6(g) consists of a ghost signal and the true field as implied by (18.14).

As was shown earlier, the distance between the source and observation points imposes an upper bound on the subsignal duration. Consequently, considering the nearest possible source and observation locations in the two-sphere setting, the choice of $T_s < (R_c - 2R_s)/c$ permits the elimination of the ghost signal with proper time gating. This condition implies that $u_l(\mathbf{r}, t) = 0$ for $t < lT_s$ as a time span of at least $(R_c - 2R_s)/c$ should elapse from the onset of the source subsignal $q_l(\mathbf{r}, t)$ at $t = (l-1)T_s$ for any field to propagate to the observer sphere. Also, $u_l(\mathbf{r}, t) = \tilde{u}_l(\mathbf{r}, t)$ for $t \geq lT_s$ since the ghost signals vanish throughout the observer sphere for $t \geq -(R_c - 2R_s)/c + lT_s$ by (18.14) and the observer and source spheres are assumed to be nonoverlapping, i.e., $R_c > 2R_s$. Therefore, choosing $T_s < (R_c - 2R_s)/c$ ensures that

$$u_l(\mathbf{r}, t) = \begin{cases} 0 & \text{for } t < lT_s \\ \tilde{u}_l(\mathbf{r}, t) & \text{for } t \geq lT_s \end{cases} \tag{18.20}$$

Plane-Wave Time-Domain Algorithms

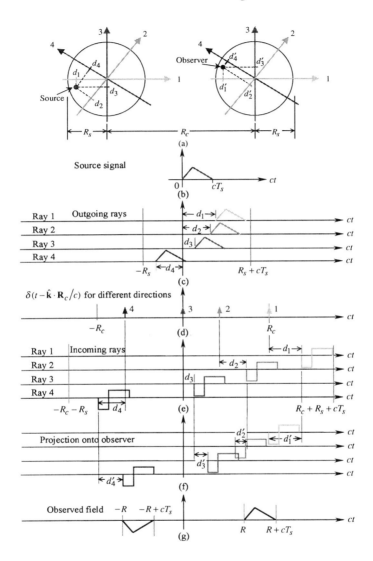

Figure 18.6 The signals associated with the three-stage PWTD algorithm. (a) The source–observation configuration and four selected ray directions. (b) The source signal. (c) The outgoing rays propagating along the selected ray directions. Note that for any source location, the outgoing rays are confined to the interval $[-R_s, R_s + cT_s]$. (d) The $\delta\left(t - \hat{\mathbf{k}} \cdot \mathbf{R}_c\right)$ term of the translation functions associated with the selected ray directions. (e) Incoming rays obtained after translation operation. (f) Projection of the incoming rays onto the observer location. (g) The observed field.

18.3.2 Implementation Issues

Several remarks regarding the implementation of the three-stage PWTD algorithm are in order.

1. *Spatial integration.* The integration over the source domain in (18.17) should be carried out using appropriate quadrature rules. The same holds true when testing the fields, as in (18.6). Since the choice of integration rule depends on the actual discretization function used for representing the source distribution or the testing function, this issue will not be further discussed here.

2. *Spectral integration.* To numerically evaluate the fields via (18.15), the integration over the unit sphere has to be performed using an appropriate quadrature rule that requires sampling of the integrand only in a small number of directions. Three basic observations allow us to achieve arbitrary accuracy when numerically evaluating (18.15). First, the integrand in (18.15), excluding the translation function, can be interpreted as the time-dependent radiation pattern of a source distribution enclosed in a sphere of radius $2R_s$ [25]. Therefore, assuming that $q_l(\mathbf{r}, t)$ is temporarily bandlimited to some $\omega_s > \omega_{max}$ (see part 3 below), this part of the integrand can be represented in terms of spherical harmonics $Y_{km}(\theta, \phi)$ as

$$g(\hat{\mathbf{k}}, \mathbf{r}, t) = \delta\left[t - \hat{\mathbf{k}} \cdot (\mathbf{r} - \mathbf{r}_o)/c\right] *$$
$$\int_S d\mathbf{r}' \delta\left[t + \hat{\mathbf{k}} \cdot (\mathbf{r}' - \mathbf{r}_s)/c\right] * q_l(\mathbf{r}', t)$$
$$= \int_S d\mathbf{r}' \delta\left[t + \hat{\mathbf{k}} \cdot [(\mathbf{r}' - \mathbf{r}_s) - (\mathbf{r} - \mathbf{r}_o)]/c\right] * q_l(\mathbf{r}', t) \quad (18.21)$$
$$= \sum_{k=0}^{K} \sum_{m=-k}^{k} g_{km}(\mathbf{r}, t) Y_{km}(\theta, \phi)$$

where $K = \lceil \chi_1 2 R_s \omega_s / c \rceil$ and $\chi_1 > 1$ is an excess bandwidth factor [26, 27] that ensures rapid convergence of the series in (18.21). Secondly, the translation function is only a function of the angle θ' between $\hat{\mathbf{k}}$ and $\mathbf{R}_c$, and can be expressed in terms of associated Legendre polynomials in θ' or the spherical harmonics in (θ, ϕ) as

$$\mathcal{T}(\hat{\mathbf{k}}, \mathbf{R}_c, t) =$$
$$= \begin{cases} -\frac{\partial_t}{16\pi^2 R_c} \sum_{k=0}^{\infty} (2k+1) P_k\left(ct/R_c\right) P_k\left(\cos\theta'\right) & |t| \leq R_c/c \\ 0 & \text{elsewhere} \end{cases}$$
$$= \begin{cases} \sum_{k=0}^{\infty} \sum_{m=-k}^{k} \mathcal{T}_{km}(\mathbf{R}_c, t) Y_{km}(\theta, \phi) & |t| \leq R_c/c \\ 0 & \text{elsewhere} \end{cases}$$
(18.22)

Finally, due to the orthogonality of the spherical harmonics, the terms in (18.22), for which $k > K$, do not contribute to the final result when integrating $g(\hat{\mathbf{k}}, \mathbf{r}, t) * \mathcal{T}(\hat{\mathbf{k}}, \mathbf{R}_c, t)$ over the unit sphere. Hence, the summation in (18.22) can be truncated at $k = K$, and, in all previous expressions, the translation function $\mathcal{T}(\hat{\mathbf{k}}, \mathbf{R}_c, t)$ can be replaced by its truncated version $\bar{\mathcal{T}}(\hat{\mathbf{k}}, \mathbf{R}_c, t)$ given by

$$\bar{\mathcal{T}}(\hat{\mathbf{k}}, \mathbf{R}_c, t) =$$
$$= \begin{cases} -\frac{\partial_t}{16\pi^2 R_c} \sum_{k=0}^{K} (2k+1) P_k\left(ct/R_c\right) P_k\left(\cos\theta'\right) & |t| \leq R_c/c \\ 0 & \text{elsewhere} \end{cases}$$
(18.23)

As illustrated in Figure 18.7, with this truncation, the translation function becomes nonlocalized and spans a duration of $2R_c/c$. Note that, as the harmonics with $k > K$ do not contribute to the final result, they can be tapered off with a smooth windowing instead of the hard truncation in (18.23) so that the translation function spans a shorter duration. This also implies that the translation function is not unique. This point will be further exploited in Sections 18.5 and 18.6 to develop more efficient algorithms.

It is now clear that the integrand in (18.15) can be written as a product of two functions, each of which is expressible in terms of spherical harmonics $Y_{km}(\theta, \phi)$, $k = 0, \ldots, K$, $m = -k, \ldots, k$. Such an integral can be evaluated exactly by using a $(2K + 1)$-point trapezoidal rule in the ϕ direction and a $(K + 1)$-point Gauss-Legendre quadrature in the θ direction yielding the following expression for $\tilde{u}_l(\mathbf{r}, t)$:

$$\tilde{u}_l(\mathbf{r}, t) = \sum_{p=0}^{K} \sum_{q=-K}^{K} w_{pq} \, \delta\left[t - \hat{\mathbf{k}}_{pq} \cdot (\mathbf{r} - \mathbf{r}_o)/c\right] * \bar{\mathcal{T}}(\hat{\mathbf{k}}_{pq}.\mathbf{R}_c, t)$$
$$* \int_S d\mathbf{r}' \, \delta\left[t + \hat{\mathbf{k}}_{pq} \cdot (\mathbf{r}' - \mathbf{r}_s)/c\right] * q_l(\mathbf{r}', t)$$
(18.24)

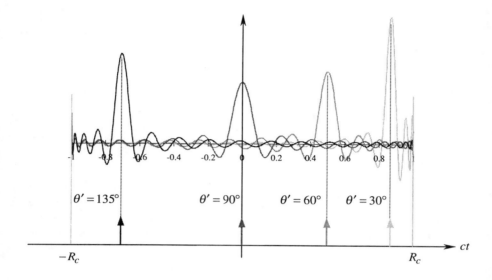

Figure 18.7 Effect of bandlimiting the translation function. The lower part illustrates the Dirac functions expressed by the infinite series in (18.22), and the upper part illustrates the truncated series of (18.23) for $K = 30$.

where

$$w_{pq} = \frac{4\pi(1 - \cos^2 \theta_p)}{(2K + 1)\left[(K + 1)P_K(\cos \theta_p)\right]^2}$$

$$\hat{\mathbf{k}}_{pq} = \hat{\mathbf{x}} \sin \theta_p \cos \phi_q + \hat{\mathbf{y}} \sin \theta_p \sin \phi_q + \hat{\mathbf{z}} \cos \theta_p \quad (18.25)$$

$$\phi_q = q2\pi/(2K + 1)$$

θ_p is the $(p + 1)^{th}$ zero of $P_{K+1}(\cos \theta)$

Although this choice of sampling points seems to depend on the coordinate system, this is not true owing to the fact that under a coordinate system rotation a spherical harmonic of degree K is transformed into a linear combination of harmonics of the same degree [28, 29]. Hence, as already implied by the second equality in (18.22), sampling points can be chosen with respect to any coordinate system independent of $\mathbf{R}_c$.

3. *Subsignal temporal representation.* Finally, in the derivation of the above quadrature rule, it was assumed that the subsignals are temporally bandlimited. This assumption contradicts the requirement that each subsignal has to be time limited as well. However, the whole signal that is bandlimited to ω_{max} can

be broken up into subsignals that are both bandlimited to $\omega_s = \chi_0 \omega_{max}$ with $\chi_2 > 1$ and *approximately* time limited to a given interval length by using proper local interpolation functions as follows.

Since $q(\mathbf{r}, t)$ is temporally bandlimited (i.e., the temporal spectrum of $q(\mathbf{r}, t)$ vanishes for $\omega > \omega_{max}$) it can be sampled and locally interpolated using temporally bandlimited and approximately time limited functions as

$$q(\mathbf{r}, t) \cong \sum_{k=1}^{N_t} q(\mathbf{r}, k\Delta_t) \psi_k(t) \qquad (18.26)$$

where Δ_t is the time step size and $\psi_k(t)$ is a bandlimited interpolant [30]. Many choices for the interpolation function exist; however, a near optimal one—a variant of the approximate prolate spheroidal (APS) functions introduced by Knab [31]—is given by

$$\psi_k(t) = \frac{\omega_+}{\omega_s} \frac{\sin(\omega_+(t - k\Delta_t))}{\omega_+(t - k\Delta_t)} \\ \times \frac{\sinh\left(\omega_- p_t \Delta_t \sqrt{1 - [(t - k\Delta_t)/p_t\Delta_t]^2}\right)}{\sinh(\omega_- p_t \Delta_t)\sqrt{1 - [(t - k\Delta_t)/p_t\Delta_t]^2}} \qquad (18.27)$$

where

$$\omega_s = \pi/\Delta_t = \chi_0 \omega_{max} \qquad (18.28)$$

In the above, $\chi_0 > 1$ is the oversampling ratio, $\omega_{\pm} = (\omega_s \pm \omega_{max})/2$, and p_t is an integer that defines the approximate duration of the interpolation function. In practice, a truncated version of $\psi_k(t)$, obtained by setting $\psi_k(t) = 0$ for $|t - k\Delta_t| > p_t \Delta_t$, is used. The relative interpolation error ε_t introduced by this truncation can be shown to be bounded as [31]

$$|\varepsilon_t| \leq \frac{1}{\sinh(\omega_- p_t \Delta_t)} \qquad (18.29)$$

which decreases exponentially fast with increasing χ_0 for a fixed p_t. Hence, (18.26) permits local interpolation in terms of $2p_t$ samples.

From (18.26), it follows that the source signal $q(\mathbf{r}, t)$ can be broken up into subsignals $q_l(\mathbf{r}, t)$, as in (18.7), given by

$$q_l(\mathbf{r}, t) = \sum_{k=l M_t}^{(l+1)M_t - 1} q(\mathbf{r}, k\Delta_t) \psi_k(t) \qquad (18.30)$$

Each subsignal $q_l(\mathbf{r}, t)$ is defined in terms of M_t samples of the signal $q(\mathbf{r}, t)$ but spans $M_t' = M_t + 2p_t$ time steps. In other words, whereas each subsignal

$q_l(\mathbf{r}, t)$ is formed from samples of $q(\mathbf{r}, t)$ in an interval of length $T_s = M_t \Delta_t$, the duration of each $q_l(\mathbf{r}, t)$ is $T'_s = M'_t \Delta_t$, and adjacent subsignals overlap by $2p_t$ samples as illustrated in Figure 18.2(b). Obviously, the total number of time samples equals $N_t = LM_t$. Since the interpolation function $\psi_k(t)$ is bandlimited to ω_s, so is each subsignal.

From the above discussions, it should be obvious that the accuracy of the three-stage PWTD algorithm is solely determined by the choice of χ_1 and χ_0. To verify this and to determine the level of accuracy that can be achieved using the proposed algorithm, several numerical experiments were conducted. The results of one such experiment are depicted in Figure 18.8. In this experiment, eight point sources were distributed over a unit sphere centered at $(x, y, z) = (-5, -5, -5)$ meters, and an observation sphere of unit radius was centered at the origin. The time signature of each source was a Gaussian pulse given by $e^{-(t-7.75\sigma)^2/(2\sigma^2)}$ with $\sigma = 3.183$ ns, and the wave speed of the medium was assumed to be $c = 3 \times 10^8$ m/s. The radiated fields were evaluated via the three-stage PWTD algorithm throughout a 6m × 6m region in the $x = y$ plane centered around the observation sphere. Then, the calculated fields were compared with those obtained analytically. The energy norm of the error thus obtained is plotted in Figure 18.8 for four cases, in which the parameters used for forming the subsignals as well as the truncation limit for the translation functions were estimated to yield L2 errors of order 10^{-4}, 10^{-7}, 10^{-10}, and 10^{-14}. As seen in the figure, the targeted level of accuracy has been achieved within the observer sphere for each case.

18.4 IMPLEMENTATION OF THE PWTD-ENHANCED MOT SCHEMES

The previous section described a PWTD algorithm that permits the plane wave–based evaluation of transient fields due to a known source distribution. Incorporation of this algorithm into the classical MOT framework results in source reconstruction schemes whose costs scale much more favorably than that of the conventional MOT method. The most naive way to combine the PWTD and MOT schemes is to divide the scatterer into equally sized subscatterers and to evaluate interactions between nearby and distant subscatterers using the classical MOT and PWTD algorithms, respectively. This yields a "two-level" scheme with $O(N_t N_s^{1.5} \log N_s)$ complexity. A more efficient "multilevel" algorithm with $O(N_t N_s \log^2 N_s)$ complexity is obtained by grouping very distant subscatterers into even larger entities before invoking the PWTD scheme (Figure 18.9). In this section, the two-level algorithm will be described first and its multilevel extension will be outlined next.

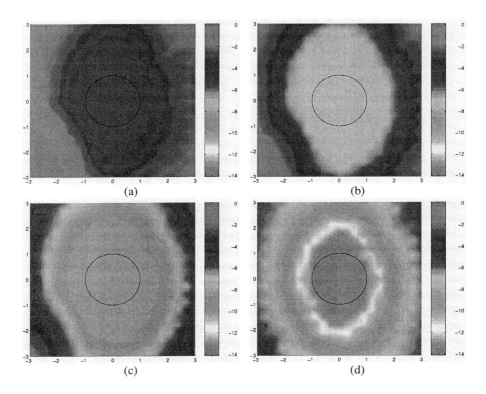

Figure 18.8 Energy norm of the error calculated by the three-stage PWTD algorithm around the observation sphere. The error estimates prior to the calculation were of order (a) 10^{-4}, (b) 10^{-7}, (c) 10^{-10}, (d) 10^{-14}.

18.4.1 A Two-Level PWTD-Enhanced MOT Algorithm

The basic idea supporting the two-level PWTD-enhanced MOT scheme is very simple. To rapidly evaluate the sum appearing on the right-hand side of (18.5), the scatterer is subdivided into a large number of subscatterers. Next, all contributions to this sum that arise from spatial basis and testing functions residing on nearby subscatterers are evaluated directly (i.e., as in (18.5)). All other contributions are evaluated using the PWTD scheme.

These ideas are formalized using the following definitions. Consider a fictitious cubical volume enclosing the scatterer that is subdivided into many equally sized smaller boxes, each of which fits into a circumscribing sphere of radius R_s as

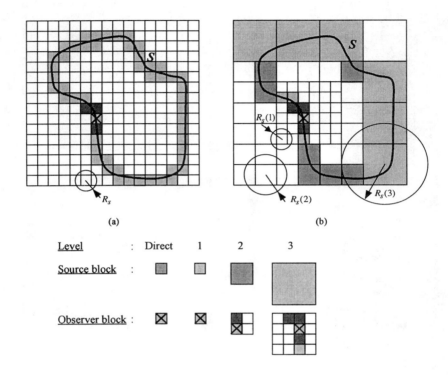

Figure 18.9 Cross-sectional view of the (a) two-level and (b) multilevel partitioning of the scatterer surface when the observer is located in the box with an ×.

illustrated in Figure 18.9(a). Let N_g denote the total number of nonempty boxes. Henceforth, the collection of spatial basis functions that reside within a nonempty box is termed a group and the average number of spatial basis functions per group is denoted by $M_s = N_s/N_g$. It can be verified that, for a surface scatterer, $M_s \propto (R_s \omega_{max}/c)^2$. Each group pair (γ, γ'), $\gamma, \gamma' = 1, \ldots, N_g$, is identified as either a "near field" or a "far field" pair depending on whether their group centers are separated by less than or more than a preset distance $R_{c,min}$, respectively. It is assumed that $R_{c,min} = \xi R_s$, where ξ typically varies between 3 and 6. This implies that every group participates in only a small number of near field pairs, and therefore that the total number of near field pairs is proportional to N_g.

In order to facilitate a ghost-free evaluation of fields involving far field source and observer groups using the PWTD algorithm, appropriate subsignal durations must be defined. To this end, the fundamental subsignal duration T_s is defined as $T_s = M_t \Delta_t$, where $M_t = \min_{\gamma,\gamma'} \{\lfloor (R_{c,\gamma\gamma'} - 2R_s)/(c\Delta_t) \rfloor \}$ and $R_{c,\gamma\gamma'}$ denotes the distance between the centers of groups γ and γ'. This T_s is easily identified as the maximum duration of a subsignal that can be translated without incurring any ghost signals between the nearest groups classified as a far field pair. Subsignal durations that feature in the evaluation of fields related to other far field pairs (γ, γ') are represented as $T_{s,\gamma\gamma'} = M_{t,\gamma\gamma'} \Delta_t$. The integers $M_{t,\gamma\gamma'}$ are multiples of M_t and are computed as $M_{t,\gamma\gamma'} = M_t \lfloor (R_{c,\gamma\gamma'} - 2R_s)/(cT_s) \rfloor$; hence, all subsignal durations $T_{s,\gamma\gamma'}$ are integer multiples of the fundamental subsignal duration T_s. Note that these definitions enable one to recycle information since outgoing rays that feature in the interaction of any far field pair can be obtained by concatenating an integer number of rays corresponding to subsignals of duration T_s.

Before embarking on a systematic description of the two-level scheme, a few words are in order regarding the convolution of outgoing rays with translation functions (18.18). The maximum duration of the outgoing rays that are to be translated between groups γ and γ' is $T_{s,\gamma\gamma'} + 2R_s$, which can be shown to be proportional to $R_{c,\gamma\gamma'}/c \propto M_{t,\gamma\gamma'} \Delta_t$ with the aid of the above definitions. Also, as evident from (18.23), the duration of the translation functions is proportional to $R_{c,\gamma\gamma'}/c$—in fact, it exactly equals $2R_{c,\gamma\gamma'}/c$. This implies that both signals that are to be convolved last $O(M_{t,\gamma\gamma'})$ time steps. To evaluate this convolution using fast Fourier transform (FFT) techniques in $O(M_{t,\gamma\gamma'} \log M_{t,\gamma\gamma'})$ operations, note that, as mentioned in Section 18.3.2, the subsignals and therefore the outgoing rays are bandlimited. Consequently, their spectrum can be obtained using FFTs. However, as the translation functions are not bandlimited, their spectrum cannot be obtained by simple FFTs. Fortunately, analytic expressions for the Fourier transforms of the translation functions are available, enabling the rapid evaluation of their spectra. Using (18.23), the Fourier transform of the translation functions is

$$\mathcal{F}\left\{\bar{\mathcal{T}}(\hat{\mathbf{k}}, \mathbf{R}_{c,\gamma\gamma'}, t)\right\} = \int_{-\infty}^{+\infty} dt \, \bar{\mathcal{T}}(\hat{\mathbf{k}}, \mathbf{R}_{c,\gamma\gamma'}, t) \, e^{-j\omega t}$$

$$= -\frac{j\omega}{8\pi^2 c} \sum_{k=0}^{K} (2k+1)(-j)^k j_k \left(\omega R_{c,\gamma\gamma'}/c\right) P_k \left(\cos \theta'\right)$$

(18.31)

where $j_k(\cdot)$ is the spherical Bessel function of order k. An efficient way to construct these functions for any sphere pair (γ, γ') is to tabulate the normalized translation function

$$\tilde{T}(\theta', \Omega) = R_{c,\gamma\gamma'} \left\{ \tilde{T}(\hat{\mathbf{k}}, \mathbf{R}_{c,\gamma\gamma'}, t) \right\}$$
$$= -\frac{j\Omega}{8\pi^2} \sum_{k=0}^{K} (2k+1)(-j)^k j_k(\Omega) P_k(\cos\theta') \quad (18.32)$$

with respect to the ray angle θ' and the normalized frequency $\Omega = \omega R_{c,\gamma\gamma'}/c$. Since $\tilde{T}(\theta', \Omega)$ is bandlimited in both θ' and Ω, it can be reconstructed from judiciously selected samples in the (θ', Ω) plane. Specifically, it can be shown that $O(\max_{\gamma,\gamma'}\{M_{t,\gamma\gamma'}\}) = O(\sqrt{N_s})$ samples in Ω and $O(K)$ samples in θ' permit the reconstruction of $\tilde{T}(\theta', \Omega)$ to arbitrary precision for all far field pairs.

A two-level algorithm for the rapid evaluation of the sum on the right-hand side of (18.5) can now be prescribed. In this algorithm, the contributions to this sum due to all near and far field pairs are evaluated separately.

1. *Evaluation of the near field contributions.* At each time step, the sum

$$\sum_{k=1}^{j-1} \bar{\mathbf{Z}}_k^{\gamma\gamma'} \mathbf{Q}_{j-k}^{\gamma'} \quad (18.33)$$

is computed for all near field pairs (γ, γ'). In (18.33), $\bar{\mathbf{Z}}_k^{\gamma\gamma'}$ denotes the submatrix of the interaction matrix $\bar{\mathbf{Z}}_k$ that relates fields over group γ to sources residing in group γ', and $\mathbf{Q}_{j-k}^{\gamma'}$ is a vector comprised of the $q_{n,j-k}$ for all sources n in group γ'. For every time step, the evaluation of the sum (18.33) requires $O(M_s^2)$ operations per near field pair. Since each group interacts in a direct fashion only with a small set of near field neighbors, the overall cost of this step scales as $O(N_t N_g M_s^2) = O(N_t N_s M_s)$.

2. *Evaluation of the far field contributions.* The three-step procedure outlined in Section 18.3.1 that accounts for the far field contributions is incorporated into the MOT scheme as follows.

   (a) *Construction of outgoing rays.* For each group, a set of outgoing rays describing transient far fields that are generated by subsignals of fundamental duration T_s is constructed every M_t time steps. Each such set consists of $O(K^2) = O(R_s^2 \omega_{max}^2/c^2) = O(M_s)$ rays that are formed by contributions from all sources residing in a group. The contribution from the spatial basis function $f_n(\mathbf{r})$ to the outgoing ray propagating in the $\hat{\mathbf{k}}_{pq}$ direction is obtained by convolving the subsignal associated with $f_n(\mathbf{r})$ with

$$V_n^+(\hat{\mathbf{k}}_{pq}.t) = \int_S d\mathbf{r}' \, \delta\left[t + \hat{\mathbf{k}}_{pq} \cdot (\mathbf{r}' - \mathbf{r}_c)/c\right] f_n(\mathbf{r}') \quad (18.34)$$

where $\mathbf{r}_c$ denotes the center of the group to which $f_n(\mathbf{r})$ belongs. Since this mapping from $O(M_s)$ sources to $O(K^2)$ rays is carried out for all N_g groups a total of N_t/M_t times and since the cost of projecting a single spatial source onto a single direction scales as $O(M_t)$ per subsignal, the cost of constructing all outgoing rays for the duration of the analysis is $O(M_s K^2 N_g (N_t/M_t) M_t) = O(N_t N_s M_s)$.

(b) *Translation.* For each far field pair (γ, γ'), outgoing rays are translated (i.e., converted from outgoing to incoming rays) from group γ to γ' every $M_{t,\gamma\gamma'}$ time steps. Note that the duration of the subsignals that are translated is $T_{s,\gamma\gamma'}$. As mentioned previously, although outgoing rays describing such subsignals are not readily available, they can be formed by concatenating $M_{t,\gamma\gamma'}/M_t$ rays of fundamental duration T_s computed in step 2a. This is feasible as $M_{t,\gamma\gamma'}$ is an integer multiple of M_t by design. Each outgoing ray is translated to an incoming ray by executing the following steps:

  i. An outgoing ray described by $O(M_{t,\gamma\gamma'})$ time samples is constructed by concatenating the rays stored in step 2a. Furthermore, anticipating its convolution with a translation function, its Fourier transform is computed using an FFT, which can be accomplished in $O(M_{t,\gamma\gamma'} \log M_{t,\gamma\gamma'})$ operations.

  ii. The spectrum of the pertinent translation function is extracted from the $\tilde{\mathcal{T}}(\theta', \Omega)$ table through local interpolation in $O(M_{t,\gamma\gamma'})$ operations.

  iii. The outgoing ray spectrum is multiplied with the translation function spectrum in $O(M_{t,\gamma\gamma'})$ operations.

  iv. The result is inverse Fourier transformed back into time domain in $O(M_{t,\gamma\gamma'} \log M_{t,\gamma\gamma'})$ operations. The rays obtained through this procedure are superimposed onto incoming rays that impinge upon the receiver group from all other source groups.

Clearly, the dominant cost per direction scales as $O(M_{t,\gamma\gamma'} \log M_{t,\gamma\gamma'})$. For a single group pair, the above steps are to be repeated $N_t/M_{t,\gamma\gamma'}$ times to cover the duration of the entire transient analysis, and for all $O(K^2)$ directions, which results in a cost that scales as $O((N_t/M_{t,\gamma\gamma'})K^2 M_{t,\gamma\gamma'} \log M_{t,\gamma\gamma'}) = O(N_t M_s \log M_{t,\gamma\gamma'})$. Note that the subsignal duration $M_{t,\gamma\gamma'}$ is proportional to the distance between groups, which is bounded by the maximum linear dimension of the scatterer. For a surface scatterer, the maximum linear dimension is proportional to $\sqrt{N_s}$. Hence, in the above estimate, $\log M_{t,\gamma\gamma'}$ can safely be replaced by $\log N_s$. The total computational cost of translation between all $O(N_g^2)$ far field group pairs therefore scales as $O(N_g^2 N_t M_s \log N_s) = O(N_t N_s^2 M_s^{-1} \log N_s)$.

(c) *Projection of incoming rays onto observers.* At each time step, the field at the nth observer is formed by convolving the incoming rays with

$$\tilde{V}_n^-(\hat{\mathbf{k}}_{pq}.t) = \int_S d\mathbf{r}' \, \delta\left[t - \hat{\mathbf{k}}_{pq} \cdot (\mathbf{r}' - \mathbf{r}_c)/c\right] \tilde{f}_n(\mathbf{r}') \qquad (18.35)$$

and by performing the spherical integration (i.e., adding up the signals due to all incoming rays). This step constitutes a mapping from $O(K^2)$ directions to $O(M_s)$ sources for all groups and all time steps, and can be accomplished in $O(N_t N_g K^2 M_s) = O(N_t N_s M_s)$ operations.

Evidently, the computational complexity of each of the steps in this algorithm depends on the group size M_s. It can be verified by adding up the costs associated with steps 1 and 2 above that the overall complexity is minimized by choosing M_s proportional to $\sqrt{N_s}$. With this choice, the computational cost of performing a transient scattering analysis using the two-level algorithm scales as $O(N_t N_s^{1.5} \log N_s)$.

18.4.2 A Multilevel PWTD-Enhanced MOT Algorithm

A multilevel PWTD-enhanced MOT scheme can be constructed by casting the two-level algorithm into a divide-and-conquer framework. The procedure that permits a further reduction in computational complexity of the two-level scheme involves the multilevel aggregation of small subscatterers into larger entities before translation, inasmuch as permitted by the PWTD algorithm. This section will introduce a systematic scheme for achieving a multilevel partitioning of a scatterer along with an arsenal of multilevel notation. Also, four operations—ray interpolation, ray splicing, ray resection, and ray anterpolation—that permit the transfer of information across levels will be elucidated. Finally, the multilevel PWTD algorithm will be described in detail.

A hierarchical subdivision of the scatterer is achieved by recursively subdividing a fictitious cubical box that encloses the scatterer. Initially, this box is divided into eight boxes, each of which is further subdivided into eight smaller boxes in a recursive manner. A box that is subdivided into smaller boxes is termed the "parent" of the "child" boxes that result from the operation. The finest boxes so obtained are termed level 1 boxes, and the collection of spatial basis functions that fall into a level 1 box is said to form a level 1 group; higher level boxes and groups are defined similarly. For levels $i = 1, ..., N_l$, let $N_g(i)$ denote the number of groups (nonempty boxes), $M_s(i)$ the average number of sources in each group, $R_s(i)$ the radius of the sphere that encloses a level i box, and $K(i)$ the number of spherical harmonics used in construction of the translation functions. It then follows from the discussion of the two-level scheme that $M_s(i) \propto (R_s(i) \omega_{max}/c)^2 \propto K^2(i)$. It is assumed that level 1 boxes have linear dimensions that are proportional to the wavelength at ω_{max} [i.e., $(R_s(1) \omega_{max}/c)$ is of $O(1)$]. This in turn implies that level 1 groups contain only a small number of basis functions that is independent of the problem size [i.e., $M_s(1)$ is of $O(1)$]. It also follows that $N_g(1) \propto N_s$ and that $N_l \propto \log N_s$. Finally, note that, for a surface scatterer, $N_g(i+1) \propto N_g(i)/4$ and $M_s(i+1) \propto 4M_s(i)$, and therefore $N_g(i+1)M_s(i+1) \approx N_g(i)M_s(i) \approx N_g(1)M_s(1) \approx N_s$.

Next, a set of near and far field group pairs is constructed. Just as in the two-level algorithm, each and every source/observer (basis/testing function) combination belongs to one and only one pair. However, in contrast to the two-level PWTD algorithm, all far field pairs do not all reside at the same level: distant source/observer combinations tend to belong to higher-level far field pairs than those that reside close to one another [Figure 18.9(b)]. Consequently, cutoff separations for all levels are defined as $R_{c,min}(i) = \xi R_s(i)$. To construct the near and far field pairs, first, each group pair at the highest level whose centers are separated by more than $R_{c,min}(N_l)$ is classified as a "level N_l far field pair." Next, all level $N_l - 1$ pairs with group centers separated by more than $R_{c,min}(N_l - 1)$, and describing interactions that have not yet been accounted for by any of the level N_l far field group pairs, are classified as "level $N_l - 1$ far field pairs." This process is continued, and far field pairs are identified at each and every level, including level 1, as those pairs that are considered well separated at a given level, but that have not yet been accounted for at a higher level. The level 1 pairs with group centers separated by less than $R_{c,min}(1)$ are classified as near field pairs.

As in the two-level algorithm, the fundamental subsignal duration for level i is calculated as $T_s(i) = M_t(i)\Delta_t$ with $M_t(i) = \min_{\gamma,\gamma'} \{\lfloor (R_{c,\gamma\gamma'} - 2R_s(i))/(c\Delta_t) \rfloor\}$ where γ and γ' vary over all the level i far field groups. Also, subsignal durations for evaluation of fields involving a far field pair (γ, γ') at level i are defined as $T_{s,\gamma\gamma'}(i) = M_{t,\gamma\gamma'}(i)\Delta_t$ with $M_{t,\gamma\gamma'}(i) = M_t(i) \lfloor (R_{c,\gamma\gamma'} - 2R_s(i))/(cT_s(i)) \rfloor$.

For a given level i far field group pair (γ, γ'), the field at group γ due to $q_l(\mathbf{r}, t)$ in group γ' (and vice versa) will be calculated using the three-stage PWTD algorithm. The first stage of this algorithm calls for the evaluation of the SSTs to form the outgoing rays associated with group γ' along $O(K^2(i))$ directions. For $i = 1$, the outgoing rays can be obtained by directly evaluating $V_n^+(\hat{\mathbf{k}}_{pq}.t)$ as in (18.34). However, for higher levels, direct evaluation of the SSTs is computationally expensive. The fact that the same source information is to be used to construct outgoing rays at each level suggests that rays at level $i > 1$ can be constructed economically by reusing information already stored in level $i - 1$ rays. This is achieved by two operations termed *interpolation* and *splicing*, which are schematically illustrated in Figure 18.10(a). Interpolation is needed as more rays are to be associated with a parent group than with one of its children and splicing is used in assembling a single parent group ray from interpolated child rays. Similarly, the last stage of the PWTD algorithm calls for a projection of incoming rays onto observers. Directly projecting the incoming rays at level $i > 1$ onto the observers is more expensive than disaggregating these rays into level $i - 1$ rays and propagating the information contained in these rays through the multilevel structure until level 1 rays are projected onto the observer locations using (18.35). Two operations termed *resection* and *anterpolation*, which are complementary to the splicing and interpolation operators, construct level $i-1$ incoming rays for a group from those of its parent as illustrated in Figure 18.10(b). Implementation of these four operations is discussed next.

First, consider the interpolation and anterpolation operations. As mentioned earlier, the outgoing rays associated with a group describe the time-dependent radiation pattern of the source distribution associated with that group. Because the radiation pattern of a source distribution that is spatially bounded by a sphere of radius $R_s(i)$ and spectrally bandlimited to ω_s can be expressed in terms of $K(i)$ spherical harmonics [26], the interpolation operator increases the sampling rate and zero-pads the excess spherical spectrum introduced. Similarly, anterpolation operations call for the application of a spherical filter with uniform resolution to a set of outgoing or incoming rays. In other words, anterpolation is equivalent to truncating the spherical harmonic content and lowering the sampling rate over the sphere. If implemented as described by Jakob-Chien and Alpert [32], the application of interpolation and anterpolation operators between levels i and $i+1$ can be completed in $O(K^2(i) \log K(i))$ operations per time step and per child-parent group pair. This yields approximately $O(N_s \log K(i))$ operations per time step at level i, and an overall computational cost of order less than $O(N_t N_s \log^2 N_s)$.

The nature of the splicing and resection operations can be understood by inspecting (18.24). With reference to Figure 18.11(a), assume that an outgoing ray of a level $i+1$ box, whose center is denoted by $\mathbf{r}_c$, is to be formed. Denoting the centers of the N_c child boxes associated with this level $i+1$ box as $\mathbf{r}_{c,\xi}$, $\xi = 1, \ldots, N_c$, and the part of the surface S that lies inside each of these level i boxes as S_ξ, the expression for an outgoing ray of the parent box takes the form

$$\int_S d\mathbf{r}' \, \delta\left[t + \hat{\mathbf{k}}_{pq} \cdot (\mathbf{r}' - \mathbf{r}_c)/c\right] * q_l^{i+1}(\mathbf{r}', t)$$

$$= \sum_{\xi=1}^{N_c} \delta\left[t + \hat{\mathbf{k}}_{pq} \cdot (\mathbf{r}_{c,\xi} - \mathbf{r}_c)/c\right] \quad (18.36)$$

$$* \int_{S_\xi} d\mathbf{r}' \, \delta\left[t + \hat{\mathbf{k}}_{pq} \cdot (\mathbf{r}' - \mathbf{r}_{c,\xi})/c\right] * q_l^{i+1}(\mathbf{r}', t)$$

In (18.36), the superscript $i+1$ on $q_l(\mathbf{r}', t)$ signifies the fact that the source subsignal is of duration $T_s(i+1)$, and $\hat{\mathbf{k}}_{pq}$ denotes a ray direction at level $i+1$. Since each $q_l^{i+1}(\mathbf{r}, t)$ can be obtained by splicing two level i subsignals $q_{l'}^i(\mathbf{r}, t)$ and $q_{l'+1}^i(\mathbf{r}, t)$ for some l' as shown in Figure 18.11(b), the outgoing ray of the parent box can be

Plane-Wave Time-Domain Algorithms

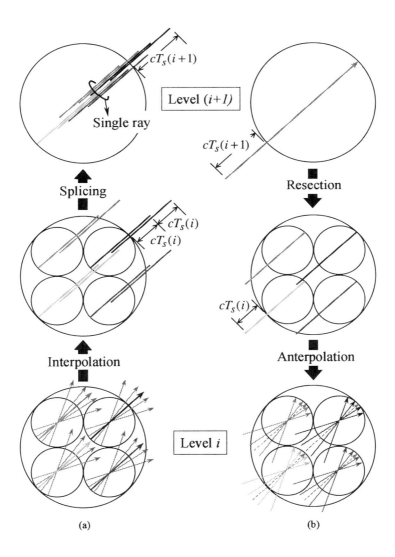

Figure 18.10 (a) Ray interpolation and ray splicing operations to form the outgoing rays of level $i + 1$ from the rays of level i. (b) Ray resection and ray anterpolation operations to obtain incoming rays of level i from the rays of level $i + 1$.

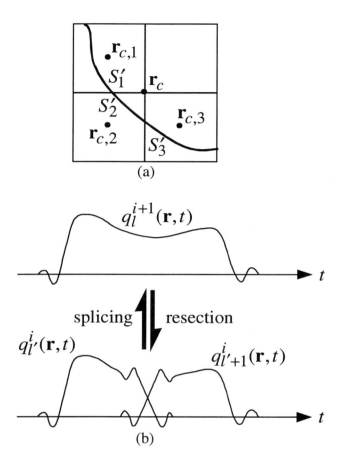

Figure 18.11 Schematic description of the splicing and resection operations.

expressed as

$$\sum_{\zeta=l'}^{l'+1} \sum_{\xi=1}^{N_c} \delta\left[t + \hat{\mathbf{k}}_{pq} \cdot (\mathbf{r}_{c,\xi} - \mathbf{r}_c)/c\right] \\ * \int_{S_\xi} d\mathbf{r}' \, \delta\left[t + \hat{\mathbf{k}}_{pq} \cdot (\mathbf{r}' - \mathbf{r}_{c,\xi})/c\right] * q_\zeta^i(\mathbf{r}', t) \quad (18.37)$$

Note that in (18.37), the integral over each S_ξ is nothing but an outgoing ray of a child box propagating along direction $\hat{\mathbf{k}}_{pq}$. Hence, (18.37) clearly indicates that once the outgoing rays of the child boxes are interpolated to level $i+1$ directions $\hat{\mathbf{k}}_{pq}$, they can be time-advanced by an amount $s_\xi(\hat{\mathbf{k}}_{pq}) = \hat{\mathbf{k}}_{pq} \cdot (\mathbf{r}_{c,\xi} - \mathbf{r}_c)/c$ and spliced together to form the outgoing rays of the parent box. In a completely complementary manner, it can be shown that the incoming rays of a level i group can be obtained by resecting the incoming rays of its parent box as depicted in Figure 18.11(b), delaying these rays by an amount $s_\xi(\hat{\mathbf{k}}_{pq})$, and anterpolating the resulting rays to level i directions.

With all requisite tools for a multilevel scheme defined, the fast multilevel evaluation of the sum in (18.5) proceeds as follows:

1. *Evaluation of the near field contributions.* As in the two-level scheme, all interactions between source/observer combinations that belong to near field pairs are accounted for classically. As mentioned earlier, all near field pairs reside on the finest level. Since each finest level group interacts with only a small number of its immediate neighbors through the near field, the computational complexity associated with the evaluation of the near field contributions per time step scales as $O(N_g(1)M_s^2(1))$. However, since $N_g(1) \propto N_s$ and $M_s(1)$ is of $O(1)$, the total cost for near field evaluation for all time steps scales as $O(N_t N_s)$.

2. *Evaluation of the far field contributions.* These contributions are still evaluated via a three-stage process reminiscent of that described for the two-level scheme. However, the independent generation of outgoing and incoming rays at all levels would result in too high a computational cost. Fortunately, these rays can be efficiently constructed by supplementing the first and the last stages by the ray interpolation, splicing, resection, and anterpolation operations. With these modifications in mind, the three-stage far field evaluation scheme can be described as follows:

    (a) *Construction of outgoing rays.* Outgoing rays for all groups at level i are constructed every $M_t(i)$ time steps. At level 1, this is accomplished by convolving the source signatures with $V_n^+(\hat{\mathbf{k}}_{pq}.t)$. Higher-level rays are constructed from those at the previous level through interpolation and splicing. Note that, since the fundamental subsignal duration at level $i+1$ is twice that at level i (Figure 18.10(a)), the levels must be traversed starting from level 1. This guarantees that level i rays are constructed before they are needed by the ray splicing operator to obtain the rays at level $i+1$. As mentioned above, the total cost of constructing outgoing rays for all levels using interpolation and splicing scales as $O(N_t N_s \log^2 N_s)$.

    (b) *Translation.* As in the two-level algorithm, rays are translated between the far field pairs (γ, γ') at level i every $M_{t,\gamma\gamma'}(i)$ time steps by concatenating rays of fundamental duration $T_s(i)$ formed in step 2a. Note that a different $\tilde{\mathcal{T}}(\theta', \Omega)$ table has to be constructed for each level since the number of harmonics $K(i)$ required in the computation of the translation functions varies across levels. The translation of rays for all level i far field pairs can be accomplished in $O\left(N_g(i)K^2(i)N_t \log M_{t,\gamma\gamma'}(i)\right) = O\left(N_s N_t \log M_s(i)\right)$ operations. When summed over all levels, this yields a complexity of $O\left(N_t N_s \sum_{i=1}^{\log N_s} \log M_s(i)\right) = O(N_t N_s \log^2 N_s)$ for translation.

(c) *Projection of incoming rays onto observers.* Starting with level $N_l - 1$, incoming rays for each group are resected and anterpolated from the incoming rays at the higher level every $M_t(i)$ time steps. The fields at the observer locations are constructed by convolving incoming rays at level 1 with $\tilde{V}_n^-(\hat{\mathbf{k}}_{pq}.t)$ and performing the spherical integration. This step, which is conceptually the transpose of forming the outgoing rays, can also be accomplished in $O(N_t N_s \log^2 N_s)$ operations.

Considering all the steps of the multilevel algorithm, it is seen that a scattering analysis using a multilevel PWTD-enhanced MOT scheme can be completed in $O(N_t N_s \log^2 N_s)$ operations.

18.5 THE WINDOWED PLANE-WAVE TIME-DOMAIN ALGORITHM

The previous section has shown that two-level and multilevel PWTD-enhanced algorithms offer a considerable reduction in computational complexity over traditional MOT schemes. As with frequency domain fast multipole methods, the computational complexity of the proposed schemes can be further reduced by using windowed translation functions [6, 8]. This additional cost reduction hinges on the observation that, if the time span of the translation functions can be shortened, the number of rays that need to be translated for the evaluation of the observed field shrinks to a constant independent of the group size for sufficiently remote source and observation groups. Shorter translation functions can be constructed either by smoothly tapering off, or windowing, the series in (18.23) for $k < K$ instead of truncating it at K or by using sampled field representations that rely on the bandlimitedness of the radiated fields, as will be done in the present section.

This section introduces windowed diagonal time-domain translation operators that permit the rapid evaluation of transient fields produced by surface-bound source distributions. It will be shown that the computational complexities associated with the solution of large-scale surface scattering problems using the proposed two-level and multilevel windowed PWTD algorithms based on these operators scale as $O(N_t N_s^{4/3} \log N_s)$ and $O(N_t N_s \log N_s)$, respectively.

18.5.1 Windowed Plane-Wave Decomposition

The immediate goal of this section is to develop a plane-wave representation of the field $u(\mathbf{r},t)$ radiated by the source distribution $q(\mathbf{r},t)$ that is more economical than the one presented in the previous section. Again, it is assumed that both the source and field are sectioned as in (18.7)–(18.9). To arrive at a windowed plane-wave representation of $u_l(\mathbf{r},t)$, we again consider the field $\tilde{u}_l(\mathbf{r},t)$ defined by (18.10). However, this time, we don't restrict the discussion to $\theta_{int} = \pi$. Here, we repeat for convenience the result in (18.13) that was derived from (18.10) subject to the constraint in (18.12):

$$\tilde{u}_l(\mathbf{r},t) = u_l(\mathbf{r},t) - \int_S d\mathbf{r}' \frac{\delta(t + R\cos\theta'_{int}(\phi',\mathbf{r},\mathbf{r}')/c)}{4\pi R} * q_l(\mathbf{r}',t) \quad (18.38)$$

Note that, as in (18.14), there is again a ghost signal represented by the second term. However, the ghost signal in (18.38) is spread out and spans a longer duration than the one in (18.14). In what follows, a scheme is derived that permits time-gating out the spread ghost signal from $\tilde{u}_l(\mathbf{r},t)$ in order to retain only the true observer field.

From (18.38) it follows that the ghost signal present in $\tilde{u}_l(\mathbf{r},t)$ vanishes after

$$\begin{aligned} t_l^{ghost} &= \frac{R}{c}\cos\theta'_{min} + (l+1)T_s \\ &< (R_c \cos\theta_{int} + 2R_s)/c + (l+1)T_s \end{aligned} \quad (18.39)$$

where $\theta'_{min} = \min[\theta'_{int}(\phi',\mathbf{r},\mathbf{r}')]$, and the upper bound follows from geometrical considerations [see Figure 18.3(b)]. The fields in the observation sphere coincide with the true fields after the ghost signal has vanished. Also, the true field does not reach the observation sphere before

$$t_l^{trans} = (R_c - 2R_s)/c + lT_s \quad (18.40)$$

Therefore, provided that $t_l^{trans} > t_l^{ghost}$, all ghost fields in the observation sphere cease to exist before the true signal arrives. In addition, assuming that $q_l(\mathbf{r},t) = 0$ outside the interval $lT_s \leq t < (l+1)T_s$, if $t_l^{trans} > (l+1)T_s$, all source activity related to the lth time interval ends before the true signal reaches any observer. In summary,

$$t_l^{trans} \geq t_l^{ghost} \Rightarrow u_l(\mathbf{r},t) = \begin{cases} 0 & t < t_l^{trans} \\ \tilde{u}_l(\mathbf{r},t) & t \geq t_l^{trans} \end{cases} \quad (18.41)$$

$$t_l^{trans} \geq (l+1)T_s \Rightarrow q_l(\mathbf{r},t) = 0, \quad t \geq t_l^{trans} \quad (18.42)$$

The above two conditions can be restated, using (18.39) and (18.40), as

$$\frac{cT_s}{R_s} \leq \frac{R_c}{R_s} - 2 \quad (18.43)$$

and

$$\frac{cT_s}{R_s} \leq \frac{R_c}{R_s}(1 - \cos\theta_{int}) - 4 \qquad (18.44)$$

It can be shown that constraint (18.12) is automatically satisfied provided that (18.44) holds for any $T_s \geq 0$.

The above two constraints are key to the development of the windowed PWTD algorithm. Equation (18.41) implies that if, for a given source and observation sphere pair (i.e., for a given R_c/R_s), a cT_s/R_s and a θ_{int} that satisfy both (18.43) and (18.44) are selected, then the field $u(\mathbf{r},t)$ can be reconstructed as a superposition of time-gated $\tilde{u}_l(\mathbf{r},t)$. The contribution of each of the time-gated $\tilde{u}_l(\mathbf{r},t)$ to the observed field can be obtained by translating the SST of the source distribution $q_l(\mathbf{r},t)$ at $t = t_l^{trans}$. It is easily recognized that the SST of $q_l(\mathbf{r},t)$ corresponds to "outgoing" rays, leaving the source sphere; similarly, it will be shown in the next subsection that after translation (i.e., after $t = t_l^{trans}$), $u_l(\mathbf{r},t)$ can be described in terms of "incoming" rays impinging upon the observation sphere. Condition (18.43) ensures that this SST can be completely constructed prior to the translation time, enabling the PWTD algorithm to be incorporated into any time marching scheme.

In practice, provided that a cT_s/R_s that satisfies (18.43) is chosen for a given R_c/R_s, θ_{int} is computed from (18.44) by enforcing the equality. This procedure minimizes θ_{int} and hence will minimize the computational cost associated with the numerical procedure for evaluating $\tilde{u}_l(\mathbf{r},t)$, as described in Section 18.6. For this choice of cT_s/R_s and θ_{int}, it follows from (18.39)–(18.41) that at $t = t_l^{trans}$, the ray traveling along $\theta = 0$ is about to enter the observation sphere, while rays traveling along directions $\theta = \theta_{int}$ have all exited the sphere [Figure 18.3(b)]. At $t = t_l^{trans}$, rays traveling at intermediate angles partially overlap with the observation sphere, but add up to a null field in its interior.

The implications of inequalities (18.43) and (18.44) are further illustrated in Figure 18.12. For a given R_c/R_s, combinations of θ_{int} and cT_s/R_s that satisfy both (18.43) and (18.44) lie to the lower right of the intersection of the curves obtained by enforcing the equalities in (18.43) and (18.44). For example, while the point $(\theta_{int}, cT_s/R_s) = (120°, 15)$ permits a ghost-free solution for $R_c/R_s = 20$, this same combination does not permit a ghost-free solution for $R_c/R_s = 10$. Note that as the two spheres approach each other, the region that satisfies both conditions collapses to the point $(\theta_{int}, cT_s/R_s) = (180°, 0)$.

18.5.2 Implementation Using Sampled Field Representations

Equation (18.10), together with constraints (18.43) and (18.44), is the basis for formulating the windowed PWTD algorithm for sampled field representations. To

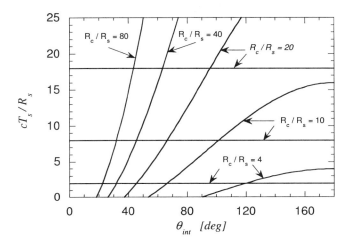

Figure 18.12 Graphical representation of the constraints (18.43) and (18.44) for different values of R_c/R_s.

derive a closed-form windowed translation operator, note that

$$\int_S d\mathbf{r}' \delta\left(t - \hat{\mathbf{k}} \cdot \mathbf{R}/c\right) * q_l(\mathbf{r}', t) \tag{18.45}$$
$$= \delta\left(t - \hat{\mathbf{k}} \cdot \tilde{\mathbf{r}}_o/c\right) * \delta\left(t - \hat{\mathbf{k}} \cdot \mathbf{R}_c/c\right) * q_l^{out}(\hat{\mathbf{k}}, t)$$

where $\tilde{\mathbf{r}}_o = \mathbf{r} - \mathbf{r}_o$ and $q_l^{out}(\hat{\mathbf{k}}, t)$ is the SST of $q_l(\mathbf{r}, t)$ as given in (18.17). Using (18.45) in (18.10) and letting $\tilde{\mathbf{r}}_s$ denote $(\mathbf{r}' - \mathbf{r}_s)$ yields

$$\tilde{u}_l(\mathbf{r}, t) = -\frac{\partial_t}{8\pi^2 c} \int_0^{2\pi} d\phi \int_0^{\theta_{int}} d\theta \sin\theta \, \delta\left(t - \hat{\mathbf{k}} \cdot \tilde{\mathbf{r}}_s/c\right)$$
$$* \delta\left(t - \hat{\mathbf{k}} \cdot \mathbf{R}_c/c\right) * \int_S d\mathbf{r}' \delta\left(t + \hat{\mathbf{k}} \cdot \tilde{\mathbf{r}}_s/c\right) * q_l(\mathbf{r}', t) \tag{18.46}$$

As mentioned earlier, the SST of the source distribution represented by the last convolution in (18.46) can be interpreted as outgoing rays leaving the source sphere. It is seen that for a point source, the SST imposes a direction-dependent shift on $q_l(\mathbf{r}, t)$ by an amount of $\hat{\mathbf{k}} \cdot \tilde{\mathbf{r}}_s/c$ which can be incorporated into (18.26). Note that the leftmost convolution in (18.46) is the same as the SST except that the direction dependency is reversed. Therefore, $\tilde{u}_l(\mathbf{r}, t)$ can be interpreted as a superposition of incoming rays projected onto the observers. However, the use of the interpolation function defined in (18.27) implies that knowledge of the field at the edge of the observation sphere requires knowledge of samples from incoming rays that reside p_t samples exterior to the sphere in all directions. Therefore, when working with sampled field representations, constraints (18.43) and (18.44) should be satisfied in terms of T'_s and $R'_s = R_s + p_t c \Delta_t$, instead of T_s and R_s. Translation times should also be computed in terms of the primed quantities.

To efficiently evaluate $\tilde{u}_l(\mathbf{r}, t)$, define

$$g_l(\hat{\mathbf{k}}, t) = \delta\left(t - \hat{\mathbf{k}} \cdot \tilde{\mathbf{r}}_o/c\right) * \int_S d\mathbf{r}' \, \delta\left(t + \hat{\mathbf{k}} \cdot \tilde{\mathbf{r}}_s/c\right) * q_l(\mathbf{r}', t)$$
$$= \int_S d\mathbf{r}' \, \delta\left(t - \hat{\mathbf{k}} \cdot (\tilde{\mathbf{r}}_o - \tilde{\mathbf{r}}_s)/c\right) * q_l(\mathbf{r}', t) \qquad (18.47)$$

The function $g_l(\hat{\mathbf{k}}, t)$ can be interpreted as the time-dependent radiation pattern of a source distribution residing in a sphere of radius $2R'_s$. Therefore, $g_l(\hat{\mathbf{k}}, t)$ is spatially quasi-bandlimited and can be reconstructed to arbitrary precision, provided it is sampled densely enough over the sphere, using the expansion [27]

$$g_l(\hat{\mathbf{k}}, t) = \sum_{n=0}^{M'} \sum_{m=-M_n}^{M_n} g_l(\hat{\mathbf{k}}_{nm}, t) \, \Omega_{nm}(\hat{\mathbf{k}}) \qquad (18.48)$$

where the functions $\Omega_{nm}(\hat{\mathbf{k}})$ represent bandlimited spherical interpolation functions. As with the temporal interpolants, many different choices for the $\Omega_{nm}(\hat{\mathbf{k}})$ exist. One near optimal choice is a combination of the Dirichlet kernel and the cylindrical APS function introduced by Bucci et al. [27], for which

$$\hat{\mathbf{k}}_{nm} = \hat{\mathbf{x}} \sin\theta_n \cos\phi_{nm} + \hat{\mathbf{y}} \sin\theta_n \sin\phi_{nm} + \hat{\mathbf{z}} \cos\theta_n \qquad (18.49)$$

$$\phi_{nm} = m2\pi/(2M_n + 1) \qquad (18.50)$$

$$\theta_n = n2\pi/(2M' + 1) \qquad (18.51)$$

$$M' = \text{Int}(\chi_2 M) \qquad (18.52)$$

$$M = \text{Int}(2\chi_1 \omega_s R'_s/c) + 1 \qquad (18.53)$$

$$M_n = \text{Int}\left(2\left[\sin\theta_n + (\chi_1 - 1)\sin^{1/3}\theta_n\right] \omega_s R'_s/c\right) + 1 \qquad (18.54)$$

$$\Omega_{nm}(\hat{\mathbf{k}}) = \begin{cases} S_0(\theta) & n = 0 \\ S_n(\theta) D_{M_n}(\phi - \phi_{nm}) \\ \quad + S_n(-\theta) D_{M_n}(\phi + \pi - \phi_{nm}) & n \neq 0 \end{cases} \qquad (18.55)$$

$$D_{M_n}(\phi) = \frac{\sin\left[(2M_n + 1)\phi/2\right]}{(2M_n + 1)\sin(\phi/2)} \quad (18.56)$$

$$S_n(\theta) = \frac{R_N(\theta - \theta_n \cdot p_s \Delta_\theta)}{R_N(0, p_s \Delta_\theta)} D_{M'}(\theta - \theta_n) \quad (18.57)$$

$$\Delta_\theta = 2\pi/(2M' + 1) \quad (18.58)$$

$$R_N(\theta \cdot p_s \Delta_\theta)$$
$$= \frac{\sinh\left[(2N+1)\sinh^{-1}\sqrt{\sin^2(p_s\Delta_\theta/2) - \sin^2(\theta/2)}\right]}{\sqrt{\sin^2(p_s\Delta_\theta/2) - \sin^2(\theta/2)}} \quad (18.59)$$

In the above, $\chi_1 > 1$ is the excess bandwidth factor, $\chi_2 > 1$ is the oversampling ratio in elevation, p_s is an integer that defines the approximate angular extent of $S_n(\theta)$, and $N = M' - M$. As with the temporal interpolation functions, $S_n(\theta)$ can also be truncated for $|\theta - \theta_n| > p_s \Delta_\theta$ yielding a relative interpolation error ε_s bounded by

$$|\varepsilon_s| \leq \frac{1}{\sinh\left[\pi p_s (1 - 1/\chi_2)\right]} \quad (18.60)$$

This error also decreases exponentially fast with increasing p_s. Hence, (18.48) permits local interpolation in elevation in terms of $2p_s$ samples.

Substituting expansion (18.48) in (18.46), rearranging the terms, and interchanging the order of summations and integrations yield

$$\tilde{u}_l(\mathbf{r}_o.t) = \sum_{n=0}^{M'} \sum_{m=-M_n}^{M_n} \delta\left(t - \hat{\mathbf{k}}_{nm} \cdot \tilde{\mathbf{r}}_o/c\right) * \mathcal{T}_{nm}(\mathbf{R}_c, t)$$
$$* \int_S d\mathbf{r}' \, \delta\left(t + \hat{\mathbf{k}}_{nm} \cdot \tilde{\mathbf{r}}_s/c\right) * q_l(\mathbf{r}', t) \quad (18.61)$$

where

$$\mathcal{T}_{nm}(\mathbf{R}_c, t) = -\frac{\partial_t}{8\pi^2 c} \int_0^{2\pi} d\phi \int_0^{\theta_{int}} d\theta \sin\theta \, \Omega_{nm}(\hat{\mathbf{k}}) \delta\left(t - \hat{\mathbf{k}} \cdot \mathbf{R}_c/c\right) \quad (18.62)$$

is the translation function. Since $\hat{\mathbf{k}} \cdot \mathbf{R}_c = R_c \cos\theta$, the integral in (18.62) can be evaluated in closed form and the translation function can be succinctly expressed as

$$\mathcal{T}_{nm}(\mathbf{R}_c, t) = -\frac{\partial_t}{4\pi R_c (2M_n + 1)} \Psi_n\left(\cos^{-1}\frac{ct}{R_c}\right) \quad (18.63)$$

in the interval $(R_c/c)\cos\theta_{int} \leq t \leq R_c/c$ with $\Psi_n(\theta)$ defined as

$$\Psi_n(\theta) = \begin{cases} S_0(\theta) & n = 0 \\ S_n(\theta) + S_n(-\theta) & n \neq 0 \end{cases} \quad (18.64)$$

Another useful expression for $\mathcal{T}_{nm}(\mathbf{R}_c, t)$ results upon expanding the spatially bandlimited and even functions $\Psi_n(\theta)$ in a cosine series as

$$\Psi_n(\theta) = \sum_{k=0}^{M'+N} a_{n,k} \cos(k\theta) \quad (18.65)$$

Substituting (18.65) into (18.63), and using the relation $T_k(x) = \cos(k\cos^{-1} x)$, where $T_k(x)$ is the kth-order Chebyshev polynomial, yields

$$\mathcal{T}_{nm}(\mathbf{R}_c, t) = -\frac{\partial_t}{4\pi R_c(2M_n + 1)} \sum_{k=0}^{M'+N} a_{n,k} T_k\left(\frac{ct}{R_c}\right) \quad (18.66)$$

in the interval $(R_c/c)\cos\theta_{int} \leq t \leq R_c/c$. Equation (18.66) shows that the translation function can be expressed as a finite-order polynomial. The sum in (18.66) can be efficiently evaluated using Clenshaw's recurrence algorithm [33].

To elucidate the properties of the translation function, the functions $\Psi_n(\theta)$ are plotted with respect to θ for $n = 0, \ldots, 10$, $M = 4$, $M' = 10$, and $p_s = 3$ in Figure 18.13(a). The corresponding time signals $\Psi_n\left(\cos^{-1}(ct/R_c)\right)$ are shown in Figure 18.13(b) as a function of the time parameter $\tau = ct/R_c$. Clearly, the duration of the translation function is $(1 - \cos\theta_{int})R_c/c$. If θ_{int} is chosen as outlined in the concluding paragraphs of Section 18.5.1(i.e., by enforcing the equality in (18.44) for a T'_s/cR'_s that satisfies constraint (18.43)), this duration equals $T'_s + 4R'_s/c$. However, a truncated version of $\Psi_n(\theta)$ may be used because $S_n(\theta)$ is vanishingly small for $|\theta - \theta_n| > p_s\Delta_\theta$. As can be seen in Figure 18.13, the translation function associated with the directions for which $\theta_n > \theta_{int} + p_s\Delta_\theta$ vanishes as $\Psi_n(\theta) \cong 0$ for $0 < \theta < \theta_{int}$. The $\Psi_n(\theta)$ associated with these directions ($n = 7, \ldots, 10$) are plotted with dash-dotted lines in Figure 18.13. For other directions, the nonvanishing portion of $\Psi_n(\theta)$ in the interval $0 < \theta < \theta_{int}$ contributes to the translation function and is plotted with a solid line in Figure 18.13. Note, however, that for these directions, the duration of the translation function may become much shorter than $T'_s + 4R'_s/c$. Also, if T'_s is fixed, constraint (18.44) dictates that, as the spheres move further apart, the number of contributing directions decreases.

The above analysis can easily be extended to source and observation spheres for which $\mathbf{R}_c$ is not aligned with the z-axis. One approach is to use interpolation functions that are windowed in both elevation and azimuth instead of the above introduced interpolants that are windowed solely in elevation. Alternatively, instead of relying on the bandlimited nature of the far field interpolation functions to bandlimit

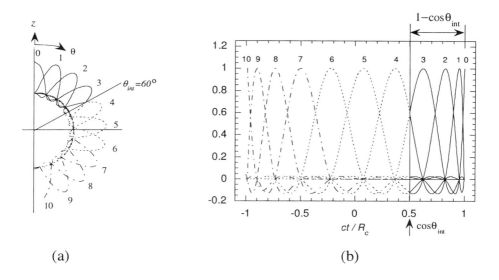

Figure 18.13 (a) Interpolation functions $\Psi_n(\theta)$ as a function of θ. (b) Time signals $\Psi_n\left(\cos^{-1}(ct/R_c)\right)$ as a function of the time parameter $\tau = ct/R_c$.

the translation operator as in the above derivation, the translation operator can be explicitly bandlimited and the integration over the sphere performed using an exact quadrature formula. This procedure, which was demonstrated in Section 18.3.2, leads to a translation function for which the angular and temporal dependences can be expressed in terms of Legendre polynomials as in (18.23). This derivation is akin to the traditional construction of windowed translation operators for frequency domain fast multipole methods [6, 8, 34].

18.6 IMPLEMENTATION OF THE WINDOWED PWTD-ENHANCED MOT SCHEMES

This section describes the practical implementation of the windowed PWTD algorithm and studies the computational complexity of MOT schemes supplemented with this algorithm in two-level and multilevel settings. It should be clear that using the windowed translation operator requires only minor changes to the algorithms described in Section 18.4. Therefore, the algorithms described there will not be repeated here. Instead, the use of the windowed PWTD for evaluating fields at multiple observers due to multiple sources residing in geometrically separate observation and source spheres will be detailed in Section 18.6.1. Then, the computational complexities of the two-level and multilevel schemes using the windowed PWTD algorithm will be briefly derived in Sections 18.6.2 and 18.6.3, respectively.

18.6.1 Sphere-to-Sphere Translation

This section outlines a sequence of operations leading to a successful implementation of the windowed PWTD algorithm for computing observer fields. The setting is again as described in Section 18.3.1 and illustrated in Figure 18.5. It is assumed that $q(\mathbf{r}, t)$ consists of M_s point sources located at $\mathbf{r}_s^j$, $j = 1, ..., M_s$, distributed throughout the source sphere and characterized by temporal signatures $f^j(t)$; that is,

$$q(\mathbf{r}, t) = \sum_{j=1}^{M_s} f^j(t)\, \delta(\mathbf{r} - \mathbf{r}_s^j) \tag{18.67}$$

It is assumed that the spectra of all $f^j(t)$ vanish for $\omega > \omega_{max}$. The field due to $q(\mathbf{r}, t)$ is to be evaluated at M_o observers located at $\mathbf{r}_o^i$, $i = 1, \cdots, M_o$, distributed throughout the observation sphere. To simplify the notation, the positions of the source and observation points relative to their respective sphere centers are denoted by $\tilde{\mathbf{r}}_s^j = \mathbf{r}_s^j - \mathbf{r}_s$ and $\tilde{\mathbf{r}}_o^i = \mathbf{r}_o^i - \mathbf{r}_o$.

Direct evaluation of the fields at all M_o observers for all N_t time steps is a computationally expensive task as its cost scales as $O(N_t M_s M_o)$, which is of $O(N_t M_s^2)$ if $M_o \propto M_s$. Alternatively, the fields at the M_o observers can be evaluated using (18.61), which, for the source density expressed by (18.67), takes the form

$$\tilde{u}_l(\mathbf{r}_o^i.t) = \sum_{n=0}^{M'} \sum_{m=-M_n}^{M_n} \delta\left(t - \hat{\mathbf{k}}_{nm} \cdot \tilde{\mathbf{r}}_o^i/c\right) * \mathcal{T}_{nm}(\mathbf{R}_c, t) \\ * \sum_{j=1}^{M_s} \delta\left(t + \hat{\mathbf{k}}_{nm} \cdot \tilde{\mathbf{r}}_s^j/c\right) * f_l^j(t) \tag{18.68}$$

where it is assumed that each source signal $f^j(t)$ is broken up into L subsignals $f_l^j(t)$, $j = 1, \ldots, M_s$. Equation (18.68) is the crux of the PWTD algorithm and indicates that the fields within the observation sphere can be constructed via a three-step process consisting of source aggregation, ray translation, and ray disaggregation. The aggregation step, which is represented by the innermost summation in (18.68), maps the source subsignals onto a set of time-dependent plane waves—henceforth termed *subrays*—propagating along the $\hat{\mathbf{k}}_{nm}$ vectors. The translation step is carried out by convolving these subrays with the translation functions given in (18.66). The disaggregation process can be viewed as the reverse of the aggregation process and maps a set of incoming rays onto observer locations.

To implement this algorithm for a given R'_s and R_c, a T'_s is selected which (1) satisfies constraint (18.43) and for which (2) $R'_s/(cT'_s)$ is of $O(1)$. If it is impossible to satisfy both of these conditions, the fields in the observer sphere should be computed using classical procedures, as the PWTD becomes less efficient. Next, a θ_{int} is computed by enforcing the equality in constraint (18.44). The following three operations are then performed for all L time intervals:

1. Compute the sampled SST of the source distribution for all ray directions [i.e., perform the rightmost convolution and carry out the innermost summation in (18.68)].

2. At $t = t_l^{trans}$, convolve each subray with the translation function on a direction-by-direction basis, and add the resulting subrays onto incoming rays which propagate through the observation sphere [i.e., perform the middle convolution in (18.68)]. While in certain cases it is advantageous to perform this convolution directly in the time domain, it is assumed here that the convolution is performed using an FFT. However, care should be exercised as the translation function is not bandlimited and cannot be sampled without aliasing. On the other hand, since each subray is bandlimited, so is the result of the convolution. In practice, the Fourier transform of the translation function is evaluated analytically at the frequency points required by the FFT. This is efficiently accomplished by locally expanding the translation function in terms of a small set of orthogonal polynomials whose Fourier transforms are well defined. Note that this operation translates each subray onto an incoming ray that propagates in the same direction, analogous to diagonal frequency domain fast multipole translation operators.

3. Evaluate the fields at the observers as the incoming rays travel across the observation sphere [i.e., perform the leftmost convolution and summations in (18.68)].

Note that each subray can be at most $(2R'_s/c + T'_s)/\Delta_t$ time steps long. Furthermore, as discussed in Section 18.5.2, the translation function associated with each ray direction is $(4R'_s/c + T'_s)/\Delta_t$ time steps long. By virtue of the choice $cT'_s \propto R'_s$, both the subray duration and the translation function length scale as $O(M'_t)$. From (18.48), it is seen that the number of ray directions D_s equals $\sum_{n=0}^{M'} (2M_n + 1)$. Using (18.52)–(18.54), it can be shown that D_s is proportional to the surface area of each sphere [i.e., $D_s \propto (R'_s/(c\Delta_t))^2$].

Since the aggregation step maps M_s source subsignals onto D_s subrays, its computational cost scales as $O(M'_t M_s D_s)$. The dominant cost in the translation step is due to the convolution and scales as $O(M'_t \log M'_t)$ if evaluated using an FFT. This operation is performed for all D_s directions, yielding a computational complexity of $O(D_s M'_t \log M'_t)$. The disaggregation step has the same complexity as the aggregation step.

For a surface scatterer, the number of sources or observers in a sphere is proportional to the surface area of the sphere [i.e., $M_s \propto (R'_s/(c\Delta_t))^2$]. This implies that $M'_t \propto \sqrt{M_s}$, and that $D_s \propto M_s$. It can be verified that the costs of the aggregation and dissaggregation processes dominate that of the translation process and that the cost of evaluating the observed fields for one subsignal using PWTD algorithm scales as $O(M'_t M_s^2)$. Hence, the cost of evaluating the fields due to all L subsignals scales as $O(N_t M_s^2)$. This cost is no less than that of the classical algorithm. Nonetheless, the PWTD scheme permits the reuse of SST information, which results in a reduction of the computational complexity when applied in an integral equation setting.

However, as noted previously, not all outgoing rays contribute to the observed fields if use is made of the windowed character of the translation operator. In fact, as is evident in Figure 18.12, the number of subrays that need to be translated shrinks to a constant as the ratio of the distance between the spheres and the sphere radii increases. Under these circumstances, D_s can be assumed constant and omitted in the above complexity estimates; hence, the computational complexity of computing the fields associated with one time interval scales as $O(M'_t M_s)$, and that of all L subsignals combined scales as $O(N_t M_s)$. Note that the aim of achieving a more efficient algorithm than the standard PWTD algorithm has been achieved through windowing.

A series of numerical experiments was conducted to validate and examine accuracy versus efficiency trade-offs in the windowed PWTD algorithm. To this end, six point sources with Gaussian time signatures $f^j(t) = \exp\left[-(t - 7.75\sigma)^2/(2\sigma^2)\right]$, $j = 1, \ldots, 6$, were distributed in the yz-plane on the surface of a spherical region in space whose center was at $(x, y, z) = (0, 0, -20)$ m and whose radius was $R_s = 1$m. This region is dubbed the source sphere. The time function was characterized by $\sigma = 2.12$ ns, which yields a pulse with a duration (full-width between half maximum) of 5 ns. For all practical purposes, this pulse can be assumed to be bandlimited to $\omega_{max} = 600\pi \times 10^6$ rad/s. The time step size Δ_t is fixed at 0.5 ns and the wave speed of the medium is chosen to be $c = 3 \times 10^8$ m/s. It was assumed that all the observers would be confined within a 1m radius around the origin, dubbed the observation sphere. For a sampled field representation, this choice of parameters yields $t_l^{trans} = 120\Delta_t + lT_s - 2p_l\Delta_t$. The radiated fields were evaluated throughout a 12m × 12m region in the yz-plane centered about the observation sphere center. The normalized error in each observer response was calculated by dividing the L_2 norm of the difference between the exact fields and those computed using the PWTD algorithm by the L_2 norm of the exact fields. Error distributions for cases, in which the parameters defining the temporal and spherical interpolation functions were chosen to yield 10^{-4}, 10^{-6}, 10^{-8}, and 10^{-10} accuracy, are plotted throughout the square observation domain in Figure 18.14. The location of the observation sphere is also depicted in these figures, and it is seen that for all four cases the desired accuracy is obtained throughout the observation sphere. Note that the structure of the error depicted in Figure 18.14 is quite different than that shown in Figure 18.8. This is due

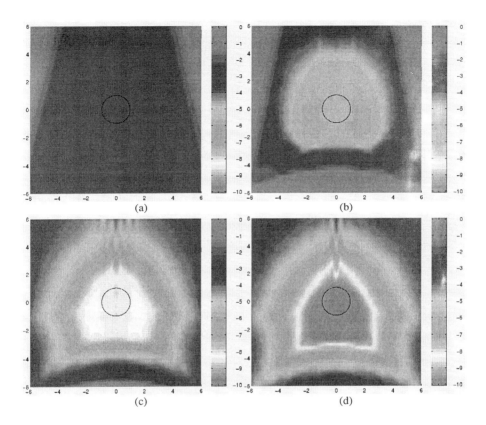

Figure 18.14 Logarithm (base 10) of normalized errors observed in a region enclosing the observation sphere (black circle) for several cases with estimated accuracy levels of (a) 10^{-4}, (b) 10^{-6}, (c) 10^{-8}, and (d) 10^{-10}.

to the fact that the algorithm there relied on a Whittaker-type expansion rather than a finite-cone representation of the field as done here.

18.6.2 A Two-Level Windowed PWTD-Enhanced MOT Algorithm

Now that the applicability of the windowed PWTD algorithm has been verified for a pair of source and observation groups, we will demonstrate that this algorithm results in lower computational complexities than those obtained by the nonwindowed PWTD algorithm when applied to the analysis of scattering from large surfaces using time-domain integral-equation schemes like MOT.

The first step in forming the two-level scheme is to subdivide the scatterer into subscatterers or groups and to identify near field and far field pairs. This step was described in detail in Section 18.4.1 and illustrated in Figure 18.9. The next step is to evaluate the near field interactions. Clearly, this process is independent of how the PWTD algorithm is implemented. Hence, its computational complexity will still scale as $O(N_t N_s M_s)$. The far field interactions are evaluated through the aggregation, translation, and disaggregation stages for each far field pair as described in the previous section. Of these three stages, only the translation process will be different from that of the nonwindowed two-level scheme. The aggregation and disaggregation stages will not change since—although only a limited number of rays will be translated from one group to the other—the rays propagating in all directions still need to be formed for each group. This is due to the fact that an outgoing ray of, say, group γ might not be translated to group γ' but is needed for translation to another group. Similarly, a group might receive only a limited number of incoming rays from a particular group but the contributions from other interactions might produce incoming rays in all directions. Therefore, the cost scaling of the aggregation and disaggregation stages will be again $O(N_t N_s M_s)$ as in the nonwindowed case.

The cost scaling of the translation process, however, will be lower than the $O(N_t N_s^2 M_s^{-1} \log N_s)$ complexity of the nonwindowed two-level scheme, because in the present algorithm not all plane waves need to be translated from one group to another. Indeed, as the interacting groups move further apart, the number of plane wave components that participate in translation shrinks to a constant.[1] This yields a reduction in complexity by a factor proportional to M_s. That is, the cost of the translation stage now scales as $O(N_t N_s^2 M_s^{-2} \log N_s)$, and the total cost associated with the computation of the fields at all the observers asymptotically scales as

$$c_1(N_t N_s M_s) + c_2(N_t N_s^2 M_s^{-2} \log N_s) \qquad (18.69)$$

where c_1 and c_2 are machine and program dependent constants. It can be verified that for the algorithm described above, the optimal number of unknowns per group is $M_s \propto N_s^{1/3}$, and the total cost of the field computation scales as

$$C_T \propto N_t N_s^{4/3} \log N_s \qquad (18.70)$$

18.6.3 A Multilevel Windowed PWTD-Enhanced MOT Algorithm

It was shown in the previous section that the only difference between the windowed and nonwindowed two-level schemes was the translation process. This is, in essence,

---

[1] Note that, for source and observation spheres that are not separated by a distance that is very large compared to the sphere radius, the number of rays that need to be translated becomes larger than the constant assumed in the derivation of (18.69). However, it can be shown through a more rigorous derivation that this effect does not alter the derived complexity estimate.

also true for the multilevel scheme. However, the source of computational savings in the multilevel scheme is quite different than that in the two-level case. In the two-level scheme, the savings stemmed from the observation that the number of rays that need to be translated shrinks to a constant as the distance between the source and observer groups increase. In a multilevel setting, as the ratio of the maximum group separation to the group size remains constant, the same observation does not help reduce the computational burden. Here, the key observation is that as the group size increases, the temporal extent of the translation function remains a constant. Indeed, when the truncated version of $\Psi_n(\theta)$ is used in forming the translation functions (see Section 18.5.2), the length of these windowed translation functions scales as $(R_c/c)\, p_s \Delta_\theta\, \theta_{int}$ for sufficiently large scatterers. Since, in a multilevel setting, R_c scales as R_s, Δ_θ scales as $1/R_s$, and p_s and θ_{int} remain constant for all levels, the translation function length is of $O(1)$. Therefore, directly convolving each subray with the translation function in the time domain will be more efficient than using an FFT, reducing the cost per subray translation from $O\left(M'_t(i)\log M'_t(i)\right)$ to $O\left(M'_t(i)\right)$. Hence, the cost of performing all the translations in a scattering analysis reduces to $O(N_t N_s \log N_s)$. Unfortunately, this reduction alone is not sufficient to reduce the complexity of the whole scattering analysis because, as elucidated in Section 18.4.2, the cost of performing the interpolation and anterpolation operations in the multilevel setting also scale as $O(N_t N_s \log^2 N_s)$. This cost scaling can also be reduced to $O(N_t N_s \log N_s)$ by using local interpolation/anterpolation schemes for radiated fields over the sphere [7, 27]. This reduction comes at the expense of a modest (constant) increase in the number of directions at each level due to oversampling. Hence, the total computational complexity of a scattering analysis using the windowed PWTD algorithm in a multilevel framework turns out to be

$$O(N_t N_s \log N_s) \qquad (18.71)$$

18.7 SUMMARY

This chapter introduced two PWTD algorithms and outlined their incorporation into existing MOT algorithms to considerably reduce the computational complexity associated with the evaluation of RTBIs and the analysis of transient surface scattering phenomena. This reduction in cost is achieved through expansion of transient wave fields in terms of a propagating plane wave basis that is characterized by a diagonal translation operator. The translation function derived for the (nonwindowed) PWTD algorithm turns out to be the Fourier transform of the regular part of the frequency domain, fast multipole method, translation function, and it gives rise to an anticausal ghost signal. Properly sectioning the source signal into subsignals and translating each subsignal separately isolates the ghost signals, which are then time-gated out. The applications of the proposed plane wave expansions on a subsignal basis are formulated as three-stage PWTD algorithms that facilitate ghost-free evaluation of transient fields. It is illustrated both theoretically and experimentally that these algorithms permit the reconstruction of transient fields to arbitrary precision. Incorporation of the (nonwindowed) PWTD algorithm into the classical MOT scheme in a two-level setting permits the analysis of surface scattering phenomena in $O(N_t N_s^{1.5} \log N_s)$ operations as opposed to $O(N_t N_s^2)$ operations required by the classical method. The ideas underlying the two-level algorithm are extended into a multilevel framework thereby enabling a further reduction of the computational complexity to $O(N_t N_s \log^2 N_s)$. Employing the windowed PWTD algorithm further reduces the complexities of the two-level and multilevel schemes to $O(N_t N_s^{4/3} \log N_s)$ and $O(N_t N_s \log N_s)$, respectively. The efficacy of these schemes will be demonstrated in the next chapter.

REFERENCES

1. A. Devaney and G. C. Sherman, "Plane-wave representations for scalar wave fields," *SIAM Review*, vol. 15, no. 4, pp. 765–786, 1973.

2. M. Tygel and P. Hubral, *Transient Waves in Layered Media*, Volume 26 of *Methods in Geochemistry and Geophysics*, Amsterdam: Elsevier Science Publishers BV, 1987.

3. E. Whittaker, "On the partial differential equations of mathematical physics," *Mathemaische Annalen*, vol. 57, pp. 333–355, 1902.

4. R. Coifman, V. Rokhlin, and S. Wandzura, "The fast multipole method for the wave equation: A pedestrian prescription," *IEEE Antennas Propagat. Mag.*, vol. 35, pp. 7–12, 1993.

5. E. Heyman, "Time-dependent plane-wave spectrum representations for radiation from volume source distributions," *J. Math. Phys.*, vol. 37, no. 2, pp. 658–681, 1996.

6. R. Coifman, V. Rokhlin, and S. Wandzura, "Faster single-stage multipole method for the wave equation," in *10th Annual Review of Progress in Applied Computational Electromagnetics*, vol. 1, Monterey, CA, pp. 19–24, 1994.

7. J. Song and W. Chew, "Multilevel fast-multipole algorithm for solving combined field integral equations of electromagnetic scattering," *Microwave Opt. Tech. Lett.*, vol. 10, no. 1, pp. 14–19, 1995.

8. R. L. Wagner and W. C. Chew, "A ray-propagation fast multipole algorithm," *Microwave Opt. Tech. Lett.*, vol. 7, no. 10, pp. 435–438, 1994.

9. B. Hu, W. C. Chew, E. Michielssen, and J. Zhao, "An improved fast steepest descent path algorithm," *Proc. IEEE Antennas and Propag. Int. Symp.*, vol. 3, Atlanta, GA, pp. 1542–1545, IEEE, 1998.

10. V. Jandhyala, E. Michielssen, and B. Shanker, "A combined steepest descent-fast multipole algorithm for the analysis of three-dimensional scattering by rough surfaces," *IEEE Trans. Geosci. Remote Sensing*, vol. 36, p. 738, 1997.

11. A. A. Ergin, B. Shanker, and E. Michielssen, "Fast evaluation of transient wave fields using diagonal translation operators," *J. Comp. Phys.*, vol. 146, pp. 157–180, 1998.

12. A. A. Ergin, B. Shanker, and E. Michielssen, "The plane wave time domain algorithm for the fast analysis of transient wave phenomena," *IEEE Antennas Propagat. Mag.*, vol. 41, no. 4, pp. 39–52, 1999.

13. S. M. Rao and D. R. Wilton, "Transient scattering by conducting surfaces of arbitrary shape," *IEEE Trans. Antennas Propagat.*, vol. 39, no. 1, pp. 56–61, 1991.

14. B. Rynne, "Time domain scattering from arbitrary surfaces using the electric field integral equation," *J. Electromag. Waves Applicat.*, vol. 5, no. 1, pp. 93–112, 1991.

15. A. Tijhuis, *Electromagnetic Inverse Profiling: Theory and Numerical Implementation*, Utrecht, The Netherlands: VNU Science Press BV, 1987.

16. A. A. Ergin, B. Shanker, K. Aygun, and E. Michielssen, "Computational complexity and implementation of two-level plane wave time domain algorithm for scalar wave equation," *Proc. IEEE Antennas and Propag. Int. Symp.*, vol. 2, Atlanta, GA, pp. 944–947, 1998.

17. B. Shanker, A. A. Ergin, K. Aygun, and E. Michielssen, "Computation of transient scattering from electrically large structures using the plane wave time domain algorithm," in *Proc. IEEE Antennas and Propag. Int. Symp.*, vol. 2, Atlanta, GA, pp. 948–951, 1998.

18. S. P. Walker, "Developments in time domain integral-equation modeling at Imperial College," *IEEE Antennas Propagat. Mag.*, vol. 39, no. 1, pp. 7–19, 1997.

19. S. Walker, "Scattering analysis via time-domain integral equations: Methods to reduce the scaling of cost with frequency," *IEEE Antennas Propagat. Mag.*, vol. 39, no. 5, pp. 13–20, 1997.

20. V. Rokhlin, "Diagonal forms of translation operators for the Helmholtz equation in three dimensions," *Appl. Comp. Harmonic Anal.*, vol. 1, pp. 82–93, 1993.

21. P. Clemmow, *The Plane Wave Spectrum Representation of Electromagnetic Fields*, Second Edition, IEEE/OUP series on electromagnetic wave theory, New York: IEEE Press, 1996. (First edition:Pergamon Press, 1966).

22. A. D. Yaghjian and T. B. Hansen, "Time-domain far fields," *J. Appl. Phys.*, vol. 79, no. 6, pp. 2822–2830, 1996.

23. T. B. Hansen and A. D. Yaghjian, *Plane-Wave Theory of Time Domain Fields: Near Field Scanning Applications*, New York: IEEE Press, 1999.

24. C. Chapman, "Generalized radon transforms and slant stacks," *Geophys. J. R. Astron. Soc.*, vol. 66, no. 2, pp. 445–453, 1981.

25. A. Shlivinski, E. Heyman, and R. Kastner, "Antenna characterization in the time domain," *IEEE Trans. Antennas Propagat.*, vol. 45, no. 7, pp. 1140–1149, 1997.

26. O. M. Bucci and G. Franceschetti, "On the spatial bandwidth of scattered fields," *IEEE Trans. Antennas Propagat.*, vol. 35, no. 12, pp. 1445–1455, 1987.

27. O. M. Bucci, C. Gennarelli, and C. Savarese, "Optimal interpolation of radiated fields over a sphere," *IEEE Trans. Antennas Propagat.*, vol. 39, no. 11, pp. 1633–1643, 1991.

28. J. R. Driscoll and J. D. M. Healy, "Asymptotically fast algorithms for spherical and related transforms," *30th Annual Symposium on Foundations of Computer Science*, Research Triangle Park, NC, pp. 344–349, IEEE Computer Society Press, 1989.

29. J. A. Stratton, *Electromagnetic Theory*, (International series in physics), New York: McGraw-Hill, 1941.

30. S. Huestis, "Interpolation formulas for oversampled band-limited functions," *SIAM Review*, vol. 34, no. 3, pp. 477–481, 1992.

31. J. J. Knab, "Interpolation of band-limited functions using the approximate prolate series," *IEEE Trans. Information Theory*, vol. 25, no. 6, pp. 717–720, 1979.

32. R. Jakob-Chien and B. K. Alpert, "A fast spherical filter with uniform resolution," *J. Comp. Phys.*, vol. 136, pp. 580–584, 1997.

33. W. H. Press, S. A. Teukolsky, W. T. Vetterling, and B. P. Flannery, *Numerical Recipes in Fortran 77: The Art of Scientific Computing*, Volume 1 of *Fortran Numerical Recipes*; Cambridge: Cambridge University Press, 1996.

34. V. Rokhlin, "Sparse diagonal forms for translation operators for the helmholtz equation in two dimensions," Tech. Rep. YALEU/DCS/RR-1095, Yale University, 1995.

19

Plane-Wave Time-Domain Algorithm Enhanced Time-Domain Integral Equation Solvers

Balasubramaniam Shanker, A. Arif Ergin, Kemal Aygün, and Eric Michielssen

19.1 INTRODUCTION

Numerical methods for analyzing electromagnetic transients find widespread engineering applications ranging from the analysis of broadband scattering to the design of modern antennas to the study of nonlinear phenomena and more. These methods typically are based either on differential equations (DEs) [1, 2] or integral equations (IEs) [3–5]. Historically, DE methods have been favored over their IE counterparts (the vast majority of which are marching-on-in-time (MOT) methods [3]) as the latter often were found to be unstable [6] and highly expensive in application [7]. IE-based techniques, however, offer unmistakable advantages over DE-based methods when applied to the analysis of homogeneous/surface scatterers. First, IE solvers only require a discretization of the scatterer surface rather than a volume enclosing the latter, which results in a sharp decrease in the number of unknowns when compared to DE methods. Second, IE techniques automatically impose the radiation condition, hence, there is no need for (approximate local) absorbing boundary conditions that are required in the truncation of finite grids used by DE methods. Their intrinsic qualities not withstanding, time-domain integral equation (TDIE)-based techniques have not enjoyed widespread application, even in the study of surface scattering phenomena. This is in large part due to two principal reasons: (1) many MOT schemes have been shown prone to late time instabilities, and (2) the cost associated with classical MOT schemes scale unfavorably with problem size.

Over the past several years a considerable research effort has been expended on eliminating the aforementioned drawbacks of MOT solvers. Many a study has deepened our understanding of the origins of MOT instabilities and new methods have been proposed for confronting them (see [6, 8–12] and references therein). A series of recent papers by Walker's group demonstrates that MOT schemes for solving magnetic field integral equations (MFIEs) can be stabilized for "all practical purposes" by relying on accurate spatial integration rules and implicit time stepping methods [12, 13]. Likewise, it has been shown that the MOT schemes for solving the electric field integral equation (EFIE) can be stabilized albeit with a little more difficulty [11, 14–16].

While improvements in ensuring the accuracy and stability of the time marching schemes are necessary, alleviating the computational complexity of MOT schemes is imperative if these schemes are to be used for large scale analysis. It is well known that the cost of analyzing transient scattering for N_t time steps from an object whose surface current is discretized in terms of N_s spatial unknowns scales as $\mathcal{O}(N_t N_s^2)$. As a result, MOT solvers quickly swamp the available computational resources when applied to large scale scattering problems. Recently, Walker proposed a scheme for amortizing the cost of MOT scheme by discounting noninteracting portions of the scatterer; however, while this scheme has been used for analyzing scattering from a class of interesting scatterers, it is heuristic in nature. The problem of reducing the computational complexity of time domain solvers is very similar to the those that plagued the widespread applicability of classical method of moments to large-scale scattering and radiation problems. These problems have largely been overcome by the fast multipole method (FMM) that achieves significant reduction in complexity by expressing radiated fields in terms of plane waves [17, 18]. Our recent efforts have focused on developing an analogous scheme for solving time domain problems. Indeed, recently we introduced the plane-wave time-domain (PWTD) algorithm [19] that exploits similar ideas to arrive at a reduced complexity scheme for scalar fields. Since then, this scheme has been applied to solving a host of practical problems ranging from transient acoustic [20] and electromagnetic scattering analysis [21] to EMI/EMC analysis [22].

An accidental by-product of our efforts to develop novel computation schemes for large-scale analysis was the development of a time domain combined field integral equation (CFIE), whose solution is free of corruption by resonant modes when it is used for analyzing transient scattering from closed bodies. The presence of resonance modes was investigated earlier to determine whether they would contribute to instabilities that were observed while solving either the EFIE or the MFIE [6, 10]. Indeed, it is eminently possible that the MOT schemes for solving these equations would be unstable if the singularity expansion method poles that characterize the resolvent of the corresponding integral equation drifted into the right-half plane by virtue of numerical approximations. Poles describing interior resonances are prime candidates as they reside on the imaginary axis. Implicit schemes for solving the MFIE are virtually always stable and tend to avoid undesirable pole displacements (unfortunately the same is not true of EFIE). However, the presence of cavity modes adversely affects the accuracy of the solutions. While in theory cavity modes are never excited upon external illumination, practical numerical schemes will develop solutions that correspond to these perturbed solutions [23]. So while these solutions are not necessarily unstable, they are inaccurate.

The twin objectives of this chapter are as follows:

1. To develop a CFIE and demonstrate that using this equation to analyze transient scattering from closed bodies yields an accurate solution;

2. To elucidate the PWTD-scheme as it applies to the solution of the vector wave equations, cast it into a framework wherein it can be incorporated into existing MOT schemes with facile ease, and demonstrate that the PWTD-augmented MOT scheme is efficient both in terms of memory and computational complexity. Sufficient details are presented to facilitate implementation.

The proposed objectives are met by analyzing transient scattering from electrically large perfect electrically conducting (PEC) bodies that reside in free space, and all results are obtained using an implicit time marching scheme. This chapter is organized as follows: Section 19.2 will describe the EFIE, MFIE and CFIE, and the MOT procedure that is used to solve these equations. Section 19.3 details the PWTD algorithm, its practical implementation within the context of an MOT solver, and the theoretical computational complexity of the resulting scheme. Several numerical results are presented in Section 19.4. These serve to demonstrate the accuracy of the CFIE, and the applicability of the PWTD scheme for large scale scattering. Theoretical complexity estimates are validated via numerical experiments. Finally, Section 19.5 summarizes ideas presented in this chapter.

19.2 FORMULATION

In this section, three surface IEs for analyzing transient electromagnetic scattering phenomena—namely, an EFIE, an MFIE, and a CFIE—are introduced. It is argued that contrary to the EFIE and the MFIE, the CFIE cannot support interior cavity modes. An MOT scheme for solving these equations is presented, together with a brief discussion on the stability properties of these equations

19.2.1 Integral Equations

Consider a closed PEC body (Figure 19.1) with surface S residing in free space. In what follows, $\hat{\mathbf{n}}$ denotes an outward pointing position dependent normal to S, and S_- and S_+ are hypothetical surfaces that are conformal to, but residing just inside and outside S, respectively. An impressed field with electric and magnetic components $\{\mathbf{E}^i(\mathbf{r},t), \mathbf{H}^i(\mathbf{r},t)\}$ that is temporally bandlimited to f_{max} impinges on S. The surface current $\mathbf{J}(\mathbf{r},t)$ that is induced on S by virtue of this interaction generates a scattered field $\{\mathbf{E}^s(\mathbf{r},t), \mathbf{H}^s(\mathbf{r},t)\}$ that is fully characterized by the vector potential $\mathbf{A}(\mathbf{r},t)$:

$$\mathbf{A}(\mathbf{r},t) = \frac{\mu_0}{4\pi} \int_S dS' \frac{\mathbf{J}(\mathbf{r}',\tau)}{R} \tag{19.1}$$

where $R = |\mathbf{r}-\mathbf{r}'|$, $\tau = t - R/c$ denotes the retarded time, c is the speed of light, and μ_0 is the permeability of free space. In what follows, it is assumed that the incident field does not interact with S for $t \leq 0$ [i.e., $\mathbf{J}(\mathbf{r},t) = 0$ for $t \leq 0$].

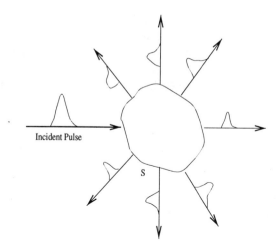

Figure 19.1 Generic scattering problem description.

A time domain EFIE can be constructed by expressing the scattered electric field $\mathbf{E}^s(\mathbf{r},t)$ in terms of $\mathbf{A}(\mathbf{r},t)$ as

$$\mathbf{E}^s(\mathbf{r},t) = -\int_0^t dt' \left(\partial_{t'}^2 \mathcal{I} - c^2 \nabla\nabla\right) \cdot \mathbf{A}(\mathbf{r},t') \tag{19.2}$$

where $\mathcal{I}$ is the identity dyad. Enforcing the total electric field tangential to S to vanish (the same condition holds on both S_- and S_+) yields

$$\hat{\mathbf{n}} \times \hat{\mathbf{n}} \times \mathbf{E}^i(\mathbf{r},t) = -\hat{\mathbf{n}} \times \hat{\mathbf{n}} \times \mathbf{E}^s(\mathbf{r},t) \quad \forall \mathbf{r} \in S, S_-, S_+ \tag{19.3}$$

Combining (19.1)–(19.3), and denoting $\mathcal{V}_e \{\mathbf{E}^i(\mathbf{r},t), \mathbf{H}^i(\mathbf{r},t)\} = \hat{\mathbf{n}} \times \hat{\mathbf{n}} \times \mathbf{E}^i(\mathbf{r},t)$ results in the following EFIE:

$$\mathcal{V}_e \{\mathbf{E}^i(\mathbf{r},t), \mathbf{H}^i(\mathbf{r},t)\} = \mathcal{L}_e \{\mathbf{J}(\mathbf{r},t)\} \quad \forall \mathbf{r} \in S, S_-, S_+ \quad (19.4a)$$

where

$$\mathcal{L}_e \{\mathbf{J}(\mathbf{r},t)\} = \hat{\mathbf{n}} \times \hat{\mathbf{n}} \times \left\{ \frac{\mu_0}{4\pi} \int_0^t dt' \int_S dS' \left(\partial_{t'}^2 \mathcal{I} - c^2 \nabla \nabla \right) \cdot \frac{\mathbf{J}(\mathbf{r}', t' - R/c)}{R} \right\} \quad (19.4b)$$

Equation (19.4) not only holds true for closed S, but also for open structures.

A time domain MFIE can be constructed by expressing the scattered magnetic field $\mathbf{H}^s(\mathbf{r},t)$ in terms of $\mathbf{A}(\mathbf{r},t)$ as

$$\mathbf{H}^s(\mathbf{r},t) = \frac{1}{\mu_0} \nabla \times \mathbf{A}(\mathbf{r},t) \quad (19.5)$$

and by enforcing the condition that the total magnetic field tangential to S_- vanishes; that is,

$$\hat{\mathbf{n}} \times \mathbf{H}^i(\mathbf{r},t) = -\hat{\mathbf{n}} \times \mathbf{H}^s(\mathbf{r},t) \quad \forall \mathbf{r} \in S_- \quad (19.6)$$

Using (19.1), (19.5), and (19.6), and denoting $\mathcal{V}_h \{\mathbf{E}^i(\mathbf{r},t), \mathbf{H}^i(\mathbf{r},t)\} = \hat{\mathbf{n}} \times \mathbf{H}^i(\mathbf{r},t)$, the following MFIE can be derived [24]:

$$\mathcal{V}_h \{\mathbf{E}^i(\mathbf{r},t), \mathbf{H}^i(\mathbf{r},t)\} = \mathcal{L}_h \{\mathbf{J}(\mathbf{r},t)\} \quad \forall \mathbf{r} \in S_- \quad (19.7a)$$

where

$$\mathcal{L}_h \{\mathbf{J}(\mathbf{r},t)\} = -\frac{1}{4\pi} \hat{\mathbf{n}} \times \int_S dS' \left[\frac{1}{cR^2} \partial_\tau \mathbf{J}(\mathbf{r}', \tau) + \frac{1}{R^3} \mathbf{J}(\mathbf{r}', \tau) \right] \times (\mathbf{r} - \mathbf{r}') \quad (19.7b)$$

where $\partial_\tau \mathbf{J}(\mathbf{r}', \tau) \Longrightarrow \partial \mathbf{J}(\mathbf{r}', t)/\partial t|_{t=\tau}$.

The singularity expansion method shows that solutions to the homogeneous time domain EFIE and MFIE are characterized by the poles of the resolvent of $\mathcal{L}_e$ and $\mathcal{L}_h$ [23], respectively. Of course, some of these poles lie close to the right half-plane, and those that are on the imaginary axis correspond to the frequencies of the interior cavity modes that the EFIE and MFIE support. It has been established that in theory the incident field does not couple to the interior modes [23]. Unfortunately, because of two important effects, the situation changes drastically when (19.4) and (19.7) are solved numerically. First, inaccuracies inherent to a numerical scheme will result in shifting of the system poles, some of which may end up in the right half-plane, which is problematic as this leads to instabilities. It has been argued that as inaccuracies are more substantial at higher frequencies, the movement of the poles associated with these frequencies will be larger [6, 25]. The size of the time step that is chosen for simulation is thus crucial. Smaller time steps (that are used in explicit schemes) model higher frequencies than larger ones (that are used in implicit schemes). Thus, explicit MOT schemes lead to larger pole displacement which renders these techniques virtually always unstable. In contrast, Bluck and Walker [12] experimentally demonstrate that implicit schemes are for all practical purposes stable, which implies that the pole displacement into the right half plane due to these schemes is substantially smaller. Second, in a numerical framework, the incident field does couple to the perturbed interior modes that are supported by the EFIE and MFIE. While this does not necessarily lead to unstable behavior when implicit methods are used, it does follow that the solution can be corrupted by the presence of perturbed cavity modes. The actual level of excitation of these modes depends heavily on the details of the implementation. Bluck and Walker [12] report schemes that appear rather insensitive though not totally immune to the excitation of these resonance. In contrast, as will be demonstrated in Section 19.4, our flat-panel triangular-patch implicit MOT implementation does pick up resonances, albeit often in only minute quantities. However, we conjecture that any EFIE/MFIE implementation will pick up these modes provided that the incident pulse contains sufficient energy in the frequency band near these resonances.

To combat this resonance problem, a time domain CFIE is constructed in analogy to its frequency domain counterpart by combining the EFIE and MFIE as

$$-\beta/\eta_0 \hat{n} \times \hat{n} \times \mathbf{E}^t(\mathbf{r},t) + \hat{n} \times \mathbf{H}^t(\mathbf{r},t) = 0 \quad \forall \mathbf{r} \in S_- \quad (19.8)$$

where the superscript t is used to denote the total field (i.e., sum of the incident and the scattered fields), and $\eta_0 = \sqrt{\mu_0/\epsilon_0}$ is the intrinsic impedance of the free space, introduced in (19.8) for scaling purposes, and β is a positive (real) constant that is greater than zero. Denoting $\mathcal{V}_c \{\mathbf{E}^i(\mathbf{r},t), \mathbf{H}^i(\mathbf{r},t)\} = -\beta/\eta_0 \mathcal{V}_e \{\mathbf{E}^i(\mathbf{r},t), \mathbf{H}^i(\mathbf{r},t)\} + \mathcal{V}_h \{\mathbf{E}^i(\mathbf{r},t), \mathbf{H}^i(\mathbf{r},t)\}$, this CFIE reads

$$\mathcal{V}_c \{\mathbf{E}^i(\mathbf{r},t), \mathbf{H}^i(\mathbf{r},t)\} = \mathcal{L}_c \{\mathbf{J}(\mathbf{r},t)\} \quad \forall \mathbf{r} \in S_- \quad (19.9a)$$

where

$$\mathcal{L}_c \{\mathbf{J}(\mathbf{r},t)\} = -\beta/\eta_0 \mathcal{L}_e \{\mathbf{J}(\mathbf{r},t)\} + \mathcal{L}_h \{\mathbf{J}(\mathbf{r},t)\} \quad (19.9b)$$

As pointed out before, when numerically solving (19.4) and (19.7), poles corresponding to the resonance frequencies of the cavity formed by S can be excited. On the other hand, our numerical results indicate that, for all structures tested, the solution to the CFIE is free of cavity modes. As with the frequency domain CFIE [23], $\beta > 0$ resulted in a resonant free solution.

An earlier investigation into the use of the CFIE was presented by Vechinski and Rao in [26]. There, the authors used

$$-\beta/\eta_0 \hat{n} \times \hat{n} \times \partial_t \mathbf{E}^i(\mathbf{r},t) + \hat{n} \times \partial_t \mathbf{H}^i(\mathbf{r},t) = \partial_t \mathcal{L}_c \{\mathbf{J}(\mathbf{r},t)\} \quad \forall \mathbf{r} \in S_- \quad (19.10)$$

for the analysis of 2D scattering instead of (19.9). The purpose of the analysis presented in [26] was to determine whether the late–time instabilities commonly encountered in MOT schemes that rely on explicit time stepping can be overcome by using a combined field formulation. In this regard, no benefit from using a CFIE was observed. This can again be explained using the arguments put forth by Rynne and Smith [6]. A CFIE formulation only eliminates the poles that lie on the imaginary axis corresponding to the interior cavity modes, and does not affect the location of the poles describing the exterior problem. As the MOT scheme prescribed in [26] relies on an explicit time-marching scheme, it is conjectured that the poles of the resolvent of $\partial_t \mathcal{L}_c$ that lie close to the imaginary axis easily shift into the right half-plane, thereby generating instabilities. As a result, little benefit is observed from using a resonance suppressing integral equation method. Finally, experiments indicate that our implicit time stepping scheme for solving (19.9) yields a more accurate solution than that of (19.10), based upon otherwise very similar implementation choices; it is for this reason that in this work (19.9) was selected over (19.10).

19.2.2 Marching-on-in-Time Formulation

This section develops a classical MOT-based scheme [6, 27, 28] for solving the integral equations (19.4a), (19.7a), (19.9a). Throughout this section, a subscript $q = e, h, c$ is used to represent quantities associated with the EFIE, MFIE, and CFIE, respectively. As a first step towards solving these equations using an MOT scheme, spatial and temporal variations of the current $\mathbf{J}(\mathbf{r}, t)$ are represented in terms of basis functions $\mathbf{S}_n(\mathbf{r})$, $n = 1, \cdots, N_s$, and $T_j(t)$, $j = 1, \cdots, N_t$, as

$$\mathbf{J}(\mathbf{r}, t) = \sum_{j=1}^{N_t} \sum_{n=1}^{N_s} I_{n,j} \mathbf{S}_n(\mathbf{r}) T_j(t) \tag{19.11}$$

Here, $I_{n,j}$ is the weight associated with the space-time basis function $\mathbf{S}_n(\mathbf{r}) T_j(t)$. Denoting the surface area of the scatterer by S_a, the number of spatial basis functions is chosen as $N_s \propto S_a f_{max}^2 / c^2$ to ensure a discretization dense enough to represent the current at f_{max}. Furthermore, assuming that $\mathbf{J}(\mathbf{r}, t)$ resides on S for $0 < t < T_t$ and becomes vanishingly small thereafter, the number of temporal samples is chosen as $N_t \propto T_t f_{max}$.

Rao-Wilton-Glisson functions, which in the past have been used extensively in both frequency and time domain analysis [27, 29], are chosen to model the spatial variation of the current. To this end, S is approximated in terms of flat triangular panels, and one basis function $\mathbf{S}_n(\mathbf{r})$ associated with each edge joining two triangles:

$$\mathbf{S}_n(\mathbf{r}) = \begin{cases} \dfrac{l_n}{2 A_n^+} \boldsymbol{\rho}_n^+(\mathbf{r}) & \text{for } \mathbf{r} \in \Gamma_n^+ \\ \dfrac{l_n}{2 A_n^-} \boldsymbol{\rho}_n^-(\mathbf{r}) & \text{for } \mathbf{r} \in \Gamma_n^- \\ 0 & \text{elsewhere} \end{cases} \tag{19.12}$$

where l_n is the length of the common edge between the triangles Γ_n^+ and Γ_n^-, $A_n^\pm$ is the area of the triangle $\Gamma_n^\pm$, and $\boldsymbol{\rho}_n^\pm(\mathbf{r})$ is the position vector with respect to the free vertex of the corresponding triangle [27]. Linearly interpolating (triangular) functions are used to represent the temporal variation of the current, i.e., $T_j(t) = T(t - j\Delta_t)$, where $\Delta_t = T_t/N_t$ is the time step size, and $T(t) = 1$ for $t = 0$ and linearly interpolates to zero at $t = \pm\Delta_t$.

Substituting (19.11) in (19.4), (19.7), (19.9), and using a spatial Galerkin testing procedure at $t_j = j\Delta_t$, leads to a set of equations that can be represented in matrix form as

$$\mathcal{Z}_q^0 \mathcal{I}_j = \mathcal{F}_q^j - \sum_{i=1}^{j-1} \mathcal{Z}_q^i \mathcal{I}_{j-i} \tag{19.13a}$$

where $\mathcal{I}_j$ is an array of the weights $I_{n,j}$, $n = 1, \cdots, N_s$,

$$\mathcal{F}_{q,m}^j = \langle \mathbf{S}_m(\mathbf{r}), \mathcal{V}_q \{ \mathbf{E}^i(\mathbf{r},t), \mathbf{H}^i(\mathbf{r},t) \} \rangle \bigg|_{t=t_j} \tag{19.13b}$$

and

$$\mathcal{Z}_{q,mn}^i = \langle \mathbf{S}_m(\mathbf{r}), \mathcal{L}_q \{ \mathbf{S}_n(\mathbf{r}) T_{j-i}(t) \} \rangle \bigg|_{t=t_j} \tag{19.13c}$$

In the above equations $\langle \cdot, \cdot \rangle$ denotes the standard inner product [27]. Equation (19.13a) constitutes the basis of the classical MOT scheme; it relates the currents on the surface at t_j to those at $t \leq t_{j-1}$, and hence permits the recursive computation of currents for all times.

Until recently, all MOT schemes were prone to late time instabilities. A vast body of literature on the analysis of, and remedies for addressing, these instabilities exists (see [6, 14, 25, 27] and references therein). Rynne and Smith [6] provide an insightful explanation of the origin of these instabilities and suggest averaging schemes to stabilize the MOT procedure. Their analysis hinges on the location and movement of the poles of the resolvent of the integral operator $\mathcal{L}_q$ [23]. More recently, Dodson et al. observed that use of an implicit time stepping procedure stabilizes the MOT procedure "for all practical purposes" [13]. It has also been suggested that the stability of MOT schemes can be further improved by relying on backward temporal differencing and higher-order temporal basis functions [14].

In our implementation, all inner products are evaluated by using seven-point Gaussian quadrature rules over triangular domains [29]. This testing procedure leads to an implicit scheme, even for the time step size suggested in [27]. In our implementation, the time step size chosen is independent of the spatial discretization and is $\Delta_t = \alpha/(10 f_{max})$, where $\alpha \geq 1$. The resulting scheme is termed implicit because $\mathcal{Z}_q^0$ is not diagonal. However, as this matrix is highly sparse, a nonstationary iterative solver such as QMR (Quasi Minimal Residual) [30] can be used to efficiently solve for $\mathcal{I}_j$ each and every time step. The additional cost incurred from the use of such a solver is insignificant when compared to the overall cost. In addition to implicit time stepping, accurate spatial and temporal integration rules are used for computing $\mathcal{L}_q\{\mathbf{J}(\mathbf{r},t)\}$ [12, 31].

The dominant cost in the construction of (19.13a) involves the vector sum that appears on its right-hand side. The cost of evaluating this sum scales as $\mathcal{O}(N_s^2)$; indeed the reaction at any testing point comprises of contributions from all N_s source basis functions. Since this sum is evaluated for all N_t time steps, the total cost of this analysis scales as $\mathcal{O}(N_t N_s^2)$. In the next section, a succinct derivation of the two-level PWTD algorithm that results in a reduction of this cost will be presented and details regarding its implementation into existing MOT codes will be outlined.

19.3 PLANE-WAVE TIME-DOMAIN ALGORITHM

From the proceeding discussion, it is apparent that computing interactions between individual basis functions, as in traditional MOT schemes, leads to a computationally inefficient algorithm. This suggests that one of the keys to developing a reduced-complexity algorithm is constructing schemes that permit the computation of interactions in a group-wise manner. To this end, assume that the scatterer can be enclosed in a fictitious cubical box, which is further subdivided in many smaller equal-sized cubes or boxes (Figure 19.2). The radius of a sphere circumscribing each cube is denoted by R_s, and the set of basis functions that are contained in a box is called a group. Then, all nonempty boxes are identified and numbered from $\alpha = 1, \cdots, N_g$. Next, a pair of boxes (α, α') is classified as either a "near field" or "far field" pair, depending on a separation criterion based on the distance between the box centers. In our analysis, this distance was chosen to be $R_{c,min} < \gamma R_s$, where $4 \leq \gamma \leq 6$. Since $R_{c,min} = \mathcal{O}(R_s)$, the number of near field pairs is proportional to the total number of nonempty boxes N_g. The interactions between the basis functions that reside in a near field box pair are computed using the classical MOT scheme. However, the interactions between those that reside in a far field box pair are computed in a group-wise manner using the PWTD algorithm. The foundations of the PWTD algorithm lie in expressing the field at a point due to a sufficiently separated source distribution as a superposition of plane waves. Such a representation, as in the FMM, has been extremely effective in reducing the computational complexity of the frequency domain IE solvers. Consequently, the analysis in the first part of this section will focus on the development of such a scheme in the time domain. Next, the incorporation of this algorithm into an MOT scheme in a two-level setting is elucidated. Finally, it is shown that the resulting PWTD-enhanced MOT algorithm has a computational complexity of $\mathcal{O}(N_t N_s^{3/2} \log N_s)$.

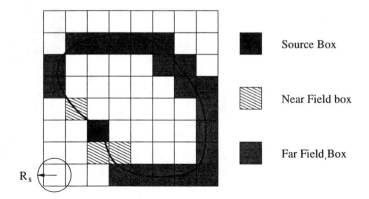

Figure 19.2 Division of the geometry into near and far field blocks.

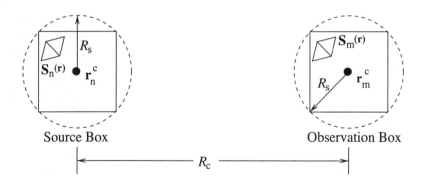

Figure 19.3 Two far field boxes (α, α') with basis functions $\mathbf{S}_n(\mathbf{r})$ and $\mathbf{S}_m(\mathbf{r})$.

19.3.1 Plane Wave Representations

Consider a far field box pair as shown in Figure 19.3, one of which contains the basis function $\mathbf{S}_n(\mathbf{r})$ and is dubbed the source box, the other contains the basis function $\mathbf{S}_m(\mathbf{r})$ and is called the observation box. The centers of these boxes are located at $\mathbf{r}_n^c$ and $\mathbf{r}_m^c$, respectively. The vector connecting the centers of these boxes is denoted by $\mathbf{R}_c = \mathbf{r}_m^c - \mathbf{r}_n^c$ and $R_c = |\mathbf{R}_c|$. For the purpose of this exposition, let the current associated with the source basis function be characterized as

$$\mathbf{J}_n(\mathbf{r}, t) = \sum_{j=1}^{N_t} I_{n,j} \mathbf{S}_n(\mathbf{r}) T_j(t) = \mathbf{S}_n(\mathbf{r}) f_n(t) \qquad (19.14)$$

The time signature $f_n(t)$ is divided into L consecutive subsignals $f_{n,l}(t)$, each of duration $T_s = (M_t + 1)\Delta_t$ (with $LM_t = N_t$) occupying a time slice $T_{l,start} \leq t \leq T_{l,stop}$ for $l = 1, \cdots, L$, where $T_{l,start} = (l-1)M_t\Delta_t$, and $T_{l,stop} = lM_t\Delta_t + \Delta_t$. In keeping with this division, the current source $\mathbf{J}_n(\mathbf{r},t)$ can be rewritten as

$$\mathbf{J}_n(\mathbf{r},t) = \sum_{l=1}^{L} \mathbf{J}_{n,l}(\mathbf{r},t) \tag{19.15a}$$

where

$$\mathbf{J}_{n,l}(\mathbf{r},t) = \mathbf{S}_n(\mathbf{r})f_{n,l}(t); \quad f_{n,l}(t) = \sum_{j=(l-1)M_t+1}^{lM_t} I_{n,j}T_j(t) \tag{19.15b}$$

Then, the vector potential associated with one subsignal is

$$\mathbf{A}_{n,l}(\mathbf{r},t) = \frac{\mu_0}{4\pi} \int_{S_n} dS' \frac{\mathbf{S}_n(\mathbf{r}')f_{n,l}(t - R/c)}{R} \tag{19.16}$$

Alternatively, motivated by frequency domain fast multipole methods that rely heavily on plane wave expansions, consider the field

$$\tilde{\mathbf{A}}_{n,l}(\mathbf{r},t) = -\frac{\mu_0 \partial_t}{8\pi^2 c} \int d^2\Omega \int_{S_n} dS' \, \mathbf{S}_n(\mathbf{r}')\delta\left(t - \hat{\mathbf{k}}\cdot(\mathbf{r}-\mathbf{r}')/c\right) * f_{n,l}(t) \tag{19.17a}$$

where $\hat{\mathbf{k}} = \hat{\mathbf{x}}\sin\theta\cos\phi + \hat{\mathbf{y}}\sin\theta\sin\phi + \hat{\mathbf{z}}\cos\theta$, * denotes a temporal convolution, and

$$\int d^2\Omega \doteq \int_0^{2\pi} d\phi \int_0^{\pi} d\theta \sin\theta \tag{19.17b}$$

To relate $\tilde{\mathbf{A}}_{n,l}(\mathbf{r},t)$ to the vector potential $\mathbf{A}_{n,l}(\mathbf{r},t)$, the spectral integral in (19.17a) is evaluated by interchanging the order of integrations and transforming the spatial variables to a new coordinate system in which $\hat{\mathbf{z}}'$ is parallel to $\mathbf{R}$. Defining θ' and ϕ' as the angular coordinates in this new system, (19.17a) reduces to [32, 33]

$$\begin{aligned}\tilde{\mathbf{A}}_{n,l}(\mathbf{r},t) &= -\frac{\mu_0 \partial_t}{8\pi^2 c} \int_{S_n} dS' \int d^2\Omega' \mathbf{S}_n(\mathbf{r}') f_{n,l}\left(t - \frac{R}{c}\cos\theta'\right) \\ &= -\frac{\mu_0 \partial_t}{8\pi^2} \int_{S_n} dS' \int_0^{2\pi} d\phi \int_{-R/c}^{R/c} d\tau \frac{\mathbf{S}_n(\mathbf{r}') f_{n,l}(t-\tau)}{R} \\ &= \frac{\mu_0}{4\pi} \int_{S_n} dS' \frac{\mathbf{S}_n(\mathbf{r}') f_{n,l}(t - R/c)}{R} \\ &\quad - \frac{\mu_0}{4\pi} \int_{S_n} dS' \frac{\mathbf{S}_n(\mathbf{r}') f_{n,l}(t + R/c)}{R}\end{aligned} \tag{19.18}$$

Comparing (19.16) and (19.18), it is apparent that the evaluation of (19.17a) yields the true vector potential (first term) and an anticausal signal (second term), which will henceforth be referred to as the ghost signal. As the ghost signal appears at the observer before the true signal arrives, it is possible to time-gate $\tilde{\mathbf{A}}_{n,l}(\mathbf{r}, t)$ and recover the true vector potential provided that certain conditions are met.

To develop a scheme for time-gating $\tilde{\mathbf{A}}_{n,l}(\mathbf{r}, t)$, the following observations are in order. With reference to (19.18), the true vector potential radiated by the subsignal $f_{n,l}(t)$ reaches the observer no sooner than $t = T_{l,start} + \min\{R\}/c$, and the ghost signal vanishes after $t = T_{l,stop} - \min\{R\}/c$. Hence, choosing the duration of the signal $T_s < \min\{R\}/c$ implies that $f_{n,l}(t)$ vanishes before the true signal reaches the observer and that the ghost and true fields never overlap in time. These observations can now be generalized for an arbitrary distribution of sources and observers residing in their respective spheres. For this configuration, the arrival of the true signal can be no sooner than $(R_c - 2R_s)/c$ following the onset of a source signal at $t = T_{l,start}$. Hence, the choice $T_s < (R_c - 2R_s)/c$ dictates that $\mathbf{A}_{n,l}(\mathbf{r}, t) = 0$ for $t < T_{l,stop}$ and $\mathbf{A}_{n,l}(\mathbf{r}, t) = \tilde{\mathbf{A}}_{n,l}(\mathbf{r}, t)$ for $t \geq T_{l,stop}$ as the ghost signal vanishes for $t > -(R_c - 2R_s)/c + T_{l,stop}$. Summarizing these observations, the choice of $T_s < (R_c - 2R_s)/c$ ensures that

$$\mathbf{A}_{n,l}(\mathbf{r}, t) = \begin{cases} 0 & \text{for } t < T_{l,stop} \\ \tilde{\mathbf{A}}_{n,l}(\mathbf{r}, t) & \text{for } t \geq T_{l,stop} \end{cases} \quad (19.19)$$

It follows that both the electric and magnetic fields can be evaluated from (19.17a). Indeed, using (19.2) and (19.5) and the fact that $\nabla \iff -\partial_t \hat{\mathbf{k}}/c$ holds for plane wave basis, the fields in the observation sphere due to the lth source time slice for $t > T_{l,stop}$ can be written as

$$\mathbf{E}_{n,l}^s(\mathbf{r}, t) = \frac{\eta_0}{8\pi^2 c^2} \int_{T_{l,stop}}^{t} dt' \, \partial_{t'}^3 \int d^2\Omega \, \left(\mathcal{I} - \hat{\mathbf{k}}\hat{\mathbf{k}}\right) \\ \cdot \int_{S_n} dS' \, \mathbf{S}_n(\mathbf{r}')\delta\left(t' - \hat{\mathbf{k}} \cdot (\mathbf{r} - \mathbf{r}')/c\right) * f_{n,l}(t') \quad (19.20a)$$

$$\mathbf{H}_{n,l}^s(\mathbf{r}, t) = \frac{1}{8\pi^2 c^2} \int_{T_{l,stop}}^{t} dt' \, \partial_{t'}^3 \int d^2\Omega \, \hat{\mathbf{k}} \\ \times \int_{S_n} dS' \, \mathbf{S}_n(\mathbf{r}')\delta\left(t' - \hat{\mathbf{k}} \cdot (\mathbf{r} - \mathbf{r}')/c\right) * f_{n,l}(t') \quad (19.20b)$$

Equations (19.20) indicate that the scattered electric and magnetic fields can be constructed using the transverse components of the vector potential. It should also be noted that the time integrals in (19.20) should not be cancelled by one of the time derivatives that follow as this will introduce an undesirable static ghost signal.

To explore the merits of a plane wave representation in constructing a fast algorithm, note that $\tilde{\mathbf{A}}(\mathbf{r},t)$ tested with a basis function $\mathbf{S}_m(\mathbf{r})$ can be written as

$$\langle \mathbf{S}_m(\mathbf{r}), \tilde{\mathbf{A}}(\mathbf{r},t) \rangle = -\frac{\mu_0 \partial_t}{8\pi^2 c} \int d^2\Omega \left[\int_{S_m} dS \mathbf{S}_m(\mathbf{r}) \delta\left(t - \hat{\mathbf{k}} \cdot (\mathbf{r} - \mathbf{r}_m^c)/c\right) \right]^T$$
$$* \delta\left(t - \hat{\mathbf{k}} \cdot \mathbf{R}_c/c\right) * \left[\int_{S_n} dS' \, \mathbf{S}_n(\mathbf{r}') \delta\left(t + \hat{\mathbf{k}} \cdot (\mathbf{r}' - \mathbf{r}_n^c)/c\right) \right] * f_{n,l}(t) \tag{19.21}$$

for $t > T_{l,stop}$, where the superscript T is used to denote a transpose. Similarly, using (19.4), (19.7), (19.9), and (19.20), it can be shown that the inner products of the fields $\mathbf{E}_{n,l}^s(\mathbf{r},t)$ and $\mathbf{H}_{n,l}^s(\mathbf{r},t)$ with $\mathbf{S}_m(\mathbf{r})$ are zero for $t < T_{l,stop}$, and for $t \geq T_{l,stop}$:

$$\langle \mathbf{S}_m(\mathbf{r}), \mathcal{L}_e \{\mathbf{J}_{n,l}(\mathbf{r},t)\} \rangle = \frac{\eta_0}{8\pi^2 c^2} \int_{T_{l,stop}}^{t} dt' \int d^2\Omega \left[\mathcal{S}_m^-\left(\hat{\mathbf{k}}, t', \hat{\mathbf{k}}\right) \right]^T$$
$$* \mathcal{T}\left(\hat{\mathbf{k}}, t'\right) * \left[\mathcal{S}_n^+\left(\hat{\mathbf{k}}, t', \hat{\mathbf{k}}\right) \right] * f_{n,l}(t') \tag{19.22a}$$

$$\langle \mathbf{S}_m(\mathbf{r}), \mathcal{L}_h \{\mathbf{J}_{n,l}(\mathbf{r},t)\} \rangle = \frac{1}{8\pi^2 c^2} \int_{T_{l,stop}}^{t} dt' \int d^2\Omega \left[\mathcal{S}_m^-\left(\hat{\mathbf{k}}, t', \hat{\mathbf{n}}\right) \right]^T$$
$$* \mathcal{T}\left(\hat{\mathbf{k}}, t'\right) * \left[\mathcal{S}_n^+\left(\hat{\mathbf{k}}, t', \hat{\mathbf{k}}\right) \right] * f_{n,l}(t') \tag{19.22b}$$

$$\langle \mathbf{S}_m(\mathbf{r}), \mathcal{L}_c \{\mathbf{J}_{n,l}(\mathbf{r},t)\} \rangle = -\frac{\beta}{\eta_0} \langle \mathbf{S}_m(\mathbf{r}), \mathcal{L}_e \{\mathbf{J}_{n,l}(\mathbf{r},t)\} \rangle$$
$$+ \langle \mathbf{S}_m(\mathbf{r}), \mathcal{L}_h \{\mathbf{J}_{n,l}(\mathbf{r},t)\} \rangle \tag{19.22c}$$

In the above equations, $\mathcal{T}(\hat{\mathbf{k}},t)$ denotes the translation function

$$\mathcal{T}(\hat{\mathbf{k}},t) = \partial_t^3 \delta\left(t - \hat{\mathbf{k}} \cdot \mathbf{R}_c/c\right) \tag{19.23a}$$

and

$$\mathcal{S}_o^\pm\left(\hat{\mathbf{k}}, t, \hat{\mathbf{v}}\right) = \int_{S_o} dS' \hat{\mathbf{v}} \times \mathbf{S}_o(\mathbf{r}') \delta\left(t \pm \hat{\mathbf{k}} \cdot (\mathbf{r}' - \mathbf{r}_o^c)/c\right) \tag{19.23b}$$

where $o = \{m, n\}$.

Computing all interactions requires numerical implementation of (19.22). In practice, as all spherical integrals are computed using quadrature rules, it is necessary to temporally bandlimit the current densities. This follows naturally as the excitation is assumed to be bandlimited to f_{max}. Thus, the current density can be locally interpolated using temporally bandlimited and approximately time-limited functions such that the time signature of the current for the lth time slice can be written as

$$f_{n,l}(t) = \sum_{j=(l-1)M_t+1}^{lM_t} I_{n,j} \Psi_j(t) \tag{19.24}$$

where $\Psi_j(t) = \Psi(t - j\Delta_t)$ is an interpolant bandlimited to $f_s > f_{max}$. In this study, $\Psi_j(t)$ is chosen to be a variant of the approximate prolate spheroidal functions [34]. Timelimiting these functions to a duration of $(2p + 1)\Delta_t$ introduces an error that decreases exponentially with increasing p. As a result, $f_{n,l}(t)$ can be considered bandlimited and spans a duration $M'_t\Delta_t = (M_t + 2p)\Delta_t$. Evaluation of (19.22) is done by sampling on the surface of the sphere. As the far field is bandlimited, it is possible to show that the number of samples needed to completely characterize the field is of $\mathcal{O}(M^2)$, where

$$M = \text{Int}(4\pi\chi f_s R_s/c) + 1 \tag{19.25}$$

and χ is an oversampling factor [35, 36]. Note that the translation function (19.23a) can be succinctly expressed as $\mathcal{T}(\hat{\mathbf{k}}, t) = \hat{\mathcal{T}}(\hat{\mathbf{k}}, t, \infty)$, where [31]

$$\hat{\mathcal{T}}(\hat{\mathbf{k}}, t, \hat{M}) = \frac{c\partial_t^3}{2R_c} \sum_{\nu=0}^{\hat{M}} (2\nu + 1) P_\nu \left(\frac{ct}{R_c}\right) P_\nu(\cos\theta') \quad \text{for } |t| \leq \frac{R_c}{c} \tag{19.26}$$

and $P_\nu(x)$ are Legendre polynomials of degree ν. In keeping with the bandlimitedness of the fields, the upper limit in the above summation can be truncated to $\hat{M} = M$ [37]. Consequently, (19.22) can be evaluated numerically using

$$\langle \mathbf{S}_m(\mathbf{r}), \mathcal{L}_e\{\mathbf{J}_{n,l}(\mathbf{r}, t)\}\rangle = \frac{\eta_0}{8\pi^2 c^2} \int_{T_{l,stop}}^{t} dt' \sum_{k=0}^{M} \sum_{p=-M}^{M} w_{kp} \left[\mathcal{S}_m^-\left(\hat{\mathbf{k}}_{kp}, t', \hat{\mathbf{k}}_{kp}\right)\right]^T$$
$$* \hat{\mathcal{T}}\left(\hat{\mathbf{k}}_{kp}, t', M\right) * \left[\mathcal{S}_n^+\left(\hat{\mathbf{k}}_{kp}, t', \hat{\mathbf{k}}_{kp}\right)\right] * f_{n,l}(t') \tag{19.27a}$$

$$\langle \mathbf{S}_m(\mathbf{r}), \mathcal{L}_h\{\mathbf{J}_{n,l}(\mathbf{r}, t)\}\rangle = \frac{1}{8\pi^2 c^2} \int_{T_{l,stop}}^{t} dt' \sum_{k=0}^{M} \sum_{p=-M}^{M} w_{kp} \left[\mathcal{S}_m^-\left(\hat{\mathbf{k}}_{kp}, t', \hat{\mathbf{n}}\right)\right]^T$$
$$* \hat{\mathcal{T}}\left(\hat{\mathbf{k}}_{kp}, t', M\right) * \left[\mathcal{S}_n^+\left(\hat{\mathbf{k}}_{kp}, t', \hat{\mathbf{k}}_{kp}\right)\right] * f_{n,l}(t') \tag{19.27b}$$

The corresponding equation for the $\langle \mathbf{S}_m(\mathbf{r}), \mathcal{L}_c\{\mathbf{J}_{n,l}(\mathbf{r}, t)\}\rangle$ can be obtained using (19.22c), and the time integral in the above equation is performed using standard integration rules [38]. In the above equations, w_{kp} are the quadrature weights [37] given by

$$w_{kp} = \frac{4\pi(1 - \cos^2\theta_k)}{(2M + 1)\left[(M + 1)P_M(\cos\theta_k)\right]^2} \tag{19.28}$$

$$\hat{\mathbf{k}}_{kp} = \hat{\mathbf{x}}\sin\theta_k\cos\phi_p + \hat{\mathbf{y}}\sin\theta_k\sin\phi_p + \hat{\mathbf{z}}\cos\theta_k \tag{19.29}$$

$$\phi_p = 2\pi p/(2M + 1) \tag{19.30}$$

θ_k is the $(k + 1)^{th}$ zero of $P_{M+1}(\cos\theta)$ \hfill (19.31)

To gain insight into (19.27) note that the rightmost convolution maps the source distribution onto a set of plane waves which will henceforth be referred to as "outgoing rays." In the literature, this mapping is also known as the Slant Stack Transform (SST) [33]. The center convolution "translates" these to "incoming rays," which impinge on the observation sphere. Finally, via the last convolution and the spectral integration the incoming rays are mapped onto the observers. The reconstruction of transient field using this three-stage process of aggregation, translation, and disaggregation, is reminiscent of the popular frequency domain FMM.

19.3.2 Implementation of Two-Level PWTD Enhanced MOT Solvers

The ideas outlined in the preceding section give rise to a procedure via which the interaction between basis functions residing in any far field box pair (α, α') can be computed as a superposition of transient plane waves. To complete the PWTD algorithmic prescription, these ideas are now systematized such that they can be efficiently applied in conjunction with an MOT scheme. With this in mind, a fundamental subsignal duration $T_s = (M_t + 2p)\Delta_t$ is defined, where

$$M_t = \min_{(\alpha, \alpha')} \left\{ \left\lfloor \frac{R_{c,\alpha\alpha'} - 2R_s}{c\Delta_t} \right\rfloor \right\} - 2p \qquad (19.32)$$

and $R_{c,\alpha\alpha'}$ is the distance between the centers of the boxes (α, α'). Such a definition stems from the fact that T_s corresponds to the duration of the longest possible subsignal that can be translated "ghost-free" between the closest far field pair. For all other far field pairs, the subsignal lengths are chosen to be an integer multiple of the fundamental duration [i.e., $T_{s,\alpha\alpha'} = (M_{t,\alpha\alpha'} + 2p)\Delta_t$ where $M_{t,\alpha\alpha'} = M_t \lfloor (R_{c,\alpha\alpha'} - 2R_s - 2pc\Delta_t)/M_t \rfloor$]. The rationale behind this choice will soon become transparent.

The task of computing the current distribution at each time step is divided into (1) evaluating near field interactions using the usual MOT scheme, and (2) computing far field interactions using the PWTD algorithm.

(1) *Near Field Evaluation*: At each time step, the sum

$$\sum_{i=1}^{j-1} \mathcal{Z}^i_{q,mn} \mathcal{I}_{j-i,n} \quad \forall\, m \in \alpha \text{ and } \forall\, n \in \alpha' \qquad (19.33)$$

is computed for all near field interaction pairs (α, α').

(2) *Far Field Evaluation*: To take all the far field interactions into account, the algorithm follows the three stage process that was alluded to in the previous section.

   (a) The first task is the construction of outgoing rays for all boxes. This involves computing $\mathcal{S}_n^+(\hat{\mathbf{k}}_{kp}, t, \hat{\mathbf{k}}_{kp})$ for all ray directions for a fundamental subsignal comprising of $M_t + 2p$ samples and duration T_s. Note that, once this information has been computed, it can be reused for different interaction pairs. Furthermore, the number of time slices for which one needs to store these outgoing rays is proportional to the largest linear dimension of the scatterer.

(b) The next step is to translate the outgoing rays from a source group α' to an observer group α. This is done every $M_{t,\alpha\alpha'}$ time steps. As $M_{t,\alpha\alpha'}$ is an integer multiple of M_t, the rays to be translated can be formed by concatenating an appropriate number of outgoing rays from group α'. Since the length of the translation function is $2R_{c,\alpha\alpha'}/(c\Delta_t) = \mathcal{O}(M_{t,\alpha\alpha'})$ time steps, the convolution of the outgoing ray with the translation function can be efficiently accomplished using fast fourier transforms (FFTs) since the length of the outgoing ray to be translated is of the same order. Unfortunately, as the translation function is not bandlimited, simple FFTs cannot be used to transform it into the Fourier domain. However, this hurdle can be surmounted as an analytical expression for the Fourier transform of the translation function is available. For a far field pair (α, α'), the Fourier transform of the translation function is

$$\check{\mathcal{T}}(\hat{\mathbf{k}}, \omega) = \int_{-\infty}^{\infty} dt\, e^{-j\omega t} \hat{\mathcal{T}}(\hat{\mathbf{k}}, t, \hat{M})$$

$$= \int_{-\infty}^{\infty} dt\, e^{-j\Omega_f ct/R_{c,\alpha\alpha'}} \hat{\mathcal{T}}(\hat{\mathbf{k}}, t, \hat{M}) \qquad (19.34)$$

$$= -\frac{jc^3}{R_{c,\alpha\alpha'}^3} \tilde{\mathcal{T}}(\hat{\mathbf{k}}, \Omega_f)$$

where $\tilde{\mathcal{T}}(\hat{\mathbf{k}}, \Omega_f) = \Omega_f^3 \sum_{\nu=0}^{\hat{M}}(2\nu+1)(-j)^\nu j_\nu(\Omega_f) P_\nu(\cos\theta)$, $\Omega_f = \omega R_{c,\alpha\alpha'}/c$ is the normalized frequency, and $j_\nu(\cdot)$ is a spherical Bessel function of the νth order [31]. This equation also indicates that the translation function for an arbitrary sphere pair can be constructed from a function $\tilde{\mathcal{T}}(\hat{\mathbf{k}}, \Omega_f)$ that is bandlimited in both the Ω_f and θ. Hence, this function can be sampled at a discrete set of points, and the translation function for any far field sphere pair reconstructed by interpolating through these samples. After convolving the outgoing rays with the translation function, the resulting rays are then superimposed on to the incoming rays of the observer group. It should be noted that as the evaluation of $\mathcal{S}_n^+(\hat{\mathbf{k}}_{kp}, t, \hat{\mathbf{k}}_{kp})$ yields two real signals, namely, the $\hat{\theta}$ and $\hat{\phi}$ components of the field which are to be translated, this operation is most efficiently performed using one complex FFT [38].

(c) Finally, the rays entering all the spheres are projected on to the observers. This is done via the leftmost convolution in (19.27).

It should be pointed out that the error incurred in computing the fields via a plane wave expansion method can be controlled to arbitrary precision [19, 37].

19.3.3 Complexity Analysis

To analyze the computational complexity of the two-level PWTD-enhanced MOT solver described above, assume that there are N_g nonempty boxes, each containing approximately $M_s = N_s/N_g$ unknowns. In what follows, implementation-dependent constants are denoted as C_i, $i = 0, 1, 2, 3$. The cost of a PWTD-MOT analysis is comprised of near and far field components. The computation of all near field interactions for the duration of the analysis requires

$$C_{NF} = C_0 N_g M_s^2 N_t \qquad (19.35a)$$

operations. The cost of evaluating all far field interactions consists of those incurred in constructing outgoing rays (C_{FF}^1), translating the latter onto incoming rays (C_{FF}^2), and projecting incoming rays onto observers (C_{FF}^3). Constructing outgoing rays involves the projection of all current elements in the N_g nonempty boxes onto $\mathcal{O}(M^2) = \mathcal{O}(M_s)$ plane waves for all N_t time steps. Projecting incoming rays onto observers involves a very similar set of operations. Hence, the cost associated with these operations is

$$C_{FF}^{1,3} = C_{1,3} N_g M_s^2 N_t \qquad (19.35b)$$

The cost of translating one ray between groups (α, α') scales as $\mathcal{O}(M_{t,\alpha\alpha'} \log M_{t,\alpha\alpha'})$. As this operation has to be performed for all $N_t/M_{t,\alpha\alpha'}$ time slices and $\mathcal{O}(M^2) = \mathcal{O}(M_s)$ directions, the cost of translating information between a given pair of spheres for the duration of the analysis scales as $\mathcal{O}(N_t M^2 \log M_{t,\alpha\alpha'})$. Since $M_{t,\alpha\alpha'}$ is bounded by the maximum linear dimension of the scatterer ($\sqrt{N_s}$), the cost of translations for all N_g^2 far field group interactions is

$$C_{FF}^2 = C_2 N_t N_s^2 / M_s \log N_s \qquad (19.35c)$$

The total cost associated with the PWTD-enhanced MOT analysis is

$$C_T = C_{NF} + C_{FF}^1 + C_{FF}^2 + C_{FF}^3 \qquad (19.36)$$

Minimizing C_T with respect to M_s reveals that the optimal number of unknowns per group grows as $M_s \propto N_s^{1/2}$, and that C_T scales as $\mathcal{O}(N_t N_s^{3/2} \log N_s)$.

19.4 NUMERICAL RESULTS

This section presents numerical results that serve to demonstrate the twin objectives stated in the introduction; (1) validate and demonstrate the accuracy of the CFIE in analyzing scattering from closed bodies, (2) validate and (3) demonstrate the efficacy of the PWTD-augmented MOT schemes. To demonstrate the first, transient scattering from electrically small structures is analyzed. Time-domain far field signatures computed using the EFIE/MFIE/CFIE codes are used to construct radar cross-section (RCS) data at a set of frequencies. These are then compared against RCS data obtained from a frequency domain CFIE code (Fast Illinois Solver Code (FISC)). The efficacy of the PWTD-augmented MOT scheme is demonstrated by analyzing transient scattering from electrically large scatterers. Again, similar data is generated and compared against those obtained using FISC. In all the examples presented herein, the incident field is a modulated Gaussian plane wave parameterized as

$$\mathbf{E}^i(\mathbf{r}, t) = \hat{\mathbf{p}} \cos\left(2\pi f_0 \left[t - \mathbf{r} \cdot \hat{\mathbf{k}}/c\right]\right) \exp\left[-\frac{(ct - \mathbf{r} \cdot \hat{\mathbf{k}} - ct_p)^2}{2\sigma^2 c^2}\right] \quad (19.37)$$

where f_0 is the pulse's center frequency, $\hat{\mathbf{k}}$ and $\hat{\mathbf{p}}$ denote its direction of travel and polarization, $\sigma = 6/(2\pi f_{bw})$, and $t_p = 3.5\sigma$. The parameter f_{bw} will be referred to as the "bandwidth" of the signal. It is to be noted that the power in the incident pulse is down by 160 dB at $f = f_0 \pm f_{bw}$ relative to f_0. Also, the details on the geometries of some of the scatterers that are used here to illustrate the capabilities of this code can be obtained from [39].

19.4.1 Efficacy of the CFIE

To validate the time domain CFIE against the time domain MFIE and EFIE, consider a cube of dimensions 1m × 1m × 1m shown in the inset in Figure 19.4. A modulated Gaussian plane wave with $f_0 = 100$ MHz and $f_{bw} = 40$ MHz, traveling along $\hat{\mathbf{k}} = -\hat{\mathbf{z}}$ with $\hat{\mathbf{p}} = \hat{\mathbf{x}}$ is incident on the cube. The cube is discretized into 450 spatial unknowns. It is ensured that $f = f_0 + f_{bw}$ is less than 150 MHz, which is the frequency of the first resonant mode of the cube. In Figure 19.4 the magnitude of the current at a point on the cube's upper surface, computed using $\beta = 0.0$ (MFIE), $0.25, 1.0, 4.0, \infty$ (EFIE), is plotted against time. It is seen that the temporal signatures of the current computed using all values of β agree well with each other. However, unlike the currents computed using the CFIE and MFIE, whose magnitude keeps decreasing with time, those obtained using the EFIE stabilize at a value three orders of magnitude below the peak. This behavior of the EFIE has also been observed by other researchers [40].

Having ascertained that the numerical implementation of the time domain CFIE does yield a solution that coincides with that of the MFIE and the EFIE when no resonant modes are excited, we next examine the CFIE's performance when the spectrum of the incident pulse encompasses the body's resonance frequencies. To this end, consider a sphere of radius 1m that is discretized using 2,793 spatial unknowns, illuminated by a $\hat{\mathbf{p}} = \hat{\mathbf{x}}$ polarized modulated Gaussian pulse with $\hat{\mathbf{k}} = -\hat{\mathbf{z}}$. The

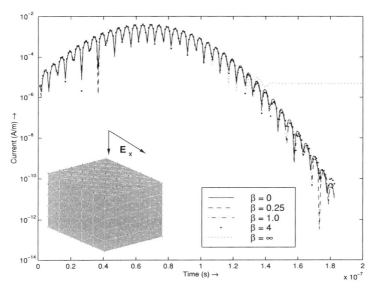

Figure 19.4 Comparison of the currents observed on a cube as a function of time. The currents are computed using the time domain CFIE code for $\beta = 0$(MFIE), $0.25, 1.0, 4.0$, and ∞(EFIE).

pulse has a bandwidth of $f_{bw} = 150$ MHz, and a center frequency of $f_0 = 200$ MHz. The above choice of center frequency and bandwidth is such that the $\text{TM}_{m,2,1}$ (184 MHz), $\text{TE}_{m,1,1}$ (214 MHz), and $\text{TM}_{m,3,1}$ (237 MHz) modes are excited. While the sphere theoretically also resonates at $f = 131, 275, 289, 292, 333, 340$ MHz, etc., these modes are barely excited as the power in the incident pulse at these frequencies is down by at least 30 dB from its peak at f_0. The magnitudes of the current at a point on the sphere with $\{\theta = 65°, \phi = 72°\}$ obtained using the EFIE, MFIE, and CFIE ($\beta = 0.25$) codes are compared in Figure 19.5(a). It is apparent that the solution to the MFIE exhibits a characteristic ringing, whereas that of the CFIE dies down. As mentioned earlier, since the spectrum of the incident pulse contains resonance frequencies characteristic of the body, it was necessary to use four-point temporal averaging [6] in addition to implicit time stepping to stabilize the solution to the EFIE. As a consequence of using the averaging process, which suppresses resonance effects to some extent, the EFIE solution does not exhibit the same features as that of the MFIE. However, both are still quite different from that obtained using the CFIE. The differences between the currents are further highlighted by examining their Fourier transforms. Figure 19.5(b) compares the Fourier transforms of the current obtained using the CFIE, MFIE, and EFIE with those obtained analytically (using Mie series). As is apparent, the CFIE and analytical solutions agree very well with each other whereas the the others do not; indeed, the variation of the results obtained using the CFIE, MFIE, and the EFIE from the analytical solutions is 0.7%, 7%, and 18%, respectively. The MFIE results are significantly different from the CFIE in the vicinity of of the above mentioned resonance frequencies. Also, the Fourier transform of the current obtained using the EFIE is slightly smaller than that obtained using the CFIE and is distorted at the ends of the spectrum, both of which are consequences of temporal averaging. Examination of Figure 19.5(c), where the difference between the Fourier transforms of the current obtained using the MFIE and CFIE are plotted, reveals the presence of the principal resonant modes more clearly.

In what follows, the RCS patterns obtained using the MFIE and CFIE codes are compared against those obtained using FISC. The RCS patterns obtained using the EFIE codes are not shown as it is well known that, while the currents on the surface computed using the EFIE are corrupted by interior modes, the scattered far fields obtained from them are not [24]. Figure 19.6 compares the RCS pattern in the x–z plane computed using the time domain CFIE and MFIE codes, FISC, and Mie series solutions at two different frequencies chosen either towards the end of the spectrum or close to a resonance. As is seen in these figures, the time domain CFIE faithfully reproduces analytical results as well as those obtained from FISC while the MFIE does not. It should be noted that the CFIE results agree reasonably well with those from FISC and Mie methods at 120 MHz in spite of the fact that at this frequency the power in the incident field is down by 45 dB from its peak value. These results are not surprising, as the existence of nonphysical resonant currents in the solution to the MFIE will cause errors to propagate in any MOT scheme. Thus, when the RCS pattern is extracted from the far field signature, these errors are most conspicuous at the ends of the band. It should be noted that MFIE, CFIE, FISC, and Mie results at any frequency f agree well with each other as long as f is not close to any of the resonance frequencies.

Next, scattering from a cone-sphere is studied. The cone is 1m long, the radius of the half sphere attached to the cone is 0.235m, and the cone-sphere is discretized with 1,656 unknowns. The incident field is a modulated Gaussian pulse with center frequency $f_0 = 400$ MHz and bandwidth $f_{bw} = 350$ MHz; it is $\hat{p} = \hat{z}$ polarized, and is traveling in the $\hat{k} = -\hat{x}$ direction. The RCS patterns in the x–z plane obtained from the time domain MFIE and CFIE codes are compared against those obtained from FISC for two different frequencies, as shown in Figure 19.7. As before, while the results obtained from the CFIE code ($\beta = 1.0$) agree very well with those from FISC for both frequencies, those obtained from the MFIE code do not.

In the next two examples the scatterer being analyzed is an almond with a maximum height of 0.0575m, maximum width of 1.15m, and length of 3m, and is discretized using 1,104 spatial unknowns. The incident field has a center frequency of $f_0 = 200$ MHz, and bandwidth of $f_{bw} = 150$ MHz. In the first of these two examples, the

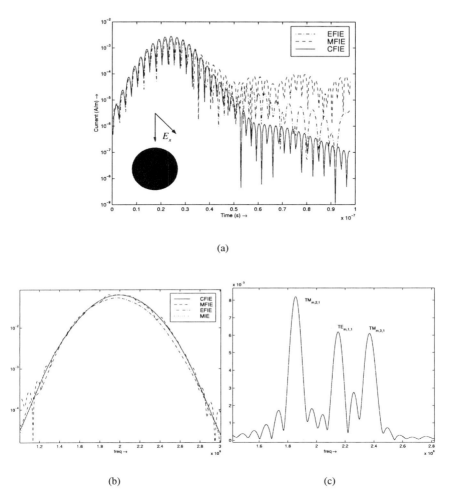

Figure 19.5 (a) Comparison of the current observed at a point on a sphere computed using MFIE, EFIE, and CFIE ($\beta = 0.25$). (b) Comparison of the absolute value of the Fourier transform of these currents. (c) The absolute value of the Fourier transform of the difference between the currents computed using the time domain CFIE and MFIE.

incident wave propagates along $\hat{\mathbf{k}} = -\hat{\mathbf{z}}$, and is $\hat{\mathbf{p}} = \hat{\mathbf{y}}$ polarized. The RCS in the y–z is computed and representative results are shown in Figure 19.8(a) and (b). Figure 19.8(c,d) compares the RCS data in the x–y due to an incident field that propagates in along $\hat{\mathbf{k}} = -\hat{\mathbf{x}}$ and is $\hat{\mathbf{p}} = \hat{\mathbf{y}}$ polarized. As is evident from these results, the

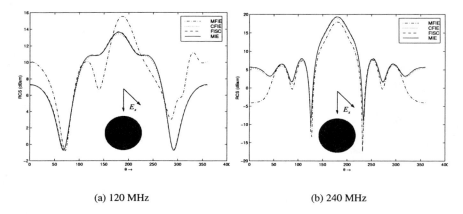

Figure 19.6 The radar scattering cross-section of a sphere in the x–z plane extracted from the time domain CFIE and MFIE is compared to that obtained from FISC for two different frequencies. The incident wave propagates along $\hat{\mathbf{k}} = -\hat{\mathbf{z}}$ and is $\hat{\mathbf{p}} = \hat{\mathbf{x}}$ polarized.

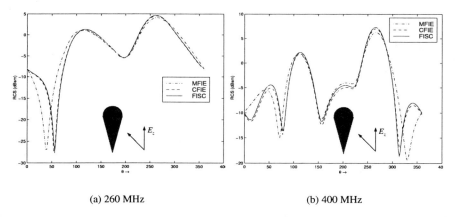

Figure 19.7 The radar scattering cross-section of a cone-sphere in the z–x plane, extracted from the time domain CFIE and MFIE is compared to that obtained from FISC for a set of frequencies. The incident wave is $\hat{\mathbf{p}} = \hat{\mathbf{z}}$ polarized and is traveling in the $\hat{\mathbf{k}} = -\hat{\mathbf{x}}$ direction.

CFIE code ($\beta = 1.0$) agrees very well with those obtained from FISC, whereas those obtained using the MFIE show enormous deviations. It should also be noted that it is possible to extract meaningful results from the CFIE time domain data at 120 MHz in spite of the fact that at this frequency the power of the incident pulse is down by about 46 dB with respect to its peak value.

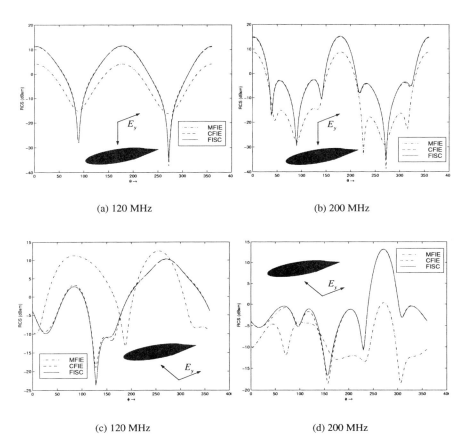

Figure 19.8 The radar scattering cross-section of an almond in: (a) and (b) the z–y plane, extracted from the time domain CFIE and MFIE is compared to that obtained from FISC for a set of frequencies. The incident wave is $\hat{\mathbf{p}} = \hat{\mathbf{y}}$ polarized and is traveling in the $\hat{\mathbf{k}} = -\hat{\mathbf{z}}$ direction; (c) and (d) in the x–y plane, extracted from the time domain CFIE and MFIE is compared to that obtained from FISC for a set of frequencies. The incident wave is $\hat{\mathbf{p}} = \hat{\mathbf{y}}$ polarized and is traveling in the $\hat{\mathbf{k}} = -\hat{\mathbf{x}}$ direction.

19.4.2 Validating the PWTD-Augmented MOT Solver

Thus far, we have established that the use of the CFIE is imperative if transient scattering from closed bodies are to be analyzed. Next, it is necessary to verify that the PWTD-enhanced and classical MOT solvers yield identical results, and validate these schemes against measurements. To this end, three objects are analyzed. First, a rectangular plate of dimensions 2m × 15m that resides in the x–y plane. A $\hat{\mathbf{p}} = \hat{\mathbf{x}}$

polarized pulse traveling in the $\hat{\mathbf{k}} = -\hat{\mathbf{z}}$ direction, with center frequency $f_0 = 0$ MHz and bandwidth $f_{bw} = 150$ MHz is incident on the plate. The current on the plate is represented using 2,170 spatial basis functions and solved for using the EFIE. The magnitudes of the current at $(0.2, 2.0, 0.0)$ on the plate's surface, computed using both the PWTD-enhanced and classical MOT schemes are compared in Figure 19.9(a). Similarly, the temporal far field signatures of the field scattering along $+\hat{\mathbf{z}}$, computed using both methods, are compared in Figure 19.9(b). Obviously, our PWTD-enhanced MOT solver yields results that agree very well with those obtained using the classical scheme.

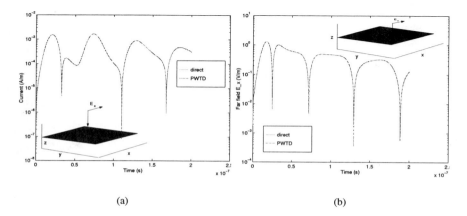

Figure 19.9 Transient scattering from a plate analyzed using the EFIE. The plate is discretized with 2,170 unknowns. The plate measures 2m × 15m. The incident field is $\hat{\mathbf{k}} = -\hat{\mathbf{z}}$ directed and $\hat{\mathbf{p}} = \hat{\mathbf{x}}$ polarized. (a) Current at a location on the plate. (b) The backscattered far field E_x.

Next, an almond that fits into a rectangular box of size 5.00m × 1.92m × 0.48m is excited by a pulse traveling along $\hat{\mathbf{k}} = -\hat{\mathbf{z}}$ and polarized along $\hat{\mathbf{p}} = \hat{\mathbf{x}}$. The center frequency and bandwidth of the pulse are $f_0 = 0$ MHz and $f_{bw} = 404$ MHz, respectively, and $t_p = 6.0\sigma$. The current on the almond is represented using 4,680 spatial basis functions and solved for using an MFIE. Figure 19.10 compares the magnitude of the current at $(0.01, 0.58, 0.03)$ on the almond and the far field signature of the $+\hat{\mathbf{z}}$ traveling scattered field obtained using PWTD-enhanced and classical MOT codes; again, good agreement between both sets of results is observed.

Finally, the resonant frequencies of a rectangular PEC box loaded with a wire are calculated [see Figure 19.11(a)]. The wire is the 0.8–*mm* inner conductor of a $50-\Omega$ coaxial line with a $50-\Omega$ source resistance, connected to the top of the box. The wire is terminated at the bottom using a $47-\Omega$ resistor. For this structure, the power delivered by the source as a function of frequency was measured and reported in

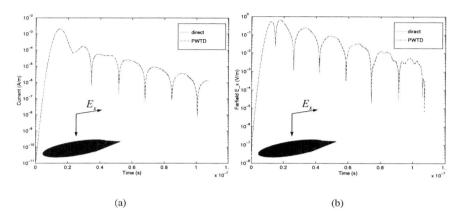

Figure 19.10 Transient scattering from an almond, which is discretized using 4,680 spatial basis functions, analyzed using the MFIE. The incident field propagates along $\hat{\mathbf{k}} = -\hat{\mathbf{z}}$ and is $\hat{\mathbf{p}} = \hat{\mathbf{x}}$ polarized. (a) Current at a location on the almond. (b) The backscattered far field E_x.

[41]. The same problem is analyzed using the PWTD-enhanced MOT scheme. The geometry is discretized using 5,800 spatial unknowns. A delta-gap voltage source is placed in the wire/box junction to model the excitation specified by

$$V^i(\mathbf{r}, t) = V_0 \cos\left(2\pi f_0 \left[t - t_p\right]\right) \exp\left[-\frac{(t - t_p)^2}{2\sigma^2}\right] \quad (19.38)$$

with $V_0 = 1\text{V}$, $f_0 = 1.15$ GHz, and $\sigma = 9.55 \times 10^{-10}$s. The $47 - \Omega$ resistor is modeled as a lumped resistive element in this scheme. The simulation was run for 10,000 time steps and, as can be verified from Figure 19.11(b), the location of the all the resonant frequencies are predicted correctly.

19.4.3 Efficacy of the PWTD-Augmented MOT Scheme for Large-Scale Analysis

Via the above numerical simulations, it has been verified that the solutions obtained using the PWTD-augmented MOT scheme agree with those obtained using the classical MOT scheme. Next, the PWTD-accelerated MOT codes is applied to the analysis of scattering from electrically large objects, and is further validated by comparing their bistatic RCS, extracted at a number of frequency points, against RCS data computed using a frequency domain solver. In all the examples that follow, the time domain CFIE is used [28]. First, transient scattering from a sphere of unit radius is analyzed. The sphere is illuminated by a modulated Gaussian pulse traveling in the $\hat{\mathbf{k}} = -\hat{\mathbf{z}}$ direction and polarized along $\hat{\mathbf{p}} = \hat{\mathbf{x}}$. The center frequency and bandwidth

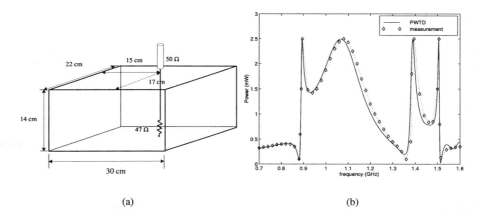

Figure 19.11 Comparison of the resonance frequencies of a rectangular box load with a wire obtained using the PWTD-augmented MOT algorithm against those obtained using measurements.

of the pulse are $f_0 = 400$ MHz and $f_{bw} = 200$ MHz, respectively, and the sphere is discretized using 9,414 spatial basis functions. Figure 19.12 compares the RCS pattern in the x–z plane obtained using the PWTD-enhanced MOT scheme and FISC for a range of frequencies within the band of excitation. Specifically, the RCS is compared at 280 and 420 MHz. Results obtained from the time and frequency domain codes are in good agreement with each other, even at 280 MHz, where the power in the incident pulse is down by 65 dB from its peak at f_0. In this and subsequent examples, comparisons at frequencies where the power is down by at least 30 dB are made to highlight the fact that meaningful results can be obtained at these points through the use of the CFIE. Indeed, if either the MFIE or the EFIE were used, errors induced by nonphysical currents at resonance frequencies would propagate in any MOT scheme and would be most visible at the ends of the band [28].

Next, transient scattering from an almond that fits into a rectangular box of dimensions 5.0m × 1.92m × 0.64m is analyzed. A pulse traveling along $\hat{\mathbf{k}} = -\hat{\mathbf{z}}$ and polarized along $\hat{\mathbf{p}} = \hat{\mathbf{x}}$ excites the almond. The center frequency and bandwidth of the pulse are $f_0 = 303.4$ MHz and $f_{bw} = 200$ MHz, respectively, and the almond is discretized using 10,620 spatial basis functions. RCS patterns in the x–z plane are extracted from temporal far field signatures at 210.2 and 322.0 MHz, and compared against FISC data in Figure 19.13; again, both RCS patterns agree well with each other. It should also be pointed out that at 210.2 MHz the power in the incident pulse is down by about 30 dB from its peak at f_0.

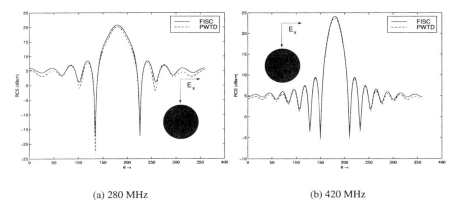

(a) 280 MHz   (b) 420 MHz

Figure 19.12 Radar cross-section of a sphere in the $z-x$ plane extracted from the time domain CFIE result is compared to that obtained from FISC for two different frequencies. The incident wave propagates along $\hat{\mathbf{k}} = -\hat{\mathbf{z}}$ and is $\hat{\mathbf{p}} = \hat{\mathbf{x}}$ polarized, and the sphere is discretized using 9,414 spatial basis functions.

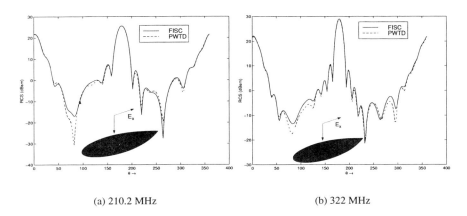

(a) 210.2 MHz   (b) 322 MHz

Figure 19.13 Radar cross-section of an almond in the $z-x$ plane extracted from the time domain CFIE result is compared to that obtained from FISC for two different frequencies. The incident wave propagates along $\hat{\mathbf{k}} = -\hat{\mathbf{z}}$ and is $\hat{\mathbf{p}} = \hat{\mathbf{x}}$ polarized, and the almond is discretized using 10,620 spatial basis functions.

Next, scattering from a cone-sphere is studied. The cone is 1m long and the radius of the half sphere attached to the cone is 0.235m. The incident pulse has a center frequency of $f_0 = 800$ MHz and bandwidth $f_{bw} = 750$ MHz, is $\hat{\mathbf{p}} = \hat{\mathbf{z}}$ polarized, and travels along the $\hat{\mathbf{k}} = -\hat{\mathbf{y}}$ direction. This cone-sphere is discretized using 11,412 spatial basis functions. The RCS patterns in the y–z plane obtained from the time domain PWTD-enhanced MOT code is compared against that obtained from FISC at 500 and 900 MHz, as shown in Figure 19.14. Examination of Figure 19.14 shows that both the FISC and the PWTD results agree very well with each other, even at 500 MHz where the power in the incident pulse is down by 25 dB from that at 800 MHz.

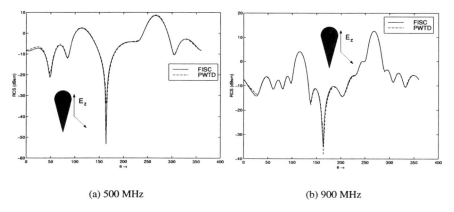

(a) 500 MHz   (b) 900 MHz

Figure 19.14 Radar cross-section of a cone-sphere in the z–y plane extracted from the time domain CFIE result is compared to that obtained from FISC for two different frequencies. The incident wave propagates along $\hat{\mathbf{k}} = -\hat{\mathbf{y}}$ and is $\hat{\mathbf{p}} = \hat{\mathbf{z}}$ polarized, and the cone-sphere is discretized using 11,412 spatial basis functions.

All examples analyzed thus far involved relatively simple geometries. Next, scattering from a VFY 218 aircraft, discretized using 9,747 spatial basis functions is analyzed. The incident field travels in the $\hat{\mathbf{k}} = -\hat{\mathbf{y}}$ direction and is polarized along $\hat{\mathbf{p}} = \hat{\mathbf{x}}$, has a center frequency of $f_0 = 100$ MHz, and bandwidth of $f_{bw} = 60$ MHz. Figure 19.15 compares the RCS patterns in the x–y plane at 75 and 105 MHz, where the pulse's power at these frequencies is down by 27, and 1.07 dB with respect to its peak value. Again, the results of the time domain code replicate all of the RCS nulls and peaks computed using the frequency domain solver.

Finally, scattering from an almond discretized with 29,700 spatial basis functions is analyzed. This simulation pushes the limits of the two-level PWTD algorithm insofar as computational resources are concerned. This almond fits inside a rectangular box of dimensions 6.00m × 2.31m × 0.77m, and is illuminated by a $\hat{\mathbf{p}} = \hat{\mathbf{x}}$ polarized Gaussian pulse traveling along $\hat{\mathbf{k}} = -\hat{\mathbf{z}}$, with a center frequency of $f_0 = 500$ MHz and a bandwidth of $f_{bw} = 300$ MHz. Figure 19.16 compares the RCS pattern in the x–z plane computed using the time and frequency domain codes at 400 and 520 MHz. At 400 MHz, power in the incident field is down by 27 dB from its peak. As

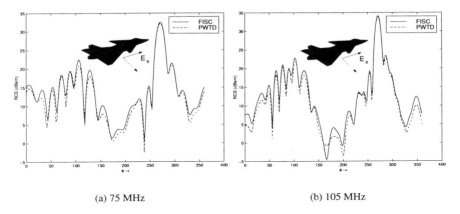

(a) 75 MHz (b) 105 MHz

Figure 19.15 Radar cross-section of an aircraft (VFY 218) in the x–y plane extracted from the time domain CFIE result is compared to that obtained from FISC for two different frequencies. The incident wave propagates along $\hat{k} = -\hat{y}$ and is $\hat{p} = \hat{x}$ polarized, and the VFY 218 is discretized using 9,747 spatial basis functions.

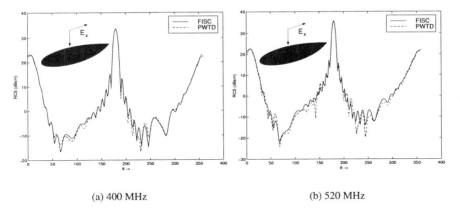

(a) 400 MHz (b) 520 MHz

Figure 19.16 Radar cross-section of an almond in the z–x plane extracted from the time domain CFIE result is compared to that obtained from FISC for two different frequencies. The incident wave propagates along $\hat{k} = -\hat{z}$ and is $\hat{p} = \hat{x}$ polarized, and the almond is discretized using 29,700 spatial basis functions.

is apparent in Figure 19.16, the RCS patterns computed by both the time domain and frequency domain codes are in very good agreement with each other.

Finally, the predicted computational complexity of both the PWTD-enhanced and classical MOT schemes is verified. All results presented above were obtained using

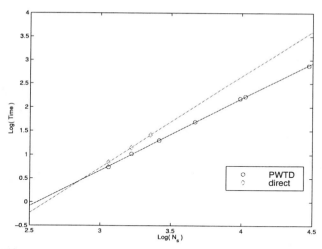

Figure 19.17 Comparison of the computational complexity of the classical and PWTD augmented MOT schemes.

an SGI Origin2000 with peak performance rated at 360 Mflops. In Figure 19.17, the logarithm of the CPU time required to compute the interactions at one time point is plotted against the $\log N_s$. This graph reveals that the computational cost of a PWTD-MOT algorithm scales as $\mathcal{O}(N_t N_s^{1.507} \log N_s)$, which is very close to the theoretically predicted scaling law. Also, it should be noted that the break-even point, or the number of unknowns where it becomes more advantageous to use the PWTD-enhanced MOT schemes as opposed to a classical MOT solver, is as low as $N_s = 700$.

19.5 SUMMARY

This chapter presented a two-level PWTD-enhanced MOT algorithm that permits the fast EFIE-, MFIE-, and CFIE-based analysis of transient electromagnetic scattering phenomena. The computational complexity of this algorithm scales as $\mathcal{O}(N_t N_s^{3/2} \log N_s)$, as opposed to $\mathcal{O}(N_t N_s^2)$ for the conventional MOT algorithm. Numerous simulations that were conducted during the course of this study demonstrate the usefulness of these solvers in characterizing broadband scattering from large objects.

REFERENCES

1. A. Taflove, *Computational Electrodynamics: The Finite Difference Time Domain Method*, Norwood, MA: Artech House, 1995.

2. K. S. Kunz and R. J. Luebbers, *The Finite Difference Time Domain Method for Electromagnetics*, Boca Raton, FL: CRC Press, 1993.

3. C. L. Bennet, "A Technique for Computing Approximate Impulse Response for Conducting Bodies," Ph.D. Dissertation, Purdue University, 1968.

4. A. J. Poggio and E. K. Miller, *Computer Techniques for Electromagnetics*, Oxford, U.K.: Pergamon Press, 1973, ch. 4..

5. R. Mittra, *Transient Electromagnetic Fields*, L. B. Felsen (ed.), New York: Springer Verlag, 1976, ch. 2.

6. B. P. Rynne and P. D. Smith, "Stability of time marching algorthms for the electric field integral equations," *J. Electromagn. Waves Applicat.*, vol. 12, pp. 1181–1205, 1990.

7. E. K. Miller, "A selective survey of computational electromagnetics," *IEEE Trans. Antennas Propagat.*, vol. 36, pp. 1281–1305, 1988.

8. A. G. Tijhuis, *Electromagnetic Inverse Profiling*, Utrecht, The Netherlands: VNU Science Press, 1987.

9. D. A. Vechinski and S. M. Rao, "A stable procedure to calculate the transient scattering by conducting surfaces of arbitrary shape," *IEEE Trans. Antennas Propagat.*, vol. 40, pp. 661–665, 1992.

10. B. P. Rynne, "Stability and convergence of time marching methods in scattering problems," *Int. J. Appl. Math.*, vol. 35, pp. 297–310, 1985.

11. W. Pinello, A. Ruehli, and A. Cangellaris, "Stabilization of time domain solutions of the EFIE based on the partial element equivalent circuit models," in *Proceedings of the IEEE Antennas and Propagation Society International Symposium*, vol. 3, pp. 966–969, 1997.

12. M. J. Bluck and S. P. Walker, "Time-domain BIE analysis of large three dimensional electromagnetic scattering problems," *IEEE Trans. Antennas Propagat.*, vol. 45, pp. 894–901, 1997.

13. S. Dodson, S. P. Walker, and M. J. Bluck, "Implicitness and stability of time domain integral equation scattering analysis," *Appl. Comp. Electromag. Soc. J.*, vol. 13, no. 3, pp. 291–301, 1998.

14. G. Manara, A. Monorchio, and R. Reggiannini, "A space-time discretization criterion for a stable time-marching solution of the electric field integral equation," *IEEE Trans. Antennas Propagat.*, vol. 45, pp. 527–532, 1997.

15. S. M. Rao and T. K. Sarkar, "An efficient method to evaluate the time-domain scattering from arbitrarily shaped conducting bodies," *Microwave Opt. Technol. Lett.*, vol. 17, pp. 321–325, 1998.

16. K. Aygün, A. A. Ergin, B. Shanker, S. E. Fisher, and E. Michielssen, "Transient analysis of thin wire antennas mounted on three-dimensional perfectly conducting bodies," in *14th Annual Review of Progress in Applied Computational Electromagnetics*, vol. 2, pp. 838–844, 1998.

17. R. Coifman, V. Rokhlin, and S. Wandzura, "The fast multipole method for the wave equation: A pedestrian prescription," *IEEE Antennas Propagat. Mag.*, vol. 35, pp. 7–12, 1993.

18. J. M. Song, C. C. Lu, and W. C. Chew, "MLFMA for electromagnetic scattering by large complex objects," *IEEE Trans. Antennas Propagat.*, vol. 45, pp. 1488–1493, 1997.

19. A. A. Ergin, B. Shanker, and E. Michielssen, "Fast evaluation of transient wave fields using diagonal translation operators," *J. Comp. Phys.*, vol. 146, pp. 157–180, 1998.

20. A. A. Ergin, B. Shanker, and E. Michielssen, "Fast transient analysis of acoustic wave scattering from rigid bodies using a two-level plane wave time domain algorithm," *J. Acoustical Soc. Am.*, vol. 106, pp. 2405–2416, 1999.

21. B. Shanker, A. A. Ergin, K. Aygün, and E. Michielssen, "Analysis of transient electromagnetic scattering phenomena using a two-level plane wave time domain algorithm," *IEEE Trans. Antennas Propagat.*, vol. 48, pp. 510–523, 2000.

22. K. Aygün, M. Lu, B. Shanker, and E. Michielssen, "Analysis of PCB level EMI phenomena using an adaptive low-frequency plane wave time domain algorithm," in *Proceedings of the IEEE International Symposium on Electromagnetic Compatability*, vol. 1, pp. 295–300, 2000.

23. D. S. Jones, *Methods in Electromagnetic Wave Propagation*, Oxford, U.K.: Oxford Science Publications, 1994.

24. J. R. Mautz and R. F. Harrington, "H-field, E-field, and combined field solutions for conducting bodies of revolution," *AEÜ*, vol. 32, pp. 157–164, 1978.

25. P. D. Smith, "Instabilities in time marching methods for scattering: Cause and rectification," *Electromagnetics*, vol. 10, pp. 439–451, 1990.

26. D. A. Vechinski and S. M. Rao, "Transient scattering from dielectric cylinders: E–field, h–field, and combined field solutions," *Radio Science*, vol. 27, pp. 611–622, 1992.

27. S. M. Rao and D. R. Wilton, "Transient scattering by conducting surfaces of arbitrary shape," *IEEE Trans. Antennas Propagat.*, vol. 39, pp. 56–61, 1991.

28. B. Shanker, A. A. Ergin, K. Aygün, and E. Michielssen, "Analysis of transient electromagnetic scattering from closed surfaces using the combined field integral equation," *IEEE Trans. Antennas Propagat.*, vol. 48, pp. 1064–1074, 2000.

29. S. M. Rao, D. R. Wilton, and A. W. Glisson, "Electromagnetic scattering by surfaces of arbitrary shape," *IEEE Trans. Antennas Propagat.*, vol. 30, pp. 408–418, 1982.

30. Y. Saad, *Iterative Methods for Sparse Linear Systems*, New York: PWS Publishing Company, 1996.

31. M. Abramowitz and I. A. Stegun, *Handbook of Mathematical Functions*, New York: Dover, 1972.

32. M. Tygel and P. Hubral, *Transient Waves in Layered Media*, Amsterdam: Elsevier, 1987.

33. E. Heyman, "Time-depenent plane-wave spectrum representations for radiation from volume source distributions," *J. Math. Phys.*, vol. 37, pp. 658–681, 1996.

34. J. J. Knab, "Interpolation of bandlimited functions using the approximate prolate series," *IEEE Trans. Information Theory*, vol. 25, pp. 717–720, 1979.

35. O. M. Bucci and G. Franceschitti, "On the spatial bandwidth of scattered fields," *IEEE Trans. Antennas Propagat.*, vol. 35, pp. 1445–1455, 1987.

36. O. M. Bucci, C. Gennarelli, and C. Savarse, "Optimal interpolation of radiated fields over a sphere," *IEEE Trans. Antennas Propagat.*, vol. 39, pp. 1633–1643, 1991.

37. A. A. Ergin, B. Shanker, and E. Michielssen, "The plane wave time domain algorithm for the fast analysis of transient wave phenomena," *IEEE Antennas Propag. Mag.*, vol. 41, no. 4, pp. 39–52, 1999.

38. W. H. Press, S. A. Teukolsky, W. T. Vetterling, and B. P. Flannery, *Numerical Recipies in Fortran 77*, Cambridge: Cambridge University Press, 1992.

39. A. C. Woo, H. T. G. Wang, M. J. Schuh, and M. L. Sanders, "Benchmark radar targets for the validation of computational electromagnetics programs," *IEEE Antennas Propagat. Mag.*, vol. 35, pp. 84–89, 1993.

40. P. J. Davies, "A stability analysis of a time marching scheme for the general surface electric field integral equation," *Appl. Numer. Math.*, vol. 27, pp. 35–57, 1998.

41. M. Li, K. Ma, D. M. Hockanson, J. L. Dwerniak, T. G. Hubing, and T. P. Doren, "Numerical and experimental corroboration of an FDTD thin-slot model for slots near the corners of shielding enclosures," *IEEE Trans. Electromagn. Compat.*, vol. 39, pp. 225–232, 1997.

About the Authors

Kemal Aygun received his B.S. degree from the Middle East Technical University, Ankara, Turkey, in 1995, and his M.S. degree from the University of Illinois at Urbana-Champaign, in 1997, both in electrical engineering. He is currently pursuing his Ph.D. degree at UIUC. He also received the 1999 and 2000 Computational Science and Engineering Fellowship. His e-mail address is aygun@decwa.ece.uiuc.edu.

Siyuan Chen is a manufacturing development engineer in the Lightwave Division of Agilent Technologies, Inc. He received his Ph.D. degree in 2000 from the University of Illinois at Urbana-Champaign. His current interests include computational electromagnetics and microwave measurement system design. His e-mail address is siyuan_chen@agilent.com.

Weng Cho Chew is Founder Professor, College of Engineering, and a professor of electrical and computer engineering, University of Illinois at Urbana-Champaign. He is also the director of the Center for Computational Electromagnetics and the Electromagnetics Laboratory. He earned his B.S., M.S., and Ph.D. degrees from the Massachusetts Institute of Technology in 1976, 1978, and 1980, respectively. He has worked as a department manager at Schlumberger-Doll Research and joined the faculty at the University of Illinois in 1985. He is interested in fast algorithms for scattering and inverse scattering computations and electromagnetics for various applications. In addition, he has written a book entitled *Waves and Fields in Inhomogeneous Media* and published over 230 journal papers and over 300 conference papers. He is an IEEE Fellow and was an NSF Presidential Young Investigator. He received the 2000 IEEE Graduate Teaching Award and the 2001 UIUC Campus Award for Excellence in Graduate and Professional Teaching. He coauthored a paper that won the Schelkulnoff Prize Paper award in 2001. His e-mail address is w-chew@uiuc.edu.

Tie Jun Cui is a research scientist at the Center for Computational Electromagnetics, Department of Electrical and Computer Engineering, University of Illinois at Urbana-Champaign. He received his B.S., M.S., and Ph.D. degrees in electrical engineering from Xidian University in 1987, 1990, and 1993, respectively. His research interests include wave propagation, scattering, inverse scattering, landmine detection, geophysical subsurface sensing, fast algorithms, and intergrated circuit simulations. His e-mail address is tiecui@uiuc.edu.

Kalyan Donepudi is a graduate student in the Department of Electrical and Computer Engineering at the University of Illinois at Urbana-Champaign. His research interests are fast methods, high-frequency scattering, and higher-order solutions. His e-mail address is donepudi@students.uiuc.edu.

A. Arif Ergin is an assistant professor at the Gebze Institute of Technology, Kocaeli, Turkey. He received his Ph.D. degree from the University of Illinois at Urbana-Champaign in 2000. His current research interests include computational methods for analyzing wave propagation and scattering in acoustics and electromagnetics, antenna design and measurements, and electromagnetic compatibility. His e-mail address is aergin@penta.gyte.edu.tr.

Andrew D. Greenwood received his B.S. (summa cum laude) and M.S. degrees in electrical engineering from Brigham Young University, Provo, Utah, in 1993 and 1995, respectively. He received his Ph.D. degree in electrical engineering from the University of Illinois at Urbana-Champaign in 1998. He received the Air Force Palace Knight Fellowship and is an IEEE member. He is currently with the Air Force Research Laboratory, Directed Energy Directorate, Kirtland AFB, New Mexico. His research interest is numerical techniques for electromagnetic problems. His e-mail address is Andrew.Greenwood@kirtland.af.mil.

Vikram Jandhyala is an assistant professor in the Department of Electrical Engineering, University of Washington, Seattle, Washington. He received his Ph.D. degree from the University of Illinois at Urbana-Champaign in 1998 and worked as a research and development engineer developing fast electromagnetic solvers at Ansoft, Pittsburgh, from 1998 to 2000. His research interests include computational electromagnetics, integral equations, fast algorithms, and field solvers for high-speed circuits. He is a recipient of the NSF Career Award (2001), Outstanding Graduate Research Award at UIUC (1998), and IEEE Microwave Graduate Fellowship (1997). His e-mail address is jandhyala@ee.washington.edu.

Dan Jiao is currently working toward a Ph.D. degree in electrical engineering at the University of Illinois at Urbana-Champaign. She received the 2000 Raj Mittra Outstanding Research Award presented by the Department of Electrical and Computer Engineering, University of Illinois at Urbana-Champaign. She has published over 20 papers in refereed journals. Her current research interests include fast computational methods in electromagnetics and time-domain numerical techniques. Her e-mail address is danjiao@uiuc.edu.

Jian-Ming Jin is a professor of electrical and computer engineering and an associate director of the Center for Computational Electromagnetics at the University of Illinois at Urbana-Champaign. He has authored *The Finite Element Method in Electromag-*

netics and *Electromagnetic Analysis and Design in Magnetic Resonance Imaging*, and coauthored *Computation of Special Functions*. He has served as an associate editor of *Radio Science* and *IEEE Transactions on Antennas and Propagation*. He received the 1994 National Science Foundation Young Investigator Award and the 1995 Office of Naval Research Young Investigator Award. He also received the 1997 and 2000 Xerox Research Awards from the University of Illinois and was appointed the first Henry Magnuski Outstanding Young Scholar in the Department of Electrical and Computer Engineering in 1998. He was a Distinguished Visiting Professor in the Air Force Research Laboratory in 1999 and was elected an IEEE Fellow in 2000. His e-mail address is j-jin1@uiuc.edu.

Gang Kang received his Ph.D. degree from Peking University, Beijing, China, in 1998. From June 1998 to December 1999, he was a research associate in the Center for Computational Electromagnetics, University of Illinois at Urbana-Champaign. He is currently a research associate in the Department of Electrical Engineering, University of Utah, Salt Lake City, Utah. His interests include numerical techniques for electromagnetics and their applications in bioelectromagnetics, scattering, and antenna design. His e-mail address is gkang@ee.utah.edu.

Feng Ling is a senior engineer/scientist at the Semiconductor Products Section, Motorola, Inc., Tempe, Arizona. He received his B.S. and M.S. degrees in electrical engineering from Nanjing University of Science and Technology in 1993 and 1996, respectively, and the Ph.D. degree in electrical engineering from the University of Illinois at Urbana-Champaign in 2000. His research interests include computational electromagnetics and modeling of integrated passive components and interconnects for RF applications. His e-mail address is fengling@ieee.org.

Jian Liu received his B.S. and M.S. degrees in electrical engineering from the University of Science and Technology of China in 1995 and 1998, respectively. Since 1998, he has been working in the Center of Computational Electromagnetics at the University of Illinois at Urbana-Champaign as a research assistant. His research interests include the application of finite element and moment methods to electromagnetics problems. His e-mail address is jianliu@uiuc.edu.

Cai-Cheng Lu is an assistant professor in the Department of Electrical and Computer Engineering at the University of Kentucky. He received his Ph.D. degree in 1995 from the Department of Electrical and Computer Engineering, University of Illinois at Urbana-Champaign. His research interests are wave scattering, microwave circuit simulation, and antenna analysis. He is especially experienced in fast algorithms for computational electromagnetics. He is a recipient of the NSF Career Award as well as the ONR Young Investigator Award. His e-mail address is cclu@engr.uky.edu.

Eric Michielssen is an associate professor in the Department of Electrical and Computer Engineering, University of Illinois at Urbana-Champaign, and an associate director of the Center for Computational Electromagnetics. He received his Ph.D. degree from the University of Illinois at Urbana-Champaign in 1992 and has been on the UIUC faculty ever since. His research interests include all aspects of theoretical and applied computational electromagnetics, and specifically fast algorithms for solving frequency/time domain integral equations and robust electromagnetic synthesis techniques. He received the NSF Career Award in 1995, Applied Computational Electromagnetics Valued Service Award in 1998, the International Union of Radio Science Henry G. Booker Fellowship in 1999, Issac Koga Gold Medal in 1999, and Xerox Research Award from UIUC in 2001. His e-mail address is emichiel@uiuc.edu.

Kaladhar Radhakrishnan is a senior design engineer for Intel Corporation. He received his Ph.D. degree in 1999 from the University of Illinois at Urbana-Champaign. As part of his doctoral work, he developed efficient numerical algorithms for the analysis of waveguiding structures. His current area of work involves designing microprocessor packages to meet power delivery and signal integrity requirements.

Balasubramaniam Shanker is an assistant professor in the Department of Electrical and Computer Engineering at Iowa State University. He received his B.Tech degree from the Indian Institute of Technology, Madras, India, in 1989, and M.S. and Ph.D. degrees from Pennsylvania State University in 1992 and 1993, respectively. His research interests include all aspects of computational electromagnetics with an emphasis on fast time and frequency domain algorithms. His e-mail address is shanker@iastate.edu.

Jiming Song is a staff engineer/scientist in the Digital DNA Research Laboratory of the Semiconductor Products Sector of Motorola in Tempe, Arizona. He received his B.S. and M.S. degrees in physics from Nanjing University in China, and his Ph.D. degree in electrical engineering from Michigan State University. From 1993 to 2000, he worked as a postdoctoral research associate, research scientist, and visiting assistant professor at the University of Illinois at Urbana-Champaign. From 1996 to 2000, he also worked as a research scientist at SAIC-DEMACO. His research interests include modeling and simulation of interconnect and passive components, wave scattering using fast algorithms, wave interaction with inhomogeneous media, electromagnetic simulations for foliage penetration applications, and transient electromagnetic field. His e-mail address is Jiming.Song@motorola.com.

About the Authors

Fernando L. Teixeira is an assistant professor in the Department of Electrical Engineering at The Ohio State University, where he teaches courses in electromagnetics and wireless propagation. He is also affiliated with the ElectroScience Laboratory at Ohio State. He received his Ph.D. degree from the University of Illinois at Urbana-Champaign and did postdoctoral work at the Massachusetts Institute of Technology. His current research interests include computational electromagnetics, electromagnetic wave interaction with complex materials, and scattering.

Sanjay Velamparambil is a research scientist at the Center for Computational Electromagnetics, Department of Electrical and Computer Engineering, University of Illinois, Urbana-Champaign. He obtained his M. Eng. and Ph.D. degrees from the Indian Institute of Science, Bangalore. His research interests are fast algorithms and parallel computing. His e-mail address is sanjay@sunchew.ece.uiuc.edu.

Karl Warnick is an assistant professor in the Department of Electrical and Computer Engineering at Brigham Young University, Provo, Utah. He received the B.S. degree in 1994 and his Ph.D. degree in 1997, both from Brigham Young. His research interests are electromagnetics and numerical simulation methods. His e-mail address is warnick@ee.byu.edu.

Junsheng Zhao received his B.S. degree from Shandong University, China, in 1985, his M. Eng. degree from the Second Academy of the Ministry of the Astronautics Industry of China (now China Aerospace Corporation), in 1988, and his Ph.D. degree from Tsinghua University, China, in 1995, all in electrical engineering. He is currently a research scientist at the Department of Electrical and Computer Engineering, University of Illinois at Urbana-Champaign. His research interests include fast algorithms for computational electromagnetics, microwave integrated circuits, and ferrite devices. He coauthored a paper that won the Schelkulnoff Prize Paper award in 2001. His e-mail address is zhao@sunchew.ece.uiuc.edu.

Index

$\overline{\alpha}$ matrices, 72, 73
$\overline{\beta}$ matrices, 72, 73

2D buried dielectric cylinders, 354–60
   CG-FFT algorithm, 359–60
   cyclic convolution and correlation, 357–58
   Galerkin's method, 356–57
   illustrated, 355
   integral equation, 354–56
2D electronic scattering, 40
2D FMM, 39–53
   additional theorem for Bessel functions, 42–44
   alternative derivation of diagonalized translator, 49–50
   bandwidth of radiation pattern, 51–52
   diagonalization of translation operator, 46–48
   error control, 52–53
   Green's function and, 44–46
   MOM problem, 40–42
   physical interpretation of aggregation, translation, disaggregation, 50–51
   summary, 49
   See also Fast multipole method (FMM)

2D LF-MLFMA, 153–73
   CFIE for PEC structures, 160–73
   CFIE for TE polarization, 166–73
   CFIE for TM polarization, 160–65
   computational complexity of, 158–60
   core equation, applying, 164–65, 168
   core equation, quasi-static case, 157
   CPU time for setup, 160, 161
   CPU time per iteration, 160, 161
   for expediting matrix-vector multiplication, 164
   memory requirements, 160, 162
   nonuniformly normalized, 158
   number of levels, 159
   total workload per iteration, 160
   uniformly normalized, 156–58
   See also Low frequency MLFMA (LF-MLFMA)
2D MLFMA, 152, 153–56
   CFIE for TE polarization, 166–73
   CFIE for TM polarization, 160–65
   CG method bistatic RCS comparison, 173
   diagonalized dynamic, 155–56
   dynamic, applying to CFIE for PEC structures, 160–73
   undiagonalized dynamic, 153–55
   See also MLFMA

2D spectral convergence theory, 212–56
   circular cylinder (TE), 222–25
   circular cylinder (TM), 213–22
   flat strip (edge error), 239–41
   flat strip (TE), 234–39
   flat strip (TM), 225–34
   high-order basis functions, 247–55
   rectangular cavity, 242–47
   summary, 255–56
   See also Spectral convergence theory
3D buried conducting plates, 360–64
   CG-FFT algorithm, 363–64
   Galerkin's method, 361–63
   illustrated, 360
   integral equation, 361
3D buried dielectric objects, 364–70
   CG-FFT algorithm, 369–70
   Galerkin's method, 366–69
   illustrated, 365
   integral equation, 364–66
   mesh, 367
3D FMM
   applying in matrix-vector multiply, 81
   error analysis, 85–92
   introduction to, 77–78
   undiagonalized dynamic, 152
   See also Fast multipole method (FMM)
3D LF-MLFMA, 152–53, 174–97
   applied to solve matrix equation, 179
   based on RWG basis, 189–97
   computational complexity, 180–81
   core equation for, 177–80
   floating-point overflows, 178–79
   memory requirements, 181
   nonuniformly normalized, 179
   normalized core equation, 178
   relative errors, 179, 180
   rotation of translation matrices for, 184–89
   uniformly normalized, 179
   See also Low frequency MLFMA (LF-MLFMA)
3D MLFMA, 83–85, 152
   diagonalized dynamic, core equation, 176–77
   dynamic, general formulations, 174–75
   error analysis, 85–92
   implementing, 83
   introduction to, 77–78
   number of levels in, 83
   static, core equation, 181–84
   static, rotation of translation matrices for, 184–89
   undiagonalized dynamic, core equation, 174–75
   See also MLFMA
3D spectral convergence theory, 256–61
   flat plate, 256–58
   rooftop basis functions, 258–61
   See also Error analysis; Spectral convergence theory

Absorbing boundary condition (ABC), 284, 310, 542
   adaptive (AABC), 577
   conformal, 310
   in FDTD, 487
   hybrid FEM/ABC technique, 577–82
Adaptive absorbing boundary condition (AABC), 577
   convergence geometries, 599
   hybrid FEM/AABC technique, 590–602
   iterations, 601
   RMS error vs. number of iterations, 600
Adaptive integral method (AIM), 730, 759–64
   to accelerate matrix-vector multiplication, 759
   accuracy, 762
   complexity, 762, 764
   CPU time, 762
   matrix elements calculated by, 763
   memory requirement, 761, 762
   MOM results vs., 762
Adjoint operator, 369
Aggregation
   MLFMA process, 56–57, 62, 63
   parallel MLFMA phase, 128–30
   physical interpretation of, 50–51
   sphere-to-sphere translation, 851
Air-to-PML interface, 552, 553
Ampere's Law, 4, 306, 309, 310

Angular resolutions, 305
Annular-ring power divider, 745
Antenna analysis problems, 499
Antenna radome modeling, 519–21
    effects calculation, 520
    numerical methods, 519–20
    See also Radomes
Antennas
    cavity-backed microstrip
        patch, 718–19
    corrugated horn, 560, 561, 562, 563
    inverted-L, 711
    loop, 708, 710, 712
    metallic, 702–8
    microstrip, 718–24
Anterpolation, 59–61
    3D MLFMA and, 84
    matrices, 61
    MLFMA and, 60–61
    operation, 838
Application-independent kernel, 120, 136
Approximate Green's function, 619
Approximate planar waveguide
        models, 745
Approximate prolate spheroidal (APS)
        function, 99–101, 829
    cylindrical, 846
    defined, 100
    interpolation formula, 100
    results, 101
Approximation methods, 7–8
Arnoldi method, 467
Asymmetric spare matrix equation, 462
Asymptotic error estimates, 208–10
    defined, 208
    key parameter, 208
    static approximation, 209
Asymptotic methods, 6–7
Asymptotic waveform evaluation. *See*
    AWE method
Automatic target recognition, 2
Auxiliary cells, 500
Average kernel time, 142
AWE method, 700–702, 746
    analysis of dielectric scatters, 713–18
    analysis of metallic antennas, 702–8
    analysis of metallic scatterers, 708–13
    analysis of microstrip
        antennas, 718–24
    applications, 700
    applied to MOM solution of EFIE, 700
    for broadband calculations, 699–725
    for calculation of frequency
        response, 749
    combined with FEM/BIE method, 718
    CPU time comparison, 749
    defined, 699–700
    dispersive substrates and, 721
    efficiency, 702
    implementation, 700
    summary, 725
    usefulness demonstration, 707
Axisymmetric problems, 541–70
    BOR with appendages, 566–69
    CAP formulation, 542–43
    conclusion, 569
    cylindrical PML, 552–54
    far-field calculations, 551–52
    formulation, 542–52
    introduction to, 541–42
    numerical results, 554–66
    problem definition, 543
    solutions for equations, 549–51
    variational formulation, 543–49

Backscatter RCS, 575, 604, 611
    of dielectric cube, 718
    of dielectric sphere, 717
    for microstrip antenna arrays, 767
    of rectangular dielectric box, 521
    for rectangular microstrip patch
        antenna, 743
    See also Radar cross-section (RCS)
Bandwidth
    of radiation function, 61
    of radiation pattern, 51–52
Basis functions, 212
    Fourier transforms of, 227
    "half" basis, 500
    higher-order, 247–55, 271–74, 637
    loop, 428–29, 430
    loop-star, 276
    loop-tree, 189, 190, 193, 195, 276,
        372, 428

Basis functions (continued)
  low-order, 252, 257, 638
  mixed-order, 675
  number removed, 502
  piecewise polynomial, 216, 229
  rooftop, 258–61
  RWG, 109, 189–97, 426–28
  source, 868
  star, 430
  third-order, 688
  tree, 430
  types of, 205
  vector, 596, 597, 641
  zeroth-order, 662, 685, 738
Bessel functions, 225
  additional theorem for, 42–44
  asymptotic behaviors of, 156
  Fourier coefficients involving, 47
  integral representation of, 50
  normalized, 156
  recursion relations for derivatives of, 222
  spherical, 87, 178, 183, 273, 833
Biconjugate gradient method (BCG), 263, 666, 755
Biconjugate gradient stabilized (BICGSTAB), 645
Bi-Lanczos algorithm, 465, 467
  complexity, 468
  defined, 467
  iteration vectors, 467
  number of iterations, 468
Bistatic RCS
  of 80 sphere, 144
  2D MLFMA and plain CG comparison, 173
  approximation to monostatic RCS, 94–96
  calculating, 94, 102
  of coated PEC sphere, 447, 448
  of coated sphere, 591
  computed with loop-tree basis, 436
  computed with RWG basis, 427
  copolarized, 800
  of different frequencies, 426
  E-plane, 599, 646, 650, 652
  as function of known density, 645

  as function of level exited for calculating RCS with interpolation, 102
  number of points, 102
  pattern, 646
  patterns of conducting cube, 649
  of PEC cube, 660
  of PEC sphere, 660
  RMS error in, 647, 689, 690
  of sphere coated with two dielectric layers, 592
  for sphere with 1m radius, 105
  of vertical/horizontal polarization incidences, 515
  of VFY218 at 8 GHz, 113
  of VFY218 at 500 MHz, 146
  See also Radar cross-section (RCS)
Blackman-Harris pulse, 292
Block diagonal preconditioner, 92–93
Block matrices, 498
Bodies of revolution (BOR), 541
  with appendages, 566–69
  dyadic Green's function in presence of, 566
  example, with small appendages, 568
  scattering problem, 566
Boundary conditions
  absorbing (ABC), 284, 310, 542, 577–82
  adaptive absorbing (AABC), 577, 590–602
  for conducting/dielectric bodies, 204
  for electric and magnetic fields, 475
  at final junction, 476
  finite difference formulation, 465–67
  FISC, 104–5
  impedance, 104–5
  at interior junctions, 476
  local, 298
  satisfied by transverse electric field components, 465
Boundary element method (BEM), 206
Boundary integral equation (BIE), 577
  dense matrices generated by, 589
  exact representation, 582
  for exterior fields, 584
  hybrid FEM/BIE technique, 582–90

Boundary integral method/modal approach (BIM/MODE), 677
Boundary-value problem, 584
Branch line coupler, 744

Cartesian PML, 307, 312
 analysis, 320–23
 anisotropic medium, 321
 constitutive tensor, 320
 extensions, 339
 inserting before box walls, 322
 obtaining, 315
 See also Perfectly matched layers (PMLs)
Cascaded microstrip radial stub, 762
 current distributions, 766
 S-parameters, 766
"Casualty trick," 822
Cauchy principal values, 432, 440
Cavity-backed microstrip patch antenna, 718–19
 illustrated, 719
 input impedance vs. frequency, 722
 input impedance vs. frequency (residing on dispersive substrate), 724
 input impedance vs. frequency (with probe feed), 723
 monostatic VV-polarized RCS vs. frequency, 720
 problem, 718
Cavity scattering, higher-order FEM application to, 664–72
CG-FFT algorithm, 355, 752–58
 2D buried dielectric cylinders, 359–60
 3D buried conducting plates, 363–64
 3D buried dielectric objects, 369–70
 convergence, 376
 defined, 752
 schemes, 752–54
 uniform discretization requirement, 754
 use of, 348, 376, 379
 validity of, 379
CGNE, 263
Chebyshev interpolation, 736
Circuit modeling, 499
Circular cavity, 669
 monostatic RCS of, 669, 671
 RCS calculation information, 672
Circular cylinder, 212
 geometry, 533
 plane wave scattering problem, 213
Circular cylinder (TE), 222–25
 condition number estimates, 266–67
 quadrature error, 223–25
 spectral error, 222–23
 See also 2D spectral convergence theory
Circular cylinder (TM), 213–22
 condition number estimates, 264–66
 current error, 218–20
 internal resonance, 221–22
 quadrature error, 217–18
 scattering amplitude error, 220–21
 spectral error, 215–17
 See also 2D spectral convergence theory
Clebsch-Gordon coefficients, 186
Clenshaw's recurrence algorithm, 848
Closed-form windowed translation operator, 845
Combined field integral equation (CFIE), 16, 77, 111, 502–3, 506–8, 587, 643, 708, 860
 accuracy, 875–76
 for closed conducting object, 709
 for closed surfaces, 111
 current observed at a point comparison, 879
 discretization, 79
 efficacy of, 876–81
 frequency domain, 864, 876
 for hybrid surface-volume integral equations, 506–7
 for large grid disparities, 506
 matrix elements, 82
 MOM applied to, 79
 multiregion, 781–82
 radar scattering cross-section comparison, 880
 RCS patterns obtained using, 878
 for TE polarization, 166–73
 time domain, 864, 878, 879, 883, 885, 886, 887

Combined field integral equation
(CFIE) (continued)
for TM polarization, 160–65
Complex frequency hopping (CFH), 746
Complex multipole beam approach
(CMBA), 151
Complex space, 285–86
Complex space coordinates, 284–87
frequency domain analysis, 284–87
time domain analysis, 287
See also Perfectly matched layers
(PMLs)
Complex-space nabla operator, 332
Complex stretching
approach, 287
variables, 285
Computational box, 125
Computational complexity
2D LF-MLFMA, 158–60
3D LF-MLFMA, 180–81
classical and PWTD augmented MOT
scheme comparison, 888
MLFMA, 61–63
MOT scheme, 818, 849
SDFMM, 788–90
Computational electromagnetics, 1
as analysis tool, 425
defined, 2
in electrical prospection setting, 425
error analysis approaches and, 211
high-order methods in, 637–94
hybridization in, 575–628
Computer-aided design (CAD) tools, 729
Concave PML, 310, 313, 325
cylindrical, 302, 324, 326
spherical, 329
See also Perfectly matched layers
(PMLs)
Condition number estimates, 263–77
circular cylinder (TE), 266–67
circular cylinder (TM), 264–66
expansion and testing functions
and, 266
flat plate, 274–75
flat strip (TE), 268–69
flat strip (TM), 267–68
higher-order basis functions, 271–74

low-frequency breakdown, 275–77
preconditioners, 275
rectangular cavity, 269–71
for TM polarization, 270
See also Iterative solution methods
Conductivity-object function, 400
Conformal PML, 310–17
anisotropic, in coordinate system, 315
application examples, 315
built over parallel surfaces, 337
defined, 310
equations inside, 312–13
Maxwellian, 313
obtaining, 312, 337
spherical coordinates, 310
See also Perfectly matched layers
(PMLs)
Conjugate gradient (CG), 388, 666
2D MLFMA bistatic RCS
comparison, 173
complex object function solved
by, 389
Hermitian matrices, 262–63
iterative solver, 221
MLFMA with, 103
See also CG-FFT
Conjugate gradient method (CGM), 119
Continuity condition
advantage, 502
enforcing, 501–2
Convex PML, 326, 327, 329
Corporate-fed microstrip antenna array
geometry, 756
monostatic RCS of, 758
radiation pattern, 757, 772
Corrugated horn antennas, 560
diagram, 561
photo, 562
radiation pattern, 563, 564, 565
Coupled azimuth potential (CAP)
formulation, 542, 552
CPU time
for 2D LF-MLFMA setup, 160, 161
AIM, 762
AWE method, 749
for filling translation matrices, 97
FISC, scaling, 110–12

Index

FISC requirement, as function of surface area, 110
LF-MLFMA, 182
metallic antennas, 707
metallic scatterers, 713
microstrip antennas, 721
SDFMM, 796
Curl-free mode, 276, 277
Current error
  circular cylinder (TE), 222–23
  circular cylinder (TM), 218–20
  flat strip (TE), 236–38
  flat strip (TM), 231–32
  higher-order basis functions, 251–54
Curvilinear discretization, 738
Curvilinear mesh, 504–6
Curvilinear PML, 299–317
  conformal (doubly curved) PML, 310–17
  cylindrical PML-FDTD, 299–305
  illustrated, 338
  Maxwellian PML, 307–10
  spherical PML-FDTD, 305–7
  See also Perfectly matched layers (PMLs)
Curvilinear quadrilateral, 504–5, 508
Curvilinear triangular patches, 639, 640, 683, 738
Cyclic convolution and correlation, 357–58
Cylindrical coordinate system, 43
Cylindrical harmonics, 69–70
Cylindrical mode expansion, 213, 218
Cylindrical PML, 320–23, 552–54
  analysis, 323–27
  concave, 302, 324, 326
  convex, 326, 327
  discrete, 303
  eight-layer, 303, 304
  grid termination, 303
  parameter definitions, 553–54
  reflection levels, 304
  systematic error reduction, 554
  See also Perfectly matched layers (PMLs)
Cylindrical PML-FDTD, 299–305
Cylindrical PML-PLRC-FDTD, 332–36

Maxwellian formulation, 334–36
split-field formulation, 332–34
time-stepping scheme, 333
unsplit formulation, 336

Darboux-Dupin, 310
Debye model, 288, 289, 719
DEC Personal Workstation
  CPU timings (metallic antennas), 707
  CPU timings (metallic scatterers), 713
  CPU timings (microstrip antennas), 721
Delta function, 205
DE methods, 859
Denominator functions, 329
Dense matrix, 185
Diagonalized factorization, 46
Dielectric objects
  3D, mesh of, 530
  3D buried, 364–70
  buried, 417
  free-standing inhomogeneous, 593–94
  high-contrast, 417–18
  shell, cross-sectional view, 529
Dielectric scatter analysis, 713–18
  formulation, 714–17
  numerical examples, 717–18
  See also AWE method
Dielectric substrate, 719
Dielectric volume region, 492
Differential equations (DEs), 859
Differential equation solvers, 9–12
  convergence rate, 10–12
  defined, 9–10
  fast direct, 10
  Krylov subspace method, 12
  multigrid method, 12
  second-order accurate scheme, 10
  sparse matrices, 10
  spectral Lanczos decomposition method (SLDM), 12
  spectral method, 12
Diffraction tomographic algorithms. *See* DT algorithm (2D objects); DT algorithm (3D objects)
Dipole array
  beam pointing error, 537

Dipole array (continued)
  normalized radiation comparison, 535
  radiation, in presence of three
    hemisphere radomes, 536
  See also Antenna radome modeling
Dirac delta function, 41, 321, 759, 818
Dirichlet condition, 547
Dirichlet kernel, 846
Disaggregation
  MLFMA, 58–59
  parallel MLFMA phase, 132
  physical interpretation of, 50–51
  sphere-to-sphere translation, 851
  See also Aggregation
Discrete complex image method
    (DCIM), 730
  applying, 732
  FMM with, 767
  for Green's function
    evaluation, 735–36
  with guided-mode extraction, 734, 735
  MLFMA with, 767–70
  surface-wave contribution, 730
Discrete Fourier transform
    (DFT), 357, 358, 360
Discretized operator
  flat strip (TE), 235
  flat strip (TM), 227–29
Discretized vector wave equation, 462
Dispersive medium models, 288–91
Distorted Born approximation, 386
Distorted Born iterative method
    (DBIM), 349
  multiple-frequency, 349, 392–96
  single-frequency, 349, 382–92
Distributed tree, 144
Divergence theorem, 580
Domain decomposition, 133–34, 143
  influence on memory requirement for
    geometric data structures, 146
  MLFMA induced, 143
  tree construction and, 133–34
Down-interpolation, 59
DT algorithm (2D objects), 396–406
  with 64 transmitter/64 receiver
    locations, 405, 406

permittivity and conductivity
  profiles, 400
reconstruction examples, 404–6
reconstruction results comparison, 406
setup configuration, 398
transmitter/receiver locations, 404
See also Fast inverse scattering
  methods
DT algorithm (3D objects), 407–18
  electric-dipole transmitter and
    receiver, 407–8
  inversion algorithms, 409–14
  inversion results, 414–18
  magnetic-dipole transmitter and
    receiver, 408–9
  setup configuration, 407
  See also Fast inverse scattering
    methods
Duality theorem, 470
Duffy's transform method, 508, 644
Dyadic Green's function, 88, 352, 353,
    373, 604
  3D, 389
  addition theorem application to, 656
  in Cartesian coordinates, 88
  electric-field, 352
  free-space, 585, 595, 607, 615,
    617, 703
  magnetic-field, 353, 375
  in presence of BOR, 566
  radial component, 91
  relative error, 89
  spectral, 351–52, 353
  transverse component, 91
  vector, 585
  See also Green's functions
Dynamic MLFMA, 120, 174–77
Dynamic stability criterion, 319

Edge vectors, 738
Eigenfunctions, 256, 257
  low spatial frequency, 215
  maximum, for TE polarization, 270
Eigenvalues, 465
  curl-free, 257
  difference, 245
  of discretized operator, 257–58

Index

divergence-free, 258, 274
of EFIE, 222, 243
estimates, 263
extremal, 274
of high order, 228
minimal, estimation, 273
of moment matrix, 215, 228, 264
for open cavity, 248
resonant, 244
scaling of, 277
smallest cavity mode, 245, 246–47
zero spatial frequency, 228
Electrical logging tools, 426
Electric currents, 367
Electric dipole
excitation vectors, 437–38
horizontal (HED), 731
Electric-dipole transmitter/
    receiver, 407–8, 411, 413
horizontal receiver arrangement, 416
polarization impact and, 416
reconstruction results
    (1 to 100 MHz), 416
reconstruction results
    (1.5 to 300 MHz), 415
reconstruction results (high-contrast
    object), 417
upper frequency to 10 MHz, 417
upper frequency to 30 MHz, 417
Electric field integral equation
    (EFIE), 77, 78–80, 204, 586–87,
    642, 688, 860–61, 876–79
for conducting objects, 78
current observed at a point
    comparison, 879
direct discretization, 224
discretized, Fourier
    representation, 227
EFIE-I, 714, 716
EFIE-O, 714, 716
eigenvalues, 222, 243
erroneous solution, 587
FMM formulation solution for, 81
homogeneous time domain, 863
incident field, 212
interior resonance problem, 708
matrix elements, 81

for perfectly conducting surfaces, 793
in PMCHW formulation, 714
RCS patterns obtained using, 878
scalar integral equation, 204
SDFMM-based solution, 793
solutions at resonant frequencies, 139
for TE polarization, 166, 223, 234
time domain, 862
for TM polarization, 162, 234
transient scattering from plate
    analyzed using, 882
See also Integral equations
Electric fields
boundary conditions for, 475
discretized vector wave equation, 462
for FDTD, 638
for FEM, 638
as fore vector, 470
incident, 616
intensity, 703
mixed-potential form, 738
for MOM, 638
near Luneburg lens, 560, 568
near-zone, 551
produced by magnetic current, 439
scattered, 364–65
tangential, 466
total, 490
transverse components for, 464, 477
See also Magnetic fields
Electric field volume integral
    equation, 490, 491
Electromagnetic analysis
defined, 1
introduction to, 1–3
Electromagnetic Code Consortium
    (EMCC), 599
benchmark geometry, 646, 690
benchmark targets, 554–55
Electromagnetic compatibility (EMC)
    problems, 152
Electromagnetic compatibility and
    electromagnetic interference
    (EMC/EMI), 2
Electromagnetic (EM) scattering, 347
EMCC dart object
geometry, 515

EMCC dart object (continued)
  radar cross-section, 516
  See also Electromagnetic Code Consortium (EMCC)
Empirical error analysis methods, 211
E-plane bistatic RCS, 646, 650, 652
Error analysis, 85–92, 203–78
  2D spectral convergence theory, 212–56
  3D spectral convergence theory, 256–61
  approaches to, 208–11
  asymptotic error estimates, 208–10
  computational electromagnetics and, 211
  empirical methods, 211
  iterative solution methods, 261–77
  local interpolation, 92
  numerical integral, 92
  scalar Green's function, 86–88
  spectral convergence theory, 211–12
  surface integral equations, 203–78
  for TE polarization, 236–37
  for truncation error, 85, 86–91
Error analysis (continued)
  vector Green's function, 88–91
Euclidean metric tensor, 336
Excitation vectors, 435, 498
  electric dipole, 437–38
  elements, 434
  frequency scaling properties, 434
  magnetic dipole, 436–37
  scale, 435
Expansion function, 229–30
Extended Galerkin's method, 498
Extinction theorem, 14, 438–39

Faraday's law, 305, 308
  in cylindrical system, 308
  on complex space, 308, 309
  generalized, 305–6
  in spherical coordinates, 309
Far field contributions,
    evaluation, 834–36, 841–42
  outgoing ray construction, 834–35, 841
  projection, 836, 842
  translation, 835, 841–42

Fast forward scattering methods, 354–70
  2D buried dielectric cylinders, 354–60
  3D buried conducting plates, 360–64
  3D buried dielectric objects, 364–70
  detection with, 370–80
Fast Fourier transform (FFT), 348
  methods, 637
  solvers, 22
  summations calculation, 358
  See also CG-FFT
Fast-frequency-sweep calculation, 746–52
Fast Hankel transform (FHT), 730
Fast Illinois Solver Code (FISC), 103–14
  boundary conditions, 104–5
  capabilities, 104–5
  complexity and accuracy, 106–9
  CPU time scaling, 110–12
  defined, 103
  integral equations, 104
  radar scattering cross-section comparison, 880
  results, 112–14
  scaling of memory requirements, 109–10
  summary, 112–14
  total CPU time, 111
Fast inhomogeneous plane wave algorithm, 151
Fast inverse scattering methods, 380–418
  defined, 382
  DT algorithm (2D objects), 396–406
  DT algorithm (3D objects), 407–18
  multifrequency DBIM algorithm, 392–96
  single-frequency DBIM algorithm, 382–92
Fast multipole method (FMM), 39–53, 67–74, 151–52, 860
  in 2D, 39–53
  in 3D, 77–83
  DCIM combined with, 767
  error control, 52–53
  error sources, 52
  groups, 42
  group theory and, 67–74
  introduction to, 39–53

for matrix-vector multiplication
  speedup, 764–67
matrix-vector product computation
  with, 589
representation of matrix elements in
  MOM, 124–25
for solving VIE, 513–14
steepest-descent (SDFMM), 781–810
success, 67
VSIE solution by, 513–14
See also MLFMA
FE-BI method, 664
FEM/AABC technique, 590–602
  comparison between
    computations, 601, 602
  computational information, 601
  defined, 593
  formulation, 593–98
  numerical results, 598–602
  See also Hybrid techniques
FEM/ABC technique, 577–82
  advantage, 590
  constraints, 590–92
  finite element analysis, 578–82
  ideal, 592–93
  numerical results, 582
  problem technique, 578
  solution approximation, 590
  See also Hybrid techniques
FEM/BIE technique, 582–90
  application of MLFMA, 589–90
  arbitrary shaped truncation surface
    and, 593
  combined with AWE method, 718–19
  defined, 718
  formulation, 584–88
  numerical results, 590
  See also Hybrid techniques
FEM/SBR technique, 602–14
  analysis, 609–10
  approximation and, 609
  defined, 602–4
  effects, 610
  formulation, 604–8
  monostatic RCS comparison, 613
  numerical results, 610–14
  RCS computed using, 611–14

reciprocity theorem, 608, 609
results comparison, 612
scattered field calculation, 608–9
See also Hybrid techniques
Field distribution
  in rooms without poles, 531
  in rooms with poles, 532
Field intensity plots, 478
Filtering operator, 127
Finite difference formulation, 461–67
  for approximating spatial
    derivatives, 470
  boundary conditions, 465–67
  to discretize operators, 470
  error introduced in, 463
Finite difference method (FDM), 541
Finite-difference time-domain (FDTD)
    method, 8, 730
  2D solution, 328
  absorbing boundary condition, 487
  cylindrical grid, 325
  cylindrical staggered-grid, 301
  defined, 8
  implementation, 287
  implementation issues, 283
  as PDE-based algorithm, 283
  popularity, 9, 283
  simulation, 292, 318
  solution dispersion error, 582
  update, incorporation into, 291–93
  uses, 283
Finite element analysis
  of complex axisymmetric
    problems, 541–70
  FEM/ABC technique, 578–82
Finite element and boundary integral
  (FE-BI) methods, 104
Finite element method
    (FEM), 8, 77, 104, 730
  advantages, 581
  arbitrary geometries modeling, 576
  for BOR scattering problems, 541–42
  for bounded problem with finite
    solution domain, 582
  cavity modeling by, 599
  computational advantage, 542
  discretization, 582

Finite element method (FEM) (continued)
    electric fields for, 638
    for electrodynamics, 425
    equations, solving, 549–51
    equations, symmetry properties, 548
    expansion, 546, 547
    formulation approach, 578
    handling inhomogeneous and anisotropic materials, 581
    higher-order, 658–72
    hybrid FEM/AABC technique, 590–602
    hybrid FEM/ABC technique, 577–82
    hybrid FEM/BIE technique, 582–90
    hybrid FEM/SBR technique, 602–14
    implementation issues, 283
    magnetic fields for, 638
    matrices, 581, 588, 593
    matrix, sparsity pattern, 550
    matrix equation, 665
    mesh, 554, 664
    mesh truncation, 569
    for open-region problem, 576
    variational problem solution, 580
Flat plate, 256–58
    condition number estimates, 274–75
    condition number of moment matrix, 275
    See also 3D spectral convergence theory
Flat strip (edge error), 239–41
    scattering amplitude error, 239–40
    See also 2D spectral convergence theory
Flat strip (TE), 234–39
    condition number estimates, 268–69
    current error, 236–38
    discretized operator, 235
    quadrature error, 238–39
    spectral error, 235–36
    See also 2D spectral convergence theory
Flat strip (TM), 225–34
    condition number estimates, 267–68
    current error, 231–32
    discretized operator, 227–29
    quadrature error, 230
    scattering amplitude error, 232–34
    spectral error, 229–30
    See also 2D spectral convergence theory
Floating-point underflows, 178–79
Fore vector, 470
Foster's reactance/susceptance theorem, 195
Fourier series
    convolutional property of, 47
    Parseval's theorem for, 47
Fourier transforms, 47, 64
    approximate generalized, 402
    of basis functions, 227
    discrete (DFT), 357, 358, 360
    fast (FFT), 22, 348, 358, 637
    of impulse train, 64–65
    inverse, 401
    Lagrange polynomial, 249
    of local interpolator, 66
    product, of testing and expansion functions, 216
    of sinc function, 65
    spatial, 399, 409–10
    of testing function, 217
    of translation functions, 833
Fréchet derivative operators, 392
Free-space Green's function, 204, 487, 585, 595, 607, 643, 664, 703
Free-space scattering problems, 767
Free-space wave impedance, 642, 664, 719
Free-space wavenumber, 642, 664, 719
Frequency-dependence analysis, 434
Frequency domain analysis, 284–87
Frequency normalization, 435
Frontal algorithm, 598
Full-wave analysis, 729–72
    AIM, 759–64
    CG-FFT method, 752–58
    fast frequency-sweep calculation, 746–52
    Green's functions for multilayer media, 730–37
    introduction, 729–30
    MLFMA, 764–70
    MOM solution, 737–46

of multilayer microstrip
problems, 229–772
summary, 770–72
Galerkin-based MOM, 652, 688
with RWG basis function, 652
second-order, 690
Galerkin-Petrov method, 206
Galerkin's method, 8, 16, 206, 507, 618, 643, 680, 740
2D buried dielectric cylinders, 356–57
3D buried conducting plates, 361–63
3D buried dielectric objects, 366–69
approximate integral solution, 139
extended, 498
higher-order, 652, 653
MLFMA with, 77–78
Gaunt coefficient, 175
Gauss elimination method, 499
Gaussian distribution, 791
Gaussian elimination, 452, 667
Gaussian-Legendre quadrature rule, 92, 827
Gaussian quadratures, 81, 218, 499, 688
higher-degree, 686, 688
points, 686–87
rule, 684
Gauss-Legendre quadrature rules, 644
Gauss' theorem, 13, 686, 687
Gedanken experiment, 17
Generalized minimal residual (GMRES), 263, 645
Generalized pencil-of-function (GPOF) method, 733
Geometrical optics (GO), 555
Ghost boxes, 135
Ghost signal, 822, 824, 843, 870
defined, 822
static, 870
Global interpolation, 63–65
Global loop basis function, 191, 193
Gradient vectors, 738
Gratings, 802
D-, 802, 803, 804
doubly periodic, 802
illustrated, 802
P-, 802, 803

periodic and quasi-random, 802–4
rough surface, 804
S-, 802, 803, 804
Green's functions, 9, 19–20, 23, 98, 349–54
in 2D, 19, 40
in 3D, 19–20, 782
approximate, 619
closed-form, 733
diagonal factorization, 25
discrete, 363, 369
dyadic, 88, 352, 353, 373, 585, 595, 607, 686
efficient evaluation of, 730
electric-field, 351, 352, 353
of electromagnetic problem, 349
expanding gradient of, 434
factorization of, 54
far field expansion, 207
free space, 204, 487, 585, 595, 607, 643, 664
guided-mode contributions, 733
half-space, 373, 374
as Helmholtz equation solution, 71–73
for incident field, 351–52
inefficient factorization of, 44–46
inhomogeneous medium, 385
in integral equation, 19, 350–51
magnetic-field dyadic, 353, 375
magnitude of, 735
for multilayer media, 730–37
oscillatory nature of, 20
plane-wave representation of, 73–74
PML, 322
for point source, 302
reduced 2D case, 353–54
scalar, 81, 85, 86–88, 497, 686, 784
for scalar potential, 732
for scattered field, 352–53
singularity of, 215
spectral dyadic, 351–52, 353
symmetry property, 72
truncation error in, 86–91
types of, 349
vector, 88–91
for vector potential, 732
Green's operator, 61

Ground-penetrating radar
 (GPR), 348, 349, 371
Groups
 defined, 68
 example, 68
 properties, 68
 representation of, 68–71
Group theory, 67–74
Guided-mode extraction, 734, 735

Hankel functions, 41, 225, 785
 additional theorem for, 42, 44
 argument approximation, 53
 as matrix products, 54
 recursion relations for derivatives
  of, 222
 role in matrix equation, 41
 spherical, 86–87, 98, 135
Harmonics functions, 185
Helmholtz decomposition, 426, 433
Helmholtz equation, 10
 convergence of solving, 11
 in free space, 67
 Green's function solution, 71–73
 solutions, 69, 70, 71–73
 solving, 11
 source-free, 71, 72
Helmholtz kernel, 20
Helmholtz operator, 69, 71, 213
Hermitian matrices, 262–63, 393
Hertzian dipole, 195
 input reactance of, 198
 series LC network and, 195
Higher-order basis functions, 247–55,
  271–74, 637
 condition number estimates, 271–74
 convergence of, 737
 current error, 251–54
 interpolation nodes for, 640–41
 interpolatory polynomials, 251
 orthonormal polynomials, 248–50
 scattering amplitude error, 254–55
 vector, 640
 See also 2D spectral convergence
  theory
Higher-order FEM, 658–72
 application to cavity scattering, 664–72

defined, 658–61
dispersion error and, 664
efficiency, 669
element types, 661
hierarchical basis function
 elements, 661
interpolatory basis function
 elements, 661
modeling accuracy and, 664
tetrahedral elements, 661–64
Higher-order MLFMA, 638–52
 edge length and, 652
 formulation, 639–45
 numerical examples, 645–52
 point-based, 652–58
Higher-order MOM, 638–52
 convergence behavior of, 741, 742
 formulation, 639–45
 Galerkin-based, 652
 numerical examples, 645–52
Higher-order tetrahedral
 elements, 661–64
High-order methods, 637–94
 aspects, 638
 higher-order FEM, 658–72
 mixed-order prism elements, 672–80
 MLFMA, 638–52
 MOM, 638–52
 numerical results, 688–93
 point-based grid-robust higher-order
  bases, 680–88
 point-based MLFMA, 652–58
 summary, 693–94
Hilbert space, 231
Hodge operators, 340, 341
Homogeneous material coating, 487
Horizontal electric dipole (HED), 731
Hybridization, 575–628
 defined, 575
 introduction to, 575–77
 need for, 575
 summary, 621–28
Hybrid PMLs, 341
Hybrid techniques
 classes, 576–77, 626
 defined, 575
 FEM/ABC, 577–82

FEM/BIE, 582–90
FEM/SBR, 602–14
MOM/SBR, 614–21
numerical methods, 576–77
numerical method with asymptotic
  method, 577
Hyperbolic grid generation technique, 316

Impedance boundary conditions
  (IBC), 104–5
Impedance matrix localization (IML)
  method, 22, 151
Implicit mode-matching scheme, 480
Impulse train, 64
Incident fields
  direct, 619
  indirect, 610, 619
  secondary, 619
Incoming rays, 824
Indoor radio wave propagation
    simulation, 518–19
  defined, 518–19
  results, 531–32
Inhomogeneous wave equation, 461
Input reactance, 198
Integral equations, 859
  2D buried dielectric cylinders, 354–56
  3D buried conducting plates, 361
  3D buried dielectric objects, 364–66
  combined field (CFIE), 16, 77, 79,
    160–73, 781–82
  derivation, 13
  discretization, 492, 495–99
  electric field (EFIE), 77, 78–79, 223,
    234, 354, 587, 882
  FISC, 104
  formulation, 138–40
  Green's function in, 350–51
  internal resonance problem, 15–18
  magnetic field (MFIE), 79, 502,
    586–87, 860–61, 876–79
  MOM and, 78–80
  pertinent equation, 23
  PWTD-enhanced TDIE
    solvers, 862–65
  scalar, 204
  surface, 13–15, 203–78

testing, 206
time-domain (TDIE), 859–60
volume, 19
Integral equation solvers, 13–22
  fast, 21–22
  FFT-based, 22
  Green's function, 19–20
  IML method, 22
  MOM, 20–21
  PWTD method, 22
  See also MLFMA
Integrated circuits (ICs), 729
Interaction lists
  defined, 126–27
  forward, 131
  inverse, 130, 131
Interior current error, 231
Interior resonance problem, 587, 708
Internal resonance
  circular cylinder (TM), 221–22
  defined, 221
  numerical, 221
Internal resonance problem, 15–18
  defined, 15
  Gedanken experiment, 17
  penetrable scatterer case, 17
  PMCHWT formulation, 18
Interpolation, 59–61
  error, 63–67
  global (exact), 63–65, 85
  Lagrange polynomial, 99
  local (exponentially
    accurate), 65–67, 85, 847
  matrices, 60, 61
  MLFMA and, 60–61
  nodes for higher-order basis
    functions, 640–41
  operation, 837, 838, 839
  optimal, 99–101
  of translation matrix, 99–101
Interpolation functions
  illustrated, 849
  temporal, 847
Interpolation operator, 127
Interpolation points
  additional requirement for, 685
  Gaussian quadrature rule and, 684

Interpolation points (continued)
  quadrature points and, 683
  total number of, 685
Interpolatory polynomials, 251
  Lagrange, 251
  RMS error at nodes for, 253
Introduction to, 699, 700
Inverse Fourier transform, 401
Inversion algorithms, 409–14
  partial-information, 414
  results, 414–18
Iteration count estimates, 262–63
Iteration vectors, 467, 468
Iterative solution methods, 261–77
  condition number estimates, 263–77
  iteration count estimates, 262–63

Kramers-Kronig relations, 319, 320, 321
Kronecker delta function, 682
Krylov subspace
  iteration, 262
  method, 12, 467, 484

Lagrange bases, 640, 641, 663
Lagrange expansion, relative error
  estimate, 254
Lagrange interpolation points, 682
Lagrange interpolator, 682, 684
Lagrange polynomial interpolation, 99
  defined, 99
  higher-order basis functions, 640
  results, 101
Lagrange polynomials
  Fourier transform, 249
  as interpolatory basis, 251
  interpolatory expansion, 252
Laplace kernel, 20
Laplace transform, 400
Laplacian operators, 10, 11
Large-scale computing, 92–103
  approximation of bistatic RCS to
monostatic RCS, 94–96
  block diagonal preconditioner, 92–93
  initial guess, 93–94
  interpolation of translation matrix in
    MLFMA, 96–102
  MLFMA for calculating radiation
    fields, 102–3

Legendre expansion, relative error
  estimate, 253
Legendre polynomials, 80, 98, 826,
    849, 872
  applied to flat strip, 273
  orthogonality of, 89
  orthonormal set, 249
  recurrence relations, 89
Linear discretization density, 257
Load balancing, 135–36
Local interpolation, 65–67
  defined, 65
  error in, 92
  Fourier transform, 66
  illustrated, 66
  implementation, 85
  radiation patterns, 85
  See also Interpolation
Locally complete decomposition, 133
Local reflector error, 305
Loop-antenna model, 372–74
  assumptions, 372
  comparison, 375
  defined, 372
  magnetic-dipole model
    comparison, 375
  time-domain reflected, scattered, total
    magnetic fields, 378
Loop antennas, 708, 710
  on finite ground plane, 710, 712
  S-parameters, 712
Loop basis function, 428–29
  divergence-free, 429
  illustrated, 430
  as linear combination of RWG basis
    functions, 430
Loop-star basis, 276
  loop space, 276
  MOM matrix dimension and, 432
Loop-tree basis, 189, 372
  application, 190
  bistatic RCS computed with, 436
  divergence-free loop space, 276
  for expanding unknown electric/
    magnetic currents, 432–33
  formation, 449
  impedance matrix of, 190, 453

LF-MLFMA based on, 193
  for low-frequency cases, 195
  MOM matrix dimension and, 432
  NOUR scheme for, 440, 449, 451
  number-of-unknowns reduction
    scheme, 449–58
  PMCHWT formulation using, 428
  purpose of, 449
  as superpositions of RWG-type basis
    functions, 193
Lorentzian dispersive medium, 289
Lower-upper-triangular decomposition
  (LUD), 21
Low-frequency breakdown, 275–77
  demonstration, 426
  of MOM, 276
  problem, 426
Low-frequency MLFMA
  (LF-MLFMA), 151–99
  2D, 153–73
  3D, 152–53, 174–97
  based on integration points, 193
  based on loop-tree basis, 193
  CPU time, 182
  defined, 152
  matrix-vector multiplication expedited
    by, 193
  memory requirements, 152, 182
  setup time, 181
  six-level, 180, 190
  workload per iteration, 152
  See also MLFMA
Low-frequency scattering, 425–58
  general equations and frequency
    normalization, 432–36
  interpretations, 436–40
  introduction to, 425–28
  from multibody, 440–58
  number-of-unknowns reduction
    scheme (loop-tree basis), 449–58
  number-of-unknowns reduction
    scheme (RWG basis), 443–49
  PMCHWT formulation for multibody
    problem, 440–43
  from single penetrable, 428–40
LU decomposition, 680
Lumped circuit model, 706

Luneburg lens, 559
  diameter, 566
  as directional radiator, 566
  electric field near, 560, 568
  normalized bistatic scattering
    from, 559
  radiation, excited by Hertzian
    dipole, 567

Magnetic dipole, excitation
  vectors, 437–38
Magnetic-dipole model, 374–75
  defined, 374
  time-domain reflected, scattered, total
    magnetic fields, 378
Magnetic-dipole transmitter/
    receiver, 408–9, 413
  upper frequency to 10 MHz, 417
  upper frequency to 30 MHz, 417
Magnetic field integral equation
  (MFIE), 79, 502, 586–87, 642,
  860–61, 876–79
  accuracy, 688
  for closed conducting objects, 709
  current observed at a point
    comparison, 879
  erroneous solution, 587
  homogeneous time domain, 863
  interior resonance problem, 708
  MFIE-I, 716
  MFIE-O, 716
  in PMCHW formulation, 714
  radar scattering cross-section
    comparison, 880
  solutions at resonant frequencies, 139
  for TE polarization, 166
  time domain, 863, 878, 879
  for TM polarization, 162
  transient scattering from almond
    using, 883
  vector Green's function for, 89
Magnetic fields
  boundary conditions for, 475
  for FDTD, 638
  for FEM, 638
  as fore vectors, 470
  incident, 687

Magnetic fields (continued)
  for MOM, 638
  transverse components
    of, 464, 470, 471, 472
  transverse vector wave equation
    for, 464
  See also Electric fields
Magnetic field volume integral
    equation, 491
Magnetic resonance imaging (MRI)
    system, 706
Marching-on-in-time (MOT)
    scheme, 817–19, 859, 864, 865
  classical, 865
  computational complexity, 818, 849
  cost, 819, 860
  defined, 818
  explicit, 864
  flat-panel triangular-patch
    implicit, 864
  late time instabilities, 860, 866
  multilevel PWTD-enhanced
    algorithm, 836–40
  stabilizing, 866
  two-level PWTD-enhanced
    algorithm, 831–36
  windowed PWTD-enhanced, 849–55
Matrix equation solver, 191
Matrix-vector multiply, 23
  accelerating with AIM, 759
  applying FMM in, 81
  as bottleneck in iterative solver, 41
  calculated by MLFMA, 655
  effected, 63
  evaluation phases, 127
  far interactions of, 656
  MLFMA, memory requirements/CPU
    time, 659
  one-level, 24
  sweeps, 83–84
  two-level, 24
Maxwellian PML, 293–94, 328
  benefits, 294
  bianisotropic constitutive
    parameters, 297–98
  boundary conditions satisfaction, 294
  causality violation, 319

complex-space PML vs., 341
conformal, 313
in cylindrical and spherical
    coordinates, 307–10
defined, 293
for different coordinate systems, 339
formulation, 295–98
See also Perfectly matched layers
    (PMLs)
Maxwell's equations, 4
  in anisotropic medium of constitutive
    parameters, 310
  in bianisotropic and dispersive
    media, 296
  complete solution to, 1
  computer simulations with, 2
  for conducting/dielectric bodies, 204
  in free space, 67
  in inhomogeneous media, 298
  metric and topological structure
    of, 339–41
  natural Helmholtz decomposition
    of, 426
  numerical methods for solving, 8
  solutions, 5
  solving, 1, 2, 26
  source-free, 284–85
Maxwell's theory, 1
  history, 4
  physicists regard for, 4
Memory requirements
  2D LF-MLFMA, 160, 162
  3D LF-MLFMA, 181
  adaptive integral method
    (AIM), 761, 762
  domain decomposition influence
    on, 146
  FISC estimated, 109–10
  FISC scaling, 109
  MLFMA, 85, 657, 659
  on Linux Cluster Orion, 646
  point-based higher-order
    MLFMA, 657
  SDFMM, 788, 796, 799
  TRIMOM+ScaleME, 141
Meshes
  of 3D dielectric object, 530

Index   917

convergence tests, 515
curvilinear, 504–6
density, 650
of dielectric sphere, 523
of dielectric spherical shell, 524
extending, 499–500
FEM, 554, 569, 664
generating, 492–94
hexahedron, 493, 494, 520
RCS comparison of, 524
RCS solved using different, 524
termination, 499–501
tetrahedron, 493, 528
triangle-tetrahedron, 494
type choices, 493
Message Passing Interface (MPI). *See* MPI
Metallic antenna analysis, 702–8
 formulation, 703–6
 numerical examples, 706–8
 See also AWE method
Metallic scatter analysis, 708–13
 formulation, 708–13
 numerical examples, 713
 See also AWE method
Method of moments (MOM), 8, 20–21,
  347, 425, 488, 730
 2D, 40–42, 611
 analysis, 617–18, 737–46
 applied to CFIE, 79
 defined, 20–21, 638, 700
 electric fields for, 638
 FISC and, 103
 FMM representation of matrix
  elements in, 124–25
 formulation, 686–88, 807–8
 Galerkin-based, 652, 688, 690
 higher-order, 638–52
 hybrid MOM/SBR technique, 614–21
 integral equations and, 78–80
 linear equation generation, 151
 low-frequency breakdown, 276
 magnetic fields for, 638
 matrix, 428, 432, 655
 matrix elements calculated by, 770
 matrix equation, 93
 matrix-vector products, 782
 for near-neighbor calculations, 63

 protrusion characterization with, 620
 scaled 2D, 611
 scatterer size and, 615
 scattering solution using, 22
 solving by, 617
 surface integral equations and, 204–6
Method of stationary phase, 6–7
Method of steepest descent, 6–7
Metric tensor
 complex, 337
 Euclidean, 336
 original, 337
Micro electro-mechanical sensors
  (MEMS), 2
Microstrip antenna analysis, 718–24
 formulation, 718–20
 numerical examples, 720–21
 See also AWE method
Microstrip antenna arrays on finite
  substrates analysis, 804–9
 integral equation formulation and
  SDFMM solution, 805–6
 introduction to, 804–5
 MOM formulation, 807–8
 simulation results, 808–9
 See also SDFMM
Microstrip double-stub, 750
Microstrip taper, 483, 484
Microwave thermal effect simulation, 519
 defined, 519
 example, 533
 temperature distribution, 534
Mie scattering solution, 5, 645
Mie series, 144
 calculating, 86
 near-field comparison of VIE and, 529
 RMS error, 106, 107, 108
 solutions, 435, 518
Mixed-order basis functions, 675
Mixed-order prism elements, 672–80
 advantages, 675
 efficiency, 675–77
 higher-order tetrahedral elements
  vs., 672, 675
 second/zeroth-order triangular prism
  element, 672–75
 See also High-order methods

Mixed potential formulation, 498
MLFMA, 53–67
   translation disaggregation, 58–59
   2D, 152, 153–56
   3D, 83–85, 152
   aggregation procedure illustration, 57
   aggregation process, 56–57, 62, 63
   anterpolation, 59–60
   applying to basis-based
      sources, 653, 657, 658
   applying to point-based
      sources, 653–58
   benefits, 151
   with biconjugate gradient (BCG), 103
   combined with Galerkin's method, 78
   complexity, 771
   computational complexity, 61–63
   with conjugate gradient (CG), 103
   with DCIM, 767–70
   defined, 21, 53–54
   development of, 152
   dynamic, 120, 152
   error analysis in, 85–92
   extension, 22
   for far-neighbor calculations, 63
   FEM/BIE technique and, 589–90
   for free-space scattering problems, 767
   higher-order, 638–52
   implemented to calculate bistatic
      RCS, 92
   implemented with FFT and
      1D FMM, 77
   implemented with signature function,
      interpolation and filtering, 77
   intergroup calculations, 97
   interpolation, 59–61
   interpolation of translation matrix
      in, 96–101
   introduction to, 39
   kernel, 645
   matrix elements calculated by, 770
   matrix-vector multiply calculated
      by, 655
   memory complexity of, 61
   memory requirements, 85, 657, 659
   motivation for, 53–54
   nine-level, 142
   parallelization, 119–48
   performance for hybrid VSIE, 514
   point-point interactions calculated
      by, 655, 657
   for radiation field calculation, 102–3
   radiation patterns, 57
   simplified view, 22–26
   static, 152, 181–89
   summary, 55–56
   translation-disaggregation procedure
      illustration, 58
   translation matrix, 92
   tree structure, 25
   at very low frequencies, 151–99
   See also Fast multipole method
      (FMM)
Model-based parameter estimation
   (MBPE), 699, 746
Mode matching method of lines (M30L)
   technique, 478
Moment matrix, 214
   condition number of, 265, 268, 269,
      271, 272, 274, 275
   eigenvalues of, 215, 228, 257, 264
   eigenvectors of, 257
   for rooftop basis, 259
MOM/SBR technique, 614–21
   accuracy and capability, 620
   approximation in, 619
   effects, 619
   effects not included in, 620
   formulation, 615–18
   hybrid solution, 622, 623
   iterative improvement, 619–20
   monostatic echo-width
      comparison, 624, 625
   numerical results, 620–21
   scattered field calculation, 618–19
   See also Hybrid techniques
Monostatic echo-width
   comparison, 624, 625
Monostatic RCS
   of almond with crack, 614
   approximation of bistatic RCS
      to, 94–96
   calculating, 93, 95
   of circular cavity, 669, 671

of coated double cone, 528
of coated trapezoidal plate, 603
of conducting cylinder, 583
of corporate-fed microstrip antenna array, 758
hybrid FEM/SBR and MOM comparison, 613
of inhomogeneous cylinder, 447, 448
of inhomogeneous PEC-dielectric cylinder, 449
iterative solutions of, 93, 94
NASA almond, 651, 692
of offset bend cavity, 681
of open cavity, 648
of rectangular cavity, 668, 670, 676, 680
RMS error in, 669, 670, 677, 678
for two polarizations, 599
of two small plates, 622
of VFY218 at 100 MHz, 96, 107, 108
of VFY218 with mushroom-shaped protrusion, 627
VV-polarized, 720
See also Radar cross-section (RCS)
Monte Carlo simulation, 796
copolarized bistatic scattering coefficient, 797
cross-polarized bistatic scattering coefficient, 798
Morton ordering, 132
MOT solvers, 860
MPI model, 120–21, 136
defined, 121
schematic, 121
Multibody problem, 441
Multifrequency DBIM
algorithm, 349, 392–96
for 2D objects, 392–96
defined, 392
first implementation illustration, 396
implementation illustrations, 394
implementations, 394–96
iterations, 395
second implementation illustration, 397
setup configuration, 393

single-frequency DBIM comparison, 395
See also Distorted Born iterative method (DBIM); Fast inverse scattering methods
Multifrontal algorithm, 598
Multigrid method, 12
Multilayer media
Green's function for, 730–37
interpolation scheme, 737
with source and field points, 731
Multilevel fast multipole algorithm. *See* MLFMA
Multilevel matrix decomposition algorithm (MLMDA), 151
Multilevel PWTD-enhanced MOT algorithm, 836–42
anterpolation operation, 838
defined, 836
evaluation of far field contributions, 841–42
evaluation of near field contributions, 841
far field pair, 837
far field pairs, 837
interpolation and splicing operations, 837, 839
last stage of, 837
outgoing ray construction, 841
projection, 842
resection operation, 838
subsignal duration, 837
translation, 841–42
Multilevel windowed PWTD-enhanced MOT algorithm, 854–55
Multipole amplitude, 157
Multipole cylindrical harmonics, 157

NASA almond, 646, 651, 692, 713
defined, 646
RCS frequency response, 715
Navier-Stokes equation, 5
N-degree polynomials, 684–85
Near field contributions, evaluation of, 834, 841–42
Near-neighbor preconditioners, 275
NEPAL, 21

N-junction problem, 474–77
  defined, 474–75
  structure illustration, 475
  waveguide cross-section, 475
  See also Waveguide discontinuities
Normalized radiation
  dipole array, comparison, 535
  dipole array, in presence of three
    hemisphere radomes, 536
NOUR matrix, 428, 445
NOUR schemes, 440, 447
  construction for loop-tree basis, 451
  construction rules, 454
  loop merges into loop and, 454
  loop merges into tree and, 454
  for loop-tree basis, 440, 449
  for magnetic current, 454
  need for, 454
  PEC body and, 454
  for RWG basis, 440, 443, 447
  tree merges into loop and, 454
  tree merges into tree and, 454
  valid, 449, 454
  verification, 455
Number-of-unknowns reduction scheme
  (loop-tree basis), 449–58
Number-of-unknowns reduction scheme
  (RWG basis), 443–49
  unknowns become redundant, 444
  unknowns become redundant and form
    a junction, 446
Numerical experiments, 138–44
  integral equation formulation, 138–40
  large-scale problems, 142–44
  notations, 138
  processor scaling, 142
  single processor performance, 141
  TRIMOM+ScaleME code, 140
  See also Parallelization
Numerical methods, 203
  behavior, 204
  convergence behavior, 203
  error estimates and, 211
Numerical mode matching (NMM)
  method, 12, 455
Numerical reflections, 300
Nyquist criterion, 210

Nyquist theorem, 60, 63
Nyström approach, 680

Object functions, 413, 414
Optimal interpolation, 99–101
Organization, this book, xx–xxii
Orion, 646
Orthonormal polynomials, 248–50
Overlap-gap-coupled microstrip filter
  geometry, 752
  S-parameters, 753

Padé approximate, 746, 748
Padé via Lanczos (PVL), 746
  defined, 746
  submatrices, 746
Parallel-coupled bandpass filter, 762
  current distributions, 765
  S-parameters, 765
Parallelization, 119–48
  implementation issues, 132–36
  mathematical preliminaries, 122–25
  MPI model, 120–21
  numerical experiments, 138–44
  ScaleME, 136–38
  ScaleME-2, 144–47
Parallel MLFMA, 125–32
  aggregation phase, 128–30
  algorithm, 127–32
  computational box, 125
  disaggregation phase, 132
  distance between boxes, 126
  interaction list, 126–27
  near neighbor list, 126
  translation phase, 130–32
  See also MLFMA
Parseval's theorem, 47, 53
Partial differential equations
  (PDEs), 8, 283
Partial-information algorithms, 414, 417
Penetrable 3-D scatter, 433
Perfect electrical conductor
  (PEC), 104, 204, 544, 861
  boundary, 302
  CFIE for, 160–73
  coated sphere, 447, 448
  cube, bistatic RCS, 660
  cylinder, 302

fast computation of RCS of, 700
scattering determination from, 426
sphere, bistatic RCS, 660
Perfectly matched layers
  (PMLs), 283–341, 604
  advances in theory of, 283–341
  Cartesian, 307, 312, 315, 320–23, 339
  as change on metric of space, 336–39
  complex-space, 296, 322, 323
  concave, 302, 310, 324, 325, 326, 329
  conformal, 310–17
  convex, 324, 326, 327, 329
  curvilinear, 299–317
  cylindrical, 299–305, 323–27, 552–54
  defined, 284
  for dispersive media, 288–93
  eight-layer cylindrical, 303, 304
  extension to anisotropic
    media, 294–98
  for FEM mesh, 569
  frequency-domain formulations, 341
  Green's function, 322
  Hodge operators in, 341
  hybrid, 341
  for inhomogeneous area, 298–99
  interfaces, 309, 310, 312
  layer thickness, 287
  lossy, 552
  Maxwellian, 293–298, 328, 339–41
  non-Maxwellian formulation, 295
  performance, 284
  quasi-, 330–31, 339
  reflection errors, 554
  reflectionless absorption, 284, 285
  source-free Maxwell's equations
    in, 284–85
  spectral properties, 317
  spherical, 315, 317, 327–30
  spurious reflection from, 286
  stability issues, 317–31
  thickness, 303
  via complex space
    coordinates, 284–87
Perfect magnetic conductor (PMC), 544
Periodic and quasi-random
    gratings, 802–4
Periodic sinc function, 217, 228

Permittivity-object function, 400
Petrov-Galerkin method. *See* Method of
    moments (MOM)
Piecewise constant, 205, 240
Piecewise linear, 205
Piecewise linear recursive convolution
    (PLRC) algorithm, 333
Piecewise polynomial basis, 216, 229
  product of Fourier transforms for, 216
  relative spectral error for, 229, 235
Plane waves, 70–71
  decomposition, 820–25
  representation of Green's
    function, 73–74
  representations, 868–73
  windowed, decomposition, 843–44
Plane-wave time-domain (PWTD)
    algorithms, 815–56, 860
  accuracy, 830
  defined, 22
  definitions of vectors used in, 823
  effect of bandlimiting translation
    function, 828
  enhanced TDIE solvers, 867–75
  foundations, 867
  implementation issues, 826–30
  incorporation into MOT scheme, 816
  introduction to, 815–17
  MOT method and, 817–19
  MOT scheme implementation, 830–42
  multilevel, enhanced MOT, 836–42
  nonwindowed, 816
  plane wave decomposition, 820–25
  signals associated with, 824, 825
  spatial integration, 826
  spectral integration, 826–28
  subsignal temporal
    representation, 828–30
  summary, 855–56
  three-stage, 824–30
  two-level, enhanced MOT, 831–36
  types of, 816
  windowed, 816, 842–55
  within two-level (multilevel)
    framework, 816
  See also PWTD-enhanced TDIE
    solvers

PMCHW formulation
    defined, 714
    EFIE in, 714
    MFIE in, 714
PMCHWT formulation, 18, 428, 792
    of dielectric surfaces, 793
    internal resonance problem, 18
    loop-tree basis, 428
    modification, 440
    for multibody problem, 440–43
    SDFMM-based solution, 793–94
PML-FDTD
    2D algorithm, 326
    configurations for stability test, 327
    cylindrical, 299–305
    generalized schemes, 331–41
    maximum residual amplitude, 303
    spherical, 305–7
    spherical-grid 3D algorithm, 329–30
    See also Perfectly matched layers
        (PMLs)
PML-FDTD for dispersive media, 288–93
    accuracy, 292
    dispersive medium models, 288–91
    incorporated into FDTD
        update, 291–93
    Sommerfeld solution vs., 293
    time domain analysis, 288
    See also Perfectly matched layers
        (PMLs)
Point-based grid-robust higher-order
        bases, 680–88
    defined, 682
    implementation, 682
    Lagrange interpolation points, 682
    MOM formulation, 686–88
    vector basis functions, 682–85
    See also High-order methods
Point-based higher-order MLFMA, 652–58
    advantage, 658
    complexity analysis, 657
    defined, 653–54
    formulation, 654–57
    memory requirements, 657
    number of levels, 657
    numerical results, 658
    See also High-order methods

Posteriori-based error bounds, 210
Pulse function, 216, 362
PWTD-accelerated MOT codes, 883
PWTD-augmented MOT solver, 881–88
    efficacy, for large-scale
        analysis, 883–88
    resonance frequencies comparison, 884
    validating, 881–83
PWTD-enhanced MOT
        algorithm, 867–75
    complexity analysis, 875
    defined, 867
    plane wave representations, 868–73
    two-level enhanced MOT solver
        implementation, 873–75
PWTD-enhanced TDIE solvers, 859–88
    efficacy for large-scale
        analysis, 883–88
    formulation, 861–67
    integral equations, 862–65
    introduction to, 859–61
    numerical results, 875–88
    PWTD algorithm, 867–75
    summary, 888
    validating, 881–83

Quadratic shape function, 738
Quadrature error
    circular cylinder (TE), 223–25
    circular cylinder (TM), 217–18
    flat strip (TE), 238–39
    flat strip (TM), 230
    hypersingular term of EFIE, 225
    spectral shift due to, 245
Quadrature points, 684
Quadrature rules, 212, 217
    efficient, 784
    information about, 685
    number of points in, 683
    numerical, 225
    smoothing error term and, 218
Quadrature weights, 785, 872
Quantum well grating analysis, 798–804
    introduction and formulation, 798–802
    periodic and quasi-random, 802–4
    random rough surface couplers, 804
    See also SDFMM

Quantum well infrared photodetectors
    (QWIPs), 798
  absorption improvement, 800
  back-illuminated structure, 800
  grating structure, 802
  schematic cross-sectional view, 801
  unrestrained n-type, 798
Quasi-Minimum Residual (QMR)
    method, 666, 866–67
Quasi-planar surface
  defined, 782
  side view, 783
  top view, 783
Quasi-planar surface field
    evaluation, 782–88
  scalar case, 782–86
  vector case, 787–88
  See also SDFMM
Quasi-PML, 330–31, 339
  features, 331
  practical applications, 331
  See also Perfectly-matched layers
    (PML)
Quasi-TEM mode, 480

Radar absorbing materials (RAMs), 576
Radar cross-section (RCS), 2
  of aircraft extracted from time domain
    CFIE, 887
  of almond extracted from time domain
    CFIE, 885, 887
  backscatter, 521, 575, 604, 611, 717,
    718, 743, 767
  bistatic. See Bistatic RCS
  calculated, 645
  calculation, 576
  CEM codes for calculating, 599
  of coated conducting sphere, 527
  of coated sphere, 517, 518
  of coated sphere calculated by hybrid
    integral equation, 526
  comparison of different meshes/
    dielectric constants, 524
  of cone-sphere extracted from time
    domain CFIE, 886
  of dielectric plate, 522
  fast computation for PEC objects, 700

  frequency response, 715
  HH-polarized, in x-y plane, 603
  of H-polarized incidence, 517
  of metallic conesphere, 558
  of metallic cylinder with four
    wings, 570
  of metallic double ogive, 557
  of metallic ogive, 556
  monostatic. See Monostatic RCS
  reducing, 575–76
  relative RMS error, 208
  solved using different meshes, 524
  solved using different spheres, 523
  of sphere extracted from time domain
    CFIE, 885
  for two layers of spherical dielectric
    shells, 525
  of V-polarized incidence, 517
  VV-polarized, in x-y plane, 603
Radiation fields
  diagonal forms of, 122–24
  MLFMA calculation, 102–3
Radiation patterns
  of 8 ´ N microstrip corporate-fed
    microstrip array, 757
  bandwidth of, 51–52
  of corporate-fed microstrip antenna
    array, 772
  from corrugated horn
    antennas, 560, 563, 564, 565
  in local interpolation, 85
  MLFMA, 57
  up-sampled, 62
  using SDFMM, 808, 809
Radomes, 519
  effects calculation, 520
  sharp tips, 520
  three dielectric, 521
  three-layer problem, 520
  types of, 520
  See also Antenna radome modeling
Raising operator, 72
Random rough surface couplers, 804
Random rough surfaces, 791–98
  with Gaussian distribution, 791
  integral equation formulations, 792–93
  model development, 791

Random rough surfaces (continued)
  scattering from, 791–98
  SDFMM-based solutions, 793–95
  simulation results, 795–98
  See also SDFMM
Rao-Wilton-Glisson (RWG) basis.
    See RWG basis function
RATMA, 21
Ray interpolation and splicing
    operations, 837, 839
Rayleigh-Ritz procedure, 665
Reciprocity theorem, 568–69, 608,
    609, 618
Rectangular cavity, 242–47
  condition number estimates, 269–71
  dimensions, 243
  illustrated, 679
  monostatic RCS of, 668, 670, 676, 680
  near-resonant case, 245
  numerical results, 246–47
  resonant case, 243–44
  spectral error, 245–46
  See also 2D spectral convergence
    theory
Reduced-order model, 771
Reflection errors
  local, 305
  in logarithmic scale, 317
Relative root mean square (RMS)
    error, 206, 231
  maximum, 208
  for plane wave incident angle, 232
  RCS, 208
  surface current, 220, 224
Resection operation, 838
  nature of, 838
  schematic description, 840
Resistive sheets (R-card), 104
Retarded-time boundary integrals
    (RTBIs), 815
Reverse Cuthill-McKee (RCM)
    ordering, 549
RF coils, 707
Riemann zeta function, 217
Right-hand side (RHS) vectors, 740–41
Rooftop basis functions, 258–61, 361, 754
  defined, 259

  moment matrix elements, 259
  spectral convergence rate, 260
  spectral error, 261
  triangular, 704
  See also 3D spectral convergence
    theory
Root mean square (RMS) error, 688
  bistatic RCS, 647, 689, 690
  bistatic RCS vs. angle of rotation, 691
  current, 219
  defined, 645
  as function of rotating angle, 690
  in monostatic RCS, 669, 670, 677, 678
  relative, 206, 208
  from series of defective meshes, 690
RWG basis function, 109, 190, 426,
    617, 738
  3D LF-MLFMA based on, 189–97
  3D version, 503, 504
  basis, 195
  bistatic RCS computation with, 427
  defined, 104
  definition, 428
  divergence of, 429
  extended to higher-order, 638
  for high-frequency cases, 195, 427
  illustrated, 429
  "localized," 426, 428
  loop-tree basis functions as
    superimpositions of, 193
  for modeling spatial variant of
    current, 865
  NOUR scheme for, 440, 443, 447
  number-of-unknowns reduction
    scheme, 443–49
  on regular triangular mesh, 274
  for three edges, 639
  translation to rectangular grids, 760
  for triangular discretization, 737
  for triangular patches, 638
  two adjacent triangles forming
    domain of, 503
  vector plot, 705

Sampling error, 245
Scalar Green's function, 81, 85
  relative truncation error of, 90, 91

truncation error in, 86–88
Scalar wave functions, 178
ScaleME, 136–38
    defined, 120
    embedding, 137
    host, 136
    interfaces, 137
    parallel host interface, 137
    performance, 137
    processor scaling, 142
    sequential host interface, 137
    single processor performance, 141
    SPMD paradigm, 136
    TRIMOM, 140
    user-callable functions, 137
    using, 122
ScaleME-2, 144–47
    defined, 144
    implementation, 146–47
Scattered fields
    by crack, 611
    direct, 619
    electric, 364–65
    FEM/SBR technique
        calculation, 608–9
    Green's function for, 352–53
    indirect, 610, 619
    from lossless sphere, 438
    medium and geometry parameters for
        computing, 437
    MOM/SBR technique
        calculation, 618–19
    reflected by large body, 609
    spatial-time domain, 380, 383, 384
    total, 618
    volume-surface integral equation, 492
Scattering amplitude, 207
    accuracy, 210
    bistatic, 220
    forward, 232, 254
    numerical results, 234
    stationary, 240
    total, 241
Scattering amplitude error
    circular cylinder (TE), 222–23
    circular cylinder (TM), 220–21
    first order, 240

flat strip (edge error), 239–40
flat strip (TM), 232–34
higher-order basis functions, 254–55
relative, 207, 241
Scattering approximation, 231, 233
Scattering parameters
    for microstrip taper, 483
    for step discontinuity, 482
Scattering resonances
    distance between, 246
    internal, 266
    low-order, 246
    of open cavity, 244
    study of, 242–43
SDFMM, 781–810
    3D Green's function and, 782
    for accelerating Green's functions
        computations, 787
    analysis of microstrip antenna arrays
        on finite substrates, 804–9
    application, 782
    approach validation, 795
    computational complexity
        estimates, 788–90
    conclusion, 810
    CPU time, 796
    decomposition schematic, 786
    defined, 781–82
    efficient application of, 808
    expansion for electromagnetic
        fields, 808
    field evaluation on quasi-planar
        surfaces, 782–88
    introduction to, 781–82
    matrix-vector product times, 799
    memory requirements, 788, 796, 799
    multilevel, 786
    number of levels, 788
    overall cost, 789
    parent block, 790
    quantum well grating
        analysis, 798–804
    radiation patterns obtained
        using, 808, 809
    scattering from random rough
        surfaces, 791–98
    spectral angles required by, 788

SDFMM (continued)
  traversals, 790
  tree, ascending, 789
  tree, descending, 789
  two-level implementation, 782
Second-order spatial interpolation
    scheme, 463
Second/zeroth-order triangular prism
    element, 672–75
  curved element, 673
  defined, 672
  interpolation points, 674
  order of transformation, 672
  rectilinear element, 673
  zeroth-order vector basis
      functions, 673
  See also Mixed-order prism elements
SGI Origin 2000, 650
Shooting-and-bouncing-ray (SBR)
    method
  defined, 577
  field scattered by large body without
      protrusion calculation, 618
  as high-frequency method, 606
  hybrid FEM/SBR technique, 602–14
  hybrid MOM/SBR technique, 614–21
  for incident electric field
      calculation, 616
Silvester-Lagrange polynomials, 739
Silvester polynomials, 642
Single-frequency DBIM algorithm, 349
  with 16 transmitters/
      16 receivers, 390–91
  with 32 transmitters/32 receivers, 392
  distorted Born approximation, 386
  fast convergence, 386
  iterations, 389, 391
  key step, 387
  multifrequency DBIM
      comparison, 395
  setup configuration, 385
  stability, 391
  See also Distorted Born iterative
method (DBIM); Fast inverse scattering
    methods
Single junction problem, 471–74
  case 1, 472–73

case 2, 473–74
illustrated, 472
See also Waveguide discontinuities
Single program multiple data (SPMD)
    paradigm, 121, 136
Singular integral processing, 508–13
Singularity expansion method, 863
Singularity extraction method, 687
Singular value decomposition (SVD), 680
Slant stack transform (SST), 824
  defined, 873
  direction-dependent shift, 845
  information reuse, 852
  "outgoing" rays, 844
  of source distribution, 845
  translating, 844
Smoothing errors, 245, 266
Sobolev norms, 207, 210
Sobolev space, 231
Sommerfeld half-plane problem
  illustration, 5
  solution, 6
Sommerfeld identity, 733
Sommerfeld integral, 399, 411, 412
Sommerfeld radiation
    condition, 585, 604–5, 615
Source-to-field vector, 511
Sparse matrix equations, 464
  complexity and storage issues, 468–69
  generalized formal solutions, 465
  Krylov-subspace-based method for
      solving, 467
  solution to, 467–69
Spatial Fourier transform, 399
  4D, 409–10
  of object function, 410
  See also Fourier transforms
Spatial Galerkin testing procedure, 866
Spatial-time domain scattered
    fields, 380, 383, 384
Spectral convergence theory, 211–12
  2D, 212–56
  3D, 256–61
  basis, 211
  circular cylinder (TE), 222–25
  circular cylinder (TM), 213–22
  defined, 211

flat strip (edge error), 239–41
flat strip (TE), 234–39
flat strip (TM), 225–34
higher-order basis functions, 247–55
rectangular cavity, 242–47
summary, 255–56
Spectral decomposition procedure, 468
Spectral error
    circular cylinder (TE), 222–23
    circular cylinder (TM), 215–17
    flat strip (TE), 235–36
    flat strip (TM), 229–30
    order determination, 260
    for piecewise polynomial basis functions, 229, 235
    rectangular cavity, 245–46
    relative, 222
    rooftop basis functions, 261
    total, 223
Spectral Lanczos decomposition method (SLDM), 12
    operations, 474
    solving matrix function with, 468
Sphere-to-sphere translation, 850–53
    aggregation step, 851
    disaggregation step, 851
    error distribution, 852
    logarithm of normalized errors, 853
    operations, 851
    time steps, 851, 863
    See also Windowed PWTD algorithm
Spherical harmonics, 827
    arbitrary rotation of, 185
    orthogonality of, 827
Spherical PML, 315, 317
    analysis, 327–30
    concave, 329
    convex, 329
    dynamic stability, 327
    See also Perfectly matched layers (PMLs)
Spherical PML-FDTD, 305–7
Spiral inductor
    current distribution, 747
    geometry, 747
    S-parameters, 747
Splicing operation, 837, 838, 839

nature of, 838
ray, 839
schematic description of, 840
Square integral domain, 510
Square matrix, 700
Stability criterion, 319
Star basis function, 430
    curl-free/divergence-free and, 432
    illustrated, 431
Static MLFMA, 152, 181–89
    core equation, 181–84
    relative errors, 184, 191, 192
    rotation of translation matrices for, 184–89
    See also MLFMA
Steepest-descent fast multipole method. *See* SDFMM
Steepest-descent path (SDP), 730, 784
    integration cost, 790
    integration rules, 797
    number of integration points, 790
Stirling's formula, 98
Stokes' theorem, 440
Subarrays, 850, 851
Subsignal temporal representation, 828–30
Surface discretization, 495
Surface integral equations (SIE), 13–15, 488, 700
    advantages, 488
    error analysis, 203–78
    error measures, 206–8
    formulation, 488
    grid convergence property, 515
    MOM and, 204–6
    solving, 205
Surface patches, 493
Surface scattering problem, 817
Synthetic aperture radar (SAR) images, 699

Taylor expansion, 68, 701
Temperature distribution, 534
TE polarization, 209
    CFIE for, 166–73
    EFIE for, 166, 234
    error analysis for, 236–37

TE polarization (continued)
  growth rate for, 270
  incident and scattered magnetic
      fields, 166
  induced current, 166
  integral operator, 205
  maximum condition number, 270
  maximum eigenvalue for, 270
  MFIE for, 166
Testing and expansion functions, 216
Tetrahedral elements
  curvilinear, 661
  for fixed-size cavity, 668
  higher-order, 661–64
  quadratic curvilinear, 662
  rectangular cavity modeled with, 668
Tetrahedron mesh, 493, 528
Theorem of quasioptimality, 209
Thin dielectric sheets (TDS), 104, 105
Third-order basis function, 688
Three-component, node-based
    formulation, 542
Three-stage PWTD algorithm, 824–30
  accuracy, 830
  definition of vectors used in, 823
  implementation issues, 826–30
  radiated field evaluation, 830
  signals associated with, 824, 825
  spatial integration, 826
  spectral integration, 826–28
  subsignal temporal
      representation, 828–30
  See also Plane-wave time-domain
      (PWTD) algorithms
Time domain analysis
  complex space coordinates, 287
  PML-FDTD for dispersive media, 288
Time-domain integral equation
    (TDIE), 859–60
Time-domain scattered magnetic
    fields, 378
  for different buried depths, 382
  for different conducting plate
      sizes, 380
  for different earth conductivity, 381
Time-gating scheme, 870

TM polarization
  CFIE for, 160–65
  condition number estimate of, 270
  EFIE for, 162, 234
  incident electric fields, 160
  maximum condition number, 270
  MFIE for, 162
  normalized induced current, 172
  surface currents due to, 209–10
Transient scattering
  from an almond analyzed using
      MFIE, 883
  from plate analyzed using EFIE, 882
Translation
  $\overline{\alpha}$, 58
  $\overline{\beta}$, 59
  in cylindrical coordinate system, 43
  functions, 848
  MLFMA, 58–59
  normalized, 185
  parallel MLFMA phase, 130–32
  physical interpretation of, 50–51
Translational addition theorem, 49
Translation matrices, 92
  along z axis, 186
  calculation of, 98
  compressed representation, 134
  CPU time for filling, 97
  interpolation of, in MLFMA, 96–101
  nonzero, 185
  normalized, 184
  rotation, for 3D LF-MLFMA/3D static
      MLFMA, 184–89
  scaling of, 134–35
  spectrum, 97–98
Translation operator
  diagonalization of, 46–48
  diagonalized, 71, 134
  matrix operator representation, 69
  memory required at each level, 145
  plane-wave representation of, 74
  representation in subspaces, 69
Transpose-free quasi-minimal residual
    (TFQMR)
  residual error stopping criterion
      for, 795
  use of, 808

Tree basis function, 430
  curl-free/divergence-free and, 432
  defined on two patches, 432
  structure on plate, 431
  structure on sphere, 431
Trees
  construction of, 133–34
  distributed, 134
  dummy, 136
  skeleton of, 134
  storage, 132–33
Tree space matrix, 276, 277
Triangular patches
  current density, 740
  curvilinear, 639, 640, 683, 738
  RWG basis function for, 638
  volume cells, 502–4
  zeroth-order basis function on, 738
Triangular roof-top function, 704
Tridiagonal matrix, 468
TRIMOM+ScaleME, 140
  defined, 140
  domain decomposition, 143
  speed/memory requirements comparison, 141
Truncation error, 52–53
  error analysis for, 85
  relative, 90
  in scalar Green's function, 86–88
  in vector Green's function, 88–91
Tschebysceff sampling (TS) series, 100
Two-level PWTD-enhanced MOT algorithm, 831–36
  construction of outgoing rays, 834–35
  defined, 831
  evaluation of far field contributions, 834–36
  evaluation of near field contributions, 834
  projection, 836
  translation, 835
Two-level PWTD-enhanced MOT solvers
  complexity analysis, 875
  current distribution computation, 873–74
  far field evaluation, 873–74
  implementation, 873–75

near field evaluation, 873
  total cost, 875
Two-level windowed PWTD-enhanced MOT algorithm, 853–54
Two-port asymmetric antenna, 751
TX-RX, 407
  electric-dipole, 407–8, 411, 413
  fixed offset, 411
  magnetic-dipole, 408–9, 413
  placed in cart, 412
  in practical setup of measurement system, 411

Unified theory, 336–41
  hybrid PMLs, 341
  metric and topological structure of Maxwell's equations, 339–41
  PML as change on metric of space, 336–39
Unitary basis vectors, 639
Unknown elimination schemes, 453

Vandermonde determinant, 685
Vector basis functions, 596, 597, 641, 682–85
  associated with shared edge, 662
  first-order, 740
  higher-order, 662
  higher-order interpolatory, 739
  information about, 685
  unified expression, 641
  zeroth-order, 673
Vector Green's function
  for MFIE, 89
  relative truncation error of, 90
  truncation error in, 88–91
Vector interpolator, 682
Vector wave equation, 543–44
  discretized, for electric field, 462
  for magnetic field, 464
  transverse, 464
  transverse components, 463
Very early time electromagnetic (VETEM) systems, 370–71
  buried target detection, 349
  configurations, 371
  defined, 348, 370
  detection in very lossy earth, 379

Very early time electromagnetic
(VETEM) systems (continued)
development motivation, 370–71
elements, 348, 371
frequency spectrum, 372, 377
illustrated, 371
loop-antenna model, 372–74
magnetic-dipole model, 374–75
numerical modeling, 371–75
numerical simulation, 348
simulated input signal, 377
simulation results, 375–80
spatial-time domain scattered fields
and, 380, 383, 384
time-domain reflected magnetic fields
and, 378, 379
time-domain scattered magnetic fields
and, 378, 380, 381, 382
VFY218 airplane, 621
bistatic RCS at 8 GHz, 113
bistatic RCS at 500 MHz, 146
illustrated, 626
monostatic RCS at
100 MHz, 96, 107, 108
with mushroom-shaped protrusion,
monostatic RCS, 627
Volume cells
auxiliary, 500
curvilinear, 512
curvilinear hexadedron, 504–6
curvilinear quadrangle, 504–6
in dielectrics, 500
external face of, 493
hexahedron, 499
quadrangle, 499
shapes, 493, 502–6
tetrahedron, 502–4
triangle patch, 502–4
Volume integral equation
(VIE), 19, 488, 489–91
applied to two-layer problem, 517
cell shapes, 502–6
discretization, 495–99
electric field, 490, 491
enforcing continuity condition, 501–2
FMM for solving, 513–14

grid convergence property, 515
magnetic field, 491
mesh generating, 492–94
mesh termination, 499–501
near-field comparison of Mie series
and, 529
numerical solution, 492–506
numerical treatment, 488
results by, 515–17
unknown vector, 499
Volume internal discretization, 495
Volume Jacobian, 503, 504
Volume-surface integral equation
(VSIE), 491–506
for antenna radome modeling, 519–21
applications, 518–37
CFIE for, 506–7
defined, 488, 492
for indoor radio wave propagation
simulation, 518–19
for microwave thermal effect
simulation, 519
MLFMA performance for, 514
for modeling large sized
structures, 518
numerical examples, 514–18
RCS of coated sphere calculated
by, 526
results by, 517–18
scattered field, 491
solution by FMM, 513–14

Waveguide discontinuities, 469–77
microstrip, 484–85
microstrip taper, 480
n-junction problem, 474–77
problems, 469
shielded microstrip line with step, 474
single junction problem, 471–74
step, 482
three-dimensional, 480
Waveguide junctions
with current source, 472
transverse electric field
components, 471
Waveguiding structures

conclusions, 480–84
discontinuities, 469–77
efficient analysis of, 461–84
finite difference formulation, 462–67
introduction to, 461–62
modeling, 461
numerical characterization, 461
numerical examples, 477–80
sparse matrix equation solution and, 467–69
Whitney's elements, 662
Whittaker-type field expansion, 816
Wiener-Hopf technique, 6
Windowed PWTD algorithm, 842–55
constraints, 844
defined, 816
for field evaluation at multiple observers, 849–50
formulation basis, 844–45
implementation, 849–55
implementation with sampled field representations, 844–49

multilevel windowed PWTD-enhanced MOT, 854–55
plane-wave decomposition, 843–44
sphere-to-sphere translation, 850–53
two-level windowed PWTD-enhanced MOT, 853–54
See also Plane-wave time-domain (PWTD) algorithms
Window function, 229
Wire spiral inductor
inductance, 197
input susceptance, 196
structure, 196

XPATCH, 7, 621

Yee scheme, 10

Zeroth-order basis
functions, 662, 685, 738
on flat triangular patch, 738
multiplying by interpolatory polynomial functions, 662

Recent Titles in the Artech House Antennas and Propagation Library

Thomas Milligan, Series Editor

Adaptive Array Measurements in Communications, M. A. Halim

Advances in Computational Electrodynamics: The Finite-Difference Time-Domain Method, Allen Taflove, editor

AWAS for Windows: Analysis of Wire Antennas and Scatterers, Software and User's Manual, A. R. Djordjević, et al.

Antenna Engineering Using Physical Optics: Practical CAD Techniques and Software, Leo Diaz and Thomas Milligan

Applications of Neural Networks in Electromagnetics, Christos Christodoulou and Michael Georgiopoulos

CAD of Microstrip Antennas for Wireless Applications, Robert A. Sainati

The CG-FFT Method: Application of Signal Processing Techniques to Electromagnetics, Manuel F. Cátedra, et al.

Computational Electrodynamics: The Finite-Difference Time-Domain Method, Second Edition, Allen Taflove and Susan C. Hagness

Fresnel Zones in Wireless Links, Zone Plate Lenses and Antennas, Hristo D. Hristov

Iterative and Self-Adaptive Finite-Elements in Electromagnetic Modeling, Magdalena Salazar-Palma, et al.

Measurement of Mobile Antenna Systems, Hiroyuki Arai

Mobile Antenna Systems Handbook, Second Edition, K. Fujimoto and J. R. James, editors

Microstrip Antenna Design Handbook, Ramesh Garg, et al.

Radiowave Propagation and Antennas for Personal Communications, Second Edition, Kazimierz Siwiak

For further information on these and other Artech House titles, including previously considered out-of-print books now available through our In-Print-Forever® (IPF®) program, contact:

Artech House
685 Canton Street
Norwood, MA 02062
Phone: 781-769-9750
Fax: 781-769-6334
e-mail: artech@artechhouse.com

Artech House
46 Gillingham Street
London SW1V 1AH UK
Phone: +44 (0)20 7596-8750
Fax: +44 (0)20 7630 0166
e-mail: artech-uk@artechhouse.com

Find us on the World Wide Web at: www.artechhouse.com